Philosophy of Technology

BLACKWELL PHILOSOPHY ANTHOLOGIES

Each volume in this outstanding series provides an authoritative and comprehensive collection of the essential primary readings from philosophy's main fields of study. Designed to complement the *Blackwell Companions to Philosophy* series, each volume represents an unparalleled resource in its own right, and will provide the ideal platform for course use.

1. Cottingham: *Western Philosophy: An Anthology*
2. Cahoone: *From Modernism to Postmodernism: An Anthology* (expanded second edition)
3. LaFollette: *Ethics in Practice: An Anthology* (second edition)
4. Goodin and Pettit: *Contemporary Political Philosophy: An Anthology*
5. Eze: *African Philosophy: An Anthology*
6. McNeill and Feldman: *Continental Philosophy: An Anthology*
7. Kim and Sosa: *Metaphysics: An Anthology*
8. Lycan: *Mind and Cognition: An Anthology* (second edition)
9. Kuhse and Singer: *Bioethics: An Anthology*
10. Cummins and Cummins: *Minds, Brains, and Computers – The Foundations of Cognitive Science: An Anthology*
11. Sosa and Kim: *Epistemology: An Anthology*
12. Kearney and Rasmussen: *Continental Aesthetics – Romanticism to Postmodernism: An Anthology*
13. Martinich and Sosa: *Analytic Philosophy: An Anthology*
14. Jacquette: *Philosophy of Logic: An Anthology*
15. Jacquette: *Philosophy of Mathematics: An Anthology*
16. Harris, Pratt, and Waters: *American Philosophies: An Anthology*
17. Emmanuel and Goold: *Modern Philosophy – From Descartes to Nietzsche: An Anthology*
18. Scharff and Dusek: *Philosophy of Technology – The Technological Condition: An Anthology*
19. Light and Rolston: *Environmental Ethics: An Anthology*
20. Taliaferro and Griffiths: *Philosophy of Religion: An Anthology*
21. Lamarque and Olsen: *Aesthetics and the Philosophy of Art – The Analytic Tradition: An Anthology*
22. John and Lopes: *Philosophy of Literature – Contemporary and Classic Readings: An Anthology*
23. Cudd and Andreasen: *Feminist Theory: A Philosophical Anthology*

Philosophy of Technology

The Technological Condition

An Anthology

Edited by

Robert C. Scharff and Val Dusek

Blackwell Publishing

BLACKWELL PUBLISHING
350 Main Street, Malden, MA 02148-5020, USA
9600 Garsington Road, Oxford OX4 2DQ, UK
550 Swanston Street, Carlton, Victoria 3053, Australia

First published 2003 by Blackwell Publishing Ltd

6 2007

Library of Congress Cataloging-in-Publication Data

Philosophy of technology : the technological condition : an anthology / edited by Robert C. Scharff and Val Dusek
p. cm.—(Blackwell philosophy anthologies ; 18)
Includes bibliographical references and index.
ISBN 978-0-631-22218-7 (alk. paper)—ISBN 978-0-631-22219-4 (pb. : alk. paper)
1. Technology—Philosophy. I. Scharff, Robert C. II. Dusek, Val, 1941– III. Series.
T14 .P534 2002
601—dc21
2001052652

A catalogue record for this title is available from the British Library.

Set in 9 on 11 pt Ehrhardt
by Kolam Information Services Pvt. Ltd, Pondicherry, India
Pr inted and bound in India
by Replika Press Pvt. Ltd

The publisher's policy is to use permanent paper from mills that operate a sustainable forestry policy, and which has been manufactured from pulp processed using acid-free and elementary chlorine-free practices. Furthermore, the publisher ensures that the text paper and cover board used have met acceptable environmental accreditation standards.

For further information on
Blackwell Publishing, visit our website:
www.blackwellpublishing.com

Contents

General Introduction: Philosophy and the Technological Condition

This collection grew out of the editors' experience as teachers of the philosophy of technology. Although a number of well-stocked anthologies were available, no matter which one we chose, we found ourselves repeatedly supplementing its selections. Most especially, we felt we were forever facing two philosophical lacunae. First, most anthologies contained very little material from classical sources (e.g., Aristotle, Bacon, Comte, Marx) in which technology, or basic concepts that contribute to our current ways of conceiving technological practices, are discussed. Second, in many cases, the major portion of the text focused on specific technological problems and case studies, with the result – truth be told – that the selections were often philosophically thin. In our view, especially when it comes to the philosophical consideration of technology, structuring an anthology after the familiar model of the "applied ethics" reader is likely to have unfortunate pedagogical consequences. In the typical application of this model, one starts with familiar, extra-philosophically identifiable "problems," samples the variety of "values" or "standards" in terms of which it has been claimed these problems can be handled, and then more or less leaves it up to the instructor to explain how philosophy somehow gets involved in testing the "selection" and "justification" of these values or standards.

Regarding the philosophy of technology, however, we feel that this model gets things strategically backwards in important ways. One unintended consequence of its use is that it can leave students, especially those who have not had much previous exposure to philosophy, with the impression that philosophy mostly happens at the level of a "debate" among a smorgasbord of competing sets of values that themselves are somehow simply found, or "given" as logical or sociological options. This serves to confirm the popular non-philosophical conception of philosophy as a "belief system" that one already has or should pick out and thereafter "defend." The whole idea of philosophy as a process of inquiry, or as critical self-discovery, or as involving a reflective struggle with inherited orientations, is thus muted or occluded. Moreover, as some of the authors below complain, the problems-model also has the effect of privileging one very familiar but perhaps not so innocent outlook regarding technological problems – namely, the idea that technology itself is not a problem, that it simply provides us with a collection of instrumental means, and that the main task is to decide what ends it should serve. To a significant number of philosophers of technology, this allegedly "neutral" interpretation of technology should itself be identified as a topic to be carefully questioned.

The second gap we found in the available texts is a widespread failure to consider the question of the relation between contemporary technology and modern science. As pressing and immediate as the issues of, say, technology transfer, medical patients' rights, and biotechnology in agriculture clearly are, debates that stay at the level of these issues often silently perpetuate long-standing, deeply held, but now hotly contested assumptions about the nature of science, about the technological applications of science, and even about the proper place of science and technology within the larger scope of human affairs. For example, is knowledge essentially connected to a drive for power, as Bacon claimed and Foucault still insists? Is technology primarily to be understood as "applied modern science," or is the ancient human concern for "making" already implicated in the very development of science itself, as (in very different ways) Comte, Marx, Heidegger, Mumford, Arendt, and various sorts of pragmatists maintain? And should we expect, or do we even have a choice about, technological practices increasingly coming to define the nature and axiological direction of human life? Such questions simply cannot be addressed adequately if they are permitted to arise only between the lines of selections focused primarily on issues of how to control, modify, or conceptually clarify this or that specific political, ethical, aesthetic, or engineering problem.

With these concerns in mind, then, we have structured our anthology in such a way that, with or without sharing our reasoning above, instructors have the option of making historical, metaphysical, and epistemological issues just as prominent as ethical, political, aesthetic, and engineering problems. Because we envision this text as useful for anything from introductory undergraduate courses to graduate seminars, our selections vary considerably in length and difficulty, and we have elected to place most of our introductory material at the beginning of the sections rather than all together in one opening essay. Here, we shall confine ourselves to a brief explanation of the general plan of the six main parts of the text.

The purpose of part I is to provide a forum for some familiar voices in the Western philosophical tradition whose views about the relation between knowledge and its applications have played an important role in setting up the inherited context within which contemporary philosophy of technology takes its bearings. Some consideration is also given to the question of why – in comparison to other philosophical topics – a philosophy specifically of technology is so relatively recent in origin.

Part II contains contemporary readings that especially emphasize and critically assess the basic assumptions handed down to us from the nineteenth century about science, the relation between modern science and technology, and philosophy's proper treatment of both. We have divided this part into two sections. The first section provides a kind of mini-history of the rise and decline of logical positivist, or Vienna Circle, philosophy of science, together with the emergence of various postpositivist criticisms and alternatives. Our intention is to highlight the ways in which these alternatives all tend to stress the importance of precisely the social, cultural, and historical context of scientific practice that positivistic philosophy of science urges us to ignore. The readings in the second section illustrate how stressing or ignoring this context directly affects one's conception of the nature of and the relation between the philosophies of science and of technology.

The readings in part III illustrate how many recent efforts to define technology still tend, sometimes in spite of themselves, to reflect the older, more traditional assumptions about what science is and how philosophy should approach it. In addition, these selections make it plain that any attempt to define technology also tends to take a stand on two controversial topics – namely, whether and how modern science has transformed "prescientific" technologies, and whether technology is essentially "applied science."

Part IV reprints Martin Heidegger's essay, "The Question Concerning Technology," and a sampling of responses to it. Heidegger's essay presents what is probably the single most influential – though by no means most popular – position in the field. Many of the issues discussed in the sections that follow, especially in Part VI, are framed in a way that reflects some species of agreement or disagreement with his views.

In part V, the readings raise a cluster of general issues concerning technology's proper role in mediating our relations with the natural world. One section considers the question of whether human beings are essentially just "tool-users" and thus most themselves when they are engaged in technological activities. A second section raises the issue of whether, as some writers have argued, the influence of technology in our lives is so strong and pervasive that it actually functions as a virtually autonomous force and makes all optimistic talk of "controlling" it seem naïve. The essays in the third section bring the issues of human nature and technological power together in relation to the widely debated ecological question of the legitimacy of the famous (or perhaps infamous and even male-gendered) Baconian imperative that encourages us to think of "knowledge" primarily is giving us the "power" to "control" our natural surroundings.

Part VI focuses on issues that arise when technology is viewed, not so much as an expression of human nature or as an instrument for controlling our natural surroundings, but rather as defining a specific and (at least in some parts of the world) increasingly dominant kind of social practice. The essays in the first section all ask, as one author puts it, what it is like to "be-with" technology, such that it mediates most of our relations not just with nature but also among ourselves. In "Technology and Cyberspace," the second section, several authors consider the puzzling issue of whether the computer revolution promises, like it or not, to alter our basic notions of who we are and what it is to experience "reality." A third section brings into focus a question implicit in numerous other readings, namely, what are the ramifications for the future of political democracy of our ever more predominantly technological forms of social practice?

Finally, we add a note of grateful acknowledgment. We would like to express our thanks to the publishers and other copyright holders who gave us reprint permission, and to the virtual army of persons who have encouraged and advised us in putting this text together. Among them are (with apologies to those we have inadvertently omitted) Thomas Achen, Babette Babich, Robert Crease, Fred Dallmayr, Trish Glazebrook, Donna Haraway, Patrick Heelan, Michael Heim, Don Ihde, Theodore Kisiel, David Kolb, Carolyn Merchant, Joseph Rouse, Hans Seigfried, Timm Triplett, Kenneth Westphal, Michael Zimmerman, and the six anonymous reviewers for Blackwell Publishers. Special thanks are due also to Larry Hickman for volunteering to revise "Doing and Making in a Democracy: Dewey's Experience of Technology," to Andrew Feenberg for revising "Democratic Rationalization [originally, "Subversive Rationality"]: Technology, Power, and Freedom [originally, "Democracy"]"; and to John McDermott for writing, on very short notice, a retrospective on his "Technology: The Opiate of the Intellectuals." We are also grateful for the continuing support and patience of Jeffrey Dean, our Blackwell editor, and Beth Remmes, who saw the manuscript through the press. Let us add that we are painfully aware that in this rapidly growing field, it is impossible for anyone to maintain a working knowledge of "everything important" that might be suitable for a reader such as ours. We therefore welcome all criticisms and suggestions about possible sins of omission as well as commission. And, of course, we ask that those we have thanked above be held blameless for this final product.

R. C. S. and R. V. D.
September 2001

Part I

The Historical Background

Introduction

At first glance, it may seem surprising that until recently, philosophers have not devoted much time to the question of technology. One might have thought that greater attention would at least have come to be paid to this phenomenon in the modern period, when advances in natural and biological science increasingly and obviously made technology a central and dominant feature of society and culture. Yet the fact is that even today – in both the North American and British mainstream of analytic philosophy and to a somewhat lesser extent among those influenced by late-nineteenth- and twentieth-century postpositivist, continental European sources – the philosophy of technology is still widely regarded as not much more than a small and not particularly prestigious area of specialization.

In part, the reasons for this secondary status for the philosophy of technology are reflected in the general features of modern intellectual history. In the Anglo-American empiricist, French enlightenment, and European positivist traditions, technology is widely depicted as an unproblematically beneficial force for human progress. For these traditions, technology needs only the proper association with modern science to fulfill its promise; hence the genuinely philosophical issues lie primarily in the epistemology of science, which explains how genuine knowledge is to be obtained, and in ethics, which determines what that knowledge is for. With epistemology and ethics thus focused on the two central issues of what we can know and what we should do, technology falls through the cracks, understood as just the relatively neutral means for employing scientific knowledge to bring about the ideal relations in the natural and social world that ethical decisions prescribe. It is true that for the romantic and post-Hegelian "continental" traditions, this judgment must be qualified slightly, for in these traditions, there is less inclination to conceive all knowledge according to the model of science or to conceive of science as an essentially progressive force. Yet science itself (especially natural science) is often viewed by them in strictly instrumentalist terms, and technology is widely understood as simply applied science – with the difference being that the cultural implications of all this are more likely to be conceived in critical and pessimistic terms, not in the progressive or even utopian terms characteristic of the empiricist and positivist traditions.

To fully understand the philosophical neglect of technology, however, one must go back to Ancient Greek thought and to the manner in which figures like Plato and Aristotle drew their distinctions between theoretical and practical understanding. There is no question, of course, that the ancients took the distinction seriously. It is known, for instance, that Plato's teacher, Socrates (c.470–399 BCE), often discussed this distinction. He insisted that the "craft" knowledge of farmers, shoemakers, and bakers, as well as physicians, is genuine knowledge. Socrates' point, however, is one of criticism rather than defense. Craft knowledge consists primarily in a kind of technical understanding, limited to its concern with the pursuit of particular trades or practices. Unfortunately, those who possess such knowledge (and especially those who achieve worldly success because of it) are often misled into thinking they possess wisdom about life in general. Socrates is thus at pains to argue that practically oriented craft knowledge is in fact quite different from the knowledge of the good life that has always been the concern of the religious seers and poets. Plato (c.470–c.347 BCE), too, employs this general Socratic distinction of craft knowledge vs. knowledge of life; moreover, his dialogues are full of images of actual technological devices. The water clock, the astronomical orrery, and the mechanical puppet-show all figure prominently as metaphors and models for several of his myths – for example, cosmic creation in the *Timaeus*, the last judgment in the *Republic*, the history of the cosmos in the *Statesman*, and the shadow play of puppet-objects in the myth of the cave in the *Republic*'s account of the triumph of reason over sensual experience in genuine philosophical learning.

What Plato also makes explicit, however, is that the Socratic distinction between technical or craft knowledge on the one hand and knowledge of the good life on the other is fundamentally a distinction between two unequal phenomena. Craft knowledge is ordinary, lower, sense-experientially-based understanding focused on specific practical affairs. Knowledge of the good life and of the ultimate nature of things – i.e., the "wisdom" that philosophers "love" – is a higher, theoretical, and genuinely rational knowledge to which the former kind of knowledge is rightfully and ultimately beholden. Thus, for example, in the *Republic*, Plato envisions the education of the philosopher king as involving extensive training in pure mathematics (including theoretical astronomy and music theory) as the proper background for further and still higher training in philosophical dialectics. And in the *Gorgias*, he shows how technical understanding (e.g., of rhetoric) is useless, or worse, when it is cut off from the deeper knowledge of what rhetoric is truly "good" for.

It is this higher, genuinely rational understanding of the essential nature of things that Plato identifies as the concern of the philosophers; and it is this hierarchical conception of theoretical over practical understanding the he (and in a somewhat differently interpreted way, Aristotle) bequeathed to the Western tradition. Moreover, in their enthusiastic preference for the rational and theoretical over the practical and sense-dependent, many later Platonists, neoplatonists, and some say even Plato himself (in the controversial reports of his allegedly "unwritten" doctrines and the "Lecture on the Good" given at his Academy) came to identify numbers with the ideal, timeless form of philosophical knowledge. The distinction between mathematical knowledge and philosophical knowledge thereby came to be blurred, and it is perhaps not too much of a stretch to suggest that one can hear the echo of this ancient preference for mathematical metaphor in the later Western conceptions of technocracy, or rule by scientists and technologists. In any case, Plato did advocate the rule of the "wise," by which he meant those trained in philosophy, where philosophy is understood as the love of a knowledge that is "like" that of the mathematical scientists but with an additional concern for cultivating a rational vision of the ultimate principles of all things.

Aristotle (384–322 BCE) made just a strong distinction between higher and lower understanding and was just as convinced as Plato that the highest kind of human life is one of rational contemplation of the "highest things." In contrast to and criticism of Plato, however, Aristotle argued that the distinction between practical-technical (and artistic) understanding on the one hand and scientific-philosophical understanding on the other is really not a distinction involving possession of two kinds of "theories." Plato's Socrates allegedly claims that moral virtue is a kind of knowledge; but this, countered Aristotle, cannot be – if for no other reason than that this idea of moral virtue cannot account for our familiar experience of the weakness of the will. Too obviously, it is possible to have knowledge of what to do but fail to do it. For Aristotle, moral virtue must therefore be conceived as a kind of practical reasoning achieved through exposure to experienced teachers, the building of good character, and the formation of the proper habits of activity.

Aristotle also rejected Plato's tendency to overuse mathematical imagery in depicting not only philosophical and scientific knowledge but also practical and political reasoning. In the *Republic*, for example, Plato presents abstract mathematical knowledge as a preliminary to political practice. His *Philebus* even entertains the notion of a "science of normative measure" (perhaps inspired by the mathematical theory of utility of the leading mathematician Eudoxus, who had joined the Academy). In contrast, Aristotle claims that different disciplines have different degrees of rigor appropriate to them. It is wrong to demand mathematical exactness of, say, ethics or politics. The largely implicit practical wisdom required of the citizen or politician, as well as the expertise of artist and craftsperson, is concerned with generalizing about particular situations and individual things; and this sort of expertise must be carefully distinguished from the explicit theoretical knowledge of the scientist or philosopher, who is concerned with the truly universal and essential in all situations and all things.

For all their differences, however, Plato and Aristotle both developed hierarchical conceptions of knowledge that make philosophical or scientific understanding of the universal and essential superior. Evidence of this line of thinking can be seen in the fact that the ancient Greeks did not conceive of technological change and economic production in the modern terms of efficiency and progress. Practical techniques were judged as analogous to and as facilitating for our own purposes the cosmic and natural processes about which the human spirit above all seeks knowledge and with which we are in any case involved. The Roman aqueducts, for example, may seem "overbuilt" by modern standards, but that is because they are designed not just to carry water but to do so in perpetuity, like rivers and streams. Hence for the ancients, what we might consider "merely" aesthetic considerations were taken to be a necessary measure of changes in the crafts at least as much as utility or efficiency, because these considerations received their sanction from physics and metaphysics. Also in keeping with a supposed natural order of things, the "higher" arts of economic production as well as architecture and sculpture were thought more suited to men, whereas manual labor and hands-on crafts were widely understood as lowly activities fit mostly for women and slaves.

In this Greek-pagan orientation, then, there runs a pervasive sense of our being destined to live in harmony with an awesomely comprehensive cosmos to which we are never closer than when we strive to contemplate its first principles. In contrast, Christianity tends to encourage an outlook that fosters the idea of our separation from and superiority over nature. The Christian conception of the material universe as created from nothing by an all-knowing, rational God seems somehow to make that universe both less mysterious than Greek cosmology (see Emmanuel Mesthene's "Technology and Wisdom" in part VI, chapter 53) and more remote from our true being. The theological interpretation of history as a progress toward our salvation paved the way for the later notions of linear scientific and technological progress. At the same time, the struggle for self-purification against natural and material forces (introduced in the monastic orders) implicitly increased the dignity of the idea of work, and this imagery would later be used to suggest the possibility of technological and scientific revolutions. All such developments, however, had to await the transfer of these views to a non-religious context. Only in the modern period did the human control of nature and the essential beneficence of applying scientific knowledge to technology become ruling ideals. The remaining selections in Part One provide a sample of some of the major variations on these modern themes.

Francis Bacon (1561–1626) famously claims that knowledge is power – that is, that through knowledge of nature and its technological applications, humans can achieve a purity of mind and behavior that was lost after the Fall in the Garden of Eden. Thus, in *The New Atlantis*, Bacon envisages a utopia in which the workers at "Saloman's House" study the resources of the island and the world to improve health and welfare of the inhabitants. Plato's philosopher kings have now been transformed into proto-scientists and technologists who guide the nation. Other selections reveal Bacon's colorful and (as discussed later, in part V, chapter 38) sometimes disturbingly gendered ways of depicting the acquisition of scientific knowledge and what sort of power it is that such knowledge allegedly gives us.

A sophisticated and nuanced version of the secularized doctrine of linear historical progress is found in "The Idea for a Universal History from a Cosmopolitan Point of View," by Immanuel Kant (1724–1804). All "animals," Kant suggests, are destined to fulfill their natural purposes, and in the case of human beings, this means the development of scientific and ethical rationality. The idea of human progress toward world government and perpetual peace that Rousseau ridiculed is presented here as empowered by the "unsociable sociability" of humans. Though written in an admittedly somewhat speculative vein, Kant's essay nevertheless expresses the modern outlook of many who suppose its truth even when they are silent about it. For Kant's portrayal of rational progress seems to justify the enlightement doctrines about our elevation above and over against nature. At the same time, it points forward to the less qualified historical claims of Hegel, Comte, and Marx.

The theme of what would later be called technocracy truly came into its own in the early 1800s with the writings of Henri de Saint-Simon and Auguste Comte (1760–1825). Comte began his career as Saint-Simon's assistant, and he came to exert a powerful if often unacknowledged influence over a wide and diverse range of other thinkers by clarifying and extending Saint-Simon's rather sketchy and disorganized ideas into a "System of Positive Philosophy." The first half of his system focuses on the epistemology of science (his *Cours de philosophie positive*); the last half develops his ideas on the social and political reorganization that Comte assumes successful science will make possible (his *Système de politique positive*); At the heart of his system lies Comte's philosophy of history and its Law of Three Stages. This law depicts humanity as moving (first intellectually and then in action) through three phases of development, utilizing three "methods" of philosophizing – namely, the theological (or fictive), metaphysical (or abstractly speculative), and scientific (or "positive"). Comte founded and named sociology as the science of society, and in his later work was quite explicit about the need to replace traditional religion with a "Religion of Humanity," so that scientists, social scientists, and those who apply their knowledge, not the priests of the Catholic Church, would rule. A few positivist churches were actually founded and still exist in England and Latin America (Brazil's flag contains Comte's slogan, "Order and Progress"). More importantly, however, Comte's notion of the "priestly" role of scientists and technologists possesses – albeit in a less explicit form – a much greater worldwide significance in twentieth-century industrial societies, East and West.

Marx ridiculed the Saint-Simonian and Comtean blueprints for a technocratic utopia, calling them "recipes for the cookshops of the future." Yet Marx's collaborator, Engels, displayed much greater sympathy toward their doctrines, and through him, Saint-Simonian phrases – for example, "the administration of things, not of men," "society as one vast factory," and "artists as engineers of the soul" – found their way into the Soviet Marxism of both Lenin and Stalin. (Is it mere coincidence that the planners of Disney World are called "Imagineers"?)

In his *Discourse on the Sciences and the Arts*, Jean-Jacques Rousseau (1712–78) opposes the prevalent enlightenment optimism concerning science and its applications. Rousseau shocked his contemporaries by challenging their complacent progressivism. Where they saw only scientific progress and the promise of what Comte called "social reorganization" leading to world peace and human happiness, Rousseau perceived progress of the sciences and arts as leading instead to decline and decadence. Where *philosophes* such as Voltaire, d'Alembert, and Condorcet anticipated Saint-Simon and Comte in assuming that scientific progress of necessity leads to moral and political progress, Rousseau claimed instead that the virtue and vigor of the barbarian nations is destroyed by the spread of civilization. Rousseau's sensational anti-Baconian manifesto presaged and nourished the romantic critique of the industrial revolution in subsequent generations in Germany and England.

Karl Marx (1818–83) may have ridiculed the Saint-Simonian and Comtean conceptions of scientific and technological progress. Yet he was, after Bacon, the modern philosopher who made technology most central to his system. Marx's vision of human history combines elements of the Baconians and enlightenment *philosophes* with elements of the German and British romantics. Like Bacon and the enlightenment thinkers, Marx is optimistic concerning the development of science and technology as well as about their benefiting humanity in the long run, with the eventual establishment of communism. He shares, however, the pessimism of Rousseau and the romantics concerning the oppression and alienation produced by science and technology, especially in relation to private property, in the present and short run.

Marx's account of the role of technology in social change varies from writing to writing. *The Poverty of Philosophy* contains his famous quip, "the hand-mill gives you society with the feudal lord; the steam-mill, society with the industrial capitalist." His brief and highly influential "Preface to a Contribution to a Critique of Political Economy" identifies the "base" of society as constituted by technological forces and social power relations of production. Many orthodox Marxists have interpreted these passages as proposing a technological determinism. (Ironically, many strongly anti-Marxist technocrats in the late twentieth century espouse a conception of contemporary post-industrial society that is just as technologically deterministic. For contributions to the non-Marxist debate concerning technological determinism and autonomy, see the later selections by Ellul, Winner, and Heilbronner.) In the much more detailed discussions of *Capital*, however, Marx claims that class structure and class struggle control what sort of technology is developed. In these passages, some readers have seen a social determinism which can be used to criticize the idea of technological determinism. Many twentieth-century non-Soviet Marxists – both Chinese Maoists and European Marxist humanists – have gone on to argue that the very idea of technological determinism is a product of a technocratic capitalist ideology (see, e.g., the Habermas selection). One key issue in this dispute over whether technology determines social relations or social power determines technological developments is the question of the relative weight of the technological division of labor (job roles determined by technology) vs. the social division of labor (job roles and technology determined by the desire of owners and managers to control the workers.) That Marx and Engels embrace the latter view can be seen, for example, in the *The German Ideology*, where they describe a communist utopia in which one could be at once fisherman, hunter, and intellectual, or in Engels's later writings, where he rejects the anarchists' claim that technology by itself enforces certain types of labor discipline and hierarchy.

In any case, through all of these Marx–Engels selections there runs another theme that has also become a major topic of debate. According to Marx, human beings modify their environment with tools because this is our nature, and insofar as we do so, we exhibit our difference from other animals (see, e.g., *The German Ideology*). In his lifelong collaboration with Engels, Marx developed this conception of the human species into a full-blown account of human evolution from the apes (see "The Part Played by Labor in the Transition from Ape to Man"). This evolutionary conception of humans as essentially tool-making animals has been vigorously criticized by later philosophers of technology, even including humanistic Marxists (see Mumford and Arendt in part V, chapters 29 and 30). Arendt's argument is especially challenging, for it contains an immanent criticism, namely that Marx's own concept of labor betrays a deep ambivalence between understanding technological labor, on the one hand, as involving creative world-construction and, on the other hand, as inseparably linked to degrading oppression.

1

On Dialectic and "Technē"

Plato

From the *Republic*

[...]

Next, I said, compare the effect of education and of 514 the lack of it on our nature to an experience like this: Imagine human beings living in an underground, cavelike dwelling, with an entrance a long way up, which is both open to the light and as wide as the cave itself. They've been there since childhood, fixed in the same place, with their necks and legs fettered, able to see only in front of them, because their bonds prevent them from turning their heads around. Light is provided by a fire burning far above and behind them. Also behind b them, but on higher ground, there is a path stretching between them and the fire. Imagine that along this path a low wall has been built, like the screen in front of puppeteers above which they show their puppets.

I'm imagining it.

Then also imagine that there are people along the wall, carrying all kinds of artifacts that project above it – statues of people and other animals, made out of stone, wood, and every material. c And, as you'd expect, some of the carriers are 515 talking, and some are silent.

It's a strange image you're describing, and strange prisoners.

They're like us. Do you suppose, first of all, that these prisoners see anything of themselves and one another besides the shadows that the fire casts on the wall in front of them?

How could they, if they have to keep their heads motionless throughout life? b

What about the things being carried along the wall? Isn't the same true of them?

Of course.

And if they could talk to one another, don't you think they'd suppose that the names they used applied to the things they see passing before them?[1]

They'd have to.

And what if their prison also had an echo from the wall facing them? Don't you think they'd believe that the shadows passing in front of them were talking whenever one of the carriers passing along the wall was doing so?

I certainly do.

Then the prisoners would in every way believe that the truth is nothing other than the shadows of c those artifacts.

They must surely believe that.

Consider, then, what being released from their bonds and cured of their ignorance would naturally be like. When one of them was freed and suddenly compelled to stand up, turn his head, walk, and look up toward the light, he'd be pained and dazzled and unable to see the things whose shadows he'd seen before. What do you think he'd say, if we told him that what he'd seen before d was inconsequential, but that now – because he is a bit closer to the things that are and is turned towards things that are more – he sees more correctly? Or, to put it another way, if we pointed to each of the things passing by, asked him what each

Selections from Plato, *Republic* VII, trans. G. M. A. Grube and C. D. C. Reeve, pp. 186–206, 210–12. Indianapolis: Hackett, 1992.

of them is, and compelled him to answer, don't you think he'd be at a loss and that he'd believe that the things he saw earlier were truer than the ones he was now being shown?

Much truer.

And if someone compelled him to look at the e light itself, wouldn't his eyes hurt, and wouldn't he turn around and flee towards the things he's able to see, believing that they're really clearer than the ones he's being shown?

He would.

And if someone dragged him away from there by force, up the rough, steep path, and didn't let him go until he had dragged him into the sunlight, wouldn't he be pained and irritated at being treated 516 that way? And when he came into the light, with the sun filling his eyes, wouldn't he be unable to see a single one of the things now said to be true?

He would be unable to see them, at least at first.

I suppose, then, that he'd need time to get adjusted before he could see things in the world above. At first, he'd see shadows most easily, then images of men and other things in water, then the things themselves. Of these, he'd be able to study the things in the sky and the sky itself more easily at night, looking at the light of the stars and the b moon, than during the day, looking at the sun and the light of the sun.

Of course.

Finally, I suppose, he'd be able to see the sun, not images of it in water or some alien place, but the sun itself, in its own place, and be able to study it.

Necessarily so.

And at this point he would infer and conclude that the sun provides the seasons and the years, governs everything in the visible world, and is in c some way the cause of all the things that he used to see.

It's clear that would be his next step.

What about when he reminds himself of his first dwelling place, his fellow prisoners, and what passed for wisdom there? Don't you think that he'd count himself happy for the change and pity the others?

Certainly.

And if there had been any honors, praises, or prizes among them for the one who was sharpest at identifying the shadows as they passed by and who best remembered which usually came earlier, d which later, and which simultaneously, and who could thus best divine the future, do you think that our man would desire these rewards or envy those among the prisoners who were honored and held power? Instead, wouldn't he feel, with Homer, that he'd much prefer to "work the earth as a serf to another, one without possessions,"[2] and go through any sufferings, rather than share their opinions and live as they do?

I suppose he would rather suffer anything than live like that. e

Consider this too. If this man went down into the cave again and sat down in his same seat, wouldn't his eyes – coming suddenly out of the sun like that – be filled with darkness?

They certainly would.

And before his eyes had recovered – and the adjustment would not be quick – while his vision was still dim, if he had to compete again with the perpetual prisoners in recognizing the shadows, wouldn't he invite ridicule? Wouldn't it be said of 517 him that he'd returned from his upward journey with his eyesight ruined and that it isn't worthwhile even to try to travel upward? And, as for anyone who tried to free them and lead them upward, if they could somehow get their hands on him, wouldn't they kill him?

They certainly would.

This whole image, Glaucon, must be fitted together with what we said before. The visible realm b should be likened to the prison dwelling, and the light of the fire inside it to the power of the sun. And if you interpret the upward journey and the study of things above as the upward journey of the soul to the intelligible realm, you'll grasp what I hope to convey, since that is what you wanted to hear about. Whether it's true or not, only the god knows. But this is how I see it: In the knowable realm, the form of the good is the last thing to be seen, and it is reached only with difficulty. Once one has seen it, however, one must conclude that it is the cause of all that is correct and beautiful in anything, that it produces both light and its source c in the visible realm, and that in the intelligible realm it controls and provides truth and understanding, so that anyone who is to act sensibly in private or public must see it.

I have the same thought, at least as far as I'm able.

Come, then, share with me this thought also: It isn't surprising that the ones who get to this point are unwilling to occupy themselves with human affairs and that their souls are always pressing upwards, eager to spend their time above, for, after all, this is surely what we'd expect, if indeed things fit the image I described before. d

It is.

What about what happens when someone turns from divine study to the evils of human life? Do you think it's surprising, since his sight is still dim, and he hasn't yet become accustomed to the darkness around him, that he behaves awkwardly and appears completely ridiculous if he's compelled, either in the courts or elsewhere, to contend about the shadows of justice or the statues of which they are the shadows and to dispute about the way these things are understood by people who have never e seen justice itself?

That's not surprising at all.

518 No, it isn't. But anyone with any understanding would remember that the eyes may be confused in two ways and from two causes, namely, when they've come from the light into the darkness *and* when they've come from the darkness into the light. Realizing that the same applies to the soul, when someone sees a soul disturbed and unable to see something, he won't laugh mindlessly, but he'll take into consideration whether it has come from a brighter life and is dimmed through not having yet become accustomed to the dark or whether it has come from greater ignorance into greater light and is dazzled by the increased brilliance. Then he'll declare the first soul happy in its experience b and life, and he'll pity the latter – but even if he chose to make fun of it, at least he'd be less ridiculous than if he laughed at a soul that has come from the light above.

What you say is very reasonable.

If that's true, then here's what we must think about these matters: Education isn't what some people declare it to be, namely, putting knowledge c into souls that lack it, like putting sight into blind eyes.

They do say that.

But our present discussion, on the other hand, shows that the power to learn is present in everyone's soul and that the instrument with which each learns is like an eye that cannot be turned around from darkness to light without turning the whole body. This instrument cannot be turned around from that which is coming into being without turning the whole soul until it is able to study that which is and the brightest thing that d is, namely, the one we call the good. Isn't that right?

Yes.

Then education is the craft concerned with doing this very thing, this turning around, and with how the soul can most easily and effectively be made to do it. It isn't the craft of putting sight into the soul. Education takes for granted that sight is there but that it isn't turned the right way or looking where it ought to look, and it tries to redirect it appropriately.

So it seems.

Now, it looks as though the other so-called virtues of the soul are akin to those of the body, for they really aren't there beforehand but are added later by habit and practice. However, the e virtue of reason seems to belong above all to something more divine,[3] which never loses its power but is either useful and beneficial or useless and harmful, depending on the way it is turned. Or have you never noticed this about people who are said to be 519 vicious but clever, how keen the vision of their little souls is and how sharply it distinguishes the things it is turned towards? This shows that its sight isn't inferior but rather is forced to serve evil ends, so that the sharper it sees, the more evil it accomplishes.

Absolutely.

However, if a nature of this sort had been hammered at from childhood and freed from the bonds of kinship with becoming, which have been fastened to it by feasting, greed, and other such pleasures and which, like leaden weights, pull its vision downwards – if, being rid of these, it turned to look b at true things, then I say that the same soul of the same person would see these most sharply, just as it now does the things it is presently turned towards.

Probably so.

And what about the uneducated who have no experience of truth? Isn't it likely – indeed, doesn't it follow necessarily from what was said before – that they will never adequately govern a city? But neither would those who've been allowed to spend their whole lives being educated. The former would fail because they don't have a single goal at which all their actions, public and private, inevitably aim; the c latter would fail because they'd refuse to act, thinking that they had settled while still alive in the faraway Isles of the Blessed.[4]

That's true.

It is our task as founders, then, to compel the best natures to reach the study we said before is the most important, namely, to make the ascent and see the good. But when they've made it and looked sufficiently, we mustn't allow them to do what d they're allowed to do today.

What's that?

To stay there and refuse to go down again to the prisoners in the cave and share their labors and honors, whether they are of less worth or of greater.

Then are we to do them an injustice by making them live a worse life when they could live a better one?

You are forgetting again that it isn't the law's e concern to make any one class in the city outstandingly happy but to contrive to spread happiness throughout the city by bringing the citizens into harmony with each other through persuasion or compulsion and by making them share with each other the benefits that each class can confer on the 520 community.[5] The law produces such people in the city, not in order to allow them to turn in whatever direction they want, but to make use of them to bind the city together.

That's true, I had forgotten.

Observe, then, Glaucon, that we won't be doing an injustice to those who've become philosophers in our city and that what we'll say to them, when we compel them to guard and care for the others, will be just. We'll say: "When people like you come b to be in other cities, they're justified in not sharing in their city's labors, for they've grown there spontaneously, against the will of the constitution. And what grows of its own accord and owes no debt for its upbringing has justice on its side when it isn't keen to pay anyone for that upbringing. But we've made you kings in our city and leaders of the swarm, as it were, both for yourselves and for the rest of the city. You're better and more completely c educated than the others and are better able to share in both types of life.[6] Therefore each of you in turn must go down to live in the common dwelling place of the others and grow accustomed to seeing in the dark. When you are used to it, you'll see vastly better than the people there. And because you've seen the truth about fine, just, and good things, you'll know each image for what it is and also that of which it is the image. Thus, for you and for us, the city will be governed, not like the majority of cities nowadays, by people who fight over shadows and struggle against one another in order to rule – as if that were a great good – but by d people who are awake rather than dreaming,[7] for the truth is surely this: A city whose prospective rulers are least eager to rule must of necessity be most free from civil war, whereas a city with the opposite kind of rulers is governed in the opposite way."

Absolutely.

Then do you think that those we've nurtured will disobey us and refuse to share the labors of the city, each in turn, while living the greater part of their time with one another in the pure realm?

It isn't possible, for we'll be giving just orders to e just people. Each of them will certainly go to rule as to something compulsory, however, which is exactly the opposite of what's done by those who now rule in each city. This is how it is. If you can find a way of life that's better than ruling for the prospective rulers, your well-governed city will become a possibility, for only in it will the truly rich rule – not those who are rich in gold but those who are rich in the 521 wealth that the happy must have, namely, a good and rational life. But if beggars hungry for private goods go into public life, thinking that the good is there for the seizing, then the well-governed city is impossible, for then ruling is something fought over, and this civil and domestic war destroys these people and the rest of the city as well.

That's very true.

Can you name any life that despises political rule besides that of the true philosopher? b

No, by god, I can't.

But surely it is those who are not lovers of ruling who must rule, for if they don't, the lovers of it, who are rivals, will fight over it.

Of course.

Then who will you compel to become guardians of the city, if not those who have the best understanding of what matters for good government and who have other honors than political ones, and a better life as well?

No one.

Do you want us to consider now how such people will come to be in our city and how – just as some are said to have gone up from Hades to the c gods – we'll lead them up to the light?

[...]

Then it would be appropriate, Glaucon, to le- 525 gislate this subject for those who are going to share in the highest offices in the city and to persuade them to turn to calculation and take it up, not as laymen do, but staying with it until they reach the study of the natures of the numbers by means of c understanding itself, nor like tradesmen and retailers, for the sake of buying and selling, but for the sake of war and for ease in turning the soul around, away from becoming and towards truth and being.

Well put.

Moreover, it strikes me, now that it has been mentioned, how sophisticated the subject of calculation is and in how many ways it is useful for our d purposes, provided that one practices it for the sake of knowing rather than trading.

How is it useful?

In the very way we were talking about. It leads the soul forcibly upward and compels it to discuss the numbers themselves, never permitting anyone to propose for discussion numbers attached to visible or tangible bodies. You know what those who are clever in these matters are like: If, in the course of the argument, someone tries to divide the e one itself, they laugh and won't permit it. If you divide it, they multiply it, taking care that one thing never be found to be many parts rather than one.

That's very true.

Then what do you think would happen, Glaucon, 526 if someone were to ask them: "What kind of numbers are you talking about, in which the one is as you assume it to be, each one equal to every other, without the least difference and containing no internal parts?"

I think they'd answer that they are talking about those numbers that can be grasped only in thought and can't be dealt with in any other way.

b Then do you see that it's likely that this subject really is compulsory for us, since it apparently compels the soul to use understanding itself on the truth itself?

Indeed, it most certainly does do that.

And what about those who are naturally good at calculation or reasoning? Have you already noticed that they're naturally sharp, so to speak, in all subjects, and that those who are slow at it, if they're educated and exercised in it, even if they're benefited in no other way, nonetheless improve and become generally sharper than they were?

That's true.

Moreover, I don't think you'll easily find subjects c that are harder to learn or practice than this.

No, indeed.

Then, for all these reasons, this subject isn't to be neglected, and the best natures must be educated in it.

I agree.

Let that, then, be one of our subjects. Second, let's consider whether the subject that comes next is also appropriate for our purposes.

What subject is that? Do you mean geometry?

That's the very one I had in mind.

d Insofar as it pertains to war, it's obviously appropriate, for when it comes to setting up camp, occupying a region, concentrating troops, deploying them, or with regard to any of the other formations an army adopts in battle or on the march, it makes all the difference whether someone is a geometer or not.

But, for things like that, even a little geometry – or calculation for that matter – would suffice. What we need to consider is whether the greater and more advanced part of it tends to make it easier to see the form of the good. And we say that anything has that tendency if it compels the soul e to turn itself around towards the region in which lies the happiest of the things that are, the one the soul must see at any cost.

You're right.

Therefore, if geometry compels the soul to study being, it's appropriate, but if it compels it to study becoming, it's inappropriate.

So we've said, at any rate.

Now, no one with even a little experience of geometry will dispute that this science is entirely 527 the opposite of what is said about it in the accounts of its practitioners.

How do you mean?

They give ridiculous accounts of it, though they can't help it, for they speak like practical men, and all their accounts refer to doing things. They talk of "squaring," "applying," "adding," and the like, whereas the entire subject is pursued for the sake of knowledge. b

Absolutely.

And mustn't we also agree on a further point?

What is that?

That it is knowledge of what always is, not of what comes into being and passes away.

That's easy to agree to, for geometry *is* knowledge of what always is.

Then it draws the soul towards truth and produces philosophic thought by directing upwards what we now wrongly direct downwards.

As far as anything possibly can.

Then as far as *we* possibly can, we must require those in your fine city not to neglect geometry in any c way, for even its by-products are not insignificant.

What are they?

The ones concerned with war that you mentioned. But we also surely know that, when it comes to better understanding any subject, there is a world of difference between someone who has grasped geometry and someone who hasn't.

Yes, by god, a world of difference.

Then shall we set this down as a second subject for the young?

Let's do so, he said.

And what about astronomy? Shall we make it the third? Or do you disagree? d

That's fine with me, for a better awareness of the seasons, months, and years is no less appropriate for a general than for a farmer or navigator.

You amuse me: You're like someone who's afraid that the majority will think he is prescribing useless subjects. It's no easy task – indeed it's very difficult – to realize that in every soul there is an instrument that is purified and rekindled by such subjects when e it has been blinded and destroyed by other ways of life, an instrument that it is more important to preserve than ten thousand eyes, since only with it can the truth be seen. Those who share your belief that this is so will think you're speaking incredibly well, while those who've never been aware of it will probably think you're talking nonsense, since they see no benefit worth mentioning in these subjects. So decide right now which group you're address- 528 ing. Or are your arguments for neither of them but mostly for your own sake – though you won't begrudge anyone else whatever benefit he's able to get from them?

The latter: I want to speak, question, and answer mostly for my own sake.

Then let's fall back to our earlier position, for we were wrong just now about the subject that comes after geometry.

What was our error?

After plane surfaces, we went on to revolving solids before dealing with solids by themselves. But the right thing to do is to take up the third b dimension right after the second. And this, I suppose, consists of cubes and of whatever shares in depth.

You're right, Socrates, but this subject hasn't been developed yet.

There are two reasons for that: First, because no city values it, this difficult subject is little researched. Second, the researchers need a director, for, without one, they won't discover anything. To begin with, such a director is hard to find, and, then, c even if he could be found, those who currently do research in this field would be too arrogant to follow him. If an entire city helped him to supervise it, however, and took the lead in valuing it, then he would be followed. And, if the subject was consistently and vigorously pursued, it would soon be developed. Even now, when it isn't valued and is held in contempt by the majority and is pursued by researchers who are unable to give an account of its usefulness, nevertheless, in spite of all these handicaps, the force of its charm has caused it to develop somewhat, so that it wouldn't be surprising if it were further developed even as things stand.

The subject *has* outstanding charm. But explain d more clearly what you were saying just now. The subject that deals with plane surfaces you took to be geometry.

Yes.

And at first you put astronomy after it, but later you went back on that.

In my haste to go through them all, I've only progressed more slowly. The subject dealing with the dimension of depth was next. But because it is in a ridiculous state, I passed it by and spoke of astronomy (which deals with the motion of things having depth) after geometry. e

That's right.

Let's then put astronomy as the fourth subject, on the assumption that solid geometry will be available if a city takes it up.

That seems reasonable. And since you reproached me before for praising astronomy in a vulgar manner, I'll now praise it your way, for I think it's clear to everyone that astronomy compels the soul to look upward and leads it from things 529 here to things there.

It may be obvious to everyone except me, but that's not my view about it.

Then what *is* your view?

As it's practiced today by those who teach philosophy, it makes the soul look very much downward.

How do you mean?

In my opinion, your conception of "higher studies" is a good deal too generous, for if someone were to study something by leaning his head back and studying ornaments on a ceiling, it looks as though you'd say he's studying not with his eyes but with his understanding. Perhaps you're right, b and I'm foolish, but I can't conceive of any subject making the soul look upward except one concerned with that which is, and that which is is invisible. If anyone attempts to learn something about sensible things, whether by gaping upward or squinting downward, I'd claim – since there's no knowledge of such things – that he never learns anything and that, even if he studies lying on his back on the ground or floating on it in the sea, his soul is c looking not up but down.

You're right to reproach me, and I've been justly punished, but what did you mean when you said that astronomy must be learned in a different way from the way in which it is learned at present if it is to be a useful subject for our purposes?

It's like this: We should consider the ornaments that brighten the sky to be the most beautiful and most exact of visible things, seeing that they're embroidered on a visible surface. But we should consider their motions to fall far short of the true d ones – motions that are really fast or slow as measured in true numbers, that trace out true geometrical figures, that are all in relation to one another, and that are the true motions of the things carried along in them. And these, of course, must be grasped by reason and thought, not by sight. Or do you think otherwise?

Not at all.

Therefore, we should use the embroidery in the sky as a model in the study of these other things.[8] If someone experienced in geometry were to come upon plans very carefully drawn and worked out e by Daedalus or some other craftsman or artist, he'd consider them to be very finely executed, but he'd think it ridiculous to examine them seriously in 530 order to find the truth in them about the equal, the double, or any other ratio.

How could it be anything other than ridiculous?

Then don't you think that a real astronomer will feel the same when he looks at the motions of the stars? He'll believe that the craftsman of the heavens arranged them and all that's in them in the finest way possible for such things. But as for the ratio of night to day, of days to a month, of a month to a year, or of the motions of the stars to any of them or to each other, don't you think he'll consider it strange b to believe that they're always the same and never deviate anywhere at all or to try in any sort of way to grasp the truth about them, since they're connected to bodies and visible?

That's my opinion anyway, now that I hear it from you.

Then if, by really taking part in astronomy, we're to make the naturally intelligent part of the soul useful instead of useless, let's study astronomy by means of problems, as we do geometry, and c leave the things in the sky alone.

The task you're prescribing is a lot harder than anything now attempted in astronomy.

And I suppose that, if we are to be of any benefit as lawgivers, our prescriptions for the other subjects will be of the same kind. But have you any other appropriate subject to suggest?

Not offhand.

Well, there isn't just one form of motion but d several. Perhaps a wise person could list them all, but there are two that are evident even to us.

What are they?

Besides the one we've discussed, there is also its counterpart.

What's that?

It's likely that, as the eyes fasten on astronomical motions, so the ears fasten on harmonic ones, and that the sciences of astronomy and harmonics are closely akin. This is what the Pythagoreans[9] say, Glaucon, and we agree, don't we?

We do.

Therefore, since the subject is so huge, shouldn't we ask them what they have to say about harmonic e motions and whether there is anything else besides them, all the while keeping our own goal squarely in view?

What's that?

That those whom we are rearing should never try to learn anything incomplete, anything that doesn't reach the end that everything should reach – the end we mentioned just now in the case of astronomy. Or don't you know that people do something similar in harmonics? Measuring 531 audible consonances and sounds against one another, they labor in vain, just like present-day astronomers.

Yes, by the gods, and pretty ridiculous they are too. They talk about something they call a "dense interval" or quartertone[10] – putting their ears to their instruments like someone trying to overhear what the neighbors are saying. And some say that they hear a tone in between and that *it* is the shortest interval by which they must measure, while others argue that this tone sounds the same as a quarter tone. Both put ears before understanding. b

You mean those excellent fellows who torment their strings, torturing them, and stretching them on pegs. I won't draw out the analogy by speaking of blows with the plectrum or the accusations or denials and boastings on the part of the strings; instead I'll cut it short by saying that these aren't the people I'm talking about. The ones I mean are the ones we just said we were going to question about harmonics, for they do the same as the astronomers. They seek out the numbers that are to be found in these audible harmonies, but they do c not make the ascent to problems. They don't investigate, for example, which numbers are in harmony and which aren't or what the explanation is of each.

But that would be a superhuman task.

Yet it's useful in the search for the beautiful and the good. But pursued for any other purpose, it's useless.

Probably so.

Moreover, I take it that, if inquiry into all the subjects we've mentioned brings out their association and relationship with one another and draws conclusions about their kinship, it does contribute d something to our goal and isn't labor in vain, but that otherwise it is in vain.

I, too, divine that this is true. But you're still talking about a very big task, Socrates.

Do you mean the prelude, or what? Or don't you know that all these subjects are merely preludes to the song itself that must also be learned? Surely you don't think that people who are clever in these e matters are dialecticians.

No, by god, I don't. Although I have met a few exceptions.

But did it ever seem to you that those who can neither give nor follow an account know anything at all of the things we say they must know?

My answer to that is also no.

532 Then isn't this at last, Glaucon, the song that dialectic sings? It is intelligible, but it is imitated by the power of sight. We said that sight tries at last to look at the animals themselves, the stars themselves, and, in the end, at the sun itself.[11] In the same way, whenever someone tries through argument and apart from all sense perceptions to find the being itself of each thing and doesn't give up b until he grasps the good itself with understanding itself, he reaches the end of the intelligible, just as the other reached the end of the visible.

Absolutely.

And what about this journey? Don't you call it dialectic?

I do.

Then the release from bonds and the turning around from shadows to statues and the light of the fire and, then, the way up out of the cave to the sunlight and, there, the continuing inability to look at the animals, the plants, and the light of the sun, c but the newly acquired ability to look at divine images in water and shadows of the things that are, rather than, as before, merely at shadows of statues thrown by another source of light that is itself a shadow in relation to the sun – all this business of the crafts we've mentioned has the power to awaken the best part of the soul and lead it upward to the study of the best among the things that are, just as, before, the clearest thing in d the body was led to the brightest thing in the bodily and visible realm.

I accept that this is so, even though it seems very hard to accept in one way and hard not to accept in another. All the same, since we'll have to return to these things often in the future, rather than having to hear them just once now, let's assume that what you've said is so and turn to the song itself, discussing it in the same way as we did the prelude. So tell us the way in which the power of dialectic works, what forms it is divided into, and what e paths it follows, for these lead at last, it seems, towards that place which is a rest from the road, so to speak, and an end of journeying for the one who reaches it.

You won't be able to follow me any longer, 533 Glaucon, even though there is no lack of eagerness on my part to lead you, for you would no longer be seeing an image of what we're describing, but the truth itself. At any rate, that's how it seems to me. That it is really so is not worth insisting on any further. But that there is some such thing to be seen, *that* is something we must insist on. Isn't that so?

Of course.

And mustn't we also insist that the power of dialectic could reveal it only to someone experienced in the subjects we've described and that it cannot reveal it in any other way?

That too is worth insisting on.

At any rate, no one will dispute it when we say that there is no other inquiry that systematically attempts b to grasp with respect to each thing itself what the being of it is, for all the other crafts are concerned with human opinions and desires, with growing or construction, or with the care of growing or constructed things. And as for the rest, I mean geometry and the subjects that follow it, we described them as to some extent grasping what is, for we saw that, while they do dream about what is, they are unable to command a waking view of it as long as they make use of hypotheses that they leave untouched and that they cannot give any account of. What mechanism c could possibly turn any agreement into knowledge when it begins with something unknown and puts together the conclusion and the steps in between from what is unknown?

None.

Therefore, dialectic is the only inquiry that travels this road, doing away with hypotheses and proceeding to the first principle itself, so as to be d secure. And when the eye of the soul is really buried in a sort of barbaric bog,[12] dialectic gently pulls it out and leads it upwards, using the crafts we described to help it and cooperate with it in turning the soul around. From force of habit, we've often called these crafts sciences or kinds of knowledge, but they need another name, clearer than opinion, darker than knowledge. We called them thought somewhere before.[13] But I presume that

we won't dispute about a name when we have so
e many more important matters to investigate.

Of course not.

It will therefore be enough to call the first section knowledge, the second thought, the third belief, and the fourth imaging, just as we did before. The last two together we call opinion, the
534 other two, intellect.[14] Opinion is concerned with becoming, intellect with being. And as being is to becoming, so intellect is to opinion, and as intellect is to opinion, so knowledge is to belief and thought to imaging. But as for the ratios between the things these are set over and the division of either the opinable or the intelligible section into two, let's pass them by, Glaucon, lest they involve us in arguments many times longer than the ones we've already gone through.

I agree with you about the others in any case,
b insofar as I'm able to follow.

Then, do you call someone who is able to give an account of the being of each thing dialectical? But insofar as he's unable to give an account of something, either to himself or to another, do you deny that he has any understanding of it?

How could I do anything else?

Then the same applies to the good. Unless someone can distinguish in an account the form
c of the good from everything else, can survive all refutation, as if in a battle, striving to judge things not in accordance with opinion but in accordance with being, and can come through all this with his account still intact, you'll say that he doesn't know the good itself or any other good. And if he gets hold of some image of it, you'll say that it's through opinion, not knowledge, for he is dreaming and asleep throughout his present life, and, before he
d wakes up here, he will arrive in Hades and go to sleep forever.

Yes, by god, I'll certainly say all of that.

Then, as for those children of yours whom you're rearing and educating in theory, if you ever reared them in fact, I don't think that you'd allow them to rule in your city or be responsible for the most important things while they are as irrational as incommensurable lines.

Certainly not.

Then you'll legislate that they are to give most attention to the education that will enable them to ask and answer questions most knowledgeably?

e I'll legislate it along with you.

Then do you think that we've placed dialectic at the top of the other subjects like a coping stone and that no other subject can rightly be placed above it, but that our account of the subjects that a future ruler must learn has come to an end? 535

Probably so.

[...]

We hold from childhood certain convictions about just and fine things; we're brought up with them as with our parents, we obey and honor them. 538

Indeed, we do.

There are other ways of living, however, opposite to these and full of pleasures, that flatter the soul and attract it to themselves but which don't persuade sensible people, who continue to honor and obey the convictions of their fathers. d

That's right.

And then a questioner comes along and asks someone of this sort, "What is the fine?" And, when he answers what he has heard from the traditional lawgiver, the argument refutes him, and by refuting him often and in many places shakes him from his convictions, and makes him believe that the fine is no more fine than shameful, and the same with the just, the good, and the things he honored most. What do you think his attitude will be then to honoring and obeying his earlier convictions? e

Of necessity he won't honor or obey them in the same way.

Then, when he no longer honors and obeys those convictions and can't discover the true ones, will he be likely to adopt any other way of life than that which flatters him? 539

No, he won't.

And so, I suppose, from being law-abiding he becomes lawless.

Inevitably.

Then, as I asked before, isn't it only to be expected that this is what happens to those who take up arguments in this way, and don't they therefore deserve a lot of sympathy?

Yes, and they deserve pity too.

Then, if you don't want your thirty-year-olds to be objects of such pity, you'll have to be extremely careful about how you introduce them to arguments.

That's right.

And isn't it one lasting precaution not to let them taste arguments while they're young? I don't suppose that it has escaped your notice that, when young people get their first taste of arguments, they misuse it by treating it as a kind of game of contradiction. They imitate those who've refuted them by refuting others themselves, and, like puppies, they enjoy dragging and tearing those around them with their arguments. b

They're excessively fond of it.

Then, when they've refuted many and been refuted by them in turn, they forcefully and quickly fall into disbelieving what they believed before. And, as a result, they themselves and the c whole of philosophy are discredited in the eyes of others.

That's very true.

But an older person won't want to take part in such madness. He'll imitate someone who is willing to engage in discussion in order to look for the truth, rather than someone who plays at contradiction for sport. He'll be more sensible himself and will bring honor rather than discredit d to the philosophical way of life.

That's right.

And when we said before that those allowed to take part in arguments should be orderly and steady by nature, not as nowadays, when even the unfit are allowed to engage in them – wasn't all that also said as a precaution?

Of course.

Then if someone continuously, strenuously, and exclusively devotes himself to participation in arguments, exercising himself in them just as he did in the bodily physical training, which is their counterpart, would that be enough?

e Do you mean six years or four?

It doesn't matter. Make it five. And after that, you must make them go down into the cave again, and compel them to take command in matters of war and occupy the other offices suitable for young people, so that they won't be inferior to the others in experience. But in these, too, they must be tested to see whether they'll remain steadfast when they're pulled this way and that or shift their ground. 540

How much time do you allow for that?

Fifteen years. Then, at the age of fifty, those who've survived the tests and been successful both in practical matters and in the sciences must be led to the goal and compelled to lift up the radiant light of their souls to what itself provides light for everything. And once they've seen the good itself, they must each in turn put the city, its citizens, and themselves in order, using it as their model. Each b of them will spend most of his time with philosophy, but, when his turn comes, he must labor in politics and rule for the city's sake, not as if he were doing something fine, but rather something that has to be done. Then, having educated others like himself to take his place as guardians of the city, he will depart for the Isles of the Blessed and dwell there. And, if the Pythia agrees, the city will publicly establish memorials and sacrifices to him as a c daimon, but if not, then as a happy and divine human being.

Like a sculptor,[15] Socrates, you've produced ruling men that are completely fine.

And ruling women, too, Glaucon, for you mustn't think that what I've said applies any more to men than it does to women who are born with the appropriate natures.

That's right, if indeed they are to share everything equally with the men, as we said they should. [...]

Notes

1 Reading *parionta autous nomizein onomazein*. E.g. they would think that the name "human being" applied to the shadow of a statue of a human being.

2 *Odyssey* 11.489–90. The shade of the dead Achilles speaks these words to Odysseus, who is visiting Hades. Plato is, therefore, likening the cave dwellers to the dead.

3 See *Republic* 589d, 590d, 611b ff.

4 A place where good people are said to live in eternal happiness, normally after death.

5 See *Republic* 420b–421c, 462a–466c.

6 I.e. the practical life of ruling the city and the theoretical life of studying the good itself.

7 See *Republic* 476c–d.

8 See *Republic* 510d–511a.

9 Pythagoras of Samos (sixth century) taught a way of life (see *Republic* 600b) in which natural science became a religion. He is credited with discovering the mathematical ratios determining the principal intervals of the musical scale. He seems to have been led by this to believe that all natural phenomena are explicable in terms of numbers. He may have discovered some version of the theorem about right triangles that bears his name.

10 A dense interval is evidently the smallest difference in pitch recognized in ancient music.

11 See *Republic* 516a–b.

12 See *Republic* 519a–b.

13 See *Republic* 511d–e.

14 The reference is to *Republic* 511d–e, but there the first section is called understanding (*noēsis*) rather than knowledge (*epistēmē*). However, since we've just been told that thought (*dianoia*) is not a kind of knowledge, understanding and knowing have in

effect become identified. It is harder to explain why knowledge and thought are now referred to jointly as *noēsis*. But presumably it is because that whole section of the line is earlier referred to as the intelligible (*noēton*). See *Republic* 509d–e. To prevent misunderstanding, therefore, I have translated *noēsis* as "intellect" here.

15 See *Republic* 361d.

2

On "Technē" and "Epistēmē"

Aristotle

From *Nichomachean Ethics*

6.15 The particular virtues of thought

Then let us begin over again, and discuss these states of the soul. Let us say, then, that there are five states in which the soul grasps the truth in its affirmations or denials. These are craft, scientific knowledge, intelligence, wisdom and understanding; for belief and supposition admit of being false.

6.2 Scientific Knowledge (Epistēmē)

6.21 It is concerned with what is necessary What science is is evident from the following, if we must speak exactly and not be guided by [mere] similarities.

For we all suppose that what we know scientifically does not even admit of being otherwise; and whenever what admits of being otherwise escapes observation, we do not notice whether it is or is not, [and hence we do not know about it]. Hence what is known scientifically is by necessity. Hence it is eternal; for the things that are by unconditional necessity are all eternal, and eternal things are ingenerable and indestructible.

6.22 Its first principles cannot be scientifically known Further, every science seems to be teachable, and what is scientifically knowable is learnable. But all teaching is from what is already known, as we also say in the *Analytics*; for some teaching is through induction, some by deductive inference, [which both require previous knowledge].

Induction [reaches] the origin, i.e. the universal, while deductive inference proceeds from the universals. Hence deductive inference has origins from which it proceeds, but which are not themselves [reached] by deductive inference. Hence they are [reached] by induction.

6.23 Hence scientific knowledge requires demonstration from indemonstrable premises Scientific knowledge, then, is a demonstrative state, and has all the other features that in the *Analytics* we add to the definition. For someone has scientific knowledge when he has the appropriate sort of confidence, and the origins are known to him; for if they are not better known to him than the conclusion, he will have scientific knowledge only coincidentally.

So much for a definition of scientific knowledge.

6.3 Craft-knowledge (Technē)

6.31 Production contrasted with action What admits of being otherwise includes what is produced and what is done in action. Production and action are different; about them we rely also on

Selections from Aristotle, *Nichomachean Ethics* VI, trans. Terence Irwin, Indianapolis: Hackett, 1985, pp. 151–9; and *Metaphysics* I, 1, trans. W. D. Ross, in *The Basic Works of Aristotle*, ed. Richard McKeon, New York: Random House, 1941, pp. 689–93, Copyright © Oxford University Press.

[our] popular discussions. Hence the state involving reason and concerned with action is different from the state involving reason and concerned with production. Nor is one included in the other; for action is not production, and production is not action.

6.32 Crafts are concerned with production, not with action

6.32 Crafts are concerned with production, not with action Now building, e.g., is a craft, and is essentially a certain state involving reason concerned with production; there is no craft that is not a state involving reason concerned with production, and no such state that is not a craft. Hence a craft is the same as a state involving true reason concerned with production.

Every craft is concerned with coming to be; and the exercise of the craft is the study of how something that admits of being and not being comes to be, something whose origin is in the producer and not in the product. For a craft is not concerned with things that are or come to be by necessity; or with things that are by nature, since these have their origin in themselves.

And since production and action are different, craft must be concerned with production, not with action.

In a way craft and fortune are concerned with the same things, as Agathon says: "Craft was fond of fortune, and fortune of craft."

A craft, then, as we have said, is a state involving true reason concerned with production. Lack of craft is the contrary state involving false reason and concerned with production. Both are concerned with what admits of being otherwise.

6.4 *Intelligence (Phronesis)*

6.41 An intelligent person deliberates about living well To grasp what intelligence is we should first study the sort of people we call intelligent.

It seems proper, then, to an intelligent person to be able to deliberate finely about what is good and beneficial for himself, not about some restricted area – e.g. about what promotes health or strength – but about what promotes living well in general.

A sign of this is the fact that we call people intelligent about some [restricted area] whenever they calculate well to promote some excellent end, in an area where there is no craft. Hence where [living well] as a whole is concerned, the deliberative person will also be intelligent.

6.42 The scope of intelligence distinguishes it from scientific knowledge and from craft-knowledge Now no one deliberates about what cannot be otherwise or about what cannot be achieved by his action. Hence, if science involves demonstration, but there is no demonstration of anything whose origins admit of being otherwise, since every such thing itself admits of being otherwise; and if we cannot deliberate about what is by necessity; it follows that intelligence is not science nor yet craft-knowledge. It is not science, because what is done in action admits of being otherwise; and it is not craft-knowledge, because action and production belong to different kinds.

6.43 Definition of intelligence The remaining possibility, then, is that intelligence is a state grasping the truth, involving reason, concerned with action about what is good or bad for a human being.

It must be concerned with action, not with production For production has its end beyond it; but action does not, since its end is doing well itself, [and doing well is the concern of intelligence].

6.44 The definition is confirmed by commonly recognized features of intelligence

Commonly recognized intelligent people Hence Pericles and such people are the ones whom we regard as intelligent, because they are able to study what is good for themselves and for human beings; and we think that household managers and politicians are such people.

The recognized connection between temperance and intelligence This is also how we come to give temperance (*sōphrosunē*) its name, because we think that it preserves intelligence, (*sōzousan tēn phronēsin*). This is the sort of supposition that it preserves. For the sort of supposition that is corrupted and perverted by what is pleasant or painful is not every sort – not, e.g., the supposition that the triangle does or does not have two right angles – but suppositions about what is done in action.

For the origin of what is done in action is the goal it aims at; and if pleasure or pain has corrupted someone, it follows that the origin will not appear to him. Hence it will not be apparent that this must be the goal and cause of all his choice and action; for vice corrupts the origin.

Hence [since intelligence is what temperance preserves, and what temperance preserves is a true supposition about action], intelligence must be a state grasping the truth, involving reason, and concerned with action about human goods.

We recognize that intelligence cannot be misused... Moreover, there is virtue [or vice in the use] of craft, but not [in the use] of intelligence. Further, in a craft, someone who makes errors voluntarily is more choiceworthy; but with intelligence, as with the virtues, the reverse is true. Clearly, then, intelligence is a virtue, not craft-knowledge.

There are two parts of the soul that have reason. Intelligence is a virtue of one of them, of the part that has belief; for belief is concerned, as intelligence is, with what admits of being otherwise.

...And that it cannot be forgotten Moreover, it is not only a state involving reason. A sign of this is the fact that such a state can be forgotten, but intelligence cannot.

6.5 *Understanding (Nous)*

6.51 There must be a virtue of thought concerned with first principles Scientific knowledge is supposition about universals, things that are by necessity. Further, everything demonstrable and every science have origins, since scientific knowledge involves reason.

Hence there can be neither scientific knowledge nor craft-knowledge nor intelligence about the origins of what is scientifically known. For what is scientifically known is demonstrable, [but the origins are not]; and craft and intelligence are about what admits of being otherwise. Nor is wisdom [exclusively] about origins; for it is proper to the wise person to have a demonstration of some things.

6.52 Since no other virtue of thought grasps first principles, understanding must grasp them [The states of the soul] by which we always grasp the truth and never make mistakes, about what can or cannot be otherwise, are scientific knowledge, intelligence, wisdom and understanding. But none of the first three – intelligence, scientific knowledge, wisdom – is possible about origins. The remaining possibility, then, is that we have understanding about origins.

6.6 *Wisdom (Sophia)*

6.61 It is concerned with scientific knowledge and with understanding, not with action We ascribe wisdom in crafts to the people who have the most exact expertise in the crafts, e.g. we call Pheidias a wise stone-worker and Polycleitus a wise bronze-worker, signifying nothing else by wisdom than excellence in a craft. But we also think some people are wise in general, not wise in some [restricted] area, or in some other [specific] way, as Homer says in the *Margites*: "The gods did not make him a digger or a ploughman or wise in anything else." Clearly, then, wisdom is the most exact [form] of scientific knowledge.

Hence the wise person must not only know what is derived from the origins of a science, but also grasp the truth about the origins. Therefore wisdom is understanding plus scientific knowledge; it is scientific knowledge of the most honourable things that has received [understanding as] its coping-stone.

6.62 It must be distinguished from intelligence and political science For it would be absurd for someone to think that political science or intelligence is the most excellent science, when the best thing in the universe is not a human being [and the most excellent science must be of the best things].

Moreover, what is good and healthy for human beings and for fish is not the same, but what is white or straight is always the same. Hence everyone would say that the content of wisdom is always the same, but the content of intelligence is not. For the agent they would call intelligent is the one who studies well each question about his own [good], and he is the one to whom they would entrust such questions. Hence intelligence is also ascribed to some of the beasts, the ones that are evidently capable of forethought about their own life.

It is also evident that wisdom is not the same as political science. For if people are to say that sci- 30 ence about what is beneficial to themselves [as human beings] counts as wisdom, there will be many types of wisdom [corresponding to the different species of animals]. For if there is no one medical science about all beings, there is no one science about the good of all animals, but a different science about each specific good. [Hence there will be many types of wisdom, contrary to our assumption that it has always the same content].

And it does not matter if human beings are the best among the animals. For there are other beings 1141b of a far more divine nature than human beings; e.g., most evidently, the beings composing the universe.

What we have said makes it clear that wisdom is both scientific knowledge and understanding about what is by nature most honourable. That is why 5 people say that Anaxagoras or Thales or that sort of person is wise, but not intelligent, when they see that he is ignorant of what benefits himself. And so they say that what he knows is extraordinary, amazing, difficult and divine, but useless, because it is not human goods that he looks for.

6.7 Intelligence Compared with the Other Virtues of Thought

6.71 It is concerned with action, and hence with particulars Intelligence, by con- 10 trast, is about human concerns, about what is open to deliberation. For we say that deliberating well is the function of the intelligent person more than anyone else; but no one deliberates about what cannot be otherwise, or about what lacks a goal that is a good achievable in action. The unconditionally good deliberator is the one whose aim expresses rational calculation in pursuit of the best good for a human being that is achievable in action.

15 Nor is intelligence about universals only. It must also come to know particulars, since it is concerned with action and action is about particulars. Hence in other areas also some people who lack knowledge but have experience are better in action than others who have knowledge. For someone who knows that 20 light meats are digestible and healthy, but not which sorts of meats are light, will not produce health; the one who knows that bird meats are healthy will be better at producing health. And since intelligence is concerned with action, it must possess both [the universal and the particular knowledge] or the [particular] more [than the universal]. Here too, however, [as in medicine] there is a ruling [science].

From *Metaphysics*

Book A (I)

1 All men by nature desire to know. An indication of this is the delight we take in our senses; for 980a even apart from their usefulness they are loved for themselves; and above all others the sense of sight. For not only with a view to action, but even when we are not going to do anything, we prefer seeing 25 (one might say) to everything else. The reason is that this, most of all the senses, makes us know and brings to light many differences between things.

By nature animals are born with the faculty of sensation, and from sensation memory is produced in some of them, though not in others. And therefore the former are more intelligent and apt at learning than those which cannot remember; those 980b which are incapable of hearing sounds are intelligent though they cannot be taught, e.g. the bee, and any other race of animals that may be like it; and those which besides memory have this sense of hearing can be taught.

The animals other than man live by appearances and memories, and have but little of connected 25 experience; but the human race lives also by art and reasonings. Now from memory experience is produced in men; for the several memories of the same thing produce finally the capacity for a single experience. And experience seems pretty much like 981a science and art (technē), but really science and art come to men *through* experience; for "experience made art", as Polus says,[1] "but inexperience luck". Now art arises when from many notions gained by 5 experience one universal judgement about a class of objects is produced. For to have a judgement that when Callias was ill of this disease this did him good, and similarly in the case of Socrates and in many individual cases, is a matter of experience; but to judge that it has done good to all persons of a certain constitution, marked off in one class, when 10 they were ill of this disease, e.g. to phlegmatic or bilious people when burning with fever – this is a matter of art.

With a view to action experience seems in no respect inferior to art, and men of experience succeed even better than those who have theory without experience. (The reason is that experience is knowledge of individuals, art of universals, and actions and productions are all concerned with the individual; for the physician does not cure *man*, except in an incidental way, but Callias or Socrates or some other called by some such individual name, who happens to be a man. If, then, a man has the theory without the experience, and recognizes the universal but does not know the individual included in this, he will often fail to cure; for it is the individual that is to be cured.) But yet we think that *knowledge* and *understanding* belong to art rather than to experience, and we suppose artists to be wiser than men of experience (which implies that Wisdom depends in all cases rather on knowledge); and this because the former know the cause, but the latter do not. For men of experience know that the thing is so, but do not know why, while the others know the "why" and the cause. Hence we think also that the master-workers in each craft are more honourable and know in a truer sense and are wiser than the manual workers, because they know the causes of the things that are done (we think the manual workers are like certain lifeless things which act indeed, but act without knowing what they do, as fire burns – but while the lifeless things perform each of their functions by a natural tendency, the labourers perform them through habit); thus we view them as being wiser not in virtue of being able to act, but of having the theory for themselves and knowing the causes. And in general it is a sign of the man who knows and of the man who does not know, that the former can teach, and therefore we think art more truly knowledge than experience is; for artists can teach, and men of mere experience cannot.

Again, we do not regard any of the senses as Wisdom; yet surely these give the most authoritative knowledge of particulars. But they do not tell us the "why" of anything – e.g. why fire is hot; they only say *that* it is hot.

At first he who invented any art whatever that went beyond the common perceptions of man was naturally admired by men, not only because there was something useful in the inventions, but because he was thought wise and superior to the rest. But as more arts were invented, and some were directed to the necessities of life, others to recreation, the inventors of the latter were naturally always regarded as wiser than the inventors of the former, because their branches of knowledge did not aim at utility. Hence when all such inventions were already established, the sciences which do not aim at giving pleasure or at the necessities of life were discovered, and first in the places where men first began to have leisure. This is why the mathematical arts were founded in Egypt; for there the priestly caste was allowed to be at leisure.

We have said in the *Ethics*[2] what the difference is between art and science and the other kindred faculties; but the point of our present discussion is this, that all men suppose what is called Wisdom to deal with the first causes and the principles of things; so that, as has been said before, the man of experience is thought to be wiser than the possessors of any sense-perception whatever, the artist wiser than the men of experience, the master-worker than the mechanic, and the theoretical kinds of knowledge to be more of the nature of Wisdom than the productive. Clearly then Wisdom is knowledge about certain principles and causes.

2 Since we are seeking this knowledge, we must inquire of what kind are the causes and the principles, the knowledge of which is Wisdom. If one were to take the notions we have about the wise man, this might perhaps make the answer more evident. We suppose first, then, that the wise man knows all things, as far as possible, although he has not knowledge of each of them in detail; secondly, that he who can learn things that are difficult, and not easy for man to know, is wise (sense-perception is common to all, and therefore easy and no mark of Wisdom); again, that he who is more exact and more capable of teaching the causes is wiser, in every branch of knowledge; and that of the sciences, also, that which is desirable on its own account and for the sake of knowing it is more of the nature of Wisdom than that which is desirable on account of its results, and the superior science is more of the nature of Wisdom than the ancillary; for the wise man must not be ordered but must order, and he must not obey another, but the less wise must obey *him*.

Such and so many are the notions, then, which we have about Wisdom and the wise. Now of these characteristics that of knowing all things must belong to him who has in the highest degree universal knowledge; for he knows in a sense all the instances that fall under the universal. And these things, the most universal, are on the whole the hardest for men to know; for they are farthest from the senses. And the most exact of the sciences are those which deal most with first principles; for those which involve fewer principles are more exact than those which involve additional principles, e.g. arithmetic than geometry. But the science which investigates causes is also *instructive*, in a higher degree, for the people who instruct us are those who tell the causes of each thing. And understanding and knowledge pursued for their own sake are found most in the knowledge of that which is most knowable (for he who chooses to know for the sake of knowing will choose most readily that which is most truly knowledge, and such is the knowledge of that which is most knowable); and the first principles and the causes are most knowable; for by reason of these, and from these, all other things come to be known, and not these by means of the things subordinate to them. And the science which knows to what end each thing must be done is the most authoritative of the sciences, and more authoritative than any ancillary science; and this end is the good of that thing, and in general the supreme good in the whole of nature. Judged by all the tests we have mentioned, then, the name in question falls to the same science; this must be a science that investigates the first principles and causes; for the good, i.e. the end, is one of the causes.

That it is not a science of production is clear even from the history of the earliest philosophers. For it is owing to their wonder that men both now begin and at first began to philosophize; they wondered originally at the obvious difficulties, then advanced little by little and stated difficulties about the greater matters, e.g. about the phenomena of the moon and those of the sun and of the stars, and about the genesis of the universe. And a man who is puzzled and wonders thinks himself ignorant (whence even the lover of myth is in a sense a lover of Wisdom, for the myth is composed of wonders); therefore since they philosophized in order to escape from ignorance, evidently they were pursuing science in order to know, and not for any utilitarian end. And this is confirmed by the facts; for it was when almost all the necessities of life and the things that make for comfort and recreation had been secured, that such knowledge began to be sought. Evidently then we do not seek it for the sake of any other advantage; but as the man is free, we say, who exists for his own sake and not for another's so we pursue this as the only free science, for it alone exists for its own sake.

Hence also the possession of it might be justly regarded as beyond human power; for in many ways human nature is in bondage, so that according to Simonides "God alone can have this privilege", and it is unfitting that man should not be content to seek the knowledge that is suited to him. If, then, there is something in what the poets say, and jealousy is natural to the divine power, it would probably occur in this case above all, and all who excelled in this knowledge would be unfortunate. But the divine power cannot be jealous (nay, according to the proverb, "bards tell many a lie"), nor should any other science be thought more honourable than one of this sort. For the most divine science is also most honourable; and this science alone must be, in two ways, most divine. For the science which it would be most meet for God to have is a divine science, and so is any science that deals with divine objects; and this science alone has both these qualities; for (1) God is thought to be among the causes of all things and to be a first principle, and (2) such a science either God alone can have, or God above all others. All the sciences, indeed, are more necessary than this, but none is better.

Yet the acquisition of it must in a sense end in something which is the opposite of our original inquiries. For all men begin, as we said, by wondering that things are as they are, as they do about self-moving marionettes, or about the solstices or the incommensurability of the diagonal of a square with the side; for it seems wonderful to all who have not yet seen the reason, that there is a thing which cannot be measured even by the smallest unit. But we must end in the contrary and, according to the proverb, the better state, as is the case in these instances too when men learn the cause; for there is nothing which would surprise a geometer so much as if the diagonal turned out to be commensurable.

We have stated, then, what is the nature of the science we are searching for, and what is the mark which our search and our whole investigation must reach.

Notes

1 Cf. Pl. *Gorg*. 448 c, 462 BC.

2 1130^b 14–1141^b 8.

3

On the Idols, the Scientific Study of Nature, and the Reformation of Education

Francis Bacon

Thoughts and Conclusions on the Interpretation of Nature as a Science Productive of Works

15. The Need for a Philosophy of Invention

But it is not only the methods of demonstration that are at fault. The methods of discovery, or invention, if indeed there be any, are just as much in need of examination. And here Bacon found evidence not so much of wandering from the true path, but simply of solitude and emptiness. He was indeed dumbfounded at it. To think that no man among men should ever have had it in his heart or on his mind to direct the resources of human wit and intellect towards the arts and sciences and to pave a path towards that goal! To think that this whole endeavour should have been, should still be, left to the obscurity of tradition, the dizzy round of argument, the eddies and whirlpools of chance and mere experience! He felt driven to condone the strange practice of the Egyptians. Like other ancient peoples they deified their inventors; and if they set up images even of brute beasts in their temples, well, they had the excuse that the irrational animals have discovered almost as many of nature's operations as men have done. Men indeed have failed to use their prerogative of reason to this end. However, we must not neglect to look into such discoveries as are made.

We may consider first the simple artless mode of discovery habitual to men. All this amounts to is that everyone who makes the attempt first seeks out and peruses what others have said on the subject and then adds his own quota of thought. But it is a baseless procedure either to entrust oneself to the authority of others or to solicit, to invoke, one's own spirit to deliver oracles. Next comes the kind of discovery or research in favour with the dialecticians. It shares no more than the name with what we have in mind. The dialecticians are not concerned to seek out the principles and axioms on which arts depend; they look only for logical consistency. When exceptionally keen and persistent researchers come to bother them with their questions, the dialecticians' practice is to urge them to put their trust in, nay, to take an oath of blind loyalty to, the existent art, such as it is. Finally, there are inventions due to experience pure and simple. If it just happens, it is called chance; if it has been sought after, it is called experiment. But these are only examples of what the proverb calls the broom that has come untied.

"Thoughts and Conclusions on the Interpretation of Nature as a Science Productive of Works," trans. B. Farrington, in *The Philosophy of Francis Bacon*, Liverpool: Liverpool University Press, 1964, pp. 90–6, abridged. "The Plan of the Work 6" and "Aphorisms Concerning the Interpretation of Nature and the Kingdom of Man" from *Novum Organum*, trans. and ed. Peter Urbach and John Gibson, Chicago and La Salle: Open Court Publishing, 1994, pp. 29, 43, 53–6, 292–3, abridged. "On the Idols and on the Scientific Study of Nature" from *New Atlantis* and *The Great Instauration*, ed. Jerry Weinberge, Wheeling, IL: Harlan Davidson, Inc., rev. ed. 1989, pp. 71–4, 77, 79–83, abridged, Copyright © 1980, 1989. "On the Reformation of Education" from "Translation of the 'De Augmentis'" in *The Works of Francis Bacon*, coll. and ed. J. Spedding, R. L. Ellis, and D. D. Heath, London: Longman & Co. et al., 1860, pp. 284–91, abridged.

Those who try to discover the nature or mode of operation of anything by the repetition of random experiments are never at one stay. They alternate between puzzled inaction or giddy activity; hot on the trail one moment, covered with confusion the next; their one discovery being the need for further investigation. How could it be otherwise? To imagine that the nature of anything can be found by examining that thing in isolation, is a notion born of ignorance and inexperience. The nature one seeks may be latent in some things, but manifest and palpable in others; in some things a matter for astonishment, in others too common to notice. There is, for example, a property in bodies which makes them hold together. Water-bubbles seem to shut themselves up in little hemispherical membranes; the force by which they do this strikes us as something mysterious and ingenious. The force which holds wood or stone together we take for granted, and give them the name of solid.

Bacon's conclusion was that men should perhaps be called rather unlucky than ignorant. It is not that they have not exerted themselves; but ill-luck or fond illusions have deflected them from their course.

16. The Time Has Come for a Fresh Start

It is time an end were put to this desperation, or at least to these laments. We must decide once for all whether it would be better to abandon the endeavour and rest content with what we have, or make a serious exertion to improve our lot. The first step to this end is to set up in view the worthiness and excellence of the aim proposed, and so kindle greater enthusiasm for hard work on an exacting business. In this connection Bacon recalled how in antiquity extravagant enthusiasm led men to accord divine honours to inventors; while on those who deserved well of their fellows in civil affairs, on founders of cities and empires, Lawmakers, Liberators of their country from long-standing evils, over-throwers of tyrants, and others of this ilk, the style of Heroes only was conferred. Not for nothing, Bacon reflected, was this distinction observed in ancient times; for the benefits inventors confer extend to the whole human race, while those of civil heroes are confined to particular regions and narrower circles of human settlement. And there is this too. Inventions come without force or disturbance to bless the life of mankind, while civil changes rarely proceed without uproar and violence. If then the utility of some one particular invention so impresses men that they exalt to superhuman rank the man who is responsible for it, how much more noble would that discovery be which should contain within itself the potentiality of all particular inventions, and open up to the human spirit a path of direct and easy access to new remoter powers. Take an example from history. In olden days, when men directed their course at sea by observation of the stars, they merely skirted the shores of the old continent or ventured to traverse small land-locked seas. They had to await the discovery of a more reliable guide, the needle, before they crossed the ocean and opened up the regions of the New World. Similarly, men's discoveries in the arts and sciences up till now are such as could be made by intuition, experience, observation, thought; they concerned only things accessible to the senses. But, before men can voyage to remote and hidden regions of nature, they must first be provided with some better use and management of the human mind. Such a discovery would, without a doubt, be the noblest, the truly masculine birth of time.

Again Bacon noted that, in the Scriptures, King Solomon, though blessed with empire, gold, splendour of architecture, satellites, servants, ministers and slaves of every kind and degree, with a fleet to boot, and a glorious name and with the flattering admiration of the world, yet prided himself on none of these things. Instead he declared that *It was the glory of God to conceal a thing, the glory of a King to find it out*; as if the divine nature enjoyed the kindly innocence of such hide-and-seek, hiding only in order to be found, and with characteristic indulgence desired the human mind to join Him in this sport. And indeed it is this glory of discovery that is the true ornament of mankind. In contrast with civil business it never harmed any man, never burdened a conscience with remorse. Its blessing and reward is without ruin, wrong or wretchedness to any. For light is in itself pure and innocent; it may be wrongly used, but cannot in its nature be defiled.

Bacon next considered the nature of human ambition and found it to be of three kinds, one perhaps not worthy of the name. The first is of those men who with restless striving seek to augment their personal power in their own country. This is the vulgar and degenerate sort. The second is of those who seek to advance the position of their own country in the world; and this may be allowed to have more worth in it and less selfishness. The third is of those whose endeavour is to restore and exalt the power and dominion of man himself, of the human race, over the universe. Surely this is nobler and holier than the former two. Now the dominion of man over nature rests only on knowledge. His power of action is limited to what he knows. No force avails to break the chain of natural causation. Nature cannot be conquered but by obeying her.

This put Bacon upon thinking of examples to illustrate not simply the force of inventions but how such force is accompanied by rewards and blessings. This force is most plainly seen in those three inventions unknown to antiquity, and whose origins are still to us obscure and inglorious, to wit, Printing, Gunpowder and the Nautical Needle. These three, few in number, and not lying much out of the way, have changed the face and status of the world of men, first in learning, next in warfare, and finally in navigation. On them have followed countless changes, as a close scrutiny reveals. In fact, no empire, no school, no star seems to have exerted a greater influence on human affairs than these mechanical inventions. As for their value, the soonest way to grasp it is this. Consider the abyss which separates the life of men in some highly civilised region of Europe from that of some savage, barbarous tract of New India. So great is it that the one man might appear a god to the other, not only in respect of any service rendered but on a comparison of their ways of life. And this is the effect not of soil, not of climate, not of physique, but of the arts. Thus, in the geographical world, the old is much more civilised than the new. In the scientific world this is not so. On the contrary, recent acquisitions must be held the more important. They do not, like the old, merely exert a gentle guidance over nature's course; they have the power to conquer and subdue her, to shake her to her foundations. For the rule is that what discoveries lie on the surface exert but little force. The roots of things, where strength resides, are buried deep.

It may be that there are some on whose ear my frequent and honourable mention of practical activities makes a harsh and unpleasing sound because they are wholly given over in love and reverence to contemplation. Let them bethink themselves that they are the enemies of their own desires. For in nature practical results are not only the means to improve well-being but the guarantee of truth. The rule of religion, that a man should show his faith by his works, holds good in natural philosophy too. Science also must be known by works. It is by the witness of works, rather than by logic or even observation, that truth is revealed and established. Whence it follows that the improvement of man's mind and the improvement of his lot are one and the same thing.

Bacon drew the conclusion that all he has said about the worthiness of the end which he has marked out and measured in his mind is not exaggerated but falls short of the truth.

17. Omens Favourable to Research

But, since what has been said about the excellence of the end may be regarded as a dream, let us consider with full care what hopeful prospect there is and in what quarter it shows itself. We must not suffer ourselves to become prisoners of a vision of supreme goodness and beauty and so abandon, or impair, the strictness of our judgment. Rather we should apply the rule current in civil affairs and be suspicious on principle and look on the dark side of human prospects. Let us cast aside all slighter hopes and vigorously canvass even those that seem most firm. In this determination Bacon consulted the auspices with all due care; and here the first thing that struck him was that the business in hand, being eminently good, was manifestly of God, and in the works of His hand small beginnings draw after them great ends. Then the omens from the nature of Time were also good. All concur that truth is the daughter of Time. How pusillanimous, then, to grovel before authors but to allow to Time, the author of authors and of all authority, less than his due! Nor were his hopes drawn only from the universal character of time, but from the special prerogative of our own age. The opinion men cherish about Antiquity is ill-considered and ill-suited to the word. The term should mean the ripe age, the fullness of years, of the whole world. Now among men we expect greater knowledge of affairs and more maturity of judgment from an old man in proportion to his experience and the multitude of things he has seen, heard and pondered; so from our modern age, if it but realised its powers and would put them boldly to the trial, far greater things are to be expected than from those distant days; for the world has grown older and immeasurably increased its store of experience and observation. It ought not to go for nothing that through the long voyages and travels which are the mark of our age many things in nature have been revealed which might throw new light on natural philosophy. Nay, it would be a disgrace for mankind if the expanse of the material globe, the lands, the seas, the stars, was opened up and brought to light, while, in contrast with this enormous expansion, the bounds of the intellectual globe should be restricted to what was known to the ancients.

It is worth bearing in mind, too, that the political conditions of Europe in this age are favourable. England is stronger, France is restored to peace, Spain is exhausted, Italy and Germany are undisturbed. The balance of power is restored and, in this tranquil state of the most famous nations, there is a turning towards peace; and peace is fair weather for the sciences to flourish.[1] Nor is the state of letters unfavourable. Rather, it has many auspicious aspects. By the Art of Printing, a thing unknown to antiquity, the discoveries and thoughts of individuals are now spread abroad like a flash of lightning. Religious controversies have become a weariness of the spirit, and men are perhaps more ready to contemplate the power, wisdom, and goodness of God in His works. Still, let us assume we have to do with a man who is overwhelmed by the unanimity and duration of the world's acquiescence in the opinions of former days. If such a man considers closely he cannot fail to see that leaders of opinion are but few and that all the rest, their followers, are but ciphers. They have never given a valid assent to the general opinion, for this results from an act of independent judgment. All they have managed to do is to make the step from ignorance to prejudice. If, then, the unanimity of these opinions is an illusion, so is their duration. On examination it shrinks to very narrow bounds. Suppose we allow twenty-five centuries to the recorded history of mankind. Of these scarce five can be set apart as propitious towards, and fruitful in, scientific progress, and the kind of sciences they cultivated were as far as possible removed from that natural philosophy we have in mind. Three periods only can be counted when the wheel of knowledge really turned: one among the Greeks, the second with the Romans, the last among the nations of Western Europe. All other ages have been given over to wars or other pursuits. So far as any scientific harvest is concerned they were barren wastes.

Another favourable omen is found in an understanding of the power and true nature of Chance. Chance, operating in suitable circumstances, has prompted many discoveries. This explains why, in the discovery of fire, the Prometheus of New India followed a different course from the European Prometheus. Flint is scarce in New India. Clearly in inventions which depend on the availability of suitable materials chance plays a large part; in inventions remote from daily experience, a smaller one; yet, be its role big or little, in every age it is the fertile parent of discoveries, nor have we any reason to suppose it has grown old and past bearing. Bacon accordingly opined that, since discoveries occur even when men are not looking for them and are thinking of something else, it is reasonable to expect that when men *are* thought determined and intelligent experimentalists. But, whichever group they belong to, these pretensions are only evidence of their wish to have a reputation above their fellows. In fact the divorce between the two activities, speculation and experiment, has always obtained. But if the two could be joined in a closer and holier union the prospects of a numerous and happy issue are bright indeed.

There is also this further ground for comfort. When he reviewed the infinite expenditure of brains, time, and money on objects and pursuits which, fairly judged, are useless, Bacon was certain that a small portion of this expenditure devoted to sane and solid purposes could triumph over every obstacle. Men shrink back from the multiplicity of particular facts. Yet the phenomena of the arts are easily grasped in comparison with the fictions of the mind once they break free from the control of factual evidence. Thus all the arguments adduced above urge us on to adopt a hopeful view. But the surest ground of hope is in the mistakes of the past. When the affairs of a commonwealth had been mismanaged, there was comfort in the remark: The blacker the past, the brighter the hope for the future. In philosophy, too, if the old errors are abandoned (and to be made aware of them is the first step to amendment), things will take a turn for the better. But if men had been on the right path all those ages past and yet had got no further, what hope could there be? Then it would have been clear that the difficulty lay in the material to be investigated (which is out of our control), not in the instrument, that is to say, the human mind and its management, which is ours to improve. As things are, it is plain that there are no insuperable or immovable objects in the way; simply it lies in a direction untrodden by the feet of men. It may frighten us a little by its loneliness; it offers no other threat. A new world beckons. Even if the breezes that reach us from it were of far less promise and hope, Bacon was resolved that the trial should be made. Not to try is a greater hazard than to fail. If we fail, it is the loss but of a trifling effort. Not to try is to forgo the prospect of measureless good.

The conclusion of this meditation, of what has been said and left unsaid, is that there is no lack of hope. There is hope enough both to launch the man of enterprise on the venture and to convince the deep and sober mind of the likelihood of success.

Note

1 Bacon's optimism here is facile. The Thirty Years War and the English Revolution were at hand.

From *Novum Organum*

The Plan of the Work

I have made a beginning that, I hope, is not to be despised; the fortune of mankind will give the outcome, such as men in the present state of things and of minds may perhaps be unable to grasp or measure. For the matter in hand is not just a pleasant speculation, but in truth concerns the affairs and fortunes of mankind and all the power of its works. For man is only the servant and interpreter of Nature and he only does and understands so much as he shall have observed, in fact or in thought, of the course of Nature; more than this he neither knows, nor can do. No force whatever can unfasten or break the chain of causes, and Nature is only overcome by obeying her. So it is that those two objects of mankind, *Knowledge* and *Power*, come in fact to the same thing; and the failure of works derives mostly from ignorance of causes.

[...]

Aphorisms Concerning the Interpretation of Nature and the Kingdom of Man

3

Human knowledge and human power come to the same thing, for where the cause is not known the effect cannot be produced. We can only command Nature by obeying her, and what in contemplation represents the cause, in operation stands as the rule.

[...]

38

The idols and false notions that have hitherto occupied the human understanding, and lie deep-seated there, have not only so beset men's minds that their approach to the truth becomes difficult; but even when access to it is given and conceded, they will present themselves and interfere in that very restoration (*instauratio*) of the sciences, unless men are forewarned and protect themselves against them as far as possible.

39

There are four kinds of idols besetting human minds. To help in my teaching, I have given them names. I call the first, *Idols of the Tribe;* the second, *Idols of the Cave;* the third, *Idols of the Market-place;* and the fourth, *Idols of the Theatre*.

40

The formation of notions and axioms by true *induction* is of course the proper remedy for warding off and clearing away these idols, but just to point them out is very useful. For the doctrine concerning idols is to the *Interpretation of Nature* as the doctrine of Sophistical Refutations[1] is to ordinary dialectic.

41

The *Idols of the Tribe* lie deep in human nature itself and in the very tribe or race of mankind. For it is wrongly asserted that the human sense is the measure of things.[2] It is rather the case that all our perceptions, both of our sense and of our minds, are reflections of man, not of the universe, and the human understanding is like an uneven mirror that cannot reflect truly the rays from objects, but distorts and corrupts the nature of things by mingling its own nature with it.

42

The *Idols of the Cave* are those specific to individual men.[3] For besides the errors common to human nature in general, each of us has his own private cave or den, which breaks up and falsifies the light of Nature; either because of his own distinct and individual nature, or because of what he has been taught or gained in conversation with others, or from his reading, and the authority of those whom he respects and admires; or from the different impressions [he gains from things], according as they present themselves to a mind prejudiced and already committed, or to one impartial and moderate, or the like. So that the human spirit (according to how it is distributed in individual men) is variable and always in commotion, and as it were, subject to chance. Whence Heraclitus[4] has well said that men seek knowledge in lesser worlds, and not in the greater or common world.

43

There are also idols arising from the dealings and association of men with one another, which I call *Idols of the Market-place*, because of the commerce and meeting of men there. For speech is the means of association among men; but words are applied according to common understanding. And in consequence, a wrong and inappropriate application of words obstructs the mind to a remarkable extent. Nor do the definitions or explanations with which learned men have sometimes been accustomed to defend and vindicate themselves in any way remedy the situation. Indeed, words plainly do violence to the understanding and throw everything into confusion, and lead men into innumerable empty controversies and fictions.

44

Finally, there are idols which have crept into human minds from the various dogmas of philosophies, and also from faulty laws of demonstrations. These I call *Idols of the Theatre*, because I regard all the philosophies that have been received or invented as so many stage plays creating fictitious and imaginary worlds. Nor am I only speaking of present philosophies, nor indeed only of the ancient philosophies and their sects, for numerous other plays of the same kind may yet be composed and contrived, since the most diverse errors spring sometimes from similar causes. Nor again do I mean this only in regard to universal philosophies, but also to many principles and axioms of the sciences, which have become established through tradition, credulity and neglect.

[...]

And from this an improvement of the estate of man is sure to follow, and an enlargement of his power over Nature. For man by the Fall fell both from his state of innocence and his dominion over creation. Both of these, however, can even in this life be to some extent made good; the former by religion and faith, the latter by arts and sciences. For the curse did not make creation entirely and for ever rebellious; but in virtue of that ordinance "in the sweat of thy face shalt thou eat thy bread",[5] by every kind of effort (certainly not by disputations and empty magic ceremonies), it will at length in some measure be subdued so as to provide man with his bread, that is, with the necessities of human life.

Notes

1. This is a reference to Aristotle's work *De Sophisticis Elenchis* (*On Sophistical Refutations*), in which he expounds and attempts to solve various sophistical puzzles that arise from verbal ambiguities and equivocations. Sophistical refutations were "arguments which appear to be refutations but are really fallacious and not refutations" (164a20). And in this work, Aristotle characterised dialectical arguments as ones "which, starting from generally accepted opinions, reason to establish a contradiction" (165b3).
2. This presumably alludes to Protagoras of Abdera (fifth century BC), the first and most notable of the Greek sophists, whose famous contention was "Man is the measure of all things, of things that are that they are, and of things that are not that they are not". This seems to have been intended as an expression of scepticism and relativism, though Bacon is evidently reading it in the opposite sense. We have noted before that Bacon did not regard sophist teachings as sceptical.
3. Bacon is alluding here to Plato's Myth of the Cave, from the seventh book of the *Republic*, which depicts mankind as trapped in a cave, mistaking the shadows passing across the walls of the cave for realities.
4. In the *Advancement of Learning*, I, (*Works*, III, p. 292) Bacon makes his meaning much clearer: "Heraclitus gave a just censure, saying, *Men sought truth in their own little worlds, and not in the great and common world*; for they disdain to spell and so by degrees to read in the volume of God's works; and contrariwise by continual meditation and agitation of wit do urge and as it were invocate their own spirits to divine and give oracles unto them, whereby they are deservedly de-

luded". (Heraclitus of Ephesus was a Greek philosopher who was born around 500 BC.)

5 Genesis iii, 19.

On the Idols and on the Scientific Study of Nature

For I will impart unto thee, for the love of God and men, a relation of the true state of Salomon's House. Son, to make you know the true state of Salomon's House, I will keep this order. First, I will set forth unto you the end of our foundation. Secondly, the preparations and instruments we have for our works. Thirdly, the several employments and functions whereto our fellows are assigned. And fourthly, the ordinances and rites which we observe.

"The End of our Foundation is the knowledge of Causes, and secret motions of things; and the enlarging of the bounds of Human Empire, to the effecting of all things possible.

"The Preparations and Instruments are these. We have large and deep caves of several depths: the deepest are sunk six hundred fathom; and some of them are digged and made under great hills and mountains: so that if you reckon together the depth of the hill and the depth of the cave, they are (some of them) above three miles deep. For we find that the depth of a hill, and the depth of a cave from the flat, is the same thing; both remote alike from the sun and heaven's beams, and from the open air.[1] These caves we call the Lower Region. And we use them for all coagulations, indurations,[2] refrigerations, and conservations of bodies. We use them likewise for the imitation of natural mines;[3] and the producing also of new artificial metals, by compositions and materials which we use, and lay there for many years. We use them[4] also sometimes, (which may seem strange,) for curing of some diseases, and for prolongation of life in some hermits that choose to live there, well accommodated of[5] all things necessary; and indeed live very long; by whom also we learn many things.

"We have burials in several earths, where we put divers cements, as the Chineses do their porcellain. But we have them in greater variety, and some of them more fine. We have also great variety of composts, and soils, for the making of the earth fruitful.

"We have high towers; the highest about half a mile in height; and some of them likewise set upon high mountains; so that the vantage[6] of the hill with the tower is in the highest of them three miles at least. And these places we call the Upper Region: accounting the air between the high places and the low, as a Middle Region. We use these towers, according to their several heights and situations, for insolation,[7] refrigeration, conservation; and for the view of divers meteors;[8] as winds, rain, snow, hail; and some of the fiery meteors[9] also. And upon them, in some places, are dwellings of hermits, whom we visit sometimes, and instruct what to observe.

[…]

"We have also great and spacious houses, where we imitate and demonstrate meteors; as snow, hail, rain, some artificial rains of bodies and not of water, thunders, lightnings; also generations of bodies in air; as frogs, flies, and divers others.

"We have also certain chambers, which we call Chambers of Health, where we qualify[10] the air as we think good and proper for the cure of divers diseases, and preservation of health.

"We have also fair and large baths, of several mixtures, for the cure of diseases, and the restoring of man's body from arefaction:[11] and others for the confirming[12] of it in strength of sinews, vital parts, and the very juice and substance of the body.

[…]

"We have also divers mechanical arts, which you have not; and stuffs made by them; as papers, linen, silks, tissues; dainty works of feathers of wonderful lustre; excellent dyes, and many others; and shops likewise, as well for such as are not brought into vulgar use amongst us as for those that are. For you must know that of the things before recited, many of them are grown into use throughout the kingdom; but yet if they did flow from our invention, we have of them also for patterns and principals.[13]

"We have also furnaces of great diversities, and that keep great diversity of heats; fierce and quick; strong and constant; soft and mild; blown, quiet; dry, moist; and the like. But above all, we have heats in imitation of the sun's and heavenly bodies' heats, that pass divers inequalities[14] and (as it were) orbs,[15] progresses,[16] and returns,[17] whereby we produce admirable effects. Besides, we have heats of dungs, and of bellies and maws of living creatures, and of their bloods and bodies; and of hays and herbs laid up moist; of lime unquenched;[18] and such like. Instruments also which generate heat only by motion. And farther, places for strong insolations; and again, places under the earth, which by nature or art yield heat. These divers heats we use, as the nature of the operation which we intend requireth.

[...]

"We have also engine-houses, where are prepared engines and instruments for all sorts of motions. There we imitate and practise to make swifter motions than any you have, either out of your muskets or any engine that you have; and to make them and multiply them more easily, and with small force, by wheels and other means: and to make them stronger, and more violent than yours are; exceeding your greatest cannons and basilisks.[19] We represent also ordnance and instruments of war, and engines of all kinds: and likewise new mixtures and compositions of gun-powder, wildfires burning in water, and unquenchable. Also fire-works of all variety both for pleasure and use. We imitate also flights of birds; we have some degrees[20] of flying in the air; we have ships and boats for going under water, and brooking of seas; also swimming-girdles[21] and supporters. We have divers curious[22] clocks, and other like motions of return,[23] and some perpetual motions. We imitate also motions of living creatures, by images[24] of men, beasts, birds, fishes, and serpents. We have also a great number of other various motions, strange for equality, fineness, and subtilty.[25]

"We have also a mathematical house, where are represented all instruments, as well of geometry as astronomy, exquisitely made.

"We have also houses of deceits of the senses; where we represent all manner of feats of juggling, false apparitions, impostures, and illusions; and their fallacies.[26] And surely you will easily believe that we that have so many things truly natural which induce admiration, could in a world of particulars[27] deceive the senses, if we would disguise those things and labour to make them seem more miraculous. But we do hate all impostures and lies: insomuch as[28] we have severely forbidden it to all our fellows, under pain of ignominy and fines, that they do not[29] shew any natural work or thing, adorned or swelling;[30] but only pure as it is, and without all affectation of strangeness.

"These are (my son) the riches of Salomon's House.

"For the several employments and offices of our fellows; we have twelve that sail into foreign countries, under the names of other nations, (for our own we conceal;) who bring us the books, and abstracts,[31] and patterns of experiments of all other parts. These we call Merchants of Light.

"We have three that collect the experiments which are in all books. These we call Depredators.[32]

"We have three that collect the experiments of all mechanical arts; and also of liberal sciences;[33] and also of practices which are not brought into arts.[34] These we call Mystery-men.

"We have three that try new experiments, such as themselves think good. These we call Pioners [*sic*] or Miners.

"We have three that draw the experiments of the former four into titles and tables,[35] to give the better light[36] for the drawing of observations and axioms out of them. These we call Compilers.

"We have three that bend themselves,[37] looking into the experiments of their fellows, and cast about[38] how to draw out of them things of use and practice for man's life, and knowledge as well for works as for plain demonstration[39] of causes, means of natural divinations,[40] and the easy and clear discovery of the virtues[41] and parts of bodies. These we call Dowry-men or Benefactors.

"Then after divers meetings and consults[42] of our whole number, to consider of the former labours and collections, we have three that take care, out of them, to direct new experiments, of a higher light,[43] more penetrating into nature than the former. These we call Lamps.

"We have three others that do execute the experiments so directed, and report them. These we call Inoculators.[44]

"Lastly, we have three that raise the former discoveries by experiments into greater observations, axioms, and aphorisms. These we call Interpreters of Nature.

"We have also, as you must think, novices and apprentices, that the succession of the former employed men do not fail; besides a great number of servants and attendants, men and women. And this we do also: we have consultations, which of the inventions and experiences which we have discovered shall be published, and which not: and take all an oath of secrecy, for the concealing of those which we think fit to keep secret: though some of those we do reveal sometimes to the state, and some not.

"For our ordinances and rites: we have two very long and fair galleries: in one of these we place patterns and samples of all manner of the more rare and excellent inventions: in the other we place the statua's[45] of all principal inventors. There we have the statua of your Columbus, that discovered the West Indies: also the inventor of ships: your monk[46] that was the inventor of ordnance and of gunpowder: the inventor of music: the inventor of letters: the inventor of printing: the inventor of observations of astronomy: the inventor of works in metal: the inventor of glass: the inventor of silk of the worm: the inventor of wine: the inventor of corn and bread: the inventor of sugars: and all these by more certain[47] tradition than you have. Then have we divers inventors of our own, of excellent works; which since you have not seen, it were too long to make descriptions of them; and besides, in the right understanding of those descriptions you might easily err. For upon every invention of value, we erect a statua to the inventor, and give him a liberal and honourable reward. These statua's are some of brass; some of marble and touch-stone;[48] some of cedar and other special woods gilt and adorned: some of iron; some of silver; some of gold.

"We have certain hymns and services, which we say daily, of laud,[49] and thanks to God for his marvellous works: and forms of prayers, imploring his aid and blessing for the illumination of our labours, and the turning of them into good and holy uses.

"Lastly, we have circuits or visits of divers principal cities of the kingdom; where, as it cometh to pass, we do publish such new profitable inventions as we think good. And we do also declare natural divinations of diseases, plagues, swarms of hurtful creatures, scarcity, tempests, earthquakes, great inundations, comets, temperature[50] of the year, and divers other things; and we give counsel thereupon what the people shall do for the prevention and remedy of them."

And when he had said this, he stood up; and I, as I had been taught, kneeled down; and he laid his right hand upon my head, and said; "God bless thee, my son, and God bless this relation which I have made. I give thee leave to publish it for the good of other nations; for we here are in God's bosom, a land unknown." And so he left me; having assigned a value of about two thousand ducats,[51] for a bounty[52] to me and my fellows. For they give great largesses[53] where they come upon all occasions.

Notes

1 *Thus*, caves dug downward under hills are especially deep.
2 *Hardenings.*
3 *Mineral* veins.
4 *The* caves.
5 *Well* provided with.
6 *Total* height.
7 *Exposure* to the sun.
8 *Atmospheric* phenomena.
9 *Shooting* stars and lightning.
10 *Alter*, change.
11 *Withering.*
12 *Strengthening.*
13 *Preserve* them only as examples or originals.
14 *Changes* in intensity.
15 *Planes.*
16 *Forward* courses.
17 *Return* courses.
18 *Unslaked.*
19 *Large* cannons named for a deadly serpent.
20 *Some* success.
21 *Life* preservers.
22 *Precise.*
23 *Machines* producing cyclical motion.
24 *Robots*, automata.
25 *Extraordinary* for regularity, sharpness, and complexity.
26 *How* they are false.
27 *That* could in the practical world.
28 *So* much that.
29 *An* acceptable double negative in Bacon's time.
30 *Exaggerated.*
31 *Summaries.*
32 *Those* who plunder or pillage.
33 *Liberal* arts.
34 *Unsystematic* practices.
35 *Those* who classify and list the experiments of the former.
36 *The* better to enable.
37 *Direct* their attention to.
38 *Consider.*
39 *Theoretical* demonstration.
40 *Discovering* and predicting the secrets of nature.
41 *Characteristics*, powers.
42 *Consultations.*
43 *Producing* more general knowledge.
44 *Men* who bud trees.
45 *Statues.*
46 *Roger Bacon* or Berthold Schwarz.

47 *More* trustworthy.
48 *Dark* quartz or jasper.
49 *In* praise of.
50 *Climate.*
51 *Gold* coins.
52 *Gift*, gratuity.
53 *Free* gifts.

On the Reformation of Education

...let us now review and consider with ourselves what has hitherto been done by kings and others for the increase and advancement of learning, and what has been left undone; and let us discuss the question solidly and distinctly, in a style active and masculine, without digressing or dilating. We may begin then by assuming (which will not be disputed) that all the greatest and most difficult works are overcome either by amplitude of reward, or by prudence and soundness of direction, or by conjunction of labours; whereof the first stimulates endeavour, the second removes uncertainty and error, and the third supplies the frailty of man. But of these three, prudence and soundness of direction, – that is, the pointing out and setting forth of the straight and ready way to the thing which is to be done, – must be placed first. For the cripple in the right way (as the saying is) outstrips the runner in the wrong. And Solomon observes, most aptly to the point in question, that "if the iron be blunt it requireth more strength, but wisdom is that which prevaileth;" signifying that the prudent choice of the mean is more effectual for the purpose than either the enforcement or the accumulation of endeavours. This I am induced to say, for that (not derogating from the honour of those who have been in any way deservers towards the state of learning) I observe nevertheless that most of their works and acts have had in view rather their own magnificence and memory than the progress and advancement of learning, and have rather augmented the number of learned men than raised and rectified the sciences themselves.

The works or acts which pertain to the advancement of learning are conversant about three objects; the places of learning, the books of learning, and the persons of the learned. For as water, whether it be the dew of Heaven or the springs of the earth, easily scatters and loses itself in the ground, except it be collected into some receptacle where it may by union and consort comfort and sustain itself (and for that cause the industry of man has devised aqueducts, cisterns, and pools, and likewise beautified them with various ornaments, for magnificence and state as well as for use and necessity); so this excellent liquor of knowledge, whether it descend from divine inspiration or spring from human sense, would soon perish and vanish into oblivion; if it were not preserved in books, traditions, and conferences; and especially in places appointed for such matters, as universities, colleges, and schools, where it may have both a fixed habitation and means and opportunity of increasing and collecting itself.

And first, the works which concern the *places of learning* are four; buildings, endowments with revenues, grants of franchises and privileges, and institutions and ordinances of government; all tending (for the most part) to retirement and quietness of life, and a release from cares and trouble; like the stations which Virgil prescribes for the hiving of honey bees.

Principio sedes apibus statioque petenda,
Quo neque sit ventis aditus, &c.[1]

The principal works touching *books* are two; first, libraries, which are as the shrines wherein all the relics of the ancient saints full of true virtue are preserved. Secondly, new editions of authors, with more correct impressions, more faithful translations, more profitable commentaries, more diligent annotations, and the like.

The works pertaining to the *persons of the learned* (besides the advancement and countenancing of them in general) are likewise two. The remuneration and designation of lecturers in arts already extant and invented; and the remuneration and appointment of writers and inquirers concerning those parts of learning not yet sufficiently laboured or prosecuted.

These are summarily the works and acts wherein the merits of many excellent princes and other illustrious personages towards learning have been manifested. As for the particular commemoration of any one who has deserved well of literature, I call to mind what Cicero said when, on his return from exile, he gave general thanks; "It is hard to remember all, ungrateful to pass by any."[2] Let us rather (after the advice of Scripture) look forward to that part of the race which is still to be run, than look back to that which has been passed.

First therefore, among so many noble foundations of colleges in Europe, I find it strange that they are all dedicated to professions, and none left free to the study of arts and sciences at large. For if men judge that learning should be referred to use and action, they judge well; but it is easy in this to fall into the error pointed at in the ancient fable; in which the other parts of the body found fault with the stomach, because it neither performed the office of motion as the limbs do, nor of sense, as the head does; but yet notwithstanding it is the stomach which digests and distributes the aliment to all the rest. So if any man think that Philosophy and Universality are idle and unprofitable studies, he does not consider that all arts and professions are from thence supplied with sap and strength. And this I take to be a great cause, which has so long hindered the more flourishing progress of learning; because these fundamental knowledges have been studied but in passage, and not drunk deeper of. For if you will have a tree bear more fruit than it has used to do, it is not anything you can do to the boughs, but it is the stirring of the earth, and putting richer mould about the roots, that must work it. Neither is it to be forgotten that this dedication of colleges and societies to the use only of professory learning has not only been inimical to the growth of the sciences, but has also been prejudicial to states and governments. For hence it proceeds that princes when they have to choose men for business of state find a wonderful dearth of able men around them; because there is no collegiate education designed for these purposes, where men naturally so disposed and affected might (besides other arts) give themselves especially to histories, modern languages, books of policy and civil discourse; whereby they might come better prepared and instructed to offices of state.

And because founders of Colleges do plant, and founders of Lectures do water, I must next speak of the deficiencies which I find in public lectures; wherein I especially disapprove of the smallness of the salary assigned to lecturers in arts and professions, particularly amongst ourselves. For it is very necessary to the progression of sciences that lecturers in every sort be of the most able and sufficient men; as those who are ordained not for transitory use, but for keeping up the race and succession of knowledge from age to age. This cannot be, except their condition and endowment be such that the most eminent professors may be well contented and willing to spend their whole life in that function and attendance, without caring for practice. And therefore if you will have sciences flourish, you must observe David's military law; which was, "That those who stayed with the baggage should have equal part with those who were in the action;"[3] else will the baggage be ill attended. So lecturers in sciences are as it were the keepers and guardians of the whole store and provision of learning, whence the active and militant part of the sciences is furnished; and therefore they ought to have equal entertainment and profit with the men of active life. Otherwise if the fathers in sciences be not amply and handsomely maintained, it will come to pass, as Virgil says of horses, –

Et patrum invalidi referent jejunia nati;[4]

the poor keeping of the parents will be seen in the weakliness of the children.

I will now notice another defect, wherein I should call in some alchemist to help me; one of those who advise the studious to sell their books and build furnaces, and forsaking Minerva and the Muses as barren virgins, to rely upon Vulcan. But certain it is that for depth of speculation no less than for fruit of operation in some sciences (especially natural philosophy and physic) other helps are required besides books. Wherein also the beneficence of men has not been altogether wanting; for we see spheres, globes, astrolabes, maps, and the like have been provided and prepared as assistants to astronomy and cosmography, as well as books. We see likewise that some places instituted for physic have gardens for the examination and knowledge of simples of all sorts, and are not without the use of dead bodies for anatomical observations. But these respect but a few things. In general, it may be held for certain that there will hardly be any great progress in the unravelling and unlocking of the secrets of nature, except there be a full allowance for expenses about experiments; whether they be experiments appertaining to Vulcan or Dædalus (that is, the furnace or engine), or any other kind. And therefore as secretaries and emissaries of princes are allowed to bring in bills of expenses for their diligence in exploring and unravelling plots and civil secrets, so the searchers and spies of nature must have their expenses paid, or else you will never be well informed of a great number of things most worthy to be known. For if Alexander made such a liberal assignation of money to Aristotle, to support hunters, fowlers, fishers and the like, that he might be better furnished for compiling a History of Animals; certainly much more do they deserve it, who instead of wandering in the forests of nature, make their way through the labyrinths of arts.

Another defect to be noticed (and one of great importance) is a neglect of consultation in governors of universities, and of visitation in princes or superior persons, to enter into careful account and consideration whether the readings, disputations, and other scholastic exercises anciently begun, and since continued up to our time, may be profitably kept up, or whether we should rather abolish them and substitute better. For I find it is one of your Majesty's most wise maxims; "That in all usages or precedents the times be considered wherein they first began; which, if they were disordered or ignorant, it derogates greatly from the authority of the precedents, and leaves all things for suspect." And therefore inasmuch as most of the institutions of the universities are derived from times a good deal more obscure and ignorant than our own, it is the more convenient that they be re-examined. In this kind I will give an instance or two, of things which appear the most obvious and familiar. It is a general custom (and yet I hold it to be an error) that scholars come too soon and too unripe to the study of logic and rhetoric, arts fitter for graduates than children and novices; for these two rightly taken are the gravest of sciences, being the arts of arts, the one for judgment, the other for ornament; besides they give the rule and direction how both to set forth and illustrate the subject matter. And therefore for minds empty and ignorant (and which have not yet gathered what Cicero calls "stuff"[5] or "furniture,"[6] that is matter and variety) to begin with those arts (as if one should learn to weigh or to measure or to paint the wind), works but this effect, that the virtue and faculty of those arts (which are great and universal) are almost made contemptible, and either degenerate into childish sophistry and ridiculous affectation, or at least lose not a little of their reputation. And further, the premature and untimely learning of these arts has drawn on, by consequence, the superficial and unprofitable teaching and handling of them, – a manner of teaching suited to the capacity of children. Another instance of an error which has long prevailed in universities is this; that they make too great and mischievous a divorce between invention and memory. For most of the speeches there are either entirely premeditate, and delivered in preconceived words, where nothing is left to invention; or merely extempore, where little is left to memory; whereas in common life and action there is little use of either of these separately, but rather of intermixtures of them; that is of notes or commentaries and extempore speech; and thus the exercise fits not the practice, nor the image the life. But it must ever be observed as a rule in exercises, that they be made to represent in everything (as near as may be) the real actions of life; for otherwise they will pervert the motions and faculties of the mind, and not prepare them. The truth whereof appears clearly enough when scholars come to the practice of their professions, or other offices of civil life; which when they set into, this want I speak of is soon found out by themselves, but still sooner by others. But this part, touching the amendment of the Institutions and Orders of Universities, I will conclude with a sentence taken from one of Cæsar's letters to Oppius and Balbus; "How this may be done, some means occur to me, and many may be found; I beg you therefore to take these matters into consideration."[7]

Another defect which I note ascends a little higher than the preceding. For as the progress of learning consists not a little in the wise ordering and institutions of each several university; so it would be yet much more advanced if there were a closer connexion and relationship between all the different universities of Europe than now there is. For we see there are many orders and societies which, though they be divided under distant sovereignties and territories, yet enter into and maintain among themselves a kind of contract and fraternity, insomuch that they have governors (both provincial and general) whom they all obey. And surely as nature creates brotherhood in families, and arts mechanical contract brotherhoods in societies, and the anointment of God superinduces a brotherhood in kings and bishops, and vows and regulations make a brotherhood in religious orders; so in like manner there cannot but be a noble and generous brotherhood contracted among men by learning and illumination, seeing that God himself is called "the Father of Lights."[8]

The last defect I complain of (to which I have already alluded) is that there has not been, or very rarely been, any public designation of fit men either to write or to make inquiry concerning such parts of knowledge as have not been already sufficiently laboured. To which point it will greatly conduce, if a review and *census* be made of the sciences, and account be taken what parts of them are rich and well advanced, and what poor and destitute. For the opinion of plenty is amongst the causes of want; and the great quantity of books makes a show rather of superfluity than lack; of which surcharge nevertheless the true remedy is not to destroy the old books, but to make more good ones; of such a kind that like the serpent of Moses, they may devour the serpents of the enchanters.[9]

The removal of all the defects formerly enumerated, except the last, and of the active part also of the last, which relates to the designation of writers, are truly works for a king; towards which the endeavours and industry of a private man can be but as an image in a crossway, that may point at the way but cannot go it. But the speculative part of it, which relates to the survey of knowledges to see what in each is deficient, is open likewise to private industry. Wherefore I now intend to make a general and faithful perambulation and survey of learning, with a very careful and accurate inquiry what parts thereof lie fresh and waste, and not yet improved and converted to use by the industry of man; to the end that such a plot marked out, and recorded to memory, may minister light both to public designations and voluntary endeavours. Wherein nevertheless my purpose is at this time to note only omissions and deficiencies, and not to make any redargution of errors and failures; for it is one thing to point out what parts lie untilled, and another thing to mend the manner of tillage.

In addressing myself to which task I am not ignorant how great a work I attempt, and how difficult a province I take upon me; nor again how far unequal my strength is to my will. Nevertheless I have great hope that if my extreme love to learning carry me too far, I may obtain the excuse of affection; for that "it is not granted to any man at the same time to love and to be wise."[10] But I know well I can use no other liberty of judgement, than I must leave to others; and I for my part shall be equally glad either to perform myself or to accept from others that duty of humanity, to put the wanderer on the right way: *nam qui erranti comiter monstrat viam*[11], &c. I foresee likewise that many of those things which I shall think fit to enter in this registry of mine as omitted and deficient will incur censure on different accounts; some as being already done and extant; others as savouring of curiosity, and promising very scanty fruit; others as being too difficult and almost impossible to be compassed and effected by man. For the two first I refer myself to the particulars themselves. For the last, touching impossibility, I take it that all those things are to be held possible and performable, which may be done by some persons, though not by every one; and which may be done by many together, though not by one alone; and which may be done in the succession of ages, though not in one man's life; and lastly, which may be done by public designation and expense, though not by private means and endeavour. But notwithstanding if any man will take to himself rather the saying of Solomon, "The slothful man says there is a lion in the path,"[12] than that of Virgil, *Possunt, quia posse videntur*[13], "they find it possible because they think it possible," I shall be content that my labours be esteemed but as the better sort of wishes. For as it asks some knowledge of a thing to demand a question not impertinent, so it requires some sense to make a wish not absurd.

Notes

1 Virg. Georg. iv. 8.: –First for thy bees a quiet station find, And lodge them under covert of the wind.
2 Cicero, Post Red. c. 12.
3 I Sam. xxx, 24.
4 Georg. iii. 128.
5 *Sylva.* De Orator, iii. 26.
6 *Supellex*. Orator. c. 24.
7 Cic. Ep. ad Att. ix. 8.
8 St. James, i. 17.
9 Not Moses, but Aaron. Ex. vii. 12.
10 Senecæ Proverbia.
11 Ennius, ap. Aul. Gell. xii. 4. and ap. Cic. De Officiis, i. 17.
12 Prov. xxvi. 13.
13 Virg. *Æn*. v. 231.

4

Idea for a Universal History from a Cosmopolitan Point of View

Immanuel Kant

Whatever concept one may hold, from a metaphysical point of view, concerning the freedom of the will, certainly its appearances, which are human actions, like every other natural event are determined by universal laws. However obscure their causes, history, which is concerned with narrating these appearances, permits us to hope that if we attend to the play of freedom of the human will in the large, we may be able to discern a regular movement in it, and that what seems complex and chaotic in the single individual may be seen from the standpoint of the human race as a whole to be a steady and progressive though slow evolution of its original endowment. Since the free will of man has obvious influence upon marriages, births, and deaths, they seem to be subject to no rule by which the number of them could be reckoned in advance. Yet the annual tables of them in the major countries prove that they occur according to laws as stable as [those of] the unstable weather, which we likewise cannot determine in advance, but which, in the large, maintain the growth of plants, the flow of rivers, and other natural events in an unbroken, uniform course. Individuals and even whole peoples think little on this. Each, according to his own inclination, follows his own purpose, often in opposition to others; yet each individual and people, as if following some guiding thread, go toward a natural but to each of them unknown goal; all work toward furthering it, even if they would set little store by it if they did know it.

Since men in their endeavors behave, on the whole, not just instinctively, like the brutes, nor yet like rational citizens of the world according to some agreed-on plan, no history of man conceived according to a plan seems to be possible, as it might be possible to have such a history of bees or beavers. One cannot suppress a certain indignation when one sees men's actions on the great world-stage and finds, beside the wisdom that appears here and there among individuals, everything in the large woven together from folly, childish vanity, even from childish malice and destructiveness. In the end, one does not know what to think of the human race, so conceited in its gifts. Since the philosopher cannot presuppose any [conscious] individual purpose among men in their great drama, there is no other expedient for him except to try to see if he can discover a natural purpose in this idiotic course of things human. In keeping with this purpose, it might be possible to have a history with a definite natural plan for creatures who have no plan of their own.

We wish to see if we can succeed in finding a clue to such a history; we leave it to Nature to produce the man capable of composing it. Thus Nature produced Kepler, who subjected, in an unexpected way, the eccentric paths of the planets to definite laws; and she produced Newton, who explained these laws by a universal natural cause.

From Immanuel Kant, *On History*, ed. and intro. L. W. Beck, trans. L. W. Beck, R. E. Anchor, and E. L. Fackenheim, Indianapolis and New York: Bobbs-Merrill, Inc., 1957, pp. 11–26.

First Thesis

All natural capacities of a creature are destined to evolve completely to their natural end.

Observation of both the outward form and inward structure of all animals confirms this of them. An organ that is of no use, an arrangement that does not achieve its purpose, are contradictions in the teleological theory of nature. If we give up this fundamental principle, we no longer have a lawful but an aimless course of nature, and blind chance takes the place of the guiding thread of reason.

Second Thesis

In man (as the only rational creature on earth) *those natural capacities which are directed to the use of his reason are to be fully developed only in the race, not in the individual.*

Reason in a creature is a faculty of widening the rules and purposes of the use of all its powers far beyond natural instinct; it acknowledges no limits to its projects. Reason itself does not work instinctively, but requires trial, practice, and instruction in order gradually to progress from one level of insight to another. Therefore a single man would have to live excessively long in order to learn to make full use of all his natural capacities. Since Nature has set only a short period for his life, she needs a perhaps unreckonable series of generations, each of which passes its own enlightenment to its successor in order finally to bring the seeds of enlightenment to that degree of development in our race which is completely suitable to Nature's purpose. This point of time must be, at least as an ideal, the goal of man's efforts, for otherwise his natural capacities would have to be counted as for the most part vain and aimless. This would destroy all practical principles, and Nature, whose wisdom must serve as the fundamental principle in judging all her other offspring, would thereby make man alone a contemptible plaything.

Third Thesis

Nature has willed that man should, by himself, produce everything that goes beyond the mechanical ordering of his animal existence, and that he should partake of no other happiness or perfection than that which he himself, independently of instinct, has created by his own reason.

Nature does nothing in vain, and in the use of means to her goals she is not prodigal. Her giving to man reason and the freedom of the will which depends upon it is clear indication of her purpose. Man accordingly was not to be guided by instinct, not nurtured and instructed with ready-made knowledge; rather, he should bring forth everything out of his own resources. Securing his own food, shelter, safety and defense (for which Nature gave him neither the horns of the bull, nor the claws of the lion, nor the fangs of the dog, but hands only), all amusement which can make life pleasant, insight and intelligence, finally even goodness of heart – all this should be wholly his own work. In this, Nature seems to have moved with the strictest parsimony, and to have measured her animal gifts precisely to the most stringent needs of a beginning existence, just as if she had willed that, if man ever did advance from the lowest barbarity to the highest skill and mental perfection and thereby worked himself up to happiness (so far as it is possible on earth), he alone should have the credit and should have only himself to thank – exactly as if she aimed more at his rational self-esteem than at his well-being. For along this march of human affairs, there was a host of troubles awaiting him. But it seems not to have concerned Nature that he should live well, but only that he should work himself upward so as to make himself, through his own actions, worthy of life and of well-being.

It remains strange that the earlier generations appear to carry through their toilsome labor only for the sake of the later, to prepare for them a foundation on which the later generations could erect the higher edifice which was Nature's goal, and yet that only the latest of the generations should have the good fortune to inhabit the building on which a long line of their ancestors had (unintentionally) labored without being permitted to partake of the fortune they had prepared. However puzzling this may be, it is necessary if one assumes that a species of animals should have reason, and, as a class of rational beings each of whom dies while the species is immortal, should develop their capacities to perfection.

Fourth Thesis

The means employed by Nature to bring about the development of all the capacities of men is their antagonism in society, so far as this is, in the end, the cause of a lawful order among men.

By "antagonism" I mean the unsocial sociability of men, i.e., their propensity to enter into society, bound together with a mutual opposition which constantly threatens to break up the society. Man has an inclination to associate with others, because in society he feels himself to be more than man, i.e., as more than the developed form of his natural capacities. But he also has a strong propensity to isolate himself from others, because he finds in himself at the same time the unsocial characteristic of wishing to have everything go according to his own wish. Thus he expects opposition on all sides because, in knowing himself, he knows that he, on his own part, is inclined to oppose others. This opposition it is which awakens all his powers, brings him to conquer his inclination to laziness and, propelled by vainglory, lust for power, and avarice, to achieve a rank among his fellows whom he cannot tolerate but from whom he cannot withdraw. Thus are taken the first true steps from barbarism to culture, which consists in the social worth of man; thence gradually develop all talents, and taste is refined; through continued enlightenment the beginnings are laid for a way of thought which can in time convert the coarse, natural disposition for moral discrimination into definite practical principles, and thereby change a society of men driven together by their natural feelings into a moral whole. Without those in themselves unamiable characteristics of unsociability from whence opposition springs – characteristics each man must find in his own selfish pretensions – all talents would remain hidden, unborn in an Arcadian shepherd's life, with all its concord, contentment, and mutual affection. Men, good-natured as the sheep they herd, would hardly reach a higher worth than their beasts; they would not fill the empty place in creation by achieving their end, which is rational nature. Thanks be to Nature, then, for the incompatibility, for heartless competitive vanity, for the insatiable desire to possess and to rule! Without them, all the excellent natural capacities of humanity would forever sleep, undeveloped. Man wishes concord; but Nature knows better what is good for the race; she wills discord. He wishes to live comfortably and pleasantly; Nature wills that he should be plunged from sloth and passive contentment into labor and trouble, in order that he may find means of extricating himself from them. The natural urges to this, the sources of unsociableness and mutual opposition from which so many evils arise, drive men to new exertions of their forces and thus to the manifold development of their capacities. They thereby perhaps show the ordering of a wise Creator and not the hand of an evil spirit, who bungled in his great work or spoiled it out of envy.

Fifth Thesis

The greatest problem for the human race, to the solution of which Nature drives man, is the achievement of a universal civic society which administers law among men.

The highest purpose of Nature, which is the development of all the capacities which can be achieved by mankind, is attainable only in society, and more specifically in the society with the greatest freedom. Such a society is one in which there is mutual opposition among the members, together with the most exact definition of freedom and fixing of its limits so that it may be consistent with the freedom of others. Nature demands that humankind should itself achieve this goal like all its other destined goals. Thus a society in which freedom under external laws is associated in the highest degree with irresistible power, i.e., a perfectly just civic constitution, is the highest problem Nature assigns to the human race; for Nature can achieve her other purposes for mankind only upon the solution and completion of this assignment. Need forces men, so enamored otherwise of their boundless freedom, into this state of constraint. They are forced to it by the greatest of all needs, a need they themselves occasion inasmuch as their passions keep them from living long together in wild freedom. Once in such a preserve as a civic union, these same passions subsequently do the most good. It is just the same with trees in a forest: each needs the others, since each in seeking to take the air and sunlight from others must strive upward, and thereby each realizes a beautiful, straight stature, while those that live in isolated freedom put out branches at random and grow stunted, crooked, and twisted. All culture, art which adorns mankind, and the finest social order are fruits of unsociableness, which forces itself to discipline itself and so, by a contrived art, to develop the natural seeds to perfection.

Sixth Thesis

This problem is the most difficult and the last to be solved by mankind.

The difficulty which the mere thought of this problem puts before our eyes is this. Man is an animal which, if it lives among others of its kind, requires a master. For he certainly abuses his freedom with respect to other men, and although as a reasonable being he wishes to have a law which limits the freedom of all, his selfish animal impulses tempt him, where possible, to exempt himself from them. He thus requires a master, who will break his will and force him to obey a will that is universally valid, under which each can be free. But whence does he get this master? Only from the human race. But then the master is himself an animal, and needs a master. Let him begin it as he will, it is not to be seen how he can procure a magistracy which can maintain public justice and which is itself just, whether it be a single person or a group of several elected persons. For each of them will always abuse him freedom if he has none above him to exercise force in accord with the laws. The highest master should be just in himself, and yet a man. This task is therefore the hardest of all; indeed, its complete solution is impossible, for from such crooked wood as man is made of, nothing perfectly straight can be built.[1] That it is the last problem to be solved follows also from this: it requires that there be a correct conception of a possible constitution, great experience gained in many paths of life, and – far beyond these – a good will ready to accept such a constitution. Three such things are very hard, and if they are ever to be found together, it will be very late and after many vain attempts.

Seventh Thesis

The problem of establishing a perfect civic constitution is dependent upon the problem of a lawful external relation among states and cannot be solved without a solution of the latter problem.

What is the use of working toward a lawful civic constitution among individuals, i.e., toward the creation of a common wealth? The same unsociability which drives man to this causes any single commonwealth to stand in unrestricted freedom in relation to others; consequently, each of them must expect from another precisely the evil which oppressed the individuals and forced them to enter into a lawful civic state. The friction among men, the inevitable antagonism, which is a mark of even the largest societies and political bodies, is used by Nature as a means to establish a condition of quiet and security. Through war, through the taxing and never-ending accumulation of armament, through the want which any state, even in peacetime, must suffer internally, Nature forces them to make at first inadequate and tentative attempts; finally, after devastations, revolutions, and even complete exhaustion, she brings them to that which reason could have told them at the beginning and with far less sad experience, to wit, to step from the lawless condition of savages into a league of nations. In a league of nations, even the smallest state could expect security and justice, not from its own power and by its own decrees, but only from this great league of nations (*Foedus Amphictyonum*[2]), from a united power acting according to decisions reached under the laws of their united will. However fantastical this idea may seem – and it was laughed at as fantastical by the Abbé de St. Pierre[3] and by Rousseau,[4] perhaps because they believed it was too near to realization – the necessary outcome of the destitution to which each man is brought by his fellows is to force the states to the same decision (hard though it be for them) that savage man also was reluctantly forced to take, namely, to give up their brutish freedom and to seek quiet and security under a lawful constitution.

All wars are accordingly so many attempts (not in the intention of man, but in the intention of Nature) to establish new relations among states, and through the destruction or at least the dismemberment of all of them to create new political bodies, which, again, either internally or externally, cannot maintain themselves and which must thus suffer like revolutions; until finally, through the best possible civic constitution and common agreement and legislation in external affairs, a state is created which, like a civic commonwealth, can maintain itself automatically.

[There are three questions here, which really come to one.] Would it be expected from an Epicurean concourse of efficient causes that states, like minute particles of matter in their chance contacts, should form all sorts of unions which in their turn are destroyed by new impacts, until once, finally, by chance a structure should arise which could maintain its existence – a fortunate accident that could hardly occur? Or are we not rather to suppose that Nature here follows a lawful course in gradually lifting our race from the lower levels of animality to the highest level of humanity, doing this by her own secret art, and developing in accord with her law all the original gifts of man in this apparently chaotic disorder? Or perhaps we should prefer to conclude that, from all these actions and counteractions of men in the large, absolutely nothing, at least nothing wise, is to issue? That everything should remain as it always was, that we cannot therefore tell but that discord, natural to our race, may not prepare for us a hell of evils, however civilized we may now be, by annihilating civilization and all cultural progress through barbarous devastation? (This is the fate we may well have to suffer under the rule of blind chance – which is in fact identical with lawless freedom – if there is no secret wise guidance in Nature.) These three questions, I say, mean about the same as this: Is it reasonable to assume a purposiveness in all the parts of nature and to deny it to the whole?

Purposeless savagery held back the development of the capacities of our race; but finally, through the evil into which it plunged mankind, it forced our race to renounce this condition and to enter into a civic order in which those capacities could be developed. The same is done by the barbaric freedom of established states. Through wasting the powers of the commonwealths in armaments to be used against each other, through devastation brought on by war, and even more by the necessity of holding themselves in constant readiness for war, they stunt the full development of human nature. But because of the evils which thus arise, our race is forced to find, above the (in itself healthy) opposition of states which is a consequence of their freedom, a law of equilibrium and a united power to give it effect. Thus it is forced to institute a cosmopolitan condition to secure the external safety of each state.

Such a condition is not unattended by the danger that the vitality of mankind may fall asleep; but it is at least not without a principle of balance among men's actions and counteractions, without which they might be altogether destroyed. Until this last step to a union of states is taken, which is the halfway mark in the development of mankind, human nature must suffer the cruelest hardships under the guise of external well-being; and Rousseau was not far wrong in preferring the state of savages, so long, that is, as the last stage to which the human race must climb is not attained.

To a high degree we are, through art and science, *cultured*. We are *civilized* – perhaps too much for our own good – in all sorts of social grace and decorum. But to consider ourselves as having reached *morality* – for that, much is lacking. The ideal of morality belongs to culture; its use for some simulacrum of morality in the love of honor and outward decorum constitutes mere civilization. So long as states waste their forces in vain and violent self-expansion, and thereby constantly thwart the slow efforts to improve the minds of their citizens by even withdrawing all support from them, nothing in the way of a moral order is to be expected. For such an end, a long internal working of each political body toward the education of its citizens is required. Everything good that is not based on a morally good disposition, however, is nothing but pretense and glittering misery. In such a condition the human species will no doubt remain until, in the way I have described, it works its way out of the chaotic conditions of its international relations.

Eighth Thesis

The history of mankind can be seen, in the large, as the realization of Nature's secret plan to bring forth a perfectly constituted state as the only condition in which the capacities of mankind can be fully developed, and also bring forth that external relation among states which is perfectly adequate to this end.

This is a corollary to the preceding. Everyone can see that philosophy can have her belief in a millennium, but her millenarianism is not Utopian, since the Idea can help, though only from afar, to bring the millennium to pass. The only question is: Does Nature reveal anything of a path to this end? And I say: She reveals something, but very little. This great revolution seems to require so long for its completion that the short period during which humanity has been following this course permits us to determine its path and the relation of the parts to the whole with as little certainty as we can determine, from all previous astronomical observation, the path of the sun and his host of satellites among the fixed stars. Yet, on the fundamental premise of the systematic structure of the cosmos and from the little that has been observed, we can confidently infer the reality of such a revolution.

Moreover, human nature is so constituted that we cannot be indifferent to the most remote epoch our race may come to, if only we may expect it with certainty. Such indifference is even less possible for us, since it seems that our own intelligent action may hasten this happy time for our posterity. For that reason, even faint indications of approach to it are very important to us. At present, states are in such an artificial relation to each other that none of them can neglect its internal cultural development without losing power and influence among the others. Therefore the preservation of this natural end [culture], if not progress in it, is fairly well assured by the ambitions of states. Furthermore, civic freedom can hardly be infringed without the evil consequences being felt in all walks of life, especially in commerce, where the effect is loss of power of the state in its foreign relations. But this freedom spreads by degrees. When the citizen is hindered in seeking his own welfare in his own way, so long as it is consistent with the freedom of others, the vitality of the entire enterprise is sapped, and therewith the powers of the whole are diminished. Therefore limitations on personal actions are step by step removed, and general religious freedom is permitted. Enlightenment comes gradually, with intermittent folly and caprice, as a great good which must finally save men from the selfish aggrandizement of their masters, always assuming that the latter know their own interest. This enlightenment, and with it a certain commitment of heart which the enlightened man cannot fail to make to the good he clearly understands, must step by step ascend the throne and influence the principles of government.

Although, for instance, our world rulers at present have no money left over for public education and for anything that concerns what is best in the world, since all they have is already committed to future wars, they will still find it to their own interest at least not to hinder the weak and slow, independent efforts of their peoples in this work. In the end, war itself will be seen as not only so artificial, in outcome so uncertain for both sides, in aftereffects so painful in the form of an ever-growing war debt (a new invention) that cannot be met, that it will be regarded as a most dubious undertaking. The impact of any revolution on all states on our continent, so closely knit together through commerce, will be so obvious that the other states, driven by their own danger but without any legal basis, will offer themselves as arbiters, and thus they will prepare the way for a distant international government for which there is no precedent in world history. Although this government at present exists only as a rough outline, nevertheless in all the members there is rising a feeling which each has for the preservation of the whole. This gives hope finally that after many reformative revolutions, a universal cosmopolitan condition, which Nature has as her ultimate purpose, will come into being as the womb wherein all the original capacities of the human race can develop.

Ninth Thesis

A philosophical attempt to work out a universal history according to a natural plan directed to achieving the civic union of the human race must be regarded as possible and, indeed, as contributing to this end of Nature.

It is strange and apparently silly to wish to write a history in accordance with an Idea of how the course of the world must be if it is to lead to certain rational ends. It seems that with such an Idea only a romance could be written. Nevertheless, if one may assume that Nature, even in the play of human freedom, works not without plan or purpose, this Idea could still be of use. Even if we are too blind to see the secret mechanism of its workings, this Idea may still serve as a guiding thread for presenting as a system, at least in broad outlines, what would otherwise be a planless conglomeration of human actions. For if one starts with Greek history, through which every older or contemporaneous history has been handed down or at least certified;[5] if one follows the influence of Greek history on the construction and misconstruction of the Roman state which swallowed up the Greek, then the Roman influence on the barbarians who in turn destroyed it, and so on down to our times; if one adds episodes from the national histories of other peoples insofar as they are known from the history of the enlightened nations, one will discover a regular progress in the constitution of states on our continent (which will probably give law, eventually, to all the others). If, further, one concentrates on the civic constitutions and their laws and on the relations among states, insofar as through the good they contained they served over long periods of time to elevate and adorn nations and their arts and sciences, while through the evil they contained they destroyed them, if only a germ of enlightenment was left to be further developed by this overthrow and a higher level was thus prepared – if, I say, one carries through this study, a guiding thread will be revealed. It can serve not only for clarifying the confused play of things human, and not only for the art of prophesying later political changes (a use which has already been made of history even when seen as the disconnected effect of lawless freedom), but for giving a consoling view of the future (which could not be reasonably hoped for without the presupposition of a natural plan) in which there will be exhibited in the distance how the human race finally achieves the condition in which all the seeds planted in it by Nature can fully develop and in which the destiny of the race can be fulfilled here on earth.

Such a justification of Nature – or, better, of Providence – is no unimportant reason for choosing a standpoint toward world history. For what is the good of esteeming the majesty and wisdom of Creation in the realm of brute nature and of recommending that we contemplate it, if that part of the great stage of supreme wisdom which contains the purpose of all the others – the history of mankind – must remain an unceasing reproach to it? If we are forced to turn our eyes from it in disgust, doubting that we can ever find a perfectly rational purpose in it and hoping for that only in another world?

That I would want to displace the work of practicing empirical historians with this Idea of world history, which is to some extent based upon an a priori principle, would be a misinterpretation of my intention. It is only a suggestion of what a philosophical mind (which would have to be well versed in history) could essay from another point of view. Otherwise the notorious complexity of a history of our time must naturally lead to serious doubt as to how our descendants will begin to grasp the burden of the history we shall leave to them after a few centuries. They will naturally value the history of earlier times, from which the documents may long since have disappeared, only from the point of view of what interests them, i.e., in answer to the question of what the various nations and governments have contributed to the goal of world citizenship, and what they have done to damage it. To consider this, so as to direct the ambitions of sovereigns and their agents to the only means by which their fame can be spread to later ages: this can be a minor motive for attempting such a philosophical history.

Notes

A statement in the "Short Notices" or the twelfth number of the *Gothaische Gelehrte Zeitung* of this year [1784], which no doubt was based on my conversation with a scholar who was traveling through, occasions this essay, without which that statement could not be understood.

[The notice said: "A favorite idea of Professor Kant's is that the ultimate purpose of the human race is to achieve the most perfect civic constitution, and he wishes that a philosophical historian might undertake to give us a history of humanity from this point of view, and to show to what extent humanity in various ages has approached or drawn away from this final purpose and what remains to be done in order to reach it."]

1 The role of man is very artificial. How it may be with the dwellers on other planets and their nature we do not know. If, however, we carry out well the mandate given us by Nature, we can perhaps flatter ourselves that we may claim among our neighbors in the cosmos no mean rank. Maybe among them each individual can perfectly attain his destiny in his own life. Among us, it is different; only the race can hope to attain it.

2 [An allusion to the Amphictyonic League, a league of Greek tribes originally for the protection of a religious shrine, which later gained considerable political power.]

3 [Charles-Irénée Castel, Abbé de Saint Pierre (1658–1743), in his *Projet de paix perpetuelle* (Utrecht, 1713). Trans. H. H. Bellot (London, 1927).]

4 [In his *Extrait du projet de paix perpetuelle de M. l'Abbé de St. Pierre* (1760). Trans. C. E. Vaughn, *A Lasting Peace through the Federation of Europe* (London, 1917).]

5 Only a learned public, which has lasted from its beginning to our own day, can certify ancient history. Outside it, everything else is *terra incognita*; and the history of peoples outside it can only be begun when they come into contact with it. This happened with the Jews in the time of the Ptolemies through the translation of the Bible into Greek, without which we would give little credence to their isolated narratives. From this point, when once properly fixed, we can retrace their history. And so with all other peoples. The first page of Thucydides, says Hume ("Of the Populousness of Ancient Nations," in *Essays Moral, Political, and Literary*, eds. Green and Grose, Vol. I, p. 414), is the only beginning of all real history.

5

The Nature and Importance of the Positive Philosophy

Auguste Comte

In order to explain properly the true nature and peculiar character of the positive philosophy, it is indispensable that we should first take a brief survey of the progressive growth of the human mind viewed as a whole; for no idea can be properly understood apart from its history.

In thus studying the total development of human intelligence in its different spheres of activity, from its first and simplest beginning up to our own time, I believe that I have discovered a great fundamental law, to which the mind is subjected by an invariable necessity. The truth of this law can, I think, be demonstrated both by reasoned proofs furnished by a knowledge of our mental organization, and by historical verification due to an attentive study of the past. This law consists in the fact that each of our principal conceptions, each branch of our knowledge, passes in succession through three different theoretical states: the theological or fictitious state, the metaphysical or abstract state, and the scientific or positive state. In other words, the human mind – by its very nature – makes use successively in each of its researches of three methods of philosophizing, whose characters are essentially different and even radically opposed to each other. We have first the theological method, then the metaphysical method, and finally the positive method. Hence, there are three kinds of philosophy or general systems of conceptions on the aggregate of phenomena which are mutually exclusive of each other. The first is the necessary starting point of human intelligence; the third represents its fixed and definitive state; the second is destined to serve only as a transitional method.

From Auguste Comte, *Introduction to Positive Philosophy*, ed. and trans. Frederick Ferré, Indianapolis and Cambridge, MA: Hackett, 1988, pp. 1–33. Reprinted by permission of Hackett Publishing Company, Inc.

In the theological state, the human mind directs its researches mainly toward the inner nature of beings, and toward the first and final causes of all the phenomena that it observes – in a word, toward absolute knowledge. It therefore represents these phenomena as being produced by the direct and continuous action of more or less numerous supernatural agents, whose arbitrary intervention explains all the apparent anomalies of the universe.

In the metaphysical state, which is in reality only a simple general modification of the first state, the supernatural agents are replaced by abstract forces, real entities or personified abstractions, inherent in the different beings of the world. These entities are looked upon as capable of giving rise by themselves to all the phenomena observed, each phenomenon being explained by assigning it to its corresponding entity.

Finally, in the positive state, the human mind, recognizing the impossibility of obtaining absolute truth, gives up the search after the origin and hidden causes of the universe and a knowledge of the final causes of phenomena. It endeavours now only to discover, by a well-combined use of reasoning and observation, the actual laws of phenomena – that is to say, their invariable relations of succession and likeness. The explanation of facts, thus reduced to its real terms, consists henceforth only in the connection established between different particular phenomena and some general facts, the number of which the progress of science tends more and more to diminish.

The theological system arrived at its highest form of perfection when it substituted the providential action of a single being for the varied play of the numerous independent gods which had been imagined by the primitive mind. In the same way, the last stage of the metaphysical system consisted in replacing the different special entities by the idea of a single great general entity – nature – looked upon as the sole source of all phenomena. Similarly, the ideal of the positive system, toward which it constantly tends, although in all probability it will never attain such a stage, would be reached if we could look upon all the different phenomena observable as so many particular cases of a single general fact, such as that of gravitation, for example.

This is not the place to give a special demonstration of this fundamental law of mental development, and to deduce from it its most important consequences. We shall make a direct study of it, with all the necessary details, in the part of this work relating to social phenomena.[1] I am considering it now only in order to determine precisely the true character of the positive philosophy, as opposed to the two other philosophies which have successively dominated our whole intellectual system up to these latter centuries. For the present, to avoid leaving entirely undemonstrated so important a law, the applications of which frequently occur throughout this work, I must confine myself to a rapid enumeration of the most evident general reasons that prove its exactitude.

In the first place, it is, I think, sufficient merely to enunciate such a law for its accuracy to be immediately verified by all those who are fairly well acquainted with the general history of the sciences. For there is not a single science that has today reached the positive stage, which was not in the past – as each can easily see for himself – composed mainly of metaphysical abstractions, and, going back further still, it was altogether under the sway of theological conceptions. Unfortunately, we shall have to recognize on more than one occasion in the different parts of this course, that even the most perfect sciences retain today some very evident traces of these two primitive states.

This general revolution of the human mind can, moreover, be easily verified today in a very obvious, although indirect, manner, if we consider the development of the individual intelligence. The starting point being necessarily the same in the education of the individual as in that of the race, the various principal phases of the former must reproduce the fundamental epochs of the latter. Now, does not each of us in contemplating his own history recollect that he has been successively – as regards the most important ideas – a theologian in childhood, a metaphysician in youth, and a natural philosopher in manhood? This verification of the law can easily be made by all who are on a level with their era.

But in addition to the proofs of the truth of this law furnished by direct observation of the race or the individual, I must, above all, mention in this brief summary the theoretical considerations that show its necessity.

The most important of these considerations arises from the very nature of the subject itself. It consists in the need at every epoch of having some theory to connect the facts, while, on the other hand, it was clearly impossible for the primitive human mind to form theories based on observation.

All competent thinkers agree with Bacon[2] that there can be no real knowledge except that which rests upon observed facts. This fundamental maxim is evidently indisputable if it is applied, as it ought to be, to the mature state of our intelligence. But, if we consider the origin of our knowledge, it is no less certain that the primitive human mind could not and, indeed, ought not to have thought in that way. For if, on the one hand, every positive theory must necessarily be founded upon observations, it is, on the other hand, no less true that, in order to observe, our mind has need of some theory or other. If in contemplating phenomena we did not immediately connect them with some principles, not only would it be impossible for us to combine these isolated observations and, therefore, to derive any profit from them, but we should even be entirely incapable of remembering the facts, which would for the most part remain unnoted by us.

Thus, there were two difficulties to be overcome: the human mind had to observe in order to form real theories; and yet it had to form theories of some sort before it could apply itself to a connected series of observations. The primitive human mind, therefore, found itself involved in a vicious circle, from which it would never have had any means of escaping if a natural way out of the difficulty had not fortunately been found by the spontaneous development of theological conceptions. These presented a rallying point for the efforts of the mind, and furnished materials for its activity. This is the fundamental motive which demonstrates the logical necessity for the purely theological character of primitive philosophy, apart from those important social considerations relating to the matter which I cannot even indicate now.

This necessity becomes still more evident when we regard the perfect congruity of theological philosophy with the peculiar nature of the researches on which the human mind in its infancy concentrated to so high a degree all its efforts. It is, indeed, very noticeable that the most insoluble questions – such as the inner nature of objects, or the origin and purpose of all phenomena – are precisely those which the human mind proposes to itself, in preference to all others, in its primitive state, all really soluble problems being looked upon as hardly worthy of serious thought. The reason for this is very obvious, since it is experience alone that has enabled us to estimate our abilities rightly, and, if man had not commenced by overestimating his forces, these would never have been able to acquire all the development of which they are capable. This fact is a necessity of our organization. But, be that as it may, let us picture to ourselves as far as we are able this [early] mental disposition, so universal and so prominent, and let us ask ourselves what kind of reception would have been accorded at such an epoch to the positive philosophy, supposing it to have been then formed. The highest ambition of this philosophy is to discover the laws of phenomena, and its main characteristic is precisely that of regarding as necessarily interdicted to the human reason all those sublime mysteries which theological philosophy, on the contrary, explains with such admirable facility, even to the smallest detail. [Under such circumstances, it is easy to see what the choice of primitive man would be.]

The same thing is true when we consider from a practical standpoint the nature of the pursuits with which the human mind first occupies itself. Under that aspect they offer to man the strong attraction of an unlimited control over the exterior world, which is regarded as being entirely destined for our use, while all its phenomena seem to have close and continuous relations with our existence. These chimerical hopes, these exaggerated ideas of man's importance in the universe, to which the theological philosophy gives rise, are destroyed irrevocably by the first fruits of the positive philosophy. But at the beginning they afforded an indispensable stimulus without the aid of which we cannot, indeed, conceive how the primitive human mind would have been induced to undertake any arduous labors.

We are at the present time so far removed from that early state of mind – at least as regards the majority of phenomena – that it is difficult for us to appreciate properly the force and necessity of such considerations. Human reason is now so mature that we are able to undertake laborious scientific researches without having in view any extraneous goal capable of strongly exciting the imagination, such as that which the astrologers or alchemists proposed to themselves. Our intellectual activity is sufficiently excited by the mere hope of discovering the laws of phenomena, by the simple desire of verifying or disproving a theory. This, however, could not be the case in the infancy of the human mind. Without the attractive chimeras of astrology, or the powerful deceptions of alchemy, for example, where should we have found the perseverance and ardor necessary for collecting the long series of observations and experiments which later on served as a basis for the first positive theories of these two classes of phenomena?

The need for such a stimulus to our intellectual development was keenly felt long ago by Kepler[3] in the case of astronomy, and has been justly appreciated in our own time by Berthollet[4] in chemistry.

The above considerations show us that, although the positive philosophy represents the true final state of human intelligence – that to which it has always tended more and more – it was nonetheless necessary to employ the theological philosophy at first and during many centuries, both as a method and as furnishing provisional doctrines. Because the theological philosophy was spontaneous in its character, it was the only one possible in the beginning; it was also the only one to offer a sufficient interest to our budding intelligence. It is now very easy to see that, in order to pass from this provisional form of philosophy to the final stage, the human mind was naturally obliged to adopt metaphysical methods and doctrines as a transitional form of philosophy. This last consideration is indispensable in order to complete the general sketch of the great law which I have pointed out.

It is easily seen that our understanding, [which was] compelled to progress by almost insensible steps, could not pass suddenly and without any intermediate stages from theological to positive philosophy. Theology and physics are so profoundly incompatible, their conceptions are so radically opposed in character, that, before giving up the one in order to employ the other exclusively, the human intelligence had to make use of intermediate conceptions, which, being of a hybrid character, were eminently fitted to bring about a gradual transition. That is the part played by metaphysical conceptions, and they have no other real use. By substituting, in the study of phenomena, a corresponding inseparable entity for a direct supernatural agency – although at first the former was only held to be an offshoot of the latter – man gradually accustomed himself to consider only the facts themselves. This development was caused by the concepts of metaphysical agents gradually becoming so empty through oversubtle qualification that all right-minded persons considered them to be only the abstract names of the phenomena in question. It is impossible to imagine by what other method our understanding could have passed from frankly supernatural to purely natural considerations, or, in other words, from the theological to the positive régime.

I have thus established, insofar as it is possible without entering into a special discussion, which would be out of place at the present moment, that which I conceive to be the general law of mental development. It will now be easy for us to determine precisely the exact nature of the positive philosophy. To do that is the special object of this chapter.

We have seen that the fundamental character of the positive philosophy is to consider all phenomena as subject to invariable natural laws. The exact discovery of these laws and their reduction to the least possible number constitute the goal of all our efforts; for we regard the search after what are called causes, whether first or final, as absolutely inaccessible and unmeaning. It is unnecessary to dwell much on a principle that has now become so familiar to all who have made anything like a serious study of the observational sciences. Everybody, indeed, knows that in our positive explanations, even when they are most complete, we do not pretend to explain the real causes of phenomena, as this would merely throw the difficulty further back; we try only to analyze correctly the circumstances of their production, and to connect them by normal relations of succession and similarity.

Thus, to cite the best example, we say that the general phenomena of the universe are explained – as far as they can be – by the Newtonian law of gravitation. On the one hand, this admirable theory shows us all the immense variety of astronomical facts as only a single fact looked at from different points of view, that fact being the constant tendency of all molecules towards each other in direct proportion to their masses and inversely as the squares of their distances. On the other hand, this general fact is shown to be the simple extension of an extremely familiar and, therefore, well-known phenomenon – the weight of a body at the earth's surface. As to determining what attraction and weight are in themselves, or what their causes are – these are questions which we regard as insoluble and outside the domain of the positive philosophy; we, therefore, rightly abandon them to the imagination of the theologians or the subtleties of the metaphysicians. That it is clearly impossible to solve such questions is shown by the fact that, whenever an attempt has been made to give a rational explanation of the matter, the greatest thinkers have only been able to define one of these principles by the other. Attraction is defined as nothing but universal weight, and weight is said to consist simply in terrestrial attraction. Explanations of this kind raise a smile, if put forward as furnishing us with a knowledge of "things-in-themselves" and the mode of causation of phenomena. They are, however, the only satisfactory results obtainable, for they present as identical two orders of phenomena which for so long a time were regarded as unconnected. No sensible person would nowadays seek to go beyond this.

It would be easy to multiply these examples, which will occur very frequently throughout this treatise, for at the present day all great intellectual operations are conducted in this spirit. To take a single example of this from contemporary works, I will choose the fine series of researches made by Fourier[5] on the theory of heat. This affords us an excellent verification of the preceding general remarks. In this work, the philosophical character of which is so eminently positive, the most important and most precise laws of thermal phenomena are disclosed; but the author has not once inquired into the intimate nature of heat itself, nor has he mentioned, except to point out its uselessness, the vigorous controversy between the partisans of heat as a material substance and those who make it consist in the vibrations of a universal ether. Yet, that work treats of the most important questions, several of which had never been raised – a clear proof that the human mind, by simply confining itself to researches of an entirely positive order, can find therein inexhaustible food for its highest form of activity without attacking inaccessible problems.

Having thus indicated, insofar as it was possible in this general sketch, the spirit of the positive philosophy, which the whole of this course is intended to develop, we must next consider what stage in the formation of that philosophy has now been reached and what remains to be done in order to constitute it fully.

For this purpose, we must in the first place remember that the different branches of our knowledge were not able to pass at the same rate through the three great phases of their development indicated above, and that consequently they did not arrive simultaneously at the positive state. There exists in this respect an invariable and necessary order that our various classes of conceptions have followed, and were bound to follow, in their progressive course; and the exact consideration of this order is the indispensable complement of the fundamental mental law previously enunciated. That order will form the special subject of the next chapter. At present it is sufficient to know that it conforms to the diverse nature of the phenomena, and that it is determined by their degree of generality, of simplicity, and of reciprocal independence – three considerations which, although quite distinct, lead to the same result. Thus, astronomical phenomena, being the most general, the simplest, and the most independent of all others, were the first to be subjected to positive theories; then followed in succession and for the same reasons the phenomena of terrestrial physics, properly so called, those of chemistry, and, finally, those of physiology.

It is impossible to fix the precise date of this mental revolution; we can say only that, like all other great human events, it took place continuously and at an increasing rate, especially since the labors of Aristotle and the Alexandrian school, and afterward from the introduction of natural science into the west of Europe by the Arabs. However, as it is better to fix an epoch in order to give greater precision to our ideas, I would select that of the great movement imparted to the human intellect two centuries ago by the combined influence of the precepts of Bacon, the conceptions of Descartes,[6] and the discoveries of Galileo.[7] It was then that the spirit of the positive philosophy began to assert itself in the world, in evident opposition to the theological and metaphysical spirit; for it was then that positive conceptions disengaged themselves clearly from the superstitious and scholastic alloy, which had more or less disguised the true character of all the previous scientific work.

Since that memorable epoch, the increasing influence of the positive philosophy and the decadent movement of theological and metaphysical philosophy have been extremely marked. These movements have at last become so pronounced that at the present day it is impossible for any observer acquainted with the spirit of his age to fail to recognize the final bent of the human mind toward positive studies, and the irrevocable break henceforth from those fruitless doctrines and provisional methods that were suited only to its first flight. This fundamental mental revolution will, therefore, necessarily be carried out to the fullest extent. If, then, there still remains some great conquest to be made, some important division of the intellectual domain to be invaded, we can be certain that the transformation will take place there also, as it has been carried out in all the other branches of science. It would evidently be absurd to suppose that the human mind, which is so disposed to unity of method, would yet preserve indefinitely, in the case of a single class of phenomena, its primitive mode of philosophizing, when it has once adopted for the other classes a new philosophic path of an entirely opposite character.

The whole thing reduces itself, therefore, to a simple question of fact: Does the positive philosophy, which during the last two centuries has gradually acquired so great an extension, embrace at the present day all classes of phenomena? It is evident that it does not; therefore, a great scientific work still remains to be executed in order to give the positive philosophy that universal character indispensable for its final constitution.

In the four principal categories of natural phenomena enumerated above – astronomical, physical, chemical, and physiological – we notice an important omission relating to social phenomena. Although these are implicitly comprised among physiological phenomena, yet, owing to their importance and the inherent difficulties of their study, they deserve to form a distinct class. This last order of ideas is concerned with the most special, most complicated, and most dependent of all phenomena; it has, therefore, necessarily progressed more slowly than all the preceding orders, even if we do not take into account the more special obstacles to its study which we shall consider later on. However that may be, it is evident that it has not yet been included within the domain of positive philosophy. Theological and metaphysical methods are never used now by anyone in dealing with all the other kinds of phenomena, either as a means of investigation or even as a mode of reasoning. But these discarded methods are, on the contrary, still used exclusively for both purposes in everything that concerns social phenomena, although their insufficiency in this respect has been fully felt already by all good minds, such men being tired of these empty and endless discussions between, e.g., divine right and the sovereignty of the people.

Here, then, is the great, but evidently the only, gap that has to be filled in order to finish the construction of the positive philosophy. Now that the human mind has founded celestial physics, terrestrial physics (mechanical and chemical), and organic physics (vegetable and animal), it only remains to complete the system of observational sciences by the foundation of social physics.[8] This is at the present time, under several important aspects, the greatest and most pressing of our cognitive needs, and to meet this need is, I make bold to say, the first purpose of this work, its special object.

The conceptions which I shall endeavor to present relating to the study of social phenomena, and of which I hope the present chapter has already enabled us to see the germ, cannot be expected to raise social physics at once to the degree of perfection that has been reached by the earlier branches of natural philosophy. Such a hope would be evidently chimerical, seeing that these branches still differ widely from one another in perfectness, as was, indeed, inevitable. But I aim at impressing upon this last branch of our knowledge the same positive character that already marks all the other branches. If this condition is once really fulfilled, the philosophical system of the modern world will be founded at last in its entirety; for there is no observable fact that would not then be included in one or another of the five great categories of astronomical, physical, chemical, physiological, and social phenomena. All our fundamental conceptions having thus been rendered homogeneous, philosophy will be constituted finally in the positive state. Its character will be henceforth unchangeable, and it will then have only to develop itself indefinitely, by incorporating the constantly increasing knowledge that inevitably results from new observations or more profound meditations. Having by this means acquired the character of universality which as yet it lacks, the positive philosophy, with all its natural superiority, will be able to displace entirely the theological and metaphysical philosophies. The only real property possessed by theology and metaphysics at the present day is their character of universality, and when deprived of this motive for preference they will have for our successors only a historical interest.

The first and special object of this course having been thus set forth, it is easy to comprehend its second and general aim, that which constitutes it a course of positive philosophy, and not merely a course on social physics.

The formation of social physics at last completes the system of natural sciences. It, therefore, becomes possible and even necessary to summarize these different sciences, so that they may be coordinated by presenting them as so many branches of a single trunk, instead of continuing to look upon them as only so many isolated groups. Therefore, before proceeding to the study of social phenomena, I shall successively consider, in the encyclopedic order given above, the different positive sciences already formed.

It is, I think, unnecessary to warn the reader that I do not claim to give here a series of special courses of lectures on each of the principal branches of natural philosophy. Not to speak of the enormous time that such an enterprise would take, it is clear that I cannot claim to be equipped for it, nor, I think I may add, can anyone else in the present state of human education. On the contrary, a course of the kind contemplated here requires, if it is to be understood properly, a previous series of special studies on the different sciences which will be treated therein. In the absence of this condition, it is very difficult to realize, and impossible to estimate, the philosophical reflections that will be made upon these sciences. In one word, it is a course on positive philosophy, and not on the positive sciences, that I propose to give. We shall have to consider here only each fundamental science in its relations with the whole positive system and the spirit characterizing it; that is to say, under the twofold aspect of its essential methods and its principal achievements. As to the achievements, indeed, I shall often do no more than mention them as known to specialists, though I shall try to estimate their importance.

In order to sum up the ideas relating to the twofold purpose of this course, I must call attention to the two objects – the one special, the other general – that I have in view and that, although distinct in themselves, are necessarily inseparable.

On the one hand, it would be impossible to conceive of a course of positive philosophy unless social physics had been founded first, since an essential element would then be lacking; consequently, the conceptions of such a course would not have that character of generality that ought to be their principal attribute and that distinguishes our present study from any series of special studies. On the other hand, how can we proceed with sure step to the positive study of social phenomena if the mind has not been prepared first by the thorough consideration of positive methods in the case of less complex phenomena, and furnished in addition with a knowledge of the principal laws of earlier phenomena, all of which have a more or less direct influence upon social facts?

Although all the fundamental sciences do not inspire ordinary minds with an equal interest, there is not one of them that should be neglected in such a study as we are about to undertake. As regards the welfare of the human race, all of them are certainly of equal importance when we examine them thoroughly. Besides, those whose results seem at first sight to offer only a minor practical interest are yet of the greatest importance, owing to either the greater perfection of their methods or the indispensable foundation of all the others. This is a consideration to which I shall have special occasion to refer in the next chapter.

To guard as far as possible against the misconceptions likely to arise respecting a work as novel as this, I must add a few remarks to the explanations already given. I refer especially to that universal predominance of specialism, which hasty readers might think was the tendency of this course, and which is so rightly looked upon as wholly contrary to the true spirit of the positive philosophy. These remarks, moreover, will have the more important advantage of exhibiting this spirit under a new aspect, calculated to make its general idea clearer.

In the primitive state of our knowledge, no regular division exists among intellectual labors; all the sciences are cultivated simultaneously by the same minds. This method of organizing human studies is at first inevitable and even indispensable, as I shall have occasion to show later on; but it gradually changes in proportion as the different orders of conceptions develop themselves. By a law whose necessity is evident, each branch of the scientific system gradually separates from the trunk when it has developed far enough to admit of separate cultivation – that is to say, when it has arrived at a stage in which it is capable of constituting the sole pursuit of certain minds. It is to this division of the various kinds of research among different orders of scientists that we evidently owe the development which each distinct class of human knowledge has attained in our time; but this very division renders it impossible for modern scientists to practice that simultaneous cultivation of all the sciences which was so easy and so common in antiquity. In a word, the division of intellectual labor, carried out further and further, is one of the most important and characteristic attributes of the positive philosophy.

But, while recognizing the prodigious results due to this division, and while seeing that it henceforth constitutes the true fundamental basis of the general organization of the scientific world, it is, on the other hand, impossible not to be struck by the great inconveniences which it at present produces, because of the excessive specialization of the ideas that exclusively occupy each mind. This unfortunate result, being inherent in the very principle of the division of labor, is no doubt inevitable up to a certain point. Do what we will, therefore, we shall never be able to equal the ancients in this respect, for their general superiority was due to the slight degree of development of their knowledge. Yet, I think we can, by proper means, avoid the most pernicious effects of an exaggerated specialism without doing injury to the fruitful influence of the division of labor in research. There is an urgent need to consider this question seriously, for these inconveniences, which by their very nature tend constantly to increase, are now becoming very apparent. Everyone agrees that the divisions which we establish between the various branches of natural philosophy, in order to make our labors more perfect, are at bottom artificial. In spite of this admission, we must not forget that the number of scientists who study the whole of even a single science is already very small, although such a science is, in its turn, only a part of a greater whole. The majority of scientists already confine themselves entirely to the isolated consideration of a more or less extensive section of a particular science, without concerning themselves much about the relationship between their special work and the general system of positive knowledge. Let us hasten to remedy this evil before it becomes more serious. Let us take care that the human mind does not lose its way in a mass of detail. We must not conceal from ourselves that this is the essentially weak side of our system, and that this is the point on which the partisans of theological and metaphysical philosophy may still attack the positive philosophy with some hope of success.

The true means of arresting the pernicious influence that seems to threaten the intellectual future of mankind, because of too great a specialization of individual researches, is clearly not to return to the ancient confusion of labors. This would tend to put the human mind back; and, besides, such a return has happily become impossible now. The true remedy consists, on the contrary, in perfecting the division of labor itself. All that is necessary is to create one more great speciality, consisting in the study of general scientific traits. We need a new class of properly trained scientists who, instead of devoting themselves to the special study of any particular branch of science, shall employ themselves solely in the consideration of the different positive sciences in their present state. It would be their function to determine exactly the character of each science, to discover the relations and concatenation of the sciences, and to reduce, if possible, all their chief principles to the smallest number of common principles, while always conforming to the fundamental maxims of the positive method. At the same time the other scientists, before devoting themselves to their respective specialities, should have received a previous training embracing all the general principles of positive knowledge. This would enable them henceforth to make immediate use of the light thrown on their work by the scientists devoted to the study of the sciences in general, whose results the specialists would in turn be able to rectify. That is a state of things to which the existing scientists are drawing nearer every day. If these two great conditions were once fulfilled, as they evidently can be, then the division of labor in the sciences could be carried on without any danger as far as the development of the different kinds of knowledge required. There would be a distinct class of men [always open to the critical discipline of all the other classes], whose special and permanent function would consist in connecting each new special discovery with the general system; and we should then have no cause to fear that too great an attention bestowed upon the details would ever prevent us from perceiving the whole. In a word, the modern organization of the scientific world would then be accomplished, and would be susceptible of indefinite development, while always preserving the same character.

To make the study of the universal characteristics of the sciences a distinct department of intellectual labor is merely a further extension of the same principle of division that led to the successive separation of the different sciences. As long as the different positive sciences were only slightly developed, their mutual relations were not important enough to give rise (at all events permanently) to a special discipline, nor was the need of this new study nearly as urgent as it is now. But at the present day each of the sciences has developed on its own lines to such an extent that the examination of a mutual relationship affords material for systematic and continued labor, while at the same time this new order of studies becomes indispensable to prevent the dispersion of human ideas.

Such, in my view, is the office of the positive philosophy in relation to the positive sciences, properly so called. Such, at all events, is the aim of the present work.

I have now determined, as exactly as possible in a first sketch, the general spirit of a course of positive philosophy. In order to bring out its full character, I must state concisely the principal general advantages that such a work may have – if its essential conditions are fulfilled properly – as regards intellectual progress. I will mention only four. They are fundamental qualities of the positive philosophy.

In the first place, the study of the positive philosophy, by considering the results of the activity of our intellectual faculties, furnishes us with the only really rational means of exhibiting the logical laws of the human mind, which have hitherto been sought by methods so ill calculated to reveal them.

To explain what I mean on this point I must first recall a philosophical conception of the highest importance, set forth by Blainville[9] in the fine introduction to his *Principles of Comparative Anatomy*. According to him, every active being, and especially every living being, may be studied in all its manifestations under two fundamental relations – the static and the dynamic; that is, as fitted to act and as actually acting. It is clear that all the considerations which might be presented will necessarily fall under the one or the other of these heads. Let us apply this luminous fundamental maxim to the study of intellectual functions.

If these functions are regarded from a static point of view, their study can consist only in determining the organic conditions on which they depend; it thus forms an essential part of anatomy and physiology. When considering the question from a dynamic point of view, we have merely to study the actual march of the human intellect, in practice, by examining the procedures used by it in order to acquire a knowledge of the various sciences; this constitutes essentially the general object of the positive philosophy as I have already defined it in this chapter. In brief, we must look upon all scientific theories as so many great logical facts; and it is only by a thorough observation of these facts that we can rise to the knowledge of logical laws.

These are evidently the only two general methods, complementary to each other, by the use of which we are able to arrive at any really rational ideas concerning intellectual phenomena. We see that in no case is there room for that illusory psychology – the last transformation of theology – the revival of which attempts are being made so vainly at the present day. This theory, while ignoring and discarding the physiological study of our intellectual organs and the observation of the rational methods that actually direct our various scientific researches, claims that it can discover the fundamental laws of the human mind, by contemplating it in itself, without paying any attention to either the causes or the effects of its activity.

The preponderance of the positive philosophy has been growing steadily since Bacon's time. It has today acquired, indirectly, so great a hold over even those minds that are the least familiar with its immense development that the metaphysicians devoted to the study of the intellect could only hope to check the decadence of their pretended science by presenting their doctrines as also being founded upon the observation of facts. In order to do this, they have recently attempted to distinguish, by a very singular subtlety, two kinds of observation of equal importance, the one exterior, the other interior, the latter being devoted solely to the study of intellectual phenomena. To enter into a special discussion of this fundamental sophism would be out of place here. I must be content with indicating the principal consideration which proves clearly that this pretended direct contemplation of the mind by itself is a pure illusion.

It was thought until quite recently that vision was explained by saying that the luminous action of bodies produces on the retina actual images representing exterior forms and colors. To this the physiologists have reasonably objected that, if the luminous impressions produced real images on the retina, we should need another eye to see them. Is not this reasoning still more applicable in the present instance?

It is clear that, by an inevitable necessity, the human mind can observe all phenomena directly, except its own. Otherwise, by whom would the observation be made? As far as moral phenomena are concerned, it may be granted that it is possible for a man to observe the passions that animate him, for the anatomical reason that the organs which are their seat are distinct from those whose functions are devoted to observation. Everyone has had occasion to notice this fact for himself, but such observations would evidently never possess much scientific value. The best way of knowing the passions will always be to observe them from the outside; for a person in any state of extreme passion – that is to say, in precisely the state that it is most essential to examine – would necessarily be incapacitated for observing himself. But in the case of intellectual phenomena, to observe them in this manner while they are taking place is clearly out of the question. The thinking individual cannot cut himself in two – one of the parts reasoning, while the other is looking on. Since in this case the organ observed and the observing organ are identical, how could any observation be made?

The principle of this so-called psychological method is therefore entirely worthless. Besides, consider to what thoroughly contradictory proceedings it immediately leads! On the one hand, you are recommended to isolate yourself as far as possible from the outer world, and you must especially give up all intellectual work; for if you were engaged in making only the simplest calculation, what would become of the interior observation? On the other hand, after having, by means of due precautions, at last attained this perfect state of intellectual slumber, you must then occupy yourself in contemplating the operations that will be taking place in a mind supposed to be blank! Our descendants will no doubt see such pretensions ridiculed on the stage some day.

The results of such a strange procedure are in thorough accordance with the principle. For the last two thousand years metaphysicians have been cultivating psychology in this manner, and yet they have not been able to agree on one single intelligible and sound proposition. They are, even at the present day, divided into a multitude of schools that are incessantly disputing on the first elements of their doctrines. In fact, interior observation gives rise to almost as many divergent opinions as there are so-called observers.

The true scientists – the men devoted to the positive sciences – are still calling in vain on these psychologists to cite a single real discovery, great or small, due to this much-vaunted method. It does not follow that all their labors have been absolutely fruitless as regards the general progress of our knowledge, and we must remember the valuable service that they rendered in sustaining the activity of human intelligence at a time when it could not find a more substantial ailment. But their writings consist largely of that which an illustrious positive philosopher, M. Cuvier,[10] has well called "metaphors mistaken for reasoning." We may safely affirm that any true notions they present have been obtained, not by their pretended method, but by observations on the progress of the human mind – observations to which the development of the sciences has from time to time given birth. And even these ideas, so scanty in number, although proclaimed with so much emphasis, and due only to the unfaithfulness of the psychologists to their pretended methods, are generally either greatly exaggerated or very incomplete, and they are very inferior to the remarks that scientists have already unostentatiously made upon the methods which they employ. It would be easy to cite some striking examples of this, if I did not fear that I should be prolonging the discussion of the point too much: take, for instance, the treatment of the theory of [algebraical] signs [by metaphysicians and geometers respectively].

The considerations relating to logical science which I have just indicated become still more evident when we deal with the art of logic.

For when we want not only to know what the positive method consists in, but also to have such a clear and deep knowledge of it as to be able to use it effectively, we must consider it in action; we must study the various great applications of the method that the human mind has made and already verified. In a word, it is only by a philosophical examination of the sciences that we can attain the desired result. The method does not admit of being studied apart from the researches on which it is employed; or, at all events, it is only a lifeless study, incapable of fertilizing the mind that resorts to it. Looking at it in that abstract way, the only real information that you can give about it amounts to no more than a few general propositions, so vague that they can have no influence on mental habits. When we have thoroughly established as a logical thesis that all our knowledge must be founded upon observation, that we must proceed sometimes from facts to principles, at other times from principles to facts, and some other similar aphorisms, we still know the method far less clearly than he who, even without any philosophical purpose in view, has studied at all completely a single positive science. It is because they have failed to recognize this essential fact that our psychologists have been led to take their reveries for science, in the belief that they understood the positive method because they have read the precepts of Bacon or the discourse of Descartes.

I do not know if, in the future, it will become possible to construct by a priori reasoning a genuine course on method, wholly independent of the philosophical study of the sciences; but I am quite convinced that it cannot be done at present, for the great logical methods cannot yet be explained with sufficient precision apart from their applications. I venture to add, moreover, that, even if such an enterprise could be carried out eventually, which is conceivable, it would nevertheless be only through the study of regular applications of scientific methods that we could succeed in forming a good system of intellectual habits; this is, however, the essential object to be gained by studying method. There is no need to insist further just now on a subject that will recur frequently throughout this work and in regard to which I shall present some new considerations in the next chapter.

The first great direct result of the positive philosophy is then the manifestation by experience of the laws that our intellectual functions follow in their operations and, consequently, a precise knowledge of the general rules that are suitable for our guidance in the investigation of truth.

A second consequence, of no less importance and of much more urgent concern, which must immediately result from the establishment of the positive philosophy as defined in this chapter, is the general recasting of our educational system.

Competent judges are already unanimous in recognizing the necessity of replacing our European education, which is still essentially theological, metaphysical, and literary, by a positive education in accordance with the spirit of our time and adapted to the needs of modern civilization. Various attempts have been made in increasing number during the last hundred years, and especially during recent years, to spread and augment, without ceasing, instruction of a positive kind. Such attempts, which the different European governments have always eagerly encouraged and often initiated, are a sufficient testimony that the spontaneous feeling of this necessity is everywhere growing. But, while supporting these useful enterprises as much as possible, we must not conceal the fact that in the present state of our ideas they are not at all capable of attaining their principal object – namely, the fundamental regeneration of general education. The exclusive speciality, the too rigid isolation, which still characterizes our way of conceiving and of cultivating the sciences, has necessarily a marked influence upon the mode of teaching them. An intelligent person who wishes at the present day to study the principal branches of natural philosophy, in order to acquire a general system of positive ideas, is obliged to study each separate science in the same way and with the same amount of detail as if he wished to become an astronomical or chemical specialist, etc. This renders such an education almost impossible and necessarily very imperfect, even in the case of the most intelligent minds placed in the most favorable circumstances. Such a mode of proceeding would, therefore, be wholly chimerical as regards general education; and yet, an essential requirement of the latter is a complete body of positive conceptions on all the great classes of natural phenomena. It is such a general survey, on a more or less extended scale, which must henceforth constitute, even among the mass of the people, the permanent basis of all human combinations; it must, in short, constitute the mental framework of our descendants. In order that natural philosophy may be able to complete the already partially accomplished regeneration of our intellectual system, it is therefore indispensable that the different sciences of which it is composed – regarding them as the different branches of a single trunk – should first be reduced to what constitutes their essence – that is, to their principal methods and most important results. It is in this way only that the teaching of the sciences can become the basis of a new general and really rational education for our people. Of course, each individual, after receiving this general education, will have to supplement it by such special education as he may require, in which he will study one or other of the special sciences. But the essential consideration which I wish to point out here is that all these special studies, even if all of them were toilsomely compiled, would necessarily be insufficient really to renew our educational system, if they did not rest on the preliminary basis of this general education which is itself the direct result of the positive philosophy as defined in this discourse.

The special study of the general traits of the sciences is not only destined to reorganize education, but it will also contribute to the particular progress of the different positive sciences. This constitutes the third fundamental property that I have to point out.

The divisions that we establish between the sciences, although not arbitrary as some people suppose, are yet essentially artificial. In reality, the subject of all our researches is one; we divide it only so that we may, by separating the difficulties, resolve them more easily. And so it not infrequently happens that these established divisions are a hindrance, and that questions arise which need to be treated by combining the points of view of several sciences. This cannot be done easily when scientists are so addicted to specialization. Hence, the problems are left unsolved for a much longer time than would otherwise be necessary. Such an inconvenience must make itself especially felt in the case of the more essential doctrines of each positive science. Very striking examples of this fact could be cited easily, and I shall carefully call attention to them as they occur in the course of this work.

I could cite a very memorable example of this from the past, in the case of the admirable conception of Descartes relating to analytical geometry. This fundamental discovery, which has changed the aspect of mathematical science and in which we should see the true germ of all the great subsequent progress, is it not simply the result of establishing a closer connection between two sciences that had hitherto been regarded from separate standpoints. But the case will be even more decisive if we consider some questions that are still under discussion.

I will take the case, in chemistry, of the important doctrine of definite proportions. It is certain that the memorable discussion which has been raised in our own time, relating to the fundamental principle of this theory, cannot yet be considered, in spite of appearances, as irrevocably terminated. For this is not, in my opinion, a simple question of chemistry. I venture to assert that, in order to settle the point definitively – that is, to determine whether it is a law of nature that atoms necessarily combine together in fixed proportions – it will be indispensable to unite the chemical with the physiological point of view. This is shown by the fact that, even in the opinion of the illustrious chemists who have most powerfully contributed to the formation of this doctrine, the utmost that can be said is that it is always verified in the composition of inorganic bodies; but it is no less constantly at fault in the case of organic compounds, to which up to the present it seems quite impossible to extend the doctrine. Now, before erecting the theory into a truly fundamental principle, ought not this immense exception to be considered first? Does it not belong to the same general characteristic of all organic bodies, that in none of their phenomena can we make use of invariable numbers? However that may be, an entirely new order of considerations, belonging equally to chemistry and physiology, is evidently necessary in order to decide finally, in some way or other, this great question of natural philosophy.

I think it will be well to consider here a second example of the same kind, which since it relates to a subject of much more limited scope, shows even more conclusively the special importance of the positive philosophy in the solution of questions that need the combination of several sciences. This example, which I also take from chemistry, is the still controverted question as to whether, in the present state of our knowledge, nitrogen should be regarded as an element or a compound. The illustrious Berzelius[11] [differing from almost all living chemists] believes it to be a compound; and his reasons, of a purely chemical nature, successfully balance those of present-day chemists. But what I want particularly to point out is that Berzelius, as he admits himself – and a most instructive admission it is – was greatly influenced by the physiological observation that animals that feed on non-nitrogenous matter contain in their tissues just as much nitrogen as the carnivorous animals. It is therefore quite clear that, in order to decide whether nitrogen is or is not an element, we must necessarily call in the aid of physiology, and combine with chemical considerations, properly so called, a series of new researches on the relationships between the composition of living bodies and the nature of their food.

It would be superfluous now to go on multiplying examples of these complex problems, which can be solved only by the ultimate combination of several sciences that are at present cultivated in a wholly independent manner. Those which I have just cited are sufficient to show in a general way the importance of the function that the positive philosophy will perform in perfecting each of the natural sciences, for it is directly destined to organize in a permanent manner combinations of this kind, which could not be formed suitably without its aid.

I must draw attention to a fourth and last fundamental property of that which I have called the positive philosophy, and which no doubt deserves our notice more than any other property, for it is today the most important one from a practical point of view. We may look upon the positive philosophy as constituting the only solid basis of the social reorganization that must terminate the crisis in which the most civilized nations have found themselves for so long. The last part of this course will be specially devoted to establish and develop this proposition. But the general sketch of my great subject which I have undertaken to give in this chapter would lack one of its most characteristic elements if I failed to call attention here to such an essential consideration.

It may be thought that I am making too ambitious a claim for the positive philosophy. But a few very simple reflections will suffice to justify it.

There is no need to prove to readers of this work that the world is governed and overturned by ideas, or, in other words, that the whole social mechanism rests finally on opinions. They know, above all, that the great political and moral crisis of existing societies is due at bottom to intellectual anarchy. Our gravest evil consists, indeed, in this profound divergence that now exists among all minds, with regard to all the fundamental maxims whose fixity is the first condition of a true social order. As long as individual minds are not unanimously agreed upon a certain number of general ideas capable of forming a common social doctrine, we cannot disguise the fact that the nations will necessarily remain in an essentially revolutionary state, in spite of all the political palliatives that may be adopted. Such a condition of things really admits only of provisional institutions. It is equally certain that, if this general agreement upon first principles can once be obtained, the appropriate institutions will necessarily follow, without giving rise to any grave shock; for the greater part of the disorder will have been already dissipated by the mere fact of the agreement. All those, therefore, who feel the importance of a truly normal state of things should direct their attention mainly to this point.

And now, from the lofty standpoint to which the various considerations indicated in this chapter have step by step raised us, it is easy both to characterize clearly the present state of society as regards its inner spirit, and to deduce therefrom the means by which that state can be changed essentially. Returning to the fundamental law enunciated at the commencement of this chapter,

I think we may sum up exactly all the observations relating to the existing situation of society, by the simple statement that the actual confusion of men's minds is at bottom due to the simultaneous employment of three radically incompatible philosophies – the theological, the metaphysical, and the positive. It is quite clear that, if any one of these three philosophies really obtained a complete and universal preponderance, a fixed social order would result, whereas the existing evil consists above all in the absence of any true organization. It is the existence of these three opposite philosophies that absolutely prevents all agreement on any essential point. Now, if this opinion be correct, all that is necessary is to know which of the three philosophies can and must prevail by the nature of things; every sensible man should next endeavor to work for the triumph of that philosophy, whatever his particular opinions may have been before the question was analyzed. The question being once reduced to these simple terms, the issue cannot long remain doubtful, because it is evident for all kinds of reasons, some of the principal of which have been indicated in this chapter, that the positive philosophy is alone destined to prevail in the ordinary course of things. It alone has been making constant progress for many centuries, while its antagonists have been as constantly in a state of decay. Whether this is a good or a bad thing matters little; the general fact cannot be denied, and that is sufficient. We may deplore the fact, but we are unable to destroy it; nor, consequently, can we neglect it, on pain of giving ourselves up to illusory speculations. This general revolution of the human mind is at the present time almost entirely accomplished. Nothing more remains to be done, as I have already explained, than to complete the positive philosophy by including in it the study of social phenomena, and then to sum them up in a single body of homogeneous doctrine. When these two tasks have made sufficient progress, the final triumph of the positive philosophy will take place spontaneously, and will reestablish order in society. The marked preference which almost all minds, from the highest to the lowest, show at the present day for positive knowledge, as contrasted with vague and mystical conceptions, augers well for the reception that awaits this philosophy when it shall have acquired the only quality that it still lacks – a character of suitable generality.

To sum up the matter: the theological and metaphysical philosophies are now disputing with each other the task of reorganizing society, although the task is really too hard for their united efforts; it is between these schools only that any struggle still exists in this respect. The positive philosophy has, up to the present, intervened in the contest only in order to criticize both schools; and it has accomplished this task so well as to discredit them entirely. Let us put it in a condition to play an active part, without paying any further attention to debates that have become useless. We must complete the vast intellectual operation commenced by Bacon, Descartes, and Galileo, by furnishing the positive philosophy with the system of general ideas that is destined to prevail henceforth, and for an indefinite future, among the human race. The revolutionary crisis which harasses civilized peoples will then be at an end.

Such are the four principal advantages that will follow from the establishment of the positive philosophy. I have thought it well to mention them at once, because they supplement the general definition that I have tried to give of it.

Before concluding, I desire to caution the reader briefly against an erroneous anticipation which he might form as to the nature of the present work.

In saying that the aim of the positive philosophy was to sum up, in a single body of homogeneous doctrine, the aggregate of acquired knowledge relating to the different orders of natural phenomena, I did not mean that we should proceed to the general study of these phenomena by looking upon them all as so many different effects of a single principle, as reducible to one sole law. Although I must treat this question specially in the next chapter, I think it necessary to say so much at once, in order to avoid unfounded objections that might otherwise be raised. I refer to those critics who might jump to the conclusion that this course is one of those attempts at universal explanation by a single law, which one sees made daily by men who are entire strangers to scientific methods and knowledge. Nothing of that kind is intended here; and the development of this course will furnish the best proof of it to all those whom the explanations contained in this chapter might have left in any doubt on the subject.

It is my deep personal conviction that these attempts at the universal explanation of all phenomena by a single law are highly chimerical, even when they are made by the most competent minds. I believe that the resources of the human mind are too feeble, and the universe is too complicated, to admit of our ever attaining such scientific perfection; and I also think that a very exaggerated idea is generally formed of the advantages to be derived from it, even were it attainable. In any case, it seems to me evident that, considering the present state of our knowledge, we are yet a long way from the time when any such attempt might reasonably be expected to succeed. It seems to me that we could hope to arrive at it only by connecting all natural phenomena with the most general positive law with which we are acquainted – the law of gravitation – which already links all astronomical phenomena to some of the phenomena of terrestrial physics. Laplace[12] has effectively brought forward a conception by which chemical phenomena would be regarded as purely simple molecular effects of Newtonian attraction, modified by the figure and mutual position of the atoms. This conception would probably always remain an open question, owing to the absence of any essential data respecting the intimate constitution of bodies; and it is almost certain that the difficulty of applying the idea would be so great that we should still be obliged to retain, as an artificial aid, the division which at present is regarded as natural between astronomy and chemistry. Accordingly, Laplace only presented this idea as a mere philosophical game which is incapable of really exercising any useful influence on the progress of chemical science. The case is really stronger, however, for even if we supposed this insurmountable difficulty overcome, we should still not have attained scientific unity, since it would be necessary next to connect the same law of gravitation with the whole of physiology; and this would certainly not be the least difficult part of the task. Yet, the hypothesis which we have just been discussing would be, on the whole, the most favorable to this much-desired unity.

I have no need to go further into details in order to convince the reader that the object of this course is by no means to present all natural phenomena as being at bottom identical, apart from the variety of circumstances. The positive philosophy would no doubt be more perfect if this were possible. But this condition is not at all necessary, either for its systematic formation or for the realization of the great and happy consequences which we have seen that it is destined to produce. The only indispensable unity for those purposes is that of method, which can and evidently must be, and is already largely established. As to the scientific product, it is not necessary that it should be unified; it is sufficient if it be homogeneous. It is, therefore, from the double standpoint of unity of method and homogeneity of scientific propositions that the different classes of positive theories will be considered in the present work. While trying to diminish as far as possible the number of general laws necessary for the positive explanation of natural phenomena – which is the real philosophic purpose of all science – we shall think it rash ever to hope, even in the most distant future, to reduce these laws rigorously to a single one.

I have attempted in this chapter to determine, as exactly as I could, the aim, the spirit, and the influence of the positive philosophy. I have, therefore, indicated the goal toward which my labors have always tended, and always will tend unceasingly, in this course or elsewhere. No one is more profoundly convinced than myself of the inadequacy of my intellectual powers, even if they were far superior to what they are, to undertake such a vast and noble work. But, although the task is too great for a single mind or a single lifetime, yet one man can state the problem clearly, and that is all I am ambitious of doing.

Having thus expounded the true aim of this course, by setting the point of view from which I shall consider the various principal branches of natural philosophy, I shall in the next chapter complete these general preliminaries by explaining the plan I have adopted – that is to say, by determining the encyclopedic order that should be established among the several classes of natural phenomena and, consequently, among the corresponding positive sciences.

Notes

1 Readers who desire to have a fuller explanation of this subject, without delay, may consult with advantage three articles entitled "Philosophical Considerations on the Sciences and Men of Science," which I published in November, 1825, in a journal called the *Producteur* (numbers seven, eight, and ten), and especially the first part of my *System of Positive Polity*, addressed in April, 1824, to the Academy of Sciences, where I placed on record for the first time my discovery of this law. [This note appears in the original text. All other notes have been added by the editor.]

2 Francis Bacon (1561–1626), English philosopher largely responsible for laying the modern foundations of experimentalism in science and resolute empiricism in philosophy.
3 Johann Kepler (1571–1630), German astronomer and mathematician, one of the principal founders of modern astronomy through the mathematical formulation of the laws of planetary motion.
4 Claude Louis Berthollet (1748–1822), French chemist who, with Antoine Lavoisier (1743–1794) reformed modern chemical nomenclature and thus helped to found the modern science of chemistry.
5 Jean Baptise Joseph Fourier (1768–1830), French mathematician and physicist. Probably in the private audience to which these remarks were first addressed.
6 René Descartes (1596–1650), French philosopher and mathematician, largely responsible for shaping the problems of modern philosophy and for emphasizing the rational, mathematical, and theoretical aspects of science and philosophy.
7 Galileo Galilei (1564–1642), Italian physicist and astronomer, whose combination of inductive with deductive ways of thinking (uniting Bacon and Descartes, as it were) founded the methodology of modern science, and whose discoveries in various fields provided tremendous impetus to early modern science.
8 Comte invented the term "sociology," meant to designate the rigorous study of social phenomena according to the precepts of positive philosophy. But since here Comte is clearly attempting to show the parallels between the various fields of science, his expression "social physics" will be retained.
9 Henri Marie Ducrotay de Blainville (1778–1850), French naturalist. Probably in the private audience to which these remarks were first addressed.
10 Georges L. C. F. D. Cuvier (1769–1832), French naturalist and founder of the science of comparative anatomy. An opponent of the evolutionary theories of Jean Baptiste Lamarck (1744–1829).
11 Jöns Jakob Berzelius (1779–1848), Swedish chemist, discoverer of several new elements and notable contributor to atomic theory after John Dalton (1766–1844).
12 Pierre Simon Laplace (1749–1827), French astronomer and mathematician, author of *Mécanique céleste* (1799–1825) and particularly noted for his conviction that all the phenomena of the universe can in principle be explained and predicted in terms of the laws of classical mechanics alone.

6

On the Sciences and Arts

Jean-Jacques Rousseau

Final Reply [by] J. J. Rousseau Of Geneva[1]

> Ne, dum tacemus, non verecundiae sed diffidentiae causa tacere videamur.[2]
>
> *Cyprian,* Against Demetrianus

It is with extreme reluctance that I amuse idle Readers who care very little about the truth with my disputes. But the manner in which it has just been attacked forces me to spring to its defense once again, so that my silence is not taken by the multitude as consent, nor by Philosophers as disdain.

I must be repetitious. I know that very well, and the public will not forgive me for it. But the wise will say: this man does not need to seek new reasons continually. That is a proof of the solidity of his own.[3]

Since those who attack me never fail to stray from the question and leave out the essential distinctions I included, I must always begin by taking them back to them. Here, then, is a summary of the propositions I affirmed and will continue to affirm as long as I consult no other interest than that of the truth.

Originally "Final Reply" in *Collected Writing of Jean-Jacques Rousseau*, vol. 1, ed. Roger D. Masters and Christopher Kelly, Hanover, NH and London: University Press of New England, 1992, pp. 110–99.

The Sciences are the masterpiece of genius and reason. The spirit of imitation produced the fine Arts and experience has perfected them. We are indebted to the mechanical arts for a great number of useful inventions which have added to the pleasures and conveniences of life. These are truths about which I surely agree wholeheartedly. But now let's consider all this knowledge in relation to morals.[4]

If celestial intellects cultivated the sciences, only good would result. I say the same of great men, who are destined to guide others. A learned and virtuous Socrates was the pride of humanity. But the vices of ordinary men poison the most sublime knowledge and make it pernicious for Nations. The wicked derive many harmful things from it. The good derive little more. If no one other than Socrates had prided himself on Philosophy in Athens, the blood of a just man would not have cried out for revenge against the fatherland of the Sciences and Arts.[5]

One question to examine is whether it would be advantageous for men to have Science, assuming that what they call by that name in fact deserves it. But it is folly to pretend that the chimeras of Philosophy, the errors and lies of Philosophers can ever be good for anything. Will we always be the dupe of words? And won't we ever understand that study, knowledge, learning, and Philosophy are only vain semblances constructed by human pride and very unworthy of the pompous names it gives them?

To the extent that the taste for these foolish things spreads in a nation, it loses the taste for solid virtues. For it costs less to distinguish oneself by babble than by good morals, as soon as one is dispensed from being a good man provided one is a pleasant man.

The more the interior is corrupted, the more the exterior is composed.[6] In this way the cultivation of Letters imperceptibly engenders politeness. Taste is also born from the same source. Public approbation being the first reward for literary works, it is natural for those preoccupied by them to reflect on the ways to please. And it is these reflections which, in the long run, form style, purify taste, and spread the graces and urbanity everywhere. All these things will be, if you will, the supplement of virtue. But it will never be possible to say that they are virtue, and they will rarely be associated with it. There will always be this difference, that the person who makes himself useful labors for others, and the one who thinks only of making himself pleasing labors only for himself. The flatterer, for example, spares no effort to please, and yet he does only evil.

The vanity and idleness that have engendered our sciences have also engendered luxury. The taste for luxury always accompanies that of Letters, and the taste for Letters often accompanies that for luxury.[7] All these things are rather faithful companions, because they are all the work of the same vices.

If experience were not in accord with these demonstrated propositions, it would be necessary to seek the particular causes of that contrary result. But the first idea of these propositions is itself born from a long meditation about experience. And to see to what extent it confirms them, it is necessary only to open the annals of the world.

The first men were very ignorant. How would anyone dare to say they were corrupt in times when the sources of corruption were not yet open?

Across the obscurity of ancient times and the rusticity of ancient Peoples, one perceives very great virtues in several of them, especially a severity of morals that is an infallible mark of their purity, good faith, hospitality, justice, and – what is very important – great horror for debauchery,[8] the fertile mother of all the other vices. Virtue is therefore not incompatible with ignorance.

It is not always its companion, either, for several very ignorant peoples were very vicious. Ignorance is an obstacle to neither good nor evil. It is only the natural state of man.[9]

One cannot say as much about science. All learned Peoples have been corrupt, and that is already a terrible prejudice against it. But since comparisons from People to People are difficult, since a great number of objects must be taken into consideration, and since they always lack exactness in some respect, it is much more certain to follow the history of the same People and compare the progress of its knowledge with the revolutions of its morals. Now the result of this examination is that the beautiful time, the time of virtue for each People was that of its ignorance. And to the extent to which it has become learned, Artistic, and Philosophical, it has lost its morals and its probity. It has redescended in this respect to the rank of ignorant and vicious Nations which are the shame of humanity. If one wishes to persist stubbornly in seeking out the differences, I can recognize one, and this is it: It is that all barbarous Peoples, even those who are without virtue, nonetheless always honor virtue, whereas by dint of progress, learned and Philosophical Peoples finally come to ridicule and scorn it. It is when a nation has once reached this point that corruption can be said to be at its peak and there is no hope for remedies.

That is the summary of the things I asserted, and for which I believe I gave proofs. Let us look now at the summary of the Doctrine opposed to me.

"Men are naturally evil. They were that way before the formation of societies. And every place where the sciences have not carried their flame, peoples – abandoned to the *faculties of instinct* alone, reduced with the lions and bears to a purely animal life – have remained immersed in barbarity and wretchedness.

"Greece alone in ancient times thought and *elevated itself by the mind* to all that can make a People praiseworthy. Philosophers formed its morals and gave it laws.

"Sparta, it's true, was poor and ignorant by institution and by choice. But its laws had great defects, its Citizens a great tendency to allow themselves to be corrupted. Its glory had little solidity, and it soon lost its institutions, its laws, and its morals.

"Athens and Rome degenerated too. One yielded to the success of Macedonia. The other succumbed under the weight of its own greatness, because the laws of a small city were not made to govern the whole world. If it has happened sometimes that the glory of great Empires has not long survived that of letters, it is because the Empire was at its peak when letters were cultivated there, and it is the fate of human things not to last long in the same state. By granting, then, that the alteration of laws and morals influenced these great events, one is not forced to agree that the Sciences and Arts contributed to it. And on the contrary, it can be observed that the progress of letters and their decline is always in exact proportion with the success and fall of empires.

"This truth is confirmed by the experience of recent times, when one sees in a vast and powerful Monarchy the prosperity of the State, the cultivation of the Sciences and Arts, and warlike virtue cooperating simultaneously for the glory and greatness of the Empire.

"Our morals are the best there can be. Several vices have been proscribed among us. Those that remain belong to humanity, and the sciences have no part in them.

"Luxury has nothing in common with them either. Thus the disorders it can cause must not be attributed to them. Besides, luxury is necessary in large States. It does more good than harm. It is useful in occupying idle Citizens and providing bread for the poor.

"Politeness ought to be counted among the virtues rather than among the vices. It prevents men from showing themselves as they are, a very necessary precaution to make them tolerable to one another.

"The Sciences have rarely attained the goal they set, but at least they aim for it. We progress by slow steps in knowledge of the truth, which doesn't prevent us from making some progress.

"Finally, even if it were true that the Sciences and Arts enfeeble courage, aren't the infinite goods they procure for us still preferable to the barbarous and fierce virtue that makes humanity tremble?" I skip the useless and pompous review of these goods. And to start on this last point with an admission suited to forestall much verbiage, I declare once and for all that if something can compensate for the ruin of morals, I am ready to concede that the Sciences do more good than harm. Let us turn now to the rest.

I could without much risk assume all this as proven, since in so many boldly advanced assertions, there are very few that touch on the heart of the question, fewer still from which one can draw any valid conclusion against my sentiment, and since most of them even provide new arguments in my favor if my cause needed some.

Indeed, 1. If men are wicked by nature, it can happen, if you will, that the sciences will produce some good in their hands. But it is very certain that they will do much more harm. Madmen should not be given weapons.

2. If the sciences rarely attain their goal, there will always be much more time lost than time well used. And if it were true that we had found the best methods, most of our labors would still be as ridiculous as those of a man who, very sure he follows the plumb line precisely, would like to drive a well to the center of the earth.

3. We must not be so afraid of the purely animal life, nor consider it as the worst state into which we can fall. For it is still better to resemble a sheep than a fallen Angel.

4. Greece owed its morals and its laws to Philosophers and Legislators. I acknowledge that. I have already said a hundred times that it is good for there to be Philosophers provided that the People doesn't get mixed up in being Philosophers.

5. Not daring to assert that Sparta didn't have good laws, one blames the laws of Sparta for having had great defects. So that in order to twist around my reproaches to learned peoples for having always been corrupt, ignorant Peoples are reproached for not having attained perfection.

6. The progress of letters is always proportional to the greatness of Empires. So be it. I see that one always speaks to me of success and greatness. I was talking about morals and virtue.

7. Our morals are the best that wicked men like ourselves can have. That may be. We have proscribed several vices. I don't disagree about that. I don't accuse the men of this century of having all the vices. They have only those of cowardly souls. They are merely imposters and rascals. As for the vices that presuppose courage and firmness, I think they are incapable of them.

8. Luxury may be necessary to provide bread for the poor. But if there were no luxury, there would not be any poor people.[10] It keeps idle Citizens occupied. And why are there idle Citizens? When agriculture held a place of honor, there was neither misery nor idleness, and there were many fewer vices.

9. I see they take very much to heart this issue of luxury, which they pretend, however, to want to separate from that of the Sciences and Arts. I will agree, then, since they wish for it so absolutely, that luxury serves to support States as Caryatids serve to hold up the palaces they decorate, or rather like those beams with which rotted buildings are supported and which often end up toppling them. Wise and prudent men, get out of any house that is propped up.

This may show how easy it would be for me to turn around in my favor most of the things with which people claim to oppose me. But to speak frankly, I don't find them well enough proved to have the courage to take advantage of them.

It is asserted that the first men were wicked, from which it follows that man is naturally wicked.[11] This is not an assertion of slight importance. It seems to me it was worth the trouble of being proved. The Annals of all the peoples one dares to cite as proof are much more favorable to the opposite assumption. And there would have to be much testimony to oblige me to believe an absurdity. Before those dreadful words *thine* and *mine* were invented, before there were any of that cruel and brutal species of man called masters and of that other species of roguish and lying men called slaves; before there were men abominable enough to dare have superfluities while other men die of hunger; before mutual dependance forced them all to become imposters, jealous, and traitors; I very much wish someone would explain to me what those vices, those crimes could have been with which they are reproached so emphatically. I am assured that people have long since been disabused of the chimera of the golden Age. Why not add that people have long since been disabused of the chimera of virtue?

Notes

Rousseau originally intended this reply to Charles Bordes's "Discourse on the Advantages of the Sciences and Arts," published in the *Mercury* in April 1752, to close the debate over his *First Discourse*. Although he ultimately published another defense ("Letter [to Lecat]", [*Collected Writings*, I] pp. 175–179) and, after Bordes published a rejoinder in 1753, started the manuscript of a "Second Letter to Bordes" (pp. 182–185), the "Final Reply" is one of the clearest statements of his paradoxical and often misunderstood position. See Pléiade, II, 1270–1271, 1283 and compare *Confessions*, Book VIII (Pléiade, I, 366). [Ed.]

1 In 1761, Rousseau explained that he had used the words "Citizen of Geneva" only "on those works that I think will do honor" to the city (*La Nouvelle Héloise*, Second Preface [Pléiade, II, 27]); hence, this self-identification indicates the importance he attached to the "Final Reply." Such an interpretation is not contradicted by the fact that, on a copy of his text intended for a corrected edition of his *Collected Writings*, Rousseau later deleted the words "of Geneva" (Plèiade, I, 1270). These corrections are subsequent to the condemnation of Rousseau's work by Geneva in 1762 and include a similar deletion on the title page of the *First Discourse* (Pléiade, I, 1240). [Ed.]

2 "We must no longer remain silent for fear that silence seem to be dictated by weakness rather than by discretion" St. Cyprian (c. 210–258) was bishop of Carthage. [Ed.]

3 There are very reliable truths that appear at first glance to be absurdities, and that will always be viewed as such by most people. Go say to a man of the People that the sun is closer to us in winter than in summer, or that it has set before we stop seeing it, and he will laugh at you. The same is true of the sentiment I hold. The most superficial men have always been the the quickest to oppose me. The true Philosophers are in less of a hurry. And if I have the honor of having made some proselytes, it is only among the latter. Before explaining myself, I meditated on my subject at length and deeply, and I tried to consider all aspects of it. I doubt that any of my adversaries can say as much. At least I don't perceive in their writings any of those luminous truths that are no less striking in their obviousness than in their novelty, and that are always the fruit and proof of an adequate meditation. I dare say that they have never raised a reasonable objection that I did not anticipate and to which I did not reply in advance. That is why I am always compelled to restate the same things.

4 *Knowledge makes men gentle*, says that famous philosopher whose work – always profound and sometimes sublime – exudes everywhere love of humanity. [Editor's note: Plutarch, "On Tranquillity of Mind," 4; *Moralia*, 466D; Loeb Classical Library, trans. W. C. Helmbold (London: Heinemann, 1939), VI, 177.] In these few words and, which is rare, without declamation, he wrote the most solid statement ever made in favor of letters. It is true, knowledge does make men gentle. But gentleness, which is the most appealing of the virtues, is sometimes also a weakness of the soul. Virtue is not always gentle. It knows how to arm itself appropriately with severity against vice; it is inflamed with indignation against crime.

And the just knows no way to pardon the wicked.

A king of Lacedemonia gave a very wise reply to those who praised in his presence the extreme goodness of his Colleague Charillus. *And how can he be good*, he said to them, *if he doesn't know how to be terrible to the wicked?* [Editor's note: Plutarch, "On Envy and Hate," 5, *Moralia*, 537D; Loeb Classical Library, trans. Philip H. de Lacey and Benedict Einarson (London: Heinemann, 1959), VII, 101.

This same text had been cited, without an attribution to a Spartan king, by Montaigne, *Essays*, II, xii, *ad finem*. In the manuscript of corrections (see note 1 above), Rousseau added: *quos malos boni oderint, bonos oportet esse* ("That good men hate the wicked is the proof of their goodness"). Cf. Plutarch, "On Moral Virtue," 12; *Moralia*, 451E (Loeb ed., VI, 81).] Brutus was not a gentle man. Who would have the impudence to say that he was not virtuous? On the contrary, there are cowardly and pusillanimous souls that have neither fire nor warmth, and that are gentle only through indifference about good and evil. Such is the gentleness that is inspired in Peoples by the taste for Letters.

5 It cost Socrates his life to say precisely the same things I am saying. In the proceedings that were instituted against him, one of his accusers pleaded for Artists, another for Orators, the third for Poets, all for the supposed cause of the Gods. The Poets, the Artists, the Fanatics, the Rhetoricians triumphed; and Socrates perished. I am very afraid I have given my century too much credit in asserting that Socrates would not have drunk the hemlock now. [Editor's notes: Note that, as in the text of the *First Discourse*, Rousseau speaks of "artists" where Plato had referred to "artisans": cf. *Apology*, 23e and *First Discourse*, p. 10 and Introduction, note 10. In the manuscript corrected when preparing republication of his works, referring to the legal prosecution to which he had been subject, Rousseau added the note: "It will be remarked that I said this as early as the year 1752" (Pléiade, III, 1272; Intégrale, II, 142).]

6 I never attend a presentation of a Comedy by Molière without admiring the delicacy of the spectators. A word that is a little loose, an expression that is coarse rather than obscene, everything wounds their chaste ears. And I have no doubt whatever that the most corrupt are always the most scandalized. Yet if the morals of Molière's century were compared with those of ours, is there anyone who believes that the result will be in favor of ours? Once the imagination has been sullied, everything becomes a subject of scandal for it. When nothing good is left but the exterior, all efforts are redoubled to preserve it.

7 Somewhere the luxury of Asiatic peoples has been used to contradict me, by the same manner of reasoning which uses the vices of ignorant peoples to contradict me. But through a misfortune that pursues my adversaries, they are mistaken even about facts that prove nothing against me. I know that the peoples of the Orient are not less ignorant than we are. But that doesn't prevent them from being as vain and from writing almost as many books. The Turks, those who cultivate Letters the least of them all, counted five hundred eighty classical Poets among them toward the middle of the last century.

8 I have no scheme to pay court to women. I consent to their honoring me with the epithet Pedant, so dreaded by all our gallant Philosophers. I am coarse, sullen, impolite on principle, and want no supporters. Therefore I am going to speak the truth just as I please.

Man and woman are made to love one another and unite. But beyond this legitimate union, all commerce of love between them is a dreadful source of disorders in society and morals. It is certain that women alone could restore honor and probity among us. But they disdain to take from the hands of virtue an empire they wish to owe only to their charms. Thus, they do only evil, and often receive themselves the punishment for this preference. It is hard to conceive how, in such a pure Religion, chastity could have become a base and monkish virtue, capable of making ridiculous any man and I daresay almost any woman who would dare to pride themselves on it. Whereas among the Pagans, this same virtue was universally honored, regarded as suited to great men, and admired in their most illustrious heroes. I can name three of them who will not be inferior to any other and who – without Religion being involved – all gave memorable examples of continence: Cyrus, Alexander, and Scipio the Younger. [Editor's note: On Cyrus and Alexander, see Plutarch, "Of Curiosity," *Moralia*, 521F–522A (Loeb ed., VI, 509). On Scipio, Rousseau seems to refer to a saying attributed to Scipio the Elder when he captured Carthage as a man of twenty-five; Plutarch, "Sayings of the Romans," *Moralia*, 196B (Loeb ed., III, 163).] Of all the rare objects in the King's Collection, I would like to see only the silver shield that was given to the latter by the Peoples of Spain; on which they engraved the triumph of his virtue. This was the Romans' way of subjugating Peoples, as much by the veneration due to their morals as by the effort of their arms. This was how the city of the Falisci was subjugated and Pyrrhus the victor chased out of Italy.

I remember having read somewhere a rather good reply by the Poet Dryden to some young English Lord, who reproached him because in one of his Tragedies, Cleomenes enjoyed chatting tête-à-tête with his love rather than formulating some enterprise worthy of his love. When I am near a beautiful woman, the young Lord said to him, I make better use of my time. I believe it, replied Dryden, but you also have to admit that you aren't a Hero.

9 I can't help laughing when I see I don't know how many very learned men who honor me with their criticism always raising in objection the vices of a multitude of ignorant Peoples, as though that had something to do with the question. Does it follow from the fact that science necessarily engenders vice that ignorance necessarily engenders virtue? This manner of arguing may be good for Rhetoricians or for the children by whom I was refuted in my country. [Editor's note: There was a refutation of Rousseau, in the form of a dialogue in Latin, at the graduation ceremonies of the College of Geneva on May 23, 1751. Actually, however, it appears that the

sixteen-year-old Jean-Alphonse Turrettini praised Rousseau's position, and the condemnation was presented by the Pastor Jacob Vernes (Pléiade, III, 1273; Intégrale, II, 143).] But Philosophers ought to reason in another way.

10 Luxury feeds a hundred poor people in our cities and causes a hundred thousand to die in our countryside. The money that circulates between the hands of the rich and the Artists in order to provide for their superfluities is lost for the subsistence of the Farmer. And the latter has no clothes precisely because the former have braid on theirs. The waste of materials that go into food for people alone suffices to make luxury odious to humanity. My adversaries are most fortunate that the culpable delicacy of our language prevents me from offering details about this that would make them blush for the cause that they dare defend. Gravy is necessary for our cooking; that is why so many sick people lack broth. We must have liquors on our table; that is why the peasant drinks only water. We must powder our wigs; that is why so many poor people have no bread.

11 This note is for Philosophers. I advise others to skip over it.

If man is wicked by his nature, it is clear that the Sciences will only make him worse. Thus their cause is lost by this assumption alone. But it is necessary to note well that although man is naturally good, as I believe and as I have the happiness to feel, it does not follow from this that the sciences are salutary for him. For every situation that places a people in the position to cultivate them necessarily announces the beginning of corruption which the sciences quickly accelerate. Then the vice of the constitution does all the harm that nature could have done, and bad prejudices take the place of bad inclinations.

References

Collected Writings, I: Jean-Jacques Rousseau. *Rousseau Judge of Jean-Jacques, Dialogues. Collected Writings of Rousseau*, Vol. 1. Edited by Roger D. Masters and Christopher Kelly. Hanover, N.H.: University Press of New England, 1991.

Pléiade: Jean-Jacques Rousseau. *Oeuvres complètes*. Vols. 1–5. Paris: NRF-Editions de la Pléiade, 1959ff.

7

Capitalism and the Modern Labor Process

Karl Marx and Friedrich Engels

From Karl Marx, *Capital: A Critique of Political Economy*

Chapter VII: The labour-process

The labour-process or the production of use-values Labour is, in the first place, a process in which both man and Nature participate, and in which man of his own accord starts, regulates, and controls the material re-actions between himself and Nature. He opposes himself to Nature as one of her own forces, setting in motion arms and legs, head and hands, the natural forces of his body, in order to appropriate Nature's productions in a form adapted to his own wants. By thus acting on the external world and changing it, he at the same time changes his own nature. He develops his slumbering powers and compels them to act in obedience to his sway. We are not now dealing with those primitive instinctive forms of labour that remind us of the mere animal. An immeasurable interval of time separates the state of things in which a man brings his labour-power to market for sale as a commodity, from that state in which human labour was still in its first instinctive stage. We pre-suppose labour in a form that stamps it as exclusively human. A spider conducts operations that resemble those of a weaver, and a bee puts to shame many an architect in the construction of her cells. But what distinguishes the worst architect from the best of bees is this, that the architect raises his structure in imagination before he erects it in reality. At the end of every labour-process, we get a result that already existed in the imagination of the labourer at its commencement. He not only effects a change of form in the material on which he works, but he also realises a purpose of his own that gives the law to his modus operandi, and to which he must subordinate his will. And this subordination is no mere momentary act. Besides the exertion of the bodily organs, the process demands that, during the whole operation, the workman's will be steadily in consonance with his purpose. This means close attention. The less he is attracted by the nature of the work, and the mode in which it is carried on, and the less, therefore, he enjoys it as something which gives play to his bodily and mental powers, the more close his attention is forced to be.

The elementary factors of the labour-process are 1, the personal activity of man, *i.e.*, work itself, 2, the subject of that work, and 3, its instruments.

Karl Marx, "Chapter VII: The Labour-Process and the Process of Producing Surplus Value," in *Capital: A Critique of Political Economy*, ed. Frederick Engels, New York: International Publishers, 1967, pp. 173–7, abridged; Preface to *A Contribution to the Critique of Political Economy*, New York: International Publishers, 1970, pp. 20–2, abridged; Karl Marx and Friedrich Engels, "Materialist Method," from *The German Ideology*, ed. C. J. Arthur, New York: International Publishers, 1970, pp. 42–3, 46–9, abridged; Friedrich Engels, "The Part Played by Labour in the Transition from Ape to Man," from *Dialectics of Nature*, 2nd ed., trans. Clemens Dutt, Moscow: Progress Publishers, 1954, pp. 170–83; Friedrich Engels, "On Authority," from Karl Marx and Friedrich Engels, *Basic Writings on Politics and Philosophy*, ed. Lewis S. Feuer, Garden City, NY: The Doubleday Broadway Publishing Group, 1959, pp. 481–5, abridged.

The soil (and this, economically speaking, includes water) in the virgin state in which it supplies[1] man with necessaries or the means of subsistence ready to hand, exists independently of him, and is the universal subject of human labour. All those things which labour merely separates from immediate connexion with their environment, are subjects of labour spontaneously provided by Nature. Such are fish which we catch and take from their element, water, timber which we fell in the virgin forest, and ores which we extract from their veins. If, on the other hand, the subject of labour has, so to say, been filtered through previous labour, we call it raw material; such is ore already extracted and ready for washing. All raw material is the subject of labour, but not every subject of labour is raw material: it can only become so, after it has undergone some alteration by means of labour.

An instrument of labour is a thing, or a complex of things, which the labourer interposes between himself and the subject of his labour, and which serves as the conductor of his activity. He makes use of the mechanical, physical, and chemical properties of some substances in order to make other substances subservient to his aims.[2] Leaving out of consideration such ready-made means of subsistence as fruits, in gathering which a man's own limbs serve as the instruments of his labour, the first thing of which the labourer possesses himself is not the subject of labour but its instrument. Thus Nature becomes one of the organs of his activity, one that he annexes to his own bodily organs, adding stature to himself in spite of the Bible. As the earth is his original larder, so too it is his original tool house. It supplies him, for instance, with stones for throwing, grinding, pressing, cutting, &c. The earth itself is an instrument of labour, but when used as such in agriculture implies a whole series of other instruments and a comparatively high development of labour.[3] No sooner does labour undergo the least development, than it requires specially prepared instruments. Thus in the oldest caves we find stone implements and weapons. In the earliest period of human history domesticated animals, *i.e.*, animals which have been bred for the purpose, and have undergone modifications by means of labour, play the chief part as instruments of labour along with specially prepared stones, wood, bones, and shells.[4] The use and fabrication of instruments of labour, although existing in the germ among certain species of animals, is specifically characteristic of the human labour-process, and Franklin therefore defines man as a tool-making animal. Relics of bygone instruments of labour possess the same importance for the investigation of extinct economic forms of society, as do fossil bones for the determination of extinct species of animals. It is not the articles made, but how they are made, and by what instruments, that enables us to distinguish different economic epochs.[5] Instruments of labour not only supply a standard of the degree of development to which human labour has attained, but they are also indicators of the social conditions under which that labour is carried on. Among the instruments of labour, those of a mechanical nature, which, taken as a whole, we may call the bone and muscles of production, offer much more decided characteristics of a given epoch of production, than those which, like pipes, tubs, baskets, jars, &c., serve only to hold the materials for labour, which latter class, we may in a general way, call the vascular system of production. The latter first begins to play an important part in the chemical industries.

In a wider sense we may include among the instruments of labour, in addition to those things that are used for directly transferring labour to its subject, and which therefore, in one way or another, serve as conductors of activity, all such objects as are necessary for carrying on the labour- process. These do not enter directly into the process, but without them it is either impossible for it to take place at all, or possible only to a partial extent. Once more we find the earth to be a universal instrument of this sort, for it furnishes a locus standi to the labourer and a field of employment for his activity. Among instruments that are the result of previous labour and also belong to this class, we find workshops, canals, roads, and so forth.

In the labour-process, therefore, man's activity, with the help of the instruments of labour, effects an alteration, designed from the commencement, in the material worked upon. The process disappears in the product; the latter is a use-value, Nature's material adapted by a change of form to the wants of man. Labour has incorporated itself with its subject: the former is materialised, the latter transformed. That which in the labourer appeared as movement, now appears in the product as a fixed quality without motion. The blacksmith forges and the product is a forging.

If we examine the whole process from the point of view of its result, the product, it is plain that both the instruments and the subject of labour, are means of production,[6] and that the labour itself is productive labour.[7]

Though a use-value, in the form of a product, issues from the labour-process, yet other use-values, products of previous labour, enter into it as means of production. The same use-value is both the product of a previous process, and a means of production in a later process. Products are therefore not only results, but also essential conditions of labour.

With the exception of the extractive industries, in which the material for labour is provided immediately by Nature, such as mining, hunting, fishing, and agriculture (so far as the latter is confined to breaking up virgin soil), all branches of industry manipulate raw material, objects already filtered through labour, already products of labour. Such is seed in agriculture. Animals and plants, which we are accustomed to consider as products of Nature, are in their present form, not only products of, say last year's labour, but the result of a gradual transformation, continued through many generations, under man's superintendence, and by means of his labour. But in the great majority of cases, instruments of labour show even to the most superficial observer, traces of the labour of past ages.

[...]

Notes

1 "The earth's spontaneous productions being in small quantity, and quite independent of man, appear, as it were, to be furnished by Nature, in the same way as a small sum is given to a young man, in order to put him in a way of industry, and of making his fortune." (James Steuart: "Principles of Polit. Econ." edit. Dublin, 1770, v. I, p. 116.)

2 "Reason is just as cunning as she is powerful. Her cunning consists principally in her mediating activity, which, by causing objects to act and re-act on each other in accordance with their own nature, in this way, without any direct interference in the process, carries out reason's intentions." (Hegel: "Enzyklopädie, Erster Theil, Die Logik," Berlin, 1840, p. 382.)

3 In his otherwise miserable work ("Théorie de l'Econ. Polit." Paris, 1815), Ganilh enumerates in a striking manner in opposition to the "Physiocrats" the long series of previous processes necessary before agriculture properly so called can commence.

4 Turgot in his "Réflexions sur la Formation et la Distribution des Richesses" (1766) brings well into prominence the importance of domesticated animals to early civilisation.

5 The least important commodities of all for the technological comparison of different epochs of production are articles of luxury, in the strict meaning of the term. However little our written histories up to this time notice the development of material production, which is the basis of all social life, and therefore of all real history, yet prehistoric times have been classified in accordance with the results, not of so-called historical, but of materialistic investigations. These periods have been divided, to correspond with the materials from which their implements and weapons were made, viz., into the stone, the bronze, and the iron ages.

6 It appears paradoxical to assert, that uncaught fish, for instance, are a means of production in the fishing industry. But hitherto no one has discovered the art of catching fish in waters that contain none.

7 This method of determining, from the standpoint of the labour-process alone, what is productive labour, is by no means directly applicable to the case of the capitalist process of production.

From Karl Marx, *A Contribution to the Critique of Political Economy*

Preface

[...]

The general conclusion at which I arrived and which, once reached, became the guiding principle of my studies can be summarised as follows. In the social production of their existence, men inevitably enter into definite relations, which are independent of their will, namely relations of production appropriate to a given stage in the development of their material forces of production. The totality of these relations of production constitutes the economic structure of society, the real foundation, on which arises a legal and political superstructure and to which correspond definite forms of social consciousness. The mode of production of material life conditions the general process of social, political and intellectual life. It is not the consciousness of men that determines their existence, but their social existence that determines their consciousness. At a certain stage of development, the material productive forces of society come into conflict with the existing relations of production or – this merely expresses the same thing in legal terms – with the property relations within the framework of which they have operated hitherto. From forms of development of the productive forces these relations turn into their fetters. Then begins an era of social revolution. The changes in the economic foundation lead sooner or later to the transformation of the whole immense superstructure. In studying such transformations it is always necessary to distinguish between the material transformation of the economic conditions of production, which can be determined with the precision of natural science, and the legal, political, religious, artistic or philosophic – in short, ideological forms in which men become conscious of this conflict and fight it out. Just as one does not judge an individual by what he thinks about himself, so one cannot judge such a period of transformation by its consciousness, but, on the contrary, this consciousness must be explained from the contradictions of material life, from the conflict existing between the social forces of production and the relations of production. No social order is ever destroyed before all the productive forces for which it is sufficient have been developed, and new superior relations of production never replace older ones before the material conditions for their existence have matured within the framework of the old society. Mankind thus inevitably sets itself only such tasks as it is able to solve, since closer examination will always show that the problem itself arises only when the material conditions for its solution are already present or at least in the course of formation. In broad outline, the Asiatic, ancient, feudal and modern bourgeois modes of production may be designated as epochs marking progress in the economic development of society. The bourgeois mode of production is the last antagonistic form of the social process of production – antagonistic not in the sense of individual antagonism but of an antagonism that emanates from the individuals' social conditions of existence – but the productive forces developing within bourgeois society create also the material conditions for a solution of this antagonism. The prehistory of human society accordingly closes with this social formation.

From Karl Marx and Friedrich Engels, *The German Ideology*

First premises of materialist method

The premises from which we begin are not arbitrary ones, not dogmas, but real premises from which abstraction can only be made in the imagination. They are the real individuals, their activity and the material conditions under which they live, both those which they find already existing and those produced by their activity. These premises can thus be verified in a purely empirical way.

The first premise of all human history is, of course, the existence of living human individuals. Thus the first fact to be established is the physical organisation of these individuals and their consequent relation to the rest of nature. Of course, we cannot here go either into the actual physical nature of man, or into the natural conditions in which man finds himself – geological, oreohydrographical, climatic and so on. The writing of history must always set out from these natural bases and their modification in the course of history through the action of men.

Men can be distinguished from animals by consciousness, by religion or anything else you like. They themselves begin to distinguish themselves from animals as soon as they begin to *produce* their means of subsistence, a step which is conditioned by their physical organisation. By producing their means of subsistence men are indirectly producing their actual material life.

The way in which men produce their means of subsistence depends first of all on the nature of the actual means of subsistence they find in existence and have to reproduce. This mode of production must not be considered simply as being the production of the physical existence of the individuals. Rather it is a definite form of activity of these individuals, a definite form of expressing their life, a definite *mode of life* on their part. As individuals express their life, so they are. What they are, therefore, coincides with their production, both with *what* they produce and with *how* they produce. The nature of individuals thus depends on the material conditions determining their production.

This production only makes its appearance with the *increase of population*. In its turn this presupposes the *intercourse* [*Verkehr*][1] of individuals with one another. The form of this intercourse is again determined by production.

The relations of different nations among themselves depend upon the extent to which each has developed its productive forces, the division of labour and internal intercourse. This statement is generally recognised. But not only the relation of one nation to others, but also the whole internal structure of the nation itself depends on the stage of development reached by its production and its internal and external intercourse. How far the productive forces of a nation are developed is shown most manifestly by the degree to which the division of labour has been carried. Each new productive force, insofar as it is not merely a quantitative extension of productive forces already known (for instance the bringing into cultivation of fresh land), causes a further development of the division of labour.

The division of labour inside a nation leads at first to the separation of industrial and commercial from agricultural labour, and hence to the separation of *town* and *country* and to the conflict of their interests. Its further development leads to the separation of commercial from industrial labour. At the same time through the division of labour inside these various branches there develop various divisions among the individuals co-operating in definite kinds of labour. The relative position of these individual groups is determined by the methods employed in agriculture, industry and commerce (patriarchalism, slavery, estates, classes). These same conditions are to be seen (given a more developed intercourse) in the relations of different nations to one another.

The various stages of development in the division of labour are just so many different forms of ownership, i.e. the existing stage in the division of labour determines also the relations of individuals to one another with reference to the material, instrument, and product of labour.

[...]

The fact is, therefore, that definite individuals who are productively active in a definite way enter into these definite social and political relations. Empirical observation must in each separate instance bring out empirically, and without any mystification and speculation, the connection of the social and political structure with production. The social structure and the State are continually evolving out of the life-process of definite individuals, but of individuals, not as they may appear in their own or other people's imagination, but as they *really* are; i.e. as they operate, produce materially, and hence as they work under definite material limits, presuppositions and conditions independent of their will.

The production of ideas, of conceptions, of consciousness, is at first directly interwoven with the material activity and the material intercourse of men, the language of real life. Conceiving, thinking, the mental intercourse of men, appear at this stage as the direct efflux of their material behaviour. The same applies to mental production as expressed in the language of politics, laws, morality, religion, metaphysics, etc. of a people. Men are the producers of their conceptions, ideas, etc. – real, active men, as they are conditioned by a definite development of their productive forces and of the intercourse corresponding to these, up to its furthest forms. Consciousness can never be anything else than conscious existence, and the existence of men is their actual life-process. If in all ideology men and their circumstances appear upside-down as in a *camera obscura*, this phenomenon arises just as much from their historical life-process as the inversion of objects on the retina does from their physical life-process.

In direct contrast to German philosophy which descends from heaven to earth, here we ascend from earth to heaven. That is to say, we do not set out from what men say, imagine, conceive, nor from men as narrated, thought of, imagined, conceived, in order to arrive at men in the flesh. We set out from real, active men, and on the basis of their real life-process we demonstrate the development of the ideological reflexes and echoes of this life-process. The phantoms formed in the human brain are also, necessarily, sublimates of their material life-process, which is empirically verifiable and bound to material premises. Morality, religion, metaphysics, all the rest of ideology and their corresponding forms of consciousness, thus no longer retain the semblance of independence. They have no history, no development; but men, developing their material production and their material intercourse, alter, along with this their real existence, their thinking and the products of their thinking. Life is not determined by consciousness, but consciousness by life. In the first method of approach the starting-point is consciousness taken as the living individual; in the second method, which conforms to real life, it is the real living individuals themselves, and consciousness is considered solely as *their* consciousness.

[...]

History: fundamental conditions

...we must begin by stating the first premise of all human existence and, therefore, of all history, the premise, namely, that men must be in a position to live in order to be able to "make history". But life involves before everything else eating and drinking, a habitation, clothing and many other things. The first historical act is thus the production of the means to satisfy these needs, the production of material life itself. And indeed this is an historical act, a fundamental condition of all history, which today, as thousands of years ago, must daily and hourly be fulfilled merely in order to sustain human life. Even when the sensuous world is reduced to a minimum, to a stick as with Saint Bruno [Bauer], it presupposes the action of producing the stick. Therefore in any interpretation of history one has first of all to observe this fundamental fact in all its significance and all its implications and to accord it its due importance. It is well known that the Germans have never done this, and they have never, therefore, had an *earthly* basis for history and consequently never an historian. The French and the English, even if they have conceived the relation of this fact with so-called history only in an extremely one-sided fashion, particularly as long as they remained in the toils of political ideology, have nevertheless made the first attempts to give the writing of history a materialistic basis by being the first to write histories of civil society, of commerce and industry.

The second point is that the satisfaction of the first need (the action of satisfying, and the instrument of satisfaction which has been acquired) leads to new needs; and this production of new needs is the first historical act.

Note

1 In *The German Ideology* the word "*Verkehr*" is used in a very wide sense, encompassing the material and spiritual intercourse of separate individuals, social groups and entire countries. Marx and Engels show that material intercourse, and above all the intercourse of men with each other in the production process, is the basis of every other form of intercourse.

The terms "*Verkehrstorm*" (form of intercourse), "*Verkehrsweise*" (mode of intercourse) and "*Verkehrsverhaltnisse*" (relations, or conditions, of intercourse) which we encounter in *The German Ideology* are used by Marx and Engels to express the concept "relations of production" which during that period was taking shape in their mind.

The ordinary dictionary meanings of "*Verkehr*" are traffic, intercourse, commerce. In this translation the word "*Verkehr*" has been mostly rendered as "intercourse" and occasionally as "association" or "commerce". – Ed.

From Friedrick Engels, *Dialectics of Nature*

The part played by labour in the transition from ape to man

Labour is the source of all wealth, the political economists assert. And it really is the source – next to nature, which supplies it with the material that it converts into wealth. But it is even infinitely more than this. It is the prime basic condition for all human existence, and this to such an extent that, in a sense, we have to say that labour created man himself.

Many hundreds of thousands of years ago, during an epoch, not yet definitely determinable, of that period of the earth's history known to geologists as the Tertiary period, most likely towards the end of it, a particularly highly-developed race of anthropoid apes lived somewhere in the tropical zone – probably on a great continent that has now sunk to the bottom of the Indian Ocean. Darwin has given us an approximate description of these ancestors of ours. They were completely covered with hair, they had beards and pointed ears, and they lived in bands in the trees.

Climbing assigns different functions to the hands and the feet, and when their mode of life involved locomotion on level ground, these apes gradually got out of the habit of using their hands [in walking – *Tr.*] and adopted a more and more erect posture. This was *the decisive step in the transition from ape to man*.

All extant anthropoid apes can stand erect and move about on their feet alone, but only in case of urgent need and in a very clumsy way. Their natural gait is in a half-erect posture and includes the use of the hands. The majority rest the knuckles of the fist on the ground and, with legs drawn up, swing the body through their long arms, much as a cripple moves on crutches. In general, all the transition stages from walking on all fours to walking on two legs are still to be observed among the apes today. The latter gait, however, has never become more than a makeshift for any of them.

It stands to reason that if erect gait among our hairy ancestors became first the rule and then, in time, a necessity, other diverse functions must, in the meantime, have devolved upon the hands. Already among the apes there is some difference in the way the hands and the feet are employed. In climbing, as mentioned above, the hands and feet have different uses. The hands are used mainly for gathering and holding food in the same way as the fore paws of the lower mammals are used. Many apes use their hands to build themselves nests in the trees, or even to construct roofs between the branches to protect themselves against the weather, as the chimpanzee, for example, does. With their hands they grasp sticks to defend themselves against enemies, and with their hands they bombard their enemies with fruits and stones. In captivity they use their hands for a number of simple operations copied from human beings. It is in this that one sees the great gulf between the undeveloped hand of even the most man-like apes and the human hand that has been highly perfected by hundreds of thousands of years of labour. The number and general arrangement of the bones and muscles are the same in both hands, but the hand of the lowest savage can perform hundreds of operations that no simian hand can imitate – no simian hand has ever fashioned even the crudest stone knife.

The first operations for which our ancestors gradually learned to adapt their hands during the many thousands of years of transition from ape to man could have been only very simple ones. The lowest savages, even those in whom regression to a more animal-like condition with a simultaneous physical degeneration can be assumed, are nevertheless far superior to these transitional beings. Before the first flint could be fashioned into a knife by human hands, a period of time probably elapsed in comparison with which the historical period known to us appears insignificant. But the decisive step had been taken, *the hand had become free* and could henceforth attain ever greater dexterity; the greater flexibility thus acquired was inherited and increased from generation to generation.

Thus the hand is not only the organ of labour, *it is also the product of labour*. Labour, adaptation to ever new operations, the inheritance of muscles, ligaments, and, over longer periods of time, bones that had undergone special development and the ever-renewed employment of this inherited finesse in new, more and more complicated operations, have given the human hand the high degree of perfection required to conjure into being the pictures of a Raphael, the statues of a Thorwaldsen, the music of a Paganini.

But the hand did not exist alone, it was only one member of an integral, highly complex organism. And what benefited the hand, benefited also the whole body it served; and this in two ways.

In the first place, the body benefited from the law of correlation of growth, as Darwin called it. This law states that the specialised forms of separate parts of an organic being are always bound up with certain forms of other parts that apparently have no connection with them. Thus all animals that have red blood cells without cell nuclei, and in which the head is attached to the first vertebra by means of a double articulation (condyles), also without exception possess lacteal glands for suckling their young. Similarly, cloven hoofs in mammals are regularly associated with the possession of a multiple stomach for rumination. Changes in certain forms involve changes in the form of other parts of the body, although we cannot explain the connection. Perfectly white cats with blue eyes are always, or almost always, deaf. The gradually increasing perfection of the human hand, and the commensurate adaptation of the feet for erect gait, have undoubtedly, by virtue of such correlation, reacted on other parts of the organism. However, this action has not as yet been sufficiently investigated for us to be able to do more here than to state the fact in general terms.

Much more important is the direct, demonstrable influence of the development of the hand on the rest of the organism. It has already been noted that our simian ancestors were gregarious; it is obviously impossible to seek the derivation of man, the most social of all animals, from non-gregarious immediate ancestors. Mastery over nature began with the development of the hand, with labour, and widened man's horizon at every new advance. He was continually discovering new, hitherto unknown properties in natural objects. On the other hand, the development of labour necessarily helped to bring the members of society closer together by increasing cases of mutual support and joint activity, and by making clear the advantage of this joint activity to each individual. In short, men in the making arrived at the point where *they had something to say* to each other. Necessity created the organ; the undeveloped larynx of the ape was slowly but surely transformed by modulation to produce constantly more developed modulation, and the organs of the mouth gradually learned to pronounce one articulate sound after another.

Comparison with animals proves that this explanation of the origin of language from and in the process of labour is the only correct one. The little that even the most highly-developed animals need to communicate to each other does not require articulate speech. In a state of nature, no animal feels handicapped by its inability to speak or to understand human speech. It is quite different when it has been tamed by man. The dog and the horse, by association with man, have developed such a good ear for articulate speech that they easily learn to understand any language within their range of concept. Moreover they have acquired the capacity for feelings such as affection for man, gratitude, etc., which were previously foreign to them. Anyone who has had much to do with such animals will hardly be able to escape the conviction that in many cases they *now* feel their inability to speak as a defect, although, unfortunately, it is one that can no longer be remedied because their vocal organs are too specialised in a definite direction. However, where vocal organs exist, within certain limits even this inability disappears. The buccal organs of birds are as different from those of man as they can be, yet birds are the only animals that can learn to speak; and it is the bird with the most hideous voice, the parrot, that speaks best of all. Let no one object that the parrot does not understand what it says. It is true that for the sheer pleasure of talking and associating with human beings, the parrot will chatter for hours at a stretch, continually repeating its whole vocabulary. But within the limits of its range of concepts it can also learn to understand what it is saying. Teach a parrot swear words in such a way that it gets an idea of their meaning (one of the great amusements of sailors returning from the tropics); tease it and you will soon discover that it knows how to use its swear words just as correctly as a Berlin costermonger. The same is true of begging for titbits.

First labour, after it and then with it speech – these were the two most essential stimuli under the influence of which the brain of the ape gradually changed into that of man, which for all its similarity is far larger and more perfect. Hand in hand with the development of the brain went the development of its most immediate instruments – the senses. Just as the gradual development of speech is inevitably accompanied by a corresponding refinement of the organ of hearing, so the development of the brain as a whole is accompanied by a refinement of all the senses. The eagle sees much farther than man, but the human eye discerns considerably more in things than does the eye of the eagle. The dog has a far keener sense of smell than man, but it does not distinguish a hundredth part of the odours that for man are definite signs denoting different things. And the sense of touch, which the ape hardly possesses in its crudest initial form, has been developed only side by side with the development of the human hand itself, through the medium of labour.

The reaction on labour and speech of the development of the brain and its attendant senses, of the increasing clarity of consciousness, power of abstraction and of conclusion, gave both labour and speech an ever-renewed impulse to further development. This development did not reach its conclusion when man finally became distinct from the ape, but on the whole made further powerful progress, its degree and direction varying among different peoples and at different times, and here and there even being interrupted by local or temporary regression. This further development has been strongly urged forward, on the one hand, and guided along more definite directions, on the other, by a new element which came into play with the appearance of fully-fledged man, namely, *society*.

Hundreds of thousands of years – of no greater significance in the history of the earth than one second in the life of man[1] – certainly elapsed before human society arose out of a troupe of tree-climbing monkeys. Yet it did finally appear. And what do we find once more as the characteristic difference between the troupe of monkeys and human society? *Labour*. The ape herd was satisfied to browse over the feeding area determined for it by geographical conditions or the resistance of neighbouring herds; it undertook migrations and struggles to win new feeding grounds, but it was incapable of extracting from them more than they offered in their natural state, except that it unconsciously fertilised the soil with its own excrement. As soon as all possible feeding grounds were occupied, there could be no further increase in the ape population; the number of animals could at best remain stationary. But all animals waste a great deal of food, and, in addition, destroy in the germ the next generation of the food supply. Unlike the hunter, the wolf does not spare the doe which would provide it with the young the next year; the goats in Greece, that eat away the young bushes before they grow to maturity, have eaten bare all the mountains of the country. This "predatory economy" of animals plays an important part in the gradual transformation of species by forcing them to adapt themselves to other than the usual food, thanks to which their blood acquires a different chemical composition and the whole physical constitution gradually alters, while species that have remained unadapted die out. There is no doubt that this predatory economy contributed powerfully to the transition of our ancestors from ape to man. In a race of apes that far surpassed all others in intelligence and adaptability, this predatory economy must have led to a continual increase in the number of plants used for food and to the consumption of more and more edible parts of food plants. In short, food became more and more varied, as did also the substances entering the body with it, substances that were the chemical premises for the transition to man. But all that was not yet labour in the proper sense of the word. Labour begins with the making of tools. And what are the most ancient tools that we find – the most ancient judging by the heirlooms of prehistoric man that have been discovered, and by the mode of life of the earliest historical peoples and of the rawest of contemporary savages? They are hunting and fishing implements, the former at the same time serving as weapons. But hunting and fishing presuppose the transition from an exclusively vegetable diet to the concomitant use of meat, and this is another important step in the process of transition from ape to man. A *meat diet* contained in an almost ready state the most essential ingredients required by the organism for its metabolism. By shortening the time required for digestion, it also shortened the other vegetative bodily processes that correspond to those of plant life, and thus gained further time, material and desire for the active manifestation of animal life proper. And the farther man in the making moved from the vegetable kingdom the higher he rose above the animal. Just as becoming accustomed to a vegetable diet side by side with meat converted wild cats and dogs into the servants of man, so also adaptation to a meat diet, side by side with a vegetable diet, greatly contributed towards giving bodily strength and independence to man in the making. The meat diet, however, had its greatest effect on the brain, which now received a far richer flow of the materials necessary for its nourishment and development, and which, therefore, could develop more rapidly and perfectly from generation to generation. With all due respect to the vegetarians man did not come into existence without a meat diet, and if the latter, among all peoples known to us, has led to cannibalism at some time or other (the forefathers of the Berliners, the Weletabians or Wilzians, used to eat their parents as late as the tenth century), that is of no consequence to us today.

The meat diet led to two new advances of decisive importance – the harnessing of fire and the domestication of animals. The first still further shortened the digestive process, as it provided the mouth with food already, as it were, half-digested; the second made meat more copious by opening up a new, more regular source of supply in addition to hunting, and moreover provided, in milk and its products, a new article of food at least as valuable as meat in its composition. Thus both these advances were, in themselves, new means for the emancipation of man. It would lead us too far afield to dwell here in detail on their indirect effects notwithstanding the great importance they have had for the development of man and society.

Just as man learned to consume everything edible, he also learned to live in any climate. He spread over the whole of the habitable world, being the only animal fully able to do so of its own accord. The other animals that have become accustomed to all climates – domestic animals and vermin – did not become so independently, but only in the wake of man. And the transition from the uniformly hot climate of the original home of man to colder regions, where the year was divided into summer and winter, created new requirements – shelter and clothing as protection against cold and damp, and hence new spheres of labour, new forms of activity, which further and further separated man from the animal.

By the combined functioning of hands, speech organs and brain, not only in each individual but also in society, men became capable of executing more and more complicated operations, and were able to set themselves, and achieve, higher and higher aims. The work of each generation itself became different, more perfect and more diversified. Agriculture was added to hunting and cattle raising; then came spinning, weaving, metalworking, pottery and navigation. Along with trade and industry, art and science finally appeared. Tribes developed into nations and states. Law and politics arose, and with them that fantastic reflection of human things in the human mind – religion. In the face of all these images, which appeared in the first place to be products of the mind and seemed to dominate human societies, the more modest productions of the working hand retreated into the background, the more so since the mind that planned the labour was able, at a very early stage in the development of society (for example, already in the primitive family), to have the labour that had been planned carried out by other hands than its own. All merit for the swift advance of civilisation was ascribed to the mind, to the development and activity of the brain. Men became accustomed to explain their actions as arising out of thoughts instead of their needs (which in any case are reflected and perceived in the mind); and so in the course of time there emerged that idealistic world outlook which, especially since the fall of the world of antiquity, has dominated men's minds. It still rules them to such a degree that even the most materialistic natural scientists of the Darwinian school are still unable to form any clear idea of the origin of man, because under this ideological influence they do not recognise the part that has been played therein by labour.

Animals, as has already been pointed out, change the environment by their activities in the same way, even if not to the same extent, as man does, and these changes, as we have seen, in turn react upon and change those who made them. In nature nothing takes place in isolation. Everything affects and is affected by every other thing, and it is mostly because this manifold motion and interaction is forgotten that our natural scientists are prevented from gaining a clear insight into the simplest things. We have seen how goats have prevented the regeneration of forests in Greece; on the island of St. Helena, goats and pigs brought by the first arrivals have succeeded in exterminating its old vegetation almost completely, and so have prepared the ground for the spreading of plants brought by later sailors and colonists. But animals exert a lasting effect on their environment unintentionally and, as far as the animals themselves are concerned, accidentally. The further removed men are from animals, however, the more their effect on nature assumes the character of premeditated, planned action directed towards definite preconceived ends. The animal destroys the vegetation of a locality without realising what it is doing. Man destroys it in order to sow field crops on the soil thus released, or to plant trees or vines which he knows will yield many times the amount planted. He transfers useful plants and domestic animals from one country to another and thus changes the flora and fauna of whole continents. More than this. Through artificial breeding both plants and animals are so changed by the hand of man that they become unrecognisable. The wild plants from which our grain varieties originated are still being sought in vain. There is still some dispute about the wild animals from which our very different breeds of dogs or our equally numerous breeds of horses are descended.

It goes without saying that it would not occur to us to dispute the ability of animals to act in a planned, premeditated fashion. On the contrary, a planned mode of action exists in embryo wherever protoplasm, living albumen, exists and reacts, that is, carries out definite, even if extremely simple, movements as a result of definite external stimuli. Such reaction takes place even where there is yet no cell at all, far less a nerve cell. There is something of the planned action in the way insect-eating plants capture their prey, although they do it quite unconsciously. In animals the capacity for conscious, planned action is proportional to the development of the nervous system, and among mammals it attains a fairly high level. While fox-hunting in England one can daily observe how unerringly the fox makes use of its excellent knowledge of the locality in order to elude its pursuers, and how well it knows and turns to account all favourable features of the ground that cause the scent to be lost. Among our domestic animals, more highly developed thanks to association with man, one can constantly observe acts of cunning on exactly the same level as those of children. For, just as the development history of the human embryo in the mother's womb is only an abbreviated repetition of the history, extending over millions of years, of the bodily evolution of our animal ancestors, starting from the worm, so the mental development of the human child is only a still more abbreviated repetition of the intellectual development of these same ancestors, at least of the later ones. But all the planned action of all animals has never succeeded in impressing the stamp of their will upon the earth. That was left for man.

In short, the animal merely *uses* its environment, and brings about changes in it simply by its presence; man by his changes makes it serve his ends, *masters* it. This is the final, essential distinction between man and other animals, and once again it is labour that brings about this distinction.[2]

Let us not, however, flatter ourselves overmuch on account of our human victories over nature. For each such victory nature takes its revenge on us. Each victory, it is true, in the first place brings about the results we expected, but in the second and third places it has quite different, unforeseen effects which only too often cancel the first. The people who, in Mesopotamia, Greece, Asia Minor and elsewhere, destroyed the forests to obtain cultivable land, never dreamed that by removing along with the forests the collecting centres and reservoirs of moisture they were laying the basis for the present forlorn state of those countries. When the Italians of the Alps used up the pine forests on the southern slopes, so carefully cherished on the northern slopes, they had no inkling that by doing so they were cutting at the roots of the dairy industry in their region; they had still less inkling that they were thereby depriving their mountain springs of water for the greater part of the year, and making it possible for them to pour still more furious torrents on the plains during the rainy seasons. Those who spread the potato in Europe were not aware that with these farinaceous tubers they were at the same time spreading scrofula. Thus at every step we are reminded that we by no means rule over nature like a conqueror over a foreign people, like someone standing outside nature – but that we, with flesh, blood and brain, belong to nature, and exist in its midst, and that all our mastery of it consists in the fact that we have the advantage over all other creatures of being able to learn its laws and apply them correctly.

And, in fact, with every day that passes we are acquiring a better understanding of these laws and getting to perceive both the more immediate and the more remote consequences of our interference with the traditional course of nature. In particular, after the mighty advances made by the natural sciences in the present century, we are more than ever in a position to realise, and hence to control, even the more remote natural consequences of at least our day-to-day production activities. But the more this progresses the more will men not only feel but also know their oneness with nature, and the more impossible will become the senseless and unnatural idea of a contrast between mind and matter, man and nature, soul and body, such as arose after the decline of classical antiquity in Europe and obtained its highest elaboration in Christianity.

It required the labour of thousands of years for us to learn a little of how to calculate the more remote *natural* effects of our actions in the field of production, but it has been still more difficult in regard to the more remote *social* effects of these actions. We mentioned the potato and the resulting spread of scrofula. But what is scrofula compared to the effect which the reduction of the workers to a potato diet had on the living conditions of the masses of the people in whole countries, or compared to the famine the potato blight brought to Ireland in 1847, which consigned to the grave a million Irishmen, nourished solely or almost exclusively on potatoes, and forced the emigration overseas of two million more? When the Arabs learned to distil spirits, it never entered their heads that by so doing they were creating one of the chief weapons for the annihilation of the aborigines of the then still undiscovered American continent. And when afterwards Columbus discovered this America, he did not know that by doing so he was laying the basis for the Negro slave trade and giving a new lease of life to slavery, which in Europe had long ago been done away with. The men who in the seventeenth and eighteenth centuries laboured to create the steam-engine had no idea that they were preparing the instrument which more than any other was to revolutionise social relations throughout the world. Especially in Europe, by concentrating wealth in the hands of a minority and dispossessing the huge majority, this instrument was destined at first to give social and political domination to the bourgeoisie, but later, to give rise to a class struggle between bourgeoisie and proletariat which can end only in the overthrow of the bourgeoisie and the abolition of all class antagonisms. But in this sphere, too, by long and often cruel experience and by collecting and analysing historical material, we are gradually learning to get a clear view of the indirect, more remote social effects of our production activity, and so are afforded an opportunity to control and regulate these effects as well.

This regulation, however, requires something more than mere knowledge. It requires a complete revolution in our hitherto existing mode of production, and simultaneously a revolution in our whole contemporary social order.

All hitherto existing modes of production have aimed merely at achieving the most immediately and directly useful effect of labour. The further consequences, which appear only later and become effective through gradual repetition and accumulation, were totally neglected. The original common ownership of land corresponded, on the one hand, to a level of development of human beings in which their horizon was restricted in general to what lay immediately available, and presupposed, on the other hand, a certain superfluity of land that would allow some latitude for correcting the possible bad results of this primeval type of economy. When this surplus land was exhausted, common ownership also declined. All higher forms of production, however, led to the division of the population into different classes and thereby to the antagonism of ruling and oppressed classes. Thus the interests of the ruling class became the driving factor of production, since production was no longer restricted to providing the barest means of subsistence for the oppressed people. This has been put into effect most completely in the capitalist mode of production prevailing today in Western Europe. The individual capitalists, who dominate production and exchange, are able to concern themselves only with the most immediate useful effect of their actions. Indeed, even this useful effect – inasmuch as it is a question of the usefulness of the article that is produced or exchanged – retreats far into the background, and the sole incentive becomes the profit to be made on selling.[3]

Classical political economy, the social science of the bourgeoisie, in the main examines only social effects of human actions in the fields of production and exchange that are actually intended. This fully corresponds to the social organisation of which it is the theoretical expression. As individual capitalists are engaged in production and exchange for the sake of the immediate profit, only the nearest, most immediate results must first be taken into account. As long as the individual manufacturer or merchant sells a manufactured or purchased commodity with the usual coveted profit, he is satisfied and does not concern himself with what afterwards becomes of the commodity and its purchasers. The same thing applies to the natural effects of the same actions. What cared the Spanish planters in Cuba, who burned down forests on the slopes of the mountains and obtained from the ashes sufficient fertiliser for *one* generation of very highly profitable coffee trees – what cared they that the heavy tropical rainfall afterwards washed away the unprotected upper stratum of the soil, leaving behind only bare rock! In relation to nature, as to society, the present mode of production is predominantly concerned only about the immediate, the most tangible result; and then surprise is expressed that the more remote effects of actions directed to this end turn out to be quite different, are mostly quite the opposite in character; that the harmony of supply and demand is transformed into the very reverse opposite, as shown by the course of each ten years' industrial cycle – even Germany has had a little preliminary experience of it in the "crash"; that private ownership based on one's own labour must of necessity develop into the expropriation of the workers, while all wealth becomes more and more concentrated in the hands of non-workers; that [...][4]

Notes

1 A leading authority in this respect, Sir William Thomson, has calculated that *little more than a hundred million years* could have elapsed since the time when the earth had cooled sufficiently for plants and animals to be able to live on it. [*Note by Engels.*]

2 In the margin of the manuscript is written in pencil: "Ennoblement." – *Ed.*

3 The MS ends here. What follows was written on a separate sheet of paper with a note in a different hand to the effect that it was the last page of the first draft. – *Ed.*

4 Here the manuscript breaks off. – *Ed.*

From Karl Marx and Friedrich Engels, *Basic Writings on Politics and Philosophy*

Engels: On authority

A number of socialists have latterly launched a regular crusade against what they call the *principle* of authority. It suffices to tell them that this or that act is *authoritarian* for it to be condemned. This summary mode of procedure is being abused to such an extent that it has become necessary to look into the matter somewhat more closely. Authority, in the sense in which the word is used here, means the imposition of the will of another upon ours; on the other hand, authority presupposes subordination. Now since these two words sound bad and the relationship which they represent is disagreeable to the subordinated party, the question is to ascertain whether there is any way of dispensing with it, whether – given the conditions of present-day society – we could not create another social system, in which this authority would be given no scope any longer and would consequently have to disappear. On examining the economic, industrial, and agricultural conditions which form the basis of present-day bourgeois society, we find that they tend more and more to replace isolated action by combined action of individuals. Modern industry with its big factories and mills, where hundreds of workers supervise complicated machines driven by steam, has superseded the small workshops of the separate producers; the carriages and wagons of the highways have been replaced by railway trains, just as the small schooners and sailing feluccas have been by steamboats. Even agriculture falls increasingly under the dominion of the machine and of steam, which slowly but relentlessly put in the place of the small proprietors big capitalists, who with the aid of hired workers cultivate vast stretches of land. Everywhere combined action, the complication of processes dependent upon each other, displaces independent action by individuals. But whoever mentions combined action speaks of organization; now is it possible to have organization without authority?

Supposing a social revolution dethroned the capitalists, who now exercise their authority over the production and circulation of wealth. Supposing, to adopt entirely the point of view of the anti-authoritarians, that the land and the instruments of labor had become the collective property of the workers who use them. Will authority have disappeared or will it only have changed its form? Let us see.

Let us take by way of example a cotton-spinning mill. The cotton must pass through at least six successive operations before it is reduced to the state of thread, and these operations take place for the most part in different rooms. Furthermore, keeping the machines going requires an engineer to look after the steam engine, mechanics to make the current repairs, and many other laborers, whose business it is to transfer the products from one room to another, and so forth. All these workers, men, women, and children, are obliged to begin and finish their work at the hours fixed by the authority of the steam, which cares nothing for individual autonomy. The workers must, therefore, first come to an understanding on the hours of work; and these hours, once they are fixed, must be observed by all, without any exception. Thereafter particular questions arise in each room and at every moment concerning the mode of production, distribution of materials, etc., which must be settled at once on pain of seeing all production immediately stopped; whether they are settled by decision of a delegate placed at the head of each branch of labor or, if possible, by a majority vote, the will of the single individual will always have to subordinate itself, which means that questions are settled in an authoritarian way. The automatic machinery of a big factory is much more despotic than the small capitalists who employ workers ever have been. At least with regard to the hours of work one may write upon the portals of these factories: *Lasciate ogni autonomia, voi che entrate!*[1] If man, by dint of his knowledge and inventive genius, has subdued the forces of nature, the latter avenge themselves upon him by subjecting him, in so far as he employs them, to a veritable despotism, independent of all social organization. Wanting to abolish authority in large-scale industry is tantamount to wanting to abolish industry itself, to destroy the power loom in order to return to the spinning wheel.

Let us take another example – the railway. Here, too, the co-operation of an infinite number of individuals is absolutely necessary, and this co-operation must be practiced during precisely fixed hours, so that no accidents may happen. Here, too, the first condition of the job is a dominant will that settles all subordinate questions, whether this will is represented by a single delegate or a committee charged with the execution of the resolutions of the majority of persons interested. In either case there is very pronounced authority. Moreover, what would happen to the first train dispatched if the authority of the railway employees over the honorable passengers were abolished?

But the necessity of authority, and of imperious authority at that, will nowhere be found more evident than on board a ship on the high seas. There, in time of danger, the lives of all depend on the instantaneous and absolute obedience of all to the will of one.

When I submitted arguments like these to the most rabid anti-authoritarians, the only answer they were able to give me was the following: Yes, that's true, but here it is not a case of authority which we confer on our delegates, *but of a commission entrusted!* These gentlemen think that when they have changed the names of things they have changed the things themselves. This is how these profound thinkers mock at the whole world.

We have thus seen that, on the one hand, a certain authority, no matter how delegated, and, on the other hand, a certain subordination are things which, independent of all social organization, are imposed upon us together with the material conditions under which we produce and make products circulate.

We have seen, besides, that the material conditions of production and circulation inevitably develop with large-scale industry and large-scale agriculture, and increasingly tend to enlarge the scope of this authority. Hence it is absurd to speak of the principle of authority as being absolutely evil and of the principle of autonomy as being absolutely good. Authority and autonomy are relative things, whose spheres vary with the various phases of the development of society. If the autonomists confined themselves to saying that the social organization of the future would restrict authority solely to the limits within which the conditions of production render it inevitable, we could understand each other; but they are blind to all facts that make the thing necessary and they passionately fight the word.

Why do the anti-authoritarians not confine themselves to crying out against political authority, the state? All socialists are agreed that the political state, and with it political authority, will disappear as a result of the coming social revolution, that is, that public functions will lose their political character and be transformed into the simple administrative functions of watching over the true interests of society. But the anti-authoritarians demand that the authoritarian political state be abolished at one stroke, even before the social conditions that gave birth to it have been destroyed. They demand that the first act of the social revolution shall be the abolition of authority. Have these gentlemen ever seen a revolution? A revolution is certainly the most authoritarian thing there is; it is the act whereby one part of the population imposes its will upon the other part by means of rifles, bayonets, and cannon – authoritarian means, if such there be at all; and if the victorious party does not want to have fought in vain, it must maintain this rule by means of the terror which its arms inspire in the reactionaries. Would the Paris Commune have lasted a single day if it had not made use of this authority of the armed people against the bourgeois? Should we not, on the contrary, reproach it for not having used it freely enough?

Therefore, either one of two things: either the anti-authoritarians don't know what they are talking about, in which case they are creating nothing but confusion, or they do know, and in that case they are betraying the movement of the proletariat. In either case they serve the reaction.

Note

1 [Leave, ye that enter in, all autonomy behind!]

Part II

Philosophy, Modern Science, and Technology

Positivist and Postpositivist Philosophies of Science

Introduction

Perhaps in keeping with the popular idea that philosophy simply begins by critically evaluating some marked-off "field" of inquiry (e.g., ethics, art, mind, law, language), current investigations of technology, too, often define themselves by identifying technological topics and practices that seem to raise philosophical "issues." As the selections in Parts II and III suggest, however, this approach tends to result in inquiries that remain silent both about what counts as "technology" in the first place and about the crucial question of the relation between technology and science. The latter failure is especially significant, because many of the current debates over technology rest upon oversimplified or outdated conceptions of science. Optimistic and "pro-technology" writings often presuppose fairly crude, positivistic visions of science and oversimplified conceptions of the objectivity of scientific knowledge. Social constructivist and "critical" writings on technology often assume equally crude, antipositivistic conceptions of science and tend to reduce scientific knowledge claims to mere functions of societal or cultural "consensus." With the decline of the influence of positivism on the philosophy of science in recent decades, however, there have emerged a number of sophisticated reinterpretations of science and scientific knowledge. After the first two selections, the readings in this section are representative of these recent developments; and it will become plain in the sections that follow that these post- and non-positivist philosophies of science promise to open up new and more nuanced possibilities for the evaluation of technology and its place in culture.

During the first two-thirds of the twentieth century, logical positivism and its more qualified successor, logical empiricism, dominated the philosophy of science, particularly in England and North America. Viewed epistemologically, logical positivism famously and explicitly combines Bacon's empiricism with twentieth-century developments in mathematical logic. In addition, because this kind of epistemology appears to entail hostility toward speculative theorizing and is often accompanied by an emotivist position in ethics, it has been widely assumed that logical positivists and empiricists do not espouse any general historical theory concerning scientific and general intellectual progress. Indeed, several members of the movement, following John Stuart Mill, explicitly reject the classical positivism of Comte, first, for failing to develop a formal, rational reconstruction of the scientific method and second, for embracing a speculative philosophy of history based on an empirically suspect theory of "three stages" of intellectual development. The second reading selection, Carl Hempel and Paul Oppenheim's classic "deductive-nomological" account of scientific explanation, exemplifies the greater epistemological concern displayed by all the twentieth-century positivists – in this case, in its somewhat more chastened logical empiricist version. Formal, syntactic reconstruction of scientific theories and a strict separation of the purely logical from the purely observational parts of science characterize this approach. In the first selection, however, the twentieth-century positivists' allegedly anti-speculative and purely epistemic stance is shown to be something of a myth. Like Comte, only silently, the Vienna Circle manifesto, "The Scientific Conception of the World," exhibits the same militant opposition to traditional religious and metaphysical reasoning, the same conviction of the superiority of scientific rationality, and the same assumption that social and political progress depend upon the acquisition of scientific knowledge. The influence of these assumptions in the debates over the use and importance of technology will, of course, emerge in many of the later selections. For the moment, it might be noted that both Comte and Marx, for all the differences in their projected political programs, share with twentieth-century positivists the same optimistic and progressivistic picture of a world improved through the application of scientific knowledge. Marx, we recall, envisioned the rise of a "scientific" socialism.

Beginning in the late 1950s and early 1960s, new approaches to the philosophy of science arose in Anglo-American philosophy, influenced by Wittgenstein's later writings and by burgeoning interest in the history of science. Stephen Toulmin's *Foresight and Understanding* (1961) is representative of this development. Drawing on R. G. Collingwood's theory of "absolute presuppositions" as the basis of philosophical systems and on Wittgenstein's analyses of the uses of language, Toulmin argues that scientific theories should be understood in terms of Collingwood-like "ideals of natural order" – that is, global presuppositions concerning how natural processes work when undisturbed by external complications. What science "explains," then, is not the natural processes themselves but only deviations from them. For these global, or idealizing background presuppositions, Toulmin introduces the notion of a "paradigm" – a term soon to be made famous in Thomas Kuhn's account of scientific revolutions as involving paradigm shifts.

Where the writings of Toulmin, Kuhn, and other early postpositivist philosophers are still explicitly marked by polemics against their predecessors, the feminist philosophy of science that began to emerge about ten years later more often simply starts out from where these thinkers left off. For if Toulmin, Kuhn, and others are right – that is, if scientific theories are neither direct, Baconian-empiricist reports of the facts nor purely neutral, Hempelian-like formal machines for making scientific predictions – if, instead, they are explanatory systems grounded in idealizing presuppositions and global paradigms that are often neither explicit nor articulated, then it is important to ask what kinds of attitudes, worldviews, and biases are likely to enter even the best of scientific theories and methodologies. In particular, feminist philosophers have argued for the existence of a "masculine" bias toward understanding scientific research – one that privileges the detached, experimental manipulation of objects and thereby fosters the neglect of aspects of nature that might nevertheless be revealed to science through other approaches. Nancy Tuana's recent survey of feminist philosophy of science identifies the influence of such masculine orientations, and describes feminist alternatives on topics ranging from primate research and theories of human evolution to general attitudes toward the scientific epistemology.

It is worth remembering that before logical positivism came to dominate North American philosophy of science, an indigenous philosophy of pragmatism had already developed a richer alternative conception of the scientific method. The classical American pragmatists (e.g., James, Peirce, and Dewey) did not share the positivists' Baconian/British Empiricist conception of experience, which they regarded as poverty-stricken. "British Empiricism," James famously quipped, "is not very empirical." For these pragmatists, "experience" can be something as broad and complex as the reflective discovery of a mental habit or social dilemma, or something as narrow and measurable as the passive registration of sense data. Not surprisingly, given this experiential pluralism, they displayed much greater tolerance to the idea that what the "scientific method" is and does cannot be reduced to one standard description.

With the waning of the influence of classical American pragmatism, especially after World War II, only its more narrowly instrumentalist, naturalistic, and manipulative strains were taken seriously and absorbed into American logical empiricism, which was much more inclined to view all scientific inquiry on the model of nineteenth-century experiments in physics. Pragmatism's richer social and contextual notions of scientific research were largely ignored by the Anglo-American mainstream until the early 1980s, when postpositivists like Richard Rorty and Hilary Putnam began to draw again upon the work of James and Dewey in order to express their growing displeasure with the narrow and dated character of the positivist epistemologies they had inherited.

In the meantime, in Germany and much of continental Europe, logical empiricism enjoyed a much shorter and less dominant run. By the first two decades of the twentieth century, an influential alternative approach to the understanding of the methods of science had already been developed. Drawing first on Dilthey's argument for separate epistemologies for "natural" and "human" science, and later on Husserl's phenomenological conception of irreducibly different "regions" of reality and on Heidegger's conception of philosophy as hermeneutics, or the interpretation of the lived-through "meaning" of whatever we encounter, "hermeneutical" philosophy of science was originally offered specifically as an alternative to the positivists' formal, "rationally reconstructive" explications of science and scientific theory. Such an alternative, it was argued, was necessary especially in the social and human sciences, where the focus is often on unique individuals and where even generalizing accounts of human phenomena are more concerned with understanding what actions "mean" to people than with finding what "causes" them or making predictions about future actions. Eventually, however, the scope of hermeneutics was widened to accommodate a reconsideration of all science, not just human science, and by the end of the twentieth century, a hermeneutics of natural science and of science generally had been extensively developed. Patrick Heelan and Jay Schulkin's article gives a particularly lucid exposition of these developments, as well as a very helpful comparison of the hermeneutical conception of science with that of pragmatism.

The social and contextual aspects of science are also stressed by the French philosopher and sociologist of science, Bruno Latour, who complains that the very idea of an opposition or separation of "science" from its "society" involves a false dichotomy. There is as much society within science as without. In his earlier work, Latour presented a kind of anthropology of scientific laboratories, but more recently he has moved toward developing a generalized actor-network theory. Latour's work, highly influential in science studies in the United States, popularizes (often without crediting) Michel Foucault's power-network interpretations of human activities and the deconstructive techniques of Jacques Derrida. He employs the latter to undermine numerous oppositions assumed in traditional studies of science and technology – for example, that between science and technology. Using Gaston Bachelard's term, "technoscience," and transforming the subtle and difficult prose of Derrida into a jokey and cartoon-filled presentation, he makes his case by recounting scientific and technological case studies. Latour's approach, like that of the social constructivists excerpted elsewhere in this anthology, has been criticized in the so-called science wars as rejecting the reality of the objects of science. Latour wittily rejects this accusation as a red herring, and insightfully suggests that the contemporary science wars replay in a crude fashion the debates of Plato's *Gorgias* between Truth and political power.

Finally, the post-Kuhnian, pragmatic, and hermeneutic approaches to science raise questions concerning the relation of Western science to traditional or indigenous science in the non-Western world. If there are no clear demarcations between science and non-science, as both the logical positivists and Karl Popper insisted, then indigenous knowledge cannot be sharply demarcated from science, and science itself might better come to be seen as a particularly powerful form of "local knowledge," involving tacit skills, traditions, and rituals. Sandra Harding's "Dysfunctional Universality Claims?" succinctly raises a number of issues concerning the claim of Western science to universality and superiority.

Taken together with the previous readings, Harding's piece makes it plain that the new approaches to science have numerous implications for the philosophy of technology. For example, so long as science is taken to be a presuppositionless and completely detached recounting of objective atomic facts and technology is presumed to be simply applied science, the latter is easily conceived as a benign and neutral means of taking advantage of the latest research to make whatever changes to our physical and social world that we decide upon. In reaction to this happy but deeply questionable picture, political activists against particular technological projects (e.g., logging, advertised as simply making use of a "renewable resource") and postmodern literary and anthropological accounts of science have sometimes come to perceive science itself as an arbitrary and subjective construct. Yet neither the positivist nor these antipositivist and constructivist accounts shed much real light on the relation between science and technology. Postpositivistic, pragmatic, and hermeneutic approaches to science, all of which in various ways emphasize the cultural influences that help shape but do not entirely determine the development of technology, promise something better.

8

The Scientific Conception of the World: The Vienna Circle

Rudolf Carnap, Hans Hahn, and Otto Neurath

Dedicated to Moritz Schlick[1]

Preface

At the beginning of 1929 Moritz Schlick received a very tempting call to Bonn. After some vacillation he decided to remain in Vienna. On this occasion, for the first time it became clear to him and us that there is such a thing as the "Vienna Circle" of the scientific conception of the world, which goes on developing this mode of thought in a collaborative effort. This circle has no rigid organization; it consists of people of an equal and basic scientific attitude; each individual endeavours to fit in, each puts common ties in the foreground, none wishes to disturb the links through idiosyncrasies. In many cases one can deputise for another, the work of one can be carried on by another.

The Vienna Circle aims at making contact with those similarly oriented and at influencing those who stand further off. Collaboration in the Ernst Mach Society is the expression of this endeavour; Schlick is the chairman of this society and several members of Schlick's circle belong to the committee.

On 15–16 September 1929, the Ernst Mach Society, with the Society for Empirical Philosophy (Berlin), will hold a conference in Prague, on the epistemology of the exact sciences, in conjunction with the conference of the German Physical Society and the German Association of Mathematicians which will take place there at the same time. Besides technical questions, questions of principle are to be discussed. It was decided that on the occasion of this conference the present pamphlet on the Vienna Circle of the scientific conception of the world was to be published. It is to be handed to Schlick in October 1929 when he returns from his visiting professorship at Stanford University, California, as token of gratitude and joy at his remaining in Vienna. The second part of the pamphlet contains a bibliography compiled in collaboration with those concerned. It is to give a survey of the area of problems in which those who belong to, or are near to, the Vienna Circle are working.

Vienna, August 1929

For the Ernst Mach Society

Hans Hahn
Otto Neurath *Rudolf Carnap*

1 The Vienna Circle of the Scientific Conception of the World

1.1 Historical background

Many assert that metaphysical and theologizing thought is again on the increase today, not only in life but also in science. Is this a general phenomenon or merely a change restricted to certain circles? The assertion itself is easily confirmed if one looks at the topics of university courses and at the titles of

From *Otto Neurath: Empiricism and Sociology*, ed. Marie Neurath and R. S. Cohen, Dordrecht: D. Reidel Publishing Co., 1973, pp. 1–20, abridged. Reprinted by permission of Kluwer Academic Publishers, Dordrecht, The Netherlands.

philosophic publications. But likewise the opposite spirit of enlightenment and *anti-metaphysical factual research* is growing stronger today, in that it is becoming conscious of its existence and task. In some circles the mode of thought grounded in experience and averse to speculation is stronger than ever, being strengthened precisely by the new opposition that has arisen.

In the research work of all branches of empirical science this *spirit of a scientific conception of the world* is alive. However only a very few leading thinkers give it systematic thought or advocate its principles, and but rarely are they in a position to assemble a circle of like-minded colleagues around them. We find anti-metaphysical endeavours especially in England, where the tradition of the great empiricists is still alive; the investigations of Russell and Whitehead on logic and the analysis of reality have won international significance. In the U.S.A. these endeavours take on the most varied forms; in a certain sense James belongs to this group too. The new Russia definitely is seeking for a scientific world conception, even if partly leaning on older materialistic currents. On the continent of Europe, a concentration of productive work in the direction of a scientific world conception is to be found especially in Berlin (Reichenbach, Petzoldt, Grelling, Dubislav and others) and in Vienna.

That Vienna was specially suitable ground for this development is historically understandable. In the second half of the nineteenth century, liberalism was long the dominant political current. Its world of ideas stems from the enlightenment, from empiricism, utilitarianism and the free trade movement of England. In Vienna's liberal movement, scholars of world renown occupied leading positions. Here an anti-metaphysical spirit was cultivated, for instance, by men like Theodor Gomperz who translated the works of J. S. Mill, Suess, Jodl and others.

Thanks to this spirit of enlightenment, Vienna has been leading in a scientifically oriented people's education. With the collaboration of Victor Adler and Friedrich Jodl, the society for popular education was founded and carried forth; "popular university courses" and the "people's college" were set up by the well-known historian Ludo Hartmann whose anti-metaphysical attitude and materialist conception of history expressed itself in all his actions. The same spirit also inspired the movement of the "Free School" which was the forerunner of today's school reform.

In this liberal atmosphere lived Ernst Mach (born 1838) who was in Vienna as student and as *privatdozent* (1861–64). He returned to Vienna only at an advanced age when a special chair of the philosophy of the inductive sciences was created for him (1895). He was especially intent on cleansing empirical science, and in the first place, physics, of metaphysical notions. We recall his critique of absolute space which made him a forerunner of Einstein, his struggle against the metaphysics of the thing-in-itself and of the concept of substance, and his investigations of the construction of scientific concepts from ultimate elements, namely sense data. In some points the development of science has not vindicated his views, for instance in his opposition to atomic theory and in his expectation that physics would be advanced through the physiology of the senses. The essential points of his conception however were of positive use in the further development of science. Mach's chair was later occupied by Ludwig Boltzmann (1902–06) who held decidedly empiricist views.

The activity of the physicists Mach and Boltzmann in a philosophical professorship makes it conceivable that there was a lively dominant interest in the epistemological and logical problems that are linked with the foundations of physics. These problems concerning foundations also led toward a renewal of logic. The path towards these objectives had also been cleared in Vienna from quite a different quarter by Franz Brentano (during 1874–80 professor of philosophy in the theological faculty, and later lecturer in the philosophical faculty). As a Catholic priest Brentano understood scholasticism; he started directly from the scholastic logic and from Leibniz's endeavours to reform logic, while leaving aside Kant and the idealist system-builders. Brentano and his students time and again showed their understanding of men like Bolzano (*Wissenschaftslehre*, 1837) and others who were working toward a rigorous new foundation of logic. In particular Alois Höfler (1853–1922) put this side of Brentano's philosophy in the foreground before a forum in which, through Mach's and Boltzmann's influence, the adherents of the scientific world conception were strongly represented. In the Philosophical Society at the University of Vienna numerous discussions took place under Höfler's direction, concerning questions of the foundation of physics and allied epistemological and logical problems. The Philosophical Society published *Prefaces and Introductions to Classical Works on Mechanics* (1899), as well as the individual papers of Bolzano (edited by Höfler and Hahn, 1914 and 1921). In Brentano's Viennese circle there was the young Alexius von Meinong (1870–82, later professor in Graz), whose theory of objects (1907) has certainly some affinity to modern theories of concepts and whose pupil Ernst Mally (Graz) also worked in the field of logistics. The early writings of Hans Pichler (1909) also belong to these circles.

Roughly at the same time as Mach, his contemporary and friend Josef Popper-Lynkeus worked in Vienna. Beside his physical and technical achievements we mention his large-scale, if unsystematic philosophical reflections (1899) and his rational economic plan (*A General Peacetime Labour Draft*, 1878). He consciously served the spirit of enlightenment, as is also evident from his book on Voltaire. His rejection of metaphysics was shared by many other Viennese sociologists, for example Rudolf Goldscheid. It is remarkable that in the field of political economy, too, there was in Vienna a strictly scientific method, used by the marginal utility school (Carl Menger, 1871); this method took root in England, France and Scandinavia, but not in Germany. Marxist theory likewise was cultivated and extended with special emphasis in Vienna (Otto Bauer, Rudolf Hilferding, Max Adler and others).

These influences from various sides had the result, especially since 1900, that there was in Vienna a sizeable number of people who frequently and assiduously discussed more general problems in close connection with empirical sciences. Above all these were epistemological and methodological problems of physics, for instance Poincaré's conventionalism, Duhem's conception of the aim and structure of physical theories (his translator was the Viennese Friedrich Adler, a follower of Mach, at that time *privatdozent* in Zürich); also questions about the foundations of mathematics, problems of axiomatics, logistic and the like. The following were the main strands from the history of science and philosophy that came together here, marked by those of their representatives whose works were mainly read and discussed:

(1) Positivism and empiricism: Hume, Enlightenment, Comte, J. S. Mill, Richard Avenarius, Mach.
(2) Foundations, aims and methods of empirical science (hypotheses in physics, geometry, etc.): Helmholtz, Riemann, Mach, Poincaré, Enriques, Duhem, Boltzmann, Einstein.
(3) Logistic and its application to reality: Leibniz, Peano, Frege, Schröder, Russell, Whitehead, Wittgenstein.
(4) Axiomatics: Pasch, Peano, Vailati, Pieri, Hilbert.
(5) Hedonism and positivist sociology: Epicurus, Hume, Bentham, J. S. Mill, Comte, Feuerbach, Marx, Spencer, Müller-Lyer, Popper-Lynkeus, Carl Menger (the elder).

1.2 The circle around Schlick

In 1922 Moritz Schlick was called from Kiel to Vienna. His activities fitted well into the historical development of the Viennese scientific atmosphere. Himself originally a physicist, he awakened to new life the tradition that had been started by Mach and Boltzmann and, in a certain sense, carried on by the anti-metaphysically inclined Adolf Stöhr. (In Vienna successively: Mach, Boltzmann, Stöhr, Schlick; in Prague: Mach, Einstein, Philipp Frank.)

Around Schlick, there gathered in the course of time a circle whose members united various endeavours in the direction of a scientific conception of the world. This concentration produced a fruitful mutual inspiration. Not one of the members is a so-called "pure" philosopher; all of them have done work in a special field of science. Moreover they come from different branches of science and originally from different philosophic attitudes. But over the years a growing uniformity appeared; this too was a result of the specifically scientific attitude: "What can be said at all, can be said clearly" (Wittgenstein); if there are differences of opinion, it is in the end possible to agree, and therefore agreement is demanded. It became increasingly clearer that a position not only free from metaphysics, but opposed to metaphysics was the common goal of all.

The attitudes toward questions of life also showed a noteworthy agreement, although these questions were not in the foreground of themes discussed within the Circle. For these attitudes are more closely related to the scientific world-conception than it might at first glance appear from a purely theoretical point of view. For instance, endeavours toward a new organization of economic and social relations, toward the unification of mankind, toward a reform of school and education, all show an inner link with the scientific world-conception; it appears that these endeavours are welcomed and regarded with sympathy by the members of the Circle, some of whom indeed actively further them.

The Vienna Circle does not confine itself to collective work as a closed group. It is also trying to make contact with the living movements of the present, so far as they are well disposed toward the scientific world-conception and turn away from metaphysics and theology. The Ernst Mach Society is today the place from which the Circle speaks to a wider public. This society, as stated in its program, wishes to "further and disseminate the scientific world-conception. It will organize lectures and publications about the present position of the scientific world-conception, in order to demonstrate the significance of exact research for the social sciences and the natural sciences. In this way intellectual tools should be formed for modern empiricism, tools that are also needed in forming public and private life." By the choice of its name, the society wishes to describe its basic orientation: science free of metaphysics. This, however, does not mean that the society declares itself in programmatic agreement with the individual doctrines of Mach. The Vienna Circle believes that in collaborating with the Ernst Mach Society it fulfils a demand of the day: we have to fashion intellectual tools for everyday life, for the daily life of the scholar but also for the daily life of all those who in some way join in working at the conscious reshaping of life. The vitality that shows itself in the efforts for a rational transformation of the social and economic order, permeates the movement for a scientific world-conception too. It is typical of the present situation in Vienna that when the Ernst Mach Society was founded in November 1928, Schlick was chosen chairman; round him the common work in the field of the scientific world-conception had concentrated most strongly.

Schlick and Philipp Frank jointly edit the collection of *Monographs on the Scientific World-Conception* [*Schriften zur wissenschaftlichen Weltauffassung*] in which members of the Vienna Circle preponderate.

2 The Scientific World Conception

The scientific world conception is characterized not so much by theses of its own, but rather by its basic attitude, its points of view and direction of research. The goal ahead is *unified science*. The endeavour is to link and harmonize the achievements of individual investigators in their various fields of science. From this aim follows the emphasis on *collective efforts*, and also the emphasis on what can be grasped intersubjectively; from this springs the search for a neutral system of formulae, for a symbolism freed from the slag of historical languages; and also the search for a total system of concepts. Neatness and clarity are striven for, and dark distances and unfathomable depths rejected. In science there are no "depths"; there is surface everywhere: all experience forms a complex network, which cannot always be surveyed and can often be grasped only in parts. Everything is accessible to man; and man is the measure of all things. Here is an affinity with the Sophists, not with the Platonists; with the Epicureans, not with the Pythagoreans; with all those who stand for earthly being and the here and now. The scientific world-conception knows *no unsolvable riddle*. Clarification of the traditional philosophical problems leads us partly to unmask them as pseudo-problems, and partly to transform them into empirical problems and thereby subject them to the judgment of experimental science. The task of philosophical work lies in this clarification of problems and assertions, not in the propounding of special "philosophical" pronouncements. The method of this clarification is that of *logical analysis*; of it, Russell says (*Our Knowledge of the External World*, p. 4) that it "has gradually crept into philosophy through the critical scrutiny of mathematics... It represents, I believe, the same kind of advance as was introduced into physics by Galileo: the substitution of piecemeal, detailed and verifiable results for large untested generalities recommended only by a certain appeal to imagination."[2]

It is *the method of logical analysis* that essentially distinguishes recent empiricism and positivism from the earlier version that was more biological-psychological in its orientation. If someone asserts "there is a God", "the primary basis of the world is the unconscious", "there is an entelechy which is the leading principle in the living organism", we do not say to him: "what you say is false"; but we ask him: "what do you mean by these statements?" Then it appears that there is a sharp boundary between two kinds of statements. To one belong statements as they are made by empirical science; their meaning can be determined by logical analysis or, more precisely, through reduction to the simplest statements about the empirically given. The other statements, to which belong those cited above, reveal themselves as empty of meaning if one takes them in the way that metaphysicians intend. One can, of course, often re-interpret them as empirical statements; but then they lose the content of feeling which is usually essential to the metaphysician. The metaphysician and the theologian believe, thereby misunderstanding themselves, that their statements say something, or that they denote a state of affairs. Analysis, however, shows that these statements say nothing but merely express a certain mood and spirit. To express such feelings for life can be a significant task. But the proper medium for doing so is art, for instance lyric poetry or music. It is dangerous to choose the linguistic garb of a theory instead: a theoretical content is simulated where none exists. If a metaphysician or theologian wants to retain the usual medium of language, then he must himself realise and bring out clearly that he is giving not description but expression, not theory or communication of knowledge, but poetry or myth. If a mystic asserts that he has experiences that lie above and beyond all concepts, one cannot deny this. But the mystic cannot talk about it, for talking implies capture by concepts and reduction to scientifically classifiable states of affairs.

The scientific world-conception rejects metaphysical philosophy. But how can we explain the wrong paths of metaphysics? This question may be posed from several points of view: psychological, sociological and logical. Research in a psychological direction is still in its early stages; the beginnings of more penetrating explanation may perhaps be seen in the investigations of Freudian psychoanalysis. The state of sociological investigation is similar; we may mention the theory of the "ideological superstructure"; here the field remains open to worthwhile further research.

More advanced is the clarification of *the logical origins of metaphysical aberration*, especially through the works of Russell and Wittgenstein. In metaphysical theory, and even in the very form of the questions, there are two basic logical mistakes: too narrow a tie to the form of *traditional languages* and a confusion about the logical achievement of thought. Ordinary language for instance uses the same part of speech, the substantive, for things ("apple") as well as as for qualities ("hardness"), relations ("friendship"), and processes ("sleep"); therefore it misleads one into a thing-like conception of functional concepts (hypostasis, substantialization). One can quote countless similar examples of linguistic misleading, that have been equally fatal to philosophers.

The second basic error of metaphysics consists in the notion that *thinking* can either lead to knowledge out of its own resources without using any empirical material; or at least arrive at new contents by an inference from given states of affair. Logical investigation, however, leads to the result that all thought and inference consists of nothing but a transition from statements to other statements that contain nothing that was not already in the former (tautological transformation). It is therefore not possible to develop a metaphysic from "pure thought".

In such a way logical analysis overcomes not only metaphysics in the proper, classical sense of the word, especially scholastic metaphysics and that of the systems of German idealism, but also the hidden metaphysics of Kantian and modern *apriorism*. The scientific world-conception knows no unconditionally valid knowledge derived from pure reason, no "synthetic judgments a priori" of the kind that lie at the basis of Kantian epistemology and even more of all pre- and post-Kantian ontology and metaphysics. The judgments of arithmetic, geometry, and certain fundamental principles of physics, that Kant took as examples of a priori knowledge will be discussed later. It is precisely in the rejection of the possibility of synthetic knowledge a priori that the basic thesis of modern empiricism lies. The scientific world-conception knows only empirical statements about things of all kinds, and analytic statements of logic and mathematics.

In rejecting overt metaphysics and the concealed variety of apriorism, all adherents of the scientific world-conception are at one. Beyond this, the Vienna Circle maintain the view that the statements of (critical) *realism* and *idealism* about the reality or non-reality of the external world and other minds are of a metaphysical character, because they are open to the same objections as are the statements of the old metaphysics: they are meaningless, because unverifiable and without content. For us, *something is "real" through being incorporated into the total structure of experience.*

Intuition which is especially emphasised by metaphysicians as a source of knowledge, is not rejected as such by the scientific world-conception. However, rational justification has to pursue all intuitive knowledge step by step. The seeker is allowed any method; but what has been found must stand up to testing. The view which attributes to intuition a superior and more penetrating power of knowing, capable of leading beyond the contents of sense experience and not to be confined by the shackles of conceptual thought – this view is rejected.

We have characterised the *scientific world-conception* essentially by *two features. First* it is *empiricist and positivist*: there is knowledge only from experience, which rests on what is immediately given. This sets the limits for the content of legitimate science. *Second*, the scientific world-conception is marked by application of a certain method, namely *logical analysis*. The aim of scientific effort is to reach the goal, unified science, by applying logical analysis to the empirical material. Since the meaning of every statement of science must be statable by reduction to a statement about the given, likewise the meaning of any concept, whatever branch of science it may belong to, must be statable by step-wise reduction to other concepts, down to the concepts of the lowest level which refer directly to the given. If such an analysis were carried through for all concepts, they would thus be ordered into a reductive system, a "constitutive system". Investigations towards such a constitutive system, the "constitutive theory", thus form the framework within which logical analysis is applied by the scientific world-conception. Such investigations show very soon that traditional Aristotelian scholastic logic is quite inadequate for this purpose. Only modern symbolic logic ("logistic") succeeds in gaining the required precision of concept definitions and of statements, and in formalizing the intuitive process of inference of ordinary thought, that is to bring it into a rigorous automatically controlled form by means of a symbolic mechanism. Investigations into constitutive theory show that the lowest layers of the constitutive system contain concepts of the experience and qualities of the individual psyche; in the layer above are physical objects; from these are constituted other minds and lastly the objects of social science. The arrangement of the concepts of the various branches of science into the constitutive system can already be discerned in outline today, but much remains to be done in detail. With the proof of the possibility and the outline of the shape of the total system of concepts, the relation of all statements to the given and with it the general structure of *unified science* become recognizable too.

A scientific description can contain only the *structure* (form of order) of objects, not their "essence". What unites men in language are structural formulae; in them the content of the common knowledge of men presents itself. Subjectively experienced qualities – redness, pleasure – are as such only experiences, not knowledge; physical optics admits only what is in principle understandable by a blind man too.

3 Fields of Problems

3.1 Foundations of arithmetic

In the writings and discussions of the Vienna Circle many different problems are treated, stemming from various branches of science. Attempts are made to arrange the various lines of problems systematically and thereby to clarify the situation.

The problems concerning the foundations of arithmetic have become of special historical significance for the development of the scientific world-conception because they gave impulse to the development of a new logic. After the very fruitful developments of mathematics in the 18th and 19th century during which more attention was given to the wealth of new results than to subtle examination of their conceptual foundations, this examination became unavoidable if mathematics were not to lose the traditionally celebrated certainty of its structure. This examination became even more urgent when certain contradictions, the "paradoxes of set theory", arose. It was soon recognized that these were not just difficulties in a special part of mathematics, but rather they were general logical contradictions, "antinomies", which pointed to essential mistakes in the foundations of traditional logic. The task of eliminating these contradictions gave a very strong impulse to the further development of logic. Here efforts for *clarification of the concept of number* met with those for an internal *reform of logic*. Since Leibniz and Lambert, the idea had come up again and again to master reality through a greater precision of concepts and inferential processes, and to obtain this precision by means of a symbolism fashioned after mathematics. After Boole, Venn and others, especially Frege (1884), Schröder (1890) and Peano (1895) worked on this problem. On the basis of these preparatory efforts *Whitehead* and *Russell* (1910) were able to establish a coherent system of logic in symbolic form ("logistic"), not only avoiding the contradictions of traditional logic, but far exceeding that logic in intellectual wealth and practical applicability. From this logical system they derived the concepts of arithmetic and analysis, thereby giving mathematics a secure foundation in logic.

Certain difficulties however remained in this attempt at overcoming the foundation crisis of arithmetic (and set theory) and have so far not found a definitively satisfactory solution. At present three different views confront each other in this field; besides the "logicism" of Russell and Whitehead, there is Hilbert's "formalism" which regards arithmetic as a playing with formulae according to certain rules, and Brouwer's "intutionism" according to which arithmetic knowledge rests on a not further reducible intuition of duality and unity [*Zwei-einheit*]. The debates are followed with great interest in the Vienna Circle. Where the decision will lead in the end cannot yet be foreseen; in any case, it will also imply a decision about the structure of logic; hence the importance of this problem for the scientific world-conception. Some hold that the three views are not so far apart as it seems. They surmise that essential features of all three will come closer in the course of future development and probably, using the far-reaching ideas of Wittgenstein, will be united in the ultimate solution.

The conception of mathematics as tautological in character, which is based on the investigations of Russell and Wittgenstein, is also held by the Vienna Circle. It is to be noted that this conception is opposed not only to apriorism and intuitionism, but also to the older empiricism (for instance of J. S. Mill), which tried to derive mathematics and logic in an experimental-inductive manner as it were.

Connected with the problems of arithmetic and logic are the investigations into the nature of the *axiomatic method* in general (concepts of completeness, independence, monomorphism, unambiguity and so on) and on the establishment of axiom-systems for certain branches of mathematics.

3.2 Foundations of physics

Originally the Vienna Circle's strongest interest was in the method of empirical science. Inspired by ideas of Mach, Poincaré, and Duhem, the problems of mastering reality through scientific systems, especially through *systems of hypotheses and axioms*, were discussed. A system of axioms, cut loose from all empirical application, can at first be regarded as a system of implicit definitions; that is to say, the concepts that appear in the axioms are fixed, or as it were defined, not from their content but only from their mutual relations through the axioms. Such a system of axioms attains a meaning for reality only by the addition of further definitions, namely the "coordinating definitions", which state what objects of reality are to be regarded as members of the system of axioms. The development of empirical science, which is to represent reality by means of as uniform and simple a net of concepts and judgments as possible, can now proceed in one of two ways, as history shows. The changes imposed by new experience can be made either in the axioms or in the coordinating definitions. Here we touch the problem of conventions, particularly treated by Poincaré.

The methodological problem of the application of axiom systems to reality may in principle arise for any branch of science. That these investigations have thus far been fruitful almost solely for physics, however, can be understood from the present stage of historical development of science: in regard to precision and refinement of concepts, physics is far ahead of the other branches of science.

Epistemological analysis of the leading concepts of natural science has freed them more and more from *metaphysical admixtures* which had clung to them from ancient time. In particular, Helmholtz, Mach, Einstein, and others have cleansed the concepts of *space, time, substance, causality, and probability*. The doctrines of absolute space and time have been overcome by the theory of relativity; space and time are no longer absolute containers but only ordering manifolds for elementary processes. Material substance has been dissolved by atomic theory and field theory. Causality was divested of the anthropomorphic character of "influence" or "necessary connection" and reduced to a relation among conditions, a functional coordination. Further, in place of the many laws of nature which were considered to be strictly valid, statistical laws have appeared; following the quantum theory there is even doubt whether the concept of strictly causal lawfulness is applicable to phenomena in very small space-time regions. The concept of probability is reduced to the empirically graspable concept of relative frequency.

Through the application of the *axiomatic method* to these problems, the empirical components always separate from the merely conventional ones, the content of statements from definitions. No room remains for a priori synthetic judgments. That knowledge of the world is possible rests not on human reason impressing its form on the material, but on the material being ordered in a certain way. The kind and degree of this order cannot be known beforehand. The world might be ordered much more strictly than it is; but it might equally be ordered much less without jeopardising the possibility of knowledge. Only step by step can the advancing research of empirical science teach us in what degree the world is regular. The method of induction, the inference from yesterday to tomorrow, from here to there, is of course only valid if regularity exists. But this method does not rest on some a priori presupposition of this regularity. It may be applied wherever it leads to fruitful results, whether or not it be adequately founded; it never yields certainty. However, epistemological reflection demands that an inductive inference should be given significance only insofar as it can be tested empirically. The scientific world-conception will not condemn the success of a piece of research because it has been gathered by means that are inadequate, logically unclear or empirically unfounded. But it will always strive at testing with clarified aids, and demand an indirect or direct reduction to experience.

3.3 Foundations of geometry

Among the questions about the foundations of physics, the problem of *physical space* has received special significance in recent decades. The investigations of Gauss (1816), Bolyai (1823), Lobatchevski (1835) and others led to *non-Euclidean geometry*, to a realisation that the hitherto dominant classical geometric system of Euclid was only one of an infinite set of systems, all of equal logical merit. This raised the question, which of these geometries was that of actual space. Gauss had wanted to resolve this question by measuring the angles of a large triangle. This made *physical geometry* into an empirical science, a branch of physics. The problems were further studied particularly by Riemann (1868), Helmholtz (1868) and Poincaré (1904). Poincaré especially emphasized the link of physical geometry with all other branches of physics: the question concerning the nature of actual space can be answered only in connection with a total system of physics. Einstein then found such a total system, which answered the question in favour of a certain non-Euclidean system.

Through this development, physical geometry became more and more clearly separated from pure *mathematical* geometry. The latter gradually became more and more formalized through further development of logical analysis. First it was arithmetized, that is, interpreted as the theory of a certain number system. Next it was axiomatized, that is, represented by means of a system of axioms that conceives the geometrical elements (points, etc.) as undefined objects, and fixes only their mutual relations. Finally geometry was logicized, namely represented as a theory of certain structural relations. Thus geometry became the most important field of application for the axiomatic method and for the general theory of relations. In this way, it gave the strongest impulse to the development of the two methods which in turn became so important for the development of logic itself, and thereby again for the scientific world-conception.

The relations between mathematical and physical geometry naturally led to the problem of the application of axiom systems to reality which, as mentioned, played a big role in the more general investigations about the foundations of physics.

3.4 Problems of the foundations of biology and psychology

Metaphysicians have always been fond of singling out biology as a special field. This came out in the doctrine of a special life force, the theory of *vitalism*. The modern representatives of this theory endeavour to bring it from the unclear, confused form of the past into a conceptually clear formulation. In place of the life force, we have "dominants" (Reinke, 1899) or "entelechies" (Driesch, 1905). Since these concepts do not satisfy the requirement of reducibility to the given, the scientific world-conception rejects them as metaphysical. The same holds true of so-called "psychovitalism" which puts forward an intervention of the soul, a "role of leadership of the mental in the material". If, however, one digs out of this metaphysical vitalism the empirically graspable kernel, there remains the thesis that the processes of organic nature proceed according to laws that cannot be reduced to physical laws. A more precise analysis shows that this thesis is equivalent to the assertion that certain fields of reality are not subject to a uniform and pervasive regularity.

It is understandable that the scientific world-conception can show more definite confirmation for its views in those fields which have already achieved conceptual precision than in others: in physics more than in psychology. The linguistic forms which we still use in psychology today have their origin in certain ancient metaphysical notions of the soul. The formation of concepts in psychology is made difficult by these defects of language: metaphysical burdens and logical incongruities. Moreover there are certain factual difficulties. The result is that hitherto most of the concepts used in psychology are inadequately defined; of some, it is not known whether they have meaning or only simulate meaning through usage. So, in this field nearly everything in the way of epistemological analysis still remains to be done; of course, analysis here is more difficult than in physics. The attempt of behaviorist psychology to grasp the psychic through the behavior of bodies, which is at a level accessible to perception, is, in its principled attitude, close to the scientific world-conception.

3.5 *Foundations of the social sciences*

As we have specially considered with respect to physics and mathematics, every branch of science is led to recognise that, sooner or later in its development, it must conduct an epistemological examination of its foundations, a logical analysis of its concepts. So too with the social sciences, and in the first place with history and economics. For about a hundred years, a process of elimination of metaphysical admixtures has been operating in these fields. Of course the purification has not yet reached the same degree as in physics; on the other hand, the task of cleansing is less urgent perhaps. For it seems that even in the heyday of metaphysics and theology, the metaphysical strain was not particularly strong here; maybe this is because the concepts in this field, such as war and peace, import and export, are closer to direct perception than concepts like atom and ether. It is not too difficult to drop concepts like "folk spirit" and instead to choose, as our object, groups of individuals of a certain kind. Scholars from the most diverse trends, such as Quesnay, Adam Smith, Ricardo, Comte, Marx, Menger, Walras, Müller-Lyer, have worked in the sense of the empiricist, anti-metaphysical attitude. The object of history and economics are people, things and their arrangement.

4 Retrospect and Prospect

The modern scientific world-conception has developed from work on the problems just mentioned. We have seen how in physics, the endeavours to gain tangible results, at first even with inadequate or still insufficiently clarified scientific tools, found itself forced more and more into methodological investigations. Out of this developed the method of forming hypotheses and, further, the axiomatic method and logical analysis; there by concept formation gained greater clarity and strength. The same methodological problems were met also in the development of foundations research in physical geometry, mathematical geometry and arithmetic, as we have seen. It is mainly from all these sources that the problems arise with which representatives of the scientific world-conception particularly concern themselves at present. Of course it is still clearly noticeable from which of the various problem areas the individual members of the Vienna Circle come. This often results in differences in lines of interests and points of view, which in turn lead to differences in conception. But it is characteristic that an endeavour toward precise formulation, application of an exact logical language and symbolism, and accurate differentiation between the theoretical content of a thesis and its mere attendant notions, diminish the separation. Step by step the common fund of conceptions is increased, forming the nucleus of a scientific world-conception around which the outer layers gather with stronger subjective divergence.

Looking back we now see clearly what is the *essence of the new scientific world-conception* in contrast with traditional philosophy. No special "philosophic assertions" are established, assertions are merely clarified; and at that assertions of empirical science, as we have seen when we discussed the various problem areas. Some representatives of the scientific world-conception no longer want to use the term "philosophy" for their work at all, so as to emphasise the contrast with the philosophy of (metaphysical) systems even more strongly. Whichever term may be used to describe such investigations, this much is certain: *there is no such thing as philosophy as a basic or universal science alongside or above the various fields of the one empirical science*; there is no way to genuine knowledge other than the way of experience; there is no way to genuine knowledge other than the way of experience; there is no realm of ideas that stands over or beyond experience. Nevertheless the work of "philosophic" or "foundational" investigations remains important in accord with the scientific world-conception. For the logical clarification of scientific concepts, statements and methods liberates one from inhibiting prejudices. Logical and epistemological analysis does not wish to set barriers to scientific enquiry; on the contrary, analysis provides science with as complete a range of formal possibilities as is possible, from which to select what best fits each empirical finding (example: non-Euclidean geometries and the theory of relativity).

The representatives of the scientific world-conception resolutely stand on the ground of simple human experience. They confidently approach the task of removing the metaphysical and theological debris of millennia. Or, as some have it: returning, after a metaphysical interlude, to a unified picture of this world which had, in a sense, been at the basis of magical beliefs, free from theology, in the earliest times.

The increase of metaphysical and theologizing leanings which shows itself today in many associations and sects, in books and journals, in talks and university lectures, seems to be based on the fierce social and economic struggles of the present: one group of combatants, holding fast to traditional social forms, cultivates traditional attitudes of metaphysics and theology whose content has long since been superseded; while the other group, especially in central Europe, faces modern times, rejects these views and takes its stand on the ground of empirical science. This development is connected with that of the modern process of production, which is becoming ever more rigorously mechanised and leaves ever less room for metaphysical ideas. It is also connected with the disappointment of broad masses of people with the attitude of those who preach traditional metaphysical and theological doctrines. So it is that in many countries the masses now reject these doctrines much more consciously than ever before, and along with their socialist attitudes tend to lean towards a down-to-earth empiricist view. In previous times, *materialism* was the expression of this view; meanwhile, however, modern empiricism has shed a number of inadequacies and has taken a strong shape in the *scientific world-conception*.

Thus, the scientific world-conception is close to the life of the present. Certainly it is threatened with hard struggles and hostility. Nevertheless there are many who do not despair but, in view of the present sociological situation, look forward with hope to the course of events to come. Of course not every single adherent of the scientific world-conception will be a fighter. Some, glad of solitude, will lead a withdrawn existence on the icy slopes of logic; some may even disdain mingling with the masses and regret the "trivialized" form that these matters inevitably take on spreading. However, their achievements too will take a place among the historic developments. We witness the spirit of the scientific world-conception penetrating in growing measure the forms of personal and public life, in education, upbringing, architecture, and the shaping of economic and social life according to rational principles. *The scientific world-conception serves life, and life receives it.*

References

1 [The pamphlet *Wissenschaftliche Weltauffassung, Der Wiener Kreis* does not give an author's name on the title page – unless one considers "Der Wiener Kreis" as author, being printed in smaller type. This pamphlet is the product of teamwork; Neurath did the writing, Hahn and Carnap edited the text with him; other members of the Circle were asked for their comments and contributions. (H. Feigl mentions F. Waismann and himself, see: "Wiener Kreis in America" in *Perspectives in American History, II*, 1968.) See also H. Neider's remarks in his contribution to our first chapter; he was a witness, as I was myself. (The publisher, Artur Wolf, also published the first colour book of the Social and Economic Museum in Vienna.) Carnap and Hahn's widow gave us their permission to include the pamphlet among Otto Neurath's writings. In fact, the name Wiener Kreis (Vienna Circle) was invented and suggested by Neurath. (See the Neurath–Carnap correspondence in a later volume in this series.) – M. N.]

2 [Note: In his text, Russell wrote about "logical atomism", not specifically of "logical analysis" – Trans.].

9

Studies in the Logic of Explanation

Carl G. Hempel and Paul Oppenheim

§1 *Introduction*

To explain the phenomena in the world of our experience, to answer the question "why?" rather than only the question "what?", is one of the foremost objectives of all rational inquiry; and especially, scientific research in its various branches strives to go beyond a mere description of its subject matter by providing an explanation of the phenomena it investigates. While there is rather general agreement about this chief objective of science, there exists considerable difference of opinion as to the function and the essential characteristics of scientific explanation. In the present essay, an attempt will be made to shed some light on these issues by means of an elementary survey of the basic pattern of scientific explanation and a subsequent more rigorous analysis of the concept of law and of the logical structure of explanatory arguments. ...

Part I Elementary Survey of Scientific Explanation

§2 *Some illustrations*

A mercury thermometer is rapidly immersed in hot water; there occurs a temporary drop of the mercury column, which is then followed by a swift rise. How is this phenomenon to be explained? The increase in temperature affects at first only the glass tube of the thermometer; it expands and thus provides a larger space for the mercury inside, whose surface therefore drops. As soon as by heat conduction the rise in temperature reaches the mercury, however, the latter expands, and as its coefficient of expansion is considerably larger than that of glass, a rise of the mercury level results. – This account consists of statements of two kinds. Those of the first kind indicate certain conditions which are realized prior to, or at the same time as, the phenomenon to be explained; we shall refer to them briefly as antecedent conditions. In our illustration, the antecedent conditions include, among others, the fact that the thermometer consists of a glass tube which is partly filled with mercury, and that it is immersed into hot water. The statements of the second kind express certain general laws; in our case, these include the laws of the thermic expansion of mercury and of glass, and a statement about the small thermic conductivity of glass. The two sets of statements, if adequately and completely formulated, explain the phenomenon under consideration: They entail the consequence that the mercury will first drop, then rise. Thus, the event under discussion is explained by subsuming it under general laws, i.e., by showing that it occurred in accordance with those laws, by virtue of the realization of certain specified antecedent conditions.

Consider another illustration. To an observer in a row boat, that part of an oar which is under water appears to be bent upwards. The phenomenon is explained by means of general laws – mainly the law of refraction and the law that water is an optically denser medium than air – and by refer-

From *Philosophy of Science* 15/2 (1948): 135–75, abridged. Reprinted by permission of The University of Chicago Press.

ence to certain antecedent conditions – especially the facts that part of the oar is in the water, part in the air, and that the oar is practically a straight piece of wood. – Thus, here again, the question "*Why* does the phenomenon happen?" is construed as meaning "according to what general laws, and by virtue of what antecedent conditions does the phenomenon occur?"

So far, we have considered exclusively the explanation of particular events occurring at a certain time and place. But the question "Why?" may be raised also in regard to general laws. Thus, in our last illustration, the question might be asked: Why does the propagation of light conform to the law of refraction? Classical physics answers in terms of the undulatory theory of light, i.e. by stating that the propagation of light is a wave phenomenon of a certain general type, and that all wave phenomena of that type satisfy the law of refraction. Thus, the explanation of a general regularity consists in subsuming it under another, more comprehensive regularity, under a more general law. – Similarly, the validity of Galileo's law for the free fall of bodies near the earth's surface can be explained by deducing it from a more comprehensive set of laws, namely Newton's laws of motion and his law of gravitation, together with some statements about particular facts, namely the mass and the radius of the earth.

§3 *The basic pattern of scientific explanation*

From the preceding sample cases let us now abstract some general characteristics of scientific explanation. We divide an explanation into two major constituents, the explanandum and the explanans.[1] By the explanandum, we understand the sentence describing the phenomenon to be explained (not that phenomenon itself); by the explanans, the class of those sentences which are adduced to account for the phenomenon. As was noted before, the explanans falls into two subclasses; one of these contains certain sentences $C_1, C_2, \ldots, C_k$ which state specific antecedent conditions; the other is a set of sentences $L_1, L_2, \ldots L_r$ which represent general laws.

If a proposed explanation is to be sound, its constituents have to satisfy certain conditions of adequacy, which may be divided into logical and empirical conditions. For the following discussion, it will be sufficient to formulate these requirements in a slightly vague manner; in Part III, a more rigorous anlysis and a more precise restatement of these criteria will be presented.

I *Logical conditions of adequacy*

(R1) The explanandum must be a logical consequence of the explanans; in other words, the explanandum must be logically deducible from the information contained in the explanans, for otherwise, the explanans would not constitute adequate grounds for the explanandum.

(R2) The explanans must contain general laws, and these must actually be required for the derivation of the explanandum. – We shall not make it a necessary condition for a sound explanation, however, that the explanans must contain at least one statement which is not a law; for, to mention just one reason, we would surely want to consider as an explanation the derivation of the general regularities governing the motion of double stars from the laws of celestial mechanics, even though all the statements in the explanans are general laws.

(R3) The explanans must have empirical content; i.e., it must be capable, at least in principle, of test by experiment or observation. – This condition is implicit in (R1); for since the explanandum is assumed to describe some empirical phenomenon, it follows from (R1) that the explanans entails at least one consequence of empirical character, and this fact confers upon it testability and empirical content. But the point deserves special mention because, as will be seen in §4, certain arguments which have been offered as explanations in the natural and in the social sciences violate this requirement.

II *Empirical condition of adequacy*

(R4) The sentences constituting the explanans must be true.

That in a sound explanation, the statements constituting the explanans have to satisfy some condition of factual correctness is obvious. But it might seem more appropriate to stipulate that the explanans has to be highly confirmed by all the relevant evidence available rather than that it should be true. This stipulation however, leads to awkward consequences. Suppose that a certain phenomenon was explained at an earlier stage of science, by means of an explanans which was well supported by the evidence then at hand, but which had been highly disconfirmed by more recent empirical findings. In such a case, we would have to say that originally the explanatory account was a correct explanation, but that it ceased to be one later, when unfavorable evidence was discovered. This does not appear to accord with sound common usage, which directs us to say that on the basis of the limited initial evidence, the truth of the explanans, and thus the soundness of the explanation, had been quite probable, but that the ampler evidence now available made it highly probable that the explanans was not true, and hence that the account in question was not – and had never been – a correct explanation. (A similar point will be made and illustrated, with respect to the requirement of truth for laws, in the beginning of §6.)

Some of the characteristics of an explanation which have been indicated so far may be summarized in the following schema:

Logical deduction	$C_1, C_2, \ldots, C_k$	Statements of antecedent conditions	Explanans
	$L_1, L_2, \ldots, L_r$	General Laws	
	E	Description of the empirical phenomenon to be explained	Explanandum

Let us note here that the same formal analysis, including the four necessary conditions, applies to scientific prediction as well as to explanation. The difference between the two is of a pragmatic character. If E is given, i.e. if we know that the phenomenon described by E has occurred, and a suitable set of statements $C_1, C_2, \ldots, C_k$, $L_1, L_2, \ldots, L_r$ is provided afterwards, we speak of an explanation of the phenomenon in question. If the latter statements are given and E is derived prior to the occurrence of the phenomenon it describes, we speak of a prediction. It may be said, therefore, that an explanation is not fully adequate unless its explanans, if taken account of in time, could have served as a basis for predicting the phenomenon under consideration.[2] – Consequently, whatever will be said in this article concerning the logical characteristics of explanation or prediction will be applicable to either, even if only one of them should be mentioned.

It is this potential predictive force which gives scientific explanation its importance: only to the extent that we are able to explain empirical facts can we attain the major objective of scientific research, namely not merely to record the phenomena of our experience, but to learn from them, by basing upon them theoretical generalizations which enable us to anticipate new occurrences and to control, at least to some extent, the changes in our environment.

Many explanations which are customarily offered, especially in pre-scientific discourse, lack this predictive character, however. Thus, it may be explained that a car turned over on the road "because" one of its tires blew out while the car was travelling at high speed. Clearly, on the basis of just this information, the accident could not have been predicted, for the explanans provides no explicit general laws by means of which the prediction might be effected, nor does it state adequately the antecedent conditions which would be needed for the prediction. – The same point may be illustrated by reference to W. S. Jevons's view that every explanation consists in pointing out a resemblance between facts, and that in some cases this process may require no reference to laws at all and "may involve nothing more than a single identity, as when we explain the appearance of shooting stars by showing that they are identical with portions of a comet".[3] But clearly, this identity does not provide an explanation of the phenomenon of shooting stars unless we presuppose the laws governing the development of heat and light as the effect of friction. The observation of similarities has explanatory value only if it involves at least tacit reference to general laws.

In some cases, incomplete explanatory arguments of the kind here illustrated suppress parts of the explanans simply as "obvious"; in other cases, they seem to involve the assumption that while the missing parts are not obvious, the incomplete explanans could at least, with appropriate effort, be so supplemented as to make a strict derivation of the explanandum possible. This assumption may be justifiable in some cases, as when we say that a lump of sugar disappeared "because" it was put into hot tea, but it is surely not satisfied in many other cases. Thus, when certain peculiarities in the work of an artist are explained as outgrowths of a specific type of neurosis, this observation may contain significant clues, but in general it does not afford a sufficient basis for a potential prediction of those peculiarities. In cases of this kind, an incomplete explanation may at best be considered as indicating some positive correlation between the antecedent conditions adduced and the type of phenomenon to be explained, and as pointing out a direction in which further research might be carried on in order to complete the explanatory account.

The type of explanation which has been considered here so far is often referred to as causal explanation. If E describes a particular event, then the antecedent circumstances described in the sentences $C_1, C_2, \ldots, C_k$ may be said jointly to "cause" that event, in the sense that there are certain empirical regularities, expressed by the laws $L_1, L_2, \ldots, L_r$, which imply that whenever conditions of the kind indicated by $C_1, C_2, \ldots, C_k$ occur, an event of the kind described in E will take place. Statements such as $L_1, L_2, \ldots, L_r$, which assert general and unexceptional connections between specified characteristics of events, are customarily called causal, or deterministic, laws. They are to be distinguished from the so-called statistical laws which assert that in the long run, an explicitly stated percentage of all cases satisfying a given set of conditions are accompanied by an event of a certain specified kind. Certain cases of scientific explanation involve "subsumption" of the explanandum under a set of laws of which at least some are statistical in character. Analysis of the peculiar logical structure of that type of subsumption involves difficult special problems. The present essay will be restricted to an examination of the causal type of explanation, which has retained its significance in large segments of contemporary science, and even in some areas where a more adequate account calls for reference to statistical laws.[4]

§4 *Explanation in the non-physical sciences. Motivational and teleological approaches*

Our characterization of scientific explanation is so far based on a study of cases taken from the physical sciences. But the general principles thus obtained apply also outside this area.[5] Thus, various types of behavior in laboratory animals and in human subjects are explained in psychology by subsumption under laws or even general theories of learning or conditioning; and while frequently, the regularities invoked cannot be stated with the same generality and precision as in physics or chemistry, it is clear, at least, that the general character of those explanations conforms to our earlier characterization.

Let us now consider an illustration involving sociological and economic factors. In the fall of 1946, there occurred at the cotton exchanges of the United States a price drop which was so severe that the exchanges in New York, New Orleans, and Chicago had to suspend their activities temporarily. In an attempt to explain this occurrence, newspapers traced it back to a large-scale speculator in New Orleans who had feared his holdings were too large and had therefore begun to liquidate his stocks; smaller speculators had then followed his example in a panic and had thus touched off the critical decline. Without attempting to assess the merits of the argument, let us note that the explanation here suggested again involves statements about antecedent conditions and the assumption of general regularities. The former include the facts that the first speculator had large stocks of cotton, that there were smaller speculators with considerable holdings, that there existed the institution of the cotton exchanges with their specific mode of operation, etc. The general regularities referred to are – as often in semi-popular explanations – not explicitly mentioned; but there is obviously implied some form of the law of supply and demand to account for the drop in cotton prices in terms of the greatly increased supply under conditions of practically unchanged demand; besides, reliance is necessary on certain regularities in the behavior of individuals who are trying to preserve or improve their economic position. Such laws cannot be formulated at present with satisfactory precision and generality, and therefore, the suggested explanation is surely incomplete, but its intention is unmistakably to account for the phenomenon by integrating it into a general pattern of economic and socio-psychological regularities.

We turn to an explanatory argument taken from the field of linguistics.[6] In Northern France, there exist a large variety of words synonymous with the English "bee," whereas in Southern France, essentially only one such word is in existence. For this discrepancy, the explanation has been suggested that in the Latin epoch, the South of France used the word "apicula", the North the word "apis". The latter, because of a process of phonologic decay in Northern France, became the monosyllabic word "é"; and monosyllables tend to be eliminated, especially if they contain few consonantic elements, for they are apt to give rise to misunderstandings. Thus, to avoid confusion, other words were selected. But "apicula", which was reduced to "abelho", remained clear enough and was retained, and finally it even entered into the standard language, in the form "abeille". While the explanation here described is incomplete in the sense characterized in the previous section, it clearly exhibits reference to specific antecedent conditions as well as to general laws.[7]

While illustrations of this kind tend to support the view that explanation in biology, psychology, and the social sciences has the same structure as in the physical sciences, the opinion is rather widely held that in many instances, the causal type of explanation is essentially inadequate in fields other than physics and chemistry, and especially in the study of purposive behavior. Let us examine briefly some of the reasons which have been adduced in support of this view.

One of the most familiar among them is the idea that events involving the activities of humans singly or in groups have a peculiar uniqueness and irrepeatability which makes them inaccessible to causal explanation because the latter, which its reliance upon uniformities, presupposes repeatability of the phenomena under consideration. This argument which, incidentally, has also been used in support of the contention that the experimental method is inapplicable in psychology and the social sciences, involves a misunderstanding of the logical character of causal explanation. Every individual event, in the physical sciences no less than in psychology or the social sciences, is unique in the sense that it, with all its peculiar characteristics, does not repeat itself. Nevertheless, individual events may conform to, and thus be explainable by means of, general laws of the causal type. For all that a causal law asserts is that any event of a specified kind, i.e. any event having certain specified characteristics, is accompanied by another event which in turn has certain specified characteristics; for example, that in any event involving friction, heat is developed. And all that is needed for the testability and applicability of such laws is the recurrence of events with the antecedent characteristics, i.e. the repetition of those characteristics, but not of their individual instances. Thus, the argument is inconclusive. It gives occasion, however, to emphasize an important point concerning our earlier analysis: When we spoke of the explanation of a single event, the term "event" referred to the occurrence of some more or less complex characteristic in a specific spatio-temporal location or in a certain individual object, and not to *all* the characteristics of that object, or to all that goes on in that space-time region.

A second argument that should be mentioned here[8] contends that the establishment of scientific generalizations – and thus of explanatory principles – for human behavior is impossible because the reactions of an individual in a given situation depend not only upon that situation, but also upon the previous history of the individual. – But surely, there is no *a priori* reason why generalizations should not be attainable which take into account this dependence of behavior on the past history of the agent. That indeed the given argument "proves" too much, and is therefore a *non sequitur*, is made evident by the existence of certain physical phenomena, such as magnetic hysteresis and elastic fatigue, in which the magnitude of a specific physical effect depends upon the past history of the system involved, and for which nevertheless certain general regularities have been established.

A third argument insists that the explanation of any phenomenon involving purposive behavior calls for reference to motivations and thus for teleological rather than causal analysis. Thus, for example, a fuller statement of the suggested explanation for the break in the cotton prices would have to indicate the large-scale speculator's motivations as one of the factors determining the event in question. Thus, we have to refer to goals sought, and this, so the argument runs, introduces a type of explanation alien to the physical sciences. Unquestionably, many of the – frequently incomplete –

explanations which are offered for human actions involve reference to goals and motives; but does this make them essentially different from the causal explanations of physics and chemistry? One difference which suggests itself lies in the circumstance that in motivated behavior, the future appears to affect the present in a manner which is not found in the causal explanations of the physical sciences. But clearly, when the action of a person is motivated, say, by the desire to reach a certain objective, then it is not the as yet unrealized future event of attaining the goal which can be said to determine his present behavior, for indeed the goal may never be actually reached; rather – to put it in crude terms – it is (a) his desire, present before the action, to attain that particular objective, and (b) his belief, likewise present before the action, that such and such a course of action is most likely to have the desired effect. The determining motives and beliefs, therefore, have to be classified among the antecedent conditions of a motivational explanation, and there is no formal difference on this account between motivational and causal explanation.

Neither does the fact that motives are not accessible to direct observation by an outside observer constitute an essential difference between the two kinds of explanation; for also the determining factors adduced in physical explanations are very frequently inaccessible to direct observation. This is the case, for instance, when opposite electric charges are adduced in explanation of the mutual attraction of two metal spheres. The presence of those charges, while eluding all direct observation, can be ascertained by various kinds of indirect test, and that is sufficient to guarantee the empirical character of the explanatory statement. Similarly, the presence of certain motivations may be ascertainable only by indirect methods, which may include reference to linguistic utterances of the subject in question, slips of the pen or of the tongue, etc.; but as long as these methods are "operationally determined" with reasonable clarity and precision, there is no essential difference in this respect between motivational explanation and causal explanation in physics.

A potential danger of explanation by motives lies in the fact that the method lends itself to the facile construction of ex-post-facto accounts without predictive force. It is a widespread tendency to "explain" an action by ascribing it to motives conjectured only after the action has taken place. While this procedure is not in itself objectionable, its soundness requires that (1) the motivational assumptions in question be capable of test, and (2) that suitable general laws be available to lend explanatory power to the assumed motives. Disregard of these requirements frequently deprives alleged motivational explanations of their cognitive significance.

The explanation of an action in terms of the motives of the agent is sometimes considered as a special kind of teleological explanation. As was pointed out above, motivational explanation, if adequately formulated, conforms to the conditions for causal explanation, so that the term "teleological" is a misnomer if it is meant to imply either a non-causal character of the explanation or a peculiar determination of the present by the future. If this is borne in mind, however, the term "teleological" may be viewed, in this context, as referring to causal explanations in which some of the antecedent conditions are motives of the agent whose actions are to be explained.[9]

Teleological explanations of this kind have to be distinguished from a much more sweeping type, which has been claimed by certain schools of thought to be indispensable especially in biology. It consists in explaining characteristics of an organism by reference to certain ends or purposes which the characteristics are said to serve. In contradistinction to the cases examined before, the ends are not assumed here to be consciously or subconsciously pursued by the organism in question. Thus, for the phenomenon of mimicry, the explanation is sometimes offered that it serves the purpose of protecting the animals endowed with it from detection by its pursuers and thus tends to preserve the species. – Before teleological hypotheses of this kind can be appraised as to their potential explanatory power, their meaning has to be clarified. If they are intended somehow to express the idea that the purposes they refer to are inherent in the design of the universe, then clearly they are not capable of empirical test and thus violate the requirement (R3) stated in §3. In certain cases, however, assertions about the purposes of biological characteristics may be translatable into statements in non-teleological terminology which assert that those characteristics function in a specific manner which is essential to keeping the organism alive or to preserving the species.[10] An attempt to state precisely what is meant by this latter assertion – or by the similar one that without those characteristics, and other things being equal, the organism or the species would not survive – encounters considerable difficulties. But these need not be discussed here. For even if we assume that biological statements in teleological form can be adequately translated into descriptive statements about the life-preserving function of certain biological characteristics, it is clear that (1) the use of the concept of purpose is not essential in these contexts, since the term "purpose" can be completely eliminated from the statements in question, and (2) teleological assumptions, while now endowed with empirical content, cannot serve as explanatory principles in the customary contexts. Thus, e.g., the fact that a given species of butterflies displays a particular kind of coloring cannot be inferred from – and therefore cannot be explained by means of – the statement that this type of coloring has the effect of protecting the butterflies from detection by pursuing birds, nor can the presence of red corpuscles in the human blood be inferred from the statement that those corpuscles have a specific function in assimilating oxygen and that this function is essential for the maintenance of life.

One of the reasons for the perseverance of teleological considerations in biology probably lies in the fruitfulness of the teleological approach as a heuristic device: Biological research which was psychologically motivated by a teleological orientation, by an interest in purposes in nature, has frequently led to important results which can be stated in non-teleological terminology and which increase our scientific knowledge of the causal connections between biological phenomena.

Another aspect that lends appeal to teleological considerations is their anthropomorphic character. A teleological explanation tends to make us feel that we really "understand" the phenomenon in question, because it is accounted for in terms of purposes, with which we are familiar from our own experience of purposive behavior. But it is important to distinguish here understanding in the psychological sense of a feeling of empathic familiarity from understanding in the theoretical, or cognitive, sense of exhibiting the phenomenon to be explained as a special case of some general regularity. The frequent insistence that explanation means the reduction of something unfamiliar to ideas or experiences already familiar to us is indeed misleading. For while some scientific explanations do have this psychological effect, it is by no means universal: The free fall of a physical body may well be said to be a more familiar phenomenon than the law of gravitation, by means of which it can be explained; and surely the basic ideas of the theory of relativity will appear to many to be far less familiar than the phenomena for which the theory accounts.

"Familiarity" of the explicans is not only not necessary for a sound explanation – as we have just tried to show –, but it is not sufficient either. This is shown by the many cases in which a proposed explicans sounds suggestively familiar, but upon closer inspection proves to be a mere metaphor, or an account lacking testability, or a set of statements which includes no general laws and therefore lacks explanatory power. A case in point is the neovitalistic attempt to explain biological phenomena by reference to an entelechy or vital force. The crucial point here is not – as it is sometimes made out to be – that entelechies cannot be seen or otherwise directly observed; for that is true also of gravitational fields, and yet, reference to such fields is essential in the explanation of various physical phenomena. The decisive difference between the two cases is that the physical explanation provides (1) methods of testing, albeit indirectly, assertions about gravitational fields, and (2) general laws concerning the strength of gravitational fields, and the behavior of objects moving in them. Explanations by entelechies satisfy the analogue of neither of these two conditions. Failure to satisfy the first condition represents a violation of (R3); it renders all statements about entelechies inaccessible to empirical test and thus devoid of empirical meaning. Failure to comply with the second condition involves a violation of (R2). It deprives the concept of entelechy of all explanatory import; for explanatory power never resides in a concept, but always in the general laws in which it functions. Therefore, notwithstanding the flavor of familiarity of the metaphor it invokes, the neovitalistic approach cannot provide theoretical understanding.

The preceding observations about familiarity and understanding can be applied, in a similar manner, to the view held by some scholars that the explanation, or the understanding, of human actions requires an empathic understanding of the personalities of the agents.[11] This understanding of another person in terms of one's own psychological functioning may prove a useful heuristic device in the search for general psychological principles which might provide a theoretical explanation; but the existence of empathy on the part of the scientist is neither a necessary nor a sufficient condition for the explanation, or the scientific understanding, of any human action. It is not necessary, for the behavior of psychotics or of people belonging to a culture very different from that of the scientist may sometimes be explainable and predictable in terms of general principles even though the scientist who establishes or applies those principles may not be able to understand his subjects empathically. And empathy is not sufficient to guarantee a sound explanation, for a strong feeling of empathy may exist even in cases where we completely misjudge a given personality. Moreover, as the late Dr. Zilsel has pointed out, empathy leads with ease to incompatible results; thus, when the population of a town has long been subjected to heavy bombing attacks, we can understand, in the empathic sense, that its morale should have broken down completely, but we can understand with the same ease also that it should have developed a defiant spirit of resistance. Arguments of this kind often appear quite convincing; but they are of an *ex post facto* character and lack cognitive significance unless they are supplemented by testable explanatory principles in the form of laws or theories.

Familiarity of the explanans, therefore, no matter whether it is achieved through the use of teleological terminology, through neovitalistic metaphors, or through other means, is no indication of the cognitive import and the predictive force of a proposed explanation. Besides, the extent to which an idea will be considered as familiar varies from person to person and from time to time, and a psychological factor of this kind certainly cannot serve as a standard in assessing the worth of a proposed explanation. The decisive requirement for every sound explanation remains that it subsume the explanandum under general laws.

[...]

Part III Logical Analysis of Law and Explanation

§6 *Problems of the concept of general law*

From our general survey of the characteristics of scientific explanation, we now turn to a closer examination of its logical structure. The explanation of a phenomenon, we noted, consists in its subsumption under laws or under a theory. But what is a law, what is a theory? While the meaning of these concepts seems intuitively clear, an attempt to construct adequate explicit definitions for them encounters considerable difficulties. In the present section, some basic problems of the concept of law will be described and analyzed; in the next section, we intend to propose, on the basis of the suggestions thus obtained, definitions of law and of explanation for a formalized model language of a simple logical structure.

The concept of law will be construed here so as to apply to true statements only. The apparently plausible alternative procedure of requiring high confirmation rather than truth of a law seems to be inadequate: It would lead to a relativized concept of law, which would be expressed by the phrase "sentence S is a law relatively to the evidence E". This does not seem to accord with the meaning customarily assigned to the concept of law in science and in methodological inquiry. Thus, for example, we would not say that Bode's general formula for the distance of the planets from the sun was a law relatively to the astronomical evidence available in the 1770s, when Bode propounded it, and that it ceased to be a law after the discovery of Neptune and the determination of its distance from the sun; rather, we would say that the limited original evidence had given a high probability to the assumption that the formula was a law, whereas more recent additional information reduced that probability so much as to make it practically certain that Bode's formula is not generally true, and hence not a law.[12]

Apart from being true, a law will have to satisfy a number of additional conditions. These can be studied independently of the factual requirement of truth, for they refer, as it were, to all logically possible laws, no matter whether factually true or false. Adopting a convenient term proposed by Goodman,[13] we will say that a sentence is lawlike if it has all the characteristics of a general law, with the possible exception of truth. Hence, every law is a lawlike sentence, but not conversely.

Our problem of analyzing the concept of law thus reduces to that of explicating the meaning of "lawlike sentence". We shall construe the class of lawlike sentences as including analytic general statements, such as "A rose is a rose", as well as the lawlike sentences of empirical science, which have empirical content.[14] It will not be necessary to require that each lawlike sentence permissible in explanatory contexts be of the second kind; rather, our definition of explanation will be so constructed as to guarantee the factual character of the totality of the laws – though not of every single one of them – which function in an explanation of an empirical fact.

What are the characteristics of lawlike sentences? First of all, lawlike sentences are statements of universal form, such as "All robins' eggs are greenish-blue", "All metals are conductors of electricity", "At constant pressure, any gas expands with increasing temperature". As these examples illustrate, a lawlike sentence usually is not only of universal, but also of conditional form; it makes an assertion to the effect that universally, if a certain set of conditions, *C*, is realized, then another specified set of conditions, *E*, is realized as well. The standard form for the symbolic expression of a lawlike sentence is therefore the universal conditional. However, since any conditional statement can be transformed into a non-conditional one, conditional form will not be considered as essential for a lawlike sentence, while universal character will be held indispensable.

But the requirement of universal form is not sufficient to characterize lawlike sentences. Suppose, for example, that a certain basket, *b*, contains at a certain time *t* a number of red apples and nothing else.[15] Then the statement

(S_1) Every apple in basket *b* at time *t* is red

is both true and of universal form. Yet the sentence does not qualify as a law; we would refuse, for example, to explain by subsumption under it the fact that a particular apple chosen at random from the basket is red. What distinguishes S_1 from a lawlike sentence? Two points suggest themselves, which will be considered in turn, namely, finite scope, and reference to a specified object.

First, the sentence S_1 makes, in effect, an assertion about a finite number of objects only, and this seems irreconcilable with the claim to universality which is commonly associated with the notion of law.[16] But are not Kepler's laws considered as lawlike although they refer to a finite set of planets only? And might we not even be willing to consider as lawlike a sentence such as the following?

(S_2) All the sixteen ice cubes in the freezing tray of this refrigerator have a temperature of less than 10 degrees centigrade.

This point might well be granted; but there is an essential difference between S_1 on the one hand and Kepler's laws as well as S_2 on the other: The latter, while finite in scope, are known to be consequences of more comprehensive laws whose scope is not limited, while for S_1 this is not the case.

Adopting a procedure recently suggested by Reichenbach,[17] we will therefore distinguish between fundamental and derivative laws. A statement will be called a derivative law if it is of universal character and follows from some fundamental laws. The concept of fundamental law requires further clarification; so far, we may say that fundamental laws, and similarly fundamental lawlike sentences, should satisfy a certain condition of non-limitation of scope.

It would be excessive, however, to deny the status of fundamental lawlike sentence to all statements which, in effect, make an assertion about a finite class of objects only, for that would rule out also a sentence such as "All robins' eggs are greenish-blue", since presumably the class of all robins' eggs – past, present, and future – is finite. But again, there is an essential difference between this sentence and, say, S_1. It requires empirical knowledge to establish the finiteness of the class of robins' eggs, whereas, when the sentence S_1 is construed in a manner which renders it intuitively unlawlike, the terms "basket *b*" and "apple" are understood so as to imply finiteness of the class of apples in the basket at time *t*. Thus, so to speak, the meaning of its constitutive terms alone – without additional factual information – entails that S_1 has a finite scope. – Fundamental laws, then, will have to be construed so as to satisfy what we have called a condition of non-limited scope; our formulation of that condition however, which refers to what is entailed by "the meaning" of certain expressions, is too vague and will have to be revised later. Let us note in passing that the stipulation here envisaged would bar from the class of fundamental lawlike sentences also such undesirable candidates as "All uranic objects are spherical", where "uranic" means the property of being the planet Uranus; indeed, while this sentence has universal form, it fails to satisfy the condition of non-limited scope.

In our search for a general characterization of lawlike sentences, we now turn to a second clue which is provided by the sentence S_1. In addition to violating the condition of non-limited scope, this sentence has the peculiarity of making reference to a particular object, the basket *b*; and this, too, seems to violate the universal character of a law.[18] The restriction which seems indicated here, should however again be applied to fundamental lawlike sentences only; for a true general statement about the free fall of physical bodies on the moon, while referring to a particular object, would still constitute a law, albeit a derivative one.

It seems reasonable to stipulate, therefore, that a fundamental lawlike sentence must be of universal form and must contain no essential – i.e., uneliminable – occurrences of designations for particular objects. But this is not sufficient; indeed, just at this point, a particularly serious difficulty presents itself. Consider the sentence

(S_3) Everything that is either an apple in basket *b* at time *t* or a sample of ferric oxide is red.

If we use a special expression, say "*x* is ferple", as synonymous with "*x* is either an apple in *b* at *t* or a sample of ferric oxide", then the content of S_3 can be expressed in the form

(S_4) Everything that is ferple is red.

The statement thus obtained is of universal form and contains no designations of particular objects, and it also satisfies the condition of non-limited scope; yet clearly, S_4 can qualify as a fundamental lawlike sentence no more than can S_3.

As long as "ferple" is a defined term of our language, the difficulty can readily be met by stipulating that after elimination of defined terms, a fundamental lawlike sentence must not contain essential occurrences of designations for particular objects. But this way out is of no avail when "ferple", or another term of the kind illustrated by it, is a primitive predicate of the language under consideration. This reflection indicates that certain restrictions have to be imposed upon those predicates – i.e., terms for properties or relations, – which may occur in fundamental lawlike sentences.[19]

More specifically, the idea suggests itself of permitting a predicate in a fundamental lawlike sentence only if it is purely universal, or, as we shall say, purely qualitative, in character; in other words, if a statement of its meaning does not require reference to any one particular object or spatio-temporal location. Thus, the terms "soft", "green", "warmer than", "as long as", "liquid", "electrically charged", "female", "father of" are purely qualitative predicates, while "taller than the Eiffel Tower", "medieval", "lunar", "arctic", "Ming" are not.[20]

Exclusion from fundamental lawlike sentences of predicates which are not purely qualitative would at the same time ensure satisfaction of the condition of non-limited scope; for the meaning of a purely qualitative predicate does not require a finite extension; and indeed, all the sentences considered above which violate the condition of non-limited scope make explicit or implicit reference to specific objects.

The stipulation just proposed suffers, however, from the vagueness of the concept of purely qualitative predicate. The question whether indication of the meaning of a given predicate in English does or does not require reference to some one specific object does not always permit an unequivocal answer since English as a natural language does not provide explicit definitions or other clear explications of meaning for its terms. It seems therefore reasonable to attempt definition of the concept of law not with respect to English or any other natural language, but rather with respect to a formalized language – let us call it a model language, *L*, – which is governed by a well-determined system of logical rules, and in which every term either is characterized as primitive or is introduced by an explicit definition in terms of the primitives.

This reference to a well-determined system is customary in logical research and is indeed quite natural in the context of any attempt to develop precise criteria for certain logical distinctions. But it does not by itself suffice to overcome the specific difficulty under discussion. For while it is now readily possible to characterize as not purely qualitative all those among the defined predicates in *L* whose definiens contains an essential occurrence of some individual name, our problem remains open for the primitives of the language, whose meanings are not determined by definitions within the language, but rather by semantical rules of interpretation. For we want to permit the interpretation of the primitives of *L* by means of such attributes as blue, hard, solid, warmer, but not by the properties of being a descendant of Napoleon, or an arctic animal, or a Greek statue; and the difficulty is precisely that of stating rigorous criteria for the distinction between the permissible and the non-permissible interpretations. Thus the problem of setting up an adequate definition for purely qualitative attributes now arises again; namely for the concepts of the metalanguage in which the semantical interpretation of the primitives is formulated. We may postpone an encounter with the difficulty by presupposing formalization of the semantical meta-language, the meta-meta-language, and so forth; but somewhere, we will have to stop at a non-formalized meta-language, and for it a characterization of purely qualitative predicates will be needed and will present much the same problems as non-formalized English, with which we began. The characterization of a purely qualitative predicate as one whose meaning can be made explicit without reference to any one particular object points to the intended meaning but does not explicate it precisely, and the problem of an adequate definition of purely qualitative predicates remains open.

There can be little doubt, however, that there exists a large number of property and relation terms which would be rather generally recognized as purely qualitative in the sense here pointed out, and as permissible in the formulation of fundamental lawlike sentences; some examples have been given above, and the list could be readily enlarged. When we speak of purely qualitative predicates, we shall henceforth have in mind predicates of this kind.

In the following section, a model language *L* of a rather simple logical structure will be described, whose primitives will be assumed to be qualitative in the sense just indicated. For this language, the concepts of law and explanation will then be defined in a manner which takes into account the general observations set forth in the present section.

[...]

Notes

1 These two expressions, derived from the Latin *explanare*, were adopted in preference to the perhaps more customary terms "explicandum" and "explicans" in order to reserve the latter for use in the context of explication of meaning, or analysis. On explication in this sense, cf. Carnap, [Concepts], p. 513. – Abbreviated titles in brackets refer to the bibliography at the end of this article.

2 The logical similarity of explanation and prediction, and the fact that one is directed towards past occurrences, the other towards future ones, is well expressed in the terms "postdictability" and "predictability" used by Reichenbach in [Quantum Mechanics], p. 13.

3 [Principles], p. 533.

4 The account given above of the general characteristics of explanation and prediction in science is by no means novel; it merely summarizes and states explicitly some fundamental points which have been recognized by many scientists and methodologists.

Thus, e.g., Mill says: "An individual fact is said to be explained by pointing out its cause, that is, by stating the law or laws of causation of which its production is an instance", and "a law of uniformity in nature is said to be explained when another law or laws are pointed out, of which that law itself is but a case, and from which it could be deduced." ([Logic], Book III, Chapter XII, section 1). Similarly, Jevons, whose general characterization of explanation was critically discussed above, stresses that "the most important process of explanation consists in showing that an observed fact is one case of a general law or tendency." ([Principles], p. 533). Ducasse states the same point as follows: "Explanation essentially consists in the offering of a hypothesis of fact, standing to the fact to be explained as case of antecedent to case of consequent of some already known law of connection." ([Explanation], pp. 150–51). A lucid analysis of the fundamental structure of explanation and prediction was given by Popper in [Forschung], section 12, and, in an improved version, in his work [Society], especially in Chapter 25 and in note 7 referring to that chapter. – For a recent characterization of explanation as subsumption under general theories, cf., for example, Hull's concise discussion in [Principles], chapter I. A clear elementary examination of certain aspects of explanation is given in Hospers, [Explanation], and a concise survey of many of the essentials of scientific explanation which are considered in the first two parts of the present study may be found in Feigl, [Operationism], pp. 284 ff.

5 On the subject of explanation in the social sciences, especially in history, cf. also the following publications, which may serve to supplement and amplify the brief discussion to be presented here: Hempel, [Laws]; Popper, [Society]; White, [Explanation]; and the articles *Cause* and *Understanding* in Beard and Hook, [Terminology].

6 The illustration is taken from Bonfante, [Semantics], section 3.

7 While in each of the last two illustrations, certain regularities are unquestionably relied upon in the explanatory argument, it is not possible to argue convincingly that the intended laws, which at present cannot all be stated explicitly, are of a causal rather than a statistical character. It is quite possible that most or all of the regularities which will be discovered as sociology develops will be of a statistical type. Cf., on this point, the suggestive observations by Zilsel in [Empiricism] section 8, and [Laws]. This issue does not affect, however, the main point we wish to make here, namely that in the social no less than in the physical sciences, subsumption under general

regularities is indispensable for the explanation and the theoretical understanding of any phenomenon.

8 Cf., for example, F. H. Knight's presentation of this argument in [Limitations], pp. 251–52.

9 For a detailed logical analysis of the character and the function of the motivation concept in psychological theory, see Koch, [Motivation]. – A stimulating discussion of teleological behavior from the standpoint of contemporary physics and biology is contained in the article [Teleology] by Rosenblueth, Wiener and Bigelow. The authors propose an interpretation of the concept of purpose which is free from metaphysical connotations, and they stress the importance of the concept thus obtained for a behavioristic analysis of machines and living organisms. While our formulations above intentionally use the crude terminology frequently applied in philosophical arguments concerning the applicability of causal explanation to purposive behavior, the analysis presented in the article referred to is couched in behavioristic terms and avoids reference to "motives" and the like.

10 An analysis of teleological statements in biology along these lines may be found in Woodger, [Principles], especially pp. 432 ff; essentially the same interpretation is advocated by Kaufmann in [Methodology], chapter 8.

11 For a more detailed discussion of this view on the basis of the general principles outlined above, cf. Zilsel, [Empiricism], sections 7 and 8, and Hempel, [Laws], section 6.

12 The requirement of truth for laws has the consequence that a given empirical statement *S* can never be definitely known to be a law; for the sentence affirming the truth of *S* is logically equivalent with *S* and is therefore capable only of acquiring a more or less high probability, or degree of confirmation, relatively to the experimental evidence available at any given time. On this point, cf. Carnap, [Remarks]. – For an excellent non-technical exposition of the semantical concept of truth, which is here applied, the reader is referred to Tarski, [Truth].

13 [Counterfactuals]. p. 125.

14 This procedure was suggested by Goodman's approach in [Counterfactuals]. – Reichenbach, in a detailed examination of the concept of law, similarly construes his concept of nomological statement as including both analytic and synthetic sentences; cf. [Logic], chapter VIII.

15 The difficulty illustrated by this example was stated concisely by Langford ([Review]), who referred to it as the problem of distinguishing between universals of fact and causal universals. For further discussion and illustration of this point, see also Chisholm [Conditional], especially pp. 301f. – A systematic analysis of the problem was given by Goodman in [Counterfactuals], especially part III. – While not concerned with the specific point under discussion, the detailed examination of counterfactual conditionals and their relation to laws of nature, in Chapter VIII of Lewis's work [Analysis], contains important observations on several of the issues raised in the present section.

16 The view that laws should be construed as not being limited to a finite domain has been expressed, among others, by Popper ([Forschung], section 13) and by Reichenbach ([Logic], p. 369).

17 [Logic], p. 361. – Our terminology as well as the definitions to be proposed later for the two types of law do not coincide with Reichenbach's, however.

18 In physics, the idea that a law should not refer to any particular object has found its expression in the maxim that the general laws of physics should contain no reference to specific space-time points, and that spatio-temporal coordinates should occur in them only in the form of differences or differentials.

19 The point illustrated by the sentences S_3 and S_4 above was made by Goodman, who has also emphasized the need to impose certain restrictions upon the predicates whose occurrence is to be permissible in lawlike sentences. These predicates are essentially the same as those which Goodman calls projectible. Goodman has suggested that the problems of establishing precise criteria for projectibility, of interpreting counterfactual conditionals, and of defining the concept of law are so intimately related as to be virtually aspects of a single problem. (Cf. his articles [Query] and [Counterfactuals].) One suggestion for an analysis of projectibility has recently been made by Carnap in [Application]. Goodman's note [Infirmities] contains critical observations on Carnap's proposals.

20 That laws, in addition to being of universal form, must contain only purely universal predicates was clearly argued by Popper ([Forschung], section 14, 15). – Our alternative expression "purely qualitative predicate" was chosen in analogy to Carnap's term "purely qualitative property" (cf. [Application]). – The above characterization of purely universal predicates seems preferable to a simpler and perhaps more customary one, to the effect that a statement of the meaning of the predicate must require no reference to particular objects. For this formulation might be too exclusive since it could be argued that stating the meaning of such purely qualitative terms as "blue" or "hot" requires illustrative reference to some particular object which has the quality in question. The essential point is that no one specific object has to be chosen; any one in the logically unlimited set of blue or of hot objects will do. In explicating the meaning of "taller than the Eiffel Tower", "being an apple in basket *b* at time *t*", "medieval", etc., however, reference has to be made to one specific object or to some one in a limited set of objects.

Bibliography

Throughout the article, the abbreviated titles in brackets are used for reference

Beard, Charles A., and Hook, Sidney. [Terminology] Problems of terminology in historical writing. Chapter IV of *Theory and practice in historical study: A report of the Committee on Historiography*. Social Science Research Council, New York, 1946.

Bonfante, G. [Semantics] Semantics, language. An article in P. L. Harriman, ed., *The encyclopedia of psychology*. Philosophical Library, New York, 1946.

Carnap, Rudolf. 1942. [Concepts] The two concepts of probability. *Philosophy and phenomenological research*, vol. 5 (1945), pp. 513–532.

——. [Remarks] Remarks on induction and truth. *Philosophy and phenomenological research*, vol. 6 (1946), pp. 590–602.

——. [Application] On the application of inductive logic. *Philosophy and phenomenological research*, vol. 8 (1947), pp. 133–147.

Chisholm, Roderick M. [Conditional] The contrary-to-fact conditional. *Mind*, vol. 55 (1946), pp. 289–307.

Ducasse, C. J. [Explanation] Explanation, mechanism, and teleology. *The journal of philosophy*, vol. 22 (1925), pp. 150–155.

Feigl, Herbert. [Operationism] Operationism and scientific method. *Psychological review*, vol. 52 (1945), pp. 250–259 and 284–288.

Goodman, Nelson. [Query] A query on confirmation. *The journal of philosophy*, vol. 43 (1946), pp. 383–385.

——. [Counterfactuals]. The problem of counterfactual conditionals. *The journal of philosophy*, vol. 44 (1947), pp. 113–128.

——. [Infirmities] On infirmities of confirmation theory. *Philosophy and phenomenological research*, vol. 8 (1947), pp. 149–151.

Hempel, Carl G. [Laws] The function of general laws in history. *The journal of philosophy*, vol. 39 (1942), pp. 35–48.

Hospers, John. [Explanation] On explanation. *The journal of philosophy*, vol. 43 (1946), pp. 337–356.

Hull, Clark L. [Variables] The problem of intervening variables in molar behavior theory. *Psychological review*, vol. 50 (1943), pp. 273–291.

——. [Principles] *Principles of behavior*. New York, 1943.

Jevons, W. Stanley. [Principles] *The principles of science*. London, 1924. (1st ed. 1874).

Kaufmann, Felix. [Methodology] *Methodology of the social sciences*. New York, 1944.

Knight, Frank H. [Limitations] The limitations of scientific method in economics. In Tugwell, R., ed., *The trend of economics*. New York, 1924.

Koch, Sigmund. [Motivation] The logical character of the motivation concept. *Psychological review*, vol. 48 (1941). Part I: pp. 15–38, Part II: pp. 127–154.

Langford, C. H. [Review] Review in *The journal of symbolic logic*, vol. 6 (1941), pp. 67–68.

Lewis, C.I. [Analysis] *An analysis of knowledge and valuation*. La Salle, Ill., 1946.

Mill, John Stuart. [Logic] *A system of Logic*. London, 1843.

Popper, Karl. [forschung] *Logik der Forschung*. Wien, 1935.

——. [Society] *The open society and its enemies*. London, 1945.

Reichenbach, Hans. [Logic] *Elements of symbolic logic*. New York, 1947.

——. [Quantum mechanics] *Philosophic foundations of quantum mechanics*. University of California Press, 1944.

Rosenblueth, A., Wiener, N., and Bigelow, J. [Teleology] Behavior, purpose, and teleology. *Philosophy of science*, vol. 10 (1943), pp. 18–24.

Tarski, Alfred. [Truth] The semantical conception of truth, and the foundations of semantics. *Philosophy and phenomenological research*, vol. 4 (1944), pp. 341–376.

White, Morton G. [Explanation] Historical explanation. *Mind*, vol. 52 (1943), pp. 212–229.

Woodger, J. H. [Principles] *Biological principles*. New York, 1929.

Zilsel, Edgar. [Empiricism] Problems of empiricism. In *International encyclopedia of unified science*, vol. II, no. 8. University of Chicago Press, 1941.

——. [Laws] Physics and the problem of historico-sociological laws. *Philosophy of science*, vol. 8 (1941), pp. 567–579.

10

Ideals of Natural Order

Stephen Toulmin

What is a phenomenon? How do scientists tell when an event has to be recognized as a "phenomenon"; and how do they know what sort of a phenomenon it is? The predictivist view of explanation distracts our attention from this question, and that is a pity. For it suggests that, when it comes to applying our theories, all events are on a par – in the same way that all tides, sunrises, and eclipses are to the forecaster. If we have a technique for predicting high-tides or eclipses at all, it must apply equally to all such events; and why (one might begin by asking) should it be any different with explanation?

There is, in fact, an important difference here. A prognosticator may forecast all events of a given type equally, but for the scientist a phenomenon is not just *any* event of the sort he is interested in – it is (as the lexicographers rightly say) "an event... whose cause is in question", and particularly one which is "highly unexpected". Further, if a phenomenon is an unexpected event, this indicates, not that the scientist neglected or simply failed to predict it, but rather that he had certain prior expectations, which *made* the event unexpected.

So far as the prognosticator is concerned, the course of Nature need consist only of "one damn thing after another". He himself is not going to be caught napping, for he has discovered a way of telling what is going to happen next; but this is not to say that he understands what is happening. The scientist is in a very different position. He *begins* with the conviction that things are not just happening (not even just-happening-regularly) but rather that some fixed set of laws or patterns or mechanisms accounts for Nature's following the course it does, and that his understanding of these should guide his expectations. Furthermore, he has the beginnings of an idea what these laws and mechanisms are, so he does not (and should not) approach Nature devoid of all prejudices and prior beliefs. Rather, he is looking for evidence which will show him how to trim and shape his ideas further, so that they will more adequately fit the Nature with which he wrestles.

This is what makes "phenomena" important for him. The games-player improves his sporting techniques most quickly by playing against opponents who are just *one* degree his superior. The scientist, likewise, is on the look-out for events which are not yet *quite* intelligible, but which could probably be mastered as a result of some intellectual step which he has power to take. So long as everything proceeds according to his prior expectations, he has no opportunity to improve on his theories. He must look out for deviations that are not yet explained, but promise to be explicable.

"Deviations" – as soon as one begins to characterize phenomena, the very ink in one's pen becomes saturated with revealing words like "deviation", "anomaly", and "irregularity". All these imply quite clearly that we know of a straight, smooth, regular course of events which would be intelligible and rational and natural in a way that the "phenomenon" is not. And this is just the conclusion we are now prepared for: the scientist's prior expectations are governed by certain rational

From Stephen Toulmin, *Foresight and Understanding*, Bloomington: Indiana University Press, 1961, pp. 44–61.

ideas or conceptions of the regular order of Nature. Things which happen according to these ideas he finds unmysterious; the cause or explanation of an event comes in question (i.e. it becomes a phenomenon) through seemingly deviating from this regular way; its classification among the different sorts of phenomenon (e.g. "anomalous refraction") is decided by contrasting it with the regular, intelligible case; and, before the scientist can be satisfied, he must find some way of applying or extending or modifying his prior ideas about Nature so as to bring the deviant event into the fold. Let us now look at some representative cases in which this intellectual procedure is displayed, so as to show something of the function which "ideals of natural order" have in the development and application of scientific theory.

We may at this stage look back once again into the history of science, this time turning our attention to the seventeenth century. That period saw drastic changes in several branches of science, including two quite fundamental reorientations, which will be our chief topics in this chapter and the next. To begin with, let me illustrate the points I have been making by reference to the internal reordering within the science of dynamics, through which Newton's basic conceptions finally displaced those of Aristotle. In the next chapter we shall look at some changes which began seriously only at the end of the seventeenth century, and affected, not the internal organization of one science, but rather the mutual relations between two different sciences – physiology and matter-theory.

In each case a purely chronological account can be given of the experiments and publications and empirical discoveries of the scientists involved; but the intellectual changes which took place in their thought are intelligible only if we go deeper, and attempt to recognize the fundamental patterns of expectation at stake in the disputes. Happenings of sorts which earlier men had accepted as the natural course of events now came (we shall see) to be regarded as complex and anomalous; while others, which had earlier appeared exceptional, anomalous, or even inconceivable, came to be treated as perfect instances of the natural order. But let us get down to the cases.

First, consider the seventeenth-century revolution in dynamics. To bring out clearly the central change this involved, we must begin by looking at the popular caricature of pre-Galilean theories of motion, which can ultimately be traced back to Aristotle. "Men's ideas about dynamics before Galileo," this caricature suggests, "rest upon a simple mistake. Aristotle was a philosopher, or at best a naturalist, rather than a true scientist: he may have been skilled at collecting specimens and miscellaneous information, but he was bad at explaining things; and he put forward certain clearly mistaken views about the ways in which the motion of a body is related to the forces acting on it. The benighted man asserted that the effect of a given force acting continuously upon a given body was to keep it in motion at a constant speed; whereas we have now looked and seen that a constant force produces not a constant speed but a constant acceleration. Aristotle's successors, having an exaggerated idea of his intellectual capacities, trusted to his words rather than to their own eyes, and only the work of that obstinately common-sensical genius Galileo – who refused to allow himself to be befuddled by mere words, and insisted on submitting even the most august and authoritative doctrines to the test of experience – led to this chimaera being blown away into the oblivion where it properly belonged."

So stated, this may be less a caricature than the caricature of a caricature; though in less blatant forms, or in part, or by implication, one comes across this view often enough. Still, the picture implicit in this account, both of Aristotelian mechanics and of Galileo's own contribution to our thought, embodies a collection of anachronisms and legends exceptional even for the history of science – a subject in which the George Washingtons have for too long been chopping down their fathers' cherry trees. What one must protest against is not only the intrinsic unlikelihood that a man of Aristotle's capacities could have fallen for so elementary a blunder; but even more, the way in which this caricature degrades a fascinating episode into a prosaic one.

What, then, is wrong? To begin with, this picture gives Aristotle credit for attempting to do something he never seems to have envisaged. It treats him as putting forward a mathematical relationship of the sort familiar from modern dynamical theory. The relationship in question could be written either in words, as

Force varies as Weight times Speed

or alternatively in symbolic shorthand, as

$$F \propto W \times V$$

But this can be read into Aristotle's works only through an anachronism. We scarcely encounter this sort of mathematical equation before the sixteenth century A.D. – not just because the notation employed had yet to be developed, but because the very ideas implicit in the use of such equations were worked out only in the years immediately preceding 1600.

Of course, if we accept this equation as an expression of Aristotle's view, and interpret it in modern terms, we shall find it sadly mistaken. For nowadays it would be natural to take the symbol for speed as meaning "instantaneous velocity", and the symbol for force in its standard Newtonian sense – both of them notions formulated with complete clarity only in 1687. At once objections arise. The term "weight" now appears entirely out of place, and should presumably be replaced by the term "mass"; and even so, the ratio of the force acting on a body to its mass surely determines not its velocity but its acceleration. Yet the question ought to be asked: are we taking Aristotle in a sense which he ever intended? If we read things into him, it will not be surprising if we end up by finding him seriously at fault.

How else, then, can Aristotle's thesis be taken? In general, his practice in the *Physics* is to put forward, not precise equations, but at most ratios or proportionalities relating (say) the lengths of time different bodies will take to go the same distances when different degrees of effort are exerted upon them. He presents these examples as concerned with *tasks*: posing his questions in the form: "If such-and-such a task takes such-and-such a time, how long will such-and-such another task take?" – e.g. if one man can shift a given body 100 yards by himself in one hour, how large a body can two men jointly shift through the same distance in the same time? Aristotle concludes that, within limits, the amount a body can be displaced by a given effort will vary in inverse proportion to the size of the body to be moved; and also, that a given body can be displaced in a set time through a distance directly proportional to the effort available.

Of course (he allows) beyond certain limits this sort of ratio does not apply: a body may be so large that it can be shifted only by a team of men, and will not respond at all to one man working single-handed – he cites the instance of a team of men moving a ship. And he further remarks, with equal truth, that the effect one can achieve by a given effort depends entirely on the resistances to be overcome. A team of men pulling a ship will take longer to go from one point to another across rough ground than to move it the same distance over smoother ground. As a first approximation, and lacking any better definition of "resistance", Aristotle accordingly puts forward the further proportionality: that the distance travelled in a given time will vary inversely as the strength of the resistance offered to motion.

Three things need saying about these ratios of Aristotle's, before we look at the dynamical innovations of the seventeenth century. The first is this: Aristotle concentrated his attention on the motion of bodies against appreciable resistance, and on the length of time required for a complete change of position from one place to another. For a variety of reasons, he never really tackled the problem of defining "velocity" in the case when one considers progressively shorter and shorter periods of time – i.e. instantaneous velocity. Nor was he prepared to pay serious attention to the question how bodies would move if all resisting agencies were effectively or completely removed. As things turned out, his hesitations were unfortunate; yet his reasons for hesitating are understandable, and in their way laudable. Though he was a philosopher – and so, in some people's eyes, bound to have had his head in the clouds and his feet off the ground – Aristotle was always unwilling to be drawn into discussing impossible or extreme examples. Leaving aside free fall for the moment as a special case, all the motions we observe going on close around us happen as they do (he saw) through a more-or-less complete balance between two sets of forces: those tending to maintain the motion and those tending to resist it. In real life, too, a body always takes a definite time to go a definite distance. So the question of instantaneous velocity would have struck him as over-abstract; and he felt the same way about the idea of a completely unresisted motion, which he dismissed as unreal. In point of fact (I suppose) he was right. Even in the interstellar void, where the obstacles to the motion of a body are for practical purposes entirely negligible, there do nevertheless remain some minute, if intermittent, resistances.

In the second place: if we pay attention directly to the kinds of motion Aristotle himself thought typical, we shall find that his rough proportionalities retain a respected place even in twentieth-century physics. Interpreted not as rival laws of nature to Newton's, but as generalizations about familiar experience, many of the things he said are entirely true. One can even represent him as having spoken more wisely than he knew. For, where he argued only for rough, qualitative ratios connecting gross measures of distances and time, contemporary physics actually recognizes an exact mathematical equation corresponding closely to them – though, of course, one which relates instantaneous variables of a kind Aristotle himself never employed.

This equation is known as "Stokes' Law". It relates the speed at which a body will move when placed in a resisting medium, such as a liquid, to the force acting on it and the thickness (viscosity) of the medium. According to Stokes, the body's speed under those circumstances will be directly proportional to the force moving it, and inversely proportional to the liquid's viscosity. Suppose we take a billiard ball and drop it through liquids of different viscosities in turn – water and honey and mercury: in each case it will accelerate for a moment, and then move steadily down at a limiting (terminal) speed determined by the viscosity of the liquid in question. If the impressed force is doubled, the speed of fall will be doubled: if one liquid is twice as viscous as another, the billiard ball will travel at only half the speed.

The third point combines these two previous ones. The fact is that Aristotle based his analysis on one particular explanatory conception or *paradigm*, which he formulated by considering examples of a standard type; and he used these examples as objects of comparison when trying to understand and explain *any* kind of motion. If you want to understand the motion of a body (in his view) you should think of it as you would think of a horse-and-cart: i.e. you should look for two factors – the external agency (the horse) keeping the body (the cart) in motion, and the resistances (the roughness of the road and the friction of the cart) tending to bring the motion to a stop. Explaining the phenomenon means recognizing that the body is moving at the rate appropriate to an object of its weight, when subjected to just that particular balance of force and resistance. Steady motion under a balance of actions and resistances is the natural thing to expect. Anything which can be shown to exemplify this balance will thereby be explained.

In the case of bodies moving against a sufficiently slight resistance, as we all know, Aristotle's analysis ceases to apply. If you drop a billiard ball through air instead of through water or treacle, it will go on accelerating for a long time: under normal terrestrial conditions, it could never fall far enough to reach the "terminal velocity" at which Stokes' Law would begin to apply. The factor of paramount importance in this case will for once be the initial period of acceleration, and that was something to which Aristotle paid very little attention. If he had thought more about the problem of acceleration, indeed, he might have seen the need for something more sophisticated than his simple proportionalities.

As things turned out, Strato, the very first of Aristotle's followers to take an active interest in mechanics, turned his attention at once to this very phenomenon. Yet, for many reasons – some of them intellectual, some of them historical – neither he nor his ancient successors made any great progress beyond Aristotle's ratios. It was left to the Oxford mathematicians of the early fourteenth century to add an adequate definition of acceleration to Aristotle's previous accounts of speed, and so to pave the way for the work of Stevin and Galileo and Newton.

So much for the background: what, then, did happen in dynamics during the seventeenth century? Certainly the popular caricature is wrong in one respect: men did not suddenly become aware that Aristotle's views about motion were false, whereas their predecessors had trusted blindly in their truth. Aristotle himself stated his ratios as applying only within certain limits, and John Philoponos (around A.D. 500) made it absolutely clear that projectiles and freely falling bodies could be explained only by bringing in some radically new conception. The problem was, *how* to remedy matters.

In retrospect we can see that the paradigm at the heart of Aristotle's analysis had to be abandoned and replaced by another, which placed proper importance on acceleration. Yet this was not easy: men were accustomed to think of motion as a balance between force and resistance, as much on the basis of everyday experience as through "blind trust in Aristotle's authority". They took the necessary steps hesitantly, a bit at a time, and in the face of their inherited common-sense. The most radical single step was taken by Galileo, yet even he stopped short of the conclusion which is generally credited to him.

There is nothing uniquely natural or rational, Galileo rightly insisted, in a terrestrial body coming to rest when outside forces are removed: rest and uniform motion alike, he argues, are "natural" for a body on the Earth. Let us only approach gradually towards the extreme case of zero resistance, which Aristotle had denounced as impossible, and we shall recognize this. Think of a ship (say) on a calm sea, and imagine the resistances to motion progressively reduced, until we could neglect them entirely. If that were to happen, said Galileo, the ship would retain its original motion without change. If it had originally been at rest, it would remain at rest until some outside force started it moving; while, if it were originally moving, it would go on travelling along the same course at the same speed until it met an obstacle. Continuous, steady motion could therefore be just as natural and self-explanatory as rest, and outside resistances alone could bring terrestrial bodies to a halt.

By this step, Galileo went a long way towards the classical Newtonian view, but he did not go the whole way. True, he had exchanged Aristotle's paradigm of natural motion – the horse-and-cart being pulled along against resistances at a constant speed – for a very different one. For Aristotle, all continuous terrestrial motion was a "phenomenon", or departure from the regular order of things, and he would have asked: "What is to keep Galileo's imaginary ship moving?" Galileo, however, now demanded only that we account for *changes* in the motion of bodies. His ship could move for ever without a motive force.

Now this result looks, at first sight, very like our modern "law of inertia". Yet Galileo's paradigm was no more identical with our own than Aristotle's had been. For what he envisaged as his ideal case was a ship moving unflaggingly across the ocean along a Great Circle track, for lack of any external force to speed it up or slow it down. He saw that uniform motion could be quite as natural as rest; but this "uniform motion" took place along a closed horizontal track circling the centre of the Earth; and Galileo took such circular motion as entirely natural and self-explanatory. He does not seem to have regarded the ship as constrained by its own weight from flying off the Earth on a tangent – the image which can clearly be found in Newton.

Indeed, if Galileo's imagined ship *had* taken off from the sea and disappeared off into space along a Euclidean straight line, he would have been no less surprised – in fact, *more* surprised – than us. We should have one possible hypothesis at hand to explain this amazing event – namely, that the action of gravity on the ship had been suspended, so that it was no longer constrained to remain in contact with the Earth's surface and could fly off along its natural path. For Galileo, however, this option was not yet available: in his eyes, some active force alone could have obliged the ship to travel in a perfectly rectilinear path, instead of cruising of its own accord round its natural Great Circle track.

When we turn to Newton we find that the ideal of natural motion has changed yet again. The fundamental example is completely idealized. From now on, a body's motion is treated as self-explanatory only when it is free from all forces, even including its own weight. Galileo could explain his conception of "inertia" by referring to real objects – ships moving on the sea. Newton started his theory by offering us a completely abstract example as the paradigm – namely, a body moving at uniform speed in a Euclidean straight line – and this, as Aristotle would have retorted, is the last thing we should ever encounter in the real world. But, then, Newton does not have to claim that, as a matter of fact, any actual body moves exactly as his first law specifies. He is providing us, rather, with a criterion for telling in what respects a body's motion calls for explanation; and what impressed forces we must bring to light if we are to succeed in explaining it. Only if a body ever were left completely to itself would it move steadily along a straight line, and no real body ever actually is placed in this extreme position. This is, for Newton, simply a dynamical ideal, the sole kind of motion which would be self-explanatory, free of all complexity, calling for no further comment – if it ever happened.

It should be clear, by now, why I present Newton's first law of motion or principle of inertia as an "ideal of natural order" – one of those standards of rationality and intelligibility which (as I see it) lie at the heart of scientific theory. At their deepest point, the seventeenth-century changes in dynamics, which had been brewing ever since the early 1300's, involved the replacement of Aristotle's common-sensical paradigm by Newton's new, idealized one. From some angles, this could look like a regression: from now on it was necessary, for theoretical purposes, to relate familiar everyday happenings to idealized, imaginary states-of-affairs that never in practice occur – ideals to which even the motions of the planets can only approximate. Yet the change paid dividends. Once this new theoretical ideal was accepted, the single hypothesis of universal gravitation brought into an intelligible pattern a dozen classes of happenings, many of which had previously been entirely unexplained; and, in the resulting theory, Newton could display a whole new range of relationships and necessities as part of the intelligible order of Nature.

This example has illustrated how the idea of explanation is tied up with our prior patterns of expectation, which in turn reflect our ideas about the order of Nature. To sum up: any dynamical theory involves some explicit or implicit reference to a standard case or "paradigm". This paradigm specifies the manner in which, in the natural course of events, bodies may be expected to move. By comparing the motion of any actual body with this standard example, we can discover what, if anything, needs to be regarded as a "phenomenon". If the motion under examination turns out to be a phenomenon – i.e. "an event...whose cause is in question" as being "highly unexpected" – the theory must indicate how we are to set about accounting for it. (In Newton's theory, this is the prime task of the second law of motion.) By bringing to light causes of the appropriate kind, e.g. Newtonian "forces", we may reconcile the phenomenon to the theory; and if this can be done we shall have achieved our "explanation". Every step of the procedure – from the initial identification of "phenomena" requiring explanation to the final decision that our explanation is satisfactory – is governed and directed by the fundamental conceptions of the theory.

No wonder that the replacement of one ideal of natural motion by another represents so profound a change in dynamics. Men who accept different ideals and paradigms have really no common theoretical terms in which to discuss their problems fruitfully. They will not even *have* the same problem: events which are "phenomena" in one man's eyes will be passed over by the other as "perfectly natural". These ideals have something "absolute" about them, like the "basic presuppositions" of science about which R. G. Collingwood wrote.

If that is so, the problem at once arises: how do we know which presuppositions to adopt? Certainly, explanatory paradigms and ideals of natural order are not "true" or "false", in any naive sense. Rather, they "take us further (or less far)", and are theoretically more or less "fruitful". At a first, everyday level of analysis, Aristotle's paradigm of uniform, resisted motion had genuine merits. But a complete mathematical theory of dynamics required a different ideal. It was no good first taking uniform, resisted motion as one's paradigm, and supposing that one could later explain how bodies would move in the absence of resistances by cancelling out the counteracting forces: that way inevitably led to the unhelpful conclusion that a completely unresisted motion was inconceivable – since the attempt to describe it in everyday terms entangles one in contradictions. (Suppose you reduce the resistances finally to zero, then, in Aristotle's ratio of motive force to resistance, the denominator becomes zero; and you are landed in all the difficulties which spring from "dividing by nought".) On the contrary: it was necessary to proceed in the opposite direction. One must first start by taking entirely unresisted motion as one's ideal of perfectly simple and natural motion; and only later introduce resistances – showing how, as they are progressively allowed for, the uniform acceleration produced by a single force gives way to the uniform terminal speed of a horse-and-cart.

Changes in our ideals of natural order may sometimes be justifiable, but they do have to be justified positively. In due course uniform rectilinear motion became as natural and self-explanatory to Newton's successors as rest had been for Aristotle. Yet neither view of inertia was self-evidently correct: each must be known by its fruits. So its tenure as the fundamental ideal of dynamics was conditional, and provisional. For just as long as we continue to operate with the fundamental notions of the Newtonian theory, his principle of inertia keeps its place in physics. Yet, at the most refined level of analysis, it has already lost its authority. As one consequence of the twentieth-century changeover to relativity physics, the conception of "natural motion" expressed in Newton's first law has again had to be reconsidered. The implications of the resulting amendments in our ideas may have been less drastic than those which flowed from the seventeenth-century revolution; yet – at the theoretical level – the change has been none the less profound.

Before we go on to our second example, let us return to a less rarefied atmosphere. The general point I am making does not apply only to abstract and highly developed sciences, such as dynamics. We use similar patterns of thought in the common affairs of daily life; and, in a sense, the task of science is to extend, improve on, and refine the patterns of expectation we display every day. There is a continual interplay between the two fields.

Suppose, for example, that we look out of the window, into the street. One car travels steadily down the road, comes into sight, passes our window, and goes on out of sight again: it may well escape our attention. Another car comes down the road haltingly, perhaps jerking and backfiring, perhaps only stopping dead and starting up again several times: our attention is immediately arrested, and we begin to ask questions – "Why is it behaving like that?" From this example it is only a step to the case of a practical astronomer, for whom the continued motion round its orbit of the planet Jupiter is no mystery: but for whom questions would immediately arise if the planet were suddenly to fly off along a tangent to its orbit and out into space: "What made it do that?" And from this it is only one further step to the mathematicians' point of view, according to which, if left to itself, Jupiter ought to travel, not in a closed orbit, but in a straight line – so that even its normal, elliptical path demands explanation.

All the same: though the form of this thought-pattern is preserved, its content changes drastically, and one popular epigram about explanation is falsified in the process. For it is often said that "explanation" consists in relating things with which we are unfamiliar (and which so need explaining) to others which are familiar to us (and so stand in no need of explanation). At a certain level this epigram has a point. If you are explaining something *to somebody* – what might be called an explanation *ad hominem* – it is sensible to start from things he knows about and understands, and to relate the things he finds mysterious back to those which he finds intelligible. This is one of the purposes of "models" in the physical sciences. The beginner in electricity is helped to understand the relations between voltage, current, and resistance by having the flow of electricity in a wire compared with the flow of water down a tube: "Don't you see? Voltage is like the head of water in the system, resistance is like the narrowness of a pipe, and the current of water or electricity depends in each case on both factors."

Scientific discoveries, however, do not consist in arguments which are plausible *ad hominem*, but rather in explanations which will stand on their own feet. In these explanations, the relation between the "familiar" and the "unfamiliar" may be reversed. Revert for a moment to Newtonian dynamics: the ideal of inertial motion which underlies Newtonian explanations can hardly be described as *familiar*. (Aristotle would laugh at that suggestion.) If we were to insist on accounting for the "unfamiliar" in terms of the "familiar", instead of *vice versa*, we should never be able to shake ourselves loose of Aristotelian dynamics. Aristotle's paradigm is familiar in a way that Newton's never can be; and the Newtonian programme of treating the motion of horses and carts as being something highly complex, which can be understood only by starting from planets and projectiles and then allowing for a multiplicity of interfering forces – remains rather paradoxical to the commonsense mind.

What are the lessons of this first example? In ordinary life explanation may, perhaps, consist in "relating the unfamiliar to the familiar". But, as science develops, this turns into "relating the anomalous to the accepted", and so in due course into "relating the phenomena to our paradigms". This is inevitable. Which things are familiar and which unfamiliar is a relative matter. (A man who lived in a desert might find the idea of "the head of water" a difficult one to grasp, and be more mystified by hydraulics than by electricity.) On the other hand, whether an event is "anomalous" or not need not be so personal a question. It can be discussed rationally – still more, if we go to the length of labelling the event as a "phenomenon" and implying that it needs to be squared with theory. For then our standard must be, not what is familiar, but rather what is intelligible and reasonable in the course of Nature. And where we are led once we recognize this distinction, it has been the aim of this chapter to show.

11

Revaluing Science: Starting from the Practices of Women

Nancy Tuana

Introduction

Work in the social studies of science in the last twenty years has undermined the belief common to positivist models of science that value-neutrality is both a hallmark and goal of scientific knowledge. The ideal of a value-free science was linked to the tenet that neither the individual beliefs or desires of a scientist nor the social values of a scientific community are relevant to the production of knowledge, and models of scientific method were constructed with the goal of factoring out such contaminating influences. The rapid militarization of science in the United States since the 1970s and the current rise of influence of venture capital in charting the direction of scientific research have made it increasingly difficult to draw any clear lines between a "pure," disinterested science, and a goal-oriented, transformative "applied" science. Questions in the philosophy of science have shifted from the "pure" epistemological question "How do we know?" to questions that reflect the locations of science within society and the relationships between power and knowledge: "Why do we know what we know?" "Why don't we know what we don't know?" "Who benefits or is disadvantaged from knowing what we know?" "Who benefits or is disadvantaged from what we don't know?" "Why is science practiced in the way that it is and who is advantaged or disadvantaged by this approach?" "How might the practice of science be different?"

Feminist theorists of science have been active participants in this research program. Our work has added an important dimension to discourses concerning the value-neutrality of science by focusing attention onto the dynamics of gender and oppression in the theories and methods of science.[1] One of the central insights of feminist science studies has been the increased awareness of the ways in which social locations, locations that include political and ethical dimensions, are gendered. Through this attention to gender we have contributed to the transformation of the traditional question "How do we know?" in numerous ways, including investigating whether traditional models of rationality and of the scientific method have been gender biased, that is, have privileged traits viewed as masculine and denigrated those perceived to be feminine; documenting the ways in which scientific theories have reinforced sexist and/or racist biases: delineating the ways in which men in dominant groups have benefited (and been hindered) by the questions asked and avoided in science; and analyzing the impact of the exclusion, as well as the inclusion, of women in science.

An important resource for feminist investigations of science has been the practices of women scientists. Many feminist theorists, particularly those who embrace a feminist standpoint epistemology, have argued that the distinctive experiences of women in a gender-stratified society provide an important resource, a resource typically

From *Feminism, Science, and the Philosophy of Science*, ed. L. H. Nelson and J. Nelson, Dordrecht: Kluwer, 1996, pp. 17–35, abridged.

overlooked by nonfeminist theorists, that, in the words of Sandra Harding, enables "feminism to produce empirically more accurate descriptions and theoretically richer explanations than does conventional research" (Harding, 1991, p. 119). One of my goals in this essay is to illustrate the ways in which the experiences of women, particularly women scientists, provide a resource for feminist critiques of the ideal of value-neutrality in science.

Women's differences, both their differences from men and their differences from one another, can highlight overlooked or minimized aspects of the knowledge process in science. I will here limit my analysis to three of these, each of which is relevant to transformations of the traditional epistemological question "How do we know?" and the rejection of the ideal of value-neutrality in science:

(1) replacing the traditional model of the knower as a detached, disinterested individual with the dynamic model of engaged, committed individuals in communities;
(2) recognition of the epistemic value of affective processes;
(3) examination of the role of embodiment in the knowledge process.

Individuals in Communities

Descartes envisioned himself alone in this study, attempting to put aside all he had learned from authority and all the beliefs he had unquestioningly inherited from his culture, as well as endeavoring to suppress the needs of his body. Descartes believed that only after he had removed all such influences from his rational processes would he be capable of pursuing his method for gaining true knowledge, alone and unencumbered by others.

Although Descartes was hardly an empiricist, it is the Cartesian subject that is designed to hold the subject position in S-knows-that-p models of knowledge. This is a model of knowledge that aims ideally at removing all individual traces of the knowing subject. Both perception and cognition are assumed to be invariant from knower to knower – at least in the ideal case. All other factors such as personal beliefs, desires, and bodily configurations are deemed irrelevant at best, contaminating at worst. Based on this picture of rationality, much of modern epistemology has been focused on the ways in which variations between knowers could be filtered out.

This model of the knowing subject is in tension with the feminist acknowledgement of the fact that as humans we are always in relations of interdependence and that these relationships are crucial not simply for personal satisfaction, but also for moral, political, and scientific deliberation. In the words of Seyla Benhabib, "the self only becomes an I in a community of other selves who are also I's. Every act of self-reference expresses simultaneously the uniqueness and difference of the self as well as the commonality among selves" (Benhabib, 1987, p. 94).

A careful study of the actual practice of science also discloses a different model of the knowing subject, one that necessitates a rejection of the model of the isolated knower and replaces it with a dynamic model of individuals in communities. An examination of the complexity of the communities relevant to the production of knowledge in science also reveals that the production of good science does not require disinterested, dispassionate scientists. As Sandra Harding has convincingly argued, objectivity does not require neutrality.[2] A scientist's social locations can be epistemically significant to her or his practice of science. The ideal of a pure science, a science uninfluenced by values, and the scientist as a neutral recorder of facts are myths, ones that can be rejected without abandoning objectivity.

Changing the subject of evolution

The development of "woman, the gatherer" theories of human evolution has been the subject of much discussion in feminist science literature because this example is an excellent illustration of not only the inescapable fact of value within the construction of scientific theories, but also the potential epistemic significance of the various communities, including political communities, in which the knower participates.[3]

Feminist discussions of the epistemological significance of being part of the feminist community have found the science of primatology to be particularly relevant due to the fact that its stories of human evolution arise out of origin myths, that is, accounts of the origins of the family, of the sexes and their roles, as well as of the divisions between human and nonhuman animals. In other words, accounts of human evolution wear the metaphysics of their authors "on their sleeves" and thus provide clear accounts of the ways participation in alternative communities can be epistemologically significant.

To understand the androcentrism of traditional "man, the hunter" accounts of evolution, we need only attend to the respective roles of women and men. "Man, the hunter" theories of human evolution attribute the evolution of *Homo sapiens* to those activities and behaviors engaged in and exhibited by male ancestors. Males, the explanation goes, having the important and dangerous task of hunting big animals to provide the central food source, invented not only tools but also a social organization, including the development of language, that enabled them to do so most successfully. Hunting behavior is posited as the rudimentary beginnings of social and political organization. "In a very real sense our intellect, interests, emotions, and basic social life – all are evolutionary products of the success of the hunting adaptation. ...The biology, psychology, and customs that separate us from apes – all these we owe to the hunters of time past" (Washburn and Lancaster, 1976, pp. 293, 303).

Such accounts do not omit women, but place them firmly "at home." While men are out hunting, women are taking care of hearth and children, dependent upon the men for sustenance and protection. Note the assumptions embedded in this account. Only male activities are depicted as skilled or socially oriented. Women's actions are represented as biologically oriented and based on "nature." This definition of woman's functions as natural curtails any analysis of them, such as their relation to the physical and social environments or the role they might play in determining other social arrangements. Men are depicted as actively transforming their nature, while women are portrayed as constrained by it.

The alternative origin stories told by feminist primatologists transform women from a passive, sexual resource for males to active agents and creators. The work of Linda Marie Fedigan, Sarah Blaffer Hardy, Lila Leibowitz, Sally Linton Slocum, Barbara Smuts, Shirley Strum, Nancy Tanner, and Adrienne Zihlman, among others, began in the 1970s to transform the complexion of accounts of the nature of woman and man. One key to understanding the explosion of alternative images of women's nature lies in the woman's movement of the 1960s that contested the definitions of woman as the second sex, definitions that simultaneously relegated her role to the private realm of family while designating the public realm of culture and politics as that which makes one fully human.

Feminist attention to perceptions of women's roles and the linkage of woman and nature provided the basis for a rethinking of evolution for a number of scientists. The anthropologist Sally Linton Slocum, for example, in her 1970 essay "Woman the Gatherer: Male Bias in Anthropology" [Slocum 1971] identified ways in which females were being obscured within evolutionary theories by the association of their actions with nature and began to question the assumption that women's actions were unimportant because they were derived from instinct and thus not relevant to the evolutionary process. Slocum's position was in turn developed by the paleoanthropological research of Adrienne Zihlman and Nancy Tanner. This shift of attention was the result not of any biological difference between women and men scientists, but because women scientists were more likely to be affected by and participate in the feminist community – a community that had been actively exposing the history and the impact of the androcentric bias of associating women with nature and men with culture, as well as working to revalue the socially defined work of women, including childcare and housework.[4] This political awareness arising from the influences of the feminist community changed the focus of attention for researchers like Slocum, Tanner, and Zihlman and contributed to the construction of alternative questions.

But it would be inaccurate to see the accounts of these scientists as influenced only by their participation in communities that were redefining woman's nature. These women were also influenced by their membership in scientific communities and the then current theories of evolution. The point is that accounts by women primatologists, particularly feminist primatologists, while marked both by their gender and their politics as they attempt to carve a role for women out of the standard narrative of evolution, nevertheless evolve out of and are influenced by the accepted narratives and standards of evidence of their scientific communities.

Nor should alternative evolutionary accounts such as "woman, the gatherer" be seen simply as feminist "correctives," that is, as an ideological image imposed onto the data. I will argue that this alternative model of evolution arose in response to changes within the scientific community, provided more accurate accounts of the evidence, and was therefore the result of better science. But this is not incompatible with saying that the model emerged from the practice of feminist scientists who, because of the impact of their communities, attended differently to the data. To say that the practice of science is marked by gender and by politics is not the same as claiming that it arises out of wishful thinking or ideological concerns. A scientific theory can provide consistent methods for obtaining reliable knowledge, yet be influenced by certain values or interests. Objectivity and neutrality are not the same thing.

[...]

Zihlman's creation of an alternative to the androcentric "man, the hunter" theory was made possible by the knowledge she gained from the communities of which she was a member, in this case both scientific and nonscientific. However, it is not a coincidence that the "woman, the gatherer" hypothesis was initiated by the work of Sally Linton Slocum, and developed by Nancy Tanner and Adrienne Zihlman. Being a feminist scientist can affect one's practice of science. In the words of Lynn Hankinson Nelson, "it makes a difference to one's observations, appraisals of theories, and one's own theorizing, if one *recognizes* androcentric and sexist assumptions, categories, or questions and if one questions the inevitability of male dominance and/or the universality of hierarchical dominance relationships. In short, it makes a difference if one is working from a feminist perspective" (Nelson, 1990, p. 224). But Zihlman's participation within particular scientific communities was also a crucial factor in the development of her research. The point is that a scientist is simultaneously a member of a number of different epistemic communities and subcommunities. The values and beliefs of these various communities often interact in complex ways over the course of the knowledge process. Fully understanding the development of knowledge then requires an appreciation of the interactive effects of all relevant communities and an understanding of the underlying presuppositions, metaphysical as well as aesthetic and moral values, of each community's system of beliefs.

Engaged Knowers

Acknowledging that social values enter into the practice of science problematizes the traditional model of the knower as detached, disinterested, and autonomous. Both the individualism as well as the goal of neutrality posited by traditional accounts of knowledge must be questioned. Many feminist theorists of science contend that women's relative absence from the practice of science is not due simply to institutional barriers such as limited access to advanced science training, but is also an aspect of a model of the scientist that privileges traits that have historically been associated with masculinity (autonomous, detached, disinterested) and suppress those traditionally associated with femininity (dependent, connected, engaged).[5] In other words, despite its professed neutrality, the positivist model of knowledge, like all models, arises out of a tradition and is imprinted with the values of that tradition. Neither science nor our models of science correspond to the neutrality ideal.

Feminist studies of science thus reveal the myopia of traditional individualist accounts of the knowing subject. On the traditional S-knows-that-p model of knowledge, we need have no knowledge of S. Knowers, while envisioned as distinct individuals, are not seen as *dinstinctive*. Neither the body nor any "subjective" aspect of an individual's mental activity is seen as affecting the proper pursuit of knowledge. This model of knowledge is linked to the belief in a universal faculty of reason common to all potential knowers. Whether it be Descartes' ability to apprehend clear and distinct ideas or the positivist vision of a deductive logic, knowing capacities are invariant (though not all equally developed). S-knows-that-p models thus embrace the vision of the generic "man" – a sameness that removes the threat of allegedly biased or partial perspectives.

Feminist investigations of science are resulting in what Helen Longino labels the strategy of "changing the subject" of knowledge. We are finding that S-knows-that-p models of knowledge are inadequate to the actual practice of science. The conception of the subject of knowledge as "generic" and hence not itself a subject of study does not fit the epistemic importance of differences between subjects. Such a model, for example, does not account for the epistemic role of the complex relationships between agents of knowledge as evidenced in examples like that of "woman, the gatherer" theories of human evolution. Equally problematic, this model overlooks the epistemic significance of subjective aspects of the relationship between scientists and the subject of inquiry, such as the scientist's commitments, desires, and interests. It also ignores the fact and nature of a scientist's embodiment. In this section I examine the role of what has traditionally been labeled "the passions" in the knowledge process in science, reserving the question of the role of the body for the final section of the essay.

A feeling for the organism

Evelyn Fox Keller offers a portrait of the geneticist Barbara McClintock that provides a very different image of the scientist than that of the disinterested, detached observer. McClintock describes herself as having developed a close relationship with the objects of her investigation. "I start with the seedling [of maize]. I don't want to leave it. I don't feel I really know the story if I don't watch the plant all the way along. So I know every plant in the field. I know them intimately, and I find it a great pleasure to know them" (Keller, 1983, p. 198). McClintock viewed the complexity of nature as being beyond full human comprehension. "Organisms can do all types of things; they do fantastic things. They do everything that we do, and they do it better, more efficiently, more marvelously....Trying to make everything fit into set dogma won't work.... There's no such thing as a central dogma into which everything will fit" (Keller, 1983, p. 179). In holding to this belief – a metaphysical value – McClintock deviated from the positivist assumption – yet another metaphysical value – that there were underlying regularities of nature, the laws of nature, that were discrete and individually knowable by humans. This difference in basic values contributed to McClintock's commitment of developing a close relationship with the material she was studying, for only by listening carefully can one "let the material tell you." "I feel that much of the work [in science] is done because one wants to impose an answer on it. They have the answer ready, and they [know what they] want the material to tell them. [Anything else it tells them] they don't really recognize as there, or they think it's a mistake and throw it out....*If you'd only just let the material tell you*" (Keller, 1983, p. 179).

[...]

Knowing other people

Feminist studies of science, particularly the detailed studies of the practices of women scientists, have served as an important resource for feminist epistemologists. Influenced by examples like that of McClintock, many feminists are developing epistemologies that include the tenet that subjectivity is an important and indispensable component of the process of gaining knowledge.[6] But successfully doing so requires offering alternative models of knowledge. Lorraine Code has offered such an alternative, the model of "knowing other people." While S-knows-that-p models of knowledge are based on what Code calls ordinary knowledge of medium-sized objects in the immediate environment – the red book, the open door – Code's model is based on the centrality of our relationships with others. "Developmentally, recognizing other people, learning what can be expected of them, is both one of the first and one of the most essential kinds of knowledge a child acquires" (Code, 1991, p. 37). Code presents this model as an *addition* to the S-knows-that-p epistemologies that perhaps work for simple objects in simple settings. She argues that the latter model is not sufficient for more complex instances in which knowledge requires constant learning, is open to interpretation at various levels, admits of degree, and is not primarily propositional. For such cases, a standard of knowledge modeled after our knowledge of other people would be more accurate.

Code, influenced by examples like that of McClintock, argues for a remapping of the epistemic terrain. A model that posits knowing other people as a paradigmatic kind of knowing challenges the desirability or even possibility of the disinterested and dislocated view from nowhere. Code's model of knowing other people is a dynamic, interactive model. It is a vision of a process of coming to know, "knowing other people in relationships requires constant learning: how to be with them, respond to them, and act toward them" (Code, 1993, p. 33). It is a model of knowledge that admits of degree, that is not fixed or complete, that is not primarily propositional, and is acquired interactively.

Code's model embraces the subjective components of the knowledge process illustrated in McClintock's method. McClintock's desire to develop an intimate relationship with the subject of her study, to take the time to listen, to get right down there, to develop a feeling for the organism, are all aspects of the knowing process accounted for by Code's model. "Rocks, cells, and scientists are located in multiple relations to one another, all of which are open to analysis and critique. Singling out and privileging the asymmetrical observer-observed relation is but one possibility" (Code, 1991, p. 164). Code's alternative model, unlike S-knows-that-p models, embraces McClintock's metaphysical belief that nature, like other people, is far too complex to allow for complete and universal knowledge. For McClintock, our knowledge of nature will always be partial, always changing, always in process – just as is our knowledge of people. This is not a critique or belittlement of our knowledge capacities, but rather a recognition and appreciation of the extraordinary complexity and continual evolution of both nature and of people. Such recognition leads to a model of knowledge that embraces the importance of empathy and imagination as a resource for "letting the material tell you." It is a model that, while acknowledging the importance of categories and theories, does not privilege them over and above the importance of listening attentively and responsibly to the stories told to us – accounting for the differences rather than imposing a model upon the world.

Code's point and one that is shared by many feminist epistemologists is that the traditional image of the dispassionate scientist removed from her or his object of study has blinded us to the complexity of the possible relationships between subjects and objects. Code argues that McClintock gained her knowledge *because* of her engaged relationship with the object of her study. That is, McClintock's fascination with the maize is epistemically significant. She is drawn to it not just to predict the genetic patterns, but because she desires a full understanding of the organism in all its stages. When she refers to her study of maize, she does so with affection – "these were my friends."

Code posits nothing like an essential femininity that entails that all and only women will embrace an engaged style of knowledge production. She argues rather that McClintock's femaleness is one aspect of the complex conjunction of subjective factors at play in her practice of science. Code's goal and the goal of other feminists is to open epistemology and science education and practice to the importance of such subjective features and to argue that S-knows-that-p models of scientific knowledge are inadequate to the full complexity of knowing. Code's intention is to reclaim subjective components of the knowledge process, components often defined as "feminine" and suppressed from traditional accounts. The aim is not to create a "feminine" science, but "to make a space in scientific research for suppressed practices and values that, coincidentally or otherwise, are commonly associated with 'the feminine'" (Code, 1991, p. 152).

Embodied Knowers

The feminist rejection of the supposedly "generic" knower thus requires that attention be paid to the characteristics and situation of the knower as an important part of the knowledge process. As illustrated in my example of Zihlman's practice of science, the various communities of which one is a part, including one's political beliefs, can be epistemically significant to the knowledge process. As we see with McClintock, a knower's emotive capacities and her or his openness to their relevance to the knowing process, can also be epistemically significant. This is the content of Code's claim that a person's gender can be epistemically significant. In contemporary Western culture, one who is female is more likely than one who is male to be socialized in such a way as to make her more proficient in and accepting of the usefulness of emotions such as empathy and imagination. Just as a feminist is more likely to question the categorization of female activities as "natural" and male activities as "cultural," a woman in contemporary Western culture is more likely to be accepting of and skilled in the employment of emotions in the knowledge process.

But an additional aspect of the knowing subject that is epistemically significant is the fact of and nature of their embodiment. The model of the generic knower has traditionally rejected the relevance of our bodily differences. Attention to the body calls attention to the specificities and partiality of human knowledge, as well as reminding us of the importance of acknowledging the body, and its variations, in the knowledge process. Once we admit the body into our theories of knowledge, we must also recognize its variations; we must, for example, examine the ways in which bodies are "sexed."[7]

Vision

Traditional models of knowledge privilege vision over the other senses. The association of knowledge and vision provides a model of knowledge as disembodied. Vision, perceived as the most detached of the senses, is employed in such a way as to conceal the action of the body. The world appears to my gaze without any apparent movement or action on my part. The action of the body disappears into the background and with it as a model of knowledge, the philosopher places the world at a distance from the observer, thereby dematerializing knowledge. The perceived scene, as well as the perceiver, is to be physically unaffected by the gaze.

There have been many studies that have examined the ways in which this conception of vision has shaped traditional Western conceptions of knowledge.[8] The construction of an image of reason based on metaphors of vision has led to the notion of a "mind's eye" and a conception of knowledge in which the world is separated from the observer who sees it and thereby gains knowledge of it, without in any way contaminating it or being affected by it.

But these disembodied images of vision are possible only by "forgetting" the fact of our embodiment. What we are capable of seeing and what we attend to are part of our location within the world. Let me begin by using a very different case study than those I've so far employed. Let us think about frogs and dogs.[9]

There are many ways to remember the significance of the situatedness of vision and thereby inhibit the tendency to use visual metaphors to construct allegedly generic images of reason. One of these is to reflect upon the significance of the specificities of human vision. A frog's visual cortex is different from ours. Neural response is linked to small objects in rapid, erratic motion. Objects at rest elicit little neural response and large objects evoke a qualitatively different response than small ones. Although this makes sense for frogs, let us imagine, along with Katherine Hayles, that a frog is presented with Newton's law of motion:

> The first law, you recall, says that an object at rest remains so unless acted upon by a force. Encoded into the formulation is the assumption that the object stays the same; the new element is the force. This presupposition, so obvious from a human point of view, would be almost unthinkable from a frog's perspective, since for the frog moving objects are processed in an entirely different way than stationary ones. Newton's first law further states, as a corollary, that an object moving in a straight line continues to move so unless compelled to change by forces acting upon it. The proposition would certainly not follow as a corollary for the frog, for variation of motion rather than continuation counts in his perceptual scheme. Moreover, it ignores the *size* of the object, which from a frog's point of view is crucial to how information about movement is processed (Hayles, 1993, p. 28).

The point is that bodily differences in perceptual organs and neural patterns organize perception in highly specific, in this case species specific, ways. Far from being the neutral receptor or static mirroring of the visual metaphors informing traditional accounts of knowledge, observation is a dynamic process of organization in which our bodily being plays a central role.

The image of disembodied vision is similarly discounted by imagining a walk with one's dog. Haraway reminds us of the lessons that can be learned from such a walk. "I learned in part walking with my dog and wondering how the world looks without a fovea and very few retinal cells for colour vision, but with a huge neural processing and sensory area for smells....[that] all eyes, including our own organic ones, are active perceptual systems, building in translations and specific *ways* of seeing, that is ways of life" (Haraway, 1991, p. 190).

Although the walk with your dog may remind you of human emphasis on color and shape over a canine attention to smell, it may also remind you, depending on your focus of attention, of the essential and intimate connection of vision with our kinaesthetic sense and our sense of touch. As you walk through a meadow you may meditate upon the way in which your two eyes integrate to produce a unified vision of your dog, and as you reach out to pet her be reminded of the ways in which vision is woven together with motility and touch. Such a walk can impress upon us the realization that the image of vision as disembodied is a circumscribed perspective, and that its emphasis has been the result of complex factors. That is, it is an example of a partial, situated knowledge.

When we consider the human specificities of vision, those mandated by our bodies as well as by the social contexts which shape our experiences of it, we are reminded that the privileging of an image of vision which views it as passive, detached, and disinterested is itself a partial and biased perspective. As any loving parent who looks into the eyes of her or his six month old child or any lover who gazes into the eyes of the person she or he loves know, vision can also be a way in which we actively connect and interact with other people. It can be a way in which we express feelings and negotiate our relationships. Such vision is active, engaged, and reciprocal. An emphasis on vision as passive, detached, and disinterested is a situated vision, one that arises out of particular social situations and values. We are reminded again that vision, as well as objectivity, is not about neutrality, but is embedded in particular and specific embodiments.

A recognition of the epistemic importance of our embodiment requires a conception of knowledge as embodied, in which the emphasis on vision as the primary source of knowledge is replaced by an appreciation of the multiplicity of senses involved in the process of knowledge and an understanding of the ways in which faculties such as empathy, intuition, and reason enter into and interact in this process.

[...]

Conclusion

As a final example I would like to quickly mention the ways in which this study, and others like it, serve as case studies of the very model of knowledge I am here professing. Much of feminist scholarship and practice over the last two decades has been devoted to revaluing the importance of interpersonal relationships. Many feminist political theorists have argued that we must revalue the so-called "private realm" of relationships. Psychoanalytic feminists have called attention to the centrality of interpersonal relationships in the development of our personalities, our genders, and our desires. Feminist ethicists have offered and examined an ethics of care in which moral action is intimately linked to our relationships with others. Add to this the fact that women's prescribed social role of primary caretakers of children, of the elderly, and of the ill, contributes to a heightened sensitivity to the fact and importance of our essential relationality and our embodiment, and it should be no surprise that it is feminist philosophers of science and epistemologists who are vociferously rejecting the Cartesian model of the isolated knowing subject and replacing it with models that emphasize the centrality of our relationships with others to the process of knowing.[10]

I and many other feminists came to positions like these *because* of our participation in feminist communities. This obviously was not the only epistemically significant community; we are also philosophers, historians, sociologists, and scientists. Nor does it mean that *only* feminists will hold such views. There are many theorists of science who do not participate in feminist communities who argue for versions of the above tenets. But a difference of feminist analyses is the persistent attention to gender as a variable of analysis. This is how our feminism is epistemologically significant.

What feminist epistemologists have realized is that it is a mistake to ask for a value-free science. As illustrated in the example of McClintock's "feeling for the organism," the development of knowledge, including scientific knowledge, is affectively influenced. And as the example of primatology illustrates, we cannot treat politics as *inherently* distorting the practice of science. Scientific research, as well as all cognitive endeavors, begins with metaphysical and methodological commitments. It arises out of and is conditioned by our participation in various epistemic communities. Each of us, in being part of a community and a number of subcommunities, participates in an evolving conceptual scheme that makes intersubjective experience possible, influence our interests and desires, and also sets the standards of what constitutes evidence.

The acceptance of the essentially relational nature of knowledge and the inseparability of subjective and objective components of knowledge does not result in relativism, though it does require an abandonment of the traditional "view-from- nowhere" conception of objectivity. This alternative notion of objectivity has been the research program of many feminist philosophers of science ... (see, e.g., Harding, Longino, Nelson). Although I will refer you to their work for the details of feminist accounts of objectivity, let me call attention to yet another way the development of feminist epistemologies are compatible with the model offered. Although there are significant differences between feminist epistemologies, one common tenet is the emphasis on diversity within the scientific community to ensure objectivity. To cite just one of many possible examples, consider Helen Longino's claim that

> ...because background assumptions can be and most frequently are invisible to the members of the scientific community for which they are background and because unreflective acceptance of such assumptions can come to define what it is to be a member of such a community

> (thus making criticism impossible), effective criticism of background assumptions requires the presence and expression of alternative points of view. This sort of account allows us to see how social values and interests can become enshrined in otherwise acceptable research programs (i.e., research programs that strive for empirical adequacy and engage in criticism). As long as representatives of alternative points of view are not included in the community, shared values will not be identified as shaping observation or reasoning (Longino, 1993, pp. 111–12).

Once again we see the impact of the politics of feminism upon the development of feminist epistemology, for a central emphasis of feminism has been the importance of inclusion of previously excluded groups and viewpoints. Earlier feminist accounts focused on the impact of including women and attention to gender upon society, scholarly methods, politics, and so on. The last decade has intensified this commitment as feminists have become aware of the differences between women and have acknowledged the ways in which attention to such factors as class, race, and sexuality, as well as gender, reveals previously hidden assumptions and opens up new research programs.

Feminist philosophers of science have thus actively developed research programs consistent with the values and commitments we express in the rest of our lives. In this sense we are creating "feminist sciences," the doing of science from the politics of feminism. We also acknowledge the need for science to be open to diverse groups of individuals and to have these groups engage in what Longino calls "an interactive dialogic community" (Longino, 1993, p. 113). This is not a simple pluralism, but one in which critical interchange between communities is highly valued. This, of course, does not mean that "anything goes." Although scientific standards are not seen as unchanging or unresponsive to such critical interaction, they do provide standards for acceptability. The "woman, the gatherer" model in human evolution studies arises out of a feminist political agenda yet meets the standards set by the field in which it is proposed. And this is important. Only if these alternative models receive a hearing within the scientific community will they ever secure serious attention.

A value implicit in this vision of science is that the best form of science will be that which is the product of the most inclusive scientific community. This suggests that the problem of developing a new science is the problem of creating a new social and political reality.

Notes

1 A relatively recent correction of contemporary feminist theory in general and feminist philosophy of science in particular is that theories that do not attend to the interactions of various forms of oppression, including class, race, and sexuality, distort the nature of gender oppressions. For an important contribution to this discourse in relation to science studies see Harding, 1993.
2 Harding, 1992. See also Proctor, 1991.
3 Accounts of "woman, the gatherer" theories can be found in Haraway, 1989; Longino, 1990; and Nelson, 1990. My analysis here is thoroughly influenced by Haraway's *Primate Visions*.
4 For a more detailed argument in support of this position see Haraway, 1989.
5 See Keller, 1984 and 1992.
6 See Code, 1991 and 1993; Jaggar, 1989; Keller, 1984; Longino, 1990; Nelson, 1990; and Rose, 1983 and 1994.
7 Although I do not have the time to develop this point, I feel it is too important not to mention and to urge readers to explore the work done on this topic in the writings of feminists such as Rosi Braidotti, Elizabeth Grosz, and Luce Irigaray.
8 See Code, 1991; Keller and Grontkowski, 1983; Jonas, 1966; Leder, 1990; Merleau-Ponty, 1962; and Rorty, 1979.
9 My account here is influenced by Haraway, 1989 and Hayles, 1993, dogs and frogs respectively.
10 I am *not* claiming that feminists are the only theorists developing such a model, but that there is an epistemic link between this model of the subject of knowledge and the politics of feminism.

References

Benhabib, Seyla: 1987, "The Generalized and the Concrete Other: The Kohlberg-Gilligan Controversy and Feminist Theory," in Seyla Benhabib and Drucilla Cornell (eds), *Feminism as Critique: On the Politics of Gender*. University of Minnesota Press, Minneapolis.

Code, Lorraine: 1991, *What Can She Know? Feminist Theory and the Construction of Knowledge*. Cornell University Press, Ithaca.

Code, Lorraine: 1993, "Taking Subjectivity into Account," in Linda Alcoff and Elizabeth Potter (eds), *Feminist Epistemologies*. Routledge, New York, pp. 15–48.

Haraway, Donna: 1989, *Primate Visions: Gender, Race and Nature in the World of Modern Science*. Routledge, New York.

Haraway, Donna: 1991, "Situated Knowledges: The Science Question in Feminism and the Privilege of Partial Perspective," in *Simians, Cyborgs, and Women: The Reinvention of Nature*. Routledge, New York.

Harding, Sandra: 1991, *Whose Science? Whose Knowledge?* Cornell University Press, Ithaca.

Harding, Sandra: 1992, "After the Neutrality Ideal: Science, Politics, and 'Strong Objectivity'," *Social Research*, 59, 567–87.

Harding, Sandra: 1993, *The "Racial" Economy of Science: Toward a Democratic Future*. Indiana University Press, Bloomington.

Hayles, N. Katherine: 1993, "Constrained Constructivism: Locating Scientific Inquiry in the Theater of Representation," in George Levine (ed.), *Realism and Representation*. University of Wisconsin Press, Madison.

Jaggar, Alison: 1989, "Love and Knowledge: Emotion in Feminist Epistemology," in Susan Bordo and Alison Jaggar (eds), *Gender/Body/Knowledge*. Rutgers University Press, New Brunswick.

Jonas, Hans: 1966, *The Phenomenon of Life*. Chicago University Press, Chicago.

Keefer, Chester: 1994, "Skillful Listening," *Cortland Forum*, June, 74.

Keller, Evelyn Fox: 1983, *A Feeling for the Organism*. W. H. Freeman and Company, New York.

Keller, Evelyn Fox: 1984, *Reflections on Gender and Science*. Yale University Press, New Haven.

Keller, Evelyn Fox: 1992, *Secrets of Life, Secrets of Death: Essays on Language, Gender and Science*. Routledge, New York.

Keller, Evelyn Fox and Christine R. Grontkowski: 1983, "The Mind's Eye," in Sandra Harding and Merrill B. Hintikka (eds), *Discovering Reality*. D. Reidel, Dordrecht, pp. 207–224.

Leder, Drew: 1990, *The Absent Body*. University of Chicago Press, Chicago.

Longino, Helen: 1990, *Science as Social Knowledge*. Princeton University Press, Princeton.

Longino, Helen: 1993, "Subjects, Power, and Knowledge: Description and Prescription in Feminist Philosophies of Science," in Linda Alcoff and Elizabeth Potter (eds), *Feminist Epistemologies*. Routledge, New York, pp. 101–120.

Merleau-Ponty, Maurice: 1962, *Phenomenology of Perception*, trans. Colin Smith. Routledge and Kegan Paul, London.

Nelson, Lynn Hankinson: 1990, *Who Knows: From Quine to a Feminist Empiricism*. Temple University Press, Philadelphia.

Proctor, Robert N.: 1991, *Value-Free Science? Purity and Power in Modern Knowledge*. Harvard University Press, Cambridge.

Rose, Hilary: 1983, "Hand, Brain and Heart: A Feminist Epistemology for the Natural Sciences," *Signs*, **9**, 73–90.

Rose, Hilary: 1994, *Love, Power, and Knowledge: Towards a Feminist Transformation of the Sciences*. Indiana University Press, Bloomington.

Rorty, Richard: 1979, *Philosophy and the Mirror of Nature*. Princeton University Press, Princeton.

Slocum, Sally Linton: 1971, "Woman the Gatherer: Male Bias in Anthropology," in Sue-Ellen Jacobs (ed.), *Women in Perspective: A Guide for Cross Cultural Studies*. University of Illinois Press, Urbana.

Washburn, Sherwood L. and C. S. Lancaster: 1976, "Evolution of Hunting," in R. B. Lee and I. De Vore (eds), *Kalahari Hunter–Gatherers*. Harvard University Press, Cambridge.

12

Do You Believe in Reality? News from the Trenches of the Science Wars

Bruno Latour

"I have a question for you," he said, taking out of his pocket a crumpled piece of paper on which he had scribbled a few key words. He took a breath: "Do you believe in reality?"

"But of course!" I laughed. "What a question! Is reality something we have to believe in?"

He had asked me to meet him for a private discussion in a place I found as bizarre as the question: by the lake near the chalet, in this strange imitation of a Swiss resort located in the tropical mountains of Teresopolis in Brazil. Has reality truly become something people have to believe in, I wondered, the answer to a serious question asked in a hushed and embarrassed tone? Is reality something like God, the topic of a confession reached after a long and intimate discussion? Are there people on earth who *don't* believe in reality?

When I noticed that he was relieved by my quick and laughing answer, I was even more baffled, since his relief proved clearly enough that he had anticipated a *negative* reply, something like "Of course not! Do you think I am that naive?" This was not a joke, then: he really was concerned, and his query had been in earnest.

"I have two more questions," he added, sounding more relaxed. "Do we know more than we used to?"

"But of course! A thousand times more!"

From Bruno Latour, *Pandora's Hope: Essays on the Reality of Science Studies*, pp. 1–23, Cambridge, MA: Harvard University Press, 1999.

"But is science cumulative?" he continued with some anxiety, as if he did not want to be won over too fast.

"I guess so," I replied, "although I am less positive on this one, since the sciences also forget so much, so much of their past and so much of their bygone research programs – but, on the whole, let's say yes. Why are you asking me these questions? Who do you think I am?"

I had to switch interpretations fast enough to comprehend both the monster he was seeing me as when he raised these questions and his touching openness of mind in daring to address such a monster privately. It must have taken courage for him to meet with one of these creatures that threatened, in his view, the whole establishment of science, one of these people from a mysterious field called "science studies," of which he had never before met a flesh-and-blood representative but which – at least so he had been told – was another threat to science in a country, America, where scientific inquiry had never had a completely secure foothold.

He was a highly respected psychologist, and we had both been invited by the Wenner-Grenn Foundation to a gathering made up of two-thirds scientists and one-third "science students." This division itself, announced by the organizers, baffled me. How could we be pitted *against* the scientists? That we are studying a subject matter does not mean that we are attacking it. Are biologists anti-life, astronomers anti-stars, immunologists anti-antibodies? Besides, I had taught for twenty years in scientific schools, I wrote regularly in scientific journals, I and my colleagues lived on contract research carried out on behalf of many groups of scientists in industry and in the academy. Was I not part of the French scientific establishment? I was a bit vexed to be excluded so casually. Of course I am just a philosopher, but what would my friends in science studies say? Most of them have been trained in the sciences, and several of them, at least, pride themselves on *extending* the scientific outlook to science itself. They could be labeled as members of another discipline or another subfield, but certainly not as "anti-scientists" meeting halfway with scientists, as if the two groups were opposing armies conferring under a flag of truce before returning to the battlefield!

I could not get over the strangeness of the question posed by this man I considered a colleague, yes, a colleague (and who has since become a good friend). If science studies has achieved anything, I thought, surely it has *added* reality to science, not withdrawn any from it. Instead of the stuffed scientists hanging on the walls of the armchair philosophers of science of the past, we have portrayed lively characters, immersed in their laboratories, full of passion, loaded with instruments, steeped in know-how, closely connected to a larger and more vibrant milieu. Instead of the pale and bloodless objectivity of science, we have all shown, it seemed to me, that the many nonhumans mixed into our collective life through laboratory practice have a history, flexibility, culture, blood – in short, all the characteristics that were denied to them by the humanists on the other side of the campus. Indeed, I naively thought, if scientists have a faithful ally, it is we, the "science students" who have managed over the years to interest scores of literary folk in science and technology, readers who were convinced, until science studies came along, that "science does not think" as Heidegger, one of their masters, had said.

The psychologist's suspicion struck me as deeply unfair, since he did not seem to understand that in this guerrilla warfare being conducted in the no-man's-land between the "two cultures," *we were the ones* being attacked by militants, activists, sociologists, philosophers, and technophobes of all hues, precisely because of our interest in the inner workings of scientific facts. Who loves the sciences, I asked myself, more than this tiny scientific tribe that has learned to open up facts, machines, and theories with all their roots, blood vessels, networks, rhizomes, and tendrils? Who believes more in the objectivity of science than those who claim that it can be turned into an object of inquiry?

Then I realized that I was wrong. What I would call "adding realism to science" was actually seen, by the scientists at this gathering, as a threat to the calling of science, as a way of decreasing its stake in truth and their claims to certainty. How has this misunderstanding come about? How could I have lived long enough to be asked in all seriousness this incredible question: "Do you believe in reality?" The distance between what I thought we had achieved in science studies and what was implied by this question was so vast that I needed to retrace my steps a bit. And so this book was born.

The Strange Invention of an "Outside" World

There is no natural situation on earth in which someone could be asked this strangest of all questions: "Do you believe in reality?" To ask such a question one has to become so *distant* from reality that the fear of *losing* it entirely becomes plausible – and this fear itself has an intellectual history that should at least be sketched. Without this detour we would never be able to fathom the extent of the misunderstanding between my colleague and me, or to measure the extraordinary form of radical realism that science studies has been uncovering.

I remembered that my colleague's question was not so new. My compatriot Descartes had raised it against himself when asking how an isolated mind could be *absolutely* as opposed to relatively sure of anything about the outside world. Of course, he framed his question in a way that made it impossible to give the only reasonable answer, which we in science studies have slowly rediscovered three centuries later: that we are *relatively* sure of many of the things with which we are daily engaged through the practice of our laboratories. By Descartes's time this sturdy relativism, based on the number of *relations* established with the world, was already in the past, a once-passable path now lost in a thicket of brambles. Descartes was asking for absolute certainty from a brain-in-a-vat, a certainty that was not needed when the brain (or the mind) was firmly attached to its body and the body thoroughly involved n its normal ecology. As in Curt Siodmak's novel *Donovan's Brain*, absolute certainty is the sort of neurotic fantasy that only a surgically removed mind would look for after it had lost everything else. Like a heart taken out of a young woman who has just died in an accident and soon to be transplanted into someone else's thorax thousands of miles away, Descartes's mind requires artificial life-support to keep it viable. Only a mind put in the strangest position, looking at a world *from the inside out* and linked to the outside by nothing but the tenuous connection of the *gaze*, will throb in the constant fear of losing reality; only such a bodiless observer will desperately look for some absolute life-supporting survival kit.

For Descartes the only route by which his mind-in-a-vat could reestablish some reasonably sure connection with the outside world was through God. My friend the psychologist was thus right to phrase his query using the same formula I had learned in Sunday school: "Do you believe in reality?" – "Credo in unum Deum," or rather, "Credo in unam realitam," as my friend Donna Haraway kept chanting in Teresopolis! After Descartes, however, many people thought that going through God to reach the world was a bit expensive and far-fetched. They looked for a shortcut. They wondered whether the world could *directly* send us enough information to produce a stable image of itself in our minds.

But in asking this question the empiricists kept going along the same path. They did not retrace their steps. They never plugged the wriggling and squiggling brain back into its withering body. They were still dealing with a mind looking through the gaze at a lost outside world. They simply tried to train it to recognize patterns. God was out, to be sure, but the *tabula rasa* of the empiricists was as disconnected as the mind in Descartes's times. The brain-in-a-vat simply exchanged one survival kit for another. Bombarded by a world reduced to meaningless stimuli, it was supposed to extract from these stimuli everything it needed to recompose the world's shapes and stories. The result was like a badly connected TV set, and no amount of tuning made this precursor of neural nets produce more than a fuzzy set of blurry lines, with white points falling like snow. No shape was recognizable. Absolute certainty was lost, so precarious were the connections of the senses to a world that was pushed ever further outside. There was too much static to get any clear picture.

The solution came, but in the form of a catastrophe from which we are only now beginning to extricate ourselves. Instead of retracing their steps and taking the other path at the forgotten fork in the road, philosophers abandoned even the claim to absolute certainty, and settled instead on a makeshift solution that preserved at least some access to an outside reality. Since the empiricists' associative neural net was unable to offer clear pictures of the lost world, this must prove, they said, that the mind (still in a vat) extracts *from itself* everything it needs to form shapes and stories. Everything, that is, except the reality itself. Instead of the fuzzy lines on the poorly tuned TV set, we got the fixed tuning grid, molding the confused static, dots, and lines of the empiricist channel into a steady picture held in place by the mindset's predesigned categories. Kant's *a priori* started this extravagant form of constructivism, which neither Descartes, with his detour through God, nor Hume, with his shortcut to associated stimuli, would ever have dreamed of.

Now, with the Konigsberg broadcast, everything was ruled by the mind itself and reality came in simply to say that it was there, indeed, and not imaginary! For the banquet of reality, the mind provided the food, and the inaccessible things-in-themselves to which the world had been reduced simply dropped by to say "We are here, what you eat is not dust," but otherwise remained mute and stoic guests. If we abandon absolute certainty, Kant said, we can at least retrieve universality as long as we remain inside the restricted sphere of science, to which the world outside contributes decisively but minimally. The rest of the quest for the absolute is to be found in morality, another *a priori* certainty that the mind-in-the-vat extracts from its own wiring. Under the name of a "Copernican Revolution" Kant invented this science-fiction nightmare: the outside world now turns around the mind-in-the-vat, which dictates most of that world's laws, laws it has extracted from itself without help from anyone else. A crippled despot now ruled the world of reality. This philosophy was thought, strangely enough, to be the deepest of all, because it had at once managed to abandon the quest for absolute certainty and to retain it under the banner of "universal *a priori*s," a clever sleight of hand that hid the lost path even deeper in the thickets.

Do we really have to swallow these unsavory pellets of textbook philosophy to understand the psychologist's question? I am afraid so, because otherwise the innovations of science studies will remain invisible. The worst is yet to come. Kant had invented a form of constructivism in which the mind-in-the-vat built everything by itself but not entirely without constraints: what it learned from itself had to be universal and could be elicited only by some experiential contact with a reality out there, a reality reduced to its barest minimum, but there nonetheless. For Kant there was still something that revolved around the crippled despot, a green planet around this pathetic sun. It would not be long before people realized that this "transcendental Ego," as Kant named it, was a fiction, a line in the sand, a negotiating position in a complicated settlement to avoid the complete loss of the world or the complete abandonment of the quest for absolute certainty. It was soon replaced by a more reasonable candidate, *society*. Instead of a mythical Mind giving shape to reality, carving it, cutting it, ordering it, it was now the prejudices, categories, and paradigms of a group of people living together that determined the representations of every one of those people. This new definition, however, in spite of the use of the word "social," had only a superficial resemblance to the realism to which we science students have become attached, and which I will outline over the course of this book.

First, this replacement of the despotic Ego with the sacred "society" did not retrace the philosophers' steps but went even *further* in distancing the individual's vision, now a "view of the world," from the definitely lost outside world. Between the two, society interposed its filters; its paraphernalia of biases, theories, cultures, traditions, and standpoints became an opaque window. Nothing of the world could pass through so many intermediaries and reach the individual mind. People were now locked not only into the prison of their own categories but into that of their social groups as well. Second, this "society" itself was just a series of minds-in-a-vat, many minds and many vats to be sure, but each of them still composed of that strangest of beasts: a detached mind gazing at an outside world. Some improvement! If prisoners were no longer in isolated cells, they were now confined to the same dormitory, the same collective mentality. Third, the next shift, from one Ego to multiple cultures, jeopardized the only good thing about Kant, that is, the universality of the *a priori* categories, the only bit of ersatz absolute certainty he had been able to retain. Everyone was not locked in the same prison any more; now there were *many* prisons, incommensurable, unconnected. Not only was the mind disconnected from the world, but each collective mind, each culture was disconnected from the others. More and more progress in a philosophy dreamed up, it seems, by prison wardens.

But there was a fourth reason, even more dramatic, even sadder, that made this shift to "society" a catastrophe following fast on the heels of the Kantian revolution. The claims to knowledge of all these poor minds, prisoners in their long rows of vats, were now made part of an even more bizarre history, were now associated with an even more ancient threat, *the fear of mob rule*. If my friend's voice quivered as he asked me "Do you believe in reality?" it was not only because he feared that all connection with the outside world might be lost, but above all because he worried that I might answer, "Reality depends on whatever the mob thinks is right at any given time." It is the resonance of these two fears, the *loss* of any certain access to reality and the *invasion* by the mob, that makes his question at once so unfair and so serious.

But before we disentangle this second threat, let me finish with the first one. The sad story, unfortunately, does not end here. However incredible it seems, it is possible to go even further along the wrong path, always thinking that a more radical solution will solve the problems accumulated from the past decision. One solution, or more exactly another clever sleight of hand, is to become so very pleased with the loss of absolute certainty and universal *a prioris* that one rejoices in abandoning them. Every defect of the former position is now taken to be its best quality. Yes, we have lost the world. Yes, we are forever prisoners of language. No, we will never regain certainty. No, we will never get beyond our biases. Yes, we will forever be stuck within our own selfish standpoint. Bravo! Encore! The prisoners are now gagging even those who ask them to look out their cell windows; they will "deconstruct," as they say – which means destroy in slow motion – anyone who reminds them that there was a time when they were free and when their language bore a connection with the world.

Who can avoid hearing the cry of despair that echoes deep down, carefully repressed, meticulously denied, in these paradoxical claims for a joyous, jubilant, free construction of narratives and stories by people forever in chains? But even if there *were* people who could say such things with a blissful and light heart (their existence is as uncertain to me as that of the Loch Ness monster, or, for that matter, as uncertain as that of the real world would be to these mythical creatures), how could we avoid noticing that we have not moved an inch since Descartes? That the mind is still in its vat, excised from the rest, disconnected, and contemplating (now with a blind gaze) the world (now lost in darkness) from the very same bubbling glassware? Such people may be able to smile smugly instead of trembling with fear, but they are still descending further and further along the spiraling curves of the same hell. At the end of this chapter we will meet these gloating prisoners again.

In our century, though, a second solution has been proposed, one that has occupied many bright minds. This solution consists of taking only a *part* of the mind out of the vat and then doing the obvious thing, that is, offering it a body again and putting the reassembled aggregate back into relation with a world that is no longer a spectacle at which we gaze but a lived, self-evident, and unreflexive extension of ourselves. In appearance, the progress is immense, and the descent into damnation suspended, since we no longer have a mind dealing with an outside world, but a lived world to which a semi-conscious and intentional body is now attached.

Unfortunately, however, in order to succeed, this emergency operation must chop the mind into even smaller pieces. The real world, the one known by science, is left entirely to itself. Phenomenology deals only with the world-for-a-human-consciousness. It will teach us a lot about how we never distance ourselves from what we see, how we never gaze at a distant spectacle, how we are always immersed in the world's rich and lived texture, but, alas, this knowledge will be of no use in accounting for how things really are, since we will never be able to escape from the narrow focus of human intentionality. Instead of exploring the ways we can shift from standpoint to standpoint, we will always be fixed in the human one. We will hear much talk about the real, fleshy, pre-reflexive lived world, but this will not be enough to cover the noise of the second ring of prison doors slamming even more tightly shut behind us. For all its claims to overcoming the distance between subject and object – as if this distinction were something that could be overcome! as if it had not been devised so as *not* to be overcome! – phenomenology leaves us with the most dramatic split in this whole sad story: a world of science left entirely to itself, entirely cold, absolutely inhuman; and a rich lived world of intentional stances entirely limited to humans, absolutely divorced from what things are in and for themselves. A slight pause on the way down before sliding even further in the same direction.

Why not choose the opposite solution and forget the mind-in-a-vat altogether? Why not let the "outside world" invade the scene, break the glassware, spill the bubbling liquid, and turn the mind into a brain, into a neuronal machine sitting inside a Darwinian animal struggling for its life? Would that not solve all the problems and reverse the fatal downward spiral? Instead of the complex "life-world" of the phenomenologists, why not study the adaptation of humans, as naturalists have studied all other aspects of "life"? If science can invade everything, it surely can put an end to Descartes's long-lasting fallacy and make the mind a wriggling and squiggling part of nature. This would certainly please my friend the psychologist – or would it? No, because the ingredients that make up this "nature," this hegemonic and all-encompassing nature, which would now include the human species, are the *very same ones* that have constituted the spectacle of a world viewed from inside by a brain-in-a-vat. Inhuman, reductionist, causal, law-like, certain, objective, cold, unanimous, absolute – all these expressions do not pertain to nature *as such*, but to nature viewed through the deforming prism of the glass vessel!

If there is something unattainable, it is the dream of treating nature as a homogeneous unity in order to unify the different views the sciences have of it! This would require us to ignore too many controversies, too much history, too much unfinished business, too many loose ends. If phenomenology abandoned science to its destiny by limiting it to human intention, the opposite move, studying humans as "natural phenomena," would be even worse: it would abandon the rich and controversial human history of science – and for what? The averaged-out orthodoxy of a few neurophilosophers? A blind Darwinian process that would limit the mind's activity to a struggle for survival to "fit" with a reality whose true nature would escape us forever? No, no, we can surely do better, we can surely stop the downward slide and retrace our steps, retaining both the history of humans' involvement in the making of scientific facts and the sciences' involvement in the making of human history.

Unfortunately, we can't do this, not yet. We are prevented from returning to the lost crossroads and taking the other path by the dangerous bogeyman I mentioned earlier. It is the threat of mob rule that stops us, the same threat that made my friend's voice quake and quiver.

The Fear of Mob Rule

As I said, two fears lay behind my friend's strange question. The first one, the fear of a mind-in-a-vat losing its connection to a world outside, has a shorter history than the second, which stems from this truism: if reason does not rule, then mere force will take over. So great is this threat that any and every political expedient is used with impunity against those who are deemed to advocate force against reason. But where does this striking opposition between the camp of reason and the camp of force come from? It comes from an old and venerable debate, one that probably occurs in many places but that is staged most clearly and influentially in Plato's *Gorgias*. In this dialog,...Socrates, the true scientist, confronts Callicles, another of those monsters who must be interviewed in order to expose their nonsense, this time not on the shores of a Brazilian lake but in the agora in Athens. He tells Callicles: "You've failed to notice *how much power geometrical equality has among gods and men*, and this neglect of geometry has led you to believe that one should try to gain a *disproportionate* share of things" (508a).[1]

Callicles is an expert at disproportion, no doubt about that. "I think," he boasts in a preview of Social Darwinism, "we only have to look at nature to find evidence that it is right for better to have a greater share than worse...The superior person shall dominate the inferior person and have more than him" (483c–d). Might makes Right, Callicles frankly admits. But, as we shall see at the end of this book, there is a little snag. As both of the two protagonists are quick to point out, there are at least two sorts of Mights to consider: that of Callicles and that of the Athenian mob. "What else do you think I've been saying?" Callicles asks. "Law consists of the statements made by an assembly of slaves and assorted other forms of human debris who could be completely *discounted if it weren't for the fact they do have physical strength at their disposal*" (489c). So the question is not simply the opposition of force and reason, Might and Right, but the Might of the solitary patrician against the superior force of the crowd. How can the combined forces of the people of Athens be nullified? "Here's your position, then," Socrates ironizes: "a single clever person is almost bound to be *superior to ten thousand fools;* political power should be his and they should be his subjects; and it is appropriate for someone with political power to have more than his subjects" (490a). When Callicles speaks of brute force, what he means is an inherited moral force superior to that of ten thousand brutes.

But is it fair for Socrates to practice irony on Callicles? What sort of disproportion is Socrates himself setting in motion? What sort of power is he trying to wield? The Might that Socrates sides with is the *power of reason*, "the power of geometrical equality," the force which "rules over gods and men," which he knows, which Callicles and the mob ignore. As we shall see, there is a second little snag here, because there are two forces of reason, one directed against Callicles, the ideal foil, and the other directed sideways, aimed at reversing the balance of power between Socrates and all the other Athenians. Socrates is also looking for a force able to nullify that of "ten thousand fools." He too tries to get the biggest share. His success at reversing the balance of forces is so extraordinary that he boasts, at the end of the *Gorgias*, of being "the only real statesman of Athens," the only winner of the biggest share of all, an eternity of glory that will be awarded to him by Rhadamantes, Aeacus, and Minos, who preside over the tribunal of hell! He ridicules all the famous Athenian politicians, Pericles included, and he alone, equipped with "the power of geometrical equality," will rule over the citizens of the city even beyond death. One of the first of many in the long literary history of mad scientists.

"As if your slapdash history of modern philosophy is not enough," the reader may complain, "do you also have to drag us all the way back to the Greeks just to account for the question asked by your psychologist in Brazil?" I am afraid both of these detours were necessary, because only now can the two threads, the two threats, be tied together to explain my friend's worries. Only after these digressions can my position, I hope, be clarified at last.

Why, in the first place, did we even need the idea of an *outside world* looked at through a gaze from the very uncomfortable observation post of a mind-in-a-vat? This has puzzled me ever since I started in the field of science studies almost twenty-five years ago. How could it be so important to maintain this awkward position, in spite of all the cramps it gave philosophers, instead of doing the obvious: retracing our steps, pruning back the brambles hiding the lost fork in the road, and firmly walking on the other, forgotten path? And why burden this solitary mind with the impossible task of finding absolute certainty instead of plugging it into the connections that would provide it with all the relative certainties it needed to know and to act? Why shout out of both sides of our mouths these two contradictory orders: "Be absolutely disconnected!" "Find absolute proof that you are connected!" Who could untangle such an impossible double bind? No wonder so many philosophers wound up in asylums. In order to justify such a self-inflicted, maniacal torture, we would have to be pursuing a loftier goal, and such indeed has been the case. This is the place where the two threads connect: it is in order to avoid the inhuman crowd that we need to rely on another inhuman resource, the objective object untouched by human hands.

To avoid the threat of a mob rule that would make everything lowly, monstrous, and inhuman, we have to depend on something that has no human origin, no trace of humanity, something that is purely, blindly, and coldly outside of the City. The idea of a completely *outside* world dreamed up by epistemologists is the only way, in the eyes of moralists, to avoid falling prey to mob rule. *Only inhumanity will quash inhumanity*. But how is it possible to imagine an outside world? Has anyone ever seen such a bizarre oddity? No problem. We will make the world into a spectacle seen *from* the inside.

To obtain such a contrast, we will imagine that there is a mind-in-a-vat that is totally disconnected from the world and accesses it only through one narrow, artificial conduit. This minimal link, psychologists are confident, will be enough to keep the world outside, to keep the mind informed, provided we later manage to rig up some absolute means of getting certainty back – no mean feat, as it turns out. But this way we will achieve our overarching agenda: *to keep the crowds at bay*. It is because we want to fend off the irascible mob that we need a world that is totally outside – while remaining accessible! – and it is in order to reach this impossible goal that we came up with the extraordinary invention of a mind-in-a-vat disconnected from everything else, striving for absolute truth, and, alas, failing to get it. As we can see in Figure 1, *epistemology, morality, politics, and psychology go hand in hand and are aiming at the same settlement*.

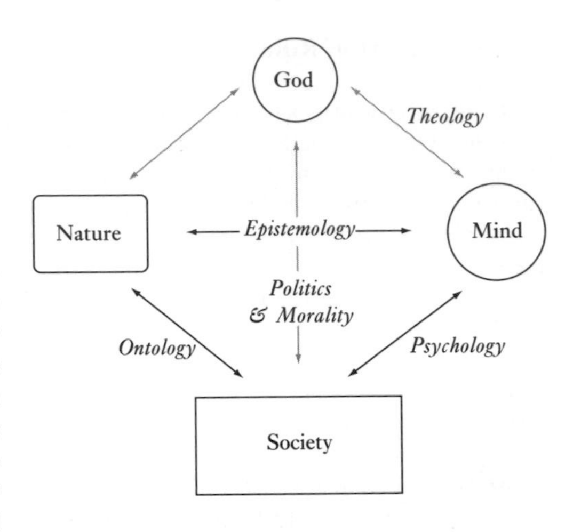

Figure 1. The modernist settlement. For science studies there is no sense in talking independently of epistemology, ontology, psychology, and politics – not to mention theology. In short: "out there," "nature"; "in there," the mind; "down there," the social; "up there," God. We do not claim that these spheres are cut off from one another, but rather that they all pertain to the same settlement, a settlement that can be replaced by several alternative ones.

This is...the reason the reality of science studies is so difficult to locate. Behind the cold epistemological question – can our representations capture with some certainty stable features of the world out there? – the second, more burning anxiety is always lurking: can we find a way to fend off the people? Conversely, behind any definition of the "social" is the same worry: will we still be able to use objective reality to shut the mob's too many mouths?

My friend's question, on the shore of the lake, shaded by the chalet's roof from the tropical noontime sun in this austral winter, becomes clear at last: "Do you believe in reality?" means "Are you willing to accept this settlement of epistemology, morality, politics, and psychology?" – to which the quick and laughing answer is, obviously: "*No!* Of course not! Who do you think I am? How could I believe reality to be the answer to a question of belief asked by a brain-in-a-vat terrified of losing contact with an outside world because it is even more terrified of being invaded by a social world stigmatized as inhuman?" Reality is an object of belief only for those who have started down this impossible cascade of settlements, always tumbling into a worse and more radical solution. Let them clean up their own mess and accept the responsibility for their own sins. My trajectory has always been different. "Let the dead bury the dead," and, please, listen for one minute to what we have to say on our own account, instead of trying to shut us up by putting in our mouths the words that Plato, all those centuries ago, placed in the mouths of Socrates and Callicles to keep the people silent.

Science studies, as I see it, has made two related discoveries that were very slow in coming because of the power of the settlement that I have now exposed – as well as for a few other reasons I will explain later. This joint discovery is that *neither the object nor the social* has the *inhuman* character that Socrates' and Callicles' melodramatic show required. When we say there is no outside world, this does not mean that we deny its existence, but, on the contrary, that we refuse to grant it the ahistorical, isolated, inhuman, cold, objective existence that it was given *only* to combat the crowd. When we say that science is social, the word social for us does not bear the stigma of the "human debris," of the "unruly mob" that Socrates and Callicles were so quick to invoke in order to justify the search for a force strong enough to reverse the power of "ten thousand fools."

Neither of these two monstrous forms of inhumanity – the mob "down there," the objective world "out there" – interests us very much. And thus we have no need for a mind- or brain-in-a-vat, that crippled despot constantly fearful of losing either "access" to the world or its "superior force" against the people. We long neither for the absolute certainty of a contact with the world nor for the absolute certainty of a transcendent force against the unruly mob. We do not *lack* certainty, because we never dreamed of *dominating* the people. For us there is no inhumanity to be quashed with another inhumanity. Humans and nonhumans are enough for us. We do not need a social world to break the back of objective reality, nor an objective reality to silence the mob. It is quite simple, even though it may sound incredible in these times of the science wars: we are *not* at war.

As soon as we refuse to engage the scientific disciplines in this dispute about who should hold sway over the people, the lost crossroads is rediscovered, and there is no major difficulty in treading along the neglected path. Realism now returns in force, as will be made obvious, I hope, in later chapters, which should look like milestones along the route to a more "realistic realism." My argument in this book recapitulates the halting "two steps forward, one step back" advance of science studies along this long-forgotten pathway.

We started when we first began to talk about scientific *practice* and thus offered a more realistic account of science-in-the-making, grounding it firmly in laboratory sites, experiments, and groups of colleagues.... Facts, we found, were clearly fabricated. Then realism gushed forth again when, instead of talking about objects and objectivity, we began to speak of *nonhumans* that were socialized through the laboratory and with which scientists and engineers began to swap properties.... Pasteur makes his microbes while the microbes "make their Pasteur"; [more generally,] humans and nonhumans fold[...] into each other, forming constantly changing collectives. Whereas objects had been made cold, asocial, and distant for political reasons, we found that nonhumans were close, hot, and easier to enroll and to enlist, adding more and more reality to the many struggles in which scientists and engineers had engaged.

But realism became even more abundant when nonhumans began to have a *history*, too, and were allowed the multiplicity of interpretations, the flexibility, the complexity that had been reserved, until then, for humans....Through a series of counter-Copernican revolutions, Kant's nightmarish fantasy slowly lost its pervasive dominance over the philosophy of science. There was again a clear sense in which we could say that words have *reference* to the world and that science grasps the things themselves.... Naïveté was back at last, a naïveté appropriate for those who had never understood how the world could be "outside" in the first place. We have yet to provide a real alternative to that fateful distinction between construction and reality; I attempt to provide one here with the notion of "factish." [The word] is a combination of the words "fact" and "fetish," in which the work of fabrication has been twice added, canceling the twin effects of belief and knowledge.

Instead of the three poles – a reality "out there," a mind "in there," and a mob "down there" – we have finally arrived at a sense of what I call a *collective*. As the...*Gorgias*...demonstrates, Socrates has defined this collective very well before switching to his bellicose collusion with Callicles: "The expert's opinion is that co-operation, love, order, discipline, and justice *bind* heaven and earth, gods and men. That's why they call the universe an *ordered whole*, my friend, rather than a disorderly mess or an *unruly shambles*" (507e–508a).

Yes, we live in a hybrid world made up at once of gods, people, stars, electrons, nuclear plants, and markets, and it is our duty to turn it into either an "unruly shambles" or an "ordered whole," a *cosmos* as the Greek text puts it, undertaking what Isabelle Stengers gives the beautiful name of cosmopolitics (Stengers 1996). Once there is no longer a mind-in-a-vat looking through the gaze at an outside world, the search for absolute certainty becomes less urgent, and thus there is no great difficulty in reconnecting with the relativism, the relations, the relativity on which the sciences have always thrived. Once the social realm no longer bears these stigmata branded upon it by those who want to silence the mob, there is no great difficulty in recognizing the human character of scientific practice, its lively history, its many connections with the rest of the collective. Realism comes back like blood through the many vessels now reattached by the clever hands of the surgeons – there is no longer any need for a survival kit. After following this route, no one would even think of asking the bizarre question "Do you believe in reality?" – at least not of asking *us!*

The Originality of Science Studies

Nevertheless, my friend the psychologist would still be entitled to pose another, more serious query: "Why is it that, in spite of what you claim your field has achieved, I was *tempted* to ask you my silly question *as if* it were a worthwhile one? Why is it that in spite of all these philosophies you zigzagged me through, I still doubt the radical realism you advocate? I can't avoid the nasty feeling that there is a science war going on. In the end, are you a friend of science or its enemy?"

Three different phenomena explain, to me at least, why the novelty of "science studies" cannot be registered so easily. The first is that we are situated, as I said, in the no-man's-land between the two cultures, much like the fields between the Siegfried and Maginot lines in which French and German soldiers grew cabbages and turnips during the "phony war" in 1940. Scientists always stomp around meetings talking about "bridging the two-culture gap," but when scores of people from outside the sciences begin to build just that bridge they recoil in horror and want to impose the strangest of all gags on free speech since Socrates: only scientists should speak about science!

Just imagine if that slogan were generalized: only politicians should speak about politics, businessmen about business; or even worse: only rats will speak about rats, frogs about frogs, electrons about electrons! Speech implies by definition the risk of misunderstanding across the huge gaps between different species. If scientists want to bridge the two-culture divide for good, they will have to get used to a lot of noise and, yes, more than a little bit of nonsense. After all, the humanists and the literati do not make such a fuss about the many absurdities uttered by the team of scientists building the bridge from the other end. More seriously, bridging the gap cannot mean extending the unquestionable *results* of science in order to stop the "human debris" from behaving irrationally. Such an attempt can at best be called pedagogy, at worst propaganda. This cannot pass for the cosmopolitics that would require the collective to socialize into its midst the humans, the nonhumans, and the gods together. Bridging the two-culture gap cannot mean lending a helping hand to Socrates' and Plato's dreams of utter control.

But where does the two-culture debate itself originate? In a division of labor between the two sides of the campus. One camp deems the sciences accurate only when they have been purged of any contamination by subjectivity, politics, or passion; the other camp, spread out much more widely, deems humanity, morality, subjectivity, or rights worthwhile only when they have been protected from any contact with science, technology, and objectivity. We in science studies fight against these two purges, against both purifications at once, and this is what makes us traitors to both camps. We tell the scientists that *the more connected a science* is to the rest of the collective, *the better* it is, the more accurate, the more verifiable, the more solid... – and this runs against all the conditioned reflexes of epistemologists. When we tell them that the social world is good for science's health, they hear us as saying that Callicles' mobs are coming to ransack their laboratories.

But, against the other camp, we tell the humanists that *the more nonhumans share existence with humans, the more humane* a collective is – and this too runs against what they have been trained for years to believe. When we try to focus their attention on solid facts and hard mechanisms, when we say that objects are good for the subjects' health because objects have none of the inhuman characteristics they fear so much, they scream that the iron hand of objectivity is turning frail and pliable souls into reified machines. But we keep defecting and counter-defecting from both sides, and we insist and insist again that there is a social history of things and a "thingy" history of humans, but that neither "the social" nor "the objective world" plays the role assigned to it by Socrates and Callicles in their grotesque melodrama.

If anything, and here we can be rightly accused of a slight lack of symmetry, "science students" fight the humanists who are trying to invent a human world purged of nonhumans *much more* than we combat the epistemologists who are trying to purify the sciences of any contamination by the social. Why? Because scientists spend only a fraction of their time purifying their sciences and, frankly, do not give a damn about the philosophers of science coming to their rescue, while the humanists spend all their time on and take very seriously the task of freeing the human subjects from the dangers of objectification and reification. Good scientists enlist in the science wars only in their spare time or when they are retired or have run out of grant money, but the others are up in arms day and night and even get granting agencies to join in their battle. This is what makes us so angry about the suspicion of our scientist colleagues. They don't seem to be able to differentiate friends from foes anymore. Some are pursuing the vain dream of an autonomous and isolated science, Socrates' way, while we are pointing out the very means they need to reconnect the facts to the realities without which the existence of the sciences cannot be sustained. Who first offered us this treasure trove of knowledge? The scientists themselves!

I find this blindness all the more bizarre because, in the last twenty years, many scientific disciplines have joined us, crowding into the tiny no-man's-land between the two lines. This is the second reason "science studies" is so contentious. By mistake, it is caught in the middle of another dispute, this one *within* the sciences themselves. On one side there are what could be called the "cold war disciplines," which still look superficially like the Science of the past, autonomous and detached from the collective; on the other side there are strange imbroglios of politics, science, technology, markets, values, ethics, facts, which cannot easily be captured by the word Science with a capital S.

If there is some plausibility in the assertion that cosmology does not have the slightest connection with society – although even that is wrong, as Plato reminds us so tellingly – it is hard to say the same of neuropsychology, sociobiology, primatology, computer sciences, marketing, soil science, cryptology, genome mapping, or fuzzy logic, to name just a few of these active zones, a few of the "disorderly messes" as Socrates would call them. On the one hand we have a model that still applies the earlier slogan – the less connected a science the better – while on the other we have many disciplines, uncertain of their exact status, striving to apply the old model, unable to reinstate it, and not yet prepared to mutter something like what we have been saying all along: "Relax, calm down, the more connected a science is the better. Being part of a collective will not deprive you of the nonhumans you socialize so well. It will only deprive you of the polemical kind of objectivity that has no other use than as a weapon for waging a political war *against* politics."

To put it even more bluntly, science studies has become a hostage in a huge shift from Science to what we could call Research (or Science No. 2, as I will call it...). While Science had certainty, coldness, aloofness, objectivity, distance, and necessity, Research appears to have all the opposite characteristics: it is uncertain; open-ended; immersed in many lowly problems of money, instruments, and know-how; unable to differentiate as yet between hot and cold, subjective and objective, human and nonhuman. If Science thrived by behaving as if it were totally disconnected from the collective, Research is best seen as a *collective experimentation* about what humans and nonhumans together are able to swallow or to withstand. It seems to me that the second model is wiser than the former. No longer do we have to choose between Right and Might, because there is now a third party in the dispute, that is, the collective; no longer do we have to decide between Science and Anti-Science, because here too there is a third party – *the same* third party, the collective.

Research is this zone into which humans and nonhumans are thrown, in which has been practiced, over the ages, the most extraordinary collective experiment to distinguish, in real time, between "cosmos" and "unruly shambles" with no one, neither the scientists nor the "science students," knowing in advance what the provisional answer will be. Maybe science studies is anti-Science, after all, but in that case it is wholeheartedly *for* Research, and, in the future, when the spirit of the times will have taken a firmer grip on public opinion, it will be in the same camp as all of the active scientists, leaving on the other side only a few disgruntled cold-war physicists still wishing to help Socrates shut the mouths of the "ten thousand fools" with an unquestionable and indisputable absolute truth coming from nowhere. The opposite of relativism, we should never forget, is called absolutism (Bloor [1976] 1991).

I am being a bit disingenuous, I know – because there is a third reason that makes it hard to believe that science studies could have so many goodies to offer. By an unfortunate coincidence, or maybe through a strange case of Darwinian mimicry in the ecology of the social sciences, or – who knows? – through some case of mutual contamination, science studies bears a superficial resemblance to those prisoners locked in their cells whom we left, a few pages ago, in their slow descent from Kant to hell and smiling smugly all the way down, since they claim no longer to care about the ability of language to refer to reality. When we talk about hybrids and imbroglios, mediations, practice, networks, relativism, relations, provisional answers, partial connections, humans and nonhumans, "disorderly messes," it may sound as if we, too, are marching along the same path, in a hurried flight from truth and reason, fragmenting into ever smaller pieces the categories that keep the human mind forever removed from the presence of reality. And yet – there is no need to paper it over – just as there is a fight inside the scientific disciplines between the model of Science and the model of Research, there is a fight in the social sciences and the humanities between two opposite models, one that can loosely be called postmodern and the other that I have called nonmodern. Everything the first takes to be a justification for more absence, more debunking, more negation, more deconstruction, the second takes as a proof of presence, deployment, affirmation, and construction.

The cause of the radical differences as well as of the passing resemblances is not difficult to ferret out. Postmodernism, as the name indicates, is descended from the series of settlements that have defined modernity. It has inherited from these the disconnected mind-in-the-vat's quest for absolute truth, the debate between Might and Right, the radical distinction between science and politics, Kant's constructivism, and the critical urge that goes with it, but it has *stopped* believing it is possible to carry out this implausible program successfully. In this disappointment it shows good common sense, and that is something to say in its favor. But it has not retraced the path of modernity all the way back to the various bifurcations that started this impossible project in the first place. It feels the same nostalgia as modernism, except that it tries to take on, as positive features, the overwhelming failures of the rationalist project. Hence its apology on behalf of Callicles and the Sophists, its rejoicing in virtual reality, its debunking of "master narratives," its claim that it is good to be stuck inside one's own standpoint, its overemphasis on reflexivity, its maddening efforts to write texts that do not carry any risk of presence.

Science studies, as I see it, has been engaged in a very different nonmodern task. For us, modernity has never been the order of the day. Reality and morality have never been lacking. The fight for or against absolute truth, for or against multiple standpoints, for or against social construction, for or against presence, has never been the important one. The program of debunking, exposing, avoiding being taken in, steals energy from the task that has always seemed much more important to the collective of people, things, and gods, namely, the task of sorting out the "cosmos" from an "unruly shambles." We are aiming at a *politics of things*, not at the bygone dispute about whether or not words refer to the world. Of course they do! You might as well ask me if I believe in Mom and apple pie or, for that matter, if I believe in reality!

Are you still unconvinced, my friend? Still uncertain if we are fish or fowl, friends or foes? I must confess that it takes more than a small act of faith to accept this portrayal of our work in such a light, but since you asked your question with such an open mind, I thought you deserved to be answered with the same frankness. It is true that it is a bit difficult to locate us in the middle of the two-culture divide, in the midst of the epochal shift from Science to Research, torn between the postmodern and the nonmodern predicament. I hope you are convinced, at least, that there is no deliberate obfuscation in our position, but that being faithful to your own scientific work in these troubled times is just damned difficult. In my view, your work and that of your many colleagues, your effort to establish facts, has been taken hostage in a tired old dispute about how best to control the people. We believe the sciences deserve better than this kidnapping by Science.

Contrary to what you may have thought when you asked me for this private conversation, far from being the ones who have limited science to "mere social construction" by the frantic disorderly mob invented to satisfy Callicles' and Socrates' urge for power, we in science studies may be *the first to have found a way to free the sciences from politics* – the politics of reason, that old settlement among epistemology, morality, psychology, and theology. We may be the first to have freed nonhumans from the politics of objectivity and humans from the politics of subjectification. The disciplines themselves, the facts and the artifacts with their beautiful roots, their delicate articulations, their many tendrils, and their fragile networks remain, for the most part, to be investigated and described. I try my best, in the pages that follow, to untangle a few of them. Far from the rumblings of the science wars in which neither you nor I want to fight (well, maybe I won't mind firing a few shots!), facts and artifacts can be part of many other conversations, much less bellicose, much more productive, and, yes, much friendlier.

I have to admit I am being disingenuous again. In opening the black box of scientific facts, we knew we would be opening Pandora's box. There was no way to avoid it. It was tightly sealed as long as it remained in the two-culture no-man's-land, buried among the cabbages and the turnips, blissfully ignored by the humanists trying to avoid all the dangers of objectification and by the epistemologists trying to fend off all the ills carried by the unruly mob. Now that it has been opened, with plagues and curses, sins and ills whirling around, there is only one thing to do, and that is to go even deeper, all the way down into the almost-empty box, in order to retrieve what, according to the venerable legend, has been left at the bottom – yes, *hope*. It is much too deep for me on my own; are you willing to help me reach it? May I give you a hand?

Note

1 I use the recent translation by Robin Waterfield (Oxford: Oxford University Press, 1994).

References

Bloor, D. (1976) 1991. *Knowledge and Social Imagery*, 2nd ed., with new foreword. Chicago: University of Chicago Press.

Stengers, I. 1996. *Cosmopolitiques*, tome 1: *La Guerre des Sciences.* Paris: La Découverte et les Empêcheurs de penser en rond.

13

Hermeneutical Philosophy and Pragmatism: A Philosophy of Science

Patrick A. Heelan and Jay Schulkin

1 Introduction

The philosophy of science rooted in the works of Descartes, Hume, and Comte, took a great detour in the twentieth century (Babich 1994b). Beginning with a strong historical voice in P. Duhem, E. Mach, and G. Bachelard, often focussed on the theme of scientific creativity, this voice became muted at the mid-century following the rise and dominance of positivism and analysis. But with the close of the century, logical positivism is dead, and the historical voice has returned, joined now by those in the social studies of science, to confront the anti-historicism and reductionism of the prevailing tradition and its dismissal of interest in what N. R. Hanson (Hanson 1961) called patterns of discovery. One would expect then a revival of interest in the cultural processes that shepherd scientific inquiry in making theories and testing them within the socio-historical context in which they function to give meaning to the empirical world, and the modalities under which people, scientists and non-scientists, come to an understanding of the mission of science. These are areas crucial today for interdisciplinary cooperation and for public trust and understanding of one of our greatest institutions.

From *Synthese* 115 (1998): 269–302, with some notes and references cut. Reprinted by permission of Kluwer Academic Publishers, Dordrecht, The Netherlands.

2 Elements of Consensus

Debate and criticism of the past few decades seem to have brought about widespread agreement on certain theses. Aware that the formulae do not mean quite the same for all, there is nevertheless strong support among diverse philosophers in opposition to residual positivism and to non-linguistic cognitive givens (knowledge is not built of such bricks). There is support for contextuality and semantic networks (whatever we know is relative to a particular background language and culture providing the relevant categories of human experience), and for an instrumentalistic orientation toward formalistic theories;[1] mathematical models, being abstract and divorced from the world of which they purport to speak, provide not *what* is known to be in the world but *means through which* the world becomes understandable to human knowers equipped to interact purposefully with it. With some uneasiness, there is a very general agreement that, in general, theories or semantic networks are properly used in social historical contexts to explore experiential horizons. In the experimental sciences theories are characteristically used with special instruments and technologies which provide research data that from the start are laden with scientific meanings – they are, it is said, "theory-laden", but more about this below. There is a new focus on the diverse perspectives of different local scientific communities, embodied in their historical institutions and cultural artifacts. We focus on these themes in so far as they affect inquiry into scientific inquiry, but they also provide the context of inquiry into any form of inquiry.

In keeping with this transformed background, we can say with a certain presumption of broad agreement that perceptual experience is not to be taken as a curtain that cuts human inquirers off from a real world of generally imperceptible entities, but the milieu within which by interaction we come to understand the world of experience itself and the furniture it contains. There is general agreement that the entities of science are to be counted among the furniture of the world, but whether they are perceptual entities is still disputed. Also there is widespread agreement that experience is active, embodied, and engaged with public cultural realities, and that it is not passive, nor merely the private content of individual minds. Experience is ever more deeply penetrated by theoretical understanding from which people learn to adapt with growing success and flexibility to a changing environment. The experiential emphasis of all inquiry rests then on current and prospective shared knowledge of diverse perspectives through language, action, perception, and culture. This new background has changed the focus of philosophical interest in scientific inquiry away from the context of justification and prediction to the context of (let us call it) scientific culture.[2] This is the context of discovery, interpretation, laboratory practices, historical change, and the influence of science-based technologies on general culture.

So far, despite their different languages and origins, *pragmatism* and *hermeneutic philosophy* have each something to say that addresses these issues, each aiming in its own usage to speak "with truth" about the same contemporary lifeworld we all actively share.

3 Pragmatism

Pragmatism was a reaction on the one hand against the philosophical idealism of the Hegelians and Neo-kantians and on the other against the dogmatic authority of cultural, mostly religious, elites who claimed to possess privileged knowledge of the world stemming from a transcendent or supernatural source. It developed from a keen sense of and respect for the processes of human inquiry particularly those of the natural sciences and adopted as its root metaphor science as the potent instrument of human evolutionary and cultural adaptability. However its fundamental principle was broader than just science: all processes of human inquiry occur within the domain of human experience as it struggles to cope with the world around it, and human language is designed to express this struggle. It is this dimension that makes it a philosophy.

Pragmatism proposed that all language was grounded in the adaptability of human experience to the social and wordly conditions of life. Thus, language purportedly about other-worldly agencies, immaterial forms, or religio-mythic figures was taken to be implicitly and metaphorically language about human life. The most potent tool to critique, correct, and enlarge language and to promote the goal of human adaptability by control of the environment was scientific inquiry. Science, however, was more than a method, but how much more was a matter of dispute among pragmatists. Did it confer "objectivity"[3] on scientific entities or theories? Or show them to be "real" outside of the context of human history and culture? These themes will be taken up below.

Our principal concern in this paper is with the classical pragmatists: C. S. Peirce, W. James, G. H. Mead, C. I. Lewis, and J. Dewey. They reconfigured the rules and maxims of classical inquiry. In their judgment, these were aimed at providing on the one hand, as it were, permanent exhibits in a Science Museum of Knowledge, and on the other, topics of discourse for parlor intellectuals. These were not proper functions of serious inquiry. Inquiry was at the heart of everyday activity and it occurred at the level of ordinary people. Science only gave it precision.

For pragmatism, all human inquiry was tied exclusively to experience, and experience was active, never passive, and science was continually opening up new areas of experience for exploration. Philosophy of science then was centered on the activity of scientific research and science's aim to enrich human experience with new goals worth pursuing.

Peirce pre-eminently among the classical pragmatists believed he understood the nature of scientific inquiry. Science was no disembodied study of the world, it needed environments for empirical work, such as the laboratory for interdisciplinary research in psychophysics which he set up at Johns Hopkins University which was one of the first of its kind in America. He also worked for the U.S. Coast and Geodetic Survey. This dedication to empirical work tended to give him a laboratory frame of mind even in philosophy, replete with tests, methods and ideas.

However, he understood earlier than most of his colleagues the meaning of symbolic computational systems and their dependence in use on principles of interpretation. His familiarity with formal logic and its history grounded his logic of inquiry. The logic of inquiry comprises three modes or stages: *abduction* or the genesis of an idea, *induction* informed by specific instances, and *deduction* of logical consequences from general principles. Of these, abduction is the most interesting from the point of view of hermeneutics.

Peirce was a critical realist, contrary to Rorty's revisionistic views he did not emphasize the "glassy essence" and "the world well lost". Peirce understood that our theories are constrained by the things we encounter in the environments with which we are trying to cope, and that our ideas as a consequence are shaped by these environments, since adaptation requires us to forge a plan of action and implement strategies based on a coherent representation of these worldly environments.

With respect to perception, Peirce says that seeing is never (simply) seeing, it is always relative to a background, a "seeing as". To this, Dewey and Mead would add the sense of a rich social milieu, constituting what A. Schutz called "the lifeworld" (Schutz 1973). The pragmatists went to great lengths to distinguish their sense of experience from that of 17th and 18th century empiricism. The empiricism of Locke and Hume was passive and about the association of sensations. For James as for other pragmatists, experience was active, structured by ideas and tested by actions, and aimed functionally at broadening horizons by informing them with new ideas; experience was not just a passive reception of the world narrowly focussed, say, on sense data or other "givens".

Pragmatists took inquiry to stem from a breakdown of a form of life that had been replete with functional coherence. Frustrated with the doubt and insecurity that follows this breakdown, inquiry begins as the search to recover what was lost. For Peirce, abduction with its associated inductive and deductive activities was the generative process of new ideas that aimed to restore a secure and settled form of life and a stable pattern of action.

Inquiry has at its heart two components that are often in conflict with one another as history and practice show, these are *commitment* and *correction*. The spirit of commitment tends narrowly to be satisfied with what has been accepted as "funded knowledge" or the permanent accomplishments of the disciplines. The spirit of correction prompts inquiry to look beyond the current state of "funded knowledge" and to cross borders among the disciplines. Both of these are also characteristics of hermeneutical inquiry.

Dewey shared the view that the origin of inquiry is in the precariousness of human existence and the countless searches to recapture life's lost equilibrium. In the pursuit of inquiry, theories were to guide actions, and feedback from actions was to correct theories. In Dewey's naturalistic evolutionary vision of life, the twin cardinal poles of human action were strife and resolution.

James' radical empiricism (James 1912/1996) is a kind of functionalism based on the search for adaptive behaviors through instrumental action, that is, on inquiry as the search to secure stability amidst ever changing circumstances (see Parrott and Schulkin 1993).

Although Dewey's categories of experience were often construed exclusively in terms of appetitive and consummatory experience as if human behavior was driven just to promote material satisfaction, esthetic experience as a way of balancing the books against too crass an interpretation of human life also constituted a serious focus of Dewey's interests (see Dewey 1958).

Besides the factors mentioned above as initiating inquiry, Peirce and other pragmatists also included *musement* or *wonder*. This is the kind of exploration of an issue essential to discerning what about it is to be understood as relevant. Musement has a certain resonance in Husserl's and Heidegger's turn towards the exploration of the "pre-predicative" state of human understanding prior to conventional categorization and predication.

A central message of classical pragmatism is that inquiry is a mode of action, of doing, in contrast with the cognitively sterile view that answers are to be researched in a museum of settled knowledge. Pursuing this metaphor, since science and philosophy are not museum pieces, they share, as it were, a common interactive space and can influence each other across disciplinary boundaries without one absorbing or being reduced to the other.

Pragmatism influenced the philosophy of science through the works of N. R. Hanson, E. Nagel, W. Sellars, W.v.O. Quine, H. Putnam, N. Goodman, I. Sheffler, among others. Hanson (1958) defended the thesis that scientific inquiry is theory-laden, replete with theoretical meaning. Abduction for him was the process of creating a hypothesis from the fall-out of disputed hypotheses, within an investigatory form of life rooted in action guided by empirical consequences.

Sellars (1963) contributed a new and broad perspective which established the context dependence of all experience. All experience was understood as contextual either within the world's pragmatic "manifest image" which is the world as experienced perceptually (or in quasi-personal terms), or within the theoretical "scientific image", which is the world as explained in terms of "non-perceptual" – i.e., "theoretical" – scientific entities. Sellars departed from the pragmatic tradition by prioritizing the non-perceptual and mathematical-theoretical. In this respect he was more Kantian than pragmatist.

Another important contributor, Quine (1960), saw that, although the positivists' version of sense data had satisfied Mach and Carnap and the earlier generation of positivistically inclined philosophers, the positivists' view was incoherent and would not suffice for the confirmation of theory. He also came to see that science and metaphysics could not be separated as the positivists had envisioned. While the influence of pragmatism is seen in his support for historically changing scientific frameworks, Quine, like many other pragmatists, could not break the tie to a world of basic objective empirically given units of knowledge.

After Quine, some philosophers of science turned, like Quine himself, to philosophical behaviorism; others, like Chomsky and Fodor, turned to cognitivism, and the Churchlands, to eliminative materialis. Thereafter the perspective that the classical pragmatists gave to scientific inquiry became muted. It found limited expression in the discussion of (what T. S. Kuhn called) "paradigms" and in theories about the social constitution of knowledge. But the sense of rationality that mattered most to classical pragmatists, the grounding of a philosophy of human inquiry in human experience, seemed to flitter away.

One reason for this was ambiguity about the meaning and role of theory. Philosophers of science, seeing physics as the privileged exemplar of science, took theory to be a mathematical model tested against observations, while pragmatists, seeing experimental praxis as the privileged exemplar of science, took theory to be descriptive of scientific entities as these were perceived in laboratory praxis. The biological and social sciences today tend to use "theory" in the latter sense, while the physical sciences continue to exploit the mathematical imagination in search of new theoretical models, or in their terms, simply theories. One of the problems that hermeneutical method and philosophy will address is the diversity in the meaning given to theories and the usage of the term.

4 Hermeneutical Philosophy

Although hermeneutical philosophy also comes from a scientific background, it does not come from the quantitative sciences, but rather from the side of (what the Germans call) the *Geisteswissenschaften* and their characteristic method, hermeneutics. This is the tradition of humanistic scholarship in scriptural studies, history, art, philology and literature, and the humanistically oriented social sciences. Its characteristic method, hermeneutics, is oriented towards meaning not power, and towards the things – signs, symbols, actions – that can be construed as having meaning. These are, for example, the relics of past events, social institutions, religious myths and rituals, cosmological and natural phenomena, cultural artifacts not just in the domain of the arts, but most particularly, the spoken word and written texts. Biblical texts were among the earliest subject matter of modern hermeneutical science. All of the things just mentioned enter into public awareness endowed with some – though usually ambiguous – meaning. Many of these symbolic vehicles of meaning are clearly human artifacts, but others are natural or cosmological phenomena that seem at first sight to belong totally to a non-human realm. On deeper scrutiny, however, the meanings they carry also turn out to have a constitutive hermeneutic dimension capable of calling into question any hard and fast distinction between nature and culture.

Just as contemporary philosophy of science went through an early positivist phase that was overcome, so hermeneutical method had an early quasi-positive phase that too had to be overcome. In its early phase, hermeneutics assumed that each natural thing and event, each product of human making, each text and cultural object, had a specific original (though possibly hidden or disguised) meaning that was given to it by the cause (author, founder, ruler, creator, etc.) that brought it into being or gave it a charge. In its early phase the science of hermeneutics aimed precisely at uncovering these original meanings. Under the positivist influence, however, the meanings in question were taken to be private mental entities, not verifiable empirical entities, and therefore not objects suitable for scientific research.

The process of hermeneutics in this early positivist phase was modelled on the metaphor of a "conduit" through which "messages" originating at a distant or currently non-existent source were passed on to a living receiver. The "conduit" could be inscriptions, architectural remains, or other artifacts, but most commonly it was a text inscribed on some durable material such as papyrus, paper, leather, or stone. As the "message" "travelled" in space and time from its sender/source to a distant human receiver/interpreter, there was a concern with the possible distortion of the "message" due to the difference between the linguistic and cultural environments of the receiver and sender. As long as the "message" to be communicated was thought to be the original meaning imposed on the inscription by the sender, the receiver/interpreter was enjoined to practice the art of re-living in imagination the cultural and historical world of the source and to re-construct the "message" out of empathy with the sender. This view of hermeneutics was to change dramatically due to the writings of W. Dilthey, E. Husserl, H-G. Gadamer, M. Heidegger, M. Merleau-Ponty, P. Ricoeur, and others.

For hermeneutical method to be recognized as a valid tool of modern scientific method universally applicable even to the natural sciences, a new understanding of the notion of meaning and hermeneutics had to take place. Some salient aspects of this new understanding are briefly summarized in the paragraphs below. We shall not defend them here since they have been defended elsewhere. But most of the extensive literature on such topics is found outside the philosophy of science corpus in what is called "continental philosophy". Some of the positions laid out in these propositions have indeed entered analytic philosophy in its recent postmodern phase, but more as a kind of Feyerabendian *bricolage* than as a consequence of any basic revision of viewpoint.

5 Meaning

Meaning is not a private mental entity but a shared social entity embodied in language (understood always to include other language-like inscriptions, whether passive, like road signs, or active, like performances) and a cultural environment embodying community purposes.

Meanings are not fully complete unless incorporated in a linguistic utterance used to affirm or deny some content that finds itself *fulfilled in public experience*.

Perception relates to the perceptual field of the lifeworld. By "perceptual experience", we do not mean simply sensory inputs, but the public recognition of the existence of objects in the space(s) and time(s) of the lifeworld that are understood and categorized through sensory and bodily interactions, and which are the referents of a public ostensive/descriptive language.

Fulfilled and perceptual meanings are not just private mental representations of something, a referent, but are in fact by intent identical with the referent that is presented in experience and give access to the *ontic* and *ontological*[4] character of that referent under the aspect of what is in truth on this occasion given to understanding. They include *but are not exhausted by* whatever can be reached by a reflective and hermeneutical study of the constitution of fulfilled meanings. Husserl, for instance, typically focused on how "objects" (contents) of knowledge are "constituted" (presented to communal knowers) within "noetic" contexts of meaning (directed by a communal vector of inquiry). Heidegger referred to such objects as "ontic beings" disclosed perspectivally to the "circumspective care of the human inquirer as *Da-sein*,[5] his term for the human being as coping with the lifeworld and immersed in the "ontological" history of Being. Merleau-Ponty studied them as the "flesh" (or embodiment) of things revealed by perception through the forms of embodied human life in the world (Merleau-Ponty 1962).

To the extent that language and other public expressive signs are the only means through which we articulate our public world and come to understand one another, the meanings that these signs convey are construals of human cultural communities and cannot be attributed to non-human sources except by metaphor. Taking meanings to be cultural does not mean that there is no truth, but that the truth possessed, even scientific truth, is always mediated by human language and culture which are a part of human history.

Knowledge is handed down by the medium of linguistic and expressive inscriptions and the cultural forms of life in which they find fulfillment. Phrases, however, that once meant one thing come to mean another with the passage of time, for language and culture change. As historians of science know, this is as true for natural science as it is for literature and politics. Of special interest then are the circumstances of continuity and change in the historical transmission of scientific meanings via the media of language, mathematics, laboratory praxes, and the culture of the scientific community. Meanings originating at one (linguistic, historical, cultural, geographic) local site are received/interpreted/fulfilled at a distant local site as different meanings. These latter are adopted from traditions of interpretation, or constructed or re-constructed in keeping with the responsibilities, constraints, and presumptions of rational hermeneutical inquiry (see below, the *hermeneutical circle*) which require that each find fulfillment in local experience at the reception site. One of these constraints is the extent of the linguistic and cultural resources available to the distant reader. One of the presumptions is that there is no single meaning that is relevant to and fulfillable by all readers of such a text; there are many, and they depend on the linguistic and cultural resources as well as on the cultural ambience of the reader (Nickles 1995). Like a hammer or any piece of equipment, a text can be used successfully for several meaningful cultural purposes. As in the case of the hammer, for each useful purpose there are criteria as to how well it performs for this purpose in human life. The uses are not arbitrary, for nothing but nonsense would be gained by arbitrary use, but this does not imply that there is just a single correct meaningful use. Once again, as in the case of the hammer, there may be a priority of users set by a firm cultural tradition – hammers for construction workers, scientific results for scientific research communities – but no one use need go unchallenged either by logic or by experience nor should any one use become the sole property of just one interested group.

Rational hermeneutic inquiry acknowledges the existence of *traditions of interpretation* that give to today's readers and inquirers a culturally privileged version (shaped to the goals of the linguistic and cultural environment of the community with special "ownership" rights in the subject matter) of past sources. Within the sciences such traditions of interpretation are at the basis of what Kuhn called "paradigms".

In addition to meanings construed on the basis of a common tradition of interpretation (with its presumption of continuity), other meanings can be legitimate that are independent from any presumption of the existence of a continuity of meaning with the source through a common tradition of life, action, and interpretation. Such *discontinuities of meaning* within the sciences are exemplified by what Kuhn called "revolutions" in which old "paradigms" are replaced by new ones. In the work of hermeneutics, however, a radically new meaning need not expel the old, because each, though different, may be a valid historical and cultural perspective. Indeed, despite some sense of discomfort, we often find in the sciences the old flourishing side by side with the radically new, quantum mechanics with Newtonian mechanics (though these are formally incompatible with one another), statistical thermodynamics with phenomenological thermodynamics, an so on. Each through its own empirical processes of testing and measurement is in dialogue with confirming or disconfirming data.

In summary and for the purposes of this paper, hermeneutic method is the strategy according to which a living inquirer goes about the task of finding or constructing a contemporary meaning for a non-present or, at least, non-understood source event, natural or artificial; if the source event is a human artifact, say, a text or an action, the hermeneutic process of the interpreter has to bridge the difference between the linguistic/cultural environments of the source and the interpretation. This strategy is the *hermeneutical circle*. It is also called a *method*, though results are not guaranteed; but it is an orderly process and to that extent *methodological*. Interpretative work of this kind is clearly historical, cultural, and anthropological, multidisciplinary in character and in need of a philosophical foundation which hermeneutical philosophy (to be taken up below) tries to provide. In this work lies the significance and power of hermeneutic method and hermeneutic philosophy for the history and philosophy of science. And not just for these, but also for understanding how quantitative empirical methods function in science to give meaning to empirical contents, in particular, how measurement equipment plays a double role creating both theoretical and cultural meanings, and how theory-laden data depend on the successful public self presentation of the (so-called) "theoretical entity" as a public cultural entity. Such a self presentation is an essential part of laboratory measurement and is an ordinary part of the culture of scientific research. But a theoretical entity, say, a magnetic field, can also manifest its presence as a cultural object to a wider public through an essential role it happens to play in a public standard technology, say, a magnetic compass.

Hermeneutical philosophy and *pragmatism* have their root metaphors in different kinds of scholarship, the former in humanistic scholarship and the latter in the quantitative sciences. Such a difference prompts us to ask whether there is as a consequence a serious risk of systematic misunderstanding or a significant obstacle to translation from one to the other.

As a prelude to our attempt to give an answer to this question we need to consider the nature of *philosophical inquiry*.

6 Philosophical Inquiry

Inquiry for Husserl and Heidegger – as incidentally also for Peirce and Dewey – begins when some real expectation based in experience fails and we are curious why, and look for an answer that will enable us to fulfill our failed expectation, or failing this, to go around the problem or alternatively to transform the context re-assessing if need be our goals. For Husserl, eidetic phenomenological analysis explores the invariant boundaries of an imagined experience that is subjected to imagined variations of approach. Both see inquiry as connected with a breakdown of intelligibility – for Heidegger when action fails in the world, for Husserl when the noetic structure of the imagination fails. Husserl's approach is more logical, conceptual, and abstract while Heidegger's is more existential, historical, and action oriented. It is then to Heidegger's philosophy that we will turn almost exclusively for an account of hermeneutical philosophy.

For the limited purpose of this inquiry into scientific inquiry, it is useful to enter Heidegger's philosophy through *Being and Time*, first drawing down from his analysis of equipment to lay out the methodology of the hermeneutical circle (Heidegger 1927/1996, 357–364). Here he illustrates the *genesis and process of any inquiry* by focusing on what happens when in the middle of a task, a tool, say, a hammer, breaks. A hammer, a tree, a text, an atom, all are recognized by their characteristic function in the lifeworld; each may fail on some occasion – perhaps even systematically – to fulfil its expected function, and such a failure would initiate a process of inquiry.

To start with the philosophical background: the breakdown of the task initiates inquiry by calling on the deep structure of pre-theoretical pre-categorial understanding[6] of Being which is the lifeworld (*Vorhabe*). The human inquirer is *Da-sein*, *Da-sein* is "existence", the embodied understanding of Being, a "there-ness" (*Da*) in the domain of Being (*Sein*). Inquiry is awakened when *Da-sein* poses a directed question (*Vorsicht*) which, like all directed questions, already implicitly contains an outline of a search and discovery strategy that aims at finding a solution to the problem or an answer to the query. The question so construed in this case is not yet articulated; only later will it achieve an adequate expression in apophantic discourse as (what philosophers of science call) an "explanation". There follows an active dialogue between *Vorsicht* and *Vorhabe*, accompanied by actions seeking practical fulfillment in the awareness that the sought-for understanding (*die Sache selbst*) has presented itself and made itself manifest to the inquirer (*Vorgriff*). If on first trial the sought-for understanding is absent, something nevertheless has been learnt, and the search resumes, dipping again into the available resources of *Vorhabe*, *Vorsicht* and *Vorgriff*. This circle of inquiry, called the "hermeneutical circle" – sometimes called the "hermeneutical spiral" – is repeated until a solution presents itself within a new cultural praxis in the lifeworld. Only at that time is it in order to express the solution linguistically in statements, that is, in apophantic form.

Let us now apply the hermeneutical circle to the problem of the broken hammer. In order to finish the job on hand, we draw on our background understanding of the lifeworld (*Vorhabe*), and ask ourselves, perhaps for the first time in our lives, what kind of thing is a hammer, what specifications does it have. The question itself (*Vorsicht*) suggests strategies for searching and finding. A dialogue is initiated that has theoretical and practical dimensions. First we aim at defining the theoretical specifications of a hammer, then we look for or construct something that fulfills these specifications, and when we have it, we try it out (*Vorgriff*). Does it work? If it does the job – if our first attempt at defining the theoretical specifications is fulfilled in practical experience – we are aware of the presence of what we were looking for, something with which to finish this job. We may still need a new hammer, but for the moment the job can go on. If, however, the trial fails, a thought may intervene, "We may have to revise our goals!" But No! We revise the theoretical specifications in the light of the previous outcome (a revised *Vorsicht*) and try again, modifying the conditions of the experimental trial if necessary (a revised *Vorgriff*). This phase may be repeated several times until we have a theory and practice that works. If we experience nothing but failure, we re-assess our options (returning to the *Vorhabe* for other suggestions), e.g., getting the job done in a different way, hiring a carpenter, or turning to a different technology, or finally, we could just fold our tent for the time being.

This philosophico-methodological process of inquiry is *hermeneutical* because it is a search for a theoretical – in fact, explanatory – meaning to be fulfilled in a new cultural praxis in the lifeworld. The process has a cyclic pattern which is repeated over and over, from the general background understanding of the lifeworld (*Vorhabe*) to proposed theory (*Vorsicht*) to trial experience (*Vorgriff*) or summarily, the circle of background, theory, and experience. This process is called the *hermeneutical circle*.[7] The term "*hermeneutical spiral*" is sometimes used to indicate both the cycling and the progressive character of the process. Every rational inquiry then moves in a forward spiral aiming at fulfillment in a solution made present in experience as governed by a conscious if revisable goal.

7 Theoretical Understanding in a Hermeneutical Perspective

During his analysis, Heidegger (focusing his discussion initially on equipment and the like) makes a special and highly critical point about *theoretical understanding*. Since the characteristic goal of all scientific or scholarly inquiry is theoretical understanding, it is important to understand what theory does. Theory, as in the case of a hammer, a tree, a text, or an atom is always connected with something as fulfilling a social or cultural function. Theory-making arises then out of some public need and the desire to learn how to fulfill that need. He would remind us that, when presented with an individual piece of equipment, say, a hammer, we must realize on the one hand that the theory of a hammer does not assign to it an exclusive or "objective" essence, for that which can function as a hammer can function on occasion in other ways too, as a door stop, nutcracker, etc. On the other hand, old shoes and wooden mallets can also on occasion function for the hammering of nails. All individual tools or equipment are (as Heidegger says) no more than a *mere resource* unless they are in actual use or designated for use, when they are *dedicated (or designated) resources*. Equipment is a dedicated resource when it is pragmatically related to the fulfillment of its role within a cultural function-as-meant. The distinction is significant because only dedicated resources belong to the furniture of the lifeworld and so have ontic status.

These distinctions are reflected in the use of words. The sentence, "I want a hammer", can be used in a theory-laden context where the sentence refers to the structure that makes hammering possible, or in a praxis-laden context where the sentence refers to something that is in actual use in construction or designated for such use. Words and sentences about tools or equipment take on different meanings according to whether they are used in one or other of these contexts.

Returning to the cultural praxis-laden meaning of the hammer, what is it? It is what ties a thing – the hammer – to construction or building projects. This is different from its theory-laden meaning for this latter relates to its specifications as a tool and "explains" the thing *qua* hammer by specifying the conditions under which it can be the host of the cultural meaning of a hammer. There are then two meanings in dialogue, a *theory-laden meaning* and a *cultural praxis-laden meaning*. The *theory-laden* meaning makes sense only if the individual hammer is *praxis-laden* within the function of construction. If perchance one were to look for a theory to explain construction-praxes, one would look beyond (hammer-)theory, toward a theory at the level of architectural and engineering practices.

The same kind of analysis holds for the social sciences. There is, however, a difference in efficacy: in the natural sciences, theory or explanatory meaning generally confers predictive power and effective control over some physical aspect of the environment possessed of the cultural meaning which the theory explains; in the social sciences, theory is no less meaningful but it is less powerful since cultural aspects of the environment are less easy to manipulate.

Despite the fact then that (hammer-)theory "explains" (hammering) praxis, they belong to different perspectives and there is no one-to-one correlation between the two. The (hammer-)theory-ladenness of a thing is just a mere possibility of serving as a real hammer (it could alternatively serve as a nutcracker); the (hammering-)praxis-ladenness of the thing in the context of construction could be served by means other than the use of hammers. Let us call the theoretical principles of construction, "architecture/engineering theory".

Then, (hammer-)theory is not an essential part of architecture/engineering theory, nor is (hammering-)praxis an essential part of construction praxis, although each may be a contingent part of the other. Theoretical inquiry into hammers and hammering, however, can inaugurate a dialogue with the higher cultural function (construction) and lead to the design of new science-based (designer) construction techniques that dispense with hammers using, say, ready-made clip-together plastic units.

When such new science-based technologies are added to the lifeworld, scientific terms can be introduced into everyday descriptive language with new (non-theoretical) practical lifeworld meanings. A set of new lifeworld entities are thereby naturalized, the effect of which has the power to transform the old cultural environment. This happened, for example, in the *quattrocento* during the Italian Renaissance when perceptual space was subjected to universal measurement by mathematical perspective and in the process was transformed from a variable geometry space into a Euclidean space thus preparing the way for the Copernican revolution (see Heelan 1983/1988, Part I).

In any case, to be theory-laden means to "explain" – to prescribe the "infrastructural" conditions – why something *can* play a particular socio-cultural role, but it does not explain whether, or if so, why it is in fact playing that role or has been designated to play that role. "To be theory-laden" then always implies an unexpressed cultural or lifeworld hypothetical. Otherwise "to be theory-laden" implies no more than "to be a mere resource" disjoint from the lifeworld – and this no more entitles a thing to be included in the furniture of the world than every old shoe under the category of hammers.

What kind of entity then is a hammer as a dedicated resource? It is a public cultural reality, a physical reality constituted by a socio-cultural meaning (in this case, the open cultural perspective of construction and its practical underpinnings in architecture and engineering) to which the individual hammer has been dedicated by a social choice. It has a specific theory-laden meaning that conceals (renders tacit/implicit) but does not replace (say, by a reductive move) its higher socio-cultural meaning as described above. On such empirical entities (where the "natural" and the "artificial" are not distinguished) a new kind of "empiricism" can be built. Some might see in pragmatism a fulfillment of this goal – but more about this below.

Now, to the extent that nothing (or almost nothing) in our experience – a hammer, a tree, a text, or an atom – is without a cultural purpose, everything in our experience bears some resemblance, however distant, to a tool or instrument. This deep structure of pre-theoretical pre-categorical understanding belongs to the human inquirer's (*Dasein's*) background understanding of the lifeworld (*Vorhabe*) as described above. Everything then while actually praxis-laden in a cultural sense, is also involved in the possibility of becoming theory-laden in an explanatory sense *even prior to actual scientific inquiry, because such is the deep structure (Vorhabe) of the human inquirer*.

These conclusions have important consequences for understanding the role of *measurement equipment in the quantitative sciences*. The data – or better, the "raw data" or "proto-data" – that are produced belong *hypothetically* to the perspective of measurement but *affirmatively* to the perspective of some public cultural reality within which the data display "in truth" not just themselves but the public cultural entity of which they are profiles. By saying that the raw or proto-data belong "hypothetically" to the measurement context, we mean that they do so only if they are recognized as meaningful in a cultural forum. This forum could be, for example, scientific research strategy and the research "narratives" that Rouse (1996, 27 and in chap. 9) speaks about. But it could also extend to technological applications, media, entertainment, or other aspects of general culture. Only in such fora are the data real – given "in truth" as *die Sache selbst*. There they can witness to the presence of individual scientific entities, say, electrons or atoms, as public cultural realities, dedicated resources, part of the furniture of the world, and incidentally perceptual objects (see Heelan 1983/1988 and 1989). As such the data and the scientific entities they exhibit are only implicitly theory-laden. They emulate the relationship between individual hammers and construction projects. If they are not recognized as dedicated resources within an appropriate cultural context, they are not data at all, but functionally meaningless marks (non-entities, junk, etc.) with all the indeterminacy of (positivism's) *sense data*.

Heidegger embodied this conclusion in his choice of the Greek term, *alethia* (literally "uncovering") for truth (see Heidegger 1927/1996, 213f). It signalled a change in the notion of truth from the classical model of full transparency to the human mind, towards one of only partial, practical, or contextual transparency. Polanyi says the same in different terms: the *explicit meaning* conceals a *tacit meaning* (see Polanyi 1964, x–xi). Let us reflect on what this means historically.

People everywhere and always have lived in a socially linguistically represented action-oriented world in which what a thing is must be derived from what it comes to mean within human life. This is what Husserl, Heidegger, and Schutz called "the lifeworld" (see Husserl 1954/1970, Heidegger 1927/1996, Schutz 1973), and for which W. Sellars coined the term "manifest image of the world" (Sellars 1963, 6). Within this perspective, many things are first grasped as having fixed essences dedicated (as it were) "by nature" to a single function. Such was the opinion of Plato and Aristotle, Aquinas and Descartes, Bacon and Newton, and it is a view still held by many philosophers and scientists today. However, with the advent of modernity, the world changed adopting as its defining characteristic an *inquiring theorizing scientific spirit*. Open season for scientific inquiry was declared on whatever is given in human experience, not just hammers, trees, texts, or atoms, but also political society, perception, food, athletics, emotions, religion, love itself, all present themselves as possible subjects for scientific studies. From each of these studies there is a latent theory to be derived that explains something in its individual socio-cultural meaning *all things being equal* by a set of explanatory theoretical parameters. As in the case of the hammer, these explanatory parameters address just one aspect of the thing in question, the *explanandum* or cultural meaning to be explained, forgetful of the other aspects the thing may have. But in so doing the theorizing process reveals the extent to which the *explanandum* can be taken over and assumed by other things or artifacts than the exemplars studied. Theory shows that what makes this or any individual hammer to be a hammer – or what makes this or any individual thing in human experience to be what it is perceived to be – is *not a defining essence but a movable contextual set of properties that can be found or engineered in many different ways in many different physical hosts*. To the extent then that explanatory theorizing scientific inquiry is successful, it turns the objects of its inquiry – whether natural or cultural – into mere replaceable resources to serve a variety of cultural lifeworld functions.

Everything that has been said about theory stands whether theory is interpreted 1. as a mathematical model connected technically with laboratory practice (the way most physicists look on theory) or 2. as a defining set of normative scientific concepts exemplified culturally in experimental practice (the way biologists and sociologists use theory). Heidegger also intended his critique of theory to apply 3. to the predication of all abstract, universal, quantitative and non-quantitative concepts to the perceptual domain, therefore also (among others) to the domains of natural and social sciences; they too constitute a theory about what is perceived, for they simplify experience by focusing just on one "as-such" aspect; they are then also subject to the conclusions stated above.

Whether then one interprets *theory* according to 1. or 2. above, one is faced with a duality of (explanatory) theory-laden and (cultural) praxis-laden contexts at the heart of all scientific culture. Moreover, this is also true for 3. Theory-laden and praxis-laden contexts are relatively independent of one another. They generate two loosely coordinated perspectival languages that are not isomorphic to one another but have only many-to-one and one-to-many correspondences; these are partially ordered by statement inclusion within a complemented non-distributive lattice (or *context logic*).

Human beings receive great benefits from scientific theories, not just in such tradition-bound domains as agriculture, health care, and the organization of work, but in every domain, for there is no domain that cannot be transformed by the products of science, from air travel to childbirth, from how we perceive to how we control our emotions. But there is a price to be paid for such changes. They affect the way cultural life teaches people to be human and communicates to them the sense of the wholeness, integrity, and goodness of the world, the self, and human communities. Changing the traditional vehicles by means of which these core meanings are maintained and handed on inevitably changes how people regard themselves, their neighbors, and the world around, with consequent risks of cultural instability in all three areas. Whether and with what consequences science is changing our culture is the domain of sociology, politics, and cultural anthropology (see Geertz 1973; and 1983). For such reasons hermeneutical philosophy must also become a salient feature of the philosophy of science, even of the natural sciences.

8 Hermeneutic Philosophy and Classical Pragmatism; Common Themes

How close is the hermeneutical philosophy of science outlined above to a pragmatism that draws its inspiration from the works of Peirce, Dewey, James, C. I. Lewis, and Mead? Pragmatism, like hermeneutic philosophy, turns to the lived world of inquiry but, unlike the latter with its root metaphor in the humanities, it derives its inspiration from the quantitative sciences. We have two idioms and two perspectives, a pragmatist one and a hermeneutic one. How much is *intertranslatable without residue*?[8] We shall next try to express the common themes, and put down separately what resists translation without residue because peculiar to just one of the perspectives.

We listed at the start of this paper common ground elements which, though coming from diverse sources and often understood in different ways, have nevertheless acquired widespread support within the larger philosophy of science community. One would expect to find then that pragmatism and hermeneutical philosophy have some measure of intertranslatability.

Pragmatists emphasize the process of inquiry and experience, as does the hermeneutical tradition. Both are rooted in a critique of scientific culture and the scientism of its current philosophy; the pragmatists focus on the quantitative and model-building sciences, the hermeneutical philosophers the cultural meaning-oriented sciences. Both are self-reflective. While one emphasizes meaning and the other action, these turn out to be in great part mirror reflections of each other. Both traditions acknowledge that inquiry is initiated by the breakdown of habits and beliefs that normally serve as guides to thinking, and is directed toward the search for new hypotheses that cross old disciplinary boundaries and the construction of ways to test these in human experience. Each emphasizes the expanded role of perception replete with cultural meaning which for Peirce and Dewey is a part of habitual knowledge. For Heidegger, there is the additional problem (to be taken up below), the inauthenticity of habitual knowledge.

In particular, all three see themselves as grounded in a lifeworld that is historical, concern-filled, social, and technological of which we have an embodied understanding. Both see the interpretation of experience as evolving through actions that are understandable and are themselves interpretations – hermeneutic philosophy calls them "existential interpretations" of experience; they have an affinity with artistic "performances", a concern of pragmatism. Both see science as the creation not just of theoretical meanings but more significantly of cultural meanings. While agreeing about the instrumental character of scientific theory, both pragmatism and hermeneutic philosophy naturalize such objects in the lifeworld regarding them therefore as perceptual objects though they may offer different justifications for this move (see Bourgeois 1996). There is then much in common between pragmatism and hermeneutic philosophy that is roughly intertranslatable.

9 Themes Peculiar to Pragmatism

Classical pragmatists, rooted in a sense of nature, and very much taken by the power of evolutionary biology, construed human problem solving on a continuum with animal problem solving. All knowledge is greatly oriented to ways of adapting to dilemmas whether local or global, animals adapting almost exclusively to local problems, and humans to a confluence of both local and global problems. The root metaphor of past evolutionary continuity of humans with animals is a powerful influence on pragmatists whatever they are discussing, whether it is the philosophy of science or esthetics. Classical pragmatists then place the roots of problem solving in animal life where the body is the medium in which the world is received and represented. This body is not passive but replete with perceptual and cognitive strategies, and the horizons of both human and animal understanding are expanded through bodily action in the world. Minds and bodies are not separated substances, and the continuum of problem solving abilities is not broken at the juncture between animals and humans.

Classical pragmatists adopted (what they took to be) the epistemology of the natural sciences; in this sense, they were naturalized epistemologists, but they were not overtly reductionistic as later variants of this position became. Their approach to the evolution of problem-solving was creative. After all, their philosophy of science was rooted in the capacities of human perception to affirm even what positivists called "theoretical entities" as things in the lifeworld.

They erred, in the case of James, in rendering truth cheap; in the case of Dewey in favoring technological domination; and for Peirce, in a representational theory which legislated that truth will be determined by whatever science says in the long run, an unhelpful and sometimes misleading view. However, the saving grace of classical pragmatism is its wide notion of inquiry, and the prominent role given to perception in knowledge acquisition.

10 Themes Peculiar to Hermeneutic Philosophy

Hermeneutic philosophy seeks a level of understanding beyond pragmatism which is rather the condition of possibility of the pragmatist's world in which theory and practice merge in action. This is (what Heidegger called) the *ontological* dimension, an understanding at the level of the human inquirer's pre-predicative and pre-categorized[9] active involvement with the lifeworld. The human inquirer, *Da-sein*, unlike the pragmatist's more commonsensical inquirer, is the human individual thrown into the world at a certain time and place, conscious of having a temporal destiny, and sharing the historicity of the human community's involvement with its most fundamental concern, Being; Da-sein's existence is understanding historical Being.

Although Heidegger sees theoretical or "calculative" thinking as necessary, he sees it as ever in danger of dissolving the dialectical distinction between cultural goals and their theoretical explanation, substituting the latter for the former. Such for him is the danger of inauthenticity with the reduction of (what he calls) "meditative" thinking to "calculative" thinking. The core stratum of meditative thinking, the deepest subject of human inquiry, Heidegger finds, neither in theory, nor indeed in practice, but in ontological Being. At this level, there are no articulated theories – neither quantitative, nor descriptive of scientific cultural entities, nor theories providing abstract categories of the perceptual world. But there, at their source in *Da-sein*, led by cultural as well as explanatory desires for the human good, is the upward movement toward the articulation of human understanding. This upward movement is met by a downward movement that seeks fulfillment in the self-manifestation of "*die Sache selbst*" thus closing the hermeneutical circle. Ontological Being then justifies the search for – as well as the subsequent discovery of – entities that (in the hermeneutical sense) are both theory – and praxis-laden. The art of such discovery is learned in daily life (say, from broken hammers), and it is brought to a fine art within the sciences where it can be learned only from apprenticeship to an expert community of research scientists. *Such a conclusion offers to provide an epistemic hermeneutical justification of pragmatism's – and science's – abductive methods.*

Hermeneutic philosophy's approach to scientific inquiry is also subtly different from pragmatism's in another respect, because its thinking is shaped predominantly by a different *strenge Wissenschaft*, one guided by literary, moral, social, political, historical, and religious interests. All *wissenschaftliche* inquiry takes place within the lifeworld and is characterized by a *theoretical* thrust that necessarily results in a distinction between the theory-laden mere resource of the explanatory or abstract dimension, and the praxis-laden dedicated resource of the public cultural domain. In humanistic scholarship explicit categories often introduce no more than play items making suggestions to the imagination without going so far as to cause them to be revealed in the present and actual lifeworld. Scientific research can benefit both from the imagination of the humanistic disciplines and the extraordinary potential for changing the lifeworld that the scientific disciplines possess. What may begin as a work of literary imagination – say, to discover black holes in our back yard or intelligent creatures in the galaxy – can become a scientific program and, if successful, it could change forever, not just human imagination, but the human lifeworld into which future generations will be born.

Hermeneutical activity, then, in contrast with pragmatism's more earthy abductive methods, is fundamentally metaphorical, multi-valued, and flexible until definite orientations are chosen for assertive expression with a consequent specification of meaning. Where an inquiry seeks new categories for the subject matter at hand, metaphors are first borrowed from areas better understood, to serve the inquiring imagination (as *Vorsicht*) in the hermeneutical circle of discovery. The history of science is full of such examples, from billiard balls, elastic bands, ethers, mechanical devices, and molecular bench models, to computer simulations, harmonic oscillators, and ten-dimensional spaces. Nor is it possible to come to understand modern physics or biology without passing through stages of metaphor. In the context of discovery, an initial multiplicity of metaphorical meanings is sifted by a cyclical interplay among the trio of background, theory, and experience that is called the "hermeneutical circle". In this process a proto-theory (or bunch of proto-theories) are fashioned and refashioned into something usable as a potential finite-dimensional model that eventually comes to express a certain way of meaningfully grasping experience within the cultural context of the particular inquiry. The outcome then usually gets handed on as the literal meaning of the discovery.

Since it is the case that scientists value literal expression over the metaphorical and make the (disputed) claim that theoretical language is normatively literal, the question arises: how do metaphorical expressions become literal? They take on literal meanings when authoritative conventions are made by those who can effectively establish "ownership" of a domain of knowledge, "ownership" in this case normatively implying acknowledged expertise. Metaphor becomes literal expression then only when there exists an authoritative community of experts (see Dear 1995, 23).

There is, however, always a suspicion of inauthenticity in literal usage, since literal use is always defined by a community whose credentials for expertise and legitimacy can – and sometimes should – be scrutinized. Moreover, uniformity of meaning tends to go hand in hand with forgetfulness of the diversity of meaning resources implicit in their hermeneutical origin and pragmatic use. Again, theoretical models are precise while life-world situations are always imprecise, and so theories always function in a cultural milieu that, being praxis-laden, does not need or support unlimited univocity or precision.

There is another aspect of hermeneutical thought of which pragmatists on the whole have little cognizance; this is the hermeneutic concern with "facticity". This refers to the hermeneutical fact that each individual human inquirer enters into life at a definite place and time as a member of a definite network of communities and cultures and pursues thereafter a unique life trajectory until death. It is within this limiting framework that the individual human inquirer, whether philosopher or scientist, comes to experience the circumspective (concern-filled) onticity of the temporal life-world.[10] The temporality of human existence puts limits to among others the achievement of authenticity. While scientists and philosophers were very much aware of the insecurity of habitual human knowledge seeking stability amid insecurity, Heidegger saw a danger in the search for security as indicative of a desire for a kind of existence that is contradicted by every individual's life-towards-death. If the human inquirer, for instance, has a tendency to objectify whatever is expressed by nouns in human language, such as ideas, theories, things, and functions, as if these existed independently of human culture, only rarely do human inquirers have the opportunity to delve deeply into the originary roots and metaphors in the imagination that feed the meanings of these words.

A similar temporal tendency towards inauthenticity affects community cultural paradigms and institutions leading to forgetfulness of their historical origins in a common metaphor-filled life. So old forms of life are fated to be dropped and ever to be replaced by new ones. This too is the story of science whether one organizes the story according to "paradigms" and their overthrow by "revolutions" or in some other way. "Scientific revolutions", whether Kuhnian or not, share this revolutionary character with all institutionalized forms of knowledge. Within the hermeneutical context, this means that there is no guarantee of historical convergence toward a final scientific account or any other account.

11 Implications for the Philosophy of Science

Pragmatism and hermeneutic philosophy both treat science (or in general, all scholarship) as a *form of human culture* which approaches the world in the spirit of active inquiry in which background, theory, and praxis work together in orderly succession. Pragmatism tends to see them as working together seamlessly in solving problems and helping humans to adapt to changing ontic environments. Hermeneutic philosophy pries them apart to study the contribution each makes to the generation of meaning; among them the lifeworld has priority as ontological background, because *there* is to be found living human understanding, and from it springs the intellectual desires and frustrations that stimulate inquiry and serve human life. Established cultural praxes have a certain priority because theoretical inquiry is instigated in order to improve, re-establish, or adapt these practices to human needs; they thus furnish *explananda*. The search for the *explanans* is then the search for the relevant model or theoretical infrastructure of an existing cultural praxis and it proceeds by the process of the hermeneutical circle (ontological background, to theory, to trial, and back). When a solution is found, a new cultural context is inaugurated in which, all things being equal, the *explanandum* becomes subject to manipulation through its model-theoretical infrastructure. The historical character of the relationship between *explanadum* and *explanans* gives point to Hume's concern that the occult powers that cause the observed regularity in nature may not be reliable in the long run (see Dear 1995, 18).

In science, the *explanans* is apt to contain new entities, "theoretical entities". These are revealed in experimental practice as parts of the explanatory infrastructure, the *explanans*, of the *explanandum*. Exploration of the *explanans* constitutes a new scientific research program to which these entities belong essentially. Turning to the laboratory, the reference of the *explanans* is in the first place to the measurement setup, its design and engineering; it is to this environment that the new "theoretical entities" primarily and essentially belong: they belong only contingently, if at all, to the constitution of the *explanandum* (see above).

Both pragmatism and hermeneutic philosophy have a stake in answering the question: What is theory and what is its role in culture? But the tension to clarify the role of theory is strong in hermeneutical philosophy and weak in pragmatism. The hermeneutical answer to this question should dispel the view widespread even among some pragmatists that the old cultural account is capable of being replaced by a single new scientifico-cultural account, this last as we have shown is beset by logical problems which will be mentioned all too briefly below.

Since all scientific and scholarly inquiry embarks on a project whose goal is to construct an explanatory theory about a starting point that is anchored in the cultural life of people, the discovery process is always constrained by the condition that a meaningful coordinated relationship to public culture of common as well as scientific experience be maintained throughout the inquiry. Common sense then always retains a certain authority. For this reason, the social study of science and its cultural anthropology are of keen interest both to pragmatism and to a hermeneutical philosophy of science. From the perspective of hermeneutical philosophy, these studies need to be guided – as indeed all inquiry should be guided – by the hermeneutical circle of ontological background, theory, and practice discussed in this paper. The pragmatist circle in which cultural background, theory, and praxis are seamlessly joined fails to address some deeper interpretative issues.

Since hermeneutic philosophy sees the effects of technologies on the historicity of scientific culture and the lifeworld, an important task of hermeneutic philosophy is to inquire into the coherence implicit within the multiplicity of cultural and theoretical perspectives that arise in sociohistorical and technological contexts. Has it a logical structure? Is it, say, the logical structure of context-logic?

While we do not ask of a philosophy that it contribute to the successful practice of science, science continually throws up metaphysical questions that divide the scientific community and constrain or limit its energies in a world of temporal inquirers, historical institutions, and finite resources. Increased clarity about the mission and process of scientific inquiry can be helpful without intruding on the actual practice of science.

For example, the philosophical principles laid out in this paper could conceivably throw new light on some special issues that arguably need the dialactical character of theory and cultural praxis for a solution. Among them, for instance, are conceptions of the universalizing theoretical role of individual experiments (see Dear 1995, 21–25), and of *measurement* as the locus of the interface between scientific theory making and its practical cultural object. Other deeply philosophical problems in natural science relate to Wigner's problem of the "unreasonable effectiveness" of mathematics, the Bohr–Heisenberg problem of the distinction between large and small, the praxis-laden meaning of theoretical non-commutability in quantum physics, a theory/praxis approach to the quantum paradoxes of space and time and to the possibility of a unified quantum relativity, and the praxis-laden stratification of natural science under broken theoretical symmetries. In the social, psychological, and neural sciences, confusion between the theoretical and the cultural abounds, in great part because these sciences have not yet a way of coping with the contextual dependency that empirical data in these fields have on conscious or unconscious cultural influences.

Hermeneutical philosophy and pragmatism both support the principle that scientific entities, even those not perceptible to the unaided senses, function (with the help of science-based technologies) as naturalized citizens of the lifeworld. In both philosophies, such entities arguably become public cultural entities by their ability to carry and fulfill perceptual meanings within the lifeworld. In this case scientific entities become a common possession of society integrating science with the lifeworld, a highly desirable situation for a democratic society.

Within a political society such as ours that is so deeply aware of the enormous needs for and benefits from theoretical knowledge, but that nevertheless has its own independent set of practical goals, it is highly desirable for scientific institutions to be able to give a better account of their public role as the principal producer of theoretical knowledge. It may be a mistake to suppose that the current questioning of the agenda of scientific culture by public agencies and media indicates a "flight from rationality" or a lapse of public confidence in scientific knowledge, rather than a general failure of public institutions including scientific ones to present a believable culturally attuned and historical account of science.

Returning to where we began in this paper. The import of historical, social, and political studies of the practice of science first got serious public attention through the work of the late Thomas S. Kuhn, who gave *paradigms* and *scientific revolutions* their names. He failed, however, as most others have failed too, to give a good philosophical account of historical theory change and the cultural construction of scientific entities (see Hoyningen-Huene 1993, Dear 1995, Shapin 1996). Hermeneutical philosophy and a hermeneutically sensitive pragmatism could fare better. What is needed in addition is a philosophy of scientific discovery and change to help bring the history and philosophy of science together with the social and cultural studies of science. Such an accomplishment would provide elements for a better scholarly and public appreciation of what is certainly one of the greatest institutions of our society.

A final word and conclusion: the principles laid out above stress the fact that so-called "theoretical entities" are naturalizable in the lifeworld through science-based technologies and thereby become public cultural and perceptual entities. These principles which span both pragmatism and hermeneutical philosophy can found a *new empiricism* and restore kinship with the scientific revolution that began, not with Descartes or Boyle, but with the empiricism of Galileo.

Notes

1 For the purposes of this paper unless otherwise indicated by the context, we take "theories" and "categories" to be reflectively defined abstract objects such as scholarship and science provide; they may be mathematically defined objects or kinds of perceptual objects.

2 In this essay, we follow Pickering (1995) in taking *scientific culture* to cover not just a field of knowledge – "scientific facts and theories" – but in addition "to denote the 'made things' of science, in which [we] include skills and social relations, machines and instruments…" (p. 3). We interpret this to include the external relations of the active scientific community to other communities, for instance, in terms of technological use and cultural transformations.

3 By "objective" knowledge – the quotes are just to draw attention to the need for clarification – we mean, knowledge that purports to represent reality according to a correspondence theory of truth, that is, as it is independently of and undeformed by cultural biases.

4 The terms "ontic" and "ontological" are used in Heidegger's sense; "ontic" applying to things distinctly categorized in the lifeworld, "ontological" signifying the background of Being in which *Da-sein* lives antecedently to all descriptive categories.

5 By *Da-sein* is meant the human inquirer, individual, historical, thrown into the world at a certain time and place, and yet *Da-sein* shares in the destiny of the human community's involvement with Being. Heidegger speaks of *Da-sein* as to-understand-Being, to-be-there-amid-Being, to-have-been-thrown-into-Being, etc.

6 Readers may find some difficulty with the meaning of this language. It attempts to direct thinking to the conditions of possibility of every articulation of human experience in the lived understanding of the lifeworld, antecedent (in principle) to the formation of perceptual kinds (of things, events, etc.) and their representations in language or language-like signs. I have translated *Vorhabe* as *lifeworld background, Vorsicht* as *proposed theory*, and *Vorgriff* as *looked-for fulfillment in experience or experiment*. To anticipate what comes later in this essay: the pre-theoretical pre-predicative understanding of the world is taken as the condition of possibility of what in Dewey's thought is understood as the intelligent habits presupposed by our everyday behavior; this is the pragmatist's world.

7 Some readers are confused by the word "circle" taking it to mean "return to the same starting point", but that is not what is meant here. The "circle" of hermeneutics indicates the repetition of a methodological cycle, not a return to the same contents but to progressively transformed contents.

8 By "[inter]translatable without residue" is meant: there is nothing lost or gained in the translation and as a consequence there is a minimum risk of misinterpretation when a translation is made.

9 By "pre-categorized" is meant: when language is used ostensively, that is, before descriptive terms are reflexively given abstract definitions. By "pre-predicative" is meant: before such terms are used to make assertions (by predicating them of presentations in human experience). The degree of specificity of categorial definitions will vary with subject matter, but in each case it will be an abstract content and theory-like.

10 *Facticity* is partially reflected in pragmatism's concern for the unresolved conflicts of the human condition, but this is not a fundamental theme that deeply colors the characteristic optimism of classical pragmatism.

References

Babich, B. E.: 1994b, "Philosophies of Science: Mach, Duhem, Bachelard", in Richard Kearney (ed.), *Twentieth Century Continental Philosophy, Routledge Encyclopedia of Twentieth Century Philosophy*, Volume VIII, Routledge, New York, pp. 175–221.

Bourgeois, P.: 1996, "Merleau-Ponty, Scientific Method, and Pragmatism", *Journal of Speculative Philosophy* **10**, 120–127.

Dear, P.: 1995, *Discipline and Experience: The Mathematical Way in the Scientific Revolution*, University of Chicago Press, Chicago.

Dewey, J.: 1958, *Art as Experience*, Putnam, New York.

Geertz, C.: 1973, *The Interpretation of Culture*, Basic Books, New York.

Geertz, C.: 1983, *Local Knowledge*, Basic Books, New York.

Hanson, N. R.: 1958, *Patterns of Discovery*, Cambridge University Press, Cambridge.

Heelan, P. A.: 1989, "After Experiment: Research and Reality", *Amer. Phil. Qrtly* **26**, 297–308.

Heelan, P. A.: 1983/1988, *Space-Perception and the Philosophy of Science*, University of California Press, Berkeley and Los Angeles.

Heidegger, M.: 1927/1996, *Being and Time*, trans. by Joan Stambaugh, Albany NY, SUNY Press. Page references in the text and notes are to the German 1927 edition and are given in the margins of this and the Macquarrie/Robinson translation.

Hoyningen-Huene, P.: 1993, *Reconstructing Scientific Revolutions: Thomas S. Kuhn's Philosophy of Science*, University of Chicago Press, Chicago.

Husserl, E.: 1954/1970, *Crisis of European Sciences and Transcendental Phenomenology*, Northwestern University Press, Evanston.

James, W. 1912/1996, *Essays in Radical Empiricism*, University of Nebraska Press, Lincoln and London.

Merleau-Ponty, M.: 1962, *The Phenomenology of Perception*, trans. by Colin Smith, Routledge and Kegan Paul, London.

Nickles, T.: 1995, "Philosophy of Science and History of Science", *Osiris* **10**, 139–163.

Parrott, W. J. and Schulkin, J.: 1993, "Neuropsychology and the Cognitive Nature of the Emotions", *Cognition and Emotion* **7**, 43–59.

Pickering, A.: 1995, *The Mangle of Practice: Time, Agency and Science*, Chicago University Press, Chicago.

Polanyi, M.: 1964, *Personal Knowledge*, Harper and Row, New York.

Quine, W.: 1960, *Word and Object*. MIT Press, Cambridge MA.

Rouse, J.: 1996, *Engaging Science*, Cornell University Press, Ithaca and London.

Schutz, A.: 1973, *The Problem of Social Reality*, Vol. 1 of *Collected Papers*, Martinus Nijhoff, The Hague.

Sellars, W.: 1963, *Science, Perception, and Reality*, Routledge, London.

Shapin, S.: 1996, *The Scientific Revolution*, University of Chicago Press, Chicago.

14

Dysfunctional Universality Claims? Scientific, Epistemological, and Political Issues

Sandra Harding

1. Universality and modernity. The ability to produce a uniquely universal science is commonly thought to be a distinctive mark of modernity. Such a science should have uniquely valid standards of rationality, objectivity, method, and what counts as nature's order and as knowers. Such purported features of modern science have been called on to legitimate as uniquely socially progressive European models of government, law, education, social policy, and even ethics. From this perspective, modern Europe and its diasporas in the Americas, Australia, and elsewhere provide a uniquely desirable model of human achievements, social relations, and standards of living.

Other cultures' accounts of themselves and the world around them, like those of premodern Europe, cannot really be considered in the same category as modern sciences, this account holds. They should not even be called science traditions since there is one and only one possible true account of nature, and that is the one modern science has been struggling to piece together.[1] [S]uch other accounts are to be regarded only as "ethnosciences," "folk thought," "precursors of modern science," belief systems "with scientific elements," or – worse – savage and barbaric thought, witchcraft, superstitions, magic, and products of prelogical mentalities.

Moreover, models of objectivity, rationality, scientific method, and their ability to advance modernity have also invariably been defined in terms of idealizations not only of Europeans, but also of masculinity, as...feminist critiques...have shown. The desirable virility of European civilization is often signified by the progressiveness of its modern sciences, and the desirability of dominant models of manliness by their links to modernity, its rationality and social progressiveness. The thinking and behavior of women of European descent, too, are to be assigned to the premodern, according to this kind of conventional thinking.[2] Thus, the meanings of claims to modern science's universality are constructed in opposition to meanings of womanliness and cultural otherness.

The power of these meanings is indicated by their immunity to empirical evidence to the contrary, as feminist and postcolonial science and technology theorists have pointed out. No matter how effective other cultures' (including "women's cultures"') knowledge traditions are, were, or might have been for enabling effective interaction with natural worlds, they are not counted as real sciences. No matter how much modern sciences might have incorporated elements of other cultures' concepts and theories about nature, their mathematical and empirical techniques, and even whole bodies of their accumulated navigational, medical, pharmacological, climatological, agricultural, manufacturing, or other effective knowledge enabling prediction, and control of nature, these other bodies of knowledge are not counted as "real science" until incorporated into European knowledge systems. And no matter how poor at explanation, prediction, and control European sciences are – for example, with respect to social causes

From Sandra Harding, *Is Science Multicultural? Postcolonialisms, Feminisms, and Epistemologies*, Bloomington: Indiana University Press, 1998, pp. 165–225.

of environmental destruction, or the causes of patterns of carcinogens or contagious diseases – these inadequacies do not count against European sciences' purportedly unique universal validity. ...[3] How is this pattern of prevailing thought to be explained in light of the fact that success at prediction and control of nature are always claimed to be the most reliable hallmarks of a real science?

When assumptions are used to define the unique identity and value of a people, its civilization, and its history in the way the universal science ideal and its supporting internalist epistemology tradition apparently do, critical examination of them obviously is going to draw forth deep psychic anxieties and resistances. When assumptions carry this much moral and political weight, exposing them to contrary evidence is not going to be a comfortable process.

Four unity-of-science claims. In the early twentieth century, the unity of science thesis became an important form in which the universality claim was widely defended. This thesis overtly makes three claims: there exists just one world, one and only one possible true account of it ("one truth"), and one unique science that can piece together the one account that will accurately reflect the truth about that one world. What this thesis means – what methodological, metaphysical, or other features could constitute the unity of physics, chemistry, biology, psychology, economics, and so on – are issues about which philosophers and scientists have been concerned to make sense.[4] However, as political theorists are aware, a fourth claim is also assumed in these universality/unity arguments: that there is a distinctive universal human "class" – some distinctive group of humans – who should be taken as exemplars of the uniquely or admirably human to whom the truth about the world could become evident. For early modern scientists and philosophers, such a group was those members of the new educated classes whose minds were trained to reflect the order of nature that God's mind had created, as God's mind had also created human minds "in his own image." God's mind, human minds, and nature's order were assumed to be congruent or homologous. Scientists and the educated classes that could see the truth and importance of scientific accounts represented the universal class that could learn to detect the one possible true account of nature's order. For nineteenth- and twentieth-century Marxists, the proletariat represented this universal class. This class alone, since its labor transformed nature into provisions for everyday human life to exist, had the potential to become the unique representative of distinctively human knowers. This class alone had the potential to detect the real relations of nature and social life beneath the distorted appearances produced by class society. Some forms of feminism have flirted with a similar kind of transvaluation of gender that considers the possibility that women are the uniquely human gender: if it made sense in sexist society to imagine men as the model of the uniquely human, then perhaps it is reasonable to consider how in many respects women's characteristics – their claimed altruism, pacifism, sensitivity to others' needs, or some other putative virtue – are more reasonably regarded as uniquely valuable models of the human, capable of producing less distorted understandings of natural and social orders. And some African Americans have claimed that the suffering, compassion, or some other characteristic of African Americans under the horrible conditions of slavery uniquely equips them to understand natural and social orders in ways unavailable to those who have not had such experiences.

There are important insights behind such claims.[5] In the contemporary world of multicultural, postcolonial, and (more complex and diverse) feminist politics and social theories, however, faith has declined in the possibility and desirability of such a universal class – Enlightenment, proletarian, feminine, or culturally distinctive in any other way. In these worlds in which we all live (whether or not we acknowledge the effects post-World War II emancipatory social movements have had), who could such a distinctively human, universal class be? What group could democratically gain assent to their own abilities uniquely to represent accurately universal human interests and the one true natural and social order such interests supposedly could reveal in the face of other groups' different but also valuable cultural conditions for the production of knowledge for them and their survival? In contemporary life, many kinds of important differences between humans – biological and, more importantly, social, economic, political, psychic, and otherwise cultural – are recognized as resources for producing effective knowledge and advancing democratic social relations.[6]

The universality/unity ideal is no mere philosopher's notion; in one form or another it has been one of the most central and enduring values of otherwise conflicting conceptual and political tendencies in modernity's social theories. However, it is now attracting critical attention from many groups around the globe which claim that for them it has had primarily bad scientific, epistemological, and political effects.

2. Science and democracy: Allies or enemies? Does the ideal of a single, universally valid science decrease global democracy? Does the goal of global democracy advance or obstruct the plausibility of such an ideal? In the late nineteenth and early twentieth centuries, defenders of the universality ideal hoped that it could serve as a powerful antidote to the tides of racialist and nationalist partisan conflict that again and again had resulted in violence and even genocide. For them, appreciation of the universality of science and its standards of rationality and objectivity could only support and advance democratic social relations. Today the universality ideal's defenders see in the unique standards of scientific rationality and objectivity the main hope for restoring what they think of as the fair and orderly social relations now being disrupted by the claims and demands of multiculturalism, feminism and "relativism" in the post-Kuhnian social studies of science. In effect, these defenders of the universality ideal fear and do not find plausible the analyses produced by the three distinctive schools of science and technology studies that have emerged since World War II... .

For many feminist, multiculturalist, and post-Kuhnian science studies theorists, however, the universality ideal increasingly appears as a force for maintaining inequality and obstructing democratic tendencies, and for obstructing the growth of knowledge. For these groups, claims to the transcultural truth of modern sciences' representations of nature, and of only those of modern science, function to mask the ways that modern sciences and their representations of nature's order tend to distribute the cognitive and social benefits of scientific and technological changes disproportionately to those already positioned to take advantage of them, and the costs primarily to those least able to resist them. Moreover, universality claims legitimate the devaluation and even destruction of knowledge traditions that have enabled women, the poor, and less powerful cultures to interact effectively with their environments. The unique universality claims also have bad epistemological and scientific effects, in addition to their political consequences. In a variety of ways, they function to increase the production of systematic ignorance. From this perspective the universality claims are epistemologically, scientifically, and politically dysfunctional.

Of course the familiar "universalist" response to such claims is to insist that the "anti-universalists" are confused. Lamentable as the worsening situation of women and racially and culturally disadvantaged groups may be ("if this *is* the situation," they demur), such arguments only address the applications and technologies of science, and nobody denies that these often are shaped by anti-as well as pro-democratic politics. Of course politics can misuse and abuse applications of sciences and technologies – "Think of Lysenkoism! Think of Nazi science!" they argue. Real sciences could not possibly have any political consequences. "Real sciences" simply provide pure information about nature's order, according to this older view. One can note that such a position also blocks the argument that science, its rationality and objectivity, support or advance democratic social relations, or have any other good social effects, however. If science were culturally neutral, then it could not have any social effects at all. Of course this recognition motivated the invention of "positivism" in Auguste Comte's proposal that the pure information that sciences produce is politically neutral, and the only *positive* social effects of science are to be found in its distinctive method.

However, this attempt to disassociate modern sciences from their effects has become increasingly difficult to defend. Earlier chapters reviewed the evidence against this "pure science" claim when they focussed on the explicit mission-directed character of so much valuable scientific research, from Galileo's work in the Venice armory and Pasteur's concerns with public health, to contemporary medical, economic, and military-directed scientific and technological projects.[7] Such histories clarify that mission-directed research should not always be conceptualized as an obstacle to the growth of scientific knowledge; obviously, it can produce valuable information about and explanations of nature's regularities, whether or not one approves of the "missions." Sometimes the cultural interests and values that constitute a scientific project, its conceptual framework, methods, and purpose are politically relatively uncontroversial; sometimes not. After all, medical research to discover the causes of cancer or of AIDS, and research to establish space satellites for military surveillance have all produced reliable and valuable information about nature's regularities regardless of how one evaluates the social desirability of such projects. How are the technologies and applications of a science to be regarded as completely separate from that science's information when the information is produced specifically *for* such technologies and applications? Obviously, to start with, the patterns of knowledge and ignorance that a science or collection of sciences generates bear a close relation to the culturally local overt purposes and unarticulated interests in such information. If such science is conceptualized at its cognitive core in ways suitable to culturally local medical or military purposes – whether or not individual scientists are aware of such a match – in what sense is it "pure"?

Another kind of evidence against the purity-of-science thesis, however, showed that technologies of scientific research themselves contribute to constituting what can count as legitimate scientific knowledge, and that those research technologies themselves have always been constituted through social, economic, and political processes that have social, economic, and political consequences. Such research technologies are part of scientific methods, yet modern science's methods were long cited as value neutral and thus responsible for science's unique universality. Consequently, philosophers of science have been forced to reexamine what it is about scientific methods that makes modern science so effective if it is not their social and political neutrality.[8]

Thus, the assumption that the universality ideal can only advance both the growth of knowledge and a democratic social order has come under increased suspicion from a variety of sources.

3. The epistemological status of universality/unity claims. The universality/unity claims express ideals, not the results of empirical observation. The question of this chapter is, therefore, whether they should retain their status as ideals. However, many of their defenders seem to think they are scientific claims; that the history and present achievements of science somehow prove that it is uniquely universally valid and unified.

One stream of common sense captured by the older histories and philosophies of science seems to conjoin everyday experience and historical evidence to make the universality thesis appear obviously supported by empirical evidence. In this way of thinking, our observing, perceiving bodies-with-rational-minds and their environments seem to fit together in ways that enable modern sciences to gain greater and greater accuracy in their predictions and control of nature. How else could we account for the many magnificent achievements of modern medicine, the ability of humans to walk on the moon, or the possibility of composing these sentences on this computer? Modern sciences must be providing more and more pieces of the one coherent account that reflects or uniquely corresponds to how the world is for these technological achievements to be possible. Doesn't the history of such successes constitute incontrovertible evidence that there is one nature, one truth about it, and one science that captures that truth? How could scientific predictions increasingly achieve such accuracy were this not the case?

On the other hand, before the pull of this way of thinking becomes too irresistible, we can recollect that common sense and the history of science also tell us that there is something wrong with it. We all know that scientific claims can never be regarded as once and for all proved (or disproved). They always must be left only tentatively confirmed by observation and reasoning since new evidence continually shows how familiar scientific ways of thinking have limitations that were not earlier visible. Every scientific project makes "background assumptions" about properties of the instruments, theories of vision, what the relevant variable local conditions are, the cultural neutrality of the relevant conceptual framework, and much more that can at another time be thrown into doubt. Perhaps even more importantly, any given scientific framework eventually outlives its usefulness in advancing the growth of knowledge. At moments of revolutionary scientific change, when a new framework promises to replace the older one, old data is repositioned within a different conceptual scheme. Most of the observations made in Ptolemaic astronomy were repositioned within Copernican astronomy. Do they provide "the same information" within these conflicting conceptual frameworks? Well, yes and no. Of course many observations of the moon, the sun, and the planets made prior to Copernicus and Galileo were retained in the heliocentric account. In one sense we see "the same heavens" that cultures did a millennium and more ago. On the other hand, what patterns such observations form, how such patterns are to be explained, and what such observations, patterns, and explanations mean to different groups of scientists and their diverse audiences (intended and unintended) – these differ immensely between the two theories. In important respects, the pre-Galilean observations do not provide "the same information" for heliocentric theories.[9]

In recognizing the importance of this kind of understanding of the history of science, one need not hold that new conceptual frameworks are fully incommensurable with the older ones, or that ones from different disciplines leave their practitioners unable to communicate with each other or to work together on scientific projects, as both "incommensurabilists" such as Thomas Kuhn and also his critics claimed must be the case when one gives up these kinds of universalist assumptions. Temporary, local *universalizing* strategies are devised in the face of the de facto unavailability of reliable universality claims. Historical and sociological evidence shows how scientists continually make effective, "good-enough" translations – pidgin languages – and technical equivalences to get from one conceptual terrain to another and to enable them to work together effectively. Thus, their joint scientific projects can draw on otherwise disunified, heterogeneous, local scientific practices and cultures. For example, Peter Galison examines the "trading zones" scientists create between their otherwise disunified work. When H-bomb designers, logicians, aerodynamical engineers, and statisticians sat down together in the 1940s and 1950s to construct computer-simulated realities, they brought to such interactions very different notions of "randomness," "experiment," and other terms.

> While the mathematician thinks about the best definition of "random" quite differently from the physicist, in the cauldron of those early days of computer simulations a notion of "random enough for present purposes" emerged, borrowing from several cultures, yet belonging exclusively to none.[10]

Others have pointed to the diversity of such translation devices developed not only in modern science traditions but also in the scientific and technological traditions of other cultures. Modern sciences are like other scientific and technological traditions in that they have developed just such means of establishing effective continuity in scientific and technological projects by peoples with disparate, heterogeneous kinds of knowledge. Historians, philosophers, and ethnographers of sciences and technologies are still struggling to come up with a way to represent adequately the immense innovativeness of scientific and technological workers in developing these complexes of conceptual, technical, institutional, rhetorical, practical, and other kinds of devices.[11]

Such analyses of the strategies scientists use to communicate and work together across their heterogeneous cultures have not yet come to inform common sense. Nevertheless, reflection enables anyone to realize that in order for the growth of scientific knowledge to have occurred, scientific change must be more open to the value of alternative models of nature's order than the "one world, one truth, one science" ideal (and one culturally distinctive kind of ideal knower) can suggest. The ideal could not accurately be reflecting the history of science as that is widely understood after more than three decades of the post-Kuhnian, feminist, and postcolonial science studies. These schools of science studies delineated the resources provided for the growth of knowledge by assuming many human worlds, many truths, many sciences, and many culturally diverse knowers.

Thus, the universal science ideal appears to be inconsistent with the best history of science and with the best contemporary scientific practices. If the results of scientific research cannot and should not be protected from historical change, then it is not clear what it could mean to claim them to be universally valid, let alone uniquely so. On the other hand, before leaping gleefully to the presumed final defeat of this particular conceptual framework and the social programs in which it was embedded, we must note that the defenses of the localness, heterogeneity, and disunity of science also should not be taken to be transcultural truths, fully supported by incontrovertible evidence from within the sciences and from the history and sociology of science. The argument here, instead, is only that the universality arguments seem to block our understanding of the history and necessary practices of the sciences. One could say that both the universality and anti-universality arguments are strategies for keeping understandings of human knowledge of nature in balance when confronted with tides that threaten to pull too strongly toward insistence either on unique universality or only on incommensurable localness in knowledge systems.

There is another kind of defense of the unique universality ideal that must be considered. Are there weaker forms of it that might be more plausibly defended?

Weak forms of the universality claims. i. Cultural plurality of scientists attests to universal validity of modern science. Let us set aside first a popular defense of the universality thesis that consists in pointing to how modern science's creators and users today come from many conflicting cultures, yet all agree to scientific claims. The individuals who have made modern sciences have come from

Great Britain, Japan, Germany, India, Denmark, and many other nations; they have held diverse religious, political, and other kinds of cultural beliefs; yet they have been able to agree to scientific claims though there might well be little else to which they could agree. Something about modern science must be universally valid for it to emerge from such culturally diverse peoples in the first place, and then to continue to gain assent from more and more culturally diverse peoples around the globe.

This popular argument must be taken seriously since it is just the one that the Vienna Circle scientist/philosophers made in other forms: in a world of partisan conflict, Rudolf Carnap and others argued in Europe in the late 1930s, pursuit of the universally valid standards of scientific rationality provided the only imaginable hope for achieving peaceful resolutions to social problems. From this perspective, Nazism and the Holocaust were the consequence of too little scientific rationality. Such icons as this argument make it difficult to resist the appeal of "one nature, one truth, one science."

One can gain a useful perspective on such an issue by nothing that European scientists have all agreed to claims that were, and for many people from other cultures still are, embedded in other, non-European worldviews. The number "0," to take one example, has distinctive cultural meanings for many peoples. It represents the beginning or origin of all that follows. From nothing emerges everything. Thus, it is often regarded as different in quality from all the numbers that follow. It is not in the same cultural category as 1, 2, 3 and the others. Indeed, the number one often has such distinctive properties in many cultures: from homogenous unity emerges the heterogeneous multiplicity of the world we know. Moreover, zero has been invented or discovered independently in several different cultures, in each of which it has occupied a distinctive place in the prevailing cultural discourses. However, it, like all the other numbers, is presumed in modern scientific thought to have no culturally distinctive meanings at all, thought even in European history prior to the modern era zero also carried different meanings for different groups. Is the cultural neutrality presumed for numbers in the modern European worldview itself a culturally distinctive feature of that worldview? Chapter 4 examined why it is illuminating to understand how such assumptions and ideals of cultural neutrality are themselves culturally distinctive.[12]

To take another example, now acupuncture has been integrated into European medical practices for the control of chronic pain. Of course what has been extracted from the Asian systems of thought and practice from which it has been borrowed is just those elements that will fit into modern biomedicine. Discarded are the rest of the Asian beliefs about bodies that get in or out of "balance," and how to keep them in balance, that do not coherently fit into modern biomedicine's ontology and epistemology.[13] Yet we can ask: if these elements of other cultures' knowledge systems have become scientific once they have been adopted into the culturally distinctive ontology, epistemology, psychic structures, and political economy of European sciences, why weren't they just as scientific before, when they were embedded in different culturally distinctive interests, discourses, and ways of organizing the production of knowledge? After all, zero and acupuncture "worked" both theoretically and practically in those contexts no less than they do in this one. And why should we regard those larger bodies of culturally distinctive beliefs of other cultures as only containing obstacles to the growth of knowledge about the natural world when those cultural values and interests produced the thinking and practices of zero and acupuncture that have proved so valuable far beyond their cultures of origin?

Such reflections suggest that we need a different way of thinking about the fact that the claims of modern sciences, like those of other cultures, "hold true" – or, can be useful – far distant from their sites of original observation, and for phenomena that may be described and explained very differently in other cultures. They enable accurate prediction and control of nature. Evidently claims emerging from other knowledge traditions also can be regarded as universal or universally valid in this sense no less than modern scientific claims. And we can ask how such knowledge is different from the ways Ptolemaic astronomy and Aristotelian physics, in spite of their culturally local meanings and referents, predict accurately a great deal that Copernican astronomy and Newtonian physics respectively also predict. Don't modern sciences in such cases agree to the claims of knowledge traditions that are not modern European ones? It no longer appears reasonable to argue that while some cultures manage to stumble upon empirically adequate claims, only modern science's claims are produced by culturally neutral, and thus universally valid, methods.

ii. Metaphysical, methodological, and reasoning unity. Attempts to support the unity-of-science thesis and its universality ideal have been worked out in other forms by philosophers of science. John Dewey meant by the unity of science only a "scientific attitude" – a weak form of only methodological unity. For Rudolf Carnap, however, it was a linguistic unity that was important. The laws of biology and physics both "could be expressed in terms of everyday physical terms and procedures."[14] He did not mean that physical laws could replace biological ones, or that the entities of physics and biology would turn out to be one and the same. Otto Neurath's understanding of what the "unification" of science entailed was much like Carnap's.

On the other hand, Victor Lenzen thought that the laws of different fields of physics, and of physics and biology could be integrated. To Hilary Putnam and Paul Oppenheim, the domains of science could be reduced one to another, physics forming a base for the familiar hierarchically structured pyramid of unified sciences. And in the 1990s, physicist Steven Weinberg argued for a metaphysical unity of science that justifies U.S. government funding of the superconducting super collider (in his home state of Texas). Such a project would enable modern science finally to identify the fundamental constituents of the universe.[15] Such a metaphysical unity of the sciences produces "fundamental facts" that must be part of any subsequent science worthy of the name.[16]

Philosopher Ian Hacking points out that the idea of unity has always emphasized or blended two distinct ideas: singleness and integrated harmony. These have been weighted in different proportions in defense of diverse kinds of metaphysical unity, methodological unity, and logical or "styles of reasoning" unities. Here is where the unique universality ideal (singleness) can be seen to be embedded in the unity (integrated harmony) arguments. Surveying diverse arguments for just metaphysical and methodological unities, Hacking identifies "a metaphysical sentiment, three metaphysical theses, three practical precepts, and two logical maximums" each of which weights differently unity as singleness and unity as harmony.[17] Moreover, there evidently are as many styles of scientific reasoning as there are of effective human reasoning more generally. Crombie identifies six important ones, each of which brings its own standards of adequacy.[18]

Thus, these post-Kuhnian science and technology studies arrive at the same assessment that the postcolonial and feminist science theorists do: science cannot plausibly be understood as one single kind of thing at all. "There is no set of features peculiar to all the sciences, and possessed only by sciences. There is no necessary and sufficient condition for being a science."[19] In the face of such disunity, many different kinds of techniques serve as "unifiers," as Hacking refers to them. Mathematics is the earliest recognized to have such a function. However, it turns out that there is no "one thing" that is mathematics since it, like the sciences to which it gives an appearance of unity, is a motley collection of principles and practices, as he and other historians of mathematics point out.[20] And numerous other such unifiers are to be found among scientific instruments, techniques, attitudes...all those inventive strategies that occur in the "trading zones" within which scientists in the modern West, like other indigenous knowers, work to communicate across the diverse cultural and natural conditions that separate them.

Thus, common sense, the history of science, and observation of contemporary scientific practice do not support either the inherent singleness or "integrated harmony" of modern science. Moreover, earlier chapters explored the scientific, epistemological, and political advantages of understanding science as necessarily and desirably plural. Let us briefly review those findings and then turn to consider some techniques pointed out in the postcolonial and feminist accounts through which the illusion of the unity and universality of modern sciences has been historically established.

4. Producing illusions of unity and of universality. Some of the most important cognitive, technical resources of modern sciences are to be found in their distinctively local features, as post-Kuhnian, postcolonial and feminist science studies showed in earlier chapters. Of course nature's order – "reality" – has a great deal to do with what sciences – modern or other – come up with as the best results of research. "Nature" is a major player in producing sciences' most reliable and widely accepted claims, as is the case also with much of the belief of everyday life. People who do not pay attention to known regularities of nature, or who are disinterested in charting what turn out to be important regularities of nature, are among those who tend to live shorter lives as they are felled by avoidable "accidents" and other threats to life and limb. But nature does not have *everything* to do with even the best representations of such regularities. Scientific claims are not mere reflections of nature's order such that no traces of social values, interests, and inquiry processes are visible in such claims – let alone in the patterns of such claims that the sciences of different eras and different cultures tend to produce. As earlier chapters explored, it turns out that nothing in science can be protected from cultural influence – not its methods, its research technologies, its conceptions of nature's fundamental ordering principles, its other concepts, metaphors, models, narrative structures, or even formal languages that all play crucial roles in advancing the growth of knowledge – that is, its research questions and the consequent distinctive patterns of systematic knowledge and systematic ignorance that it generates. Many socially constituted theories about nature can be *consistent* with nature's order, but none are uniquely *congruent* with it; none uniquely correspond to it. Indeed, we should not want to protect sciences from all cultural influences, for many of these are precisely responsible for their great successes and are necessary for their continued successes in the future.[21] This is where the universality ideal turns out to be costly to the growth of scientific knowledge.

Indeed, it was precisely the use of local resources that enabled modern sciences to emerge in Europe according to postcolonial accounts. European expansion and the growth of modern sciences in Europe were causally linked in that each contributed important resources to the success of the other. Without the knowledge of those daunting aspects of nature's order that Europeans encountered in their "voyages of discovery," knowledge provided by the emerging modern sciences, Europe could not have successfully developed the imperial and colonial relations that permitted it to achieve global leadership. Modern sciences helped Europe shift from being just one of a number of cultures around the globe that were living through the beginning of the end of feudalism in the late Middle Ages to becoming the single most important center of the early or proto-capitalist global economic and political relations that it would spread around the globe. Moreover, without the diverse resources provided by European expansion, modern sciences would have had a far more difficult time emerging – perhaps they would not have emerged in Europe.[22]

Through this process, modern sciences developed distinctive patterns of systematic knowledge and systematic ignorance, traces of which remain visible in contemporary modern science. Early modern science produced information about those aspects of nature's order that European societies needed in order successfully to expand into the Americas, around the Cape of Good Hope, into Asia and eventually Australia, New Zealand, and Africa (and, now, out into space stations and onto other planets in our solar system). As some of the development theorists have argued, the "science and technology transfer" involved in post-World War II so-called development projects that overtly were intended to bring the "underdeveloped countries" up to the standard of living of the "developed countries" in fact have continued the colonial process begun five centuries earlier. These expansionist projects shaped what modern sciences would and would not know about nature's order.[23]

I have identified four respects in which this kind of postcolonial account as well as the post-Kuhnian and feminist accounts argue for the advantages as well as the limitations that local resources provide for the growth of scientific and technological knowledge for every culture's knowledge projects. There it was pointed out that different cultures are located in different parts of heterogeneous nature's order; their environments are always local ones whether restricted to a Pacific isle or the trajectory between Spain and the Caribbean, or Cape Canaveral and the moon. Cultures are interested in whatever they count as their own environments, but even in "the same" environment, they will tend to have different interests generating different questions about the world around them. For example, on the shores of the Atlantic, one culture will be interested in fishing, another in coastal trading, a third in oil and minerals lying beneath the ocean's floor, a fourth in possibilities for transporting slaves, sugar, and rum back and forth across it, a fifth in using the ocean as a toxic dumping site, and so forth. Such culturally local interests lead to different patterns of knowledge and ignorance about local environments.

Moreover these patterns are organized and produced through culturally distinctive discursive resources. Both advancing and limiting a culture's patterns of knowledge are metaphors, models, and narratives about, for example, the Garden of Eden, peaceable kingdoms, wild and unruly nature, nature as a machine, or as a product of God's mind, as a computer, a spaceship, and a lifeboat; about noble, innocent, childlike, animal-like, primitive, or evil natives; about manly, heroic explorers, conquistadors, and natural philosophers, or ones that purportedly represented the admirable national temperaments of Spain, England, France, or other European nations. These kinds of discursive resources represent a distinctive European legacy rather than, for instance, one that could be found in an Islamic or Native American culture. Finally, the production of knowledge is organized in the distinctive ways that different cultures tend to organize social activities more generally. The characteristic ways that work, travel, conquest, and other social relations were organized in fifteenth- through twentieth-century societies of Europe and its diasporas shaped how the work of science and of European expansion were organized. The "voyages of discovery" were distinctively European ways of organizing the production of these parts of modern sciences' knowledge. Though these four kinds of local resources have been described here as if they were completely separate, in daily practice they are partly interlocked and shape each other.

Similarly, as earlier chapters explored, post-Kuhnian and feminist histories, sociologies, ethnographies, and cultural studies of modern sciences produced in the last thirty years have charted precisely the ways that scientists have used local resources to generate such new theories and interpretations. Thus, the postcolonial studies converge with northern science and technology studies, and with feminist components of both, in highlighting the strengths as well as the limitations of sciences' uses of local resources. Modern sciences are "local knowledge systems" no less than are the science and technology knowledge systems of other cultures. Of course modern sciences in many respects are much more powerful than other cultures' knowledge systems, though other cultures' knowledge systems also have their relative strengths over modern sciences, for their locations in nature, interests, discursive resources, and ways of organizing the production of knowledge enable them to learn patterns in nature's order that are not visible from modern science's perspective. But no matter how global the successes of modern science's predictions of nature's order, they can never achieve a unique universality in the sense of being culture free, or destined to persist through history with their meanings and conceptual contexts unchanged. Further, significant parts of the knowledge systems of other cultures also have been able to achieve effective prediction far from the original cultural location of their production.

If it is not useful to think of modern sciences as single or harmoniously integrated in the various senses reviewed above, how is it that we have all been fooled about this? Noted above were various strategies of translation, "trading zones," mathematical and other unifiers that have permitted modern scientists to think and communicate across their disparate projects. Postcolonial histories of science and technology enable us to identify several practices of European expansion that have also contributed to this effect. First, as such expansion turned the world into a laboratory for emerging European sciences, Europeans could test the hypotheses they developed over vastly larger and more diverse natural terrains than could other cultures. European expansion gained access for European sciences to a far greater diversity in nature's order than was available to cultures not so engaged in expansion and, in some cases, whose trade routes and other travels Europeans curtailed. Not all the sciences benefitted equally from this aspect of European expansion, of course. Astronomy, physics, and chemistry benefitted less than did cartography, geology, geography, climatology, and many kinds of biology, though they still did benefit as they addressed the challenges of expansion and its effects on European economies, and, more indirectly, as they became funded through riches gained from the Americas.[24] Agricultural, pharmacological, and medical sciences immensely benefitted from expansion, as did such social sciences as linguistics and anthropology. Of course any culture engaged in such expansionist projects could also have developed their systematic knowledge about natural and social worlds. Indeed, they would have to have done so in order to succeed at travelling through or settling in unfamiliar environments and climates, and cohabiting with the indigenes they encountered. The internalist epistemology of modern science has provided no resources for understanding the effects on European sciences of these expansionist projects.

Second, European sciences could forage in other cultures for elements of those cultures' "ethnosciences" to incorporate into European sciences. It was not just "hypotheses" about nature's order that Europeans came up with all by themselves that were tested in the course of expansionist projects. Native informants taught Europeans about the local flora and fauna and how to use them, minerals and ores and how to extract them, climates and how to survive them, diseases and other threats to health and how to avoid them, pharmacological remedies, agricultural, fishing and engineering practices, land and sea routes, and much of the rest of the knowledge traditions developed and stored in local cultures. Incorporated into the European sciences were those parts of this local knowledge that fit into the prevailing European conceptual frameworks. Those parts that did not fit were ignored or rejected. As the historians point out, not all the resources for modern sciences that the Europeans encountered in other cultures' knowledge traditions were initially perceived to be such, however. Europeans encountered mathematical notions they could not use till decades and even centuries later, and pharmacological knowledge that only now are the northern pharmaceutical companies systematically interested in gathering.[25] A history of "unborrowed knowledge" can provide an illuminating accompaniment to the histories of borrowed knowledge.

Moreover, the Europeans could combine knowledge gathered through observation or foraging from one part of the globe with knowledge so gained elsewhere to create kinds of knowledge that could not emerge from fewer sites in nature and in culture. This, too, is a unifying strategy. For example, Linnaeus's categories were designed to accommodate species from many different parts of the world; Darwin's hypotheses came to him as a result of thinking back and forth between what he had learned at different sites in his travels. Conceptual frameworks designed to explain the relation between observations made at different sites around the globe contributed to the idea that universal sciences were in the making.

Third, at the same time, European expansion suppressed or destroyed – both intentionally and unintentionally – competitive local knowledge systems. Some cultures were wiped out by diseases inadvertently carried by the Europeans; they were infected before there was a chance for them to be conquered, as one historian puts the point.[26] Others were destroyed by conquest. In both cases, the cultures took to the grave their repositories of knowledge about nature and social relations.

Even when the indigenes survived their first encounters with Europeans, their local knowledge traditions were often destroyed nevertheless, both intentionally and unintentionally. For example, the British set out to destroy the Indian textile industry, and succeeded in doing so, in order to sell their British-made textiles in the Indian market. In the United States, Native Americans were neither permitted to speak their native languages in the government schools nor to develop their traditional repositories of knowledge there. Again, the British did not permit Indians to learn the mathematics that had been created by Indian mathematicians.[27] Land upon which local knowledge traditions depended was appropriated by Europeans, turned to "scientific" agriculture, forestry, or other profit production for Europeans, and often environmentally impoverished. In such ways the basis for local knowledge traditions was removed from local access and often destroyed.[28] So the suppression of other cultures' knowledge traditions also contributed to producing the illusion that only European sciences were and could be universal ones.

A fourth way the illusion of unique success was created has been through the dissemination of a predatory conceptual framework for and by European sciences. This conceptual framework spread through the societies Europeans encountered as a central feature of the imposition or adoption of European culture. What is meant by a "predatory conceptual framework"? One way this occurred was through the persistent substitution of abstract, transcultural and ahistorical concepts of nature and processes of gaining knowledge for concrete, locally situated, and historical ones. The former were claimed unique to modern sciences and responsible for their successes, and the latter devalued as merely characteristic of "folk science." For example, features of local environments become aspects of omnipresent "nature" to be explained adequately only by universally valid laws of nature.[29]

There is nothing wrong with abstractions and generalizations in themselves. The point is rather that such abstract concepts always must in fact be accompanied by local knowledge about how to apply such concepts – when and where they are relevant, how to revise and extend them. Yet such an abstract, universalizing conceptual framework devalues this very local knowledge that it needs in order to complete our understanding of it as empirical knowledge – how it relates to the world around us. One could say that the abstract and universal perpetually depend upon and reproduce the "premodern" forms of local knowledge required for the "universal" to be regarded as empirically relevant. It is not that modern science actually replaces its premodern predecessor; rather, it insists on its continual production as a devalued form of knowledge.[30] Moreover, such "foreign" concepts consistently have been used to legitimate the authority of powerful groups over economically and politically vulnerable ones.

Most effective in establishing the impression of universality is simply insisting upon it as an empirical fact; there is one and only one kind of "right" or "real" science, and that is the kind practiced in modern Europe. This replication in modern science of the monologic voice of the Judeo-Christian God is buttressed with various supporting theses – scientific accounts are value free, nature is value free, no kinds of interventions in or uses of nature are forbidden either by nature or by science, and so on. However, as reviewed above and in earlier chapters, such claims are not themselves the results of scientific or historical investigations, nor is there anything logically necessary about them. They are, instead, articles of faith, as is the insistence in modern science on its own "solo performances," that are so well suited to the belief in nature's "monovocality."

Thus, the appearance of universality is created not by any internal epistemic features of modern sciences, as the universality ideal assumes. Instead, it is produced by the kind of hard scientific work reported in the post-Kuhnian accounts and by contingencies of history and political strategies, obscured by, or fit into, a dogmatic conceptual framework that is persistently rhetorically elaborated. This will sound like a harsh judgment. Yet it is important to state as clearly as possible these findings of postcolonial science studies that are strongly supported also by the other two schools of science studies examined here. We need better ways to conceptualize the successes and limitations of modern sciences than are provided by the universality ideal of the internalist scientific epistemology. The ideal neither fits the facts of state-of-the-art history, sociology, and philosophy of science, nor does it make sense in light of what is now understood about how human knowledge of the nature's order must be gathered, preserved, and expanded.

5. The local-global continuum: An alternative conceptual frame. If the successes of sciences – modern or other – cannot be attributed to their internal epistemic features – such as a uniquely universally valid metaphysics, methodology, language, or standards of objectivity and rationality – what does account for them? Apparently there is no distinguishing feature of a science, as we saw earlier, but we still might usefully ask about causes of, or influences on, variations in the powers of different science traditions.

One strategy of many science and technology observers has been to try to bring into visibility the set of practices through which different modern scientific projects have maintained valuable tensions between the local and the global. Knowledge systems, any knowledge systems, always are constituted initially through a set of local conditions. However, the most widely successful ones, such as many parts of modern sciences, manage to travel effectively to become useful in other sets of local conditions – parts of nature, interests, discursive resources, ways of organizing the production of knowledge – that are different in significant respects from those that initially produced them. Without claiming a universality for them that we now can see is historically and conceptually misleading, how could we usefully think about valuable tensions between the local and this movability, or ability to travel, that has characterized parts of modern sciences in particular, but also parts of other knowledge systems (e.g., the concept zero and acupuncture)?

"Technoscience," proposed by Bruno Latour, is one term that has proved useful for drawing attention to the value of maintaining certain kinds of tensions between the local and the global in modern scientific practices.[31] Other science observers have focussed on different sets of components of such complexes that enable them to maintain the local/global tensions. As Helen Watson-Verran and David Turnbull point out, this is an area of study still emerging since no single analysis so far proposed quite captures all of the heterogeneous practices modern technosciences have developed.

> Though scientific culture is now being more frequently recognized as deeply heterogeneous (see, e.g., Law, 1991; Pickering, 1992), there is, at present, no term in general usage that adequately captures the amalgam of places, bodies, voices, skills, practices, technical devices, theories, social strategies, and collective work that together constitute technoscientific knowledge/practices. Foucault's epistemes; Kuhn's paradigms; Callon, Law, and Latour's actor networks; Hacking's self-vindicating constellations; Fujimura and Star's standardized packages and boundary objects, and Knorr-Cetina's reconfigurations – each embraces some of the range of possible components but none seems sufficiently all-encompassing.[32]

Watson-Verran and Turnbull's work is especially interesting because of the way it links European and other cultures' scientific practices in these respects. They are concerned to show how other scientific and technology traditions besides those of modern science have achieved similar kinds of balances between the local and the global in their most successful projects. They develop Deleuze and Guattari's term "assemblage" as a more satisfactory way to capture the set of technoscientific "power practices" that enable cultures in different ways to maintain that crucial tension between the local and the global.[33] Watson-Verran and Turnbull show how medieval European cathedral builders, the Anasazi, the Inca, Australian aborigines, and Pacific navigators, like modern scientists, develop "social strategies and technical devices" that enable them to create "equivalences and connections whereby otherwise heterogeneous and isolated knowledges are enabled to move in space and time from the local site and moment of their production and application to other places and times."[34] This is not the place to explore their interesting account in greater detail. Rather, they provide a good example of the postcolonial arguments showing that it is not internal epistemological features but diverse combinations of technical and social strategies that enable both some modern and "indigenous" technoscientific traditions to become more successful than others. We do not need the notion of universal validity under consideration in this chapter, for other frameworks are becoming available that can preserve what was valuable in the universal science framework without the severe costs of the latter's historical and conceptual inadequacies.

It is time to summarize those costs for the sciences, for epistemologies, and for democratic social relations.

6. Dysfunctional universality claims: Scientific, epistemological, and political costs. To conclude, whatever the remaining benefits of supporting the universality ideal, the sources and arguments reviewed here show that its costs are significant. Let us look first at scientific and epistemological and then political costs of maintaining this ideal.

For one, the unique universality thesis supports the legitimacy of appeals to the authority of a single, monolithic "science" to support individual scientific claims, rather than each having to stand on its own – to "face the tribunal of observation" without the crutch of the general authority of modern science. Feminist and postcolonial critics have pointed this out again and again. Hypotheses about women's psychologies, reproductive systems, or physical abilities have achieved legitimacy when they can claim to be scientific ones, whether or not rigorous testing of such hypotheses has occurred.[35] The empirical reliability of agriculture and forestry principles developed for European environments has been presumed superior to local practices in African, Indian, and other environments when the former can claim the status of modern scientific principles rather than only the ones local farmers and peasants have developed and improved for generations.[36] In such cases individual scientific claims have not had to face the empirical tests that are demanded of claims that cannot appeal to modern science for their legitimacy.

This argument has been voiced far more widely. As philosopher John Dupre puts the point, "The political power of science rests in considerable part on the assumption that it is a unified whole." If science is disunified, then "particular appeals to the authority of science must stand on their own merits."[37] The unity-of-science thesis and its unique universality claim encourage what Dupre refers to as the "unity of scientism."[38] Without the crutch of such dogmas as the universality claim, many purportedly viable scientific claims would have to face much more rigorous tests of empirical and theoretical adequacy.

Secondly, the universality claim legitimates resistance to the most valuable criticisms of contemporary science. Feminist theorists frequently are challenged either to show the ideological bias in physics, or admit the irrelevance of any other kind of evidence to support their claims of androcentric assumptions in the constitution of scientific claims.[39] Criticisms of modern sciences that cannot be recognized as coming from within the sciences can be devalued or ignored without the kind of consideration that criticisms from within the sciences would receive by those who hold the unique universality ideal. Yet it is precisely the fact that they come from what is perceived to be outside the sciences that makes such critiques especially valuable. It is only by starting from outside the dominant conceptual frameworks that such frameworks can themselves come into sharp focus, as the arguments in earlier chapters for standpoint epistemologies and their standards for "strong objectivity" pointed out.

This issue has been central in the postcolonial accounts also. The global authority of a claimed uniquely unitary science, especially one associated with increasingly widespread eurocentric ideals of modernity, progress, unique human potential, and manliness defined in eurocentric terms, conspires to silence what are potentially the most viable alternatives to modern sciences' claims and concerns. The universality claim makes it difficult to see the limitations to modern sciences' institutions, cultures, and practices that accompany their strengths. Moreover, the universality claim works against the overt, valuable claim of modern sciences and their philosophies that it is vigorous criticism that most advances the growth of knowledge. Instead of encouraging such criticism, the universality thesis suppresses it. The way the universality ideal tends to immunize sciences against their most telling criticisms points to the problematic assumption of a single, unified "class" of knowers to whom responsibility for discovering nature's order should be assigned. Women, non-Europeans, and activists/scientists among such groups, who "bring their special-interest politics into science," have never been considered appropriate members of such a class of knowers by those who have benefitted most from scientific and technological change.

Third, the universality thesis is dysfunctional for the growth of scientific knowledge in another related respect. It has the effect of decreasing valuable forms of cognitive diversity, as the postcolonial critics in particular have argued. There is no evidence that the kinds of sciences favored in the modern North today will remain the most useful ones in the future either for other cultures or for the heirs of modern European cultures. Indeed, the arguments here, and many others explored earlier in this book, point to ways in which the ontologies and methodologies of modern sciences, and the interests and discursive resources that shape them, are not the most useful ones for many scientific research projects today. The universality ideal functions to delegitimate any but the scientific problems found interesting in the modern West.

Fourth, the strongest form of the universality ideal has raised distinctive obstacles to our understanding of certain kinds of ways the world is arranged and changes over time. It does this by promoting only narrow conceptions of both nature and science. As long as physics is assigned the status of the model for all sciences, whether on historical, ontological, methodological, logical, or other grounds, modern sciences will unnecessarily generate a certain kind of distinctive pattern of ignorance about the world around and in us. When physics, especially the narrowest conceptions of it, is permitted to set the standards for what counts as nature and what counts as scientific accounts, our knowledge will tend to focus disproportionately on discrete, isolated, short-term, and "purely physical" aspects of the world around us. (Here, the phrase "nature's order" starts to look suspiciously narrow.) It blocks our ability to get into focus the social elements – institutions, practices, languages, meanings – in what are often presented as purportedly merely natural, scientific, and technological changes. It makes it especially hard to see those that are distant, broadscale, and long-term.

Moreover, it blocks our ability to grasp systematic patterns of ignorance that any preferred pattern of knowledge will also generate. The universality ideal encourages the unfortunate tendency to internalize the benefits of scientific and technological change and externalize their costs. The benefits tend to be seen as the consequence of internal features of the epistemology of modern sciences, and the costs as the consequence only of misapplications of scientific knowledge or of their technologies, but not of scientific processes themselves.

Finally, such a model in the natural sciences also promotes the production of systematic ignorance in the social sciences. There are the social sciences that overtly model themselves on the natural sciences: physicalist psychologies, rational choice theorists in economics and international relations, and positivistic sociologies, for example. But there are also the social sciences that conceptualize their projects in such single-minded opposition to the naturalistic models prevailing in their research areas that they cannot get at the ways that more global forces shape or are the consequences of the social phenomena that they study. They get stuck in the local as a reaction to naturalistic social sciences' devaluations of the local. In such cases universalism's conceptual world is advanced in unarticulated forms.

Last but not least, there are the political costs. Feminist and postcolonial theorists especially have argued that the bad political effects of modern sciences' universality ideal are in part a consequence of the scientific and epistemological costs of this ideal. We have seen in this chapter and earlier ones how the universality thesis supports the devaluation of forms of knowledge-seeking that have proved valuable in non-western and premodern cultures, and in devalued subcultures in the West (and elsewhere), such as women's cultures. Indeed, modern sciences have ended up, unintentionally usually, partners with the worst genocidal social projects when they lend legitimation to the destruction of other peoples and the cultures that sustain them, not just their knowledge systems, as the inevitable costs of "human" progress. The centrality of European sciences and technologies to the further development of the world's least advantaged peoples in the name of human "development" is one place where such bad effects of the universality ideal can be seen. Here the universality thesis legitimates continuing to move access to nature's resources from those who are already the most economically and politically vulnerable to those who are already the best positioned to take advantage of such access. The universality thesis elevates to a desirable ideal models of the distinctively rational, progressive, civilized, and human that are constructed in opposition to, in terms of their distance from, the non-European, the economically frugal, and the feminine. Indeed, the universality thesis elevates authoritarianism – the necessity and desirability of acknowledging the legitimacy of just one true account of the world – to a necessity for the distinctively rational, progressive, civilized, and human.

The philosophies of modern sciences have always claimed that such modern knowledge-seeking contributes to democratic social relations. One can find such assessments throughout the evaluations of modern sciences from the Baconian New Science Movement in the seventeenth century through the advent of Comte's positivism in the nineteenth century, the logical positivist philosophies of the 1930s and 1940s, to today's debates about the appropriate projects for sciences and technologies after the Cold War. However, the explorations of the scientific and epistemological dysfunctionality of the universality thesis support long-voiced arguments that there is another, conflicting story to be told about the relationship between modern sciences and democratic social relations. Other conceptual frameworks can do the historical, empirical, and theoretical work that was provided by the universality ideal without invoking the latter's scientific and epistemological dysfunctionality or its ethnocentric, antidemocratic politics.[40]

Notes

1 As noted in earlier chapters, there are also postcolonial reasons to resist subsuming all cultures' traditions of systematic knowledge about themselves and the world around them under what the West has in the last century or so referred to as "science." (Even in the West, the term is a recent one, since Galileo, Newton, and Boyle's work was referred to as "natural philosophy.") After all, why should other cultures' projects have to be named in European terms in order to be taken seriously by Europeans?

2 Compare, e.g., Susan Bordo, *The Flight to Objectivity: Essays on Cartesianism and Culture* (Albany: State University of New York Press, 1987); Alison Jaggar, "Love and Knowledge: Emotions in Feminist Epistemology," in *Gender/Body/Knowledge*, ed. Susan Bordo and Alison Jaggar (New Brunswick: Rutgers University Press, 1989); Evelyn Fox Keller, *Reflections on Gender and Science* (New Haven: Yale University Press, 1984); Genevieve Lloyd, *The Man of Reason: "Male" and "Female" in Western Philosophy* (Minneapolis: University of Minnesota Press, 1984); Phyllis Rooney, "Recent Work in Feminist Discussions of Reason," *American Philosophical Quarterly* 31:1, 1–21.

3 Compare, e.g., Susantha Goonatilake, "The Voyages of Discovery and the Loss and Rediscovery of the 'Other's' Knowledge," *Impact of Science on Society* no. 167 (1992), 241–64; R. K. Kochhar, "Science in British India," parts I and II, *Current Science* 63:11 (1992–93), 689–94; and 64:1 (1992–93) 55–62 (India); Joseph Needham, *The Grand Titration: Science and Society in East and West* (Toronto: University of Toronto Press, 1969); Patrick Petitjean et al., eds., *Science and Empires: Historical Studies about Scientific Development and European Expansion* (Dordrecht: Kluwer, 1992); Ziauddin Sardar, ed., *The Revenge of Athena: Science, Exploitation, and the Third World* (London: Mansell, 1988).

4 For illuminating recent accounts of the history and current scientific support for these unity of science claims, see John Dupre, *The Disorder of Things: Metaphysical Foundations for the Disunity of Science* (Cambridge: Harvard University Press, 1993); and *The Disunity of Science*, ed. Peter Galison and David Stump (Stanford: Stanford University Press, 1996). For one exploration of problems with the unity of science claims from a postcolonial perspective, see David J. Hess, *Science and Technology in a Multicultural World* (New York: Columbia University Press,

1995). For an older harbinger of these arguments, see Patrick Suppes, "The Plurality of Science," *Philosophy of Science Association 1978*, vol. 2, ed. P. Asquith and I. Hacking (East Lansing: Philosophy of Science Association, 1978). Chapter 4 above, "Cultures as Toolboxes for Sciences and Technologies," examined what it is about nature and social relations that insures that science and technology inevitably and desirably must be plural. Note that the universality claim in its unity of science form, as well as in other forms, asserts the uniquely maximal reliability of scientific claims (often expressed in terms of their truth, or in terms of the "fact that science works") and the unique validity of sciences' logic of research and explanation that produced them. The focus here will be primarily on the validity claim since it is sciences' logic of research and explanation that is thought responsible for its production of empirically reliable claims.

5 Such insights are the beginning of the development of standpoint epistemologies – only the beginning, not the end, since these insights express "identity epistemologies" while standpoint epistemologies center not socially unmediated experience but distinctive kinds of critically and dialogically achieved discourses as generators of knowledge. See chapters 8 and 9, and, e.g., Patricia Hill Collins, *Black Feminist Thought: Knowledge, Consciousness, and the Politics of Empowerment* (New York: Routledge, 1991).

6 Thanks to Val Plumwood for pointing out to me this fourth assumption in the unity of science thesis.

7 Mario Biagioli, *Galileo Courtier* (Cambridge: Harvard University Press, 1993); Bruno Latour, *The Pasteurization of France* (Cambridge: Harvard University Press, 1988); Robert Proctor, *Cancer Wars: How Politics Shapes What We Know and Don't Know about Cancer* (Boston: Basic, 1995).

8 Compare, e.g., John A. Schuster and Richard R. Yeo, eds., *The Politics and Rhetoric of Scientific Method: Historical Studies* (Dordrecht: Reidel, 1986); Steven Shapin and Simon Schaffer, *Leviathan and the Air Pump* (Princeton: Princeton University Press, 1985).

9 This was a major point of Thomas Kuhn's *The Structure of Scientific Revolutions*, 2d ed. (Chicago: University of Chicago Press, 1970) and the outpouring of subsequent histories, sociologies, ethnographies, and philosophies of science and technology that followed it. For the demise of the idea of "crucial experiments," see also Sandra Harding, ed., *Can Theories Be Refuted? Essays on the Duhem-Quine Thesis* (Dordrecht: Kluwer/Reidel, 1976).

10 Galison, "Introduction," in *Disunity*, ed. Galison and Stump, 14–15.

11 Compare, e.g., Helen Watson-Verran and David Turnbull, "Science and Other Indigenous Knowledge Systems," in *Handbook of Science and Technology Studies*, ed. S. Jasanoff, G. Markle, T. Pinch, and J. Petersen (Thousand Oaks, Calif.: Sage, 1995), 115–39.

12 Compare David Bloor, *Knowledge and Social Imagery* (London: Routledge and Kegan Paul, 1977); George Gheverghese Joseph, *The Crest of the Peacock: Non-European Roots of Mathematics* (New York: I. B. Tauris, 1991); Sal Restivo, *Mathematics in Society and History: Sociological Inquiries* (Dordrecht: Kluwer, 1992). Arthur B. Powell and Marilyn Frankenstein, eds., *Ethnomathematics: Challenging Eurocentrism in Mathematics Education* (Albany: State University of New York Press, 1997).

13 Compare Ted J. Kaptchuk, *The Web That Has No Weaver: Understanding Chinese Medicine* (New York: Congdon and Weed, 1983).

14 Galison, *Disunity*, 5.

15 Ibid., 3–8.

16 Weinberg's original argument is in *Dreams of a Final Theory* (New York: Pantheon, 1993). A later, more measured, statement can be found in his discussion of Maxwell's equations in "Sokal's Hoax," *New York Review of Books* (8 Aug. 1996), 13.

17 Ian Hacking, "The Disunities of the Sciences," in *Disunity*, ed. Galison and Stump, 52.

18 A. C. Crombie, *Styles of Scientific Thinking in the European Tradition* (London: Duckworth, 1994); Hacking, in *Disunity*, ed. Galison and Stump, 52.

19 Hacking, in *Disunity*, ed. Galison and Stump, 68.

20 Bloor, *Knowledge*; Morris Kline, *Mathematics: The Loss of Certainty* (New York: Oxford, 1980); Restivo, *Mathematics*.

21 These issues were discussed in earlier chapters. In addition to the citations against the unique universality of modern science provided above, see, e.g., N. Katherine Hayles, "Constrained Constructivism: Locating Scientific Inquiry in the Theater of Representation," in *Realism and Representation*, ed. George Levine (Madison: University of Wisconsin Press, 1993); Bas Van Fraassen and Jill Sigman, "Interpretation in Science and in the Arts," in *Realism*, ed. Levine.

22 See the notes to earlier chapters, e.g., J. M. Blaut, *1492: The Debate on Colonialism, Eurocentrism, and History* (Trenton, N.J.: Africa World Press, 1992); Lucille Brockway, *Science and Colonial Expansion: The Role of the British Royal Botanical Gardens* (New York: Academic, 1979); Goonatilake, "Voyage"; James E. McClellan, *Colonialism and Science: Saint Domingue in the Old Regime* (Baltimore: Johns Hopkins University Press, 1992); Petitjean, *Empires*; Nathan Reingold and Marc Rothenberg, eds., *Scientific Colonialism* (Washington, D.C.: Smithsonian Institution Press, 1987); *Revenge*, ed. Sardar.

23 See Wolfgang Sachs, ed., *The Development Dictionary: A Guide to Knowledge as Power* (Atlantic Highlands, N.J.: Zed, 1992; and many of the essays in *Revenge*, ed. Sardar. See Third World Network,

"Modern Science in Crisis: A Third World Response," in *Revenge*, ed. Sardar, as a separate monograph published by the Third World Network (Penang, Malaysia, 1988), and reprinted in Sandra Harding, ed., *The "Racial" Economy of Science: Toward a Democratic Future* (Bloomington: Indiana University Press, 1993).

24 See Blaut, *Debate*; Boris Hessen, *The Economic Roots of Newton's Principia* (New York: Howard Fertig, 1970).

25 Joseph, *Crest*.

26 J. M. Blaut, *The Colonizer's Model of the World: Geographical Diffusionism and Eurocentric History* (New York: Guilford Press, 1993).

27 Michael Adas, *Machines as the Measure of Man* (Ithaca: Cornell University Press, 1989).

28 Susantha Goonatilake, *Aborted Discovery: Science and Creativity in the Third World* (London: Zed, 1984); Ashis Nandy, ed., *Science, Hegemony, and Violence: A Requiem for Modernity* (Delhi: Oxford University Press, 1990); Sachs, *Development*; Vandana Shiva, *Staying Alive: Women, Ecology, and Development* (London: Zed, 1989); *Revenge*, ed. Sardar.

29 Compare Tom Patterson's arguments that the concept "nature" has a class history. It was persistently introduced by protocapitalist "outside experts" (often groups in their own society) in their struggles with peasants and/or farmers over who would have the power to decide how land was to be used. Tom Patterson, "Nature: The Shadow of Civilization," (forthcoming), and his *Inventing Western Civilization* (New York: Monthly Review Press, 1997). See also the issues about "nature" raised in chapters 5 and 6.

30 See chapter 7 for related discussion of the dependency of modern, masculinized forms of knowledge on purportedly premodern women's forms. Maria Mies makes similar arguments in *Patriarchy and Accumulation on a World Scale: Women in the International Division of Labor* (Atlantic Highlands, N. J.: Zed, 1986).

31 Bruno Latour, *Science in Action* (Cambridge: Harvard University Press, 1987).

32 Watson-Verran and Turnbull, "Science," 117.

33 Gilles Deleuze and Félix Guattari, *A Thousand Plateaus: Capitalism and Schizophrenia* (Minneapolis: University of Minnesota Press, 1987).

34 David Turnbull, "Local Knowledge and Comparative Scientific Traditions," *Knowledge and Policy* 6:3/4 (1993), 29.

35 See, e.g., Carolyn Wood Sherif, "Bias in Psychology," in *Feminism and Methodology*, ed. Sandra Harding (Bloomington: Indiana University Press, 1987); Ruth Hubbard, *The Politics of Women's Biology* (New Brunswick: Rutgers University Press, 1990); Anne Fausto-Sterling, *Myths of Gender: Biological Theories about Women and Men*, 2d ed. (New York: Basic, 1994).

36 See, e.g., Thomas A. Bass, *Camping with the Prince, and Other Tales of Science in Africa* (Boston: Houghton Mifflin, 1990), and Vandana Shiva, *Staying Alive: Women, Ecology, and Development* (London: Zed, 1989).

37 John Dupre, "Metaphysical Disorder and Scientific Disunity," in *Disunity*, ed. Galison and Stump, 115. See also Dupre, *Disorder*.

38 Dupre, "Metaphysical," 115.

39 Dupre, "Metaphysical," also makes this point about resistance to feminist criticisms (116). Cf. Sandra Harding, *The Science Question in Feminism* (Ithaca: Cornell University Press, 1986).

40 Similar criticisms have been made of the idea that the sciences do and should make truth claims. For one review of the issues, see "Are Truth Claims Dysfunctional?" by Sandra Harding, in *Philosophy of Language: The Big Questions*, ed. Andrea Nye (New York: Blackwell, 1998).

References

Law, John, ed. 1991. *A Sociology of Monsters? Essays on Power, Technology, and Domination*. London: Routledge.

Pickering, Andrew. 1992. "Objectivity and the Mangle of Practice." *Annals of Scholarship* 8:3, ed. Alan Megill.

The Task of a Philosophy of Technology

Introduction

In spite of the relatively recent emergence of philosophies of technology, an impressive diversity of approaches has already developed. In general, one can at least say that, not surprisingly, analytic philosophies of technology tend to reflect the characteristics of the predominantly empiricist-positivist tradition they inherit. Hence, for example, given this tradition's well-established suspicion of speculative systems and extra-scientific claims, its philosophers of technology tend to look first toward actual, or real-world technological issues and problems and to eschew evaluations of anything like technology "as such." Also, given their tradition's long-standing preference for scientific, or at least science-like models of knowledge, analytic philosophers of technology usually take the scientific basis of modern technology for granted and concentrate on the ethical evaluation of the application of scientific knowledge in technology and by technologists. Continental philosophies of technology, on the other hand, tend to reflect their tradition's long-standing suspicion of enlightenment conceptions of reason and of the scientistic and utopian attitude toward technology to which these conceptions have often led. As a result, continental philosophies of technology frequently display considerable tolerance for holistic and extra-scientific evaluations of technological phenomena, and they rarely make a point of sharply distinguishing questions of the logic and facts of technology from questions of its value and valuation. All of these generalizations, however, are fairly high-level abstractions, and none of them capture adequately the plurality of actual positions. Moreover, as the following selections make plain, not every philosophy of technology (e.g., those inspired by classical pragmatism or by recent feminism) is easily classified under either the analytic or continental label.

Analytical philosophy of technology is exemplified by the selection from Mario Bunge. A former physicist and disciple of Karl Popper whose writings reveal a substantial commitment to general systems theory, Bunge has been a vocal participant in the so-called Science Wars alluded to in the previous section. He is a passionate opponent of Romanticism and of anti-technological attitudes in philosophy generally and is also a severe critic of social constructivist and hermeneutical approaches to technology specifically. In order to show that technology, when properly conceptualized, is not by nature "soulless, aphilosophical, or even antithetical to philosophy," Bunge describes the relations between technology and philosophy in terms of inputs and outputs (which is itself, of course, a technology-influenced terminology). On the output side, he notes that technology supplies system-theoretical ontologies (i.e., conceptual systems of the nature of scientifically knowable objects like the one Bunge himself has produced in a multi-volume treatise). Technology, he adds, has also, and less fortunately, given us the philosophy of pragmatism. As a disciple of Popper, Bunge is critical of pragmatism, at least as he understands it, but he admits it is clearly one of the major philosophies of the modern world. (For a much more subtle and favorable estimation of pragmatism, see the article by Heelan and Schulkin in the previous section, chapter 13.)

In the selections from various introductory notes to his famous *The Technological Society*, Jacques Ellul makes it plain that his approach to technology does not proceed by way of empirical descriptions of technological problems, techniques, and practitioners. Indeed, he suggests that such an approach will never arrive at an adequate conception of what technology is and how it functions. Only a characterization of "the real nature of the technological phenomenon" as a whole can shed light on its actual and pervasive – and, he thinks, also fundamentally dangerous – effect in the contemporary world. One especially provocative aspect of Ellul's approach, however, is that although it clearly exemplifies the global, or holistic outlook one expects from continental philosophers, he explicitly denies that this makes it either speculative or evaluative. His approach, he insists, is entirely "descriptive." Not only does he claim to deliberately avoid offering ethical and aesthetic evaluations of technology, he accuses those who read him as promoting a negative or pessimistic picture of technology as themselves simply reacting on the basis of their own prior (and extra-descriptive, "metaphysical") value commitments. Moreover, he argues that when critics accuse him of going beyond mere description in referring to "technology" itself as if it were a real phenomenon instead of, at best, a sociological abstraction, that simply reveals their commitment (common especially among non-continental philosophers) to methodological individualism. Philosophical "description" of technology cannot mean focusing only on individuals and their practices, because society is not simply the sum of actions of individuals, but has a collective reality. Without a proper account of this extra-individual character of the technological phenomenon, one will never understand its "deterministic" power in contemporary life and will therefore underestimate the extent to which we are currently deprived of our freedom by it.

Kristen Shrader-Frechette's well-known entry from the *Encyclopedia of Ethics* presents a clear summary of what is perhaps the most popular philosophical treatment of technology – indeed, one that is widely supposed to represent philosophy of technology's only task – namely, that of ethical evaluation. As Shrader-Frechette points out, there is considerable overlap between philosophers and non-philosophers in this area. For example, questions of the political and social responsibility of engineers and scientists, as well as risk-benefit analyses of technological projects and systems, are major concerns of policy scientists as well as (especially) analytic philosophers of technology.

In his essay, Hans Jonas resembles Ellul in presenting an unabashedly holistic account of the irreversibility and inevitability of technological change; but unlike Ellul, he combines this account with an appeal to shoulder the "cosmic task" of establishing ethical imperatives responsive to this change. Jonas distinguishes between the "formal dynamics" and the "substantive content" of technology. Formally, he argues, modern technology differs from premodern technology insofar as the former is "an enterprise and process," where the latter was more of "a possession and a state." Jonas stresses the fact that because modern technology is driven by consciously developed plans and ideas, its innovations tend to build upon one another sequentially and spread rapidly across the globe. In this way, a concept of technology as involving genuine progress – a concept in which invention and change are understood as bringing about conditions of life that are superior to those of the present or past – replaces the older idea of using technology to reach an accommodation with a static and stable natural order. Today, observes Jonas, the older "unilinear" idea of knowable but fixed ends and accommodating means, according to which good theory always precedes successful practice, has been replaced by a "circular" one. Science and technology have become inseparably intertwined (cf. Latour's reference to "technoscience"), and technological innovation is now just as likely to suggest new goals as advances in scientific knowledge. Jonas sees the inherent "restlessness" of modern science and technology as leading to the disastrous situation where the sheer process of production and alteration of objects and objectives itself becomes the end of life, thus threatening any substantial and extra-technoscientific idea of what we are like and what life is for. Hence, our most urgent philosophical need is for an ethics of averting disaster – an ethics that encourages a world in which diverse images of humanity and the quality of life legitimately contend, and people in power are as little beholden as possible to the interests generated by technology. Yet one must ask, says Jonas, echoing the problem Plato's philosopher king faces in the *Republic*, Book 7 (see chapter 1), what the role of the philosopher can be in such a world, and one must consider the inevitable compromises that a well-meaning person will have to make in order to be effectively involved in public policy-making.

15

Philosophical Inputs and Outputs of Technology

Mario Bunge

Technology is often considered soulless, aphilosophical, or even antithetical to philosophy. This paper contends that such an image of technology is erroneous and that:

1 Far from being aphilosophical, let alone antiphilosophical, technology is permeated with some of the philosophy it has inherited from pure science along with scientific methods and theories – as exemplified by its reliance on the philosophical principle that we can get some knowledge of reality through experience and reason, and even improve on it.

2 Far from being philosophically passive or sterile, technology puts forth a number of philosophically significant theories, such as automata theory, and important (though perhaps mistaken) philosophical views, such as pragmatism.

3 Far from being ethically neutral, like pure science, technology is involved with ethics and wavers between good and evil.

In other words, this paper proposes the thesis that technology has a philosophical input and a philosophical output and, moreover, part of the latter controls the former. If this is true, then technology is not cut off from culture nor is it a detachable part of culture; technology is instead a major organ of contemporary culture. This being so, the philosopher must pay it far more attention than before; he should build a fully developed philosophy of technology related to but distinct from the philosophy of science.

From *The History of Philosophy and Technology*, ed. George Bugliarello and Dean B. Doner, Urbana: University of Illinois Press, 1979, pp. 262–81, abridged. Reprinted by permission of George Bugliarello.

Tasks of the Philosophy of Technology

The concern of the philosophy of technology – one of the underdeveloped areas of philosophy – is the investigation of the philosophy inherent in technology as well as of the philosophical ideas suggested by the technological process. Some of the typical problems in the philosophy of technology are these: (a) Which characteristics does technological knowledge share with scientific knowledge, and which are exclusive of the former? (b) In what does the ontology of artifacts differ from that of natural objects? (c) What distinguishes a technological forecast from a scientific forecast? (d) How are rule of thumb, technological rule, and scientific law related? (e) Which philosophical principles play a heuristic, and which a blocking, role in technological research? (f) Does pragmatism account for the theoretical richness of technology? (g) What are the value systems and the ethical norms of technology? (h) What are the conceptual relations between technology and the other branches of contemporary culture?

Where are we to search for the philosophical components of technology? Clearly not among the products of technology – cars, drugs, healed patients, or victims of technological warfare – which are about the only technological items the anti-technological philosopher is acquainted with. We must search for philosophy among the ideas of technology – in technological research and in the planning of research and development. We are likely to find them here, as philosophy is found in every department of mature thinking. Indeed mature thinking is always guided (or misguided) and controlled (or exhilarated) by methodological rules as well as by epistemological, ontological, and ethical principles. Just think of the problems posed by the design of any new product. Is the relevant scientific knowledge reliable, and is it likely to be sufficient? Will the new product be radically new – that is, will it exhibit new emergent properties – or will it be just a rearrangement of existing components? Shall we design the product so as to maximize performance, social usefulness, profit, or what?

Since the philosophical components of technology must be searched for among technological ideas, we had better start by recalling what the loci of these ideas are. Moreover, since there is some uncertainty about what "technology" includes, we should enumerate the branches of technology as we understand it.

Branches of Contemporary Technology

We take technology to be that field of research and action that aims at the control or transformation of reality whether natural or social. (Pure science, if it is experimental, also controls and transforms reality but does so only on a small scale and in order to know it, not as an end in itself. Whereas science elicits changes in order to know, technology knows in order to elicit changes.) We discern the following branches of technology;

Material	Physical (civil, electrical, nuclear, and space engineering) Chemical engineering Biochemical (pharmacology) Biological (agronomy, medicine)
Social	Psychological (education, psychology, psychiatry) Psychosociological (industrial, commercial, and war psychologies) Sociological (politology, jurisprudence, city planning) Economic (management science, operations research) Warfare (military science)
Conceptual	Computer sciences
General	Automata theory, information theory, linear system theory, control theory, optimization theory, and so forth

This list is not exhaustive, and some technologists may feel ill at ease with the bed fellows I have chosen for them. The list is intended to be only a partial extensional definition of "technology." It includes the miscellany I have called "general technology" because its theories can be applied almost everywhere regardless of the kind of system; we shall see later in the paper that it constitutes the great contribution of technology to metaphysics. On the other hand, the list does not include futurology, because the latter is just long-term planning and hence is part of social technology.

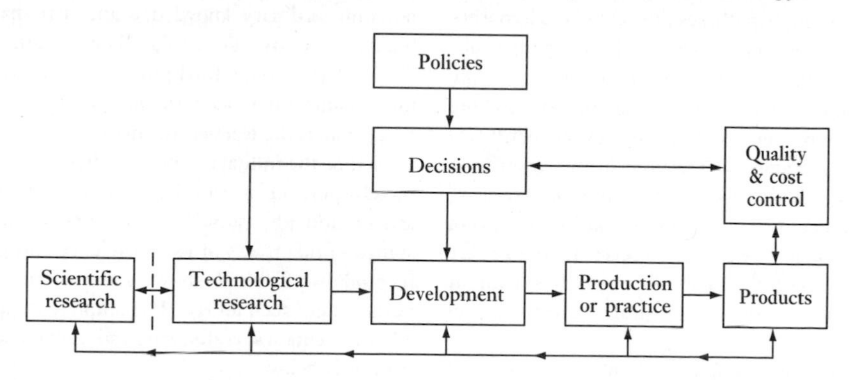

Figure 1. Flow Diagram of Technological Process. The first stage, scientific research, is occasionally missing or completed at a scientific institution – hence the dotted vertical line. The end product of a technological process need not be an industrial good or a service; it may be a rationally organized institution, a mass of docile consumers of material or ideological goods, a throng of grateful if fleeced patients, or a war cemetery.

Let us now locate the areas of maximal conceptual density regardless of subject matter: there we must cast our net. To this end we must take a brief look at the technological process.

Technological Research and Policy

A technological process exhibits the stages shown in Figure 1.

Most technological ideas are found in two of the stages or aspects of a technological process: policy and decision making (largely in the hands of management) and research (in the hands of investigators). In any high-grade technological process, such as one taking place in a petroleum refinery, in a hospital, or in an army, managers as well as technological investigators (but not technicians and blue- and white-collar workers) employ a number of sophisticated conceptual tools – belonging, for example, to organic chemistry or operations research. If they are innovative or creative, policy makers and investigators will try out or even invent new theories or procedures. In sum, technology is not alien to theory, nor is it just an application of pure science; it has a creative component, which is particularly visible in the design of technological policies and in technological research.

Consider technological research for a moment. Methodologically, it is no different from scientific research. In either case, a research cycle looks schematically like this: (1) spotting the problem; (2) trying to solve the problem with available theoretical or empirical knowledge; (3) if that attempt fails, inventing hypotheses or even whole hypothetico-deductive systems capable of solving the problem; (4) finding a solution to the problem with the help of the new conceptual system; (5) checking the solution, for example by experiment; (6) making the required corrections in the hypotheses or even in the formulation of the original problem. Besides being methodologically alike, both kinds of research are goal-oriented; however, their goals are different. The goal of scientific research is truth for its own sake; that of technological research is useful truth.

The conceptual side of technology is neglected or even ignored by those who equate technology with its practice or even with its material outputs. (Curiously enough, not only idealist philosophers but also pragmatists ignore the conceptual richness of technology. Hence neither of them can be expected to give a correct account of the philosophy inherent in technology.) We must distinguish the various stages or aspects of the technological process and focus on technological research, as well as on the design of technological policies, if we are to discover the philosophical components of technology.

Before we face our specific problem we shall make one more preliminary investigation – this time into the conceptual relations among technology and a few other branches of culture, both alive and dead.

Near Neighbors of Technology

Nothing, especially not technology, comes out of nothing. Hence nothing, especially not technology, can be understood in isolation from its kin and neighbors. Modern technology grows out of the very soil it fertilizes, industrial civilization and modern culture. The distinction between civilization and culture is particularly useful for understanding the nature of technology. One can have some modern industry without modern culture, provided one imports technological know-how and does not expect great technological innovations. One can have scraps of modern culture without modern industry – provided one is willing to put up with a one-sided and rickety culture. No creative technology, however, is possible outside modern civilization (which includes modern industry) and modern culture (which of course includes modern technology).

In particular, modern technology presupposes not only ordinary knowledge and artisanal skills but also scientific knowledge, hence mathematics. Technology is not a final product, either; it shades into technical practice – the practice of the general practitioner, the teacher, the manager, the financial expert, or the military expert. Things are not completely pure in or around technology; besides its artistic and philosophical components, one occasionally finds traces of pseudoscience and pseudotechnology. Table 1 shows some of the nearest neighbors of technology. To complete the picture, add mathematics, crafts, arts, and humanities, as in Figure 2, below.

Having sketched a map of technology and having listed some of its neighbors very schematically, we are now in a position to try to explore the philosophy inherent in technological research and policy making.

Table 1. The nearest neighbors of technology.

PROTOSCIENCE	SCIENCE	TECHNOLOGY	TECHNICAL PRACTICE	PSEUDOTECHNOLOGY
Ancient & medieval physics & astronomy	Modern physics & astronomy	Physical engineering	Engineering practice	Astrology
Ancient & medieval mineralogy & part of alchemy	Chemistry	Chemical engineering	Chemical engineering practice	Alchemy
Ancient & medieval natural history	Biology	Agronomy, medicine	Agrotechnical & medical practices	Homeopathy, chiropractic, Lysenkoism
Philosophy of mind (partly)	Psychology	Psychopathology	Drug & behavior therapy	Psychoanalysis, graphology
	Economics	Economic & financial planning	Economic management	Economic miraclemanship
		Computer science	Computation & control	GIGO computeering

The Epistemology of Technology

Technology shares with pure science a number of epistemological assumptions. We mention only the following: (1) there is an external world; (2) the external world can be known, if only partially; (3) every piece of knowledge of the external world can be improved upon if only we care to. These assumptions belong to epistemological realism. The classical technologist was not only a realist but usually a naive realist, in that he took our representations of reality for more or less accurate pictures of it. The modern technologist, involved as he is with constructing sophisticated mathematical models of things and processes, is still a realist but a critical one. He realizes that our scientific and technological theories are not pictures but symbolic representations that fail to cover every detail (and sometimes the very essence) of their referents. He knows that those theories are over-simplifications and also that they contain many concepts – like the proverbial massless piston – which lack real counterparts.

However, the critical realism of technology is tempered and distorted by a strong instrumentalist or pragmatist attitude, the normal attitude among people intent on obtaining practical results. This attitude is obvious from the technologist's way of dealing with both reality and the knowledge of it. For him, reality, the object of pure science, is the sum total of resources (natural and human), and factual knowledge, the aim of pure science, is chiefly a means.

In other words, whereas for the scientist an object of study is a *Ding an sich*, for the technologist it is a *Ding für uns*. Whereas to the scientist knowledge is an ultimate goal, to the technologist it is an intermediate goal, something to be achieved only in order to be used as a means for attaining a practical goal. It is no wonder that instrumentalism (pragmatism, operationalism) has such a great appeal both to technologists and to those who mistake technology for pure science.

Because of his pragmatic attitudes, the technologist will tend to disregard any sector of nature that is not or does not promise to become a resource. For the same reason he is prone to push aside any sector of culture unlikely to be instrumental for achieving his goals. This is just as well as long as he is open-minded enough to tolerate whatever he disregards.

The pragmatic attitude toward knowledge is reflected, in particular, in the way the technologist treats the concept of truth. Although in practice he adopts the correspondence conception of truth as adequacy of the intellect or mind to the thing, he will care for true data, hypotheses, and theories only as long as they are conducive to the desired outcomes. He will often prefer a simple half-truth to a complex truth. He must, because he is always in a hurry to get useful results. Besides, any error made in neglecting some factor (or some decimal figure) is likely to be overshadowed by unpredictable disturbances his real system may undergo. Unlike the physicist, the chemist, or the biologist, he cannot protect his systems against shocks other than by building shock-absorbing mechanisms into them. For similar reasons, the technologist cannot prefer deep but involved theories when superficial ones will do. However, unless he is a pseudotechnologist, he will not shy away from complex and deep theories if they promise success. (For example, he will employ the quantum theory of solids in the design of solid state components and genetics in obtaining improved varieties of corn.) The technologist, in sum, will adopt a mixture of critical realism and pragmatism, varying these ingredients according to his needs. He will seem to confirm first one and then another epistemology, while actually all he intends to do is to maximize his own efficiency regardless of philosophical loyalties.

The technologist's opportunistic conception of truth is just one – although major – epistemological component of technology. We shall now cite two specific items of epistemology that have taken part in technological developments, one in education, the other in artificial intelligence. It is well known that Pestalozzi's educational techniques were based on the slogan of British empiricism, "No concept without a percept." Likewise the philosophical basis of Dewey's educational techniques was the pragmatist thesis, "No concept without an action." The philosophy underlying artificial intelligence studies contains one major ontological hypothesis, "Whatever behaves like an intelligent being is intelligent," and a batch of epistemological hypotheses, among them "Every perception is the acceptance of an external stimulus" and "Some spatial patterns are perceptible and discrete."

There is more to the epistemology of technology, but we must hurry on to the metaphysics of technology.

The Metaphysics of Technology

Technology inherits some of the metaphysics of science and has in turn produced some remarkable metaphysics of its own. We shall list without discussion a few examples of each.

Here are some of the metaphysical hypotheses inherent in both scientific and technological research:

1 *The world is composed of things*, that is, it is not simple, and it is not made of ideas or of shades of ideas. (Were this not so, we could not get things done by cleverly manipulating things – people among them. Mere wishes or incantations would suffice.)
2 *Things get together in systems* (composed of things in more or less close interaction), and *some systems are fairly well isolated from others.* (Otherwise we would not be able to assemble and dismantle things, nor would we be capable of acting upon anything without at the same time disturbing everything else.)
3 *All things, all facts, all processes, whether in nature or in society, fit into objective stable patterns (laws).* Some of these laws are deterministic, others are stochastic, and all are objective. (Otherwise we would not need to know any laws in order to transform nature and society: ordinary knowledge would have sufficed to bring forth modern technology.)
4 *Nothing comes out of nothing and nothing goes over into nothingness.* There are antecedents or causes for everything, and whatever is the case leaves some trace or other. If this were not so, there should be no need to work and no worries about energy.
5 *Determination is often multiple and probabilistic rather than simple or linear.* (If this were not so, we would be unable to attain most goals through different means, and there would be no point in searching for optimal means or in calculating probabilities of success.)

So much for the metaphysics that takes part in technological research and policy making. Now let us look at some of the metaphysical outputs of contemporary technology. While some of them are loose though important theses, others are full-blown ontological theories. Among the former we point out the following:

1 With the help of technology man can alter certain natural processes in a deliberate and planned fashion.
2 Thanks to technology man can create or wipe out entire natural kinds, thus increasing the variety of reality in some respects and decreasing it in others.
3 Because artifacts are under intelligent control or are endowed with control mechanisms which have not emerged spontaneously in a

process of natural evolution, they constitute a distinct ontic level characterized by properties and laws of its own – whence the need for elaborating a technological ontology besides the ontologies of natural and of social science.

As for the metaphysical theories evolved by contemporary technology, they belong in what I have called general technology. They are high grade (though mathematically often simple) general theories such as automata theory, the general theory of machines, general network theory, linear system theory, information theory, control theory, and optimization theory. They qualify not only as technological (or scientific) theories but also as ontological theories for the following reasons. First, they are concerned with generic traits of entire genera (rather than species) of systems: they are cross-disciplinary theories. (Think of the variety of applications of automata theory and control theory.) Second, those theories are stuff-free (independent of any kind of material), hence independent of any particular physical or chemical law. (They focus on structure and behavior rather than on specific composition and mechanism.) Third, those theories are untestable without further effort, if only because they issue no predictions. (They can be made to issue projections and thus become testable upon conjoining them with items of specific information concerning the concrete systems they are applied to.)

In sum, whether they like it or not technologists have built a conceptual building which houses all of the metaphysics of science plus some distinctly technological metaphysics. Metaphysics, banned from philosophy departments, is alive and well in the schools of advanced technology.

[...]

The Value Orientation of Technology

To the scientist all concrete objects are equally worthy of study and devoid of value. Not so to the technologist: he partitions reality into resources, artifacts, and the rest – the set of useless things. He values artifacts more than resources and these in turn more than the rest. His, then, is not a value-free cosmology but one resembling the value-laden ontology of the primitive and archaic cultures. One example should suffice to bring this point home.

Let P and Q be two components or properties of a certain system of technological interest. Assume that, far from being mutually independent, Q interferes with or inhibits P. If P is desirable (in the eyes of the technologist) then Q will often be called an impurity. Unless the impurity is necessary to obtain a third desirable item R (such as conductivity, fluorescence, or a given color), the technologist will regard Q as a disvaluable item to be minimized or neutralized. To the scientist Q may be interesting or uninteresting in some respects, but never disvaluable.

This value orientation of technological knowledge and action contrasts with the value neutrality of pure science. True, social science does not ignore values but attempts to account for them. However, to pure science nothing is either pure or impure in an axiological sense, not even pollutants. In pure science valuation bears not on the objects of study but on the research tools (e.g., measurement techniques) and outcomes (e.g., theories). One lunar theory may be better (truer) than another, but the moon is neither good nor bad. That is not so for the space scientist and the politician behind him. Whereas the technologist evaluates everything, the scientist *qua* scientist evaluates only his own activity and its outcomes. He approaches even valuation in a value-free fashion.

The value orientation of technology gives the philosopher a splendid opportunity to analyze the valuation process in concrete cases rather than setting up a priori (or else conventional) "value tables." It can even inspire him to build realistic value theories, where valuation appears as a human activity, largely rational, done in the light of definite antecedent knowledge and definite desiderata. As a matter of fact, technology has already had an impact on value theory; utility theory (the theory of subjective value), though originally proposed as a psychological theory, has recently been revived and elaborated in response to the needs of managers. One may also think of a theory of objective value even more closely in tune with technology – one defining value as the degree of satisfaction of an objective need.

We turn now to a few other instances of the impact of technology upon philosophy.

Technology as a Source of Inspiration for the Philosophy of History

We have seen that technology is both a consumer and a producer of philosophical ideas. In addition, it can inspire or suggest interesting new developments in the philosophy of action, in particular ethics, legal philosophy, and the philosophy of history. Let us look into the last.

A number of historians are applying mathematics to problems in history. Here are a few examples of the mathematization of history: (a) cleansing historical data (such as chronologies) with the help of mathematical statistics; (b) finding historical trends or quasi-laws in a number of socioeconomic variables (notably by the French historians of the *Annales: Economies, Sociétés, Civilisations*); (c) building mathematical models of certain historical processes, such as the expansion and decline of empires; and (d) studying certain historical events and processes in the light of decision theory. This last approach, suggested partly by management science, is legitimate with reference to deliberate decisions affecting the life of entire communities. The passing of important new legislation, the launching of a war, the call to a nationwide strike, and the outbreak of a planned revolution are occasions for the application of decision theory. Indeed, all the necessary components are there or can be conjectured: the decision makers who are supposed to maximize their expected utilities, the goals, the utilities of them, the means or courses of action considered by the decision makers, and the probability of attaining a given goal with a certain means.

The philosophy of history can acquire a whole new dimension in the light of decision theory, provided, of course, it is not employed to resurrect the great hero theory of history. Certainly, important areas of historiography, such as the anonymous history studied by historical demography, historical geography, and economic history, remain beyond the decision-theory approach. However, in an increasingly technological society, rational (but, alas, often wicked) action, based on carefully designed policies, plays an increasingly important role and can therefore be partly understood with the help of decision theory.

Technology as a Source of Inspiration for Ethics and Legal Philosophy

Other fields of the philosophy of action that technology can fertilize are ethics and legal philosophy, by teaching them to spell out norms as grounded rules or even as conclusions of arguments. Thus, instead of issuing blind commands of the form "Do *x*," or blind ethical norms of the form "You ought to do *x*," the technologist will proceed as follows. He will propose and test grounded rules of the form "Do *x* in order to get *y*," on the basis of the knowledge that doing *x* does in fact bring about *y* either invariably or with a certain probability. By stating explicitly the ground for a rule of action, one kills three birds with one stone: (a) one breaks the fact/norm barrier, (b) one transforms moral decision making into a rational activity, and (c) one dispenses with the logic of norms.

This proposal, even if feasible, does not allow us to build a value-free ethics. This would be impossible, because moral decision making is as value-oriented as technological policy design. What technology can teach us is, rather, to render values explicit so as to be able to examine them critically instead of receiving them uncritically. In other words, it is impossible to translate a normative sentence into a value-free declarative sentence without loss. On the other hand, it is possible to spell out a norm into a pair law sentence-value sentence, in this way: "Do *x*" or "You ought to do *x*" may be construed as short for "There is a *y* such that *x* brings about *y* and (you value *y* or there is a *z* such that not doing *x* brings about *z* and you disvalue *z*)." The command (or the norm) and its expansion, though not logically equivalent, are related in that the former is just an abbreviation of the latter.

For example, "Do not cheat" can be expanded into "(Any) cheating does (some) harm and you do not want to do any harm." But the same norm can also be expanded into "(Any) cheating jeopardizes your credit and you want to keep your credit in good standing." This ambiguity is to be blamed on the norm itself and not on its rational translation. In any event, a norm, when grounded and formulated in the declarative mode, appears as a consequence of a set of premises. And at least one of these premises is a law statement while at least one other is a value judgment. Consequently, the handling of norms requires only ordinary logic (instead of the logic of norms) and value theory. In other words, we can reconstruct normative science without norms, but with values.

(Superficially, ordinary logic would seem to suffice. Thus, in the case of the injunction not to cheat because it causes harm and harm is undesirable, we would seem to have just an instance of *modus tollens*, namely: $C \rightarrow H \;\&\; \neg H \therefore \neg C$. However, the H occurring in the first premise differs from that occurring in the second: the latter is not really H but rather "H is valuable." Likewise, the conclusion is a value statement. A task of value theory is to compute the value of the conclusion in terms of the values occurring in the premises. But we cannot go into this here.)

What holds for ethics holds for legal philosophy: here, too, norms are profitably expanded into complex statements or construed as consequences of sets of premises. For example, "Murderers must be put away" somehow follows from "Murder endangers the social structure and we value the social structure." However, the same norm also follows from premises in a different field, e.g., "Murderers are sick people and it is disvaluable to leave sick people at large," as well as from premises inspired in still other value systems. The advantage of such expansions is obvious: they force the law giver to lay bare the grounds of positive law – which is often cruel, unfair, or even absurd – and invite him to ground legal technology on sociology and psychology.

In sum, technology suggests that we replace every authoritarian set of imperatives with a grounded set of rules – rules based on laws and value judgments. In this way, whatever was implicit or even concealed can be analyzed, criticized, reconstructed, and systematized. Technology can thus act as a methodological model for the normative sciences, in particular ethics. Unfortunately, far from having served as a moral model, technology is in need of some ethical bridling. This deserves another section.

The Dubious Morals of Technology

Knowing is a good in itself. (Even knowing how to inflict pain may be valuable, as it can assist us in avoiding the act of inflicting pain.) However, there are ways and means of knowing, and some of them may be morally objectionable, such as torturing and killing people in order to find out more about fear. Hence scientific research gets somewhat involved with ethics. In practice, a few rather obvious strictures usually suffice to keep it unsoiled. There are, of course, uncertainty zones, but they can be bounded. For example, in research into fear mild torturing might be condoned provided it is done with the free consent of the experimental subject and it can be safely predicted that it will not be traumatic. In short, pure science needs only a mild external ethical control. As a matter of fact, scientific research has built into it an ethical code of honesty, responsibility, and hard work that can inspire other human activities.

Things are different in technology. Here not only some of the means and ways of knowing may be impure, but also the entire technological process may be morally objectionable for aiming exclusively at evil practical goals. For instance, it is wicked to conduct research into forest defoliation, the poisoning of water reservoirs, the maiming of civilians, the manipulating of consumers or voters, and the like, because the knowledge gained in research of this kind is likely to be used for evil purposes and unlikely to serve good purposes. It is not just a matter of an unexpected evil use of a piece of neutral knowledge, as is the case with the misuse of a pair of scissors: the technique of evil doing is evil itself. The few valuable items it may deliver are by far outnumbered by its negative output. Try to find a good use for the stocks of lethal germs accumulated for chemical warfare, for example, or for plans for the rational organization of an extermination camp.

Technology can then be either a blessing or a curse. That it is always a blessing, if not in the short run then in the long run, is a tenet which has been preached by a number of progressive philosophers since the dawn of the modern period. Other philosophers claimed instead that technology is a curse, but they did so for the wrong reasons – because they were against social progress and cultural expansion. It is only very recently that most of us have come to realize that technology itself can in fact be wicked and must therefore be checked. We have learned that, while accelerating advance in some respects (such as the size of the GNP), technology is also accelerating our decline in other respects (such as the quality of life) and is even jeopardizing the very existence of the biosphere.

Of course, there is nothing unavoidable about the evils of technology. Except for isolated cases of unexpected bad side effects, technology could be all good instead of being half-saintly and half-devilish. It is up to the policy makers to have the technological investigator produce good or evil technological items. It is up to the technologist to take orders or to disobey them. In any event, technology is by its nature morally committed one way or another, and it needs some ethical bridling.

The Ethics of Technology

Every human activity is either explicitly controlled or criticizable by some behavior code which is partly legal and partly moral. In particular, the technological process has usually been guided (or misguided) by the following maxims:

1 Man is separate from and more valuable than nature.
2 Man has a right (or even the duty) to subdue nature to his own (private or social) benefit.
3 Man has no responsibility toward nature: he may be the keeper (or even the prison warden) of his brother, but he is not the nanny of nature.
4 The ultimate task of technology is the fullest exploitation of natural and human resources (the unlimited increase in GNP) at the lowest cost without regard for anything else.
5 Technologists and technicians are morally irresponsible; they are to carry on their task without being distracted by any ethical or aesthetic scruples. The latter are the exclusive responsibility of the policy makers.

These maxims constitute the core of the ethics of the technology that has prevailed heretofore in all industrial societies, regardless of the type of social organization. Certainly those maxims are not justified by technology itself: rather, they justify boundless exploitation of the natural and social resources. Moreover, they have not evolved within technology or science but within certain religions, ideologies, and philosophies.

In recent years we have come to distrust these maxims or even reject them altogether because we have started to realize that they condone the dark side of technology. As yet, we have not offered an alternative ethical code. It is high time we attempted to build alternative ethics of technology, ones with different desiderata and based on our improved knowledge of both nature and society, which were largely unknown at the time the old code was formulated, toward the beginning of the seventeenth century. If we wish to keep most of modern technology while minimizing its evil components and negative side effects, we must design and enforce an ethical code for technology that covers every technological process and its repercussions at both the individual and the social levels. Such a code should consist of the following components: (1) *An individual ethical code* for the technologist *qua* investigator. This should include the ethics of science, namely the set of ethical norms securing the search for truth and its dissemination. It should also take into account the peculiar moral problems faced by the technologist bent on attaining noncognitive goals. These additional norms should emphasize the personal responsibility of the technologist in his professional work and his duty to decline taking part in any project aiming for antisocial goals. Such moral imperatives, or rather grounded rules, should be consistent with (2) *a social ethical code* for technological policy making, research, and development of practices, disallowing the pursuit of unworthy goals and limiting any technological processes that, while pursuing worthy goals, interfere severely with further desiderata. This social ethical code should be inspired by the overall needs and desiderata of society rather than being dictated by any privileged group within it. Otherwise it would be unfair, and it might not be enforceable.

Such a two-tiered ethical code would make impossible, or at least reprehensible, the "Dr. Jekyll-Mr. Hyde" type of scientist who deserves both the Nobel prize for his contributions to elementary particle research and a hanging verdict for designing diabolical new means of mass murder. There would be no toleration of double ethical standards today if there were not two ethical codes, one for the pure scientist and the other for the impure technologist. If we are to keep technology in check, we need a single ethics of technology covering its whole wide spectrum, from knowledge to action.

Conclusion: The Centrality of Technology

Nobody denies that technology is central to industrial civilization. What is sometimes denied is that technology forms an essential part of modern intellectual culture. Indeed, it is often held that technology is alien or even inimical to culture. This is a mistake, one which betrays a total ignorance of the intellectual richness of the technological process, in particular of the innovating one. The mistake has obnoxious consequences, for it perpetuates the training of scholars with a traditional (preindustrial) cast of mind and conceptual equipment, contemptuous and afraid of whatever they do not understand about modern life. When they wield power in governmental or educational institutions, such people try to isolate the technologist as a skillful barbarian who must be kept in his modest place as the provider of material comfort. By behaving in this way, those scholars in fact deepen the gaps among the various subcultures and miss the chance of contributing to steering the course of technology along a path beneficial to society as a whole.

Like every other culture, ours is a complex system of heterogeneous interacting components. Some of them are already past their creative prime, others are blossoming, while still others are just budding. The creative components of our culture are some of the humanities, mathematics, science (natural and social), technology, and the arts. Modern technology is both an essential component and the youngest of all. Perhaps this is why we do not fully realize how central it is to our culture. In fact, instead of being an isolated component, technology interacts strongly with every other branch of culture. (On the other hand, art hardly interacts at all with mathematics.) Moreover, technology and the humanities (in particular philosophy) are the only components of living culture that interact vigorously with all the other components (see Figure 2). In particular, technology interacts fairly strongly with several branches of systematic philosophy: logic, epistemology, metaphysics, value theory, and ethics.

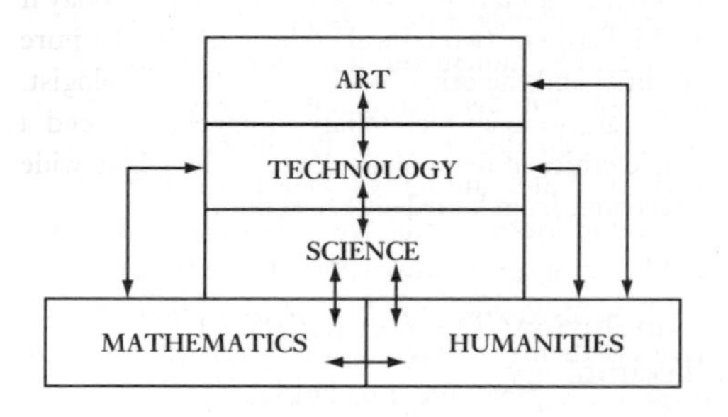

Figure 2. Flow Diagram of the System of Contemporary Culture. The noncreative components have been discarded.

Not only does technology interact with every other living sector of contemporary culture, in particular philosophy, but it overlaps partially with some of them. Thus, architecture and industrial design are at the intersection of technology and art; much of physics and chemistry is as much engineering as it is science; applied genetics is hardly distinguishable from pure genetics; and even some of metaphysics is at the intersection of technology and philosophy, as was discussed above.

Like science, technology consumes, produces, and circulates philosophical goods. Some of these are the same as those activated by science; others are peculiar to technology. Thus, because of its emphasis on usefulness, the epistemology of technology has a pragmatist streak and is therefore coarser than the epistemology of scientific research. On the other hand, the metaphysics and the ethics of technology are richer than those of science.

Because of the conceptual richness of technological processes, and because of the multiple contacts between technology and the other creative components of modern culture, technology is central to that culture. We cannot ignore the organic integration of technology with the rest of modern culture if we wish to improve the health and even save the life of our culture. We cannot afford to ignore the nature of technology, let alone despise it, if we want to gain full control over technology in order to check its dark side. We must then build up all the disciplines dealing with technology, not least of them the philosophy of technology – the more so since it is often mistaken for the philosophy of science. The history, sociology, and psychology of technology tell us much about technologies and technologists, but only the philosophy of technology makes it its business to tell us what the methodological, epistemological, metaphysical, and ethical pennants of technology look like.

16

On the Aims of a Philosophy of Technology

Jacques Ellul

Author's Preface to the French Edition of *The Technological Society* [1954]

Let us, first of all, clear up certain misunderstandings that inevitably arise in any discussion of technique.

It is not the business of this book to describe the various techniques which, taken together, make up the technological society. It would take a whole library to describe the countless technical means invented by man; and such an undertaking would be of little value. Moreover, quite enough elementary works describing the various techniques are already available. I shall frequently allude to some of these techniques on the assumption that their applications or their mechanics are familiar to the reader.

I do not intend to draw up a balance sheet, positive or negative, of what has been so far accomplished by means of these techniques, or to compare their advantages and disadvantages. I shall not repeat what has so often been stated, that through technology the work week has been materially shortened, that living standards have risen, and so forth; or, on the other side of the ledger, that the worker has encountered many difficulties in adapting to the machine. Indeed, no one is capable of making a true and itemized account of the total effect of existing techniques. Only fragmentary and superficial surveys are possible.

Finally, it is not my intention to make ethical or aesthetic judgments on technique. A human being is, of course, human and not a mere photographic plate, so that his own point of view inevitably appears. But this does not preclude a deeper objectivity. The sign of it will be that worshippers of technique will no doubt find this work pessimistic and haters of technique will find it optimistic.

I have attempted simply to present, by means of a comprehensive analysis, a concrete and fundamental interpretation of technique.

That is the sole object of this book.

Note to the Reader [1963]

I think the task of the reader will be lightened if at the outset I attempt a definition of *technique*. The whole first chapter is devoted to making clear what constitutes technique in the present-day world, but as a preliminary there must be a simple idea, a definition.

The term *technique*, as I use it, does not mean machines, technology, or this or that procedure for attaining an end. In our technological society, *technique* is the *totality of methods rationally arrived at and having absolute efficiency* (for a given stage of development) in *every* field of human activity. Its characteristics are new; the technique of the present has no common measure with that of the past.

This definition is not a theoretical construct. It is arrived at by examining each activity and observing the facts of what modern man calls technique in

From Jacques Ellul, *The Technological Society*, trans. John Wilkinson, New York: Knopf and London: Jonathan Cape, 1964, pp. xxxv–xxxvi, xxv–xxxiii. Reprinted by permission of Alfred A. Knopf, a Division of Random House, Inc., and Jonathan Cape Ltd, London.

general, as well as by investigating the different areas in which specialists declare they have a technique.

In the course of this work, the word *technique* will be used with varying emphasis on one or another aspect of this definition. At one point, the emphasis may be on rationality, at another on efficiency or procedure, but the over-all definition will remain the same.

Finally, we shall be looking at technique in its sociological aspect; that is, we shall consider the effect of technique on social relationships, political structures, economic phenomena. Technique is not an isolated fact in society (as the term *technology* would lead us to believe) but is related to every factor in the life of modern man; it affects social facts as well as all others. Thus technique itself is a sociological phenomenon, and it is in this light that we shall study it.

Author's Foreword to the Revised American Edition [1964]

At the beginning I must try to make clear the direction and aim of this book. Although descriptive, it is not without purpose. I do not limit myself to describing my findings with cold objectivity in the manner of a research worker reporting what he sees under a microscope. I am keenly aware that I am myself involved in technological civilization, and that its history is also my own. I may be compared rather with a physician or physicist who is describing a group situation in which he is himself involved. The physician in an epidemic, the physicist exposed to radioactivity: in such situations the mind may remain cold and lucid, and the method objective, but there is inevitably a profound tension of the whole being.

Although I have deliberately not gone beyond description, the reader may perhaps receive an impression of pessimism. I am neither by nature, nor doctrinally, a pessimist, nor have I pessimistic prejudices. I am concerned only with knowing whether things are so or not. The reader tempted to brand me a pessimist should begin to examine his own conscience, and ask himself what causes him to make such a judgment. For behind this judgment, I believe, will always be found previous metaphysical value judgments, such as: "Man is free"; "Man is lord of creation"; "Man has always overcome challenges" (so why not this one too?); "Man is good." Or again: "Progress is always positive"; "Man has an eternal soul, and so cannot be put in jeopardy." Those who hold such convictions will say that my description of technological civilization is incorrect and pessimistic. I ask only that the reader place himself on the factual level and address himself to these questions: "Are the facts analyzed here false?" "Is the analysis inaccurate?" "Are the conclusions unwarranted?" "Are there substantial gaps and omissions?" It will not do for him to challenge factual analysis on the basis of his own ethical or metaphysical presuppositions.

The reader deserves and has my assurance that I have not set out to prove anything. I do not seek to show, say, that man is determined, or that technique is bad, or anything else of the kind.

Two other factors may lead the reader to the feeling of pessimism. It may be that he feels a rigorous determinism is here described that leaves no room for effective individual action, or that he cannot find any solution for the problems raised in the book. These two factors must now engage our attention.

As to the rigorous determinism, I should explain that I have tried to perform a work of sociological reflection, involving analysis of large groups of people and of major trends, but not of individual actions. I do not deny the existence of individual action or of some inner sphere of freedom. I merely hold that these are not discernible at the most general level of analysis, and that the individual's acts or ideas do not *here and now* exert any influence on social, political, or economic mechanisms. By making this statement, I explicitly take a partisan position in a dispute between schools of sociology. To me the sociological does not consist of the addition and combination of individual actions. I believe that there is a collective sociological reality, which is independent of the individual. As I see it, individual decisions are always made within the framework of this sociological reality, itself preexistent and more or less determinative. I have simply endeavored to describe technique as a sociological reality. We are dealing with collective mechanisms, with relationships among collective movements, and with modifications of political or economic structures. It should not be surprising, therefore, that no reference is made to the separate, independent initiative of individuals. It is not possible for me to treat the individual sphere. But I do not deny that it exists. I do not maintain that the individual is more determined today than he has been in the past; rather, that he is differently determined. Primitive man, hemmed in by prohibitions, taboos, and rites, was, of course, socially determined. But it is an illusion – unfortunately very widespread – to think that because we have broken through the prohibitions, taboos, and rites that bound primitive man, we have become free. We are conditioned by something new: technological civilization. I make no reference to a past period of history in which men were allegedly free, happy, and independent. The determinisms of the past no longer concern us; they are finished and done with. If I do refer to the past, it is only to emphasize that present determinants did not exist in the past, and men did not have to grapple with them then. The men of classical antiquity could not have found a solution to our present determinisms, and it is useless to look into the works of Plato or Aristotle for an answer to the problem of freedom.

Keeping in mind that sociological mechanisms are always significant determinants – of more or less significance – for the individual, I would maintain that we have moved from one set of determinants to another. The pressure of these mechanisms is today very great; they operate in increasingly wide areas and penetrate more and more deeply into human existence. Therein lies the specifically modern problem.

This determinism has, however, another aspect. There will be a temptation to use the word *fatalism* in connection with the phenomena described in this book. The reader may be inclined to say that, if everything happens as stated in the book, man is entirely helpless – helpless either to preserve his personal freedom or to change the course of events. Once again, I think the question is badly put. I would reverse the terms and say: if man – if each one of us – abdicates his responsibilities with regard to values; if each of us limits himself to leading a trivial existence in a technological civilization, with greater adaptation and increasing success as his sole objectives; if we do not even consider the possibility of making a stand against these determinants, then everything *will* happen as I have described it, and the determinants *will* be transformed into inevitabilities. But, in describing sociological currents, I obviously cannot take into account the contingent decisions of this or that individual, even if these decisions could modify the course of social development. For these decisions are not visible, and if they are truly personal, they cannot be foreseen. I have tried to describe the technical phenomenon as it exists at present and to indicate its *probable* evolution. Fatalism is not involved; it is rather a question of probability, and I have indicated what I think to be its most likely development.

What is the basis for this most likely eventuality? I would say that it lies in social, economic, and political phenomena, and in certain chains of events and sequences. If we may not speak of laws, we may, at any rate, speak of repetitions. If we may not speak of mechanisms in the strict sense of the word, we may speak of interdependencies. There is a certain logic (though not a formal logic) in economic phenomena which makes certain forecasts possible. This is true of sociology and, to a lesser degree, of politics. There is a certain logic in the evolution of institutions which is easily discernible. It is possible, without resorting to imagination or science fiction, to describe the path that a social body or institutional complex will follow. An extrapolation is perfectly proper and scientific when it is made with care. Such an extrapolation is what we have attempted. But it never represents more than a probability, and may be proved false by events.

External factors could change the course of history. The probable development I describe might be forestalled by the emergence of new phenomena. I give three examples – widely different, and deliberately so – of possible disturbing phenomena:

1) If a general war breaks out, and if there are any survivors, the destruction will be so enormous, and the conditions of survival so different, that a technological society will no longer exist.

2) If an increasing number of people become fully aware of the threat the technological world poses to man's personal and spiritual life, and if they determine to assert their freedom by upsetting the course of this evolution, my forecast will be invalidated.

3) If God decides to intervene, man's freedom may be saved by a change in the direction of history or in the nature of man.

But in sociological analysis these possibilities cannot be considered. The last two lie outside the field of sociology, and confront us with an upheaval so vast that its consequences cannot be assessed. But *sociological* analysis does not permit consideration of these possibilities. In addition, the first two possibilities offer no analyzable fact on which to base any attempt at projection. They have no place in an inquiry into facts; I cannot deny that they may occur, but I cannot take them rationally into account. I am in the position of a physician who must diagnose a disease and guess its probable course, but who recognizes that God may work a miracle, that the patient may have an unexpected constitutional reaction, or that the patient – suffering from tuberculosis – may die unexpectedly of a heart attack. The reader must always keep in mind the implicit presupposition that *if* man does not pull himself together and assert himself (or if some other unpredictable but decisive phenomenon does not intervene), *then* things will go the way I describe.

The reader may be pessimistic on yet another score. In this study no solution is put forward to the problems raised. Questions are asked, but not answered. I have indeed deliberately refrained from providing solutions. One reason is that the solutions would necessarily be theoretical and abstract, since they are nowhere apparent in existing facts. I do not say that no solutions will be found; I merely aver that in the present social situation there is not even a beginning of a solution, no breach in the system of technical necessity. Any solutions I might propose would be idealistic and fanciful. In a sense, it would even be dishonest to suggest solutions: the reader might think them real rather than merely literary. I am acquainted with the "solutions" offered by Emmanuel Mounier, Pierre Teilhard de Chardin, Ragnor Frisch, Jean Fourastié, Georges Friedmann, and others. Unfortunately, all these belong to the realm of fancy and have no bearing on reality. I cannot rationally consider them in analyzing the present situation.

However, I will not make a final judgment on tomorrow before it arrives. I do not presume to put chains around man. But I do insist that a distinction be made between diagnosis and treatment. Before a remedy can be found, it is first necessary to make a detailed study of the disease and the patient, to do laboratory research, and to isolate the virus. It is necessary to establish criteria that will make it possible to recognize the disease when it occurs, and to describe the patient's symptoms at each stage of his illness. This preliminary work is indispensable for eventual discovery and application of a remedy.

By this comparison I do not mean to suggest that technique is a disease of the body social, but rather to indicate a working procedure. Technique presents man with multiple problems. As long as the first stage of analysis is incomplete, as long as the problems are not correctly stated, it is useless to proffer solutions. And, before we can pose the problems correctly, we must have an exact description of the phenomena involved. As far as I know, there is no over-all and exact description of the facts which would make it possible to formulate the problems correctly.

The existing works on the subject either are limited to a single aspect of the problem – the effect of motion pictures on the nervous system, for example – or else propose solutions without the requisite preliminary study. I offer these pages as a first effort in laying the necessary ground; much more work will have to follow before we can see what man's true response is to the challenge before him.

But this must not lead the reader to say to himself: "All right, here is some information on the problem, and other sociologists, economists, philosophers, and theologians will carry on the work, so I have simply got to wait." This will not do, for the challenge is not to scholars and university professors, but to all of us. At stake is our very life, and we shall need all the energy, inventiveness, imagination, goodness, and strength we can muster to triumph in our predicament. While waiting for the specialists to get on with their work on behalf of society, each of us, in his own life, must seek ways of resisting and transcending technological determinants. Each man must make this effort in every area of life, in his profession and in his social, religious, and family relationships.

In my conception, freedom is not an immutable fact graven in nature and on the heart of man. It is not inherent in man or in society, and it is meaningless to write it into law. The mathematical, physical, biological, sociological, and psychological sciences reveal nothing but necessities and determinisms on all sides. As a matter of fact, reality is itself a combination of determinisms, and freedom consists in overcoming and transcending these determinisms. Freedom is completely without meaning unless it is related to necessity, unless it represents victory over necessity. To say that freedom is graven in the nature of man, is to say that man is free because he obeys his nature, or, to put it another way, because he is conditioned by his nature. This is nonsense. We must not think of the problem in terms of a choice between being determined and being free. We must look at it dialectically, and say that man is indeed determined, but that it is open to him to overcome necessity, and that this *act* is freedom. Freedom is not static but dynamic; not a vested interest, but a prize continually to be won. The moment man stops and resigns himself, he becomes subject to determinism. He is most enslaved when he thinks he is comfortably settled in freedom.

In the modern world, the most dangerous form of determinism is the technological phenomenon. It is not a question of getting rid of it, but, by an act of freedom, of transcending it. How is this to be done? I do not yet know. That is why this book is an appeal to the individual's sense of responsibility. The first step in the quest, the first act of freedom, is to become aware of the necessity. The very fact that man can see, measure, and analyze the determinisms that press on him means that he can face them and, by so doing, act as a free man. If man were to say: "These are not necessities; I am free because of technique, or despite technique," this would prove that he is totally determined. However, by grasping the real nature of the technological phenomenon, and the extent to which it is robbing him of freedom, he confronts the blind mechanisms as a conscious being.

At the beginning of this foreword I stated that this book has a purpose. That purpose is to arouse the reader to an awareness of technological necessity and what it means. It is a call to the sleeper to awake.

17

Technology and Ethics

Kristin Shrader-Frechette

Aristotle (384–322 B.C.) pointed out in Book III of the *Nicomachean Ethics* that one can deliberate only about what is within one's power to do. Technologies such as gene splicing and nuclear fission were not within the power of the Greeks, so there was no ethical deliberation about them until centuries later. Throughout history, technology (knowledge associated with the industrial arts, applied sciences, and various forms of engineering) has opened new possibilities for actions. As a result, it has also raised new ethical questions.

Most of these questions have not generated new ethical concepts; instead they have expanded the scope of existing ones. For example, because hazardous technologies threaten those who live nearby, ethicists have expanded the notion of "equal treatment" to include "geographical equity," equal treatment of persons located different distances from dangerous facilities.

Because new developments force the expansion of ethical concepts, those who investigate technology and ethics need both technical and philosophical skills. To assess the ethical desirability of using biological (versus chemical) pest control, for example, one must know the relevant biology and chemistry, as well as the economic constraints on the choice. Although such factual knowledge does not determine the ethical decision, it constrains it in important and unavoidable ways.

Since policymakers evaluate virtually all technologies, at least in part, by methods such as benefit-cost or benefit-risk analysis, knowledge of economics is essential for informed discussions of technology and ethics. Philosophers investigate both the ethical constraints on developing or implementing particular technologies and the ethical acceptability of various economic and policy methods used to evaluate technology.

Originally "Technology" in *Encyclopedia of Ethics*, vol. 2, ed. Lawrence C. Becker and Charlotte B. Becker, New York: Garland Publishing, 1992, pp. 1231–4.

Philosophical questions about technology and ethics generally fall into one of at least five categories. These are (1) conceptual or *metaethical* questions; (2) *general normative* questions; (3) *particular normative* questions about specific technologies; (4) questions about the ethical *consequences* of technological developments; and (5) questions about the ethical justifiability of various *methods* of technology assessment.

Examples of (1) are: "how ought one to define 'free, informed consent' to risks imposed by sophisticated technologies?" or "how ought one to define 'equal protection' from such risks?" Examples of (2) are: "does one have a right, as Alan Gewirth argues, not to be caused to contract cancer?" or "are there duties to future generations potentially harmed by various technologies?" Examples of (3) are: "should commercial nuclear power licensees, contrary to the Price–Anderson Act, be subject to strict and full liability?" or "should the US continue to export banned pesticides to developing nations?" Examples of (4) are: "would development of a plutonium-based energy technology threaten civil liberties?" or "would deregulation of the airline industry result in less safe air travel?" Examples of (5) are: "does benefit-cost analysis ignore noneconomic components of human welfare?" or "do Bayesian methods of technology assessment ignore the well-being of minorities likely to be harmed by a technological development?"

The leading philosophical issues concerning technology and ethics are the following:

How to Define Technological Risk

Engineers and technical experts tend to define "technological risk" as a probability of physical harm, usually as "average annual probability of fatality." Philosophers and other humanistic critics claim both that technological risk cannot be defined purely quantitatively, and that it includes more than physical harm. Instead they argue that technology often threatens other goods, such as civil liberties, personal autonomy, or rights such as due process.

The technologists argue for a quantitative definition of risk, claiming that we need a common denominator for evaluating diverse technological hazards. They also claim that it is imposible to evaluate nonquantitative notions, such as the technological threat to democracy. Those who oppose the quantitative definition argue not only that it excludes qualitative factors (like equity of risk distribution) affecting welfare, but also that the nonquantitative factors are sometimes more important than the quantitative ones. Hence they argue, for example, that an equitably distributed technological risk could be more desirable than a quantitatively smaller one (in terms of probability of fatality) that is inequitably distributed.

How to Evaluate Technologies in the Face of Uncertainty

Whether one technological risk is quantitatively greater than another, in terms of average annual probability of fatality, however, is often difficult to determine. Most evaluations of technology are conducted in the face of probabilistic uncertainty about the magnitude of potential hazards. Typically this uncertainty ranges from two to four orders of magnitude. It arises because the developments most needing evaluation, e.g., biotechnology, are new. We have limited experience with them and hence limited data about their accident frequency.

How should one evaluate technologies whose level of risk is uncertain? According to John Harsanyi, the majority position is that, in such situations, one should either use subjective probabilities of experts, or assume that all uncertain events are equally probable. The desirable technological choice is then the one having the highest "average expected utility," as measured by the probability and utility of the outcomes associated with each choice. Critics of the majority position, like John Rawls, maintain that it has all the flaws of utilitarianism. It fails, they say, to take adequate account of minorities likely to be harmed by high-consequence, low-probability risks. Rawls argues instead that we should use a maximin rule in situations of probabilistic uncertainty. Such a rule, like the difference principle, would direct us to avoid the outcome having the worst possible consequences, regardless of its alleged probability.

Critics of the Rawlsian position claim that it is irrational to choose so as to avoid worst-case technological accidents. They claim that taking small chances with technology often brings great economic benefits for everyone. Opponents of the majority Bayesian position respond, however, that such benefits are neither assured nor worth the risk, and that the subjective probabilities of experts often exhibit an "overconfidence bias" that there will be no serious accidents or negative health effects from a given technology.

Technological Threats to Due Process

Ethicists also charge that technology threatens due-process rights. To the extent that hazardous technologies cause (what Judith Thomson calls) "incompensable risks," like death, due process is impossible because the victim cannot be compensated.

One of the most controversial due-process debates concerns commercial nuclear fission for generating electricity. Current United States law limits liability of the nuclear licensees to less than three percent of total possible losses in a catastrophic accident. Critics maintain that this law (the Price–Anderson Act) violates citizens' due-process rights. Defenders argue that it is needed to protect the industry from possible bankruptcy, and that a catastrophic nuclear accident is unlikely. Critics respond that if a catastrophic nuclear accident is unlikely, then industry needs no protection from bankruptcy caused by such an event.

How Safe Is Safe Enough?

Because a zero-risk society is impossible, philosophers and policy makers debate both how much risk is acceptable and how it ought to be distributed. The distribution controversies raise all the classical problems associated with utilitarian *versus* egalitarian ethical schemes. Conflicts over how much technological risk is acceptable typically raise issues of whether the public has certain welfare rights, like the right to breathe clean air. The controversies also focus on how much economic progress can be traded for the negative health consequences of technology-induced risks.

Philosophers are particularly divided about how to evaluate numerous negligible risks, from a variety of technologies, that together pose a serious hazard. Small cancer risks that are singly harmless, but cumulatively and synergistically harmful, provide a good example of such cases. They raise the classical ethical problem of the contributor's dilemma. This dilemma occurs because the benefit of avoiding imposing a single small technological risk is imperceptible, although the cumulative benefit of everyone's doing so is great. Some philosophers view such small risks as ethically insignificant, while others claim they are important. Those in the latter group argue that agents are responsible for the effects of *sets* of acts (that together cause harm) of which their individual act is only one member.

Consent to Risk

The sophistication of many technologies, from genetically engineered organisms to the latest nuclear weapons, makes it questionable whether many individuals understand them. If they do not, then it is likewise questionable whether persons are able to give free, informed consent to the risks that they impose. Critics of some contemporary technologies point out that those persons most likely to take technological risks (e.g., blue-collar workers in chemical or radiation-related industries) are precisely those who are least able to give free, informed consent to them. This is because they are often persons with limited education and no alternative job skills.

Those who claim that both workers and the public have given consent to technological risks use notions like "compensating wage differential" to defend their position. They say that, since workers in hazardous technologies receive correspondingly higher pay because of the greater risks that they face, they are compensated. Likewise they maintain that accepting a risky job constitutes a form of consent. They also claim that society's acceptance of the economic benefits created by hazardous technologies constitutes implicit acceptance of the technologies.

In response, more conservative ethicists argue both that economic analysis does not show the existence of a compensating wage differential in all cases, and that mere acceptance of a job in a risky technology does not constitute consent to the hazard, especially if the worker has no other realistic employment alternatives. They also argue that acceptance of the benefits of hazardous technologies does not constitute acceptance of the technologies themselves since many people are inadequately informed about such risks.

Bibliography

Baier, Kurt, and Nicholas Rescher, eds. *Values and the Future: The Impact of Technological Change on American Values*. New York: Free Press, 1969. Philosophical analysis of problems associated with technological development.

Durbin, Paul T., ed. *Philosophy and Technology*. Boston: Kluwer, 1985–. Annual publication of the Society for Philosophy and Technology. Essays, many of which deal with ethics and technology, written from a variety of philosophical perspectives.

Ellul, Jacques. *The Technological Society*. Translated by John Wilkinson. New York: Knopf, 1964. General overview of problems associated with technology; widely viewed as anti-technology.

——. *The Technological System*. Translated by Joachim Neugroschel. New York: Continuum, 1980. General overview of problems associated with technology; widely viewed as anti-technology.

Goodpaster, Kenneth, and Kenneth Sayre, eds. *Ethics and the Problems of the 21st Century*. South Bend, Ind.: University of Notre Dame Press, 1979. Original philosophical essays on various problems associated with technological development.

Harsanyi, John C. "Can the Maximin Principle Serve as a Basis for Morality? A Critique of John Rawls' Theory." *American Political Science Review* 59, no. 2 (1975): 594–605.

Jonas, Hans. *The Imperative of Responsibility*. Chicago: University of Chicago Press, 1985. Criticism of contemporary technological society and proposals for reform.

Kasperson, Roger, and Mimi Berberian, eds. *Equity Issues in Radioactive Waste Management*. Boston: Oelgeschlager, 1983. Descriptive analysis of equity issues related to radwaste management.

MacLean, Douglas. *Values at Risk*. Totowa, N.J.: Rowman and Allanheld, 1984. Consideration of the ethical problems posed by issues of nuclear security and deterrence.

MacLean, Douglas, and Peter Brown, eds. *Energy and the Future*. Totowa, N.J.: Rowman and Allanheld, 1983. Essays on ethical issues associated with energy technologies.

Rescher, Nicholas. *Unpopular Essays on Technological Progress*. Pittsburgh, Pa.: University of Pittsburgh Press, 1980. Widely ranging essays on a variety of topics related to technology.

Shrader-Frechette, Kristin. *Nuclear Power and Public Policy*. Boston: Reidel, 1983. Criticism of nuclear technology on grounds that it violates various ethical principles.

——. *Science Policy, Ethics, and Economic Methodology*. Boston: Reidel, 1984. Analysis of problems with benefit-cost methods and proposals for amending them.

——. *Risk Analysis and Scientific Method*. Boston: Reidel, 1984. Analysis of problems associated with assessment of technological risk assessment and proposals for amending them.

Thomson, Judith Jarvis. *Rights, Restitution, and Risk*. Cambridge: Harvard University Press, 1986. Treats a variety of ethical issues related to technological risk.

18

Toward a Philosophy of Technology

Hans Jonas

Are there philosophical aspects to technology? Of course there are, as there are to all things of importance in human endeavor and destiny. Modern technology touches on almost everything vital to man's existence – material, mental, and spiritual. Indeed, what of man is *not* involved? The way he lives his life and looks at objects, his intercourse with the world and with his peers, his powers and modes of action, kinds of goals, states and changes of society, objectives and forms of politics (including warfare no less than welfare), the sense and quality of life, even man's fate and that of his environment: all these are involved in the technological enterprise as it extends in magnitude and depth. The mere enumeration suggests a staggering host of potentially philosophic themes.

To put it bluntly: if there is a philosophy of science, language, history, and art; if there is social, political, and moral philosophy; philosophy of thought and of action, of reason and passion, of decision and value – all facets of the inclusive philosophy of man – how then could there not be a philosophy of technology, the focal fact of modern life? And at that a philosophy so spacious that it can house portions from all the other branches of philosophy? It is almost a truism, but at the same time so immense a proposition that its challenge staggers the mind. Economy and modesty require that we select, for a beginning, the most obvious from the multitude of aspects that invite philosophical attention.

The old but useful distinction of "form" and "matter" allows us to distinguish between these two major themes: (1) the *formal dynamics* of technology as a continuing collective enterprise, which advances by its own "laws of motion"; and (2) the *substantive content* of technology in terms of the things it puts into human use, the powers it confers, the novel objectives it opens up or dictates, and the altered manner of human action by which these objectives are realized.

The first theme considers technology as an abstract whole of movement; the second considers its concrete uses and their impact on our world and our lives. The formal approach will try to grasp the pervasive "process properties" by which modern technology propels itself – through our agency, to be sure – into ever-succeeding and superceding novelty. The material approach will look at the species of novelties themselves, their taxonomy, as it were, and try to make out how the world furnished with them looks. A third, overarching theme is the *moral* side of technology as a burden on human responsibility, especially its long-term effects on the global condition of man and environment. This – my own main preoccupation over the past years – will only be touched upon.

I The Formal Dynamics of Technology

First some observations about technology's form as an abstract whole of movement. We are concerned with characteristics of *modern* technology and therefore ask first what distinguishes it *formally* from all previous technology. One major distinction is that modern technology is an enterprise and process, whereas earlier technology was a possession and a state. If we roughly describe technology as comprising the use of artificial implements for the business of life, together with their original invention, improvement, and occasional additions, such a tranquil description will do for most of technology through mankind's career (with which it is coeval), but not for modern technology. In the past, generally speaking, a given inventory of tools and procedures used to be fairly constant, tending toward a mutually adjusting, stable equilibrium of ends and means, which – once established – represented for lengthy periods an unchallenged optimum of technical competence.

To be sure, revolutions occurred, but more by accident than by design. The agricultural revolution, the metallurgical revolution that led from the neolithic to the iron age, the rise of cities, and such developments, *happened* rather than were consciously created. Their pace was so slow that only in the time-contraction of historical retrospect do they appear to be "revolutions" (with the misleading connotation that their contemporaries experienced them as such). Even where the change was sudden, as with the introduction first of the chariot, then of armed horsemen into warfare – a violent, if short-lived, revolution indeed – the innovation did not originate from within the military art of the advanced societies that it affected, but was thrust on it from outside by the (much less civilized) peoples of Central Asia. Instead of spreading through the technological universe of their time, other technical breakthroughs, like Phoenician purple-dying, Byzantine "greek fire," Chinese porcelain and silk, and Damascene steel-tempering, remained jealously guarded monopolies of the inventor communities. Still others, like the hydraulic and steam playthings of Alexandrian mechanics, or compass and gunpowder of the Chinese, passed unnoticed in their serious technological potentials.[1]

On the whole (not counting rare upheavals), the great classical civilizations had comparatively early reached a point of technological saturation – the aforementioned "optimum" in equilibrium of means with acknowledged needs and goals – and had little cause later to go beyond it. From there on, convention reigned supreme. From pottery to monumental architecture, from food growing to shipbuilding, from textiles to engines of war, from time measuring to stargazing: tools, techniques, and objectives remained essentially the same over long times; improvements were sporadic and unplanned. Progress therefore – if it occurred at all[2] – was by inconspicuous increments to a universally high level that still excites our admiration and, in historical fact, was more liable to regression than to surpassing. The former at least was the more noted phenomenon, deplored by the epigones with a nostalgic remembrance of a better past (as in the declining Roman world). More important, there was, even in the best and most vigorous times, no proclaimed *idea* of a future of *constant progress* in the arts. Most important, there was never a deliberate method of going about it like "research," the willingness to undergo the risks of trying unorthodox paths, exchanging information widely about the experience, and so on. Least of all was there a "natural science" as a growing body of theory to guide such semi-theoretical, prepractical activities, plus their social institutionalization. In routines as well as panoply of instruments, accomplished as they were for the purposes they served, the "arts" seemed as settled as those purposes themselves.[3]

Traits of modern technology

The exact opposite of this picture holds for modern technology, and this is its first philosophical aspect. Let us begin with some manifest traits.

1. Every new step in whatever direction of whatever technological field tends *not* to approach an equilibrium or saturation point in the process of fitting means to ends (nor is it meant to), but, on the contrary, to give rise, if successful, to further steps in all kinds of direction and with a fluidity of the ends themselves. "Tends to" becomes a compelling "is bound to" with any major or important step (this almost being its criterion); and the innovators themselves expect, beyond the accomplishment, each time, of their immediate task, the constant future repetition of their inventive activity.

2. Every technical innovation is sure to spread quickly through the technological world community, as also do theoretical discoveries in the sciences. The spreading is in terms of knowledge and of practical adoption, the first (and its speed) guaranteed by the universal intercommunication that is itself part of the technological complex, the second enforced by the pressure of competition.

3. The relation of means to ends is not unilinear but circular. Familiar ends of long-standing may find better satisfaction by new technologies whose genesis they had inspired. But equally – and in-

creasingly typical – new technologies may suggest, create, even impose new ends, never before conceived, simply by offering their feasibility. (Who had ever wished to have in his living room the Philharmonic orchestra, or open heart surgery, or a helicopter defoliating a Vietnam forest? or to drink his coffee from a disposable plastic cup? or to have artificial insemination, test-tube babies, and host pregnancies? or to see clones of himself and others walking about?) Technology thus adds to the very objectives of human desires, including objectives for technology itself. The last point indicates the dialectics or circularity of the case: once incorporated into the socioeconomic demand diet, ends first gratuitously (perhaps accidentally) generated by technological invention become necessities of life and set technology the task of further perfecting the means of realizing them.

4. Progress, therefore, is not just an ideological gloss on modern technology, and not at all a mere option offered by it, but an inherent drive which acts willy-nilly in the formal automatics of its *modus operandi* as it interacts with society. "Progress" is here not a value term but purely descriptive. We may resent the fact and despise its fruits and yet must go along with it, for – short of a stop by the fiat of total political power, or by a sustained general strike of its clients or some internal collapse of their societies, or by self-destruction through its works (the last, alas, the least unlikely of these) – the juggernaut moves on relentlessly, spawning its always mutated progeny by coping with the challenges and lures of the now. But while not a value term, "progress" here is not a neutral term either, for which we could simply substitute "change." For it is in the nature of the case, or a law of the series, that a later stage is always, in terms of technology itself, *superior* to the preceding stage.[4] Thus we have here a case of the entropy-defying sort (organic evolution is another), where the internal motion of a system, left to itself and not interfered with, leads to ever "higher," not "lower" states of itself. Such at least is the present evidence.[5] If Napoleon once said, "Politics is destiny," we may well say today, "Technology is destiny."

These points go some way to explicate the initial statement that modern technology, unlike traditional, is an enterprise and not a possession, a process and not a state, a dynamic thrust and not a set of implements and skills. And they already adumbrate certain "laws of motion" for this restless phenomenon. What we have described, let us remember, were formal traits which as yet say little about the contents of the enterprise. We ask two questions of this descriptive picture: *why* is this so, that is, what *causes* the restlessness of modern technology; what is the nature of the thrust? And, what is the philosophical import of the facts so explained?

The nature of restless technology

As we would expect in such a complex phenomenon, the motive forces are many, and some causal hints appeared already in the descriptive account. We have mentioned *pressure of competition* – for profit, but also for power, security, and so forth – as one perpetual mover in the universal appropriation of technical improvements. It is equally operative in their origination, that is, in the process of invention itself, nowadays dependent on constant outside subsidy and even goal-setting: potent interests see to both. War, or the threat of it, has proved an especially powerful agent. The less dramatic, but no less compelling, everyday agents are legion. To keep one's head above the water is their common principle (somewhat paradoxical, in view of an abundance already far surpassing what former ages would have lived with happily ever after). Of pressures other than the competitive ones, we must mention those of population growth and of impending exhaustion of natural resources. Since both phenomena are themselves already by-products of technology (the first by way of medical improvements, the second by the voracity of industry), they offer a good example of the more general truth that to a considerable extent technology itself begets the problems which it is then called upon to overcome by a new forward jump. (The Green Revolution and the development of synthetic substitute materials or of alternate sources of energy come under this heading.) These compulsive pressures for progress, then, would operate even for a technology in a noncompetitive, for example, a socialist setting.

A motive force more autonomous and spontaneous than these almost mechanical pushes with their "sink or swim" imperative would be the pull of the quasi-utopian *vision* of an ever better life, whether vulgarly conceived or nobly, once technology had proved the open-ended capacity for procuring the conditions for it: perceived possibility whetting the appetite ("the American dream," "the revolution of rising expectations").

This less palpable factor is more difficult to appraise, but its playing a role is undeniable. Its deliberate fostering and manipulation by the dream merchants of the industrial-mercantile complex is yet another matter and somewhat taints the spontaneity of the motive, as it also degrades the quality of the dream. It is also moot to what extent the vision itself is *post hoc* rather than *ante hoc*, that is, instilled by the dazzling feats of a technological progress already underway and thus more a response to than a motor of it.

Groping in these obscure regions of motivation, one may as well descend, for an explanation of the dynamism as such, into the Spenglerian mystery of a "Faustian soul" innate in Western culture, that drives it, nonrationally, to infinite novelty and unplumbed possibilities for their own sake; or into the Heideggerian depths of a fateful, metaphysical decision of the will for boundless power over the world of things – a decision equally peculiar to the Western mind: speculative intuitions which do strike a resonance in us, but are beyond proof and disproof.

Surfacing once more, we may also look at the very sober, functional facts of industrialism as such, of production and distribution, output maximization, managerial and labor aspects, which even apart from competitive pressure provide their own incentives for technical progress. Similar observations apply to the requirements of *rule* or control in the vast and populous states of our time, those giant territorial super-organisms which for their very cohesion depend on advanced technology (for example, in information, communication, and transportation, not to speak of weaponry) and thus have a stake in its promotion: the more so, the more centralized they are. This holds for socialist systems no less than for free-market societies. May we conclude from this that even a communist world state, freed from external rivals as well as from internal free-market competition, might still have to push technology ahead for purposes of control on this colossal scale? Marxism, in any case, has its own inbuilt commitment to technological progress beyond necessity. But even disregarding all dynamics of these conjectural kinds, the most monolithic case imaginable would, at any rate, still be exposed to those noncompetitive, natural pressures like population growth and dwindling resources that beset industrialism as such. Thus, it seems, the compulsive element of technological progress may not be bound to its original breeding ground, the capitalist system. Perhaps the odds for an eventual stabilization look somewhat better in a socialist system, provided it is worldwide – and possibly totalitarian in the bargain. As it is, the pluralism we are thankful for ensures the constancy of compulsive advance.

We could go on unravelling the causal skein and would be sure to find many more strands. But none nor all of them, much as they explain, would go to the heart of the matter. For all of them have one premise in common without which they could not operate for long: the premise that there *can* be indefinite progress because there *is* always something new and better to find. The, by no means obvious, givenness of this objective condition is also the pragmatic conviction of the performers in the technological drama; but without its being true, the conviction would help as little as the dream of the alchemists. Unlike theirs, it is backed up by an impressive record of past successes, and for many this is sufficient ground for their belief. (Perhaps holding or not holding it does not even greatly matter.) What makes it more than a sanguine belief, however, is an underlying and well-grounded, theoretical view of the nature of things and of human cognition, according to which they do not set a limit to novelty of discovery and invention, indeed, that they of themselves will at each point offer another opening for the as yet unknown and undone. The corollary conviction, then, is that a technology tailored to a nature and to a knowledge of this indefinite potential ensures its indefinitely continued conversion into the practical powers, each step of it begetting the next, with never a cutoff from internal exhaustion of possibilities.

Only habituation dulls our wonder at this wholly unprecedented belief in virtual "infinity." And by all our present comprehension of reality, the belief is most likely true – at least enough of it to keep the road for innovative technology in the wake of advancing science open for a long time ahead. Unless we understand this ontologic-epistomological premise, we have not understood the inmost agent of technological dynamics, on which the working of all the adventitious causal factors is contingent in the long run.

Let us remember that the virtual infinitude of advance we here seek to explain is in essence different from the always avowed perfectibility of every human accomplishment. Even the undisputed master of his craft always had to admit as possible that he might be surpassed in skill or tools or materials; and no excellence of product ever foreclosed that it might still be bettered, just as today's champion runner must know that his time may one day be beaten. But these are improvements within a given genus, not different in kind from what went before, and they must accrue in diminishing fractions. Clearly, the phenomenon of an exponentially growing *generic* innovation is qualitatively different.

Science as a source of restlessness

The answer lies in the interaction of *science* and *technology* that is the hallmark of modern progress, and thus ultimately in the kind of nature which modern science progressively discloses. For it is here, in the movement of *knowledge*, where relevant novelty first and constantly occurs. This is itself a novelty. To Newtonian physics, nature appeared simple, almost crude, running its show with a few kinds of basic entities and forces by a few universal laws, and the application of those well-known laws to an ever greater variety of composite phenomena promised ever widening knowledge indeed, but no real surprises. Since the mid-nineteenth century, this minimalistic and somehow finished picture of nature has changed with breathtaking acceleration. In a reciprocal interplay with the growing subtlety of exploration (instrumental and conceptual), nature itself stands forth as ever more subtle. The progress of probing makes the object grow richer in modes of operation, not sparer as classical mechanics had expected. And instead of narrowing the margin of the still-undiscovered, science now surprises itself with unlocking dimension after dimension of new depths. The very essence of matter has turned from a blunt, irreducible ultimate to an always reopened challenge for further penetration. No one can say whether this will go on forever, but a suspicion of intrinsic infinity in the very being of things obtrudes itself and therewith an anticipation of unending inquiry of the sort where succeeding steps will not find the same old story again (Descartes' "matter in motion"), but always add new twists to it. If then the art of technology is correlative to the knowledge of nature, technology too acquires from this source that potential of infinity for its innovative advance.

But it is not just that indefinite scientific progress offers the *option* of indefinite technological progress, to be exercised or not as other interests see fit. Rather the cognitive process itself moves by interaction with the technological, and in the most internally vital sense: for its own *theoretical* purpose, science must generate an increasingly sophisticated and physically formidable technology as its tool. What it finds with this help initiates new departures in the practical sphere, and the latter as a whole, that is, technology at work provides with its experiences a large-scale laboratory for science again, a breeding ground for new questions, and so on in an unending cycle. In brief, a mutual feedback operates between science and technology; each requires and propels the other; and as matters now stand, they can only live together or must die together. For the dynamics of technology, with which we are here concerned, this means that (all external promptings apart) an agent of restlessness is implanted in it by its functionally integral bond with science. As long, therefore, as the cognitive impulse lasts, technology is sure to move ahead with it. The cognitive impulse, in its turn, culturally vulnerable in itself, liable to lag or to grow conservative with a treasured canon – that theoretical eros itself no longer lives on the delicate appetite for truth alone, but is spurred on by its hardier offspring, technology, which communicates to it impulsions from the broadest arena of struggling, insistent life. Intellectual curiosity is seconded by interminably self-renewing practical aim.

I am conscious of the conjectural character of some of these thoughts. The revolutions in science over the last fifty years or so are a fact, and so are the revolutionary style they imparted to technology and the reciprocity between the two concurrent streams (nuclear physics is a good example). But whether those scientific revolutions, which hold primacy in the whole syndrome, will be typical for science henceforth – something like a law of motion for its future – or represent only a singular phase in its longer run, is unsure. To the extent, then, that our forecast of incessant novelty for technology was predicated on a guess concerning the future of science, even concerning the nature of things, it is hypothetical, as such extrapolations are bound to be. But even if the recent past did not usher in a state of permanent revolution for science, and the life of theory settles down again to a more sedate pace, the scope for technological innovation will not easily shrink; and what may no longer be a revolution in science, may still revolutionize our lives in its practical impact through technology. "Infinity" being too large a word anyway, let us say that present signs of potential and of incentives point to an indefinite perpetuation and fertility of the technological momentum.

The philosophical implications

It remains to draw philosophical conclusions from our findings, at least to pinpoint aspects of philosophical interest. Some preceding remarks have already been straying into philosophy of science in the technical sense. Of broader issues, two will be ample to provide food for further thought beyond the limitations of this paper. One concerns the status of knowledge in the human scheme, the other the status of technology itself as a human goal, or its tendency to become that from being a means, in a dialectical inversion of the means-end order itself.

Concerning knowledge, it is obvious that the time-honored division of theory and practice has vanished for both sides. The thirst for pure knowledge may persist undiminished, but the involvement of knowing at the heights with doing in the lowlands of life, mediated by technology, has become inextricable; and the aristocratic self-sufficiency of knowing for its own (and the knower's) sake has gone. Nobility has been exchanged for utility. With the possible exception of philosophy, which still can do with paper and pen and tossing thoughts around among peers, all knowledge has become thus tainted, or elevated if you will, whether utility is intended or not. The technological syndrome, in other words, has brought about a thorough *socializing* of the theoretical realm, enlisting it in the service of common need. What used to be the freest of human choices, an extravagance snatched from the pressure of the world – the esoteric life of thought – has become part of the great public play of necessities and a prime necessity in the action of the play.[6] Remotest abstraction has become enmeshed with nearest concreteness. What this pragmatic functionalization of the once highest indulgence in impractical pursuits portends for the image of man, for the restructuring of a hallowed hierarchy of values, for the idea of "wisdom," and so on, is surely a subject for philosophical pondering.

Concerning technology itself, its actual role in modern life (as distinct from the purely instrumental definition of technology as such) has made the relation of means and ends equivocal all the way up from the daily living to the very vocation of man. There could be no question in former technology that its role was that of humble servant – pride of workmanship and esthetic embellishment of the useful notwithstanding. The Promethean enterprise of modern technology speaks a different language. The word "enterprise" gives the clue, and its unendingness another. We have mentioned that the effect of its innovations is disequilibrating rather than equilibrating with respect to the balance of wants and supply, always breeding its own new wants. This in itself compels the constant attention of the best minds, engaging the full capital of human ingenuity for meeting challenge after challenge and seizing the new chances. It is psychologically natural for that degree of engagement to be invested with the dignity of dominant purpose. Not only does technology dominate our lives in fact, it nourishes also a belief in its being of predominant worth. The sheer grandeur of the enterprise and its seeming infinity inspire enthusiasm and fire ambition. Thus, in addition to spawning new ends (worthy or frivolous) from the mere invention of means, technology as a grand venture tends to establish *itself* as the transcendent end. At least the suggestion is there and casts its spell on the modern mind. At its most modest, it means elevating *homo faber* to the essential aspect of man; at its most extravagant, it means elevating *power* to the position of his dominant and interminable goal. To become ever more masters of the world, to advance from power to power, even if only collectively and perhaps no longer by choice, *can* now be seen to be the chief vocation of mankind. Surely, this again poses philosophical questions that may well lead unto the uncertain grounds of metaphysics or of faith.

I here break off, arbitrarily, the formal account of the technological movement in general, which as yet has told us little of what the enterprise is about. To this subject I now turn, that is, to the new kinds of powers and objectives that technology opens to modern man and the consequently altered quality of human action itself.

II The Material Works of Technology

Technology is a species of power, and we can ask questions about how and on what object any power is exercised. Adopting Aristotle's rule in *de anima* that for understanding a faculty one should begin with its objects, we start from them too – "objects" meaning both the visible *things* technology generates and puts into human use, and the *objectives* they serve. The objects of modern technology are first everything that had always been an object of human artifice and labor: food, clothing, shelter, implements, transportation – all the material necessities and comforts of life. The technological intervention changed at first not the product but its production, in speed, ease, and quantity. However, this is true only of the very first stage of the industrial revolution with which large-scale scientific technology began. For example, the cloth for the steam-driven looms of Lancashire remained the same. Even then, one significant new product was added to the traditional list – the machines themselves, which required an entire new industry with further subsidiary industries to build them. These novel entities, machines – at first capital goods only, not consumer goods – had from the beginning their own impact on man's symbiosis with nature by being consumers themselves. For example: steam-powered water pumps facilitated coal mining, required in turn extra coal for firing their boilers, more coal for the foundries and forges that made those boilers, more for the mining of the requisite iron ore, more for its transportation to the foundries, more – both coal and iron – for the rails and locomotives made in these same foundries, more for the conveyance of the foundries' product to the pitheads and return, and finally more for the distribution of the more abundant coal to the users outside this cycle, among which were increasingly still more machines spawned by the increased availability of coal. Lest it be forgotten over this long chain, we have been speaking of James Watt's modest steam engine for pumping water out of mine shafts. This syndrome of self-proliferation – by no means a linear chain but an intricate web of reciprocity – has been part of modern technology ever since. To generalize, technology exponentially increases man's drain on nature's resources (of substances and of energy), not only through the multiplication of the final goods for consumption, but also, and perhaps more so, through the production and operation of its own mechanical means. And with these means – machines – it introduced a new category of goods, not for consumption, added to the furniture of our world. That is, among the objects of technology a prominent class is that of technological apparatus itself.

Soon other features also changed the initial picture of a merely mechanized production of familiar commodities. The final products reaching the consumer ceased to be the same, even if still serving the same age-old needs; new needs, or desires, were added by commodities of entirely new kinds which changed the habits of life. Of such commodities, machines themselves became increasingly part of the consumer's daily life to be used directly by himself, as an article not of production but of consumption. My survey can be brief as the facts are familiar.

New kinds of commodities

When I said that the cloth of the mechanized looms of Lancashire remained the same, everyone will have thought of today's synthetic fibre textiles for which the statement surely no longer holds. This is fairly recent, but the general phenomenon starts much earlier, in the synthetic dyes and fertilizers with which the chemical industry – the first to be wholly a fruit of science – began. The original rationale of these technological feats was substitution of artificial for natural materials (for reasons of scarcity or cost), with as nearly as possible the same properties for effective use. But we need only think of plastics to realize that art progressed from substitutes to the creation of really new substances with properties not so found in any natural one, raw or processed, thereby also initiating uses not thought of before and giving rise to new classes of objects to serve them. In chemical (molecular) engineering, man does more than in mechanical (molar) engineering which constructs machinery from natural materials; his intervention is deeper, redesigning the infra-patterns of nature, making substances to specification by arbitrary disposition of molecules. And this, be it noted, is done deductively from the bottom, from the thoroughly analyzed last elements, that is, in a real *via compositiva* after the completed *via resolutiva*, very different from the long-known empirical practice of coaxing substances into new properties, as in metal alloys from the bronze age on. Artificiality or creative engineering with abstract construction invades the heart of matter. This, in molecular biology, points to further, awesome potentialities.

With the sophistication of molecular alchemy we are ahead of our story. Even in straightforward hardware engineering, right in the first blush of the mechanical revolution, the objects of use that came out of the factories did not really remain the same, even where the objectives did. Take the old objective of travel. Railroads and ocean liners are relevantly different from the stage coach and from the sailing ship, not merely in construction and efficiency but in the very feel of the user, making travel a different experience altogether, something one may do for its own sake. Airplanes, finally, leave behind any similarity with former conveyances, except the purpose of getting from here to there, with no experience of what lies in between. And these instrumental objects occupy a prominent, even obtrusive place in our world, far beyond anything wagons and boats ever did. Also they are constantly subject to improvement of design, with obsolescence rather than wear determining their life span.

Or take the oldest, most static of artifacts: human habitation. The multistoried office building of steel, concrete, and glass is a qualitatively different entity from the wood, brick, and stone structures of old. With all that goes into it besides the structures as such – the plumbing and wiring, the elevators, the lighting, heating, and cooling systems – it embodies the end products of a whole spectrum of technologies and far-flung industries, where only at the remote sources human hands still meet with primary materials, no longer recognizable in the final result. The ultimate customer inhabiting the product is ensconced in a shell of thoroughly derivative artifacts (perhaps relieved by a nice piece of driftwood). This transformation into utter artificiality is generally, and increasingly, the effect of technology on the human environment, down to the items of daily use. Only in agriculture has the product so far escaped this transformation by the changed modes of its production. We still eat the meat and rice of our ancestors.[7]

Then, speaking of the commodities that technology injects into private use, there are machines themselves, those very devices of its own running, originally confined to the economic sphere. This unprecedented novum in the records of individual living started late in the nineteenth century and has since grown to a pervading mass phenomenon in the Western world. The prime example, of course, is the automobile, but we must add to it the whole gamut of household appliances – refrigerators, washers, dryers, vacuum cleaners – by now more common in the lifestyle of the general population than running water or central heating were one hundred years ago. Add lawn mowers and other power tools for home and garden: we are mechanized in our daily chores and recreations (including the toys of our children) with every expectation that new gadgets will continue to arrive.

These paraphernalia are machines in the precise sense that they perform work and consume energy, and their moving parts are of the familiar magnitudes of our perceptual world. But an additional and profoundly different category of technical apparatus was dropped into the lap of the private citizen, not labor-saving and work-performing, partly not even utilitarian, but – with minimal energy input – catering to the senses and the mind: telephone, radio, television, tape recorders, calculators, record players – all the domestic terminals of the electronics industry, the latest arrival on the technological scene. Not only by their insubstantial, mind-addressed output, also by the subvisible, not literally "mechanical" physics of their functioning do these devices differ in kind from all the macroscopic, bodily moving machinery of the classical type. Before inspecting this momentous turn from power engineering, the hallmark of the first industrial revolution, to communication engineering, which almost amounts to a second industrial-technological revolution, we must take a look at its natural base: electricity.

In the march of technology to ever greater artificiality, abstraction, and subtlety, the unlocking of electricity marks a decisive step. Here is a universal force of nature which yet does not naturally appear to man (except in lightning). It is not a datum of uncontrived experience. Its very "appearance" had to wait for science, which contrived the experience for it. Here, then, a technology depended on science for the mere providing of its "object," the entity itself it would deal with – the first case where theory alone, not ordinary experience, wholly preceded practice (repeated later in the case of nuclear energy). And what sort of entity! Heat and steam are familiar objects of sensuous experience, their force bodily displayed in nature; the matter of chemistry is still the concrete, corporeal stuff mankind had always known. But electricity is an abstract object, disembodied, immaterial, unseen; in its usable form, it is entirely an artifact, generated in a subtle transformation from grosser forms of energy (ultimately from heat via motion). Its theory indeed had to be essentially complete before utilization could begin.

Revolutionary as electrical technology was in itself, its purpose was at first the by now conventional one of the industrial revolution in general: to supply motive power for the propulsion of machines. Its advantages lay in the unique versatility of the new force, the ease of its transmission, transformation and distribution – an unsubstantial commodity, no bulk, no weight, instantaneously delivered at the point of consumption. Nothing like it had ever existed before in man's traffic with matter, space, and time. It made possible the spread of mechanization to every home; this alone was a tremendous boost to the technological tide, at the same time hooking private lives into centralized public networks and thus making them dependent on the functioning of a total system as never before, in fact, for every moment. Remember, you cannot hoard electricity as you can coal and oil, or flour and sugar for that matter.

But something much more unorthodox was to follow. As we all know, the discovery of the universe of electromagnetics caused a revolution in theoretical physics that is still underway. Without it, there would be no relativity theory, no quantum mechanics, no nuclear and subnuclear physics. It also caused a revolution in technology beyond what it contributed, as we noted, to its classical program. The revolution consisted in the passage from electrical to electronic technology which signifies a new level of abstraction in means and ends. It is the difference between power and communication engineering. Its object, the most impalpable of all, is information. Cognitive instruments had been known before – sextant, compass, clock, telescope, microscope, thermometer, all of them for information and not for work. At one time, they were called "philosophical" or "metaphysical" instruments. By the same general criterion, amusing as it may seem, the new electronic information devices, too, could be classed as "philosophical instruments." But those earlier cognitive devices, except the clock, were inert and passive, not generating information actively, as the new instrumentalities do.

Theoretically as well as practically, electronics signifies a genuinely new phase of the scientific-technological revolution. Compared with the sophistication of its theory as well as the delicacy of its apparatus, everything which came before seems crude, almost natural. To appreciate the point, take the man-made satellites now in orbit. In one sense, they are indeed an imitation of celestial mechanics – Newton's laws finally verified by cosmic experiment: astronomy, for millennia the most purely contemplative of the physical sciences, turned into a practical art! Yet, amazing as it is, the astronomic imitation, with all the unleashing of forces and the finesse of techniques that went into it, is the least interesting aspect of those entities. In that respect, they still fall within the terms and feats of classical mechanics (except for the remote-control course corrections).

Their true interest lies in the instruments they carry through the voids of space and in what these do, their measuring, recording, analyzing, computing, their receiving, processing, and transmitting abstract information and even images over cosmic distances. There is nothing in all nature which even remotely foreshadows the kind of things that now ride the heavenly spheres. Man's imitative practical astronomy merely provides the vehicle for something else with which he sovereignly passes beyond all the models and usages of known nature.[8] That the advent of man portended, in its inner secret of mind and will, a cosmic event was known to religion and philosophy: now it manifests itself as such by fact of things and acts in the visible universe. Electronics indeed creates a range of objects imitating nothing and progressively added to by pure invention.

And no less invented are the ends they serve. Power engineering and chemistry for the most part still answered to the natural needs of man: for food, clothing, shelter, locomotion, and so forth. Communication engineering answers to needs of information and control solely created by the civilization that made this technology possible and, once started, imperative. The novelty of the means continues to engender no less novel ends – both becoming as necessary to the functioning of the civilization that spawned them as they would have been pointless for any former one. The world they help to constitute and which needs computers for its very running is no longer nature supplemented, imitated, improved, transformed, the original habitat made more habitable. In the pervasive mentalization of physical relationships it is a *trans-nature* of human making, but with this inherent paradox: that it threatens the obsolescence of man himself, as increasing automation ousts him from the places of work where he formerly proved his humanhood. And there is a further threat: its strain on nature herself may reach a breaking point.

The last stage of the revolution?

That sentence would make a good dramatic ending. But it is not the end of the story. There may be in the offing another, conceivably the last, stage of the technological revolution, after the mechanical, chemical, electrical, electronic stages we have surveyed, and the nuclear we omitted. All these were based on physics and had to do with what man can put to his use. What about biology? And what about the user himself? Are we, perhaps, on the verge of a technology, based on biological knowledge and wielding an engineering art which, this time, has man himself for its object? This has become a theoretical possibility with the advent of molecular biology and its understanding of genetic programming; and it has been rendered morally possible by the metaphysical neutralizing of man. But the latter, while giving us the license to do as we wish, at the same time denies us the guidance for knowing what to wish. Since the same evolutionary doctrine of which genetics is a cornerstone has deprived us of a valid image of man, the actual techniques, when they are ready, may find us strangely unready for their responsible use. The anti-essentialism of prevailing theory, which knows only of *de facto* outcomes of evolutionary accident and of no valid essences that would give sanction to them, surrenders our being to a freedom without norms. Thus the technological call of the new microbiology is the twofold one of physical feasibility and metaphysical admissibility. Assuming the genetic mechanism to be completely analyzed and its script finally decoded, we can set about rewriting the text. Biologists vary in their estimates of how close we are to the capability; few seem to doubt the right to use it. Judging by the rhetoric of its prophets, the idea of taking our evolution into our own hands is intoxicating even to many scientists.

In any case, the idea of making over man is no longer fantastic, nor interdicted by an inviolable taboo. If and when *that* revolution occurs, if technological power is really going to tinker with the elemental keys on which life will have to play its melody in generations of men to come (perhaps the only such melody in the universe), then a reflection on what is humanly desirable and what should determine the choice – a reflection, in short, on the image of man, becomes an imperative more urgent than any ever inflicted on the understanding of mortal man. Philosophy, it must be confessed, is sadly unprepared for this, its first cosmic task.

III Toward an Ethics of Technology

The last topic has moved naturally from the descriptive and analytic plane, on which the objects of technology are displayed for inspection, onto the evaluative plane where their ethical challenge poses itself for decision. The particular case forced the transition so directly because there the (as yet hypothetical) technological object was man directly. But once removed, man is involved in all the other objects of technology, as these singly and jointly remake the worldly frame of his life, in both the narrower and the wider of its senses: that of the artificial frame of civilization in which social man leads his life proximately, and that of the natural terrestrial environment in which this artifact is embedded and on which it ultimately depends.

Again, because of the magnitude of technological effects on both these vital environments in their totality, both the quality of human life and its very preservation in the future are at stake in the rampage of technology. In short, certainly the "image" of man, and possibly the survival of the species (or of much of it), are in jeopardy. This would summon man's duty to his cause even if the jeopardy were not of his own making. But it is, and, in addition to his ageless obligation to meet the threat of things, he bears for the first time the responsibility of prime agent in the threatening disposition of things. Hence nothing is more natural than the passage from the objects to the ethics of technology, from the things made to the duties of their makers and users.

A similar experience of inevitable passage from analysis of fact to ethical significance, let us remember, befell us toward the end of the first section. As in the case of the matter, so also in the case of the form of the technological dynamics, the image of man appeared at stake. In view of the quasi-automatic compulsion of those dynamics, with their perspective of indefinite progression, every existential and moral question that the objects of technology raise assumes the curiously eschatological quality with which we are becoming familiar from the extrapolating guesses of futurology. But apart from thus raising all challenges of present particular matter to the higher powers of future exponential magnification, the despotic dynamics of the technological movement as such, sweeping its captive movers along in its breathless momentum, poses its own questions to man's axiological conception of himself. Thus, form and matter of technology alike enter into the dimension of ethics.

The questions raised for ethics by the objects of technology are defined by the major areas of their impact and thus fall into such fields of knowledge as ecology (with all its biospheric subdivisions of land, sea, and air), demography economics, biomedical and behavioral sciences (even the psychology of mind pollution by television), and so forth. Not even a sketch of the substantive problems, let alone of ethical policies for dealing with them, can here be attempted. Clearly, for a normative rationale of the latter, ethical theory must plumb the very foundations of value, obligation, and the human good.

The same holds of the different kind of questions raised for ethics by the sheer fact of the formal dynamics of technology. But here, a question of another order is added to the straightforward ethical questions of both kinds, subjecting any resolution of them to a pragmatic proviso of harrowing uncertainty. Given the mastery of the creation over its creators, which yet does not abrogate their responsibility nor silence their vital interest, what are the chances and what are the means of gaining *control* of the process, so that the results of any ethical (or even purely prudential) insights can be translated into effective action? How in short can man's freedom prevail against the determinism he has created for himself? On this most clouded question, whereby hangs not only the effectuality or futility of the ethical search which the facts invite (assuming it to be blessed with *theoretical* success!), but perhaps the future of mankind itself, I will make a few concluding, but – alas – inconclusive, remarks. They are intended to touch on the whole ethical enterprise.

Problematic preconditions of an effective ethics

First, a look at the novel state of determinism. Prima facie, it would seem that the greater and more varied powers bequeathed by technology have expanded the range of choices and hence increased human freedom. For economics, for example, the argument has been made[9] that the uniform compulsion which scarcity and subsistence previously imposed on economic behavior with a virtual denial of alternatives (and hence – conjoined with the universal "maximization" motive of capitalist market competition – gave classical economics at least the appearance of a deterministic "science") has given way to a latitude of indeterminacy. The plenty and powers provided by industrial technology allow a pluralism of choosable alternatives (hence disallow scientific prediction). We are not here concerned with the status of economics as a science. But as to the altered state of things alleged in the argument. I submit that the change means rather that one, relatively homogeneous determinism (thus relatively easy to formalize into a law) has been supplanted by another, more complex, multifarious determinism, namely, that exercised by the human artifact itself upon its creator and user. We, abstractly speaking the possessors of those powers, are concretely subject to their emancipated dynamics and the sheer momentum of our own multitude, the vehicle of those dynamics.

I have spoken elsewhere[10] of the "new realm of necessity" set up, like a second nature, by the feedbacks of our achievements. The almighty we, or Man personified is, alas, an abstraction. *Man* may have become more powerful; *men* very probably the opposite, enmeshed as they are in more dependencies than ever before. What ideal Man now can do is not the same as what real men permit or dictate to be done. And here I am thinking not only of the immanent dynamism, almost automatism, of the impersonal technological complex I have invoked so far, but also of the pathology of its client society. Its compulsions, I fear, are at least as great as were those of unconquered nature. Talk of the blind forces of nature! Are those of the sorcerer's creation less blind? They differ indeed in the serial shape of their causality: the action of nature's forces is cyclical, with periodical recurrence of the same, while that of the technological forces is linear, progressive, cumulative, thus replacing the curse of constant toil with the threat of maturing crisis and possible catastrophe. Apart from this significant vector difference, I seriously wonder whether the tyranny of fate has not become greater, the latitude of spontaneity smaller; and whether man has not actually been weakened in his decision-making capacity by his accretion of collective strength.

However, in speaking, as I have just done, of "his" decision-making capacity, I have been guilty of the same abstraction I had earlier criticized in the use of the term "man." Actually, the subject of the statement was no real or representative individual but Hobbes' "Artificiall Man," "that great Leviathan, called a Common-Wealth," or the "large horse" to which Socrates likened the city, "which because of its great size tends to be sluggish and needs stirring by a gadfly." Now, the chances of there being such gadflies among the numbers of the commonwealth are today no worse nor better than they have ever been, and in fact they are around and stinging in our field of concern. In that respect, the free spontaneity of personal insight, judgment, and responsible action by speech can be trusted as an ineradicable (if also incalculable) endowment of humanity, and smallness of number is in itself no impediment to shaking public complacency. The problem, however, is not so much complacency or apathy as the counterforces of active, and anything but complacent, interests and the complicity with them of all of us in our daily consumer existence. These interests themselves are factors in the determinism which technology has set up in the space of its sway. The question, then, is that of the possible chances of unselfish insight in the arena of (by nature) selfish *power*, and more particularly; of one long-range, interloping insight against the short-range goals of many incumbent powers. Is there hope that wisdom itself can become power? This renews the thorny old subject of Plato's philosopher-king and – with that inclusion of realism which the utopian Plato did not lack – of the role of myth, not knowledge, in the education of the guardians. Applied to our topic: the *knowledge* of objective dangers and of values endangered, as well as of the technical remedies, is beginning to be there and to be disseminated: but to make it prevail in the marketplace is a matter less of the rational dissemination of truth than of public relations techniques, persuasion, indoctrination, and manipulation, also of unholy alliances, perhaps even conspiracy. The philosopher's descent into the cave may well have to go all the way to "if you can't lick them, join them."

That is so not merely because of the active resistance of special interests but because of the optical illusion of the near and the far which condemns the long-range view to impotence against the enticement and threats of the nearby: it is this incurable shortsightedness of animal-human nature more than ill will that makes it difficult to move even those who have no special axe to grind, but still are in countless ways, as we all are, beneficiaries of the untamed system and so have something dear in the present to lose with the inevitable cost of its taming. The taskmaster, I fear, will have to be actual pain beginning to strike, when the far has moved close to the skin and has vulgar optics on its side. Even then, one may resort to palliatives of the hour. In any event, one should try as much as one can to forestall the advent of emergency with its high tax of suffering or, at the least, prepare for it. This is where the scientist can redeem his role in the technological estate.

The incipient knowledge about technological danger trends must be developed, coordinated, systematized, and the full force of computer-aided projection techniques be deployed to determine priorities of action, so as to inform preventive efforts wherever they can be elicited, to minimize the necessary sacrifices, and at the worst to preplan the saving measures which the terror of beginning calamity will eventually make people willing to accept. Even now, hardly a decade after the first stirrings of "environmental" consciousness, much of the requisite knowledge, plus the rational persuasion, is available inside and outside academia for any well-meaning powerholder to draw upon. To this, we – the growing band of concerned intellectuals – ought persistently to contribute our bit of competence and passion.

But the real problem is to get the well-meaning into power and have that power as little as possible beholden to the interests which the technological colossus generates on its path. It is the problem of the philosopher-king compounded by the greater magnitude and complexity (also sophistication) of the forces to contend with. Ethically, it becomes a problem of playing the game by its impure rules. For the servant of truth to join in it means to sacrifice some of his time-honored role: he may have to turn apostle or agitator or political operator. This raises moral questions beyond those which technology itself poses, that of sanctioning immoral means for a surpassing end, of giving unto Caesar so as to promote what is not Caesar's. It is the grave question of moral casuistry, or of Dostoevsky's Grand Inquisitor, or of regarding cherished liberties as no longer affordable luxuries (which may well bring the anxious friend of mankind into odious political company) – questions one excusably hesitates to touch but in the further tide of things may not be permitted to evade.

What is, prior to joining the fray, the role of philosophy, that is, of a philosophically grounded ethical knowledge, in all this? The somber note of the last remarks responded to the quasi-apocalyptic prospects of the technological tide, where stark issues of planetary survival loom ahead. There, no philosophical ethics is needed to tell us that disaster must be averted. Mainly, this is the case of the ecological dangers. But there are other, noncatastrophic things afoot in technology where not the existence but the image of man is at stake. They are with us now and will accompany us and be joined by others at every new turn technology may take. Mainly, they are in the biomedical, behavioral, and social fields. They lack the stark simplicity of the survival issue, and there is none of the (at least declaratory) unanimity on them which the spectre of extreme crisis commands. It is here where a philosophical ethics or theory of values has its task. Whether its voice will be listened to in the dispute on policies is not for it to ask; perhaps it cannot even muster an authoritative voice with which to speak – a house divided, as philosophy is. But the philosopher must try for normative knowledge, and if his labors fall predictably short of producing a compelling axiomatics, at least his clarifications can counteract rashness and make people pause for a thoughtful view.

Where not existence but "quality" of life is in question, there is room for honest dissent on goals, time for theory to ponder them, and freedom from the tyranny of the lifeboat situation. Here, philosophy can have its try and its say. Not so on the extremity of the survival issue. The philosopher, to be sure, will also strive for a theoretical grounding of the very proposition that there ought to be men on earth, and that present generations are obligated to the existence of future ones. But such esoteric, ultimate validation of the perpetuity imperative for the species – whether obtainable or not to the satisfaction of reason – is happily not needed for consensus in the face of ultimate threat. Agreement in favor of life is pretheoretical, instinctive, and universal. Averting disaster takes precedence over everything else, including pursuit of the good, and suspends otherwise inviolable prohibitions and rules. All moral standards for individual or group behavior, even demands for individual sacrifice of life, are premised on the continued existence of human life. As I have said elsewhere,[11] "No rules can be devised for the waiving of rules in extremities. As with the famous shipwreck examples of ethical theory, the less said about it, the better."

Never before was there cause for considering the contingency that all mankind may find itself in a lifeboat, but this is exactly what we face when the viability of the planet is at stake. Once the situation becomes desperate, then what there is to do for salvaging it must be done, so that there be life – which "then," after the storm has been weathered, can again be adorned by ethical conduct. The moral inference to be drawn from this lurid eventuality of a moral pause is that we must never allow a lifeboat situation for humanity to arise.[12] One part of the ethics of technology is precisely to guard the space in which any ethics can operate. For the rest, it must grapple with the cross-currents of value in the complexity of life.

A final word on the question of determinism versus freedom which our presentation of the technological syndrome has raised. The best hope of man rests in his most trouble some gift: the spontaneity of human acting which confounds all prediction. As the late Hannah Arendt never tired of stressing: the continuing arrival of newborn individuals in the world assures ever-new beginnings. We should expect to be surprised and to see our predictions come to naught. But those predictions themselves, with their warning voice, can have a vital share in provoking and informing the spontaneity that is going to confound them.

Notes

1 But as serious an actuality as the Chinese plough "wandered" slowly westward with little traces of its route and finally caused a major, highly beneficial revolution in medieval European agriculture, which almost no one deemed worth recording when it happened (cf. Paul Leser, *Entstehung und Verbreitung des Pfluges*, Münster, 1931; reprint: The International Secretariate for Research on the History of Agricultural Implements, Brede-Lingby, Denmark, 1971).

2 Progress did, in fact, occur even at the heights of classical civilizations. The Roman arch and vault, for example, were distinct engineering advances over the horizontal entablature and flat ceiling of Greek (and Egyptian) architecture, permitting spanning feats and thereby construction objectives not contemplated before (stone bridges, aqueducts, the vast baths and other public halls of Imperial Rome). But materials, tools, and techniques were still the same, the role of human labor and crafts remained unaltered, stonecutting and brickbaking went on as before. An existing technology was enlarged in its scope of performance, but none of its means or even goals made obsolete. [Ed.]

3 One meaning of "classical" is that those civilizations had somehow implicitly "defined" themselves and neither encouraged nor even allowed to pass beyond their innate terms. The – more or less – achieved "equilibrium" was their very pride. [Ed.]

4 This only seems to be but is not a value statement, as the reflection on, for example, an ever more destructive atom bomb shows. [Ed.]

5 There may conceivably be internal degenerative factors – such as the overloading of finite information-processing capacity – that may bring the (exponential) movement to a halt or even make the system fall apart. We don't know yet. [Ed.]

6 There is a paradoxical side effect to this change of roles. That very science which forfeited its place in the domain of leisure to become a busy toiler in the field of common needs, creates by its toils a growing domain of leisure for the masses, who reap this with the other fruits of technology as an additional (and no less novel) article of forced consumption. Hence leisure, from a privilege of the few, has become a problem for the many to cope with. Science, not idle, provides for the needs of this idleness too: no small part of technology is spent on filling the leisure-time gap which technology itself has made a fact of life. [Ed.]

7 Not so, objects my colleague Robert Heilbroner in a letter to me; "I'm sorry to tell you that meat and rice are both *profoundly* influenced by technology. Not even they are left untouched." Correct, but they are at least generically the same (their really profound changes lie far back in the original breeding of domesticated strains from wild ones – as in the case of all cereal plants under cultivation). I am speaking here of an order of transformation in which the results bear no resemblance to the natural materials at their source, nor to any naturally occurring state of them. [Ed.]

8 Note also that in radio technology, the medium of action is nothing material, like wires conducting currents, but the entirely immaterial electromagnetic "field", i.e., space itself. The symbolic picture of "waves" is the last remaining link to the forms of our perceptual world. [Ed.]

9 I here loosely refer to Adolph Lowe, "The Normative Roots of Economic Values," in Sidney Hook, ed., *Human Values and Economic Policy* (New York: New York University Press, 1967) and, more perhaps, to the many discussions I had with Lowe over the years. For my side of the argument, see "Economic Knowledge and the Critique of Goals," in R. L. Heilbroner, ed., *Economic Means and Social Ends* (Englewood Cliffs, N.J.: Prentice-Hall, 1969), reprinted in Hans Jonas, *Philosophical Essays* (Englewood Cliffs, N.J.: Prentice-Hall, 1969), reprinted in Hans Jonas, *Philosophical Essays* (Englewood Cliffs, N.J.: Prentice-Hall, 1974).

10 "The Practical Uses of Theory," *Social Research* 26 (1959), reprinted in Hans Jonas, *The Phenomenon of Life* (New York, 1966). The reference is to pp. 209–10 in the latter edition.

11 "Philosophical Reflections on Experimenting with Human Subjects," in Paul A. Freund, ed., *Experimentation with Human Subjects* (New York: George Braziller, 1970), reprinted in Hans Jonas, *Philosophical Essays*. The reference is to pp. 124–25 in the latter edition.

12 For a comprehensive view of the demands which such a situation or even its approach would make on our social and political values, see Geoffrey Vickers, *Freedom in a Rocking Boat* (London, 1970).

Part III

Defining Technology

Introduction

Efforts to define "technology" face at least three serious difficulties. First, as several previous essays have suggested, there is the problem of trying to assign technology a general "nature" at all – a problem that seems all the harder to solve because any proposed definition faces the problem of explaining away the already numerous competing alternatives. Second, there is the question of how, or even whether to distinguish prescientific from modern scientific technology. And third, there are the special difficulties associated with criticizing the widely popular but deeply problematic characterizations of technology as equipment or as applied science.

Stephen J. Kline suggests that we think of technology first and foremost as involving both "sociotechnical systems of manufacture" and "sociotechnical systems of use." Although it is certainly common to think of technology in terms of its products (i.e., hardware or artifacts), or its facilitating knowledge and techniques, Kline argues that it is only in terms of the *purposes* furthered by those social systems that these products, know-how, and techniques can be fully understood. In the end, he concludes, it is and always has been these systems, not merely their products or procedures, that "form the physical bases of all human societies past and present." In this way, Kline separates himself from those who conceive technology primarily as the mere neutral means to whatever are our chosen ends, as well as from those who see technology primarily in its relation to modern science. It is, of course, an open question whether even those who might otherwise be sympathetic with his general conclusion would also accept his apparently "materialistic" construal of the idea of a social system.

For Arnold Gehlen, technology is best understood in connection with questions about human nature rather than simply in light of the social systems in which it functions. He thus interprets technology in terms of a "philosophical anthropology," a field in which he has been a leading figure. In the continental tradition, philosophical anthropology is understood to be a kind of twentieth-century transformation of "rational psychology," i.e., that branch of traditional metaphysics concerned with the nature of the soul. Today, its task is to integrate into a coherent system what we have come to know about the biological and psychological as well as the social aspects of human beings. Anglo-American philosophers do not officially recognize this species of philosophy, and they tend to get nervous whenever it is suggested that something they do has roots in traditional metaphysics. Nevertheless, many of the same themes addressed by philosophical anthropologists figure prominently in analytic discussions of the philosophical implications of cognitive science and evolutionary biology for theories of human nature.

Like Kant (see chapter 4) and many others, Gehlen sees human beings as lacking the reliable instincts and physical prowess so important to survival for other animals, and he thinks of us as compensating for these lacks by our capacity for reason and creativity. Anticipating the discussions in the first section of Part V, Gehlen makes his theory of human nature the basis for his account of technology. In his view, technology helps us construct a safe, artificial physical environment in which to propagate and to work out an appropriate social, cultural, and political organization. For the first time in history, he concludes, we are witness to the rise of an ideology that expresses values capable of receiving universal assent – namely, those of peace, equality, and material and spiritual development. This is because we realize that with these values, it is possible to work out in a successful and "mutually enforcing" way our twin propensities for control of nature and human propagation.

Trevor Pinch and Wiebe Bijker defend a social constructionist account of technology. Such accounts have become increasingly influential in science and technology studies. Social constructionism (also "constructivism") first emerged in postpositivistic studies of science, where the formal and purely rational conceptions of scientific procedure were recontextualized to show that science is best understood as an actual human practice – one in which defining and accepting data, discoveries, and explanations are all shown to involve a process of negotiation among real persons and social groups. Social constructionists insist that they remain neutral with regard to the truth-status of scientific claims; but this neutrality has been interpreted in two very different ways. On the one hand, some take it to be a methodological stance of "bracketing," in which science as a socio-cultural practice is simply studied without the need to take a stand on the objectivity of its results or the reality of its subject matter. On the other hand, some regard social constructionist neutrality as a substantive philosophical position, according to which the so-called facts and objects of science are all socially constructed and thus not to be treated as "real" in the sense of present in an objective world, independently of humans. This latter construal of neutrality, especially, has been attacked, whether rightly or wrongly, for its allegedly anti-realist account of science (see Latour's remarks about the science wars, Part II, chapter 12).

In their accounts of technology, social constructionists usually draw special attention to the actual processes of invention. They complain that the holistic characterizations of technology by such thinkers as Heidegger, Ellul, Marcuse, and Mumford wrongly incline us to think of it in extra-empirical, deterministic terms, or to picture it as an autonomous, even out-of-control force. They urge instead that we examine the fine structure of particular technologies and the groups of inventors and consumers involved in the development and acceptance of these technologies. What results do these groups desire or demand? they ask. What tensions does this produce, and what satisfactory resolution to these tensions is possible? From addressing these questions, they argue, one can arrive at a picture of both the social negotiations that determine the final form of artifacts and also the social consensus that accepts it as a desired product.

Langdon Winner admits that social constructivism has methodological advantages, but he criticizes it for focusing excessively on the invention of technologies and neglecting their social impact. In this, he argues, it is just transposing to technology studies its earlier tendency to investigate the discovery process in science without enough concern for the problem of its cultural impact. Social constructivism also shares the weakness of all pluralistic methodologies in political sociology. By treating the various groups of inventors, consumers, and persons in government and business as equal participants in the negotiations over technology, it inevitably ignores the extent to which more powerful groups and interests may call the tune. More seriously and generally, it neglects what is not "visible" in the negotiations – namely, groups that have been excluded, and dominant orientations and attitudes (e.g., of inclusion and exclusion) shared by the visible participants. From a social constructivist perspective, then, questions about the problematic character of modern technological orientation generally – questions central to thinkers such as Heidegger, Marx, and Ellul – cannot even arise. Finally, Winner objects to social constructivism's value-neutrality. One result of the constructivist stance is that the criticisms of technology by ecologists, feminists, and other advocates of social justice are all flattened out in a social relativism that merely notes "neutrally" how, yes, these are some of the possible evaluations of technology. In this, says Winner, social constructivists share the outlook of postmodernism – not only in the rejection of any "grand theories" of technology but also in the tendency to make evaluative judgments about technology mere functions of their time and location. Of course, as the exchange between Winner and Elam (Part VI, chapter 52) suggests, whether one agrees with Winner depends in part on whether one accepts the viability of grand theories of society and of objective values.

19

What is Technology

Stephen J. Kline

In the late 20th century, there is only one thing most people agree about concerning technology – it is important. It is discussed almost as much as the weather, and sometimes it seems, with as little effect.

But what is "technology?" If we look with even a little care, we find this same word is being used to represent things, actions, processes, methods and systems. "Technology" is also used symbolically as an epithet, for important working procedures, and to represent progress. This much conflict within the usage of one of our central terms won't do; it can lead only to chaos. Even more important, the current vague use of the word "technology" hides from view two central concepts, and a central pattern of human behavior that we must have to make sense of our views of many critical questions in the current world including how we understand innovation, how we can communicate across Snow's culture gap, and how we understand the way in which we humans make our living on the planet.

We cannot get on with our work in STS studies even reasonably well until we "unpack" the word "technology" – take apart the various usages and agree on names for each of the important concepts so that we can understand one another at least adequately. As Seneca told us two millennia ago, "When the words are corrupt, the mind is also."

Perhaps the commonest usage of "technology" is to denote *manufactured articles* – things made by humans that do not occur naturally on earth, for example: refrigerators, eyeglasses, atom bombs, paints, automobiles, pianos, paper, rubber, glass, aspirin, penicillin, airplanes, copying machines, furniture, roads, rifles, printing presses, boots, bicycles, and on and on. In the late 20th century, the list is very long. Engineers often call manufactured articles "hardware"; anthropologists usually call them "artifacts." Since the phrase "manufactured articles" is awkward, we might use either the word "hardware" or the word "artifact."

USAGE 1: HARDWARE (OR ARTIFACTS): Possible denotation: non-natural objects, of all kinds, manufactured by humans.

The next most common usage of 'technology' is the process of *manufacturing hardware*. Usually this usage includes the manufacturing equipment, and sometimes it includes in addition the people who operate the equipment. In either case this is a truncated usage in a very important sense. What is usually being implied, in the important sense of this usage, must be much more than just the machinery and the people. It is what I will call a *sociotechnical system of manufacture*. For example, a complete system for manufacturing airplanes (or pianos, bicycles, eyeglasses, atom bombs, aspirin, blue jeans, etc.). This full usage is essential for many reasons that will be illustrated shortly.

USAGE 2: SOCIOTECHNICAL SYSTEM OF MANUFACTURE. Possible denotation: All the elements needed to manufacture a particular kind of hardware, the complete working system including its inputs: people; machinery; resources;

From *Bulletin of Science, Technology & Society* 1 (1985): 215–18, New York, Pergamon Press, Copyright © 1985 STS Press. Reprinted by permission of Sage Publications, Inc.

processes; and legal, economic, political and physical environment.

A third common usage of the word "technology" is *technique*, *methodology* or "*know-how*." In a famous polemic Ellul uses the word "technology" to denote any form of rationalized methodology. Others, such as Brooks, suggest "technology" be used to denote the knowledge needed within a sociotechnical system of manufacture. These are even more truncated uses of the term "technology," and in some instances this truncation has caused significant confusions, as in some interpretations of Ellul's that occured in part owing to the translation of the term "*La Technique*" from the French. Others appear to use "technology" to denote methodology for accomplishing any given task, as in, "we have the technology to do the job." In order to avoid these potential confusions, we might call these various functions: "knowledge," "technique," "know-how," "methodology" (or "sociotechnical system") as appropriate.

USAGE 3: The information, skills, processes, and procedures for accomplishing tasks: Possible denotation: KNOWLEDGE, TECHNIQUE, KNOW-HOW, OR METHODOLOGY in the usual sense of these words.

When we include the many variations and shadings of the three usages discussed above, they constitute together the common usages of the term "technology." However, there is a fourth related concept that has no common name, but which is essential to understanding the human implications of "technology" in the ways intended by much public discussion. I will call this fourth concept *sociotechnical systems of use*. Such systems form the basis of what we do with the hardware after we have manufactured it.

For example, we embody automobiles in a system of roads, gas stations, laws for ownership and operation, rules of the road, etc., and use the combined system (the autos plus all the rest) to extend the human capacity for moving ourselves and our possessions about – transport. We manufacture violins, pianos, drums, guitars, and other musical instruments. We then embody them in orchestras and bands to extend the ways in which we can make music. We build microscopes, telescopes, CAT-scanners, thermometers, and other instruments and utilize them in systems to extend our ability to sense various aspects of the world around us. We make rifles, pistols, grenades, atom bombs, and other weapons and diffuse them into armies, navies and air forces which are systems for extending our capacity to kill, to oppress other peoples, and to protect ourselves from being oppressed by others. And, we do all this in order to perform tasks which individual humans cannot perform without such systems.

USAGE 4: A SOCIOTECHNICAL SYSTEM OF USE is a system using combinations of hardware and people (and usually other elements) to accomplish tasks that humans cannot perform unaided by such systems – to extend human capacities.

I am making a point of saying "these systems" and not just "hardware" because nearly always we need more than just the hardware to create these extensions of human capacities. Even if a single human is using a musical instrument, there is the need for knowledge of music and ingrained neuromuscular skills, from long practice, to make even adequate music. To create a band, a system of transport, an army, or a football team takes many more elements. The central point is that we have learned to vastly extend our muscular, sensing, and mental capacities through the use of sociotechnical systems of manufacture *and* use.

These extensions of human capacities by use of sociotechnical systems are both quantitative and qualitative. Using autos or trains we can move over the land much faster than we can unaided by such systems of transport. Using sociotechnical systems built to exploit airplanes, we can fly, a function hardly any of us can perform unaided. Using various weapons in armies, navies or air forces, we can vastly extend our ability to kill and oppress. Using telescopes and microscopes, X-rays and other hardware in appropriate systems, we can see far beyond the scope of the naked eye. Using engines and motors in appropriate systems, we can create manifold extensions of our muscle power. In a relatively recent set of systems, we are extending our memory and data manipulating capacities by use of computer systems.

Without sociotechnical system of use, the manufacture of hardware would have no purpose. Taken together, sociotechnical systems of manufacture and sociotechnical systems of use form the physical bases of all human societies past and present. The human consequences of this statement are so profound that they need a book for complete elaboration, but I will give a few illustrations here.

The pattern of creating hardware in special sociotechnical systems of manufacture and diffusing the hardware into other sociotechnical systems of use in order to extend our human capacities is not a product of the "high-tech age." On the contrary, the pattern was first adopted by our evolutionary ancestors two species before we became homo sapiens, roughly two million years ago according to the best current evidence from paleoanthropology. We humans have been making our living on earth by use of this pattern for so long a time, that it has materially affected our evolutionary path.

Other animals use sociotechnical systems (beavers, ants, bees, and prairie dogs, to mention only a few). However, we humans are the only species that purposefully makes innovations in our sociotechnical systems in order to (hopefully) improve their functioning. This characteristic of purposeful innovation in sociotechnical systems distinguishes humans from other animals as least as clearly as any single characteristic.

The history of changes in sociotechnical systems is a history that accelerates with major eras each roughly a period of ten times shorter than the preceding period for the past million or two years. The changes look much like a growth function. There is a sharp break in the rate of acceleration in the extension of human powers via sociotechnical systems about 1840 as both Lienhard (1979) and Kline (1977) have documented independently.

Without sociotechnical systems, we humans might not exist as a species, and if we did, we would be relatively powerless, few in number and of little import on the planet. Using the extensions that become possible with current sociotechnical systems, we have in a large measure become the lords of the planet. If we are to exercise the powers of lordship well, we will certainly need to be clear on the source of those powers and the processes through which they are exercised.

Few topics are more basic to STS studies than an understanding of the nature of sociotechnical systems and the pattern in which we humans use them to create the physical bases for our societies past and present.

References

Stephen J. Kline (1977). Toward the Understanding of Technology in Society: Part One – The Basic Pattern. *Mechanical Engineering* 99/1: 40–5.

John H. Lienhard (1979). The Rate of Technological Improvement Before and After the 1830s. *Technology and Culture* 20.

20

A Philosophical-Anthropological Perspective on Technology

Arnold Gehlen

I

Ernst Kapp in his *Philosophie der Technik* (1877) was probably the first to note the connection between man's organic shortcomings and his inventive intelligence. Shortly thereafter the philosopher of language Ludwig Noiré in *Das Werkzeug* [The Tool] (1880) wrote: "Then man became released from nature, for he became his own creator, created his own organs, became a tool maker, a *tool making animal*."[1] Along with Paul Alsberg (*Das Menschheitsrätsel* [The Riddle of Mankind], 1922), José Ortega y Gasset (*Vom Menschen also utopischem Wesen*, 1951),[2] and Werner Sombart, we agree that the necessity for technology derives from man's organ deficiencies. My *Der Mensch* (1940), using Herder's concept of the deficient being, described how man in any natural, uncultivated environment is not able to survive because of a lack of specialized organs and instincts, and how consequently he has to create the conditions for his physical survival by intelligently altering existing environmental conditions. The use of weapons, fire, and hunting techniques thus belong to the behavioral patterns designed to preserve the species, so that the word "technology" [*Technik*] must refer to both the actual tools and those skills needed to create and use them which make it possible for this instinct-poor and defenseless creature "to preserve himself."

From *Research in Philosophy and Technology* 6 (1983): 205–16, abridged. Trans. Dorthe Thrane Rogers and Carl Mitcham, from "Anthropologische Ansicht der Technik," in *Technik im technischen Zeitalter*, ed. Hans Freyer, Johannes Chr. Papalekas, and Georg Weippert, Düsseldorf: J. Schilling, 1965, pp. 101–18. Reprinted with permission from Elsevier Science.

One must begin, then, with an interpretation of the tool, and Kapp's concept of "organ projection" already provides some help, making it possible to distinguish the principles of organic relief [*Organentlastung*], organic substitution or replacement [*Organersatzes*], and organic strengthening or improvement [*Organüberbeitung*], either in themselves or through their interactions. It is worth noting, though, that one must avoid the assumption that man-made tools always have a model in nature, because the bow and arrow, wheel, or even knife can only be understood as free, constructive inventions developed through experimentation. The Chinese ascribe the invention of the knot to their first mythical emperor.

The principle of organic substitution manifests itself in the tendency to replace organic by inorganic material in artifacts. In this way the replacement of wooden weapons and utensils by metal ones defines an entire era (the bronze age), and in the contemporary world one can point to the chemistry of synthetics. Of even greater impact, however, is another development which dates back less than two hundred years – namely the replacement of the organic power of man and beast with inorganic or stored and recovered solar energies: coal, oil, electricity, nuclear energy.

Wind power can be productively utilized only at sea, and the definitive break with the ancient view of life did not come until the fifteenth century, when naval technology and superiority became crucial to world domination. On land, however, human beings still had to make do largely with organic power and the work of men and beasts. Apart from the inventions of the printing press and gunpowder, both of which were highly successful, there was little technological progress – which explains, first, the enormous stability of the agrarian period in human history and, second, the backwardness of land warfare. If we disregard firearms, Napoleon waged war the same way Caesar did: with troops marching over enormous distances, followed by endless supply lines plied by beasts of burden, following roads which often had to be constructed for this very purpose; and in the end battles were decided by close combat engagements, which explains the great importance of the cavalry.

The discovery of uses for coal and oil, plus the use of electricity and nuclear energy, resulted in an immeasurable increase in the available energy sources, the potential development of which is beyond assessment.

II

Hans Freyer's *Über das Dominantwerden technischer Kategorien in der Lebenswelt der industriellen Gesellschaft*[3] pointed out that the technology of the industrial age does not fit the description which applies to all technology prior to the eighteenth century – namely, that man fabricates appropriate means and tools to attain his goals in life, and then improves these until they serve his purposes. The meaning of technology, according to this formula, rests with its subordination to the successful achievement of what man wants or wills. According to Freyer, such an idea is outdated. It applied to a state of technological development which was primitive by comparison with the present. Instead, Freyer maintains that technology now creates a kind of abstract ability, and it is not until later that the question arises of what one "wants" to do with the existing means.

This thought no doubt applies first and foremost to the above mentioned immeasurable increase in energy production potential. Under such circumstances, as Freyer argues, the technological spirit becomes an absolute and is no longer tied to set goals. Mentally, the desire for controlling and interfering with nature, and for progress, has become so manifest as to create the impression that we set preposterous goals, such as landing on the moon, in order to solve the extremely difficult technical problems which only arise out of this venture. At the same time, an increasing proportion of whatever technical intelligence is available must go into dealing with those unintentionally created side effects, nuisances, and major problems of this uncontrolled race into the future – from automobile exhaust and traffic jams to the expansive and dark question of feeding the world's population, the gigantic growth of which was a secondary result of technological progress, with the food-supply problem being a tertiary effect – or with projects such as how to control that purely emotional manipulation of information which entire nations are exposed to as a consequence of the completely autonomous worldwide dissemination of information.

It is evident, then, that just as one cannot precisely define "art," it is difficult to determine exactly what "technology" is, especially when one includes organic surgical interventions of a medical-therapeutic sort – not to mention the fairly predictable effects of attempts at mental manipulation, in which one cannot always clearly distinguish between advertisement, propaganda, and education. It is therefore necessary to stick with a rather general definition, and above all to keep in mind that this concept includes still others.

Even if we start with the narrow field of mechanics and machine technology, the traditional idea that with regard to their effects we are dealing with "applied natural science" is incorrect. This particular technology is related to the natural sciences, on the one hand, and to industrial machine production, on the other, in a complicated relationship of mutual influences. And these three spheres, of course, along with the entire sphere of information, can be seen as a *superstructure*, the existence of which constitutes the primary difference between our culture and all previous cultures.

The definition of technology as applied science does not take into account the interaction between the various components of this superstructure. For instance, progress in the natural sciences depends on technology, on devices for sensing and measuring, as well as the entire sphere of technical procedures within the special world of the various types of apparatus. A physics experiment today usually requires at least one machine built especially for that particular purpose, to produce the expected data; and the researcher who is examining some problematic phenomena must come up with plans for non-existing equipment to produce observable natural processes which will confirm or falsify the hypothesis in question. Here we are dealing with machines which, in the service of the natural sciences, are not producing goods but phenomena. This cooperation between technology and natural science is witnessed to by the wonders which, in the form of sensing and computing devices, are being launched into "outer space."

To this reciprocal interaction must be added that of industrial application, unless these projects are not themselves promoted by the interaction in the first place. Thus the production engineer becomes the link between technology and industrial production. For their part, large industrial companies with their large-scale resources also contribute to scientific research. They set up their own laboratories, where millions are spent on research not targeted for specific purposes, on top of which such industrial organizations even grant money to other institutions for scientific investigation. The amounts spent in these two ways far exceed the state budgets for university education in the Federal Republic of Germany. In particular cases the long-standing interaction of these forces is difficult to determine or retrace, and it is impossible to provide a summary picture of its extent. However, the concept of the "superstructure" should be kept in mind, since later on it will have to be part of our discussion.

Thus it is not surprising that the question has arisen about whether this no longer surveyable wave of progress can be said to have a master anymore, or whether, as suggested by Frederick Jonas's "Aspekte des Entwicklungsproblems in Industriestaaten" [Aspects of the problem of development in industrial countries][4] we are not dealing with an age without a subject term or primary actor. That is, perhaps the idea of a uniform structure which can be governed or for which a "leading spirit"[5] can be imagined is an illusion even in the area of one particular sector such as communications, not to mention industrial-technological development as a whole. Rational behavior, says Jonas, no longer signifies a control of relationships, but the optimal reaction to facts which are continuously changing due to the boundless stream of new occurrences – i.e., optimal reaction to the unexpected.

Then of course the strange question arises as to whether or not it is possible, instead of referring to a subject term or actor, to speak of the passive beneficiary of this rapid development of progress. And such a beneficiary actually seems to be the human race, admittedly in a purely biological, passionless sense. Between 1650 and 1850 it roughly doubled in numbers; from 1850 to 1950 it doubled again; and by 2000 it will have doubled once more – a rather depressing state of affairs.

III

If one views technological development to date in the context of human liberation and the expansion of power, then a law formulated by Hermann Schmidt appears persuasive.[6] Schmidt sees the *objectification of human* work through technological means as a process begun in previous ages but concluding in our own. This process passes through three phases: in the first, that of the tool, the necessary physical power comes from work while a required intellectual contribution is rendered by the subject. To this stage belong the above-mentioned distinctions between different types of organ projection. The second phase, according to Schmidt, is that of work and power machinery, the operation of which brings about the technological objectification of human physical power. And in the third phase of automation even the intellectual contribution of the subject is dispensed with by technological means, until the situation is reached in which the automaton can act independent of any physical or intellectual debt to us.

This law expresses an inner technological development which is not wholly dependent on the intentions of mankind. On the contrary, this law operates, so to speak, behind the back, and extends throughout the whole of human cultural history. Furthermore, the law does not allow for any technological development beyond the stage of complete automation, for there are no more areas of human capability which can be objectified or projected into the environment. The technological control loop [*Regelkreis*], the basic structure of the third stage, Schmidt sees as an objectified [stimulus-response] circle of human action [*Handlungskreis*], which can best be illustrated by the working hand and the controlling eye in which the actual and the ideal state of the system are always in harmonious relation to one another. The speech-hearing [stimulus-response] circle, which is very important, is also described as such a self-contained circle of human action.

We can make the following remarks with regard to the history of this model. In 1940 Friedrich von Weizsäcker's book *Der Gestaltkreis* appeared, along with my own *Der Mensch*. Today the unusually long life of my book can probably be attributed in part to the fact that, for the first time, it described man as what one would now call a feedback system [*rückgekoppeltes System*], namely a system which reacts to its own products. Essential help in this analysis was provided by the descriptions of so-called sensorimotor processes in *Mental Development in the Child and the Race* (1895) by J. M. Baldwin (1861–1934), who gave an account of "walking" as a movement which produces the incentive for its own continuation.

Schmidt, on the other hand, reports that "from approximately 1940 on insofar as the federal patent office had an easy overview of the state of technology, control loop technology could no longer be considered the center of technological development," and yet in that same year the Berlin physiologists W. Trendelenburg and K. Kramer confirmed a number of organic-somatic control loops. "Every one of our deliberate actions," says Schmidt, "necessarily takes place in the form of such a self-contained circle of human action in which we are reacting to ourselves *via* the respective results of our actions."

IV

The question of the consequences of this externalization of our functions for man's self-understanding can, for the moment, probably only be discussed and not actually answered. When the feedback circle of human action, which is one of the most basic human phenomena, is externalized, at the same time that the technological control loop is "desubjectified," this of course does not constitute progress in the technological production of organic systems. After all, both circles, the subjectively possessed and the externalized, are isomorphic – i.e., each "element" is replaced by a totally different one, while the relationship of the parts within the whole remains the same in both cases.

Earlier, Schmidt had pointed out the importance of the supposition that the tendency to objectification which in the past development of the human race had taken place unconsciously and instinctively was now being purposefully and deliberately directed so that human life was being relieved of its instinctiveness, at least in this respect. As late as 1964 he expected an acceleration in the continuation of our individuation since we, in cybernetics, are working on "the expansion and development of our psycho-physical existence." Obviously we are here also dealing with basic moral convictions, such as belief in the value of rationality, although this has never been tested by solving problems of contemporary dimensions.

The following consideration, however, is somewhat more obvious: Up till now, thinking in physics has abstracted from the circular relation of thought and experimentation the idea of the object as a "nature in itself," thus opening up the famous epistemological gap between subject and object – a gap which, in prescientific life, had always been camouflaged primarily through the linguistic dramatization of the external world. However, in feedback machines man objectifies not a "picture" of nature but the circle of human action, i.e., a *relationship* to nature. Yet here there are grounds for the supposition, which we shall first limit to the organic field, that nature itself in many respects operates according to the model of a dynamic feedback circle and that thus, to use an old formula, what goes on in nature, "in itself" actually appears in man "in and for itself" (in and for nature). This signifies an intimacy with nature much like that which was outlined ages ago merely linguistically and therefore ineffectively. Now, however, it also includes practice, so that nature in itself seems almost to meet halfway the interventions and directions which we have been carrying out in medical and biological operations for a long time. This is an obvious point which, interpreted the other way around, will have important implications if interventions in the social structure are ever going to appear "natural."

From what has just been said one must distinguish real technological progress from the way it has been realized in the automation of machine production, in space programs, etc. – which still consists of man's self-relief and goes so far as to be a relief from dependence on the conditions essential to life in an earth-environment. Insofar as this also includes control of the external world, both areas are related.

"Cybernetics" is the name of a comprehensive science which is at best in an embryonic state, and which can only be taken seriously in some specializations such as automation technology. In early formulations of its general theory, concepts such as those of information, game theory, homeostasis, planning, prediction, automata, learning machines, etc., are combined in various ways. The danger exists that the maturation of this science will be short-circuted politically, since it is already being used as a conceptual instrument for disqualifying certain social attitudes (or to say the same thing in a different way: for legitimizing certain claims of competence). There are interpretations of cybernetics which, with respect to so-called "alternative" views about how technology should be used, make the latter out to be simply examples of cultural backwardness or a preoccupation with pleasure, emotional childishness, a need for dependence, etc.; i.e., the scientific project serves as a "Progressive" substitute for Marxist ideology.[7] It is worth noting that someone in the field, such as Valentin Braitenberg (Naples) can say: "Cybernetics is no more a social reality than it is an academic one. There are no university departments of cybernetics, and there is not even a uniform definition of what is to be understood by it."[8]

V

Cybernetics as a super-science, therefore, for the present, can still only be considered a future possibility; society cannot yet be programed. Yet the question of the *presently discernible* human impact of the modern, technologically-created life-environment can be dealt with, and in this connection we have on previous occasions already quoted the words of Max Weber from 1908 to the effect that "modern civilization has already changed almost beyond recognition the spiritual countenance of mankind" and that it will continue to do so.[9]

From the beginning the effect of the transformation of a rural agrarian society into an over-built urban system was apparent everywhere. It brought about, as we all know, a radical uprooting and intimidation of the people involved, as has been described by anthropologists who study similar processes in the colonizing of primitive peoples. The whole traditional pattern of thinking and feeling of those affected is upset, and apart from the arousal of new needs it also brings about numerous shock reactions ranging from neuroses, mass hysteria and state of panic, to indifference and apathy. Commonly a mass of reflection and thought is also stirred up, which must be considered some sort of "search organization" looking for supportive guidance systems. But the adaptive attempts of populaces which have thus been totally "redirected" are never successful if they are based on consciousness alone. It take generations of growing accustomed to the altered circumstances, which seems to have taken place to a certain extent in North America, Central Europe, and the Soviet Union.

There is no longer any doubt concerning man's increasing *existential dependence* on his new constructed environment; i.e., physical survival presupposes undisturbed functioning of energy plants, water supply systems, communication and information systems, chemical industries, etc. If man's dependence on nature was severe in earlier days, then it has increased in today's self-created cultural environment; for although an individual with his natural organic resources and what he might make with his own hands could hold out alone for a while in nature, he could not get through three days in a non-functioning technological-industrial system. This may be expressed, as has been done by Ernst Forstshoff (*Die Daseinsvorsorge als Aufgabe der modernen Verwaltung*) [Care-for-Being as a task of modern governance][10] in the distinction between self-contained living spaces and real ones, i.e., the ones in which life actually takes place. A self-contained living space which belongs to the individual no longer exists for an increasing number of people.

This is not a purely theoretical problem. It may be expanded by the question of reserve or standby living space in the case of disaster, and this was the direction of the important "shelter debate" in the American press in October of 1961. In order to deal with the case of nuclear attack, emergency shelters were sold with room for one family; they were equipped with food, water, oxygen, medicine, etc., and were supposed to offer some chance of survival at least for a while. Debate focused on whether, in the case of a real attack, for defense against those who would storm such shelters because they themselves did not have any, weapons should be brought into these private dugouts. This debate was carried on in good American fashion, from all conceivable angles and with full disclosure of names. With unprecedented sharpness, the debate raised the question of "self-contained living spaces."

From this perspective it is no exaggeration to say, as Hannah Arendt has, that the apparatuses which we once handled freely have now started to become part of our biological make-up to such an extent that it looks as if mankind no longer belongs to a species of mammals, but has instead turned into some kind of shellfish. If we conjoin this thought with the above-mentioned view that an instinct-like, unconscious process has propelled the various epochs of technology, this would lead to an acknowledgement of Heisenberg's idea, as quoted by Arendt, to the effect that "technology in fact no longer appears 'as the product of a conscious human effort to enlarge material power, but rather like a biological development of mankind in which the innate structures of the human organism are transformed in an ever-increasing measure to the environment of man'" in a biological process which is no longer subject to human control.[11]

Along this line it is well to remember that Henri Bergson's hope for an alliance between "Mechanics and Mysticism" (last chapter of *The Two Sources of Morality and Religion*) has not been fulfilled. The industrial world has brought about no "mystical" golden age of spirituality, and the following prophecy of his did not come true: "The material barrier then has well nigh vanished. Tomorrow the way will be clear...[so that] the summons of the hero [may] come."[12]

Instead we see our culture, which rests on technology, exhibiting the same effects that have always been observed to follow great spiritual exertion and epoch-making discovery. Whether we are dealing with the human movement onto the land and the rapid spread of agricultural settlements; or with monotheism and the neutralization of the external world and a consequent surrender to inventive practices; or with the first large-scale utilization of a non-organic power source (the wind), ocean navigation, and the white man's massive land claims to sparsely populated areas overseas – human life in a biological sense always made its way into those free spaces as into open niches. The normal course always consisted in a multiplication of the number of people. The same happens again now as we watch with apprehension the population explosion brought about in this age of technology: the doubling of the world population in shorter and shorter intervals (1650–1850, 1850–1950; 1950–2000).

VI

We cannot resist the temptation to extend the previous lines of reflection into the future, a course which might lead to the following picture.

With the increasing existential dependence of the individual, it becomes likely that in Max Weber's sense there will follow a spiritual adaptation to this condition. At first the *need* of free space for creating a real living arrangement and for an expansive lifestyle will decrease. Then, conversely, *generalized* interests will move into the center of social concerns and thus also into the experiential center of the individual. Such interests include, e.g., numerous civil rights, as well as basic desires for health care and job security. This development leads to a general welfare system whose goal is that not just every immediate need but also any imaginable one which might arise in a crisis situation, as long as it can be expressed in generalized terms, will be transformed into an individual demand to be fulfilled with the aid of the large-scale means of large industrial societies.

When these processes have become established, it will then be likely that the individual will *feel* only such needs as have a chance of collective fulfillment or which are guaranteed by law and are thus for the common good. In this way the individual will be collectivized from within and absorbed into large organizational states. The individual will only give out such fragments of desire, so to speak, as the large group will accept, as can be thrown into the common pot.

This will above all severely restrict from within the powers of the founding entrepreneur in the nineteenth-century sense; he will be compared in his historical context with the great slave-related initiatives of feudal times. The determined individualist and all subjectivists will take on the part of eccentrics, or society will develop a need for them and hand out eccentric roles. In part those who overthrow literary taboos even now exhibit a theoretical, impractical free space extending the mission of "progressive" literature. Those symptoms should be noted, however, which point to a growing and strongly felt popularity of criminals who, like the Carbonari or the Mafia, appear to be liberating heroes warring against the tyranny of circumstances. The attraction of detective stories can be explained partly by the fact that the detective and the criminal are seen as two alien entrepreneurs set apart from the whole of society – they are newstyle practical heroes.

A spiritual adaptation to the age of technology can be followed like concentric circles inward from the areas of contact with the external world all the way to the interior of man. One may start with Hans Freyer's observation about how originally neutral words of everyday speech take on a purely technical meaning, as with "*beanspruchen*" (require) or "*schalten*" (govern/direct); and everyday speech takes in words originating in the area of technology as "*Leerlauf*" (neutral) or "*auslosen*" (release). We should alas mention those dynamic but abstract patterns of speech which are approaching us "from all sides" and which we are being forced "to learn to live with."

The matter-of-factness with which we take the concept of "organization" and an expectation of technically correct functioning and apply it to social contexts leads to a change not just in the contents but in the very structure of consciousness. Thus one speaks of the "*Aufbau*" (structure) of a distribution and marketing system, where the word "*Aufbau*" has long since lost its original architectural meaning; the term is even extended so that one can refer to "personality structure." Then there is the widespread desire for internal rationalization of a business which is described as an effort to eliminate "areas of friction" and follow a "smooth course." Along this line there is also the strangely matter-of-fact way in which we reject the complexities of things. If someone says, "That must be standardized," it hardly draws any objection. Consequently, everything conceivable is being planned, including the "structure" of entire countries or forms of government, with everything being copied like blueprints.

The ambiguous effects of worldwide news services, to which we are exposed daily in both word and picture, lead still further into the interior of man. On the one hand, the news brings about novel and totally unprecedented psychic possibilities, increased concentration of experience and a supply of stimuli – in short, a process which little by little seems to result in a rising demand for education and entertainment. People are being fed continuously by radio, the press, and television; they are more reflexive; and insofar as they think more they alas to some extent become more insecure, at the very time they are becoming more committed. But such conditions are inseparably interwoven with disadvantages of *secondary* experience; our entire conception of the world is transmitted to us electronically and goes beyond the range of immediate communication. Between us and the world electromagnetic information processes are at work and within them a host of anonymous transforming minds. In the case of great spectacular political events, the facts are so obscure that they seem to be arranged around an imaginary center.

In this dubious expansion of consciousness one invariably develops abstract centers of stabilization in consciousness, especially that infamous area of "public opinion," with its curious Don Quixote-like mixture of subjective certainty with objective bias. High degrees of ignorance of the world can then be quite effectively communicated by means of current technology.

At this point in our effort to outline the imaginative core of contemporary consciousness, the possibility presents itself that perhaps consciousness can adapt to thinking machines. Although a good mechanical translation of one of Holderlin's odes into French would be impossible today, it is conceivable that in his creations, some future lyric poet might subconsciously supply the basis for such a translation.

When discussing the process of the mechanization of consciousness one should never forget that it is counter-balanced to the degree that our intellectual life is renewed. It might be that a very well formed, biographically sound, and in the highest sense original, art could not find its way into the intensified media; it would thus circulate in a non-public manner in the way centers of tradition in the late classical period shifted to remote provincial areas. Again, it is a well-attested fact that the monumental experiences of people in government seem to make them reticent to appear on television or to speak over the radio; so one does not know with how much or how little wisdom the world is being governed.

This is an idea which reaches to the deepest levels. The ideological difference between today and other epochs is that the values of equality, peace, and development are accepted *without opposition* throughout the world. Never before has there been a system of belief without opposition; today, from Washington to Peking, it is impossible to speak against this system. Nevertheless, inequality among peoples remains an irrefutable fact; there is always a war going on somewhere; and the capacity of industrial society to increase specific massive populations is simply unbelievable. The present culture is the most dishonest that ever existed, while it has advanced much farther toward objective recognition of reality than any before it. If someone asks what causes universal agreement in regard to such values as equality, peace, and development, the answer is that in such values the operational needs of the age of technology have found common cause with the biological imperatives generated by the population explosion. Control of nature, both in technology and in the basic organization of society, and human propagation with social control catalyzed by this propagation, have reached a high level of mutual reinforcement.

Notes

All notes except no. 7 are supplied by the translators, sometimes on the basis of textual hints.

1 The italicized words are in English in the original.

2 This is the German translation of *Ideas y creencias* [Thoughts and beliefs], first published in 1940. For a readily available text in English that makes the same argument see Ortega's "Thoughts on Technology" in Carl Mitcham and Robert Mackey, eds., *Philosophy and Technology* (New York: Free Press, 1972).

3 This is a 15-page pamphlet (Mainz: Akademie der Wissenschaften und der Literatur, 1960), the title of which can be rendered as follows: "Concerning the developing domination of technical categories in the life-world of industrial society."

4 This essay is published in *Schmollers Jahrbuch* 83, no. 3 (1963): 315–330.

5 In English in the original.

6 Reference is being made here and following to Schmidt's two articles "Die Entwicklung der Technik als Phase der Wandlung des Menschen" [The development of technology as a phase of human transformation], *VDI-Zeitschrift* 96, no. 5 (1954): 118–122, and "Beginn und Aufstieg der Kybernetik" [Start and advancement of cybernetics], *VDI-Zeitschrift* 106, no. 17 (1964): 749–753. For another summary and discussion of Schmidt's theories see Simon Moser, "Toward a Metaphysics of Technology," *Philosophy Today* 15, no. 2 (Summer 1971), esp. pp. 139–143.

7 See O. W. Haseloff, in *Bergedorfer Gesprache*, 10th meeting (1963), and Haseloff's admission "that a kind of social reorganization is looking here for a name."

8 Source not given by Gehlen.

9 Max Weber, "Erhebungen über Auslese und Anpassung (Berufswahl und Berufsschicksal) der Arbeiterschaft der geschlossenen Grossindustrie," a written manuscript from 1908, published in Marianne Weber, ed., *Gesammelte Aufsätze zur Soziologie und Sozialpolitik* (Tübingen: Mohr-Siebeck, 1924), pp. 1–60, under the title "Methodologische Einleitung für die Erhebungen des Vereins für Sozialpolitik über Auslese und Anpassung (Berufswahlen und Berufsschicksal) der Arbeiterschaft der geschlossenen Grossindustrie."

10 This may be a faulty reference to Ernst Forsthoff's pamphlet, *Die Daseinsvorsorge und die Kommunen* [Care-for-Being and communities] (Munich: Sigillum, 1958).

11 Hannah Arendt, *The Human Condition* (Garden City, NY: Doubleday Anchor, n.d. [first published, 1958], p. 133. The quotation from Heisenberg is from his *Das naturbild der heutigen Physik* (Hamburg: Rowohlt, 1955), pp. 14–15.

12 Henri Bergson, *The Two Sources of Mortality and Religion* (Garden City, NY: Doubleday Anchor, n.d.), p. 312. Bergson's book was originally published in French in 1932.

21

The Social Construction of Facts and Artifacts

Trevor J. Pinch and Wiebe E. Bijker

One of the most striking features of the growth of "science studies" in recent years has been the separation of science from technology. Sociological studies of new knowledge in science abound, as do studies of technological innovation, but thus far there has been little attempt to bring such bodies of work together.[1] It may well be the case that science and technology are essentially different and that different approaches to their study are warranted. However, until the attempt to treat them within the same analytical endeavor has been undertaken, we cannot be sure of this.

It is the contention of this chapter that the study of science and the study of technology should, and indeed can, benefit from each other. In particular we argue that the social constructivist view that is prevalent within the sociology of science and also emerging within the sociology of technology provides a useful starting point. We set out the constitutive questions that such a unified social constructivist approach must address analytically and empirically.

This chapter falls into three main sections. In the first part we outline various strands of argumentation and review bodies of literature that we consider to be relevant to our goals. We then discuss the two specific approaches from which our integrated viewpoint has developed: the "Empirical Programme of Relativism" (Collins 1981c) and a social constructivist approach to the study of technology (Bijker et al. 1984). In the third part we bring these two approaches together and give some empirical examples. We conclude by summarizing our provisional findings and by indicating the directions in which we believe the program can most usefully be pursued.

Originally "The Social Construction of Facts and Artifacts: Or How the Sociology of Science and the Sociology of Technology Might Benefit Each Other" in *The Social Construction of Technological Systems: New Directions in the Sociology and History of Technology*, ed. Wiebe E. Bijker, Thomas P. Hughes, and Trevor J. Pinch, Cambridge, MA: The MIT Press, 1987, pp. 17–50, 349–72, abridged.

Some Relevant Literature

In this section we draw attention to three bodies of literature in science and technology studies. The three areas discussed are the sociology of science, the science-technology relationship, and technology studies. We take each in turn.

Sociology of Science

It is not our intention to review in any depth developments in this field as a whole.[2] We are concerned here with only the recent emergence of the sociology of scientific *knowledge*.[3] Studies in this area take the actual content of scientific ideas, theories, and experiments as the subject of analysis. This contrasts with earlier work in the sociology of science, which was concerned with science as an institution and the study of scientists' norms, career patterns, and reward structures.[4] One major – if not *the* major – development in the field in the last decade has been the extension of the sociology of knowledge into the arena of the "hard sciences." The need for such a "strong programme" has been outlined by Bloor: Its central tenets are that, in investigating the causes of beliefs, sociologists should be impartial to the truth or falsity of the beliefs, and that such beliefs should be explained symmetrically (Bloor 1973). In other words, differing explanations should not be sought for what is taken to be a scientific "truth" (for example, the existence of x-rays) and a scientific "falsehood" (for example, the existence of n-rays). Within such a program all knowledge and all knowledge claims are to be treated as being socially constructed; that is, explanations for the genesis, acceptance, and rejection of knowledge claims are sought in the domain of the social world rather than in the natural world.[5]

This approach has generated a vigorous program of empirical research, and it is now possible to understand the processes of the construction of scientific knowledge in a variety of locations and contexts. For instance, one group of researchers has concentrated their attention on the study of the laboratory bench.[6] Another has chosen the scientific controversy as the location for their research and have thereby focused on the social construction of scientific knowledge among a wider community of scientists.[7] As well as in hard sciences, such as physics and biology, the approach has been shown to be fruitful in the study of fringe science[8] and in the study of public-science debates, such as lead pollution.[9]

Although there are the usual differences of opinion among researchers as to the best place to locate such research (for instance, the laboratory, the controversy, or the scientific paper) and although there are differences as to the most appropriate methodological strategy to pursue,[10] there is widespread agreement that scientific knowledge can be, and indeed has been, shown to be thoroughly socially constituted. These approaches, which we refer to as "social constructivist," mark an important new development in the sociology of science. The treatment of scientific knowledge as a social construction implies that there is nothing espistemologically special about the nature of scientific knowledge: It is merely one in a whole series of knowledge cultures (including, for instance, the knowledge systems pertaining to "primitive" tribes) (Barnes 1974; Collins and Pinch 1982). Of course, the successes and failures of certain knowledge cultures still need to be explained, but this is to be seen as a sociological task, not an epistemological one.

The sociology of scientific knowledge promises much for other areas of "science studies." For example, it has been argued that the new work has relevance for the history of science (Shapin 1982), philosophy of science (Nickles 1982), and science policy (Healey 1982; Collins 1983b). The social constructivist view not only seems to be gaining ground as an important body of work in its own right but also shows every potential of wider application. It is this body of work that forms one of the pillars of our own approach to the study of science and technology.

Science-Technology Relationship

The literature on the relationship between science and technology, unlike that already referred to, is rather heterogeneous and includes contributions from a variety of disciplinary perspectives. We do not claim to present anything other than a partial review, reflecting our own particular interests.

One theme that has been pursued by philosophers is the attempt to separate technology from science on analytical grounds. In doing so, philosophers tend to posit overidealized distinctions, such as that science is about the discovery of truth whereas technology is about the application of truth. Indeed, the literature on the philosophy of technology is rather disappointing (Johnston 1984). We prefer to suspend judgment on it until philosophers propose more realistic models of both science and technology.

Another line of investigation into the nature of the science-technology relationship has been carried out by innovation researchers. They have attempted to investigate empirically the degree to which technological innovation incorporates, or originates from, basic science. A corollary of this approach has been the work of some scholars who have looked for relationships in the other direction; that is, they have argued that pure science is indebted to developments in technology.[11] The results of the empirical investigations of the dependence of technology on science have been rather frustrating. It has been difficult to specify the interdependence. For example, Project Hindsight, funded by the US Defense Department, found that most technological growth came from mission-oriented projects and engineering R&D, rather than from pure science (Sherwin and Isenson 1966, 1967). These results were to some extent supported by a later British study (Langrish et al.

1972). On the other hand, Project TRACES, funded by the NSF in response to Project Hindsight, found that most technological development stemmed from basic research (Illinois Institute of Technology, 1968). All these studies have been criticized for lack of methodological rigor, and one must be cautious in drawing any firm conclusions from such work (Kreilkamp 1971; Mowery and Rosenberg 1979). Most researchers today seem willing to agree that technological innovation takes place in a wide range of circumstances and historical epochs and that the import that can be attached to basic science therefore probably varies considerably.[12] Certainly the view prevalent in the "bad old days" (Barnes 1982a) – that science discovers and technology applies – will no longer suffice. Simplistic models and generalizations have been abandoned. As Layton remarked in a recent review:

> Science and technology have become intermixed. Modern technology involves scientists who "do" technology and technologists who function as scientists.... The old view that basic sciences generate all the knowledge which technologists then apply will simply not help in understanding contemporary technology. (Layton 1977, p. 210)

Researchers concerned with measuring the exact interdependence of science and technology seem to have asked the wrong question because they have assumed that science and technology are well-defined monolithic structures. In short, they have not grasped that science and technology are themselves socially produced in a variety of social circumstances (Mayr 1976). It does seem, however, that there is now a move toward a more sociological conception of the science-technology relationship. For instance, Layton writes:

> The divisions between science and technology are not between the abstract functions of knowing and doing. Rather they are social. (Layton 1977, p. 209)

Barnes has recently described this change of thinking:

> I start with the major reorientation in our thinking about the science-technology relationship which has occurred in recent years....We recognize science and technology to be on a par with each other. Both sets of practitioners creatively extend and develop their existing culture; but both also take up and exploit some part of the culture of the other....They are in fact enmeshed in a symbiotic relationship. (Barnes 1982a, p. 166)

Although Barnes may be overly optimistic in claiming that a "major reorientation" has occurred, it can be seen that a social constructivist view of science and technology fits well with his conception of the science-technology relationship. Scientists and technologists can be regarded as constructing their respective bodies of knowledge and techniques with each drawing on the resources of the other when and where such resources can profitably be exploited. In other words, both science and technology are socially constructed cultures and bring to bear whatever cultural resources are appropriate for the purposes at hand. In his view the boundary between science and technology is, in particular instances, a matter for social negotiation and represents no underlying distinction. It then makes little sense to treat the science-technology relationship in a general unidirectional way. Although we do not pursue this issue further in this chapter, the social construction of the science-technology relationship is clearly a matter deserving further empirical investigation.

Technology Studies

Our discussion of technology studies work is even more schematic. There is a large amount of writing that falls under the rubric of "technology studies." It is convenient to divide the literature into three parts: innovation studies, history of technology, and sociology of technology. We discuss each in turn.

Most innovation studies have been carried out by economists looking for the conditions for success in innovation. Factors researched include various aspects of the innovating firm (for example, size of R&D effort, management strength, and marketing capability) along with macroeconomic factors pertaining to the economy as a whole.[13] This literature is in some ways reminiscent of the early days in the sociology of science, when scientific knowledge was treated like a "black box" (Whitley 1972) and, for the purpose of such studies, scientists might as well have produced meat pies. Similarly, in the economic analysis of technological innovation everything is included that might be expected to influence innovation, except any discussion of the technology itself. As Layton notes:

> What is needed is an understanding of technology from inside, both as a body of knowledge and as a social system. Instead, technology is often treated as a "black box" whose contents and behaviour may be assumed to be common knowledge. (Layton 1977, p. 198)

Only recently have economists started to look into this black box.[14]

The failure to take into account the content of technological innovations results in the widespread use of simple linear models to describe the process of innovation. The number of developmental steps assumed in these models seems to be rather arbitrary....[15] Although such studies have undoubtedly contributed much to our understanding of the conditions for economic success in technological innovation, because they ignore the technological content they cannot be used as the basis for a social constructivist view of technology.[16]

This criticism cannot be leveled at the history of technology, where there are many finely crafted studies of the development of particular technologies. However, for the purposes of a sociology of technology, this work presents two kinds of problem. The first is that descriptive historiography is endemic in this field. Few scholars (but there are some notable exceptions) seem concerned with generalizing beyond historical instances, and it is difficult to discern any overall patterns on which to build a theory of technology (Staudenmaier 1983, 1985). This is not to say that such studies might not be useful building blocks for a social constructivist view of technology – merely that these historians have not yet demonstrated that they are doing sociology of knowledge in a different guise.[17]

The second problem concerns the asymmetric focus of the analysis. For example, it has been claimed that in twenty-five volumes of *Technology and Culture* only nine articles were devoted to the study of failed technological innovations (Staudenmaier 1985). This contributes to the implicit adoption of a linear structure of technological development, which suggests that

> the whole history of technological development had followed an orderly or rational path, as though today's world was the precise goal toward which all decisions, made since the beginning of history, were consciously directed. (Ferguson 1974, p. 19)

This preference for successful innovations seems to lead scholars to assume that the success of an artifact is an explanation of its subsequent development. Historians of technology often seem content to rely on the manifest success of the artifact as evidence that there is no further explanatory work to be done. For example, many histories of synthetic plastics start by describing the "technically sweet" characteristics of Bakelite; these features are then used implicitly to position Bakelite at the starting point of the glorious development of the field:

> God said: "let Baekeland be" and all was plastics! (Kaufman 1963, p. 61)

However, a more detailed study of the developments of plastic and varnish chemistry, following the publication of the Bakelite process in 1909 (Baekeland 1909a, b), shows that Bakelite was at first hardly recognized as the marvelous synthetic resin that it later proved to be.[18] And this situation did not change much for some ten years. During the First World War the market prospects for synthetic plastics actually grew worse. However, the dumping of war supplies of phenol (used in the manufacture of Bakelite) in 1918 changed all this (Haynes 1954, pp. 137–138) and made it possible to keep the price sufficiently low to compete with (semi-) natural resins, such as celluloid.[19] One can speculate over whether Bakelite would have acquired its prominence if it had not profited from that phenol dumping. In any case it is clear that a historical account founded on the retrospective success of the artifact leaves much untold.

Given our intention of building a sociology of technology that treats technological knowledge in the same symmetric, impartial manner that scientific facts are treated within the sociology of scientific knowledge, it would seem that much of the historical material does not go far enough. The success of an artifact is precisely what needs to be explained. For a sociological theory of technology it should be the *explanandum*, not the *explanans*.

Our account would not be complete, however, without mentioning some recent developments, especially in the American history of technology. These show the emergence of a growing number of theoretical themes on which research is focused (Staudenmaier 1985; Hughes 1979). For example, the systems approach to technology,[20] consideration of the effect of labor relations on technological development,[21] and detailed studies of some not-so-successful inventions[22] seem to herald departures from the "old" history of technology. Such work promises to be valuable for a sociological analysis of technology, and we return to some of it later.

The final body of work we wish to discuss is what might be described as "sociology of technology."[23] There have been some limited attempts in recent years to launch such a sociology, using ideas developed in the history and sociology of science – studies by, for example, Johnston (1972) and Dosi (1982), who advocate the description of technological knowledge in terms of Kuhnian paradigms.[24] Such approaches certainly appear to be more promising than standard descriptive historiography, but it is not clear whether or not these authors share our understanding of technological artifacts as social constructs. For example, neither Johnston nor Dosi considers explicitly the need for a symmetric sociological explanation that treats successful and failed artifacts in an equivalent way. Indeed, by locating their discussion at the level of technological paradigms, we are not sure how the artifacts themselves are to be approached. As neither author has yet produced an empirical study using Kuhnian ideas, it is difficult to evaluate how the Kuhnian terms may be utilized.[25] Certainly this has been a pressing problem in the sociology of science, where it has not always been possible to give Kuhn's terms a clear empirical reference.

The possibilities of a more radical social constructivist view of technology have been touched on by Mulkay (1979a). He argues that the success and efficacy of technology could pose a special problem for the social constructivist view of *scientific knowledge*. The argument Mulkay wishes to counter is that the practical effectiveness of technology somehow demonstrates the privileged epistemology of science and thereby exempts it from sociological explanation. Mulkay opposes this view, rightly in our opinion, by pointing out the problem of the "science discovers, technology applies" notion implicit in such claims. In a second argument against this position, Mulkay notes (following Mario Bunge (1966)) that it is possible for a false or partly false theory to be used as the basis for successful practical application: The success of the technology would not then have anything to say about the "truth" of the scientific knowledge on which it was based. We find this second point not entirely satisfactory. We would rather stress that the truth or falsity of scientific knowledge is irrelevant to sociological analysis of belief: To retreat to the argument that science may be wrong but good technology can still be based on it is missing this point. Furthermore, the success of technology is still left unexplained within such an argument. The only effective way to deal with these difficulties is to adopt a perspective that attempts to show that technology, as well as science, can be understood as a social construct.

Mulkay seems to be reluctant to take this step because, as he points out, "there are very few studies...which consider how the technical meaning of hard technology is socially constructed" (Mulkay 1979a, p. 77). This situation however, is starting to change: A number of such studies have recently emerged. For example, Michel Callon, in a pioneering study, has shown the effectiveness of focusing on technological controversies. He draws on an extensive case study of the electric vehicle in France (1960–75) to demonstrate that almost everything is negotiable: what is certain and what is not; who is a scientist and who is a technologist; what is technological and what is social; and who can participate in the controversy (Callon 1980a, b, 1981, and Bijker et al. 1985). David Noble's study of the introduction of numerically controlled machine tools can also be regarded as an important contribution to a social constructivist view of technology (Noble 1984). Noble's explanatory goals come from a rather different (Marxist) tradition,[26] and his study has much to recommend it: He considers the development of both a successful and a failed technology and gives a symmetric account of both developments. Another intriguing study in this tradition is Lazonick's account (1979) of the introduction of the self-acting mule: He shows that aspects of this technical development can be understood in terms of the relations of production rather than any inner logic of technological development. The work undertaken by Bijker, Bönig, and Van Oost is another attempt to show how the socially constructed character of the content of some technological artifacts might be approached empirically: Six case studies were carried out, using historical sources.[27]

In summary, then, we can say that the predominant traditions in technology studies – innovation studies and the history of technology – do not yet provide much encouragement for our program. There are exceptions, however, and some recent studies in the sociology of technology present promising starts on which a unified approach could be built. We now give a more extensive account of how these ideas may be synthesized.

EPOR and SCOT

In this part we outline in more detail the concepts and methods that we wish to employ. We start by describing the "Empirical Programme of Relativism" as it was developed in the sociology of scientific knowledge. We then go on to discuss in more detail the approach taken by Bijker and his collaborators in the sociology of technology.

The Empirical Programme of Relativism (EPOR)

The EPOR is an approach that has produced several studies demonstrating the social construction of scientific knowledge in the "hard" sciences. This tradition of research has emerged from recent sociology of scientific knowledge. Its main characteristics, which distinguish it from other approaches in the same area, are the focus on the empirical study of contemporary scientific developments and the study, in particular, of scientific controversies.[28]

Three stages in the explanatory aims of the EPOR can be identified. In the *first stage* the interpretative flexibility of scientific findings is displayed; in other words, it is shown that scientific findings are open to more than one interpretation. This shifts the focus for the explanation of scientific developments from the natural world to the social world. Although this interpretative flexibility can be recovered in certain circumstances, it remains the case that such flexibility soon disappears in science; that is, a scientific consensus as to what the "truth" is in any particular instance usually emerges. Social mechanisms that limit interpretative flexibility and thus allow scientific controversies to be terminated are described in the *second stage*. A *third stage*, which has not yet been carried through in any study of contemporary science, is to relate such "closure mechanisms" to the wider social-cultural milieu. If all three stages were to be addressed in a single study, as Collins writes, "the impact of society on knowledge 'produced' at the laboratory bench would then have been followed through in the hardest possible case" (Collins 1981c, p. 7).

The EPOR represents a continuing effort by sociologists to understand the content of the natural sciences in terms of social construction. Various parts of the program are better researched than others. The third stage of the program has not yet even been addressed, but there are many excellent studies exploring the first stage. Most current research is aimed at elucidating the closure mechanisms whereby consensus emerges (the second stage). Many studies within the EPOR have been most fruitfully located in the area of scientific controversy. Controversies offer a methodological advantage in the comparative ease with which they reveal the interpretative flexibility of scientific results. Interviews conducted with scientists engaged in a controversy usually reveal strong and differing opinions over scientific findings. As such flexibility soon vanishes from science, it is difficult to recover from the textual sources with which historians usually work. Collins has highlighted the importance of the "controversy group" in science by his use of the term "core set" (Collins 1981b). These are the scientists most intimately involved in a controversial research topic. Because the core set is defined in relation to knowledge production in science (the core set constructs scientific knowledge), some of the empirical problems encountered in the identification of groups in science by purely sociometric means can be overcome. And studying the core set has another methodological advantage, in that the resulting consensus can be monitored. In other words, the group of scientists who experiment and theorize at the research frontiers and who become embroiled in scientific controversy will also reflect the growing consensus as to the outcome of that controversy. The same group of core set scientists can then be studied in both the first and second stages of the EPOR. For the purposes of the third stage, the notion of a core set may be too limited.

The Social Construction of Technology (SCOT)

Before outlining some of the concepts found to be fruitful by Bijker and his collaborators in their studies in the sociology of technology, we should point out an imbalance between the two approaches (EPOR and SCOT) we are considering. The EPOR is part of a flourishing tradition in the sociology of scientific knowledge: It is a well-established program supported by much empirical research. In contrast, the sociology of technology is an embryonic field with no well-established traditions of research, and the approach we draw on specifically (SCOT) is only in its early empirical stages, although clearly gaining momentum.[29]

In SCOT the developmental process of a technological artifact is described as an alternation of variation and selection.[30] This results in a "multidirectional" model, in contrast with the linear models used explicitly in many innovation studies and implicitly in much history of technology. Such a multidirectional view is essential to any social constructivist account of technology. Of course, with historical hindsight, it is possible to collapse the multidirectional model on to a simpler linear model; but this misses the thrust of our argument that the "successful" stages in the development are not the only possible ones.

[...]

The wider context

Finally, we come to the third stage of our research program. The task here in the area of technology would seem to be the same as for science – to relate the content of a technological artifact to the wider sociopolitical milieu. This aspect has not yet been demonstrated for the science case,[31] at least not in contemporaneous sociological studies.[32] However, the SCOT method of describing technological artifacts by focusing on the meanings given to them by relevant social groups seems to suggest a way forward. Obviously, the sociocultural and political situation of a social group shapes its norms and values, which in turn influence the meaning given to an artifact. Because we have shown how different meanings can constitute different lines of development, SCOT's descriptive model seems to offer an operationalization of the relationship between the wider milieu and the actual content of technology. To follow this line of analysis, see Bijker 1985.

Conclusion

In this chapter we have been concerned with outlining an integrated social constructivist approach to the empirical study of science and technology. We reviewed several relevant bodies of literature and strands of argument. We indicated that the social constructivist approach is a flourishing tradition within the sociology of science and that it shows every promise of wider application. We reviewed the literature on the science-technology relationship and showed that here, too, the social constructivist approach is starting to bear fruit. And we reviewed some of the main traditions in technology studies. We argued that innovation studies and much of the history of technology are unsuitable for our sociological purposes. We discussed some recent work in the sociology of technology and noted encouraging signs that a new wave of social constructivist case studies is beginning to emerge.

We then outlined in more detail the two approaches – one in the sociology of scientific knowledge (EPOR) and one in the field of sociology of technology (SCOT) – on which we base our integrated perspective. Finally, we indicated the similarity of the explanatory goals of the two approaches and illustrated these goals with some examples drawn from technology. In particular, we have seen that the concepts of interpretative flexibility and closure mechanism and the notion of social group can be given empirical reference in the social study of technology.

As we have noted throughout this chapter, the sociology of technology is still underdeveloped, in comparison with the sociology of scientific knowledge. It would be a shame if the advances made in the latter field could not be used to throw light on the study of technology. On the other hand, in our studies of technology it appeared to be fruitful to include several social groups in the analysis, and there are some indications that this method may also bear fruit in studies of science. Thus our integrated approach to the social study of science and technology indicates how the sociology of science and the sociology of technology might benefit each other.

But there is another reason, and perhaps an even more important one, to argue for such an integrated approach. And this brings us to a question that some readers might have expected to be dealt with in the first paragraph of this chapter, namely, the question of how to distinguish science from technology. We think that it is rather unfruitful to make such an a priori distinction. Instead, it seems worthwhile to start with commonsense notions of science and technology and to study them in an integrated way, as we have proposed. Whatever interesting differences may exist will gain contrast within such a program. This would constitute another concrete result of the integrated study of the social construction of facts and artifacts.

Notes

This chapter is a shortened and updated version of Pinch and Bijker (1984).

We are grateful to Henk van den Belt, Ernst Homburg, Donald MacKenzie, and Steve Woolgar for comments on an earlier draft of this chapter. We would like to thank the Stiftung Volkswagen, Federal Republic of Germany, the Twente University of Technology, The Netherlands, and the UK SSRC (under grant G/00123/0072/1) for financial support.

1 The science technology divorce seems to have resulted not so much from the lack of overall analytical goals within "science studies" but more from the contingent demands of carrying out empirical work in these areas. To give an example, the new sociology of scientific knowledge, which attempts to take into account the actual content of scientific knowledge, can best be carried out by researchers who have some training in the science they study, or at least by those who are familiar with an extensive body of technical literature (indeed, many researchers are ex-natural scientists). Having gained such expertise, the researchers tend to stay within the domain where that expertise can best be deployed. Similarly, R&D studies and innovation studies, in which the analysis centers on the firm and the marketplace, have tended to demand the specialized competence of economists. Such disparate bodies of work do not easily lead to a more integrated conception of science and technology. One notable exception is Ravetz (1971). This is one of the few works of recent science studies in which both science and technology and their differences are explored within a common framework.

2 A comprehensive review can be found in Mulkay and Milič (1980).

3 For a recent review of the sociology of scientific knowledge, see Collins (1983c).

4 For a discussion of the earlier work (largely associated with Robert Merton and his students), see Whitley (1972).

5 For more discussion, see Barnes (1974), Mulkay (1979b), Collins (1983c), and Barnes and Edge (1982). The origins of this approach can be found in Fleck (1935).

6 See, for example, Latour and Woolgar (1979), Knorr-Cetina (1981), Lynch (1985), and Woolgar (1982).

7 See, for example, Collins (1975), Wynne (1976), Pinch (1977, 1986), Pickering (1984), and the studies by Pickering, Harvey, Collins, Travis, and Pinch in Collins (1981a).

8 Collins and Pinch (1979, 1982).

9 Robbins and Johnston (1976). For a similar analysis of public science controversies, see Gillespie et al. (1979) and McCrea and Markle (1984).

10 Some of the most recent debates can be found in Knorr-Cetina and Mulkay (1983).

11 The *locus classicus* is the study by Hessen (1931).

12 See, for example, de Solla Price (1969), Jevons (1976), and Mayr (1976).

13 See, for example, Schumpeter (1928, 1942), Schmookler (1966, 1972), Freeman (1974, 1977), and Scholz (1977).

14 See, for example, Rosenberg (1982), Nelson and Winter (1977, 1982), and Dosi (1982, 1984). A study that preceded these is Rosenberg and Vincenti (1978).

15 Adapted from Uhlmann (1978), p. 45.

16 For another critique of these linear models, see Kline (1985).

17 Shapin writes that "a proper perspective of the uses of science might reveal that sociology of knowledge and history of technology have more in common than is usually thought" (1980, p. 132). Although we are sympathetic to Shapin's argument, we think the time is now ripe for asking more searching questions of historical studies.

18 Manuals describing resinous materials do mention Bakelite but not with the amount of attention that, retrospectively, we would think to be justified. Professor Max Bottler, for example, devotes only one page to Bakelite in his 228-page book on resins and the resin industry (Bottler 1924). Even when Bottler concentrates in another book on the *synthetic* resinous materials, Bakelite does not receive an indisputable "first place." Only half of the book is devoted to phenol/formaldehyde condensation products, and roughly half of that part is devoted to Bakelite (Bottler 1919). See also Matthis (1920).

19 For an account of other aspects of Bakelite's success, see Bijker et al. (1985).

20 See, for example, Constant (1980), Hughes (1983), and Hanieski (1973).

21 See, for example, Noble (1979), Smith (1977), and Lazonick (1979).

22 See, for example, Vincenti (1986).

23 There is an American tradition in the sociology of technology. See, for example, Gilfillan (1935), Ogburn (1945), Ogburn and Meyers Nimkoff (1955), and Westrum (1983). A fairly comprehensive view of the present state of the art in German sociology of technology can be obtained from Jokisch (1982). Several studies in the sociology of technology that attempt to break with the traditional approach can be found in Krohn et al. (1978).

24 Dosi uses the concept of technological trajectory, developed by Nelson and Winter (1977); see also Van den Belt and Rip (Bijker et al. 1985). Other approaches to technology based on Kuhn's idea of

the community structure of science are mentioned by Bijker et al. (1985). See also Constant (Bijker et al. 1985) and the collection edited by Laudan (1984).

25 One is reminded of the first blush of Kuhnian studies in the sociology of science. It was hoped that Kuhn's "paradigm" concept might be straightforwardly employed by sociologists in their studies of science. Indeed there were a number of studies in which attempts were made to identify phases in science, such as preparadigmatic, normal, and revolutionary. It soon became apparent, however, that Kuhn's terms were loosely formulated, could be subject to a variety of interpretations, and did not lend themselves to operationalization in any straightforward manner. See, for example, the inconclusive discussion over whether a Kuhnian analysis applies to psychology in Palermo (1973). A notable exception is Barnes's contribution to the discussion of Kuhn's work (Barnes 1982b).

26 For a valuable review of Marxist work in this area, see MacKenzie (1984).

27 For a provisional report of this study, see Bijker et al. (1984). The five artifacts that are studied are Bakelite, fluorescent lighting, the safety bicycle, the Sulzer loom, and the transistor. See also Bijker et al. (1985).

28 Work that might be classified as falling within the EPOR has been carried out primarily by Collins, Pinch, and Travis at the Science Studies Centre, University of Bath, and by Harvey and Pickering at the Science Studies Unit, University of Edinburgh. See, for example, the references in note 7.

29 See, for example, Bijker and Pinch (1983), Bijker (1984), and Bijker et al. (1985). Studies by Van den Belt (1985), Schot (1985, 1986), Jelsma and Smit (1986), and Elzen (1985, 1986) are also based on SCOT.

30 Constant (1980) used a similar evolutionary approach. Both Constant's model and our model seem to arise out of the work in evolutionary epistemology; see, for example, Toulmin (1972) and Campbell (1974). Elster (1983) gives a review of evolutionary models of technical change. See also Van den Belt and Rip (Bijker et al. 1985).

31 A model of such a "stage 3" explanation is offered by Collins (1983a).

32 Historical studies that address the third stage may be a useful guide here. See, for example, MacKenzie (1978), Shapin (1979, 1984), and Shapin and Schaffer (1985).

References

Baekeland, L. H. 1909a. "On soluble, fusible, resinous condensation products of phenols and formaldehyde." *Journal of Industrial and Engineering Chemistry* 1:345–349.

Baekeland, L. H. 1909b. "The synthesis, constitution, and use of Bakelite." *Journal of Industrial and Engineering Chemistry* 1:149–161.

Barnes, B. 1974. *Scientific Knowledge and Sociological Theory*. London: Routledge and Kegan Paul.

Barnes, B. 1982a. "The science-technology relationship: A model and a query." *Social Studies of Science* 12: 166–172.

Barnes, B. 1982b. *T. S. Kuhn and Social Science*. London: Macmillan.

Barnes, B., and Edge, D., eds. 1982. *Science in Context*. Milton Keynes: Open University Press.

Bijker, W. E. 1984. "Collectifs technologiques et styles technologiques: Eléments pour un modèle explicatif de la construction sociale des artefacts techniques," in *Travailleur collectif et relations science-production*, J. H. Jacot, ed. Paris: Editions du CNRS, 113–120.

Bijker, W. E., and Pinch, T. J. 1983. "La construction sociale de faits et d'artefacts: Impératifs stratégiques et méthodologiques pour une approche unifiée de l'étude des sciences et de la technique." Paper presented to L'atelier de recherche (III) sur les problèmes stratégiques et méthodologiques en milieu scientifique et technique. Paris, March.

Bijker, W. E., Bonig, J., and van Oost, E. C. J. 1984. "The social construction of technological artefacts." Paper presented to the EASST Conference. A shorter version of this paper is published in *Zeitschrift für Wissenschaftsforschung*, special issue 2, 3:39–52 (1984).

Bijker, W. E., Hughes, T. P., and Pinch, T. J., eds. 1985. *The Social Construction of Technological Systems: New Directions in the Sociology and History of Technology*. Cambridge, Mass.: MIT Press.

Bloor, D. 1973. "Wittgenstein and Manheim on the sociology of mathematics." *Studies in the History and Philosophy of Science* 4:173–191.

Bottler, M. 1924. *Harze und Harzinustrie*. Leipzig: Max Janecke.

Bunge, M. 1966. "Technology as applied science." *Technology and Culture* 7:329–347.

Callon, M. 1980a. "The state and technical innovation: A case study of the electrical vehicle in France." *Research Policy* 9:358–376.

Callon, M. 1980b. "Struggles and negotiations to define what is problematic and what is not: The sociologic of translation," in *The Social Process of Scientific Investigation*, K. Knorr, R. Krohn, and R. Whitley, eds. Dordrecht and Boston: Reidel, vol. 4, 197–219.

Callon, M. 1981. "Pour une sociologie des controverses technologiques." *Fundamenta Scientiae* 2:381–399.

Campbell, D. T. 1974. "Evolutionary epistemology," in *The Philosophy of Karl Popper, The Library of Living Philosophers*, P. A. Schlipp, ed. La Salle, Ill.: Open Court, vol. 14–1, 413–463.

Collins, H. M. 1975. "The seven sexes: A study in the sociology of a phenomenon, or the replication of experiments in physics." *Sociology* 9:205–224.

Collins, H. M., ed. 1981a. "Knowledge and controversy." *Social Studies of Science* 11:3–158.

Collins, H. M. 1981b. "The place of the core-set in modern science: Social contingency with methodological propriety in science." *History of Science* 19:6–19.

Collins, H. M. 1981c. "Stages in the empirical programme of relativism." *Social Studies of Science* 11:3–10.

Collins, H. M. 1983a. "An empirical relativist programme in the sociology of scientific knowledge," in *Science Observed: Perspectives on the Social Study of Science*, K. D. Knorr-Cetina and M. J. Mulkay, eds. Beverly Hills: Sage, 85–113.

Collins, H. M. 1983b. "Scientific knowledge and science policy: Some foreseeable implications." *EASST Newsletter*, November, 2:5–8.

Collins, H. M. 1983c. "The sociology of scientific knowledge: Studies of contemporary science." *Annual Review of Sociology* 9:265–285.

Collins, H. M., and Pinch, T. J. 1979. "The construction of the paranormal: Nothing unscientific is happening," in *On the Margins of Science: The Social Construction of Rejected Knowledge*, R. Wallis, ed. Keele: University of Keele, 237–270.

Collins, H. M., and Pinch, T. J. 1982. *Frames of Meaning: The Social Construction of Extraordinary Science*. London: Routledge and Kegan Paul.

Constant, E. W., II. 1980. *The Origins of the Turbojet Revolution*. Baltimore: Johns Hopkins University Press.

de Solla Price, D. J. 1969. "The structure of publication in science and technology," in *Factors in the Transfer of Technology*, W. H. Gruber and D. G. Marquis, eds. Cambridge, Mass.: MIT Press, 91–104.

Dosi, G. 1982. "Technological paradigms and technological trajectories: A suggested interpretation of the determinants and directions of technical change." *Research Policy* 11: 147–162.

Dosi, G. 1984. *Technical Change and Industrial Transformation*. London: Macmillan.

Elster, J. 1983. *Explaining Technical Change*. Cambridge: Cambridge University Press.

Elzen, B. 1985. "De ultracentrifuge: op zoek naar patronen in technologische ontwikkeling door een vergelijkin van twee case-studies." *Jaarboek voor de Geschiedenis van Bedrif en Techniek* 2:250–278.

Elzen, B. 1986. "Two ultracentrifuges: A comparative study of the social construction of artefacts." *Social Studies of Science* 16:621–662.

Ferguson, E. 1974. "Toward a discipline of the history of technology." *Technology and Culture* 15:13–30.

Fleck, L. 1935. *Entstehung und Entwicklung einer wissenschaftlichen Tatsache: Einfürhung in die Lehre vom Denkstil und Denkkollektiv*. Basel: Benno Schwabe. Reprinted by Suhrkamp, Frankfurt am Main, 1980. English translation: *The Genesis and Development of a Scientific Fact*, Chicago: The University of Chicago Press, 1979.

Freeman, C. 1974. *The Economics of Industrial Innovation*. Harmondsworth: Penguin. Reprinted by Frances Pinter, London, 1982.

Freeman, C. 1977. "Economics of research and development," in *Science, Technology and Society. A Cross-Disciplinary Perspective*, I. Spiegel-Rösing and D. de Solla Price, eds. London and Beverly Hills: Sage, 223–275.

Gilfillan, S. G. 1935. *The Sociology of Invention*. Cambridge, Mass.: MIT Press.

Gillespie, B., Eva, D., and Johnston, R. 1979. "Carcinogenic risk assessment in the United States and Great Britain: The case of aldrin/dieldrin." *Social Studies of Science* 9: 265–301.

Hanieski, J. F. 1973. "The airplane as an economic variable: Aspects of technological change in aeronautics, 1903–1955." *Technology and Culture* 14:535–552.

Haynes, W. 1954. *American Chemical Industry*, Vol. 2. New York: Van Nostrand.

Healey, P. 1982. "The research funding organization as a focus for science studies." Paper presented to the Science Studies Conference, Oxford, September.

Hessen, B. 1931. "The social and economic roots of Newton's *Principia*," in *Science at the Crossroads*, by N. I. Bukharin, A. F. Joffe, M. Rubinstein, B. Zavadovsky, E. Colman, N. I. Vavilov, W. Th. Mitkewich, and B. Hessen. London: Frank Cass, 147–212.

Hughes, T. P. 1979. "Emerging themes in the history of technology." *Technology and Culture* 20:697–711.

Hughes, T. P. 1983. *Networks of Power: Electrification in Western Society, 1880–1930*. Baltimore: Johns Hopkins University Press.

Illinois Institute of Technology. 1968. *Technology in Retrospect and Critical Events in Science (TRACES)*. Chicago: ITT Research Institute.

Jelsma, J., and Smit, W. A. 1986. "Risks of recombinant DNA research: From uncertainty to certainty," in *Impact Assessment Today*, H. A. Becker and A. L. Porter, eds. Van Arkel: Utrecht, 715–741.

Jevons, F. R. 1976. "The interaction of science and technology today, or, is science the mother of invention?" *Technology and Culture* 17:729–742.

Johnston, R. 1972. "The internal structure of technology," in *The Sociology of Science*, P. Halmos, ed. Keele: University of Keele, 117–130.

Johnston, R. 1984. "Controlling technology: An issue for the social studies of science." *Social Studies of Science* 14:97–112.

Jokisch, R., ed. 1982. *Technikosoziologie*. Frankfurt am Main: Suhrkamp.

Kaufman, M. 1963. *The First Century of Plastics: Celluloid and Its Sequel*. London: The Plastics Institute.

Kline, S. 1985. "Research, invention, innovation, and production: Models and reality." *Research Management* 28(4):36–45.

Knorr-Cetina, K. D. 1981. *The Manufacture of Knowledge: An Essay on the Constructivist and Contextual Nature of Science*. Oxford: Pergamon.

Knorr-Cetina, K. D., and Mulkay, M. J., eds. 1983. *Science Observed: Perspectives on the Social Study of Science*. London and Beverly Hills: Sage.

Kreilkamp, K. 1971. "Hindsight and the real world of science policy." *Science Studies* 1:43–66.

Krohn, W., Layton, E. T., and Weingart, P., eds. 1978. *The Dynamics of Science and Technology, Sociology of the Sciences Yearbook*, vol. 2. Dordrecht: Reidel.

Langrish, J., Gibbons, M., Evans, W. G., and Jevons, F. R. 1972. *Wealth from Knowledge*. London: Macmillan.

Latour, B. and Woolgar, S. 1979. *Laboratory Life. The Social Construction of Scientific Facts*. London and Beverly Hills: Sage; second edition published as *Laboratory Life. The Construction of Scientific Facts*, Princeton, N.J.: Princeton University Press, 1986.

Laudan, R., ed. 1984. *The Nature of Technological Knowledge: Are Models of Scientific Change Relevant?* Dordrecht: Reidel.

Layton, E. 1977. "Conditions of technological development," in *Science, Technology and Society: A Cross-Disciplinary Perspective*, I. Spiegel-Rösing and D. de Solla Price, eds. London and Beverly Hills: Sage, 197–222.

Lazonick, W. 1979. "Industrial relations and technical change: The case of the self-acting mule." *Cambridge Journal of Economics* 3:231–262.

Lynch, M. 1985. *Art and Artefact in Laboratory Science: A Study of Shop Work and Shop Talk in a Research Laboratory*. London: Routledge and Kegan Paul.

MacKenzie, D. 1978. "Statistical theory and social interest: A case study." *Social Studies of Science* 8:35–83.

MacKenzie, D. 1984. "Marx and the machine." *Technology and Culture* 25:473–502.

Matthis, A. R. c. 1920. *Insulating Varnishes in Electrotechnics*. London: John Heywood.

Mayr, O. 1976. "The science-technology relationship as a historiographic problem." *Technology and Culture* 17: 663–673.

McCrea, F. B., and Markle, G. E. 1984. "The estrogen replacement controversy in the USA and UK: Different answers to the same question?" *Social Studies of Science* 14: 1–26.

Mowery, D. C., and Rosenberg, N. 1979. "The influence of market demand upon innovation: A critical review of some recent empirical studies." *Research Policy* 8:103–153.

Mulkay, M. J. 1979a. "Knowledge and utility: Implications for the sociology of knowledge." *Social Studies of Science* 9:63–80.

Mulkay, M. J. 1979b. *Science and the Sociology of Knowledge*. London: Allen and Unwin.

Mulkay, M. J., and Milič, V. 1980. "The sociology of science in East and West." *Current Sociology*, Winter, 28:1–342.

Nelson, R. R., and Winter, S. G. 1977. "In search of a useful theory of innovation." *Research Policy* 6:36–76.

Nelson, R. R., and Winter, S. G. 1982. *An Evolutionary Theory of Economic Change*. Cambridge, Mass.: The Belknap Press of Harvard University Press.

Nickles, T. 1982. "How discovery is important to cognitive studies of science." Paper presented to the Philosophy of Science Association Meeting, Philadelphia, October.

Noble, D. F. 1979. "Social choice in machine design: The case of automatically controlled machine tools," in *Case Studies on the Labour Process*, A. Zimbalist, ed. New York: Monthly Review Press, 18–50.

Noble, D. F. 1984. *Forces of Production: A Social History of Industrial Automation*. New York: Knopf.

Ogburn, W. F. 1945. *The Social Effects of Aviation*. Boston: Houghton Mifflin.

Ogburn, W. F., and Meyers Nimkoff, F. 1955. *Technology and the Changing Family*. Boston: Houghton Mifflin.

Palermo, D. S. 1973. "Is a scientific revolution taking place in psychology?" *Science Studies* 3:211–244.

Pickering, A. 1984. *Constructing Quarks – A Sociological History of Particle Physics*. Chicago and Edinburgh: University of Chicago Press and Edinburgh University Press.

Pinch, T. J. 1977. "What does a proof do if it does not prove? A study of the social conditions and metaphysical divisions leading to David Bohm and John von Neumann failing to communicate in quantum physics," in *The Social Production of Scientific Knowledge*, E. Mendelsohn, P. Weingart, and R. Whitley, eds. Dordrecht: Reidel, 171–215.

Pinch, T. J. 1986. *Confronting Nature: The Sociology of Solar-Neutrino Detection*. Dordrecht: Reidel.

Pinch, T. J., and Bijker, W. E. 1984. "The social construction of facts and artefacts: Or how the sociology of science and the sociology of technology might benefit each other." *Social Studies of Science* 14:399–441. Also published in Serbo-Croatian as "Društveno Proizvodenje Činjenica I Tvorevina: O Cjelovitom Pristupu Izučavanju Znanosti I Tehnologije," *Gledišta, časopis za društvenu kritiku i teoriju*, March–April, 25: 21–57 (1984).

Ravetz, J. R. 1971. *Scientific Knowledge and Its Social Problems*. Oxford: Oxford University Press.

Robbins, D., and Johnston, R. 1976. "The role of cognitive and occupational differentiation in scientific controversies." *Social Studies of Science* 6:349–368.

Rosenberg, N. 1982. *Inside the Black Box: Technology and Economics*. Cambridge: Cambridge University Press.

Rosenberg, N., and Vincenti, W. G. 1978. *The Britannia Bridge: The Generation and Diffusion of Knowledge*. Cambridge, Mass.: MIT Press.

Schmookler, J. 1966. *Invention and Economic Growth*. Cambridge, Mass.: Harvard University Press.

Schmookler, J. 1972. *Patents, Invention and Economic Change, Data and Selected Essays*. Z. Griliches and L. Hurwicz, eds. Cambridge, Mass.: Harvard University Press.

Scholz, L. with a contribution by G. von L. Uhlmann, 1977. *Technik-Indikatoren, Ansätze zur Messung des Standes der Technik in der industriellen Produktion*. Berlin and München: Duncker & Humblot.

Schot, J. 1985. "De ontwikkeling van de techniek als een variatieen selectieproces. De meekrapteelt en-bereiding in het licht van een alternatieve techniekopvatting." Master's thesis, Erasmus University of Rotterdam, unpublished.

Schot, J. 1986. "De meekrapnijverheid: de ontwikkeling van de techniek als een proces van variatie en selectie," in *Jaarboek voor de Geschiedenis van Bedrijf en Techniek*, E. S. A. Bloemen, W. E. Bijker, W. van den Brocke, et al., eds. Utrecht: Stichting, vol. 3, 43–62.

Schumpeter, J. 1971 [1928]. "The instability of capitalism," in *The Economics of Technological Change*, N. Rosenberg, ed. Harmondsworth: Penguin, 13–42.

Schumpeter, J. 1942. *Capitalism, Socialism and Democracy*. New York: Harper & Row. Reprinted 1974 by Unwin University Books, London.

Shapin, S. 1979. "The politics of observation: Cerebral anatomy and social interests in the Edinburgh phrenology disputes," in *On the Margins of Science: The Social Construction of Rejected Knowledge*, R. Wallis, ed. Keele: University of Keele, 139–178.

Shapin, S. 1980. "Social uses of science," in *The Ferment of Knowledge*, G. S. Rousseau and R. Porter, eds. Cambridge: Cambridge University Press, 93–139.

Shapin, S. 1982. "History of science and its sociological reconstructions." *History of Science* 20:157–211.

Shapin, S. 1984. "Pump and circumstance: Robert Boyle's literary technology." *Social Studies of Science* 14: 481–520.

Shapin, S., and Schaffer, S. 1985. *Leviathan and the Air-Pump: Hobbes, Boyle and the Experimental Life*. Princeton: Princeton University Press.

Sherwin, C. S., and Isenson, R. S. 1966. *First Interim Report on Project Hindsight: Summary*. Washington, D.C.: Office of the Director of Defense Research and Engineering.

Sherwin, C. S., and Isenson, R. S. 1967. "Project hindsight: A Defense Department study of the utility of research." *Science* 156: 1571–1577.

Smith, M. Roe. 1977. *Harpers Ferry Armory and the New Technology: The Challenge of Change*. Ithaca, N.Y.: Cornell University Press.

Staudenmaier, J. M., SJ. 1983. "What SHOT hath wrought and what SHOT hath not: Reflections on 25 years of the history of technology." Paper presented to the Twenty-fifth Annual Meeting of SHOT.

Staudenmaier, J. M., SJ. 1985. *Technology's Storytellers: Reweaving the Human Fabric*. Cambridge, Mass.: MIT Press.

Toulmin, S. 1972. *Human Understanding*, vol. 1. Oxford: Oxford University Press.

Uhlmann, L. 1978. *Der Innovationsprozess in westeurpäischen Industrieländern. Band 2: Den Ablauf industriellen Innovationsprozesses*. Berlin and München: Duncker and Humblot.

van den Belt, H. 1985. "A. W. Hofman en de Franse Octrooiprocessen rond anilinerood: demarcatie als sociale constructie." *Jaarboek voor de Geschiedenis van Bedrijf en Techniek* 2:64–86.

Vincenti, W. G. 1986. "The Davis wing and the problem of airfoil design: Uncertainty and growth in engineering knowledge." *Technology and Culture* 26.

Westrum, R. 1983. "What happened to the old sociology of technology?" Paper presented to the Eighth Annual Meeting of the Society for Social Studies of Science, Blacksburg, Virginia, November.

Whitley, R. D. 1972. "Black boxism and the sociology of science: A discussion of the major developments in the field," in *The Sociology of Science*, P. Halmos, ed. Keele: University of Keele, 62–92.

Woolgar, S. 1982. "Laboratory studies: A comment on the state of the art." *Social Studies of Science* 12:481–498.

Wynne, B. 1976. "C. G. Barkla and the J phenomenon: A case study of the treatment of deviance in physics." *Social Studies of Science* 6:307–347.

22

Social Constructivism: Opening the Black Box and Finding it Empty

Langdon Winner

What do philosophers need to know about technology? I mean, of course, philosophers who want to think and write about technology in fruitful ways. What kind of knowledge do we need to have? And how much?

Perhaps it is enough simply to have lived in a society in which a wide variety of technologies are in common use. Drawing upon an everyday understanding of such matters, one can move on to develop general perspectives and theories that may enable us to answer important questions about technology in general. I expect that many philosophers who have written on this topic have used an approach of roughly this kind. The problem is that one's grasp may be superficial, failing to do justice to the phenomena one wants to explain and interpret. One may seize upon a limited range of vaguely understood examples of technical applications – a dam on a river, a robot in a factory, or some other typification – and try to wring universal implications from a sample that is perhaps too small to carry the weight placed upon it.

An alternative would be to focus one's attention more carefully, becoming expert in the technical knowledge of a specific field, attaining the deeper understanding of, say, a worker, engineer or technical professional. Even that may prove limiting, however, because the experience available in one field of practice may not be useful in comprehending the origins, character and consequences of technical practices in other domains. The sheer multiplicity of technologies in modern society poses serious difficulties for anyone who seeks an overarching grasp of human experience in a technological society.

Yet another strategy might be to study particular varieties of technology in a scholarly mode, drawing upon existing histories and contemporary social studies of technological change as one's base of understanding. And one might make the effort to expand this base of knowledge by contributing research of one's own. Noel Mostert's wonderful book *Supership* is such a work; a philosophical reflection upon the world of oil tankers in which the author takes care to examine details of the construction, economic context, and daily operation of these enormous vessels (Mostert, 1974).

The use of colourful, substantive cases like Mostert's suggests an interesting question. *Where* does one go to learn what one needs to know to write confidently about philosophy and technology? For Mostert, it meant not only going to the library to study the history, engineering and economics of supertankers, but also living on a tanker himself during several voyages. An identifying mark of the different philosophical approaches to technology can be found in the typical locations writers prefer to visit, if only in their minds. It is common for many Marxist thinkers, for example, to want to return to the scene of the crimes described by Marx himself; namely, to the industrial factory, to

From *Science as Culture* 16 (1993): 427–52, plates removed. Reprinted by permission of Carfax Publishing Ltd. Originally "Upon Opening the Black Box and Finding it Empty: Social Constructivism and the Philosophy of Technology," Presidential Address delivered to the Biennial Conference of the Society for Philosophy and Technology, Mayaguez, Puerto Rico, March 1991, and published in the conference proceedings.

the shop floor, noticing the social relations and productive forces displayed there. By the same token, many feminist writers have turned their attention to technologies in the home, office and hospital – places where technological designs and policies have historically affected the lives of women.

The list of typical locations in which a detailed understanding of technology might be gained is very large indeed. Where should a philosopher go to learn about technology? To a research and development laboratory? A farm? An electrical power plant? Communications centre? Airport? Arsenal? Construction site? Offices of an agency that funds research? A toxic waste dump? Automated theme park? A school where computers are being introduced? What does one's understanding of a specific location and specific varieties of technical apparatus, knowledge and practice contribute to one's ability to talk in penetrating, reliable ways about modern technology in general?

As studies in philosophy and technology mature, as I hope they will, it will be increasingly important for us to think critically about the origins and relative quality of the knowledge we draw upon as we address the key questions. There are bound to be disagreements about which strategies of enquiry are the best ones to follow. But it seems perfectly clear that faced with the enormously diverse kinds of technology in the world, philosophers must somehow gain a well-developed understanding of at least a representative slice of them.

For those of us engaged in studies of philosophy and technology, this need is all the more crucial just now because those working in others of the various subdisciplines of contemporary science and technology studies are at work on roughly the same turf on which philosophers commonly situate their enquiries. This flurry of activity in the social sciences poses both an opportunity and a peril. The opportunity is that one can enter into discussion with persons who employ other approaches, learning from their results, sharing ideas about similar topics. The peril is that philosophers may find themselves outflanked by these developments because the rich, empirical detail of historical and social science studies of technology can make the abstract speculations of philosophers appear vacuous and armchair-bound by comparison.

My purpose here is to look briefly at some recent work in the lively cross-disciplinary field of science and technology studies and ask: how well does it help orient our understanding of the place of technology in human affairs? The particular school of thought I shall briefly examine is one currently fashionable among historians and sociologists who study technology and society. Its most common label is the "social construction of technology" or, simply, "social constructivism". It is of interest not only for the specific features of its approach to the study of technology and society, but also for the way it regards past and present philosophical enquiries in technology and philosophy. To ignore the central claims of this important school of thought, to fail to examine its basic notions, would be to overlook an important challenge.

Among the names of those involved in this project are a number of Europeans and Americans: H. M. Collins, Trevor Pinch, Wiebe Bijker, Donald MacKenzie, Steven Woolgar, Bruno Latour, Michel Callon, Thomas Hughes, and John Law. These and other scholars of similar persuasion are now very active doing research, publishing articles, building academic programmes. They are also openly proselytizing and even self-consciously imperial in their hopes for establishing this approach. It is clear they would like to establish social constructivism as the dominant research strategy and intellectual agenda within science and technology studies for many years to come.[1]

The Dynamics of Change

An important aim of the social constructivist mode of enquiry is to look carefully at the inner workings of real technologies and their histories to see what is actually taking place. It recommends that rather than employ such broad-gauged notions as "technological determinism" or "technological imperatives", scholars need to talk more precisely about the dynamics of technological change. Rather than try to explain things through such loosely conceived notions as "the trajectory of a technical field" or "technical momentum", we need to look very closely at the artefacts and varieties of technical knowledge in question and at the social actors whose activities affect their development. In that light their preferred locations for research so far have been contemporary research and development laboratories as well as the archives that contain records of R&D accomplishments of the past.

The plea frequently voiced by the social constructivists is that we open "the black box" of historical and contemporary technology to see what is there (Pinch and Bijker, 1987). The term "black box" in both technical and social science parlance is a device or system that, for convenience, is described solely in terms of its inputs and outputs. One need not understand anything about what goes on inside such black boxes. One simply brackets them as instruments that perform certain valuable functions.[2]

In my view, the social constructivists are correct in criticizing writers in the social sciences and humanities who have often looked upon technological developments as black boxes while neglecting any comprehensive account of their structures, workings and social origins. To find more precise, detailed descriptions and explanations of the dynamics of technical change is a goal well worth pursuing.

As they go about opening the black box, the historians and sociologists in this school of thought follow methodological guidelines established during the past two decades within sociology of science, in particular an approach that studies the sociology of scientific knowledge (Collins, 1983). In this mode of analysis there is a strong tendency to regard technology as the lesser relative of science. Because science deals with the fundamentals of human knowledge, it is considered the more elevated and significant topic. In that light, for both historians and sociologists, the "turn to technology" is sometimes portrayed as a kind of intellectual slumming (Woolgar, 1991). There is even some doubt that sociologists of scientific knowledge will benefit greatly from studying such grubby technological matters at all.

The attitude of the sociologists of science towards technology calls to mind the story about a man who is taking a truckload of penguins to their new home at the city zoo. Unfortunately, in the middle of his journey the truck breaks down. Worried that the penguins will suffer in the afternoon heat, the driver flags down a passing motorist and asks him to take the birds to the zoo while the truck is being fixed. The other fellow amicably agrees and the truck driver gives him $50 for any expenses he might incur on the errand. A couple of hours later the truck has been repaired and the driver heads toward the zoo. But what should he see but the other car, still filled with penguins, heading in the opposite direction. The truck driver turns around, flags down the motorist and exclaims, "What are you doing? I thought I asked you to take these penguins to the zoo!" "I did," the man replies, "but we had a little money left over so we thought we'd go to the beach."

That, roughly speaking, is how sociologists of science regard prospects for social studies of technology. They see this as a new field in which to apply a powerful but as yet under utilized research apparatus that had been successful in studies of the sociology of scientific knowledge.

From that vantage point most past and contemporary work in the philosophy of technology is greeted with scorn. As Pinch and Bijker conclude in their widely-cited survey article: "Philosophers tend to posit over-idealized distinctions, such as that science is about the discovery of truth whereas technology is about the application of truth. Indeed, the literature on the philosophy of technology is rather disappointing. We prefer to suspend judgement on it until philosophers propose more realistic models of both science and technology" (Pinch and Bijker, 1987, p. 19).[3]

In quest of "more realistic" models of their own, social constructivists employ a methodological posture – "the empirical programme of relativism" – commonly used in the sociology of science. As with its studies of the workings of science, this approach requires that all claims to scientific knowledge must be bracketed, not judged true or false by some independent standard. One assumes that scientific disputes are settled not through access to a firmer grasp upon objective reality, but through processes in which knowledge is socially constructed. Thus, "truth" can be seen to emerge through a variety of social activities in which different social groups contend to establish their knowledge claims.

Adapting this stance to the study of technology requires some modification. What social analysts do in this new focus is to study the "interpretive flexibility" of technical artefacts and their uses. One begins by noticing that people in different situations interpret the meaning of a particular machine or design of an instrument in different ways. People may use the same kind of artefact for widely different purposes. The meanings attached to a particular artefact and its uses can vary widely as well. In this way of seeing, sociologists and historians must locate the "relevant social groups" involved in the development of a particular technological device or system or process. They must pay attention to the variety of interpretations of what a particular technological entity in a process of development means, and how people act in different ways to achieve their purposes within that process.

For example, in their study of the development of the modern bicycle in nineteenth-century England, Pinch and Bijker draw a map of a wide variety of social groups who had different needs for a two-wheeled, human-powered vehicle. Some groups stressed speed; others emphasized safety; others wanted comfort and so forth. The task of scholarship is to show exactly how these various needs and interests are expressed as a wide range of technical problems and an even wider range of proposed solutions. From there one moves on to explain how processes of technological development achieve "closure", where the relevant social groups manage to agree "This is a bicycle" or "This is television". At such points of closure, the fundamental process of innovation ceases, at least for the time being. The social analysts' next step is to study the extension of this closure into society as a whole (Pinch and Bijker, 1987, pp. 40–6).

I want to emphasize that social constructivism is by no means an entirely unified viewpoint. There are some important differences among its leading practitioners.[4] For some who work in this perspective the conventional distinction between technology and society has finally broken down altogether. In the approach of Michel Callon and Bruno Latour, for example, we find the methodological premiss (eventually upheld as a basic social truth) that the modern world is composed of actor networks in which the significant social actors include both living persons and non-living technological entities. Others like Trevor Pinch and Wiebe Bijker prefer to maintain the notion that society is an environment or context in which technologies develop. But despite such differences of emphasis, the basic disposition and viewpoint of social constructivism is fairly consistent.

As a way of studying the dynamics of technological change, this approach does offer some interesting advantages. It offers clear, step-by-step guidance for doing case studies of technological innovation. One can present this method to graduate students who need a rigid conceptual framework to get started, and expect them to come up with empirical studies of how particular technologies are "socially constructed". Indeed, the social constructivists promise to deliver a veritable goldmine of those most highly valued of academic treasures: case studies. One locates a field of technological development that one finds interesting. Then one identifies the relevant social groups or design constituencies involved in a process of technological choice. One goes on to study exactly how the social actors express their needs and interests and how those come to be defined as technical problems to be solved, and how those problems are answered with not one, but several possible solutions. One notices that a particular device or technique or system can be interpreted in a number of different ways. One remains sceptical of claims about which particular person or group discovered or invented it. One looks for areas of conflict and cooperation, agreement, disagreement, and possible consensus. And one watches for that moment of closure where the struggle to define the shape of a technical artefact comes to an end through negotiation, or the machinations of the most powerful actors.

Using this approach, social constructivists in history and sociology have begun to produce a range of detailed studies of the development of specific technical devices and systems. They have studied the development of Bakelite, missile guidance systems, electric vehicles, expert systems in computer science, networks of electrical power generation and distribution, and several other corners of technological development.[5] Research results usually indicate that technological innovation is a multi-centred, complex process, not the unilinear progression depicted in many earlier writings. Along the way, social constructivists have helped debunk the idea that new technologies spring full-born from the work of "great men". Another useful contribution of this approach is to reveal the spectrum of possible technological choices, alternatives and branching points within patterns sometimes thought to be necessary. Social constructivist interpretations of technology emphasize contingency and choice rather than forces of necessity in the history of technology.

Although they are not alone in doing so, the social constructivists have been quite helpful in calling into question the sometimes highly arbitrary distinctions between the social sphere and the technical sphere. In my view, the ability to break down such arbitrary distinctions opens up some interesting possibilities for those who want to understand the place of technology in human experience. For that reason alone, the literature in the new sociology of technology is well worth a philosopher's attention.

As they proceed with their work, social constructivists are eager to call attention to the inadequacies of their predecessors, identifying their accomplishments as a clear advance over earlier ways of thinking about technology and society. Theirs is said to be a more rigorous, methodologically refined and clearsighted vision of technology and society than that which came before.

What are the significant points of comparison? Among the cast of characters one would certainly have to include the whole range of thinkers who have written about the origins and significance of modern technology. Among those explicitly or implicitly criticized are sociologists of technology like William Ogburn, historians of technology like Lynn White, and a variety of economists who have written on the economic correlates of innovation. Not far in the background are the likes of Lewis Mumford, Jacques Ellul, Ivan Illich, members of the Frankfurt school of critical theory, and any number of Marxist social theorists, not to mention Marx and Engels themselves.

As they refer to earlier generations of sociologists, the social constructivists often appear to be saying: "Yes, these were, indeed, great thinkers, but they were wrong and we are right". Whether or not this judgement comes to be accepted by the scholarly community as a whole, only time will tell. But the aspirations of social constructivism are fairly evident. Part of what is going on here is a socially constructed social construction of knowledge that seeks to depict earlier and contemporary approaches as outmoded or dead. Clearly, one of the ways in which this approach can be said to be "more complex" than previous ones has something to do with the Oedipus complex.

Before we join the swelling applause for social constructivism and anoint this school as the cutting edge in technology studies, we must pause to ask whether or not their approach does amount to an improvement over other approaches. Before we forget our Marx, or our Mumford, Ellul or Heidegger, it important to notice what one gives up as well as what one gains in choosing this intellectual path to the study of technology and human affairs.

Technology and Human Experience?

I hope I've made clear the aspects of this work that I find valuable: its conceptual rigour, its concern for specifics, its attempt to provide empirical models of technological change that better reveal the actual course of events. But as I read the works of the social constructivists and ponder the character of their research programme, I am increasingly struck by the narrowness of this perspective. Advances along this line of enquiry take place at a significant cost: a willingness to disregard important questions about technology and human experience, questions very much alive in other theoretical approaches.

The most obvious lack in social constructionist writing is an almost total disregard for the social consequences of technical choice. This is a social theory and method geared to explaining how technologies arise, how they are shaped through various kinds of social interaction. One tries to show why it is that particular devices, designs and social constituencies are the ones that prevail within the range of alternatives available at a given time. But the consequences of prevailing are seldom a focus of study. What the introduction of new artefacts has done for people's sense of self, for the texture of human communities, for qualities of everyday living, and for the broader distribution of power in society – these are not matters of explicit concern.[6]

The commitment to studying the origins of technology rather than the consequences of technological choices stems in part from the belief – a woefully mistaken one in my view – that the consequences or effects or "impacts" of technological change have already been studied to death by earlier generations of humanists and social scientists. As Donald MacKenzie and Judy Wajcman put the matter, the urgent but neglected question is, "What has shaped the technology that is having 'effects'? What has caused and is causing the technological change whose 'impact' we are experiencing?" (MacKenzie and Wajcman, 1985, p. 2).

Another reason for social constructivists' turn away from the study of consequences, in my reading, springs from the basic orientation of the social constructivists: namely, an application of ideas and methods employed in sociology of science to what they regard as a new and less important field of enquiry – namely, technology. In the sociology of science the primary issues are ones that have to do with the origins of knowledge about natural phenomena. On translating this approach to the study of technology, the focus tends to become the closest corresponding phenomenon the sociologist can identify: namely, the origins and dynamics of technological innovation.

In a peculiar way, then, this is a sociology of technology that has little concern for the ways in which technologies transform personal experience and social relations. The object of fascination is social construction of technical artefacts and processes. But why such innovations matter in the broader context is no longer of any great concern.

A second variety of narrowness can be seen in the social constructivists' favoured conception of social process. Here, as I've noted, one usually finds a field of what are called "relevant social actors" who are engaged in a process of defining technical problems, seeking solutions, and having their solutions adopted as authoritative within prevailing patterns of social use. As a student of politics examining this approach, I am struck by the ways in which it echoes the conceptual and theoretical commitments of theories of political pluralism and of bureaucratic politics. Proposed as a way of understanding the workings of modern democracy, pluralist theories point to the complex interactions of interest groups within society as a whole, and within and around particular organizations. Decisions and policies emerge as a vector outcome of the combined pushes and pulls within an essentially pluralist framework.

But there is an annoying question for political pluralism that can be posed for social constructivism as well. Who says what are "relevant" social groups and social interests? What about groups which have no voice but which nevertheless will be affected by the results of technological change? What of groups which have been suppressed or deliberately excluded? How does one account for potentially important choices that never surface as matters for debate and choice?

As critics of pluralist theory in political science have argued, it is important to notice not only which decisions are made and how, but also which decisions never land on the agenda at all; which possibilities are relegated to the sphere of non-decisions (Bachrach, 1980). By noticing which issues are never (or seldom) articulated or legitimized, by observing which groups are consistently excluded from power, one begins to understand the enduring social structures upon which more obvious kinds of political behaviour rest. Failing to do this, social scientists offer an account of politics and society that is implicitly conservative, an account that attends to the needs and machinations of the powerful as if they were all that mattered.

The corresponding problem for social constructivism is that its ways of modelling the relationship between social interests and technological innovation will conceal as much as they reveal. Looking at contemporary research and development in manufacturing technology in the United States, for example, it is remarkable how thoroughly the interests and perspectives of labour have simply been eliminated as a focus of any serious concerns. In research models of computer-integrated manufacturing (CIM), the traditional roles of blue collar workers are simply no longer present. Can research in the social construction of technology succeed if its map of the relevant social groups does not indicate which social groups have finally been sandbagged out of the laboratories and which social voices effectively silenced?

As a programme of enquiry, social constructivism is careful to avoid the technological version of the "Whig theory of history", in which the past is read as a sequence of steps leading inevitably to the accomplishments of today. But although social constructivism escapes the bind of Whig history, it seems not to have noticed the problem of elitism, the ways in which even a broad, multi-centred spectrum of technical possibilities is skewed in ways that favour some social interests while excluding others. While this approach rejects the "great man theory" of technological development, it still attends to the needs and problems of the powerful persons and groups: those with the resources to enter the game and define its terms. Although it succeeds in finding contingency rather than necessity in the course of technological change, it seems so far to have little to say about the deep-seated political biases that can underlie the spectrum of choices that surface for "relevant" social actors.

This point leads to my third problem with social constructivism: namely, that it disregards the possibility that there may be dynamics evident in technological change behind those revealed by studying the immediate needs, interests, problems, and solutions of specific groups and social actors. One of the key claims in philosophical writings is that if one looks closely, one sees basic conditions that underlie the busy social activities of technology-making. Marxists, for example, argue that a key condition is the phenomenon of social class. In this view, the structural relationships between classes are fundamental conditions that underlie all economic institutions, government policies and technological choices.

Other thinkers have pointed to a basic metaphysical disposition that establishes the split between human beings and nature and the attitude of mastery and domination that characterizes modern technics, whatever its particular forms may be.

Others still have pointed to the form of underdimensioned rationality that plays itself out in all modern technological projects.

The possibility that the ebb and flow of social interaction among social groups may reflect other, more deeply seated processes in society is not an idea that the social constructivists choose to explore. They usually find it sufficient to gather evidence of social activities most clearly connected to technological change. In so far as there exist deeper cultural, intellectual or economic origins of social choices about technology or deeper issues surrounding these choices, the social constructivists do not seek to reveal them.

The notion of "autonomous technology", for example, they reject as a now discredited determinism, eclipsed by their models of a dynamic, multicentred process of social selection. But in more subtle versions of "autonomous technology", determinism is not the central issue at all. As people pursue their interests, socially constructing technologies that succeed at some level of practice, they sometimes undermine what are or ought to be key concerns at another level. Each technically embodied affirmation may also count as a betrayal, perhaps even self-betrayal. The same devices that have brought wonderful conveniences in transportation and communication have also tended to erode community. In the maxim of theologian Richard Penniman, "They got what they wanted, but they lost what they had".

In that light, the interesting questions have nothing at all to do with any alleged self-generating properties of modern technology. Instead they have to do with the often painful ironies of technical choice. Although the social constructivists are energetic researchers, they always seem not to be careful readers. Thus, they simply overlook aspects of philosophical discussion about autonomous technology that do not fit their preferred conceptual straw man: technological determinism.

Evaluation, Morality and Political Principle

A fourth and final quality of this mode of enquiry that deserves comment is one to which I've already alluded: namely, its lack of and, indeed, apparent disdain for anything resembling an evaluative stance, or any particular moral or political principles, that might help people judge the possibilities that technologies present. The empirical programme of relativism in the sociology of science becomes the methodology of interpretive flexibility in the new sociology of technology. Rather than attribute any particular meaning to a technical device or its uses, social research tries to understand how it is that some people see a developing artefact in one way while others see it quite differently.

This strategy seems to me well worth using – up to a point. It helps to reveal the broad range of demands and desires that are packed into technical developments of various kinds. Some welcomed the modern safety bicycle with its balloon tyres and foot brakes because it was fast and stable; others liked it because it presented fewer riding hazards than its predecessors; and so on. The premiss of interpretive flexibility works especially well in cases where social consensus is achievable; where all or most parties can say at the end of the process, "Thank God we came together around this set of design features". In that way the underlying leitmotif of the sociologist's composition is still (implicitly) that of progress; kudos all around. But what about circumstances in which there are serious disagreements about the design or use of an artefact or technological system? How will the social analyst evaluate the terms of the disagreement?

As regards the analysis of scientific knowledge, the epistemological programme of relativism in the sociology of science remains neutral as regards judgements about whether or not the proclaimed discoveries or theories of scientists are true or not. Extrapolating to technology, social constructivists choose to remain agnostic as regards the ultimate good or ill attached to particular technical accomplishments. As a feature of a purely descriptive, explanatory project in sociology, this may make sense. A researcher may even suggest that at some later point and in a different setting it may be possible to offer well-considered judgements about values associated with a particular technology. But in fact, researchers in the social construction of technology programme have neither made such promises nor, to my knowledge, taken such steps. As far as I can tell, they have no theoretical or practical position on technology and human well-being at all. In fact, to announce such a position seems forbidden on methodological grounds. And since purity of social science methodology is of such pre-eminent concern, it is likely that social constructivists will continue their research without taking a stand on the larger questions about technology and the human condition that matter most in modern history.

In this way, the methodological bracketing of questions about interests and interpretations amounts to a political stance which regards the status quo and its ills and injustices with precious equanimity. Interpretive flexibility soon becomes moral and political indifference. In my view, the frequency with which technology looms as a crucial issue for commitment in modern society makes this posture an extremely vain and unhelpful one. Sometimes it matters what a thing is, what name it has and how people judge its properties. For example, was the structure in Iraq that was photographed during the Gulf War of 1991 a baby food factory or chemical weapons plant? It is true that some people claimed the building was one thing while others said it was something else. But noticing the diversity and flexibility of interpretations in such cases is of little help. Ultimately, one has to decide what one is dealing with and why it matters.

But the methodological posture of social constructivism is characteristically unwilling to engage in argument about the aspects of technology that now weigh heavily in key debates about the place of technology in human affairs. Such concerns are now deleted from historical accounts of how technologies arise, as well as from contemporary descriptions of technological and social change. There is, similarly, no willingness to examine the underlying patterns that characterize the quality of life in modern technological societies. There is also no desire to weigh arguments about right and wrong involved in particular social choices in energy, transportation, weaponry, manufacturing, agriculture, computing and the like. Even less is there any effort to evaluate patterns of life in technological societies taken as a whole. All the emphasis is focused upon specific cases and how they illuminate a standard, often repeated hypothesis: namely, that technologies are socially constructed.

To give an example of how the constructivist orientation seeks to sidestep questions that require moral and political argument, I want to look briefly at an article written by one of the leading proponents of this view, Steven Woolgar, in a recent edition of the journal *Science, Technology and Human Values*. Woolgar examines "the turn to technology", deploying familiar constructivist moves to build his point of view. Along the way he takes on an argument that I made several years ago about the politics of technology. The case in point involves some bridges on the Long Island Expressway built decades ago by the powerful New York planner, Robert Moses. My claim is that Moses deliberately built the overpasses on the Expressway fairly low to the ground so that buses would not be able to pass under them. I see this as an expression of Moses' desire to separate different social class and racial groups in New York City. Blacks and poor tended not to have cars and would have to take buses if they wanted to get to places like Jones Beach on Long Island. Thus, the height of the bridge was a political statement – inequality in built form – that became an enduring part of both the physical infrastructure and social landscape of New York (Winner, 1986).

Woolgar looks at this finding with amusement. In the style of social constructivism he notes that the bridges could be flexibly interpreted in a variety of ways. He says that Langdon Winner can interpret them as political artefacts "if you like". But, he says, "the interpretist response espouses a measure of impartiality by proposing that analysis deals with the ways in which readings are done, without prejudice to their relative truth" (Woolgar, 1991, p. 41).

Woolgar is certainly justified in asking "what is it that makes one reading of the text (technology) more persuasive than another?" However, he is wrong to suggest that the issue is simply not decidable. I agree that all structures, including Moses' bridges, can be interpreted in a variety of different ways; in fact, my analysis presupposes exactly that. What makes the conclusion that Moses' bridges are inegalitarian political artefacts a strongly defensible proposition is not difficult to grasp. It can be seen in the role that bridges play in the social and political history of a particular community at a particular time, as well as in the personal history of a power broker notorious in his willingness to use all possible means, including public works projects, to shape social patterns to match his vision of what was desirable. To avoid this conclusion through the use of postmodernist interpretive irony is, in my view, politically naïve.

In situations in which there are admittedly a variety of points of view that matter in making choices about technology, I believe it is necessary for social theorists to go beyond what positivists used to call "value neutrality" and what social constructivists resurrect as "interpretive flexibility". One must move on to offer coherent arguments about which ends, principles and conditions deserve not only our attention, but also our commitment. At that point one ceases interpreting interpretations of interpretations and, for better or worse, takes a stand on choices to develop and/or limit the technologies available to humankind.

Power holders who have technological megaprojects in mind could well find comfort in a vision like that now offered by the social constructivists. Unlike the enquiries of previous generations of critical social thinkers, social constructivism provides no solid, systematic standpoint or core of moral concerns from which to criticize or oppose any particular pattern of technical development. Neither does it show any desire to move beyond elaborate descriptions, interpretations and explanations to discuss what ought to be done. Robert Moses, for example, might well have applauded such an approach. For it implicitly affirms what he eventually came to believe: that what matters in the end is simply the exercise of raw power.

A Remarkably Hollow Box

My conclusion is, then, that although the social constructivists have opened the black box and shown a colourful array of social factors, processes and images therein, the box they reveal is still a remarkably hollow one. Yes, they regularly succeed in tracking a great deal of intense activity around technological developments of various kinds. They also show us the fascinating dynamics of conflict, disagreement and consensus formation that surround some choices of great importance. But as they survey the evidence, they offer no judgement on what it all means, other than to notice that some technological projects succeed and others fail, that new forms of power arise while other forms decline.

Unlike other approaches – those of Marx, Ellul, Heidegger, Mumford and Illich, for example – this perspective does not explore or in any way call into question the basic commitments and projects of modern technological society. The attitude of the social constructivists seems to be that it is enough to provide clearer, well-nuanced explanations of technological development. As compared to any of the major philosophical discussions of technology, there is something very important missing here; namely, a general position on the social and technological patterns under study.

In contrast, the corresponding enquiries of traditional Marxists have always shown a concern for the condition of the working class and the world's downtrodden, expressing suspicion of the manipulations of capital and a hope that the dynamics of history would produce human liberation.

With liberal theorists, similarly, there is a fundamental conviction that expanding technology and economic growth will eventually make everyone relatively wealthy.

With Heideggerians one always has the sense that there might some day be a "turning" within the history of being to save humanity from the perils of modernity.

With Lewis Mumford there is always an underlying hope that the abstract, mechanistic obsessions of the modern age would be replaced by a more humane, organic sense of technical possibilities.

With Jacques Ellul there remains the possibility that even as the technological system reaches its maturity, humanity will renew its covenant with a forgiving God.

What are the corresponding prospects envisioned by social constructivism? The answer is by no means clear. Up to this point the dreams and projects of the social constructivists have been primarily academic ones, carefully sanitized of any critical standpoint that might contribute to substantive debates about the political and environmental dimensions of technological choice.

Perhaps the helpful insight they want to offer is simply that choices are available, that the course of technological development is not fore-ordained by outside forces, but instead a product of complex social interactions. If that is the point of their enquiries, then constructivists are now repeating it *ad nauseam*. Alas, this increasingly redundant theme has not been incorporated into anything like a programme for positive change or a theoretical perspective that anticipates anything better than the current course of events. We do not find, for example, arguments by social constructivists to justify expanded democratic participation in key technological choices. Neither are there suggestions to illuminate processes of technological design in ways that might serve the ends of freedom and justice. Indeed, many social constructivists appear much more concerned to gaze at themselves within that endlessly enchanting hall of mirrors – sociological reflexivity.

In the manner in which it now presents itself, social constructivism offers a very limited purchase on the issues that surround technology. In its own distinctive manner, the accomplishments it recommends are largely technical ones, ways of enriching increasingly specialized sociological and historical research. As such, social constructivism now appears content to define itself as a narrow academic subfield – innovation studies. At present it shows no inclination to reach further, to fashion conceptual links to the larger questions about technology and the human condition that have engaged social and political thinkers throughout the nineteenth and twentieth centuries.

The intellectual vogue of social constructivism arises at a crucial time. In the late twentieth century a great many people – scholars and ordinary citizens alike – have begun to realize that the key question is not how technology is constructed, but how to come to terms with ways in which our technology-centred world might be reconstructed. Faced with a variety of social and environmental ills, there is growing recognition that what is needed is a process of redirecting our technological systems and projects in ways inspired by democratic and ecological principles. How that reconstruction might occur is an open question, one ripe for widespread study, debate and action. I believe it to be the great challenge for cross-disciplinary thinking during the next several decades. How tragic it would be to find that at the moment of greatest challenge, many leading scholars of technology and society had retreated into a blasé, depoliticized scholasticism.

Fortunately, this need not happen. It turns out that the very questions that the social constructivists typically ignore are the ones that a good number of contemporary philosophers, political theorists and social activists are still interested in posing. While there is much we can learn from the new sociology of technology, there is also much in this way of thinking that needs to be criticized, reformulated and refocused around a clearer understanding of what the aims of our thinking ought to be.

In sum, the search for a meaningful theory of technology has by no means achieved "closure". It must begin anew.

Notes

1 The best statement of the general aims and approaches of this school of thought is presented in Bijker, Hughes and Pinch, eds (1987).

2 The use of "black boxes" is a common engineering practice, a way of abbreviating complex technical processes so that the work of design can proceed. Textbooks, lectures and problems sets employed in engineering education are also filled with "black boxes" that students are taught to view as convenient "plug-ins" for problem-solving.

3 Pinch and Bijker show little awareness of the literature in philosophy and technology, past or present. That does not prevent them from delivering a peremptory judgement on the matter.

4 I do not wish to conflate works on the sociology and history of technology that have important distinguishing traits. Each published study by the scholars I mention here would well stand on its own merits alone. Some of the writers I have mentioned may even object to being classified within this category at all. Nevertheless, there has been a concerted push to affirm social constructivism as a coherent mode of analysis and to include or exclude writers according to their degree of adherence to this new canonical standard. It is that push to which I am responding here.

5 Probably the most complete application of this approach to date is Donald MacKenzie's study of missile guidance systems (MacKenzie, 1990).

6 A notable exception is the work of Ruth Schwartz Cowan, a historian sometimes included in social constructivist conferences and anthologies. Cowan's work is steadfast in its desire to show the connection between specific technological choices and how social life is affected as a consequence (Cowan, 1983).

References

Bachrach, P. (1980) *The Theory of Democratic Elitism: A Critique*. Lanham, MD: University Press of America.

Bijker, W. E., Hughes, T. P. and Pinch, T., eds (1987) *The Social Construction of Technological Systems: New Directions in the Sociology and History of Technology*. Cambridge, MA: MIT Press.

Collins, H. M. (1983) "The sociology of scientific knowledge: studies of contemporary science", *Annual Review of Sociology* 9: 265–85.

Cowan, R. S. (1983) *More Work for Mother: The Ironies of Household Technology from the Open Hearth to the Microwave*. New York: Basic Books. London: Free Association Books, 1989.

MacKenzie, D. A. (1990) *Inventing Accuracy: A Historical Sociology of Nuclear Missile Guidance.* Cambridge, MA: MIT Press.

MacKenzie, D. A. and Wajcman, J. (1985) *The Social Shaping of Technology: How the Refrigerator Got Its Hum.* Milton Keynes: Open University Press.

Mostert, N. (1974) *Supership.* New York: Alfred A. Knopf.

Pinch, T. and Bijker, W. E. (1987) "The social construction of facts and artifacts: or how the sociology of science and the sociology of technology might benefit each other", in W. E. Bijker, T. P. Hughes and T. Pinch, eds, 1987, pp. 17–50.

Winner, L. (1986), "Do artifacts have politics?", in *The Whale and the Reactor.* Chicago: University of Chicago Press.

Woolgar, S. (1991), "The turn to technology in social studies of science", *Science, Technology and Human Values* 16: 20–50.

Heidegger on Technology

Introduction

Most philosophers of technology would probably agree that, for good or ill, Martin Heidegger's interpretation of technology, its meaning in Western history, and its role in contemporary human affairs is probably the single most influential position in the field. The selections in this part, first from Heidegger himself and then from a sampling of his readers, introduce some lines of his influence that also anticipate discussions in subsequent sections.

Heidegger's consideration of technology spans some forty years, but his core position is found in the article reprinted here. It treats technology (and its association with science) in a way that opens up what Heidegger initially called his "question of the meaning of Being." In his view, the scientifically informed technology that increasingly dominates the world is not something fundamentally new or even modern. Rather, it fulfills Western philosophy's oldest desire for knowledge of what is real as that expressed in the pre-Socratics, Plato, and Aristotle. For all the specific satisfactions this fulfillment obviously brings, however, Heidegger questions whether it can ever make us feel genuinely "at home." From the start, he was convinced that a science-driven, technologically informed "understanding of Being" – in other words, our primary, operative sense of what it means for something to be "real" – is increasingly experienced as more constraining than illuminating of our encounters with each other and with our surroundings. In *Being and Time*, he proposes to raise the Being-question "again," in hopes of opening up our thinking to a less hegemonic and more pluralistic conception of the ways that things can "be."

After *Being and Time*, however, Heidegger grew increasingly dissatisfied with several features of his initial way of posing his question. He came to believe that – quite aside from what he himself may have learned in the process – it was misleading to have conceived his inquiry as one that must start with an analysis of specifically "human" Being and then turn to the task of "overcoming" Western metaphysics. In characterizing his intentions this way, *Being and Time* created the impression that if we would only (1) make our own Being the primary topic and (2) view traditional philosophy as the culprit, then (3) we might somehow achieve a standpoint that is "after," or radically freed up from this tradition. On all three counts, however, this impression is wrong. As the later Heidegger sees it, neither something about ourselves, nor opposition to traditional metaphysics, nor the dream of a post-traditional philosophy define the direction of his inquiry. Rather, "thinking" must locate itself at and within the "site" or "clearing" where our relationships with things and people take place.

"The Question Concerning Technology" is one of many works in which Heidegger recasts his original project in terms of this "thinking." As its opening lines show, the discussion begins neither with ourselves nor with technological things but with their *relationship*. What it is like, Heidegger asks, to be "in the midst" of a technological existence? Presently, typically, we tend to be "chained" to technology; but analysis of precisely this condition can show the way to open up a "free" relationship with technology instead. The analysis has four parts. First, Heidegger distinguishes a "correct" from the "true" understanding of technology. A correct understanding interprets technology the way many readers interpreted *Being and Time*, that is, in terms of what it is "for us." Interpreted this way, technology seems obviously to be a human activity that provides the means to our ends. This "instrumentalist" interpretation is not wrong, but it fails to account for its own "uncanny" correctness. Hence, it only opens up a deeper question – namely, what is technology, "essentially," such that "the instrumental" is already understood to define the very atmosphere in which (or "site" where) means, ends, and the "will to mastery" typically predominate. Second, with this question in mind, Heidegger considers instrumentality in terms of the notion of "for what end(s)?" and – through analyses of the traditional concepts of "cause" and of "*technē*" as a practical art involving a kind of bringing-forth (*poiēsis*) – he suggests that with technology comes a distinctive mode of disclosiveness, or revealedness, that is, a kind of ontological truth (*alētheia*).

Third, Heidegger develops the idea that technological truth needs to be treated not just as a disclosure but as a "sent" disclosure. Considered in this way, technological truth can be seen to account for the way we characteristically find ourselves already "put in the midst" of our busy instrumental circumstances. In *Being and Time*, Heidegger analyzed practical and theoretical being-in-the-world, respectively, in terms of relatively simple tasks and a critique of traditional epistemology. In the technology essay, however, he describes these activities in the more specifically contemporary terms of technoscientific practice and theory. Our activities, the things we encounter and deal with, and even we ourselves all seem to happen together in a "world" where everything is "enframed" as part of a stockpile of available materials and personnel – what Heidegger calls a "standing-reserve" (*Bestand*), always ready for technological purposes. Enframing (*Gestell*), then, is the "essence" of the technological – that is, the disclosing of meaning which "gives" the instrumentally useful its "instrumental" sense. Heidegger goes on to consider the current hegemony of this gift of disclosure – that is, the way it infuses the world with a pervasive sense of the disposable usefulness of everything so that it tends to hide both other (i.e., non-instrumental) possibilities and also itself as precisely this sent, or occurrent enframing.

Finally, Heidegger urges us to "thoughtfully reflect" upon the "eventuation" (*Ereignis*) of this enframing, instead of "falling away" from it into the ever more frantic pursuit of instrumental means to technoscientifically defined ends. Becoming captive to this pursuit, he says, is the pervasive "danger" of our age. By reflecting upon the very occurrence of this technoscientific enframing, however, we may come to recognize a "saving power" – the other possibility into which enframing places us – namely, the possibility of opening up a "free relation with technology." Such a relation, he concludes, would be one in which technology is "decisively confronted" and in which technological engagements do not close us off from non-instrumental possibilities. Heidegger leaves us with the provocative suggestion that such a relation – one not wholly captivated by technoscientific activities but also one that does not fancifully imagine being transported beyond or outside of these activities – would necessarily involve a transformation of the very site where all of this occurs. A free relation with technology would thus have to happen, he says, "in a realm that is, on the one hand, akin to the essence of technology and, on the other, fundamentally different from it."

The other essays in this section each react quite differently to Heidegger's holistic, or "global" interpretation of technology relations as always taking place within an enframing "realm" of available things and personnel. Robert Scharff argues that, suprisingly, Comte and Heidegger actually share this interpretation. Both regard technoscience as nothing short of the "culmination" of the Western intellectual tradition; and both see themselves as thinking about the essential character of this culmination from *within* the experienced situation it defines, not from some imagined vantage point outside of it. The question, then, is how to understand their radically different reactions to the increasingly hegemonic reach of technoscience – the implication being, of course, that understanding the difference between Comte and Heidegger might also help illuminate current debates. Scharff suggests that the difference in their reactions is not a function of any disagreements about particular technologies or particular points of epistemology. The real contrast is between an upbeat (and in some moods even utopian) Comte who is unable to even imagine an era "beyond" the technoscientific and a Heidegger who finds this same era deeply, experientially, and ontologically dissatisfying. Comte conceives the future in a way that anticipates the army of technological optimists yet to come – namely, in terms of how to encourage more of the same under ever better conditions. Heidegger, however, faces the problem – described in the final pages of "The Question Concerning Technology" and echoed by numerous technological critics and pessimists – of finding a way to think the "enframing" character of technoscience which both acknowledges his inevitably continuing to exist "in the midst" of this event and yet also does justice to all those experiences that do not fit "comfortably" within its enframing sense of what is real.

Don Ihde's paper is partly a helpful commentary on "The Question Concerning Technology" and partly a statement of preference for Heidegger's earlier analyses of practical and theoretical relations in *Being and Time*. Foreshadowing his own "phenomenology of technics" (see Part VI, chapter 43), Ihde argues that these earlier analyses actually offer a less "reductive" and more promising picture of our options in a technological era. He stresses the "positive tone" of *Being and Time*'s analysis of tool use. He praises what he regards as its more nuanced and pluralistic treatment of "ready-to-hand" relations, and he complains about the thinner conception of the instrumentality in the "Technology" essay. Ihde suggests that one problem with the latter essay is that its umbrella notion of a standing-reserve (of useful material and personnel) seems to be derived entirely from a "metaphysics of scientifically organized technology," whereas Heidegger's earlier writings leave open a richer notion of the useful that derives as much from prescientific technology and the arts as from technoscience.

Albert Borgmann's "Focal Things and Practices" is excerpted from his well-known *Technology and the Character of Contemporary Life* (1986), in which he argues for a "reform of technology" based on an elaboration and correction of Heidegger's account. Locating himself between technological determinists like Ellul and instrumentalists who view technology as merely a collection of neutral means to freely chosen ends, Borgmann agrees with Heidegger about the danger of modern technology. The world of modern industrial technology does indeed threaten to reduce our relations to the instrumental/utilitarian; and Heidegger is undoubtedly right that only by "raising the rule of technology from its anonymity" (i.e., think the enframing itself) might we learn to let something extra-technological – what Borgmann, renaming Heidegger's idea of "things thinging" and "shining forth," calls "focal things" – enrich and bring greater purpose to our lives.

Borgmann argues, however, that what Heidegger says about such enriching experiences needs to be improved in two directions. First, he rejects Heidegger's "misleading and dispiriting" tendency to appeal only to "the simple things of yesterday" as potentially enriching. Second, he complains that Heidegger fails to adequately trace out the way focal concerns shape our social relations as well as our relations with things. Borgmann analyzes activities like playing music, running, gardening, and the family meal as at least potentially providing occasions for the sort of "gathering and directing" of our concerns he has in mind. Serious runners, for example, will enhance their activity by using the latest and best gear, but only to the extent that this does not transform the activity itself into something no human runner could accomplish. So, too, the family meal might again become "a focal event par excellence" – an event that brings scattered family members together, provides an occasion for cooperative preparation of carefully chosen foods, for reenacting cultural traditions and social practices, and for intimate sharing of doubts and pleasures in conversation. Borgmann admits that focal events today are relatively isolated and in constant danger of being overtaken by the press of technological life (running comes to mean 30 minutes on the treadmill, and meals, a quick Big Mac or microwaved TV dinner, eaten on the fly, with or without company). He insists, moreover, that technology itself has never produced anything which might serve a focal concern. Nevertheless, there is nothing to prevent us from cultivating such concerns, glorying their plurality, and "celebrating the social union that is fostered by that plurality." Borgmann even suggests that a reversal of the current priorities of technological practice and focal concern might lead to the development of a new conception of the Good Life that is directed by focal practices and merely enhanced by technological means.

In their response to Borgmann, Hubert Dreyfus and Charles Spinosa join him in affirming Heidegger's interpretation of the "danger" inherent in technology, but they object to Borgmann's further claim that technology itself cannot be a source of focal concerns. The problem, in their view, is Borgmann's reliance on Heidegger's earlier rather than later thought. At first, Heidegger understood technology as simply the last, greatest expression of "subjectivity" – that is, of a world full of individual human selves, each treating everything as an "object" of their desires and so as something to be controlled and/or consumed. Acceptance of this view forces Borgmann to conceive all focal concerns in terms of resistance to technological practices and thus to conceive these concerns as necessarily non-technological. For Dreyfus and Spinosa, however, Heidegger's later notion of developing a "free relation" to technology offers a more positive possibility. In their postmodern construal, which they claim to validate with reference to Heidegger's own writings, technological practices do indeed tend to fragment, or "disaggregate" our lives so that we lose any sense of our having single, unified self-identities; and the danger is indeed that this disaggregation threatens our ability to "disclose the world" in any way other than instrumentally and arbitrarily. Yet Borgmann is wrong to see only danger here. Dreyfus and Spinosa are not convinced – and do not think the later Heidegger still holds – that all disaggregation has to be viewed negatively. They reject the idea that world-disclosing must ultimately be conceived in terms of some overarching personal or communal unity. In a provocative interpretation of Heidegger's notion of technological enframing's "saving power," they argue that entering into a free relation with technology really means "freeing us from having a total fixed identity so that we may experience ourselves as multiple identities disclosing multiple worlds." In other words, Dreyfus and Spinosa see a potential benefit to be gained from the technology's undermining of the notion of a subject-self with a single, unifying self-identity. Living with technology – which we will be doing in any event – appears to require living in a plurality of local worlds, with a plurality of focal skills, with at most a "poly-identity that is neither the identity of an arbitrary desiring subject nor the rudderless adaptability of a resource." Dwelling in some of these local worlds, of course, can be expected to be short-lived – but no less enjoyable for that. Dreyfus and Spinosa mention, for example, Web-surfing, "zipping around an autobahn cloverleaf," and participating in the mercurial creative period of a rock band, with each musician then going his or her own way. Their essay closes with a rejection – again, as they see it, in Heidegger's name – of Borgmann's resurrection of the old idea of humanity eventually forming a "community of communities." The only integrity we can expect our lives to have, they conclude, is the openness and flexibility to belong to and move around among many local worlds.

For Andrew Feenberg, neither the Heideggerianism of Borgmann nor that of Dreyfus and Spinosa can rescue Heidegger from the inadequacies of his own thinking about technology; moreover, both interpretations introduce additional problems of their own. Although Feenberg does not specifically mention Dreyfus and Spinosa, his rejection of what he calls "idealist" readings of Heidegger's analysis strongly suggests that he has views like theirs in mind. If we take Heidegger at his word that the essence of technology can only be understood through our technological engagement with the world, says Feenberg, then the question immediately arises whether this engagement is merely an "attitude" or is "embedded in the actual design of modern technological devices." If one answers that it is an attitude, then it follows that Heidegger's so-called "free relation" with technology would amount to nothing more than a change in our outlook that leaves the device-filled world as we find it. When Dreyfus and Spinosa say that *we* are "active world-disclosers," that *we* might become more "in tune" with technology, that technology solicits us to change *our* sense of personal identity, that *we* might move from one "world" to another by employing different sets of skills – all this language does at least create the impression that they, in spite of their use of Heidegger's language about the "thinging of *things*," understand themselves primarily to be recommending the adoption of a new outlook toward those things.

If, however, one takes the other alternative and answers that technological thinking is too deeply engaged with the "actual design of technological devices" for any mere change of attitude to transform our engagements with them, then Feenberg wants to know how the engagements could be changed at all. His critique of Borgmann ends with the observation that it is difficult to tell whether Borgmann has this latter problem, or Dreyfus and Spinosa's, or both.

According to Feenberg, the ultimate responsibility for these difficulties lies in Heidegger's own flawed way of posing the issues. Two related criticisms seem central to Feenberg's analysis. First, there is Heidegger's demand that we consider only the "essence" of technology; and second, there is the resultant "abstractness" of Heidegger's thinking about it. Although he does not discuss it, Feenberg appears to interpret "essence" in a fairly traditional way, as designating the fundamental, universal, unchanging characteristic(s) that define what something is. On the basis of this interpretation, he finds cause to complain that a Heideggerian analysis of technology remains remote from our actual engagements with it, and that from this distance it seems to vacillate unavoidably between recommending a change of attitude and merely denouncing in very general terms how things are. Either way, Feenberg complains, the whole variety of very real differences in our actual relations with and modification of technological devices slips through the cracks. He gives as an extended example the ways in which the production and use of the computer have mutually affected each other, so that it is simply false to imagine that some "inner, techno-logic" was spelling itself out as we witnessed it.

Feenberg's critique of Heidegger is in some ways representative of those influenced by critical social theory and Neo-Marxism. In the end, for these critics, the main problem with Heidegger's philosophy of technology is that it leaves us without any concrete sense of how social and political change in a technological world might actually be achieved. Hence, Feenberg does not believe Heidegger's own characterization of his thinking as located "in the midst" of our technological engagements and as struggling to develop a free relation with technology that would alter the very "realm" in which these engagements take place. Instead, because Heidegger called this an "ontological" question and (except for the embarrassment of his own participation in the Nazi movement) refused to offer a specific sociopolitical program, Feenberg concludes that Heidegger's critique of technology's dangers – in spite of its insightful identification of technological excess – can only give us the useless advice that we should somehow "liberate" ourselves from technological engagements. The question of whether this interpretation is correct – and whether this makes any difference to the concrete question of how technological engagements might be delimited and transformed – has become a central issue in the debates over the importance of Heidegger's work.

23

The Question Concerning Technology

Martin Heidegger

In what follows we shall be *questioning* concerning technology. Questioning builds a way. We would be advised, therefore, above all to pay heed to the way, and not to fix our attention on isolated sentences and topics. The way is one of thinking. All ways of thinking, more or less perceptibly, lead through language in a manner that is extraordinary. We shall be questioning concerning *technology*, and in so doing we should like to prepare a free relationship to it. The relationship will be free if it opens our human existence to the essence of technology. When we can respond to this essence, we shall be able to experience the technological within its own bounds.

Technology is not equivalent to the essence of technology. When we are seeking the essence of "tree," we have to become aware that what pervades every tree, as tree, is not itself a tree that can be encountered among all the other trees.

Likewise, the essence of technology is by no means anything technological. Thus we shall never experience our relationship to the essence of technology so long as we merely represent and pursue the technological, put up with it, or evade it. Everywhere we remain unfree and chained to technology, whether we passionately affirm or deny it. But we are delivered over to it in the worst possible way when we regard it as something neutral; for this conception of it, to which today we particularly like to pay homage, makes us utterly blind to the essence of technology.

According to ancient doctrine, the essence of a thing is considered to be *what* the thing is. We ask the question concerning technology when we ask what it is. Everyone knows the two statements that answer our question. One says: Technology is a means to an end. The other says: Technology is a human activity. The two definitions of technology belong together. For to posit ends and procure and utilize the means to them is a human activity. The manufacture and utilization of equipment, tools, and machines, the manufactured and used things themselves, and the needs and ends that they serve, all belong to what technology is. The whole complex of these contrivances is technology. Technology itself is a contrivance – in Latin, an *instrumentum*.

The current conception of technology, according to which it is a means and a human activity, can therefore be called the instrumental and anthropological definition of technology.

Who would ever deny that it is correct? It is in obvious conformity with what we are envisaging when we talk about technology. The instrumental definition of technology is indeed so uncannily correct that it even holds for modern technology, of which, in other respects, we maintain with some justification that it is, in contrast to the older handicraft technology, something completely different and therefore new. Even the power plant

From Martin Heidegger, *Basic Writings*, rev. ed., New York: HarperCollins, Inc., 1993, pp. 311–41, with the translation altered slightly from the version in Heidegger's *The Question Concerning Technology and Other Essays*, trans. William Lovitt, New York: Harper & Row, 1977. The German text appears in Martin Heidegger, *Vorträge und Aufsätze*, Pfullingen: Günther Neske Verlag, 1954, pp. 13–44, and in the same publisher's "Opuscula" series under the title *Die Technik und die Kehre*, 1962, pp. 5–36.

with its turbines and generators is a man-made means to an end established by man. Even the jet aircraft and the high-frequency apparatus are means to ends. A radar station is of course less simple than a weather vane. To be sure, the construction of a high-frequency apparatus requires the interlocking of various processes of technical-industrial production. And certainly a sawmill in a secluded valley of the Black Forest is a primitive means compared with the hydroelectric plant on the Rhine River.

But this much remains correct: Modern technology too is a means to an end. This is why the instrumental conception of technology conditions every attempt to bring man into the right relation to technology. Everything depends on our manipulating technology in the proper manner as a means. We will, as we say, "get" technology "intelligently in hand." We will master it. The will to mastery becomes all the more urgent the more technology threatens to slip from human control.

But suppose now that technology were no mere means: how would it stand with the will to master it? Yet we said, did we not, that the instrumental definition of technology is correct? To be sure. The correct always fixes upon something pertinent in whatever is under consideration. However, in order to be correct, this fixing by no means needs to uncover the thing in question in its essence. Only at the point where such an uncovering happens does the true propriate. For that reason the merely correct is not yet the true. Only the true brings us into a free relationship with that which concerns us from its essence. Accordingly, the correct instrumental definition of technology still does not show us technology's essence. In order that we may arrive at this, or at least come close to it, we must seek the true by way of the correct. We must ask: What is the instrumental itself? Within what do such things as means and end belong? A means is that whereby something is effected and thus attained. Whatever has an effect as its consequence is called a cause. But not only that by means of which something else is effected is a cause. The end that determines the kind of means to be used may also be considered a cause. Wherever ends are pursued and means are employed, wherever instrumentality reigns, there reigns causality.

For centuries philosophy has taught that there are four causes: (1) the *causa materialis*, the material, the matter out of which, for example, a silver chalice is made; (2) the *causa formalis*, the form, the shape into which the material enters; (3) the *causa finalis*, the end, for example, the sacrificial rite in relation to which the required chalice is determined as to its form and matter; (4) the *causa efficiens*, which brings about the effect that is the finished, actual chalice, in this instance, the silversmith. What technology is, when represented as a means, discloses itself when we trace instrumentality back to fourfold causality.

But suppose that causality, for its part, is veiled in darkness with respect to what it is? Certainly for centuries we have acted as though the doctrine of the four causes had fallen from heaven as a truth as clear as daylight. But it might be that the time has come to ask: Why are there only four causes? In relation to the aforementioned four, what does "cause" really mean? From whence does it come that the causal character of the four causes is so unifiedly determined that they belong together?

So long as we do not allow ourselves to go into these questions, causality, and with it instrumentality, and with this the accepted definition of technology, remain obscure and groundless.

For a long time we have been accustomed to representing cause as that which brings something about. In this connection, to bring about means to obtain results, effects. The *causa efficiens*, but one among the four causes, sets the standard for all causality. This goes so far that we no longer even count the *causa finalis*, telic finality, as causality. *Causa, casus*, belongs to the verb *cadere*, to fall, and means that which brings it about that something turns out as a result in such and such a way. The doctrine of the four causes goes back to Aristotle. But everything that later ages seek in Greek thought under the conception and rubric "causality" in the realm of Greek thought and for Greek thought *per se* has simply nothing at all to do with bringing about and effecting. What we call cause [*Ursache*] and the Romans call *causa* is called *aition* by the Greeks, that to which something else is indebted [*das, was ein anderes verschuldet*]. The four causes are the ways, all belonging at once to each other, of being responsible for something else. An example can clarify this.

Silver is that out of which the silver chalice is made. As this matter (*hyle*), it is co-responsible for the chalice. The chalice is indebted to, i.e., owes thanks to, the silver for that of which it consists. But the sacrificial vessel is indebted not only to the silver. As a chalice, that which is indebted to the silver appears in the aspect of a chalice, and not in that of a brooch or a ring. Thus the sacred vessel is at the same time indebted to the aspect (*eidos*) of chaliceness. Both the silver into which the aspect is admitted as chalice and the aspect in which the silver appears are in their respective ways co-responsible for the sacrificial vessel.

But there remains yet a third something that is above all responsible for the sacrificial vessel. It is that which in advance confines the chalice within the realm of consecration and bestowal. Through this the chalice is circumscribed as sacrificial vessel. Circumscribing gives bounds to the thing. With the bounds the thing does not stop; rather, from within them it begins to be what after production it will be. That which gives bounds, that which completes, in this sense is called in Greek *telos*, which is all too often translated as "aim" and "purpose," and so misinterpreted. The *telos* is responsible for what as matter and what as aspect are together co-responsible for the sacrificial vessel.

Finally, there is a fourth participant in the responsibility for the finished sacrificial vessel's lying before us ready for use, i.e., the silversmith – but not at all because he, in working, brings about the finished sacrificial chalice as if it were the effect of a making; the silversmith is not a *causa efficiens*.

The Aristotelian doctrine neither knows the cause that is named by this term, nor uses a Greek word that would correspond to it.

The silversmith considers carefully and gathers together the three aforementioned ways of being responsible and indebted. To consider carefully [*überlegen*] is in Greek *legein*, *logos*. *Legein* is rooted in *apophainesthai*, to bring forward into appearance. The silversmith is co-responsible as that from which the sacred vessel's being brought forth and subsistence take and retain their first departure. The three previously mentioned ways of being responsible owe thanks to the pondering of the silversmith for the "that" and the "how" of their coming into appearance and into play for the production of the sacrificial vessel.

Thus four ways of owing hold sway in the sacrificial vessel that lies ready before us. They differ from one another, yet they belong together. What unites them from the beginning? In what does this playing in unison of the four ways of being responsible play? What is the source of the unity of the four causes? What, after all, does this owing and being responsible mean, thought as the Greeks thought it?

Today we are too easily inclined either to understand being responsible and being indebted moralistically as a lapse, or else to construe them in terms of effecting. In either case we bar from ourselves the way to the primal meaning of that which is later called causality. So long as this way is not opened up to us we shall also fail to see what instrumentality, which is based on causality, properly is.

In order to guard against such misinterpretations of being responsible and being indebted, let us clarify the four ways of being responsible in terms of that for which they are responsible. According to our example, they are responsible for the silver chalice's lying ready before us as a sacrificial vessel. Lying before and lying ready (*hypokeisthai*) characterize the presencing of something that is present. The four ways of being responsible bring something into appearance. They let it come forth into presencing [*Anwesen*]. They set it free to that place and so start it on its way, namely, into its complete arrival. The principal characteristic of being responsible is this starting something on its way into arrival that being responsible is an occasioning or an inducing to go forward [*Ver-an-lassen*]. On the basis of a look at what the Greeks experienced in being responsible, in *aitia*, we now give this verb "to occasion" a more inclusive meaning, so that it now is the name for the essence of causality thought as the Greeks thought it. The common and narrower meaning of "occasion," in contrast, is nothing more than a colliding and releasing; it means a kind of secondary cause within the whole of causality.

But in what, then, does the playing in unison of the four ways of occasioning play? These let what is not yet present arrive into presencing. Accordingly, they are unifiedly governed by a bringing that brings what presences into appearance. Plato tells us what this bringing is in a sentence from the *Symposium* (205b): *hē gar toi ek tou mē ontos eis to on ionti hotōioun aitia pasa esti poiēsis*. "Every occasion for whatever passes beyond the nonpresent and goes forward into presencing is *poiēsis*, bringing-forth [*Her-vor-bringen*]."

It is of utmost importance that we think bringing-forth in its full scope and at the same time in the sense in which the Greeks thought it. Not only handicraft manufacture, not only artistic and poetical bringing into appearance and concrete imagery, is a bringing-forth, *poiēsis*. *Physis*, also, the arising of something from out of itself, is a bringing-forth, *poiēsis*. *Physis* is indeed *poiēsis* in the highest sense. For what presences by means of *physis* has the irruption belonging to bringing-forth, e.g., the bursting of a blossom into bloom, in itself (*en heautōi*). In contrast, what is brought forth by the artisan, or the artist, e.g., the silver chalice, has the irruption belonging to bringing-forth, not in itself, but in another (*en allōi*), in the craftsman or artist.

The modes of occasioning, the four causes, are at play, then, within bringing-forth. Through bringing-forth the growing things of nature as well as whatever is completed through the crafts and the arts come at any given time to their appearance.

But how does bringing-forth happen, be it in nature or in handicraft and art? What is the bringing-forth in which the fourfold way of occasioning plays? Occasioning has to do with the presencing [*Anwesen*] of that which at any given time comes to appearance in bringing-forth. Bringing-forth brings out of concealment into unconcealment. Bringing-forth propriates only insofar as something concealed comes into unconcealment. This coming rests and moves freely within what we call revealing [*das Entbergen*]. The Greeks have the word *alētheia* for revealing. The Romans translate this with *veritas*. We say "truth" and usually understand it as correctness of representation.

But where have we strayed to? We are questioning concerning technology, and we have arrived now at *alētheia*, at revealing. What has the essence of technology to do with revealing? The answer: everything. For every bringing-forth is grounded in revealing. Bringing-forth, indeed, gathers within itself the four modes of occasioning – causality – and rules them throughout. Within its domain belong end and means as well as instrumentality. Instrumentality is considered to be the fundamental characteristic of technology. If we inquire step by step into what technology, represented as means, actually is, then we shall arrive at revealing. The possibility of all productive manufacturing lies in revealing.

Technology is therefore no mere means. Technology is a way of revealing. If we give heed to this, then another whole realm for the essence of technology will open itself up to us. It is the realm of revealing, i.e., of truth.

This prospect strikes us as strange. Indeed, it should do so, as persistently as possible and with so much urgency that we will finally take seriously the simple question of what the name "technology" means. The word stems from the Greek. *Technikon* means that which belongs to *technē*. We must observe two things with respect to the meaning of this word. One is that *technē* is the name not only for the activities and skills of the craftsman but also for the arts of the mind and the fine arts. *Technē* belongs to bringing-forth, to *poiēsis*; it is something poetic.

The other thing that we should observe with regard to *technē* is even more important. From earliest times until Plato the word *technē* is linked with the word *epistēmē*. Both words are terms for knowing in the widest sense. They mean to be entirely at home in something, to understand and be expert in it. Such knowing provides an opening up. As an opening up it is a revealing. Aristotle, in a discussion of special importance (*Nicomachean Ethics*, Bk. VI, chaps. 3 and 4), distinguishes between *epistēmē* and *technē* and indeed with respect to what and how they reveal. *Technē* is a mode of *alētheuein*. It reveals whatever does not bring itself forth and does not yet lie here before us, whatever can look and turn out now one way and now another. Whoever builds a house or a ship or forges a sacrificial chalice reveals what is to be brought forth, according to the terms of the four modes of occasioning. This revealing gathers together in advance the aspect and the matter of ship or house, with a view to the finished thing envisaged as completed, and from this gathering determines the manner of its construction. Thus what is decisive in *technē* does not at all lie in making and manipulating, nor in the using of means, but rather in the revealing mentioned before. It is as revealing, and not as manufacturing, that *technē* is a bringing-forth.

Thus the clue to what the word *technē* means and to how the Greeks defined it leads us into the same context that opened itself to us when we pursued the question of what instrumentality as such in truth might be.

Technology is a mode of revealing. Technology comes to presence in the realm where revealing and unconcealment take place, where *alētheia*, truth, happens.

In opposition to this definition of the essential domain of technology, one can object that it indeed holds for Greek thought and that at best it might apply to the techniques of the handicraftsman, but that it simply does not fit modern machine-powered technology. And it is precisely the latter and it alone that is the disturbing thing, that moves us to ask the question concerning technology *per se*. It is said that modern technology is something incomparably different from all earlier technologies because it is based on modern physics as an exact science. Meanwhile, we have come to understand more clearly that the reverse holds true as well: modern physics, as experimental, is dependent upon technical apparatus and upon progress in the building of apparatus. The establishing of this mutual relationship between technology and physics is correct. But it remains a merely historiological establishing of facts and says nothing about that in which this mutual relationship is grounded. The decisive question still remains: Of what essence is modern technology that it thinks of putting exact science to use?

What is modern technology? It too is a revealing. Only when we allow our attention to rest on this fundamental characteristic does that which is new in modern technology show itself to us.

And yet, the revealing that holds sway throughout modern technology does not unfold into a bringing-forth in the sense of *poiēsis*. The revealing that rules in modern technology is a challenging [*Herausfordern*], which puts to nature the unreasonable demand that it supply energy which can be extracted and stored as such. But does this not hold true for the old windmill as well? No. Its sails do indeed turn in the wind; they are left entirely to the wind's blowing. But the windmill does not unlock energy from the air currents in order to store it.

In contrast, a tract of land is challenged in the hauling out of coal and ore. The earth now reveals itself as a coal mining district, the soil as a mineral deposit. The field that the peasant formerly cultivated and set in order appears differently than it did when to set in order still meant to take care of and maintain. The work of the peasant does not challenge the soil of the field. In sowing grain it places seed in the keeping of the forces of growth and watches over its increase. But meanwhile even the cultivation of the field has come under the grip of another kind of setting-in-order, which *sets upon* nature. It sets upon it in the sense of challenging it. Agriculture is now the mechanized food industry. Air is now set upon to yield nitrogen, the earth to yield ore, ore to yield uranium, for example; uranium is set upon to yield atomic energy, which can be unleashed either for destructive or for peaceful purposes.

This setting-upon that challenges the energies of nature is an expediting, and in two ways. It expedites in that it unlocks and exposes. Yet that expediting is always itself directed from the beginning toward furthering something else, i.e., toward driving on to the maximum yield at the minimum expense. The coal that has been hauled out in some mining district has not been produced in order that it may simply be at hand somewhere or other. It is being stored; that is, it is on call, ready to deliver the sun's warmth that is stored in it. The sun's warmth is challenged forth for heat, which in turn is ordered to deliver steam whose pressure turns the wheels that keep a factory running.

The hydroelectric plant is set into the current of the Rhine. It sets the Rhine to supplying its hydraulic pressure, which then sets the turbines turning. This turning sets those machines in motion whose thrust sets going the electric current for which the long-distance power station and its network of cables are set up to dispatch electricity. In the context of the interlocking processes pertaining to the orderly disposition of electrical energy, even the Rhine itself appears to be something at our command. The hydro-electric plant is not built into the Rhine River as was the old wooden bridge that joined bank with bank for hundreds of years. Rather, the river is dammed up into the power plant. What the river is now, namely, a water-power supplier, derives from the essence of the power station. In order that we may even remotely consider the monstrousness that reigns here, let us ponder for a moment the contrast that is spoken by the two titles: "The Rhine," as dammed up into the *power* works, and "The Rhine," as uttered by the *art*-work, in Hölderlin's hymn by that name. But, it will be replied, the Rhine is still a river in the landscape, is it not? Perhaps. But how? In no other way than as an object on call for inspection by a tour group ordered there by the vacation industry.

The revealing that rules throughout modern technology has the character of a setting-upon, in the sense of a challenging-forth. Such challenging happens in that the energy concealed in nature is unlocked, what is unlocked is transformed, what is transformed is stored up, what is stored up is in turn distributed, and what is distributed is switched about ever anew. Unlocking, transforming, storing, distributing, and switching about are ways of revealing. But the revealing never simply comes to an end. Neither does it run off into the indeterminate. The revealing reveals to itself its own manifoldly interlocking paths, through regulating their course. This regulating itself is, for its part, everywhere secured. Regulating and securing even become the chief characteristics of the revealing that challenges.

What kind of unconcealment is it, then, that is peculiar to that which results from this setting-upon that challenges? Everywhere everything is ordered to stand by, to be immediately on hand, indeed to stand there just so that it may be on call for a further ordering. Whatever is ordered about in this way has its own standing. We call it the standing-reserve [*Bestand*]. The word expresses here something more, and something more essential, than mere "stock." The word "standing-reserve" assumes the rank of an inclusive rubric. It designates nothing less than the way in which everything presences that is wrought upon by the revealing that challenges. Whatever stands by in the sense of standing-reserve no longer stands over against us as object.

Yet an airliner that stands on the runway is surely an object. Certainly. We can represent the machine so. But then it conceals itself as to what and how it is. Revealed, it stands on the taxi strip only as standing-reserve, inasmuch as it is ordered to insure the possibility of transportation. For this it must be in its whole structure and in every one of its constituent parts itself on call for duty, i.e., ready for takeoff. (Here it would be appropriate to discuss Hegel's definition of the machine as an autonomous tool. When applied to the tools of the craftsman, his characterization is correct. Characterized in this way, however, the machine is not thought at all from the essence of technology within which it belongs. Seen in terms of the standing-reserve, the machine is completely non-autonomous, for it has its standing only on the basis of the ordering of the orderable.)

The fact that now, wherever we try to point to modern technology as the revealing that challenges, the words "setting-upon," "ordering," "standing-reserve," obtrude and accumulate in a dry, monotonous, and therefore oppressive way – this fact has its basis in what is now coming to utterance.

Who accomplishes the challenging setting-upon through which what we call the actual is revealed as standing-reserve? Obviously, man. To what extent is man capable of such a revealing? Man can indeed conceive, fashion, and carry through this or that in one way or another. But man does not have control over unconcealment itself, in which at any given time the actual shows itself or withdraws. The fact that it has been showing itself in the light of Ideas ever since the time of Plato, Plato did not bring about. The thinker only responded to what addressed itself to him.

Only to the extent that man for his part is already challenged to exploit the energies of nature can this revealing that orders happen. If man is challenged, ordered, to do this, then does not man himself belong even more originally than nature within the standing-reserve? The current talk about human resources, about the supply of patients for a clinic, gives evidence of this. The forester who measures the felled timber in the woods and who to all appearances walks the forest path in the same way his grandfather did is today ordered by the industry that produces commercial woods, whether he knows it or not. He is made subordinate to the orderability of cellulose, which for its part is challenged forth by the need for paper, which is then delivered to newspapers and illustrated magazines. The latter, in their turn, set public opinion to swallowing what is printed, so that a set configuration of opinion becomes available on demand. Yet precisely because man is challenged more originally than are the energies of nature, i.e., into the process of ordering, he never is transformed into mere standing-reserve. Since man drives technology forward, he takes part in ordering as a way of revealing. But the unconcealment itself, within which ordering unfolds, is never a human handiwork, any more than is the realm man traverses every time he as a subject relates to an object.

Where and how does this revealing happen if it is no mere handiwork of man? We need not look far. We need only apprehend in an unbiased way that which has already claimed man so decisively that he can only be man at any given time as the one so claimed. Wherever man opens his eyes and ears, unlocks his heart, and gives himself over to meditating and striving, shaping and working, entreating and thanking, he finds himself everywhere already brought into the unconcealed. The unconcealment of the unconcealed has already propriated whenever it calls man forth into the modes of revealing allotted to him. When man, in his way, from within unconcealment reveals that which presences, he merely responds to the call of unconcealment, even when he contradicts it. Thus when man, investigating, observing, pursues nature as an area of his own conceiving, he has already been claimed by a way of revealing that challenges him to approach nature as an object of research, until even the object disappears into the objectlessness of standing-reserve.

Modern technology, as a revealing that orders, is thus no mere human doing. Therefore we must take the challenging that sets upon man to order the actual as standing-reserve in accordance with the way it shows itself. That challenging gathers man into ordering. This gathering concentrates man upon ordering the actual as standing-reserve.

That which primordially unfolds the mountains into mountain ranges and pervades them in their folded contiguity is the gathering that we call *Gebirg* [mountain chain].

That original gathering from which unfold the ways in which we have feelings of one kind or another we name *Gemüt* [disposition].

We now name the challenging claim that gathers man with a view to ordering the self-revealing as standing-reserve: *Ge-stell* [enframing].

We dare to use this word in a sense that has been thoroughly unfamiliar up to now.

According to ordinary usage, the word *Gestell* [frame] means some kind of apparatus, e.g., a bookrack. *Gestell* is also the name for a skeleton. And the employment of the word *Gestell* [enframing] that is now required of us seems equally eerie, not to speak of the arbitrariness with which words of a mature language are so misused. Can anything be more strange? Surely not. Yet this strangeness is an old custom of thought. And indeed thinkers follow this custom precisely at the point where it is a matter of thinking that which is highest. We, late born, are no longer in a position to appreciate the significance of Plato's daring to use the word *eidos* for that which in everything and in each particular thing endures as present. For *eidos*, in the common speech, meant the outward aspect [*Ansicht*] that a visible thing offers to the physical eye. Plato exacts of this word, however, something utterly extraordinary: that it name what precisely is not and never will be perceivable with physical eyes. But even this is by no means the full extent of what is extraordinary here. For *idea* names not only the nonsensuous aspect of what is physically visible. Aspect (*idea*) names and also is that which constitutes the essence in the audible, the tasteable, the tactile, in everything that is in any way accessible. Compared with the demands that Plato makes on language and thought in this and in other instances, the use of the word *Gestell* as the name for the essence of modern technology, which we are venturing, is almost harmless. Even so, the usage now required remains something exacting and is open to misinterpretation.

Enframing means the gathering together of the setting-upon that sets upon man, i.e., challenges him forth, to reveal the actual, in the mode of ordering, as standing-reserve. Enframing means the way of revealing that holds sway in the essence of modern technology and that is itself nothing technological. On the other hand, all those things that are so familiar to us and are standard parts of assembly, such as rods, pistons, and chassis, belong to the technological. The assembly itself, however, together with the aforementioned stockparts, fall within the sphere of technological activity. Such activity always merely responds to the challenge of enframing, but it never comprises enframing itself or brings it about.

The word *stellen* [to set] in the name *Ge-stell* [enframing] does not only mean challenging. At the same time it should preserve the suggestion of another *Stellen* from which it stems, namely that producing and presenting [*Her-und Dar-stellen*], which, in the sense of *poiēsis*, lets what presences come forth into unconcealment. This producing that brings forth, e.g., erecting a statue in the temple precinct, and the ordering that challenges now under consideration are indeed fundamentally different, and yet they remain related in their essence. Both are ways of revealing, of *alētheia*. In enframing, the unconcealment propriates in conformity with which the work of modern technology reveals the actual as standing-reserve. This work is therefore neither only a human activity nor a mere means within such activity. The merely instrumental, merely anthropological definition of technology is therefore in principle untenable. And it may not be rounded out by being referred back to some metaphysical or religious explanation that undergirds it.

It remains true nonetheless that man in the technological age is, in a particularly striking way, challenged forth into revealing. Such revealing concerns nature, above all, as the chief storehouse of the standing energy reserve. Accordingly, man's ordering attitude and behavior display themselves first in the rise of modern physics as an exact science. Modern science's way of representing pursues and entraps nature as a calculable coherence of forces. Modern physics is not experimental physics because it applies apparatus to the questioning of nature. The reverse is true. Because physics, indeed already as pure theory, sets nature up to exhibit itself as a coherence of forces calculable in advance, it orders its experiments precisely for the purpose of asking whether and how nature reports itself when set up in this way.

But, after all, mathematical science arose almost two centuries before technology. How, then, could it have already been set upon by modern technology and placed in its service? The facts testify to the contrary. Surely technology got under way only when it could be supported by exact physical science. Reckoned chronologically, this is correct. Thought historically, it does not hit upon the truth.

The modern physical theory of nature prepares the way not simply for technology but for the essence of modern technology. For such gathering-together, which challenges man to reveal by way of ordering, already holds sway in physics. But in it that gathering does not yet come expressly to the fore. Modern physics is the herald of enframing, a herald whose provenance is still unknown. The essence of modern technology has for a long time been concealed, even where power machinery has been invented, where electrical technology is in full swing, and where atomic technology is well under way.

All coming to presence, not only modern technology, keeps itself everywhere concealed to the last. Nevertheless, it remains, with respect to its holding sway, that which precedes all: the earliest. The Greek thinkers already knew of this when they said: That which is earlier with regard to its rise into dominance becomes manifest to us men only later. That which is primally early shows itself only ultimately to men. Therefore, in the realm of thinking, a painstaking effort to think through still more primally what was primally thought is not the absurd wish to revive what is past, but rather the sober readiness to be astounded before the coming of the dawn.

Chronologically speaking, modern physical science begins in the seventeenth century. In contrast, machine-power technology develops only in the second half of the eighteenth century. But modern technology, which for chronological reckoning is the later, is, from the point of view of the essence holding sway within it, historically earlier.

If modern physics must resign itself ever increasingly to the fact that its realm of representation remains inscrutable and incapable of being visualized, this resignation is not dictated by any committee of researchers. It is challenged forth by the rule of enframing, which demands that nature be orderable as standing-reserve. Hence physics, in its retreat from the kind of representation that turns only to objects, which has been the sole standard until recently, will never be able to renounce this one thing: that nature report itself in some way or other that is identifiable through calculation and that it remain orderable as a system of information. This system is then determined by a causality that has changed once again. Causality now displays neither the character of the occasioning that brings forth nor the nature of the *causa efficiens*, let alone that of the *causa formalis*. It seems as though causality is shrinking into a reporting – a reporting challenged forth – of standing-reserves that must be guaranteed either simultaneously or in sequence. To this shrinking would correspond the process of growing resignation that Heisenberg's lecture depicts in so impressive a manner.[1]

Because the essence of modern technology lies in enframing, modern technology must employ exact physical science. Through its so doing the deceptive appearance arises that modern technology is applied physical science. This illusion can maintain itself precisely insofar as neither the essential provenance of modern science nor indeed the essence of modern technology is adequately sought in our questioning.

We are questioning concerning technology in order to bring to light our relationship to its essence. The essence of modern technology shows itself in what we call enframing. But simply to point to this is still in no way to answer the question concerning technology, if to answer means to respond, in the sense of correspond, to the essence of what is being asked about.

Where do we find ourselves if now we think one step further regarding what enframing itself actually is? It is nothing technological, nothing on the order of a machine. It is the way in which the actual reveals itself as standing-reserve. Again we ask: Does such revealing happen somewhere beyond all human doing? No. But neither does it happen exclusively *in* man, or definitively *through* man.

Enframing is the gathering together which belongs to that setting-upon which challenges man and puts him in position to reveal the actual, in the mode of ordering, as standing-reserve. As the one who is challenged forth in this way, man stands within the essential realm of enframing. He can never take up a relationship to it only subsequently. Thus the question as to how we are to arrive at a relationship to the essence of technology, asked in this way, always comes too late. But never too late comes the question as to whether we actually experience ourselves as the ones whose activities everywhere, public and private, are challenged forth by enframing. Above all, never too late comes the question as to whether and how we actually admit ourselves into that wherein enframing itself essentially unfolds.

The essence of modern technology starts man upon the way of that revealing through which the actual everywhere, more or less distinctly, becomes standing-reserve. "To start upon a way" means "to send" in our ordinary language. We shall call the sending that gathers [*versammelnde Schicken*], that first starts man upon a way of revealing, *destining* [*Geschick*]. It is from this destining that the essence of all history [*Geschichte*] is determined. History is neither simply the object of written chronicle nor merely the process of human activity. That activity first becomes history as something destined.[2] And it is only the destining into objectifying representation that makes the historical accessible as an object for historiography, i.e., for a science, and on this basis makes possible the current equating of the historical with that which is chronicled.

Enframing, as a challenging-forth into ordering, sends into a way of revealing. Enframing is an ordaining of destining, as is every way of revealing. Bringing-forth, *poiēsis*, is also a destining in this sense.

Always the unconcealment of that which is goes upon a way of revealing. Always the destining of revealing holds complete sway over men. But that destining is never a fate that compels. For man becomes truly free only insofar as he belongs to the realm of destining and so becomes one who listens, though not one who simply obeys.

The essence of freedom is *originally* not connected with the will or even with the causality of human willing.

Freedom governs the free space in the sense of the cleared, that is to say, the revealed. To the occurrence of revealing, i.e., of truth, freedom stands in the closest and most intimate kinship. All revealing belongs within a harboring and a concealing. But that which frees – the mystery – is concealed and always concealing itself. All revealing comes out of the free, goes into the free, and brings into the free. The freedom of the free consists neither in unfettered arbitrariness nor in the constraint of mere laws. Freedom is that which conceals in a way that opens to light, in whose clearing shimmers the veil that hides the essential occurrence of all truth and lets the veil appear as what veils. Freedom is the realm of the destining that at any given time starts a revealing on its way.

The essence of modern technology lies in enframing. Enframing belongs within the destining of revealing. These sentences express something different from the talk that we hear more frequently, to the effect that technology is the fate of our age, where "fate" means the inevitableness of an unalterable course.

But when we consider the essence of technology we experience enframing as a destining of revealing. In this way we are already sojourning within the free space of destining, a destining that in no way confines us to a stultified compulsion to push on blindly with technology or, what comes to the same, to rebel helplessly against it and curse it as the work of the devil. Quite to the contrary, when we once open ourselves expressly to the *essence* of technology we find ourselves unexpectedly taken into a freeing claim.

The essence of technology lies in enframing. Its holding sway belongs within destining. Since destining at any given time starts man on a way of revealing, man, thus under way, is continually approaching the brink of the possibility of pursuing and promulgating nothing but what is revealed in ordering, and of deriving all his standards on this basis. Through this the other possibility is blocked – that man might rather be admitted sooner and ever more primally to the essence of what is unconcealed and to its unconcealment, in order that he might experience as his essence the requisite belonging to revealing.

Placed between these possibilities, man is endangered by destining. The destining of revealing is as such, in every one of its modes, and therefore necessarily, *danger*.

In whatever way the destining of revealing may hold sway, the unconcealment in which everything that is shows itself at any given time harbors the danger that man may misconstrue the unconcealed and misinterpret it. Thus where everything that presences exhibits itself in the light of a cause-effect coherence, even God, for representational thinking, can lose all that is exalted and holy, the mysteriousness of his distance. In the light of causality, God can sink to the level of a cause, of *causa efficiens*. He then becomes even in theology the God of the philosophers, namely, of those who define the unconcealed and the concealed in terms of the causality of making, without ever considering the essential provenance of this causality.

In a similar way the unconcealment in accordance with which nature presents itself as a calculable complex of the effects of forces can indeed permit correct determinations; but precisely through these successes the danger may remain that in the midst of all that is correct the true will withdraw.

The destining of revealing is in itself not just any danger, but *the* danger.

Yet when destining reigns in the mode of enframing, it is the supreme danger. This danger attests itself to us in two ways. As soon as what is unconcealed no longer concerns man even as object, but exclusively as standing-reserve, and man in the midst of objectlessness is nothing but the orderer of the standing-reserve, then he comes to the very brink of a precipitous fall; that is, he comes to the point where he himself will have to be taken as standing-reserve. Meanwhile, man, precisely as the one so threatened, exalts himself and postures as lord of the earth. In this way the illusion comes to prevail that everything man encounters exists only insofar as it is his construct. This illusion gives rise in turn to one final delusion: it seems as though man everywhere and always encounters only himself. Heisenberg has with complete correctness pointed out that the actual must present itself to contemporary man in this way.[3] *In truth, however, precisely nowhere does man today any longer encounter himself, i.e., his essence.* Man stands so decisively in subservience to the challenging-forth of enframing that he does not grasp enframing as a claim, that he fails to see himself as the one spoken to, and hence also fails in every way to hear in what respect he ek-sists, in terms of his essence, in a realm where he is addressed, so that he *can never* encounter only himself.

But enframing does not simply endanger man in his relationship to himself and to everything that is. As a destining, it banishes man into the kind of revealing that is an ordering. Where this ordering holds sway, it drives out every other possibility of revealing. Above all, enframing conceals that revealing which, in the sense of *poiēsis*, lets what presences come forth into appearance. As compared with that other revealing, the setting-upon that challenges forth thrusts man into a relation to whatever is that is at once antithetical and rigorously ordered. Where enframing holds sway, regulating and securing of the standing-reserve mark all revealing. They no longer even let their own fundamental characteristic appear, namely, this revealing as such.

Thus the challenging-enframing not only conceals a former way of revealing (bringing-forth) but also conceals revealing itself and with it that wherein unconcealment, i.e., truth, propriates.

Enframing blocks the shining-forth and holding sway of truth. The destining that sends into ordering is consequently the extreme danger. What is dangerous is not technology. Technology is not demonic; but its essence is mysterious. The essence of technology, as a destining of revealing, is the danger. The transformed meaning of the word "enframing" will perhaps become somewhat more familiar to us now if we think enframing in the sense of destining and danger.

The threat to man does not come in the first instance from the potentially lethal machines and apparatus of technology. The actual threat has already afflicted man in his essence. The rule of enframing threatens man with the possibility that it could be denied to him to enter into a more original revealing and hence to experience the call of a more primal truth.

Thus where enframing reigns, there is *danger* in the highest sense.

But where danger is, grows
The saving power also.

Let us think carefully about these words of Hölderlin.[4] What does it mean to "save"? Usually we think that it means only to seize hold of a thing threatened by ruin in order to secure it in its former continuance. But the verb "to save" says more. "To save" is to fetch something home into its essence, in order to bring the essence for the first time into its proper appearing. If the essence of technology, enframing, is the extreme danger, if there is truth in Hölderlin's words, then the rule of enframing cannot exhaust itself solely in blocking all lighting-up of every revealing, all appearing of truth. Rather, precisely the essence of technology must harbor in itself the growth of the saving power. But in that case, might not an adequate look into what enframing is, as a destining of revealing, bring the upsurgence of the saving power into appearance?

In what respect does the saving power grow also there where the danger is? Where something grows, there it takes root, from thence it thrives. Both happen concealedly and quietly and in their own time. But according to the words of the poet we have no right whatsoever to expect that there where the danger is we should be able to lay hold of the saving power immediately and without preparation. Therefore we must consider now, in advance, in what respect the saving power does most profoundly take root and thence thrive even where the extreme danger lies – in the holding sway of enframing. In order to consider this it is necessary, as a last step upon our way, to look with yet clearer eyes into the danger. Accordingly, we must once more question concerning technology.

For we have said that in technology's essence roots and thrives the saving power.

But how shall we behold the saving power in the essence of technology so long as we do not consider in what sense of "essence" it is that enframing properly is the essence of technology?

Thus far we have understood "essence" in its current meaning. In the academic language of philosophy "essence" means *what* something is; in Latin, *quid*. *Quidditas*, whatness, provides the answer to the question concerning essence. For example, what pertains to all kinds of trees – oaks, beeches, birches, firs – is the same "treeness." Under this inclusive genus – the "universal" – fall all actual and possible trees. Is then the essence of technology, enframing, the common genus for everything technological? If this were the case then the steam turbine, the radio transmitter, and the cyclotron would each be an enframing. But the word "enframing" does not mean here a tool or any kind of apparatus. Still less does it mean the general concept of such resources. The machines and apparatus are no more cases and kinds of enframing than are the man at the switchboard and the engineer in the drafting room. Each of these in its own way indeed belongs as stockpart, available resource, or executor, within enframing; but enframing is never the essence of technology in the sense of a genus. Enframing is a way of revealing that is a destining, namely, the way that challenges forth. The revealing that brings forth (*poiēsis*) is also a way that has the character of destining. But these ways are not kinds that, arrayed beside one another, fall under the concept of revealing. Revealing is that destining which, ever suddenly and inexplicably to all thinking, apportions itself into the revealing that brings forth and the revealing that challenges, and which allots itself to man. The revealing that challenges has its origin as a destining in bringing-forth. But at the same time enframing, in a way characteristic of a destining, blocks *poiēsis*.

Thus enframing, as a destining of revealing, is indeed the essence of technology, but never in the sense of genus and *essentia*. If we pay heed to this, something astounding strikes us: it is technology itself that makes the demand on us to think in another way what is usually understood by "essence." But in what way?

If we speak of the "essence of a house" and the "essence of a state" we do not mean a generic type; rather we mean the ways in which house and state hold sway, administer themselves, develop and decay – the way they "essentially unfold" [*wesen*]. Johann Peter Hebel in a poem, "Ghost on Kanderer Street," for which Goethe had a special fondness, uses the old word *die Weserei*. It means the city hall, inasmuch as there the life of the community gathers and village existence is constantly in play, i.e., essentially unfolds. It is from the verb *wesen* that the noun is derived. *Wesen* understood as a verb is the same as *währen* [to last or endure], not only in terms of meaning, but also in terms of the phonetic formation of the word. Socrates and Plato already think the essence of something as what it is that unfolds essentially, in the sense of what endures. But they think what endures is what remains permanently (*dei on*). And they find what endures permanently in what persists throughout all that happens, in what remains. That which remains they discover, in turn, in the aspect (*eidos, idea*), for example, the Idea "house."

The Idea "house" displays what anything is that is fashioned as a house. Particular, real, and possible houses, in contrast, are changing and transitory derivatives of the Idea and thus belong to what does not endure.

But it can never in any way be established that enduring is based solely on what Plato thinks as *idea* and Aristotle thinks as *to ti ēn einai* (that which any particular thing has always been), or what metaphysics in its most varied interpretations thinks as *essentia*.

All unfolding endures. But is enduring only permanent enduring? Does the essence of technology endure in the sense of the permanent enduring of an Idea that hovers over everything technological, thus making it seem that by technology we mean some mythological abstraction? The way in which technology unfolds lets itself be seen only on the basis of that permanent enduring in which enframing propriates as a destining of revealing. Goethe once uses the mysterious word *fortgewähren* [to grant continuously] in place of *fortwähren* [to endure continuously].[5] He hears *währen* [to endure] and *gewähren* [to grant] here in one unarticulated accord. And if we now ponder more carefully than we did before what it is that properly endures and perhaps alone endures, we may venture to say: *Only what is granted endures. What endures primally out of the earliest beginning is what grants.*

As the essencing of technology, enframing is what endures. Does enframing hold sway at all in the sense of granting? No doubt the question seems a horrendous blunder. For according to everything that has been said, enframing is rather a destining that gathers together into the revealing that challenges forth. Challenging is anything but a granting. So it seems, so long as we do not notice that the challenging-forth into the ordering of the actual as standing-reserve remains a destining that starts man upon a way of revealing. As this destining, the essential unfolding of technology gives man entry into something which, of himself, he can neither invent nor in any way make. For there is no such thing as a man who exists singly and solely on his own.

But if this destining, enframing, is the extreme danger, not only for man's essential unfolding, but for all revealing as such, should this destining still be called a granting? Yes, most emphatically, if in this destining the saving power is said to grow. Every destining of revealing propriates from a granting and as such a granting. For it is granting that first conveys to man that share in revealing that the propriative event of revealing needs. So needed and used, man is given to belong to the propriative event of truth. The granting that sends one way or another into revealing is as such the saving power. For the saving power lets man see and enter into the highest dignity of his essence. This dignity lies in keeping watch over the unconcealment – and with it, from the first, the concealment – of all essential unfolding on this earth. It is precisely in enframing, which threatens to sweep man away into ordering as the ostensibly sole way of revealing, and so thrusts man into the danger of the surrender of his free essence – it is precisely in this extreme danger that the innermost indestructible belongingness of man within grating may come to light, provided that we, for our part, begin to pay heed to the essence of technology.

Thus the essential unfolding of technology harbors in itself what we least suspect, the possible rise of the saving power.

Everything, then, depends upon this: that we ponder this rising and that, recollecting, we watch over it. How can this happen? Above all through our catching sight of the essential unfolding in technology, instead of merely gaping at the technological. So long as we represent technology as an instrument, we remain transfixed in the will to master it. We press on past the essence of technology.

When, however, we ask how the instrumental unfolds essentially as a kind of causality, then we experience this essential unfolding as the destining of a revealing.

When we consider, finally, that the essential unfolding of the essence of technology propriates in the granting that needs and uses man so that he may share in revealing, then the following becomes clear:

The essence of technology is in a lofty sense ambiguous. Such ambiguity points to the mystery of all revealing, i.e., of truth.

On the one hand, enframing challenges forth into the frenziedness of ordering that blocks every view into the propriative event of revealing and so radically endangers the relation to the essence of truth.

On the other hand, enframing propriates for its part in the granting that lets man endure – as yet inexperienced, but perhaps more experienced in the future – that he may be the one who is needed and used for the safekeeping of the essence of truth. Thus the rising of the saving power appears.

The irresistibility of ordering and the restraint of the saving power draw past each other like the paths of two stars in the course of the heavens. But precisely this, their passing by, is the hidden side of their nearness.

When we look into the ambiguous essence of technology, we behold the constellation, the stellar course of the mystery.

The question concerning technology is the question concerning the constellation in which revealing and concealing, in which the essential unfolding of truth propriates.

But what help is it to us to look into the constellation of truth? We look into the danger and see the growth of the saving power.

Through this we are not yet saved. But we are thereupon summoned to hope in the growing light of the saving power. How can this happen? Here and now and in little things, that we may foster the saving power in its increase. This includes holding always before our eyes the extreme danger.

The essential unfolding of technology threatens revealing, threatens it with the possibility that all revealing will be consumed in ordering and that everything will present itself only in the unconcealment of standing-reserve. Human activity can never directly counter this danger. Human achievement alone can never banish it. But human reflection can ponder the fact that all saving power must be of a higher essence than what is endangered, though at the same time kindred to it.

But might there not perhaps be a more primally granted revealing that could bring the saving power into its first shining-forth in the midst of the danger that in the technological age rather conceals than shows itself?

There was a time when it was not technology alone that bore the name *technē*. Once the revealing that brings forth truth into the splendor of radiant appearance was also called *technē*.

There was a time when the bringing-forth of the true into the beautiful was called *technē*. The *poiēsis* of the fine arts was also called *technē*.

At the outset of the destining of the West, in Greece, the arts soared to the supreme height of the revealing granted them. They illuminated the presence [*Gegenwart*] of the gods and the dialogue of divine and human destinings. And art was called simply *technē*. It was a single, manifold revealing. It was pious, *promos*, i.e., yielding to the holding sway and the safekeeping of truth.

The arts were not derived from the artistic. Artworks were not enjoyed aesthetically. Art was not a sector of cultural activity.

What was art – perhaps only for that brief but magnificent age? Why did art bear the modest name *technē*? Because it was a revealing that brought forth and made present, and therefore belonged within *poiēsis*. It was finally that revealing which holds complete sway in all the fine arts, in poetry, and in everything poetical that obtained *poiēsis* as its proper name.

The same poet from whom we heard the words

> But where danger is, grows
> The saving power also...

says to us:

> ...poetically man dwells on this earth.

The poetical brings the true into the splendor of what Plato in the *Phaedrus* calls *to ekphanestaton*, that which shines forth most purely. The poetical thoroughly pervades every art, every revealing of essential unfolding into the beautiful.

Could it be that the fine arts are called to poetic revealing? Could it be that revealing lays claim to the arts most primally, so that they for their part may expressly foster the growth of the saving power, may awaken and found anew our vision of, and trust in, that which grants?

Whether art may be granted this highest possibility of its essence in the midst of the extreme danger, no one can tell. Yet we can be astounded. Before what? Before this other possibility: that the frenziedness of technology may entrench itself everywhere to such an extent that someday, throughout everything technological, the essence of technology may unfold essentially in the propriative event of truth.

Because the essence of technology is nothing technological, essential reflection upon technology and decisive confrontation with it must happen in a realm that is, on the one hand, akin to the essence of technology and, on the other, fundamentally different from it.

Such a realm is art. But certainly only if reflection upon art, for its part, does not shut its eyes to the constellation of truth, concerning which we are *questioning*.

Thus questioning, we bear witness to the crisis that in our sheer preoccupation with technology we do not yet experience the essential unfolding of technology, that in our sheer aesthetic-mindedness we no longer guard and preserve the essential unfolding of art. Yet the more questioningly we ponder the essence of technology, the more mysterious the essence of art becomes.

The closer we come to the danger, the more brightly do the ways into the saving power begin to shine and the more questioning we become. For questioning is the piety of thought.

Notes

1 W. Heisenberg, "Das Naturbild in der heutigen Physik," in *Die Künste im technischen Zeitalter* (Munich, 1954), pp. 43ff. [See also W. Heisenberg, *Physics and Philosophy: The Revolution in Modern Science* (New York: Harper & Row, 1958). – Ed.]

2 See "On the Essence of Truth" (1930), first edition 1943, pp. 16ff.

3 "Das Naturbild," pp. 60ff.

4 From "Patmos." Cf. *Friedrich Hölderlin Poems and Fragments*, trans. Michael Hamburger (Ann Arbor: The University of Michigan Press, 1966), pp. 462–63. – Ed.

5 "Die Wahlverwandtschaften," pt. 2, chap. 10, in the novel *Die wunderlichen Nachbarskinder*.

24

On Philosophy's "Ending" in Technoscience: Heidegger vs. Comte

Robert C. Scharff

I Introduction

The comparison of Comte and Heidegger developed in this essay begins by considering two remarkably similar features of their thinking, in order subsequently to highlight the one crucial issue over which they are in complete disagreement. On the one hand, first, both Comte and Heidegger see the technologized science (i.e., technoscience) that dominates the current age as nothing short of the culmination of the Western intellectual tradition; moreover, second, both of them hold that one can only understand this culminating event by becoming fully aware of the historically determinate character (i.e., historicity) of all thinking. On the other hand, while neither of them see the "dominance" of technoscience as entailing complete suppression of other (nonscientific) possibilities, for Comte all such other possibilities are regressive, whereas for Heidegger at least some of them would constitute our "saving grace." I am interested in this comparison because it involves what Heidegger calls, in pre-*Being and Time* language, the question of what it is to "*be* historical" and how this primal fact about us matters to philosophical (or more technically, post-philosophical) thinking.

What the later Heidegger understands by the task of "thinking" at the technoscientific "end of philosophy" is, of course, the subject of widespread disagreement.[1] For some of his critics, Heidegger's conception of post-philosophical *Denken* makes him anti-technological; for others, it holds open the positive possibility of a transformed and liberating relation to technology.[2] For all their differences, however, a common thread runs through these interpretations. They are formulated by persons who see themselves *first* as critics of current philosophical practice and advocates with an agenda of what is to be done about it and only *then* as readers of Heidegger. There is certainly no crime in this; and it is easy to share some of their sentiments. Yet it may also be useful to recognize what this approach obscures. In their eagerness to speak as critics and advocates, they all tend to place at the periphery the topic Heidegger most wants to address. To put the point quickly, when Heidegger reflects on "the task of thinking" at the technoscientific end of philosophy, he is not primarily concerned either with the end itself, or with criticisms of it, or with what might emerge once we get past it. His main topic, in other words, is neither metaphysics and technoscience on the one hand nor the supposed features of a post-philosophical era on the other. For him, the primary issue is what kind of thinking might be possible, as he says, "*at* the end." It is this thinking – or for the moment, more precisely and modestly, just its "point of departure" and "task" – toward which he directs his attention.[3]

In this essay, then, I propose to directly consider Heidegger's own attempt to think "at" philosophy's end, instead of reading him through the agendas of his famous interpreters. And it is in terms of this purpose that, in spite of appearances, Auguste Comte, the notorious nineteenth-century positivist who openly and happily concludes that our age of technologized science will never "end," becomes relevant. I shall argue that if we come to understand why Comte was unable to even conceive of the modern era's ending or being surpassed, this can actually shed light on Heidegger's claim that thinking "at" such an end is today what is most needed.

My discussion has three parts. First, I review briefly Comte's famous three-stage law, rejecting the usual practice of construing it as primarily either an empirical-naturalistic or a speculative-historical theory. Rather, I show that for Comte, the law serves above all as the basis of his reflective defense of the claim that the Western philosophical tradition is culminating in our scientific era – a defense that is laid out in such a way that it makes the very idea of a further stage literally unthinkable. Heidegger, too, argues that the Western tradition marks out a path by which the cosmology of the ancients fulfills itself in modern science. Hence in my second part, I show that the affinities between Comte and Heidegger are much greater than one might have expected. At the same time, however, while it is true that Comte and Heidegger do both tend to see the rise of modern science as a consummatory event *for the Western tradition*, their ideas of the significance of this event *for the future* are radically different. Unlike Comte, Heidegger finds this event deeply, experientially, and ontologically dissatisfying. Hence, whereas Comte is content to define philosophy's future scientistically, Heidegger realizes that he must try to work out an extra-philosophical way to "think" this dissatisfaction.[4]

In part III, I pursue further this contrast between the satisfied, optimistic Comte and the dissatisfied, historically burdened Heidegger. My point will be to emphasize what I believe is Heidegger's primary but often neglected motivation. In 1962, Heidegger characterized "Das Ende der Philosophie und die Aufgabe des Denkens" (EP) as an effort to "persist in questioning" through an "immanent critique" of *Sein und Zeit* (SZ). I take this to mean that he is above all moved, from the start to finish of his path of thinking, by a concern to follow out an experiential/ontological dissatisfaction with his "culminating" tradition that originates in the decade before SZ. My comparison with Comte is intended to put the focus squarely on this motivating dissatisfaction, so that Heidegger's trajectory is less likely to be construed as a function either of the agendas of his most famous readers, or of the various techniques and approaches – whether, e.g., existential, phenomenological, Kantian, "destructive" or otherwise – that Heidegger attempts to enlist for his purposes at various stages of his thinking.

II Science and Comte's Three-Stage Law

According to Comte's famous three-stage law, "human intelligence in all of its inquiries... successively makes use of three essentially different methods of philosophizing," so that "each branch of our knowledge... necessarily passes through three different theoretical stages: the theological or fictive, the metaphysical or abstract, and the scientific or positive."[5] Today, Comte's law is usually dismissed either as being empirically false or as fostering a presumptuous metanarrative about World History. Yet both of these criticisms hide more than they reveal. Comte's conceptions of "science," and of scientific thinking as an intellectual "stage," are far more sophisticated than logical positivism's; and his law is in the first instance neither an empirical thesis nor part of a speculative philosophy of history. Unlike later positivists, Comte does not reduce theology and metaphysics to failed efforts to be science. In fact, for him, theology, metaphysics, and science are not primarily "knowledge systems" at all, but rather *global "approaches" to our surroundings* – what Comte calls "ways of philosophizing." So, for him, thinking *occurs* in three ways: theologically, metaphysically, or scientifically. Each kind of thinking arises from a distinctive sense of experiential encounter that is articulated in a distinctive sort of "speculation" (i.e., theorizing). That all thinking follows a "method" simply means that whatever its specific rules/criteria, it manifests one of the three global outlooks.

For Comte, famously, **theology** constitutes a necessary intellectual "childhood." Both historically and as individuals, we initially lack both reliable theories about our surroundings, which must be based on previous observation, and fruitful observation, which needs guidance from established theory. We would thus have remained forever mired in a "vicious circle," had it not been for the "spontaneous conceptualizations" of primitive peoples. In an important way, then, their early fetishist or animistic conceptualizations deserve an epistemic respect not due the more sophisticated polytheisms and monotheisms that follow. For fetishism is a direct, experience-based, creative, practical-minded response to surprising and disturbing encounters with our otherwise routine and predictable surroundings. It aims to restore sense to a temporarily disrupted existence – so that one can get on with life. Of course, says Comte, such early feeling- and imagination-driven accounts "overstimulate" the mind by promising answers to life's unanswerable ultimate mysteries. Yet such feeling-based, imaginative speculation does teach us how to theorize.[6] So, the theological era, especially the earliest part of it, is philosophically important – and not just for being a time of the first feeble stirrings of Reason, as the later positivists would have it. Most crucially, it is a period when all the human faculties achieve ("without prior deliberation") their initial expression. It is the period when it becomes clear that we have these faculties, and that each of them has a role to play (albeit in retrospect, not yet the right one) in knowledge. Moreover, by giving us guidance for ritual and social interaction, theological speculation grounds our original form of universal praxis. As we might say now, Comte thus construes prayer and ritual as the first technology – i.e., as a first global effort to accommodate and restore the disrupted natural relations that initially stimulated speculation.

Especially given this praise for primitive forms of theology, it is important to note that, for Comte, it is not practical but theoretical dissatisfaction that drives the mind past the first stage and toward metaphysics. Readers of Heidegger should find it suggestive that Comte sees this development both as intellectually unavoidable and as something of a betrayal of an admirably concrete and intimate sort of originary relatedness to our natural and social surroundings. For the unsystematic and disparate schemes of fetishism and polytheism give way to monotheism, not out of any hope for greater human flourishing, but because reason demands a more systematic account of the universal cosmic necessity that it is thought must somehow underlie the multiple activities of the god(s). In this very effort, finally monotheism demonstrates how cognitively unsatisfying all theology must remain. For if there is such a cosmic necessity, then philosophy's fundamental subject is the laws of this necessity and not the fact that the god(s) happen to make use of them.[7] With the emergence of this realization, thinking turns metaphysical.

Not surprisingly, Comte's concept of **metaphysics** foreshadows later positivism's. As expressions of intellectual "adolescence," metaphysical systems are both an advance over theology and yet ultimately a roadblock to genuine natural knowledge. Taken as a whole, the metaphysical era is a basically unstable but necessary time of transition. It has earlier and later forms, but no substages. Though nature and nature's laws are always its central concern, earlier metaphysical systems still tend to see natural phenomena as moved by God's agency; later systems make the forces of "nature itself" the powers behind everything. To this extent, Comte's depiction of metaphysics seems positivist in the familiar, later sense. His way of estimating its philosophical worth, however, is suggestively different. In the first place, he is as outspoken about the value of this intellectual stage for human progress as he is about its limitations. For him, it is just as important to praise the metaphysical era as the period of reason's liberation from feelings and religious authority as it is to condemn it as a time of epistemic fixation on doctrines of inner essences, hidden causes, and a priori principles. Moreover, if like the later positivists he objects to the unscientific character of metaphysical speculation, this is not what Comte sees as its most serious fault. Far more disturbing to him is its tendency to become enamored of reason's sheer logical power. The problem always runs the same course. Starting from supposedly self-evident conviction(s), metaphysical minds create competing dogmatic systems. Since each system is logically consistent, and Reason is the final court of appeal, all disputes are endless and unresolvable. Yet this cognitive result is not the worst of it. Given that their first commitment is always to the allegedly "self-evident," metaphysical thinkers are never more than secondarily focused on practical concerns. All of them wind up being so inattentive to what is actually observable that it is pride and not truth which is primarily at stake in choosing among their systems.[8] Ultimately, then, "pure" liberated reason provides us with neither genuine science nor a better life, and the presumption that our mere possession of reason somehow elevates us above the rest of nature is hollow. For Comte, the main problem with metaphysics is not that it succumbs to abstract thinking. Its real problem is that in glorifying the "life of reason," it promotes infatuation with analysis, argument, unified systems, and formal rules – with the inevitable practical result, in the end, that force replaces intellect whenever changing hearts and minds is the goal.

Eventually, then, the mind faces the prospect of "maturity." Metaphysical thinking, reluctantly realizing that it will always remain abstract and merely logical, transforms itself into **scientific**, or positive, reasoning when it becomes clear that although reason should not be a slave to feelings or to alleged revelations, it is unfit to be its own authority. Three modern figures are historically crucial (and developmentally representative) in this transition. *Bacon* pioneers the move toward "observation," away from old intellectual "idols." *Descartes* turns philosophy toward epistemology, in recognition that if reason is to properly serve observation, its primary concern must lie with method. And regarding the proper fusion of intellect with experience, *Galileo* exemplifies how nature's laws are to be "discovered" and Bacon again illuminates the idea that applied knowledge is practical "power." Field by field, scientific naturalism (which explains mechanistically *how* things work) replaces metaphysical naturalism (which can only conjecture teleologically *why* things work). Studies of observable phenomena replace the old search for answers to life's ultimate mysteries; and these studies organize themselves hierarchically – from mathematics, the simplest and most abstract, to sociology, the most complex and, *philosophically* speaking, most important. Here again, Comte gives life, not epistemology, the last word. For sociology gains its special significance not, as he puts it, "objectively," from criteria of reason, but "subjectively," from the needs of feeling, social sensibility, and altruistic love. It is understanding social behavior that will lead ultimately to the establishment of truly peaceful, prosperous societies.[9] Positive/scientific knowledge is thus the basis of the third and final form of universal praxis – namely, a truly global technology that effectuates a control over material nature and a means of "social reorganization" that will bring into being the harmonious condition humanity has desired since its earliest experiences of cosmic disruption.

The point of this quick summary of Comte's conception of the transition from theological to scientific thinking is to make clear that the usual critiques of his three-stage law as either failing empirical warrant or as merely expressing historicist ideology miss its basic function as the *measure of Comte's own outlook and philosophical self-image*. Comte displays in this regard a historical-critical reflectiveness no later positivist would tolerate. From Mill and Mach to Reichenbach and Ayer, Comte's way of mixing historical and "subjective" considerations into his philosophical self-description would seem to undermine in principle positivism's supposedly "objective" orientation. To Comte, however, this criticism is confused, even self-deceiving. In his view no philosophy, not even positivism, succeeds in placing itself beyond its inheritance. The ahistorical attitude of the later positivists would strike him as ignorance, not enlightenment.[10] In fact, since science develops *from* theology and metaphysics, it is impossible even to recognize its "superiority" and "maturity" without understanding its indebtedness to and transformation of the thinking of the two earlier stages. Comte's reflective application of the three-stage law *to his own orientation* makes his positivism deeply different in spirit from that of its later proponents. But it also, I think, brings him some way toward Heidegger – above all in his identification of the positive stage as the "consummation" of Western intellectual development.

III Technoscience as the "Consummation" of Philosophy

Stated in Comte's own terms, his primary employment of the three-stage law takes the form of a historico-critical account of why one must now be a positivist. Restated in Heideggerian terms, Comte is committed to facing the reflective question of securing the proper access (*Zugang*) to philosophical inquiry, where this inquiry is understood as originating at the "ending" of the Western tradition. Or is he? Let us give conventional accounts their due. Comte does picture the Western tradition in three "stages," not multiple "epochs." He does say that cognitive activity precedes its practical applications; and he does "explain" historical development in terms of teleological covering law of progress. Even if his positivism also possesses surprisingly unpositivist features, how much less like Heideggerian *Denken* could Comte's positive philosophizing be?

Yet we have all learned from Heidegger that questions about other thinkers posed in this way belong to the discourse of the traditional metaphysics of presence. To ask about the "propositions" Comte asserts and "positions" he takes is to have nothing of "one's own" at stake in the asking – and so nothing to acquire or "appropriate" (*aneignet*) with one's answers. So, I follow

Heidegger's example instead and ask: Toward what is Comte's work "on the way"? In other words, I submit Comte's *work* to what Heidegger called a "destructive" retrieval.[11] A "critical dismantling" of Comte's explicit pronouncements about his three-stage law can give us an originary "foreconception" of that hitherto unsatisfactorily articulated concern that is both at the heart of Comte's whole project and also still a vital issue today. This concern is: How is it today to *be* historical – as a philosopher and in one's own thinking?

With such a destructive retrieval in mind, we see easily how much is missed if Comte's law is construed as an objective theory of what once was and now is. True enough, Comte himself sometimes treats the law this way; but to see only this is to be ignorant of his own clear intentions. As noted above, Comte does not share the typical positivist vision of theology and metaphysics as mere expressions of superstition and nonsense. Nor does he see us as living "in" the scientific age, with theology and metaphysics effectively left behind us. In just these unpositivist ideas, we find his reasons for historico-critical reflection "on" science. He seeks two results. First, he intends to cultivate *in his own thinking* an awareness of positive philosophy's kinship with and debts to (not just superiority over) tradition. Second, he wants to monitor his own living-through and thinking-with an emerging transformation of prescientific inheritances that seems so far only partially visible in the successes of the natural and less "complex" sciences. In later positivism, sympathy for science *displaces* all concern for prescientific tradition. Why, we should ask, is Comte, in contrast, so insistently careful to *think in terms of* what he sees as our historically "cumulative" and inherited circumstances? What does Comte "already understand that he is unable to make known to us"?[12] From his texts, we can develop a foreconception of what is "underway" here.

Perhaps most suggestively, there is Comte's notion of time's "philosophical order" – an idea in terms of which, at the very start of his career, he expresses his concern for the current relevance of the intellectual past to the emerging third stage. For us positivists, he says, the

> chronological order of epochs is emphatically not their philosophical order. Rather than say the past, the present, and the future, we should say the past, the future, and the present. Indeed, it is only when we have conceived the future by means of the past that we can profitably return to the present so as to grasp its true character.[13]

This passage is wonderfully destructible. In other words, it merits a triple reading – one that juxtaposes what Comte **asserts**, what he **intends**, and what he must somehow already **understand**. Of course, any attempt to "assert" The Temporal Order would be, in fine Cartesian style, a mistaken effort to fix the truth about time as if from Nowhere. Yet Comte explicitly depicts his circumstances differently. He characterizes them as the circumstances of someone who already "conceives the future by means of the past" – and moreover intends this conceptual previsioning as being for the sake of a "return to the [unsatisfactorily lived-through] present." By contrast, later positivists engage in a familiar modernist double favoring of "the" present. First, they inflate it, over past and future, into a *focal Now* in which the mind is liberated from "former" thoughts and "future" guesswork, so it can focus on "today's" job of epistemic analysis and reconstruction. Second, their formal reconstructions are then themselves construed as if they were taking shape in a kind of timelessly *ideal Now* where epistemologists all possess the proper "criteria of rationality." In the usual accounts, Comte is just a somewhat more candid member of this band of good epistemologists – one who actually knows that, as Habermas puts it, "the real job . . . of positivism . . . [is] to justif[y] the sciences' scientistic belief in themselves by construing the history of the species as the history of the realization of the positive spirit."[14]

This interpretation, however, is quite wrong. Like a good Heideggerian hermeneut, Comte sees his thinking as originating in a lived, not a focal and idealized, present. It is, he says, a matter of "grasping the true character" of an *emerging* scientific era, and it is *his own reasoning* that he wishes to bring under this observation. His concern, then, is not the prescientific tradition as "the" past – i.e., as something he is "now" choosing to bring before us again so he can criticize it. Rather, Comte understands himself to be considering, let us call it, *the lingering of inherited ways "in" current practice*. Comte treats his theologico-metaphysical inheritance as making itself known to him in an experienced lack of fit between emergent scientific research and a received, still typically prescientific cluster of epistemic ideas that distort the natural sciences and contest the very idea of a human, "sociological" one.[15] Thus, when he *speaks* of this tension between emergent practice and inherited conception, and *intends* to resolve the tension in favor of emergent practice, Comte articulates an *understanding* of his aim in terms of a "philosophically ordered" time. Comte is, as Heidegger would say, "on the way" here toward philosophizing in terms of a mode of temporalization which is *not* "chronological."

A destructive retrieval of Comte's work also sheds new light on his famous remark that "From science comes prevision; [and] from prevision comes action."[16] On the standard view, this is reducible to Baconian sloganeering – or perhaps to an anticipation of the earlier Foucault. All genuine "theories" are *scientific* theories that predict the natural "order," and applied theories constitute a technology of control that enables us to organize material and social life. In other words, "knowledge is power." This, however, is precisely not Comte's view. "Prevision" is not for him an exclusively scientific concern. All reasoning, even in fetishism, aims to find theories that enable us to reestablish the peaceful and predictable relations with our surroundings that are felt to precede our experiences of its disruption. At every stage, this aim is pursued by whatever method maximally improves upon the earlier ones. Hence, all that science really does is actually fulfill this oldest aim by successfully transforming its theological and metaphysical approaches.

Destructively retrieved, then, what Comte's three-stage narrative reveals is that to envisage history's "cumulative" meaning and time's "philosophical order," one must already understand human existence in a way that does not coincide with the definitions typical of any of the three stages. We who are living into the positive era are not, of course, either theological or metaphysical beings; but for Comte, neither are we fundamentally scientific. Whatever else may be open to us, we are at bottom already "practical" – in Comte's silently operative sense that we start with a globally benign sense of relation to the surroundings and live in the expectation that this condition can always be sustained or restored if properly "previsioned." Thus for Comte, it is just as wrong to say with the later positivists that all real philosophy is scientific, anti-religious, and anti-metaphysical as it is to say with the traditionalists that philosophy's origin is always mystic feeling or wonder. At different points in our history, the imaginative response to mystic feelings, the intellectual articulation of wonder, and the positive regard for observation have all been vital to our reasoning, and all are equally moved by the desire for prevision and for the satisfaction of natural and interpersonal need. For Comte, as I said above, modern technology is simply the third kind of applied prevision. It is to science what worship and contemplation are, respectively, to theology and metaphysics – namely, an action-oriented application of a "comprehensive system" – only this time, we really get what we understand ourselves to have wanted from the start.

In a crucial respect, then, Comte's story anticipates Heidegger's argument that while it is "chronologically correct" to say modern technology needs the "prior support" of the positive sciences, this statement cannot illuminate the "truth" about technology. Construing the three ways of thinking in terms of time considered in its "philosophical" instead of "chronological" order, Comte has them originate in a practical need for a "relation of harmony between human beings and their surroundings." So for him, modern technology, like all technology when properly thought *for the future and in terms of the past* is, just as Heidegger says, "earlier" than science "from the point of view of the essence that holds sway within it."[17] As with Heidegger's treatment of onto-theological epochs in the Western tradition, Comte does not dream of leaping beyond all previous eras, nor does he offer a metanarrative "logic" that would join the stages, substages, or systems within them – and this is not because he lacks the requisite pioneering spirit or philosophical imagination, but because he thinks the very idea of such leaps and logics is inappropriate.[18] Employed reflectively, Comte's law is not a template for reckoning with the tradition from outside. "Progress" from fetishism to science is a matter of bringing to fruition a possibility *in* positive science. It is – to use Heidegger's phrasing for characterizing the end of metaphysics – a completing or "gathering" rather than "perfecting" of this possibility. And just as Heidegger has us "receive our very being" in taking up this possibility, Comte envisions human beings as becoming ever more like themselves. In other words, we enact an originary relatedness that experientially "engages" us before we reason, that is lived through before it is articulated, and that is always as much a matter of feeling and imagination (i.e., embodiment) as it is of reason. In this way, as Heidegger says, "the oldest of the old follows behind us in our thinking and yet it comes to meet us" in life from out ahead.[19]

Destructively retrieved, then, Comte's "cumulative" story of the tradition offers us a Janus face. In what he *says*, we hear the familiar song of scientism and its technological power. In what he *intends* by his story, however, there lies something more promising. Embedded in the usual talk of progress, there is an articulation of Comte's general understanding of what (in Heidegger's phrase) gets *disclosed* in the three kinds of philosophizing. In Comte's narrative, all thinking, in all three stages, originates in one very specific sense of disclosiveness – namely, the revealing of our surroundings as regular, regularizable, and responsive to human need. This sense of disclosiveness is understood as "older" than both modern science and any technology, ancient or modern. And since Comte construes science through this ruling disclosiveness and not in terms of its epistemic surface, he stays open to a proper recognition of the priority of the question of technology over that of science. Linking "prevision" to the unfolding process itself instead of to modern science or to its type of technology, he lets prevision exhibit something of the "alēthic" character that Heidegger finds in both classical technē and modern technology. But the parallel goes even farther. For in the same general way that Heidegger contrasts Aristotle's "poietic" technē with the more intrusive technology of modern science, Comte distinguishes early theology from today's science by depicting the latter as less inclined to find/restore order than to demand/make it. In this way Comte's account, like Heidegger's, sees modern scientific activity as no longer a responsive "bringing-forth" of what does not otherwise bring itself to presence, but as an aggressive "setting-up that challenges" whatever is encountered.[20]

To sum up this retrieval, Comte's positivism seems to be *on the way*, as Heidegger says, toward thinking modern technoscience "in its essence." By treating it, not as something now being resolutely chosen over past options, but rather as basically the successful articulation of a venerable disclosive process, the Comtean account even implies that this process itself is something more than "just" a human activity. In this sense, one might venture to suggest that Comte's historico-critically reflective account of the scientific culmination of Western philosophizing is already on its way toward Heidegger's later question, not of Being, but of Being's presencing or eventuation.

IV Conclusion: Toward a Thinking "at" the End

Yet even if my retrieval of Comte is so far warranted, we should be on guard against too much enthusiasm for our newfound kinship. Foreconceptions, after all, are not only *of* something that is underway; they are also *for* someone to follow out. Too much stress on what Comte must *understand* at the expense of what he actually *says* would usurp instead of appropriate his work.[21] So far, I have construed generously what is underway in Comte; but my ultimate aim is to highlight, by contrast, his most glaring divergence from Heidegger. For unlike Heidegger (and, presumably, unlike ourselves), on the question of what if anything comes *after* science, Comte is has nothing to say. Indeed for him, there is simply nothing to ask. That the third stage is the final stage seems established by the very fact that science is a *culminating occurrence* – i.e., an end*ing* in which knowledge is henceforth only "relative" to available evidence, never "absolutely" guaranteed by feeling, faith, or reason, and always ready to restore the natural and social order we expect.

We should not, I think, run past this point too quickly. It is not just that Comte stops with his arguments for a third stage. The question is, if Comte does not in fact think past the age of science, why does he fail to consider even the possibility? The answer, I think, does not lie in any theoretical or ideological convictions. It lies in the fact that *Comte has no factical-experiential incentive to move beyond his internalist vision of the third stage*. This, then, we should probably grant him: "For" Comte's story, science may really *be* a successful gathering and completing of a disclosive process; and this completion may really *be* understood as an eventuation rather than an event; and technology may really *be* its especially intrusive mode of disclosing; and we may really *belong* to this eventuation, rather than the reverse. What we, considering these matters today, need to notice, however, is that to the Comte who tells this story, none of these ontological factors are ever explicitly identified – let alone are they ever elaborated and made the basis of any inquiry into something like Heidegger's "distressing difficulty . . . of an unsuitably thought relation between *Being and human being*."[22] Here we must remember that Comte speaks from an understanding and at a time when positive science (to say nothing of its "applications") is still more promise than actualization, when naturalistic models for human-historical study are still more projected than deployed, and when prescientific thinking is still mostly seen as ill-fated rather than meaningless. In this milieu, he simply has no experiential basis for asking whether the positive stage might, in its eventual unfolding, mark out an essentially oppressive and occlusive ontological site. Thus, Comte is able to think science *as* a culmination, but he cannot think *at* its culmination – and so he cannot ask whether, as Heidegger puts it, "the world civilization just now beginning might one day overcome its technological-scientific-industrial character as the sole criterion of our world sojourn."[23] One cannot think of overcoming what has not yet "arrived."

By contrast, Heidegger finds this culmination – and specifically the experienced spread of its intrusive disclosure – profoundly unsatisfying. From start to finish, Heidegger's rethinking of the Western tradition is informed by an issue Comte experiences no need to explore. For the twentieth-century Heidegger, but not for the nineteenth-century Comte, the fundamental question is: *How does it come to pass that today so much is encountered which seems ontologically "out of place" wherever knowing and acting occur in their usual senses?* With no need to ask this question, Comte is free to consider the future "scientistically" and to assign positivism the job of articulating and defending it. But Heidegger must explore the possibility of a non-metaphysical "thinking" that might articulate his ontological dissatisfaction with the future so understood. Contrasting this thinking with the culminating (metaphysical) "philosophy," he asks,

> is the end of philosophy . . . already the complete actualization of all the possibilities in which the thinking of philosophy became set? Or is there a *first* possibility for thinking apart from the *last* possibility (of philosophy's dissolution into the technologized sciences), a first possibility from which the thinking of philosophy would have to start, but which as philosophy it could nevertheless not expressly experience or take up?[24]

In the foreconception I have worked out above, Comte's tale of the Western tradition's ending articulates what Heidegger calls (with some conviction that this cannot literally be the case) philosophy's "last possibility." In my view, Comte's formulation of this possibility – framed in the midst of and as an emerging "positive spirit" – sheds light on how there might be for us, today, a "first possibility" that there could not be for him. Comte's Janus face, however, is crucial here. When we critically dismantle Comte's actual pronouncements, it is evident how they threaten to cover over what is foreconceptually most promising. Drawing on Heidegger's early language, we might say that the point must be to explicitly recognize this regressive tendency in Comte's pronouncements in order to learn how, for us, it could just as well serve

> to manifest the *history of our very Dasein* – history not as the totality of public events but as the *mode of happening of this Dasein*. That such happening is possible and reigns in this way for thousands of years manifests a particular mode of Dasein's being, a specific tendency toward . . . declining [*Verfallen*], from which it does not escape [and which] first really comes into its own *when Dasein rebels against this tendency*.[25]

The aim of my destructive retrieval of Comte's historical-critical reflection on the meaning of the three-stage law for his own thinking has been to stress its promise – both *against* Comte's own frequently regressive portrayals of his results, and also *ahead* of all those "rebellious" later positivists who remain blind to the promise of this reflective thinking because they refuse to see any value in it all.

If Comte's tradition-bound gesturing and later positivism's more insistently scientistic rebelliousness are manifestations of "decline," then by their "destruction" we may come to see how to strengthen Comte's deeper and more promising intentions *against* these tendencies.[26] The aim must be to discern Comte's real goal all the more clearly *in* the way its pursuit is continually undermined and diverted. If we engage in what Heidegger calls a "radial reflection in and from" the site of this diversion, we might ourselves "secure" (albeit not permanently "transcend") this site as a point of departure against our own tendency toward falling, or "declining" understanding.[27]

Here, then, the real distance between Heidegger and Comte stands out clearly. Both may rightly be said to be thinking out of a concern with what it means to be historical, but only Heidegger makes *being historical itself* an explicit theme. Comte does insist that ideas cannot be understood apart from their history, and he does believe that historical-critical reflection on *being* a positivist is current philosophy's premier issue. But the record shows that he is unable to take these points to heart. By contrast, a central feature of SZ's characterization of its point of departure is that it articulates Heidegger's own experience of raising the Being-question and confronting "in" that very act of questioning a series of traditional prejudices that block it. He finds himself with a double experience of "*Perplexed, I raise again the question of Being*..."/"*It's not worth asking because*...." And this, for him, is a sufficient "beginning" – an interplay between inherited past and possible future, mediated by the dissatisfactions of present experience.

My aim in this essay has been to focus attention on this "reflective" feature of Heidegger's thinking by juxtaposing him with someone who, on the one hand, has a similarly "consummatory" understanding of technologized science and who is similarly convinced that philosophizing is an especially disciplined way of working out what it is to "be" historical, but someone who, on the other hand, also has no cause to make *this very point of departure itself* an issue. Viewed in this way, Heidegger's pervasive ontological dissatisfaction with the same "culminating eventuation" that leaves Comte happy eventually leads him to say, in effect, that we must turn the Comtean position inside out. We must, that is, explicitly "make known" to ourselves an eventuation (*Ereignis*) Comte himself only silently "understood." "We may venture the step back out of philosophy into the thinking of Being," says Heidegger, "as soon as we become at home in the provenance of thinking."[28] This remark, as the title of the work in which it is made suggests, comes "*out of the experience* of thinking." It is thus offered by someone who has already made an issue out of *not* being at home – someone who was unwilling, already a decade before SZ, to simply assume that he knew how to *be* a philosopher, or at first, even how to articulate this as a problem.

Abbreviations

Comte's Works

CPP *Cours de philosophie positive*, 6 vols. (Paris: Bachelier, 1830–42) [latest reprint, in two volumes, *Philosophie premièr: Cours de philosophie positive, leçons 1 à 45*, ed. Michel Serres, François Dagonet, and Allal Sinaceur, and *Physique sociale: Cours de philosophie positive, leçons 46–60*, ed. Jean-Paul Enthoven. Paris: Hermann, 1975]. Subsequent French editions have various paginations for the same 60 "Lessons." My citations are by volume, lesson, and first-edition pagination, and bracketed pages to the Frederick Ferré translation [F], *Introduction to Positive Philosophy* (Indianapolis, IN: Hackett, 1988).

DEP *Discours [préliminaire] sur l'esprit positif*, printed separately; also as introduction (same pagination) to *Traité philosophique d'astronomie populaire* (Paris: Carilian-Goeury and Victor Dalmont, 1844) [trans. Edward Spencer Beesly, *A Discourse on the Positive Spirit* (London: William Reeves, 1903)].

SPP *Système de politique positive, ou traité de sociologie, instituant la religion de l'humanité*, 4 vols. (Paris: L. Mathias, 1851–4) [trans. J. H. Bridges et al., *System of Positive Polity*, 4 vols. (London: Longmans, Green, 1875–7)]. Six "early essays on social philosophy" in "Appendice général," Vol. 4, cited as "SPP4a" to accommodate the separate French pagination.

Heidegger's Works

SZ *Sein und Zeit*, 10th ed. (Tübingen: Max Niemeyer, 1963) [*Being and Time*, trans. John Macquarrie and Edward Robinson (New York: Harper & Row; 1962); and Joan Stambaugh (Albany: State University of New York Press, 1996)].

EP "Das Ende der Philosophie und die Aufgabe des Denkens" [trans. Joan Stambaugh, alterations by David Farrell Krell in *Basic Writings*, rev. ed., ed. D. F. Krell (New York: HarperCollins, 1993), 431–49].

FT "Die Frage nach der Technik," in *Vorträge und Aufsätze* (Pfullingen: Günther Neske, 1954), 5–36 [trans. William Lovitt, alterations by David Farrell Krell, in *Basic Writings*, pp. 311–41].

SD *Zur Sache des Denkens* (Tübingen: Max Niemeyer, 1969) [*On Time and Being*, trans. Joan Stambaugh (New York: Harper & Row, 1972)]. Contains EP (SD German/English pagination given in brackets).

Notes

1 For Jacques Derrida, Heidegger's critique of the philosophical tradition never becomes radical enough, so that to some extent his thinking always "remains within metaphysics and only mak[es] explicit its principles" ("*Ousia* and *Grammē*: Note on a Note from *Being and Time,*" in *Margins of Philosophy*, trans. Alan Bass [Chicago: University of Chicago Press, 1982], 48). For Hans-Georg Gadamer, Heidegger's attempt to mark an end to "philosophy" appears "burdened with violence," and Gadamer pointedly identifies his own universal hermeneutic as having "the truth of a corrective" against those misguided thinkers who "make radical inferences from everything" and assume "the role of prophet, Cassandra, preacher, or ... know-it-all" ("The History of Philosophy," in *Heidegger's Ways*, trans. John W. Stanley et al. [Albany: State University of New York Press, 1994], 165; *Truth and Method*, 2nd ed., trans. Joel Weinsheimer and Donald G. Marshall [New York: Continuum, 1989], xxxvii–xxxviii), and *Philosophical Apprenticeships*, trans. Robert R. Sullivan [Cambridge, MA: MIT Press, 1985], 186–9). According to Jürgen Habermas, late Heideggerian *Denken* is the expression of his mystical and politically motivated post-war *Ursprungsphilosophie (The Philosophical Discourse of Modernity: Twelve Lectures*, trans. Frederick Lawrence [Cambridge, MA: MIT Press, 1987], 152–60). And, to name just one more critic, Richard Rorty argues that although Heidegger himself ultimately fails to deconstruct the history of philosophy, he does manage "to further encapsulate and isolate it, thus enabling *us* to circumvent it" ("Deconstruction and Circumvention," in his *Collected Papers, Vol. 2: Essays on Heidegger and Others* [Cambridge: Cambridge University Press, 1991], 105, my emphasis).

2 For the anti-technological interpretation, see, e.g., Andrew Feenberg, *Questioning Technology* (London: Routledge, 1999), in this volume, ch. 28. For the transformed relation construal, see, e.g., Hubert L. Dreyfus, "Heidegger on Gaining a Free Relation to Technology," in *Technology and the Politics of Knowledge*, ed. Andrew Feenberg (Bloomington: Indiana University Press, 1995), 97–107; and Hubert Dreyfus and Charles Spinosa, "Highway Bridges and Feasts: Heidegger and Borgmann on How to Affirm Technology," in this volume, ch. 27.

3 EP, 431 (61/55).

4 Though my focus will necessarily be on Heidegger's later work, where tradition is interpreted most explicitly in terms of its ending in modern [techno]science, I will have occasion to note that the attitude of (ontological) dissatisfaction is just as central to his earlier work as well.

5 CPP1 (1), 1 (F, 1–2) and SPP4a, 77/547, respectively. For detailed analyses of the three stages, see my *Comte After Positivism* (Cambridge: Cambridge University Press, 1995), 73–91; and "Comte, Philosophy, and the Question of Its History," *Philosophical Topics* 19/2 (1991), 184–99.

6 "[I]t is experience alone that has enabled us to estimate our abilities rightly, and, if man had not commenced by overestimating his forces, these would never have been able to acquire all the development of which they are capable" CPP 1 (1), 10 (F, 5).

7 I note here that Comte, like Heidegger, understands the Western tradition in its dominant intellectual mood as onto-theological – yet also regards the Greek concern for Being, not Judeo-Christian variations on this theme, as the grounding experience of this mood. Comte, moreover, sees theology's influence on tradition as derivative and, in terms of its underlying impulses, relatively short-lived. Long before most people are ready to abandon God talk, he says, the "metaphysical" stage is already latent in the advanced theologian's efforts to reason logically, demand conceptual clarity, and engage in disputation.

There are other interesting parallels here between Comte and Heidegger on religion. Neither wants to claim that religious faith/experience, as such, play a central role in the dominant Western tradition, apparently for similar reasons. Comte actually admires the act of believing faith for its felt intensity, its preoccupation with the concrete and experiential, and its way of informing the whole of one's life. For these reasons, he admires fetishism above any other theological outlook. Indeed, the spirit of fetishism is as central to his conception of positivist culture as to the positive spirit. As Juliette Grange rightly notes, the grand subjective synthesis of reason and feeling toward which our age aspires is called a "new fetishism" (*La philosophie d'Auguste Comte: Science, politique, religion* [Paris: Presses Universitaires de France, 1996], 17–19). For no amount of scientific progress will ever produce an "objective" or absolute synthesis, either of the natural or social order. Hence, our sense of the unity of nature and social harmony will always be only "subjective" or "fictional"; and it is in fetishism that we find the purest expression of both the struggle to "subordinate the subjective to the objective" – *and* that recognition of the "fundamental preponderance of the heart over the intellect" which mature positivism is finally in a position to appreciate fully (SPP3, 82–122/68–101, quoted from 121/100 and 120/99). On how Heidegger's religious struggles fuel his originary sense of religious experience as (in Kisiel's words) a "phenomenological paradigm" and his lifelong sense of the remoteness of metaphysicalized theology from "factical" religious life, see, e.g., Theodore Kisiel, *The Genesis of Heidegger's "Being and Time"* (Berkeley: University of California Press; 1993), esp. Ch. 2, quote from 80; of Thomas

Sheehan, "Reading a Life: Heidegger and Hard Times," in *The Cambridge Companion to Heidegger*, ed. Charles Guignon (Cambridge: Cambridge University Press, 1993), 70–96.

8 In Comte's view, this weakness applies equally to systems of thought from the empiricist and from the rationalist traditions. For whether one's a priori commitment is to a substantive principle or to a methodological rule (e.g., "one always encounters nature via impressions or sense data"), the "system of thought" that results is just as closed and unreceptive to genuine "observation." (William James, for example, who knew Comte's positivism well, often remarked that classical empiricism is not very empirical.)

9 Comte's lifelong "subjective" (i.e., ethico-political) interest in fostering the establishment of sociology is only one of several indications that his conception of the hierarchy of the sciences is not reductivist, in the manner of logical positivism. See Peter T. Manicas, *A History and Philosophy of the Social Sciences* (Oxford: Blackwell, 1987), 60–2.

10 Actually, those who imagine themselves reasoning as if from nowhere, presuppositionless except for whatever methodological tools they bring with them to keep their thoughts legitimated, are in Comte's scheme acting more like metaphysicians than scientific thinkers. Of course for many purposes, the substitution of purely formal-systematic (*dogmatique*) accounts for historical ones, Comte admits, is a constant and even admirable "tendency" of the human mind. Yet ultimately, "an idea cannot be properly understood except through its history" CPP (1), 3 (F, 1); and those who construct such formalisms must never forget what is suppressed and smoothed over in the constructing CPP (2), 77–80 (F, 46–8). See also *Comte After Positivism*, 98–105.

11 Heidegger first worked out this kind of acquisitive or appropriative reading of one's predecessors in the decade before SZ, especially in connection with his analysis of Wilhelm Dilthey's attempt to establish the foundations for genuinely human, as opposed to physical/material sciences. See my "Heidegger's 'Appropriation' of Dilthey before *Being and Time*," *Journal of the History of Philosophy* 35/1 (1997), 99–121.

12 SZ, 396. The subject of Heidegger's quote is, of course, Nietzsche (in his *Advantage and Disadvantage of History for Life*).

13 SPP4a, 100/563.

14 Jürgen Habermas, *Knowledge and Human Interests*, trans. Jeremy J. Shapiro (Boston: Beacon Press, 1971), 72.

15 See, e.g., CPP1 (1), 31–56 (F, 18–33); and SPP1, 1–6 ("Préface"), 2–3/ix–xiii, 1–3.

16 CPP1 (2), 63 (F, 38); DEP, 45/72; SPP1, 1–7, 321, 701–5/1–5, 257, 566–70. In the "Dedication" to his *Système*, Comte adds altruistic "Love" to the "Order" and "Progress" that the *Cours* envisions as promoted, respectively, by scientific knowledge and its application – for although "we grow tired of thinking, and even of acting; we never tire of loving" (SPP1, 1/1).

17 Cf. FT, 21–2 (326–7) and, e.g., SPP4, 28ff./23ff.

18 E.g., CPP (2), 82–4 (F, 49–50), where Comte affirms both pedagogical and epistemic advantages to be gained by replacing historical with formal accounts of the rise of science – yet stresses the necessarily "artificial" character of formal accounts and the danger that they will seriously distort the activity being systematized. Cf., on Heidegger, especially "The Age of the World Picture" (trans. William Lovitt, in *The Question Concerning Technology and Other Essays* [New York: Harper & Row, 1977], 115–54) and his remarks on Hegel in the "Protokoll" of the seminar on his "Time and Being" lecture, SD, 51–3 (48–50). It is helpful to know that, for Comte, it is not Hegel but Kant whose "reflective" rather than "constitutive" remarks on historical development he praised (*Comte After Positivism*, 54n.21).

19 *Aus der Erfahrung des Denkens* (Pfullingen: Günther Neske, 1954), 19 ("The Thinker as Poet," in *Poetry, Language, and Thought*, trans. Albert Hofstadter [New York: Harper & Row, 1971], 10).

20 My phrasing, of course, follows Heidegger's in FT, 10–16 (316–22). This point is of considerable importance for distinguishing Comte from other positivists as someone worth "retrieving"; for insofar as he still acknowledges fundamentally different ancient and modern modes of technicity, he cannot yet be counted as an advocate of either instrumental rationality or Nature viewed as mere standing-reserve. This contrast is clear in Comte's remark that even now, fetishism's more spontaneous and responsive kind of speculation is still needed by scientists today whenever they confront situations where both plausible theories and reliable data are lacking (SPP3, 82–3/68–9). In the *Cours*, Comte says this is the standard condition of the fledgling science, sociology (CPP4 [48], 412–21).

21 Indeed, my general view is that Comte is already too much of a Cartesian by inheritance to become the historicocritically reflective thinker he starts out to be. Hence the standard interpretations of Comte's work may not be very hermeneutical; but they were inevitable (see *Comte After Positivism*, Ch. 4, esp. 118–27).

22 This is Heidegger's phrasing for that project which runs from the early "hermeneutics of facticity" to EP's "immanent criticism" of SZ, in the 1956 "Zusatz" to *Der Ursprung des Kunstwerkes* (Stuttgart: Reclam, 1960), 100 (trans. Albert Hofstadter in *Basic Writings*, 211).

23 EP, 437 (67/60).

24 EP, 435 (65/59).

25 *Prolegomena zur Geschichte des Zeitbegriffs: Gesamtausgabe 20* (Frankfurt/Main: Vittorio Klostermann,

1979), 180 (*History of the Concept of Time: Prolegomena*, trans. Theodore Kisiel [Bloomington: Indiana University Press, 1985], 129–30), emphasis altered.

26 "Securing hermeneutical intentions against Verfallensein" is precisely the point of SZ's "repetitive" procedures, but the idea precedes SZ by several years. See, e.g., what Heidegger says on "ruinance" in the final section of the Winter Semester 1921–2 lecture course entitled "Vier formal-anzeigende Charaktere der Ruinanz" (*Gesamtausgabe 61* [Frankfurt/Main: Vittorio Klostermann, 1985], 140–55); and cf. Th. C. W. Oudemans, "Heidegger: Reading Against the Grain," in *Reading Heidegger from the Start*, ed. Theodore Kisiel and John van Buren (Albany: State University of New York Press, 1994), 39–45.

27 The language is of course SZ's, but the expressed sentiment – namely, that interpreting past forms of philosophical inquiry, when done perspicaciously, can simultaneously be an exercise in philosophical self-understanding – is explicitly announced at least as early as the famous 1922 manuscript, "Phänomenologische Interpretation zu Aristoteles (Anzeige der hermeneutischen Situation)," first published in *Dilthey-Jahrbuch* 6 (1989), 239.

28 *Aus der Erfahrung des Denkens*, 19 [10].

25

Heidegger's Philosophy of Technology

Don Ihde

Among the few philosophers to date to have taken technology seriously, it should be apparent that Martin Heidegger is a pioneer in this field. He was among the first to raise technology to a central concern for philosophy and he was among the first to see in it a genuine ontological issue. This is the case in spite of the dominant and sometimes superficial interpretations of Heidegger which see in him only a negative attitude to technology.

It will be the aim of this essay to examine some of Heidegger's main theses concerning technology and to elucidate the strategies which motivate these. To make the task manageable, I have chosen to limit myself to his 1954 lecture, "The Question Concerning Technology", and the earlier foundational work, *Being and Time* (1927).[1] And as an interpretative device, I shall read these two works retrospectively. That is, I shall isolate what emerge as the principal themes concerning technology from the lecture and then show how they reflect and are anticipated by *Being and Time*.

In so doing, I shall show how Heidegger's philosophy of technology is directly phenomenological in the sense of exhibiting the *existential foundations* of the technological enterprise. This type of phenomenology, already apparent in *Being and Time*, gives a certain priority to what I shall call the *praxical dimension* of human existence and it continues to be a key to the later work on technology.

From Don Ihde, *Technics and Praxis*, Dordrecht: Reidel Publishing, 1979, pp. 103–29. Reprinted by permission of Kluwer Academic Publishers, Dordrecht, The Netherlands.

It is my own conviction that Heidegger's philosophy of technology is one of the most penetrating to date. By examining the ontological grounds of technics, Heidegger has begun to lift technology out of its subjectivistic and merely instrumentalist interpretations and made of it a primary philosophical question. But this is not to say that this first work concerning technology is also the last word. There are implicit limitations in the Heideggerian program which lay the basis for the current misinterpretations of Heidegger and for which Heidegger himself must be blamed. I shall point to some of these along the way.

Finally, although I shall not develop this theme in substance, I do wish to point up the resultant internal need within the Heideggerian program concerning technology for the emergence of an "aesthetic" as the counterfoil to the limitations of technology as the Heidegger sees them. The *poiesis* which characterizes much of the "late" Heidegger's work arises directly as a response to technology. I shall point to how this is the case within the exposition.

I The Question Concerning Technology

The question referred to by the title of the lecture is one which takes recognizable phenomenological shape quite immediately. The query is into the *essence* of technology in its *relationship* with human *existence*.

> We shall be questioning concerning *technology*, and in so doing we should like to prepare a

> free relationship to it. The relationship will be free if it opens our human existence to the essence of technology. When we can respond to this essence we shall be able to experience the technological within its own bounds. (T 287)

The analysis is to make the phenomenon of technology stand out such that its horizon, limit, is bared, but this in relationship to human existence. These are the typical marks of Heidegger's version of phenomenology in which the intentional arc of *human–existence relation–world* are interpreted existentially such that intentionality is best described as an *existential* intentionality.

To uncover the phenomenon, it must be freed from its layers of less adequate interpretation which, again in typical fashion, Heidegger attributes to a "subjectivistic" understanding, here called the instrumental and anthropological definitions of technology.

> One says: Technology is a means to an end. The other says: Technology is a human activity. The two definitions of technology belong together. For to posit ends and procure and utilize the means to them is a human activity. The manufacture and utilization of equipment, tools, and machines, the manufactured and used things themselves, and the needs and ends that they serve, all belong to what technology is. The whole complex of these contrivances is technology. Technology itself is a contrivance – in Latin, an *instrumentum*.
>
> The current conception of technology, according to which it is a means and a human activity, can therefore be called the instrumental and anthropological definition of technology. (T 288)

Such a definition implies that technology is merely an invention of a "subject" and functions as a mere, neutral instrument. The definition, Heidegger characterizes, is *correct*. But, then, in a move directly reflective of his earlier analysis of logical or propositional truth in relation to truth as disclosure, he notes that what is correct is not yet *true*.

Correctness turns out to be "true" in a very limited sense, true with respect to some aspect or part of a larger whole. The whole, however, is more than which contains parts, it is ultimately the set of conditions of possibility which found the parts.

> The correct always fixes upon something pertinent in whatever is under consideration. However, in order to be correct, this fixing by no means needs to uncover the thing in question in its essence. Only at the point where such an uncovering happens does the true come to pass. For that reason the merely correct is not yet the true. (T 289)

The phenomenological form of the argument here is that correctness is not in itself untrue, but limited or inadequate, and may be characterized as a partial truth. But the catch is that unless it is seen for precisely this it can be taken for more than a partial truth in which case it now covers over the larger or more basic truth which founds it. It then becomes *functionally* untrue by concealing its origin. Moreover, it is only by comprehending the whole which founds correctness that it can be seen as partial. Thus what is involved in taking correctness for truth is like a fallacy of taking a part for the whole. But it is also more than that in that comprehension of the whole is a necessary condition for recognizing what is a part.

Heidegger's strategy becomes clearer if it is seen that his overall theory of truth is, in effect, a complex field theory. Truth is *alēthia*, translated as "unconcealedness", brought to presence within some opening which itself has a structure. Beings or entities thus *appear* only against, from and within a background or opening, a framework. But the opening or clearing within which they take the shapes they assume, is itself structured. Overall this structure has as an invariant feature a concealing–revealing ratio. Thus one may say that it always has some selectivity factor as an essential feature.

Understood in this way, it becomes clear that beings as such are never simply *given*: they appear or come to presence in some definite way which is dependent upon the total field of revealing in which they are situated. Preliminarily it is important to note that the field or opening in which things are "gathered" is, in a sense, given. It is given historically as an epoch of Being. This is to say that the Heideggerian notion of truth has something like a "civilizational given" as a *variable*. It is what is taken for granted by the humans who inhabit such a "world". Variables given in this sense are particular shapes of the invariant revealing–concealing structure of truth.

What is usually missed concerning this complex field theory of truth is its phenomenological role in

Heidegger. The phenomenology of truth isolates the invariance of truth as the revealing–concealing structure itself, the ratio of gathered presences to what is not revealed. Thus any particular variant is but a variant upon this overall structure. Most interpreters have missed this and failed to see that Heidegger's use of the Greeks, for example, serves as a contrasting variation upon the contemporary scene in order to point up the specific features of our epoch of truth. The interpreters often miss the counterpart characterization of the Greek modes of concealedness to which Heidegger also refers.

Thus, in Heidegger's sense, one must see beyond correctness if one is to attain truth since correctness is grounded upon some framework which makes it what it is. The process by which this penetration is accomplished is familiar from *Being and Time* as well. One begins with what is called the *ontic* in *Being and Time*. But then, by what I shall call an act of *inversion*, Heidegger seeks through it an *ontological* condition. It is only through the ontic that the ontological can be understood, but the ontological dimension is in turn the field of the conditions of possibility which founds the ontic.

It is precisely this strategy which Heidegger applies to technology. The anthropological–instrumental definition of technology is functionally ontic, correct but partial, limited to a subjectivistic set of conditions. Heidegger inverts this definition by asking a question which belongs to the transcendental tradition of philosophy; what are the set of conditions of possibility which make technology possible? Technology, as Heidegger sees it, is not only ontic, but ontological.

At first such a move seems strange, but placed within the Heideggerian theory of truth, it begins to make sense in the following way. The things of technology (instruments) and the activities (of subjects) which engage them, appear as they do only against the background and founding stratum of some kind of framework. Technology in its ontological sense is not just the collection of things and activities, but a *mode of truth* or a field within which things and activities may appear as they do. Technology is thus elevated to an ontological dimension.

> *Technē* is a mode of *alētheuein*. It reveals whatever does not bring itself forth and does not yet lie here before us, whatever can look and turn out now one way and now another. (T 295)...Thus what is decisive in *technē* does not lie at all in making and manipulating nor in the using of means, but rather in the revealing mentioned before. It is as revealing, and not as manufacturing, that *technē* is a bringing forth ...Technology is therefore no mere means. Technology is a way of revealing. (T 294)

Technology as a mode of truth assumes the overall shape of Heidegger's truth theory. "Technology is a mode of revealing. Technology comes to presence in the realm where revealing and unconcealment take place, where *alēthia*, truth, happens". (T 295)

A mode of truth as a variant upon the revealing–concealing invariant carries with it certain characteristics. A few of these are important to note with respect to the specific characteristics of technological truth. I have called what Heidegger sometimes calls "epochs of Being", civilizational givens. These are something like deeply held, dynamic but enduring traditions, historical but no more easily thrown over than one's own deepest character or personality. Thus for an individual it is possible to say that he stands in or stands over against that which precedes him. "...The coming to presence of technology gives man entry into something which, of himself, he can neither invent nor in any way make. For there is no such thing as a man who exists singly and solely on his own". (T 313) Secondly, these civilizational givens make a claim upon those who inhabit them such that some response is necessary (although variations might range from sheer rebellion to willing acceptance). And, thirdly, they have a *telos* or inherent direction which Heidegger terms a *destiny*. But, as will be noted, a destiny is not a strict determination, it is more like a direction of growth and decay. "...We do not mean a generic type; rather we mean the ways in which house and state hold sway, administer themselves, develop and decay". (T 312) Technology, ontologically, is what characterizes the variant of this epoch of Being, thus penetration of its essence or shape becomes a central philosophical concern if we are to understand our era and prepare a response to it. Again, technology is elevated to a seldom seen philosophical importance in Heidegger's sense.

Now every shape of truth as a variant upon the revealing–concealing ratio has a certain definiteness to it. It has an essence or structure which is not merely its genus but is the particular form of its set of possibilities which found what we take as contemporary technics. The name for this shape of technological truth, Heidegger calls *Ge-stell*. *Ge-stell* means in ordinary German, frame or apparatus, skeleton and in Heidegger's use, *enframing*.

With *Ge-stell* the essence of technology is named: "we now name that challenging claim which gathers man thither to order the self-revealing as standing-reserve: Ge-stell". (T 301) This is Heidegger's ontological definition of technology. It has the features previously mentioned of being a civilizational variant into which humans have moved; of being a mode of revealing which serves as the set of possibilities by which technology ontically appears as it does; of making a call or claim upon humans for some necessary response; and it has a telos or destiny as a direction of development.

By introducing *Ge-stell* at this point, I have leapt over Heidegger's development in the lecture, but I have done so in order to display what may now be called his ontology of technology, the elements of which have been mentioned. What yet remains is to examine this notion is such a way that it can be seen to account for the major features of technology in its contemporary sense and to note more specifically Heidegger's claim that technology can be thought of as the *primary* mode of truth for the contemporary era. To accomplish this task I shall turn to some more specific aspects of each of the structural features of technology as Heidegger exhibits them.

Technology is a mode of revealing. Revealing is a coming to presence within a framework. Already at this level one can detect the emergent value given to *praxis* by Heidegger. In typical fashion he reverts to etymological expositions upon Greek thought which stands at the origin of our epoch of Being. *Technē*, Heidegger points out, is originally thought of as broader than "technique" in the contemporary thought. "*Technē* is the name not only for the activities and skills of the craftsman, but also for the arts of the mind and the fine arts. *Technē* belongs to bringing forth, to *poiēsis*; it is something poetic." (T 294) *Poiēsis* is both making and bringing forth, but bringing forth is presencing and thus is a *praxical* truth. Here is already the seed for the primacy of the praxical which characterizes Heidegger's phenomenology, but at this point is is only important to see that *technē*, as with the ancients, is linked to *epistemē* as a mode of truth as bringing to presence. *Technē* reveals or brings to presence something which is possible. "What has the essence of technology to do with revealing? The answer: everything. For every bringing-forth is grounded in revealing." (T 294)

But what is revealed? Technological revealing takes its particular shape from its field of possibilities, its framework. And its framework is a particular form of the human taking up a relation to a world through some existential intentionality. There is thus some particular presumed shape to world and some particular activity which responds to that shape of the world.

The world in its technological shape, is the set of conditions which Heidegger defines as world taken as *standing reserve (Bestand)*. This is to say that the world, revealed technologically, is taken in a certain way, as a field of energy or power which can be captured and stored. "The revealing that rules in modern technology is a challenging, which puts to nature the unreasonable demand that it supply energy which can be extracted and stored as such." (T 296) This makes world a field as standing-reserve.

This view has certain consequences, for example, "The earth now reveals itself as a coal mining district, the soil as a mineral deposit," (T 296) which is to say that nature appears as a certain potential for human use. This is a *variant* upon how nature may be viewed. It stands in contrast to those civilizational variants which, for instance, regard the earth as mother and to which one does not even put a plow. Thus one may say equivalently that the technologically viewed world is a variant upon civilizational possibilities or that it is a historical *transformation* upon how nature is taken.

Heidegger argues that such an understanding of the world is a condition of the possibility for our taking up the kinds of technologies which we actually develop now. He emphasizes the transformational features of this enterprise. Thus not only is it the case that the earth may be viewed as a resource, but what was previously taken as the dominance of nature over man becomes inverted so that man dominates nature through technology. "In the context of the interlocking processes pertaining to the orderly disposition of electrical energy, even the Rhine appears to be something at our command...the river is dammed up into the power plant. What the river is now, namely a water-power supplier, derives from the essence of the power station." (T 297) Technology, in this sense, is both the condition of the possibility of the shape of world in the contemporary sense, and the transformation of nature itself as it is taken into technology.

Phenomenologically, for every variant noematic condition there is a corresponding noetic condition. Thus if world is viewed as standing-reserve, the basic way in which the world is perceived, there must also be a correlated human response. That, too, takes particular shape in a technological epoch. The activities of humans in response to world as standing-reserve are those of revealing that world's possibilities, characterized by Heidegger as "unlocking, transforming, storing, distributing, and switching about." (T 298) Man is taken into the process of *ordering*: "Precisely because man is challenged more originally than are the energies of nature, i.e., into the process of ordering, he never is transformed into mere standing-reserve. Since man drives technology forward, he takes part in ordering as a way of revealing." (T 299–300)

Here, once again as in *Being and Time*, there begins to emerge the primacy of praxis which characterizes Heidegger's version of phenomenology. And it is here that I shall begin to make the most specific connection with Heidegger's famous "tool analysis" which serves as the model for his philosophy of technology. The common view of technology, related to what Heidgger calls the instrumental and anthropological view, holds that modern technology is a child of modern science. Technology is a mere tool of science or, at best, an applied science. Heidegger inverts this view and claims that modern science is essentially the child of technology. The strategy by which he seeks to show this is a reflection of the same functional inversion employed in *Being and Time*. This inversion of science and technology calls for careful examination.

There are two correlated ideas which appear at the beginning of the strategy which bear initial note. First, Heidegger grants that the contemporary dominant view of technology seeks to strongly differentiate between scientific technology and the older handwork technology. Heidegger does not deny that there are differences, but he plays these down. For instance, in granting correctness (not truth) to the instrumental view of technology, he notes that this view can bring both handwork and scientific technology under the same rubric as "means" or as instrumental towards ends. Here the difference between technologies is merely a matter of relative complexity. (T 288–9) Secondly, the constant emphasis upon technology as *poiēsis* and as *technē*, a making in the ancient broad sense, tends to play down a difference between ancient and modern technology. But thirdly, and most profoundly, the difference is played down strategically because the essence of technology is not itself technological, but is existential. What Heidegger does grant is that modern technology allows the secret grounds of technology as enframing to emerge more clearly, allows what was long latent and originary to be made more explicit.

Correlated with this downplay of an essential difference between ancient and modern technology is the necessary admission that modern technology is chronologically later than modern science.

> Chronologically speaking, modern physical science begins in the seventeenth century. In contrast, machine-power technology develops only in the second half of the eighteenth century. But modern technology, which for chronological reckoning is the later, is, from the point of view of the essence holding sway within it, historically earlier. (T 304)

(Here one must recall the difference between history as a destiny and historiology as a chronicle developed in *Being and Time*.) The essence of technology is not chronologically prior, but it is historically, ontologically, prior to modern science itself. It is from this inversion that Heidegger makes his claim that the technological epoch is what characterizes the contemporary era. The claim is clearly reflective of his earlier explicit claims regarding the primacy of the praxical.

In the lecture the inversion first takes explicit shape regarding science and its instruments.

> It is said that modern technology is something incomparably different from all earlier technologies because it is based on modern physics as an exact science. Meanwhile we have come to understand more clearly *that the reverse holds true as well*: modern physics, as experimental, is *dependent* upon technical apparatus and upon progress in the building of apparatus. (T295–6; italics mine).

This is to say that modern science is *embodied* technologically. One might very well say that one basic difference between modern science and its ancient counterpart is precisely its increasingly technological embodiment in instruments.

But if science is embodied in instruments as a necessary condition for its investigations, this is not yet to say that technology is its origin. Yet that is the claim Heidegger ultimately makes. The form the argument takes is essentially that it is first necessary to view nature as a storehouse or standing-reserve towards which man's ordering behavior can be directed. This provides the condition of the possibility for a calculative modern science.

> Modern science's way of representing pursues and entraps nature as a calculable coherence of forces. Modern physics is not experimental physics because it applies apparatus to the questioning of nature. The reverse is true. Because physics, indeed already as pure theory, sets nature up to exhibit itself as a coherence of forces calculable in advance, it orders its experiments precisely for the purpose of asking whether and how nature reports itself when set up in this way. (T 303)

Thus, hidden behind modern physics is the spirit of technology, technology in its ontological sense as world-taken-as-standing-reserve. Its firstness, however, only gradually becomes clear. Such conditions are not necessarily first known, they only gradually come clear. Historiologically, then, modern science does play a role. It begins to announce what lies behind science as technology comes to presence.

> The modern physical theory of nature prepares the way not simply for technology but for the essence of modern technology. For such gathering-together, which challenges man to reveal by way of ordering, already holds sway in physics. But in it that gathering does not yet come expressly to the fore. Modern physics is the herald of enframing, a herald whose origin is still unknown. (T 303)

But the origin does gradually become clear, the origin which is technology as ontologically interpreted. "All coming to presence, not only modern technology, keeps itself everywhere concealed until the last. Nevertheless, it remains with respect to its holding sway, that which precedes all: the earliest." (T 303)

Technology as enframing, *Ge-stell*, as originally, is the condition of the possibility of modern science. In Heidegger's terms this is the primacy of technology.

> Because the essence of modern technology lies in enframing, modern technology must employ exact physical science. Through its so doing the deceptive illusion arises that modern technology is applied physical science. This illusion can maintain itself only so long as neither the essential origin of modern science nor indeed the essence of modern technology is adequately found out through questioning. (T 304–5)

Here the inversion is complete; technology is the source of science, technology as enframing is the origin of the scientific view of the world as standing-reserve.

Enframing is both the condition of the possibility of modern science and the field of possibilities within which it moves. Enframing is the *ontological horizon* of modern science such that what occurs within it appears as it does through its types of ordering. Such is the shape of the contemporary variant so far as world is concerned.

For the limited purposes here I shall consider that the exposition of technology as grounded in enframing, the world which appears as a standing-reserve, completes the *noematic* analysis of the phenomenological program. World is that which both stands before humans and that into which they are "thrown" in Heidegger's earlier language. Thus they must necessarily enter into some kind of relationship with this world. In the context of the contemporary era the dominant mode of revealing of world is technological, thus the *noetic* analysis would have as its task the unfolding of the range of possible responses to the essence of technology as enframing.

I have already noted that the normative response is what is called by Heidegger, the *ordering* of the world (unlocking, transforming, storing, etc.). On the surface it then appears that the human response to the world seen as enframed is the activity of calculatively ordering the disposition of resources. Thus just as nature appears, within enframing, as standing-reserve, so the human task appears as a kind of command of nature through technological means.

What is normative, however, is merely symptomatic of the essence of technology as enframing. It is indicative of the core or central destiny (telos) of world under the guise of technology. It is with the notion of destiny that Heidegger undertakes what must be considered the noetic analysis of enframing. Here, again, the standard moves of a phenomenological program appear:

(1) The noematic (world) correlate appears and is defined or described essentially. "The essence of modern technology shows itself in what we call enframing." (T 305) "It is the way in which the real reveals itself as standing-reserve." (T 305) (2)

Then the question of a relationship to this essence is taken up, the noetic correlate. "We are questioning concerning technology in order to bring to light our relationship to its essence." (T 305) (3) This relationship is characterized by Heidegger as a mode of "being-in" as follows:

> Enframing is the gathering together which belongs to that setting-upon which challenges man and puts him in position to reveal the real, in the mode of ordering, as standing-reserve. As the one who is challenged forth in this way, man *stands within* the essential realm of enframing. (T 305) (italics mine)

In short, the response or relationship of man to the essence of technology will be in terms of the way enframing appears. And this selectivity of a way of seeing the world contains a direction or destiny. "We shall call the sending that gathers, that first starts man upon a way of revealing, *destining*." (T305–6)

Destining, in Heidegger's terms, is not described as a determination. It is rather a telos, a direction, which at best may be said to set a framework and provide a set of conditions as an inclination. "But that destining is never a fate that compels. For man becomes truly free only insofar as he belongs to the realm of destining and so becomes one who listens, though not one who simply obeys." (T 306)

It is at this juncture that Heidegger makes a strategic phenomenological move. To recognize and identify the essence of technology, to comprehend it, is to have located it or to take note of it *as bounded*, as having a horizon. Thus by the same move that grasps technology in its essence, the possibility of becoming free occurs.

> But when we consider the essence of technology we experience enframing as a destining of revealing. In this way we are already sojourning within the open space of destining, a destining that in no way confines us to a stultifying compulsion to push on blindly with technology, or, what comes to the same, to rebel helplessly against it and curse it as a work of the devil. Quite to the contrary, when we once open ourselves expressly to the *essence* of technology, we find ourselves unexpectedly taken into a freeing claim. (T 307)

What this amounts to, in the Heideggerian program, is to have recognized that the relationship to technology is not technological, but is an existential relationship and hence circumscribed by all the features which characterize existentiality. And to characterize the human response to technology, now located and limited, is to recognize that technology is (a) not neutral, "We are delivered over to it in the worst possible way when we regard it as something neutral; for this conception of it, to which today we are particularly like to do homage, makes us utterly blind to the essence of technology," (T 288); (b) is ambiguous, "the essence of technology is in a lofty sense ambiguous," "(T 314), and (c) is mysterious, "technology is not demonic, but its essence is mysterious." (T 309) But all of these are characterizations of existential intentionality with respect to the truth structure of concealing–revealing.

I have indicated that Heidegger's theory of truth is a *complex* field theory. It is complex because the structure of revealing is inextricably bound to concealing – indeed, bounded by concealing which is its horizon. "All revealing belongs within a harboring and a concealing. But that which frees – the mystery – is concealed and always concealing itself...Freedom is that which conceals in a way that opens to light...Freedom is the realm of the destining that at any given time starts a revealing on its way." (T 306) I shall not here go into the complexity of the ratio of concealing to revealing which marks Heidegger's theory of truth, but it is important to note its result for a human relationship to the essence of technology.

Heidegger characterizes a range of possible responses to technology. These range from blind obedience to equally blind rebellion. But he also allows for a free (authentic?) relationship which faces technology in its essence. But because there is such a range, there is also danger: "Placed between these possibilities, man is endangered by destining. The destining of revealing is as such, in every one of its modes, and therefore necessarily, danger." (T 307)

But what is the danger? The answer is essentially the same as the previously noted danger of taking correctness as truth, the danger of taking a part for the whole. "In whatever way the destining of revealing may hold sway, the unconcealment in which everything that is shows itself at any given time harbors the danger that man may *misconstrue the unconcealed and misinterpret it*." (T 307, italics mine)

A misinterpretation, for Heidegger, contains elements which reflect the errors possible in taking correctness for truth. They revolve around his version of mistaking the part for the founding whole. Thus, unless it is recognized that technological revealing is also a concealing (and it is from concealing that the origin of freedom arises), it can be mistaken for the totality. Technology, by its very status as a mode of revealing, may harbor this temptation.

> The coming to presence of technology threatens revealing, threatens it with the possibility that *all* revealing will be consumed in ordering and that everything will present itself *only* in the unconcealedness of standing-reserve. (T 315, italics mine)

Noematically, this is the implicit claim of ultimate truth, world must appear *totally* or ultimately as standing-reserve. Noetically, the same index for danger can occur. By reflexively taking account of their place within world, humans face the danger that they can also be taken as standing-reserve.

> When destining reigns in the mode of enframing, it is the supreme danger. This danger attests itself to us in two ways. As soon as what is unconcealed no longer concerns man even as object, but exclusively as standing-reserve, then he comes to the very brink of a precipitous fall, that is, he comes to the point where he himself will have to be taken as standing-reserve. (T 308)

If world becomes totally perceived as standing-reserve, then reflexively, humanity itself may come to perceive itself as the same.

In one respect this is to note that the technological mode of truth is "reductionistic." But it is reductionistic in a special Heideggerian sense because it is not that something can be "added to" this mode of revealing which will correct it – although it appears that Heidegger himself opts for something like this alternative as a solution. Rather one mode of revealing can only be changed by in effect being replaced. Its "reductionism" is a reductionism of disregarding the concealed, the horizon of all unconcealedness or revealing.

Although I am not particularly concerned here with Heidegger's response to the danger of technology, but rather concerned with its explanatory scope, it is perhaps well to note that his response was never well formed. In the technology essay the response was, in fact, a form of remedy for "reductionism". It contained two primary steps. The first remains continuous with what may be called "phenomenological therapy". This therapy is to address the critical question to technology, as to any truth claim, and seek to limit its hubris towards totality. Critical questioning, in Heidegger's sense, calls us back to noting the structure of the invariant *concealing-unconcealing* which limits every totality. This is the perennial philosophical task:

> Because the essence of technology is nothing technological, essential reflection upon technology and decisive confrontation with it must happen in a realm that is, on the one hand, akin to the essence of technology, and, on the other, fundamentally different from it...For questioning is the piety of thought. (T 317)

The other dimension of Heidegger's response is or may be seen as an attempt to broaden and enrich technological revealing. And the enrichment, he sees, comes from a similar activity which is in its own right praxical and poetic; the enrichment is to come through a basic revival of *technē as art*. This move is familiar throughout the Heideggerian emphasis upon the primal thinking of the poet, but it is rarely appreciated as the similar-dissimilar counterpart of *technē* as technological. Art *is* technological as *technē*, but its mode of revealing opens new ways of "saying Being" as Heidegger puts it, thus is fundamentally different from *technē* as technology.

> What was art – perhaps only for that brief but magnificent age? Why did art bear the modest name *technē*? Because it was a revealing that brought forth and made present, and therefore belonged within *poiēsis*. It was finally that revealing which holds complete sway in all the fine arts, in poetry, and in everything poetical that obtained *poiēsis* as its proper name...poetically dwells man upon this earth. (T 316)

Technology and art belong to the danger and possible salvation of the same epoch of Being.

II Being and Time

I now turn to a brief examination of the famous "tool analysis" of *Being and Time* as the full anticipation of the themes concerning technology in the lecture. It is first important to note the context and role that the "tool analysis" plays in the overall Heideggerian strategy. The analysis occurs as the vehicle by which the worldhood of the world is to be made phenomenologically apparent. World, Heidegger contends, does not just appear and neither can it be accounted for by adding up and classifying the entities within it. Such a strategy always already contains a hidden interpretation and is thus ontologically naive. Heidegger's counter-strategy is to attempt to locate what is ontological *through* a phenomenological analysis of what first appears as ontical.

This strategy appears clearly at the end of the "tool analysis" where the ontological relationship with world appears *through* the ready-to-hand:

> As the Being of something ready-to-hand, an involvement is itself discovered only on the basis of the prior discovery of a totality of involvements...In this totality of involvements which has been discovered beforehand, there lurks an ontological relationship with the world. (BT 118)

Here is already a glimpse of Heidegger's assertion of the way a whole or totality, although hidden and latent, *precedes* any individual or part as the condition of possibility for the part to appear as it does while the part, in turn, is the proximal means by which the totality itself is discovered. The ontological is discovered (literally, dis-covered) through the ontic.

Ultimately, what is ontological, however, must also be noted. Heidegger's ontology is thoroughly phenomenological, although phenomenological in the specific existential sense which Heidegger gives to the intentional arc. A phenomenological ontology is one which correlates in a unified concept three distinguishing notions. These appear in Heidegger as:

Dasein-being in-World

They are clear adaptations from the Husserlian notion of intentionality which "consciousness" is always *of* something to which the act of consciousness refers. The intentional arc in Husserl is thus: Ego-cognizing-World. It should be noted preliminarily that the interpretation in the Husserlian context is one which dominantly sticks to a more traditional perceptual and cognitional characterization of the arc as "mental".

Functionally, the intentional arc remains operative in *Being and Time* but it is no longer interpreted cognitionally – it is rather existentialized such that what turns out to be basic or primary is what I shall call the *praxical*. But it remains important to recognize that the ultimate structure of Heidegger's ontology is the arc: *Dasein-being in-World*.

Heidegger's transformation of intentionality into a praxical base may be seen in two complementary ways. It is, on the one hand, a deepening of the understanding of intentionality. It is to have noted that *all* so-called "conscious" activities are equally intentional, including such phenomena as moods and emotion and, what is more, bodily movement, such that the human being as a totality is "being-in" an environment or world. It is true that Husserl recognized this, but he continued to interpret intentionality as if it were "mental" instead of existential. Heidegger's tactic is one of simply cutting through the traditional mentalistic language and speaking of human existence as correlated with a world. But the second way in which Heidegger's transformation of Husserlian phenomenology may be seen is by way of seeing it as an *inversion* of Husserlian priorities. Again, Husserl already saw that the phenomenological aim undercut much theory and aimed at what became known in the literature as the "pre-theoretical" stratum of phenomena. Heidegger not only absorbs this notion, he inverts it in *Being and Time* such that a praxical engagement with entities becomes primary over the assumed theoretical-cognitive engagement which actually characterizes all Husserl's descriptions.

This "anti-Husserl" theme in *Being and Time* is not unfamiliar, but in this context the inversion concerning praxis and theory may be seen as the anticipation of the later explicit theme which makes technology the origin of science. I shall put the exposition in the context of its proper phenomenological strategy.

Heidegger wishes to penetrate the stratum of latent, hidden, but familiar relations with the world which characterize what he calls *everydayness*. Such a stratum constitutes, according to Heidegger, the base and limits within which subsequent specifications may be made. "The theme of our analytic is to be Being-in-the-world, and accordingly the very world itself; and these are to be considered within the horizon of average everydayness – the kind of Being which is *closest* to Dasein." (BT 94)

As already noted, the analysis takes place first in its *noematic* or world-correlate step which seeks to uncover the "worldhood of the world". The everyday world is the experienced environment (world-as-environment). It is through the familiar, but hidden environment that clues to the World as such are to be found.

When Heidegger turns, then, to a phenomenological analysis of this everyday environment, he argues that what is first or primary is the *praxical*. We have dealings first with things which we put to use.

> The kind of dealing which is closest to us is...not a bare perceptual cognition, but rather that kind of concern which manipulates things and puts them to use; and this has its own kind of "knowledge"...Such entities are not thereby objects for knowing the "world" theoretically they are simply what gets used, what gets produced, and so forth. (BT 95)

Heidegger argues that to take "things" interpreted as bare entities with properties is already to have presupposed an ontology prior to the actual investigation of human engagement with the environment.

It is from this argument that Heidegger constructs two different ways of relating to entities with the environment. These two ways of relating are well known as the distinction between the "ready-to-hand" (*Zuhandenheit*) and the "present-at-hand" (*Vorhandenheit*). It must be noted that both are qualitatively different relations to entities within the environment.

Heidegger's inversion of Husserl is one which makes a strong contrast between the "present-at-hand" relation and the "ready-to-hand" relation. The first is one in which entities (beings) appear as "just there" and as having certain qualities or predicates. They are "theoretically determined". Contrarily, the "ready-to-hand" belongs to the stratum of productive use or other forms of active engagement which characterize *praxis*. And Heidegger's strategy in *Being and Time* is to show that these are not merely two alternate modes of relation, but that one is founded upon the other, in this case the "present-at-hand" upon the "ready-to-hand". This is, in effect, an action theory of ontology.

Interestingly, what prevents the contemporary era from seeing the primacy of praxis, Heidegger contends, may be laid to the door of Greek philosophy. "The Greeks had an appropriate term of 'Things': *pragmata* – that is to say, that which one has to do with one's concernful dealings (*praxis*). But ontologically, the specifically 'pragmatic' character of the *pragmata* is just what the Greeks left in obscurity; they thought of these 'proximally' as 'mere Things.'" (BT 96–7)

I have already noted in the first section of this essay that, for Heidegger, whatever appears does so in terms of a whole. The same occurs with respect to "tools" which are what most interpreters of Heidegger on this point seem to think Heidegger is talking about. Phenomenologically, however, it should be noted that the analysis Heidegger undertakes is effectively a relational analysis in which the distinguishing features of the intentional arc are what are being described. Thus one will find that the "tool analysis" begins with the noematic correlate, the context and entity as it occurs phenomenologically. Later, and reflexively referred back to, is its noetic correlate, the mode of engagement which is entered into by the human exister, Dasein. The totality, Dasein-relating in the mode of the ready-to-hand with entities within the world, determines what shows itself overall.

The noematic description of the analysis begins typically with the phenomenological observation that no entity (whether in the mode of ready-to-hand or present-at-hand for that matter) occurs except in a context and against a background. Thus a "tool" shows itself *only* as already in a context, an equipmental context.

> Taken strictly, there "is" no such thing as *an* equipment. To the Being of any equipment there always belongs a totality of equipment, in which it can be this equipment that it is. Equipment is essentially "something-in-order-to"....A totality of equipment is constituted by various ways of the "in-order-to," such as serviceability, conduciveness, usability, manipulability. (BT 97)

This context in which equipment occurs has, moreover, a variable structure. "In the 'in-order-to' as a structure there lies an *assignment* or *reference* of something to something." (BT 97) This is to say that any given piece of equipment is what it is in an equipmental context and that it appears in such and such a way relative to that context. The homely illustrations Heidegger employs (ink pens belonging to the context of the desk, writing paper, etc.) show both the way in which an individual "tool" belongs to a context and how the context is variable. But it is noteworthy that even at this first level, the whole is what determines the part. "Out of this the "arrangement" emerges, and it is in this that any "individual" item of equipment shows itself. *Before* it does so, a totality of equipment has already been discovered." (BT 98). Here is the model of how world is "already discovered" in hidden and latent form through the use of a piece of equipment.

What emerges from this analysis is a description of equipmentally intentional structures which Heidegger calls the ready-to-hand. It is the equipmental (noematic) context which is the condition for the manifestation of a "tool" as ready-to-hand.

> The kind of Being which equipment possesses – in which it manifests itself in its own right – we call "readiness-to-hand". Only because equipment has *this* "Being-in-itself" and does not merely occur, is it manipulable in the broadest sense and at our disposal. (BT 98)

What is more, it is from this structure, that Heidegger contends one can detect a kind of *praxical knowledge* which is distinct from what we ordinarily think of as theoretical knowledge. A simply predicative knowledge of things described by properties misses this stratum. "no matter how sharply we just *look* at the 'outward appearance' of Things in whatever form this takes, we cannot discover anything ready-to-hand." (BT 98)

Contrarily, it is only in use that the distinctive characteristics of the ready-to-hand emerge. "When we deal with them by using them and manipulating them, this activity is not a blind one; it has its own kind of sight, by which our manipulation is guided and from which it acquires its specific Thingly character." (BT 98) Here the turn is made to the noetic correlate. The sight which emerges in active use, noetically, is also a field characteristic of human engagement, *circumspection*. "The sight with which they thus accommodate themselves is *circumspection*...action has *its own* kind of sight." (BT 98–9)

Heidegger sets off in strongest terms the difference between this praxical sight and a theoretical observation. The latter would focus its gaze *upon* the "tool" and thus make of it an object having such and such properties – but this precisely hides the distinctive character of the entity in use. It is the peculiar manifestation of the tool in use which is the secret to praxical sight. The tool in use appears, not as an object to be seen, but recedes or withdraws.

> The peculiarity of what is proximally ready-to-hand is that, in its readiness-to-hand, it must, as it were, withdraw in order to be ready-to-hand quite authentically. That with which our everyday dealings proximally dwell is not the tools themselves. On the contrary, that with which we concern ourselves primarily is the work. (BT 99)

Here is an essential insight concerning the ready-to-hand. The entity in praxical use "withdraws" or is taken into a manifestation which is partially "transparent". This is one reason why the ready-to-hand may be so easily overlooked and also a reason for the inappropriateness of a predicate analysis. But it is, however, a phenomenologically positive feature of the appearance. It is, moreover, thoroughly in keeping with the intentionality analysis being presupposed by Heidegger. The human user refers *through* the tool-equipment towards world in which the work or result appears. A Thing in the mode of ready-to-hand is radically different from a Thing in the mode of being "just there" or present-at-hand.

Although a full characterization of the mode of the present-at-hand is not called for in this essay, its relationship with the mode of the ready-to-hand is. It might be thought that the two modes could merely be variants upon concern with the world, but this is not the use to which Heidegger puts his distinction. Rather, he argues that one is the condition for the other, that readiness-to-hand *precedes* presence-at-hand and it is this argument which is both the inversion of Husserlian phenomenology and the source of what later becomes the primacy of technology in relation to science.

The themes which arise in this argument are precisely those which arise concerning technology in the later lecture. First, readiness-to-hand is a mode of disclosure. It is through the ready-to-hand that the environment appears as a "world". Praxis discovers Nature through the ready-to-hand. Heidegger's analysis traces this discovery not merely from a subject, but intersubjectively and on through wider and wider reaches until Nature is seen in a certain way:

> Any work with which one concerns oneself is ready-to-hand not only in the domestic world of the workshop but also in the *public* world. Along with the public world, the *environing Nature* is discovered and is accessible to everyone. In roads, streets, bridges, buildings, our

> concern discovers Nature as having some definite direction. A covered railway platform takes account of bad weather; an installation for public lighting takes account of the darkness...In a clock account is taken of some definite constellation in the world-system-...When we make use of the clock-equipment, which is proximally and inconspicuously ready-to-hand, the environing Nature is ready-to-hand along with it. (BT 100–1)

Here one sees the anticipation in *Being and Time* of the way in which the founding totality is seen through a mode of disclosure. The ready-to-hand discovers world, but only implicitly because world lies "behind" the partial withdrawal of the equipment in its use.

> Our concernful absorption in whatever work-world lies closest to us, has a function of discovering; and it is essential to this function that, depending upon the way in which we are absorbed, those entities within-the-world which are brought along in the work and with it...remain discoverable in varying degrees of explicitness and with a varying circumspective penetration. (BT 101)

Second, what is ultimately revealed is the world as a whole. "The context of equipment is lit up, not as something never seen before, but as a totality constantly sighted beforehand in circumspection. With this totality, however, the world announces itself." (BT 105)

Third, once disclosed, world is seen to be that in which Dasein already was, that in which Dasein has its relation of being-in.

> "The world is therefore something 'wherein' Dasein as an entity already *was*, and if in any manner it explicitly comes away from anything, it can never do more than come back to the world. Being-in-the-world, according to our Interpretation hitherto, amounts to a non-thematic circumspection absorption in references or assignments constitutive for the readiness-to-hand of a totality of equipment." (BT 106–7)

Each of these themes is isomorphic with their later reiteration under the aegis of technology as the current world epoch. The connection with technology has been anticipated in the primacy of the ready-to-hand announced in *Being and Time*. Moreover, the connection between ready-to-hand and world occurs by use of Heidegger's inversion which takes a specific, but peculiar turn in *Being and Time*.

What is peculiar about the mode of the ready-to-hand is precisely the way in which the entities, equipment, manifest themselves by paradoxically withdrawing in use. This partial transparency in use functions to conceal the very context in which the equipment occurs. In noting this, Heidegger is considerably subtle in his phenomenological tactics, but, simultaneously, he begins to employ what I shall call the *negative turn* to isolate the structural characteristic he is interested in displaying.

Equipment in use appears as partially transparent, as hidden from *direct* observation. To show this, Heidegger inverts the situation and contends that the equipmental context (which is the first index for world) appears through *negativity* when the equipment somehow *fails* to function.

There are two reasons for this negative turn. The first is tactical with respect to presence-at-hand. Heidegger argues that the mode of relationship which is theoretical, the present-at-hand, *cannot* discover either equipment or an equipmental context. One does not uncover the praxical at all by adding predicates to an object. A "tool" is not a bare physical entity to which one may add "values", nor is its serviceability or usability seen by a bare perceptual cognition. Thus the negative turn functions, in part, to short-circuit the temptation to give an account of the ready-to-hand in terms of a theoretical metaphysics. Regarding equipment, "we discover its unusability, however, not by looking at it and establishing its properties, but rather by the circumspection of the dealings in which we use it." (BT 102)

The second reason functions as a positive phenomenological tactic by making what must be described as the partial transparency of equipment in use appear *indirectly*. Thus by this variation – no different in function than a Husserlian fantasy variation – Heidegger displays this feature of the ready-to-hand by noting that a piece of equipment which malfunctions, is unusable or even missing serves to indirectly light up its genuine function. But in the process, the negative appearance must be characterized in partial thing-like terms: conspicuousness, obtrusiveness, obstinancy. "When its unusability is thus discovered, equipment becomes conspicuous. This *conspicuousness* presents the ready-to-hand equipment as in a certain un-readiness-to-hand...When we notice what is un-ready-to-hand, that which is ready-to-hand enters the mode of *obtrusiveness*." (BT 102–3)

This is to say that a malfunctioning piece of equipment emerges from its functional transparency and becomes a "thing" which just lies there. Indeed, it is from this *negative* characterization that Heidegger derives the origin of the present-at-hand!

> Anything which is un-ready-to-hand in this way is disturbing to us, and enables us to see the *obstinancy* of that with which we must concern ourselves in the first instance before we do anything else. With this obstinancy, the presence-at-hand of the ready-to-hand makes itself known in a new way as the Being of that which still lies before us and calls for our attending to it. (BT 103–4)

Presence-at-hand is, in this way, dependent upon the primacy of the ready-to-hand. "The modes of conspicuousness, obtrusiveness, and obstinancy all have the function of bringing to the fore the characteristic of presence-at-hand in what is ready-to-hand." (BT 104)

Now, once emergent *from* the ready-to-hand, the mode of presence-at-hand can attain its own relative autonomy. It becomes possible to attend to things predicatively, theoretically. But at the same time presence-at-hand has been derived from its praxical base. This derivative character of the present at-hand carries with it, at first, the interpretation which casts it negatively as a *deficient* mode of concern. "It [equipment] reveals itself as something just present-at-hand and no more, which cannot be budged without the thing that is missing. The helpless way in which we stand before it is a deficient mode of concern, and as such it uncovers the Being-just-present-at-hand-and-no-more of something ready-to-hand." (BT 103)

I take it that this inversion is strongly indicative of both the primacy of technology and of praxis in Heidegger's later phenomenology, but it is also penultimate with respect to the ultimate strategic use to which the negative turn is put. The purpose of the analysis is to get at the world which belongs to the ready-to-hand and the inversion is but one step along that way. What equipmental negativity ultimately reveals is the latent context to which it belongs, the "world" inhabited by concern.

> *When an assignment has been disturbed* – when something is unusable for some purpose – then the assignment becomes explicit...When an assignment to some particular "towards this" has been thus circumspectly aroused, we catch sight of the "towards this" itself, and along with it everything connected with the work – the whole workshop – as that wherein concern always dwells. (BT 105)

It may now be seen that the basic strategic and functional elements which characterize the philosophy of technology found in "The Question Concerning Technology" were present in the much earlier opus, *Being and Time*, although they are not specifically identified with technology as such there. Nevertheless, *praxis* in *Being and Time* functions as the basic existential stratum *through* which world is revealed and as the basic realm of action *from* which sciences may arise (as processes of theoretically developing presence-at-hand).

The emphasis upon praxis as existentially basic is what characterizes the Heideggerian inversion of Husserlian phenomenology. Thus it may be said with more than a touch of correct-parallelism that Heidegger is to Husserl what Marx was to Hegel.

III Technology as Emergent Theme

In this retrospective reading of Heidegger on technology, I have admittedly stressed those elements which are isomorphic between *Being and Time* and the technology lecture. These isomorphisms are basic to his philosophy of technology, but there are two related anomalies concerning praxis and technology in the early as compared to the later Heidegger and it is from these that I shall lay the groundwork for the concluding section of this essay.

Being and Time does not specifically raise the question of technology, although it may easily be seen that the praxical dimension of the ready-to-hand could become interpreted as the condition of the possibility for technology. What is missing in an explicit sense in *Being and Time* is the specific characterization of world taken as standing-reserve. There is a hint of this, to be sure, in that Nature becomes *available* to the ready-to-hand. "So in the environment certain entities become accessible which are always ready-to-hand, but which, in themselves, do not need to be produced. Hammer, tongs, and needle, refer in themselves to steel, iron, metal, mineral, wood, in that they consist of these. In equipment that is used,

'Nature' is discovered along with it by that use – the 'Nature' we find in natural products." (BT 100). Here we have an anticipation of the idea of standing-reserve in a particular interpretation of Nature which is linked to readiness-to-hand.

Heidegger contrasts this concept of Nature with that which he finds in science, which in *Being and Time* is essentially an abstract nature derived from a theoretical interpretaion of the present-at-hand. In *Being and Time* he makes the *contrast* between the Nature of the ready-to-hand and the Nature of the present-at-hand as strong as possible. The Nature of the ready-to-hand "is not to be understood as that which is *just* present-at-hand, nor as a *power of Nature*." (BT 100, first italic mine). The Nature of the ready-to-hand does anticipate the notion of standing-reserve, "the wood is a forest of timber, the mountain a quarry of rock; the river is water-power, the wind is wind 'in the sails'. As the 'environment' is discovered, the 'Nature' thus discovered is encountered too." (BT 100) But this ready-to-hand Nature contrasts with the "*just there*" Nature of the present-at-hand. "If its kind of Being as ready-to-hand is disregarded, this 'Nature' itself can be discovered and defined simply in its pure presence-at-hand...when this happens, the Nature which 'stirs and strives', which assails us and enthralls us as landscape, remains hidden. The botanist's plants are not the flowers of the hedgerow; the 'source' which the geographer establishes for a river is not the 'springhead in the dale.' " (BT 100)

What Heidegger has not yet discovered in *Being and Time* is the profound link between contemporary science and technology. The "science" of *Being and Time* is essentially a metaphysical and even contemplative science. It is a science derived from what may now be seen to be the ancient Greek ideal of speculation and deduction. It is not yet the science which is necessarily *embodied* in instrumentation; nor is it the science which is in the service of technology as calculative standing-reserve of the lecture.

Thus the latent technics of *Being and Time* remain either innocuous or even positive. The "tool analysis" has often been noted to be highly selective in one respect. Heidegger chooses as examples equipment which is used "in hand", technologies which are directly employed in work projects, technologies which extend human capacities often in terms of *handiwork*. This selectivity, I shall note, colors the entire analysis and is one element of a certain Heideggerian inadequacy of interpretation regarding technics. But first, in this context, this selectivity gives a certain tone of positivity to the ready-to-hand which is lacking in the contrasting "abstractness" of the present-at-hand.

If the first contrast between the lecture and *Being and Time* revolves around the notion of standing-reserve as the essence of technology, a second anomaly revolves around what may be called the "disappearance of the object" which functions differently in the early and later publications. In a sense, the *object* is what appears or is constituted *by* metaphysically based science in *Being and Time*. That which just stands there and which can be made the theme for presence-at-hand is the object. The object, which is characterized by predicates, is the noema of science in the view of *Being and Time*. Contrarily, equipment in use withdraws and is neither objectified nor does it appear as directly present at all.

The negative tone which permeates the Heideggerian analysis of presence-at-hand, however, is directed at its reductionism. The object is "abstract", reduced, not the full and rich Thing (the springhead in the dale) which is experienced in daily life. The object is the reduced noema of scientific contemplation. It is derived from and set aside from the full existentiality of *praxis*.

By the time of the technology lecture, however, the object has also disappeared from science. Under the concept of the standing-reserve, "whatever stands by in the sense of standing-reserve no longer stands over against us as object." (T 298) Here objects *and equipment* are, in effect, absorbed into the new totality. "Then in terms of the standing-reserve, the machine is completely unautonomous, for it has its standing only from the ordering of the orderable." (T 298–9) Nature, already noted as taken into technology as standing-reserve, is now accompanied by "tools" as well. The technological world is one in which the noematic correlate is simply standing-reserve and the noetically normative response is that of ordering this reserve.

Now many critics of Heidegger see in these moves simply a Heideggerian preference for the Romantic themes of much past German philosophy – and it must be admitted that Heidegger is not blameless for offering occasion for such criticism. The implicit problem of *Being and Time* is the reductionism of the sciences of the present-at-hand in that the object reduces and loses the full sense of existentiality. Symptomatically, nature as that which "stirs and strives," as the "springhead in the dale" is lost. In the technology lecture it would seem that what is reduced and lost is the "toolshop" itself, and with it the direct expressivity which characterized the ready-to-hand of *Being and Time*. I will not deny that Heidegger provides clues himself for such an interpretation – but it seems to me that this misses much of the thrust of the Heideggerian philosophy of technology.

There is another side to the interpretation of technology which does emerge from this surface negativity. This may be seen in the transpositions which occur between *Being and Time* and the lecture. The lecture does not make anything of the distinction between the ready-to-hand and presence-at-hand, but it does elevate to the fore a strong and comprehensive concept of technology. And it seems to me that this concept is one which *combines* certain features of both the present-at-hand and the ready-to-hand in such a way that we may speak of a unique *scientific technology*. Thus, in spite of his playing down of a distinction between traditional handiwork technology and contemporary technology, Heidegger in effect recognizes the uniqueness of the latter. Indeed, the clue to the combination is not far from the surface.

If one returns to the contrast between presence-at-hand and the ready-to-hand of *Being and Time*, not only does one note the essentially positive tone which permeates the discussion of the ready-to-hand, but sees in it certain base constants which re-emerge with a different evaluation in the technology lecture. Recall several of these key features: (a) world is revealed *through* the equipmental context; (b) the equipmental context is the condition of the possibility of specific "tools" being what they are; (c) noetically, engagement of the environment through readiness-to-hand reveals existential intentionality as *concern* (which is an index of Care in *Being and Time*), and (d) concern takes account of the context holistically as circumspection. This *praxical* dimension is where the *essence* of Dasein is shown and effected. Now each of these elements remains constant with the later technological interpretation of the contemporary world in Heidegger, but the earlier clearly positive tone coloring these elements is transformed into the ambiguous sense of *danger* which characterizes the technological world.

Contrarily, the brief characterizations of presence-at-hand in the "tool analysis" are often marked by partially negative characterizations. The present-at-hand originates by means of (a) deficient mode of concern (BT 103) and is characterized as a matter of entities appearing under the guise of (b) bare perceptual cognitions (BT 95), (c) just looking (BT 98), and (d) as abstract reductions interpreted as a world-stuff (BT 101). Positively, (a) present-at-hand may be elevated into a kind of knowledge (science) which knows the world theoretically (BT 95), (b) which can be thematically ascertained (BT 106), but which appears accordingly as the ultimately reduced *functions* of the theoretically constituted world. "By reason of their Being-just-present-at-hand-and-no-more, these latter entities can have their 'properties' defined mathematically in 'functional concepts.' Ontologically, such concepts are possible only in relation to entities whose Being has the character of pure substantiality. Functional concepts are never possible except as formalized substantial concepts." (BT 122)

Now the concept of *technology* which pervades the lecture clearly *combines* elements from both sides of the earlier contrasting modes of relation. It remains the case that only through concern with the world, through what remains the praxical, is humanity effected in its essence. And it is only because it is effected in its essence that technology can be considered dangerous. "The threat to man does not come in the first instance from the potentially lethal machines and apparatus of technology. The actual threat has already afflicted man in his essence." (T 309) But what is now taken into the very way in which world is perceived are the previously negatively characterized "reductions" whereby world becomes mere standing-reserve.

I have indicated that latently the "nature" of the ready-to-hand already anticipates the notion of standing-reserve. Taking account of nature in such a way that the "wood is a forest of timber" is already to be open to a world taken as standing-reserve, but this is a necessary and not sufficient condition. What makes it sufficient is the *addition* of thematically and systematically taking "nature" into a calculative and *universal* view of nature as standing-reserve. But this is the metaphysics of what may be characterized as a scientific or theoretically organized technology and not that of any simple handiwork technology. Thus in some sense, the illuminating distinctions of the ready-to-hand and the present-at-hand of *Being and Time* collapse in the later work and become unified.

One result of this collapse is the elimination of any purely contemplative science. There can be no "just looking" in what should more correctly now be called a technological science. The Greek ideal is what is lost – and if Heidegger is correct, then those who think they are remaining true to this ideal are merely naive and open to being used by technological culture. As with the non-neutrality of technology, there can now be no neutrality to science.

Ironically, a compatible way of interpreting this collapse of readiness-to-hand and presence-at-hand in the later Heidegger is to see that the science latent within presence-at-hand, in contemporary technological science, has become an *existentialized* science. That is why it can be thought of as effecting humanity in its essence. I shall not speculate concerning how this might literally be the case in contemporary genetic engineering, however tempting such an excursis might be, but it is in such examples that one might see how humanity itself becomes standing-reserve in the Heideggerian sense.

Technology, then, becomes the combined powers of what was earlier both readiness-to-hand and presence-at-hand. Humanity is effected essentially because science itself is technological in its contemporary sense and operates in the praxical dimension. But in these transpositions the earlier positive tone given to the praxical also disappears and is replaced with the characterizations of technological culture as "dangerous", "ambiguous", "mysterious", and as harboring even a certain "monstrousness". It is from such characterizations that Heidegger's critical attitude towards technology provides material for an interpretation which sees him as dominantly pessimistic regarding humanity's future.

While Heidegger is hardly alone in this attitude, such an interpretation misses what provides not only an opening to a different hope, but the recasting of a different set of distinctions which were never fully developed in the Heideggerian corpus. I have already noted that Heidegger's hope against any totalizing closure concerning humanity lay in technics as *art*. There is a very good strategic reason for this choice.

First, art is a technics, akin thus to the concern which is exhibited in all praxical dealings with the world. It is thus already related to technology. "...Confrontation with [technology] must happen in a realm that is, on the one hand, akin to the essence of technology..." (T 317) And art is also "theoretical" in that it does not simply take the world as that which is to be used. Its "contemplative" attitude is thus akin to science in the earlier sense of the present-at-hand of *Being and Time*. It is interesting to note that "the botanist's plants are not the flowers of the hedgerow" (BT 100) not only contrasts with the reduced objects of a theoretically dominated presence-at-hand, but neither are they the use-sources for a sheerly praxical world. There is here a hint of a new contrast, a contrast between the now combined ready-to-hand-present-at-hand existential intentionality and the poetic being-towards-the world of Heidegger's "poetic dwelling".

Strategically, however, if artful praxis is akin to technological science in its technics and its possibility of thematic distance, its difference may also be noted. "Confrontation with [technology] must happen in a realm that is...on the other [hand], fundamentally different from it." (T 317) The difference lies in its *proliferation* of possibilities. Art is essentially anti-reductive in its imaginative fecundity. Its "worlds" are effectively endless.

I am thus suggesting that in terms of Heidegger's systematic concern with praxical, now technological humanity, artful praxis is not some simple addition to the current epoch of Being, but is the strategic counter-balance to what Heidegger fears is the threat of closure. There is thus an internal need for the turn to poetics, from Heidegger's point of view, as a response to the age of technology as the current epoch of Being.

Note

1 Quotations from "The Question Concerning Technology" are from *Martin Heidegger: Basic Writings*, edited by David Krell (Harper and Row, Publishers, 1977) and will be listed in the text simply as (T, pp.). Similarly, those quotations from *Being and Time* are from the John Macquarrie and Edward Robinson translation (Harper and Row, Publishers, 1962) and are listed in the text here as (BT, pp.).

26

Focal Things and Practices

Albert Borgmann

23 Focal Things and Practices

To see that the force of nature can be encountered analogously in many other places, we must develop the general notions of focal things and practices. This is the first point of this chapter. The Latin word *focus*, its meaning and etymology, are our best guides to this task. But once we have learned tentatively to recognize the instances of focal things and practices in our midst, we must acknowledge their scattered and inconspicuous character too. Their hidden splendor comes to light when we consider Heidegger's reflections on simple and eminent things. But an inappropriate nostalgia clings to Heidegger's account. It can be dispelled, so I will argue, when we remember and realize more fully that the technological environment heightens rather than denies the radiance of genuine focal things and when we learn to understand that focal things require a practice to prosper within. These points I will try to give substance in the subsequent parts of this chapter by calling attention to the focal concerns of running and of the culture of the table.

The Latin word *focus* means hearth. ... [According to what I call the device paradigm,] the hearth or fireplace, a thing, [appears] as the counterpart to the central heating plant, a device. It [has been] pointed out that in a pretechnological house the fireplace constituted a center of warmth, of light, and of daily practices. For the Romans the *focus* was holy, the place where the housegods resided. In ancient Greece, a baby was truly joined to the family and household when it was carried about the hearth and placed before it. The union of a Roman marriage was sanctified at the hearth. And at least in the early periods the dead were buried by the hearth. The family ate by the hearth and made sacrifices to the housegods before and after the meal. The hearth sustained, ordered, and centered house and family.[1] Reflections of the hearth's significance can yet be seen in the fireplace of many American homes. The fireplace often has a central location in the house. Its fire is now symbolical since it rarely furnishes sufficient warmth. But the radiance, the sounds, and the fragrance of living fire consuming logs that are split, stacked, and felt in their grain have retained their force. There are no longer images of the ancestral gods placed by the fire; but there often are pictures of loved ones on or above the mantel, precious things of the family's history, or a clock, measuring time.[2]

The symbolical center of the house, the living room with the fireplace, often seems forbidding in comparison with the real center, the kitchen with its inviting smells and sounds. Accordingly, the architect Jeremiah Eck has rearranged homes to give them back a hearth, "a place of warmth and activity" that encompasses cooking, eating, and living and so is central to the house whether it literally has a fireplace or not.[3] Thus we can satisfy, he says, "the need for a place of focus in our family lives."[4]

Originally "Focal Things and Practices" and "Wealth and the Good Life" from Albert Borgmann, *Technology and the Character of Everyday Life*, Chicago: University of Chicago Press, 1984, pp. 196–226, 285–9, abridged.

"Focus," in English, is now a technical term of geometry and optics. Johannes Kepler was the first so to use it, and he probably drew on the then already current sense of focus as the "burning point of lens or mirror."[5] Correspondingly, an optic or geometric focus is a point where lines or rays converge or from which they diverge in a regular or lawful way. Hence "focus" is used as a verb in optics to denote moving an object in relation to a lens or modifying a combination of lenses in relation to an object so that a clear and well-defined image is produced.

These technical senses of "focus" have happily converged with the original one in ordinary language. Figuratively they suggest that a focus gathers the relations of its context and radiates into its surroundings and informs them. To focus on something or to bring it into focus is to make it central, clear, and articulate. It is in the context of these historical and living senses of "focus" that I want to speak of focal things and practices. Wilderness on this continent, it now appears, is a focal thing. It provides a center of orientation; when we bring the surrounding technology into it, our relations to technology become clarified and well-defined. But just how strong its gathering and radiating force is requires further reflection. And surely there will be other focal things and practices: music, gardening, the culture of the table, or running.

We might in a tentative way be able to see these things as focal; what we see more clearly and readily is how inconspicuous, homely, and dispersed they are. This is in stark contrast to the focal things of pretechnological times, the Greek temple or the medieval cathedral that we have mentioned before. Martin Heidegger was deeply impressed by the orienting force of the Greek temple. For him, the temple not only gave a center of meaning to its world but had orienting power in the strong sense of first originating or establishing the world, of disclosing the world's essential dimensions and criteria.[6] Whether the thesis so extremely put is defensible or not, the Greek temple was certainly more than a self-sufficient architectural sculpture, more than a jewel of well-articulated and harmoniously balanced elements, more, even, than a shrine for the image of the goddess or the god. As Vincent Scully has shown, a temple or a temple precinct gathered and disclosed the land in which they were situated. The divinity of land and sea was focused in the temple.[7]

To see the work of art as the focus and origin of the world's meaning was a pivotal discovery for Heidegger. He had begun in the modern tradition of Western philosophy where ... the sense of reality is to be grasped by determining the antecedent and controlling conditions of all there is (the *Bedingungen der Möglichkeit* as Immanuel Kant has it). Heidegger wanted to outdo this tradition in the radicality of his search for the fundamental conditions of being. Perhaps it was the relentlessness of his pursuit that disclosed the ultimate futility of it. At any rate, when the universal conditions are explicated in a suitably general and encompassing way, what truly matters still hangs in the balance because everything depends on how the conditions come to be actualized and instantiated.[8] The preoccupation with antecedent conditions not only leaves this question unanswered; it may even make it inaccessible by leaving the impression that, once the general and fundamental matters are determined, nothing of consequence remains to be considered. Heidegger's early work, however, already contained the seeds of its overcoming. In his determination to grasp reality in its concreteness, Heidegger had found and stressed the inexorable and unsurpassable givenness of human existence, and he had provided analyses of its pretechnological wholeness and its technological distraction though the significance of these descriptions for technology had remained concealed to him.[9] And then he discovered that the unique event of significance in the singular work of art, in the prophet's proclamation, and in the political deed was crucial. This insight was worked out in detail with regard to the artwork. But in an epilogue to the essay that develops this point, Heidegger recognized that the insight comes too late. To be sure, our time has brought forth admirable works of art. "But," Heidegger insists, "the question remains: is art still an essential and necessary way in which that truth happens which is decisive for historical existence, or is art no longer of this character?"[10]

Heidegger began to see technology (in his more or less substantive sense) as the force that has eclipsed the focusing powers of pretechnological times. Technology becomes for him ... the final phase of a long metaphysical development. The philosophical concern with the conditions of the possibility of whatever is now itself seen as a move into the oblivion of what finally matters. But how are we to recover orientation in the oblivious and distracted era of technology when the great embodiments of meaning, the works of art, have lost their focusing power? Amidst the complication of conditions, of the *Bedingungen*, we must uncover the simplicity of things, of the *Dinge*.[11] A jug, an earthen vessel from which we pour wine, is such a thing. It teaches us what it is to hold, to offer, to pour, and to give. In its clay, it gathers for us the earth as it does in containing the wine that has grown from the soil. It gathers the sky whose rain and sun are present in the wine. It refreshes and animates us in our mortality. And in the libation it acknowledges and calls on the divinities. In these ways the thing (in agreement with its etymologically original meaning) gathers and discloses what Heidegger calls the fourfold, the interplay of the crucial dimensions of earth and sky, mortals and divinities.[12] A thing, in Heidegger's eminent sense, is a focus; to speak of focal things is to emphasize the central point twice.

Still, Heidegger's account is but a suggestion fraught with difficulties. When Heidegger described the focusing power of the jug, he might have been thinking of a rural setting where wine jugs embody in their material, form, and craft a long and local tradition; where at noon one goes down to the cellar to draw a jug of table wine whose vintage one knows well; where at the noon meal the wine is thoughtfully poured and gratefully received.[13] Under such circumstances, there might be a gathering and disclosure of the fourfold, one that is for the most part understood and in the background and may come to the fore on festive occasions. But all of this seems as remote to most of us and as muted in its focusing power as the Parthenon or the Cathedral of Chartres. How can so simple a thing as a jug provide that turning point in our relation to technology to which Heidegger is looking forward? Heidegger's proposal for a reform of technology is even more programmatic and terse than his analysis of technology.[14] Both, however, are capable of fruitful development.[15] Two points in Heidegger's consideration of the turn of technology must particularly be noted. The first serves to remind us of arguments already developed which must be kept in mind if we are to make room for focal things and practices. Heidegger says, broadly paraphrased, that the orienting force of simple things will come to the fore only as the rule of technology is raised from its anonymity, is disclosed as the orthodoxy that heretofore has been taken for granted and allowed to remain invisible.[16] As long as we overlook the tightly patterned character of technology and believe that we live in a world of endlessly open and rich opportunities, as long as we ignore the definite ways in which we, acting technologically, have worked out the promise of technology and remain vaguely enthralled by that promise, so long simple things and practices will seem burdensome, confining, and drab. But if we recognize the central vacuity of advanced technology, that emptiness can become the opening for focal things. It works both ways, of course. When we see a focal concern of ours threatened by technology, our sight for the liabilities of mature technology is sharpened.

A second point of Heidegger's is one that we must develop now. The things that gather the fourfold, Heidegger says, are inconspicuous and humble. And when we look at his litany of things, we also see that they are scattered and of yesterday: jug and bench, footbridge and plow, tree and pond, brook and hill, heron and deer, horse and bull, mirror and clasp, book and picture, crown and cross.[17] That focal things and practices are inconspicuous is certainly true; they flourish at the margins of public attention. And they have suffered a diaspora; this too must be accepted, at least for now. That is not to say that a hidden center of these dispersed focuses may not emerge some day to unite them and bring them home. But it would clearly be a forced growth to proclaim such a unity now. A reform of technology that issues from focal concerns will be radical not in imposing a new and unified master plan on the technological universe but in discovering those sources of strength that will nourish principled and confident beginnings, measures, i.e., which will neither rival nor deny technology.

But there are two ways in which we must go beyond Heidegger. One step in the first direction has already been taken. It led us to see ... that the simple things of yesterday attain a new splendor in today's technological context. The suggestion in Heidegger's reflections that we have to seek out pretechnological enclaves to encounter focal things is misleading and dispiriting. Rather we must see any such enclave itself as a focal thing heightened by its technological context. The turn to things cannot be a setting aside and even less an escape from technology but a kind of affirmation of it. The second move beyond Heidegger is in the direction of practice, into the social and, later, the political situation of focal things.[18] Though Heidegger assigns humans their place in the fourfold when he depicts the jug in which the fourfold is focused, we scarcely see the hand that holds the jug, and far less do we see of the social setting in which the pouring of the wine comes to pass. In his consideration of another thing, a bridge, Heidegger notes the human ways and works that are gathered and directed by the bridge.[19] But these remarks too present practices from the viewpoint of the focal thing. What must be shown is that focal things can prosper in human practices only. Before we can build a bridge, Heidegger suggests, we must be able to dwell.[20] But what does that mean concretely?

The consideration of the wilderness has disclosed a center that stands in a fruitful counterposition to technology. The wilderness is beyond the procurement of technology, and our response to it takes us past consumption. But it also teaches us to accept and to appropriate technology. We must now try to discover if such centers of orientation can be found in greater proximity and intimacy to the technological everyday life. And I believe they can be found if we follow up the hints that we have gathered from and against Heidegger, the suggestions that focal things seem humble and scattered but attain splendor in technology if we grasp technology properly, and that focal things require a practice for their welfare. Running and the culture of the table are such focal things and practices. We have all been touched by them in one way or another. If we have not participated in a vigorous or competitive run, we have certainly taken walks; we have felt with surprise, perhaps, the pleasure of touching the earth, of feeling the wind, smelling the rain, of having the blood course through our bodies more steadily. In the preparation of a meal we have enjoyed the simple tasks of washing leaves and cutting bread; we have felt the force and generosity of being served a good wine and homemade bread. Such experiences have been particularly vivid when we came upon them after much sitting and watching indoors, after a surfeit of readily available snacks and drinks. To encounter a few simple things was liberating and invigorating. The normal clutter and distraction fall away when, as the poet says,

> there, in limpid brightness shine,
> on the table, bread and wine.[21]

If such experiences are deeply touching, they are fleeting as well. There seems to be no thought or discourse that would shelter and nurture such events; not in politics certainly, nor in philosophy where the prevailing idiom sanctions and applies equally to lounging and walking, to Twinkies, and to bread, the staff of life. But the reflective care of the good life has not withered away. It has left the profession of philosophy and sprung up among practical people. In fact, there is a tradition in this country of persons who are engaged by life in its concreteness and simplicity and who are so filled with this engagement that they have reached for the pen to become witnesses and teachers, speakers of deictic discourse. Melville and Thoreau are among the great prophets of this tradition. Its present health and extent are evident from the fact that it now has no overpowering heroes but many and various more or less eminent practitioners. Their work embraces a spectrum between down-to-earth instruction and soaring speculation. The span and center of their concerns vary greatly. But they all have their mooring in the attention to tangible and bodily things and practices, and they speak with an enthusiasm that is nourished by these focal concerns. Pirsig's book is an impressive and troubling monument in this tradition, impressive in the freshness of its observations and its pedagogical skill, troubling in its ambitious and failing efforts to deal with the large philosophical issues. Norman Maclean's *A River Runs through It* can be taken as a fly-fishing manual, a virtue that pleases its author.[22] But it is a literary work of art most of all and a reflection on technology inasmuch as it presents the engaging life, both dark and bright, from which we have so recently emerged. Colin Fletcher's treatise of *The Complete Walker* is most narrowly a book of instruction about hiking and backpacking.[23] The focal significance of these things is found in the interstices of equipment and technique; and when the author explicitly engages in deictic discourse he has "an unholy awful time" with it.[24] Roger B. Swain's contemplation of gardening in *Earthly Pleasures* enlightens us in cool and graceful prose about the scientific basis and background of what we witness and undertake in our gardens.[25] Philosophical significance enters unbidden and easily in the reflections on time, purposiveness, and the familiar. Looking at these books, I see a stretch of water that extends beyond my vision, disappearing in the distance. But I can see that it is a strong and steady stream, and it may well have parts that are more magnificent than the ones I know.[26]

To discover more clearly the currents and features of this, the other and more concealed, American mainstream, I take as witnesses two books where enthusiasm suffuses instruction vigorously, Robert Farrar Capon's *The Supper of the Lamb* and George Sheehan's *Running and Being*.[27]

Both are centered on focal events, the great run and the great meal. The great run, where one exults in the strength of one's body, in the ease and the length of the stride, where nature speaks powerfully in the hills, the wind, the heat, where one takes endurance to the breaking point, and where one is finally engulfed by the good will of the spectators and the fellow runners.[28] The great meal, the long session as Capon calls it, where the guests are thoughtfully invited, the table has been carefully set, where the food is the culmination of tradition, patience, and skill and the presence of the earth's most delectable textures and tastes, where there is an invocation of divinity at the beginning and memorable conversation throughout.[29]

Such focal events are compact, and if seen only in their immediate temporal and spatial extent they are easily mistaken. They are more mistakable still when they are thought of as experiences in the subjective sense, events that have their real meaning in transporting a person into a certain mental or emotional state. Focal events, so conceived, fall under the rule of technology. For when a subjective state becomes decisive, the search for a machinery that is functionally equivalent to the traditional enactment of that state begins, and it is spurred by endeavors to find machineries that will procure the state more instantaneously, ubiquitously, more assuredly and easily. If, on the other hand, we guard focal things in their depth and integrity, then, to see them fully and truly, we must see them in context. Things that are deprived of their context become ambiguous.[30] The letter "a" by itself means nothing in particular. In the context of "table" it conveys or helps to convey a more definite meaning. But "table" it turn can mean many things. It means something more powerful in the text of Capon's book where he speaks of "The Vesting of the Table."[31] But that text must finally be seen in the context and texture of the world. To say that something becomes ambiguous is to say that it is made to say less, little, or nothing. Thus to elaborate the context of focal events is to grant them their proper eloquence.

"The distance runner," Sheehan says, "is the least of all athletes. His sport the least of all sports."[32] Running is simply to move through time and space, step-by-step. But there is splendor in that simplicity. In a car we move of course much faster, farther, and more comfortably. But we are not moving on our own power and in our own right. We cash in prior labor for present motion. Being beneficiaries of science and engineering and having worked to be able to pay for a car, gasoline, and roads, we now release what has been earned and stored and use it for transportation. But when these past efforts are consumed and consummated in my driving, I can at best take credit for what I have done. What I am doing now, driving, requires no effort, and little or no skill or discipline. I am a divided person; my achievement lies in the past, my enjoyment in the present. But in the runner, effort and joy are one; the split between means and ends, labor and leisure is healed.[33] To be sure, if I have trained conscientiously, my past efforts will bear fruit in a race. But they are not just cashed in. My strength must be risked and enacted in the race which is itself a supreme effort and an occasion to expand my skill.

This unity of achievement and enjoyment, of competence and consummation, is just one aspect of a central wholeness to which running restores us. Good running engages mind and body. Here the mind is more than an intelligence that happens to be housed in a body. Rather the mind is the sensitivity and the endurance of the body.[34] Hence running in its fullness, as Sheehan stresses over and over again, is in principle different from exercise designed to procure physical health. The difference between running and physical exercise is strikingly exhibited in one and the same issue of the *New York Times Magazine*. It contains an account by Peter Wood of how, running the New York City Marathon, he took in the city with body and mind, and it has an account by Alexandra Penney of corporate fitness programs where executives, concerned about their Coronary Risk Factor Profile, run nowhere on treadmills or ride stationary bicycles.[35] In another issue, the *Magazine* shows executives exercising their bodies while busying their dissociated minds with reading.[36] To be sure, unless a runner concentrates on bodily performance, often in an effort to run the best possible race, the mind wanders as the body runs. But as in free association we range about the future and the past, the actual and the possible, our mind, like our breathing, rhythmically gathers itself to the here and now, having spread itself to distant times and faraway places.

It is clear from these reflections that the runner is mindful of the body because the body is intimate with the world. The mind becomes relatively disembodied when the body is severed from the depth of the world, i.e., when the world is split into commodious surfaces and inaccessible machineries. Thus the unity of ends and means, of mind and body, and of body and world is one and the same. It makes itself felt in the vividness with which the runner experiences reality. "Somehow you feel more in touch," Wood says, "with the realities of a massive inner-city housing problem when you are running through it slowly enough to take in the grim details, and, surprisingly, cheered on by the remaining occupants."[37] As this last remark suggests, the wholeness that running establishes embraces the human family too. The experience of that simple event releases an equally simple and profound sympathy. It is a natural goodwill, not in need of drugs nor dependent on a common enemy. It wells up from depths that have been forgotten, and it overwhelms the runners ever and again.[38] As Wood recounts his running through streets normally besieged by crime and violence, he remarks: "But we can only be amazed today at the warmth that emanates from streets usually better known for violent crime." And his response to the spectators' enthusiasm is this: "I feel a great proximity to the crowd, rushing past at all of nine miles per hour; a great affection for them individually; a commitment to run as well as I possibly can, to acknowledge their support."[39] For George Sheehan, finally, running discloses the divine. When he runs, he wrestles with God.[40] Serious running takes us to the limits of our being. We run into threatening and seemingly unbearable pain. Sometimes, of course, the plunge into that experience gets arrested in ambition and vanity. But it can take us further to the point where in suffering our limits we experience our greatness too. This, surely, is a hopeful place to escape technology, metaphysics, and the God of the philosophers and reach out to the God of Abraham, Isaac, and Jacob.[41]

If running allows us to center our lives by taking in the world through vigor and simplicity, the culture of the table does so by joining simplicity with cosmic wealth. Humans are such complex and capable beings that they can fairly comprehend the world and, containing it, constitute a cosmos in their own right. Because we are standing so eminently over against the world, to come in touch with the world becomes for us a challenge and a momentous event. In one sense, of course, we are always already in the world, breathing the air, touching the ground, feeling the sun. But as we can in another sense withdraw from the actual and present world, contemplating what is past and to come, what is possible and remote, we celebrate correspondingly our intimacy with the world. This we do most fundamentally when in eating we take in the world in its palpable, colorful, nourishing immediacy. Truly human eating is the union of the primal and the cosmic. In the simplicity of bread and wine, of meat and vegetable, the world is gathered.

The great meal of the day, be it at noon or in the evening, is a focal event par excellence. It gathers the scattered family around the table. And on the table it gathers the most delectable things nature has brought forth. But it also recollects and presents a tradition, the immemorial experiences of the race in identifying and cultivating edible plants, in domesticating and butchering animals; it brings into focus closer relations of national or regional customs, and more intimate traditions still of family recipes and dishes. It is evident from the preceding chapters how this living texture is being rent through the procurement of food as a commodity and the replacement of the culture of the table by the food industry. Once food has become freely available, it is only consistent that the gathering of the meal is shattered and disintegrates into snacks, T.V. dinners, bites that are grabbed to be eaten; and eating itself is scattered around television shows, late and early meetings, activities, overtime work, and other business. This is increasingly the normal condition of technological eating. But it is within our power to clear a central space amid the clutter and distraction. We can begin with the simplicity of a meal that has a beginning, a middle, and an end and that breaks through the superficiality of convenience food in the simple steps of beginning with raw ingredients, preparing and transforming them, and bringing them to the table. In this way we can again become freeholders of our culture. We are disfranchised from world citizenship when the foods we eat are mere commodities. Being essentially opaque surfaces, they repel all efforts at extending our sensibility and competence into the deeper reaches of the world. A Big Mac and a Coke can overwhelm our tastebuds and accommodate our hunger. Technology is not, after all, a children's crusade but a principled and skillful enterprise of defining and satisfying human needs. Through the diversion and busyness of consumption we may have unlearned to feel constrained by the shallowness of commodities. But having gotten along for a time and quite well, it seemed, on institutional or convenience food, scales fall from our eyes when we step up to a festively set family table. The foods stand out more clearly, the fragrances are stronger, eating has once more become an occasion that engages and accepts us fully.

To understand the radiance and wealth of a festive meal we must be alive to the interplay of things and humans, of ends and means. At first a meal, once it is on the table, appears to have commodity character since it is now available before us, ready to be consumed without effort or merit. But though there is of course in any eating a moment of mere consuming, in a festive meal eating is one with an order and discipline that challenges and ennobles the participants. The great meal has its structure. It begins with a moment of reflection in which we place ourselves in the presence of the first and last things. It has a sequence of courses; it requires and sponsors memorable conversation; and all this is enacted in the discipline called table manners. They are warranted when they constitute the respectful and skilled response to the great things that are coming to pass in the meal. We can see how order and discipline have collapsed when we eat a Big Mac. In consumption there is the pointlike and inconsequential conflation of a sharply delimited human need with an equally contextless and closely fitting commodity. In a Big Mac the sequence of courses has been compacted into one object and the discipline of table manners has been reduced to grabbing and eating. The social context reaches no further than the pleasant faces and quick hands of the people who run the fast-food outlet. In a festive meal, however, the food is served, one of the most generous gestures human beings are capable of. The serving is of a piece with garnishing; garnishing is the final phase of cooking, and cooking is one with preparing the food. And if we are blessed with rural circumstances, the preparation of food draws near the harvesting and the raising of the vegetables in the garden close by. This context of activities is embodied in persons. The dish and the cook, the vegetable and the gardener tell of one another. Especially when we are guests, much of the meal's deeper context is socially and conversationally mediated. But that mediation has translucence and intelligibility because it extends into the farther and deeper recesses without break and with a bodily immediacy that we too have enacted or at least witnessed firsthand. And what seems to be a mere receiving and consuming of food is in fact the enactment of generosity and gratitude, the affirmation of mutual and perhaps religious obligations. Thus eating in a focal setting differs sharply from the social and cultural anonymity of a fast-food outlet.

The pretechnological world was engaging through and through, and not always positively. There also was ignorance, to be sure, of the final workings of God and king; but even the unknown engaged one through mystery and awe. In this web of engagement, meals already had focal character, certainly as soon as there was anything like a culture of the table.[42] Today, however, the great meal does not gather and order a web of thoroughgoing relations of engagement; within the technological setting it stands out as a place of profound calm, one in which we can leave behind the narrow concentration and one-sided strain of labor and the tiring and elusive diversity of consumption. In the technological setting, the culture of the table not only focuses our life; it is also distinguished as a place of healing, one that restores us to the depth of the world and to the wholeness of our being.

As said before, we all have had occasion to experience the profound pleasure of an invigorating walk or a festive meal. And on such occasions we may have regretted the scarcity of such events; we might have been ready to allow such events a more regular and central place in our lives. But for the most part these events remain occasional, and indeed the ones that still grace us may be slipping from our grasp. ... we have seen various aspects of this malaise, especially its connection with television. But why are we acting against our better insights and aspirations?[43] This at first seems all the more puzzling as the engagement in a focal activity is for most citizens of the technological society an instantaneous and ubiquitous possibility. On any day I can decide to run or to prepare a meal after work. Everyone has some sort of suitable equipment. At worst one has to stop on the way home to pick up this or that. It is of course technology that has opened up these very possibilities. But why are they lying fallow for the most part? There is a convergence of several factors. Labor is exhausting, especially when it is divided. When we come home, we often feel drained and crippled. Diversion and pleasurable consumption appear to be consonant with this sort of disability. They promise to untie the knots and to soothe the aches. And so they do at a shallow level of our existence. At any rate, the call for exertion and engagement seems like a cruel and unjust demand. We have sat in the easy chair, beer at hand and television before us: when we felt stirrings of ambition, we found it easy to ignore our superego.[44] But we also may have had our alibi refuted on occasion when someone to whom we could not say no prevailed on us to put on our coat and to step out into cold and windy weather to take a walk. At first our indignation grew. The discomfort was worse than we had thought. But gradually a transformation set in. Our gait became steady, our blood began to flow vigorously and wash away our tension, we smelled the rain, began thoughtfully to speak with our companion, and finally returned home settled, alert, and with a fatigue that was capable of restful sleep.

But why did such occurrences remain episodes also? The reason lies in the mistaken assumption that the shaping of our lives can be left to a series of individual decisions. Whatever goal in life we entrust to this kind of implementation we in fact surrender to erosion. Such a policy ignores both the frailty and strength of human nature. On the spur of the moment, we normally act out what has been nurtured in our daily practices as they have been shaped by the norms of our time. When we sit in our easy chair and contemplate what to do, we are firmly enmeshed in the framework of technology with our labor behind us and the blessings of our labor about us, the diversions and enrichments of consumption. This arrangement has had our lifelong allegiance, and we know it to have the approval and support of our fellows. It would take superhuman strength to stand up to this order ever and again. If we are to challenge *the rule of technology*, we can do so only through *the practice of engagement*.

The human ability to establish and commit oneself to a practice reflects our capacity to comprehend the world, to harbor it in its expanse as a context that is oriented by its focal points. To found a practice is to guard a focal concern, to shelter it against the vicissitudes of fate and our frailty. John Rawls has pointed out that there is decisive difference between the justification of a practice and of a particular action falling under it.[45] Analogously, it is one thing to decide for a focal practice and quite another to decide for a particular action that appears to have focal character.[46] Putting the matter more clearly, we must say that without a practice an engaging action or event can momentarily light up our life, but it cannot order and orient it focally. Competence, excellence, or virtue, as Aristotle first saw, come into being as an *éthos*, a settled disposition and a way of life.[47] Through a practice, Alasdair MacIntyre says accordingly, "human powers to achieve excellence, and human conceptions of the ends and goods involved, are systematically extended."[48] Through a practice we are able to accomplish what remains unattainable when aimed at in a series of individual decisions and acts.

How can a practice be established today? Here, as in the case of focal things, it is helpful to consider the foundation of pretechnological practices. In mythic times the latter were often established through the founding and consecrating act of a divine power or mythic ancestor. Such an act ... set up a sacred precinct and center that gave order to a violent and hostile world. A sacred practice, then, consisted in the regular reenactment of the founding act, and so it renewed and sustained the order of the world. Christianity came into being this way; the eucharistic meal, the Supper of the Lamb, is its central event, established with the instruction that it be reenacted. Clearly a focal practice today should have centering and orienting force as well. But it differs in important regards from its grand precursors. A mythic focal practice derived much force from the power of its opposition. The alternative to the preservation of the cosmos was chaos, social and physical disorder and collapse. It is a reduction to see mythic practices merely as coping behavior of high survival value. A myth does not just aid survival; it defines what truly human life is. Still, as in the case of pretechnological morality, economic and social factors were interwoven with mythic practices. Thus the force of brute necessity supported, though it did not define, mythic focal practices. Since a mythic focal practice united in itself the social, the economic, and the cosmic, it was naturally a prominent and public affair. It rested securely in collective memory and in the mutual expectations of the people.

This sketch, of course, fails to consider many other kinds of pretechnological practices. But it does present one important aspect of them and more particularly one that serves well as a backdrop for focal practices in a technological setting. It is evident that technology is itself a sort of practice, and it procures its own kind of order and security. Its history contains great moments of innovation, but it did not arise out of a founding event that would have focal character; nor has it ... produced focal things. Thus it is not a *focal* practice, and it has indeed, so I have urged, a debilitating tendency to scatter our attention and to clutter our surroundings. A focal practice today, then, meets no tangible or overtly hostile opposition from its context and is so deprived of the wholesome vigor that derives from such opposition. But there is of course an opposition at a more profound and more subtle level. To feel the support of that opposing force one must have experienced the subtly debilitating character of technology, and above all one must understand, explicitly or implicitly, that the peril of technology lies not in this or that of its manifestations but in *the pervasiveness and consistency of its pattern*. There are always occasions where a Big Mac, an exercycle, or a television program are unobjectionable and truly helpful answers to human needs. This makes a case-by-case appraisal of technology so inconclusive. It is when we attempt to take the measure of technologial life in its normal totality that we are distressed by its shallowness. And I believe that the more strongly we sense and the more clearly we understand the coherence and the character of technology, the more evident it becomes to us that technology must be countered by an equally patterned and social commitment, i.e., by a practice.

At this level the opposition of technology does become fruitful to focal practices. They can now be seen as restoring a depth and integrity to our lives that are in principle excluded within the paradigm of technology. MacIntyre, though his foil is the Enlightenment more than technology, captures this point by including in his definition of practice the notion of "goods internal to a practice."[49] These are one with the practice and can only be obtained through that practice. The split between means and ends is healed. In contrast "there are those goods externally and contingently attached" to a practice; and in that case there "are always alternative ways for achieving such goods, and their achievement is never to be had *only* by engaging in some particular kind of practice."[50] Thus practices (in a looser sense) that serve external goods are subvertible by technology. But MacIntyre's point needs to be clarified and extended to include or emphasize not only the essential unity of human being and a particular sort of doing but also the tangible things in which the world comes to be focused. The importance of this point has been suggested by the consideration of running and the culture of the table. There are objections to this suggestion that will be examined in the next chapter. Here I want to advance the thesis by considering Rawls's contention that a practice is defined by rules. We can take a rule as an instruction for a particular domain of life to act in a certain way under specified circumstances. How important is the particular character of the tangible setting of the rules? Though Rawls does not address this question directly he suggests in using baseball for illustration that "a peculiarly shaped piece of wood" and a kind of bag become a bat and base only within the confines defined by the rules of baseball.[51] Rules and the practice they define, we might argue in analogy to what Rawls says about their relation to particular cases, are logically prior to their tangible setting. But the opposite contention seems stronger to me. Clearly the possibilities and challenges of baseball are crucially determined by the layout and the surface of the field, the weight and resilience of the ball, the shape and size of the bat, etc. One might of course reply that there are rules that define the physical circumstances of the game. But this is to take "rule" in broader sense. Moreover it would be more accurate to say that the rules of this latter sort reflect and protect the identity of the original tangible circumstances in which the game grew up. The rules, too, that circumscribe the actions of the players can be taken as ways of securing and ordering the playful challenges that arise in the human interplay with reality. To be sure there are developments and innovations in sporting equipment. But either they quite change the nature of the sport as in pole vaulting, or they are restrained to preserve the identity of the game as in baseball.

It is certainly the purpose of a focal practice to guard in its undiminished depth and identity the thing that is central to the practice, to shield it against the technological diremption into means and end. Like values, rules and practices are recollections, anticipations, and, we can now say, guardians of the concrete things and events that finally matter. Practices protect focal things not only from technological subversion but also against human frailty. It was emphasized ... that the ultimately significant things to which we respond in deictic discourse cannot be possessed or controlled. Hence when we reach out for them, we miss them occasionally and sometimes for quite some time. Running becomes unrelieved pain and cooking a thankless chore. If in the technological mode we insisted on assured results or if more generally we estimated the value of future efforts on the basis of recent experience, focal things would vanish from our lives. A practice keeps faith with focal things and saves for them an opening in our lives. To be sure, eventually the practice needs to be empowered again by the reemergence of the great thing in its splendor. A practice that is not so revived degenerates into an empty and perhaps deadening ritual.

We can now summarize the significance of a focal practice and say that such a practice is required to counter technology in its patterned pervasiveness and to guard focal things in their depth and integrity. Countering technology through a practice is to take account of our susceptibility to technological distraction, and it is also to engage the peculiarly human strength of comprehension, i.e., the power to take in the world in its extent and significance and to respond through an enduring commitment. Practically a focal practice comes into being through resoluteness, either an explicit resolution where one vows regularly to engage in a focal activity from this day on or in a more implicit resolve that is nurtured by a focal thing in favorable circumstances and matures into a settled custom.

In considering these practical circumstances we must acknowledge a final difference between focal practices today and their eminent pretechnological predecessors. The latter, being public and prominent, commanded elaborate social and physical settings: hierarchies, offices, ceremonies, and choirs; edifices, altars, implements, and vestments. In comparison our focal practices are humble and scattered. Sometimes they can hardly be called practices, being private and limited. Often they begin as a personal regimen and mature into a routine without ever attaining the social richness that distinguishes a practice. Given the often precarious and inchoate nature of focal practices, evidently focal things and practices, for all the splendor of their simplicity and their fruitful opposition to technology, must be further clarified in their relation to our everyday world if they are to be seen as a foundation for the reform of technology.

24 Wealth and the Good Life

Strong claims have been made for focal things and practices. Focal concerns supposedly allow us to center our lives and to launch a reform of technology and so to usher in the good life that has eluded technology. At the end of the preceding chapter we have seen that focal practices today tend to be isolated and rudimentary. But these are marginal deficiencies, due to unfavorable circumstances. Surely there are central problems as well that pertain to focal practices no matter how well developed. Before we can proceed to suggestions about how technology may be reformed to make room for the good life, the most important objections regarding focal practices, the pivots of that reform, must be considered and, if possible, refuted. These disputations are not intended to furnish the impregnable defense of focal concerns, which ... is neither possible nor to be wished for. The deliberations of this chapter are rather efforts to connect the notion of a focal practice more closely with the prevailing conceptual and social situation and so to advance the standing of focal concerns in our midst. To make the technological universe hospitable to focal things turns out to be the heart of the reform of technology. What follows are first steps in this direction.

Among these, the first in turn requires us to consider the problem of the plurality of focal things and practices. It has a negative and positive aspect; negative because my devotion to a focal concern is rejected or challenged by the commitment of other people to contrary focal practices; positive because the plurality can have the character of a complementary richness in what is called a social union. The latter possibility, however, may be realized in the superficial diversity of various styles of consumption. As a counterforce to such shallowness I will consider in the first half of the present chapter the mode of developing one's faculties which is guided by the so-called Aristotelian Principle. It defines a notion of excellence which revolves about a notion of complexity. The more complex the faculties to whose cultivation we are devoted, the more excellent our life. This turns out to be an ambiguous result. Excellence so defined is no longer a counterforce to technology. On the other hand, it is compatible with a notion of engagement that seems to capture the most important aspirations of focal concerns and at the same time avoids the occasionally, perhaps essentially, constricting effects of the latter. When we measure these findings against an actual focal concern, we will see, however, that it is misguided to think of focal things as being entered in a competition with the concept of engagement and the Aristotelian Principle in a quest to reform technology. Only things that we experience as greater and other than ourselves can move us to judge and change technology in the first place.

Given this clarification of focal concerns we can without fear of misunderstanding explicate their generic features. On the basis of this generic definition of focal things and practices, an explicit definition of the reform of technology becomes possible. A reform so defined is neither the modification nor the rejection of the technological paradigm but the recognition and restraint of the pattern of technology so as to give focal concerns a central place in our lives. The remainder of this chapter provides a twofold application and elaboration of that reform proposal. First and applied to the private and personal realm, it will be seen to engender an intelligently selective attitude toward technology and a life of wealth in a well-defined sense. Second and in regard to traditional excellence and the family, the reform of technology makes possible a revival of these institutions.

First, then, we must consider the question of the plurality of focal commitments. A focal concern, it has been said, centers one's life. It is a final and dominant end which alone truly matters and fulfills and which therefore assigns all other things and activities their rank and place. But it is obvious that the ultimacy and dominance of a focal concern is contradicted by the fact that there are a number of different and apparently competing concerns. It cannot be that both running and fly-fishing matter ultimately. If one does, the other cannot. Focal practices in pretechnological times clearly possessed this dominance and exclusiveness. In the early Middle Ages, everyone went to church on Sundays and holy days, and Hubert, who went hunting, was a sinner for that reason. If focal practices were to become prominent in the life of this country, there surely would be a diversity of them. And would not sympathy require me to question other focal concerns and to win other people over to mine? Even if we heed the counsel of tolerance, the situation would remain unsettled and troubling. In reply we first must note how far removed we are from such a state of affairs and how many salutary measures would have to be taken before a prominent controversy about focal practices could arise.

But let us assume that there will be an evident plurality of focal concerns. How controversial would it be? It may be helpful to begin by considering the origins of that plurality. It became possible in the West when at the beginning of the modern era the unity of the Christian church was shattered through reform movements, scientific and geographical discoveries, and finally through the liberating forces of democracy and technology. ... we saw that in light of the new scientific laws our actual world appears as one instantiation of all that is physically possible. Similarly, within the context of the immense information and the varied practical possibilities that technology has procured, every actual concern now appears as one surrounded by alternatives. The severing of the ties between focal concerns and social and economic necessity that has been repeatedly noted is just a corollary of this phenomenon.

But we also must remember that we would not want to regain the support of cogency when testifying on behalf of a focal power. This would, on the one hand, compromise the grace and depth of such a power and, on the other, degrade us as respondents to that power. Our parents in their old age, as said before, address us not inasmuch as we are weak and helpless but insofar as we are capable of gratitude and receptive to wisdom, tradition, and mortality. In short, the new adulthood and maturity that are required of us are of a piece with the peculiar radiance and dignity that focal concerns now have. This status of the focal thing has the technological setting for a necessary condition, and it has the plurality of alternative concerns as a compatible background. Perhaps one should take "compatible" in the original and strong sense. We should be able to suffer the contradiction that the background of alternatives constitutes along with the joy that comes from our focal practice. And what we suffer is not just the implicit denial of what matters most to us; we suffer being deprived of great and unreachable things that are sometimes placed not only beyond our time and energy but outside our very comprehension. Sheehan is an eloquent witness:

> I may have difficulty comprehending the grasp that music has on its enthusiasts, but I see that as a deficiency in myself, not the music lovers. When a musician tells me Beethoven's Opus 132 is not simply an hour of music but of universal truth, is in fact a flood of beauty and wisdom, I envy him. I don't label him a nut. And being a city kid, I may be slow to appreciate the impact of nature on those raised differently, but, again, I regret that failure. And when Pablo Casals said, as he did on his ninety-fifth birthday, "I pass hours looking at a tree or a flower. And sometimes I cry at their beauty," I don't think age has finally gotten to old Pablo. I cry for myself.[52]

But can we not instead take the diversity of people's engagements in a positive way? Wilhelm von Humboldt, who is one of the authors ... of the liberal democratic notion of self-realization, has also pointed out that no one person can hope to realize all that human beings are capable of; we would in fact weaken our development if we tried. But far from being frustrated by our inevitable one-sidedness, we should embrace and develop our peculiarity and join it with those of others and through this connection experience and enjoy the fullness of humanity.[53] This is the idea of social union which Rawls has rediscovered and elaborated.[54] Clearly, it is an idea that affirms, deepens, and conjoins the notions of sympathy and tolerance.

It appears then that the plurality of focal concerns must be accepted and perhaps can even be seen in a positive light. But the latter possibility must be further pursued and taken to the point where it seems possible clearly to discern a unity underlying the plurality. We can begin with the apparent susceptibility of a social union to technological subversion. One might reply, as Rawls would, that the shallow and distracting diversity of self-realization that the consumption of commodities offers conflicts with the kind of self-development suggested by the Aristotelian Principle which is an integral part of a social union. The Principle says that "other things equal, human beings enjoy the exercise of their realized capacities (their innate or trained abilities), and this enjoyment increases the more the capacity is realized, or the greater its complexity."[55] Accordingly, people will not only prefer chess to checkers, as Rawls has it, but checkers to watching television and cooking a meal from basic ingredients to warming a frozen dinner. Rawls recognizes that the Principle is but a tendency and can be overridden. Yet he is confident that "the tendency postulated should be relatively strong and not easily counterbalanced."[56] But as we have seen, ... technology has not just counterbalanced but very nearly buried it. There is a difference, however, between technological obliteration and subversion. Technology can overcome the wilderness by brute force, but it cannot bring it (easily and obviously) under its rule and procure it as a commodity. Conversely, technology could hardly annihilate values (such as freedom, prosperity, or pleasure), but it can surely subvert them by specifying them in terms of the availability and consumption of commodities. Accordingly, the Aristotelian Principle is not impugned as a counterforce to technology if it can be technologically overrun so long as it resists technological subversion. But does it? Clearly the concept of complexity is crucial here. Rawls contends that "until we have some relatively precise theory and measure of complexity" we can intuitively grasp the nature of complexity and rank various activities by complexity in accordance with a principle of inclusiveness where "cases of greater complexity are those in which one of the activities compared includes all the skills and discriminations of the other activity and some further ones in addition."[57] Thus the computer game Defender might rank higher than fly-fishing since the former requires quicker hand-eye coordination, more intricate strategy, and evasive as well as aggressive skills. Similarly, exercising with a Nautilus might be more complex than running since the former allows one to sense and to work many more muscle groups. Theory here seems to conflict with considered judgment. Whether both can be saved in their essence and balanced in a reflective equilibrium is a question that will concern us in a moment.

Meanwhile let us note that the theory, i.e., the Aristotelian Principle, is attractive not only in helping to reconcile the variety of human endeavors within a social union but also in suggesting ways in which the variety of focal practices can be similarly united. It was pointed out ... that there is an apparent kinship among significant or focal things; and, symmetrically, there are common traits to be found among focal practices. These can be seen when we consider that although both Capon and a fast-food junkie are deeply concerned with food, Capon, I expect, would have a deeper appreciation of Sheehan's concern than the junkie's. Unlike the latter, both Capon and Sheehan practice the acquisition of skills, the fidelity to a daily discipline, the broadening of sensibility, the profound interaction of human beings, and the preservation and development of tradition. These traits we may bring together under the heading of engagement. The good life, then, is one of engagement, and engagement is variously realized by various people. Engagement would not only harmonize the variety among people but also within the life of one person. Sheehan, for instance, finds engagement not only in running but also in literature, and Capon finds it not only in the culture of the table but also in music.

Engagement is a more flexible and inclusive principle of ordering one's life, and being so it meets the critique of dominant ends that Rawls puts forward. If such an end deserves its name and is clearly specified, Rawls argues, there is a danger of "fanaticism and inhumanity" because the narrowness of the goal does violence to the breadth of human capacities.[58] There seems to be intuitive confirmation of Rawls's claim. Initially, the firm guidance that a dominant end affords in one's life is appealing, as Rawls notes.[59] Taking up some thing and practice as a focal and dominant end, one does, as Sheehan did, experience a sense of clarity and liberation. One is no longer caught in obliging other people's expectations and in struggling to balance a plethora of conflicting and confusing aims. Having centered my life in an ultimate concern, I have clear and principled answers to life's endless and distracting demands. But both Sheehan and Capon testify to the dark night of the soul that settles upon one from time to time, not when one has allowed distraction to erode the core of one's life but just when dedication to the focal thing has been vigorous and faithful.[60] And such darkness, depression, and collapse can be witnessed among people who have dedicated themselves to a cause that is more selfless and sublime than running or the culture of the table. These failures are so much more threatening if not devastating than those that occur under the guidance of inclusive ends because the former case admits of no alibi. One has dedicated oneself to one's highest aspiration and profoundest experience, and one has failed. Where to turn now? In a life of an inclusive end, disappointment here allows one to turn elsewhere for consolation. The question then is whether the collective plurality and the individual restrictiveness of focal concerns can be overcome through the notion of an inclusive end, placed in a social union of persons who shape their lives according to the Aristotelian Principle or according to the concept of engagement. This problem is best approached by connecting it with a still further problem, the question, i.e., whether there can be engagement of an essentially technological or purely mental sort. We have touched on this area in discussing complexity as a mark of excellence in human activities. It seemed that playing the computer game Defender is more excellent in this sense than fly-fishing. Moreover it, or more generally the playing of computer games, seems to satisfy the conditions of engagement. It certainly requires skill, discipline, and endurance; as the games develop technologically, more human capacities are called upon; and the computer game arcades have a social setting of their own and can lead to close human ties.[61] In fact, is not the computer game console a focal thing? It certainly seems to challenge and fulfill the player and to center the player's life. "There's not a lot of fun things in life," says one. "It's taken away my boredom. I've never been as serious about anything as Pac Man."[62]

The status of focal concerns as the basis of a reform of technology is now challenged in two ways. First, it appears that the ultimate givenness of a focal thing as something that unforethinkably addresses us in its own right is denied by the Aristotelian Principle or the concept of engagement. If the latter have independent standing and guiding force, focal things are mere complements that are chosen according to convenience. Second, the essentially metatechnological status of focal things and practices which in the abstract would be compatible with the Aristotelian Principle and the notion of engagement is denied by the apparent existence of essentially technological engagement.[63] Let us try to meet these challenges by pursuing Rawls's goal of achieving a reflective equilibrium and assume to begin with that in our considered judgment fly-fishing is more excellent than playing Defender. Can we align this judgment with the cluster of theories composed of the Aristotelian Principle, the notion of engagement and of a technological focal concern? Fly-fishing is more complex, we might say, because it requires more encompassing and discriminating knowledge. One must know in what season and at what time of day certain insects are hatching and trout are feeding. One must be able to read the water to recognize the riffles and the pools where the big rainbows are lying in wait. There are more intricate bodily skills in casting a line that involves not just the pushing of buttons and the movement of a stick but the harmonious interplay of rod, line, and fly, compensating for the wind, avoiding the willows, using hand, arm, and shoulder while maintaining one's stance in a slippery streambed. And to have a line and finally the fly settle gently on the river, as gently nearly as a real insect might, is one of the most delicate maneuvers humans are capable of. Fly-fishing also centers one's life more clearly and discriminatingly. Just as the grizzly is a symbol of the vastness and power of the open land, so the trout is a focus of the health and fertility of a drainage or even of a continent, considering the ravages of acid rain. To maintain the conditions that are conducive to big fish and to peaceful fishing is to take the measure of the world at large. In contrast, it appears, playing Defender requires a narrow range of highly sharpened skills, and it proceeds in utter indifference to the surrounding world. It is an activity that, given a sufficient store of energy and food, could proceed well underground should the natural environment have become unlivable.

The claim has been made, of course, that computer games allow one to become at home in the computer world. "We have a whole generation growing up," an educational consultant says, "who have no problem at all approaching the computer. They could become the haves."[64] "Kids are becoming masters of the computer," an astrophysicist contends. "When most grown-ups talk about computers, they fear the machines will dominate and displace. But these kids are learning to live and *play* with intelligent machines."[65] What the kids are learning to master is the enjoyment of a commodity; but with the supporting electronic and logical machinery they are as little familiar as consumers are with the substructure of the technological universe.[66]

But what of the people who devote their lives to the design and construction of computers? Surely they have an intimate and competent grasp of what characterizes our era. Tracy Kidder has provided an illuminating account of work at the leading edge of technology, the story of the design and construction of a computer.[67] Such work is among the best technology has to offer. It is challenging and skillful, requiring creativity, enormous dedication, and discipline. Clearly it engages, excites, and fulfills its practitioners. It occupies the center of their lives and enforces profound personal interactions. It is practiced, at least by a good number of the workers, as art for art's sake, without emphasis on remuneration, with seemingly little support from the firm's executives, with no hope of gaining fame in the world at large, and with diffidence or indifference regarding the uses to which the product will be put. Still it seems to me, judging by the evidence of Kidder's book, that computer design is deeply flawed as a focal practice. Some of the flaws are due to unhappy social arrangements. It is at least conceivable that the accomplishments that are rightly celebrated in Kidder's story could come about under socially more balanced and stable circumstances. A more serious flaw is the purely mental and essentially disembodied character of this kind of engagement. But this one-sidedness it has in common with writing music, poetry, and philosophy, with playing chess and reading novels.

Yet the poet in the stillness of writing and in the calm of speaking gathers and presents the world in the comprehensive and intimate ways that distinguish human beings. Through the poet's deictic discourse we come to comprehend the world more fully and are so empowered to inhabit it more appropriately in our tangible and bodily activities. Between poetry and practical engagement there is the complementary rhythm of comprehension and action, of systole and diastole. The focal significance of a mental activity should be judged, I believe, by the force and extent with which it gathers and illuminates the tangible world and our appropriation of it.

Is the design and construction of computers a focal concern by that standard? Not in the setting that Kidder presents. Work on the computer alienates most of the workers from the larger world. And the object of their endeavors to which they devote themselves as an end they know at the same time to be a means for whatever ends. They know that the intoxicating and engaging circumstances of their work have been granted them because for the company and the world at large the new computer will be a mere means. But these again are contingent circumstances. Inasmuch as computers embody and illuminate phenomena such as intelligence, organization, determinism, decidability, system, and the like, they surely have a kind of focal character, and a concern with computers in that sense is focal as well.[68] But the focal significance of work with computers seems precarious to me and requires for its health the essentially complementary concern with things in their own right. Otherwise the world is more likely lost than comprehended.

Have we reached a reflective equilibrium? It may seem as though a more precise inquiry of activities, traditionally thought to be excellent, will show them to be more complex than their more recent and technological rivals. Perhaps this welcome result is due to the fact that a more meticulous scrutiny comes closer to Rawls's "relatively precise theory and measure of complexity" which presumably would settle comparisons conclusively. But all this is semblance. What we have really done is to bring activities back to the things to which we respond in those activities. It is the dignity and greatness of a thing in its own right that give substance and guiding force to the notion of complexity. Complexity by itself and as a formal property is ... too flexible a notion to serve as a guide to the value of wild nature; and so it is as a guide to the excellence of human activities.[69] Rawls's Aristotelian Principle is not, to be sure, accidentally tied to a formal notion of complexity.[70] The thrust of *A Theory of Justice*, consistent with its allegiance to the deontological tradition, is to keep the contingent and historical world at bay.[71] Thus Rawls's theory screens out the presence of those things that alone, I believe, can orient our lives. To say this is, of course, to speak approximately and ambiguously. It is after all not finally decisive whether and how we succeed in securing an ordered and excellent life for worldlessly conceived subjects. The point is to remind or to suggest that in all significant reflection of the good life things in their own right have already graced us.

But if this is the pivot of ethics, is it not possible that a technological device or, more generally, a technological invention may someday address us as such a thing, one that, whatever its genesis, has taken on a character of its own, that challenges and fulfills us, that centers and illuminates our world? ... it is possible that such an invention will appear and that technology will give birth to a focal thing or event. But none are to be found now, and we must not allow vague promises of technological magnificence to blight the simple splendor of the things that now center and sustain our lives. At the same time we must, in a new kind of maturity and adulthood, accept the plurality of focal concerns, and we can take pleasure in the social union that is fostered by that plurality. But the diverse and complementary nature of our concerns should not be seen as the convergence of the Aristotelian Principle and human finitude. That would diminish focal things to the indifferent furniture of an abstract principle. The threat of one-sidedness that Rawls fears if focal things and practices are taken as dominant ends does not really obtain. Significant or focal things ... have an unsurpassable depth which surely distinguishes them from a dominant end in Rawls's precise sense where such an end "is clearly specified as attaining some objective goal such as political power or material wealth."[72] A dominant end in this sharp conception is more consonant with technology where gifted and ambitious people, dissatisfied with the shallowness of consumption, seek a transcendent goal and yet remain enthralled by technology in choosing a goal that has in principle procurable and controllable, i.e., measurable, character.

There remains one possibility of unity and coherence arising among the dispersed focal concerns. It appears when we remember that the variety of "focal" practices in pretechnological societies was centered about one focus proper, religious in nature. The focus proper did not unite all the subordinate engaging activities as a rule covers its applications. Rather the central focus surpassed the peripheral ones in concreteness, depth, and significance. ... there may be a hidden focus of that sort now, or one may emerge sometime. But who is to say? To the blight of the enthrallment with technology there corresponds symmetrically the impatient waiting or insistence on the great epiphany of the world's central focus. Instead we should gratefully record the present wealth of focal things and practices, take these things to heart, and work toward a republic of focal concerns.

Having secured, to some extent, a place for the plurality, concreteness and simplicity of focal concerns, we must now show more soberly and specifically how they serve as a basis for the reform of technology. And the first question is: How broad a basis will it be? I have suggested ... that there is a wide and steady, if frequently concealed, current of focal practices that runs through the history of this country. It is the other American mainstream. Its various stretches are linked by the generic features that focal things have in common, and it may be helpful to outline this kinship more formally as a set of traits that focal things and practices exhibit for the most part. These traits are not conditions that are sufficient to qualify something as focal. Nor is each of these traits necessary. Rather these features reflect general recollections and anticipations of focal concerns.

These generic features are divided between the things and practices of focal concern. But the division is not sharp since things and practices are tightly and variously interwoven. The practice of fly-fishing is centered around a definite, independent, and resplendent thing: the trout. The thing in backpacking is expansive, broadly defined, and it exists in its own right: the wilderness. In the practice of running, the thing is always and already there, Sheehan's ocean road, for instance, or the course that the New York City Marathon takes. But it lies there, inconspicuous and indistinct, till the runners bring it into relief. And the great meal and its courses must be prepared and brought forth by the cook and the host. Still we might say this about focal things in general. They are concrete, tangible, and deep, admitting of no functional equivalents; they have a tradition, structure, and rhythm of their own. They are unprocurable and finally beyond our control. They engage us in the fullness of our capacities. And they thrive in a technological setting. A focal practice, generally, is the resolute and regular dedication to a focal thing. It sponsors discipline and skill which are exercised in a unity of achievement and enjoyment, of mind, body, and the world, of myself and others, and in a social union.

This is just a summary of issues discussed before. An additional point must now be made. Focal practices are at ease with the natural sciences. Since focal things are concrete and tangible, they are at home in the possibility space that the sciences circumscribe. Because the givenness of these things is so eloquent and articulate, the scientific investigation of such things is not found to be a dissolution but an illumination of them. Correspondingly, the human being, as it is engaged and oriented by great things and is so an eminent focus itself, suffers no threat or diminishment from scientific examination. Capon and Sheehan testify to this openness. They use scientific insight gladly, easily, and often to bring out the splendor and depth of the things that matter to them. It is clear … that the reform of technology would rest on a treacherous foundation if focal things and practices violated or resented the bounds of science.

We now turn explicitly to the reform of technology. It is evident … that the reform must be one *of* and not merely one *within* the device paradigm. It is reasonable to expect that a reform of the paradigm would involve a restructuring of the device, perhaps the deletion, addition, and rearrangement of internal features. And this would lead, one might think, to the construction of different, perhaps intrinsically and necessarily benign technological devices. But I believe the device paradigm is perfect in its way, and if concrete pefections within the overall pattern are to be achieved, this will be the task of research and development scientists and engineers, not of philosophers. A reform of the paradigm is even less, of course, a dismantling of technology or of the technological universe. It is rather *the recognition and the restraint* of the paradigm. To restrain the paradigm is to restrict it to its proper sphere. Its proper sphere is the background or periphery of focal things and practices. Technology so reformed is no longer the characteristic and dominant way in which we take up with reality; rather it is a way of proceeding that we follow at certain times and up to a point, one that is left behind when we reach the threshold of our focal and final concerns. The concerns that move us to undertake a reform of the paradigm lead to reforms within the paradigm as well. Since a focal practice discloses the significance of things and the dignity of humans, it engenders a concern for the safety and well-being of things and persons. Consequently, focal concerns will stress and support the paradigm's native tendency toward safety, both locally and globally. It will concur with the efforts of consumer advocates and environmentalists, not of course to save and entrench the rule of technology but to provide a secure margin for what matters centrally.

But is this really a radical and remarkable reform proposal? Is it not indistinguishable from all the programs that are worried about the excesses of technology, about the imbalance between means and ends, about the suppression of the value question, and about the enslavement of humankind by its own invention? Would it not be fair to say that these programs have anticipated the goal of the present reform proposal, namely, to restrict technology to the status of a means and to introduce new ends? The question is simply unanswerable because it is deeply ambiguous. If by new ends we mean different commodities, then the present proposal differs sharply from traditional programs of reform. Reform must make room for focal things and practices. In a broad sense, these are the ends that technology should serve. But this broader sense of the means-ends relation is in conflict with the means-ends structure, embodied in the device paradigm. We can put the point at issue clearly, if baldly, this way. Both the common and the present reform proposals revolve about a means-ends distinction. In the common view, the distinction is placed within the device paradigm, in alignment with the machinery-commodity distinction. Thus the role of technology remains invisible and unchallenged. The present proposal is to restrict the entire paradigm, both the machinery and the commodities, to the status of a means and let focal things and practices be our ends. The conflict between these two views is easily overlooked. It is that unresolved conflict that infects the question above with ambiguity. More important, as argued repeatedly, … the sharpness, pervasiveness, and concealment of the technological means-ends relation exert a nearly irresistible pressure toward resolving the ambiguity in favor of technology. Most traditional reform proposals are finally ensnared by the device paradigm and fail to challenge the rule of technology and its debilitating consequences. Hence a radical reform, as said above, requires *the recognition* and the restraint of the device paradigm, a recognition that is guided by a focal concern. Such recognition can, as suggested in the preceding chapter, shade over into an implicit understanding though explication, it is hoped, would sharpen it.

Let me now draw out the concrete consequences of this kind of reform. I begin with particular illustrations and proceed to broader observations. Sheehan's focal concern is running, but he does not run everywhere he wants to go. To get to work he drives a car. He depends on that technological device and its entire associated machinery of production, service, resources, and roads. Clearly, one in Sheehan's position would want the car to be as perfect a technological device as possible: safe, reliable, easy to operate, free of maintenance. Since runners deeply enjoy the air, the trees, and the open spaces that grace their running, and since human vigor and health are essential to their enterprise, it would be consistent of them to want an environmentally benign car, one that is free of pollution and requires a minimum of resources for its production and operation. Since runners express themselves through running, they would not need to do so through the glitter, size, and newness of their vehicles.[73]

At the threshold of their focal concern, runners leave technology behind, technology, i.e., as a way of taking up with the world. The products of technology remain ubiquitous, of course: clothing, shoes, watches, and the roads. But technology can produce instruments as well as devices, objects that call forth engagement and allow for a more skilled and intimate contact with the world.[74] Runners appreciate shoes that are light, firm, and shock absorbing. They allow one to move faster, farther, and more fluidly. But runners would not want to have such movement procured by a motorcycle, nor would they, on the other side, want to obtain merely the physiological benefit of such bodily movement from a treadmill.

A focal practice engenders an intelligent and selective attitude toward technology. It leads to a simplification and perfection of technology in the background of one's focal concern and to a discerning use of technological products at the center of one's practice. I am not, of course, describing an evident development or state of affairs. It does appear from what little we know statistically of the runners in this country, for instance, that they lead a more engaged, discriminating, and a socially more profound life.[75] I am rather concerned to draw out the consequences that naturally follow for technology from a focal commitment and from a recognition of the device pattern. There is much diffidence, I suspect, among people whose life is centered, even in their work, around a great concern. Music is surely one of these. But at times, it seems to me, musicians confine the radiance, the rhythm, and the order of music and the ennobling competence that it requires to the hours and places of performance. The entrenchment of technology may make it seem quixotic to want to lead a fully musical life or to change the larger technological setting so that it would be more hospitable and attentive to music. Moreover, as social creatures we seek the approval of our fellows according to the prevailing standards. One may be a runner first and most of all; but one wants to prove too that one has been successful in the received sense. Proof requires at least the display, if not the consumption, of expensive commodities. Such inconsistency is regrettable, not because we just have to have reform of technology but because it is a partial disavowal of one's central concern. To have a focal thing radiate transformatively into its environment is not to exact some kind of service from it but to grant it its proper eloquence.

There is of course intuitive evidence for the thesis that a focal commitment leads to an intelligent limitation of technology. There are people who, struck by a focal concern, remove much technological clutter from their lives. In happy situations, the personal and private reforms take three directions. The first is of course to clear a central space for the focal thing, to establish an inviolate time for running, or to establish a hearth in one's home for the culture of the table. And this central clearing goes hand in hand, as just suggested, with a newly discriminating use of technology.[76] The second direction of reform is the simplification of the context that surrounds and supports the focal area. And then there is a third endeavor, that of extending the sphere of engagement as far as possible. Having experienced the depth of things and the pleasure of full-bodied competence at the center, one seeks to extend such excellence to the margins of life. "Do it yourself" is the maxim of this tendency and "self-sufficiency" its goal. But the tendencies for which these titles stand also exhibit the dangers of this third direction of reform. Engagement, however skilled and disciplined, becomes disoriented when it exhausts itself in the building, rebuilding, refinement, and maintenance of stages on which nothing is ever enacted. People finish their basements, fertilize their lawns, fix their cars. What for? The peripheral engagement suffocates the center, and festivity, joy, and humor disappear. Similarly, the striving for self-sufficiency may open up a world of close and intimate relations with things and people. But the demands of the goal draw a narrow and impermeable boundary about that world. There is no time to be a citizen of the cultural and political world at large and no possibility of assuming one's responsibility in it. The antidote to such disorientation and constriction is the appropriate acceptance of technology. In one or another area of one's life one should gratefully accept the disburdenment from daily and time-consuming chores and allow celebration and world citizenship to prosper in the time that has been gained.

What emerges here is a distinct notion of the good life or more precisely the private or personal side of one. Clearly, it will remain crippled if it cannot unfold into the world of labor and the public realm. ... To begin on the side of leisure and privacy is to acknowledge the presently dispersed and limited standing of focal powers. It is also to avail oneself of the immediate and undeniably large discretion one has in shaping one's free time and private sphere.[77] Even within these boundaries the good life that is centered on focal concerns is distinctive enough. Evidently, it is a favored and prosperous life. It possesses the time and the implements that are needed to devote oneself to a great calling. Technology provides us with the leisure, the space, the books, the instruments, the equipment, and the instruction that allow us to become equal to some great thing that has beckoned us from afar or that has come to us through a tradition. The citizen of the technological society has been spared the abysmal bitterness of knowing himself or herself to be capable of some excellence or achievement and of being at the same time worn-out by poor and endless work, with no time to spare and no possibility of acquiring the implements of one's desire. That bitterness is aggravated when one has a gifted child that is similarly deprived, and is exacerbated further through class distinctions where one sees richer but less gifted and dedicated persons showered with opportunities of excellence. There is prosperity also in knowing that one is able to engage in a focal practice with a great certainty of physical health and economic security. One can be relatively sure that the joy that one receives from a focal thing will not be overshadowed by the sudden loss of a loved one with whom that joy is shared. And one prospers not only in being engaged in a profound and living center but also in having a view of the world at large in its essential political, cultural, and scientific dimensions. Such a life is centrally prosperous, of course, in opening up a familiar world where things stand out clearly and steadily, where life has a rhythm and depth, where we encounter our fellow human beings in the fullness of their capacities, and where we know ourselves to be equal to that world in depth and strength.

This kind of prosperity is made possible by technology, and it is centered in a focal concern. Let us call it wealth to distinguish it from the prosperity that is confined to technology and that I want to call affluence. Affluence consists in the possession and consumption of the most numerous, refined, and varied commodities. This superlative formulation betrays its relative character. "Really" to be affluent is to live now and to rank close to the top of the hierarchy of inequality. All of the citizens of a typical technological society are more affluent than anyone in the Middle Ages. But this affluence, astounding when seen over time, is dimmed or even insensible at any one time for all but those who have a disproportionately large share of it. Affluence, strictly defined, has an undeniable glamour. It is the embodiment of the free, rich, and imperial life that technology has promised. So at least it appears from below whence it is seen by most people. Wealth in comparison is homely, homely in the sense of being plain and simple but homely also in allowing us to be at home in our world, intimate with its great things, and familiar with our fellow human beings. This simplicity, as said before, has its own splendor that is more sustaining than the glamour of affluence which leaves its beneficiaries, so we hear, sad and bored.[78] Wealth is a romantic notion also in that it continues and develops a tradition of concerns and of excellence that is rooted on the other side of the modern divide, i.e., of the Enlightenment. A life of wealth is certainly not romantic in the sense of constituting an uncomprehending rejection of the modern era and a utopian reform proposal.[79]

How wealth can be secured and advanced politically and economically is the topic of the next chapter. I will conclude this chapter by considering the narrower sphere of wealth and by connecting it with the traditional notions of excellence and of the family. ... I suggested that the virtues of world citizenship, of gallantry, musicianship, and charity still command an uneasy sort of allegiance and that it is natural, therefore, to measure the technological culture by these standards. Perhaps people are ready to accept the distressing results of such measurement with a rueful sort of agreement. But obviously the acceptance of the standards, if there is one, is not strong enough to engender the reforms that the pursuit of traditional excellence would demand. This, I believe, is due to the fact that the traditional virtues have for too long been uprooted from the soil that used to nourish them. Values, standards, and rules, I have urged repeatedly, are recollections and anticipations of great things and events. They provide bonds of continuity with past greatness and allow us to ready ourselves and our children for the great things we look forward to. Rules and values inform and are acted out in practices. A virtue is the practiced and accomplished faculty that makes one equal to a great event. From such considerations it is evident that the real circumstances and forces to which the traditional values, virtues, and rules used to answer are all but beyond recollection, and there is little in the technological universe that they can anticipate and ready us for. The peculiar character of technological reality has escaped the attention of the modern students of ethics.

To sketch a notion of excellence that is appropriate to technology is, in one sense, simply to present another version of the reform of technology that has been developed so far. But it is also to uncover and to strengthen ties to a tradition that the modern era has neglected to its peril. As regards world citizenship today, the problem is not confinement but the proliferation of channels of communication and of information. From the mass of available information we select by the criteria of utility and entertainment. We pay attention to information that is useful to the maintenance and advancement of technology, and we consume those news items that divert us. In the latter case the world is shredded into colorful bits of entertainment, and the distracted kind of knowledge that corresponds to that sort of information is the very opposite of the principled appropriation of the world that is meant by world citizenship.[80] The realm of technically useful information does not provide access to world citizenship either. Technical information is taken up primarily in one's work. Since most work in technology is unskilled, the demands on technical knowledge are low, and most people know little of science, engineering, economics, and politics. The people at the leading edge of technology have difficulty in absorbing and integrating the information that pertains to their field.[81] But even if the flood of technical information is appropriately channeled, as I think it can be, its mastery still constitutes knowledge of the social machinery, of the means rather than the ends of life. What is needed if we are to make the world truly and finally ours again is the recovery of a center and a standpoint from which one can tell what matters in the world and what merely clutters it up. A focal concern is that center of orientation. What is at issue here comes to the fore when we compare the simple and authentic world appropriation of someone like Mother Teresa with the shallow and vagrant omniscience of a technocrat.

Gallantry in a life of wealth is the fitness of the human body for the greatness and the playfulness of the world. Thus it has a grounding and a dignity that are lost in traditional gallantry, a loss that leaves the latter open to the technological concept of the perfect body where the body is narcissistically stylized into a glamorous something by whatever scientific means and according to the prevailing fashion. In the case of musicianship the tradition of excellence is unbroken and has expanded into jazz and popular music. What the notion of wealth can contribute to the central splendor and competence of music is to make us sensible to the confinement and the procurement of music. Confinement and procurement are aspects of the same phenomenon. The discipline and the rhythmic grace and order that characterize music are often confined, as said above, to the performance proper and are not allowed to inform the broader environment. This is because the unreformed structure of the technological universe leaves no room for such forces. Accordingly, music is allowed to conform to technology and is procured as a commodity that is widely and inconsequentially consumed. A focal concern for musicianship, then, will curtail the consumption of music and secure a more influential position for the authentic devotion to music.

Finally, one may hope that focal practices will lead to a deepening of charity and compassion. Focal practices provide a profounder commerce with reality and bring us closer to that intensity of experience where the world engages one painfully in hunger, disease, and confinement. A focal practice also discloses fellow human beings more fully and may make us more sensitive to the plight of those persons whose integrity is violated or suppressed. In short, a life of engagement may dispel the astounding callousness that insulates the citizens of the technological societies from the well-known misery in much of the world. The crucial point has been well made by Duane Elgin:

> When people deliberately choose to live closer to the level of material sufficiency, they are brought closer to the reality of material existence for a majority of persons on this planet. There is not the day-to-day insulation from material poverty that accompanies the hypnosis of a culture of affluence.[82]

The plight of the family, finally, consists ... in the absorption of its tasks and substance by technology. The reduction of the household to the family and the growing emptiness of family life leave the parents bewildered and the children without guidance. Since less and less of vital significance remains entrusted to the family, the parents have ceased to embody rightful authority and a tradition of competence, and correspondingly there is less and less legitimate reason to hold children to any kind of discipline. Parental love is deprived of tangible and serious circumstances in which to realize itself. Focal practices naturally reside in the family, and the parents are the ones who should initiate and train their children in them. Surely parental love is one of the deepest forms of sympathy. But sympathy needs enthusiasm to have substance. Families, I have found, that we are willing to call healthy, close, or warm turn out, on closer inspection, to be centered on a focal concern. And even in families that exhibit the typical looseness of structure, the diffidence of parents, and the impertinence of children, we can often discover a bond of respect and deep affection between parent and youngster, one that is secured in a common concern such as a sport and keeps the family from being scattered to the winds.

Notes

1 See *Paulys Realencyclopädie der classischen Altertumswissenschaft* (Stuttgart, 1893–1963), 15:615–17; See also Fustel de Coulanges, "The Sacred Fire," in *The Ancient City*, trans. Willard Small (Garden City, N.Y., n.d. [first published in 1864]), pp. 25–33.
2 See Kent C. Bloomer and Charles W. Moore, *Body, Memory, and Architecture* (New Haven, 1977), pp. 2–3 and 50–51.
3 See Jeremiah Eck, "Home Is Where the Hearth Is," *Quest* 3 (April 1979): 12.
4 Ibid., p. 11.
5 See *The Oxford English Dictionary*.
6 See Martin Heidegger, "The Origin of the Work of Art," in *Poetry, Language, Thought*, trans. Albert Hofstadter (New York, 1971), pp. 15–87.
7 See Vincent Scully, *The Earth, the Temple, and the Gods* (New Haven, 1962).
8 See my *The Philosophy of Language* (The Hague, 1974), pp. 126–31.
9 See Heidegger, *Being and Time*, trans. John Macquarrie and Edward Robinson (New York, 1962), pp. 95–107, 163–68, 210–24.
10 See Heidegger, "The Origin of the Work of Art," p. 80.
11 See Heidegger, "The Thing," in *Poetry, Language, Thought*, pp. 163–82. Heidegger alludes to the turn from the *Bedingungen* to the *Dinge* on p. 179 of the original, "Das Ding," in *Vorträge und Aufsätze* (Pfullingen, 1959). He alludes to the turn from technology to (focal) things in "The Question Concerning Technology," in *The Question Concerning Technology and Other Essays*, trans. William Lovitt (New York, 1977), p. 43.
12 See Heidegger, "The Thing."
13 See M. F. K. Fisher, *The Cooking of Provincial France* (New York, 1968), p. 50.
14 Though there are seeds for a reform of technology to be found in Heidegger as I want to show, Heidegger insists that "philosophy will not be able to effect an immediate transformation of the present condition of the world. Only a god can save us." See "Only a God Can Save Us: Der Spiegel's Interview with Martin Heidegger," trans. Maria P. Alter and John D. Caputo, *Philosophy Today* 20 (1976): 277.
15 I am not concerned to establish or defend the claim that my account of Heidegger or my development of his views are authoritative. It is merely a matter here of acknowledging a debt.
16 See Heidegger, "The Question Concerning Technology," p. 43; Langdon Winner makes a similar point in "The Political Philosophy of Alternative Technology," in *Technology and Man's Future*, ed. Albert H. Teich, 3d ed. (New York, 1981), pp. 369–73.
17 See Heidegger, "The Thing," pp. 180–82.
18 The need of complementing Heidegger's notion of the thing with the notion of practice was brought home to me by Hubert L. Dreyfus's essay, "Holism and Hermeneutics," *Review of Metaphysics* 34 (1980): 22–23.
19 See Heidegger, "Building Dwelling Thinking," in *Poetry, Language, Thought*, pp. 152–53.
20 Ibid., pp. 148–49.
21 Georg Trakl, quoted by Heidegger in "Language," in *Poetry, Language, Thought*, pp. 194–95 (I have taken some liberty with Hofstadter's translation).
22 See Norman Maclean, *A River Runs through It and Other Stories* (Chicago, 1976). Only the first of the three stories instructs the reader about fly fishing.

23 See Colin Fletcher, *The Complete Walker* (New York, 1971).
24 Ibid., p. 9.
25 See Roger B. Swain, *Earthly Pleasures: Tales from a Biologist's Garden* (New York, 1981).
26 Here are a few more: Wendell Berry, *Farming: A Handbook* (New York, 1970); Stephen Kiesling, *The Shell Game: Reflections on Rowing and the Pursuit of Excellence* (New York, 1982); John Richard Young, *Schooling for Young Riders* (Norman, Okla., 1970); W. Timothy Gallwey, *The Inner Game of Tennis* (New York, 1974); Ruedi Bear, *Pianta Su: Ski Like the Best* (Boston, 1976). Such books must be sharply distinguished from those that promise to teach accomplishments without effort and in no time. The latter kind of book is technological in intent and fraudulent in fact.
27 See Robert Farrar Capon, *The Supper of the Lamb: A Culinary Reflection* (Garden City, N.Y., 1969); and George Sheehan, *Running and Being: The Total Experience* (New York, 1978).
28 See Sheehan, pp. 211–20 and elsewhere.
29 See Capon, pp. 167–181.
30 See my "Mind, Body, and World," *Philosophical Forum* 8 (1976): 76–79. ...
31 See Capon, pp. 176–77.
32 See Sheehan, p. 127.
33 On the unity of achievement and enjoyment, see Alasdair MacIntyre, *After Virtue* (Notre Dame, Ind., 1981), p. 184.
34 See my "Mind, Body, and World," pp. 68–86.
35 See Peter Wood, "Seeing New York on the Run," *New York Times Magazine*, 7 October 1979; Alexandra Penney, "Health and Grooming: Shaping Up the Corporate Image," ibid.
36 See *New York Times Magazine*, 3 August 1980, pp. 20–21.
37 See Wood, p. 112.
38 See Sheehan, pp. 211–17.
39 See Wood, p. 116.
40 See Sheehan, pp. 221–31 and passim.
41 There is substantial anthropological evidence to show that running has been a profound focal practice in certain pretechnological cultures. I am unable to discuss it here. Nor have I discussed the problem, here and elsewhere touched upon, of technology and religion. The present study, I believe, has important implications for that issue, but to draw them out would require more space and circumspection than are available now. I have made attempts to provide an explication in "Christianity and the Cultural Center of Gravity," *Listening* 18 (1983): 93–102; and in "Prospects for the Theology of Technology," *Theology and Technology*, ed. Carl Mitcham and Jim Grote (Lanham, Md., 1984), pp. 305–22.
42 See M. F. K. Fisher, pp. 9–31.
43 For what social and empirical basis there is to this question see Chapter 24.
44 Some therapists advise lying down till these stirrings go away.
45 See John Rawls, "Two Concepts of Rules," *Philosophical Review* 64 (1955): 3–32.
46 Conversely, it is one thing to break a practice and quite another to omit a particular action. For we define ourselves and our lives in our practices; hence to break a practice is to jeopardize one's identity while omitting a particular action is relatively inconsequential.
47 See Aristotle's *Nicomachean Ethics*, the beginning of Book Two in particular.
48 See MacIntyre, p. 175.
49 Ibid., pp. 175–77.
50 Ibid., p. 176.
51 See Rawls, p. 25.
52 See George Sheehan, *Running and Being* (New York, 1978), p. 102.
53 See Wilhelm von Humboldt, "Ideen zu einem Versuch, die Gränzen der Wirksamkeit des Staats zu bestimmen," in *Werke*, ed. Andreas Flitner and Klaus Giel (Stuttgart, 1960–64), 1:64–69.
54 John Rawls, *A Theory of Justice* (Cambridge, Mass., 1971), pp. 520–29.
55 Ibid., p. 426.
56 Ibid., p. 429.
57 Ibid., p. 427.
58 Ibid., p. 554.
59 Ibid., p. 552.
60 See Robert Farrar Capon, *The Supper of the Lamb* (Garden City, N.Y., 1969), pp. 182–91; and Sheehan, pp. 238–40.
61 See Aaron Latham, "Video Games Star War," *New York Times Magazine*, 25 October 1981; "Invasion of the Video Creatures," *Newsweek*, 16 November 1981, pp. 90–94; "Games That Play People," *Time*, 18 January 1982, pp. 50–58.
62 See *Time*, p. 56.
63 How can there be technological engagement if technology is defined as disengaging? The answer is that technology in this essay is defined according to the device paradigm, and so defined it becomes disengaging primarily (*a*) in consumption and (*b*) in the mature phase after it has taken the ironical turn. Hence the possibility of technological engagement suggests that technology could achieve an alternative maturity.

There is already technological engagement in the sense that certain activities essentially depend on technological products as alpine skiing or bicycle racing do. But note that the technological devices do not procure but mediate engagement. The engagement is finally with slopes, snow conditions, courses, turns, etc. And there is full and skilled bodily engagement too.

64 See *Newsweek*, 16 November 1981, p. 94.
65 Ibid., p. 90.
66 Along similar lines one could show that running is more excellent than exercising with a Nautilus.
67 See Tracy Kidder, *The Soul of a New Machine* (Boston, 1981).
68 Remember Daniel Bell's remark ... that the computer is the "thing" in which the postindustrial society is coming to be symbolized. For a treatise in support of the thesis, see Douglas R. Hofstadter, *Gödel, Escher, Bach: An Eternal Golden Braid* (New York, 1979).
69 Engagement, too, taken as a formal notion or as a concept of a certain sort of worldlessly conceived human existence, fails as a guide to enduring excellence. But understood as a term that recollects and anticipates the human response to focal things, it is a helpful vocable, and I will continue so to use it.
70 Aristotle's Principle is not so dependent on complexity since it is balanced by the rightful assumption and careful explication of a definite and articulate world. ...
71 Rawls attempts to control the empirical world by admitting it into the design of the just society in an orderly sequence of four stages (see *A Theory of Justice*, pp. 195–201). This procedure does succeed in occluding the good life of focal things and practices. The apparent opening, so Rawls hopes, will be filled with the good life that springs from rationality, the Aristotelian Principle, and social union. But as we saw, the hope is not fulfilled. Instead ... the technological society emerges as the only possible but unacknowledged realization of Rawls's just society. One might argue, incidentally, that technology is not only an indispensable aid but also a guide for Rawls's theory which is designed to make justice available in the technological sense. I believe, however, that, appearances to the contrary, the formidable machinery of Rawls's theory bespeaks a commitment to fairness, openness, and compassion primarily and secondarily, at most, to technology.
72 See Rawls, p. 554.
73 On the general rise and decline of the car as a symbol of success, see Daniel Yankelovich, *New Rules: Searching for Self-Fulfillment in a World Turned Upside Down* (Toronto, 1982), pp. 36–39.
74 Although these technological instruments are translucent relative to the world and so permit engagement with the world, they still possess an opaque machinery that mediates engagement but is not itself experienced either directly or through social mediation. See also the remarks in n. 63 above.
75 See "Who Is the American Runner?" *Runner's World* 15 (December 1980): 36–42.
76 Capon's book is the most impressive document of such discriminating use of technology.
77 A point that is emphatically made by E. F. Schumacher in *Small Is Beautiful* (New York, 1973) and in *Good Work* (New York, 1979); by Duane Elgin in *Voluntary Simplicity* (New York, 1981); and by Yankelovich in *New Rules*.
78 See Roger Rosenblatt, "The Sad Truth about Big Spenders," *Time*, 8 December 1980, pp. 84 and 89.
79 On the confusions that beset romanticism in its opposition to technology, see Lewis Mumford, *Technics and Civilization* (New York, 1963), pp. 285–303.
80 See Daniel J. Boorstin, *Democracy and Its Discontents* (New York, 1975), pp. 12–25.
81 See Elgin, pp. 251–71. In believing that the mass of complex technical information poses a mortal threat to bureaucracies, Elgin, it seems to me, indulges in the unwarranted pessimism of the optimists.
82 Ibid., p. 71.

27

Heidegger and Borgmann on How to Affirm Technology

Hubert L. Dreyfus and Charles Spinosa

Albert Borgmann advances an American frontiersman's version of the question concerning technology that was pursued by Heidegger almost half a century ago among the peasants in the Black Forest. Since the *critique* of technology pioneered by these thinkers has by now become widely known, we would like to address a subsequent question with which each has also struggled. How can we relate ourselves to technology in a way that not only resists its devastation but also gives it a positive role in our lives? This is an extremely difficult question to which no one has yet given an adequate response, but it is perhaps *the* question for our generation. Through a sympathetic examination of the Borgmannian and Heidegerian alternatives, we hope we can show that Heidegger suggests a more coherent and credible answer than Borgmann's.

1 The Essence of Technology

In writing about technology, Heidegger formulates the goal we are concerned with here as that of gaining a free relation to technology – a way of living with technology that does not allow it to "warp, confuse, and lay waste our nature."[1] According to Heidegger our nature is to be world disclosers. That is, by means of our equipment and coordinated practices we human beings open coherent, distinct contexts or worlds in which we perceive, act, and think. Each such world makes possible a distinct and pervasive way in which things, people, and selves can appear and in which certain ways of acting make sense. The Heidegger of *Being and Time* called a world an understanding of being and argued that such an understanding of being is what makes it possible for us to encounter people and things as such. He considered his discovery of the ontological difference – the difference between the understanding of being and the beings that can show up given an understanding of being – his single great contribution to Western thought.

Middle Heidegger (roughly from the 1930s to 1950) added that there have been a series of total understandings of being in the West, each focused by a cultural paradigm which he called a work of art.[2] He distinguished roughly six epochs in our changing understanding of being. First things were understood on the model of wild nature as *physis*, i.e. as springing forth on their own. Then on the basis of *poeisis*, or nurturing, things were dealt with as needing to be helped to come forth. This was followed by an understanding of things as finished works, which in turn led to the understanding of all beings as *creatures* produced by a creator God. This religious world gave way to the modern one in which everything was organized to stand over against and satisfy the desires of autonomous and stable subjects. In 1950, Heidegger claimed that we were entering a final epoch which he called *the technological understanding of being*.

Originally "Highway bridges and feasts: Heidegger and Borgmann on how to affirm technology" in *Man and World* 30/2 (1997): 159–77. Reprinted by permission of Kluwer Academic Publishers, Dordrecht, The Netherlands.

But until late in his development, Heidegger was not clear as to how technology worked. He held for a long time that the danger of technology was that man was dominating everything and exploiting all beings for his own satisfaction, as if man were a subject in control and the objectification of everything were the problem. Thus, in 1940 he says:

> Man is what lies at the bottom of all beings; and that is, in modern terms, at the bottom of all objectification and representability.[3]

To test this early claim we turn to the work of Albert Borgmann since he has given us the best account of this aspect of Heidegger's thinking. Rather than doing an exegesis of Heidegger's texts, Borgmann does just what Heidegger wants his readers to do. He follows Heidegger on his path of thought, which always means finding the phenomena about which Heidegger is thinking. In *Technology and the Character of Contemporary Life*, Borgmann draws attention to the phenomenon of the technological device. Before the triumph of technological devices, people primarily engaged in practices that nurtured or crafted various things. So gardeners developed the skills and put in the effort necessary for nurturing plants, musicians acquired the skill necessary for bringing forth music, the fireplace had to be filled with wood of certain types and carefully maintained in order to provide warmth for the family. Technology, as Borgmann understands it, belongs to the last stage in the history of the understandings of being in the West. It replaces the worlds of *poiesis*, craftsmen, and Christians with a world in which subjects control objects. In such a world the things that call for and focus nurturing, craftsmanly, or praising practices are replaced by devices that offer a more and more transparent or commodious way of satisfying a desire. Thus the wood-burning fireplace as the foyer or focus of family activity is replaced by the stove and then by the furnace.

As Heidegger's thinking about technology deepened, however, he saw that even objects cannot resist the advance of technology. He came to see this in two steps. First, he saw that the nature of technology does not depend on subjects understanding and using objects. In 1946 he said that exploitation and control are not the subject's doing; "that man becomes the subject and the world the object, is a consequence of technology's nature establishing itself, and not the other way around."[4] And in his final analysis of technology, Heidegger was critical of those who, still caught in the subject/object picture, thought that technology was dangerous because it embodied instrumental reason. Modern technology, he insists, is "something completely different and therefore new."[5] The goal of technology Heidegger then tells us, is the more and more flexible and efficient ordering of resources, not as objects to satisfy our desires, but simply for the sake of ordering. He writes:

> Everywhere everything is ordered to stand by, to be immediately at hand, indeed to stand there just so that it may be on call for a further ordering. Whatever is ordered about in this way...we call...standing-reserve....Whatever stands by in the sense of standing reserve no longer stands over against us as object.[6]

Like late Heidegger, recent Borgmann sees that the direction technology is taking will eventually get rid altogether of objects. In his latest book, *Crossing the Postmodern Divide*, Borgmann takes up the difference between *modern* and *postmodern* technology. He distinguishes *modern hard* technology from *postmodern soft* technology. On Borgmann's account, modern technology, by rigidity and control, overcame the resistance of nature and succeeded in fabricating impressive structures such as railroad bridges as well as a host of standard durable devices. *Postmodern* technology, by being flexible and adaptive, produces instead a diverse array of quality goods such as high-tech athletic shoes designed specifically for each particular athletic activity.

Borgmann notes that as our postmodern society has moved from production to service industries our products have evolved from sophisticated goods to information. He further sees that this postmodern instrumental reality is giving way in its turn to the hyperreality of simulators that seek to get rid of the limitations imposed by the real world. Taken to the limit the simulator puts an improved reality completely at our disposal. Thus the limit of postmodernity, as Borgmann understands it, would be reached, not by the total objectification and exploitation of nature, but by getting rid of natural objects and replacing them with simulacra that are completely under our control. The essential feature of such hyperreality on Borgmann's account is that it is "entirely subject to my desire."[7] Thus for Borgmann the *object* disappears precisely to the extent that the *subject* gains total control. But Borgmann adds the important qualification that in gaining total control, the postmodern subject is reduced to "a point of arbitrary desires."[8] In the end, Borgmann's postmodern hyperreality would eliminate both objects and modernist subjects who have long-term identities and commitments. Nevertheless, Borgmann still remains within the field of subjectivity by maintaining that hyperreality is driven by the satisfaction of desires.

Even though he wrote almost half a century ago, Heidegger already had a similar account of the last stage of modernity. Like Borgmann he saw that information is replacing objects in our lives, and Heidegger and Borgmann would agree that information's main characteristic is that it can be easily transformed. But, whereas Borgmann sees the goal of these transformations as serving a minimal subject's desires, Heidegger claims that "both the subject and the object are sucked up as standing-reserve."[9] To see what he means by this, we can begin by examining Heidegger's half-century-old example. Heidegger describes the hydroelectric power station on the Rhine as his paradigm technological device because for him electricity is the paradigm technological stuff. He says:

> The revealing that rules throughout modern technology has the character of a setting-upon, in the sense of a challenging-forth. That challenging happens in that the energy concealed in nature is unlocked, what is unlocked is transformed, what is transformed is stored up, what is stored up is, in turn, distributed, and what is distributed is switched about *ever anew*.[10]

But we can see now that electricity is not a perfect example of technological stuff because it ends up finally turned into light, heat, or motion to satisfy some subject's desire. Heidegger's intuition is that treating everything as standing reserve or, as we might better say, resources, makes possible *endless* disaggregation, redistribution, and reaggregation *for its own sake*. As soon as he sees that information is truly endlessly transformable Heidegger switches to computer manipulation of information as his paradigm.[11]

As noted, when Heidegger says that technology is not instrumental and objectifying but "something entirely new," he means that, along with objects, subjects are eliminated by this new mode of being. Thus for Heidegger post-modern technology is not the culmination of the modern subject's controlling of objects but a new stage in the understanding of being. Heidegger, standing on Nietzsche's shoulders, gains a glimpse of this new understanding when he interprets Nietzsche as holding that the will to power is not the will to gain control for the sake of satisfying one's desires – even arbitrary ones – but the tendency in the practices to produce and maintain flexible ordering so that the fixity of even the past can be conquered; this cashes out as flexible ordering for the sake of more ordering and reordering without limit, which, according to Heidegger, Nietzsche expresses as the eternal return of the same.[12] Thanks to Nietzsche, Heidegger could sense that, when everything becomes standing reserve or resources, people and things will no longer be understood as having essences or identities or, for people, the goal of satisfying arbitrary desires, but back in 1955 he could not yet make out just how such a world would look.

Now, half a century after Heidegger wrote *The Question Concerning Technology*, the new understanding of being is becoming evident. A concrete example of this change and of an old fashioned subject's resistance to it can be seen in a recent *New York Times* article entitled: "An Era When Fluidity Has Replaced Maturity" (March 20th, 1995). The author, Michiko Kakutani, laments that "for many people...shape-shifting and metamorphosis seem to have replaced the conventional process of maturation." She then quotes a psychiatrist, Robert Jay Lifton, who notes in his book *The Protean Self* that "We are becoming fluid and many-sided. Without quite realizing it, we have been evolving a sense of self appropriate to the restlessness and flux of our time."[13] Kakutani then comments:

> Certainly signs of the flux and restlessness Mr. Lifton describes can be found everywhere one looks. On a superficial cultural level, we are surrounded by images of shape-shifting and reinvention, from sci-fi creatures who "morph" from form to form, to children's toys [she has in mind Transformers that metamorphose from people into vehicles]; from Madonna's ever expanding gallery of ready-to-wear personas to New Age mystics who claim they can "channel" other people or remember "previous" lives.[14]

In a quite different domain, in a talk at Berkeley on the difference between the modern library culture and the new information-retrieval culture,

Terry Winograd notes a series of oppositions which, when organized into a chart, show the transformation of the Modern into the Postmodern along the lines that Heidegger described. Here are a few of the oppositions that Winograd found:

Library Culture	*Information-Retrieval Culture*
Careful selection:	Access to everything:
a. quality of editions	a. inclusiveness of editions
b. perspicuous descriptions on cards to enable judgment	b. operational training on search engines to enable coping
c. authenticity of the text	c. availability of texts
Classification:	Diversification:
a. disciplinary standards	a. user friendliness
b. stable, organized, defined by specific interests	b. hypertext – following all lines of curiosity
Permanent collections:	Dynamic collections:
a. preservation of a fixed text	a. intertextual evolution
b. browsing	b. surfing the web

It is clear from these opposed lists that more has changed than the move from control of objects to flexibility of storage and access. What is being stored and accessed is no longer a fixed body of objects with fixed identities and contents. Moreover, the user seeking the information is not a subject who desires a more complete and reliable model of the world, but a protean being ready to be opened up to ever new horizons. In short, the postmodern human being is not interested in collecting but is constituted by connecting.

The perfect postmodern artifact is, thus, the Internet, and Sherry Turkle has described how the Net is changing the background practices that determine the kinds of selves we can be. In her recent book, *Life on the Screen: Identity in the Age of the Internet*, she details "the ability of the Internet to change popular understandings of identity." On the Internet, she tells us, "we are encouraged to think of ourselves as fluid, emergent, decentralized, multiplicitous, and ever in process."[15] Thus "the Internet has become a significant social laboratory for experimenting with the constructions and reconstructions of self that characterize postmodern life."[16] Precisely what sort of identity does the Net encourage us to construct?

There seem to be two answers that Turkle does not clearly distinguish. She uses as her paradigm Net experience the MUD, which is an acronym for Multi-User Dungeon – a virtual space popular with adults that has its origin in a teenagers' role-playing game. A MUD, she says, "can become a context for discovering who one is and wishes to be."[17] Thus some people explore roles in order to become more clearly and confidently themselves. The Net then functions in the old subject/object mode "to facilitate self knowledge and personal growth."[18] But, on the other hand, although Turkle continues to use the out-dated, modernist language of personal growth, she sees that the computer and the Internet promote something totally different and new. "MUDs," she tells us, "make possible the creation of an identity so fluid and multiple that it strains the limits of the notion."[19] Indeed, the MUD's disembodiment and lack of commitment enables people to be many selves without having to integrate these selves or to use them to improve a single identity. As Turkle notes:

> In MUDs you can write and revise your character's self-description whenever you wish. On some MUDs you can even create a character that "morphs" into another with the command "morph."[20]

Once we become accustomed to the age of the Net, we shall have many different skills for identity construction, and we shall move around virtual spaces and real spaces seeking ways to exercise these skills, powers, and passions as best we can. We might imagine people joining in this or that activity with a particular identity for so long as the identity and activity are exhilarating and then moving on to new identities and activities. Such people would thrive on having no home community and no home sense of self. The promise of the Net is that we will all develop sufficient skills to do one kind of work with one set of partners and then move on to do some other kind of work with other partners. The style that would govern such a society would be one of intense, but short, involvements, and everything would be done to maintain and develop the flexible disaggregation and reaggregation of various skills and faculties. Desires and their satisfaction would give way to having the thrill of the moment.

Communities of such people would not seem like communities by today's standards. They would not have a core cadre who remained in them over long periods of time. Rather, tomorrow's communities would live and die on the model of rock groups. For a while there would be an intense effort among a group of people and an enormous flowering of talent and artistry, and then that activity would get stale, and the members would go their own ways, joining other communities.[21] If you think that today's rock groups are a special case, consider how today's businesses are getting much work done by so-called hot groups. Notoriously, the Apple Macintosh was the result of the work of such a group. More and more products are appearing that have come about through such efforts. In such a world not only fixed identities but even desiring subjects would, indeed, have been sucked up as standing reserve.

2 Heidegger's Proposal

In order to explain Heidegger's positive response to technological things, we shall generalize Heidegger's description of the gathering power of mostly Black Forest things[22] by using Borgmann's American account of what he calls focal practices. We will then be in a position to see how, given their shared view of how things and their local worlds resist technology, Borgmann's understanding of technological practices as still enmeshed with subjectivity leads him to the conclusion that technological things cannot solicit focal practices, while Heidegger's account of postmodern technological practices as radically different from modern subject/object practices enables him to see a positive role for technological things, and the practices they solicit.

In "The Thing" (1949) and "Building Dwelling Thinking" (1951), Heidegger explores a kind of gathering that would enable us to resist postmodern technological practices. In these essays, he turns from the cultural gathering he explored in "The Origin of the Work of Art" (that sets up shared meaningful differences and thereby unifies an entire culture) to local gatherings that set up local worlds. Such local worlds occur around some everyday thing that temporarily brings into their own both the thing itself and those involved in the typical activity concerning the use of the thing. Heidegger calls this event a *thing thinging* and the tendency in the practices to bring things and people into their own, *appropriation*. Albert Borgmann has usefully called the practices that support this local gathering *focal practices*.[23] Heidegger's examples of things that focus such local gathering are a wine jug and an old stone bridge. Such things gather Black Forest peasant practices, but, as Borgmann has seen, the family meal acts as a focal thing when it draws on the culinary and social skills of family members and solicits fathers, mothers, husbands, wives, children, familiar warmth, good humor, and loyalty to come to the fore in their excellence, or in, as Heidegger would say, their ownmost.

Heidegger describes such focal practices in general terms by saying that when things thing they bring together earth and sky, divinities and mortals. When he speaks this way, his thinking draws on Hölderlin's difficult poetic terms of art; yet, what Heidegger means has its own coherence so long as we keep the phenomenon of a thing thinging before us. Heidegger, thinking of the taken-for-granted practices that ground situations and make them matter to us, calls them *earth*. In the example of the family meal we have borrowed from Borgmann, the grounding practices would be the traditional practices that produce, sustain, and develop the nuclear family. It is essential to the way these earthy practices operate that they make family gathering matter. For families, such dining practices are not simply options for the family to indulge in or not. They are the basis upon which all manifest options appear. To ground mattering such practices must remain in the background. Thus, Heidegger conceives of the earth as being fruitful by virtue of being withdrawing and hidden.

By *sky*, Heidegger means the disclosed or manifest stable possibilities for action that arise in focal situations.[24] When a focal situation is happening, one feels that certain actions are appropriate. At dinner, actions such as reminiscences, warm conversation, and even debate about events that have befallen family members during the day, as well as questions to draw people out are solicited. But, lecturing, impromptu combat, private jokes, and brooding silence are discouraged. What particular possibilities are relevant is determined by the situation itself.

In describing the cultural works of art that provide unified understandings of being, Heidegger was content with the categories of earth and world which map roughly on the thing's earth and sky. But when Heidegger thinks of focal practices, he also thinks in terms of *divinities*. When a focal event such as a family meal is working to the point where it has its particular integrity, one feels extraordinarily in tune with all that is happening, a special graceful ease takes over, and events seem to unfold of their own momentum – all combining to make the moment all the more centered and more a gift. A reverential sentiment arises; one feels thankful or grateful for receiving all that is brought out by this particular situation. Such sentiments are frequently manifested in practices such as toasting or in wishing others could be joining in such a moment. The older practice for expressing this sentiment was, of course, saying grace. Borgmann expresses a similar insight when, in speaking of a baseball game as attuning people, he says:

> Given such attunement, banter and laughter flow naturally across strangers and unite them into a community. When reality and community conspire this way, divinity descends on the game.[25]

Our sense that we did not and could not make the occasion a center of focal meaning by our own effort but rather that the special attunement required for such an occasion to work has to be granted to us is what Heidegger wants to capture in his claim that when a thing things the divinities must be present. How the power of the divinities will be understood will depend on the understanding of being of the culture but the phenomenon Heidegger describes is cross-cultural.

The fourth element of what Heidegger calls the fourfold is the *mortals*. By using this term, Heidegger is describing us as disclosers and he thinks that death primarily reveals our disclosive way of being to us. When he speaks of death, he does not mean demise or a medically defined death. He means an attribute of the way human practices work that causes mortals (later Heidegger's word for people who are inside a focal practice) to understand that they have no fixed identity and so must be ready to relinquish their current identity in order to assume the identity that their practices next call them into attunement with.[26] Of course, one needs an account of how such a multiplicity of identities and worlds differs from the morphing and hot groups we have just been describing. We will come back to this question shortly.

So far, following Borgmann, we have described the phenomenon of a thing thinging in its most glamorized form where we experience the family coming together as an integrated whole at a particular moment around a particular event. Heidegger calls this heightened version of a thing thinging a thing "shining forth."[27] But if we focus exclusively on the glamorized version, we can easily miss two other essential features of things that Heidegger attends to in "Building Dwelling Thinking." The first is that things thing even when we do not respond to them with full attention. For instance, when we walk off a crowded street into a cathedral, our whole demeanor changes even if we are not alert to it. We relax in its cool darkness that solicits meditativeness. Our sense of what is loud and soft changes, and we quiet our conversation. In general, we manifest and become centered in whatever reverential practices remain in our post-Christian way of life. Heidegger claims that things like bridges and town squares establish location and thereby thing even in ways more privative than our cathedral example. He seems to mean that so long as people who regularly encounter a thing are socialized to respond to it appropriately, their practices are organized around the thing, and its solicitations are taken into account even when no one notices.

Instead of cathedrals, Heidegger uses various sorts of bridges as examples of things thinging but not shining. His list of bridges includes a bridge from almost every major epoch in his history of the Western understandings of being. Heidegger's account could begin with the *physis* bridge – say some rocks or a fallen tree – which just flashes up to reward those who are alert to the offerings of nature. But he, in fact, begins his list with a bridge from the age of *poiesis*: "the river bridge near the country town [that] brings wagon and horse teams to the surrounding villages."[28] Then there is the bridge from high medieval times when being was understood as *createdness*. It "leads from the precincts of the castle to the cathedral square." Oddly enough there is no bridge from the subject/object days but Borgmann has leapt into the breach with magnificent accounts of the heroic effort involved in constructing railroad bridges, and poets, starting with Walt Whitman, have seen in the massive iron structure of the Brooklyn bridge an emblem of the imposing power and optimism of America.[29] Such a modern bridge is solid and reliable but it is rigid and locks into place the locations it connects.

After having briefly and soberly mentioned the *poiesis* bridge, Heidegger redescribes it in the style of Black Forest kitsch for which he is infamous. "The old stone bridge's humble brook crossing gives to the harvest wagon its passage from the fields into the village and carries the lumber cart from the field path to the road." Passages like this one seem to support Borgmann's contention that "an inappropriate nostalgia clings to Heidegger's account"[30] and that the things he names are "scattered and of yesterday."[31] And it is true that Heidegger distrusts typewriters,[32] phonographs, and television.[33] Borgmann finds "Heidegger's reflections that we have to seek out pretechnological enclaves to encounter focal things...misleading and dispiriting."[34]

While Borgmann shares Heidegger's distrust of technological *devices*, he, nonetheless, sees himself as different from Heidegger in that he finds a positive place for what he calls technological *instruments* in supporting traditional things and the practices they focus. He mentions the way hi-tech running shoes enhance running,[35] and one might add in the same vein that the dishwasher is a transparent technological instrument that supports, rather than interferes with or detracts from, the joys of the "great meal of the day." Still, according to Borgmann, what gets supported can never be technological devices since such devices, by satisfying our arbitrary desires as quickly and transparently as possible, cannot focus our practices and our lives but only disperse them.[36]

But if there were a way that technological devices could thing and thereby gather us, then one could be drawn into a positive relationship with them without becoming a resource engaged in this disaggregation and reaggregation of things and oneself and thereby loosing one's nature as a discloser. Precisely in response to this possibility, Heidegger, while still thinking of bridges, overcomes his Black Forest nostalgia and suggests a radical possibility unexplored by Borgmann. In reading Heidegger's list of bridges from various epochs, each of which things inconspicuously "in its own way," no one seems to have noticed the last bridge in the series. After his kitschy remarks on the humble old stone bridge, Heidegger continues: "The highway bridge is tied into the network of long-distance traffic, paced as calculated for maximum yield."[37] Clearly Heidegger is thinking of the postmodern autobahn interchange, in the middle of nowhere, connecting many highways so as to provide easy access to as many destinations as possible. Surely, one might think, Heidegger's point is that such a technological artifact could not possibly thing. Yet Heidegger continues:

> Ever differently the bridge escorts the lingering and hastening ways of men to and fro...The bridge *gathers*, as a passage that crosses, before the divinities – whether we explicitly think of, and visibly *give thanks for*, their presence, as in the figure of the saint of the bridge, or whether that divine presence is hidden or even pushed aside.[38]

Heidegger is here following out his sense that different things thing with different modes of revealing, that is, that each "*gathers* to itself in *its own way* earth and sky, divinities and mortals."[39] Figuring out what Heidegger might mean here is not a question of arcane Heidegger exegesis but an opportunity to return to the difficult question we raised at the beginning: How can we relate ourselves to technology in a positive way while resisting its devastation of our essence as world disclosers? In Heidegger's terms we must ask, How can a technological artifact like the highway bridge, dedicated as it is to optimizing options, gather the fourfold? Or, following Borgmann's sense of the phenomenon, we can ask how could a technological device like the highway bridge give one's activity a temporary focus? Granted that the highway bridge is a flexible resource, how can we get in tune with it without becoming flexible resources ourselves? How can mortals morph?

To answer this question about how we can respond to technology as disclosers or mortals, we must first get a clear picture of exactly what it is like to be turned into resources responding to each situation according to whichever of our disaggregated skills is solicited most strongly. We can get a hint of what such optimizing of disaggregated skills looks like if we think of the relations among a pack of today's teenagers. When a group of teenagers wants to get a new CD, the one with the car (with the driving skills and capacity) will be most important until they get to the store; then the one with the money (with purchasing skills and capacity) will lead; and then when they want to play the CD, the one with the CD player (with CD playing skills and capacity) will be out front. In each moment, the others will coordinate themselves to bring out maximally whatever other relevant skills (or possessions) they have such as chatting pleasantly, carrying stuff, reading maps, tuning the car radio, making wisecracks, and scouting out things that could be done for free. Consequently, they will be developing these other skills too.

If people lived their whole lives in this improvising mode, they would understand themselves only in terms of the skills that made the most sense at the moment. They would not see themselves as having a coordinated network of skills, but only in being led by chance to exercise some skill or other. Hence, they would not experience themselves as satisfying desires so much as getting along adaptably. Satisfying a desire here and there might be some small part of that.

If we now turn back to the autobahn "bridge" example, we can see the encounter with the interchange as a chance to let different skills be exercised. So on a sunny day we may encounter a interchange outside of Freiburg as we drive to a meeting in town as soliciting us to reschedule our meeting at Lake Constance. We take the appropriate exit and then use our cellular phone to make sure others do the same.

We can begin to understand how Heidegger thinks we can respond to technological things without becoming a collection of disaggregated skills, if we ask how the bridge could gather the fourfold. What is manifest like the *sky* are multiple possibilities. The interchange connects anywhere to anywhere else – strictly speaking it does not even connect two banks. All that is left of *earth* is that it matters that there are such possibilities, although it does not matter that there are these specific ones. But what about the *divinities*? Heidegger has to admit that they have been pushed aside. As one speeds around a clover leaf one has no pre-modern sense of having received a gift. Neither is there a modern sense, such as one might experience on a solid, iron railroad bridge, that human beings have here achieved a great triumph. All one is left with is a sense of flexibility and excitement. One senses how easy it would be to go anywhere. If one is in tune with technological flexibility, one feels *lucky* to be open to so many possibilities.

We can see that for Heidegger the interchange bridge is certainly not the best kind of bridge but it does have its style, and one can be sensitive to it in the way it solicits. The next question is, whether in getting in tune with the thinging of the highway bridge one is turned into a resource with no stable identity and no world that one is disclosing or whether one still has some sense of having an identity and of contributing to disclosing. This is where Heidegger's stress on our being *mortals* becomes essential. To understand oneself as mortal means to understand one's identity and world as fragile and temporary and requiring one's active engagement. In the case of the highway bridge, it means that, even while getting in tune with being a flexible resource, one does not understand oneself as being a resource all the time and everywhere. One does not always feel pressured, for instance, to optimize one's vacation possibilities by refusing to get stuck on back roads and sticking to the interstates. Rather, as one speeds along the overpass, one senses one's mortality, namely that one has other skills for bringing out other sorts of things, and therefore one is never wholly a resource.[40]

We have just described what may seem to be a paradox. We have said that even a technological thing may gather together earth, sky, mortals, and maybe even divinities, which are supposed to be the aspects of practices that gather people, equipment, and activities into local worlds, with roles, habitual practices, and a style that provide disclosers with a sense of integrity or centeredness. But technological things notoriously disperse us into a bunch of disaggregated skills with a style of flexible dispersion. So what could they gather into a local world? There is only one answer here. Neither equipment nor roles could be gathered, but the skills for treating ourselves as disaggregated skills and the world as a series of open possibilities are what are drawn together so that various dispersed skillful performances become possible.

But if we focus on the skills for dispersing alone, then the dangerous seduction of technology is enhanced. Because the word processor makes writing easy for desiring subjects and this ease in writing solicits us to enter discourses rather than produce finished works, the word processor attached to the Net solicits us to substitute it for pens and typewriters, thereby eliminating the equipment *and the skills* that were appropriate for modern subject/object practices. It takes a real commitment to focal practices based on stable subjects and objects to go on writing personal letters with a fountain pen and to insist that papers written on the word processor must reach an elegant finish. If the tendency to rely completely on the flexibility of technological devices is not resisted, we will be left with only one kind of writing implement promoting one style of practice, namely those of endless transformation and enhancement. Likewise, if we live our lives in front of our home entertainment centers where we can morph at will from being audiophiles to sports fans to distance learners, our sense of being mortals who can open various worlds and have various identities will be lost as we, indeed, become pure resources.[41]

Resistance to technological practices by cultivating focal practices is the primary solution Borgmann gives to saving ourselves from technological devastation. Borgmann cannot find anything more positive in technology – other than indulging in good running shoes and a Big Mac every now and then – because he sees technology as the highest form of subjectivity. It may fragment our identities, but it maintains us as desiring beings not world disclosers. In contrast, since Heidegger sees technology as disaggregating our identities into a contingently built up collection of skills, technological things solicit certain skills without requiring that we take ourselves as having one style of identity or another. This absence of identity may make our mode of being as world disclosers impossible for us. This would be what Heidegger calls the greatest danger. But this absence of an identity also allows us to become sensitive to the various identities we have when we are engaged in disclosing the different worlds focused by different styles of things. For, although even dispersive technological skills will always gather in some fashion as they develop, the role of mortals as active world disclosers will only be preserved if it is at least possible for the gathering of these background skills to be experienced as such. And this experience will only be possible in technology if one can shift back and forth between pre-technological identities with their style of coping and a technological style. As such disclosers we can then respond to technological things as revealing one kind of world among others. Hence, Heidegger's view of technology allows him to find a positive relation to it, but only so long as we maintain skills for disclosing other kinds of local worlds. Freeing us from having a total fixed identity so that we may experience ourselves as multiple identities disclosing multiple worlds is what Heidegger calls technology's saving power.[42]

We have seen that for Heidegger being gathered by and nurturing non-technological things makes possible being gathered by technological things. Thus, living in a plurality of local worlds is not only desirable, as Borgmann sees, but is actually necessary if we are to give a positive place to technological devices. Both thinkers must, therefore, face the question that Borgmann faces in his recent book, as to how to live in a plurality of communities of focal celebration. If we try to organize our lives so as to maximize the number of focal worlds we dwell in each day, we will find ourselves teaching, then running, then making dinner, then clearing up just in time to play chamber music. Such a controlling approach will produce a subject that is always outside the current world, planning the next. Indeed such willful organization runs against the responsiveness necessary for dwelling in local worlds at all. But if, on the other hand, one goes from world to world fully absorbed in each and then fully open to whatever thing grabs one next, one will exist either as a collection of unrelated selves or as no self at all, drifting in a disoriented way among worlds. To avoid such a morphing or empty identities, one wants a life where engaging in one focal practice leads naturally to engaging in another – a life of affiliations such that one regularly is solicited to do the next focal thing when the current one is becoming irrelevant. Borgmann has intimations of such a life:

> Musicians recognize gardeners; horse people understand artisans....The experience of this kinship...opens up a wider reality that allows one to refocus one's life when failing strength or changing circumstances withdraw a focal thing.[43]

Such a plurality of focal skills not only enables one to move from world to world; it gives one a sort of poly-identity that is neither the identity of an arbitrary desiring subject nor the rudderless adaptability of a resource.

Such a kinship of mortals opens new possibilities for relations among communities. As Borgmann says:

> People who have been captivated by music... will make music themselves, but they will not exclude the runners or condemn the writers. In fact, they may run and write themselves or have spouses or acquaintances who do. There is an interlacing of communities of celebration.[44]

Here, we suspect, we can find a positive place for technological devices. For there is room in such interconnecting worlds not only for a joyful family dinner, writing to a life-long friend, and attending the local concert but also for surfing on the Internet and happily zipping around an autobahn cloverleaf in tune with technology and glad that one is open to the possibilities of connecting with each of these worlds and many others.

But Borgmann does not end with his account of the interlacing of communities, which is where Heidegger, when he is thinking of things thinging, would end. Borgmann writes:

> To conclude matters in this way...would suppress a profound need and a crucial fact of communal celebration, namely religion. People feel a deep desire for comprehensive and comprehending orientation.[45]

Borgmann thinks that, fortunately, we postmoderns are more mature than former believers who excluded communities other than their own. Thus we can build a world that promotes both local worlds and a "community of communities" that satisfies everyone's need for comprehensiveness. To accept the view that our concerns form what Borgmann calls a *community of communities* is to embrace one, overarching understanding of being of the sort that Heidegger in his middle period hoped might once again shine forth in a unifying cultural paradigm. So we find that Borgmann, like middle Heidegger, entertains the possibility that "a hidden center of these dispersed focuses may emerge some day to unite them."[46] Moreover, such a focus would "surpass the peripheral ones in concreteness, depth, and significance."[47]

Heidegger's thinking until 1955, when he wrote "The Question Concerning Technology," was like Borgmann's current thinking in that for him preserving things was compatible with awaiting a single God.[48] Heidegger said as early as 1946 that the divinities were traces of the lost godhead.[49] But Heidegger came to think that there was an essential antagonism between a unified understanding of being and local worlds. Of course, he always realized that there would be an antagonism between the style set up by a cultural paradigm and things that could only be brought out in their ownness in a style different from the dominant cultural style. Such things would inevitably be dispersed to the margins of the culture. There, as Borgmann so well sees, they will shine in contrast to the dominant style but will have to resist being considered irrelevant or even wicked.[50] But, if there is a single understanding of being, even those things that come into their own in the dominant cultural style will be inhibited as things. Already in his "Thing" essay Heidegger goes out of his way to point out that, even though the original meaning of "thing" in German is a gathering to discuss a matter of concern to the community, in the case of the thing thinging, the gathering in question must be self contained. The focal occasion must determine which community concerns are relevant rather than the reverse.[51]

Given the way local worlds establish their own internal coherence that resists any imposition from outside there is bound to be a tension between the glorious cultural paradigm that establishes an understanding of being for a whole culture and the humble inconspicuous things. The shining of one would wash out the shining of the others. The tendency toward one unified world would impede the gathering of local worlds. Given this tension, in a late seminar Heidegger abandoned what he had considered up to then his crucial contribution to philosophy, the notion of a single understanding of being and its correlated notion of the ontological difference between being and beings. He remarks that "from the perspective of appropriation [the tendency in the practices to bring things out in their ownmost] it becomes necessary to free thinking from the ontological difference." He continues, "From the perspective of appropriation, [letting-presence] shows itself as the relation of world and thing, a relation which could in a way be understood as the relation of being and beings. But then its peculiar quality would be lost."[52] What presumably would be lost would be the self-enclosed local character of worlds focused by things thinging. It follows that, as mortal disclosers of worlds in the plural, the only integrity we can hope to achieve is our openness to dwelling in many worlds and the capacity to move among them. Only such a capacity allows us to accept Heidegger's and Borgmann's criticism of technology and still have Heidegger's genuinely positive relationship to technological things.

Notes

An earlier version of this essay was delivered as the 1996 Bugbee Lecture at the University of Montana. We would like thank Albert Borgmann, David Hoy, and Julian Young for their helpful comments on earlier drafts of this paper.

1 Martin Heidegger, *Discourse on Thinking*, trans. John M. Anderson and E. Hans Freund (New York: Harper & Row, 1966) 54.
2 Heidegger's main example of cultural paradigms are works of art, but he does allow that there can be other kinds of paradigm. Truth, or the cultural paradigm, can also establish itself through the actions of a god, a statesman, or a thinker.
3 Martin Heidegger, *Nietzsche*, Vol. 4 (New York: Harper & Row, 1982) 28.
4 Martin Heidegger, "What are Poets For?" *Poetry, Language, Thought* (New York: Harper & Row, 1971) 112.
5 Martin Heidegger, "The Question Concerning Technology," *The Question Concerning Technology and Other Essays*, trans. William Lovitt (New York: Harper & Row, 1977) 5.
6 Heidegger, "The Question Concerning Technology" 17.
7 Albert Borgmann, *Crossing the Postmodern Divide* (Chicago: University of Chicago Press, 1992) 88.
8 Borgmann, *Crossing* 108.
9 Martin Heidegger, "Science and Reflection", *The Question Concerning Technology and Other Essays* 173.
10 Heidegger, "The Question Concerning Technology" 16 (emphasis ours).
11 See Martin Heidegger, "On the Way to Language" (1959), trans. Peter D. Hertz, *On the Way to Language* (New York: Harper & Row, 1971) 132. See also Martin Heidegger, "Memorial Address" (1959), *Discourse on Thinking*, trans. John M. Anderson and E. Hans Freund (New York: Harper, 1966) 46.
12 Martin Heidegger, *What is Called Thinking*?, trans. Fred D. Wieck and J. Glenn Gray (New York: Harper & Row, 1968) 104–109.
13 Robert Jay Lifton as quoted by Michiko Kakutani, "When Fluidity Replaces Maturity", *New York Times*, 20 March 1995, C 11.
14 Michiko Kakutani, "When Fluidity Replaces Maturity."
15 Sherry Turkle, *Life on the Screen: Identity in the Age of the Internet* (New York: Simon & Schuster, 1995) 263–264.
16 Turkle, 180.
17 Turkle, 180.
18 Turkle, 185.
19 Turkle, 12.
20 Turkle, 192.
21 In his account of brief habits, Nietzsche describes a life similar to moving from one hot group to another. Brief habits are neither like long-lasting habits that produce stable identities, nor like constant improvisation. For Nietzsche, the best life occurs when one is fully committed to acting out of one brief habit until it becomes irrelevant and another takes over. See Friedrich Nietzsche, *The Gay Science*, trans. Walter Kaufmann (New York: Vintage, 1974) §295, 236–237.
22 Martin Heidegger, "The Thing," in *Poetry, Language, Thought*, 182.
23 Albert Borgmann, *Technology and the Character of Contemporary Life* (Chicago: University of Chicago Press, 1984) 196–210.
24 Martin Heidegger, "Building Dwelling Thinking", *Poetry, Language, Thought* 149.
25 Borgmann, *Crossing the Postmodern Divide* 135.
26 Heidegger, "The Thing", *Poetry, Language, Thought* 178–179.
27 Heidegger, "The Thing" 182.
28 Heidegger, "Building Dwelling Thinking" 152.
29 Borgmann, *Crossing the Postmodern Divide*, 27–34.
30 Borgmann, *Technology and the Character of Contemporary Life* 196.
31 Borgmann, *Technology and the Character of Contemporary Life* 199.
32 Martin Heidegger, *Parmenides*, trans. Andre Schuwer and Richard Rojecewicz (Bloomington: Indiana University Press, 1992) 85.
33 See Footnote #41.
34 Borgmann, *Technology and the Character of Contemporary Life* 200.
35 Borgmann, *Technology and the Character of Contemporary Life* 221.
36 In an attempt to overcome the residual nostalgia in any position that holds that technological devices can never have a centering role in a meaningful life, Robert Pirsig has argued in *Zen and the Art of Motorcycle Maintenance* that, if properly understood and maintained, technological devices can focus practices that enable us to live in harmony with technology. Although the motorcycle is a technological device, understanding and caring for it can help one to resist the modern tendency to use whatever is at hand as a commodity to satisfy one's desires and then dispose of it. But, as Borgmann points out, this saving stance of understanding and maintenance is doomed as our devices, for example computers, become more and more reliable while being constructed of such minute and complex parts that understanding and repairing them is no longer an option.
37 Heidegger, "Building Dwelling Thinking" 152.
38 Heidegger, "Building Dwelling Thinking" 152–153.
39 Heidegger, "Building Dwelling Thinking" 153.
40 If we take the case of writing implements, we can more clearly see both the positive role that can be played by technological things as well as the special danger they present to which Borgmann has made us

sensitive. Like bridges, the style of writing implements reflects their place in the history of being. The fountain pen solicits us to write to someone for whom the personality of our handwriting will make a difference. When involved in the practices that make the fountain pen seem important, we care about such matters as life plans, stable identities, character, views of the world, and so on. We are subjects dealing with other subjects. A typewriter, however, will serve us better if we are recording business matters or writing factual reports simply to convey information. A word processor hooked up to the Net with its great flexibility solicits us to select from a huge number of options in order to produce technical or scholarly papers that enter a network of conversations. And using a word processor one cannot help but feel lucky that one does not have to worry about erasing, retyping, literally cutting and pasting to move text around, and mailing the final product. But, as Borgmann points out, a device is not neutral; it affects the possibilities that show up for us. If one has a word processor and a modem, the text no longer appears to be a piece of work that one finishes and then publishes. It evolves through many drafts none of which is final. Circulating texts on the Net is the culmination of the dissolution of the finished object, where different versions (of what would have before been called a single text) are contributed to by many people. With such multiple contributions, not only is the physical work dispersed but so is the author. Such authorial dispersion is a part of the general dispersion of identity that Sherry Turkle describes.

41 Heidegger writes in "The Thing":

> Man...now receives instant information, by radio, of events which he formerly learned about only years later, if at all. The germination and growth of plants, which remained hidden throughout the seasons, is now exhibited publicly in a minute, on film. Distant sites of the most ancient cultures are shown on film as if they stood this very moment amidst today's street traffic....The peak of this abolition of every possibility of remoteness is reached by television, which will soon pervade and dominate the whole machinery of communication. (165)

42 Martin Heidegger, "The Turning," *The Question Concerning Technology and Other Essays* 43, where Heidegger claims that our turning away from a technological understanding of being will, at least initially, be a matter of turning to multiple worlds where things thing.

43 Borgmann, *Crossing the Postmodern Divide* 122.

44 Borgmann, *Crossing the Postmodern Divide* 141.

45 Borgmann, *Crossing the Postmodern Divide* 144.

46 Borgmann, *Technology and the Character of Contemporary Life* 199.

47 Borgmann, *Technology and the Character of Contemporary Life* 218.

48 Heidegger, "The Question Concerning Technology" 33–35.

49 Heidegger, "What are Poets For?" 97.

50 Borgmann, *Technology and the Character of Contemporary Life*, 212.

51 To put this in terms of meals, we can remember that in Virginia Woolf's *To the Lighthouse* arguments about politics brought in from outside almost ruin Mrs. Ramsay's family dinner which only works when the participants become so absorbed in the food that they stop paying attention to external concerns and get in tune with the actual occasion. The same thing happens in the film *Babette's Feast*. The members of an ascetic religious community go into the feast resolved to be true to their dead founder's principles and not to enjoy the food. Bickering and silence ensues until the wine and food makes them forget their founder's concerns and attunes them to the past and present relationships that are in accord with the gathering.

52 Martin Heidegger, "Summary of a Seminar on the Lecture 'Time and Being,'" *On Time and Being*, trans. Joan Stambaugh (New York: Harper Torchbook, 1972) 37.

28

Critical Evaluation of Heidegger and Borgmann

Andrew Feenberg

Heidegger's Critique of Modernity

What Heidegger called "The Question of Technology" has a peculiar status in the academy today. After World War II, the humanities and social sciences were swept by a wave of technological determinism. If technology was not praised for modernizing us, it was blamed for the crisis of our culture. Determinism provided both optimists and pessimists with a fundamental account of modernity as a unified phenomenon. This approach has now been largely abandoned for a view that admits the possibility of significant "difference," i.e. cultural variety in the reception and appropriation of modernity. Yet the breakdown of simplistic determinism has not led to the flowering of research in philosophy of technology one might have hoped for. To a considerable extent, it is the very authority of Heidegger's answer to the "Question" that has blocked new developments. If we want to acknowledge the possibility of alternative modernities, we will have to break with Heidegger.

Heidegger is no doubt the most influential philosopher of technology in this century. Of course he is many other things besides, but it is undeniable that his history of being culminates in the technological enframing. His ambition was to explain the modern world philosophically, to renew the power of reflection for our time. This project was worked out in the midst of the vast technological revolution that transformed the old European civilization, with its rural and religious roots, into a mass urban industrial order based on science and technology. Heidegger was acutely aware of this transformation which was the theme of intense philosophical and political discussion in the Germany of the 1920s and '30s (Sluga, 1993; Herf, 1984). At first he sought the political significance of "the encounter between global technology and modern man" (Heidegger, 1959: 166). The results were disastrous and he went on to purely philosophical reflection on the question of technology.

Heidegger claims that technology turns everything it touches into mere raw materials, which he calls "standing reserves" (*Bestand*) (Heidegger, 1977a). We ourselves are now incorporated into the mechanism, mobilized as objects of technique. Modern technology is based on methodical planning which itself presupposes the "enframing" (*Gestell*) of being, its conceptual and experiential reduction to a manipulable vestige of itself. He illustrates his theory with the contrast between a silver chalice made by a Greek craftsman and a modern dam on the Rhine (Heidegger, 1977a). The craftsman gathers the elements – form, matter, finality – and thereby brings out the "truth" of his materials. Modern technology "de-worlds" its materials and "summons" (*Herausfordern*) nature to submit to extrinsic demands. Instead of a world of authentic things capable of gathering a rich variety of contexts and meanings, we are left with an "objectless" heap of functions.

The contrast between art and craft on the one hand and technology on the other is rooted in an ontological distinction. Heidegger believes that art

Originally "Technology and Meaning" in Andrew Feenberg, *Questioning Technology*, New York and London: Routledge, Inc., 1999, pp. 183–99, 227–35. Reprinted by permission of Taylor & Francis.

and craft are ontological "openings" or "clearings" (*Lichtung*) through which ordered worlds are constituted. The jug gathers together nature, man, and gods in the pouring of the libation. A Greek temple lays out a space within which the city lives and grows. The poet establishes meanings that endure and bring a world to light. All these forms of *techne* let things appear as what they most profoundly are, in some sense, prior to human willing and making. For Heidegger the fundamental mystery of existence is this self-manifesting of things in an opening provided by man.

How pathetic are technological achievements compared to this! Heidegger claims that technology does not let being appear, it makes things be according to an arbitrary will. It does not *open*, it *causes*. Or at least so the West has understood itself since antiquity. The willful making that comes to fruition in technology has been the ontological model for Western metaphysics since Plato. It was there in Christian theology, which substituted the idea of divine creation as the making of the universe for the true question of being. Today it rages over the whole planet as a human deed: modern technology. But a universe ordered simply by the will has no roots and no intrinsic meaning. In such a universe, man has no special ontological place but is merely one force among others, one object of force among others. Metaphysics swallows up the metaphysician and so contradicts itself in the terrible catastrophe that is modernity. Heidegger calls for resignation and passivity (*Gelassenheit*) rather than an active program of reform which would simply constitute a further extension of modern technology. As Heidegger explained in his last interview, "Only a god can save us" from the juggernaut of progress (Heidegger, 1977b).

In what would salvation consist? This is a difficult question for Heideggerians. Michael Zimmerman has explained at length the similarities Heidegger imagined between his own thought and National Socialism. Presumably, he believed for a time that art and technique would merge anew in the Nazi state (Zimmerman, 1990: 231). If this truly represents Heidegger's view, it would strangely resemble Marcuse's position in *An Essay on Liberation* (1969) with its eschatological concept of an aesthetic revolution in technology.[1] More plausibly, Heidegger merely hoped that art would regain the power to define worlds as we detach ourselves from technology.

In a later work, the *Discourse on Thinking (Gelassenheit)*, Heidegger proposed something he called gaining a "free relation" to technology. He admits that technology is indispensable, but "We can affirm the unavoidable use of technical devices, and also deny them the right to dominate us, and so to warp, confuse, and lay waste our nature" (Heidegger, 1966: 54). If we do so, Heidegger promises, "Our relation to technology will become wonderfully simple and relaxed. We let technical devices enter our daily life, and at the same time leave them outside, that is, let them alone, as things which are nothing absolute but remain dependent upon something higher" (Heidegger, 1966: 54).

There is a good deal more along these lines in Heidegger's text. He claims that once we achieve a free relation to technology, we will stand in the presence of technology's hidden meaning. Even though we cannot know that meaning, awareness of its existence already reveals the technological enframing as an opening, dependent on man, and disclosing being. If we can receive it in that spirit, it will no longer dominate us and will leave us open to welcome a still deeper meaning than anything technology can supply (Zimmerman, 1990: 235; Dreyfus, 1995: 102).

Translated out of Heidegger's ontological language, we could restate his main point as the claim that technology is a cultural form through which everything in the modern world becomes available for control. Technology thus violates both humanity and nature at a far deeper level than war and environmental destruction. To this culture of control corresponds an inflation of the subjectivity of the controller, a narcissistic degeneration of humanity. This techno-culture leaves nothing untouched: even the homes of Heidegger's beloved Black Forest peasants are equipped with TV antennas. The functionalization of man and society is thus a destiny from which there is no escape.

Although Heidegger means his critique to cut deeper than any social or historical fact about our times, it is by no means irrelevant to a modern world armed with nuclear weapons and controlled by vast technically based organizations. These latter in particular illustrate the concept of the enframing with striking clarity. Alain Gras explores the inexorable growth of such macro-systems as the electric power and airline industries (Gras, 1993). As they apply ever more powerful technologies, absorb more and more of their environment, and plan ever further into the future, they effectively escape human control and indeed human purpose. Macro-systems take on what Thomas Hughes calls "momentum," a quasi-deterministic power to perpetuate themselves and to force other institutions to conform to their requirements (Hughes, 1987). Here we can give a clear empirical content to the concept of enframing.

Heidegger's critique of "autonomous technology" is thus not without merit. Increasingly, we lose sight of what is sacrificed in the mobilization of human beings and resources for goals that remain ultimately obscure. But there are significant ambiguities in Heidegger's approach. He warns us that the essence of technology is nothing technological, that is to say, technology cannot be understood through its usefulness, but only through our specifically technological engagement with the world. But is that engagement merely an attitude or is it embedded in the actual design of modern technological devices?

In the former case, we could achieve the "free relation" to technology which Heidegger demands without changing any of the devices we use. But that is an idealistic solution in the bad sense, and one which a generation of environmental action would seem decisively to refute. In the latter case, how is the break with "technological thinking" supposed to effect the design of actual devices? By osmosis, perhaps? But even such a vague indication is lacking in Heidegger, for whom technical design is totally indifferent. The lack of an answer to these questions leaves me in some doubt as to the supposed relevance of Heidegger's work to ecology.

Confronted with such arguments, Heidegger's defenders usually waffle on the attitude/device ambiguity. They point out that his critique of technology is not merely concerned with human attitudes but with the way being reveals itself. Again roughly translated out of Heidegger's language, this means that the modern world has a technological form in something like the sense in which, for example, the medieval world had a religious form. Form is no mere question of attitude but takes on a material life of its own: power plants are the gothic cathedrals of our time. But this interpretation of Heidegger's thought raises the expectation that he will offer criteria for a reform of technology. For example, his analysis of the tendency of modern technology to accumulate and store up nature's powers suggests the superiority of another technology that would not challenge nature in Promethean fashion.

Unfortunately, Heidegger's argument is developed at such a high level of abstraction he literally cannot discriminate between electricity and atom bombs, agricultural techniques and the Holocaust. In a 1949 lecture, he asserted: "Agriculture is now the mechanized food industry, in essence the same as the manufacturing of corpses in gas chambers and extermination camps, the same as the blockade and starvation of nations, the same as the production of hydrogen bombs." (Quoted in Rockmore (1992: 241)). All are merely different expressions of the identical enframing which we are called to transcend through the recovery of a deeper relation to being. And since Heidegger rejects technical regression while leaving no room for a better technological future, it is difficult to see in what that relation would consist beyond a mere change of attitude.

A Contemporary Critique

Technology and meaning

Heidegger holds that the restructuring of social reality by technical action is inimical to a life rich in meaning. The Heideggerian relation to being is incompatible with the overextension of technological thinking. It seems, therefore, that identification of the structural features of enframing can found a critique of modernity. I intend to test this approach through an evaluation of some key arguments in the work of Albert Borgmann, the leading American representative of philosophy of technology in the essentialist vein.[2]

Borgmann identifies the "device paradigm" as the formative principle of a technological society which aims above all at efficiency. In conformity with this paradigm, modern technology separates off the good or commodity it delivers from the contexts and means of delivery. Thus the heat of the modern furnace appears miraculously from discreet sources in contrast with the old wood stove that stands in the center of the room and is supplied by regular trips to the woodpile. The microwaved meal emerges effortlessly and instantly from its plastic wrapping at the individual's command in contrast with the laborious operations of a traditional kitchen serving the needs of a whole family.

The device paradigm offers gains in efficiency, but at the cost of distancing us from reality. Let us consider the substitution of "fast food" for the traditional family dinner. To common sense, well prepared fast food appears to supply nourishment without needless social complications. Functionally considered, eating is a technical operation that may be carried out more or less efficiently. It is a matter of ingesting calories, a means to an end, while all the ritualistic aspects of food consumption are secondary to the satisfaction of biological need. But what Borgmann calls "focal things" that gather people in meaningful activities that have value for their own sake cannot survive this functionalizing attitude.

The unity of the family, ritually reaffirmed each evening, no longer has a comparable locus of expression today. One need not claim that the rise of fast food "causes" the decline of the traditional family to recognize a significant connection. Simplifying personal access to food scatters people who need no longer construct the rituals of quotidian interaction around the necessities of daily living. Focal things require a certain effort, it is true, but without that effort, the rewards of a meaningful life are lost in the vapid disengagement of the operator of a smoothly functioning machinery (Borgmann, 1984: 204 ff).

Borgmann would willingly concede the usefulness of many devices, but the generalization of the device paradigm, its universal substitution for simpler ways, has a deadening effect. Where means and ends, contexts and commodities are strictly separated, life is drained of meaning. Individual involvement with nature and other human beings is reduced to a bare minimum, and possession and control become the highest values.

Borgmann's critique of technological society usefully concretizes themes in Heidegger. His dualism of device and meaning is also structurally similar to Habermas's distinction of work and interaction (Habermas, 1970). This dualism appears wherever the essence of technology is in question. It offers a way of theorizing the larger philosophical significance of the modernization process, and it reminds us of the existence of dimensions of human experience that are suppressed by facile scientism and the uncritical celebration of technology. Borgmann's contrast between the decontextualization of the device and the essentially contextual focal thing reprises Heidegger's distinction between modern technological enframing, and the "gathering" power of traditional craft production that draws people and nature together around a materialized site of encounter. Borgmann's solution, bounding the technical sphere to restore the centrality of meaning, is reminiscent of Habermas's strategy (although apparently not due to his influence.) It offers a more understandable response to invasive technology than anything in Heidegger.

However, Borgmann's approach suffers from both the ambiguity of Heidegger's original theory and the limitations of Habermas's. We cannot tell for sure if he is merely denouncing the modern attitude toward technology or technological design, and in the latter case, his critique is so broad it offers no criteria for constructive reform. He would probably agree with Habermas's critique of the colonization of the lifeworld, although he improves on that account by discussing the role of technology in modern social pathologies. But like Habermas, he lacks a concrete sense of the intricate connections of technology and culture beyond the few essential attributes on which his critique focuses. Since those attributes have largely negative consequences, we get no sense from the critique of the many ways in which the pursuit of meaning is intertwined with technology. And as a result, Borgmann imagines no significant restructuring of modern society around culturally distinctive technical alternatives that might preserve and enhance meaning.

But how persuasive is my objection to Borgmann's approach? After all, neither Russian nor Chinese communism, neither Islamic fundamentalism nor so-called "Asian values" have inspired a fundamentally distinctive stock of devices. Why *not* just reify the concept of technology and treat it as a singular essence? The problem with that is the existence of smaller but still significant differences which may become more important in the future rather than less so as essentialists assume. What is more, those differences often concern precisely the issues identified by Borgmann as central to a humane life. They determine our experience of education, medical care, and work, our relation to the natural environment, the functions of devices such as computers and automobiles, in ways either favorable or unfavorable to the preservation of meaning and community. Any theory of the essence of technology which forecloses the future therefore begs the question of difference in the technical sphere.

Interpreting the computer

I would like to pursue this contention further with a specific example that illustrates concretely my reasons for objecting to this approach to technology. The example I have chosen, human communication by computer, is one on which Borgmann has commented fairly extensively and which we have already discussed. While not everyone who shares the essentialist view will agree with his very negative conclusions, his position adequately represents that style of technology critique, and is therefore worth evaluating here at some length.[3]

Borgmann introduces the term "hyperintelligence" to refer to such developments as electronic mail and the Internet (Borgmann, 1992: 102 ff). Hyperintelligent communication offers unprecedented opportunities for people to interact across space and time, but, paradoxically, it also distances those it links. No longer are the individuals "commanding presences" for each other; they have become disposable experiences that can be turned on and off like water from a faucet. The person as a focal thing has become a commodity delivered by a device. This new way of relating has weakened connection and involvement while extending its range. What happens to the users of the new technology as they turn away from face-to-face contact?

> Plugged into the network of communications and computers, they seem to enjoy omniscience and omnipotence; severed from their network, they turn out to be insubstantial and disoriented, they no longer command the world as persons in their own right. Their conversation is without depth and wit; their attention is roving and vacuous; their sense of place is uncertain and fickle (Borgmann, 1992: 108).

This negative evaluation of the computer can be extended to earlier forms of mediated communication. In fact Borgmann does not hesitate to denounce the telephone as a hyperintelligent substitute for more deeply reflective written correspondence (Borgmann, 1992: 105).

There is an element of truth in this critique. On the networks, the pragmatics of personal encounter are radically simplified, reduced to the protocols of technical connection. It is easy to pass from one social contact to another, following the logic of the technical network that supports ever more rapid commutation. However, Borgmann's conclusions are too hastily drawn and simply ignore the role of social contextualizations in the appropriation of technology. A look, first, at the history of computer communication, and, second, at one of its innovative applications today refutes his overly negative evaluation. We will see that the real struggle is not between the computer and low tech alternatives, but within the realm of possibilities opened by the computer itself.

In the first place, the computer was not destined by some inner techno-logic to serve as a communications medium. ... the major networks, such as the French Teletel and the Internet, were originally conceived by technocrats and engineers as instruments for the distribution of data. In the course of the implantation of these networks, users appropriated them for unintended purposes and converted them into communications media. Soon they were flooded with messages that were considered trivial or offensive by their creators. Teletel quickly became the world's first and largest electronic singles bar (Feenberg, 1995: chap. 7). The Internet is overloaded with political debates dismissed as "trash" by unsympathetic critics. Less visible, at least to journalists, but more significant, other applications of computers to human communication gradually appeared, from business meetings to education, from discussions among medical patients, literary critics, and political activists to online journals and conferences.

How does Borgmann's critique fare in the light of this history? It seems to me there is an element of ingratitude in it. Because Borgmann takes it for granted that the computer is useful for human communication, he neither appreciates the process of making it so, nor the hermeneutic transformation it underwent in that process. He therefore also overlooks the political implications of the history sketched above. Today the networks constitute a fundamental scene of human activity. To impose a narrow regimen of data transmission to the exclusion of all human contact would surely be perceived as totalitarian in any ordinary institution. Why is it not a liberation to transcend such limitations in the virtual world that now surrounds us?

In the second place, Borgmann's critique ignores the variety of communicative interactions mediated by the networks. No doubt he is right that human experience is not enriched by much of what goes on there. But a full record of the face-to-face interactions occurring in the halls of his university would likely be no more uplifting. The problem is that we tend to judge the face-to-face at its memorable best and the computer mediated equivalent at its transcribed worst. Borgmann simply ignores more interesting uses of computers, such as the original research applications of the Internet, and teaching applications which show great promise (Harasim et al., 1995). It might surprise Borgmann to find the art of reflective letter writing reviving in these contexts.

Consider for example the discussion group on the Prodigy Medical Support Bulletin Board devoted to ALS (Amyotrophic Lateral Sclerosis or Lou Gehrig's Disease). In 1995, when I studied it, there were about 500 patients and caregivers reading exchanges in which some dozens of participants were actively engaged (Feenberg et al., 1996). Much of the conversation concerned feelings about dependency, illness, and dying. There was a long running discussion of problems of sexuality. Patients and caregivers wrote in both general and personal terms about the persistence of desire and the obstacles to satisfaction. The frankness of this discussion may owe something to the anonymity of the online environment, appropriated for very different purposes than those Borgmann criticizes. Here the very limitations of the medium open doors that might have remained closed in a face-to-face setting.

These online patient meetings have the potential for changing the accessibility, the scale, and the speed of interaction of patient groups. Face-to-face self-help groups are small and localized. With the exception of AIDS patients they have wielded no political power. If AIDS patients have been the exception, it is not because of the originality of their demands: patients with incurable illnesses have been complaining bitterly for years about the indifference of physicians and the obstacles to experimental treatment. What made the difference was that AIDS patients were "networked" politically by the gay rights movement even before they were caught up in a network of contagion (Epstein, 1996: 229). Online networks may similarly empower other patient groups by enabling them to constitute an effective technical locale out of which to act on the global medical system. In fact, Prodigy discussion participants established a list of priorities they presented to the ALS Society of America. Computer networking feeds into the rising demand by patients for more control over their own medical care. Democratic rationalization of the computer contributes to a parallel transformation of medicine.

It is difficult to see any connection between these applications of the computer and Borgmann's critique of "hyperintelligence." Is this technologically mediated process by which dying people come together despite paralyzing illness to discuss and mitigate their plight a mere instance of "technological thinking?" Certainly not. But then how would Heidegger incorporate an understanding of it into his theory, with its reproachful attitude toward modern technology in general?

Borgmann's critique of technology pursues the larger connections and social implications masked by the device paradigm. To this extent it is genuinely dereifying. But insofar as it fails to incorporate these hidden social dimensions into the concept of technology itself, it remains still partially caught in the very way of thinking it criticizes. His theory hovers uncertainly between a description of how we encounter technology and how it is designed. Technology, i.e. the real world objects so designated, both is and is not the problem, depending on whether the emphasis is on its fetish form as pure device or our subjective acceptance of that form. In neither case can we change technology "in itself." At best, we can hope to overcome our attitude toward it through a spiritual movement of some sort.[4]

The ambiguities of the computer referred to in this section are far from unique. In fact they are typical of most technologies, especially in the early phases of their development. Recognizing this malleability of technology, we can no longer rest content with globally negative theories that offer only condemnation of the present and no guidance for the future. We need a very different conceptualization that includes what I have called the secondary instrumentalizations, i.e. the integration of technologies to larger technical systems and nature, and to the symbolic orders of ethics and aesthetics, as well as their relation to the life and learning processes of workers and users, and the social organization of work and use.

The Gathering

Is there anything in Heidegger that can help us in this task? I believe there is, although we will need a "free relation" to Heidegger's thought to get at it. Recall that for Heidegger modern technology is stripped of meaning by contrast with the meaningful tradition we have lost. Even the old technical devices of the past shared in this lost meaning. For example, Heidegger shows us a jug "gathering" the contexts in which it was created and functions (Heidegger, 1971a). The concept of the thing as gathering resembles Borgmann's notion of the "focal thing." These concepts dereify the thing and activate its intrinsic value and manifold connections with the human world and nature.

Heidegger's doctrine of the thing is a puzzling combination of deep insights and idiosyncratic esotericism. In the example of the jug, Heidegger struggles to distinguish the thing as such from its representation as an object of knowledge and production. The essence of thing is not fully understandable from either point of view. Why? Because knowing and producing presuppose the thing as their object. What it is in itself escapes: the thing, Heidegger assures us, is unknown because it has never been thought. And even the suspicion of what has been overlooked disappears in the technological enframing which absolutizes knowing and producing and annihilates the thing in its essential being.

Heidegger wants to call to our attention another mode of perception that belongs to the lost past or perhaps to a future we can only dimly imagine. In that mode we share the earth with things rather than reducing them to mere resources. This is not a matter of political or moral choice. He is demanding that we recognize fully our own unsurpassable belonging to a world in which meaning guides the rituals that crystallize around things. I thus interpret Heidegger not to be imagining some radically new relation to reality, but simply calling on us to recognize our real situation as human beings. To become aware, to "assume" our human being (*Dasein*), this would be enough. Philosophy cannot make this happen but it can rethink the thing so that we at least acknowledge what has been lost in the enframing. The concept of the gathering nature of the thing is the result of this rethinking.

Heidegger's discussion of gathering is closely conected to another still more obscure concept, the "fourfold" (*Geviert*) of earth, sky, mortals, and divinities. In the pouring of wine from the jug, humans and things – land, sun, and gods – come together, united in a ritual practice. From one poetic leap to another, Heidegger arrives at the conclusion that this gathering constitutes what he means by "world," i.e., the ordered system of connections between things, tools, locations, enacted and suffered by *Dasein* (Heidegger, 1971a). This same concept of the fourfold also appears in his discussion of the work of art which, he claims, establishes a world through its power of disclosure (*Erschlossenheit*) (Heidegger, 1971c). "To be a work means to set up a world" (Heidegger, 1971c: 44). But does the thing also disclose? The answer is not clear; in this respect the line between work and thing blurs. Heidegger does say that "The thing things world" (Heidegger, 1971a: 181). Perhaps we are meant to understand that the thing has a minor disclosive power.

Heidegger's poetic notion of the "fourfold" seems to be an attempt to capture in abstract terms the essential elements of the ritual structure of the thing, the human being, and the world they inhabit. The fourfold refers to no particular system of practices and things, but reminds us of what all such systems have in common insofar as all human lives are rooted in enacted meanings of some sort (Kolb, 1986: 191). One can judge this notion in various ways. For some readers it may be evocative and profound. I must confess that I find it rather disappointing. It seems to call me out of the meanings I encounter in my own existence toward a manufactured mystery I inevitably associate with the contingent individual, Heidegger, whom I do not feel inclined to "follow." But then Heidegger generously conceded in advance that the thought of Being is easily led astray...

Rather than dwelling on these contentious matters, I would prefer to consider a more narrowly philosophical implication of Heidegger's conception of the thing. This is the break with substance metaphysics it implies. The jug is not primarily a physical object which has gathering relations. It *is* these relations and is merely released to its existence as such by production, or known in its outward appearance by representation. Heidegger writes, for example,

> Our thinking has of course long been accustomed to *understating* the nature of the thing. The consequence, in the course of Western thought, has been that the thing is represented as an unknown X to which perceptible properties are attached. From this point of view, everything *that already belongs to the gathering nature of this thing...* appear[s] as something that is afterward read into it (Heidegger, 1971b: 153).

In sum, Heidegger grasps the thing not just as a focus of practical rituals, but as *essentially* that, as constituted *qua* thing by these involvements rather than as preexisting them somehow and acquiring them later.

If we now consider the implications of this concept of the thing for technology, we encounter a paradox right from the start. Devices, Heidegger complains, race toward our goals and lack the integrity of his favorite jug or chalice. But by what rights does he make this summary judgement on the very things that surround him? Devices are things too. Modern and technological though they may be, they too focus gathering practices that bring people together with each and with "earth and sky," joining them in a world. Recall the Prodigy support network described in the previous section. One could hardly find a better illustration of the Heideggerian notion of the thing as essentially a gathering. Indeed, this modern technology suits Heidegger's definition even better than jugs and chalices.

At this point, a misunderstanding threatens. Could it be that Heidegger intends us to move from a substance metaphysics to a network metaphysics, a sort of field theory of things? Is the gathering thing a node in a network? In one sense the answer to these questions is yes, but it is a very special sense that needs to be worked out against the background of *Being and Time*. The notion of the thing as a gathering that discloses a world can be seen as a corrective to the overemphasis on the role of *Dasein* in disclosure in that earlier work. There "world" was defined not as "all that is," nor as an object of knowledge, but as the realm of everyday practice. To understand "world" in this sense, Heidegger required us to shift our point of view from the cognitive to the practical, to take the practical as ontologically significant on its own terms. From that point of view the world consisted in a network of ready-to-hand objects (*Zeug*) with *Dasein* at their center.

The problem is that Heidegger had little to say about those objects. *Dasein* opened up the world in which they dwelled, but they themselves never came forward in a positive role. The result was suspiciously like a theory of the subjective investment of a preexisting reality, an "X," with meaning, precisely what Heidegger wanted to avoid. Now he has found a way to right the balance. The disclosure takes place from out of the thing as much as from *Dasein*. Not just the Greek, but the Greek temple opens a world. Disclosure cannot be localized in man, an unfortunate conclusion drawn by some of Heidegger's readers, Sartre, for example.[5]

Note that the idea of the world as a network remains, but it is a network grasped from within from the practical standpoint. The world only reveals itself as such to a reflection that knows how to get behind cognition to a more primordial encounter with being. Such phenomenological reflection places us inside the flux of significance in which the world as network consists. This is not a collection of objective things, substances, but a lifeworld in which we actively participate and which only comes to light insofar as we understand participation as the most fundamental relation to reality.

If this interpretation is correct, it suggests a fundamental problem with Heidegger's critique of technology. As we saw in the discussion of de Certeau, modern technical networks have two sides: the strategic standpoint of the system manager and the tactical standpoint of the human beings they enroll. The first understands itself in objectivistic terms as knowledge and power; the second yields its secrets to a phenomenology of lived experience. Which standpoint does Heidegger occupy in his understanding of these networks?

The answer is obvious. Heidegger's modern technology is seen from above. This is why it lacks the pathos of gathering and disclosing. The official discourse of a technological society combines narrow functionalism with awe in the face of the technological sublime. In criticizing technology, Heidegger does not so much adopt a different standpoint as reveal the meaninglessness that haunts technological thinking. But this meaninglessness is no mistake; it results from the same type of abstraction that makes system management possible in the first place. It is curious that Heidegger adopts this view of systems, while condemning them, rather than applying his phenomenological approach to the lifeworld they support.

From the standpoint of the ordinary human being – and even system managers and philosophers are ordinary human beings in their spare time – networks are lived worlds in which humans and things participate through disclosive practices. This lifeworld of technology is the place of meaning in modern societies. Our fate is worked out here as surely as on Heidegger's forest paths. Why does Heidegger insist on adopting the managerial perspective even in the course of denouncing its hollow vision? Why doesn't he view modern technology from "within," practically, in its disclosive significance for ordinary actors?

Heidegger resisted the idea that technology could share in the disclosing power of art and things, but now this implication of his theory stares us in the face. If a Greek temple can open a space for the city, why not a modern structure? At what point in its development does architecture cease to be "art" and become "technology?" Heidegger does not seem to know. Indeed, there is even a peculiar passage in which he momentarily forgives the highway bridge for its efficiency and describes it too as "gathering" right along with the old stone bridge over the village stream (Heidegger, 1971b: 152). Surely this is right.

Of course one is not always alert to the ontological significance of highway bridges. As Borgmann shows, the inauthentic relation to devices is commonplace, and, in an improvement on Heidegger, he offers a phenomenological account of that relation. But Heidegger nowhere claims that authenticity was commonplace before modern technology intruded. On the contrary, inauthenticity has always been the average everyday mode of *Dasein*. Heidegger maintains only the existential possibility of passing from inauthenticity to authenticity, a possibility which ought to extend to our relation to modern technological devices as well as premodern craft objects. Nothing prevents us from respecting modern technology in its "finality," to use Henry Bugbee's term for the intrinsic meaningfulness of things in the world of action. Instead of passing over technical things in our haste to reach our goals, we can dwell near them, attending to them for their own sake and ours (Bugbee, 1999).

Heidegger's undeniable insight is that every making must also include a letting be, an active connection to the meanings that emerge with the thing and which we cannot "make" but only release through our productive activity. And linked to those meanings, there is also a background, a source, that necessarily remains untransformed by our making. This is Heidegger's concept of the "earth" as a reservoir of possibilities beyond human intentions. In denying these connections technological thinking defies human finitude. Neither the meanings of our lives, nor the earth, nature, can become human deeds because all deeds presuppose them (Feenberg, 1986: chap. 8). Yet I share David Rothenberg's interpretation, according to which Heidegger would also want us to recognize that our contact with the earth is technically mediated: what comes into focus as nature is not the pure immediate but what is lived at the limit of *techne* (Rothenberg, 1993: 195ff). Presumably, the same is true of meanings, which emerge on the horizon of our activities, not as their product but not in passive contemplation either. Despite occasional lapses into romanticism, this is after all the philosopher who placed readiness-to-hand at the center of *Dasein's* world.

The problem with Heidegger's critique is his unqualified claim that modern technology is essentially unable to recognize its limit. That is why he advocates liberation from it rather than reform of it. It is true that the dominant ideology, based on the strategic standpoint, leaves little room for respect for limits of any kind. But we must look beyond that ideology to the realities of modern technology and the society that depends on it. Heidegger's failure, like that of Habermas and many other thinkers in the humanistic tradition, to engage with actual technology is not to their credit but reveals the limits of a certain cultural tradition. Could it be that old disciplinary boundaries between the humanities and the sciences have determined the fundamental categories of social theory? If so, it is time to challenge the effects of those boundaries in our field, which is condemned to violate them by the very nature of its object.

Beyond those boundaries we discover that technology also "gathers" its many contexts through secondary instrumentalizations that integrate it to the surrounding world. Naturally, the results are quite different from the craft tradition Heidegger idealizes, but nostalgia is not a good guide to understanding the world today. When modern technical processes are brought into compliance with the requirements of the environment or human health, they incorporate their contexts into their very structure as truly as the jug, chalice, or bridge that Heidegger holds out as models of authenticity. Our models should be such things as reskilled work, medical practices that respect the person, architectural and urban designs that create humane living spaces, computer designs that mediate new social forms.

These promising innovations are the work of human beings enrolled in technological networks intervening in the design of the technical objects with which they are involved. This is the only meaningful "encounter between global technology and modern man." This encounter is not simply another instance of the goal oriented pursuit of efficiency, but constitutes an essential dimension of the contemporary struggle for a humane and livable world. …

Notes

1 Zimmerman does not cite a text and I have not been able to find one.
2 For another contemporary approach that complements Borgmann's, see Simpson (1995). Simpson denies that he is essentializing technology, and yet he works throughout his book with a minimum set of invariant characteristics of technology independent of the sociohistorical context (Simpson, 1995: 15–16, 182). That context is then consigned to a merely contingent level of influences, conditions, or consequences rather than being integrated to the conception of technology itself.
3 For another critique of the computer similar to Borgmann's, see Slouka (1995). The case for the defense can be found in Rheingold (1993).
4 Andrew Light has argued that I underestimate the significance of Borgmann's distinction between device and thing for an understanding of the aesthetics of everyday life. The distinction is useful for developing a critique of mass culture and could provide criteria for democratic rationalizations of the commodified environment. The story of the ALS patients told here could be interpreted in this light as an example of the creation of a meaningful community through the creative appropriation of the hyperreal technological universe Borgmann describes (Light, 1996: chap. 9). I am in general agreement with this revision of Borgmann's position, but in some doubt as to whether Borgmann himself would be open to it.
5 For Heidegger's anti-anthropological turn, see Schürmann (1990: 209 ff).

References

Borgmann, Albert (1984). *Technology and the Character of Contemporary Life*. Chicago: University of Chicago Press.

——(1992). *Crossing the Postmodern Divide*. Chicago: University of Chicago Press.

Bugbee, Henry (1999). *The Inward Morning*. Athens, Georgia: University of Georgia Press.

Dreyfus, Hubert (1995). "Heidegger on Gaining a Free Relation to Technology," in A. Feenberg and A. Hannay, eds., *Technology and the Politics of Knowledge*. Bloomington and Indianapolis: Indiana University Press.

Epstein, Steven (1996). *Impure Science: AIDS, Activism, and the Politics of Knowledge*. Berkeley: University of California Press.

Feenberg, Andrew (1986). *Lukács, Marx, and the Sources of Critical Theory*. New York: Oxford University Press.

——(1995). *Alternative Modernity: The Technical Turn in Philosophy and Social Theory*. Los Angeles: University of California Press.

Feenberg, Andrew, Licht, J., Kane, K., Moran, K., and Smith, R. (1996). "The Online Patient Meeting," *Journal of the Neurological Sciences*, no. 139.

Gras, Alain (1993). *Grandeurs et Dépendence: Sociologie des Macro-Systèmes Techniques*. Paris: Presses Universitaires de France.

Habermas, Jürgen (1970). "Technology and Science as 'Ideology,'" in *Toward a Rational Society*, trans. J. Shapiro. Boston: Beacon Press.

Harasim, Linda, Hiltz, S. R., Teles, L., and Turoff, M. (1995). *Learning Networks: A Field Guide to Teaching and Learning Online*. Cambridge, Mass.: MIT Press.

Heidegger, Martin (1959). *An Introduction to Metaphysics*, trans. R. Manheim. New York: Doubleday Anchor.

——(1966). *Discourse on Thinking*, trans. J. Anderson. New York: Harper & Row.

——(1971a). "The Thing," in A. Hofstadter, ed. and trans., *Poetry, Language, and Thought*. New York: Harper & Row.

——(1971b). "Building Dwelling Thinking," in A. Hofstadter, ed. and trans., *Poetry, Language, and Thought*. New York: Harper & Row.

——(1971c). "On the Origin of the Work of Art," in A. Hofstadter, ed. and trans., *Poetry, Language, and Thought*. New York: Harper & Row.

——(1977a). *The Question Concerning Technology*, trans. W. Lovitt. New York: Harper & Row.

——(1977b). "Only a God Can Save Us Now," trans. D. Schendler. *Graduate Faculty Philosophy Journal*, 6(1).

Herf, Jeffrey (1984). *Reactionary Modernism: Technology, Culture, and Politics in Weimar and the Third Reich*. Cambridge: Cambridge University Press.

Hughes, Thomas (1987). "The Evolution of Large Technological Systems," in W. Bijker, T. Hughes, and T. Pinch, eds., *The Social Construction of Technological Systems*. Cambridge, MA: MIT Press.

Kolb, David (1986). *The Critique of Pure Modernity: Hegel, Heidegger, and After*. Chicago: University of Chicago Press.

Light, Andrew (1996). *Nature, Class, and the Built World: Philosophical Essays between Political Ecology and Critical Technology*. Dissertation, University of California, Riverside.

Marcuse, Herbert (1969). *An Essay on Liberation*. Boston: Beacon Press.

Rheingold, Howard (1993). *The Virtual Community: Homesteading on the Electronic Frontier*. Reading, Mass.: Addison-Wesley.

Rockmore, Tom (1992). *On Heidegger's Nazism and Philosophy*. Berkeley: University of California Press.

Rothenberg, David (1993). *Hand's End: Technology and the Limits of Nature*. Los Angeles: University of California Press.

Schürmann, Reiner (1990). *Heidegger on Being and Acting: From Principles to Anarchy*. Bloomington: Indiana University Press.

Simpson, Lorenzo (1995). *Technology, Time, and the Conversations of Modernity*. New York: Routledge.

Slouka, Mark (1995). *War of the Worlds: Cyberspace and the High-Tech Assault on Reality*. New York: Basic Books.

Sluga, Hans (1993). *Heidegger's Crisis: Philosophy and Politics in Nazi Germany*. Cambridge, Mass.: Harvard University Press.

Zimmerman, Michael (1990). *Heidegger's Confrontation with Modernity: Technology, Politics, Art*. Bloomington: Indiana University Press.

PART V

Technology and Human Ends

Human Beings as "Makers" or "Tool-Users"?

Introduction

What is the place of technology in human life? Is it – or at least some form of it (e.g., its premodern or its scientifically informed version) – a fundamental expression of our nature? Not all philosophers, of course, are sympathetic to the idea that we even have a "nature," if that means some timeless essence. What we have, José Ortega y Gasset famously remarked, is a history, not a nature. Nevertheless, it is probably fair to say that even for anti-essentialists like Ortega, the general question of what role(s) technology plays or ought to play in human affairs always receives some sort of answer – even if it is only silently assumed, in the form of a background understanding about how, when, and in what fashion it should or should not be employed.

The traditional conception of human nature as that of a "rational animal" already clearly figures in the thought of Socrates and Plato, and it receives its classical formulation in Aristotle. Our capacity to "know," that is, to reason about our surroundings and to act with principled deliberation, is taken to be the primary sign of our difference from (and that often means also superiority over) other species. In one form or another, this idea dominates the Western tradition at least until the beginning of the modern period – and much farther, if one includes the instrumentalist transformations of the older, more contemplative conceptions of reason that begin with and are inspired by the rise of modern science. For some modern thinkers, however, even modeling reason more closely on scientific cognition cannot save the traditional species-genus definition. In light of the picture of our species emerging in the new biological sciences, a more radical break appears to be required from what is after all a metaphysically and theologically motivated definition of human beings that makes us into something cosmically special. Benjamin Franklin speaks for numerous writers of the early modern period in claiming that we ought better to be characterized as tool-making animals. In his essay on the role of labor in human evolution (in Part I), Engels develops this claim (following his influential contemporary, Ernst Haeckel) in arguing that *first* human beings stood up and *then* became intelligent. The erect stance freed the hands for manipulation of objects, which led to the fabrication of tools and thereafter to the enlargement of the human brain. Modern naturalistic, evolutionary conceptions of the human species thereby reverse the picture promoted by traditional theories, which typically claim that the mind or brain evolved first, followed by the evolution of the dexterous hands. In Plato, for example, we find the mythic imagery of the earliest humans as limbless heads, later given legs and arms to escape rolling into ditches. In a more serious vein, Aristole specifically rejects Anaxagoras' idea that the evolution of the hand led to the evolution of the mind; on the contrary, he insists that the hand must have evolved to serve the mind.

Even when it was no longer explicitly embraced, however, the influence of the traditional picture of our species' favored status lingered long into the modern period. For example, the famous fraud of the Piltdown Man (the phony fossil of a human skull connected to an ape's skeleton, which was in fact "planted" just before World War I) gained initial acceptance in part because it appealed to many early anthropologists, in suggesting that, Anaxagoras-like, the brain enlarged first in an ape-like body, and then the hands evolved. Against this view, Engels insists that manual activity is the source of the evolution of brain and mind. We humans are precisely those beings that manipulate our environment through labor. Engels, of course, is especially concerned to link this priority of practical, physical activity to the Marxist view of the priority of labor, not only in human evolution but in society as well. Even language, he adds, has its origins in labor, first having developed for the purposes of social cooperation in hunting.

For quite different reasons, Lewis Mumford and Hannah Arendt deny the priority of tool-making and labor in characterizing human nature. Arendt makes original use of the categories of the ancients to criticize Marx and to present her own understanding of human action. In contrast to the post-World War II Marxist humanists, she does not distinguish between a "true" Marx and the allegedly distorted interpretation by Marxists of humans as laboring animals. Marx himself, she argues, is simply ambivalent on the concept of labor. In some of his writings, it is a positive thing – a creative and thoroughly human activity. In other writings, it is a negative, onerous burden to be eliminated as far as possible. Arendt claims that to understand the issues properly, one must distinguish labor from both work and (communicative) action. "Labor" is the basic activity for the maintenance of life, from the labor of childbirth through the cleaning and maintenance of the household and the feeding of the family. In contrast, "work" involves the construction of artifacts – the introduction into the world of the fruits of a craftsperson's skill. Whereas, for example, household labor (cleaning, cooking) has no tangible, permanent, observable product, work is purposively directed toward a physical product that it creates. Finally, "action" – that is, communicative activity and in particular, political speech in the public forum – is like labor in that it produces no tangible artifact; but like work, it is purposively directed toward specific goals. According to Arendt, in the modern world of mass production, divisions of labor, and the planned obsolescence of whatever is produced, work tends to collapse into mere labor, and political action disappears along with the loss of a genuine sense of the political. For Arendt, as for Plato and Aristotle, human nature expresses its highest capacities in contemplation and communicative action, not in labor and work. But today, both contemplation and action have been expunged by the world of labor.

Mumford also opposes those who would interpret human activity primarily in terms of the use of tools and the production of artifacts, but for a different reason. He wants to emphasize the way human activity operates on and through the human body. Mumford claims that the anthropological preoccupation with tools and tool-making is in part a function of the sheer durability of the materials used to make these tools, in contrast to the relative paucity of the more fragile records of speech and ritual and of non-durable objects such as clothing. Mumford urges us to remember how much human activity was (and is) directed toward the body itself, for instance, in dance and in decoration. Moreover, the earliest human "machines" were in fact not physical artifacts but organized forces of huge numbers of laborers – what he calls "megamachines" – brought together for the purpose of building canals, pyramids, and other vast construction projects. Like many other twentieth-century thinkers, Mumford argues that it is language, or more generally symbolization, not tools and cooperative labor, that most clearly distinguish humans from other animals. For us, he argues, the tool is only significant when linked with human symbolic creativity. In recent decades, Mumford's conclusion may seem to have gained further plausibility from the discovery of considerable simple tool-use by animals. Not only chimps, but even insects and crabs appear to use twigs, pebbles, sponges, and other tools. Upon closer consideration, however, these discoveries prove little about the relative importance of tool-use for humans. Our tool-making and tool-using – like our language – is recursive and creative in a way that is not the case with other animals. Humans typically devise tools to make tools to make more tools. ...

Is there, then, *any* satisfying answer to the question of whether contemplation, or tool-using, or contemplation, or ... best represents what it is to be human? For the Dewey depicted in Larry Hickman's article, the conclusion appears to be No. The problem, it seems, is the very idea of rank ordering human capacities and the mistaken interpretations of human activity that underlie this idea. Consider first the relation between scientific "knowledge" and technological "practice." Hickman shows that for Dewey, the role of technology in human affairs will never be properly understood so long as human experience is made beholden to some species of theorizing that embraces "antecedent" and superior truths and certitudes. Whether it is the ancient metaphysics of contemplation, or some form of supernaturalism, or the "official view" of modern scientific theory as conceptually transcending the particularities of its materials and instruments – all of these at once "lofty" and "foundationalist" images of knowing have had the unjustified double effect of inflating the importance of conceptualization and denigrating material practice in such a way that the real character of both activities is seriously distorted. Dewey, notes Hickman, holds a "transactional" view of the relation between human beings and the world – which means that technology, science, and sociopolitical thought are all equally to be understood as forms of "inquiry." What makes us human is our "grappling" with surroundings, forever "experimenting" with new ways of conceiving, materially handling, or socially organizing what we experience. From this point of view, the whole idea of deciding once and for all which kind of activity is most human makes no sense. Technology, science, and sociopolitical inquiry only gain whatever "priority" they have in our affairs in light of how they affect each other in some actual practice.

Finally, in "Buddhist Economics," E. F. Schumacher offers a provocative reconsideration of the proper place of technology in our lives by identifying and then reversing many of the rarely questioned assumptions about the nature and purpose of human life upon which familiar Western accounts of economics and work depend. Drawing upon the idea of "Right Livelihood" that forms part of the Buddhist conception of the Noble Eightfold Path to spiritual liberation, Schumacher shows that by Buddhist standards, the Western economic (specifically, capitalist) conception of work is ultimately, and not surprisingly, contradictory. It appears, he says, that for employers, the optimum condition would be to have output without employees, and for employees, income without employment. Ideally, then, there should be no work. The source of the problem, argues Schumacher, is precisely this basic Western conception of work as a necessary evil, a mere means. For a Buddhist, however, work is only partially and secondarily concerned with the production of goods and services. Most importantly, work offers us the chance to develop our capacities and to overcome our ego-centeredness by cooperating in common tasks. (Although he says little about the matter here, it is clear that Buddhist teaching also has little sympathy for the Baconian idea that technoscience can give us power over our surroundings.) From this viewpoint, says Schumacher, the Western way of judging work has, unfortunately, "stood the truth on its head."

29

Tool-Users vs. Homo Sapiens *and* The Megamachine

Lewis Mumford

Chapter One: Prologue

The last century, we all realize, has witnessed a radical transformation in the entire human environment, largely as a result of the impact of the mathematical and physical sciences upon technology. This shift from an empirical, tradition-bound technics to an experimental mode has opened up such new realms as those of nuclear energy, supersonic transportation, cybernetic intelligence and instantaneous distant communication. Never since the Pyramid Age have such vast physical changes been consummated in so short a time. All these changes have, in turn, produced alterations in the human personality, while still more radical transformations, if this process continue unabated and uncorrected, loom ahead.

In terms of the currently accepted picture of the relation of man to technics, our age is passing from the primeval state of man, marked by his invention of tools and weapons for the purpose of achieving mastery over the forces of nature, to a radically different condition, in which he will have not only conquered nature, but detached himself as far as possible from the organic habitat.

With this new "megatechnics" the dominant minority will create a uniform, all-enveloping, super-planetary structure, designed for automatic operation. Instead of functioning actively as an autonomous personality, man will become a passive, purposeless, machine-conditioned animal whose proper functions, as technicians now interpret man's role, will either be fed into the machine or strictly limited and controlled for the benefit of de-personalized, collective organizations.

My purpose ... is to question both the assumptions and the predictions upon which our commitment to the present forms of technical and scientific progress, treated as if ends in themselves, has been based. I shall bring forward evidence that casts doubts upon the current theories of man's basic nature which over-rate the part that tools once played – and machines now play – in human development. I shall suggest that not only was Karl Marx in error in giving the material instruments of production the central place and directive function in human development, but that even the seemingly benign interpretation of Teilhard de Chardin reads back into the whole story of man the narrow technological rationalism of our own age, and projects into the future a final state in which all the possibilities of human development would come to an end. At that "omega-point" nothing would be left of man's autonomous original nature, except organized intelligence: a universal and omnipotent layer of abstract mind, loveless and lifeless.

"Chapter One: Prologue" from Lewis Mumford, *The Myth of the Machine: Technics and Human Development*, New York: Harcourt Brace Jovanovich, Inc., 1967, pp. 3–13. **"The Megamachine" from "The Concept of the Megamachine" in *Knowledge Among Men: Eleven Essays on Science, Culture, and Society Commemorating the 200th Anniversary of the Birth of James Smithson*, ed. Paul H. Oehser, New York: Simon & Schuster, 1966, pp. 134–41, abridged.**

Now, we cannot understand the role that technics has played in human development without a deeper insight into the historic nature of man. Yet that insight has been blurred during the last century because it has been conditioned by a social environment in which a mass of new mechanical inventions had suddenly proliferated, sweeping away ancient processes and institutions, and altering the traditional conception of both human limitations and technical possibilities.

Our predecessors mistakenly coupled their particular mode of mechanical progress with an unjustifiable sense of increasing moral superiority. But our own contemporaries, who have reason to reject this smug Victorian belief in the inevitable improvement of all other human institutions through command of the machine, nevertheless concentrate, with manic fervor, upon the continued expansion of science and technology, as if they alone magically would provide the only means of human salvation. Since our present over-commitment to technics is in part due to a radical misinterpretation of the whole course of human development, the first step toward recovering our balance is to bring under review the main stages of man's emergence from its primal beginnings onward.

Just because man's need for tools is so obvious, we must guard ourselves against over-stressing the role of stone tools hundreds of thousands of years before they became functionally differentiated and efficient. In treating tool-making as central to early man's survival, biologists and anthropologists for long underplayed, or neglected, a mass of activities in which many other species were for long more knowledgeable than man. Despite the contrary evidence put forward by R. U. Sayce, Daryll Forde, and André Leroi-Gourhan, there is still a tendency to identify tools and machines with technology: to substitute the part for the whole.

Even in describing only the material components of technics, this practice overlooks the equally vital role of containers: first hearths, pits, traps, cordage; later, baskets, bins, byres, houses, to say nothing of still later collective containers like reservoirs, canals, cities. These static components play an important part in every technology, not least in our own day, with its high-tension transformers, its giant chemical retorts, its atomic reactors.

In any adequate definition of technics, it should be plain that many insects, birds, and mammals had made far more radical innovations in the fabrication of containers, with their intricate nests and bowers, their geometric beehives, their urbanoid anthills and termitaries, their beaver lodges, than man's ancestors had achieved in the making of tools until the emergence of *Homo sapiens*. In short, if technical proficiency alone were sufficient to identify and foster intelligence, man was for long a laggard, compared with many other species. The consequences of this perception should be plain: namely, that there was nothing uniquely human in tool-making until it was modified by linguistic symbols, esthetic designs, and socially transmitted knowledge. At that point, the human brain, not just the hand, was what made a profound difference; and that brain could not possibly have been just a hand-made product, since it was already well developed in four-footed creatures like rats, which have no free-fingered hands.

More than a century ago Thomas Carlyle described man as a "tool-using animal," as if this were the one trait that elevated him above the rest of brute creation. This overweighting of tools, weapons, physical apparatus, and machines has obscured the actual path of human development. The definition of man as a tool-using animal, even when corrected to read "tool-making," would have seemed strange to Plato, who attributed man's emergence from a primitive state as much to Marsyas and Orpheus, the makers of music, as to fire-stealing Prometheus, or to Hephaestus, the blacksmith-god, the sole manual worker in the Olympic pantheon.

Yet the description of man as essentially a tool-making animal has become so firmly embedded that the mere finding of the fragments of little primate skulls in the neighborhood of chipped pebbles, as with the Australopithecines of Africa, was deemed sufficient by their finder, Dr. L. S. B. Leakey, to identify the creature as in the direct line of human ascent, despite marked physical divergences from both apes and later men. Since Leakey's sub-hominids had a brain capacity about a third of *Homo sapiens* – less indeed than some apes – the ability to chip and use crude stone tools plainly neither called for nor by itself generated man's rich cerebral equipment.

If the Australopithecines lacked the beginning of other human characteristics, their possession of tools would only prove that at least one other species outside the true genus *Homo* boasted this trait, just as parrots and magpies share the distinctly human achievement of speech, and the bower bird that for colorful decorative embellishment. No single trait, not even tool-making, is sufficient to identify man. What is specially and uniquely human is man's capacity to combine a wide variety of animal propensities into an emergent cultural entity: a human personality.

If the exact functional equivalence of tool-making with utensil-making had been appreciated by earlier investigators, it would have been plain that there was nothing notable about man's hand-made stone artifacts until far along in his development. Even a distant relative of man, the gorilla, puts together a nest of leaves for comfort in sleeping, and will throw a bridge of great fern stalks across a shallow stream, presumably to keep from wetting or scraping his feet. Five-year-old children, who can talk and read and reason, show little aptitude in using tools and still less in making them: so if tool-making were what counted, they could not yet be identified as human.

In early man we have reason to suspect the same kind of facility and the same ineptitude. When we seek for proof of man's genuine superiority to his fellow creatures, we should do well to look for a different kind of evidence than his poor stone tools alone; or rather, we should ask ourselves what activities preoccupied him during those countless years when with the same materials and the same muscular movements he later used so skillfully he might have fashioned better tools.

... [T]here was nothing specifically human in primitive technics, apart from the use and preservation of fire, until man had reconstituted his own physical organs by employing them for functions and purposes quite different from those they had originally served. Probably the first major displacement was the transformation of the quadruped's fore-limbs from specialized organs of locomotion to all-purpose tools for climbing, grasping, striking, tearing, pounding, digging, holding. Early man's hands and pebble tools played a significant part in his development, mainly because, as Du Brul has pointed out, they facilitated the preparatory functions of picking, carrying, and macerating food, and *thus liberated the mouth for speech.*

If man was indeed a tool-maker, he possessed at the beginning one primary, all-purpose tool, more important than any later assemblage: his own mind-activated body, every part of it, including those members that made clubs, hand-axes or wooden spears. To compensate for his extremely primitive working gear, early man had a much more important asset that extended his whole technical horizon: he had a far richer biological equipment than any other animal, a body not specialized for any single activity, and a brain capable of scanning a wider environment and holding all the different parts of his experience together. Precisely because of his extraordinary plasticity and sensitivity, he was able to use a larger portion of both his external environment and his internal, psychosomatic resources.

Through man's overdeveloped and incessantly active brain, he had more mental energy to tap than he needed for survival at a purely animal level; and he was accordingly under the necessity of canalizing that energy, not just into food-getting and sexual reproduction, but into modes of living that would convert this energy more directly and constructively into appropriate cultural – that is, symbolic – forms. Only by creating cultural outlets could he tap and control and fully utilize his own nature.

Cultural "work" by necessity took precedence over manual work. These new activities involved far more than the discipline of hand, muscle, and eye in making and using tools, greatly though they aided man: they likewise demanded a control over all man's natural functions, including his organs of excretion, his upsurging emotions, his promiscuous sexual activities, his tormenting and tempting dreams.

With man's persistent exploration of his own organic capabilities, nose, eyes, ears, tongue, lips, and sexual organs were given new roles to play. Even the hand was no mere horny specialized work-tool: it stroked a lover's body, held a baby close to the breast, made significant gestures, or expressed in shared ritual and ordered dance some otherwise inexpressible sentiment about life or death, a remembered past, or an anxious future. Tool-technics, in fact, is but a fragment of biotechnics: man's total equipment for life.

This gift of free neural energy already showed itself in man's primate ancestors. Dr. Alison Jolly has recently shown that brain growth in lemurs derived from their athletic playfulness, their mutual grooming, and their enhanced sociability, rather than from tool-using or food-getting habits; while man's exploratory curiosity, his imitativeness, and his idle manipulativeness, with no thought of ulterior reward, were already visible in his simian relatives. In American usage, "monkeyshines" and "monkeying" are popular identifications of that playfulness and non-utilitarian handling of objects. ... there is even reason to ask whether the standardized patterns observable in early tool-making are not in part derivable from the strictly repetitive motions of ritual, song, and dance, forms that have long existed in a state of perfection among primitive peoples, usually in far more finished style than their tools.

Only a little while ago the Dutch historian, J. Huizinga, in "Homo Ludens" brought forth a mass of evidence to suggest that play, rather than work, was the formative element in human culture: that man's most serious activity belonged to the realm of make-believe. On this showing, ritual and mimesis, sports and games and dramas, released man from his insistent animal attachments; and nothing could demonstrate this better, I would add, than those primitive ceremonies in which he played at being another kind of animal. Long before he had achieved the power to transform the natural environment, man had created a miniature environment, the symbolic field of play, in which every function of life might be re-fashioned in a strictly human style, as in a game.

So startling was the thesis of "Homo Ludens" that his shocked translator deliberately altered Huizinga's express statement, that all culture was a form of play, into the more obvious conventional notion that play is an element in culture. But the notion that man is neither *Homo sapiens* nor *Homo ludens*, but above all *Homo faber*, man the maker, had taken such firm possession of present-day Western thinkers that even Henri Bergson held it. So certain were nineteenth-century archeologists about the primacy of stone tools and weapons in the "struggle for existence" that when the first paleolithic cave paintings were discovered in Spain in 1879, they were denounced, out of hand, as an outrageous hoax, by "competent authorities" on the ground that Ice Age hunters could not have had the leisure or the mind to produce the elegant art of Altamira.

But mind was exactly what *Homo sapiens* possessed in a singular degree: mind based on the fullest use of all his bodily organs, not just his hands. In this revision of obsolete technological stereotypes, I would go even further: for I submit that at every stage man's inventions and transformations were less for the purpose of increasing the food supply or controlling nature than for utilizing his own immense organic resources and expressing his latent potentialities, in order to fulfill more adequately his superorganic demands and aspirations.

When not curbed by hostile environmental pressures, man's elaboration of symbolic culture answered a more imperative need than that for control over the environment – and, one must infer, largely predated it and for long outpaced it. Among sociologists, Leslie White deserves credit for giving due weight to this fact by his emphasis on "minding" and "symboling," though he has but recovered for the present generation the original insights of the father of anthropology, Edward Tylor.

On this reading, the evolution of language – a culmination of man's more elementary forms of expressing and transmitting meaning – was incomparably more important to further human development than the chipping of a mountain of hand-axes. Besides the relatively simple coordinations required for tool-using, the delicate interplay of the many organs needed for the creation of articulate speech was a far more striking advance. This effort must have occupied a greater part of early man's time, energy, and mental activity, since the ultimate collective product, spoken language, was infinitely more complex and sophisticated at the dawn of civilization than the Egyptian or Mesopotamian kit of tools.

To consider man, then, as primarily a tool-using animal, is to overlook the main chapters of human history. Opposed to this petrified notion, I shall develop the view that man is pre-eminently a mind-making, self-mastering, and self-designing animal; and the primary locus of all his activities lies first in his own organism, and in the social organization through which it finds fuller expression. Until man had made something of himself he could make little of the world around him.

In this process of self-discovery and self-transformation, tools, in the narrow sense, served well as subsidiary instruments, but not as the main operative agent in man's development; for technics has never till our own age dissociated itself from the larger cultural whole in which man, as man, has always functioned. The classic Greek term *"tekhne"* characteristically makes no distinction between industrial production and "fine" or symbolic art; and for the greater part of human history these aspects were inseparable, one side respecting the objective conditions and functions, the other responding to subjective needs.

At its point of origin, technics was related to the whole nature of man, and that nature played a part in every aspect of industry: thus technics, at the beginning, was broadly life-centered, not work-centered or power-centered. As in any other ecological complex, varied human interests and purposes, different organic needs, restrained the overgrowth of any single component. Though language was man's most potent symbolic expression, it flowed … from the same common source that finally produced the machine: the primeval repetitive order of ritual, a mode of order man was forced to develop, in self-protection, so as to control the tremendous overcharge of psychal energy that his large brain placed at his disposal.

So far from disparaging the role of technics, however, I shall rather demonstrate that once this basic internal organization was established, technics supported and enlarged the capacities for human expression. The discipline of tool-making and tool-using served as a timely correction, on this hypothesis, to the inordinate powers of invention that spoken language gave to man – powers that otherwise unduly inflated the ego and tempted man to substitute magical verbal formulae for efficacious work.

[...]

This survey will bring the reader to the threshold of the modern world: the sixteenth century in Western Europe. Though some of the implications of such a study cannot be fully worked out until the events of the last four centuries are re-examined and re-appraised, much that is necessary for understanding – and eventually redirecting – the course of contemporary technics will be already apparent, to a sufficiently perceptive mind, from the earliest chapters on. This widened interpretation of the past is a necessary move toward escaping the dire insufficiencies of current one-generation knowledge. If we do not take the time to review the past we shall not have sufficient insight to understand the present or command the future: for the past never leaves us, and the future is already here.

The Megamachine

… About five thousand years ago a monotechnics, devoted to the increase of power and wealth by the systematic organization of workaday activities in a rigidly mechanical pattern, came into existence. At this moment, a new conception of the nature of man arose, and with it a new stress upon the exploitation of physical energies, cosmic and human, apart from the processes of growth and reproduction, came to the fore. In Egypt, Osiris symbolizes the older, fecund, life-oriented technics: Atum-Re, the Sun God, who characteristically created the world out of his own semen without female cooperation, stands for the machine-centered one. The expansion of power, through ruthless human coercion and mechanical organization, took precedence over the nurture and enhancement of life.

The chief mark of this change was the construction of the first complex, high-powered machines; and therewith the beginning of a new regimen, accepted by all later civilized societies – though reluctantly by more archaic cultures – in which work at a single specialized task, segregated from other biological and social activities, not only occupied the entire day but increasingly engrossed the entire lifetime. That was the fundamental departure which, during the last few centuries, has led to the increasing mechanization and automation of all production. With the assemblage of the first collective machines, work, by its systematic dissociation from the rest of life, became a curse, a burden, a sacrifice, a form of punishment: and by reaction this new regimen soon awakened compensatory dreams of effortless affluence, emancipated not only from slavery but from work itself. These ancient dreams, first expressed in myth, but long delayed in realization, now dominate our own age.

The machine I refer to was never discovered in any archeological diggings for a simple reason: it was composed almost entirely of human parts. These parts were brought together in a hierarchical organization under the rule of an absolute monarch whose commands, supported by a coalition of the priesthood, the armed nobility, and the bureaucracy, secured a corpselike obedience from all the components of the machine. Let us call this archetypal collective machine – the human model for all later specialized machines – the "Megamachine." This new kind of machine was far more complex than the contemporary potter's wheel or bow-drill, and it remained the most advanced type of machine until the invention of the mechanical clock in the fourteenth century.

Only through the deliberate invention of such a high-powered machine could the colossal works of engineering that marked the Pyramid Age in both Egypt and Mesopotamia have been brought into existence, often in a single generation. This new technics came to an early climax in the Great Pyramid at Giza: that structure exhibited, as J. H. Breasted pointed out, a watchmaker's standard of exact measurement. By operating as a single mechanical unit of specialized, subdivided, interlocking parts, the 100,000 men who worked on that pyramid could generate ten thousand horsepower. This human mechanism alone made it possible to raise that colossal structure with the use of only the simplest stone and copper tools – without the aid of such otherwise indispensable machines as the wheel, the wagon, the pulley, the derrick, or the winch.

Two things must be noted about this power machine because they identify it through its whole historic course down to the present. The first is that the organizers of the machine derived their power and authority from a cosmic source. The exactitude in measurement, the abstract mechanical order, the compulsive regularity of this labor machine, sprang directly from astronomical observations and abstract scientific calculations: this inflexible, predictable order, incorporated in the calendar, was then transferred to the regimentation of the human components. By a combination of divine command and ruthless military coercion, a large population was made to endure grinding poverty and forced labor at dull repetitive tasks, in order to ensure "life, prosperity, and health" for the divine or semidivine ruler and his entourage.

The second point is that the grave social defects of the human machine – then as now – were partly offset by its superb achievements in flood control, grain production, and urban building, which plainly benefited the whole community. This laid the ground for an enlargement in every area of human culture: in monumental art, in codified law, and in systematically pursued and permanently recorded thought. Such order, such collective security and abundance as were achieved in Mesopotamia and Egypt, later in India, China, in the Andean and Mayan cultures, were never surpassed until the Megamachine was reestablished in a new form in our own time. But conceptually the machine was already detached from other human functions and purposes than the increase of mechanical power and order. With mordant symbolism, the Megamachine's ultimate products in Egypt were tombs, cemeteries, and mummies, while later in Assyria and elsewhere the chief testimonial to its dehumanized efficiency was, again typically, a waste of destroyed cities and poisoned soils.

In a word, what modern economists lately termed the Machine Age had its origin, not in the eighteenth century, but at the very outset of civilization. All its salient characteristics were present from the beginning in both the means and the ends of the collective machine. So Keynes's acute prescription of "pyramid building" as an essential means of coping with the insensate productivity of a highly mechanized technology, applies both to the earliest manifestations and the present ones; for what is a space rocket but the precise dynamic equivalent, in terms of our present-day theology and cosmology, of the static Egyptian pyramid? Both are devices for securing at an extravagant cost a passage to heaven for the favored few, while incidentally maintaining equilibrium in an economic structure threatened by its own excessive productivity.

Unfortunately, though the labor-machine lent itself to vast constructive enterprises, which no small-scale community could even contemplate, much less execute, the most conspicuous result has been achieved through military machines, in colossal acts of destruction and human extermination; acts that monotonously soil the pages of history, from the rape of Sumer to the blasting of Warsaw and Hiroshima. Sooner or later, I suggest, we must have the courage to ask ourselves: Is this association of inordinate power and productivity with equally inordinate violence and destruction a purely accidental one?

Now the misuse of Megamachines would have proved intolerable had they not also brought genuine benefits to the whole community by raising the ceiling of collective human effort and aspiration. Perhaps the most dubious of these advantages, humanly speaking, was the gain in efficiency derived from concentration upon rigorously repetitive motions in work, already indeed introduced in the grinding and polishing processes of neolithic tool-making. This inured civilized man to long spans of regular work, with possibly a higher productive efficiency per unit. But the social byproduct of this new discipline was perhaps even more significant; for some of the psychological benefits hitherto confined to religious ritual were transferred to work. The monotonous repetitive tasks imposed by the Megamachine, which in a pathological form we would associate with a compulsion neurosis, nevertheless served, I suggest, like all ritual and restrictive order, to lessen anxiety and to defend the worker himself from the often demonic promptings of the unconscious, no longer held in check by the traditions and customs of the neolithic village.

In short, mechanization and regimentation, through labor-armies, military-armies, and ultimately through the derivative modes of industrial and bureaucratic organization, supplemented and increasingly replaced religious ritual as a means of coping with anxiety and promoting psychal stability in mass populations. Orderly, repetitive work provided a daily means of self-control: a moralizing agent more pervasive, more effective, more universal than either ritual or law. This hitherto unnoticed psychological contribution was possibly more important than quantitative gains in productive efficiency, for the latter too often was offset by absolute losses in war and conquest. Unfortunately the ruling classes, which claimed immunity from manual labor, were not subject to this discipline; hence, as the historic record testifies, their disordered fantasies too often found an outlet into reality through insensate acts of destruction and extermination.

Having indicated the beginnings of this process, I must regrettably pass over the actual institutional forces that have been at work during the past five thousand years and leap, all too suddenly, into the present age, in which the ancient forms of biotechnics are being either suppressed or supplanted, and in which the extravagant enlargement of the Megamachine itself has become, with increasing compulsiveness, the condition of continued scientific and technical advance. This unconditional commitment to the Megamachine is now regarded by many as the main purpose of human existence.

But if the clues I have been attempting to expose prove helpful, many aspects of the scientific and technical transformation of the last three centuries will call for reinterpretation and judicious reconsideration. For at the very least, we are now bound to explain why the whole process of technical development has become increasingly coercive, totalitarian, and – in its direct human expression – compulsive and grimly irrational, indeed downright hostile to more spontaneous manifestations of life that cannot be fed into the machine.

Before accepting the ultimate translation of all organic processes, biological functions, and human aptitudes into an externally controllable mechanical system, increasingly automatic and self-expanding, it might be well to reexamine the ideological foundations of this whole system, with its overconcentration upon centralized power and external control. Must we not, in fact, ask ourselves if the probable destination of this system is compatible with the further development of specifically human potentialities?

Consider the alternatives now before us. If man were actually, as current theory still supposes, a creature whose manufacture and manipulation of tools played the largest formative part in his development, on what valid grounds do we now propose to strip mankind of the wide variety of autonomous activities historically associated with agriculture and manufacture, leaving the residual mass of workers with only the trivial tasks of watching buttons and dials, and responding to one-way communication and remote control? If man indeed owes his intelligence mainly to his tool-making and tool-using propensities, by what logic do we now take his tools away, so that he will become a functionless, workless being, conditioned to accept only what the Megamachine offers him: an automaton within a larger system of automation, condemned to compulsory consumption, as he was once condemned to compulsory production? What in fact will be left of human life, if one autonomous function after another is either taken over by the machine or else surgically removed – perhaps genetically altered – to fit the Megamachine?

But if the present analysis of human development in relation to technics proves sound, there is an even more fundamental criticism to be made. For we must then go on to question the basic soundness of the current scientific and educational ideology, which is now pressing to shift the locus of human activity from the organic environment, the social group, and the human personality to the Megamachine, considered as the ultimate expression of human intelligence – divorced from the limitations and qualifications of organic existence. That machine-centered metaphysics invites replacement: in both its ancient Pyramid Age form and its Nuclear Age form it is obsolete. For the prodigious advance of knowledge about man's biological origins and historic development made during the last century massively undermines this dubious underdimensioned ideology, with its specious social assumptions and "moral" imperatives, upon which the imposing fabric of science and technics, since the seventeenth century, has been based.

From our present vantage point, we can see that the inventors and controllers of the Megamachine, from the Pyramid Age onward, have in fact been haunted by delusions of omniscience and omnipotence – immediate or prospective. Those original delusions have not become less irrational, now that they have at their disposal the formidable resources of exact science and a high energy technology. The Nuclear Age conceptions of absolute power, infallible computerized intelligence, limitless expanding productivity, all culminating in a system of total control exercised by a military-scientific-industrial élite, correspond to the Bronze Age conception of Divine Kingship. Such power, to succeed on its own terms, must destroy the symbiotic cooperations between all species and communities essential to man's survival and development. Both ideologies belong to the same infantile magico–religious scheme as ritual human sacrifice. As with Captain Ahab's pursuit of Moby Dick, the scientific and technical means are entirely rational, but the ultimate ends are mad.

Living organisms, we now know, can use only limited amounts of energy, as living personalities can utilize only limited quantities of knowledge and experience. "Too much" or "too little" is equally fatal to organic existence. Even too much sophisticated abstract knowledge, insulated from feeling, from moral evaluation, from historic experience, from responsible, purposeful action, can produce a serious unbalance in both the personality and the community. Organisms, societies, human persons are nothing less than delicate devices for regulating energy and putting it at the service of life.

To the extent that our Megatechnics ignores these fundamental insights into the nature of all living organisms, it is actually pre-scientific, even when not actively irrational: a dynamic agent of arrest and regression. When the implications of this weakness are taken in, a deliberate, large-scale dismantling of the Megamachine, in all its institutional forms, must surely take place, with a redistribution of power and authority to smaller units, more open to direct human control.

If technics is to be brought back again into the service of human development, the path of advance will lead, not to the further expansion of the Megamachine, but to the deliberate cultivation of all those parts of the organic environment and the human personality that have been suppressed in order to magnify the offices of the Megamachine.

The deliberate expression and fulfillment of human potentialities requires a quite different approach from that bent solely on the control of natural forces and the modification of human capabilities in order to facilitate and expand the system of control. We know now that play and sport and ritual and dream-fantasy, no less than organized work, have exercised a formative influence upon human culture, and not least upon technics. But make-believe cannot for long be a sufficient substitute for productive work: only when play and work form part of an organic cultural whole, as in Tolstoy's picture of the mowers in *Anna Karenina*, can the many-sided requirements for full human growth be satisfied. Without serious responsible work, man progressively loses his grip on reality.

Instead of liberation *from* work being the chief contribution of mechanization and automation, I would suggest that liberation *for* work, for more educative, mind-forming, self-rewarding work, on a voluntary basis, may become the most salutary contribution of a life-centered technology. This may prove an indispensable counterbalance to universal automation: partly by protecting the displaced worker from boredom and suicidal desperation, only temporarily relievable by anesthetics, sedatives, and narcotics, partly by giving wider play to constructive impulses, autonomous functions, meaningful activities.

Relieved from abject dependence upon the Megamachine, the whole world of biotechnics would then once more become open to man; and those parts of his personality that have been crippled or paralyzed by insufficient use should again come into play, with fuller energy than ever before. Automation is indeed the proper end of a purely mechanical system; and once in its place, subordinate to other human purposes, these cunning mechanisms will serve the human community no less effectively than the reflexes, the hormones, and the autonomic nervous system – nature's earliest experiment in automation – serve the human body. But autonomy, self-direction, and self-fulfillment are the proper ends of organisms; and further technical development must aim at reestablishing this vital harmony at every stage of human growth by giving play to every part of the human personality, not merely to those functions that serve the scientific and technical requirements of the Megamachine.

I realize that in opening up these difficult questions I am not in a position to provide ready-made answers, nor do I suggest that such answers will be easy to fabricate. But it is time that our present wholesale commitment to the machine, which arises largely out of our one-sided interpretation of man's early technical development, should be replaced by a fuller picture of both human nature and the technical milieu, as both have evolved together. That is the first step toward a many-sided transformation of man's self and his work and his habitat – it will probably take many centuries to effect, even after the inertia of the forces now dominant has been overcome.

30

The "Vita Activa" and the Modern Age

Hannah Arendt

I *Vita Activa* and the Human Condition

With the term *vita activa*, I propose to designate three fundamental human activities: labor, work, and action. They are fundamental because each corresponds to one of the basic conditions under which life on earth has been given to man.

Labor is the activity which corresponds to the biological process of the human body, whose spontaneous growth, metabolism, and eventual decay are bound to the vital necessities produced and fed into the life process by labor. The human condition of labor is life itself.

Work is the activity which corresponds to the unnaturalness of human existence, which is not imbedded in, and whose mortality is not compensated by, the species' ever-recurring life cycle. Work provides an "artificial" world of things, distinctly different from all natural surroundings. Within its borders each individual life is housed, while this world itself is meant to outlast and transcend them all. The human condition of work is worldliness.

Action, the only activity that goes on directly between men without the intermediary of things or matter, corresponds to the human condition of plurality, to the fact that men, not Man, live on the earth and inhabit the world. While all aspects of the human condition are somehow related to politics, this plurality is specifically *the* condition – not only the *conditio sine qua non*, but the *conditio per quam* – of all political life. Thus the language of the Romans, perhaps the most political people we have known, used the words "to live" and "to be among men" (*inter homines esse*) or "to die" and "to cease to be among men" (*inter homines esse desinere*) as synonyms. But in its most elementary form, the human condition of action is implicit even in Genesis ("Male and female created He *them*"), if we understand that this story of man's creation is distinguished in principle from the one according to which God originally created Man (*adam*), "him" and not "them," so that the multitude of human beings becomes the result of multiplication. Action would be an unnecessary luxury, a capricious interference with general laws of behavior, if men were endlessly reproducible repetitions of the same model, whose nature or essence was the same for all and as predictable as the nature or essence of any other thing. Plurality is the condition of human action because we are all the same, that is, human, in such a way that nobody is ever the same as anyone else who ever lived, lives, or will live.

All three activities and their corresponding conditions are intimately connected with the most general condition of human existence: birth and death, natality and mortality. Labor assures not only individual survival, but the life of the species. Work and its product, the human artifact, bestow a measure of permanence and durability upon the futility of mortal life and the fleeting character of human time. Action, in so far as it engages in founding and preserving political bodies, creates the condition for remembrance, that is, for history.

From Hannah Arendt, *The Human Condition*, Chicago: University of Chicago Press, 2nd ed. 1998, pp. 7–11, 280–313, with some notes abridged and/or removed.

Labor and work, as well as action, are also rooted in natality in so far as they have the task to provide and preserve the world for, to foresee and reckon with, the constant influx of newcomers who are born into the world as strangers. However, of the three, action has the closest connection with the human condition of natality; the new beginning inherent in birth can make itself felt in the world only because the newcomer possesses the capacity of beginning something anew, that is, of acting. In this sense of initiative, an element of action, and therefore of natality, is inherent in all human activities. Moreover, since action is the political activity par excellence, natality, and not mortality, may be the central category of political, as distinguished from metaphysical, thought.

The human condition comprehends more than the conditions under which life has been given to man. Men are conditioned beings because everything they come in contact with turns immediately into a condition of their existence. The world in which the *vita activa* spends itself consists of things produced by human activities; but the things that owe their existence exclusively to men nevertheless constantly condition their human makers. In addition to the conditions under which life is given to man on earth, and partly out of them, men constantly create their own, self-made conditions, which, their human origin and their variability notwithstanding, possess the same conditioning power as natural things. Whatever touches or enters into a sustained relationship with human life immediately assumes the character of a condition of human existence. This is why men, no matter what they do, are always conditioned beings. Whatever enters the human world of its own accord or is drawn into it by human effort becomes part of the human condition. The impact of the world's reality upon human existence is felt and received as a conditioning force. The objectivity of the world – its object- or thing-character – and the human condition supplement each other; because human existence is conditioned existence, it would be impossible without things, and things would be a heap of unrelated articles, a non-world, if they were not the conditioners of human existence.

To avoid misunderstanding: the human condition is not the same as human nature, and the sum total of human activities and capabilities which correspond to the human condition does not constitute anything like human nature. For neither those we discuss here nor those we leave out, like thought and reason, and not even the most meticulous enumeration of them all, constitute essential characteristics of human existence in the sense that without them this existence would no longer be human. The most radical change in the human condition we can imagine would be an emigration of men from the earth to some other planet. Such an event, no longer totally impossible, would imply that man would have to live under man-made conditions, radically different from those the earth offers him. Neither labor nor work nor action nor, indeed, thought as we know it would then make sense any longer. Yet even these hypothetical wanderers from the earth would still be human; but the only statement we could make regarding their "nature" is that they still are conditioned beings, even though their condition is now self-made to a considerable extent.

The problem of human nature, the Augustinian *quaestio mihi factus sum* ("a question have I become for myself"), seems unanswerable in both its individual psychological sense and its general philosophical sense. It is highly unlikely that we, who can know, determine, and define the natural essences of all things surrounding us, which we are not, should ever be able to do the same for ourselves – this would be like jumping over our own shadows. Moreover, nothing entitles us to assume that man has a nature or essence in the same sense as other things. In other words, if we have a nature or essence, then surely only a god could know and define it, and the first prerequisite would be that he be able to speak about a "who" as though it were a "what."[1] The perplexity is that the modes of human cognition applicable to things with "natural" qualities, including ourselves to the limited extent that we are specimens of the most highly developed species of organic life, fail us when we raise the question: And *who* are we? This is why attempts to define human nature almost invariably end with some construction of a deity, that is, with the god of the philosophers, who, since Plato, has revealed himself upon closer inspection to be a kind of Platonic idea of man. Of course, to demask such philosophic concepts of the divine as conceptualizations of human capabilities and qualities is not a demonstration of, not even an argument for, the non-existence of God; but the fact that attempts to define the nature of man lead so easily into an idea which definitely strikes us as "superhuman" and therefore is identified with the divine may cast suspicion upon the very concept of "human nature."

On the other hand, the conditions of human existence – life itself, natality and mortality, worldliness, plurality, and the earth – can never "explain" what we are or answer the question of who we are for the simple reason that they never condition us absolutely. This has always been the opinion of philosophy, in distinction from the sciences – anthropology, psychology, biology, etc. – which also concern themselves with man. But today we may almost say that we have demonstrated even scientifically that, though we live now, and probably always will, under the earth's conditions, we are not mere earth-bound creatures. Modern natural science owes its great triumphs to having looked upon and treated earth-bound nature from a truly universal viewpoint, that is, from an Archimedean standpoint taken, wilfully and explicitly, outside the earth.

39 Introspection and the Loss of Common Sense

Introspection, as a matter of fact, not the reflection of man's mind on the state of his soul or body but the sheer cognitive concern of consciousness with its own content (and this is the essence of the Cartesian *cogitatio*, where *cogito* always means *cogito me cogitare*) must yield certainty, because here nothing is involved except what the mind has produced itself; nobody is interfering but the producer of the product, man is confronted with nothing and nobody but himself. Long before the natural and physical sciences began to wonder if man is capable of encountering, knowing, and comprehending anything except himself, modern philosophy had made sure in introspection that man concerns himself only with himself. Descartes believed that the certainty yielded by his new method of introspection is the certainty of the I-am.[2] Man, in other words, carries his certainty, the certainty of his existence, within himself; the sheer functioning of consciousness, though it cannot possibly assure a worldly reality given to the senses and to reason, confirms beyond doubt the reality of sensations and of reasoning, that is, the reality of processes which go on in the mind. These are not unlike the biological processes that go on in the body and which, when one becomes aware of them, can also convince one of its working reality. In so far as even dreams are real, since they presuppose a dreamer and a dream, the world of consciousness is real enough. The trouble is only that just as it would be impossible to infer from the awareness of bodily processes the actual shape of any body, including one's own, so it is impossible to reach out from the mere consciousness of sensations, in which one senses his senses and in which even the sensed object becomes part of sensation, into reality with its shapes, forms, colors, and constellations. The seen tree may be real enough for the sensation of vision, just as the dreamed tree is real enough for the dreamer as long as the dream lasts, but neither can ever become a real tree.

It is out of these perplexities that Descartes and Leibniz needed to prove, not the existence of God, but his goodness, the one demonstrating that no evil spirit rules the world and mocks man and the other that this world, including man, is the best of all possible worlds. The point about these exclusively modern justifications, known since Leibniz as theodicies, is that the doubt does not concern the existence of a highest being, which, on the contrary, is taken for granted, but concerns his revelation, as given in biblical tradition, and his intentions with respect to man and world, or rather the adequateness of the relationship between man and world. Of these two, the doubt that the Bible or nature contains divine revelation is a matter of course, once it has been shown that revelation as such, the disclosure of reality to the senses and of truth to reason, is no guaranty for either. Doubt of the goodness of God, however, the notion of a *Dieu trompeur*, arose out of the very experience of deception inherent in the acceptance of the new world view, a deception whose poignancy lies in its irremediable repetiveness, for no knowledge about the heliocentric nature of our planetary system can change the fact that every day the sun is seen circling the earth, rising and setting at its preordained location. Only now, when it appeared as though man, if it had not been for the accident of the telescope, might have been deceived forever, did the ways of God really become wholly inscrutable; the more man learned about the universe, the less he could understand the intentions and purposes for which he should have been created. The goodness of the God of the theodicies, therefore, is strictly the quality of a *deus ex machina*; inexplicable goodness is ultimately the only thing that saves reality in Descartes' philosophy (the coexistence of mind and extension, *res cogitans* and *res extensa*), as it saves the pre-stabilized harmony between man and world in Leibniz.

The very ingenuity of Cartesian introspection, and hence the reason why this philosophy became so all-important to the spiritual and intellectual development of the modern age, lies first in that it had used the nightmare of non-reality as a means of submerging all worldly objects into the stream of consciousness and its processes. The "seen tree" found in consciousness through introspection is no longer the tree given in sight and touch, an entity in itself with an unalterable identical shape of its own. By being processed into an object of consciousness on the same level with a merely remembered or entirely imaginary thing, it becomes part and parcel of this process itself, of that consciousness, that is, which one knows only as an ever-moving stream. Nothing perhaps could prepare our minds better for the eventual dissolution of matter into energy, of objects into a whirl of atomic occurrences, than this dissolution of objective reality into subjective states of mind or, rather, into subjective mental processes. Second, and this was of even greater relevance to the initial stages of the modern age, the Cartesian method of securing certainty against universal doubt corresponded most precisely to the most obvious conclusion to be drawn from the new physical science: though one cannot know truth as something given and disclosed, man can at least know what he makes himself. This, indeed, became the most general and most generally accepted attitude of the modern age, and it is this conviction, rather than the doubt underlying it, that propelled one generation after another for more than three hundred years into an ever-quickening pace of discovery and development.

Cartesian reason is entirely based "on the implicit assumption that the mind can only know that which it has itself produced and retains in some sense within itself." Its highest ideal must therefore be mathematical knowledge as the modern age understands it, that is, not the knowledge of ideal forms given outside the mind but of forms produced by a mind which in this particular instance does not even need the stimulation – or, rather, the irritation – of the senses by objects other than itself. This theory is certainly what Whitehead calls it, "the outcome of common-sense in retreat." For common sense, which once had been the one by which all other senses, with their intimately private sensations, were fitted into the common world, just as vision fitted man into the visible world, now became an inner faculty without any world relationship. This sense now was called common merely because it happened to be common to all. What men now have in common is not the world but the structure of their minds, and this they cannot have in common, strictly speaking; their faculty of reasoning can only happen to be the same in everybody.[3] The fact that, given the problem of two plus two we all will come out with the same answer, four, is henceforth the very model of common-sense reasoning.

Reason, in Descartes no less than in Hobbes, becomes "reckoning with consequences," the faculty of deducing and concluding, that is, of a process which man at any moment can let loose within himself. The mind of this man – to remain in the sphere of mathematics – no longer looks upon "two-and-two-are-four" as an equation in which two sides balance in a self-evident harmony, but understands the equation as the expression of a process in which two and two *become* four in order to generate further processes of addition which eventually will lead into the infinite. This faculty the modern age calls common-sense reasoning; it is the playing of the mind with itself, which comes to pass when the mind is shut off from all reality and "senses" only itself. The results of this play are compelling "truths" because the structure of one man's mind is supposed to differ no more from that of another than the shape of his body. Whatever difference there may be is a difference of mental power, which can be tested and measured like horsepower. Here the old definition of man as an *animal rationale* acquires a terrible precision: deprived of the sense through which man's five animal senses are fitted into a world common to all men, human beings are indeed no more than animals who are able to reason, "to reckon with consequences."

The perplexity inherent in the discovery of the Archimedean point was and still is that the point outside the earth was found by an earth-bound creature, who found that he himself lived not only in a different but in a topsy-turvy world the moment he tried to apply his universal world view to his actual surroundings. The Cartesian solution of this perplexity was to move the Archimedean point into man himself, to choose as ultimate point of reference the pattern of the human mind itself, which assures itself of reality and certainty within a framework of mathematical formulas which are its own products. Here the famous *reductio scientiae ad mathematicam* permits replacement of what is sensuously given by a system of mathematical equations where all real relationships are dissolved into logical relations between man-made symbols. It is this replacement which permits modern science to fulfil its "task of *producing*" the phenomena and objects it wishes to observe. And the assumption is that neither God nor an evil spirit can change the fact that two and two equal four.

40 Thought and the Modern World View

The Cartesian removal of the Archimedean point into the mind of man, while it enabled man to carry it, as it were, within himself wherever he went and thus freed him from given reality altogether – that is, from the human condition of being an inhabitant of the earth – has perhaps never been as convincing as the universal doubt from which it sprang and which it was supposed to dispel. Today, at any rate, we find in the perplexities confronting natural scientists in the midst of their greatest triumphs the same nightmares which have haunted the philosophers from the beginning of the modern age. This nightmare is present in the fact that a mathematical equation, such as of mass and energy – which originally was destined only to save the phenomena, to be in agreement with observable facts that could also be explained differently, just as the Ptolemaic and Copernican systems originally differed only in simplicity and harmony – actually lends itself to a very real conversion of mass into energy and vice versa, so that the mathematical "conversion" implicit in every equation corresponds to convertibility in reality; it is present in the weird phenomenon that the systems of non-Euclidean mathematics were found without any forethought of applicability or even empirical meaning before they gained their surprising validity in Einstein's theory; and it is even more troubling in the inevitable conclusion that "the possibility of such an application must be held open for all, even the most remote constructions of pure mathematics." If it should be true that a whole universe, or rather any number of utterly different universes will spring into existence and "prove" whatever over-all pattern the human mind has constructed, then man may indeed, for a moment, rejoice in a reassertion of the "pre-established harmony between pure mathematics and physics,"[4] between mind and matter, between man and the universe. But it will be difficult to ward off the suspicion that this mathematically preconceived world may be a dream world where every dreamed vision man himself produces has the character of reality only as long as the dream lasts. And his suspicions will be enforced when he must discover that the events and occurrences in the infinitely small, the atom, follow the same laws and regularities as in the infinitely large, the planetary systems.[5] What this seems to indicate is that if we inquire into nature from the standpoint of astronomy we receive planetary systems, while if we carry out our astronomical inquiries from the standpoint of the earth we receive geocentric, terrestrial systems.

In any event, wherever we try to transcend appearance beyond all sensual experience, even instrument-aided, in order to catch the ultimate secrets of Being, which according to our physical world view is so secretive that it never appears and still so tremendously powerful that it produces all appearance, we find that the same patterns rule the macrocosm and the microcosm alike, that we receive the same instrument readings. Here again, we may for a moment rejoice in a refound unity of the universe, only to fall prey to the suspicion that what we have found may have nothing to do with either the macrocosmos or the microcosmos, that we deal only with the patterns of our own mind, the mind which designed the instruments and put nature under its conditions in the experiment – prescribed its laws to nature, in Kant's phrase – in which case it is really as though we were in the hands of an evil spirit who mocks us and frustrates our thirst for knowledge, so that wherever we search for that which we are not, we encounter only the patterns of our own minds.

Cartesian doubt, logically the most plausible and chronologically the most immediate consequence of Galileo's discovery, was assuaged for centuries through the ingenious removal of the Archimedean point into man himself, at least so far as natural science was concerned. But the mathematization of physics, by which the absolute renunciation of the senses for the purpose of knowing was carried through, had in its last stages the unexpected and yet plausible consequence that every question man puts to nature is answered in terms of mathematical patterns to which no model can ever be adequate, since one would have to be shaped after our sense experiences.[6] At this point, the connection between thought and sense experience, inherent in the human condition, seems to take its revenge: while technology demonstrates the "truth" of modern science's most abstract concepts, it demonstrates no more than that man can always apply the results of his mind, that no matter which system he uses for the explanation of natural phenomena he will always be able to adopt it as a guiding principle for making and acting. This possibility was latent even in the beginnings of modern mathematics, when it turned out that numerical truths can be fully translated into spatial relationships. If, therefore, present-day science in its perplexity points to technical achievements to "prove" that we deal with an "authentic order" given in nature,[7] it seems it has fallen into a vicious circle, which can be formulated as follows: scientists formulate their hypotheses to arrange their experiments and then use these experiments to verify their hypotheses; during this whole enterprise, they obviously deal with a hypothetical nature.

In other words, the world of the experiment seems always capable of becoming a man-made reality, and this, while it may increase man's power of making and acting, even of creating a world, far beyond what any previous age dared to imagine in dream and phantasy, unfortunately puts man back once more – and now even more forcefully – into the prison of his own mind, into the limitations of patterns he himself created. The moment he wants what all ages before him were capable of achieving, that is, to experience the reality of what he himself is not, he will find that nature and the universe "escape him" and that a universe construed according to the behavior of nature in the experiment and in accordance with the very principles which man can translate technically into a working reality lacks all possible representation. What is new here is not that things exist of which we cannot form an image – such "things" were always known and among them, for instance, belonged the "soul" – but that the material things we see and represent and against which we had measured immaterial things for which we can form no images should likewise be "unimaginable." With the disappearance of the sensually given world, the transcendent world disappears as well, and with it the possibility of transcending the material world in concept and thought. It is therefore not surprising that the new universe is not only "practically inaccessible but not even thinkable," for "however we think it, it is wrong; not perhaps quite as meaningless as a 'triangular circle,' but much more so than a 'winged lion.' "[8]

Cartesian universal doubt has now reached the heart of physical science itself; for the escape into the mind of man himself is closed if it turns out that the modern physical universe is not only beyond presentation, which is a matter of course under the assumption that nature and Being do not reveal themselves to the senses, but is inconceivable, unthinkable in terms of pure reasoning as well.

41 The Reversal of Contemplation and Action

Perhaps the most momentous of the spiritual consequences of the discoveries of the modern age and, at the same time, the only one that could not have been avoided, since it followed closely upon the discovery of the Archimedean point and the concomitant rise of Cartesian doubt, has been the reversal of the hierarchical order between the *vita contemplativa* and the *vita activa*.

In order to understand how compelling the motives for this reversal were, it is first of all necessary to rid ourselves of the current prejudice which ascribes the development of modern science, because of its applicability, to a pragmatic desire to improve conditions and better human life on earth. It is a matter of historical record that modern technology has its origins not in the evolution of those tools man had always devised for the twofold purpose of easing his labors and erecting the human artifice, but exclusively in an altogether non-practical search for useless knowledge. Thus, the watch, one of the first modern instruments, was not invented for purposes of practical life, but exclusively for the highly "theoretical" purpose of conducting certain experiments with nature. This invention, to be sure, once its practical usefulness became apparent, changed the whole rhythm and the very physiognomy of human life; but from the standpoint of the inventors, this was a mere incident. If we had to rely only on men's so-called practical instincts, there would never have been any technology to speak of, and although today the already existing technical inventions carry a certain momentum which will probably generate improvements up to a certain point, it is not likely that our technically conditioned world could survive, let alone develop further, if we ever succeeded in convincing ourselves that man is primarily a practical being.

However that may be, the fundamental experience behind the reversal of contemplation and action was precisely that man's thirst for knowledge could be assuaged only after he had put his trust into the ingenuity of his hands. The point was not that truth and knowledge were no longer important, but that they could be won only by "action" and not by contemplation. It was an instrument, the telescope, a work of man's hands, which finally forced nature, or rather the universe, to yield its secrets. The reasons for trusting *doing* and for distrusting *contemplation* or *observation* became even more cogent after the results of the first active inquiries. After being and appearance had parted company and truth was no longer supposed to appear, to reveal and disclose itself to the mental eye of a beholder, there arose a veritable necessity to hunt for truth behind deceptive appearances. Nothing indeed could be less trustworthy for acquiring knowledge and approaching truth than passive observation or mere contemplation. In order to be certain one had to *make sure*, and in order to know one had to do. Certainty of knowledge could be reached only under a twofold condition: first, that knowledge concerned only what one had done himself – so that its ideal became mathematical knowledge, where we deal only with self-made entities of the mind – and second, that knowledge was of such a nature that it could be tested only through more doing.

Since then, scientific and philosophic truth have parted company; scientific truth not only need not be eternal, it need not even be comprehensible or adequate to human reason. It took many generations of scientists before the human mind grew bold enough to fully face this implication of modernity. If nature and the universe are products of a divine maker, and if the human mind is incapable of understanding what man has not made himself, then man cannot possibly expect to learn anything about nature that he can understand. He may be able, through ingenuity, to find out and even to imitate the devices of natural processes, but that does not mean these devices will ever make sense to him – they do not have to be intelligible. As a matter of fact, no supposedly suprarational divine revelation and no supposedly abstruse philosophic truth has ever offended human reason so glaringly as certain results of modern science. One can indeed say with Whitehead: "Heaven knows what seeming nonsense may not to-morrow be demonstrated truth."[9]

Actually, the change that took place in the seventeenth century was more radical than what a simple reversal of the established traditional order between contemplation and doing is apt to indicate. The reversal, strictly speaking, concerned only the relationship between thinking and doing, whereas contemplation, in the original sense of beholding the truth, was altogether eliminated. For thought and contemplation are not the same. Traditionally, thought was conceived as the most direct and important way to lead to the contemplation of truth. Since Plato, and probably since Socrates, thinking was understood as the inner dialogue in which one speaks with himself (*eme emautō*, to recall the idiom current in Plato's dialogues); and although this dialogue lacks all outward manifestation and even requires a more or less complete cessation of all other activities, it constitutes in itself a highly active state. Its outward inactivity is clearly separated from the passivity, the complete stillness, in which truth is finally revealed to man. If medieval scholasticism looked upon philosophy as the handmaiden of theology, it could very well have appealed to Plato and Aristotle themselves; both, albeit in a very different context, considered this dialogical thought process to be the way to prepare the soul and lead the mind to a beholding of truth beyond thought and beyond speech – a truth that is *arrhēton*, incapable of being communicated through words, as Plato put it,[10] or beyond speech, as in Aristotle.[11]

The reversal of the modern age consisted then not in raising doing to the rank of contemplating as the highest state of which human beings are capable, as though henceforth doing was the ultimate meaning for the sake of which contemplation was to be performed, just as, up to that time, all activities of the *vita activa* had been judged and justified to the extent that they made the *vita contemplativa* possible. The reversal concerned only thinking, which from then on was the handmaiden of doing as it had been the *ancilla theologiae*, the handmaiden of contemplating divine truth in medieval philosophy and the handmaiden of contemplating the truth of Being in ancient philosophy. Contemplation itself became altogether meaningless.

The radicality of this reversal is somehow obscured by another kind of reversal, with which it is frequently identified and which, since Plato, has dominated the history of Western thought. Whoever reads the Cave allegory in Plato's *Republic* in the light of Greek history will soon be aware that the *periagōgē*, the turning-about that Plato demands of the philosopher, actually amounts to a reversal of the Homeric world order. Not life after death, as in the Homeric Hades, but ordinary life on earth, is located in a "cave," in an underworld; the soul is not the shadow of the body, but the body the shadow of the soul; and the senseless, ghostlike motion ascribed by Homer to the lifeless existence of the soul after death in Hades is now ascribed to the senseless doings of men who do not leave the cave of human existence to behold the eternal ideas visible in the sky.

In this context, I am concerned only with the fact that the Platonic tradition of philosophical as well as political thought started with a reversal, and that this original reversal determined to a large extent the thought patterns into which Western philosophy almost automatically fell wherever it was not animated by a great and original philosophical impetus. Academic philosophy, as a matter of fact, has ever since been dominated by the never-ending reversals of idealism and materialism, of transcendentalism and immanentism, of realism and nominalism, of hedonism and asceticism, and so on. What matters here is the reversibility of all these systems, that they can be turned "upside down" or "downside up" at any moment in history without requiring for such reversal either historical events or changes in the structural elements involved. The concepts themselves remain the same no matter where they are placed in the various systematic orders. Once Plato had succeeded in making these structural elements and concepts reversible, reversals within the course of intellectual history no longer needed more than purely intellectual experience, and experience within the framework of conceptual thinking itself. These reversals already began with the philosophical schools in late antiquity and have remained part of the Western tradition. It is still the same tradition, the same intellectual game with paired antitheses that rules, to an extent, the famous modern reversals of spiritual hierarchies, such as Marx's turning Hegelian dialectic upside down or Nietzsche's revaluation of the sensual and natural as against the supersensual and supernatural.

The reversal we deal with here, the spiritual consequence of Galileo's discoveries, although it has frequently been interpreted in terms of the traditional reversals and hence as integral to the Western history of ideas, is of an altogether different nature. The conviction that objective truth is not given to man but that he can know only what he makes himself is not the result of skepticism but of a demonstrable discovery, and therefore does not lead to resignation but either to redoubled activity or to despair. The world loss of modern philosophy, whose introspection discovered consciousness as the inner sense with which one senses his senses and found it to be the only guaranty of reality, is different not only in degree from the age-old suspicion of the philosophers toward the world and toward the others with whom they shared the world; the philosopher no longer turns from the world of deceptive perishability to another world of eternal truth, but turns away from both and withdraws into himself. What he discovers in the region of the inner self is, again, not an image whose permanence can be beheld and contemplated, but, on the contrary, the constant movement of sensual perceptions and the no less constantly moving activity of the mind. Since the seventeenth century, philosophy has produced the best and least disputed results when it has investigated, through a supreme effort of self-inspection, the processes of the senses and of the mind. In this aspect, most of modern philosophy is indeed theory of cognition and psychology, and in the few instances where the potentialities of the Cartesian method of introspection were fully realized by men like Pascal, Kierkegaard, and Nietzsche, one is tempted to say that philosophers have experimented with their own selves no less radically and perhaps even more fearlessly than the scientists experimented with nature.

Much as we may admire the courage and respect the extraordinary ingenuity of philosophers throughout the modern age, it can hardly be denied that their influence and importance decreased as never before. It was not in the Middle Ages but in modern thinking that philosophy came to play second and even third fiddle. After Descartes based his own philosophy upon the discoveries of Galileo, philosophy has seemed condemned to be always one step behind the scientists and their ever more amazing discoveries, whose principles it has strived arduously to discover *ex post facto* and to fit into some over-all interpretation of the nature of human knowledge. As such, however, philosophy was not needed by the scientists, who – up to our time, at least – believed that they had no use for a handmaiden, let alone one who would "carry the torch in front of her gracious lady" (Kant). The philosophers became either epistemologists, worrying about an over-all theory of science which the scientists did not need, or they became, indeed, what Hegel wanted them to be, the organs of the *Zeitgeist*, the mouthpieces in which the general mood of the time was expressed with conceptual clarity. In both instances, whether they looked upon nature or upon history, they tried to understand and come to terms with what happened without them. Obviously, philosophy suffered more from modernity than any other field of human endeavor;

and it is difficult to say whether it suffered more from the almost automatic rise of activity to an altogether unexpected and unprecedented dignity or from the loss of traditional truth, that is, of the concept of truth underlying our whole tradition.

42 The Reversal within the *Vita Activa* and the Victory of *Homo Faber*

First among the activities within the *vita activa* to rise to the position formerly occupied by contemplation were the activities of making and fabricating – the prerogatives of *homo faber*. This was natural enough, since it had been an instrument and therefore man in so far as he is a toolmaker that led to the modern revolution. From then on, all scientific progress has been most intimately tied up with the ever more refined development in the manufacture of new tools and instruments. While, for instance, Galileo's experiments with the fall of heavy bodies could have been made at any time in history if men had been inclined to seek truth through experiments, Michelson's experiment with the interferometer at the end of the nineteenth century relied not merely on his "experimental genius" but "required the general advance in technology," and therefore "could not have been made earlier than it was."[12]

It is not only the paraphernalia of instruments and hence the help man had to enlist from *homo faber* to acquire knowledge that caused these activities to rise from their former humble place in the hierarchy of human capacities. Even more decisive was the element of making and fabricating present in the experiment itself, which produces its own phenomena of observation and therefore depends from the very outset upon man's productive capacities. The use of the experiment for the purpose of knowledge was already the consequence of the conviction that one can know only what he has made himself, for this conviction meant that one might learn about those things man did not make by figuring out and imitating the processes through which they had come into being. The much discussed shift of emphasis in the history of science from the old questions of "what" or "why" something is to the new question of "how" it came into being is a direct consequence of this conviction, and its answer can only be found in the experiment. The experiment repeats the natural process as though man himself were about to make nature's objects, and although in the early stages of the modern age no responsible scientist would have dreamt of the extent to which man actually is capable of "making" nature, he nevertheless from the onset approached it from the standpoint of the One who made it, and this not for practical reasons of technical applicability but exclusively for the "theoretical" reason that certainty in knowledge could not be gained otherwise: "Give me matter and I will build a world from it, that is, give me matter and I will show you how a world developed from it." These words of Kant show in a nutshell the modern blending of making and knowing, whereby it is as though a few centuries of knowing in the mode of making were needed as the apprenticeship to prepare modern man for making what he wanted to know.

Productivity and creativity, which were to become the highest ideals and even the idols of the modern age in its initial stages, are inherent standards of *homo faber*, of man as a builder and fabricator. However, there is another and perhaps even more significant element noticeable in the modern version of these faculties. The shift from the "why" and "what" to the "how" implies that the actual objects of knowledge can no longer be things or eternal motions but must be processes, and that the object of science therefore is no longer nature or the universe but the history, the story of the coming into being, of nature or life or the universe. Long before the modern age developed its unprecedented historical consciousness and the concept of history became dominant in modern philosophy, the natural sciences had developed into historical disciplines, until in the nineteenth century they added to the older disciplines of physics and chemistry, of zoology and botany, the new natural sciences of geology or history of the earth, biology or the history of life, anthropology or the history of human life, and, generally, natural history. In all these instances, development, the key concept of the historical sciences, became the central concept of the physical sciences as well. Nature, because it could be known only in processes which human ingenuity, the ingeniousness of *homo faber*, could repeat and remake in the experiment, became a process, and all particular natural things derived their significance and meaning solely from their functions in the over-all process. In the place of the concept of Being we now find the concept of Process. And whereas it is in the nature of Being to appear and thus disclose itself, it is in the nature of Process to remain invisible, to be something whose existence can only be inferred from the presence of certain phenomena. This process was originally the fabrication process which "disappears in the product," and it was based on the experience of *homo faber*, who knew that a production process necessarily precedes the actual existence of every object.

Yet while this insistence on the process of making or the insistence upon considering every thing as the result of a fabrication process is highly characteristic of *homo faber* and his sphere of experience, the exclusive emphasis the modern age placed on it at the expense of all interest in the things, the products themselves, is quite new. It actually transcends the mentality of man as a toolmaker and fabricator, for whom, on the contrary, the production process was a mere means to an end. Here, from the standpoint of *homo faber*, it was as though the means, the production process or development, was more important than the end, the finished product. The reason for this shift of emphasis is obvious: the scientist made only in order to know, not in order to produce things, and the product was a mere by-product, a side effect. Even today all true scientists will agree that the technical applicability of what they are doing is a mere by-product of their endeavor.

The full significance of this reversal of means and ends remained latent as long as the mechanistic world view, the world view of *homo faber* par excellence, was predominant. This view found its most plausible theory in the famous analogy of the relationship between nature and God with the relationship between the watch and the watchmaker. The point in our context is not so much that the eighteenth-century idea of God was obviously formed in the image of *homo faber* as that in this instance the process character of nature was still limited. Although all particular natural things had already been engulfed in the process from which they had come into being, nature as a whole was not yet a process but the more or less stable end product of a divine maker. The image of watch and watchmaker is so strikingly apposite precisely because it contains both the notion of a process character of nature in the image of the movements of the watch and the notion of its still intact object character in the image of the watch itself and its maker.

It is important at this point to remember that the specifically modern suspicion toward man's truth-receiving capacities, the mistrust of the given, and hence the new confidence in making and introspection which was inspired by the hope that in human consciousness there was a realm where knowing and producing would coincide, did not arise directly from the discovery of the Archimedean point outside the earth in the universe. They were, rather, the necessary consequences of this discovery for the discoverer himself, in so far as he was and remained an earth-bound creature. This close relationship of the modern mentality with philosophical reflection naturally implies that the victory of *homo faber* could not remain restricted to the employment of new methods in the natural sciences, the experiment and the mathematization of scientific inquiry. One of the most plausible consequences to be drawn from Cartesian doubt was to abandon the attempt to understand nature and generally to know about things not produced by man, and to turn instead exclusively to things that owed their existence to man. This kind of argument, in fact, made Vico turn his attention from natural science to history, which he thought to be the only sphere where man could obtain certain knowledge, precisely because he dealt here only with the products of human activity. The modern discovery of history and historical consciousness owed one of its greatest impulses neither to a new enthusiasm for the greatness of man, his doings and sufferings, nor to the belief that the meaning of human existence can be found in the story of mankind, but to the despair of human reason, which seemed adequate only when confronted with man-made objects.

Prior to the modern discovery of history but closely connected with it in its impulses are the seventeenth-century attempts to formulate new political philosophies or, rather, to invent the means and instruments with which to "make an artifical animal ... called a Commonwealth, or State."[13] With Hobbes as with Descartes "the prime mover was doubt,"[14] and the chosen method to establish the "art of man," by which he would make and rule his own world as "God hath made and governs the world" by the art of nature, is also introspection, "to read in himself," since this reading will show him "the similitude of the thoughts and passions of one man to the thoughts and passions of another." Here, too, the rules and standards by which to build and judge this most human of human "works of art"[15] do not lie outside of men, are not something men have in common in a worldly reality perceived by the senses or by the mind. They are, rather, inclosed in the inwardness of man, open only to introspection, so that their very validity rests on the assumption that "not ...

the objects of the passions" but the passions themselves are the same in every specimen of the species man-kind. Here again we find the image of the watch, this time applied to the human body and then used for the movements of the passions. The establishment of the Commonwealth, the human creation of "an artificial man," amounts to the building of an "automaton [an engine] that moves [itself] by springs and wheels as doth a watch."

In other words, the process which, as we saw, invaded the natural sciences through the experiment, through the attempt to imitate under artificial conditions the process of "making" by which a natural thing came into existence, serves as well or even better as the principle for doing in the realm of human affairs. For here the processes of inner life, found in the passions through introspection, can become the standards and rules for the creation of the "automatic" life of that "artificial man" who is "the great Leviathan." The results yielded by introspection, the only method likely to deliver certain knowledge, are in the nature of movements: only the objects of the senses remain as they are and endure, precede and survive, the act of sensation; only the objects of the passions are permanent and fixed to the extent that they are not devoured by the attainment of some passionate desire; only the objects of thoughts, but never thinking itself, are beyond motion and perishability. Processes, therefore, and not ideas, the models and shapes of the things to be, become the guide for the making and fabricating activities of *homo faber* in the modern age.

Hobbes's attempt to introduce the new concepts of making and reckoning into political philosophy – or, rather, his attempt to apply the newly discovered aptitudes of making to the realm of human affairs – was of the greatest importance; modern rationalism as it is currently known, with the assumed antagonism of reason and passion as its stock-in-trade, has never found a clearer and more uncompromising representative. Yet it was precisely the realm of human affairs where the new philosophy was first found wanting, because by its very nature it could not understand or even believe in reality. The idea that only what I am going to make will be real – perfectly true and legitimate in the realm of fabrication – is forever defeated by the actual course of events, where nothing happens more frequently than the totally unexpected. To act in the form of making, to reason in the form of "reckoning with consequences," means to leave out the unexpected, the event itself, since it would be unreasonable or irrational to expect what is no more than an "infinite improbability." Since, however, the event constitutes the very texture of reality within the realm of human affairs, where the "wholly improbable happens regularly," it is highly unrealistic not to reckon with it, that is, not to reckon with something with which nobody can safely reckon. The political philosophy of the modern age, whose greatest representative is still Hobbes, founders on the perplexity that modern rationalism is unreal and modern realism is irrational – which is only another way of saying that reality and human reason have parted company. Hegel's gigantic enterprise to reconcile spirit with reality (*den Geist mit der Wirklichkeit zu versöhnen*), a reconciliation that is the deepest concern of all modern theories of history, rested on the insight that modern reason foundered on the rock of reality.

The fact that modern world alienation was radical enough to extend even to the most worldly of human activities, to work and reification, the making of things and the building of a world, distinguishes modern attitudes and evaluations even more sharply from those of tradition than a mere reversal of contemplation and action, of thinking and doing, would indicate. The break with contemplation was consummated not with the elevation of man the maker to the position formerly held by man the contemplator, but with the introduction of the concept of process into making. Compared with this, the striking new arrangement of hierarchical order within the *vita activa*, where fabrication now came to occupy a rank formerly held by political action, is of minor importance. We saw before that this hierarchy had in fact, though not expressly, already been overruled in the very beginnings of political philosophy by the philosophers' deep-rooted suspicion of politics in general and action in particular.

The matter is somewhat confused because Greek political philosophy still follows the order laid down by the *polis* even when it turns against it; but in their strictly philosophical writings (to which, of course, one must turn if he wants to know their innermost thoughts), Plato as well as Aristotle tends to invert the relationship between work and action in favor of work. Thus Aristotle, in a discussion of the different kinds of cognition in his *Metaphysics*, places *dianoia* and *epistēmē praktikē*, practical insight and political science, at the lowest rank of his order, and puts above them the science of fabrication, *epistēmē poiētikē*, which immediately precedes and leads to *theōria*, the contemplation of truth.[16] And the reason for this predilection in philosophy is by no means the politically inspired suspicion of action which we mentioned before, but the philosophically much more compelling one that contemplation and fabrication (*theōria* and *poiēsis*) have an inner affinity and do not stand in the same unequivocal opposition to each other as contemplation and action. The decisive point of similarity, at least in Greek philosophy, was that contemplation, the beholding of something, was considered to be an inherent element in fabrication as well, inasmuch as the work of the craftsman was guided by the "idea," the model beheld by him before the fabrication process had started as well as after it had ended, first to tell him what to make and then to enable him to judge the finished product.

Historically, the source of this contemplation, which we find for the first time described in the Socratic school, is at least twofold. On one hand, it stands in obvious and consistent connection with the famous contention of Plato, quoted by Aristotle, that *thaumazein*, the shocked wonder at the miracle of Being, is the beginning of all philosophy.[17] It seems to me highly probable that this Platonic contention is the immediate result of an experience, perhaps the most striking one, that Socrates offered his disciples: the sight of him time and again suddenly overcome by his thoughts and thrown into a state of absorption to the point of perfect motionlessness for many hours. It seems no less plausible that this shocked wonder should be essentially speechless, that is, that its actual content should be untranslatable into words. This, at least, would explain why Plato and Aristotle, who held *thaumazein* to be the beginning of philosophy, should also agree – despite so many and such decisive disagreements – that some state of speechlessness, the essentially speechless state of contemplation, was the end of philosophy. *Theōria*, in fact, is only another word for *thaumazein*; the contemplation of truth at which the philosopher ultimately arrives is the philosophically purified speechless wonder with which he began.

There is, however, another side to this matter, which shows itself most articulately in Plato's doctrine of ideas, in its content as well as in its terminology and exemplifications. These reside in the experiences of the craftsman, who sees before his inner eye the shape of the model according to which he fabricates his object. To Plato, this model, which craftsmanship can only imitate but not create, is no product of the human mind but given to it. As such it possesses a degree of permanence and excellence which is not actualized but on the contrary spoiled in its materialization through the work of human hands. Work makes perishable and spoils the excellence of what remained eternal so long as it was the object of mere contemplation. Therefore, the proper attitude toward the models which guide work and fabrication, that is, toward Platonic ideas, is to leave them as they are and appear to the inner eye of the mind. If man only renounces his capacity for work and does not do anything, he can behold them and thus participate in their eternity. Contemplation, in this respect, is quite unlike the enraptured state of wonder with which man responds to the miracle of Being as a whole. It is and remains part and parcel of a fabrication process even though it has divorced itself from all work and all doing; in it, the beholding of the model, which now no longer is to guide any doing, is prolonged and enjoyed for its own sake.

In the tradition of philosophy, it is this second kind of contemplation that became the predominant one. Therefore the motionlessness which in the state of speechless wonder is no more than an incidental, unintended result of absorption, becomes now the condition and hence the outstanding characteristic of the *vita contemplativa*. It is not wonder that overcomes and throws man into motionlessness, but it is through the conscious cessation of activity, the activity of making, that the contemplative state is reached. If one reads medieval sources on the joys and delights of contemplation, it is as though the philosophers wanted to make sure that *homo faber* would heed the call and let his arms drop, finally realizing that his greatest desire, the desire for permanence and immortality, cannot be fulfilled by his doings, but only when he realizes that the beautiful and eternal cannot be made. In Plato's philosophy, speechless wonder, the beginning and the end of philosophy, together with the philosopher's love for the eternal and the craftsman's desire for permanence and immortality, permeate each other until they are almost indistinguishable. Yet the very fact that the philosophers' speechless wonder seemed to be an experience reserved for the few, while the craftsmen's contemplative glance was known by many, weighed heavily in favor of a contemplation primarily derived from the experiences of *homo faber*. It already weighed heavily with Plato, who drew his examples from the realm of making because they were closer to a more general human experience, and it weighed even more heavily where some kind of contemplation and meditation was required of everybody, as in medieval Christianity.

Thus it was not primarily the philosopher and philosophic speechless wonder that molded the concept and practice of contemplation and the *vita contemplativa*, but rather *homo faber* in disguise; it was man the maker and fabricator, whose job it is to do violence to nature in order to build a permanent home for himself, and who now was persuaded to renounce violence together with all activity, to leave things as they are, and to find his home in the contemplative dwelling in the neighborhood of the imperishable and eternal. *Homo faber* could be persuaded to this change of attitude because he knew contemplation and some of its delights from his own experience; he did not need a complete change of heart, a true *periagōgē*, a radical turnabout. All he had to do was let his arms drop and prolong indefinitely the act of beholding the *eidos*, the eternal shape and model he had formerly wanted to imitate and whose excellence and beauty he now knew he could only spoil through any attempt at reification.

If, therefore, the modern challenge to the priority of contemplation over every kind of activity had done no more than turn upside down the established order between making and beholding, it would still have remained in the traditional framework. This framework was forced wide open, however, when in the understanding of fabrication itself the emphasis shifted entirely away from the product and from the permanent, guiding model to the fabrication process, away from the question of what a thing is and what kind of thing was to be produced to the question of how and through which means and processes it had come into being and could be reproduced. For this implied both that contemplation was no longer believed to yield truth and that it had lost its position in the *vita activa* itself and hence within the range of ordinary human experience.

43 The Defeat of *Homo Faber* and the Principle of Happiness

If one considers only the events that led into the modern age and reflects solely upon the immediate consequences of Galileo's discovery, which must have struck the great minds of the seventeenth century with the compelling force of self-evident truth, the reversal of contemplation and fabrication, or rather the elimination of contemplation from the range of meaningful human capacities, is almost a matter of course. It seems equally plausible that this reversal should have elevated *homo faber*, the maker and fabricator, rather than man the actor or man as *animal laborans*, to the highest range of human possibilities.

And, indeed, among the outstanding characteristics of the modern age from its beginning to our own time we find the typical attitudes of *homo faber:* his instrumentalization of the world, his confidence in tools and in the productivity of the maker of artificial objects; his trust in the all-comprehensive range of the means-end category, his conviction that every issue can be solved and every human motivation reduced to the principle of utility; his sovereignty, which regards everything given as material and thinks of the whole of nature as of "an immense fabric from which we can cut out whatever we want to resew it however we like";[18] his equation of intelligence with ingenuity, that is, his contempt for all thought which cannot be considered to be "the first step ... for the fabrication of artificial objects, particularly of tools to make tools, and to vary their fabrication indefinitely";[19] finally, his matter-of-course identification of fabrication with action.

It would lead us too far afield to follow the ramifications of this mentality, and it is not necessary, for they are easily detected in the natural sciences, where the purely theoretical effort is understood to spring from the desire to create order out of "mere disorder," the "wild variety of nature," and where therefore *homo faber*'s predilection for patterns for things to be produced replaces the older notions of harmony and simplicity. It can be found in classical economics, whose highest standard is productivity and whose prejudice against non-productive activities is so strong that even Marx could justify his plea for justice for laborers only by misrepresenting the laboring, non-productive activity in terms of work and fabrication. It is most articulate, of course, in the pragmatic trends of modern philosophy, which are not only characterized by Cartesian world alienation but also by the unanimity with which English philosophy from the seventeenth century onward and French philosophy in the eighteenth century adopted the principle of utility as the key which would open all doors to the explanation of human motivation and behavior. Generally speaking, the oldest conviction of *homo faber* – that "man is the measure of all things" – advanced to the rank of a universally accepted commonplace.

What needs explanation is not the modern esteem of *homo faber* but the fact that this esteem was so quickly followed by the elevation of laboring to the highest position in the hierarchical order of the *vita activa*. This second reversal of hierarchy within the *vita activa* came about more gradually and less dramatically than either the reversal of contemplation and action in general or the reversal of action and fabrication in particular. The elevation of laboring was preceded by certain deviations and variations from the traditional mentality of *homo faber* which were highly characteristic of the modern age and which, indeed, arose almost automatically from the very nature of the events that ushered it in. What changed the mentality of *homo faber* was the central position of the concept of process in modernity. As far as *homo faber* was concerned, the modern shift of emphasis from the "what" to the "how," from the thing itself to its fabrication process, was by no means an unmixed blessing. It deprived man as maker and builder of those fixed and permanent standards and measurements which, prior to the modern age, have always served him as guides for his doing and criteria for his judgment. It is not only and perhaps not even primarily the development of commercial society that, with the triumphal victory of exchange value over use value, first introduced the principle of interchangeability, then the relativization, and finally the devaluation, of all values. For the mentality of modern man, as it was determined by the development of modern science and the concomitant unfolding of modern philosophy, it was at least as decisive that man began to consider himself part and parcel of the two superhuman, all-encompassing processes of nature and history, both of which seemed doomed to an infinite progress without ever reaching any inherent *telos* or approaching any preordained idea.

Homo faber, in other words, as he arose from the great revolution of modernity, though he was to acquire an undreamed-of ingenuity in devising instruments to measure the infinitely large and the infinitely small, was deprived of those permanent measures that precede and outlast the fabrication process and form an authentic and reliable absolute with respect to the fabricating activity. Certainly, none of the activities of the *vita activa* stood to lose as much through the elimination of contemplation from the range of meaningful human capacities as fabrication. For unlike action, which partly consists in the unchaining of processes, and unlike laboring, which follows closely the metabolic process of biological life, fabrication experiences processes, if it is aware of them at all, as mere means toward an end, that is, as something secondary and derivative. No other capacity, moreover, stood to lose as much through modern world alienation and the elevation of introspection into an omnipotent device to conquer nature as those faculties which are primarily directed toward the building of the world and the production of worldly things.

Nothing perhaps indicates clearer the ultimate failure of *homo faber* to assert himself than the rapidity with which the principle of utility, the very quintessence of his world view, was found wanting and was superseded by the principle of "the greatest happiness of the greatest number." When this happened it was manifest that the conviction of the age that man can know only what he makes himself – which seemingly was so eminently propitious to a full victory of *homo faber* – would be overruled and eventually destroyed by the even more modern principle of process, whose concepts and categories are altogether alien to the needs and ideals of *homo faber*. For the principle of utility, though its point of reference is clearly man, who uses matter to produce things, still presupposes a world of use objects by which man is surrounded and in which he moves. If this relationship between man and world is no longer secure, if worldly things are no longer primarily considered in their usefulness but as more or less incidental results of the production process which brought them into being, so that the end product of the production process is no longer a true end and the produced thing is valued not for the sake of its predetermined usage but "for its production of something else," then, obviously, the objection can be "raised that...its value is secondary only, and a world that contains no primary values can contain no secondary ones either." This radical loss of values within the restricted frame of reference of *homo faber* himself occurs almost automatically as soon as he defines himself not as the maker of objects and the builder of the human artifice who incidentally invents tools, but considers himself primarily a toolmaker and "particularly [a maker] of tools to make tools" who only incidentally also produces things. If one applies the principle of utility in this context at all, then it refers primarily not to use objects and not to usage but to the production process. Now what helps stimulate productivity and lessens pain and effort is useful. In other words, the ultimate standard of measurement is not utility and usage at all, but "happiness," that is, the amount of pain and pleasure experienced in the production or in the consumption of things.

Bentham's invention of the "pain and pleasure calculus" combined the advantage of seemingly introducing the mathematical method into the moral sciences with the even greater attraction of having found a principle which resided entirely on introspection. His "happiness," the sum total of pleasures minus pains, is as much an inner sense which senses sensations and remains unrelated to worldly objects as the Cartesian consciousness that is conscious of its own activity. Moreover, Bentham's basic assumption that what all men have in common is not the world but the sameness of their own nature, which manifests itself in the sameness of calculation and the sameness of being affected by pain and pleasure, is directly derived from the earlier philosophers of the modern age. For this philosophy, "hedonism" is even more of a misnomer than for the epicureanism of late antiquity, to which modern hedonism is only superficially related. The principle of all hedonism, as we saw before, is not pleasure but avoidance of pain, and Hume, who in contradistinction to Bentham was still a philosopher, knew quite well that he who wants to make pleasure the ultimate end of all human action is driven to admit that not pleasure but pain, not desire but fear, are his true guides. "If you...inquire, why [somebody] desires health, he will readily reply, because sickness is painful. If you push your inquiries further and desire a reason why he hates pain, it is impossible he can ever give any. This is an ultimate end, and is never referred to by any other object." The reason for this impossibility is that only pain is completely independent of any object, that only one who is in pain really senses nothing but himself; pleasure does not enjoy itself but something besides itself. Pain is the only inner sense found by introspection which can rival in independence from experienced objects the self-evident certainty of logical and arithmetical reasoning.

While this ultimate foundation of hedonism in the experience of pain is true for both its ancient and modern varieties, in the modern age it acquires an altogether different and much stronger emphasis. For here it is by no means the world, as in antiquity, that drives man into himself to escape the pains it may inflict, under which circumstance both pain and pleasure still retain a good deal of their worldly significance. Ancient world alienation in all its varieties – from stoicism to epicureanism down to hedonism and cynicism – had been inspired by a deep mistrust of the world and moved by a vehement impulse to withdraw from worldly involvement, from the trouble and pain it inflicts, into the security of an inward realm in which the self is exposed to nothing but itself. Their modern counterparts – puritanism, sensualism, and Bentham's hedonism – on the contrary, were inspired by an equally deep mistrust of man as such; they were moved by doubt of the adequacy of the human senses to receive reality, the adequacy of human reason to receive truth, and hence by the conviction of the deficiency or even depravity of human nature.

This depravity is not Christian or biblical either in origin or in content, although it was of course interpreted in terms of original sin, and it is difficult to say whether it is more harmful and repulsive when puritans denounce man's corruptness or when Benthamites brazenly hail as virtues what men always have known to be vices. While the ancients had relied upon imagination and memory, the imagination of pains from which they were free or the memory of past pleasures in situations of acute painfulness, to convince themselves of their happiness, the moderns needed the calculus of pleasure or the puritan moral bookkeeping of merits and transgressions to arrive at some illusory mathematical certainty of happiness or salvation. (These moral arithmetics are, of course, quite alien to the spirit pervading the philosophic schools of late antiquity. Moreover, one need only reflect on the rigidity of self-imposed discipline and the concomitant nobility of character, so manifest in those who had been formed by ancient stoicism or epicureanism, to become aware of the gulf by which these versions of hedonism are separated from modern puritanism, sensualism, and hedonism. For this difference, it is almost irrelevant whether the modern character is still formed by the older narrow-minded, fanatic self-righteousness or has yielded to the more recent self-centered and self-indulgent egotism with its infinite variety of futile miseries.) It seems more than doubtful that the "greatest happiness principle" would have achieved its intellectual triumphs in the English-speaking world if no more had been involved than the questionable discovery that "nature has placed mankind under the governance of two sovereign masters, pain and pleasure,"[20] or the absurd idea of establishing morals as an exact science by isolating "in the human soul that feeling which seems to be the most easily measurable."

Hidden behind this as behind other, less interesting variations of the sacredness of egoism and the all-pervasive power of self-interest, which were current to the point of being commonplace in the eighteenth and early nineteenth centuries, we find another point of reference which indeed forms a much more potent principle than any pain-pleasure calculus could ever offer, and that is the principle of life itself. What pain and pleasure, fear and desire, are actually supposed to achieve in all these systems is not happiness at all but the promotion of individual life or a guaranty of the survival of mankind. If modern egoism were the ruthless search for pleasure (called happiness) it pretends to be, it would not lack what in all truly hedonistic systems is an indispensable element of argumentation – a radical justification of suicide. This lack alone indicates that in fact we deal here with life philosophy in its most vulgar and least critical form. In the last resort, it is always life itself which is the supreme standard to which everything else is referred, and the interests of the individual as well as the interests of mankind are always equated with individual life or the life of the species as though it were a matter of course that life is the highest good.

The curious failure of *homo faber* to assert himself under conditions seemingly so extraordinarily propitious could also have been illustrated by another, philosophically even more relevant, revision of basic traditional beliefs. Hume's radical criticism of the causality principle, which prepared the way for the later adoption of the principle of evolution, has often been considered one of the origins of modern philosophy. The causality principle with its twofold central axiom – that everything that is must have a cause (*nihil sine causa*) and that the cause must be more perfect than its most perfect effect – obviously relies entirely on experiences in the realm of fabrication, where the maker is superior to his products. Seen in this context, the turning point in the intellectual history of the modern age came when the image of organic life development – where the evolution of a lower being, for instance the ape, can cause the appearance of a higher being, for instance man – appeared in the place of the image of the watchmaker who must be superior to all watches whose cause he is.

Much more is implied in this change than the mere denial of the lifeless rigidity of a mechanistic world view. It is as though in the latent seventeenth-century conflict between the two possible methods to be derived from the Galilean discovery, the method of the experiment and of making on one hand and the method of introspection on the other, the latter was to achieve a somewhat belated victory. For the only tangible object introspection yields, if it is to yield more than an entirely empty consciousness of itself, is indeed the biological process. And since this biological life, accessible in self-observation, is at the same time a metabolic process between man and nature, it is as though introspection no longer needs to get lost in the ramifications of a consciousness without reality, but has found within man – not in his mind but in his bodily processes – enough outside matter to connect him again with the outer world. The split between subject and object, inherent in human consciousness and irremediable in the Cartesian opposition of man as a *res cogitans* to a surrounding world of *res extensae*, disappears altogether in the case of a living organism, whose very survival depends upon the incorporation, the consumption, of outside matter. Naturalism, the nineteenth-century version of materialism, seemed to find in life the way to solve the problems of Cartesian philosophy and at the same time to bridge the ever-widening chasm between philosophy and science.[21]

Notes

1 Augustine, who is usually credited with having been the first to raise the so-called anthropological question in philosophy, knew this quite well. He distinguishes between the questions of "Who am I?" and "What am I?" the first being directed by man at himself ("And I directed myself at myself and said to me: You, who are you? And I answered: A man" – *tu, quis es?* [*Confessiones* x. 6]) and the second being addressed to God ("What then am I, my God? What is my nature?" – *Quid ergo sum, Deus meus? Quae natura sum?* [x. 17]). For in the "great mystery," the *grande profundum*, which man is (iv. 14), there is "something of man [*aliquid hominis*] which the spirit of man which is in him itself knoweth not. But Thou, Lord, who has made him [*fecisti eum*] knowest everything of him [*eius omnia*]" (x. 5). Thus, the most familiar of these phrases which I quoted in the text, the *quaestio mihi factus sum*, is a question raised in the presence of

God, "in whose eyes I have become a question for myself" (x. 33). In brief, the answer to the question "Who am I?" is simply: "You are a man – whatever that may be"; and the answer to the question "What am I?" can be given only by God who made man. The question about the nature of man is no less a theological question than the question about the nature of God; both can be settled only within the framework of a divinely revealed answer.

2 That the *cogito ergo sum* contains a logical error, that, as Nietzsche pointed out, it should read: *cogito, ergo cogitationes sunt*, and that therefore the mental awareness expressed in the *cogito* does not prove that I am, but only that consciousness is, is another matter and need not interest us here (see Nietzsche, *Wille zur Macht*, No. 484).

3 This transformation of common sense into an inner sense is characteristic of the whole modern age; in the German language it is indicated by the difference between the older German word *Gemeinsinn* and the more recent expression *gesunder Menschenverstand* which replaced it.

4 Hermann Minkowski, "Raum und Zeit," in Lorentz, Einstein, and Minkowski, *Das Relativitätsprinzip* (1913).

5 And this doubt is not assuaged if another coincidence is added, the coincidence between logic and reality. Logically, it seems evident indeed that "the electrons if they were to explain the sensory qualities of matter could not very well possess these sensory qualities, since in that case the question for the cause of these qualities would simply have been removed one step farther, but not solved" (Heisenberg, *Wandlungen in den Grundlagen der Naturwissenschaft* [1935], p. 66). The reason why we become suspicious is that only when "in the course of time" the scientists became aware of this logical necessity did they discover that "matter" had no qualities and therefore could no longer be called matter.

6 In the words of Erwin Schrödinger: "As our mental eye penetrates into smaller and smaller distances and shorter and shorter times, we find nature behaving so entirely differently from what we observe in visible and palpable bodies of our surrounding that *no* model shaped after our large-scale experiences can ever be 'true'" (*Science and Humanism* [1952], p. 25).

7 Heisenberg, *Wandlungen in den Grundlagen*, p. 64.

8 Schrödinger, *op. cit.*, p. 26.

9 *Science and the Modern World* [1925], p. 116.

10 In the *Seventh Letter* 341C: *rhēton gar oudamōs estin hōs alla mathēmata* ("for it is never to be expressed by words like other things we learn").

11 See esp. *Nicomachean Ethics* 1142a25 ff. and 1143a 36 ff. The current English translation distorts the meaning because it renders *logos* as "reason" or "argument."

12 Whitehead, *Science and the Modern World*, pp. 116–17.

13 Hobbes's Introduction to the *Leviathan*.

14 See Michael Oakeshott's excellent Introduction to the *Leviathan* [1651] (Blackwell's Political Texts), p. xiv.

15 *Ibid.*, p. lxiv.

16 *Metaphysics* 1025b25 ff., 1064a17 ff.

17 For Plato see *Theaetetus 155: Mala gar philosophou touto to pathos, to thaumazein; ou gar allē archē philosophias ē hautē* ("For wonder is what the philosopher endures most; for there is no other beginning of philosophy than this"). Aristotle, who at the beginning of the *Metaphysics* (982b12 ff.) seems to repeat Plato almost verbatim – "For it is owing to their wonder that men both now begin and at first began to philosophize" – actually uses this wonder in an altogether different way; to him, the actual impulse to philosophize lies in the desire "to escape ignorance."

18 Henri Bergson, *Évolution créatrice* (1948) p. 157.

19 *Ibid.*, p. 140.

20 This, of course, is the first sentence of the *Principles of Morals and Legislation* [1789].

21 The greatest representatives of modern life philosophy are Marx, Nietzsche, and Bergson, inasmuch as all three equate Life and Being. For this equation, they rely on introspection, and life is indeed the only "being" man can possibly be aware of by looking merely into himself. The difference between these and the earlier philosophers of the modern age is that life appears to be more active and more productive than consciousness, which seems to be still too closely related to contemplation and the old ideal of truth. This last stage of modern philosophy is perhaps best described as the rebellion of the philosophers against philosophy, a rebellion which, beginning with Kierkegaard and ending in existentialism, appears at first glance to emphasize action as against contemplation. Upon closer inspection, however, none of these philosophers is actually concerned with action as such. We may leave aside here Kierkegaard and his non-worldly, inward-directed acting. Nietzsche and Bergson describe action in terms of fabrication – *homo faber* instead of *homo sapiens* – just as Marx thinks of acting in terms of making and describes labor in terms of work. But their ultimate point of reference is not work and worldliness any more than action; it is life and life's fertility.

31

Doing and Making in a Democracy: Dewey's Experience of Technology

Larry Hickman

> Understanding has to be in terms of how things work and how to do things. Understanding, by its very nature, is related to action; just as information, by its very nature, is isolated from action...only...by accident.
>
> *John Dewey*[1]

Advancing a claim that was then regarded as radical, and is still widely misunderstood, John Dewey argued that most of his philosophical predecessors, even those who had claimed the methods of science as their own, had been guilty of a failure to recognize the importance of technology. He suggested that this was due in part to their prejudice against the impermanent materials utilized by artisans and craftspeople, in part to their tendency to deprecate the social classes whose members have traditionally dealt with doing and making in the practical sphere, and in part to their rejection of what he took to be the democratizing tendencies of technological methods.

Analysis of the web of technical artifacts and methods which humankind weaves and in which it lives and works was for Dewey a lifelong task.[2] His early work, between 1892 and 1898, exhibits a preoccupation with the relations between the sciences and the industrial arts, and between what were then known as "normal" and "technical" schools. His middle work, from 1899 to 1924, contains discussions of the ways in which intelligence is related to the use of technological artifacts, and of the ways in which concrete tools such as agricultural implements are related to tools that are less tangible, such as logical connectors.

Dewey's later work, from 1925 until his death in 1952, including *Experience and Nature* (1925) and *Art as Experience* (1934), developed these themes in detail. He articulated an account of the philosophical implications of technology during its classical, modern, and contemporary periods, and he anticipated many of the issues and debates which now occupy those working in the emerging field of the philosophy of technology. In this connection, chapters four and five of *Experience and Nature*, central chapters in what is regarded by many as Dewey's most important work, are devoted almost entirely to a critique of technology. Moreover, those of his later works that focus on science, education, religion, and democracy are richly furnished with examples and metaphors from the technical sphere.

It was Dewey's contention that his philosophical predecessors had for the most part mislocated technology with respect to science, metaphysics, and social thought. Plato and Aristotle, each in his own distinctive way, had attempted to relocate technology outside the work of the artisan and outside the sphere of human interaction with changeable matter. Plato, especially in the *Timaeus*, did this by establishing a kind of "grand artisan" outside the realm of nature. Aristotle did so by making nature itself the grand artisan whose task it is to establish fixed ends fit to be contemplated as ends-in-themselves, not as instruments for further ends.

What resulted was not just a perversion of technology itself, but a stunting of the growth of sci-

From *Philosophy of Technology*, ed. Paul T. Durbin, Dordrecht: Kluwer, 1989, revised by the author. Reprinted by permission of Kluwer Academic Publishers, Dordrecht, The Netherlands.

ence and social inquiry as well. *The Republic* richly documents the consequences for social thought in general and for democracy in particular of this turn against experience in its full-bodied sense. It is there that Plato relegates *technē*, the activities of the technical artisan, to the lowest rung of his sociopolitical hierarchy, and at the same time characterizes an attenuated and immaterial form of *technē*, that of the totalitarian social engineer, as the purest and most important of social activities.

It was Dewey's contention that Plato had placed the artisan at the bottom of the social hierarchy for the same reason that he had so adamantly demanded censorship of the work of the plastic and dramatic artists: the methods of *technē* are too powerful to be left in the hands of artists and craftsmen. Unhindered by the repressive legislation of the perfect guardians, the practitioners of *technē* in its concrete sense would have proved a threat to the "thinkers" of the Republic.

As for Aristotle, it was Dewey's view that the *Politics* fosters a view of the city state so constructed that its justification rests on ends "given" by nature. The activities of the practitioner of *technē* are, as they were in Plato, refined and sublimated. For Aristotle, however, the beneficiary of his transference is not a system of supernature contemplated by the philosopher king but nature itself, which becomes the grand artisan. Just as Plato had, but less perniciously so, Aristotle plundered much of the work of the artisan of its creative and social significance and relocated its content elsewhere.[3]

Dewey read the Greek attitudes toward science as part and parcel of their unfortunate attitude toward *technē*. He argued that the Greeks' abhorrence of the mutability inherent in the tasks and materials of technology had led to a science of "demonstration," to a science of contemplation, to an attempt to possess something already finished, "out there," and complete. In fact, they had invented not so much science as the idea of one. He issued the warning that when inquiry is focused in the sphere of objects esteemed for their own intrinsic qualities, whether that sphere be cast as supernatural or extranatural, as it was for Plato, or natural and immanent but complete, as it was for Aristotle, then whether that inquiry concerns itself with materials and artifacts, conceptual models of nature, or the ways in which social organization takes place, such inquiry will fail to increase our knowledge of things as they are.

As for modern science, the science of Copernicus, Galileo, Kepler, and Newton, it was Dewey's view that its advances were attributable more to what its practitioners were *doing* than to what they *thought* they were doing. It was not that what was novel in its theories did not advance its practice; it was simply that its metatheory, its metaphysics and epistemology, more often than not failed to grasp what was innovative in those first-order theories. Dewey had high praise for the new mathematical techniques of substitution, and suggested that they constituted a "system of exchange and mutual conversion carried to its limit" (LW 1:115). The objects of science thus became "amenable to transformation in virtue of reciprocal substitutions" (ibid.). But the metaphysics and epistemology of the new science were still wedded to the old ideas of a finished universe.

Dewey's claims in this regard are by no means uncontroversial. Desmond Lee, for example, in the introduction to his translation of the *Timaeus*, rejected the view that the contempt held by the Greeks for the work of the artisan discouraged experimentation and hindered the development of technology. He argued that the aristocracy of seventeenth-century England, a time and place of enormous technological development, had at least as much disdain for the artisan as had the Greeks. He further suggested that inhibitions of technological development in the classical world were not always aristocratic. The Roman contractor, whom he identified as a "fairly rough type, often a freedman,"[4] would certainly have been glad to have profited from technological development if it had been possible for him to do so. Instead, Lee suggested, the weakness of ancient metallurgy and a lack of precise instrumentation were among the inhibiting factors. But why should these technological materials and instruments not have developed? Lee suggested that there was a conceptual reason: the Greeks had tied science to philosophy, and "philosophy is concerned to understand rather than to change."[5] For Lee, the contribution of Galileo to the advancement of experimental science was that he took the technical tools and artifacts available to him, tools and artifacts that had gradually become much more sophisticated than those developed by the Greeks, and used them to "untie" science from philosophy.

Dewey repeatedly rejected any view of philosophy that had as its goal understanding without change, for he thought that understanding of any legitimate sort entails change. He also argued that hope of financial gain, even by the most "rough and ready" of contractors, is in itself insufficient to promote technological development. It may in fact thwart or divert such development. On one point at least, his position is consonant with that of Lee. They agree that experimental science requires active transaction with environing conditions. But for Dewey, those among his predecessors and those among his contemporaries who could not see that this is also true of philosophy had not fully appropriated the lessons of the scientific revolution of the seventeenth century.

Put in the sociopolitical terms by means of which he had analyzed the fate of *technē* among the Greeks of the fourth century BCE, it was Dewey's contention that the new science of the seventeenth century exhibited a surge of democratic methods and an assertion of adaptive practice. But there was a broad gulf between what the new science said it was doing and what it was actually about.

The official view of what modern science was about was still conservative and authoritarian. Its apologists continued to traffic in antecedent truths, demonstrations, and certitudes. They held fast to what we today call foundationalism and the correspondence theory of truth. Moreover, its metatheory continued in this vein long after the new science had enjoyed the prodigious successes that resulted from its practical commitment to the treatment of natural ends as instruments for further inquiry and transaction with nature, rather than as fixed objects of contemplation. Thus did the practice and first-order theories of seventeenth-century science relocate technology *de facto* in terms of its new spirit of practical experimentalism, even if its metatheory did not do so. Dewey thought that the genius of the new science was its discovery that "knowledge is an affair of *making* sure, not of grasping antecedently given sureties."[6]

Nevertheless, many of the metatheorists of seventeenth-century science, among them the most respected metaphysicians of the time, demonstrated an ignorance that its taproot was in practice, that is, in a transaction with the growing body of tools and artifacts that made the new science possible.

Contemporary historians of science and technology continue to commit this error. The following description by Daniel Boorstin of the work of Galileo the telescope-maker is illustrative of this mistake. "With no special insight into the science of optics," Boorstin writes:

> Galileo, a deft instrument-maker, had made his device by trial and error. But if Galileo had been merely a practical man, the telescope would not have been such a troublemaker.[7]

Dewey would have found in this characterization a vestige of the very mistake made by most of the early philosophers of modern science. His view was that Galileo was not so much proceeding by means of trial and error as he was "thinking" with his materials, inquiring into their possibilities in a way that had much more in common with the activities of the artists and craftsmen of Plato's Greece than the philosophers of the modern period realized. Dewey's view of what Galileo was doing is closer to the description of his activities provided us by Paolo Rossi:

> Kepler was to lay the foundation of the new optics in the *Paralipomena* of 1604, but it was to be a scientist-technician like Galileo who was to muster up the courage "to look" by using the telescope. He skillfully transformed a use-object which had progressed only "through practice," partly accepted in military circles but ignored by the official scientific establishment, into a powerful instrument of scientific exploration.[8]

Dewey argued that inquiry into materials such as that practiced by Galileo precedes and conditions inquiry of a more conceptual variety. It also informs its methodology, and terminates its activity in further concrete application. This was perhaps Dewey's most important contribution to the debates concerning the relations between science and technology. Even though the craftsman who thinks in and with materials may not translate that thought into the conceptual sphere, and conversely even though those who think by means of conceptual tools are frequently unable to bring their work to fruition in practical terms, there is nevertheless no reason to posit a methodological gap between the two enterprises.

It is only the infelicitous social prejudice regarding the media in which inquiry is undertaken, the misunderstanding of the talents and dispositions of those who direct the inquiry, and the unfortunate social and cultural boundaries assumed to exist between those modes of inquiry that perpetuate the appearance of a gap that is not in fact justified from the standpoint of methodology. In short, intelligence with respect to materials is fully the equal of intelligence with respect to those enterprises we normally consider "conceptual": for example, science and social thought. Not only are their methods basically the same, but it is only by the cooperation of each with the other that human knowledge is advanced. Dewey made this point forcefully in *Art as Experience*.

> "Any idea," he wrote, that ignores the necessary role of intelligence in production of works of art is based upon identification of thinking with use of one special kind of material, verbal signs and words. To think effectively in terms of relations of qualities is as severe a demand upon thought as to think in terms of symbols, verbal and mathematical. Indeed, since words are easily manipulated in mechanical ways, the production of a work of genuine art probably demands more intelligence than does most of the so-called thinking that goes on among those who pride themselves on being "intellectuals." (LW 10:52)

He was careful to include the "practical" or "technological" arts in this characterization. "Art," he suggested, "denotes a process of doing or making. This is as true of fine as of technological art" (LW 10:53).

It was Dewey's claim, then, that philosophy during its modern period, from the seventeenth to the nineteenth centuries, failed to locate technology properly because its allegiance was still tied to the metaphysics of contemplation, of antecedent truths, demonstration, and certitude. But his analysis did not simply take the part of the Empiricists against the Rationalists. Some modern philosophers, he pointedly reminded us, surrendered the antecedent truths of reason only to accept antecedent truths of sensation. Modern Empiricism, according to his view, committed itself to an equally egregious form of foundationalism.

For the bulk of philosophy in its modern period, nature was thought to be a vast machine. Living in the shadow of Darwin as he did, Dewey rejected the machine as leading metaphor and replaced it with the organism. But even to those who have transcended the metaphor of world-as-machine there is still the fact of machines, and the problem of how to relate to them. A machine can be contemplated as something finished, and its workings discovered and admired. Further, it can be examined as something complete but in need of occasional repair. Or it can be interacted with as something ongoing, unstable and provisional, as a tool which is utilized for enlarging transactions of self and society with environing conditions. It was Dewey's contention that the discussions of the nature of the world-as-machine in the seventeenth and eighteenth centuries were primarily focused on the first two of these attitudes. Of course each of these three possibilities involves some level of interaction with nature. But it is only with the third that there comes to be genuine transaction with nature, awareness of such transaction, and inclusion of that awareness in the metatheories of science.

In the political sphere, of course, it was a great advance over the old supernaturalist and extranaturalist views to think of the world as repairable, even if it was not yet fully open to transaction. In *Liberalism and Social Action*, Dewey praised the advances made by Bentham on just these grounds. But he also warned of treating the world-machine as *merely* examinable and repairable. He cautioned against Bentham's acceptance of humankind as "a reckoning machine" (LW 11:24). The old machine metaphors of Bentham and others neglected the fact that relations between human beings and their political environments are always "relations of ongoing affairs characterized by beginnings and endings which mark them off into unstable individuals" (LW 1:127). These individual relations are in need of continual and intelligent reevaluation and reconfiguration by means of practical inquiry.

Failure to make this conceptual shift from machine as finished though imperfect and repairable to machine as incomplete and unstable instrument has precipitated in our time a situation well described by Stuart Hampshire in a polemic against Utilitarianism, a cluster of positions against which Dewey also argued. In its emphasis on repair as opposed to transaction, Hampshire suggested, much recent thought has led to

> new abstract cruelty in politics, a dull, destructive political righteousness: mechanical, quantitative thinking, leaden academic minds setting out their moral calculation in leaden abstract prose, and more civilized and more superstitious people destroyed because of enlightened calculations that have proved wrong.[9]

It might be objected that it was during this modern period of science that the United States of America, the most influential democracy of the contemporary world, was founded. It might further be argued that among the framers of the political documents of that infant democracy were deists, practitioners of a form of religious faith that explicitly regards God as artisan and a virtually finished universe as His handiwork. But among these social experimenters were gadget-makers, mechanics, and tinkerers. Thomas Jefferson, whom Dewey greatly and publicly admired, consistently spoke of political and social experimentation in a manner that echoed his transaction with clocks, agricultural methods, and gadgets of many diverse sorts. Jefferson repeatedly referred to the government which he helped establish as an *experiment*, moreover one whose institutions and laws would be in need of recurring modification by each succeeding generation. For government, as for nature, contemplation had been replaced by examination, and that in turn by experimentation whose goal was constant attention to possibilities of adjustment and amelioration.

Jefferson's transactionist orientation to technology, to social thought, and to the broader world of his experience stands in stark contrast to the examinationist program of Descartes a little over a century earlier. L. J. Beck gives the following account of Descartes' attitude toward Galileo's telescope as exhibited in his *Dioptrique* of 1637 and his correspondence with Ferrier:

> Already at La Flèche, Descartes had probably heard of the discoveries made by Galileo through the use of the telescope. Descartes wishes to draw up a plan for the construction of an even better one, and above all of a more powerful lens. This cannot be done until, he tells us, it is known, what happens when light traverses several lenses, until the law of refraction has been established and the problem of the *linea anaclastica* solved. Then only can the plan of the various curves of the lenses be worked out. Descartes works these out but, as one can see, the unfortunate Ferrier is unable to carry out in practice the difficult requirements set by Descartes. Galileo cannot solve the problem of the *linea anaclastica*; he does not know the law of refraction, but he manages to construct an instrument which gives a substantial magnification. Kepler, slightly more theoretical, knew only of approximations to the law of refraction but his description of the telescope provided a working model for future astronomers. Descartes required the exact measurements for his lenses, and failing this, he lost interest in the whole topic.[10]

It was precisely this debate between the transactionist craftsmen-practitioners of modern science and technology and those seeking to examine its hypothetical and metaphysical foundations that was a matter of intense interest to Dewey. It provided evidence for his thesis that metaphysicians of the period had mislocated the place of technological practice. Writing of the controversy between the Cartesian school and that of Galileo and Newton, he lauded the triumph of the latter because of its emphasis on "experience." And his characterization of "experience" made extensive use of examples of inquiry in the technological sphere (LW 1:14–15). In a rather sad aside he suggested that, "We may, if sufficiently hopeful, anticipate a similar outcome in philosophy. But the date does not appear to be close at hand; we are nearer in philosophic theory to the time of Roger Bacon than to that of Newton" (LW 10:15). Dewey wanted to locate technology in a realm that is neither supernatural nor extranatural, an organic realm in which the only telic elements are those of the natural ends of objects, individuals, and events, all of which in turn may be utilized as means to further ends. It was his view that the legitimate place of technology is alongside science and social thought as one of several branches of inquiry. On his reading, technology is not inferior to its brother and sister branches, and may in some respects even be more important than they in that its unique qualities serve to inform, enhance, and promote those siblings in ways that they are incapable of reciprocating.

What are these unique qualities? I have already alluded to his commitments to what Don Ihde would later call "the historical-ontological priority of technology over science."[11] In 1925 Dewey argued the historical component of this claim when he suggested that in spite of the obvious fact that

> the sciences were born of the arts – the physical sciences of the crafts and technologies of healing, navigation, war and the working of wood, metals, leather, flax and wool; the mental sciences of the arts of political management,...it is still commonly [and erroneously] argued that technology is merely "applied science." (LW 1:105)

He further argued that modern science

> represents a generalized recognition and adoption of the point of view of the useful arts, for it proceeds by employment of a similar operative technique of manipulation and reduction. Physical science would be impossible without the appliances and procedures of separation and combination of the industrial arts. (LW 1:108)

In addressing the "ontological" component of his claim, Dewey reminded us that what is peculiar to human interaction with the world is not its enjoyment, but the necessity of grappling with it at the technological level and the knowledge, or science, which follows upon that interaction. In Dewey's words, "It was not enjoyment of the apple but the enforced penalty of labor that made man as the gods, *knowing* good and evil instead of just having and enjoying them" (LW 1:100). To contrast knowing and having, as Dewey did in this remark, is to allude to his treatment of knowledge as hypothesis, pointing to an unfinished future in which both inquiring human beings and their environments undergo alterations.

This is a view that would reemerge in Heidegger's essay, "The Question Concerning Technology." Technology is there differentiated into (1) technology as a tool of science, (2) technology as the activities of the craftsman (*technē*) and (3) technology in its ultimate sense, *aletheia* or revealing. It is this last sense of technology that is most basic to Heidegger's account: "Instrumentality is considered to be the fundamental characteristic of technology. If we inquire, step by step, into what technology, represented as means, actually is, then we shall arrive at revealing. The possibility of all productive manufacturing lies in revealing."[12] Again, "Technology is a mode of revealing. Technology comes to presence...in the realm where revealing and unconcealment take place, where aletheia, truth, happens."[13]

I have recalled Heidegger's account of the ontological priority of technology over science because of the light it sheds on Dewey's. For Dewey, it is technological instrumentality (what Heidegger calls "revealing") that characterizes the most primitive relation between the activities of men and women and the world of their experience. Such instrumentality ties together the myths that tell of the manner in which labor entered the world and the myths that constitute our most up-to-date theories of political economy.

But if technology is prior to science both historically and ontologically, it also is responsible for the prestige enjoyed by science. Dewey argued that the successes of science have been due not so much to what he called "scientific temper" as to "scientific technique." In his essay "Human Nature and Scholarship," Dewey argued that

> Scientific technique, as distinguished from the scientific temper, is concerned with the methods by which matter is manipulated. It is the source of special technologies, as in the application of electricity to daily life; it is concerned with immediate fruits of a practical kind in a sense in which *practical* has a special and technical meaning – power stations, broadcasting, lighting, the telephone, the ignition system of automobiles. (LW 11: 457)

Further,

> The inherent idealism of the scientific temper is submerged, for the mass of human beings, in the use and enjoyment of the material power and material comforts that have resulted from its technical applications. (LW 11:174)

What was Dewey's view of the location of technology with respect to epistemology? "Knowledge ceases to be a mental mirror of the universe and becomes a practical tool in the manipulation of matter" (LW 11:172). Dewey reiterated his radical position in *Experience and Nature*: "In the practice of science, knowledge is an affair of *making* sure, not of grasping antecedently given sureties" (LW 1:123).

Dewey not only viewed technology as the primary means of inquiry open to those individuals cut off from what normally goes on in laboratories, observatories, and places of special research; he suggested that technology was a special avenue of inquiry open to those individuals living in closed societies where social inquiry is suppressed. But he was neither idealist nor utopian. He knew that even in open societies there would be those who prefer appeals to tenacity, authority, or the a priori to free and open inquiry as methods of fixing their beliefs. It was with this in mind that he suggested that technology may also operate as a buffer between the forces of anti-science and those of science.

I do not think that he would have been surprised that those who now attempt to promote the teaching of a literal reading of the Genesis myth of creation do so while claiming to march under the banner of science. The advances of science propagated in the technological sphere have mooted many of the old anti-scientific arguments, or at least required that they be masked in scientific jargon. Dewey repeatedly demonstrated his conviction that the work of those who take the pluralistic values of free and open inquiry seriously will never be finished. This is a central aspect of his philosophy of education.

But if technology for Dewey forms a buffer between the forces of anti-science and those of science, it also functions as a means by which science may be appropriated by the scientifically uninformed. There are two ways in which this takes place: not only have the "fruits" of technology become ubiquitous, but the methods of science, historically and ontologically dependent on technology, may be reintroduced into the field of technological practice and use with new authority. But with these two outcomes of technology, one immediate, the other mediate, come two dangers. The first is the one indicated by José Ortega y Gasset, that technological men and women may become like the aboriginal forest or jungle dweller, just "picking" the technological fruits as if they had been supplied by a "natural" system beyond their understanding or control. The second is that technology will once again become mislocated with respect to science, that is, that it will once again suffer the deprecation it suffered during the period of classical, modern, and much of contemporary philosophy. These are dangers to social organization in general, and to democracy in particular, because they signal the truncation of the full spectrum of inquiry necessary to the transaction of human beings with their environment and the consequent knowing of things as they are and can be.

In his 1944 essay, "Democratic Faith in Education," Dewey made even more explicit his concern regarding the dangers to technology and to social inquiry. He could have been writing of our current situation when he characterized the first of these dangers as that of "laissez-faire naturalism." To those who would appeal to the forces of the "invisible hand" or "the [underfined] laws of the marketplace," Dewey had this to say:

> Technically speaking the policy known as *Laissez-faire* is one of limited application. But its limited and technical significance is one instance of a manifestation of widespread trust in the ability of impersonal forces, popularly called Nature, to do a work that has to be done by human insight, foresight and purposeful planning. (LW 15: 253)

He applied this assessment to planning in national and international affairs, suggesting that the refusal to apply the methods of the technological sciences in those areas had led to a state of imbalance and "profoundly disturbed equilibrium" (LW 15:27).

A second danger to technological and social inquiry lies in the attitudes and activities of the "humanist" who attacks technology as "inherently materialistic and as unsurping the place properly held by abstract moral precepts" (ibid.). He suggested that such moral precepts had remained abstract precisely because those defining them had divorced ends from the means by which they are to be realized.

Dewey specifically criticized the Hutchins program, which set out to separate technical training from liberal education, a situation which has both become a fact of our universities in the 1980s and 1990s and is seen by many as tending to create a permanent social wound as it becomes more widely practiced in our secondary schools. When the putrefaction of the wound is eventually discovered, the blame will likely be located at the door of "technology," and not in its proper place, namely, the failure to apply the inquiry which is characteristic of experimental science and technology at its best to all areas of human endeavor.

In that same essay Dewey mentioned a third threat to technological and social inquiry. It is consequent on the activities of contemporary Luddites, especially of the theocratic variety. They resist the application of scientific and technical methods to the field of human concerns and human affairs both because they tend to think of themselves as outside of and above nature, and because they prefer a return to the medieval prescientific doctrine of a supernatural foundation and outlook in all social and moral matters. He further suggested that this group erroneously believes that the methods of science and technology have been applied to every area of human concern – and have been found wanting.

A special variety of this third faction has become even more militant than it was at the time Dewey issued his warning. Both anti-scientific and anti-democratic, Christian fundamentalists of the extreme right have nevertheless adopted the tools of electronic technology to advance their aims. Dewey was aware that technological advances could be appropriated by authoritarian forces, and discussed this phenomenon in detail in his 1936 essay, "Religion, Science and Philosophy."

The two forces that particularly concerned him in that dangerous year, a year which saw the growing power of fascism throughout the world, were what he called "political nationalism" and "finance-capital," both important allies of the religious right in our own time. He called these movements "new religions." They, like religious fundamentalism, depend on the a priori and the revealed as a substitute for intelligent inquiry. They, like religious fundamentalism, have their

> established dogmatic creeds, their fixed rites and ceremonies, their central institutional authority, their distinction between the faithful and the unbelievers, with persecution of heretics who do not accept the true faith. (LW 11:460)

They share a further important characteristic of religions, viz., they dote on the *terminus a quo* rather than on the *terminus ad quem*, a doctrine of "original intent" rather than a careful attention to consequences. Dewey's *Experience and Nature* warns of the capture of applied science by these elements: they work to channel it toward "private and economic class purposes and privileges. When inquiry is narrowed by such motivation or interest, the consequence is in so far disastrous both to science and to human life" (LW 1:130–1). Dewey reminded us that these potential disasters are not due to the practical nature of technology, but to the defects and perversions of morality as it is embodied in institutions and the effects of such institutions upon personal disposition.

This essay would remain incomplete were I not to quote an extended passage from an address delivered by Dewey at a celebration of his eightieth birthday. There is perhaps no more eloquent characterization of the interaction between inquiry at its various technological, scientific, and political levels in all of Dewey's writings. "Democracy" is there characterized as

> a belief in the ability of human experience to generate the aims and methods by which further experience will grow in ordered richness. Every other form of moral and social faith rests on the idea that experience must be subjected at some point or other to some form of external control; to some "authority" alleged to exist outside the process of experience. Democracy is the faith that the process of experience is more important than any special result attained, so that special results achieved are of ultimate value only as they are used to enrich and order the ongoing process. Since the process of experience is capable of being educative, faith in democracy is all one with faith in experience and education. All ends and values that are cut off from the ongoing process become arrests, fixations. They strive to fixate what has been gained instead of using it to open the road and point the way to new and better experience.
>
> If one asks what is meant by experience in this connection, my reply is that it is that free interaction of individual human beings with surrounding conditions, especially the human surroundings, which develops and satisfies need and desire by increasing knowledge of things as they are. Knowledge of conditions as they are is the only solid ground for communication and sharing; all other communication means the subjection of some persons to the personal opinion of other persons. Need and desire – out of which grow purpose and direction of energy – go beyond what exists, and hence beyond knowledge, beyond science. They continually open the way into the unexplored and unattained future. (LW 14:229)

Notes

1 John Dewey, *The Collected Works of John Dewey*, ed. Jo Ann Boydston (Carbondale and Edwardsville: Southern Illinois University Press, 1967–1991); *The Later Works* (henceforth LW) 11:181.

2 Dewey devoted no single work to his analysis of technology. His treatment of the subject is diffused throughout the more than 15,000 pages of his published work. For this and other reasons, the secondary literature of this field is scant. There are, however, a few brave essays on the subject, including an important and interesting one by Webster F. Hood, "Dewey and Technology," in Paul Durbin, ed., *Research in Philosophy and Technology*, vol. 5 (Greenwich, Conn.: JAI Press, 1982), pp. 189–207.

3 As Hannah Arendt pointed out in her classic essay "The *Vita Activa* in the Modern Age," in *The Human*

Condition (Chicago: University of Chicago Press, 1958), Aristotle's *Metaphysics* had placed the sciences of fabrication above the practical sciences but below the theoretical ones. She thought that this was so because for the Greeks the contemplation of the theoretical sciences was thought to be an inherent element in fabrication, that is, what allowed the craftsman to judge the finished product. The point of Dewey's critique of the Greeks was not that they did not give *technē* some of its due, but that they thought it secondary to grander technical forces, viz., supernature for Plato, and nature for Aristotle.

4 Plato, *Timaeus and Critias*, trans. and with an introduction and appendix by Desmond Lee (New York: Penguin Books, 1983), p. 18.

5 Ibid., p. 19.

6 Dewey, *The Middle Works* 1:123. Dewey's emphasis.

7 Daniel Boorstin, *The Discoverers* (New York: Random House, 1983), p. 318.

8 Paolo Rossi, *Philosophy, Technology and the Arts in the Early Modern Era* (New York: Harper & Row, 1970), p. 35.

9 Stuart Hampshire, "Morality and Pessimism," in R. Beehler and A. Drengson, eds., *The Philosophy of Society* (London: Methuen, 1978), p. 32.

10 L. J. Beck, *The Method of Descartes* (Oxford: Clarendon Press, 1952), pp. 261–2.

11 Don Ihde, *Existential Technics* (Albany: State University of New York Press, 1983), pp. 25–46.

12 Martin Heidegger, *The Question Concerning Technology and Other Essays*, trans. and ed. William Lovitt (New York: Harper & Row, 1977), p. 12. [The essay is reprinted as chapter 23 in this volume.]

13 Ibid., p. 13.

32

Buddhist Economics

E. F. Schumacher

"Right Livelihood" is one of the requirements of the Buddha's Noble Eightfold Path. It is clear, therefore, that there must be such a thing as Buddhist economics.

Buddhist countries have often stated that they wish to remain faithful to their heritage. So Burma: "The New Burma sees no conflict between religious values and economic progress. Spiritual health and material wellbeing are not enemies: they are natural allies."[1] Or: "We can blend successfully the religious and spiritual values of our heritage with the benefits of modern technology."[2] Or: "We Burmans have a sacred duty to conform both our dreams and our acts to our faith. This we shall ever do."[3]

All the same, such countries invariably assume that they can model their economic development plans in accordance with modern economics, and they call upon modern economists from so-called advanced countries to advise them, to formulate the policies to be pursued, and to construct the grand design for development, the Five-Year Plan or whatever it may be called. No one seems to think that a Buddhist way of life would call for Buddhist economics, just as the modern materialist way of life has brought forth modern economics.

Economists themselves, like most specialists, normally suffer from a kind of metaphysical blindness, assuming that theirs is a science of absolute and invariable truths, without any presuppositions. Some go as far as to claim that economic laws are as free from "metaphysics" or "values" as the law of gravitation. We need not, however, get involved in arguments of methodology. Instead, let us take some fundamentals and see what they look like when viewed by a modern economist and a Buddhist economist.

There is universal agreement that a fundamental source of wealth is human labour. Now, the modern economist has been brought up to consider "labour" or work as little more than a necessary evil. From the point of view of the employer, it is in any case simply an item of cost, to be reduced to a minimum if it cannot be eliminated altogether, say, by automation. From the point of view of the workman, it is a "disutility"; to work is to make a sacrifice of one's leisure and comfort, and wages are a kind of compensation for the sacrifice. Hence the ideal from the point of view of the employer is to have output without employees, and the ideal from the point of view of the employee is to have income without employment.

The consequences of these attitudes both in theory and in practice are, of course, extremely far-reaching. If the ideal with regard to work is to get rid of it, every method that "reduces the work load" is a good thing. The most potent method, short of automation, is the so-called "division of labour" and the classical example is the pin factory eulogised in Adam Smith's *Wealth of Nations*. Here it is not a matter of ordinary specialisation, which mankind has practised from time immemorial, but of dividing up every complete process of production into minute parts, so that the final product can be produced at great speed without anyone having had to contribute more than a to-

From E. F. Schumacher, *Small is Beautiful: Economics as if People Mattered*, New York: Harper Trade, 1989, pp. 50–9.

tally insignificant and, in most cases, unskilled movement of his limbs.

The Buddhist point of view takes the function of work to be at least threefold: to give a man a chance to utilise and develop his faculties; to enable him to overcome his egocentredness by joining with other people in a common task; and to bring forth the goods and services needed for a becoming existence. Again, the consequences that flow from this view are endless. To organise work in such a manner that it becomes meaningless, boring, stultifying, or nerve-racking for the worker would be little short of criminal; it would indicate a greater concern with goods than with people, an evil lack of compassion and a soul-destroying degree of attachment to the most primitive side of this worldly existence. Equally, to strive for leisure as an alternative to work would be considered a complete misunderstanding of one of the basic truths of human existence, namely that work and leisure are complementary parts of the same living process and cannot be separated without destroying the joy of work and the bliss of leisure.

From the Buddhist point of view, there are therefore two types of mechanisation which must be clearly distinguished: one that enhances a man's skill and power and one that turns the work of man over to a mechanical slave, leaving man in a position of having to serve the slave. How to tell the one from the other? "The craftsman himself," says Ananda Coomaraswamy, a man equally competent to talk about the modern west as the ancient east, "can always, if allowed to, draw the delicate distinction between the machine and the tool. The carpet loom is a tool, a contrivance for holding warp threads at a stretch for the pile to be woven round them by the craftsmen's fingers; but the power loom is a machine, and its significance as a destroyer of culture lies in the fact that it does the essentially human part of the work."[4] It is clear, therefore, that Buddhist economics must be very different from the economics of modern materialism, since the Buddhist sees the essence of civilisation not in a multiplication of wants but in the purification of human character. Character, at the same time, is formed primarily by a man's work. And work, properly conducted in conditions of human dignity and freedom, blesses those who do it and equally their products. The Indian philosopher and economist J. C. Kumarappa sums the matter up as follows:

"If the nature of the work is properly appreciated and applied, it will stand in the same relation to the higher faculties as food is to the physical body. It nourishes and enlivens the higher man and urges him to produce the best he is capable of. It directs his free will along the proper course and disciplines the animal in him into progressive channels. It furnishes an excellent background for man to display his scale of values and develop his personality."[5]

If a man has no chance of obtaining work he is in a desperate position, not simply because he lacks an income but because he lacks this nourishing and enlivening factor of disciplined work which nothing can replace. A modern economist may engage in highly sophisticated calculations on whether full employment "pays" or whether it might be more "economic" to run an economy at less than full employment so as to ensure a greater mobility of labour, a better stability of wages, and so forth. His fundamental criterion of success is simply the total quantity of goods produced during a given period of time. "If the marginal urgency of goods is low," says Professor Galbraith in *The Affluent Society*, "then so is the urgency of employing the last man or the last million men in the labour force."[6] And again: "If...we can afford some unemployment in the interest of stability – a proposition, incidentally, of impeccably conservative antecedents – then we can afford to give those who are unemployed the goods that enable them to sustain their accustomed standard of living."

From a Buddhist point of view, this is standing the truth on its head by considering goods as more important than people and consumption as more important than creative activity. It means shifting the emphasis from the worker to the product of work, that is, from the human to the subhuman, a surrender to the forces of evil. The very start of Buddhist economic planning would be a planning for full employment, and the primary purpose of this would in fact be employment for everyone who needs an "outside" job: it would not be the maximisation of employment nor the maximisation of production. Women, on the whole, do not need an "outside" job, and the large-scale employment of women in offices or factories would be considered a sign of serious economic failure. In particular, to let mothers of young children work in factories while the children run wild would be as uneconomic in the eyes of a Buddhist economist as the employment of a skilled worker as a soldier in the eyes of a modern economist.

While the materialist is mainly interested in goods, the Buddhist is mainly interested in liberation. But Buddhism is "The Middle Way" and therefore in no way antagonistic to physical well-being. It is not wealth that stands in the way of liberation but the attachment to wealth; not the enjoyment of pleasurable things but the craving for them. The keynote of Buddhist economics, therefore, is simplicity and non-violence. From an economist's point of view, the marvel of the Buddhist way of life is the utter rationality of its pattern – amazingly small means leading to extraordinarily satisfactory results.

For the modern economist this is very difficult to understand. He is used to measuring the "standard of living" by the amount of annual consumption, assuming all the time that a man who consumes more is "better off" than a man who consumes less. A Buddhist economist would consider this approach excessively irrational: since consumption is merely a means to human well-being, the aim should be to obtain the maximum of well-being, with the minimum of consumption. Thus, if the purpose of clothing is a certain amount of temperature comfort and an attractive appearance, the task is to attain this purpose with the smallest possible effort, that is, with the smallest annual destruction of cloth and with the help of designs that involve the smallest possible input of toil. The less toil there is, the more time and strength is left for artistic creativity. It would be highly uneconomic, for instance, to go in for complicated tailoring, like the modern west, when a much more beautiful effect can be achieved by the skilful draping of uncut material. It would be the height of folly to make material so that it should wear out quickly and the height of barbarity to make anything ugly, shabby or mean. What has just been said about clothing applies equally to all other human requirements. The ownership and the consumption of goods is a means to an end, and Buddhist economics is the systematic study of how to attain given ends with the minimum means.

Modern economics, on the other hand, considers consumption to be the sole end and purpose of all economic activity, taking the factors of production – land, labour, and capital – as the means. The former, in short, tries to maximise human satisfactions by the optimal pattern of consumption, while the latter tries to maximise consumption by the optimal pattern of productive effort. It is easy to see that the effort needed to sustain a way of life which seeks to attain the optimal pattern of consumption is likely to be much smaller than the effort needed to sustain a drive for maximum consumption. We need not be surprised, therefore, that the pressure and strain of living is very much less in, say, Burma than it is in the United States, in spite of the fact that the amount of labour-saving machinery used in the former country is only a minute fraction of the amount used in the latter.

Simplicity and non-violence are obviously closely related. The optimal pattern of consumption, producing a high degree of human satisfaction by means of a relatively low rate of consumption, allows people to live without great pressure and strain and to fulfil the primary injunction of Buddhist teaching: "Cease to do evil; try to do good." As physical resources are everywhere limited, people satisfying their needs by means of a modest use of resources are obviously less likely to be at each other's throats than people depending upon a high rate of use. Equally, people who live in highly self-sufficient local communities are less likely to get involved in large-scale violence than people whose existence depends on world-wide systems of trade.

From the point of view of Buddhist economics, therefore, production from local resources for local needs is the most rational way of economic life, while dependence on imports from afar and the consequent need to produce for export to unknown and distant peoples is highly uneconomic and justifiable only in exceptional cases and on a small scale. Just as the modern economist would admit that a high rate of consumption of transport services between a man's home and his place of work signifies a misfortune and not a high standard of life, so the Buddhist economist would hold that to satisfy human wants from faraway sources rather than from sources nearby signifies failure rather than success. The former tends to take statistics showing an increase in the number of ton/miles per head of the population carried by a country's transport system as proof of economic progress, while to the latter – the Buddhist economist – the same statistics would indicate a highly undesirable deterioration in the *pattern* of consumption.

Another striking difference between modern economics and Buddhist economics arises over the use of natural resources. Bertrand de Jouvenel, the eminent French political philosopher, has characterised "western man" in words which may be taken as a fair description of the modern economist:

"He tends to count nothing as an expenditure, other than human effort; he does not seem to mind how much mineral matter he wastes and, far worse, how much living matter he destroys. He does not seem to realise at all that human life is a dependent part of an ecosystem of many different forms of life. As the world is ruled from towns where men are cut off from any form of life other than human, the feeling of belonging to an ecosystem is not revived. This results in a harsh and improvident treatment of things upon which we ultimately depend, such as water and trees."[7]

The teaching of the Buddha, on the other hand, enjoins a reverent and non-violent attitude not only to all sentient beings but also, with great emphasis, to trees. Every follower of the Buddha ought to plant a tree every few years and look after it until it is safely established, and the Buddhist economist can demonstrate without difficulty that the universal observation of this rule would result in a high rate of genuine economic development independent of any foreign aid. Much of the economic decay of south-east Asia (as of many other parts of the world) is undoubtedly due to a heedless and shameful neglect of trees.

Modern economics does not distinguish between renewable and non-renewable materials, as its very method is to equalise and quantify everything by means of a money price. Thus, taking various alternative fuels, like coal, oil, wood, or water-power: the only difference between them recognised by modern economics is relative cost per equivalent unit. The cheapest is automatically the one to be preferred, as to do otherwise would be irrational and "uneconomic". From a Buddhist point of view, of course, this will not do; the essential difference between non-renewable fuels like coal and oil on the one hand and renewable fuels like wood and water-power on the other cannot be simply overlooked. Non-renewable goods must be used only if they are indispensable, and then only with the greatest care and the most meticulous concern for conservation. To use them heedlessly or extravagantly is an act of violence, and while complete non-violence may not be attainable on this earth, there is nonetheless an ineluctable duty on man to aim at the ideal of non-violence in all he does.

Just as a modern European economist would not consider it a great economic achievement if all European art treasures were sold to America at attractive prices, so the Buddhist economist would insist that a population basing its economic life on non-renewable fuels is living parasitically, on capital instead of income. Such a way of life could have no permanence and could therefore be justified only as a purely temporary expedient. As the world's resources of non-renewable fuels – coal, oil and natural gas – are exceedingly unevenly distributed over the globe and undoubtedly limited in quantity, it is clear that their exploitation at an ever-increasing rate is an act of violence against nature which must almost inevitably lead to violence between men.

This fact alone might give food for thought even to those people in Buddhist countries who care nothing for the religious and spiritual values of their heritage and ardently desire to embrace the materialism of modern economics at the fastest possible speed. Before they dismiss Buddhist economics as nothing better than a nostalgic dream, they might wish to consider whether the path of economic development outlined by modern economics is likely to lead them to places where they really want to be. Towards the end of his courageous book *The Challenge of Man's Future*, Professor Harrison Brown of the California Institute of Technology gives the following appraisal:

"Thus we see that, just as industrial society is fundamentally unstable and subject to reversion to agrarian existence, so within it the conditions which offer individual freedom are unstable in their ability to avoid the conditions which impose rigid organisation and totalitarian control. Indeed, when we examine all of the foreseeable difficulties which threaten the survival of industrial civilisation, it is difficult to see how the achievement of stability and the maintenance of individual liberty can be made compatible."[8]

Even if this were dismissed as a long-term view there is the immediate question of whether "modernisation", as currently practised without regard to religious and spiritual values, is actually producing agreeable results. As far as the masses are concerned, the results appear to be disastrous – a collapse of the rural economy, a rising tide of unemployment in town and country, and the growth of a city proletariat without nourishment for either body or soul.

It is in the light of both immediate experience and long-term prospects that the study of Buddhist economics could be recommended even to those who believe that economic growth is more important than any spiritual or religious values. For it is not a question of choosing between "modern growth" and "traditional stagnation". It is a question of finding the right path of development, the Middle Way between materialist heedlessness and traditionalist immobility, in short, of finding "Right Livelihood".

Notes

1 *The New Burma* (Economics and Social Board, Government of the Union of Burma, 1954).
2 Ibid.
3 Ibid.
4 Ananda K. Coomaraswamy, *Art and Swadeshi* (Madras: Ganesh & Co. [n.d.]).
5 J. C. Kumarappa, *Economy of Permanence* (Sarva-Seva Sangh Publication, Rajghat, Kashi, 4th ed., 1958).
6 John Kenneth Galbraith, *The Affluent Society* (London: Penguin Books Ltd., 1962).
7 Richard B. Gregg, *A Philosophy of Indian Economic Development* (Ahmedabad: Navajivan Publishing House, 1954).
8 Harrison Brown, *The Challenge of Man's Future* (New York: The Viking Press, 1954).

Is Technology Autonomous?

Introduction

From the perspective of inquiries that emphasize technology's global reach and pervasive influence, the question naturally arises whether technology constitutes a power or force of its own, beyond the control of its human creators. In other words, is technology *autonomous*? Does it *determine* rather than serve our intentions and purposes? Does technology possess a logic or (more metaphorically) a will of its own?

Jacques Ellul is perhaps the most famous proponent of the autonomy thesis. In his view, the fact that people finance, invent, consume, and even seem to regulate technology – all of this is more appearance than reality. Ellul never refers to Max Weber as a source of inspiration, but his position resembles Weber's in two important respects. First, his notion of "technique" closely resembles Weber's conception of the "rationalization" of social practice. According to Weber, in our eagerness to bring everything from scientific investigation to daily human relations under rules and explicit organization, we tend increasingly to transform reason entirely into something entirely "instrumental" and means-oriented, and we thereby cease to think about the question of the ultimate ends of our practices. Moreover, with this instrumentalist focus comes a certain pervasive blindness to, as Weber puts it, the unintended consequences of human action.

Ellul's critique begins, Weber-like, by refusing to accept the idea that technology is in fact just the total collection of instrumental means. A close look at how "the technological phenomenon" actually functions demonstrates that all of the human activities we "grandly presume" are independent from and thus empowered to direct and control technology are in fact bound up with and beholden to it. Science, the political order, economics, our (vastly overestimated) mental powers – all of these are repeatedly put in service of "the technological demand" for more invention, more development, more control – all the more so at the very moment when we display the greatest vanity in assuming we can decide when and where invention, development, and control are desirable and valuable. Ellul goes on to identify several of the factors that account for our blindness to all of this. For one thing, it is characteristic for those who claim possession of the ability to control the direction of technological development to overestimate their skills. Scientists and engineers display embarrassing naïveté and shallowness in dealing with the social impact of technology. Politicians are driven by ideological assumptions rather than knowledge in their efforts to direct or regulate technical practices. And ordinary citizens and consumers are seriously uninformed about both the technical practices and social realities that dominate everyday life. Moreover, the technological system itself entrances us all – technologists, politicians, and consumers alike. Advertising and propaganda successfully channel our desires. Admired as an endlessly innovative, scientifically informed (and informing), progress-oriented phenomenon, technology comes to be admired as the primary "creative force" in our lives, and its values displace traditional morality, which is now regarded as merely something lingering "inside" our minds which has already disappeared from the "outside" world of real affairs.

Ellul's arguments seem to imply technological determinism. In simple terms, determinism is the thesis that all things and events are caused by some previous things and events. The most popular models for this thesis derive their persuasiveness from the evidence of the predictive success of classical physics, particularly Newtonian mechanics. Persuasive or not, however, determinism is never advocated for very long without evoking serious opposition from all those for whom our experience of free will is too real to be explained away. Philosophers have offered various gambits that attempt either to reconcile determinism and free will (e.g., as characterizing the physical and mental realms, respectively), or to identify equally legitimate standpoints (e.g., the scientific and the ethical, respectively) from which human action can be regarded as determined or free. Determinism, of course, comes in numerous forms. Put in very general terms, biological or genetic determinism claims that who we are is simply a function of our genetic makeup – that we are, in effect, robots guided by our genes. Economic determinism holds that all our activities are ultimately intelligible in terms of the influence of market forces. Technological determinism, then, is the thesis that technology somehow causes all other aspects of society and culture, and hence that changes in technology dictate changes in society.

Marx's Preface to *A Contribution to The Critique of Political Economy* (see Part I) has often been read as espousing technological determinism. The interpretation is difficult to sustain, however, given Marx's descriptions of the role of class struggle in the direction of technological development. In fact, it is many of the anti-Marxist defenders of a post-industrial society (e.g., Daniel Bell, Zbigniew Brzezinski, and others mentioned in McDermott's critique of Mesthene, Part VI, chapter 54) that speak the language of the single-minded technological determinist better than Marx, for they claim – and urge that we find political satisfaction in the claim – that the replacement of machine technology by information technology simply generates new classes of technocrats and service workers in a unilateral and unavoidable manner.

Robert Heilbroner's examination of technological determinism in "Do Machines Make History?" has become something of a classic. According to Heilbroner, the deterministic thesis has two parts that are often run together. About both parts he claims to be "softly" deterministic. First, there is the question of whether there has in fact been a determinate pattern, or "fixed sequence" of technological evolution. On this question, he offers a somewhat qualified Yes. He acknowledges that the linear, logical, ostensibly automatic progress of science, especially when conceived with the aid of an interpretation of technology as applied science, makes technological progress appear inevitable. Even apparently "premature" technological inventions (Heilbroner mentions Hero's toy models of a steam engine in ancient Greece and Charles Babbage's nineteenth-century calculator) are not really counterexamples; for upon closer inspection, these devices were not really feasible in their day because of limitations in other technology (here, lack of "material competence" in metalworking). On the second question, whether technology "imposes a determinate pattern" on society, Heilbroner offers a still more heavily qualified Yes. There is no doubt, he says, that the composition of the labor force and the organization of work are "influenced" by technology. Yet this influence can only seem like the operation of a full-blown causality when one looks at the effect of machines on society but fails to do the reverse. In a manner suggesting a preference for softly economic rather than technological determinism, Heilbroner closes his discussion with a list of several ways in which capitalism (or even a socialism "based on maximizing profits and minimizing costs") stimulates the development and expansion of both modern science and technology.

Critical theorists like Herbert Marcuse (chapter 35) and Jürgen Habermas (chapter 44) resist the pull of technological determinism by arguing that a certain pervasive ideology of technocracy – that is, a seriously inflated picture of the social and political importance of those who have expertise in the sciences and in technology – plays a major role in making technological autonomy appear plausible. Like Ellul, Marcuse describes the way advertising and political propaganda are utilized to control citizen-consumers; and like Weber, he stresses the tendency in contemporary society toward complete "rationalization" of all its practices. Unlike Weber, however, Marcuse does not accept this process with what in others seems to him like Stoic resignation. Instead, he argues that the so-called "rationalization" of society and culture (Marcuse is drawing here on Freud's idea of "rationalization") can be quite irrational. The key, argues Marcuse, is to recognize that science and technology are part of a "one-dimensional" universe of discourse whose moral and political agenda is set by the needs of an industrial (i.e., capitalistic) social order. The inner logic of this order – however illogical (where, e.g., "free" means acceptable to the Free World and "socialist" means everything that fails to serve the interests of private enterprise) – will continue to dominate our lives until the very success it is currently achieving through technoscientific domination of nature and human life makes the continuing effort to contain this success within outmoded institutions intolerable. Today, Marcuse argues, the whole sociopolitical order merely facilitates more productivity while suppressing the free actualization of "new dimensions of human realization" this very productivity now makes possible. Genuine rationalization would involve a very different, utopian, social arrangement. And it would have to be achieved by means of an alternative and genuinely emancipatory technology; for behind technology as presently understood, there is only the impetus for more control of human beings for more production for the greater benefit of those already benefiting the most from the present social order.

33

The "Autonomy" of the Technological Phenomenon

Jacques Ellul

[...]

An autonomous technology. This means that technology ultimately depends only on itself, it maps its own route, it is a prime and not a secondary factor, it must be regarded as an "organism" tending toward closure and self-determination: it is an end in itself. Autonomy is the very condition of technological development. This autonomy corresponds precisely to what J. Baudrillard (*Le Système des objets*) sees under the name of *functionality* when he says that "*functional* qualifies not what is adapted to an end but rather what is adapted to an order or a system." Each technological element is first adapted to the technological system, and it is in respect to this system that the element has its true functionality, far more so than in respect to a human need or a social order. And Baudrillard presents numerous examples of this autonomy, which transforms everything covered by technology into technological objects *before* being anything else: "The entire kitchen loses its culinary function and becomes a functional laboratory...an elision of prime functions for the sake of secondary functions of calculation and relation, an elision of impulses for the sake of culturality...a passage from a gestural universe of work to a gestural universe of control....The simplest mechanism elliptically replaces a sum of gestures, it becomes independent of the operator as of the material to be operated on."

Performing this function, technology endures no judgment from the outside nor any restraint. It presents itself as an intrinsic necessity. Let us recall a rather typical statement among a thousand. Professor L. Sedov, president of the Permanent Commission for the Coordination of Interplanetary Research in the USSR, has declared that no matter what difficulties or objections crop up, nothing could halt space research. "I feel that there are no forces capable today of stopping the historical processes" (October 1963). This remarkable declaration can apply to all technology. The technological system, embodied, of course, in the technicians, admits no other law, no other rule, than the technological law and rule visualized in itself and in regard to itself.[1]...

However, we must know more about this autonomy. First of all, it is the notions or hopes that are modified by technology. An important aspect of this autonomy is that technology radically modifies the objects to which it is applied while being scarcely modified in its own features (if not its forms and modalities). Let us take a simple example. We distinguish between open data and closed data. Open data relates to still unsettled questions, it has an indeterminate content, it implies the participation of the interested parties. Closed data concerns a well-defined object, it can be coded and diffused instantaneously, and, of course, it is closed to participation. Only closed data takes advantage of all the technological means, only it can be rapidly transmitted, etc.

Originally "Autonomy" in Jacques Ellul, *The Technological System*, trans. Joachim Neugroschel, New York: Continuum Publishing Corp., 1980, pp. 125–50, 335–8, abridged.

Hence, the instant that technology is applied more rigorously in coding and transmitting data, the faster it accelerates and the more the data tends to become closed, i.e., to exclude participation by everyone, despite the ideology and the moral desire one may have.

We will not take up here the problem of the relationship between technology and science and technology's relative autonomy from science, since we treated these matters in *The Technological Society*. We will merely add four things emerging from recent studies. The man who ... has investigated this most closely is Simondon. And after showing the interconnections, he concludes not so much – obviously – that there is an autonomy pure and simple of technology, but that there is a possibility for technology to keep developing for a long, long while, even without basic research:

> Even if the sciences did not advance for a certain time, the progress of the technological object toward specificity could continue. The principle of this progress is actually the way in which the object causes itself and conditions itself in its own functioning and in the responses of its functioning to utilization – the technological object, issuing from an abstract work of an organization of subensembles, is the arena for a certain number of relationships of reciprocal causality.

This text gives the precise point of the autonomy of the technological object and thereby specifies technology itself. In the same way, but going to extremes, Koyré (*Études d'histoire de la pensée scientifique*) opines that technology is independent of science and has no influence on it – which strikes me as impossible to support. J. C. Beaune, following Hall (*The Scientific Revolution*), likewise feels that science and technology have separate existences and autonomous developments, whose convergence was historically contingent; he also feels that the passage to scientific technology consisted in unifying the empirical and dispersed technologies, which I have called the passage from the technological operation to the technological phenomenon. These ideas merely take up what I wrote in 1950. Lastly, we can find numerous examples of both the correlation and the independence of technology in Closets. But they are not very significant!

The second remark: John Boli-Bennet (*Technization*), in another connection, offers a stunning analysis of the relationship between science and technology. His is the most recent analysis that I know of, after Ernest Nagel (*The Structure of Science*, 1961), Karl Popper (*The Logic of Scientific Discovery*, 1959), and Carl Hempel (*Aspects of Scientific Explanation*, 1965). There are, says Boli-Bennet, two essential characteristics of scientific knowledge. The first is the "empirical proof of error": a statement cannot be accepted as scientific knowledge if it is theoretically impossible to find empirical data in respect to which the statement is invalid. The second is intersubjectivity, a concept that has replaced scientific objectivity: a statement is scientific only if it is liable to verification or "falsification" which is not subjective and individual, but intersubjective, each scientist never being more than one subject; but each subject having a certain knowledge and a certain background can repeat the same experiment, hence arrive at the same result. In sum, a scientific statement is one that is potentially "falsifiable" on an intersubjective level.

On this basis, we can very clearly see the close relationship between science and technology, quite a different relationship from the one that observers have been hunting for years by setting up "causalities." We will come across this science/technology problem again when studying the finalities of technology. But the mutual relationship between science and technology cannot be divorced from the relationship between technology and politics. It is through, and because of, technology that science is put in the service of government and that politics is so enamored of science.

The third remark: The science/technology interpenetration has *inter alias* a radical effect that is admirably set forth by K. Pomian ("Le Malaise de la science" in *Les Terreurs de l'an 2000*, 1976): namely, the end of scientific innocence. There is no more neutral science, no more pure science. All science is implicated in the technological consequences. And the strength of Pomian's long and profound *factual* study lies in showing that there is no political implication here. As he demonstrates beyond dispute, the essential element is not the decision by politicians to use a scientific discovery in a certain way. But rather, the necessary implication of all scientific research in technology is the determining factor. It is the domination of the technological aspect over the epistemic aspect. And the factors operate in terms of one another. Militarization, nationalization, technicization are intercorrelated. In the same way, Pomian also points out that there is no good or bad use of science or technology.

The two are indissoluble, so that science, he claims, is not neutral, but ambivalent. "To believe that a methodology is neither good nor bad is to tacitly assume that human happiness and suffering are quantities with opposite signs, canceling one another. Far from it. In moral arithmetic, if there *is* an arithmetic, the sum of two opposite quantities does not equal zero." And we are gradually led to reverse the customary proposition: any *scientific* decision entails *political* consequences. "The decision to build a giant accelerator has political implications that the physicists cannot allow themselves to ignore." Pomian cites numerous present-day cases of scientists realizing the consequences of what they are doing and demanding a halt to research (and not a better political application!). Take for examples, the group working around Berg (1974) and the Conference of Asilomar (1975). In contrast, Pomian reveals the politically oriented character of the manifesto of researchers at the Pasteur Institute (the group for biological information). The object of the manifesto is not really the science/technology problem but rather a political debate in the most banal sense of the word! It is politics which is more and more induced by technology and incapable now of steering technological growth in any direction.

Lastly, we have to bring up a new analysis (1975),[2] which fairly transforms the present study of the relationship between science and technology. First of all, we have to distinguish between mathematics (which develops deductively, starting with axioms, and operates upon abstract symbols) and the physical or natural sciences (which develop on an instrumental and material basis). These latter sciences can progress only from a technological ensemble, which is itself nothing but the materialization of theoretical schemata.

Technology is both ahead of and behind science, and it is also at the very heart of science; the latter projects itself into technology and is absorbed into it, and technology is formulated in scientific theory. All science, having become experimental, depends on technology, which alone permits reproducing phenomena technologically. Now, technology abstractly reproduces nature to permit scientific experimenting. Hence, the temptation to make nature conform to theoretical models, to reduce nature to techno-scientific artificiality. "Nature is what I produce in my laboratory," says a modern physicist.

In these conditions, science becomes violence (in regard to everything it bears upon), and the technology expressing the scientific violence becomes power exclusively. Thus, we have a new correlation, which I consider fundamental, between science and technology. The scientific method itself determines technology's calling to be a technology of power. And technology, by the means it makes available to science, induces science into the process of violence (against the ecology, for instance). "The power of technology (theoretically unlimited, but impossible to utilize effectively) materializes in a technology of power. "That is the ultimate point of this relationship." Which the text summed up here calls the "Technological Baroque."

Quite obviously, an autonomy from the state and from politics does not imply that there is no interference with, or political decision-making about, technology. I will certainly not deny the existence of the famous "military-industrial complex." The state cannot help interfering. We have seen that it is tightly bound up with technology, that it is called upon by the technologies to widen its range of intervention. Hence, all the theorists, politicians, partisans, and philosophers agree on a simple view: The state decides, technology obeys. And even more, that is how it must be, it is the true recourse against technology.

In contrast, however, we have to ask who in the state intervenes, and how the state intervenes, i.e., how a decision is reached and by whom in reality, not in the idealist vision. We then learn that technicians are at the origin of political decisions. Next, we have to ask in what direction the state's decision goes. And we perceive very quickly that a remarkable conjunction occurs. The state is furnished with greater power devices by technology, and, being itself an organism of power, the state can only move in the direction of growth, it is strictly conditioned by the technologies not to make any decisions but those to increase power, its own and that of the body social.[3]

Finally, since the system is far from being fully realized, politicians sometimes intervene, taking measures about technological problems, for purely political and in no way technological reasons. The result is generally disastrous.

Those are the four points that we are going to examine rapidly.

Habermas, starting with the presupposition and the democratic ideology, vaguely poses the question: How can we reconcile technology and democracy? But since his view of the technological reality is inexact, since his discourse is purely ideological, the idea of correcting, of mending technology in the actual world of practice is purely illusory. Certainly, the first question to trouble us is: What is becoming of democracy?

Among the hundreds of articles on this topic, we can point out one by R. Lattès ("Énergie et démocratie," *Le Monde*, April 1975) as significant because, written by a scientist, it ingenuously expresses all the ideas assumed by the most unreal idealism. I will not repeat my criticism of identical positions, as set forth in my article "Propagande et démocratie" (*Revue de Science politique*, 1963). Instead, I will limit myself to underlining two particular features.

Monsieur Lattès rightly feels that for *the exercise of democracy*, all citizens must be well informed and judge with full knowledge of the facts. If parliamentary debate is to have any sense, all the deputies must be well educated and well informed. Then, regarding the problem of energy, Lattès asks seven "obvious" questions, whose answers one *must* know for any valid opinion in the energy debate. But he does not seem to realize for even an instant that this issue, paramount as its importance may be, is simply one of dozens: the risks of military policies, the multinational corporations, inflation, its causes and remedies, the ways and means of aid to the third world, etc. For each issue, the citizen would have to have a complete, serious, elaborate, and honest file. Who could fail to see the absurdity of the situation! People do not even have time to "keep up to date."

Furthermore, Lattès apparently believes that the correctly informed citizen could decide on the problem of nuclear energy beyond gut responses and panicky reactions. But (and I will develop this further on) what marks the situation is the inextricable conflict of opinions among the greatest scientists and technicians. The more informed the citizen, the *less* he can participate. Because the evaluations are perfectly contradictory. Lattès is deluding himself. But this is certainly more comforting! There is absolutely no way the citizen can decide for himself. Yet the politician is equally deprived (cf. "L'Illusion politique" in Finzi: *Il potere tecnocratico*).[4]

Thus, despite the advances made in understanding the state/technology problem, we must emphasize an opinion frequent among intellectuals: "To resolve the problems and difficulties caused by technology, we have to nationalize. We have to let the state run the whole thing." That is Closets's implicit thesis, straight through; he tries to prove that all the dangers and abuses of technology are due to its lack of direction. We have to work out a general policy of progress, set up planning agencies, reorganize, etc. But all this can be done only by the political authorities, although he does not come right out and say so. We know that this is also Galbraith's thesis.

Habermas does a superficial analysis of the relationship between technology and politics. He is content with arguments like: "the orientation of technological progress depends on public investments," hence on politics. He seems to be totally unaware of dozens of studies (including Galbraith's or mine) showing the subordination of political decisions to technological imperatives. He winds up with the elementary wish to "get hold of technology again" and "place it under the control of public opinion...reintegrate it within the consensus of the citizens." The matter is, alas, a wee bit more complicated; likewise, when he contrasts the technocratic schema with the decision-making schema. To grasp the interaction, he ought to study L. Sfez (*Critique de la décision*, 1974). And Habermas's discussion of the "pragmatic model" is along the lines of a pious hope, a wish: the process of scientification of politics, such as appears desirable to him, is a "must." But the reality of this technicization of politics actually occurs on a different model!

Habermas poses the philosophical problem honestly: The true problem is to know if, having reached a certain level of knowledge capable of bringing certain consequences, one is content to put that knowledge at the disposal of men involved in technological manipulations, or whether one wants men communicating among themselves to retake possession of that knowledge in their very language. But Habermas poses the problem outside of any reality. When reading this text, we need only ask: Who is that "one" who puts technology at the disposal of either group? Who exercises this (if you like) supreme "will"?

And Richta goes along with Galbraith! The state, they feel, returns to its true function of representing the general interest when it encourages science. "It is significant," writes Richta, "that the state intervenes most drastically in sectors in which science makes the most of itself as a productive force that, by nature, is hostile to private property and that endlessly exceeds its boundaries." The American federal government finances 65% of all basic research, the French government 64%, for the profit motive can no longer make technology advance. But we are forgetting that the state thereby becomes a technological agent itself, both integrated into the technological system, determined by its demands, and modified in its structures by its relationship to the imperative of technological growth.

[...]

Besides, given that, in any event, technology produces a specialization (which is inevitable and the very condition of its success), but also given that the technological system functions as an overall system, no technician can thus grasp the technological phenomenon. Such a grasp would require the experience of the body social, a non-technologically specialized collective organism – in other words, clearly the state. We find the same thing in the Mintz and Cohen book *America, Inc.* (1972). With enormous documentation, these authors show that the whole of American society is subject to two hundred ruling industrial firms – and for Mintz and Cohen, the sole issue is once again the supremacy of the government, which alone will permit the fight against technological abuses, against harmful effects (inequality, exploitation, etc.). It is, incidentally, once more the state that can assure technology its true place and its progress, because – they maintain – the giantism of economic ventures is one cause of blockage to technological advances (but Mintz and Cohen never raise the problem of government giantism).

Lastly (but, of course, the list is not closed), we have to recall Saint Mark's enthusiasm for having the state alone protect nature. Nationalizing and socializing nature is the way to save it – and such mastery would also make technology itself controlled, well oriented, useful, etc.

Before such a roster of authorities, one is surprised and amazed. But also confused. Just what are they talking about? That marvelous ideal organism, the incarnation of Truth and Justice, letting a sweet equality reign without suppression or repression, favoring the weak in order to equalize opportunities, representing the general interest without damaging private interests, promoting liberty for all by a happy harmony, insensitive to the pressures and struggles of interest, patient but not paternalistic, liberating while socialistic, administering without creating a bureaucracy, able to encourage new activities of regulation and concertedness without claiming to impose its law, in such a way as to allow the social actors to freely control the effects of technological progress. A state, finally, having Omnipotence, Omni-Science, without abusing them for anything in the world...

One can only pinch oneself before such a pastoral! Has anyone ever laid eyes on such a state? And if not, what guarantee, what chance do we have that it will come true? Who are the people who will staff it? Saints and martyrs? The huge, the enormous mistake of all those excellent authors is simply that they never breathe a word about this mythical state, which they entrust with so many functions.

Hitherto, the state, whatever its form, socialist or not, has been an organism of oppression, of repression, eliminating its opponents, and constituted by a political class that governs for its own benefit. Will someone explain to me in the name of whom and of what the state will be any different tomorrow – for the dictatorship of the proletariat is exactly the same thing. The marvelous state that will run technology and solve the problems is composed of men (Why should they no longer be dominated by the spirit of power?) and structures (which are more and more technological).[5] What those authors are proposing is that we hand over all power to the administrations, increase administrative power (an ineluctable growth, to be sure, but in no wise a remedy) – i.e., to transform an aleatory control into a technological organization.

In reality, not only is there no guarantee that the state will carry out its envisioned role. But, as can be demonstrated, this state, ruled by the technological imperative and no other, must unavoidably create a society that will be a hundred times more oppressive. It may be able to put order into the technological chaos, but not to control and direct it. It can only accentuate the features we are familiar with. Relying on the state (without considering the autonomy of technology and what the state will turn into under the pressure of technology) means obeying that so technological reflex of a specialist: Things are going badly in my sector, but my neighbor surely has the solution. Finally, it is interesting to note that the advocates of this position, while abominating technocracy, are summoning it with all their might. For a state qualified to dominate technology can only be made up of technicians! But we will come back to technocracy further on.

[...]

To wind up, we will cite a fact that stunningly reveals the dependence of politics and the autonomy of technology. The technological demand is dependent on technological means and not on political ideologies. For instance, Peru has immense copper resources in Cuajone. Experts are unanimous in affirming the incredible wealth of these deposits. But they are very hard to get at and extract. In 1968, Peru turned to the USSR. Soviet experts carefully examined the problem, and their highly detailed report concluded that only the United States had the technology to properly mine the deposits. These experts advised Peru to confide the work to the Americans. In early 1970, the Peruvian government was in a quandary about handing over the "Cuajone contract" after expropriating the International Petroleum Company. But what strikes me as important here is that most of the nontechnicized countries must either leave their riches unexploited or else appeal to highly technicized countries – whatever their ideological outlook may be.

Ideological imperialism is nonsense. Only the technological weight gives true superiority.

It might now be useful to focus on the idea of autonomy from economics, for misunderstandings abound. Quite clearly, one cannot *separate* technology and economy, as Simondon strikingly points out: "Thus there exists a convergence of economic constraints (decrease in raw materials, in work, in energy consumption, etc.) and of properly technological demands....But it seems that the latter would predominate in the technological evolution." Simondon shows that the areas in which the technological conditions override the economic conditions are those in which technological progress has been most rapid. The reason, he says, is that the economic causes "are not pure," they interfere with a diffuse network of motivations and preferences, which rotate or overthrow them. And it is to some extent the "pure" character of the technological phenomenon that assures its autonomy.

Hence, sociologists imperceptibly slide from the primacy (and autonomy) of economics to the primacy (and autonomy) of technology. This is not generally formalized, clearly worded, or enunciated as an overall reality; but more often, it is a subliminal thought, latently taken for granted, as it were. "It goes without saying" for most observers that technology is what determines and causes events, progress, general evolution, like an engine that runs on its own energy. Technology in the intellectual panorama plays the same part as spirituality in the Middle Ages or the idea of the individual in the nineteenth century. Observers do not proceed to any clear and total analysis, but one cannot conceive of society or history in any other way. This trend is so powerful that it crops up even in those who deny it.[6]

I must, however, add some clarification. When I first analyzed technology's autonomy from economics, certain readers saw this as a declaration of *absolute* autonomy – and their criticism was aimed at this *absolute*. Yet I had emphasized that my term did not imply an equivalence between technology and divinity. It is no use saying, "Either there is autonomy, and hence it is absolute – or it is not absolute, and hence there is no autonomy."

This kind of theoretical argument does not go very far. Everyone knows that a sovereign state today cannot do anything it pleases with its sovereignty; belonging to the "concert of nations" is a practical limit on sovereignty. Yet being sovereign, being colonized, having a government imposed by an invader, are not one and the same. Thus, I never said that technology was not dependent on anything or anyone, that it was beyond reach, etc. Obviously, it is subject to the counterthrust of political decisions, economic crises. I indicated, for example, that a government decision at odds with the law of development in technology, with the logic of the system, could halt technological progress, wipe out positive consequences, etc., but that in the conflict between politics and technology, the former would inevitably lose out, and that such a political decision, going against a technological imperative, would ultimately be ruinous for politics itself.

It is quite obvious that technology develops on the basis of a certain number of possibilities offered by the economy. And when the economic resources are lacking, technology cannot operate at its full capacity, achieving what its possibilities allow it to achieve. The relationship between technology and economy is complex. Technology is a determining factor in economic growth, but the converse is equally true. Closets shows clearly that the impact of technology on economy is ambiguous and that economic advances are not proportionately highest where there is the most technological research. Still, technology develops most rapidly in the peak sectors, and it is there too that economy follows. The relationship between the two is striking. In the United States, exports rose an average of 4% in 1967, but 58% for computers, 35% for aeronautics, 30% for telecommunications hardware. Here, the direct relationship is reestablished, but with technology being decisive for economy.

The relationship varies with the periods. It does not appear certain, first of all, that a relationship exists between the great movements of technological *invention* and the economic or social structure. The technological inventions seem like unforeseeable givens of civilization and are by no means tied to the economic level. Nor is technological invention today tied to any one country. It breaks away from those who have encouraged it and benefits countries that did not take part in the effort of scientific or technological invention. But when we leave the domain of invention and proceed to application, technology presumes the involvement of greater and greater capitals.

Can one say that industrial development is what conditions the possibility of technological growth? (Considering that industry is itself a product of technology!) Most technological research in the twentieth century, so it seems, is conditioned and stimulated when the market causes an industrial boom. However, M. Daumas (*Revue d'histoire des sciences et de leur application*, 1969), on the contrary, forcefully asserts the autonomy of technology from industry. And he maintains (which has always been my position): "There is no denying that the evolution of technologies can be understood only if placed in its original historical context; but it is all right to think that the original task of the historian of technologies consists precisely in revealing the intrinsic logic of the evolution of technologies. This evolution actually takes place with an internal logic, which is a very distinct phenomenon from the logic in the evolution of socio-economic history....Investigating this internal logic in the technological evolution is the only way for 'the technological history of the technologies' to slough off its character of data history."

With the spread and growing complexity of technological development, invention in its turn depends on already acquired technological bases (the outcome of earlier applications) and involves *more and more expensive* elements. Hence, technological invention comes to depend *also* on possibilities of economic investment. We thus perceive a mutual influence. On the one side, all modern economic growth depends on technological application, in all areas.[7] But, vice versa, the possibilities of advanced technological research and of the application of technologies depend both on the economic infrastructure and on possibilities of mobilizing economic resources. ... Negatively, the economy can thus either block technological development for lack of power or prevent technological application. The technological program is conditioned by two series of economic imperatives: in a capitalist country by the profitability of investment; and everywhere by the possibility of obtaining the funds necessary for investment.

Nevertheless, at the moment, this is less and less so, for people are coming to realize how impossible it is to calculate the profitability of investments in basic research, and they are growing more and more "convinced" that this research is essential, cannot be neglected, etc. The relationship between technological research and profitability is no longer direct. Hence, the technological applications will be highly unequal according to the economic forms and levels. The latter cause an inequality both in the intensity of technological progress and in the rapidity of access to the profits of technologies.

All this is obvious. But the importance of the economic factor notwithstanding, I will maintain the concept of technology's self-sufficience in the sense that economy can be a means of development, a condition for technological progress, or, inversely, it can be an obstacle, but never does it determine, provoke, or dominate that progress. Like political authority, an economic system that challenges the technological imperative is doomed.

It is not economic law that imposes itself on the technological phenomenon; it is the law of technology which orders and ordains, orients and modifies the economy.[8] Economics is a necessary agent. It is neither the determining factor nor the principle of orientation. Technology obeys its own determination, it realizes itself. And by so doing, it naturally employs many other, nontechnological factors. It may be blocked by their absence, but its reason for functioning and growing comes from nowhere else. Modifying a political or an economic system is perfectly ineffective today and does not alter the true condition of man, because this condition is now defined by its milieu and its technological possibilities, and because the impact of political or economic revolutions on the technological system is practically nil. At most, these troubles can hold up technological progress for a certain time; but revolutionary power changes nothing in the intrinsic law of the system.

This autonomy will get its institutional face in self-organization. That is to say, normally, the technological world will itself organize technological research, the direction of application, the distribution of funds, etc. The autonomy of the technological system must be matched by the autonomy of the institutions that are part of it, that embody it. And this, incidentally, will be the only acceptable autonomy in our society, because it will be the only one providing an ultimate justification. The basic research oriented toward technology cannot develop unless it is sufficiently autonomous! There is an excellent study on this topic by Monsieur Zuckerkandl, research director of France's National Center of Scientific Research (*Le Monde*, November 1964).

One of the effects of autonomy is that technology is becoming the principal factor in reclassifying the domains of activity, of ideological directions. Thus, in 1950, I studied the way technology is making political regimes more similar and reducing the role of ideologies: e.g., the Soviet and the American systems. Likewise, technology is causing a reclassification of public and private activities: the distinction is fading between the economic activities of these two areas. All this was taken up and demonstrated at length by Galbraith in *The New Industrial State* and by M. L. Weidenbaum in "Effets à long terme de la grande Technologie," *Analyse et prévision*, 1969. But the essential point is to see that these effects derive from the autonomy of technology.

Evidently, it is hard for the Marxists to admit that technology has become an autonomous factor, dominating the economic structure and having the same nature and effects in both a capitalist and a communist regime. The most frequently developed argument is that, without any possible doubt, technology is simply in the service of capital, that the familiar effects are due to its integration in capitalism. The technician is merely a salaried employee like the others, the ideology of efficiency is not technological but rather the reflection of the profit need. The division of labor and specialization are not products of technology, but additional ways of exploiting the working class, etc. The most complete effort at systematically demonstrating this interpretation was made by Benjamin Coriat (*Science, technique et capital*, 1976).[9] That is why I will stick to his book rather than lesser works along the same lines.

The two themes to be demonstrated bear, first of all, on the fact that the power of decision belongs to capital. It is capital that decides whether or not to use technologies; the capitalist technologies are as much technologies of production as they are technologies of controlling the exploited class; and capital uses the technologies *only* when they can procure greater profits. If the author admits that technology is not neutral, then only in the sense that it serves capitalism exclusively. The capitalist mode of production has one single goal: the valorization of capital; and by examining the contributions made by the different types of inventions to capital in its process of self-valorization, one can expose the (social) causes determining the incorporation or rejection of the various technologies. Capital utilizes only those that increase the extraction of surplus value. Likewise, the law of value defines the very space in which the technological rationality can operate.

Naturally, the author accuses Richta of dodging the law of value and the production relations *in* and *under* which technology is put to work. But the entire basis of his demonstration rests on Marx's demonstration that capital resorts to mechanization only under two conditions: (1) when the use of dead labor (accumulated in the machine) permits obtaining more surplus labor (diminishing the part of the day that the worker devotes to his own production and increasing that part which goes back to capital); (2) when the technologies allow capital to better dominate the labor process.

[...]

But the most characteristic thing about Coriat's unrealism is his living in the past. Coriat takes Taylorism and mechanization as examples, models, and the ne plus ultra of technology. We must be dreaming! Nothing fundamental has occurred; there has been no change in the technological structure since Taylor. Technology is summed up in and boils down to: the machine. We can obviously understand in these circumstances that Marx's analyses are accurate *for those facts* that are contemporary with, or very slightly subsequent to, Karl Marx. But the mistake is to claim that we are still back there. In Coriat, technology is nothing but the *industrial* application of science in terms of the production of *goods* (in the narrow sense). He blissfully declares that the technologies whose goal is not to produce goods are unemployed! And his critique of Taylorism (as if that were the present situation) corresponds to a labor situation of 1930. In other words, Coriat's "demonstration" is acceptable only for the reader who first grants total approval to the literal expression of Karl Marx's thought and who totally "pooh-poohs" the present facts about technology. Coriat remains enclosed in a problematics established on totally obliterated facts.

We would like to dwell on a further aspect of that autonomy from values and ethics.[10] Man in his hubris – above all intellectual – still believes that his mind controls technology, that he can impose any value, any meaning upon it. And the philosophers are in the forefront of this vanity. It is quite remarkable to note that the finest philosophies on the importance of technology, even the materialist philosophies, fall back upon a preeminence of man.[11] But this grand pretension is purely ideological. What is the autonomy of technology all about in regard to values and morals? One can, I feel, analyze five aspects.

First of all, technology does not progress in terms of a moral ideal, it does not seek to realize values, it does not aim at a virtue or a Good. ...

Secondly, technology does not endure any moral judgment. The technician does not tolerate any insertion of morality in his work. His work has to be free. It seems obvious that the researcher must absolutely not pose the problem of good and bad for himself, of what is permitted or prohibited in his research. His research, quite simply, *is*. And the same is true for its application. Whatever has been found is applied, quite simply. The technician applies his technology with the same independence as the researcher. Now this is the great illogic of many intellectuals. They agree on the first term, which strikes them as obvious, but they want to reintroduce judgments on good and evil, human and inhuman, etc., when they come to the second term, that the technician ought to use his technology to do good. Yet this makes no sense at all after the first term, for application coincides exactly with research. Technological invention is already the outcome of a certain behavior. The problem of behavior (on which people claim to have a value judgment) does not arise only with application. (We will study the conflict between power and values in the last part.) It is the same behavior that dictates the attitude of research (claiming it to be free) and the attitude of application. The technician who puts something to work claims to be as free as the scientist who does the research. Thus it is childish of an intellectual to bring morality into the consequences if he has rejected it in the principle. The autonomy of technology is established here chiefly by a radical division of two areas: "each for itself." Morality judges moral problems. It has nothing to do with technological problems: only the technological means and criteria are acceptable.

An absolutely engrossing study was done by an American technologist on the following theme:[12] So long as the problems are purely technological, they can always find a clear and certain solution. But once the human factor has to enter, or once these problems become too large for any direct technological handling, they seem insoluble. Confronted with these difficulties, people have been developing "social engineering." This innovation appeals to the better feelings; a whole improvement of man rests on the finer instincts and it claims that the route will be the improvement of man, albeit obtained by technologies (psychological or psychosociological technologies). Now after a certain number of examples, the author feels that this route is unsuccessful and uncertain because there are too many nontechnological factors. The only way out is to transform all the problems into a series of specifically technological questions, each receiving its solution from the adequate technology. Here, we can be sure of getting results by avoiding a mixture of types. There is no finer example affirming technological autonomy! Morality, psychology, humanism – they all get in the way. Such is the obvious verdict.

And this is reinforced by the philosophical certainty that only man can be subjected to a moral appraisal. "We are no longer in that primitive epoch when things were good or bad per se: things are only as man makes them. Everything boils down to him. Technology is nothing in itself." But in formulating this oversimplification, the intellectual fails to realize that man is dependent on technology and that, since the latter has become free of all moral judgment, the above statement would imply precisely that technology could do anything. Man does what technology allows him to do. He has thus undertaken to do anything. Maintaining that morality should not judge invention or technological operation leads to saying, unwittingly, that any human action is now beyond ethics. The autonomy of technology thus renders us amoral. Henceforth, morality will no longer be part of our domain, it will be shunted off into the void. In the eyes of scientists and technicians, morality – along with all values and what can be called humanism – is a purely private matter, having nothing to do with concrete activity (which can only be technological) and with no great interest in the seriousness of life.

Here is a small example. In March 1961, the French Minister of National Education launched a survey among students at the scientific *Grandes Écoles* (the faculties specializing in professional training) and in the preparatory classes for these schools. The questionnaire dealt with the teaching of philosophy and literature. The outcome was significant. The students were almost unanimous in denying any sense or value in philosophy. As for the teaching of French, they made a distinction: Literature was totally uninteresting; but knowledge of the language, in contrast, was useful for writing reports and describing experiments.

That is a fine illustration. The technician does not see any bearing that the study of ethics or philosophy can have on his work. Naturally, he admits that the specialists on moral problems, the philosophers, et al., can pass opinions on this work, pronounce judgments. But that is no concern of his. It is pure speculation. There are more and more works of philosophy, sociology of technology (and the theology of technology is beginning to blossom); but their only audience is within the circle of philosophers and humanists. They have no outlet whatsoever into the world of technicians, who utterly ignore all this research. And this is not simply due to specialization. These technicians live in a technological world that has become autonomous.[13]

Since technology does not support any ethical judgment, we come to the third aspect of its autonomy. It does not tolerate being halted for a moral reason. Needless to say, it is simply absurd to voice judgments of good or evil against an operation that is deemed technologically necessary. The technician quite frankly shrugs off something that strikes him as utterly fantastic; besides, we know how relative morality is. The discovery of "situational morality" is quite convenient for putting up with anything. How can we cite a variable, fleeting, constantly redefinable good in order to forbid the technician anything or stop a technological advance? The latter is at least stable, certain, evident. Technology, judging itself, is now liberated from what was once the main check on human action: beliefs (sacred, spiritual, religious) and ethics. Technology, with a theory and a system, thereby assures the freedom that it has acquired in fact. It no longer has to fear any limitation whatsoever because technology exists beyond good and evil.

For a long time, observers claimed that technology was neutral, and consequently not subject to morality. That is the situation we have just described, and the theoretician who thus described technology was merely rubber-stamping the de facto independence of technology and the technician. But this stage is already passed. The power and autonomy of technology are so well assured that now technology itself is turning into a judge of morality. A moral proposition will not be deemed valid for our time if it cannot enter the technological system and be consistent with it.[14]

The fourth aspect of this autonomy concerns legitimacy. Modern man takes for granted that anything scientific is legitimate, and, in consequence, anything technological. Today, we can no longer merely say: "Technology is a fact, we have to accept it as such, we cannot go against it." This is a serious position which reserves the possibility of judgment. But such an attitude is looked upon as pessimistic, antitechnological, and retrograde. Indeed, we must enter the technological system by acknowledging that everything occurring within it is legitimate per se. There is no exterior reference. There is no asking the question about truth (for now, truth is included in science, and the truth of praxis is technology pure and simple), or the question about good, or the question about finalities. None of these things can be discussed. The instant something is technological, it is legitimate, and any challenge is suspect. Technology has even become a power of legitimation. It is technology that now validates scientific research, as we shall demonstrate further on.

This is very remarkable, for hitherto, man has always tried to refer his actions to a superior value, which both judged and underpinned his actions, his enterprises. But this situation is vanishing for the sake of technology. Man in our society both discerns this autonomy demanded by the system (which can progress only if autonomous) and grants this system autonomy by accepting it as legitimate in itself. This autonomy is obviously not the outcome of a struggle between two personified divinities, Morality and Technology! It is man who, becoming a true believer in, and loyal supporter of, technology, views it as a supreme object. For it must be supreme if it bears its legitimacy in itself and needs nothing to justify it!

This conviction is spawned by both experience and persuasion; for the technological system contains its own technological power of legitimation, advertising. It is shallow to believe that advertising is an external addition to the system, due to the domination of technology by profit seeking. Advertising is a technology, indispensable to technological growth and meant to supply the system with its legitimacy. This legitimacy actually comes not just from the excellence that man is ready to acknowledge in technology, but by the persuasion that in fact every element of the system is good. That is why advertising had to add public relations and human relations. By no means does "the mass consumer society vote for itself," but rather, it is the technological society that integrates the individual in the technological process by means of that justification.

There is, however, a further stride to be made, and quite a normal one at that. Independent of morals and judgments, legitimate in itself, technology is becoming the creative force of new values, of a new ethics. Man cannot do without morality! Technology has destroyed all previous scales of value; it impugns the judgments coming from outside. After all, it wrecks their foundations. But being thus self-justified, it quite normally becomes justifying. What was done in the name of science was just; and now the same holds true for what is done in the name of technology. It attributes justice to human action, and man is thus spontaneously led to construct an ethics on the basis of, and in terms of, technology.[15]

This does not occur in a theoretical or systematic manner. The elaboration only comes afterwards. The technological ethics is constructed bit by bit, concretely. Technology demands a certain number of virtues from man (precision, exactness, seriousness, a realistic attitude, and, over everything else, the virtue of work) and a certain outlook on life (modesty, devotion, cooperation). Technology permits very clear value judgments (what is serious and what is not, what is effective, efficient, useful, etc.). This ethics is built up on these concrete givens; for it is primarily an experienced ethics of the behavior required for the technological system to function well. It thereby has the vast superiority over the other moralities of being truly experienced. Furthermore, it involves obvious and ineluctable sanctions (for it is the functioning of the technological system that reveals them). And this morality therefore imposes them almost self-evidently before crystallizing as a clear doctrine located far beyond the simplistic utilitarianisms of the nineteenth century.[16]

[...]

Notes

1 It is obvious – and this comment holds for all the rest of this discussion – that when I say technology "does not admit," "wants," etc., I am not personifying in any way. I am simply using an accepted rhetorical shortcut. In reality, it is the technicians on all levels who make these judgments and have this attitude; but they are so imbued, so impregnated with the technological ideology, so integrated into the system, that their vital judgments and attitudes are its direct expression. One can refer them to the system itself.

2 "Neuf thèses sur la Science and la Technique" in *Vivre et survivre* (1975). This anonymous text is probably by Groetenduijk. I have summed up the first five theses.

3 Furia, *Techniques et sociétés* (1970), leans toward the same opinion. In contrast, see U. Matz; "Die Freiheit der Wissenschaft in der technischen Welt" in *Politik und Wissenschaft* (1971). But he is actually investigating the freedom necessary *for the scientist* in a technicized state.

4 See Jacques Ellul, *The Political Illusion* (1967), and Finzi, *Il potere tecnocratico* (1977).

5 On the capacity of the state to play the role that is presumed, see Jacques Ellul, *The Technological Society*, chap. 4, and *The Political Illusion*. I will not bother repeating these demonstrations here.

6 It appears, quite oddly, in one of the most profound and rigorous thinkers of our time, Bertrand de Jouvenel; he keeps insisting that it is man who decides, and that the overall decisions are made on a political level – technology being merely secondary and subsequent. And yet his admirable book *L'Arcadie* is the best demonstration of the autonomy, the self-sufficience, of technology. This notion runs all through his book, recurring constantly, so that we wonder if the author wrote "on several levels," which are complementary but different and at times seemingly opposed to one another.

7 Of course, everyone agrees that research is the key to (economic) development and that it is therefore worth accumulating economic resources in order to achieve a greater economic advance by means of technological research. But the relation between the two is growing less and less clear. "Research and development" is a source of very great uncertainties. In France, the O.E.C.D. (Organization for Economic Cooperation and Development) has concluded: "The relations of research and development to economic growth suffer from a paradox. They are both obvious and unmeasurable.... Even excluding the money spent on military research, we are unable to bring out the correlation between the expenses of research and

development and the growth of the G.N.P." And Closets has a good formula for defining the relationship between economy and technology: One can only speak of an "economy of uncertainty." As for research and development, see the series *Analyse et prévision*, 1967 to 1970 – and the writings of Jouvenel.

8 Richta underlines an important turnabout in the Weberian school. At first, with Weber, they asserted that "one can rationalize technologically only in terms of commercial reason....The law of technological reason must always yield to the law of economic reason." But since 1960, the Weber disciples (e.g., Papalakas) have been claiming that this economic rationality is relative and that the relationship between capital and technology is reversing: "It is economic reason that must adapt to the harsh technological reality, it is technological rationality that becomes the primary dimension and that thereby dominates the principal focus of tension in society" (R. Richta, *Civilization at the Crossroads*, p. 80).

9 Also see S. Rose, *L'Idéologie de et dans la Science* (1977), a work of strict Marxist orthodoxy, which tries to prove that science is ideological. Very scholarly and very disappointing.

10 Two very good examples of this autonomy are offered, though on different premises, by G. Vahanian and by H. Orlans. G. Vahanian, *The Death of God*, shows that the "how to do" has become independent of all Christian thought and has, in fact, invaded Christianity, which is subordinated to efficiency. H. Orlans, in *Toward the Year 2000*, Daedalus, 1967, shows that "not all technological development is desirable, of course, but we cannot really see how we can prevent anything technologically possible from being realized."

11 The reader can refer to the excellent analysis of such illusions in Seligman (*A Most Notorious Victory*, 1966), who shows that the tragedy of these illusions comes from technology's having its own strength, capable of destroying the designs of man, of determining his ideologies. And, as he shows at length, this autonomy of technology makes man's autonomy "at best questionable."

12 A. M. Weinberg, "Technologie ou 'engineering' social," *Analyse et prévision* (1966).

13 Nevertheless, since 1968 we have to modify this statement slightly. Certain scientists (but no technicians as yet) are starting to ask moral questions about the legitimacy of their scientific work and its goals, however, with no results.

14 On the autonomy of technology from values, one should read the admirable pages by B. Charbonneau, *Le Chaos et le système*, particularly concerning the atomic bomb. "It is not the most monstrous tyrant that produces the bomb, but the most advanced society. And in 1944, it was not the U.S.S.R. or Nazi Germany, but an evangelical and liberal nation ruled by a president whose goal was to free the earth of fear. Who will have wanted the irreperable if ever it comes? Certainly not the scientists, who are only after knowledge, nor the technicians, who are only after power. As for the politicians, they are only after peace and justice. Unhappily, action commands. It was not Roosevelt who made the bomb: Hitler forced him, and then Stalin. But the Communists will demonstrate that the bomb is a product of capitalism. The proof is that the U.S.S.R. is exploding even more powerful bombs. Who or what is behind the bomb? Progress (science, technology, the state) left to its own devices. The U.S.S.R. was the second nation to explode the bomb because it was the second power on the globe. Marx has no more to do with this than Jesus."

15 For lengthy treatments on the contents of this ethics, see Jacques Ellul, *Le Vouloir et le faire*, vol. 1, chap. 2 (1963).

16 In regard to man, Mumford shows decisively and at length how and why the series of the most advanced technological inventions has absolutely nothing to do with man's "central historical task, the task of becoming human." If we take the most recent technological exploits – the moon landing, climate control, artificial survival, creation of life – nothing has the least relationship to the project of "becoming human." Everything obeys the internal logic of the system.

34

Do Machines Make History?

Robert L. Heilbroner

> The hand-mill gives you society with the feudal lord; the steam-mill, society with the industrial capitalist.
>
> ***Marx*, The Poverty of Philosophy**

That machines make history in some sense – that the level of technology has a direct bearing on the human drama – is of course obvious. That they do not make all of history, however that word be defined, is equally clear. The challenge, then, is to see if one can say something systematic about the matter, to see whether one can order the problem so that it becomes intellectually manageable.

To do so calls at the very beginning for a careful specification of our task. There are a number of important ways in which machines make history that will not concern us here. For example, one can study the impact of technology on the *political* course of history, evidenced most strikingly by the central role played by the technology of war. Or one can study the effect of machines on the *social* attitudes that underlie historical evolution: one thinks of the effect of radio or television on political behavior. Or one can study technology as one of the factors shaping the changeful content of life from one epoch to another: when we speak of "life" in the Middle Ages or today we define an existence much of whose texture and substance is intimately connected with the prevailing technological order.

From *Technology and Culture* 8 (1967): 335–45. Reprinted by permission of The Johns Hopkins University Press.

None of these problems will form the focus of this essay. Instead, I propose to examine the impact of technology on history in another area – an area defined by the famous quotation from Marx that stands beneath our title. The question we are interested in, then, concerns the effect of technology in determining the nature of the *socioeconomic order*. In its simplest terms the question is: did medieval technology bring about feudalism? Is industrial technology the necessary and sufficient condition for capitalism? Or, by extension, will the technology of the computer and the atom constitute the ineluctable cause of a new social order?

Even in this restricted sense, our inquiry promises to be broad and sprawling. Hence, I shall not try to attack it head-on, but to examine it in two stages:

1. If we make the assumption that the hand-mill does "give" us feudalism and the steam-mill capitalism, this places technological change in the position of a prime mover of social history. Can we then explain the "laws of motion" of technology itself? Or to put the question less grandly, can we explain why technology evolves in the sequence it does?

2. Again, taking the Marxian paradigm at face value, exactly what do we mean when we assert that the hand-mill "gives us" society with the feudal lord? Precisely how does the mode of production affect the superstructure of social relationships?

These questions will enable us to test the empirical content – or at least to see if there *is* an empirical content – in the idea of technological determinism. I do not think it will come as a surprise if I announce now that we will find *some* content, and a great deal of missing evidence, in our investigation. What will remain then will be to see if we can place the salvageable elements of the theory in historical perspective – to see, in a word, if we can explain technological determinism historically as well as explain history by technological determinism.

I

We begin with a very difficult question hardly rendered easier by the fact that there exist, to the best of my knowledge, no empirical studies on which to base our speculations. It is the question of whether there is a fixed sequence to technological development and therefore a necessitous path over which technologically developing societies must travel.

I believe there is such a sequence – that the steam-mill follows the hand-mill not by chance but because it is the next "stage" in a technical conquest of nature that follows one and only one grand avenue of advance. To put it differently, I believe that it is impossible to proceed to the age of the steam-mill until one has passed through the age of the hand-mill, and that in turn one cannot move to the age of the hydroelectric plant before one has mastered the steam-mill, nor to the nuclear power age until one has lived through that of electricity.

Before I attempt to justify so sweeping an assertion, let me make a few reservations. To begin with, I am fully conscious that not all societies are interested in developing a technology of production or in channeling to it the same quota of social energy. I am very much aware of the different pressures that different societies exert on the direction in which technology unfolds. Lastly, I am not unmindful of the difference between the discovery of a given machine and its application as a technology – for example, the invention of a steam engine (the aeolipile) by Hero of Alexandria long before its incorporation into a steam-mill. All these problems, to which we will return in our last section, refer however to the way in which technology makes its peace with the social, political, and economic institutions of the society in which it appears. They do not directly affect the contention that there exists a determinate sequence of productive technology for those societies that are interested in originating and applying such a technology.

What evidence do we have for such a view? I would put forward three suggestive pieces of evidence:

1 *The simultaneity of invention*

The phenomenon of simultaneous discovery is well known.[1] From our view, it argues that the process of discovery takes place along a well-defined frontier of knowledge rather than in grab-bag fashion. Admittedly, the concept of "simultaneity" is impressionistic,[2] but the related phenomenon of technological "clustering" again suggests that technical evolution follows a sequential and determinate rather than random course.[3]

2 *The absence of technological leaps*

All inventions and innovations, by definition, represent an advance of the art beyond existing base lines. Yet, most advances, particularly in retrospect, appear essentially incremental, evolutionary. If nature makes no sudden leaps, neither, it would appear, does technology. To make my point by exaggeration, we do do not find experiments in electricity in the year 1500, or attempts to extract power from the atom in the year 1700. On the whole, the development of the technology of production presents a fairly smooth and continuous profile rather than one of jagged peaks and discontinuities.

3 *The predictability of technology*

There is a long history of technological prediction, some of it ludicrous and some not.[4] What is interesting is that the development of technical progress has always seemed *intrinsically* predictable. This does not mean that we can lay down future timetables of technical discovery, nor does it rule out the possibility of surprises. Yet I venture to state that many scientists would be willing to make *general* predictions as to the nature of technological capability twenty-five or even fifty years ahead. This too suggests that technology follows a developmental sequence rather than arriving in a more chancy fashion.

I am aware, needless to say, that these bits of evidence do not constitute anything like a "proof" of my hypothesis. At best they establish the grounds on which a prima facie case of plausibility may be rested. But I should like now to strengthen these grounds by suggesting two deeper-seated reasons why technology *should* display a "structured" history.

The first of these is that a major constraint always operates on the technological capacity of an age, the constraint of its accumulated stock of available knowledge. The application of this knowledge may lag behind its reach; the technology of the hand-mill, for example, was by no means at the frontier of medieval technical knowledge, but technical realization can hardly precede what men generally know (although experiment may incrementally advance both technology and knowledge concurrently). Particularly from the mid-nineteenth century to the present do we sense the loosening constraints on technology stemming from successively yielding barriers of scientific knowledge – loosening constraints that result in the successive arrival of the electrical, chemical, aeronautical, electronic, nuclear, and space stages of technology.[5]

The gradual expansion of knowledge is not, however, the only order-bestowing constraint on the development of technology. A second controlling factor is the material competence of the age, its level of technical expertise. To make a steam engine, for example, requires not only some knowledge of the elastic properties of steam but the ability to cast iron cylinders of considerable dimensions with tolerable accuracy. It is one thing to produce a single steam-machine as an expensive toy, such as the machine depicted by Hero, and another to produce a machine that will produce power economically and effectively. The difficulties experienced by Watt and Boulton in achieving a fit of piston to cylinder illustrate the problems of creating a technology, in contrast with a single machine.

Yet until a metal-working technology was established – indeed, until an embryonic machine-tool industry had taken root – an industrial technology was impossible to create. Furthermore, the competence required to create such a technology does not reside alone in the ability or inability to make a particular machine (one thinks of Babbage's ill-fated calculator as an example of a machine born too soon), but in the ability of many industries to change their products or processes to "fit" a change in one key product or process.

This necessary requirement of technological congruence[6] gives us an additional cause of sequencing. For the ability of many industries to co-operate in producing the equipment needed for a "higher" stage of technology depends not alone on knowledge or sheer skill but on the division of labor and the specialization of industry. And this in turn hinges to a considerable degree on the sheer size of the stock of capital itself. Thus the slow and painful accumulation of capital, from which springs the gradual diversification of industrial function, becomes an independent regulator of the reach of technical capability.

In making this general case for a determinate pattern of technological evolution – at least insofar as that technology is concerned with production – I do not want to claim too much. I am well aware that reasoning about technical sequences is easily faulted as *post hoc ergo propter hoc*. Hence, let me leave this phase of my inquiry by suggesting no more than that the idea of a roughly ordered progression of productive technology seems logical enough to warrant further empirical investigation. To put it as concretely as possible, I do not think it is just by happenstance that the steam-mill follows, and does not precede, the hand-mill, nor is it mere fantasy in our own day when we speak of the coming of the automatic factory. In the future as in the past, the development of the technology of production seems bounded by the constraints of knowledge and capability and thus, in principle at least, open to prediction as a determinable force of the historic process.

II

The second proposition to be investigated is no less difficult than the first. It relates, we will recall, to the explicit statement that a given technology imposes certain social and political characteristics upon the society in which it is found. Is it true that, as Marx wrote in *The German Ideology*, "A certain mode of production, or industrial stage, is always combined with a certain mode of cooperation, or social stage,"[7] or as he put it in the sentence immediately preceding our hand-mill, steam-mill paradigm, "In acquiring new productive forces men change their mode of production, and in changing their mode of production they change their way of living – they change all their social relations"?

As before, we must set aside for the moment certain "cultural" aspects of the question. But if we restrict ourselves to the functional relationships directly connected with the process of production itself, I think we can indeed state that the technology of a society imposes a determinate pattern of social relations on that society.

We can, as a matter of fact, distinguish at least two such modes of influence:

1 The composition of the labor force

In order to function, a given technology must be attended by a labor force of a particular kind. Thus, the hand-mill (if we may take this as referring to late medieval technology in general) required a work force composed of skilled or semiskilled craftsmen, who were free to practice their occupations at home or in a small atelier, at times and seasons that varied considerably. By way of contrast, the steam-mill – that is, the technology of the nineteenth century – required a work force composed of semiskilled or unskilled operatives who could work only at the factory site and only at the strict time schedule enforced by turning the machinery on or off. Again, the technology of the electronic age has steadily required a higher proportion of skilled attendants; and the coming technology of automation will still further change the needed mix of skills and the locale of work, and may as well drastically lessen the requirements of labor time itself.

2 The hierarchical organization of work

Different technological apparatuses not only require different labor forces but different orders of supervision and co-ordination. The internal organization of the eighteenth-century handicraft unit, with its typical man-master relationship, presents a social configuration of a wholly different kind from that of the nineteenth-century factory with its men-manager confrontation, and this in turn differs from the internal social structure of the continuous-flow, semi-automated plant of the present. As the intricacy of the production process increases, a much more complex system of internal controls is required to maintain the system in working order.

Does this add up to the proposition that the steam-mill gives us society with the industrial capitalist? Certainly the class characteristics of a particular society are strongly implied in its functional organization. Yet it would seem wise to be very cautious before relating political effects exclusively to functional economic causes. The Soviet Union, for example, proclaims itself to be a socialist society although its technical base resembles that of old-fashioned capitalism. Had Marx written that the stream-mill gives you society with the industrial *manager*, he would have been closer to the truth.

What is less easy to decide is the degree to which the technological infrastructure is responsible for some of the sociological features of society. Is anomie, for instance, a disease of capitalism or of all industrial societies? Is the organization man a creature of monopoly capital or of all bureaucratic industry wherever found? These questions tempt us to look into the problem of the impact of technology on the existential quality of life, an area we have ruled out of bounds for this paper. Suffice it to say that superficial evidence seems to imply that the similar technologies of Russia and America are indeed giving rise to similar social phenomena of this sort.

As with the first portion of our inquiry, it seems advisable to end this section on a note of caution. There is a danger, in discussing the structure of the labor force or the nature of intrafirm organization, of assigning the sole causal efficacy to the visible presence of machinery and of overlooking the invisible influence of other factors at work. Gilfillan, for instance, writes, "engineers have committed such blunders as saying the typewriter brought women to work in offices, and with the typesetting machine made possible the great modern newspaper, forgetting that in Japan there are women office workers and great modern newspapers getting practically no help from typewriters and typesetting machines."[8] In addition, even where technology seems unquestionably to play the critical role, an independent "social" element unavoidably enters the scene in the *design* of technology, which must take into account such facts as the level of education of the work force or its relative price. In this way the machine will reflect, as much as mould, the social relationships of work.

These caveats urge us to practice what William James called a "soft determinism" with regard to the influence of the machine on social relations. Nevertheless, I would say that our cautions qualify rather than invalidate the thesis that the prevailing level of technology imposes itself powerfully on the structural organization of the productive side of society. A foreknowledge of the shape of the technical core of society fifty years hence may not allow us to describe the political attributes of that society, and may perhaps only hint at its sociological character, but assuredly it presents us with a profile of requirements, both in labor skills and in supervisory needs, that differ considerably from those of today. We cannot say whether the society of the computer will give us the latter-day capitalist or the commissar, but it seems beyond question that it will give us the technician and the bureaucrat.

III

Frequently, during our efforts thus far to demonstrate what is valid and useful in the concept of technological determinism, we have been forced to defer certain aspects of the problem until later. It is time now to turn up the rug and to examine what has been swept under it. Let us try to systematize our qualifications and objections to the basic Marxian paradigm:

1 Technological progress is itself a social activity

A theory of technological determinism must contend with the fact that the very activity of invention and innovation is an attribute of some societies and not of others. The Kalahari bushmen or the tribesmen of New Guinea, for instance, have persisted in a neolithic technology to the present day; the Arabs reached a high degree of technical proficiency in the past and have since suffered a decline; the classical Chinese developed technical expertise in some fields while unaccountably neglecting it in the area of production. What factors serve to encourage or discourage this technical thrust is a problem about which we know extremely little at the present moment.[9]

2 The course of technological advance is responsive to social direction

Whether technology advances in the area of war, the arts, agriculture, or industry depends in part on the rewards, inducements, and incentives offered by society. In this way the direction of technological advance is partially the result of social policy. For example, the system of interchangeable parts, first introduced into France and then independently into England failed to take root in either country for lack of government interest or market stimulus. Its success in America is attributable mainly to government support and to its appeal in a society without guild traditions and with high labor costs.[10] The general *level* of technology may follow an independently determined sequential path, but its areas of application certainly reflect social influences.

3 Technological change must be compatible with existing social conditions

An advance in technology not only must be congruent with the surrounding technology but must also be compatible with the existing economic and other institutions of society. For example, labor-saving machinery will not find ready acceptance in a society where labor is abundant and cheap as a factor of production. Nor would a mass production technique recommend itself to a society that did not have a mass market. Indeed, the presence of slave labor seems generally to inhibit the use of machinery and the presence of expensive labor to accelerate it.[11]

These reflections on the social forces bearing on technical progress tempt us to throw aside the whole notion of technological determinism as false or misleading.[12] Yet, to relegate technology from an undeserved position of *primum mobile* in history to that of a mediating factor, both acted upon by and acting on the body of society, is not to write off its influence but only to specify its mode of operation with greater precision. Similarly, to admit we understand very little of the cultural factors that give rise to technology does not depreciate its role but focuses our attention on that period of history when technology is clearly a major historic force, namely Western society since 1700.

IV

What is the mediating role played by technology within modern Western society? When we ask this much more modest question, the interaction of society and technology begins to clarify itself for us:

1 The rise of capitalism provided a major stimulus for the development of a technology of production

Not until the emergence of a market system organized around the principle of private property did there also emerge an institution capable of systematically guiding the inventive and innovative abilities of society to the problem of facilitating production. Hence the environment of the eighteenth and nineteenth centuries provided both a novel and an extremely effective encouragement for the development of an *industrial* technology. In addition, the slowly opening political and social framework of late mercantilist society gave rise to social aspirations for which the new technology offered the best chance of realization. It was not only the steam-mill that gave us the industrial capitalist but the rising inventor-manufacturer who gave us the steam-mill.

2 The expansion of technology within the market system took on a new "automatic" aspect

Under the burgeoning market system not alone the initiation of technical improvement but its subsequent adoption and repercussion through the economy was largely governed by market considerations. As a result, both the rise and the proliferation of technology assumed the attributes of an impersonal diffuse "force" bearing on social and economic life. This was all the more pronounced because the political control needed to buffer its disruptive consequences was seriously inhibited by the prevailing laissez-faire ideology.

3 The rise of science gave a new impetus to technology

The period of early capitalism roughly coincided with and provided a congenial setting for the development of an independent source of technological encouragement – the rise of the self-conscious activity of science. The steady expansion of scientific research, dedicated to the exploration of nature's secrets and to their harnessing for social use, provided an increasingly important stimulus for technological advance from the middle of the nineteenth century. Indeed, as the twentieth century has progressed, science has become a major historical force in its own right and is now the indispensable precondition for an effective technology.

It is for these reasons that technology takes on a special significance in the context of capitalism – or, for that matter, of a socialism based on maximizing production or minimizing costs. For in these societies, both the continuous appearance of technical advance and its diffusion throughout the society assume the attributes of autonomous process, "mysteriously" generated by society and thrust upon its members in a manner as indifferent as it is imperious. This is why, I think, the problem of technological determinism – of how machines make history – comes to us with such insistence despite the ease with which we can disprove its more extreme contentions.

Technological determinism is thus peculiarly a problem of a certain historic epoch – specifically that of high capitalism and low socialism – *in which the forces of technical change have been unleased, but when the agencies for the control or guidance of technology are still rudimentary.*

The point has relevance for the future. The surrender of society to the free play of market forces is now on the wane, but its subservience to the impetus of the scientific ethos is on the rise. The prospect before us is assuredly that of an undiminished and very likely accelerated pace of technical change. From what we can foretell about the direction of this technological advance and the structural alterations it implies, the pressures in the future will be toward a society marked by a much greater degree of organization and deliberate control. What other political, social, and existential changes the age of the computer will also bring we do not know. What seems certain, however, is that the problem of technological determinism – that is, of the impact of machines on history – will remain germane until there is forged a degree of public control over technology far greater than anything that now exists.

Notes

1 See Robert K. Merton, "Singletons and Multiples in Scientific Discovery: A Chapter in the Sociology of Science," *Proceedings* of the American Philosophical Society, CV (October 1961), 470–86.

2 See John Jewkes, David Sawers, and Richard Stillerman, *The Sources of Invention* (New York, 1960 [paperback edition]), p. 227, for a skeptical view.

3 "One can count 21 basically different means of flying, at least eight basic methods of geophysical prospecting; four ways to make uranium explosive;...20 or 30 ways to control birth....If each of these separate inventions were autonomous, i.e., without cause, how could one account for their arriving in these functional groups?" S. C. Gilfillan, "Social Implications

of Technological Advance," *Current Sociology*, I (1952), 197. See also Jacob Schmookler, "Economic Sources of Inventive Activity," *Journal of Economic History* (March 1962), pp. 1–20; and Richard Nelson, "The Economics of Invention: A Survey of the Literature," *Journal of Business*, XXXII (April 1959), 101–19.

4 Jewkes et al. (see n. 2) present a catalogue of chastening mistakes (p. 230 f.). On the other hand, for a sober predictive effort, see Francis Bello, "The 1960s: A Forecast of Technology," *Fortune*, LIX (January 1959), 74–78; and Daniel Bell, "The Study of the Future," *Public Interest*, I (Fall 1965), 119–30. Modern attempts at prediction project likely avenues of scientific advance or technological function rather than the feasibility of specific machines.

5 To be sure, the inquiry now regresses one step and forces us to ask whether there are inherent stages for the expansion of knowledge, at least insofar as it applies to nature. This is a very uncertain question. But having already risked so much, I will hazard the suggestion that the roughly parallel sequential development of scientific understanding in those few cultures that have cultivated it (mainly classical Greece, China, the high Arabian culture, and the West since the Renaissance) makes such a hypothesis possible, provided that one looks to broad outlines and not to inner detail.

6 The phrase is Richard LaPiere's in *Social Change* (New York, 1965), p. 263 f.

7 Karl Marx and Friedrich Engels, *The German Ideology* (London, 1942), p. 18.

8 Gilfillan (see n. 3), p. 202.

9 An interesting attempt to find a line of social causation is found in E. Hagen, *The Theory of Social Change* (Homewood, Ill., 1962).

10 See K. R. Gilbert, "Machine-Tools," in Charles Singer, E. J. Holmyard, A. R. Hall, and Trevor I. Williams (eds.), *A History of Technology* (Oxford, 1958), IV, chap. xiv.

11 See LaPiere (see n. 6), p. 284; also H. J. Habbakuk, *British and American Technology in the 19th Century* (Cambridge, 1962), *passim.*

12 As, for example, in A. Hansen, "The Technological Determination of History," *Quarterly Journal of Economics* (1921), pp. 76–83.

35

The New Forms of Control

Herbert Marcuse

A comfortable, smooth, reasonable, democratic unfreedom prevails in advanced industrial civilization, a token of technical progress. Indeed, what could be more rational than the suppression of individuality in the mechanization of socially necessary but painful performances; the concentration of individual enterprises in more effective, more productive corporations; the regulation of free competition among unequally equipped economic subjects; the curtailment of prerogatives and national sovereignties which impede the international organization of resources. That this technological order also involves a political and intellectual coordination may be a regrettable and yet promising development.

The rights and liberties which were such vital factors in the origins and earlier stages of industrial society yield to a higher stage of this society: they are losing their traditional rationale and content. Freedom of thought, speech, and conscience were – just as free enterprise, which they served to promote and protect – essentially *critical* ideas, designed to replace an obsolescent material and intellectual culture by a more productive and rational one. Once institutionalized, these rights and liberties shared the fate of the society of which they had become an integral part. The achievement cancels the premises.

To the degree to which freedom from want, the concrete substance of all freedom, is becoming a real possibility, the liberties which pertain to a state of lower productivity are losing their former content. Independence of thought, autonomy, and the right to political opposition are being deprived of their basic critical function in a society which seems increasingly capable of satisfying the needs of the individuals through the way in which it is organized. Such a society may justly demand acceptance of its principles and institutions, and reduce the opposition to the discussion and promotion of alternative policies *within* the status quo. In this respect, it seems to make little difference whether the increasing satisfaction of needs is accomplished by an authoritarian or a non-authoritarian system. Under the conditions of a rising standard of living, non-conformity with the system itself appears to be socially useless, and the more so when it entails tangible economic and political disadvantages and threatens the smooth operation of the whole. Indeed, at least in so far as the necessities of life are involved, there seems to be no reason why the production and distribution of goods and services should proceed through the competitive concurrence of individual liberties.

Freedom of enterprise was from the beginning not altogether a blessing. As the liberty to work or to starve, it spelled toil, insecurity, and fear for the vast majority of the population. If the individual were no longer compelled to prove himself on the market, as a free economic subject, the disappearance of this kind of freedom would be one of the greatest achievements of civilization. The technological processes of mechanization and standardization might release individual energy into a yet uncharted realm of freedom beyond necessity. The

From Herbert Marcuse, *One-Dimensional Man*, Boston: Beacon, 1964, pp. 1–18.

very structure of human existence would be altered; the individual would be liberated from the work world's imposing upon him alien needs and alien possibilities. The individual would be free to exert autonomy over a life that would be his own. If the productive apparatus could be organized and directed toward the satisfaction of the vital needs, its control might well be centralized; such control would not prevent individual autonomy, but render it possible.

This is a goal within the capabilities of advanced industrial civilization, the "end" of technological rationality. In actual fact, however, the contrary trend operates: the apparatus imposes its economic and political requirements for defense and expansion on labor time and free time, on the material and intellectual culture. By virtue of the way it has organized its technological base, contemporary industrial society tends to be totalitarian. For "totalitarian" is not only a terroristic political coordination of society, but also a non-terroristic economic-technical coordination which operates through the manipulation of needs by vested interests. It thus precludes the emergence of an effective opposition against the whole. Not only a specific form of government or party rule makes for totalitarianism, but also a specific system of production and distribution which may well be compatible with a "pluralism" of parties, newspapers, "countervailing powers," etc.

Today political power asserts itself through its power over the machine process and over the technical organization of the apparatus. The government of advanced and advancing industrial societies can maintain and secure itself only when it succeeds in mobilizing, organizing, and exploiting the technical, scientific, and mechanical productivity available to industrial civilization. And this productivity mobilizes society as a whole, above and beyond any particular individual or group interests. The brute fact that the machine's physical (only physical?) power surpasses that of the individual, and of any particular group of individuals, makes the machine the most effective political instrument in any society whose basic organization is that of the machine process. But the political trend may be reversed; essentially the power of the machine is only the stored-up and projected power of man. To the extent to which the work world is conceived of as a machine and mechanized accordingly, it becomes the *potential* basis of a new freedom for man.

Contemporary industrial civilization demonstrates that it has reached the stage at which "the free society" can no longer be adequately defined in the traditional terms of economic, political, and intellectual liberties, not because these liberties have become insignificant, but because they are too significant to be confined within the traditional forms. New modes of realization are needed, corresponding to the new capabilities of society.

Such new modes can be indicated only in negative terms because they would amount to the negation of the prevailing modes. Thus economic freedom would mean freedom *from* the economy – from being controlled by economic forces and relationships; freedom from the daily struggle for existence, from earning a living. Political freedom would mean liberation of the individuals *from* politics over which they have no effective control. Similarly, intellectual freedom would mean the restoration of individual thought now absorbed by mass communication and indoctrination, abolition of "public opinion" together with its makers. The unrealistic sound of these propositions is indicative, not of their utopian character, but of the strength of the forces which prevent their realization. The most effective and enduring form of warfare against liberation is the implanting of material and intellectual needs that perpetuate obsolete forms of the struggle for existence.

The intensity, the satisfaction and even the character of human needs, beyond the biological level, have always been preconditioned. Whether or not the possibility of doing or leaving, enjoying or destroying, possessing or rejecting something is seized as a *need* depends on whether or not it can be seen as desirable and necessary for the prevailing societal institutions and interests. In this sense, human needs are historical needs and, to the extent to which the society demands the repressive development of the individual, his needs themselves and their claim for satisfaction are subject to overriding critical standards.

We may distinguish both true and false needs. "False" are those which are superimposed upon the individual by particular social interests in his repression: the needs which perpetuate toil, aggressiveness, misery, and injustice. Their satisfaction might be most gratifying to the individual, but this happiness is not a condition which has to be maintained and protected if it serves to arrest the development of the ability (his own and others) to recognize the disease of the whole and grasp the chances of curing the disease. The result then is euphoria in unhappiness. Most of the prevailing needs to relax, to have fun, to behave and consume in accordance with the advertisements, to love and hate what others love and hate, belong to this category of false needs.

Such needs have a societal content and function which are determined by external powers over which the individual has no control; the development and satisfaction of these needs is heteronomous. No matter how much such needs may have become the individual's own, reproduced and fortified by the conditions of his existence; no matter how much he identifies himself with them and finds himself in their satisfaction, they continue to be what they were from the beginning – products of a society whose dominant interest demands repression.

The prevalence of repressive needs is an accomplished fact, accepted in ignorance and defeat, but a fact that must be undone in the interest of the happy individual as well as all those whose misery is the price of his satisfaction. The only needs that have an unqualified claim for satisfaction are the vital ones – nourishment, clothing, lodging at the attainable level of culture. The satisfaction of these needs is the prerequisite for the realization of *all* needs, of the unsublimated as well as the sublimated ones.

For any consciousness and conscience, for any experience which does not accept the prevailing societal interest as the supreme law of thought and behavior, the established universe of needs and satisfactions is a fact to be questioned – questioned in terms of truth and falsehood. These terms are historical throughout, and their objectivity is historical. The judgment of needs and their satisfaction, under the given conditions, involves standards of *priority* – standards which refer to the optimal development of the individual, of all individuals, under the optimal utilization of the material and intellectual resources available to man. The resources are calculable. "Truth" and "falsehood" of needs designate objective conditions to the extent to which the universal satisfaction of vital needs and, beyond it, the progressive alleviation of toil and poverty, are universally valid standards. But as historical standards, they do not only vary according to area and stage of development, they also can be defined only in (greater or lesser) *contradiction* to the prevailing ones. What tribunal can possibly claim the authority of decision?

In the last analysis, the question of what are true and false needs must be answered by the individuals themselves, but only in the last analysis; that is, if and when they are free to give their own answer. As long as they are kept incapable of being autonomous, as long as they are indoctrinated and manipulated (down to their very instincts), their answer to this question cannot be taken as their own. By the same token, however, no tribunal can justly arrogate to itself the right to decide which needs should be developed and satisfied. Any such tribunal is reprehensible, although our revulsion does not do away with the question: how can the people who have been the object of effective and productive domination by themselves create the conditions of freedom?

The more rational, productive, technical, and total the repressive administration of society becomes, the more unimaginable the means and ways by which the administered individuals might break their servitude and seize their own liberation. To be sure, to impose Reason upon an entire society is a paradoxical and scandalous idea – although one might dispute the righteousness of a society which ridicules this idea while making its own population into objects of total administration. All liberation depends on the consciousness of servitude, and the emergence of this consciousness is always hampered by the predominance of needs and satisfactions which, to a great extent, have become the individual's own. The process always replaces one system of preconditioning by another; the optimal goal is the replacement of false needs by true ones, the abandonment of repressive satisfaction.

The distinguishing feature of advanced industrial society is its effective suffocation of those needs which demand liberation – liberation also from that which is tolerable and rewarding and comfortable – while it sustains and absolves the destructive power and repressive function of the affluent society. Here, the social controls exact the overwhelming need for the production and consumption of waste; the need for stupefying work where it is no longer a real necessity; the need for modes of relaxation which soothe and prolong this stupefication; the need for maintaining such deceptive liberties as free competition at administered prices, a free press which censors itself, free choice between brands and gadgets.

Under the rule of a repressive whole, liberty can be made into a powerful instrument of domination. The range of choice open to the individual is not the decisive factor in determining the degree of human freedom, but *what* can be chosen and what *is* chosen by the individual. The criterion for free choice can never be an absolute one, but neither is it entirely relative. Free election of masters does not abolish the masters or the slaves. Free choice among a wide variety of goods and services does not signify freedom if these goods and services sustain social controls over a life of toil and fear – that is, if they sustain alienation. And the spontaneous reproduction of superimposed needs by the individual does not establish autonomy; it only testifies to the efficacy of the controls.

Our insistence on the depth and efficacy of these controls is open to the objection that we overrate greatly the indoctrinating power of the "media," and that by themselves the people would feel and satisfy the needs which are now imposed upon them. The objection misses the point. The preconditioning does not start with the mass production of radio and television and with the centralization of their control. The people enter this stage as preconditioned receptacles of long-standing; the decisive difference is in the flattening out of the contrast (or conflict) between the given and the possible, between the satisfied and the unsatisfied needs. Here, the so-called equalization of class distinctions reveals its ideological function. If the worker and his boss enjoy the same television program and visit the same resort places, if the typist is as attractively made up as the daughter of her employer, if the Negro owns a Cadillac, if they all read the same newspaper, then this assimilation indicates not the disappearance of classes, but the extent to which the needs and satisfactions that serve the preservation of the Establishment are shared by the underlying population.

Indeed, in the most highly developed areas of contemporary society, the transplantation of social into individual needs is so effective that the difference between them seems to be purely theoretical. Can one really distinguish between the mass media as instruments of information and entertainment, and as agents of manipulation and indoctrination? Between the automobile as nuisance and as convenience? Between the horrors and the comforts of functional architecture? Between the work for national defense and the work for corporate gain? Between the private pleasure and the commercial and political utility involved in increasing the birth rate?

We are again confronted with one of the most vexing aspects of advanced industrial civilization: the rational character of its irrationality. Its productivity and efficiency, its capacity to increase and spread comforts, to turn waste into need, and destruction into construction, the extent to which this civilization transforms the object world into an extension of man's mind and body makes the very notion of alienation questionable. The people recognize themselves in their commodities; they find their soul in their automobile, hi-fi set, split-level home, kitchen equipment. The very mechanism which ties the individual to his society has changed, and social control is anchored in the new needs which it has produced.

The prevailing forms of social control are technological in a new sense. To be sure, the technical structure and efficacy of the productive and destructive apparatus has been a major instrumentality for subjecting the population to the established social division of labor throughout the modern period. Moreover, such integration has always been accompanied by more obvious forms of compulsion: loss of livelihood, the administration of justice, the police, the armed forces. It still is. But in the contemporary period, the technological controls appear to be the very embodiment of Reason for the benefit of all social groups and interests – to such an extent that all contradiction seems irrational and all counteraction impossible.

No wonder then that, in the most advanced areas of this civilization, the social controls have been introjected to the point where even individual protest is affected at its roots. The intellectual and emotional refusal "to go along" appears neurotic and impotent. This is the socio-psychological aspect of the political event that marks the contemporary period: the passing of the historical forces which, at the preceding stage of industrial society, seemed to represent the possibility of new forms of existence.

But the term "introjection" perhaps no longer describes the way in which the individual by himself reproduces and perpetuates the external controls exercised by his society. Introjection suggests a variety of relatively spontaneous processes by which a Self (Ego) transposes the "outer" into the "inner." Thus introjection implies the existence of an inner dimension distinguished from and even antagonistic to the external exigencies – an individual consciousness and an individual unconscious *apart from* public opinion and behavior.[1] The idea of "inner freedom" here has its reality: it designates the private space in which man may become and remain "himself."

Today this private space has been invaded and whittled down by technological reality. Mass production and mass distribution claim the *entire* individual, and industrial psychology has long since ceased to be confined to the factory. The manifold processes of introjection seem to be ossified in almost mechanical reactions. The result is, not adjustment but *mimesis:* an immediate identification of the individual with *his* society and, through it, with the society as a whole.

This immediate, automatic identification (which may have been characteristic of primitive forms of association) reappears in high industrial civilization; its new "immediacy," however, is the product of a sophisticated, scientific management and organization. In this process, the "inner" dimension of the mind in which opposition to the status quo can take root is whittled down. The loss of this dimension, in which the power of negative thinking – the critical power of Reason – is at home, is the ideological counterpart to the very material process in which advanced industrial society silences and reconciles the opposition. The impact of progress turns Reason into submission to the facts of life, and to the dynamic capability of producing more and bigger facts of the same sort of life. The efficiency of the system blunts the individuals' recognition that it contains no facts which do not communicate the repressive power of the whole. If the individuals find themselves in the things which shape their life, they do so, not by giving, but by accepting the law of things – not the law of physics but the law of their society.

I have just suggested that the concept of alienation seems to become questionable when the individuals identify themselves with the existence which is imposed upon them and have in it their own development and satisfaction. This identification is not illusion but reality. However, the reality constitutes a more progressive stage of alienation. The latter has become entirely objective; the subject which is alienated is swallowed up by its alienated existence. There is only one dimension, and it is everywhere and in all forms. The achievements of progress defy ideological indictment as well as justification; before their tribunal, the "false consciousness" of their rationality becomes the true consciousness.

This absorption of ideology into reality does not, however, signify the "end of ideology." On the contrary, in a specific sense advanced industrial culture is *more* ideological than its predecessor, inasmuch as today the ideology is in the process of production itself.[2] In a provocative form, this proposition reveals the political aspects of the prevailing technological rationality. The productive apparatus and the goods and services which it produces "sell" or impose the social system as a whole. The means of mass transportation and communication, the commodities of lodging, food, and clothing, the irresistible output of the entertainment and information industry carry with them prescribed attitudes and habits, certain intellectual and emotional reactions which bind the consumers more or less pleasantly to the producers and, through the latter, to the whole. The products indoctrinate and manipulate; they promote a false consciousness which is immune against its falsehood. And as these beneficial products become available to more individuals in more social classes, the indoctrination they carry ceases to be publicity; it becomes a way of life. It is a good way of life – much better than before – and as a good way of life, it militates against qualitative change. Thus emerges a pattern of *one-dimensional thought and behavior* in which ideas, aspirations, and objectives that, by their content, transcend the established universe of discourse and action are either repelled or reduced to terms of this universe. They are redefined by the rationality of the given system and of its quantitative extension.

The trend may be related to a development in scientific method: operationalism in the physical, behaviorism in the social sciences. The common feature is a total empiricism in the treatment of concepts; their meaning is restricted to the representation of particular operations and behavior. The operational point of view is well illustrated by P. W. Bridgman's analysis of the concept of length:[3]

> We evidently know what we mean by length if we can tell what the length of any and every object is, and for the physicist nothing more is required. To find the length of an object, we have to perform certain physical operations. The concept of length is therefore fixed when the operations by which length is measured are fixed: that is, the concept of length involves as much and nothing more than the set of operations by which length is determined. In general, we mean by any concept nothing more than a set of operations; *the concept is synonymous with the corresponding set of operations.*

Bridgman has seen the wide implications of this mode of thought for the society at large:[4]

> To adopt the operational point of view involves much more than a mere restriction of the sense in which we understand "concept," but means a far-reaching change in all our habits of thought, in that we shall no longer permit ourselves to use as tools in our thinking concepts of which we cannot give an adequate account in terms of operations.

Bridgman's prediction has come true. The new mode of thought is today the predominant tendency in philosophy, psychology, sociology, and other fields. Many of the most seriously troublesome concepts are being "eliminated" by showing that no adequate account of them in terms of operations or behavior can be given. The radical empiricist onslaught ... thus provides the methodological justification for the debunking of the mind by the intellectuals – a positivism which, in its denial of the transcending elements of Reason, forms the academic counterpart of the socially required behavior.

Outside the academic establishment, the "far-reaching change in all our habits of thought" is more serious. It serves to coordinate ideas and goals with those exacted by the prevailing system, to enclose them in the system, and to repel those which are irreconcilable with the system. The reign of such a one-dimensional reality does not mean that materialism rules, and that the spiritual, metaphysical, and bohemian occupations are petering out. On the contrary, there is a great deal of "Worship together this week," "Why not try God," Zen, existentialism, and beat ways of life, etc. But such modes of protest and transcendence are no longer contradictory to the status quo and no longer negative. They are rather the ceremonial part of practical behaviorism, its harmless negation, and are quickly digested by the status quo as part of its healthy diet.

One-dimensional thought is systematically promoted by the makers of politics and their purveyors of mass information. Their universe of discourse is populated by self-validating hypotheses which, incessantly and monopolistically repeated, become hypnotic definitions or dictations. For example, "free" are the institutions which operate (and are operated on) in the countries of the Free World; other transcending modes of freedom are by definition either anarchism, communism, or propaganda. "Socialistic" are all encroachments on private enterprises not undertaken by private enterprise itself (or by government contracts), such as universal and comprehensive health insurance, or the protection of nature from all too sweeping commercialization, or the establishment of public services which may hurt private profit. This totalitarian logic of accomplished facts has its Eastern counterpart. There, freedom is the way of life instituted by a communist regime, and all other transcending modes of freedom are either capitalistic, or revisionist, or leftist sectarianism. In both camps, non-operational ideas are non-behavioral and subversive. The movement of thought is stopped at barriers which appear as the limits of Reason itself.

Such limitation of thought is certainly not new. Ascending modern rationalism, in its speculative as well as empirical form, shows a striking contrast between extreme critical radicalism in scientific and philosophic method on the one hand, and an uncritical quietism in the attitude toward established and functioning social institutions. Thus Descartes' *ego cogitans* was to leave the "great public bodies" untouched, and Hobbes held that "the present ought always to be preferred, maintained, and accounted best." Kant agreed with Locke in justifying revolution *if and when* it has succeeded in organizing the whole and in preventing subversion.

However, these accommodating concepts of Reason were always contradicted by the evident misery and injustice of the "great public bodies" and the effective, more or less conscious rebellion against them. Societal conditions existed which provoked and permitted real dissociation from the established state of affairs; a private as well as political dimension was present in which dissociation could develop into effective opposition, testing its strength and the validity of its objectives.

With the gradual closing of this dimension by the society, the self-limitation of thought assumes a larger significance. The interrelation between scientific-philosophical and societal processes, between theoretical and practical Reason, asserts itself "behind the back" of the scientists and philosophers. The society bars a whole type of oppositional operations and behavior; consequently, the concepts pertaining to them are rendered illusory or meaningless. Historical transcendence appears as metaphysical transcendence, not acceptable to science and scientific thought. The operational and behavioral point of view, practiced as a "habit of thought" at large, becomes the view of the established universe of discourse and action, needs and aspirations. The "cunning of Reason" works, as it so often did, in the interest of the powers that be. The insistence on operational and behavioral concepts turns against the efforts to free thought and behavior *from* the given reality and *for* the suppressed alternatives. Theoretical and practical Reason, academic and social behaviorism meet on common ground: that of an advanced society which makes scientific and technical progress into an instrument of domination.

"Progress" is not a neutral term; it moves toward specific ends, and these ends are defined by the possibilities of ameliorating the human condition. Advanced industrial society is approaching the stage where continued progress would demand the radical subversion of the prevailing direction and organization of progress. This stage would be reached when material production (including the necessary services) becomes automated to the extent that all vital needs can be satisfied while necessary labor time is reduced to marginal time. From this point on, technical progress would transcend the realm of necessity, where it served as the instrument of domination and exploitation which thereby limited its rationality; technology would become subject to the free play of faculties in the struggle for the pacification of nature and of society.

Such a state is envisioned in Marx's notion of the "abolition of labor." The term "pacification of existence" seems better suited to designate the historical alternative of a world which – through an international conflict which transforms and suspends the contradictions within the established societies – advances on the brink of a global war. "Pacification of existence" means the development of man's struggle with man and with nature, under conditions where the competing needs, desires, and aspirations are no longer organized by vested interests in domination and scarcity – an organization which perpetuates the destructive forms of this struggle.

Today's fight against this historical alternative finds a firm mass basis in the underlying population, and finds its ideology in the rigid orientation of thought and behavior to the given universe of facts. Validated by the accomplishments of science and technology, justified by its growing productivity, the status quo defies all transcendence. Faced with the possibility of pacification on the grounds of its technical and intellectual achievements, the mature industrial society closes itself against this alternative. Operationalism, in theory and practice, becomes the theory and practice of *containment*. Underneath its obvious dynamics, this society is a thoroughly static system of life: self-propelling in its oppressive productivity and in its beneficial coordination. Containment of technical progress goes hand in hand with its growth in the established direction. In spite of the political fetters imposed by the status quo, the more technology appears capable of creating the conditions for pacification, the more are the minds and bodies of man organized against this alternative.

The most advanced areas of industrial society exhibit throughout these two features: a trend toward consummation of technological rationality, and intensive efforts to contain this trend within the established institutions. Here is the internal contradiction of this civilization: the irrational element in its rationality. It is the token of its achievements. The industrial society which makes technology and science its own is organized for the ever-more-effective domination of man and nature, for the ever-more-effective utilization of its resources. It becomes irrational when the success of these efforts opens new dimensions of human realization. Organization for peace is different from organization for war; the institutions which served the struggle for existence cannot serve the pacification of existence. Life as an end is qualitatively different from life as a means.

Such a qualitatively new mode of existence can never be envisaged as the mere by-product of economic and political changes, as the more or less spontaneous effect of the new institutions which constitute the necessary prerequisite. Qualitative change also involves a change in the *technical* basis on which this society rests – one which sustains the economic and political institutions through which the "second nature" of man as an aggressive object of administration is stabilized. The techniques of industrialization are political techniques; as such, they prejudge the possibilities of Reason and Freedom.

To be sure, labor must precede the reduction of labor, and industrialization must precede the development of human needs and satisfactions. But as all freedom depends on the conquest of alien necessity, the realization of freedom depends on the *techniques* of this conquest. The highest productivity of labor can be used for the perpetuation of labor, and the most efficient industrialization can serve the restriction and manipulation of needs.

When this point is reached, domination – in the guise of affluence and liberty – extends to all spheres of private and public existence, integrates all authentic opposition, absorbs all alternatives. Technological rationality reveals its political character as it becomes the great vehicle of better domination, creating a truly totalitarian universe in which society and nature, mind and body are kept in a state of permanent mobilization for the defense of this universe.

Notes

1 The change in the function of the family here plays a decisive role: its "socializing" functions are increasingly taken over by outside groups and media. See my *Eros and Civilization* (Boston: Beacon Press, 1955), p. 96 ff.

2 Theodor W. Adorno, *Prismen. Kulturkritik und Gesellschaft* (Frankfurt: Suhrkamp, 1955), p. 24 f.

3 P. W. Bridgman, *The Logic of Modern Physics* (New York: Macmillan, 1928), p. 5. The operational doctrine has since been refined and qualified. Bridgman himself has extended the concept of "operation" to include the "paper-and-pencil" operations of the theorist (in Philipp J. Frank, *The Validation of Scientific Theories* [Boston: Beacon Press, 1954], Chap. II). The main impetus remains the same: it is "desirable" that the paper-and-pencil operations "be capable of eventual contact, although perhaps indirectly, with instrumental operations."

4 P. W. Bridgman, *The Logic of Modern Physics*, loc. cit., p. 31.

Technology, Ecology, and the Conquest of Nature

Introduction

As the selections in the first two sections of this part show, to consider the role technology plays (or ought to play) in human life is to make an issue of both our own nature (Are we fundamentally tool-using animals? Power-hungry? Desirous of being creative? Destined to be cogs in a mega-machine?) and also of the extent to which the development of technoscience does or might serve human purposes. Clearly, what we conclude about these issues will strongly influence our conception of what are our legitimate relations to our surroundings. To what extent is it permissible to exploit nature for our own purposes? Are there any limits to power-exercising "Baconianism"? Are there non-exploitive relations with our surroundings that now, for various reasons, are marginalized or ignored? To what extent are various relations really subject to choice? The articles in this section all attempt to bring some focus to these questions by addressing, often in the context of feminist criticism of mainstream positions, some of the ecological issues that seem to press upon us through the development of modern technologies. The first three consider the charge that technoscientific activities not only exploit but threaten the ecosphere; the last three evaluate the response of "Deep Ecology" to this perceived threat.

Carolyn Merchant's "Mining the Earth's Womb," adapted from her influential book, *The Death of Nature*, offers her controversial account of a major transition in Western history from the dominant pagan and medieval view of Mother Earth to the modern view of nature as inanimate and barren. Merchant is especially concerned to demonstrate how these two broad and general descriptions have very different "normative import," that is, encourage and even seem to justify very different sorts of human action. The transition itself as well as the normative change that comes with it can be especially well illuminated, she argues, by tracing the radical alteration in attitudes toward mining. Merchant notes how African smiths and ancient natural historians like Pliny expressed respect for the earth as a living and nurturing being and thus regarded mining as inherently a violation and defilement. What little mining did occur was undertaken with trepidation, in fear of nature's revenging herself in earthquakes and volcanic eruptions. During the Renaissance, the remnants of this pagan view that had remained into the Christian middle ages were eradicated. Merchant analyzes Georg Agricola's influential *De Re Metallica*, in which the old prohibitions against mining are turned "on their head" with arguments that pry loose the old assumption that activities like mining are disruptive from the traditional idea that what the Earth conceals is not to be acquired. Agricola argues that mining is not inherently more disruptive that catching fish or making metal tools, especially if one digs areas that are otherwise undesirable. Morever, he insists, raw materials are "neutral" in themselves. What is good or bad is our reason for making use of them. In short, what Agricola succeeds in showing – with the key example of mining – is how, in general, legitimate use and exploitation might better be distinguished by their (human) purposes than by their (allegedly "natural") essence. Such arguments, argues Merchant, set the stage for the development of a "new ethic of exploitation" – one that transforms the idea of defiling a Nurturing Mother into the new imagery of exploring and appropriating Mother Nature's treasures. Further, Merchant makes a point of showing how this change of attitude was accompanied and encouraged by changing attitudes toward women. Females and the birth process became less revered. Male medical men discredited female peasant healers, whose arts had been respected by male Renaissance hermeticists and magicians such as Paracelsus and Cornelius Agrippa. With the rise of scientific rationality came the burning of (mainly female) witches. Exit alchemy, with its belief in the "vegetative powers" of gold and silver operative in the "warm womb" of the earth. Enter Bacon, who in Merchant's view gave crucial assistance to all these changes by providing a kind of codification of the new attitudes toward both nature and women through his analogy between the experimental interrogation of nature and the seduction and conquest of woman.

For Alan Soble, however, feminists like Merchant and Sandra Harding are seriously misreading Bacon's use of metaphors of forceful seduction in his articulations of the new relation between scientist and nature. "In Defense of Bacon" alleges, for example, that it is illegitimate to construe Bacon's famous imagery of the scientist as "hounding" nature as rape. Soble argues that a close consideration of Bacon's metaphors reveals that either the imagined sexual connotations are not really there or that when they are, they are not intended to demean women. If there is any rape in Bacon's imagery, claims Soble, it is only marital rape, which would have been acceptable at the time. Soble also rejects Merchant's claim that Bacon's model for the "interrogation" of nature is the interrogation of witches by torture. Bacon nowhere actually states this connection; hence, there is no "smoking gun." In short, Soble's general strategy is to reject Merchant's appeals to the sociocultural context of references to sex and to nature as female in the quotations from Bacon that he examines. Although he does not say so, Soble appears to believe that if Bacon does not explicitly place his discussion of science in an atmosphere of gender-charged phraseology, then we should not read him as having gone beyond the particular allusions to gender in the quotes themselves. (Some years after Bacon's death, founders of the Baconian-inspired British Royal Society – among them, Henry Oldenburg, Joseph Glanvill, and Thomas Sprat – advocated the establishment of an entirely "masculine philosophy" that would eliminate anything "feminine" from their theories and activities ... .)

In her famous "Cyborg Manifesto," Donna Haraway suggests that the popular science-fiction portrayals of cyborgs as part human and part machine reflect a general blurring in our age of the traditional boundary lines not only between persons and hardware but also, thanks to biotechnology, even between species. The implication, of course, is that this change in our ideas of who we are goes hand in hand with changes in our understanding of how we relate to each other and to nature. The general process of blurring all the old species-defining lines, she argues, is a manifestation of the global dominance of an information economy, and it now issues in a condition that many call postmodern. Postmodernism and poststructuralism reject the traditional notions of a unitary, "centered" human being or society. Such notions – like those of an essential self or "subject" in Descartes, Kant, and Romanticism, as well as those of a society functioning according to some central dynamic like "labor," as in Marxism – all belong to an earlier, modern era. (On the "decentering" of the self, the psychoanalytic theories of Freud and Lacan are widely regarded as pioneering sources; the idea of a decentered social totality is indebted to, among others, the Marxist structuralist, Louis Althusser.) Essentialist feminism, according to which Woman has some unchanging (unusually inferior/subordinate) nature, argues Haraway, is just as dated and constrictive as are the older humanisms that posit an essence of Man. Drawing on the writings of multicultural feminists, Haraway proposes a view of selves and societies in which the intersection and crossing of lines of class, gender, and race result in non-unitary systems of overlapping allegiances and partial "identities." Just as boundaries between selves, ethnicities, sexes, and species are erased in science fiction's cyborgs, so also are such boundaries being erased in the rapidly developing multiethnic, transsexual, bioengineered condition of our time. It is interesting to note that Haraway's conception of the blurring of lines between human and inorganic, as well as between nature and culture, implicitly supersedes Merchant's account of the idea of nature as Nurturing Mother being replaced by another idea of nature as purposeless object of exploitation. For Haraway, both ideas are still too holistic and essentialist to reflect our present circumstances. Her article closes with the much discussed, anti-essentialist remark, "I would rather be a cyborg than a [traditionally understood] goddess."

The next three articles concentrate on the "deep-ecological" claim that an excessive estimation of our own importance – a pervasive "anthropocentrism" – has come to threaten our very universe. Arne Naess – the Norwegian philosopher whose work is in empirical linguistics and who is the founder of the deep ecology movement – grounds his conception of this movement in what he calls an "ecosophy" rather than ecology. The latter, he says, is just the scientific study of the environment as we find it, one that makes predictions about what we can expect under various conditions and offers expert advice on policy issues. But the ecological movement, he explains – in a manner that echoes Merchant – is "clearly and forcefully normative" – which means it is concerned with the "wisdom," as with scientific "knowledge." Naess envisages ecosophy as a kind of general systems theory, inspired by the naturalisms of Spinoza and Aristotle. Spinoza in particular seems to have influenced Naess; for his pantheistic, monistic rationalism treats God and the world as one substance, and mind and body as complementary aspects of a single unity. (Some of Naess's followers have noted that Einstein, too, revered Spinoza, and that in following out the full implications of General Relativity, he treated objects as singularities in the space–time continuum rather than as separate substances). Like Spinoza, Naess denies metaphysical status to presumably "separate" entities, just as he denies that it is a sufficiently "global" response to our current circumstances to fight against pollution and resource depletion. Such fights reflect a "shallow" ecological movement through which piecemeal environmental problem-solving effectively aids primarily the affluent populations of developed countries. In his later writings, Naess draws on the Buddhist conception of the unreality of the ego-self in order to argue for a conception of self-realization that involves moving beyond the idea of one's being as that of an "individual" toward the more expansive understanding of the "ecological self." According to Naess, hidden behind the more visible expressions of shallow ecological concern, there is the fact that for the majority of scientific ecologists, their experience has already begun to inspire in them the deep ecological views expressed in the article reprinted below.

Bill Devall's "The Deep Ecology Movement" makes more explicit the philosophical influences on the deep ecology movement. Using Kuhn's terminology (see the Toulmin selection, Part II, chapter 10), Devall formulates deep ecology as an alternative to the dominant (albeit now widely criticized) paradigm of economic growth, progress, and domination of nature (the "storehouse"). Devall cites numerous sources of inspiration – for example, White (see Part III) on the Judeo-Christian origins of the idea of the domination of nature, Eastern spiritual and Native American earth-centered traditions, Spinoza's critique of Descartes and Bacon, and Heidegger (see Part IV). Devall claims that to displace the traditional paradigm and promote the newly emerging "ecological consciousness," we will have to develop a new cosmology that stresses what Naess calls "biological equalitarianism," a new "ecological psychology" that attunes us to the "total intermingling" of every real thing, and a new philosophical anthropology that inspires a politics based on hunter-gatherer societies rather than on industrial capitalist societies with their stress on scarcity and acquisition.

The Australian philosopher, Ariel Salleh, criticizes deep ecology from an ecofeminist perspective. Citing both Naess and Devall, Salleh complains of the pervasive rationalistic and scientistic language in which they still philosophize. In her view, the influence on Naess of the logical empiricist and contemporary analytic philosophical traditions undermines the alternative program he tries to present. Thus, he likens his philosophical approach to general systems theory and speaks the language of policy science, data, programming, and control – apparently not realizing the way this merely reintroduces the "cultural scientism" of the very advocates of shallow ecology he ostensibly opposes. And he formulates ecosophy as a set of "intuitions" to be analytically clarified and formulated into an axiology – apparently unaware of the way this approach duplicates precisely the dualisms of norm and fact, ruler and ruled, control and submission that he supposes to be part of our problem. Above all, argues Salleh, one can see in Naess's advocacy of a new paradigm for our relations with our environment – a paradigm in which "Man" is still used generically – that deep ecologists uncritically ignore and carry forward all those features of the Western intellectual tradition whose pervasive sexism feminists have been concerned to challenge. Although Devall may have avoided Naess's scientism with his appeal to phenomenology and Heidegger, by looking for the source of our urge to dominate nature in the Judeo-Christian creation story, or in technics-run-wild as portrayed by Mumford, or in capitalism as portrayed by Marx, he too fails to recognize that patriarchy is the real cause of this urge. In contrast, claims Salleh, the convergence of the feminist and ecology movements is no accident. In a seemingly more essentialist voice than Haraway or social constructionist feminists would find acceptable, Salleh speaks of women both as intrinsically more connected with nature through menstruation, childbirth, and breast-feeding, and also, again unlike men, as much more likely to form small, intimate collectivities instead of hierarchical power structures. For Salleh, males would therefore appear to be constitutionally unable to consistently present or realize the ideals of deep ecology.

36

Mining the Earth's Womb

Carolyn Merchant

The domination of the earth through technology and the corresponding rise of the image of the world as *Machina ex Deo* were features of the Scientific Revolution of the sixteenth and seventeenth centuries. During this period, the two ideas of mechanism and the domination of nature came to be core concepts and controlling images of our modern world. An organically oriented mentality prevalent from ancient times to the Renaissance, in which the female principle played a significant positive role, was gradually undermined and replaced by a technological mindset that used female principles in an exploitative manner. As Western culture became increasingly mechanized during the 1600s, a female nurturing earth and virgin earth spirit were subdued by the machine.

The change in controlling imagery was directly related to changes in human attitudes and behavior toward the earth. Whereas the older nurturing earth image can be viewed as a cultural constraint restricting the types of socially and morally sanctioned human actions allowable with respect to the earth, the new images of mastery and domination functioned as cultural sanctions for the denudation of nature. Society needed these new images as it continued the processes of commercialism and industrialization, which depended on activities directly altering the earth – mining, drainage, deforestation, and assarting (grubbing up stumps to clear fields). The new activities utilized new technologies – lift and force pumps, cranes, windmills, geared wheels, flap valves, chains, pistons, treadmills, under- and overshot watermills, fulling mills, flywheels, bellows, excavators, bucket chains, rollers, geared and wheeled bridges, cranks, elaborate block and tackle systems, worm, spur, crown, and lantern gears, cams and eccentrics, ratchets, wrenches, presses, and screws in magnificent variation and combination.

These technological and commercial changes did not take place quickly; they developed gradually over the ancient and medieval eras, as did the accompanying environmental deterioration. Slowly, over many centuries, early Mediterranean and Greek civilization had mined and quarried the mountainsides, altered the forested landscape, and overgrazed the hills. Nevertheless, technologies were low level, people considered themselves parts of a finite cosmos, and animism and fertility cults that treated nature as sacred were numerous. Roman civilization was more pragmatic, secular, and commercial and its environmental impact more intense. Yet Roman writers such as Ovid, Seneca, Pliny, and the Stoic philosophers openly deplored mining as an abuse of their mother, the earth. With the disintegration of feudalism and the expansion of Europeans into new worlds and markets, commercial society began to have an accelerated impact on the natural environment. By the sixteenth and seventeenth centuries, the tension between technological development in the world of action and the controlling organic images in the world of the mind had become too great.

From *Machina Ex Dea: Feminist Perspectives on Technology*, ed. Joan Rothschild, Oxford: Pergamon Press, 1983, pp. 99–117. Reprinted by permission of Teachers College Press, Columbia University, New York.

The old structures were incompatible with the new activities.

Both the nurturing and domination metaphors had existed in philosophy, religion, and literature – the idea of dominion over the earth in Greek philosophy and Christian religion; that of the nurturing earth, in Greek and other pagan philosophies. But, as the economy became modernized and the Scientific Revolution proceeded, the dominion metaphor spread beyond the religious sphere and assumed ascendancy in the social and political spheres as well. These two competing images and their normative associations can be found in sixteenth-century literature, art, philosophy, and science.

The image of the earth as a living organism and nurturing mother had served as a cultural constraint restricting the actions of human beings. One does not readily slay a mother, dig into her entrails for gold, or mutilate her body, although commercial mining would soon require that. As long as the earth was considered to be alive and sensitive, it could be considered a breach of human ethical behavior to carry out destructive acts against it. For most traditional cultures, minerals and metals ripened in the uterus of the Earth Mother, mines were compared to her vagina, and metallurgy was the human hastening of the birth of the living metal in the artificial womb of the furnace – an abortion of the metals' natural growth cycle before its time. Miners offered propitiation to the deities of the soil and subterranean world, performed ceremonial sacrifices, and observed strict cleanliness, sexual abstinence, and fasting before violating the sacredness of the living earth by sinking a mine. Smiths assumed an awesome responsibility in precipitating the metal's birth through smelting, fusing, and beating it with hammer and anvil; they were often accorded the status of shaman in tribal rituals and their tools were thought to hold special powers (Eliade 1962, pp. 53–70, 79–96).

The Renaissance image of the nurturing earth still carried with it subtle ethical controls and restraints. Such imagery found in a culture's literature can play a normative role within the culture. Controlling images operate as ethical restraints or as ethical sanctions – as subtle "oughts" or "ought-nots." Thus, as the descriptive metaphors and images of nature change, a behavioral restraint can be changed into a sanction. Such a change in the image and description of nature was occurring during the course of the Scientific Revolution.

It is important to recognize the normative import of descriptive statements about nature. Contemporary philosophers of language have critically reassessed the earlier positivist distinction between the "is" of science and the "ought" of society, arguing that descriptions and norms are not opposed to one another by linguistic separation into separate "is" and "ought" statements, but are contained within each other. Descriptive statements about the world can presuppose the normative; they are then ethic-laden. A statement's normative function lies in the use itself as description. The norms may be tacit assumptions hidden within the descriptions in such a way as to act as invisible restraints or moral ought-nots. The writer or culture may not be conscious of the ethical import yet may act in accordance with its dictates. The hidden norms may become conscious or explicit when an alternative or contradiction presents itself. Because language contains a culture within itself, when language changes, a culture is also changing in important ways. By examining changes in descriptions of nature, we can then perceive something of the changes in cultural values. To be aware of the interconnectedness of descriptive and normative statements is to be able to evaluate changes in the latter by observing changes in the former (Cavell 1971, pp. 148, 165).

Not only did the image of nature as nurturing mother contain ethical implications but the organic framework itself, as a conceptual system, also carried with it an associated value system. Contemporary philosophers have argued that a given normative theory is linked with certain conceptual frameworks and not with others. The framework contains within itself certain dimensions of structural and normative variation, while denying others belonging to an alternative or rival framework (Taylor 1973).

We cannot accept a framework of explanation and yet reject its associated value judgments, because the connections to the values associated with the structure are not fortuitous. New commercial and technological innovations, however, can upset and undermine an established conceptual structure. New human and social needs can threaten associated normative constraints, thereby demanding new ones.

While the organic framework was for many centuries sufficiently integrative to override commercial development and technological innovation, the acceleration of such changes throughout Western Europe during the sixteenth and seventeenth centuries began to undermine the organic unity of the cosmos and society. Because the needs and purposes of society as a whole were changing with the commercial revolution, the values associated with the organic view of nature were no longer applicable; hence, the plausibility of the conceptual framework itself was slowly, but continuously, being threatened.

The Geocosm: The Earth As a Nurturing Mother

Not only was nature in a generalized sense seen as female, but also the earth, or geocosm, was universally viewed as a nurturing mother – sensitive, alive, and responsive to human action. The changes in imagery and attitudes relating to the earth were of enormous significance as the mechanization of nature proceeded. The nurturing earth would lose its function as a normative restraint as it changed to a dead, inanimate, physical system.

The macrocosm theory likened the cosmos to the human body, soul, and spirit with male and female reproductive components. Similarly, the geocosm theory compared the earth to the living human body, with breath, blood, sweat, and elimination systems.

For the Stoics, who flourished in Athens during the third century B.C., after the death of Aristotle, and in Rome through the first century A.D., the world itself was an intelligent organism; God and matter were synonymous. Matter was dynamic, composed of two forces: expansion and condensation – the former directed outward, the latter inward. The tension between them was the inherent force generating all substances, properties, and living forms in the cosmos and the geocosm.

Zeno of Citium (ca. 304 B.C.) and M. Tullius Cicero (106–43 B.C.) held that the world reasons, has sensation, and generates living rational beings: "The world is a living and wise being, since it produces living and wise beings" (Cicero 1775, p. 96). Every part of the universe and the earth was created for the benefit and support of another part. The earth generated and gave stability to plants, plants supported animals, and animals in turn served human beings; conversely, human skill helped to preserve these organisms. The Universe itself was created for the sake of rational beings – gods and men – but God's foresight insured the safety and preservation of all things. Humankind was given hands to transform the earth's resources and dominion over them: timber was to be used for houses and ships, soil for crops, iron for plows, and gold and silver for ornaments. Each part and imperfection existed for the sake and ultimate perfection of the whole.

The living character of the world organism meant not only that the stars and planets were alive, but that the earth too was pervaded by a force giving life and motion to the living beings on it. Lucius Seneca (4 B.C.–A.D. 65), a Roman Stoic, stated that the earth's breath nourished both the growths on its surface and the heavenly bodies above by its daily exhalations:

> How could she nourish all the different roots that sink into the soil in one place and another, had she not an abundant supply of the breath of life?...all these [heavenly bodies] draw their nourishment from materials of earth...and are sustained...by nothing else than the breath of the earth. ...Now the earth would be unable to nourish so many bodies...unless it were full of breath, which it exhales from every part of it day and night (Seneca 1910, p. 244).

The earth's springs were akin to the human blood system; its other various fluids were likened to the mucus, saliva, sweat, and other forms of lubrication in the human body, the earth being organized "...much after the plan of our bodies, in which there are both veins and arteries, the former blood vessels, the latter air vessels....So exactly alike is the resemblance to our bodies in nature's formation of the earth, that our ancestors have spoken of veins (= springs) of water." Just as the human body contained blood, marrow, mucus, saliva, tears, and lubricating fluids, so in the earth there were various fluids. Liquids that turned hard became metals, such as gold and silver, other fluids turned into stones, bitumens, and veins of sulfur. Like the human body, the earth gave forth sweat: "There is often a gathering of thin, scattered moisture like dew, which from many points flows into one spot. The dowsers call it *sweat*, because a kind of drop is either squeezed out by the pressure of the ground or raised by the heat" (Seneca 1910, pp. 126–27).

Leonardo da Vinci (1452–1519) elaborated the Greek analogy between the waters of the earth and the ebb and flow of human blood through the veins and heart:

> The water runs from the rivers to the sea and from the sea to the rivers, always making the same circuit. The water is thrust from the utmost depth of the sea to the high summits of the mountains, where, finding the veins cut, it precipitates itself and returns to the sea below, mounts once more by the branching veins and then falls back, thus going and coming between high and low, sometimes inside, sometimes outside. It acts like the blood of animals which is always moving, starting from the sea of the heart and mounting to the summit of the head (Cornford 1937, p. 330).

The earth's venous system was filled with metals and minerals. Its veins, veinlets, seams, and canals coursed through the entire earth, particularly in the mountains. Its humors flowed from the veinlets into the larger veins. The earth, like the human, even had its own elimination system. The tendency for both to break wind caused earthquakes in the case of the former and another type of quake in the latter:

> The material cause of earthquakes...is no doubt great abundance of wind, or store of gross and dry vapors, and spirits, fast shut up, and as a man would say, emprisoned in the caves, and dungeons of the earth; which wind, or vapors, seeking to be set at liberty, and to get them home to their natural lodgings, in a great fume, violently rush out, and as it were, break prison, which forcible eruption, and strong breath, causeth an earthquake (Gabriel Harvey quoted in Kendrick 1974, p. 542, spelling modernized).

Its bowels were full of channels, fire chambers, glory holes, and fissures through which fire and heat were emitted, some in the form of fiery volcanic exhalations, others as hot water springs. The most commonly used analogy, however, was between the female's reproductive and nurturing capacity and the mother earth's ability to give birth to stones and metals within its womb through its marriage with the sum.

In his *De Rerum Natura* of 1565, the Italian philosopher Bernardino Telesio referred to the marriage of the two great male and female powers: "We can see that the sky and the earth are not merely large parts of the world universe, but are of primary – even principal rank....They are like mother and father to all the others" (Telesio 1967, p. 308). The earth and the sun served as mother and father to the whole of creation: all things are "made of earth by the sun and that in the constitution of all things the earth and the sun enter respectively as mother and father." According to Giordano Bruno (1548–1600), every human being was "a citizen and servant of the world, a child of Father Sun and Mother Earth" (Bruno 1964, p. 72).

A widely held alchemical belief was the growth of the baser metals into gold in womblike matrices in the earth. The appearance of silver in lead ores or gold in silvery assays was evidence that this transformation was underway. Just as the child grew in the warmth of the female womb, so the growth of metals was fostered through the agency of heat, some places within the earth's crust being hotter and therefore hastening the maturation process. "Given to gold, silver, and the other metals [was] the vegetative powers whereby they could also reproduce themselves. For, since it was impossible for God to make anything that was not perfect, he gave to all created things, with their being, the power of multiplication." The sun acting on the earth nurtured not only the plants and animals but also "the metals, the broken sulfuric, bituminous, or nitrogenous rocks;...as well as the plants and animals – if these are not made of earth by the sun, one cannot imagine of what else or by what other agent they could be made" (Telesio 1967, p. 309).

The earth's womb was the matrix or mother not only of metals but also of living things. Paracelsus compared the earth to a female whose womb nurtured all life.

> Woman is like the earth and all the elements and in this sense she may be considered a matrix; she is the tree which grows in the earth and the child is like the fruit born of the tree.... Woman is the image of the tree. Just as the earth, its fruits, and the elements are created for the sake of the tree and in order to sustain it, so the members of woman, all her qualities, and her whole nature exist for the sake of her matrix, her womb....
>
> And yet woman in her own way is also a field of the earth and not at all different from it. She replaces it, so to speak: she is the field and the garden mold in which the child is sown and planted (Paracelsus 1951, p. 25).

The earth in the Paracelsian philosophy was the mother or matrix giving birth to plants, animals, and men.

The image of the earth as a nurse, which had appeared in the ancient world in Plato's *Timaeus* and the *Emerald Tablet* of Hermes Trismegistus, was a popular Renaissance metaphor. According to sixteenth-century alchemist Basil Valentine, all things grew in the womb of the earth, which was alive, and vital, and the nurse of all life:

> The quickening power of the earth produces all things that grow forth from it, and he who says that the earth has no life makes a statement flatly contradicted by facts. What is dead cannot produce life and growth, seeing that it is devoid of the quickening spirit.... This spirit is the life and soul that dwell in the earth, and are nourished by heavenly and sidereal influences.... This spirit is itself fed by the stars and is thereby rendered capable of imparting nutriment to all things that grow and of nursing them as a mother does her child while it is yet in the womb.... If the earth were deserted by this spirit it would be dead (Valentine 1974, p. 333).

In general, the Renaissance view was that all things were permeated by life, there being no adequate method by which to designate the inanimate from the animate. It was difficult to differentiate between living and nonliving things because of the resemblance in structures. Like plants and animals, minerals and gems were filled with small pores, tubelets, cavities, and streaks through which they seemed to nourish themselves. Crystalline salts were compared to plant forms, but criteria by which to differentiate the living from the nonliving could not successfully be formulated. This was due not only to the vitalistic framework of the period but to striking similarities between them. Minerals were thought to possess a lesser degree of the vegetative soul, because they had the capacity for medicinal action and often took the form of various parts of plants. By virtue of the vegetative soul, minerals and stones grew in the human body, in animal bodies, within trees, in the air and water, and on the earth's surface in the open country (Adams 1938, pp. 102–36).

Popular Renaissance literature was filled with hundreds of images associating nature, matter, and the earth with the female sex. The earth was alive and considered to be a beneficient, receptive, nurturing female. For most writers, there was a mingling of traditions based on ancient sources. In general, the pervasive animism of nature created a relationship of immediacy with the human being. An I-thou relationship in which nature was considered to be a person-writ-large was sufficiently prevalent that the ancient tendency to treat it as another human still existed. Such vitalistic imagery was thus so widely accepted by the Renaissance mind that it could effectively function as a restraining ethic.

In much the same way, the cultural belief-systems of many American Indian tribes had for centuries subtly guided group behavior toward nature. Smohalla of the Columbia Basin Tribes voiced the Indian objections to European attitudes in the mid-1800s:

> You ask me to plow the ground! Shall I take a knife and tear my mother's breast? Then when I die she will not take me to her bosom to rest.
>
> You ask me to dig for stone! Shall I dig under her skin for her bones? Then when I die I cannot enter her body to be born again.
>
> You ask me to cut grass and make hay and sell it, and be rich like white men! But how dare I cut off my mother's hair? (quoted in McLuhan 1971, p. 56).

In the 1960s, the Native American became a symbol in the ecology movement's search for alternatives to Western exploitative attitudes. The Indian animistic belief-system and reverence for the earth as a mother were constrasted with the Judeo-Christian heritage of dominion over nature and with capitalist practices resulting in the "tragedy of the commons" (exploitation of resources available for any person's or nation's use). But as will be seen, European culture was more complex and varied than this judgment allows. It ignores the Renaissance philosophy of the nurturing earth as well as those philosophies and social movements resistant to mainstream economic change.

Normative Constraints against the Mining of Mother Earth

If sixteenth-century descriptive statements and imagery can function as an ethical constraint and if the earth was widely viewed as a nurturing mother, did such imagery actually function as a norm against improper use of the earth? Evidence that this was indeed the case can be drawn from theories of the origins of metals and the debates about mining prevalent during the sixteenth century.

What ethical ideas were held by ancient and early modern writers on the extraction of the metals from the bowels of the living earth? The Roman compiler Pliny (A.D. 23–79), in his *Natural History*, had specifically warned against mining the depth of Mother Earth, speculating that earthquakes were an expression of her indignation at being violated in this manner:

> We trace out all the veins of the earth, and yet...are astonished that it should occasionally cleave asunder or tremble: as though, forsooth, these signs could be any other than expressions of the indignation felt by our sacred parent! We penetrate into her entrails, and seek for treasures...as though each spot we tread upon were not sufficiently bounteous and fertile for us! (Pliny 1858, vol. 6, pp. 68–69)

He went on to argue that the earth had concealed from view that which she did not wish to be disturbed, that her resources might not be exhausted by human avarice:

> For it is upon her surface, in fact, that she has presented us with these substances, equally with the cereals, bounteous and ever ready, as she is, in supplying us with all things for our benefit! It is what is concealed from our view, what is sunk far beneath her surface, objects, in fact, of no rapid formation, that urge us to our ruin, that send us to the very depth of hell....when will be the end of thus exhausting the earth, and to what point will avarice finally penetrate! (Pliny 1858, vol. 6, p. 69)

Here, then, is a striking example of the restraining force of the beneficent mother image – the living earth in her wisdom has ordained against the mining of metals by concealing them in the depths of her womb.

While mining gold led to avarice, extracting iron was the source of human cruelty in the form of war, murder, and robbery. Its use should be limited to agriculture and those activities that contributed to the "honors of more civilized life":

> For by the aid of iron we lay open the ground, we plant trees, we prepare our vineyard trees, and we force our vines each year to resume their youthful state, by cutting away their decayed branches. It is by the aid of iron that we construct houses, cleave rocks, and perform so many other useful offices of life. But it is with iron also that wars, murders, and robberies are effected,...not only hand to hand, but...by the aid of missiles and winged weapons, now launched from engines, now hurled by the human arm, and now furnished with feathery wings. Let us therefore acquit nature of a charge that here belongs to man himself (Pliny 1858, vol. 6, p. 205).

In past history, Pliny stated, there had been instances in which laws were passed to prohibit the retention of weapons and to ensure that iron was used solely for innocent purposes, such as the cultivation of fields.

In the *Metamorphoses* (A.D. 7), the Roman poet Ovid wrote of the violence done to the earth during the age of iron, when evil was let loose in the form of trickery, slyness, plotting, swindling, and violence, as men dug into the earth's entrails for iron and gold:

> The rich earth
> Was asked for more; they dug into her vitals.
> Pried out the wealth a kinder lord had hidden
> In stygian shadow, all that precious metal,
> The root of evil. They found the guilt of iron.
> And gold, more guilty still. And War
> came forth.
> (Ovid 1955, I. 137–43)

The violation of Mother Earth resulted in new forms of monsters, born of the blood of her slaughter:

> Jove struck them down
> With thunderbolts, and the bulk of
> those huge bodies
> Lay on the earth, and bled, and Mother
> earth,
> Made pregnant by that blood, brought
> forth new bodies,
> And gave them, to recall her older offspring,
> The forms of men. And this new stock
> was also
> Contemptuous of gods, and murder-hungry
> And violent. You would know they
> were sons of blood.
> (Ovid 1955, I. 155–62)

Seneca also deplored the activity of mining, although, unlike Pliny and Ovid, he did not consider it a new vice, but one that had been handed down from ancient times. "What necessity caused man, whose head points to the stars, to stoop, below, burying him in mines and plunging him in the very bowels of innermost earth to root up gold?" Not only did mining remove the earth's treasures, but it created "a sight to make [the] hair stand on end – huge rivers and vast reservoirs of sluggish waters." The defiling of the earth's waters was even then a noteworthy consequence of the quest for metals (Seneca 1910, p. 207–08).

These ancient strictures against mining were still operative during the early years of the commercial revolution when mining activities, which had lapsed after the fall of Rome, were once again revived. Ultimately, such constraints would have to be defeated by proponents of the new mercantilist philosophy.

An allegorical tale, reputedly sent to Paul Schneevogel, a professor at Leipzig about 1490–1495, expressed opposition to mining encroachments into the farmlands of Lichtenstat in Saxony, Germany, an area where the new mining activities were developing rapidly. In the following allegorical vision of an old hermit of Lichtenstat, Mother Earth is dressed in a tattered green robe and seated on the right hand of Jupiter who is represented in a court case by "glib-tongued Mercury" who charges a miner with matricide. Testimony is presented by several of nature's deities:

> Bacchus complained that his vines were uprooted and fed to the flames and his most sacred places desecrated. Ceres stated that her fields were devastated; Pluto that the blows of the miners resound like thunder through the depths of the earth, so that he could hardly reside in his own kingdom; the Naiad, that the subterranean waters were diverted and her fountains dried up; Charon that the volume of the underground waters had been so diminished that he was unable to float his boat on Acheron and carry the souls across to Pluto's realm, and the Fauns protested that the charcoal burners had destroyed whole forests to obtain fuel to smelt the miner's ores (Adams 1938, p. 172).

In his defense, the miner argued that the earth was not a real mother, but a wicked stepmother who hides and conceals the metals in her inner parts instead of making them available for human use.

In the old hermit's tale, we have a fascinating example of the relationship between images and values. The older view of nature as a kindly mother is challenged by the growing interests of the mining industry in Saxony, Bohemia, and the Harz Mountains, regions of newly found prosperity. The miner, representing these newer commercial activities, transforms the image of the nurturing mother into that of a stepmother who wickedly conceals her bounty from the deserving and needy children.

Henry Cornelius Agrippa's polemic *The Vanity of Arts and Sciences* (1530) reiterated some of the moral strictures against mining found in the ancient treatises, quoting the passage from Ovid portraying miners digging into the bowels of the earth in order to extract gold and iron. "These men," he declared, "have made the very ground more hurtful and pestiferous, by how much they are more rash and venturous than they that hazard themselves in the deep to dive for pearls." Mining thus despoiled the earth's surface, infecting it, as it were, with an epidemic disease (Agrippa 1694, pp. 81–82).

If mining were to be freed of such strictures and sanctioned as a commercial activity, the ancient arguments would have to be refuted. This task was taken up by Georg Agricola (1494–1555), who wrote the first "modern" treatise on mining. His *De Re Metallica* ("On Metals," 1556) marshaled the arguments of the detractors of mining in order to refute them and thereby promote the activity itself.

According to Agricola, people who argued against the mining of the earth for metals did so on the basis that nature herself did not wish to be discovered what she herself had concealed:

> The earth does not conceal and remove from our eyes those things which are useful and necessary to mankind, but, on the contrary, like a beneficent and kindly mother she yields in large abundance from her bounty and brings into the light of day the herbs, vegetables, grains, and fruits, and trees. The minerals, on the other hand, she buries far beneath in the depth of the ground, therefore they should not be sought (Agricola 1950, pp. 6–7).

This argument, taken directly from Pliny, reveals the normative force of the image of the earth as a nurturing mother.

A second argument of the detractors, reminiscent of Seneca and Agrippa, and based on Renaissance "ecological" concerns was the disruption of the natural environment and the pollutive effects of mining.

> But, besides this, the strongest argument of the detractors [of mining] is that the fields are devastated by mining operations, for which reason formerly Italians were warned by law that no one should dig the earth for metals and so injure their very fertile fields, their vineyards, and their olive groves. Also they argue that the woods and groves are cut down, for there is need of wood for timbers, machines, and the smelting of metals. And when the woods and groves are felled, then are exterminated the beasts and birds, many of which furnish a pleasant and agreeable food for man. Further, when the ores are washed, the water which has been used poisons the brooks and streams, and either destroys the fish or drives them away. Therefore the inhabitants of these regions, on account of the devastation of their fields, woods, groves, brooks, and rivers, find great difficulty in procuring the necessaries of life, and by reason of the destruction of the timber they are forced to greater expense in erecting buildings. Thus it is said, it is clear to all that there is greater detriment from mining than the value of the metals which the mining produces (Agricola 1950, p. 8).

Agricola may have been alluding to laws passed by the Florentines between 1420 and 1485, preventing people from dumping lime into rivers upstream from the city for the purpose of "poisoning or catching fish," as it caused severe problems for those living downstream. The laws were enacted both to preserve the trout, "a truly noble and impressive fish" and to provide Florence with "a copious and abundant supply of such fish" (Trexler 1974, p. 463).

Such ecological consciousness, however, suffered because of the failure of law enforcement, as well as because of the continuing progress of mining activities. Agricola, in his response to the detractors of mining, pointed out the congruences in the need to catch fish and to construct metal tools for the well-being of the human race. His effort can be interpreted as an attempt to liberate the activity of mining from the constraints imposed by the organic framework and the nurturing earth image, so that new values could sanction and hasten its development.

To the argument that the woods were cut down and the price of timber therefore raised, Agricola responded that most mines occurred in unproductive, gloomy areas. Where the trees were removed from more productive sites, fertile fields could be created, the profits from which would reimburse the local inhabitants for their losses in timber supplies. Where the birds and animals had been destroyed by mining operations, the profits could be used to purchase "birds without number" and "edible beasts and fish elsewhere" and refurbish the area (Agricola 1950, p. 17).

The vices associated with the metals – anger, cruelty, discord, passion for power, avarice, and lust – should be attributed instead to human conduct: "It is not the metals which are to be blamed, but the evil passions of men which become inflamed and ignited; or it is due to the blind and impious desires of their minds." Agricola's arguments are a conscious attempt to separate the older normative constraints from the image of the metals themselves so that new values can then surround them (Agricola 1950, p. 16).

Edmund Spenser's treatment of Mother Earth in the *Faerie Queene* (1595) was representative of the concurrent conflict of attitudes about mining the earth. Spenser entered fully into the sixteenth-century debates about the wisdom of mining, the two greatest sins against the earth being, according to him, avarice and lust. The arguments associating mining with avarice had appeared in the ancient texts of Pliny, Ovid, and Seneca, while during Spenser's lifetime the sermons of Johannes Mathesius, entitled *Beregpostilla, oder Sarepta* (1578), inveighed against the moral consequences of human greed for the wealth created by mining for metals (Kendrick 1974, pp. 548–53).

In Spenser's poem, Guyon presents the arguments against mining taken from Ovid and Agricola, while the description of Mammon's forge is drawn from the illustrations to the *De Re Metallica*. Gold and silver pollute the spirit and debase human values just as the mining operation itself pollutes the "purest streams" of the earth's womb:

> Then gan a cursed hand the quiet wombe
> Of his great Grandmother with steele to wound,
> And the hid treasures in her sacred tombe
> With Sacrilege to dig. Therein he found
> Fountaines of gold and silver to abound,
> Of which the matter of his huge desire
> And pompous pride eftsoones he did compound.
>
> (Spenser 1758, Bk II, Canto 7, verse 17)

The earth in Spenser's poem is passive and docile, allowing all manner of assault, violence, ill-treatment, rape by lust, and despoilment by greed. No longer a nurturer, she indiscriminately, as in Ovid's verse, supplies flesh to all life and lacking in judgment brings forth monsters and evil creatures. Her offspring fall and bite her in their own death throes. The new mining activities have altered the earth from a bountiful mother to a passive receptor of human rape (Kendrick 1974).

John Milton's *Paradise Lost* (1667) continues the Ovidian image, as Mammon leads "bands of pioneers with Spade and Pickaxe" in the wounding of the living female earth:

> ...By him first
> Men also, and by his suggestion taught.
> Ransack'd the Center, and with impious
> hands
> Rifl'd the bowels of their mother Earth
> For Treasures better hid. Soon had his crew
> Op'nd into the Hill a spacious wound
> And dig'd out ribs of Gold.
> (Milton 1975, Bk I, 11. 684–90)

Not only did mining encourage the mortal sin of avarice, it was compared by Spenser to the second great sin, human lust. Digging into the matrices and pockets of earth for metals was like mining the female flesh for pleasure. The sixteenth- and seventeenth-century imagination perceived a direct correlation between mining and digging into the nooks and crannies of a woman's body. Both mining and sex represent for Spenser the return to animality and earthly slime. In the *Faerie Queene*, lust is the basest of all human sins. The spilling of human blood, in the rush to rape the earth of her gold, taints and muddies the once fertile fields (Kendrick 1974).

The sonnets of the poet and divine John Donne (1573–1631) also played up the popular identity of mining with human lust. The poem "Love's Alchemie" begins with the sexual image, "Some that have deeper digged loves Myne than I,/say where his centrique happiness doth lie" (Donne 1957, p. 35). The Platonic lover, searching for the ideal or "centrique" experience of love, begins by digging for it within the female flesh, an act as debasing to the human being as the mining of metals is to the female earth. Happiness is not to be obtained by avarice for gold and silver, nor can the alchemical elixir be produced from base metals. Nor does ideal love result from an ascent up the hierarchical ladder from base sexual love to the love of poetry, music, and art to the highest Platonic love of the good, virtue, and God. The same equation appears in Elegie XVIII, "Love's Progress":

> Search every sphaere
> And firmament, our Cupid is not there;
> He's an infernal god and under ground,
> With Pluto dwells, where gold and fire
> abound:
> Men to such Gods, their sacrificing Coles,
> Did not in Altars lay, but pits and holes,
> Although we see Celestial bodies move
> Above the earth, the earth we Till and love:
> So we her ayres contemplate, words and heart
> And Virtues; but we love the Centrique part.
> (Donne 1957, p. 104, 11. 27–36)

Lust and love of the body do not lead to the celestial love of higher ideals; rather, physical love is associated with the pits and holes of the female body, just as the love of gold depends on the mining of Pluto's caverns within the female earth, "the earth we till and love." Love of the sexual "centrique" part of the female will not lead to the aery spiritual love of virtue. The fatal association of monetary revenue with human avarice, lust, and the female mine is driven home again in the last lines of the poem:

> Rich Nature hath in women wisely made
> Two purses, and their mouths aversely laid:
> They then, which to the lower tribute owe,
> That way which that Exchequer looks,
> must go.

Avarice and greed after money corrupted the soul, just as lust after female flesh corrupted the body.

The comparison of the female mine with the new American sources of gold, silver, and precious metals appears again in Elegie XIX, "Going to Bed." Here, however, Donne turns the image upside down and uses it to extoll the virtues of the mistress.

> License my roaving hands, and let them go,
> Before, behind, between, above, below.
> O my America! my new-found-land,
> My kingdome, safeliest when with one
> man man'd
> My Myne of precious stones, My Emperie,
> How blest am I in this discovering thee!
> (Donne 1957, p. 107, 11. 25–30)

In these lines, the comparison functions as a sanction – the search for precious gems and metals, like the sexual exploration of nature or the female, can benefit a kingdom or a man.

Moral restraints were thus clearly affiliated with the Renaissance image of the female earth and were strengthened by associations with greed, avarice, and lust. But the analogies were double-edged. If the new values connected with mining were positive, and mining was viewed as a means to improve the human condition, as they were by Agricola, then the comparison could be turned upside down. Sanctioning mining sanctioned the rape or technological exploration of the earth. The organic framework, in which the Mother-Earth image was a moral restraint against mining, was literally being undermined by the new commercial activity.

In the seventeenth century, Francis Bacon carried the new ethic a step further through metaphors that compared miners and smiths to scientists and technologists penetrating nature and shaping her on the anvil. Bacon's new man of science must not think that the "inquisition of nature is in any part interdicted or forbidden." Nature must be "bound into service" and made a "slave," put "in constraint" and "molded" by the mechanical arts. The "searchers and spies of nature" are to discover her plots and secrets (Bacon 1870, vol. 4, pp. 20, 287, 294).

This method, so readily applicable when nature is denoted by the female gender, degraded and made possible the exploitation of the natural environment. Nature's womb harbored secrets that through technology could be wrested from her grasp for use in the improvement of the human condition:

> There is therefore much ground for hoping that there are still laid up in the womb of nature many secrets of excellent use having no affinity or parallelism with anything that is now known...only by the method which we are now treating can they be speedily and suddenly and simultaneously presented and anticipated (quoted in Marsak 1964, p. 45).

The final step was to recover and sanction man's dominion over nature. Due to the Fall from the Garden of Eden (caused by the temptation of a woman), the human race lost its "dominion over creation." Before the Fall, there was no need for power or dominion, because Adam and Eve had been made sovereign over all other creatures. In this state of dominion, mankind was "like unto God." While some, accepting God's punishment, had obeyed the medieval strictures against searching too deeply into God's secrets, Bacon turned the constraints into sanctions. Only by "digging further and further into the mine of natural knowledge" could mankind recover that lost dominion. In this way, "the narrow limits of man's dominion over the universe" could be stretched "to their promised bounds" (Bacon 1870, vol. 4, p. 247, vol. 3, pp. 217, 219; Bacon 1964, p. 62).

Although a female's inquisitiveness may have caused man's fall from his god-given dominion, the relentless interrogation of another female, nature, could be used to regain it. As he argued in *The Masculine Birth of Time*, "I am come in very truth leading to you nature with all her children to bind her to your service and make her your slave." "We have no right," he asserted, "to expect nature to come to us." Instead, "Nature must be taken by the forelock, being bald behind." Delay and subtle argument "permit one only to clutch at nature, never to lay hold of her and capture her" (Bacon 1964, pp. 62, 129, 130).

Nature existed in three states – at liberty, in error, or in bondage:

> She is either free and follows her ordinary course of development as in the heavens, in the animal and vegetable creation, and in the general array of the universe; or she is driven out of her ordinary course by the perverseness, insolence, and forwardness of matter and violence of impediments, as in the case of monsters; or lastly, she is put in constraint, molded, and made as it were new by art and the hand of man; as in things artificial (Bacon 1870, vol. 4, p. 294).

The first instance was the view of nature as immanent self-development, the nature naturing herself of the Aristotelians. This was the organic view of nature as a living, growing, self-actualizing being. The second state was necessary to explain the malfunctions and monstrosities that frequently appeared and that could not have been caused by God or another higher power acting on his instruction. Since monstrosities could not be explained by the action of form or spirit, they had to be the result of matter acting perversely. Matter in Plato's *Timaeus* was recalcitrant and had to be forcefully shaped by the demiurge. Bacon frequently described matter in female imagery, as a "common harlot." "Matter is not devoid of an appetite and inclination to dissolve the world and fall back into the old chaos." It therefore must be "restrained and kept in order by the prevailing concord of things." "The vexations of art are certainly as the bonds and handcuffs of Proteus, which betray the ultimate struggles and efforts of matter" (Bacon 1870, vol. 4, pp. 320, 325, 257).

The third instance was the case of art (techne) – man operating on nature to create something new and artificial. Here, "nature takes orders from man and works under his authority." Miners and smiths should become the model for the new class of natural philosophers who would interrogate and alter nature. They had developed the two most important methods of wresting nature's secrets from her, "the one searching into the bowels of nature, the other shaping nature as on an anvil." "Why should we not divide natural philosophy into two parts, the mine and the furnace?" For "the truth of nature lies hid in certain deep mines and caves," within the earth's bosom. Bacon, like some of the practically minded alchemists, would "advise the studious to sell their books and build furnaces" and, "forsaking Minerva and the Muses as barren virgins, to rely upon Vulcan" (Bacon 1870, vol. 4, pp. 343, 287, 343, 393).

The new method of interrogation was not through abstract notions, but through the instruction of the understanding "that it may in very truth dissect nature." The instruments of the mind supply suggestions, those of the hand give motion and aid the work. "By art and the hand of man," nature can then be "forced out of her natural state and squeezed and molded." In this way, "human knowledge and human power meet as one" (Bacon 1870, vol. 4, pp. 246, 29, 247).

Here, in bold sexual imagery, is the key feature of the modern experimental method – constraint of nature in the laboratory, dissection by hand and mind, and the penetration of hidden secrets – language still used today in praising a scientist's "hard facts," "penetrating mind," or the "thrust of his argument." The constraints against mining the earth have been turned into sanctions in language that legitimates the exploitation and "rape" of nature for human good.

Scientific method, combined with mechanical technology, would create a "new organon," a new system of investigation, that unified knowledge with material power. The technological discoveries of printing, gunpowder, and the magnet in the fields of learning, warfare, and navigation "help us to think about the secrets still locked in nature's bosom." "They do not, like the old, merely exert a gentle guidance over nature's course; they have the power to conquer and subdue her, to shake her to her foundations." Under the mechanical arts, "nature betrays her secrets more fully...than when in enjoyment of her natural liberty" (Bacon 1964, pp. 96, 93, 99).

Mechanics, which gave man power over nature, consisted in motion; that is, in "the uniting or disuniting of natural bodies." Most useful were the arts that altered the materials of things – "agriculture, cookery, chemistry, dying, the manufacture of glass, enamel, sugar, gunpowder, artificial fires, paper, and the like." But in performing these operations, one was constrained to operate within the chain of causal connections; nature could "not be commanded except by being obeyed." Only by the study, interpretation, and observation of nature could these possibilities be uncovered; only by acting as the interpreter of nature could knowledge be turned into power. Of the three grades of human ambition, the most wholesome and noble was "to endeavor to establish and extend the power and dominion of the human race itself over the universe." In this way, "the human race [could] recover that right over nature which belongs to it by divine bequest" (Bacon 1870, Vol. 4, pp. 294, 257, 32, 114, 115).

By the close of the seventeenth century, a new science of mechanics in combination with the Baconian ideal of technological mastery over Nature had helped to create the modern worldview. The core of female principles that had for centuries subtly guided human behavior toward the earth had given way to a new ethic of exploitation. The nurturing earth mother was subdued by science and technology.

References

Adams, Frank D. 1938. *The birth and development of the geological sciences*. New York: Dover.

Agricola, Georg. 1950. *De re metallica*, 1556. trans. Herbert C. Hoover and Lou H. Hoover. New York: Dover.

Agrippa, Henry C. 1694. *The vanity of arts and sciences*. London. (Orig. publ. in Latin, 1530.)

Bacon, Francis. 1870. *Works*, ed. James Spedding, Robert L. Ellis, Douglas D. Heath. 14 vols. London: Longmans Green.

Bacon, Francis. 1964. *The philosophy of Francis Bacon*, ed. and trans. Benjamin Farrington. Liverpool, England: Liverpool University Press.

Bruno, Giordano. 1964. *The expulsion of the triumphant beast*, 1584, ed. and trans. Arthur D. Imerti. New Brunswick, N.J.: Rutgers University Press.

Cavell, Stanley. 1971. Must we mean what we say? *Philosophy and linguistics*, ed. Colin Lyas. London: Macmillan: 131–65.

Cicero, M. Tullius. 1775. *Of the nature of the gods*, ed. T. Francklin. London.

Cornford, Francis M. 1937. *Plato's cosmology*. New York: Liberal Arts Press.

Donne, John. 1957. *Poems of John Donne*, ed. Herbert Grierson. London: Oxford University Press.

Eliade, Mircea. 1962. *The forge and the crucible*, trans. Stephan Corrin. New York: Harper & Row.

Kendrick, Walter M. 1974. Earth of flesh, flesh of earth: Mother Earth in the *Faerie Queene*. *Renaissance Quarterly* 27:33–48.

Marsak, Leonard M., ed. 1964. *The rise of modern science in relation to society*. London: Collier-Macmillan.

McLuhan, T. C. 1971. *Touch the earth*. New York: Simon & Schuster.

Merchant, Carolyn. 1980. *The Death of Nature: Women, Ecology, and the Scientific Revolution*. New York: Harper & Row.

Milton, John. 1975. *Paradise lost*, 1667, ed., Scott Elledge. New York: W. W. Norton.

Ovid, Publius. 1955. *Metamorphoses*, A.D. 7, trans. Rolfe Humphries. Bloomington: Indiana University Press.

Paracelsus, Theophrastus. 1951. *Selected Writings*, ed. J. Jacobi. Princeton, N.J.: Princeton University Press.

Pliny. 1858. *Natural history*, ca. A.D. 23–79, trans. J. Bostock and H. T. Riley. London: Bohn.

Seneca, Lucius. 1910. *Physical science in the time of Nero: being a translation of the Quaestiones naturales of Seneca*, ca. A.D. 65, trans. John Clarke. London: Macmillan.

Spenser, Edmund. 1758. *The Faerie Queene*, 1590–95, ed. John Upton, 2 vols. London. Vol. 1.

Taylor, Charles. 1973. Neutrality in political science. *The Philosophy of social explanation*, ed. Alan Ryan. London: Oxford: 139–70.

Telesio, Bernardino. 1967. *De rerum natura iuxta propia principia*, 1587 (first published 1565). *Renaissance Philosophy* ed. and trans. Arturo B. Fallico and Herman Shapiro. New York: Modern Library.

Trexler, Richard. 1974. Measures against water pollution in fifteenth-century Florence. *Viator* 5:455–67.

Valentine, Basil. 1974. *The practica, with twelve keys*, 1678. In *The hermetic museum restored and enlarged*, trans. Arthur E. Waite. New York: Samuel Weiser.

37

A Cyborg Manifesto: Science, Technology, and Socialist-Feminism in the Late Twentieth Century

Donna Haraway

An Ironic Dream of a Common Language for Women in the Integrated Circuit

This chapter is an effort to build an ironic political myth faithful to feminism, socialism, and materialism. Perhaps more faithful as blasphemy is faithful, than as reverent worship and identification. Blasphemy has always seemed to require taking things very seriously. I know no better stance to adopt from within the secular-religious, evangelical traditions of United States politics, including the politics of socialist feminism. Blasphemy protects one from the moral majority within, while still insisting on the need for community. Blasphemy is not apostasy. Irony is about contradictions that do not resolve into larger wholes, even dialectically, about the tension of holding incompatible things together because both or all are necessary and true. Irony is about humour and serious play. It is also a rhetorical strategy and a political method, one I would like to see more honoured within socialist-feminism. At the centre of my ironic faith, my blasphemy, is the image of the cyborg.

A cyborg is a cybernetic organism, a hybrid of machine and organism, a creature of social reality as well as a creature of fiction. Social reality is lived social relations, our most important political construction, a world-changing fiction. The international women's movements have constructed "women's experience", as well as uncovered or discovered this crucial collective object. This experience is a fiction and fact of the most crucial, political kind. Liberation rests on the construction of the consciousness, the imaginative apprehension, of oppression, and so of possibility. The cyborg is a matter of fiction and lived experience that changes what counts as women's experience in the late twentieth century. This is a struggle over life and death, but the boundary between science fiction and social reality is an optical illusion.

Contemporary science fiction is full of cyborgs – creatures simultaneously animal and machine, who populate worlds ambiguously natural and crafted. Modern medicine is also full of cyborgs, of couplings between organism and machine, each conceived as coded devices, in an intimacy and with a power that was not generated in the history of sexuality. Cyborg "sex" restores some of the lovely replicative baroque of ferns and invertebrates (such nice organic prophylactics against heterosexism). Cyborg replication is uncoupled from organic reproduction. Modern production seems like a dream of cyborg colonization work, a dream that makes the nightmare of Taylorism seem idyllic. And modern war is a cyborg orgy, coded by C^3I, command-control-communication-intelligence, an $84 billion item in 1984's US defence budget. I am making an argument for the cyborg as a fiction mapping our social and bodily reality and as an imaginative resource suggesting some very fruitful couplings. Michel Foucault's biopolitics is a flaccid premonition of cyborg politics, a very open field.

By the late twentieth century, our time, a mythic time, we are all chimeras, theorized and fabricated

From Donna Haraway, *Simians, Cyborgs and Women: The Reinvention of Nature*, New York: Routledge and Institute for Social Research and Education, 1991, pp. 149–81, 243–8, 255–76, abridged.

hybrids of machine and organism; in short, we are cyborgs. The cyborg is our ontology; it gives us our politics. The cyborg is a condensed image of both imagination and material reality, the two joined centres structuring any possibility of historical transformation. In the traditions of "Western" science and politics – the tradition of racist, male-dominant capitalism; the tradition of progress; the tradition of the appropriation of nature as resource for the productions of culture; the tradition of reproduction of the self from the reflections of the other – the relation between organism and machine has been a border war. The stakes in the border war have been the territories of production, reproduction, and imagination. This chapter is an argument for *pleasure* in the confusion of boundaries and for *responsibility* in their construction. It is also an effort to contribute to socialist-feminist culture and theory in a postmodernist, non-naturalist mode and in the utopian tradition of imagining a world without gender, which is perhaps a world without genesis, but maybe also a world without end. The cyborg incarnation is outside salvation history. Nor does it mark time on an oedipal calendar, attempting to heal the terrible cleavages of gender in an oral symbiotic utopia or post-oedipal apocalypse. As Zoe Sofoulis argues in her unpublished manuscript on Jacques Lacan, Melanie Klein, and nuclear culture, *Lacklein*, the most terrible and perhaps the most promising monsters in cyborg worlds are embodied in non-oedipal narratives with a different logic of repression, which we need to understand for our survival.

The cyborg is a creature in a post-gender world; it has no truck with bisexuality, pre-oedipal symbiosis, unalienated labour, or other seductions to organic wholeness through a final appropriation of all the powers of the parts into a higher unity. In a sense, the cyborg has no origin story in the Western sense – a "final" irony since the cyborg is also the awful apocalyptic *telos* of the "West's" escalating dominations of abstract individuation, an ultimate self united at last from all dependency, a man in space. An origin story in the "Western", humanist sense depends on the myth of original unity, fullness, bliss and terror, represented by the phallic mother from whom all humans must separate, the task of individual development and of history, the twin potent myths inscribed most powerfully for us in psychoanalysis and Marxism. Hilary Klein has argued that both Marxism and psychoanalysis, in their concepts of labour and of individuation and gender formation, depend on the plot of original unity out of which difference must be produced and enlisted in a drama of escalating domination of woman/nature. The cyborg skips the step of original unity, of identification with nature in the Western sense. This is its illegitimate promise that might lead to subversion of its teleology as star wars.

The cyborg is resolutely committed to partiality, irony, intimacy, and perversity. It is oppositional, utopian, and completely without innocence. No longer structured by the polarity of public and private, the cyborg defines a technological polis based partly on a revolution of social relations in the *oikos*, the household. Nature and culture are reworked; the one can no longer be the resource for appropriation or incorporation by the other. The relationships for forming wholes from parts, including those of polarity and hierarchical domination, are at issue in the cyborg world. Unlike the hopes of Frankenstein's monster, the cyborg does not expect its father to save it through a restoration of the garden; that is, through the fabrication of a heterosexual mate, through its completion in a finished whole, a city and cosmos. The cyborg does not dream of community on the model of the organic family, this time without the oedipal project. The cyborg would not recognize the Garden of Eden; it is not made of mud and cannot dream of returning to dust. Perhaps that is why I want to see if cyborgs can subvert the apocalypse of returning to nuclear dust in the manic compulsion to name the Enemy. Cyborgs are not reverent; they do not re-member the cosmos. They are wary of holism, but needy for connection – they seem to have a natural feel for united front politics, but without the vanguard party. The main trouble with cyborgs, of course, is that they are the illegitimate offspring of militarism and patriarchal capitalism, not to mention state socialism. But illegitimate offspring are often exceedingly unfaithful to their origins. Their fathers, after all, are inessential.

I will return to the science fiction of cyborgs at the end of this chapter, but now I want to signal three crucial boundary breakdowns that make the following political-fictional (political-scientific) analysis possible. By the late twentieth century in United States scientific culture, the boundary between human and animal is thoroughly breached. The last beachheads of uniqueness have been polluted if not turned into amusement parks – language, tool use, social behaviour, mental events, nothing really convincingly settles the separation of human and animal. And many people no longer feel the need for such a separation; indeed, many branches of feminist culture affirm the pleasure of connection of human and other living creatures. Movements for animal rights are not irrational denials of human uniqueness; they are a clear-sighted recognition of connection across the discredited breach of nature and culture. Biology and evolutionary theory over the last two centuries have simultaneously produced modern organisms as objects of knowledge and reduced the line between humans and animals to a faint trace re-etched in ideological struggle or professional disputes between life and social science. Within this framework, teaching modern Christian creationism should be fought as a form of child abuse.

Biological-determinist ideology is only one position opened up in scientific culture for arguing the meanings of human animality. There is much room for radical political people to contest the meanings of the breached boundary.[1] The cyborg appears in myth precisely where the boundary between human and animal is transgressed. Far from signalling a walling off of people from other living beings, cyborgs signal disturbingly and pleasurably tight coupling. Bestiality has a new status in this cycle of marriage exchange.

The second leaky distinction is between animal-human (organism) and machine. Pre-cybernetic machines could be haunted; there was always the spectre of the ghost in the machine. This dualism structured the dialogue between materialism and idealism that was settled by a dialectical progeny, called spirit or history, according to taste. But basically machines were not self-moving, self-designing, autonomous. They could not achieve man's dream, only mock it. They were not man, an author to himself, but only a caricature of that masculinist reproductive dream. To think they were otherwise was paranoid. Now we are not so sure. Late twentieth-century machines have made thoroughly ambiguous the difference between natural and artificial, mind and body, self-developing and externally designed, and many other distinctions that used to apply to organisms and machines. Our machines are disturbingly lively, and we ourselves frighteningly inert.

Technological determination is only one ideological space opened up by the reconceptions of machine and organism as coded texts through which we engage in the play of writing and reading the world.[2] "Textualization" of everything in poststructuralist, postmodernist theory has been damned by Marxists and socialist feminists for its utopian disregard for the lived relations of domination that ground the "play" of arbitrary reading.[3] It is certainly true that postmodernist strategies, like my cyborg myth, subvert myriad organic wholes (for example, the poem, the primitive culture, the biological organism). In short, the certainty of what counts as nature – a source of insight and promise of innocence – is undermined, probably fatally. The transcendent authorization of interpretation is lost, and with it the ontology grounding "Western" epistemology. But the alternative is not cynicism or faithlessness, that is, some version of abstract existence, like the accounts of technological determinism destroying "man" by the "machine" or "meaningful political action" by the "text". Who cyborgs will be is a radical question; the answers are a matter of survival. Both chimpanzees and artefacts have politics, so why shouldn't we (de Waal, 1982; Winner, 1980)?

The third distinction is a subset of the second: the boundary between physical and non-physical is very imprecise for us. Pop physics books on the consequences of quantum theory and the indeterminacy principle are a kind of popular scientific equivalent to Harlequin romances[4] as a marker of radical change in American white heterosexuality: they get it wrong, but they are on the right subject. Modern machines are quintessentially microelectronic devices: they are everywhere and they are invisible. Modern machinery is an irreverent upstart god, mocking the Father's ubiquity and spirituality. The silicon chip is a surface for writing; it is etched in molecular scales disturbed only by atomic noise, the ultimate interference for nuclear scores. Writing, power, and technology are old partners in Western stories of the origin of civilization, but miniaturization has changed our experience of mechanism. Miniaturization has turned out to be about power; small is not so much beautiful as pre-eminently dangerous, as in cruise missiles. Contrast the TV sets of the 1950s or the news cameras of the 1970s with the TV wrist bands or hand-sized video cameras now advertised. Our best machines are made of sunshine; they are all light and clean because they are nothing but signals, electromagnetic waves, a section of a spectrum, and these machines are eminently portable, mobile – a matter of immense human pain in Detroit and Singapore. People are nowhere near so fluid, being both material and opaque. Cyborgs are ether, quintessence.

The ubiquity and invisibility of cyborgs is precisely why these sunshine-belt machines are so deadly. They are as hard to see politically as materially. They are about consciousness – or its simulation.[5] They are floating signifiers moving in pickup trucks across Europe, blocked more effectively by the witch-weavings of the displaced and so unnatural Greenham women, who read the cyborg webs of power so very well, than by the militant labour of older masculinist politics, whose natural constituency needs defence jobs. Ultimately the "hardest" science is about the realm of greatest boundary confusion, the realm of pure number, pure spirit, C^3I, cryptography, and the preservation of potent secrets. The new machines are so clean and light. Their engineers are sun-worshippers mediating a new scientific revolution associated with the night dream of post-industrial society. The diseases evoked by these clean machines are "no more" than the minuscule coding changes of an antigen in the immune system, "no more" than the experience of stress. The nimble fingers of "Oriental" women, the old fascination of little Anglo-Saxon Victorian girls with doll's houses, women's enforced attention to the small take on quite new dimensions in this world. There might be a cyborg Alice taking account of these new dimensions. Ironically, it might be the unnatural cyborg women making chips in Asia and spiral dancing in Santa Rita jail[6] whose constructed unities will guide effective oppositional strategies.

So my cyborg myth is about transgressed boundaries, potent fusions, and dangerous possibilities which progressive people might explore as one part of needed political work. One of my premises is that most American socialists and feminists see deepened dualisms of mind and body, animal and machine, idealism and materialism in the social practices, symbolic formulations, and physical artefacts associated with "high technology" and scientific culture. From *One-Dimensional Man* (Marcuse, 1964) to *The Death of Nature* (Merchant, 1980), the analytic resources developed by progressives have insisted on the necessary domination of technics and recalled us to an imagined organic body to integrate our resistance. Another of my premises is that the need for unity of people trying to resist world-wide intensification of domination has never been more acute. But a slightly perverse shift of perspective might better enable us to contest for meanings, as well as for other forms of power and pleasure in technologically mediated societies.

From one perspective, a cyborg world is about the final imposition of a grid of control on the planet, about the final abstraction embodied in a Star Wars apocalypse waged in the name of defence, about the final appropriation of women's bodies in a masculinist orgy of war (Sofia, 1984). From another perspective, a cyborg world might be about lived social and bodily realities in which people are not afraid of their joint kinship with animals and machines, not afraid of permanently partial identities and contradictory standpoints. The political struggle is to see from both perspectives at once because each reveals both dominations and possibilities unimaginable from the other vantage point. Single vision produces worse illusions than double vision or many-headed monsters. Cyborg unities are monstrous and illegitimate; in our present political circumstances, we could hardly hope for more potent myths for resistance and recoupling. I like to imagine LAG, the Livermore Action Group, as a kind of cyborg society, dedicated to realistically converting the laboratories that most fiercely embody and spew out the tools of technological apocalypse, and committed to building a political form that actually manages to hold together witches, engineers, elders, perverts, Christians, mothers, and Leninists long enough to disarm the state. Fission Impossible is the name of the affinity group in my town. (Affinity: related not by blood but by choice, the appeal of one chemical nuclear group for another, avidity.)[7]

[...]

The Informatics of Domination

In this attempt at an epistemological and political position, I would like to sketch a picture of possible unity, a picture indebted to socialist and feminist principles of design. The frame for my sketch is set by the extent and importance of rearrangements in world-wide social relations tied to science and technology. I argue for a politics rooted in claims about fundamental changes in the nature of class, race, and gender in an emerging system of world order analogous in its novelty and scope to that created by industrial capitalism; we are living through a movement from an organic, industrial society to a polymorphous, information system – from all work to all play, a deadly game. Simultaneously material and ideological, the dichotomies may be expressed in the following chart of transitions from the comfortable old hierarchical dominations to the scary new networks I have called the informatics of domination:

Representation	Simulation
Bourgeois novel, realism	Science fiction, postmodernism
Organism	Biotic component
Depth, integrity	Surface, boundary
Heat	Noise
Biology as clinical practice	Biology as inscription
Physiology	Communications engineering
Small group	Subsystem
Perfection	Optimization
Eugenics	Population Control
Decadence, *Magic Mountain*	Obsolescence, *Future Shock*
Hygiene	Stress Management
Microbiology, tuberculosis	Immunology, AIDS
Organic division of labour	Ergonomics/cybernetics of labour
Functional specialization	Modular construction
Reproduction	Replication
Organic sex role specialization	Optimal genetic strategies
Biological determinism	Evolutionary inertia, constraints
Community ecology	Ecosystem
Racial chain of being	Neo-imperialism, United Nations humanism
Scientific management in home/factory	Global factory/Electronic cottage
Family/Market/Factory	Women in the Integrated Circuit
Family wage	Comparable worth
Public/Private	Cyborg citizenship
Nature/Culture	Fields of difference
Co-operation	Communications enhancement
Freud	Lacan
Sex	Genetic engineering
Labour	Robotics
Mind	Artificial Intelligence
Second World War	Star Wars
White Capitalist Patriarchy	Informatics of Domination

This list suggests several interesting things.[8] First, the objects on the right-hand side cannot be coded as "natural", a realization that subverts naturalistic coding for the left-hand side as well. We cannot go back ideologically or materially. It's not just that "god" is dead; so is the "goddess". Or both are revivified in the worlds charged with microelectronic and biotechnological politics. In relation to objects like biotic components, one must think not in terms of essential properties, but in terms of design, boundary constraints, rates of flows, systems logics, costs of lowering constraints. Sexual reproduction is one kind of reproductive strategy among many, with costs and benefits as a function of the system environment. Ideologies of sexual reproduction can no longer reasonably call on notions of sex and sex role as organic aspects in natural objects like organisms and families. Such reasoning will be unmasked as irrational, and ironically corporate executives reading *Playboy* and anti-porn radical feminists will make strange bedfellows in jointly unmasking the irrationalism.

Likewise for race, ideologies about human diversity have to be formulated in terms of frequencies of parameters, like blood groups or intelligence scores. It is "irrational" to invoke concepts like primitive and civilized. For liberals and radicals, the search for integrated social systems gives way to a new practice called "experimental ethnography" in which an organic object dissipates in attention to the play of writing. At the level of ideology, we see translations of racism and colonialism into languages of development and under-development, rates and constraints of modernization. Any objects or persons can be reasonably thought of in terms of disassembly and reassembly; no "natural" architectures constrain system design. The financial districts in all the world's cities, as well as the export-processing and free-trade zones, proclaim this elementary fact of "late capitalism". The entire universe of objects that can be known scientifically must be formulated as problems in communications engineering (for the managers) or theories of the text (for those who would resist). Both are cyborg semiologies.

One should expect control strategies to concentrate on boundary conditions and interfaces, on rates of flow across boundaries – and not on the integrity of natural objects. "Integrity" or "sincerity" of the Western self gives way to decision procedures and expert systems. For example, control strategies applied to women's capacities to give birth to new human beings will be developed in the languages of population control and maximization of goal achievement for individual decision-makers. Control strategies will be formulated in terms of rates, costs of constraints, degrees of freedom. Human beings, like any other component or subsystem, must be localized in a system architecture whose basic modes of operation are probabilistic, statistical. No objects, spaces, or bodies are sacred in themselves; any component can be interfaced with any other if the proper standard, the proper code, can be constructed for processing signals in a common language. Exchange in this world transcends the universal translation effected by capitalist markets that Marx analysed so well. The privileged pathology affecting all kinds of components in this universe is stress – communications breakdown (Hogness, 1983). The cyborg is not subject to Foucault's biopolitics; the cyborg simulates politics, a much more potent field of operations.

This kind of analysis of scientific and cultural objects of knowledge which have appeared historically since the Second World War prepares us to notice some important inadequacies in feminist analysis which has proceeded as if the organic, hierarchical dualisms ordering discourse in "the West" since Aristotle still ruled. They have been cannibalized, or as Zoe Sofia (Sofoulis) might put it, they have been "techno-digested". The dichotomies between mind and body, animal and human, organism and machine, public and private, nature and culture, men and women, primitive and civilized are all in question ideologically. The actual situation of women is their integration/exploitation into a world system of production/reproduction and communication called the informatics of domination. The home, workplace, market, public arena, the body itself – all can be dispersed and interfaced in nearly infinite, polymorphous ways, with large consequences for women and others – consequences that themselves are very different for different people and which make potent oppositional international movements difficult to imagine and essential for survival. One important route for reconstructing socialist-feminist politics is through theory and practice addressed to the social relations of science and technology, including crucially the systems of myth and meanings structuring our imaginations. The cyborg is a kind of disassembled and reassembled, postmodern collective and personal self. This is the self feminists must code.

Communications technologies and biotechnologies are the crucial tools recrafting our bodies. These tools embody and enforce new social relations for women world-wide. Technologies and scientific discourses can be partially understood as formalizations, i.e., as frozen moments, of the fluid social interactions constituting them, but they should also be viewed as instruments for enforcing meanings. The boundary is permeable between tool and myth, instrument and concept, historical systems of social relations and historical anatomies of possible bodies, including objects of knowledge. Indeed, myth and tool mutually constitute each other.

Furthermore, communications sciences and modern biologies are constructed by a common move – *the translation of the world into a problem of coding*, a search for a common language in which all resistance to instrumental control disappears and all heterogeneity can be submitted to disassembly, reassembly, investment, and exchange.

In communications sciences, the translation of the world into a problem in coding can be illustrated by looking at cybernetic (feedback-controlled) systems theories applied to telephone technology, computer design, weapons deployment, or data base construction and maintenance. In each case, solution to the key questions rests on a theory of language and control; the key operation is determining the rates, directions, and probabilities of flow of a quantity called information. The world is subdivided by boundaries differentially permeable to information. Information is just that kind of quantifiable element (unit, basis of unity) which allows universal translation, and so unhindered instrumental power (called effective communication). The biggest threat to such power is interruption of communication. Any system breakdown is a function of stress. The fundamentals of this technology can be condensed into the metaphor C^3I, command-control-communication-intelligence, the military's symbol for its operations theory.

In modern biologies, the translation of the world into a problem in coding can be illustrated by molecular genetics, ecology, sociobiological evolutionary theory, and immunobiology. The organism has been translated into problems of genetic coding and read-out. Biotechnology, a writing technology, informs research broadly.[9] In a sense, organisms have ceased to exist as objects of knowledge, giving way to biotic components, i.e., special kinds of information-processing devices. The analogous moves in ecology could be examined by probing the history and utility of the concept of the ecosystem. Immunobiology and associated medical practices are rich exemplars of the privilege of coding and recognition systems as objects of knowledge, as constructions of bodily reality for us. Biology here is a kind of cryptography. Research is necessarily a kind of intelligence activity. Ironies abound. A stressed system goes awry; its communication processes break down; it fails to recognize the difference between self and other. Human babies with baboon hearts evoke national ethical perplexity – for animal rights activists at least as much as for the guardians of human purity. In the US gay men and intravenous drug users are the "privileged" victims of an awful immune system disease that marks (inscribes on the body) confusion of boundaries and moral pollution (Treichler, 1987).

But these excursions into communications sciences and biology have been at a rarefied level; there is a mundane, largely economic reality to support my claim that these sciences and technologies indicate fundamental transformations in the structure of the world for us. Communications technologies depend on electronics. Modern states, multinational corporations, military power, welfare state apparatuses, satellite systems, political processes, fabrication of our imaginations, labour-control systems, medical constructions of our bodies, commercial pornography, the international division of labour, and religious evangelism depend intimately upon electronics. Micro-electronics is the technical basis of simulacra; that is, of copies without originals.

Microelectronics mediates the translations of labour into robotics and word processing, sex into genetic engineering and reproductive technologies, and mind into artificial intelligence and decision procedures. The new biotechnologies concern more than human reproduction. Biology as a powerful engineering science for redesigning materials and processes has revolutionary implications for industry, perhaps most obvious today in areas of fermentation, agriculture, and energy. Communications sciences and biology are constructions of natural-technical objects of knowledge in which the difference between machine and organism is thoroughly blurred; mind, body, and tool are on very intimate terms. The "multinational" material organization of the production and reproduction of daily life and the symbolic organization of the production and reproduction of culture and imagination seem equally implicated. The boundary-maintaining images of base and superstructure, public and private, or material and ideal never seemed more feeble.

I have used Rachel Grossman's (1980) image of women in the integrated circuit to name the situation of women in a world so intimately restructured through the social relations of science and technology.[10] I used the odd circumlocution, "the social relations of science and technology", to indicate that we are not dealing with a technological determinism, but with a historical system depending upon structured relations among people. But the phrase should also indicate that science and technology provide fresh sources of power, that we need fresh sources of analysis and political action (Latour, 1984). Some of the rearrangements of race, sex, and class rooted in high-tech-facilitated social relations can make socialist-feminism more relevant to effective progressive politics.

The "Homework Economy" Outside "the Home"

The "New Industrial Revolution" is producing a new world-wide working class, as well as new sexualities and ethnicities. The extreme mobility of capital and the emerging international division of labour are intertwined with the emergence of new collectivities, and the weakening of familiar groupings. These developments are neither gender- nor race-neutral. White men in advanced industrial societies have become newly vulnerable to permanent job loss, and women are not disappearing from the job rolls at the same rates as men. It is not simply that women in Third World countries are the preferred labour force for the science-based multinationals in the export-processing sectors, particularly in electronics. The picture is more systematic and involves reproduction, sexuality, culture, consumption, and production. In the prototypical Silicon Valley, many women's lives have been structured around employment in electronics-dependent jobs, and their intimate realities include serial heterosexual monogamy, negotiating childcare, distance from extended kin or most other forms of traditional community, a high likelihood of loneliness and extreme economic vulnerability as they age. The ethnic and racial diversity of women in Silicon Valley structures a microcosm of conflicting differences in culture, family, religion, education, and language.

Richard Gordon has called this new situation the "homework economy".[11] Although he includes the phenomenon of literal homework emerging in connection with electronics assembly, Gordon intends "homework economy" to name a restructuring of work that broadly has the characteristics formerly ascribed to female jobs, jobs literally done only by women. Work is being redefined as both literally female and feminized, whether performed by men or women. To be feminized means to be made extremely vulnerable; able to be disassembled, reassembled, exploited as a reserve labour force; seen less as workers than as servers; subjected to time arrangements on and off the paid job that make a mockery of a limited work day; leading an existence that always borders on being obscene, out of place, and reducible to sex. Deskilling is an old strategy newly applicable to formerly privileged workers. However, the homework economy does not refer only to large-scale deskilling, nor does it deny that new areas of high skill are emerging, even for women and men previously excluded from skilled employment. Rather, the concept indicates that factory, home, and market are integrated on a new scale and that the places of women are crucial – and need to be analysed for differences among women and for meanings for relations between men and women in various situations.

The homework economy as a world capitalist organizational structure is made possible by (not caused by) the new technologies. The success of the attack on relatively privileged, mostly white, men's unionized jobs is tied to the power of the new communications technologies to integrate and control labour despite extensive dispersion and decentralization. The consequences of the new technologies are felt by women both in the loss of the family (male) wage (if they ever had access to this white privilege) and in the character of their own jobs, which are becoming capital-intensive; for example, office work and nursing.

The new economic and technological arrangements are also related to the collapsing welfare state and the ensuing intensification of demands on women to sustain daily life for themselves as well as for men, children, and old people. The feminization of poverty – generated by dismantling the welfare state, by the homework economy where stable jobs become the exception, and sustained by the expectation that women's wages will not be matched by a male income for the support of children – has become an urgent focus. The causes of various women-headed households are a function of race, class, or sexuality; but their increasing generality is a ground for coalitions of women on many issues. That women regularly sustain daily life partly as a function of their enforced status as mothers is hardly new; the kind of integration with the overall capitalist and progressively war-based economy is new. The particular pressure, for example, on US black women, who have achieved an escape from (barely) paid domestic service and who now hold clerical and similar jobs in large numbers, has large implications for continued enforced black poverty *with* employment. Teenage women in industrializing areas of the Third World increasingly find themselves the sole or major source of a cash wage for their families, while access to land is ever more problematic. These developments must have major consequences in the psychodynamics and politics of gender and race.

Within the framework of three major stages of capitalism (commercial/early industrial, monopoly, multinational) – tied to nationalism, imperialism, and multinationalism, and related to Jameson's three dominant aesthetic periods of realism, modernism, and postmodernism – I would argue that specific forms of families dialectically relate to forms of capital and to its political and cultural concomitants. Although lived problematically and unequally, ideal forms of these families might be schematized as (1) the patriarchal nuclear family, structured by the dichotomy between public and private and accompanied by the white bourgeois ideology of separate spheres and nineteenth-century Anglo-American bourgeois feminism; (2) the modern family mediated (or enforced) by the welfare state and institutions like the family wage, with a flowering of a-feminist heterosexual ideologies, including their radical versions represented in

Greenwich Village around the First World War; and (3) the "family" of the homework economy with its oxymoronic structure of women-headed households and its explosion of feminisms and the paradoxical intensification and erosion of gender itself. This is the context in which the projections for world-wide structural unemployment stemming from the new technologies are part of the picture of the homework economy. As robotics and related technologies put men out of work in "developed" countries and exacerbate failure to generate male jobs in Third World "development", and as the automated office becomes the rule even in labour-surplus countries, the feminization of work intensifies. Black women in the United States have long known what it looks like to face the structural underemployment ("feminization") of black men, as well as their own highly vulnerable position in the wage economy. It is no longer a secret that sexuality, reproduction, family, and community life are interwoven with this economic structure in myriad ways which have also differentiated the situations of white and black women. Many more women and men will contend with similar situations, which will make cross-gender and race alliances on issues of basic life support (with or without jobs) necessary, not just nice.

The new technologies also have a profound effect on hunger and on food production for subsistence world-wide. Rae Lessor Blumberg (1983) estimates that women produce about 50 per cent of the world's subsistence food.[12] Women are excluded generally from benefiting from the increased high-tech commodification of food and energy crops, their days are made more arduous because their responsibilities to provide food do not diminish, and their reproductive situations are made more complex. Green Revolution technologies interact with other high-tech industrial production to alter gender divisions of labour and differential gender migration patterns.

The new technologies seem deeply involved in the forms of "privatization" that Ros Petchesky (1981) has analysed, in which militarization, right-wing family ideologies and policies, and intensified definitions of corporate (and state) property as private synergistically interact.[13] The new communications technologies are fundamental to the eradication of "public life" for everyone. This facilitates the mushrooming of a permanent high-tech military establishment at the cultural and economic expense of most people, but especially of women. Technologies like video games and highly miniaturized televisions seem crucial to production of modern forms of "private life". The culture of video games is heavily orientated to individual competition and extraterrestrial warfare. High-tech, gendered imaginations are produced here, imaginations that can contemplate destruction of the planet and a sci-fi escape from its consequences. More than our imaginations is militarized; and the other realities of electronic and nuclear warfare are inescapable. These are the technologies that promise ultimate mobility and perfect exchange – and incidentally enable tourism, that perfect practice of mobility and exchange, to emerge as one of the world's largest single industries.

The new technologies affect the social relations of both sexuality and of reproduction, and not always in the same ways. The close ties of sexuality and instrumentality, of views of the body as a kind of private satisfaction- and utility-maximizing machine, are described nicely in sociobiological origin stories that stress a genetic calculus and explain the inevitable dialectic of domination of male and female gender roles.[14] These sociobiological stories depend on a high-tech view of the body as a biotic component or cybernetic communications system. Among the many transformations of reproductive situations is the medical one, where women's bodies have boundaries newly permeable to both "visualization" and "intervention". Of course, who controls the interpretation of bodily boundaries in medical hermeneutics is a major feminist issue. The speculum served as an icon of women's claiming their bodies in the 1970s; that handcraft tool is inadequate to express our needed body politics in the negotiation of reality in the practices of cyborg reproduction. Self-help is not enough. The technologies of visualization recall the important cultural practice of hunting with the camera and the deeply predatory nature of a photographic consciousness.[15] Sex, sexuality, and reproduction are central actors in high-tech myth systems structuring our imaginations of personal and social possibility.

Another critical aspect of the social relations of the new technologies is the reformulation of expectations, culture, work, and reproduction for the large scientific and technical work-force. A major social and political danger is the formation of a strongly bimodal social structure, with the masses of women and men of all ethnic groups, but especially people of colour, confined to a homework economy, illiteracy of several varieties, and general redundancy and impotence, controlled by high-tech repressive apparatuses ranging from entertainment to surveillance and disappearance. An adequate socialist-feminist politics should address women in the privileged occupational categories, and particularly in the production of science and technology that constructs scientific-technical discourses, processes, and objects.[16]

This issue is only one aspect of enquiry into the possibility of a feminist science, but it is important. What kind of constitutive role in the production of knowledge, imagination, and practice can new groups doing science have? How can these groups be allied with progressive social and political movements? What kind of political accountability can be constructed to tie women together across the scientific-technical hierarchies separating us? Might there be ways of developing feminist science/technology politics in alliance with anti-military science facility conversion action groups? Many scientific and technical workers in Silicon Valley, the high-tech cowboys included, do not want to work on military science.[17] Can these personal preferences and cultural tendencies be welded into progressive politics among this professional middle class in which women, including women of colour, are coming to be fairly numerous?

Women in the Integrated Circuit

Let me summarize the picture of women's historical locations in advanced industrial societies, as these positions have been restructured partly through the social relations of science and technology. If it was ever possible ideologically to characterize women's lives by the distinction of public and private domains – suggested by images of the division of working-class life into factory and home, of bourgeois life into market and home, and of gender existence into personal and political realms – it is now a totally misleading ideology, even to show how both terms of these dichotomies construct each other in practice and in theory. I prefer a network ideological image, suggesting the profusion of spaces and identities and the permeability of boundaries in the personal body and in the body politic. "Networking" is both a feminist practice and a multinational corporate strategy – weaving is for oppositional cyborgs.

So let me return to the earlier image of the informatics of domination and trace one vision of women's "place" in the integrated circuit, touching only a few idealized social locations seen primarily from the point of view of advanced capitalist societies: Home, Market, Paid Work Place, State, School, Clinic-Hospital, and Church. Each of these idealized spaces is logically and practically implied in every other locus, perhaps analogous to a holographic photograph. I want to suggest the impact of the social relations mediated and enforced by the new technologies in order to help formulate needed analysis and practical work. However, there is no "place" for women in these networks, only geometrics of difference and contradiction crucial to women's cyborg identities. If we learn how to read these webs of power and social life, we might learn new couplings, new coalitions. There is no way to read the following list from a standpoint of "identification", of a unitary self. The issue is dispersion. The task is to survive in the diaspora.

Home: Women-headed households, serial monogamy, flight of men, old women alone, technology of domestic work, paid homework, reemergence of home sweat-shops, home-based businesses and telecommuting, electronic cottage, urban homelessness, migration, module architecture, reinforced (simulated) nuclear family, intense domestic violence.

Market: Women's continuing consumption work, newly targeted to buy the profusion of new production from the new technologies (especially as the competitive race among industrialized and industrializing nations to avoid dangerous mass unemployment necessitates finding ever bigger new markets for ever less clearly needed commodities); bimodal buying power, coupled with advertising targeting of the numerous affluent groups and neglect of the previous mass markets; growing importance of informal markets in labour and commodities parallel to high-tech, affluent market structures; surveillance systems through electronic funds transfer; intensified market abstraction (commodification) of experience, resulting in ineffective utopian or equivalent cynical theories of community; extreme mobility (abstraction) of marketing/financing systems; interpenetration of sexual and labour markets; intensified sexualization of abstracted and alienated consumption.

Paid Work Place: Continued intense sexual and racial division of labour, but considerable growth of membership in privileged occupational categories for many white women and people of colour; impact of new technologies on women's work in clerical, service, manufacturing (especially textiles), agriculture, electronics; international restructuring of the working classes; development of new time arrangements to facilitate the homework economy (flex time, part time, over time, no time); homework and out work; increased pressures for two-tiered wage structures; significant numbers of people in cash-dependent populations world-wide with no experience or no further hope of stable employment; most labour "marginal" or "feminized".

State: Continued erosion of the welfare state; decentralizations with increased surveillance and control; citizenship by telematics; imperialism and political power broadly in the form of information rich/information poor differentiation; increased high-tech militarization increasingly opposed by many social groups; reduction of civil service jobs as a result of the growing capital intensification of office work, with implications for occupational mobility for women of colour; growing privatization of material and ideological life and culture; close integration of privatization and militarization, the high-tech forms of bourgeois capitalist personal and public life; invisibility of different social groups to each other, linked to psychological mechanisms of belief in abstract enemies.

School: Deepening coupling of high-tech capital needs and public education at all levels, differentiated by race, class, and gender; managerial classes involved in educational reform and refunding at the cost of remaining progressive educational democratic structures for children and teachers; education for mass ignorance and repression in technocratic and militarized culture; growing anti-science mystery cults in dissenting and radical political movements; continued relative scientific illiteracy among white women and people of colour; growing industrial direction of education (especially higher education) by science-based multinationals (particularly in electronics- and biotechnology-dependent companies); highly educated, numerous élites in a progressively bimodal society.

Clinic-hospital: Intensified machine–body relations; renegotiations of public metaphors which channel personal experience of the body, particularly in relation to reproduction, immune system functions, and "stress" phenomena; intensification of reproductive politics in response to world historical implications of women's unrealized, potential control of their relation to reproduction; emergence of new, historically specific diseases; struggles over meanings and means of health in environments pervaded by high technology products and processes; continuing feminization of health work; intensified struggle over state responsibility for health; continued ideological role of popular health movements as a major form of American politics.

Church: Electronic fundamentalist "super-saver" preachers solemnizing the union of electronic capital and automated fetish gods; intensified importance of churches in resisting the militarized state; central struggle over women's meanings and authority in religion; continued relevance of spirituality, intertwined with sex and health, in political struggle.

The only way to characterize the informatics of domination is as a massive intensification of insecurity and cultural impoverishment, with common failure of subsistence networks for the most vulnerable. Since much of this picture interweaves with the social relations of science and technology, the urgency of a socialist-feminist politics addressed to science and technology is plain. There is much now being done, and the grounds for political work are rich. For example, the efforts to develop forms of collective struggle for women in paid work, like SEIU's District 925,[18] should be a high priority for all of us. These efforts are profoundly tied to technical restructuring of labour processes and reformations of working classes. These efforts also are providing understanding of a more comprehensive kind of labour organization, involving community, sexuality, and family issues never privileged in the largely white male industrial unions.

The structural rearrangements related to the social relations of science and technology evoke strong ambivalence. But it is not necessary to be ultimately depressed by the implications of late twentieth-century women's relation to all aspects of work, culture, production of knowledge, sexuality, and reproduction. For excellent reasons, most Marxisms see domination best and have trouble understanding what can only look like false consciousness and people's complicity in their own domination in late capitalism. It is crucial to remember that what is lost, perhaps especially from women's points of view, is often virulent forms of oppression, nostalgically naturalized in the face of current violation. Ambivalence towards the disrupted unities mediated by high-tech culture requires not sorting consciousness into categories of "clear-sighted critique grounding a solid political epistemology" versus "manipulated false consciousness", but subtle understanding of emerging pleasures, experiences, and powers with serious potential for changing the rules of the game.

There are grounds for hope in the emerging bases for new kinds of unity across race, gender, and class, as these elementary units of socialist-feminist analysis themselves suffer protean transformations. Intensifications of hardship experienced worldwide in connection with the social relations of science and technology are severe. But what people are experiencing is not transparently clear, and we lack sufficiently subtle connections for collectively building effective theories of experience. Present efforts – Marxist, psychoanalytic, feminist, anthropological – to clarify even "our" experience are rudimentary.

I am conscious of the odd perspective provided by my historical position – a PhD in biology for an Irish Catholic girl was made possible by Sputnik's impact on US national science-education policy. I have a body and mind as much constructed by the post-Second World War arms race and cold war as by the women's movements. There are more grounds for hope in focusing on the contradictory effects of politics designed to produce loyal American technocrats, which also produced large numbers of dissidents, than in focusing on the present defeats.

The permanent partiality of feminist points of view has consequences for our expectations of forms of political organization and participation. We do not need a totality in order to work well. The feminist dream of a common language, like all dreams for a perfectly true language, of perfectly faithful naming of experience, is a totalizing and imperialist one. In that sense, dialectics too is a dream language, longing to resolve contradiction. Perhaps, ironically, we can learn from our fusions with animals and machines how not to be Man, the embodiment of Western logos. From the point of view of pleasure in these potent and taboo fusions, made inevitable by the social relations of science and technology, there might indeed be a feminist science.

Cyborgs: A Myth of Political Identity

I want to conclude with a myth about identity and boundaries which might inform late twentieth-century political imaginations. I am indebted in this story to writers like Joanna Russ, Samuel R. Delany, John Varley, James Tiptree, Jr, Octavia Butler, Monique Wittig, and Vonda McIntyre.[19] These are our story-tellers exploring what it means to be embodied in high-tech worlds. They are theorists for cyborgs. Exploring conceptions of bodily boundaries and social order, the anthropologist Mary Douglas (1966, 1970) should be credited with helping us to consciousness about how fundamental body imagery is to world view, and so to political language. French feminists like Luce Irigaray and Monique Wittig, for all their differences, know how to write the body; how to weave eroticism, cosmology, and politics from imagery of embodiment, and especially for Wittig, from imagery of fragmentation and reconstitution of bodies.[20]

American radical feminists like Susan Griffin, Audre Lorde, and Adrienne Rich have profoundly affected our political imaginations – and perhaps restricted too much what we allow as a friendly body and political language.[21] They insist on the organic, opposing it to the technological. But their symbolic systems and the related positions of ecofeminism and feminist paganism, replete with organicisms, can only be understood in Sandoval's terms as oppositional ideologies fitting the late twentieth century. They would simply bewilder anyone not preoccupied with the machines and consciousness of late capitalism. In that sense they are part of the cyborg world. But there are also great riches for feminists in explicitly embracing the possibilities inherent in the breakdown of clean distinctions between organism and machine and similar distinctions structuring the Western self. It is the simultaneity of breakdowns that cracks the matrices of domination and opens geometric possibilities. What might be learned from personal and political "technological" pollution? I look briefly at two overlapping groups of texts for their insight into the construction of a potentially helpful cyborg myth: constructions of women of colour and monstrous selves in feminist science fiction.

Earlier I suggested that "women of colour" might be understood as a cyborg identity, a potent subjectivity synthesized from fusions of outsider identities and in the complex political-historical layerings of her "biomythography", *Zami* (Lorde, 1982; King, 1987a, 1987b). There are material and cultural grids mapping this potential, Audre Lorde (1984) captures the tone in the title of her *Sister Outsider*. In my political myth, Sister Outsider is the offshore woman, whom US workers, female and feminized, are supposed to regard as the enemy preventing their solidarity, threatening their security. Onshore, inside the boundary of the United States, Sister Outsider is a potential amidst the races and ethnic identities of woman manipulated for division, competition, and exploitation in the same industries. "Women of colour" are the preferred labour force for the science-based industries, the real women for whom the world-wide sexual market, labour market, and politics of reproduction kaleidoscope into daily life. Young Korean women hired in the sex industry and in electronics assembly are recruited from high schools, educated for the integrated circuit. Literacy, especially in English, distinguishes the "cheap" female labour so attractive to the multinationals.

Contrary to orientalist stereotypes of the "oral primitive", literacy is a special mark of women of colour, acquired by US black women as well as men through a history of risking death to learn and to teach reading and writing. Writing has a special significance for all colonized groups. Writing has been crucial to the Western myth of the distinction between oral and written cultures, primitive and civilized mentalities, and more recently to the erosion of that distinction in "postmodernist" theories attacking the phallogocentrism of the West, with its worship of the monotheistic, phallic, authoritative, and singular work, the unique and perfect name.[22] Contests for the meanings of writing are a major form of contemporary political struggle. Releasing the play of writing is deadly serious. The poetry and stories of US women of colour are repeatedly about writing, about access to the power to signify; but this time that power must be neither phallic nor innocent. Cyborg writing must not be about the Fall, the imagination of a once-upon-a-time wholeness before language, before writing, before Man. Cyborg writing is about the power to survive, not on the basis of original innocence, but on the basis of seizing the tools to mark the world that marked them as other.

The tools are often stories, retold stories, versions that reverse and displace the hierarchical dualisms of naturalized identities. In retelling origin stories, cyborg authors subvert the central myths of origin of Western culture. We have all been colonized by those origin myths, with their longing for fulfilment in apocalypse. The phallogocentric origin stories most crucial for feminist cyborgs are built into the literal technologies – technologies that write the world, biotechnology and microelectronics – that have recently textualized our bodies as code problems on the grid of C^3I. Feminist cyborg stories have the task of recoding communication and intelligence to subvert command and control.

Figuratively and literally, language politics pervade the struggles of women of colour; and stories about language have a special power in the rich contemporary writing by US women of colour. For example, retellings of the story of the indigenous woman Malinche, mother of the mestizo "bastard" race of the new world, master of languages, and mistress of Cortés, carry special meaning for Chicana constructions of identity. Cherríe Moraga (1983) in *Loving in the War Years* explores the themes of identity when one never possessed the original language, never told the original story, never resided in the harmony of legitimate heterosexuality in the garden of culture, and so cannot base identity on a myth or a fall from innocence and right to natural names, mother's or father's.[23] Moraga's writing, her superb literacy, is presented in her poetry as the same kind of violation as Malinche's mastery of the conqueror's language – a violation, an illegitimate production, that allows survival. Moraga's language is not "whole"; it is self-consciously spliced, a chimera of English and Spanish, both conqueror's languages. But it is this chimeric monster, without claim to an original language before violation, that crafts the erotic, competent, potent identities of women of colour. Sister Outsider hints at the possibility of world survival not because of her innocence, but because of her ability to live on the boundaries, to write without the founding myth of original wholeness, with its inescapable apocalypse of final return to a deathly oneness that Man has imagined to be the innocent and all-powerful Mother, freed at the End from another spiral of appropriation by her son. Writing marks Moraga's body, affirms it as the body of a woman of colour, against the possibility of passing into the unmarked category of the Anglo father or into the orientalist myth of "original illiteracy" of a mother that never was. Malinche was mother here, not Eve before eating the forbidden fruit. Writing affirms Sister Outsider, not the Woman-before-the-Fall-into-Writing needed by the phallogocentric Family of Man.

Writing is pre-eminently the technology of cyborgs, etched surfaces of the late twentieth century. Cyborg politics is the struggle for language and the struggle against perfect communication, against the one code that translates all meaning perfectly, the central dogma of phallogocentrism. That is why cyborg politics insist on noise and advocate pollution, rejoicing in the illegitimate fusions of animal and machine. These are the couplings which make Man and Woman so problematic, subverting the structure of desire, the force imagined to generate language and gender, and so subverting the structure and modes of reproduction of "Western" identity, of nature and culture, of mirror and eye, slave and master, body and mind. "We" did not originally choose to be cyborgs, but choice grounds a liberal politics and epistemology that imagines the reproduction of individuals before the wider replications of "texts".

From the perspective of cyborgs, freed of the need to ground politics in "our" privileged position of the oppression that incorporates all other dominations, the innocence of the merely violated, the ground of those closer to nature, we can see powerful possibilities. Feminisms and Marxisms have run aground on Western epistemological imperatives to construct a revolutionary subject from the perspective of a hierarchy of oppressions and/or a latent position of moral superiority, innocence, and greater closeness to nature. With no available original dream of a common language or original symbiosis promising protection from hostile "masculine" separation, but written into the play of a text that has no finally privileged reading or salvation history, to recognize "oneself" as fully implicated in the world, frees us of the need to root politics in identification, vanguard parties, purity, and mothering. Stripped of identity, the bastard race teaches about the power of the margins and the importance of a mother like Malinche. Women of colour have transformed her from the evil mother of masculinist fear into the originally literate mother who teaches survival.

This is not just literary deconstruction, but liminal transformation. Every story that begins with original innocence and privileges the return to wholeness imagines the drama of life to be individuation, separation, the birth of the self, the tragedy of autonomy, the fall into writing, alienation; that is, war, tempered by imaginary respite in the bosom of the Other. These plots are ruled by a reproductive politics – rebirth without flaw, perfection, abstraction. In this plot women are imagined either better or worse off, but all agree they have less selfhood, weaker individuation, more fusion to the oral, to Mother, less at stake in masculine autonomy. But there is another route to having less at stake in masculine autonomy, a route that does not pass through Woman, Primitive, Zero, the Mirror Stage and its imaginary. It passes through women and other present-tense, illegitimate cyborgs, not of Woman born, who refuse the ideological resources of victimization so as to have a real life. These cyborgs are the people who refuse to disappear on cue, no matter how many times a "Western" commentator remarks on the sad passing of another primitive, another organic group done in by "Western" technology, by writing.[24] These real-life cyborgs (for example, the Southeast Asian village women workers in Japanese and US electronics firms described by Aihwa Ong) are actively rewriting the texts of their bodies and societies. Survival is the stakes in this play of readings.

To recapitulate, certain dualisms have been persistent in Western traditions; they have all been systemic to the logics and practices of domination of women, people of colour, nature, workers, animals – in short, domination of all constituted as others, whose task is to mirror the self. Chief among these troubling dualisms are self/other, mind/body, culture/nature, male/female, civilized/primitive, reality/appearance, whole/part, agent/resource, maker/made, active/passive, right/wrong, truth/illusion, total/partial, God/man. The self is the One who is not dominated, who knows that by the service of the other, the other is the one who holds the future, who knows that by the experience of domination, which gives the lie to the autonomy of the self. To be One is to be autonomous, to be powerful, to be God; but to be One is to be an illusion, and so to be involved in a dialectic of apocalypse with the other. Yet to be other is to be multiple, without clear boundary, frayed, insubstantial. One is too few, but two are too many.

High-tech culture challenges these dualisms in intriguing ways. It is not clear who makes and who is made in the relation between human and machine. It is not clear what is mind and what body in machines that resolve into coding practices. In so far as we know ourselves in both formal discourse (for example, biology) and in daily practice (for example, the homework economy in the integrated circuit), we find ourselves to be cyborgs, hybrids, mosaics, chimeras. Biological organisms have become biotic systems, communications devices like others. There is no fundamental, ontological separation in our formal knowledge of machine and organism, of technical and organic. The replicant Rachel in the Ridley Scott film *Blade Runner* stands as the image of a cyborg culture's fear, love, and confusion.

One consequence is that our sense of connection to our tools is heightened. The trance state experienced by many computer users has become a staple of science-fiction film and cultural jokes. Perhaps paraplegics and other severely handicapped people can (and sometimes do) have the most intense experiences of complex hybridization with other communication devices.[25] Anne McCaffrey's prefeminist *The Ship Who Sang* (1969) explored the consciousness of a cyborg, hybrid of girl's brain and complex machinery, formed after the birth of a severely handicapped child. Gender, sexuality, embodiment, skill: all were reconstituted in the story. Why should our bodies end at the skin, or include at best other beings encapsulated by skin? From the seventeenth century till now, machines could be animated – given ghostly souls to make them speak or move or to account for their orderly development and mental capacities. Or organisms could be mechanized – reduced to body understood as resource of mind. These machine/organism relationships are obsolete, unnecessary. For us, in imagination and in other practice, machines can be prosthetic devices, intimate components, friendly selves. We don't need organic holism to give impermeable wholeness, the total woman and her feminist variants (mutants?). Let me conclude this point by a very partial reading of the logic of the cyborg monsters of my second group of texts, feminist science fiction.

The cyborgs populating feminist science fiction make very problematic the statuses of man or woman, human, artefact, member of a race, individual entity, or body. Katie King clarifies how pleasure in reading these fictions is not largely based on identification. Students facing Joanna Russ for the first time, students who have learned to take modernist writers like James Joyce or Virginia Woolf without flinching, do not know what to make of *The Adventures of Alyx* or *The Female Man*, where characters refuse the reader's search for innocent wholeness while granting the wish for heroic quests, exuberant eroticism, and serious politics. *The Female Man* is the story of four versions of one genotype, all of whom meet, but even taken together do not make a whole, resolve the dilemmas of violent moral action, or remove the growing scandal of gender. The feminist science fiction of Samuel R. Delany, especially *Tales of Nevèrÿon*, mocks stories of origin by redoing the neolithic revolution, replaying the founding moves of Western civilization to subvert their plausibility. James Tiptree, Jr, an author whose fiction was regarded as particularly manly until her "true" gender was revealed, tells tales of reproduction based on nonmammalian technologies like alternation of generations of male brood pouches and male nurturing. John Varley constructs a supreme cyborg in his arch-feminist exploration of Gaea, a mad goddess-planet-trickster-old woman-technological device on whose surface an extraordinary array of postcyborg symbioses are spawned. Octavia Butler writes of an African sorceress pitting her powers of transformation against the genetic manipulations of her rival (*Wild Seed*), of time warps that bring a modern US black woman into slavery where her actions in relation to her white master-ancestor determine the possibility of her own birth (*Kindred*), and of the illegitimate insights into identity and community of an adopted cross-species child who came to know the enemy as self (*Survivor*). In *Dawn* (1987), the first instalment of a series called *Xenogenesis*, Butler tells the story of Lilith Iyapo, whose personal name recalls Adam's first and repudiated wife and whose family name marks her status as the widow of the son of Nigerian immigrants to the US. A black woman and a mother whose child is dead, Lilith mediates the transformation of humanity through genetic exchange with extra-terrestrial lovers/rescuers/destroyers/genetic engineers, who reform earth's habitats after the nuclear holocaust and coerce surviving humans into intimate fusion with them. It is a novel that interrogates reproductive, linguistic, and nuclear politics in a mythic field structured by late twentieth-century race and gender.

Because it is particularly rich in boundary transgressions, Vonda McIntyre's *Superluminal* can close this truncated catalogue of promising and dangerous monsters who help redefine the pleasures and politics of embodiment and feminist writing. In a fiction where no character is "simply" human, human status is highly problematic. Orca, a genetically altered diver, can speak with killer whales and survive deep ocean conditions, but she longs to explore space as a pilot, necessitating bionic implants jeopardizing her kinship with the divers and cetaceans. Transformations are effected by virus vectors carrying a new developmental code, by transplant surgery, by implants of microelectronic devices, by analogue doubles, and other means. Laenea becomes a pilot by accepting a heart implant and a host of other alterations allowing survival in transit at speeds exceeding that of light. Radu Dracul survives a virus-caused plague in his outerworld planet to find himself with a time sense that changes the boundaries of spatial perception for the whole species. All the characters explore the limits of language; the dream of communicating experience; and the necessity of limitation, partiality, and intimacy even in this world of protean transformation and connection. *Superluminal* stands also for the defining contradictions of a cyborg world in another sense; it embodies textually the intersection of feminist theory and colonial discourse in the science fiction I have alluded to in this chapter. This is a conjunction with a long history that many "First World" feminists have tried to repress, including myself in my readings of *Superluminal* before being called to account by Zoe Sofoulis, whose different location in the world system's informatics of domination made her acutely alert to the imperialist moment of all science fiction cultures, including women's science fiction. From an Australian feminist sensitivity, Sofoulis remembered more readily McIntyre's role as writer of the adventures of Captain Kirk and Spock in TV's *Star Trek* series than her rewriting the romance in *Superluminal*.

Monsters have always defined the limits of community in Western imaginations. The Centaurs and Amazons of ancient Greece established the limits of the centred polis of the Greek male human by their disruption of marriage and boundary pollutions of the warrior with animality and woman. Unseparated twins and hermaphrodites were the confused human material in early modern France who grounded discourse on the natural and supernatural, medical and legal, portents and diseases – all crucial to establishing modern identity.[26] The evolutionary and behavioural sciences of monkeys and apes have marked the multiple boundaries of late twentieth-century industrial identities. Cyborg monsters in feminist science fiction define quite different political possibilities and limits from those proposed by the mundane fiction of Man and Woman.

There are several consequences to taking seriously the imagery of cyborgs as other than our enemies. Our bodies, ourselves; bodies are maps of power and identity. Cyborgs are no exception. A cyborg body is not innocent; it was not born in a garden; it does not seek unitary identity and so generate antagonistic dualisms without end (or until the world ends); it takes irony for granted. One is too few, and two is only one possibility. Intense pleasure in skill, machine skill, ceases to be a sin, but an aspect of embodiment. The machine is not an *it* to be animated, worshipped, and dominated. The machine is us, our processes, an aspect of our embodiment. We can be responsible for machines; *they* do not dominate or threaten us. We are responsible for boundaries; we are they. Up till now (once upon a time), female embodiment seemed to be given, organic, necessary; and female embodiment seemed to mean skill in mothering and its metaphoric extensions. Only by being out of place could we take intense pleasure in machines, and then with excuses that this was organic activity after all, appropriate to females. Cyborgs might consider more seriously the partial, fluid, sometimes aspect of sex and sexual embodiment. Gender might not be global identity after all, even if it has profound historical breadth and depth.

The ideologically charged question of what counts as daily activity, as experience, can be approached by exploiting the cyborg image. Feminists have recently claimed that women are given to dailiness, that women more than men somehow sustain daily life, and so have a privileged epistemological position potentially. There is a compelling aspect to this claim, one that makes visible unvalued female activity and names it as the ground of life. But *the* ground of life? What about all the ignorance of women, all the exclusions and failures of knowledge and skill? What about men's access to daily competence, to knowing how to build things, to take them apart, to play? What about other embodiments? Cyborg gender is a local possibility taking a global vengeance. Race, gender, and capital require a cyborg theory of wholes and parts. There is no drive in cyborgs to produce total theory, but there is an intimate experience of boundaries, their construction and deconstruction. There is a myth system waiting to become a political language to ground one way of looking at science and technology and challenging the informatics of domination – in order to act potently.

One last image: organisms and organismic, holistic politics depend on metaphors of rebirth and invariably call on the resources of reproductive sex. I would suggest that cyborgs have more to do with regeneration and are suspicious of the reproductive matrix and of most birthing. For salamanders, regeneration after injury, such as the loss of a limb, involves regrowth of structure and restoration of function with the constant possibility of twinning or other odd topographical productions at the site of former injury. The regrown limb can be monstrous, duplicated, potent. We have all been injured, profoundly. We require regeneration, not rebirth, and the possibilities for our reconstitution include the utopian dream of the hope for a monstrous world without gender.

Cyborg imagery can help express two crucial arguments in this essay: first, the production of universal, totalizing theory is a major mistake that misses most of reality, probably always, but certainly now; and second, taking responsibility for the social relations of science and technology means refusing an anti-science metaphysics, a demonology of technology, and so means embracing the skilful task of reconstructing the boundaries of daily life, in partial connection with others, in communication with all of our parts. It is not just that science and technology are possible means of great human satisfaction, as well as a matrix of complex dominations. Cyborg imagery can suggest a way out of the maze of dualisms in which we have explained our bodies and our tools to ourselves. This is a dream not of a common language, but of a powerful infidel heteroglossia. It is an imagination of a feminist speaking in tongues to strike fear into the circuits of the super-savers of the new right. It means both building and destroying machines, identities, categories, relationships, space stories. Though both are bound in the spiral dance, I would rather be a cyborg than a goddess.

Notes

1 Useful references to left and/or feminist radical science movements and theory and to biological/biotechnical issues include: Bleier (1984, 1986), Harding (1986), Fausto-Sterling (1985), Gould (1981), Hubbard et al. (1982), Keller (1985), Lewontin et al. (1984), *Radical Science Journal* (became *Science as Culture* in 1987), 26 Freegrove Road, London N7 9RQ; *Science for the People*, 897 Main St, Cambridge, MA 02139.

2 Starting points for left and/or feminist approaches to technology and politics include: Cowan (1983), Rothschild (1983), Traweek (1988), Young and Levidow (1981, 1985), Weizenbaum (1976), Winner (1977, 1986), Zimmerman (1983), Athanasiou (1987), Cohn (1987a, 1987b), Winograd and Flores (1986), Edwards (1985), *Global Electronics Newsletter*, 867 West Dana St, #204, Mountain View, CA 94041; *Processed World*, 55 Sutter St, San Francisco, CA 94104: ISIS, Women's International Information and Communication Service, PO Box 50 (Cornavin), 1211 Geneva 2, Switzerland, and Via Santa Maria Dell'Anima 30, 00186 Rome, Italy. Fundamental approaches to modern social studies of science that do not continue the liberal mystification that it all started with Thomas Kuhn, include: Knorr-Cetina (1981), Knorr-Cetina and Mulkay (1983), Latour and Woolgar (1979), Young (1979). The 1984 Directory of the Network for the Ethnographic Study of Science, Technology, and Organizations lists a wide range of people and projects crucial to better radical analysis; available from NESSTO, PO Box 11442, Stanford, CA 94305.

3 A provocative, comprehensive argument about the politics and theories of "postmodernism" is made by Fredric Jameson (1984), who argues that postmodernism is not an option, a style among others, but a cultural dominant requiring radical reinvention of left politics from within; there is no longer any place from without that gives meaning to the comforting fiction of critical distance. Jameson also makes clear why one cannot be for or against postmodernism, an essentially moralist move. My position is that feminists (and others) need continuous cultural reinvention, postmodernist critique, and historical materialism; only a cyborg would have a chance. The old dominations of white capitalist patriarchy seem nostalgically innocent now: they normalized heterogeneity, into man and woman, white and black, for example. "Advanced capitalism" and postmodernism release heterogeneity without a norm, and we are flattened, without subjectivity, which requires depth, even unfriendly and drowning depths. It is time to write *The Death of the Clinic*. The clinic's methods required bodies and works; we have texts and surfaces. Our dominations don't work by medicalization and normalization any more; they work by networking, communications redesign, stress management. Normalization gives way to automation, utter redundancy. Michel Foucault's *Birth of the Clinic* (1963), *History of Sexuality*

(1976), and *Discipline and Punish* (1975) name a form of power at its moment of implosion. The discourse of biopolitics gives way to technobabble, the language of the spliced substantive; no noun is left whole by the multinationals. These are their names, listed from one issue of *Science*: Tech-Knowledge, Genentech, Allergen, Hybritech, Compupro, Genen-cor, Syntex, Allelix, Agrigenetics Corp., Syntro, Codon, Repligen, MicroAngelo from Scion Corp., Percom Data, Inter Systems, Cyborg Corp., Statcom Corp., Intertec. If we are imprisoned by language, then escape from that prison-house requires language poets, a kind of cultural restriction enzyme to cut the code; cyborg heteroglossia is one form of radical cultural politics. For cyborg poetry, see Perloff (1984); Fraser (1984). For feminist modernist/postmodernist "cyborg" writing, see HOW(ever), 871 Corbett Ave, San Francisco, CA 94131.

4 The US equivalent of Mills & Boon. [Ed.]

5 Baudrillard (1983). Jameson (1984, p. 66) points out that Plato's definition of the simulacrum is the copy for which there is no original, i.e., the world of advanced capitalism, of pure exchange. See *Discourse* 9 (Spring/Summer 1987) for a special issue on technology (cybernetics, ecology, and the postmodern imagination).

6 A practice at once both spiritual and political that linked guards and arrested anti-nuclear demonstrators in the Alameda County jail in California in the early 1980s. [Ed.]

7 For ethnographic accounts and political evaluations, see Epstein (forthcoming), Sturgeon (1986). Without explicit irony, adopting the spaceship earth/whole earth logo of the planet photographed from space, set off by the slogan "Love Your Mother", the May 1987 Mothers and Others Day action at the nuclear weapons testing facility in Nevada none the less took account of the tragic contradictions of views of the earth. Demonstrators applied for official permits to be on the land from officers of the Western Shoshone tribe, whose territory was invaded by the US government when it built the nuclear weapons test ground in the 1950s. Arrested for trespassing, the demonstrators argued that the police and weapons facility personnel, without authorization from the proper officials, were the trespassers. One affinity group at the women's action called themselves the Surrogate Others; and in solidarity with the creatures forced to tunnel in the same ground with the bomb, they enacted a cyborgian emergence from the constructed body of a large, non-heterosexual desert worm.

8 This chart was published in 1985. My previous efforts to understand biology as a cybernetic command-control discourse and organisms as "natural-technical objects of knowledge" were Haraway (1979, 1983, 1984). The 1979 version of this dichotomous chart appears in Haraway 1991, ch. 3; for a 1989 version, see ibid., ch. 10. The differences indicate shifts in argument.

9 For progressive analyses and action on the biotechnology debates: *GeneWatch, a Bulletin of the Committee for Responsible Genetics*, 5 Doane St, 4th Floor, Boston, MA 02109; Genetic Screening Study Group (formerly the Sociobiology Study Group of Science for the People), Cambridge, MA; Wright (1982, 1986); Yoxen (1983).

10 Starting references for "women in the integrated circuit": D'Onofrio-Flores and Pfafflin (1982), Fernandez-Kelly (1983), Fuentes and Ehrenreich (1983), Grossman (1980), Nash and Fernandez-Kelly (1983), Ong (1982), Science Policy Research Unit (1982).

11 For the "homework economy outside the home" and related arguments: Gordon (1983); Gordon and Kimball (1985); Stacey (1987); Reskin and Hartmann (1986); *Women and Poverty* (1984); Rose (1986); Collins (1982); Burr (1982); Gregory and Nussbaum (1982); Piven and Coward (1982); Microelectronics Group (1980); Stallard et al. (1983) which includes a useful organization and resource list.

12 The conjunction of the Green Revolution's social relations with biotechnologies like plant genetic engineering makes the pressures on land in the Third World increasingly intense. AID's estimates (*New York Times*, 14 October 1984) used at the 1984 World Food Day are that in Africa, women produce about 90 per cent of rural food supplies, about 60–80 per cent in Asia, and provide 40 per cent of agricultural labour in the Near East and Latin America. Blumberg charges that world organizations' agricultural politics, as well as those of multinationals and national governments in the Third World, generally ignore fundamental issues in the sexual division of labour. The present tragedy of famine in Africa might owe as much to male supremacy as to capitalism, colonialism, and rain patterns. More accurately, capitalism and racism are usually structurally male dominant. See also Blumberg (1981); Hacker (1984); Hacker and Bovit (1981); Busch and Lacy (1983); Wilfred (1982); Sachs (1983); International Fund for Agricultural Development (1985); Bird (1984).

13 See also Enloe (1983a, b).

14 For a feminist version of this logic, see Hrdy (1981). For an analysis of scientific women's story-telling practices, especially in relation to sociobiology in evolutionary debates around child abuse and infanticide, see Haraway 1991, ch. 5.

15 For the moment of transition of hunting with guns to hunting with cameras in the construction of popular meanings of nature for an American urban immigrant public, see Haraway (1984–5, 1989), Nash (1979), Sontag (1977), Preston (1984).

16 For guidance for thinking about the political/cultural/racial implications of the history of women doing science in the United States see: Haas and

Perucci (1984); Hacker (1981); Keller (1983), National Science Foundation (1988); Rossiter (1982); Schiebinger (1987); Haraway (1989).

17 Markoff and Siegel (1983). High Technology Professionals for Peace and Computer Professionals for Social Responsibility are promising organizations.

18 Service Employees International Union's office workers' organization in the US. [Ed.]

19 King (1984). An abbreviated list of feminist science fiction underlying themes of this essay: Octavia Butler, *Wild Seed, Mind of My Mind, Kindred, Survivor*, Suzy McKee Charnas, *Motherliness*; Samuel R. Delany, the Nevèrÿon series; Anne McCaffery, *The Ship Who Sang, Dinosaur Planet*; Vonda McIntyre, *Superluminal, Dreamsnake*; Joanna Russ, *Adventures of Alix, The Female Man*; James Tiptree, Jr, *Star Songs of an Old Primate, Up the Walls of the World*; John Varley, *Titan, Wizard, Demon*.

20 French feminisms contribute to cyborg heteroglossia. Burke (1981); Irigaray (1977, 1979); Marks and de Courtivron (1980); *Signs* (Autumn 1981); Wittig (1973); Duchen (1986). For English translation of some currents of francophone feminism see *Feminist Issues: A Journal of Feminist Social and Political Theory*, 1980.

21 But all these poets are very complex, not least in their treatment of themes of lying and erotic, decentred collective and personal identities. Griffin (1978), Lorde (1984), Rich (1978).

22 Derrida (1976, especially part II); Lévi-Strauss (1971, especially "The Writing Lesson"); Gates (1985); Kahn and Neumaier (1985); Ong (1982); Kramarae and Treichler (1985).

23 The sharp relation of women of colour to writing as theme and politics can be approached through: Program for "The Black Woman and the Diaspora: Hidden Connections and Extended Acknowledgments", An International Literary Conference, Michigan State University, October 1985; Evans (1984); Christian (1985); Carby (1987); Fisher (1980); *Frontiers* (1980, 1983); Kingston (1977); Lerner (1973); Giddings (1985); Moraga and Anzaldúa (1981); Morgan (1984). Anglophone European and Euro-American women have also crafted special relations to their writing as a potent sign: Gilbert and Gubar (1979), Russ (1983).

24 The convention of ideologically taming militarized high technology by publicizing its applications to speech and motion problems of the disabled/differently abled takes on a special irony in monotheistic, patriarchal, and frequently anti-semitic culture when computer-generated speech allows a boy with no voice to chant the Haftorah at his bar mitzvah. See Sussman (1986). Making the always context-relative social definitions of "ableness" particularly clear, military high-tech has a way of making human beings disabled by definition, a perverse aspect of much automated battlefield and Star Wars R&D. See Welford (1 July 1986).

25 James Clifford (1985, 1988) argues persuasively for recognition of continuous cultural reinvention, the stubborn non-disappearance of those "marked" by Western imperializing practices.

26 DuBois (1982), Daston and Park (n.d.), Park and Daston (1981). The noun *monster* shares its root with the verb *to demonstrate*.

Bibliography

Athanasiou, Tom (1987) "High-tech politics: the case of artificial intelligence", *Socialist Review* 92: 7–35.

Baudrillard, Jean (1983) *Simulations*, P. Foss, P. Patton, P. Beitchman, trans. New York: Semiotext[e].

Bird, Elizabeth (1984) "Green Revolution imperialism, I & II", papers delivered at the University of California, Santa Cruz.

Bleier, Ruth (1984) *Science and Gender: A Critique of Biology and Its Themes on Women*. New York: Pergamon.

——, ed. (1986) *Feminist Approaches to Science*. New York: Pergamon.

Blumberg, Rae Lessor (1981) *Stratification: Socioeconomic and Sexual Inequality*. Boston: Brown.

—— (1983) "A general theory of sex stratification and its application to the positions of women in today's world economy", paper delivered to Sociology Board, University of California at Santa Cruz.

Burke, Carolyn (1981) "Irigaray through the looking glass", *Feminist Studies* 7(2): 288–306.

Burr, Sara G. (1982) "Women and work", in Barbara K. Haber, ed. *The Women's Annual, 1981*. Boston: G.K. Hall.

Busch, Lawrence and Lacy, William (1983) *Science, Agriculture, and the Politics of Research*. Boulder, CO: Westview.

Butler, Octavia (1987) *Dawn*. New York: Warner.

Carby, Hazel (1987) *Reconstructing Womanhood: The Emergence of the Afro-American Woman Novelist*. New York: Oxford University Press.

Christian, Barbara (1985) *Black Feminist Criticism: Perspectives on Black Women Writers*. New York: Pergamon.

Clifford, James (1985) "On ethnographic allegory", in James Clifford and George Marcus, eds *Writing Culture: The Poetics and Politics of Ethnography*. Berkeley: University of California Press.

—— (1988) *The Predicament of Culture: Twentieth-Century Ethnography, Literature, and Art*. Cambridge, MA: Harvard University Press.

Cohn, Carol (1987a) "Nuclear language and how we learned to pat the bomb", *Bulletin of Atomic Scientists*, pp. 17–24.

—— (1987b) "Sex and death in the rational world of defense intellectuals", *Signs* 12(4): 687–718.

Collins, Patricia Hill (1982) "Third World women in America", in Barbara K. Haber, ed. *The Women's Annual, 1981*. Boston: G.K. Hall.

Cowan, Ruth Schwartz (1983) *More Work for Mother: The Ironies of Household Technology from the Open Hearth to the Microwave*. New York: Basic.

Daston, Lorraine and Park, Katherine (n.d.) "Hermaphrodites in Renaissance France", unpublished paper.

de Waal, Frans (1982) *Chimpanzee Politics: Power and Sex among the Apes*. New York: Harper & Row.

Derrida, Jacques (1976) *Of Grammatology*, G. C. Spivak, trans. and introd. Baltimore: Johns Hopkins University Press.

D'Onofrio-Flores, Pamela and Pfafflin, Sheila M., eds (1982) *Scientific–Technological Change and the Role of Women in Development*. Boulder: Westview.

Douglas, Mary (1966) *Purity and Danger*. London: Routledge & Kegan Paul.

—— (1970) *Natural Symbols*. London: Cresset Press.

DuBois, Page (1982) *Centaurs and Amazons*. Ann Arbor: University of Michigan Press.

Duchen, Claire (1986) *Feminism in France from May '68 to Mitterrand*. London: Routledge & Kegan Paul.

Edwards, Paul (1985) "Border wars: the science and politics of artificial intelligence", *Radical America* 19(6): 39–52.

Enloe, Cynthia (1983a) "Women textile workers in the militarization of Southeast Asia", in Nash and Fernandez-Kelly (1983), pp. 407–25.

Epstein, Barbara (forthcoming) *Political Protest and Cultural Revolution: Nonviolent Direct Action in the Seventies and Eighties*. Berkeley: University of California Press.

Evans, Mari, ed. (1984) *Black Women Writers: A Critical Evaluation*. Garden City, NY: Doubleday/Anchor.

Fausto-Sterling, Anne (1985) *Myths of Gender: Biological Theories about Women and Men*. New York: Basic.

Fernandez-Kelly, Maria Patricia (1983) *For We Are Sold, I and My People*. Albany: State University of New York Press.

Fisher, Dexter, ed. (1980) *The Third Woman: Minority Women Writers of the United States*. Boston: Houghton Mifflin.

Foucault, Michel (1963) *The Birth of the Clinic: An Archaeology of Medical Perception*, A.M. Smith, trans. New York: Vintage, 1975.

—— (1975) *Discipline and Punish: The Birth of the Prison*, Alan Sheridan, trans. New York: Vintage, 1979.

—— (1976) *The History of Sexuality*, Vol. 1: *An Introduction*, Robert Hurley, trans. New York: Pantheon, 1978.

Fraser, Kathleen (1984) *Something. Even Human Voices. In the Foreground, a Lake*. Berkeley, CA: Kelsey St Press.

Fuentes, Annette and Ehrenreich, Barbara (1983) *Women in the Global Factory*. Boston: South End.

Gates, Henry Louis (1985) "Writing 'race' and the difference it makes", in "*Race*", *Writing, and Difference*, special issue, *Critical Inquiry* 12(1): 1–20.

Giddings, Paula (1985) *When and Where I Enter: The Impact of Black Women on Race and Sex in America*. Toronto: Bantam.

Gilbert, Sandra M. and Gubar, Susan (1979) *The Madwoman in the Attic: The Woman Writer and the Nineteenth-Century Literary Imagination*. New Haven, CT: Yale University Press.

Gordon, Richard (1983) "The computerization of daily life, the sexual division of labor, and the homework economy", Silicon Valley Workshop conference, University of California at Santa Cruz.

—— and Kimball, Linda (1985) "High-technology, employment and the challenges of education", Silicon Valley Research Project, Working Paper, no. 1.

Gould, Stephen J. (1981) *Mismeasure of Man*. New York: Norton.

Gregory, Judith and Nussbaum, Karen (1982) "Race against time: automation of the office", *Office: Technology and People* 1: 197–236.

Griffin, Susan (1978) *Woman and Nature: The Roaring Inside Her*. New York: Harper & Row.

Grossman, Rachel (1980) "Women's place in the integrated circuit", *Radical America* 14(1): 29–50.

Haas, Violet and Perucci, Carolyn, eds (1984) *Women in Scientific and Engineering Professions*. Ann Arbor: University of Michigan Press.

Hacker, Sally (1981) "The culture of engineering: women, workplace, and machine", *Women's Studies International Quarterly* 4(3): 341–53.

—— (1984) "Doing it the hard way: ethnographic studies in the agribusiness and engineering classroom", paper delivered at the California American Studies Association, Pomona.

—— and Bovit, Liza (1981) "Agriculture to agribusiness: technical imperatives and changing roles", paper delivered at the Society for the History of Technology, Milwaukee.

Haraway, Donna J. (1979) "The biological enterprise: sex, mind, and profit from human engineering to sociobiology", *Radical History Review* 20: 206–37.

—— (1983) "Signs of dominance: from a physiology to a cybernetics of primate society", *Studies in History of Biology* 6: 129–219.

—— (1984) "Class, race, sex, scientific objects of knowledge: a socialist-feminist perspective on the social construction of productive knowledge and some political consequences", in Haas and Perucci (1984), pp. 212–29.

—— (1984–5) "Teddy bear patriarchy: taxidermy in the Garden of Eden, New York City, 1908–36", *Social Text* 11: 20–64.

—— (1989) *Primate Visions: Gender, Race, and Nature in the World of Modern Science*. New York: Routledge.

—— (1991) *Simians, Cyborgs and Women: The Reinvention of Nature*. New York: Routledge.

Harding, Sandra (1986) *The Science Question in Feminism*. Ithaca: Cornell University Press.

Hogness, E. Rusten (1983) "Why stress? A look at the making of stress, 1936–56", unpublished paper available from the author, 4437 Mill Creek Rd, Healdsburg, CA 95448.

Hrdy, Sarah Blaffer (1981) *The Woman That Never Evolved*. Cambridge, MA: Harvard University Press.

Hubbard, Ruth, Henifin, Mary Sue, and Fried, Barbara, eds (1982) *Biological Woman, the Convenient Myth*. Cambridge, MA: Schenkman.

International Fund for Agricultural Development (1985) *IFAD Experience Relating to Rural Women, 1977–84*. Rome: IFAD, 37.

Irigaray, Luce (1977) *Ce sexe qui n'en est pas un*. Paris: Minuit.

Jameson, Fredric (1984) "Post-modernism, or the cultural logic of late capitalism", *New Left Review* 146: 53–92.

Kahn, Douglas and Neumaier, Diane, eds (1985) *Cultures in Contention*. Seattle: Real Comet.

Keller, Evelyn Fox (1983) *A Feeling for the Organism*. San Francisco: Freeman.

—— (1985) *Reflections on Gender and Science*. New Haven: Yale University Press.

King, Katie (1984) "The pleasure of repetition and the limits of identification in feminist science fiction: reimaginations of the body after the cyborg", paper delivered at the California American Studies Association, Pomona.

—— (1987a) "Canons without innocence", University of California at Santa Cruz, PhD thesis.

—— (1987b) *The Passing Dreams of Choice...Once Before and After: Audre Lorde and the Apparatus of Literary Production*, book prospectus, University of Maryland at College Park.

Kingston, Maxine Hong (1977) *China Men*. New York: Knopf.

Knorr-Cetina, Karin (1981) *The Manufacture of Knowledge*. Oxford: Pergamon.

——and Mulkay, Michael, eds (1983) *Science Observed: Perspectives on the Social Study of Science*. Beverly Hills: Sage.

Kramarae, Cheris and Treichler, Paula (1985) *A Feminist Dictionary*. Boston: Pandora.

Latour, Bruno (1984) *Les microbes, guerre et paix, suivi des irréductions*. Paris: Métailié.

—— and Woolgar, Steve (1979) *Laboratory Life: The Social Construction of Scientific Facts*. Beverly Hills: Sage.

Lerner, Gerda, ed. (1973) *Black Women in White America: A Documentary History*. New York: Vintage.

Lévi-Strauss, Claude (1971) *Tristes Tropiques*, John Russell, trans. New York: Atheneum.

Lewontin, R. C., Rose, Steven, and Kamin, Leon J. (1984) *Not in Our Genes: Biology, Ideology, and Human Nature*. New York: Pantheon.

Lorde, Audre (1982) *Zami, a New Spelling of My Name*. Trumansberg, NY: Crossing, 1983.

—— (1984) *Sister Outsider*. Trumansberg, NY: Crossing.

McCaffrey, Anne (1969) *The Ship Who Sang*. New York: Ballantine.

Marcuse, Herbert (1964) *One-Dimensional Man: Studies in the Ideology of Advanced Industrial Society*. Boston: Beacon.

Markoff, John and Siegel, Lenny (1983) "Military micros", paper presented at Silicon Valley Research Project conference, University of California at Santa Cruz.

Marks, Elaine and de Courtivron, Isabelle, eds (1980) *New French Feminisms*. Amherst: University of Massachusetts Press.

Merchant, Carolyn (1980) *The Death of Nature: Women, Ecology, and the Scientific Revolution*. New York: Harper & Row.

Microelectronics Group (1980) *Microelectronics: Capitalist Technology and the Working Class*. London: CSE.

Moraga, Cherríe (1983) *Loving in the War Years: lo que nunca pasó por sus labios*. Boston: South End.

—— and Anzaldúa, Gloria, eds (1981) *This Bridge Called My Back: Writings by Radical Women of Color*. Watertown: Persephone.

Morgan, Robin, ed. (1984) *Sisterhood Is Global*. Garden City, NY: Anchor/Doubleday.

Nash, June and Fernandez-Kelly, Maria Patricia, eds (1983) *Women and Men and the International Division of Labor*. Albany: State University of New York Press.

Nash, Roderick (1979) "The exporting and importing of nature: nature-appreciation as a commodity, 1850–1980", *Perspectives in American History* 3: 517–60.

National Science Foundation (1988) *Women and Minorities in Science and Engineering*. Washington: NSF.

Ong, Walter (1982) *Orality and Literacy: The Technologizing of the Word*. New York: Methuen.

Park, Katherine and Daston, Lorraine J. (1981) "Unnatural conceptions: the study of monsters in sixteenth- and seventeenth-century France and England", *Past and Present* 92: 20–54.

Perloff, Marjorie (1984) "Dirty language and scramble systems", *Sulfur* 11: 178–83.

Petchesky, Rosalind Pollack (1981) "Abortion, anti-feminism and the rise of the New Right", *Feminist Studies* 7(2): 206–46.

Piven, Frances Fox and Coward, Richard (1982) *The New Class War: Reagan's Attack on the Welfare State and Its Consequences*. New York: Pantheon.

Preston, Douglas (1984) "Shooting in paradise", *Natural History* 93(12): 14–19.

Reskin, Barbara F. and Hartmann, Heidi, eds (1986) *Women's Work, Men's Work*. Washington: National Academy of Sciences.

Rich, Adrienne (1978) *The Dream of a Common Language*. New York: Norton.

Rose, Stephen (1986) *The American Profile Poster: Who Owns What, Who Makes How Much, Who Works Where, and Who Lives with Whom?* New York: Pantheon.

Rossiter, Margaret (1982) *Women Scientists in America*. Baltimore: Johns Hopkins University Press.

Rothschild, Joan, ed. (1983) *Machina ex Dea: Feminist Perspectives on Technology*. New York: Pergamon.

Russ, Joanna (1983) *How to Suppress Women's Writing*. Austin: University of Texas Press.

Sachs, Carolyn (1983) *The Invisible Farmers: Women in Agricultural Production*. Totowa: Rowman & Littlefield.

Schiebinger, Londa (1987) "The history and philosophy of women in science: a review essay", *Signs* 12(2): 305–32.

Science Policy Research Unit (1982) *Microelectronics and Women's Employment in Britain*. University of Sussex.

Sofia, Zoe (also Zoe Sofoulis) (1984) "Exterminating fetuses: abortion, disarmament, and the sexo-semiotics of extra-terrestrialism", *Diacritics* 14(2): 47–59.

Sontag, Susan (1977) *On Photography*. New York: Dell.

Stacey, Judity (1987) "Sexism by a subtler name? Postindustrial conditions and postfeminist consciousness", *Socialist Review* 96: 7–28.

Stallard, Karin, Ehrenreich, Barbara, and Sklar, Holly (1983) *Poverty in the American Dream*. Boston: South End.

Sturgeon, Noel (1986) "Feminism, anarchism, and non-violent direct action politics", University of California at Santa Cruz, PhD qualifying essay.

Sussman, Vic (1986) "Personal tech. Technology lends a hand", *The Washington Post Magazine*, 9 November, pp. 45–56.

Traweek, Sharon (1988) *Beamtimes and Lifetimes: The World of High Energy Physics*. Cambridge, MA: Harvard University Press.

Treichler, Paula (1987) "AIDS, homophobia, and biomedical discourse: an epidemic of signification", *October* 43: 31–70.

Weizenbaum, Joseph (1976) *Computer Power and Human Reason*. San Francisco: Freeman.

Welford, John Noble (1 July, 1986) "Pilot's helmet helps interpret high speed world", *New York Times*, pp. 21, 24.

Wilfred, Denis (1982) "Capital and agriculture, a review of Marxian problematics", *Studies in Political Economy* 7: 127–54.

Winner, Langdon (1977) *Autonomous Technology: Technics out of Control as a Theme in Political Thought*. Cambridge, MA: MIT Press.

—— (1980) "Do artifacts have politics?", *Daedalus* 109(1): 121–36.

—— (1986) *The Whale and the Reactor*. Chicago: University of Chicago Press.

Winograd, Terry and Flores, Fernando (1986) *Understanding Computers and Cognition: A New Foundation for Design*. Norwood, NJ; Ablex.

Wittig, Monique (1973) *The Lesbian Body*, David LeVay, trans. New York: Avon, 1975 (*Le corps lesbien*, 1973).

Women and Poverty, special issue (1984) *Signs* 10(2).

Wright, Susan (1982, July/August) "Recombinant DNA: the status of hazards and controls", *Environment* 24(6): 12–20, 51–53.

—— (1986) "Recombinant DNA technology and its social transformation, 1972–82", *Osiris*, 2nd series, 2: 303–60.

Young, Robert M. (1979, March) "Interpreting the production of science", *New Scientist* 29: 1026–8.

—— and Levidow, Les, eds (1981, 1985) *Science, Technology and the Labour Process*, 2 vols. London: CSE and Free Association Books.

Yoxen, Edward (1983) *The Gene Business*. New York: Harper & Row.

Zimmerman, Jan, ed. (1983) *The Technological Woman: Interfacing with Tomorrow*. New York: Praeger.

38

In Defense of Bacon

Alan Soble

> What a man had rather were true he more readily believes. Therefore he rejects difficult things from impatience of research.
>
> *Francis Bacon,* Novum organum, *book 1, 49*

Feminist science critics, in particular Sandra Harding, Carolyn Merchant, and Evelyn Fox Keller, claim that sexist sexual metaphors played an important role in the rise of modern science, and they single out the writings of Francis Bacon as an especially egregious instance. I defend Bacon.

Science and Rape

In an article printed in the *New York Times*, Sandra Harding introduced to the paper's readers one of the more shocking ideas to emerge from feminist science studies:

> Carolyn Merchant, who wrote a book called "Death of Nature," and Evelyn Keller's collection of papers called "Reflections on Gender & Science" talk about the important role that sexual metaphors played in the development of modern science. They see these notions of dominating mother nature by the good husband scientist. If we put it in the most blatant feminist terms used today, we'd talk about marital rape, the husband as scientist forcing nature to his wishes.[1]

From *A House Built on Sand: Exposing Postmodern Myths about Science*, ed. Noretta Kortege, Oxford: Oxford University Press, 1998, pp. 195–215. Reprinted by permission of Oxford University Press.

Harding asserts elsewhere, too, that sexist metaphors played an important role in the development of science.[2] But here she understates the point by referring to "marital rape" and so does not convey it in the "most blatant feminist terms," because her usual way of making the point is to talk about rape and torture in the same breath, not mentioning marriage. For example, Harding refers to "the rape and torture metaphors in the writings of Sir Francis Bacon and others (e.g., Machiavelli) enthusiastic about the new scientific method" (*SQIF*, 113). By associating rape metaphors with science, Harding expects the unsavoriness of rape to spill over onto science:

> Understanding nature as a woman indifferent to or even welcoming rape was...fundamental to the interpretations of these new conceptions of nature and inquiry...There does...appear to be reason to be concerned about the intellectual, moral, and political structures of modern science when we think about how, from its very beginning, misogynous and defensive gender politics and...scientific method have provided resources for each other. (*SQIF*, 113, 116)

I dare not hazard a guess as to how many people read Harding's article in the *Times*, how many clipped out that scandalous bit of bad publicity for science and put it on the refrigerator, or how many still have some vague idea tying science to rape. But the belief that vicious sexual metaphors were important to science has gained some currency in the academy.[3]

Contemporary Sexual Metaphors

In *Whose Science? Whose Knowledge?* Harding proposes that the "sexist and misogynistic metaphors" that have thus far "infused" science be replaced by "positive images of strong, independent women," metaphors based on "womanliness" and "female eroticism woman-designed for women" (*WSWK*, 267, 301). Harding defends her proposal by claiming that "the prevalence of such alternative metaphors" would lead to "less partial and distorted descriptions and explanations" and would "foster the growth of knowledge": "If they were to excite people's imaginations in the way that rape, torture, and other misogynistic metaphors have apparently energized generations of male science enthusiasts, there is no doubt that thought would move in new and fruitful directions" (267). What are the misogynistic metaphors that have already "energized" science and that must be replaced? In a footnote, Harding sends us to chapter 2 of *WSWK*. There we find a section entitled "The Sexual Meanings of Nature and Inquiry" (42–46), which contains merely four examples of metaphors in the writings of two philosophers (Francis Bacon and Paul Feyerabend), one scientist (Richard Feynman), and the unnamed preparers of a National Academy of Sciences (NAS) booklet, "On Being a Scientist."

In the passage from Feynman's Nobel lecture quoted by Harding, the physicist reminisces about a theory in physics as if it were a woman with whom he fell in love, a woman who has become old yet has been a good mother and left many children (*WSWK*, 43–44; *SQIF*, 120). Harding incredibly interprets this passage as "thinking of mature women as good for nothing but mothering" (*SQIF*, 112). From the NAS booklet, Harding quotes: "The laws of nature are not...waiting to be plucked like fruit from a tree. They are hidden and unyielding, and the difficulties of grasping them add greatly to the satisfaction of success" (*WSWK*, 44). Here, says Harding, one can hear "restrained but clear echoes" of sexuality. Perhaps the metaphors used by Feynman and the NAS are sexual, but they are hardly misogynistic or vicious, and I wonder why Harding put them on display.[4] These were supposed to be examples of how viciously sexist metaphors "energized" science, but they seem feeble. In fact, about her four examples Harding claims only that in Bacon's writings is there a rape metaphor.[5] But let us examine her treatment of Feyerabend first, for there are significant connections between them.

Harding quotes the closing lines of a critique of Kuhn and Lakatos by Feyerabend, who ends his technical paper with the joke that his view "changes science from a stern and demanding mistress into an attractive and yielding courtesan who tries to anticipate every wish of her lover. Of course, it is up to us to choose either a dragon or a pussy cat for our company. I do not think I need to explain my own preferences" (Feyerabend, 229).[6] Harding's complaint is not that Feyerabend employed a sexual metaphor, for in *WSWK* Harding condones "alternative" sexual images reflecting "female eroticism woman-designed for women" (267). Rather, Feyerabend's metaphor – science is either a selfish shrew who exploits men or a prostitute who waits on men hand and foot – is the wrong kind of sexual metaphor. Harding quotes the same passage in her earlier *SQIF*, giving it as an example of the attribution of gender to scientific inquiry (*SQIF*, 120).

In her view, this passage conveys, as does Feynman's, a cultural image of "manliness." Whereas Feynman's notion of manliness is "the good husband and father," Feyerabend's notion is "the sexually competitive, locker-room jock" (*SQIF*, 120). Thus science, in Feyerabend's metaphor, is a sexually passive, accommodating woman, and the scientist and the philosopher of science are the jocks she sexually pleases.[7] I do not see how portraying science as a courtesan implies that the men who visit her, scientists and philosophers, are locker-room jocks. The fancy word *courtesan*, if it implies anything at all, vaguely alludes to a debonair Hugh Hefner puffing on his pipe, not to a Terry Bradshaw swatting a bare male butt with a wet towel. (Should we homogenize men, or think of the philosopher of science as a locker-room jock wannabe?)

Harding concludes her brief discussion of Feyerabend in *SQIF* by claiming that his metaphor, coming strategically at the end of his paper, serves a pernicious purpose. He depicts "science and its theories" as "exploitable women" and the scientist as a masculine, manly man to imply to his (male) audience that his philosophical "proposal should be appreciated *because* it replicates gender politics" of a sort they find congenial (*SQIF*, 121). In *WSWK*, Harding similarly asserts that this metaphor was the way Feyerabend "recommended" his view (43).

This line of thought is not very promising. Some men readers prefer strict to submissive women; would Feyerabend's contrary preference for kittens tend to undermine for them his critique of Kuhn, because it does not match their tastes? I agree that some women reading his paper would probably not empathize with the metaphor, even if they concurred with the critique of Kuhn that preceded it. But they could, if they wished, ignore it as irrelevant to Feyerabend's arguments – at least because the metaphor comes at the end, tacked onto the arguments already made and digested. Had Bacon employed rape metaphors, Harding would be right that "it is...difficult to imagine women as an enthusiastic audience" (*SQIF*, 116). Still, had there been any women in Bacon's audience, they could have disregarded his metaphors and accepted (or rejected) the rest on its merits. For Harding to assert that the men in Feyerabend's audience would be in part persuaded by this appeal and that Feyerabend thought that he could seduce them with his "conscientious efforts at gender symbolism" (120) insults men and exaggerates the chicanery of philosophy.

Harding on Bacon

According to Harding, vicious sexual metaphors were infused into modern science at its very beginning, were instrumental in its ascent, and eventually became "a substantive part of science" (*WSWK*, 44). Harding thinks Francis Bacon (1561–1626) was crucial to this process.[8] What she says about Feyerabend, that he hoped his view would "be appreciated *because* it replicates gender politics" (*SQIF*, 121), is what she claims about Bacon, although in more extreme terms: "Francis Bacon appealed to rape metaphors to persuade his audience that experimental method is a good thing" (*WSWK*, 43; see *SQIF*, 237).[9]

This is a damning criticism. Bacon is not depicted as a negligible Feyerabend making silly jokes about science the prostitute. Rather, Harding is claiming that Bacon drew an analogy between the experimental method and rape and tried to gain advantage from it (see *SQIF*, 116). Imagine the scene that Harding implies. Bacon wants to persuade fellow scholars to study nature systematically by using experimental methods that elicit changes in nature, rather than to study nature by accumulating specimens and observing phenomena passively. So, thinking that his audience found rape desirable, attractive, permissible, or at least that it would be fun, even if despicable, Bacon champions experimentalism by drawing an analogy between it and rape. Bacon says to them: Think of doing science my way as forcing apart with your knees the slender thighs of an unwilling woman, pinning her under the weight of your body as she kicks and screams in your ears, grabbing her poor little jaw roughly with your fist to shut her mouth, and trying to thrust your penis into her dry vagina; that, boys, is what the experimental method is all about.

What did Bacon do or say to deserve such an abusive accusation? Is there any evidence that Bacon's writings contain a rape metaphor? Here is the entire text that Harding offers to support her charge:

> For you have but to hound nature in her wanderings, and you will be able when you like to lead and drive her afterwards to the same place again. Neither ought a man to make scruple of entering and penetrating into those holes and corners when the inquisition of truth is his whole object. (*WSWK*, 43)[10]

I suppose that a man who made no scruple of penetrating holes (and corners?) might be a rapist, but he also might be a foxhunter, a proctologist, or a billiard player. And I suppose that to "hound" nature could be seen as raping her. But the spirited student who storms my office and too often sits down next to me in the cafeteria, hoping for some words of wisdom – no more than that – also is hounding me.

Perhaps Harding, in reading "hound" as about rape, has in mind Robin Morgan's definition of rape:

> *Rape exists any time sexual intercourse occurs when it has not been initiated by the woman, out of her own genuine affection and desire*....How many millions of times have women had sex "willingly" with men they didn't want to have sex with?...How many times have women wished just to sleep instead or read or watch the Late Show?...Most of the decently married bedrooms across America are settings for nightly rape. (Morgan, 165–66)

The implication of Morgan's definition of rape for doing science seems to be that the search for knowledge must proceed passively, letting Nature "initiate" our congress with her; if science pressures Nature, it is raping her. In Morgan's view, a man who pesters ("hounds") his wife for sex when she prefers to watch television has committed rape if she caves in. But Bacon's audience would never have recognized this prosaic sexual phenomenon as rape. Nor would most reasonable men and women today judge it to be rape, even if we lament the fact that men's economic power sometimes gives them an advantage in marital bargaining. Morgan's definition of rape trivializes that phenomenon, and even Harding thinks the rape metaphor in Bacon is "violent," not mere pushiness (*SQIF*, 116).

Keller's work might help us discern the reasoning behind Harding's reading of Bacon's *De augmentis*. In her essay "Baconian Science," Keller repeats the first of the two sentences quoted by Harding, to illustrate her claim that for Bacon, even though "Nature may be coy," she can still "be conquered": "For you have but to follow and as it were hound nature in her wanderings, and you will be able, when you like, to lead and drive her afterwards to the same place again."[11] What "leads to [this] conquest," in Keller's view, is "not simple violation, or rape, but forceful and aggressive seduction." Just because Keller interprets this passage from *De augmentis* as a rape-free zone does not mean there is no rape image there, for Keller warns us that "the distinction between rape and conquest sometimes seems too subtle" (*REF*, 37).

Harding might have concluded, therefore, that the conquest of nature implied by "hound" is more accurately described as rape. Her rape interpretation of Bacon would then depend on eradicating the difference between rape and seduction. But Bacon recognizes this difference; he advises that science would be more successful by patiently wooing nature than by raping her: "Art...when it endeavours by much vexing of bodies to force Nature to its will and conquer and subdue her...rarely attains the particular end it aims at.... Men being too intent upon their end...struggle with Nature than woo her embraces with due observation and attention" ("Erichthonius," myth 20, *Wisdom of the Ancients*, in Robertson, 843). The seduction, in this passage at least, seems considerate and delicate, not, as Keller says, "forceful and aggressive."[12]

Something else can be gleaned from Keller. Compare the sentence Keller quoted from Bacon's *De augmentis* with the first sentence Harding attributes to Bacon. The sentence as quoted by Keller correctly includes the phrase "follow and as it were" (*Works* 4, 296), which are missing from Harding's quotation. I think there is some difference between Bacon's nuanced "follow and as it were hound" nature and Harding's rendition, the crude and unqualified "hound" her. Harding's misquotation of Bacon (really a misquotation of Merchant's quotation of Bacon) is serious, since students of Harding who read only *WSWK* will be misguided. Five years earlier, in *SQIF*, Harding quoted this *De augmentis* passage twice and almost got it right.[13] Here is one instance:

> To say "nature is rapable" – or, in Bacon's words: "For you have but to follow and as it were hound nature in her wanderings, and you will be able when you like to lead and drive her afterward to the same place again....Neither ought a man to make scruple of entering and penetrating into those holes and corners when the inquisition of truth is his whole object" – is to *recommend* that similar benefits can be gained from nature if it is conceptualized and treated like a woman resisting sexual advances. (*SQIF*, 237; ellipsis and italics are Harding's)

In repeating her accusation against Bacon, Harding does not sense that "follow and as it were" is a substantial qualification of "hound."

Furthermore, the passage's "penetrating" need not be taken as having any sexual implications.[14] And even if "penetration" is sexual (was it for Bacon and his audience?), "penetration" does not entail or suggest rape. Perhaps Harding construes the *unscrupulous* penetration of holes to be an allusion to rape.[15] But this reading makes sense only by wrenching "scruple" out of context. Bacon's point, which he repeats elsewhere, is that any scientist determined to find the truth about nature should be prepared to get his hands dirty; when truth is the goal, everything must be investigated, even if to prissy minds the methods employed and the objects studied are foul. (Think about survey researchers justifying the study of human sexual behavior.) Thus, a few lines after "scruple" in *De augmentis*, Bacon complains that "it is esteemed a kind of dishonour...for learned men to descend to inquiry or meditation upon matters mechanical" (*Works* 4, 296; see *Advancement, Works* 3, 332).

Parts of *Novum organum* and *De augmentis* (whose title begins *De dignitate*) were intended to establish the dignity, despite the dirty hands, of engaging in science to improve the human condition (see *Parasceve, Works* 4, 257–59). *Novum organum* is especially clear on this. In one aphorism, Bacon condemns "an opinion...vain and hurtful;

namely, that the dignity of the human mind is impaired by long and close intercourse with experiments."[16] Bacon returns to this theme later in *Novum organum*:

> And for things that are mean or even filthy, – things which (as Pliny says) must be introduced with an apology, – such things, no less than the most splendid and costly, must be admitted into natural history. Nor is natural history polluted thereby; for the sun enters the sewer no less than the palace, yet takes no pollution.... For whatever deserves to exist deserves also to be known, for knowledge is the image of existence; and things mean and splendid exist alike. Moreover as from certain putrid substances – musk, for instance, and civet – the sweetest odours are sometimes generated, so too from mean and sordid instances there sometimes emanates excellent light and information. But enough and more than enough of this; such fastidiousness being merely childish and effeminate. (120; see 121)

Bacon is, like Calvin, a rascal. He prefers to dissect bugs and chase snakes than play house or have an afternoon tea with Susie.

Bacon's two sentences from *De augmentis* appear once again in Harding's *SQIF*:

> Bacon uses bold sexual imagery to explain key features of the experimental method as the inquisition of nature: "For you have but to follow and as it were hound nature in her wanderings, and you will be able when you like to lead and drive her afterward to the same place again. ...Neither ought a man to make scruple of entering and penetrating into those holes and corners, when the inquisition of truth is his whole object – as your majesty has shown in your own example."... It might not be immediately obvious to the modern reader that this is Bacon's way of explaining the necessity of aggressive and controlled experiments in order to make the results of research replicable! (*SQIF*, 116; the first ellipsis is Harding's)

Harding is right about one thing: contrary to her interpretation, it is not obvious that Bacon, with the phrase "to the same place," is referring to experimental replicability.[17] Nor is Keller's reading plausible, that Bacon here asserts that nature can be "conquered" (*REF*, 36–37). William Leiss earlier suggested this reading: "Having discovered the course leading to the end result, we are able to duplicate the process at will" (59).[18] This makes sense, because Bacon's immediately preceding sentence is "From the wonders of nature is the most clear and open passage to the wonders of art," and then by way of explaining or defending this idea, that nature herself teaches us how to fabricate artificial devices, Bacon now writes "For you have but to follow."

That is, to learn how to achieve one of nature's effects (to use an anachronistic example, the overcoming [conquering] of bacterial infection), we must study how nature accomplishes it (we learn how to follow nature by pestering her in the lab). Once we discover Nature's Way (the various mechanisms of the immune system), we can then copy, modify, and rearrange its main ingredients (develop "artificial" devices, vaccinations, that elicit antibodies) to "lead" nature "to the same place again." Bacon is modifying a point that appears elsewhere in his writings and that he made just a page before in *De augmentis*:

> The artificial does not differ from the natural in form or essence, but only in the efficient....Nor matters it, provided things are put in the way to produce an effect, whether it be done by human means or otherwise. Gold is sometimes refined in the fire and sometimes found pure in the sands, nature having done the work for herself. So also the rainbow is made in the sky out of a dripping cloud; it is also made here below with a jet of water. Still therefore it is nature which governs everything.[19]

In *Cogitata et visa*, Bacon goes so far as to say that phenomena found in nature (he praises silk spun by a worm) and from which we can learn "are such as to elude and mock the imagination and thought of men" (Farrington, 96). Bacon's example in *De augmentis* of obtaining a rainbow from a spray of water is serene, even lovely, and makes it improbable that he viewed experimental manipulations as nothing but mere acts of aggression. Bacon's affirmation that nature "governs everything" – the ways of nature are responsible even for the artificial rainbow we make with a spray of water – is reason to doubt that he conceived of the relationship between science and nature principally as that between man the master and woman the dominated. At the very beginning of *Novum organum*, as well as in "The Plan" of *The Great Instauration* (*Works* 4, 32) – that is, often and in prominent places – Bacon writes that science is "the servant and interpreter of Nature" and "Nature to be commanded must be obeyed" (1, 3).[20]

In *SQIF*, Harding introduces the *De augmentis* passage by saying that it contains "bold sexual imagery,"[21] but after quoting the passage, she escalates the charge: experimentalism, the "testing of hypotheses," is "here formulated by the father of scientific method in clearly sexist metaphors" (*SQIF*, 116). Harding immediately takes the next step: "Both nature and inquiry appear conceptualized in ways modeled on rape and torture – on men's most violent and misogynous relationships to women – and this modeling is advanced as a reason to value science." Of course, if the passage contains no rape metaphor, Harding's thesis that Bacon employed a rape metaphor to recommend experimentalism falls apart. But even if the passage does contain a rape image, why think that Bacon deliberately used it to promote experimentalism? Or that the metaphor really did "energize" the new science?

Examine the two sentences Harding quotes from the 1623 *De augmentis* as they appeared in its 1605 English-language predecessor of this text, *The Advancement of Learning*: "For it is no more but by following and as it were hounding Nature in her wanderings, to be able to lead her afterwards to the same place again.…Neither ought a man to make scruple of entering into these things for inquisition of truth" (*Works*, 3, 331). Three candidate offenders, "drive," "penetrate," and "holes," are missing.[22] Which of these two texts was more momentous for Bacon's program? In the *Advancement*, written soon after the accession of James I, Bacon was surely trying to win over his king; when writing *De augmentis*, the older Bacon had been stripped of his official positions and was writing for posterity. In which situation should we expect Bacon to use harsh (or soft) language to do the persuading? This is treacherous terrain; the contrast between *Advancement* and *De augmentis* should give us pause.

In addition, in *Novum organum* and elsewhere, Bacon argues on behalf of science in terms more likely to convince his audience and "energize" science. Science will improve the human condition (81, 129; see *De augmentis, Works* 4, 297); in this way, the works of science are, for Bacon, works of love.[23] And in a different vein, at the end of book 1 of *Novum organum* (129), Bacon reminds his audience that science fulfills the biblical command (Genesis 1:26) that humans should rule the world (see Leiss, 53). Both these themes in Bacon are typical and familiar.[24] Why conjecture that Bacon also appealed to a rape metaphor, as if that were the icing on the cake of his vindication of science? Harding seems to assume that from the (purported) fact that the text contains a rape metaphor, it follows that Bacon used the metaphor to persuade his audience to embrace his philosophy. But it is dangerous to draw that conclusion solely from the presence of a rape or any weaker sexual metaphor. This mistake is similar to one made about erotica, namely, arguing solely from the presence in a text of a photographic or linguistic depiction of a certain sexual act that the text recommends or endorses the depicted act (Soble 1985, 73–74).

My mentioning erotica here is pertinent. Harding entitles the section of *SQIF* in which she first quotes Bacon "Should the History and Philosophy of Science Be X-Rated?" "This question is only slightly antic" because (previewing her comments on Bacon and Feyerabend) "we will see assumptions that…the best scientific activity and philosophical thinking about science are to be modeled on men's most misogynous relationships to women" (*SQIF*, 112). Harding thinks that science, its history and philosophy, should be rated "X" because it contains, in her view, explicit and nasty sex. But I cannot perceive, as she does, the sexually aggressive locker-room jock in Feyerabend's metaphor, nor can I perceive, as she does, the rape metaphor in Bacon.

Perhaps Harding's philosophy itself should be rated "X": she injects sex or rape where there is none to begin with. Consider the NAS metaphor: the laws of nature are "not waiting to be plucked like fruit from a tree [but] are hidden and unyielding." Harding finds here a "clear echo" of sexuality. But these few words can be read, without effort, as innocent and nonsexual; so it is Harding, like the person who feels squeamish at the sight of uncovered piano legs, who has infused the sex into them. Furthermore, if Harding is uncommonly sensitive to the nuances of language and this enables her to extract a rape metaphor out of "hound" and "holes," or if the metaphor is one that mostly or only women could sense, this would undermine Harding's claim that Bacon used a rape metaphor to persuade his (male) audience.[25]

One more example: Bacon's portrayal of inquiry as a "disclosing of the secrets of nature" (*Works* 4, 296; see *Novum organum*, 43, 89). We could construe the language of discovering, or uncovering, the hidden secrets of nature as alluding to a quest for carnal knowledge of a deeply concealed female sexuality that is not keen on being exposed (for whatever reason, be it prudish modesty, girlish self-doubt, or lazy reluctance). We might interpret such language in this way to suggest that this was the latent meaning of the philosophical claim that underneath the appearances of events are unobservable structures and forces about which we have no direct knowledge and about which we will remain ignorant unless we investigate, experimentally, their phenomenal manifestations. But the sexual metaphor, if we insist on digging it out, is tame; there is no rape and no need to twist the metaphor against its will (as Harding does) to *be* rape.

Merchant on Bacon

According to Harding, rape is Bacon's metaphor for and model of the experimental method. For the historian Carolyn Merchant, "the interrogation of witches" by torture (*DON*, 172)[26] is, instead, Bacon's "symbol for the interrogation of nature":

> Much of the imagery he used in delineating his new scientific objectives and methods...treats nature as a female to be tortured through mechanical interventions [and]...strongly suggests the interrogations of the witch trials and the mechanical devices used to torture witches. In a relevant passage, Bacon stated that the method by which nature's secrets might be discovered consisted in investigating the secrets of witchcraft by inquisition, referring to the example of James I. (*DON*, 168)

In which "relevant" passage does Bacon state such a thing? Merchant calls on our passage from *De augmentis* (*Works* 4, 296; italics and ellipses are Merchant's) to substantiate her assertion:

> *For you have but to follow and as it were hound nature in her wanderings, and you will be able when you like to lead and drive her afterward to the same place again.* Neither am I of opinion in this history of marvels, that superstitious narratives of *sorceries, witchcrafts, charms*, dreams, divinations, and the like, where there is an assurance and clear evidence of the fact, should be altogether excluded....howsoever the use and practice of such arts is to be condemned, yet from the speculation and consideration of them...a useful light may be gained, not only for a true judgment of the offenses of persons charged with such practices, *but likewise for the further disclosing of the secrets of nature. Neither ought a man to make scruple of entering and penetrating into these holes and corners, when the inquisition of truth is his whole object – as your Majesty has shown in your own example*. (*DON*, 168)[27]

Bacon does not state that nature should be tortured in the way witches are interrogated. I do not perceive any torture metaphor here. The two sentences Merchant italicizes (roughly the two in which Harding finds rape) do not deserve this interpretation, nor do the sentences that fall between.

Merchant misreads why Bacon mentions James I. In referring to his king in this passage, Bacon is not alluding to cruel methods of inquisition but is pointing out that James I was willing to get his hands dirty by studying witchcraft. What James I "show[ed] in [his] own example," says Bacon, is that everything in nature is an appropriate object for scientific study – one of Bacon's principles – not that science should torture nature as if it were a witch.[28] Thus the text provides no reason to think that "Bacon's recommendation that experimental method should characterize the new science was couched in terms of the method James I had successfully used to 'expose' witches" (Nelson, 353, n. 137).

Moreover, by 1622, when *De augmentis* was being written, James I had already changed his mind about witches and had intervened to save some of the accused from execution (Robbins, 278–79).[29] Thus Bacon could not have been appealing here to a beloved pastime of James I in devising a metaphor for experimentalism. Bacon might have had better luck appealing to a torture-the-witch metaphor in the 1605 *Advancement*, right after James's Statute of 1604, but as I have pointed out, the language of that early version of *De augmentis* is not very provocative and even less amenable to a Harding–Merchant type of reading.

If we reinsert into this passage from *De augmentis* the words that Merchant deleted (indicated in the passage by italics), we can better understand Bacon:

> Neither am I of opinion in this history of marvels, that superstitious narratives of sorceries, witchcrafts, charms, dreams, divinations, and the like, where there is an assurance and clear evidence of the fact, should be altogether

> excluded. *For it is not yet known in what cases, and how far, effects attributed to superstition participate of natural causes; and therefore* howsoever the use and practice of such arts is to be condemned, yet from the speculation and consideration of them (*if they be diligently unravelled*) a useful light may be gained, not only for a true judgment of the offenses of persons charged with such practices, but likewise for the further disclosing of the secrets of nature.

Bacon is not recommending that witches or nature be tortured. Rather, he is telling his audience to pay attention to the distinction between the context of discovery and the context of justification: regardless of the source of certain claims ("narratives"), their content might be true, and this can be known only by investigating them scientifically. This is what Harding calls the "desirable legacy" of modern philosophy of science, the notion that "we should be able to decide the validity of a knowledge claim apart from who speaks it" (*WSWK*, 269). Bacon, a true Wiccan, suggests that scientists might learn something from witches.

The passage from *De augmentis* was the first and longest quotation that Merchant used in making her case, but her best shot missed the mark. She quotes other passages, mostly scattered words and partial sentences, to round out her argument. Although I cannot deal with them all, her frequent reference to Proteus in Bacon's works does deserve discussion. Introducing one of them, Merchant claims that Bacon drew an analogy between the "inquisition" of nature and "the torture chamber" (*DON*, 169). Here are his words: "For like as a man's disposition is never well known or proved till he be crossed, nor Proteus ever changed shapes till he was *straitened* and *held fast*; so nature exhibits herself more clearly under the *trials* and *vexations* of art than when left to herself" (*De augmentis, Works* 4, 298, Merchant's italics. See also *Advancement, Works* 3, 333).

Here, force is applied to Proteus in particular. But if we take Bacon's analogy between nature and Proteus literally, the implication is that we must be smart enough to outfox nature to get a hearing for our questions. Then as we try to hold fast to nature, nature will almost always escape by changing unpredictably and uncontrollably, slithering or leaping away or disappearing as a gas, and we will not get an answer. Our attempts to bind her will be largely fruitless. The Proteus image, then, is a tribute to the sagacity and subtlety of nature (see *Novum organum*, 10, 24).

If we do not take the analogy literally, we can be content with the core of Bacon's idea, which has nothing to do with torture and everything to do with the advantages to be gained from experimental manipulations. For example, a person "left to herself" might have little opportunity to reveal the greed that lies buried in her heart. But if she is tempted by a stuffed wallet dropped on an empty street – the deliberate "vexations" arranged by a conniving social scientist – she might respond by keeping the money and not returning the wallet to its owner, thereby revealing the greedy part of her character.

Merchant quotes another Proteus passage that perhaps more strongly supports her reading: "The vexations of art are certainly as the bonds and handcuffs of Proteus, which betray the ultimate struggles and efforts of matter" (*Parasceve, Works* 4, 257; *DON*, 171). "Bonds and handcuffs" look damning when equated with "vexations of art" (experimental techniques). But these devices are being used on Proteus, the supreme Houdini, in which case we need not be so anxious about his safety (besides, Proteus is a guy). What does Bacon mean by the less than transparent "betray the ultimate struggles and efforts of matter"? In the next sentence (not quoted by Merchant), Bacon explains: "For bodies will not be destroyed or annihilated; rather than that they will turn themselves into various forms." Here Proteus stands for matter; so, no matter how much we bind matter, says Bacon, it is indestructible. Bacon is not issuing a normative claim, as if urging us to bind matter to prevent it from behaving perversely (Merchant's reading, *DON*, 171; see Bordo, 109); he is making the ontological point that no amount of binding will allow us to annihilate matter.

Bacon states the idea nicely in *Wisdom of the Ancients*:

> if any skilful Servant of Nature shall bring force to bear on matter, and shall vex it and drive it to extremities as if with the purpose of reducing it to nothing, then will matter (since annihilation or true destruction is not possible except by the omnipotence of God) finding itself in these straits, turn and transform itself into strange shapes, passing from one change to another till it has gone through the whole circle ("Proteus," myth 13 [Robertson, 838]).

Likewise, Proteus in the *Odyssey* (book 4), at the hands of Menelaus and his crew, goes through the whole cycle and is never destroyed. Note that Bacon speaks here again of the scientist as a "servant of nature."

Let us also look at Bacon's frequent use of "vex" and its congeners. Merchant italicized "vexations" in *De augmentis*, and both Leiss (59) and Keller (*REF*, 36) think "vex" conveys sexual aggression. Even though Bacon's use of "vex" is occasionally strong (e.g., in relation to imperishable matter; *Thoughts on the Nature of Things, Works* 5, 427–28), "vex" does not always or usually carry a pernicious connotation but is meant, innocuously, along the lines of his "hound" and my "pester." For example, in *Novum organum* Bacon writes:

> For even as in the business of life a man's disposition and the secret workings of his mind and affections are better discovered when he is in trouble than at other times; so likewise the secrets of nature reveal themselves more readily under the vexations of art than when they go their own way. (98)[30]

This mature and polished statement of Bacon's philosophy contains no rape, no torture, no bondage, just the thought that to know nature it is not enough to watch it; nature must be provoked into showing us its inner workings. I find nothing to complain about in this thought, especially when I consider how much of my knowledge of human nature I never would have acquired had not my family, friends, and colleagues, let alone myself, been "crossed" (*De augmentis, Works* 4, 298) and thereby goaded into exposing features of our personalities we do not ordinarily broadcast.

Keller on Bacon

Going beyond Harding and Merchant, Evelyn Fox Keller argues that Bacon's sexual images involve a more "complex sexual dialectic" (Keller 1980, 302; *REF*, 35). In Bacon's language, the scientist both aggressively dominates and is "subservient" and "responsive" to nature (*REF*, 36–37); science is both master of and obedient to nature. Keller's essay is in effect a reply to Leiss, who says about Bacon's "famous formula" – we command nature by obeying her – that "some commentators have claimed that it sounds a note of humility in man's attitude toward nature. But this interpretation…invents inconsistencies which do not really exist in Bacon's work" (58–59). I think Keller is right to find inconsistency in Bacon's metaphors but wrong in what she makes of it.

In arguing for one side of the inconsistency, that Baconian science is "aggressive" in its "conquest" of nature, Keller assembles seven passages (*REF*, 36). The first is from Bacon's *Refutation of Philosophies*: "Let us establish a chaste and lawful marriage between Mind and Nature" (Keller 1978, 413; 1980, 301; *REF*, 36).[31] The metaphor seems benign, as does that of a "chaste, holy and legal wedlock" (Keller's second passage, from *The Masculine Birth of Time*; Farrington, 72), but Keller thinks of the marital imagery in Bacon as aggressive (*REF*, 19).[32] There is no need to. Bacon uses marriage imagery promiscuously in his writings, with no hint of aggression: see the Preface to *The Great Instauration*, in which Bacon speaks about the "true and lawful marriage" he is attempting to effect "between the empirical and the rational faculty" (*Works* 4, 19) and *Novum organum*, in which Bacon criticizes those who would "deduce" Christianity from the principles of philosophy, thereby "pompously solemnising this union of sense and faith as a lawful marriage" (89). And in the two marriage passages quoted by Keller, there is no intimation of aggression at all; Bacon immediately proceeds to say (typical for him; see "The Plan" of *The Great Instauration, Works* 4, 27) that the marriage will issue in "wholesome and useful inventions…to bring relief" from "human necessities" (*Refutation*, Farrington, 131) and "will overcome the immeasurable…poverty of the human race" (*Masculine Birth*, Farrington, 72).

Thus it is difficult to agree with Keller that the marital image in Bacon "constitutes an invitation to the 'domination of nature' " (Keller 1978, 429, n. 5; *REF*, 91, n. 6) because it "sets the scientific project squarely in the midst of our unmistakably patriarchal tradition" (Keller 1978, 423). When Bacon writes in *Masculine Birth* (Keller's third passage), "I am come in very truth leading to you Nature with all her children to bind her to your service and make her your slave" (*REF*, 36, see 48; Farrington, 62), Keller sees a wedding announcement in which nature is fingered as the bride. She is a slave, since that will be her married lot. The image of binding nature as a slave is surely an ugly one, but it is not necessarily about marriage.

Keller cites four more passages that she believes exhibit the aggression toward nature in Bacon's images. In one, from the Preface to *Novum organum*, Bacon writes of trying to "penetrate further" and "find a way at length into her inner chambers" (*Works*, 4, 42), which is hardly pushy. Keller then turns to the *De augmentis* "hound...and drive" passage that I have already examined, which (contra Keller) contains no obvious conquest, sexual or otherwise. Keller also refers to Bacon's thought that more can be learned about nature by vexing her than by observing her in freedom, "left to herself."

The final passage cited by Keller is from *Cogitata et visa* (1607, published in 1653; Farrington, 57). According to Keller, Bacon says about "the discipline of scientific knowledge, and the mechanical inventions it leads to" (*REF*, 36) that they do not "merely exert a gentle guidance over nature's course; they have the power to conquer and subdue her, to shake her to her foundations."[33] This is, apparently, crude aggression. But our judgment of Bacon will be improved by realizing three things. First, Bacon is in part referring here not to mechanical devices in general but specifically to "Printing, Gunpower, and the Nautical Needle," which he thinks had more effect on "human affairs" than any "empire," "school," or "star" (Farrington, 93). Bacon makes the point hyperbolically – that these exemplary inventions are things that can conquer and shake nature to her foundations – but given his point, the sentence is a prime candidate for being read generously. Second, *Cogitata et visa* is a polished work, even though unpublished during Bacon's life, and, except for this line, is tranquil throughout. Much of *Cogitata* went right into *Novum organum*; it contains one questionable sentence that was not destined to join its sisters there and on which Keller pounces.[34] Third, in the later 1612 *Description of the Intellectual Globe*, Bacon repeats the line but softens it: he warns us against making the "subtle error," one that causes "despair," of thinking that science has "no power to make radical changes, and shake her in the foundations" (*Works* 5, 506). Both *conquer* and *subdue* are gone. In an analogous passage in the late *De augmentis*, Bacon redeems himself linguistically by advising against the "subtle error" of thinking, with "premature despair," that science has "by no means [the power] to change, transmute, or fundamentally alter nature" (*Works* 4, 294). Now the rest, *shaking* the foundations, is gone.

Guided into Bacon by Harding, Merchant, and Keller, one expects to find his work cluttered with scandalous metaphors. But Keller unveils only one clearly ugly line out of thousands of pages of Bacon's lifework, and this, "make her your slave," occurs in a tiny fragment of a manuscript written at the dawn of his philosophical career (the 1603 *Masculine Birth*, published posthumously in 1653; Farrington, 57). Furthermore, "shake her to her foundations" (*Cogitata*) is either expunged from or revised in Bacon's later writings. In "Feminism and Science," a paper widely reprinted (e.g., Boyd, Gasper, and Trout, 279–88; Harding and O'Barr, 233–46), Keller bizarrely reproduces these two lines, and only these two, to discredit the new science (1982, 598).[35] To poke through these essays and parade their meanest two lines as the truth about Bacon and his scientific philosophy, without any methodological reluctance, is uncharitable if not hostile.[36]

In arguing on behalf of the other side of the inconsistency, that Bacon's images sometimes express a different attitude, Keller mentions the well-known passages I cited in which Bacon speaks about the scientist as the servant of nature who obeys her. So Keller finds a "puzzle" in which the "ambiguities" of Bacon's images "become contradictions." Science is "aggressive yet responsive, powerful yet benign, masterful yet subservient, shrewd yet innocent" (*REF*, 37). Keller solves the problem to her own satisfaction by sensing that Bacon viewed, in *Masculine Birth*, the human mind as hermaphroditic or as sequentially bisexual. As she interprets Bacon:

> To receive God's truth, the mind must be pure and clean, submissive and open. Only then can it give birth to a masculine and virile science. That is, if the mind is pure, receptive, and submissive in its relation to God, it can be transformed by God into a forceful, potent, and virile agent in its relation to nature. Cleansed of contamination, the mind can be impregnated by God and, in that act, virilized: made potent and capable of generating virile offspring in its union with Nature. (*REF*, 38; see Keller 1980, 304, 307)

Keller's proposal, that Bacon's model of the mind is hermaphroditic or bisexual, since the mind begins as a passive female and is transformed into a virile male, does not obviously represent progress in understanding Bacon's philosophy. There is as little in the extremely brief *Masculine Birth* to support Keller's elaborate reading of Bacon as there is in *De augmentis* to support Harding's perception of rape. The fragments of *Masculine Birth* are themselves a puzzle and not to be entrusted with the task of clarifying vexatious passages in Bacon's mature works. The abundance of contrary images in Bacon's writings suggests that gender images in Bacon are *less* interesting than Keller's esoteric reading makes them out to be.[37]

Methodology

To make the case that Bacon deliberately used rape, sexual, or torture metaphors to convince his audience of the virtues of experimental science, the following (only necessary) conditions must be satisfied. The metaphors (1) should clearly be in his texts (which they are not), (2) should be located in vital places (not merely in posthumously published manuscripts), (3a) should appear in several passages (not just once) yet also (3b) not indiscriminately (see Koertge, 353–54), and (4) should not have their thrust diluted by other, contrasting images (see Landau, 49–50). If Feyerabend had likened not only science but also such disparate things as rivers, music, and champagne to a kitten, or if at one point he called science a kitten but elsewhere a stern mistress, we would not be inclined to take his metaphors seriously. Similarly, to the extent that Bacon applied his images to a wide assortment of things or used diverse, even contradictory, metaphors, it would be implausible to claim that he relied on them to vindicate science.

On the one hand, Bacon sees nature as wise and subtle, so discovering Nature's Ways requires not just diligence but shrewd intelligence. Similarly, in Bacon's world – he and his peers are gentlemen, not barroom bruisers – wary women are wooed with poetry or bribed or promised a love that will not be forthcoming. Comprehending the secrets of nature might be like uncovering the secrets of a woman, but brain, not brawn, yields the joys of science and sex. On the other hand, nature is also, for Bacon, one tough cookie, whose floods and hurricanes and famines and pestilences kill us and destroy our property. But neither image in this dualism – nature is smart but, with luck and skill, can be seduced; nature is cruel but, with luck and skill, we can avoid the worst of it – is obnoxious. Bacon's mistake was similar to Feyerabend's, who conceived of women only as kittens or stern mistresses, imagery that excessively narrows women's modes of existence. Maybe both Baconians and Feyerabendians can be prodded into avoiding these mildly sexist polarities and to think, instead or in addition, in terms of an equal relationship with an independent and capable woman of substance. But Harding has other metaphors in mind. In calling for the "prevalence" of metaphors reflecting "female eroticism woman-designed for women" and "woman-to-woman relations" (*WSWK*, 267), she leaves little or no room for rehabilitated heterosexual metaphors.

The incoherence of Bacon's images goes beyond the mastery/obedience contrast Keller exhibits. Consider, for example, how indiscriminately, in *Novum organum* alone, Bacon uses bondage imagery. In one passage, he complains that "men's powers" have been "bound up" by the "enchantments of antiquity." In another, he criticizes logical demonstrations on the grounds that they "make the world the bond-slave of human thought, and human thought the bond-slave of words." And lest we get the impression that nature is the only object of bondage, Bacon recommends that the human mind should be bound, for it will do us some good. For the sake of the improvement of knowledge, scientists should "bind themselves to two rules" and "the understanding must not...be supplied with wings, but rather hung with weights, to keep it from leaping and flying" (84, 69, 130, 104, respectively; see 20). This is good advice for those reading Bacon today, as it might prevent them from claiming too quickly that the mind is, for Bacon, an appropriate object of bondage just because the mind is sometimes or partially female and from claiming, too ideologically, that the metaphor, as patriarchal and sexually oppressive, "energized" or provided a "resource" for the new science.

Also consider that the effect of matter on the course of nature is, almost incomprehensibly, more dramatic and graphic, in Bacon's language, than what science does, or can do, to nature ("hound" and "drive"). Bacon writes in *De augmentis* that nature

> is either free, and follows her ordinary course of development;...or she is driven out of her ordinary course by the perverseness, insolence, and frowardness [*sic*] of matter, and violence of impediments; as in the case of monsters; or lastly, she is put in constraint, moulded, and made as it were new by art and the hand of man; as in things artificial. (*Works* 4, 294; see *Parasceve, Works* 4, 253)

The contrast is sharper in *Description of the Intellectual Globe*:

> For nature is either free...or again she is forced and driven quite out of her course by the perversities and insubordination of wayward and rebellious matter, and by the violence of impediments;...or lastly, she is constrained, moulded, translated, and made as it were new by art and the hand of man. (*Works* 5, 505–6)

Constraint of nature by the human hand need not be vicious, violent, a bit of torture, or perverse. A bush – as natural a piece of nature as we can imagine – can be gently pruned, thereby constrained, "made as it were new by art," watered and fed and, as a result, it will both thrive and bring us pleasure.

How should we understand Bacon's using the bearded Pan to represent nature in myth 6 of *Wisdom of the Ancients* and his using Proteus to represent matter in myth 13 and nature in *De augmentis*? (Robertson, 828–32, 838; *De augmentis, Works* 4, 298.) Similarly, what are we to make of "Atalanta," myth 25? (Robertson, 847–48.) There Bacon says, "Art remains subject to Nature, as the wife to the husband." He has reversed his purportedly favorite sexist image. Instead of the husband science dominating his woman nature in a patriarchal marriage, science is the unfortunate wife dominated by nature the man. Bacon's point, which has nothing to do with gender, is that science (Atalanta) will not win the race with nature (Hippomenes) if she allows herself to be distracted by baubles, that is, impatiently, quickly gained research results that eventually prove worthless.[38] More good advice.

I suggest, given the wide variety of Bacon's metaphors, that we not take them seriously as attempts at deliberate manipulation of his audience or as the smoke signals of his seething unconscious. And it is unbelievable that "the increasing empirical success of the scientific world view *depended* on the continued presence in it – either explicitly or tacitly – of the gender politics metaphors" (Harding 1981, 316; italics added; see also Koertge, 353–54). Bacon's metaphors are more plausibly understood as "literary embellishments" than as a "substantive part of science" (contra Harding, *WSWK*, 44; see Landau, 49). As Bacon says in *Description of the Intellectual Globe*, the metaphors are irrelevant:

> If any one dislike that arts should be called the bonds of nature, thinking that they should rather be counted as her deliverers and champions, because in some cases they enable her to fulfil her own intention by reducing obstacles to order; for my part I do not care about these refinements and elegancies of speech; all I mean is, that nature, like Proteus, is forced by art to do that which without art would not be done; call it what you will, – force and bonds, or help and perfection. (*Works* 5, 506)

As we are rightly rereading the canon through feminist lenses, let us take care lest we succumb to the "impatience of research." Otherwise, in our investigations, be they philosophical or historical, we will discover precisely what we hoped to discover, and we will project into the canon the horrors that frighten us, horrors that are not there.

> It is not the pleasure of curiosity, nor the quiet of resolution, nor the raising of the spirit, nor victory of wit, nor faculty of speech, nor lucre of profession, nor ambition of honour or fame, nor inablement for business, that are the true ends of knowledge;...but it is a restitution and reinvesting...of man to the sovereignty and power...which he had in his first state of creation....It is a discovery of all operations and possibilities of operations from immortality (if it were possible) to the meanest mechanical practice. And therefore knowledge that tendeth but to satisfaction is but as a courtesan, which is for pleasure and not for fruit or generation. (*Valerius terminus, Works* 3, 222)[39]

Notes

1 "Value-Free Research Is a Delusion," *New York Times*, October 22, 1989, E24.

2 *The Science Question in Feminism* (hereafter cited as *SQIF*), 112, 113, 116; *Whose Science? Whose Knowledge?* (hereafter cited as *WSWK*), 43, 267.

3 (1) In her review of Sandra Harding and Jean O'Barr's anthology *Sex and Scientific Inquiry*, Nancy Tuana writes: "Evelyn Fox Keller and Susan Bordo argue that the methods and epistemology of modern science are not in fact neutral, but are male-biased, grounded in metaphors of dominance and rape" (62). (2) In her review of Keller's *Reflections on Gender and Science*, Helen Longino writes: "Keller notes that Francis Bacon...first made explicit the connections between knowledge and power and...described experimentation in language appropriate to rape and seduction" (1988, 563). But as we will see, it is not Keller but Harding who thinks that Bacon conceives of the experimental method as rape. In *Science as Social Knowledge* (205), while laying out Keller's views, Longino writes that the new science's "conception of inquiry ...envisioned the seeker after knowledge as male and the object of knowledge (nature) as female, and which described the activity of inquiry in language used to describe the male pursuit of females: rape and courtship" (3). "The relationship of scientists and science to nature has been described as 'rape,' 'forced penetration,' and 'domination,'" reports Lynn Hankinson Nelson, citing Harding, Keller, and Merchant (*Who Knows: From Quine to a Feminist Empiricism*, 213, 353, n. 136). (4) Joseph Agassi, too, perpetuates this noxious idea: "Now Bacon did oppose instrumentalism, yet he did not denounce it as dogmatism – as the putting of nature in chains and the rape of nature, to use his metaphors" (92).

4 The second edition of *On Being a Scientist* does not contain the first-edition passage that Harding found objectionable. Indeed, the introduction begins in politically correct fashion by quoting Barbara McClintock.

5 Don Ihde read Harding too quickly: "The rise of Modern Science [according to Harding in *SQIF*] was...a movement into the Baconian, masculinist context of an aggression upon nature betrayed in the metaphors of science 'twisting the tail of nature' or even the use of rape metaphors which proceeded from Bacon on into very contemporary speeches by Nobel Prize acceptees" (70–71). Harding, as uncharitable as her reading of Feynman is, did not accuse him, too, of trading in rape metaphors.

6 I quote Feyerabend because Harding (*WSWK*, 43) provides the wrong "I think I do not have to explain my own preferences." In *SQIF*, Harding quotes it correctly (120).

7 Although Harding interprets Feyerabend's metaphor as being about "science and its theories" (*SQIF*, 121) and he says as much (229), it makes more sense to read it as being about Nature. We should not view Nature as a stern and demanding mistress, which it is for Popper and Lakatos: their Nature yells "False!" or "Incompatible!" when it does not like our scientific theories. Instead, Nature is an indulgent courtesan, one who lets us do whatever we want – in theory construction. It is Nature that whispers, "Anything goes, big guy," not Science.

8 For an overview of Bacon's life and philosophy, see Thomas Macaulay's 1837 essay "Francis Bacon" and John Robertson's 1905 critical reply (vii–xvi).

9 As far as I know, Harding first accused Bacon of relying on rape metaphors in her 1980 *PSA* paper "The Norms of Social Inquiry and Masculine Experience," 318. Noretta Koertge critically discussed Harding's claim in the same colloquium (353–54). In her later works, *SQIF* and *WSWK*, Harding repeats the accusation against Bacon but neither rebuts nor acknowledges Koertge's doubts. Is this a case of political fervor (or mere personal stubbornness) trouncing scholarship?

10 The two sentences are from Bacon's 1623 *Of the Dignity and Advancement of Learning, Works* 4 (hereafter cited as *De augmentis*), 296. Harding took the passage not from Bacon but from Merchant, *The Death of Nature* (hereafter cited as *DON*), 168, who did take it from *Works* 4.

11 *Reflections on Gender and Science* (hereafter cited as *REF*), 36.

12 Paolo Rossi suggests that in "Erichthonius," Bacon expressed his view that to be successful with nature, science has to "humbly beg her assistance" (101; see "humble respect," 105). Keller (*REF*, 37) quotes Rossi, but not this phrase, thereby creating the impression that he agrees with her "forceful and aggressive seduction" reading.

13 Harding quotes these two sentences from *De augmentis* three times in her two books, always informing us that her source is Merchant's *DON*, 168. Merchant includes the five missing words. In addition to failing to mark an ellipsis in the first of Bacon's two sentences, Harding makes a second mistake in *WSWK*: ellipsis points belong between the two sentences, since Harding omitted a large chunk. Any hint of rape created by the juxtaposition of these sentences in *WSWK* is therefore artificial. There are other errors (see *Works* 4, 296). In Merchant, we correctly find "these holes," but in Harding "those." Merchant and Harding write "whole object," but both are wrong; "sole object" is correct. Merchant, and Harding in *SQIF*, gives "drive her afterward," but both are wrong; in *WSWK*, Harding got "afterwards" right. Iddo Landau continues this comedy of errors. In "How Androcentric Is Western Philosophy?" (49), he accurately reproduces Harding's misquotations of Bacon and Feyerabend from *WSWK*, thereby misquoting Bacon and Feyerabend. If you want to check what you have read in today's newspaper, buy another copy (Wittgenstein).

14 Let alone, as in Merchant, "strong sexual implications" (*DON*, 168).

15 See Bordo's remark on this sentence from *De augmentis*, apparently provoked by Keller and Merchant: it illustrates "the famous Baconian imagery of sexual assault" (Bordo, 107–8).
16 Book 1, 83 (in *Works* 4); see *Cogitata et visa* (Farrington, 82). Henceforth I supply for *Novum organum* only the book 1 aphorism number.
17 Landau too quickly grants to Harding that the passage is about replicability (49).
18 Harding never mentions Leiss's reading of a sentence from Bacon that she quotes twice in *SQIF* and once again in *WSWK*. Leiss's *Domination of Nature* is included in Harding's bibliography (*SQIF*, 258), but Leiss is not listed in the index (*SQIF*, 267), nor is his book mentioned in any of *SQIF*'s innumerable footnotes.
19 *Works* 4, 294–95; see *Description of the Intellectual Globe, Works* 5, 507.
20 See 129 and *Cogitata et visa*: "No force avails to break the chain of natural causation. Nature cannot be conquered but by obeying her" (Farrington, 93). Koertge ("Methodology," 353) responded to Harding's 1980 *PSA* paper ("The Norms of Social Inquiry") by quoting five passages in Bacon that indicate "humility and respect" for nature.
21 This phrase is more than reminiscent of Merchant's description of Bacon's experimentalism: "Here, in bold sexual imagery, is the key feature of the modern experimental method" (*DON*, 171).
22 Susan Griffin is wrong to paraphrase Bacon as having written in 1609, "nature must be hounded in her wanderings before one can lead her and drive her" (16). Griffin says in her bibliographical notes (234, 258, 259) that the lines she attributes to Bacon were taken from an unpublished manuscript by Carolyn Iltis and from *The Domination of Nature* by "Leis" [*sic*]. But Leiss quotes "lead and drive" from the 1623 *De augmentis* (59).
23 *Valerius terminus, Works* 3, 217, 221–22; Preface to *The Great Instauration, Works* 4, 21; and see Farrington, 28–29.
24 But Bacon was no blind optimist: he recognized that science (and philosophy and history) done poorly would go wrong. See "Daedalus," myth 19, *Wisdom of the Ancients* (Robertson, 842–43).
25 Similarly, it will not help Harding to claim that her experiences and social location as a feminist and/or woman grant her an epistemic advantage ("standpoint epistemology") – in this case, make her an especially perceptive reader of early-seventeenth-century texts (*WSWK*, 121–33, 150–51; see my review of *WSWK*). Harding's reading of Bacon is a politically inspired reading that goes wrong; hence it subverts her claim that feminist scholarship is better because it is political. See n. 9, above.
26 Harding might have believed that Merchant found a rape metaphor in Bacon's works. After all, much later, according to Daphne Patai and Noretta Koertge, the claim that "modern science...has viewed nature as a woman who exists only to be raped" was made by Merchant (*Professing Feminism*, 123). But the only occurrence of the word *rape* in Merchant's chapter on Bacon is in quotation marks (*DON*, 171). And when Merchant talks about the "rape" of nature, she is not voicing Harding's thesis that for Bacon, the experimental method is rape, but the different thesis that the new science justified the "commercial exploration" or exploitation of the earth (e.g., mining; *DON*, 41).
27 Merchant's "a" true judgment should be "the." See n. 13, above.
28 Recall from *Novum organum*: "The sun enters the sewer no less than the palace, yet takes no pollution" (120).
29 Robbins suggests, sounding a contemporary note, that James's realization that children had been "falsely charging people as witches" was crucial to his change of mind.
30 See a variant of these lines in "The Plan" of *The Great Instauration, Works* 4, 29, which is the "vex" passage to which Keller calls attention.
31 Keller took this sentence from Leiss (25), who took it from Farrington's translation (131). *Redargutio philosophiarum* is not translated from the Latin in *Works* 3 and was not published until 1734 (Farrington, 57).
32 Londa Schiebinger thinks that the "masculine" in *Masculine Birth of Time* "was at most a tangential attack upon women" (137). Instead, it was meant to reject an Aristotelian, "passive, speculative, and effeminate philosophy." Similarly, Patai and Koertge understand "the violence of [Bacon's] images [as] directed at Aristotelians, not women" (124).
33 In her 1980 "Baconian Science" (308, n. 11) and its revision in *REF* (36), Keller claims to have taken this passage from Spedding's *Description of the Intellectual Globe, Works* 5, 506; it is not there. In her 1982 "Feminism and Science" (598, n. 22), she again says that it is from *Description of the Intellectual Globe*, but now cites p. 506 of Robertson's collection instead of Spedding's *Works*. On that page in Robertson, however, is *De augmentis*, book 5, chap. 2. (The passage is actually in Farrington's translation of *Cogitata et visa*, 93.) A clue to solving this mystery can be found in Leiss, 58, 216, n. 18; and n. 31, above.
34 When revising "Baconian Science" for inclusion in *REF*, Keller made three changes. She changed the subtitle from "A Hermaphroditic Birth" to "The Arts of Mastery and Obedience." She deleted a passage she had quoted from Bacon's *Valerius terminus* (*Works* 3, 222) and her comments on it (compare the first, 303, with the second, 37). And she changed (on the same pages) "when nature becomes divine [for Bacon]...the scientific mind becomes 'he'" to the *opposite* "when nature becomes divine [for Bacon]...the scientific mind becomes more

nearly female." Does Keller want to be held to the earlier (presumably mistaken) version? Why hold Bacon to a different standard? Note that Keller says in a later chapter in *REF* (48), "As Bacon's metaphoric ideal was the virile superman, the alchemist's ideal was the hermaphrodite.... Whereas Bacon sought domination, the alchemists asserted the necessity of allegorical, if not actual, cooperation between male and female."

35 In another essay, Keller, right after quoting the "slave" passage from Bacon, writes: "This conjunction between scientific and masculine norms has been historically functional in guaranteeing a sexual division of emotional and intellectual labor that effectively excludes most women from scientific professions" (1987, 79). Incomprehensibly, when discussing this essay, Christina Sommers (491) and Joseph Adelson (110) call the author "Elizabeth Fox-Keller." No one can read these days. (Her name is "Evelyn.")

36 Merchant mentions the "slave" passage twice in the space of two pages (169, 170); and Ruth Bleier (204–5) condemns Bacon by reproducing only the "slave" passage, which she took from Keller's "Feminism and Science."

37 Graham Hammill goes one step beyond Keller. Whereas Keller relies on the contradictory sexual images in Bacon to argue that Bacon's model of the mind is hermaphroditic or bisexual, not simply a masculine domination of female nature, Hammill relies on Bacon's purported homosexual activities with his men servants and his anally erotic penchant for enemas – as well as a few pieces of his texts – to argue that "for Bacon the anus becomes more properly the seat of knowledge" and the penetration of nature is "not an odyssey into the vagina but ...wanderlusts into the anus." The masculine birth is accomplished by "anal purging" (244, 251, 247). Bacon's imagery, we might say, expresses not an effeminate homosexuality or a woman-hating homosexuality but homosexuality nonetheless, or at least a woman-envying heterosexuality. Keller might appreciate Hammill's reading of Bacon; see her *Secrets of Life. Secrets of Death* (45–52), in which she speaks about male science as anal "production." (For a discussion of *Secrets*, see my "Gender, Objectivity, and Realism.")

38 Bacon uses the Atalanta story often, without any obfuscating marital or gender images; see *Novum organum* (70, 117) and "The Plan" of *The Great Instauration* (*Works* 4, 29).

39 The last sentence of this beautiful passage is the quotation Keller deleted from "Baconian Science" when revising that essay for *REF*; see n. 34, above.

References

Adelson, J. 1988. An academy of one's own. *The Public Interest* no. 91 (Spring): 107–10.

Agassi, J. 1989. The lark and the tortoise. *Philosophy of the Social Sciences* 19: 89–94.

Bacon, F. 1962–63. *The Works of Francis Bacon*. 14 vols. Ed. J. Spedding, R. L. Ellis, and D. D. Heath. 1857–74. Reprint, Stuttgart–Bad Canstatt: Friedrich Fromann Verlag Gunther Holzboog, 1963.

Bleier, R. 1984. *Science and Gender*. New York: Pergamon Press.

Bordo, S. 1987. *The Flight to Objectivity*. Albany: State University of New York Press.

Boyd, R., P. Gasper, and J. D. Trout, eds. 1991. *The Philosophy of Science*. Cambridge, Mass.: MIT Press.

Farrington, B. 1966. *The Philosophy of Francis Bacon*. 1964. Reprint, Chicago: University of Chicago Press.

Feyerabend, 1970. Consolations for the specialist. In *Criticism and the Growth of Knowledge*, ed. I. Lakatos and A. Musgrave, 197–230. Cambridge: Cambridge University Press.

Griffin, S. 1978. *Woman and Nature*. New York: Harper & Row.

Hammill, G. 1994. The epistemology of expurgation: Bacon and *The Masculine Birth of Time*. In *Queering the Renaissance*, ed. J. Goldberg, 236–52. Durham, N.C.: Duke University Press.

Harding, S. 1981. The norms of social inquiry and masculine experience. *PSA 1980* 2: 305–24.

——. 1986. *The Science Question in Feminism*. Ithaca, N.Y.: Cornell University Press.

——. 1989. Value-free research is a delusion. *New York Times*, October 22, E24.

——. 1991. *Whose Science? Whose Knowledge?* Ithaca, N.Y.: Cornell University Press.

Harding, S., and J. F. O'Barr, eds. 1987. *Sex and Scientific Inquiry*. Chicago: University of Chicago Press.

Ihde, D. 1993. *Philosophy of Technology: An Introduction*. New York: Paragon House.

Keller, E. F. 1978. Gender and science. *Psychoanalysis and Contemporary Thought* 1: 409–33.

——. 1980. Baconian science: a hermaphroditic birth. *Philosophical Forum* 11: 299–308.

——. 1982. Feminism and science. *Signs* 7: 589–602.

——. 1985. *Reflections on Gender and Science*. New Haven, Conn.: Yale University Press.

——. 1987. Women scientists and feminist critics of science. *Daedalus* 116 (4): 77–91.

——. 1992. *Secrets of Life. Secrets of Death: Essays on Language, Gender and Science*. New York: Routledge.

Koertge, N. 1981. Methodology, ideology and feminist critiques of science. *PSA 1980* 2: 346–59.

Landau, I. 1996. How androcentric is Western philosophy? *Philosophical Quarterly* 46 (182): 48–59.

Leiss, W. 1972. *The Domination of Nature*. New York: Braziller.

Longino, H. 1988. Review essay: science, objectivity, and feminist values. *Feminist Studies* 14: 561–74.

——. 1990. *Science as Social Knowledge*. Princeton, N.J.: Princeton University Press.

Macaulay, T. 1967. Francis Bacon. In *Critical and Historical Essays*. Vol. 2, 290–398. New York: Dutton.

MacKinnon, C. 1989. *Toward a Feminist Theory of the State*. Cambridge, Mass.: Harvard University Press.

Merchant, C. 1980. *The Death of Nature*. New York: Harper & Row.

Morgan, R. 1977. *Going Too Far*. New York: Random House.

Nelson, L. H. 1990. *Who Knows: From Quine to a Feminist Empiricism*. Philadelphia: Temple University Press.

On Being a Scientist: Responsible Conduct in Research. 1995. Washington, D.C.: National Academy [of Sciences] Press.

Patai, D., and N. Koertge. 1994. *Professing Feminism*. New York: Basic Books.

Robbins, R. H. 1959. *The Encyclopedia of Witchcraft and Demonology*. New York: Crown.

Robertson, J. M., ed. 1905. *The Philosophical Works of Francis Bacon*. London: Routledge and Sons.

Rossi, P. 1968. *Francis Bacon: From Magic to Science*. London: Routledge & Kegan Paul.

Schiebinger, L. 1989. *The Mind Has No Sex? Women in the Origins of Modern Science*. Cambridge, Mass.: Harvard University Press.

Soble, A. 1985. Pornography: defamation and the endorsement of degradation. *Social Theory and Practice* 11 (1): 61–87.

——. 1992. *Review of Whose Science? Whose Knowledge?* by S. Harding. *International Studies in the Philosophy of Science* 69: 159–62.

——. 1994. Gender, objectivity, and realism. *Monist* 77: 509–30.

Sommers, C. 1993. The feminist revelation. In *Morality and Moral Controversies*. 3rd ed., ed. J. Arthur, 479–93. Englewood Cliffs, N.J.: Prentice Hall.

Tuana, N. 1990. Review of *Sex and Scientific Inquiry*, ed. S. Harding and J. F. O'Barr. *American Philosophical Association Newsletter on Feminism and Philosophy* 89 (2): 61–62.

39

The Shallow and the Deep, Long-Range Ecology Movement

Arne Naess

Ecologically responsible policies are concerned only in part with pollution and resource depletion. There are deeper concerns which touch upon principles of diversity, complexity, autonomy, decentralization, symbiosis, egalitarianism, and classlessness.

The emergence of ecologists from their former relative obscurity marks a turning-point in our scientific communities. But their message is twisted and misused. A shallow, but presently rather powerful movement, and a deep, but less influential movement, compete for our attention. I shall make an effort to characterize the two.

1 The Shallow Ecology Movement

Fight against pollution and resource depletion. Central objective: the health and affluence of people in the developed countries.

2 The Deep Ecology Movement

(1) Rejection of the man-in-environment image in favour of *the relational, total-field image*. Organisms as knots in the biospherical net or field of intrinsic relations. An intrinsic relation between two things *A* and *B* is such that the relation belongs to the definitions or basic constitutions of *A* and *B*, so that without the relation, *A* and *B* are no longer the same things. The total-field model dissolves not only the man-in-environment concept, but every compact thing-in-milieu concept – except when talking at a superficial or preliminary level of communication.

Originally "The Shallow and the Deep, Long-Range Ecology Movement. A Summary" in *Inquiry* 16 (1973): 95–100. Reprinted with permission of Taylor & Francis, Oslo, Norway.

(2) *Biospherical egalitarianism* – in principle. The "in principle" clause is inserted because any realistic praxis necessitates some killing, exploitation, and suppression. The ecological field-worker acquires a deepseated respect, or even veneration, for ways and forms of life. He reaches an understanding from within, a kind of understanding that others reserve for fellow men and for a narrow section of ways and forms of life. To the ecological field-worker, *the equal right to live and blossom* is an intuitively clear and obvious value axiom. Its restriction to humans is an anthropocentrism with detrimental effects upon the life quality of humans themselves. This quality depends in part upon the deep pleasure and satisfaction we receive from close partnership with other forms of life. The attempt to ignore our dependence and to establish a master–slave role has contributed to the alienation of man from himself.

Ecological egalitarianism implies the reinterpretation of the future-research variable, "level of crowding", so that *general* mammalian crowding and loss of life-equality is taken seriously, not only human crowding. (Research on the high requirements of free space of certain mammals has, incidentally, suggested that theorists of human urbanism have largely underestimated human life-space requirements. Behavioural crowding symptoms [neuroses, aggressiveness, loss of traditions...] are largely the same among mammals.)

(3) *Principles of diversity and of symbiosis*. Diversity enhances the potentialities of survival, the chances of new modes of life, the richness of forms. And the so-called struggle of life, and survival of the fittest, should be interpreted in the sense of ability to coexist and cooperate in complex relationships, rather than ability to kill, exploit, and suppress. "Live and let live" is a more powerful ecological principle than "Either you or me".

The latter tends to reduce the multiplicity of kinds of forms of life, and also to create destruction within the communities of the same species. Ecologically inspired attitudes therefore favour diversity of human ways of life, of cultures, of occupations, of economies. They support the fight against economic and cultural, as much as military, invasion and domination, and they are opposed to the annihilation of seals and whales as much as to that of human tribes or cultures.

(4) *Anti-class posture*. Diversity of human ways of life is in part due to (intended or unintended) exploitation and suppression on the part of certain groups. The exploiter lives differently from the exploited, but both are adversely affected in their potentialities of self-realization. The principle of diversity does not cover differences due merely to certain attitudes or behaviours forcibly blocked or restrained. The principles of ecological egalitarianism and of symbiosis support the same anti-class posture. The ecological attitude favours the extension of all three principles to any group conflicts, including those of today between developing and developed nations. The three principles also favour extreme caution towards any over-all plans for the future, except those consistent with wide and widening classless diversity.

(5) Fight against *pollution and resource depletion*. In this fight ecologists have found powerful supporters, but sometimes to the detriment of their total stand. This happens when attention is focused on pollution and resource depletion rather than on the other points, or when projects are implemented which reduce pollution but increase evils of the other kinds. Thus, if prices of life necessities increase because of the installation of anti-pollution devices, class differences increase too. An ethics of responsibility implies that ecologists do not serve the shallow, but the deep ecological movement. That is, not only point (5), but all seven points must be considered together.

Ecologists are irreplaceable informants in any society, whatever their political colour. If well organized, they have the power to reject jobs in which they submit themselves to institutions or to planners with limited ecological perspectives. As it is now, ecologists sometimes serve masters who deliberately ignore the wider perspectives.

(6) *Complexity, not complication*. The theory of ecosystems contains an important distinction between what is complicated without any Gestalt or unifying principles – we may think of finding our way through a chaotic city – and what is complex. A multiplicity of more or less lawful, interacting factors may operate together to form a unity, a system. We make a shoe or use a map or integrate a variety of activities into a workaday pattern. Organisms, ways of life, and interactions in the biosphere in general, exhibit complexity of such an astoundingly high level as to colour the general outlook of ecologists. Such complexity makes thinking in terms of vast systems inevitable. It also makes for a keen, steady perception of the profound *human ignorance* of biospherical relationships and therefore of the effect of disturbances.

Applied to humans, the complexity-not-complication principle favours division of labour, *not fragmentation of labour*. It favours integrated actions in which the whole person is active, not mere reactions. It favours complex economies, an integrated variety of means of living. (Combinations of industrial and agricultural activity, of intellectual and manual work, of specialized and non-specialized occupations, of urban and non-urban activity, of work in city and recreation in nature with recreation in city and work in nature...)

It favours soft technique and "soft future-research", less prognosis, more clarification of possibilities. More sensitivity towards continuity and live traditions, and – most importantly – towards our state of ignorance.

The implementation of ecologically responsible policies requires in this century an exponential growth of technical skill and invention – but in new directions, directions which today are not consistently and liberally supported by the research policy organs of our nation-states.

(7) *Local autonomy and decentralization.* The vulnerability of a form of life is roughly proportional to the weight of influences from afar, from outside the local region in which that form has obtained an ecological equilibrium. This lends support to our efforts to strengthen local self-government and material and mental self-sufficiency. But these efforts presuppose an impetus towards decentralization. Pollution problems, including those of thermal pollution and recirculation of materials, also lead us in this direction, because increased local autonomy, if we are able to keep other factors constant, reduces energy consumption. (Compare an approximately self-sufficient locality with one requiring the importation of foodstuff, materials for house construction, fuel and skilled labour from other continents. The former may use only five per cent of the energy used by the latter.) Local autonomy is strengthened by a reduction in the number of links in the hierarchical chains of decision. (For example a chain consisting of local board, municipal council, highest sub-national decision-maker, a state-wide institution in a state federation, a federal national government institution, a coalition of nations, and of institutions, e.g. E.E.C. top levels, and a global institution, can be reduced to one made up of local board, nation-wide institution, and global institution.) Even if a decision follows majority rules at each step, many local interests may be dropped along the line, if it is too long.

Summing up, then, it should, first of all, be borne in mind that the norms and tendencies of the Deep Ecology movement are not derived from ecology by logic or induction. Ecological knowledge and the life-style of the ecological field-worker have *suggested, inspired, and fortified* the perspectives of the Deep Ecology movement. Many of the formulations in the above seven-point survey are rather vague generalizations, only tenable if made more precise in certain directions. But all over the world the inspiration from ecology has shown remarkable convergencies. The survey does not pretend to be more than one of the possible condensed codifications of these convergencies.

Secondly, it should be fully appreciated that the significant tenets of the Deep Ecology movement are clearly and forcefully *normative*. They express a value priority system only in part based on results (or lack of results, cf. point [6]) of scientific research. Today, ecologists try to influence policy-making bodies largely through threats, through predictions concerning pollutants and resource depletion, knowing that policy-makers accept at least certain minimum *norms* concerning health and just distribution. But it is clear that there is a vast number of people in all countries, and even a considerable number of people in power, who accept as valid the wider norms and values characteristic of the Deep Ecology movement. There are political potentials in this movement which should not be overlooked and which have little to do with pollution and resource depletion. In plotting possible futures, the norms should be freely used and elaborated.

Thirdly, in so far as ecology movements deserve our attention, they are *ecophilosophical* rather than ecological. Ecology is a *limited* science which makes *use* of scientific methods. Philosophy is the most general forum of debate on fundamentals, descriptive as well as prescriptive, and political philosophy is one of its subsections. By an *ecosophy* I mean a philosophy of ecological harmony or equilibrium. A philosophy as a kind of *sofia* wisdom, is openly normative, it contains *both* norms, rules, postulates, value priority announcements *and* hypotheses concerning the state of affairs in our universe. Wisdom is policy wisdom, prescription, not only scientific description and prediction.

The details of an ecosophy will show many variations due to significant differences concerning not only "facts" of pollution, resources, population, etc., but also value priorities. Today, however, the seven points listed provide one unified framework for ecosophical systems.

In general system theory, systems are mostly conceived in terms of casually or functionally interacting or interrelated items. An ecosophy, however, is more like a system of the kind constructed by Aristotle or Spinoza. It is expressed verbally as a set of sentences with a variety of functions, descriptive and prescriptive. The basic relation is that between subsets of premisses and subsets of conclusions, that is, the relation of derivability. The relevant notions of derivability may be classed according to rigour, with logical and mathematical deductions topping the list, but also according to how much is implicitly taken for granted. An exposition of an ecosophy must necessarily be only moderately precise considering the vast scope of relevant ecological and normative (social, political, ethical) material. At the moment, ecosophy might profitably use models of systems, rough approximations of global systematizations. It is the global character, not preciseness in detail, which distinguishes an ecosophy. It articulates and integrates the efforts of an ideal ecological team, a team comprising not only scientists from an extreme variety of disciplines, but also students of politics and active policy-makers.

Under the name of *ecologism*, various deviations from the deep movement have been championed – primarily with a one-sided stress on pollution and resource depletion, but also with a neglect of the great differences between under- and over-developed countries in favour of a vague global approach. The global approach is essential, but regional differences must largely determine policies in the coming years.

Selected Literature

Commoner, B., *The Closing Circle: Nature, Man, and Technology*, Alfred A. Knopf, New York 1971.

Ehrlich, P. R. and A. H., *Population, Resources, Environment: Issues in Human Ecology*, 2nd ed., W. H. Freeman & Co., San Francisco 1972.

Ellul, J., *The Technological Society*, English ed., Alfred A. Knopf, New York 1964.

Glacken, C. J., *Traces on the Rhodian Shore. Nature and Culture in Western Thought*, University of California Press, Berkeley 1967.

Kato, H., "The Effects of Crowding", Quality of Life Conference, Oberhausen, April 1972.

McHarg, Ian L., *Design with Nature*, 1969. Paperback 1971, Doubleday & Co., New York.

Meynaud, J., *Technocracy*, English ed., Free Press of Glencoe, Chicago 1969.

Mishan, E. J., *Technology and Growth: The Price We Pay*, Frederick A. Praeger, New York 1970.

Odum, E. P., *Fundamentals of Ecology*, 3rd ed., W. E. Saunders Co., Philadelphia 1971.

Shepard, Paul, *Man in the Landscape*, A. A. Knopf, New York.

40

The Deep Ecology Movement

Bill Devall

There are two great streams of environmentalism in the latter half of the twentieth century. One stream is reformist, attempting to control some of the worst of the air and water pollution and inefficient land use practices in industrialized nations and to save a few of the remaining pieces of wildlands as "designated wilderness areas." The other stream supports many of the reformist goals but is revolutionary, seeking a new metaphysics, epistemology, cosmology, and environmental ethics of person/planet. This paper is an intellectual archeology of the second of these streams of environmentalism, which I will call *deep ecology*.

There are several other phrases that some writers are using for the perspective I am describing in this paper. Some call it "eco-philosophy" or "foundational ecology" or the "new natural philosophy." I use "deep ecology" as the shortest label. Although I am convinced that deep ecology is radically different from the perspective of the dominant social paradigm, I do not use the phrase "radical ecology" or "revolutionary ecology" because I think those labels have such a burden of emotive associations that many people would not hear what is being said about deep ecology because of their projection of other meanings of "revolution" onto the perspective of deep ecology.

I contend that both streams of environmentalism are reactions to the successes and excesses of the implementation of the dominant social paradigm.

From *Natural Resources Journal* 20/2 (1980): 219–313, abridged.

Although reformist environmentalism treats some of the symptoms of the environmental crisis and challenges some of the assumptions of the dominant social paradigm (such as growth of the economy at any cost), deep ecology questions the fundamental premises of the dominant social paradigm. In the future, as the limits of reform are reached and environmental problems become more serious, the reform environmental movement will have to come to terms with deep ecology.

The analysis in the present paper was inspired by Arne Naess' paper on "shallow and deep, long-range" environmentalism.[1] The methods used are patterned after John Rodman's seminal critique of the resources conservation and development movement in the United States.[2] The data are the writings of a diverse group of thinkers who have been developing a theory of deep ecology, especially during the last quarter of a century. Relatively few of these writings have appeared in popular journals or in books published by mainstream publishers. I have searched these writings for common threads or themes much as Max Weber searched the sermons of Protestant ministers for themes which reflected from and back to the intellectual and social crisis of the emerging Protestant ethic and the spirit of capitalism.[3] Several questions are addressed in this paper: What are the sources of deep ecology? How do the premises of deep ecology differ from those of the dominant social paradigm? What are the areas of disagreement between reformist environmentalism and deep ecology? What is the likely future role of the deep ecology movement?

The Dominant Paradigm

A paradigm is a shorthand description of the world view, the collection of values, beliefs, habits, and norms which form the frame of reference of a collectivity of people – those who share a nation, a religion, a social class. According to one writer, a *dominant* social paradigm is the mental image of social reality that guides expectations in a society.

The dominant paradigm in North America includes the belief that "economic growth," as measured by the Gross National Product, is a measure of Progress, the belief that the primary goal of the governments of nation-states, after national defense, should be to create conditions that will increase production of commodities and satisfy material wants of citizens, and the belief that "technology can solve our problems." Nature, in this paradigm, is only a storehouse of resources which should be "developed" to satisfy ever increasing numbers of humans and ever increasing demands of humans. Science is wedded to technology, the development of *techniques* for control of natural processes (such as weather modification). Change ("planned obsolescence") is an end in itself. The new is valued over the old and the present over future generations. The goal of persons is personal satisfaction of wants and a higher standard of living as measured by possession of commodities (houses, autos, recreation vehicles, etc.).[4] Whatever its origin, this paradigm continues to be dominant, to be preached through publicity (i.e., advertising), and to be part of the world view of most citizens in North America.[5]

For some writers, the dominant social paradigm derives from Judeo-Christian origins.[6] For others, the excesses of air and water pollution, the demand for more and more centralization of political and economic power and the disregard for future generations, and the unwise use of natural resources derive from the ideology and structure of capitalism or from the Lockean view that property must be "improved" to make it valuable to the "owner" and to society.[7] For others, the dominant social paradigm derives from the "scientism" of the modern West (Europe and North America) as applied to the technique of domination.[8]

Following Thomas Kuhn's theory of the dominance of paradigms in modern science and the operation of scientists doing what he calls normal science within a paradigm, it can be argued that (1) those who subscribe to a given paradigm share a definition of what problems are and their priorities; (2) the general heuristics, or rules of the game, for approaching problems is widely agreed upon, (3) there is a definite, underlying confidence among believers of the paradigm that solutions within the paradigm do exist; and (4) those who believe the assumptions of the paradigm may argue about the validity of data, but rarely are their debates about the definition of what the problem is or whether there are solutions or not. Proposed solutions to problems arising from following the assumptions of the paradigms are evaluated as "reasonable," "realistic," or "valid" in terms of the agreed upon "rules of the game." When the data is difficult to fit to the paradigm, frequently there is dissonance disavowal, an attempt to explain away the inconsistency.[9]

It is possible for a paradigm shift to occur when a group of persons finds in comparing its data with generally accepted theory that the conclusions become "weird" when compared with expectations. In terms of the shared views of the goals, rules, and perceptions of reality in a nation, a tribe, or a religious group, for example, a charismatic leader, a social movement, or a formation of social networks of persons exploring a new social paradigm may be at the vanguard of a paradigm shift.

Reformist environmentalism in this paper refers to several social movements which are related in that the goal of all of them is to change society for "better living" without attacking the premises of the dominant social paradigm. These reform movements each defined a problem – such as need for more open space – and voluntary organizations were formed to agitate for social changes. There has also been considerable coalition building between different voluntary organizations espousing reform environmentalism. Several reformist environmental movements, including at least the following, have been active during the last century: (1) the movement to establish urban parks, designated wilderness areas, and national parks;[10] (2) the movement to mitigate the health and public safety hazards created by the technology which was applied to create the so-called industrial revolution.[11] The Union of Concerned Scientists, for example, has brought to the attention of the general public some of the hazards to public health and safety of the use of nuclear power to generate electricity; (3) the movement to develop "proper" land-use planning. This includes the city beautiful movement of the late nineteenth century and the movement to zone and plan land use such as the currently controversial attempts to zone uses along the coastal zones;[12] (4) the resources conservation and development movement symbolized by the philosophy of multiple use of Gifford Pinchot and the U.S. Forest Service;[13] (5) the "back to the land" movement of the 1960s and 1970s and the "organic farming" ideology; (6) the concern with exponential growth of human population and formation of such groups as Zero Population Growth;[14] (7) the "humane" and "animal liberation" movement directed at changing the attitudes and behavior of humans towards some other aspects of animals;[15] and (8) the "limits to growth" movement which emphasizes we should control human population and move towards a "steady-state" or "conserver society" as rapidly as possible.[16]

Sources of Deep Ecology

What I call deep ecology in this paper is premised on a gestalt of person-in-nature. The person is not above or outside of nature. The person is part of creation on-going. The person cares for and about nature, shows reverence towards and respect for nonhuman nature, loves and lives with nonhuman nature, is a person in the "earth household" and "lets being be," lets nonhuman nature follow separate evolutionary destinies. Deep ecology, unlike reform environmentalism, is not just a pragmatic, short-term social movement with a goal like stopping nuclear power or cleaning up the waterways. Deep ecology first attempts to question and present alternatives to conventional ways of thinking in the modern West. Deep ecology understands that some of the "solutions" of reform environmentalism are counter-productive. Deep ecology seeks transformation of values and social organization.

The historian Lynn White, Jr., in his influential 1967 article, "The Historical Roots of Our Ecological Crisis," provided one impetus for the current upwelling of interest in deep ecology by criticizing what he saw as the dominant Judeo-Christian view of man *versus* nature, or man at war with nature. But there are other writers, coming from diverse intellectual and spiritual disciplines, who have provided, in the cumulative impact of their work, a profound critique of the dominant social paradigm and the "single vision" of science in the modern (post-1500) West.[17]

One major stream of thought influencing the development of deep ecology has been the influx of Eastern spiritual traditions into the West which began in the 1950s with the writings of such people as Alan Watts[18] and Daisetz Suzuki.[19] Eastern traditions provided a radically different man/nature vision than that of the dominant social paradigm of the West. During the 1950s the so-called "beat poets" such as Alan Ginsberg seemed to be groping for a way through Eastern philosophy to cope with the violence, insanity, and alienation of people from people and people from nature they experienced in North America. Except for Gary Snyder, who developed into one of the most influential eco-philosophers of the 1970s, these beat poets, from the perspective of the 1970s, were naive in their understanding of both Eastern philosophy, ecology, and the philosophical traditions of the West.

During the late 1960s and 1970s, however, philosophers, scientists, and social critics have begun to compare Eastern and Western philosophic traditions as they relate to science, technology, and man/nature relations. Fritjoff Capra's *Tao of Physics*, for example, emphasizes the parallels between Eastern philosophies and the theories of twentieth century physics.[20] Joseph Needham's massive work, *Science and Civilization in China*, brought to the consciousness of the West the incredibly high level of science, technology, and civilization achieved in the East for millennia and made available to Western readers an alternative approach to science and human values.[21] More recently, Needham has suggested that modern Westerners take the philosophies of the East as a spiritual and ethical basis for modern science.[22] Works by Huston Smith, among others, have also contributed to this resurgent interest in relating the environmental crisis to the values expressed in the dominant Western paradigm. Smith and others have looked to the Eastern philosophies for spiritual-religious guidance.[23]

Several social philosophers have written brilliant critiques of Western societies but have not presented a new metaphysical basis for their philosophy nor attempted to incorporate Eastern philosophy into their analyses. Jacques Ellul wrote on *technique* and the technological society.[24] Paul Goodman discussed the question "can there be a humane technology?"[25] Herbert Marcuse analyzed "one-dimensional man" as the prototypical "modern" urbanite.[26] The works of Theodore Roszak have also had considerable impact on those thinkers interested in understanding the malaise and contradictions of modern societies by examining the premises of the dominant social paradigm.[27]

A second stream of thought contributing to deep ecology has been the re-evaluation of Native Americans (and other preliterate peoples) during the 1960s and 1970s. This is not a revival of the Romantic view of Native Americans as "noble savages" but rather an attempt to evaluate traditional religions, philosophies, and social organizations of Native Americans in objective, comparative, analytic, and critical ways.

A number of questions have been asked. How did different tribes at different times cope with changes in their natural environment (such as prolonged drought) and with technological innovation? What were the "separate realities" of Native Americans and can modern Western man *understand* and *know*, in a phenomenological sense, these "separate realities"? The experiences of Carlos Castaneda, for example, indicate it may be very difficult for modern man to develop such understanding since this requires a major perceptual shift of man/nature. Robert Ornstein concludes, "Castaneda's experience demonstrates primarily that the Western-trained intellectual, even a 'seeker' is by his culture almost completely unprepared to understand esoteric traditions."[28]

From the many sources on Native Americans which have become available during the 1970s, I quote a statement by Luther Standing Bear, an Oglala Sioux, from *Touch the Earth* to illustrate the contrast with the modern paradigm of the West:

> We do not think of the great open plains, the beautiful rolling hills, and winding streams with tangled growth, as "wild." Only to the white man was nature a "wilderness" and only to him was the land "infested" with "wild" animals and "savage" people. To us it was tame. Earth was bountiful and we were surrounded with the blessings of the Great Mystery. Not until the hairy man from the east came and with brutal frenzy heaped injustices upon us and the families we loved was it "wild" for us. When the very animals of the forest began fleeing from his approach, then it was that for us the "wild west" began.[29]

A third source of deep ecology is found in the "minority tradition" of Western religious and philosophical traditions. The philosopher George Sessions has claimed that:

> [I]n the civilized West, a tenuous thread can be drawn through the Presocratics, Theophrastus, Lucretius, St. Francis, Bruno and other neo-Platonic mystics, Spinoza, Thoreau, John Muir, Santayana, Robinson Jeffers, Aldo Leopold, Loren Eiseley, Gary Snyder, Paul Shepard, Arne Naess, and maybe that desert rat, Edward Abbey. This minority tradition, despite differences, could have provided the West with a healthy basis for a realistic portrayal of the balance and interconnectedness of three artificially separable components (God/Nature/Man) of an untimely seamless and inseparable Whole.[30]

Sessions, together with Arne Naess and Stuart Hampshire, has seen the philosopher Spinoza as providing a unique fusion of an integrated man/nature metaphysic with modern European science.[31] Spinoza's ethics is most naturally interpreted as implying biospheric egalitarianism, and science is endorsed by Spinoza as valuable primarily for contemplation of a pantheistic, sacred universe and for spiritual discipline and development. Spinoza stands out in a unique way in opposition to other 17th century philosophers – e.g., Bacon, Descartes, and Leibniz – who were at that time laying the foundations for the technocratic-industrial social paradigm and the fulfillment of the Christian imperative that man *must* dominate and control all nature. It has been claimed by several writers that the poet-philosopher Robinson Jeffers, who lived most of his life on the California coastline at Big Sur, was Spinoza's twentieth century "evangelist" and that Jeffers gave Spinoza's philosophy an explicitly ecological interpretation.[32]

Among contemporary European philosophers, the two most influential have been Alfred North Whitehead and Martin Heidegger.[33] In particular, more American philosophers, both those with an interest in ecological consciousness and those interested in contemporary philosophers, are discussing Heidegger's critique of Western philosophy and contemporary Western societies. Because Heidegger's approach to philosophy and language is so different from the language we are accustomed to in American academia, any summary of his ideas would distort the theory he is presenting. The reader is referred to the books and articles on Heidegger cited below.[34]

A fourth source of reference for the deep ecology movement has been the scientific discipline of *ecology*. For some ecology is a science of the "home," of the "relationships between," while for others ecology is a perspective. The difference is important, for ecology as a science is open for co-optation by the engineers, the "technological fixers" who want to "enhance," "manage," or "humanize" the biosphere. At the beginning of the "environmental decade" of the 1970s, two ecologists issued a warning against this approach:

> Even if we dispense with the idea that ecologists are some sort of environmental engineers and compare them to the pure physicists who provide scientific rules for engineers, do the tentative understandings we have outlined (in their article) provide a sound basis for action by those who would manage the environment? It is self-evident that they do not.... We submit that ecology as such probably cannot do what many people expect it to do; it cannot provide a set of "rules" of the kind needed to manage the environment.[35]

Donald Worster, at the conclusion of his scholarly and brilliant history of ecological thinking in the West, is of the same opinion.[36]

But ecologists do have an important task in the deep ecology movement. They can be *subversive* in their perspective. For human ecologist Paul Shepard, "the ideological status of ecology is that of a resistance movement" because its intellectual leaders such as Aldo Leopold challenge the major premises of the dominant social paradigm.[37] As Worster in his history of ecology points out:

> [A]ll science, though primarily concerned with the "Is," becomes implicated at some point with the "Ought." The continuing environmental crisis makes it obvious that man's moral visions and utopias are little more than empty enterprise when they depart too far from nature's ways. This is the major lesson we have learned from studying the effects of men's hands on environment. An ecological ethic of interdependence, man in nature may be the outcome of a dialectical relation between scientist and ethicist.[38]

A final source of inspiration for the deep, long-range ecology movement is those artists who have tried to maintain a sense of place in their work.[39] Some artists, standing against the tide of mid-century pop art, minimalist art, and conceptual art have shown remarkable clarity and objectivity in their perception of nature. This spiritual-mystical objectivism is found, for example, in the photographs of Ansel Adams.[40] For these artists, including Morris Graves, who introduced concepts of Eastern thought (including Zen Buddhism) into his art, and Larry Gray, who reveals the eloquent light of revelation of nature in his skyscapes, men reaffirm their spiritual kinship with the eternity of God in nature through art.[41]

Themes of Deep Ecology

I indicated in preceding pages that many thinkers are questioning some of the premises of the dominant social paradigm of the modern societies. They are attempting to extend on an appropriate metaphysics, epistemology, and ethics for what I call an "ecological consciousness." Some of these writers are very supportive of reformist environmental social movements, but they feel reform while necessary is not sufficient. They suggest a new paradigm is required and a new utopian vision of "right livelihood" and the "good society." Utopia stimulates our thinking concerning alternatives to present society.[42] Some persons, such as Aldo Leopold, have suggested that we begin our thinking on utopia not with a statement of "human nature" or "needs of humans" but by trying to "think like a mountain." This profound extending, "thinking like a mountain," is part and parcel of the phenomenology of ecological consciousness.[43] Deep ecology begins with Unity rather than dualism which has been the dominant theme of Western philosophy.[44]

Philosopher Henryk Skolimowski, who has written several papers on the options for the ecology movements, asserts:

> [W]e are in a period of ferment and turmoil, in which we have to challenge the limits of the analytical and empiricist comprehension of the world as we must work out a new conceptual and philosophical framework in which the multitude of new social, ethical, ecological, epistemological, and ontological problems can be accommodated and fruitfully tackled. The need for a new philosophical framework is felt by nearly everybody. It would be lamentable if professional philosophers were among the last to recognize this.[45]

Numerous other writers on deep ecology, including William Ophuls, E. F. Schumacher, George Sessions, Theodore Roszak, Paul Shepard,

Gary Snyder, and Arne Naess, have in one way or another called for a new social paradigm or a new environmental ethic. We must "think like a mountain" according to Aldo Leopold. And Roderick Nash says:

> Do rocks have rights? If the time comes when to any considerable group of us such a question is no longer ridiculous, we may be on the verge of a change of value structures that will make possible measures to cope with the growing ecologic crisis. One hopes there is enough time left.[46]

Any attempt to create artificially a "new ecological ethics" or a "new ontology of man's place in nature" out of the diverse strands of thought which make up the deep ecology movement is likely to be forced and futile. However, by explicating some of the major themes embodied in and presupposed by the intellectual movement I am calling deep ecology, some groundwork can be laid for further discussion and clarification.[47] Following the general outline of perennial philosophy, the order of the following statements summarizing deep ecology's basic principles are metaphysical-religious, psychological-epistemological, ethical, and social-economic-political. These concerns of deep ecology encompass most of reformist environmentalism's concerns but subsume them in its fundamental critique of the dominant paradigm.

According to deep ecology:

(1) *A new cosmic/ecological metaphysics which stresses the identity (I/thou) of humans with non-human nature is a necessary condition for a viable approach to building an eco-philosophy*. In deep ecology, the wholeness and integrity of person/planet together with the principle of what Arne Naess calls "biological equalitarianism" are the most important ideas. Man is an integral part of nature, not over or apart from nature. Man is a "plain citizen" of the biosphere, not its conqueror or manager. There should be a "democracy of all God's creatures" according to St. Francis; or as Spinoza said, man is a "temporary and dependent mode of the whole of God/Nature." Man flows with the system of nature rather than attempting to control all of the rest of nature. The hand of man lies lightly on the land. Man does not perfect nature, nor is man's primary duty to make nature more efficient.[48]

(2) *An objective approach to nature is required*. This approach is found, for example, in Spinoza and in the works of Spinoza's twentieth century disciple, Robinson Jeffers. Jeffers describes his orientation as a philosophy of "inhumanism" to draw a sharp and shocking contrast with the subjective anthropocentrism of the prevailing humanistic philosophy, art, and culture of the 20th century West.[49]

(3) *A new psychology is needed to integrate the metaphysics in the mind field of post-industrial society*. A major paradigm shift results from psychological changes of perception. The new paradigm requires rejection of subject/object, man/nature dualisms and will require a pervasive awareness of total intermingling of the planet earth. Psychotherapy seen as adjustment to ego-oriented society is replaced by a new ideal of psychotherapy as spiritual development.[50] The new metaphysics and psychology leads logically to a posture of biospheric egalitarianism and liberation in the sense of autonomy, psychological/emotional freedom of the individual, spiritual development for *Homo sapiens*, and the right of other species to pursue their own evolutionary destinies.[51]

(4) *There is an objective basis for environmentalism, but objective science in the new paradigm is different from the narrow, analytic conception of the "scientific method" currently popular*. Based on "ancient wisdom," science should be both objective and participatory without modern science's subject/object dualism. The main *value* of science is seen in its ancient perspective as contemplation of the cosmos and the enhancement of understanding of self and creation.[52]

(5) *There is wisdom in the stability of natural processes unchanged by human intervention*. Massive human-induced disruptions of ecosystems will be unethical and harmful to men. Design for human settlement should be with nature, not against nature.[53]

(6) *The quality of human existence and human welfare should not be measured only by quantity of products*. Technology is returned to its ancient place as an appropriate tool for human welfare, not an end in itself.[54]

(7) *Optimal human carrying capacity should be determined for the planet as a biosphere and for specific islands, valleys, and continents*. A drastic reduction of the rate of growth of population of *Homo sapiens* through humane birth control programs is required.[55]

(8) *Treating the symptoms of man/nature conflict, such as air or water pollution, may divert attention from more important issues and thus be counter-productive to "solving" the problems*. Economics must be subordinate to ecological-ethical criteria.

Economics is to be treated as a small sub branch of ecology and will assume a rightfully minor role in the new paradigm.[56]

(9) *A new philosophical anthropology will draw on data of hunting/gathering societies for principles of healthy, ecologically viable societies.* Industrial society is not the end toward which all societies should aim or try to aim.[57] Therefore, the notion of "reinhabitating the land" with hunting-gathering, and gardening as a goal and standard for post-industrial society should be seriously considered.[58]

(10) *Diversity is inherently desirable both culturally and as a principle of health and stability of ecosystems.*[59]

(11) *There should be a rapid movement toward "soft" energy paths and "appropriate technology" and toward lifestyles which will result in a drastic decrease in per capita energy consumption in advanced industrial societies while increasing appropriate energy in decentralized villages in so-called "third world" nations.*[60] Deep ecologists are committed to rapid movement to a "steady-state" or "conserver society" both from ethical principles of harmonious integration of humans with nature and from appreciation of ecological realities.[61] Integration of sophisticated, elegant, unobtrusive, ecologically sound, appropriate technology with greatly scaled down, diversified, organic, labor-intensive agriculture, hunting, and gathering is another goal.[62]

(12) *Education should have as its goal encouraging the spiritual development and personhood development of the members of a community, not just training them in occupations appropriate for oligarchic bureaucracies and for consumerism in advanced industrial societies.*[63]

(13) *More leisure as contemplation in art, dance, music, and physical skills will return play to its place as the nursery of individual fulfillment and cultural achievement.*[64]

(14) *Local autonomy and decentralization of power is preferred over centralized political control through oligarchic bureaucracies.* Even if bureaucratic modes of organization are more "efficient," other modes of organization for small scale human communities are more "effective" in terms of the principles of deep ecology.[65]

(15) *In the interim, before the steady-state economy and radically changed social structure are instituted, vast areas of the planet biospheres will be zoned "off limits" to further industrial exploitation and large-scale human settlement; these should be protected by defensive groups of people.* One ecologist has called such groups a "world wilderness police."[66]

[...]

Notes

Thanks and acknowledgment to George Sessions, Philosophy Department, Sierra College, Rocklin, California. His sympathetic support and ideas made it possible to develop and deepen many of the ideas expressed in this paper.

1 Naess, "The Shallow and the Deep, Long-Range Ecology Movement," *Inquiry* 16 (1973): 95.
2 Rodman, "Four Forms of Ecological Consciousness: Beyond Economics, Resource Conservation," Pitzer College, 1977.
3 M. Weber, *The Protestant Ethic and the Spirit of Capitalism* (1930).
4 D. Pirages & P. Erlich, *Ark II: Social Response to Environmental Imperatives* (1974), 43. See also, *The Future of the Great Plains, H. R. Doc. No.* 144, 75th Cong., 1st Sess. (1937).
5 On the history of the paradigm see V. Ferkiss, *The Future of Technological Civilization* (1974). For a critique of the "me now" consumerism of the 1970s see C. Lasch, *The Culture of Narcissism: American Life in an Age of Diminishing Expectations* (1979). See also Manager's Journal, "Monitoring America, Values of Americans," *Wall St. J.* (Oct. 2, 1978).
6 L. White, Jr., "The Historical Roots of Our Ecological Crisis," *Science* 155 (1967): 1203.
7 B. Weisberg, *Beyond Repair: The Ecology of Capitalism* (1971); England & Bluestone, "Ecology and Class Conflict," *Rev. Radical Political Econ.* 3 (1971): 31. On Locke's view of "property," see Ferkiss, *supra* note 5.
8 L. Marx, *The Machine in the Garden: Technology and the Pastoral Ideal in America* (1967); and L. Mumford, *The Pentagon of Power* (1970).
9 T. Kuhn, *The Structure of Scientific Revolutions* (2nd ed. 1970). For criticism of Kuhn, I. Lakatos & A. Musgrave, *Criticisms and the Growth of Knowledge* (1970).
10 R. Nash, *Wilderness and the American Mind* (rev. ed. 1973); Sax, "America's National Parks: Their Principles, Purposes and their Prospects," *Nat. Hist.* 35 (1976): 57.
11 B. Commoner, *The Closing Circle* (1971); J. Ridgeway, *The Politics of Ecology* (1970).
12 *Natural Resources Defense Council, Land Use Controls in the United States: A Handbook of Legal Rights of Citizens* (1977); I. McHarg, *Design with Nature* (1971).

13 Rodman, *supra* note 2; S. Hays, *Conservation and the Gospel of Efficiency* (1959); G. Pinchot, *Breaking New Ground* (1947).

14 P. Ehrlich, *The Population Bomb* (1968). See generally publications of the organization Zero Population Growth. *Population and the American Future: Report of the Commission on Population Growth and the American Future* (1972).

15 T. Regan & P. Singer, *Animal Rights and Human Obligation* (1976); P. Singer, *Animal Liberation* (1977).

16 D. L. Meadows & D. H. Meadows, *The Limits to Growth* (2nd ed. 1974); M. Mesarovic & E. Pestel, *Mankind at the Turning Point: Second Report of the Club of Rome* (1974); D. Meadows, *Alternatives to Growth* (1977). For a critique of the limits to growth model see *Models of Doom: A Critique of the Limits to Growth* (H. Cole ed. 1973); *Toward a Steady-State Economy* (H. Daly ed. 1973).

17 White, *supra* note 6.

18 A. Watts, *Psychotherapy East and West* (1975); A. Watts, *Nature, Man and Woman* (1970), A. Watts, *The Spirit of Zen: A Way of Life, Work and Art in the Far East* (1955), A. Watts, *The Essence of Alan Watts* (1977).

19 D. Suzuki, *Essays in Zen Buddhism* (1961).

20 F. Capra, *The Tao of Physics* (1975).

21 J. Needham, *Science and Civilization in China* (multi. vol. 1954–76).

22 Needham, "History and Human Values: A Chinese Perspective for World Science and Technology", *Centennial Rev.* 20 (1976): 1.

23 H. Smith, *Forgotten Truth* (1976); Smith, "Tao Now: An Ecological Testament," in *Earth Might be Fair* (I. Barbour ed. 1972).

24 J. Ellul, *The Technological Society* (1964).

25 Goodman, "Can Technology Be Humane?," in *Western Man and Environmental Ethics* (I. Barbour ed. 1973), 225.

26 H. Marcuse, *One-Dimensional Man* (1964).

27 T. Roszak, *The Making of a Counter-Culture* (1969), T. Roszak, *Where the Wasteland Ends: Politics and Transcendence in Post-Industrial Society* (1972), T. Roszak, *Unfinished Animal: The Aquarian Frontier and the Evolution of Consciousness* (1975), T. Roszak, *Person/Planet* (1978).

28 R. Ornstein, *The Mind Field* (1976), 105. The works of Carlos Castaneda have been influential. They include *The Teachings of Don Juan* (1968); *A Separate Reality* (1971); *Journey to Ixtlan* (1972); *Tales of Power* (1974).

29 L. Standing Bear, in *Touch the Earth* (T. McLuhan ed. 1971). Among the most significant and original theories of Native Americans and non-human nature see V. Deloria, *God is Red* (1975); C. Martin, *Keepers of the Game* (1978); S. Steiner, *The Vanishing White Man* (1976).

30 Sessions, "Spinoza and Jeffers on Man in Nature," *Inquiry* 20 (1977): 481; G. Sessions, "Spinoza, Perennial Philosophy and Deep Ecology," unpublished paper, Sierra College, 1979.

31 Hampshire, *Two Theories of Morality* (1977); Hampshire, *Spinoza* (1956); Naess, "Spinoza and Ecology", *Philosophia* 7 (1977): 45.

32 Sessions, *supra* note 30. See also A. Coffin, *Robinson Jeffers: Poet of Inhumanism* (1971); B. Hotchkiss, *Jeffers: The Siviastic Vision* (1975); R. Brophy, "Robinson Jeffers, Metaphysician of the West," unpublished paper, dept. of English, Long Beach State University, Long Beach, California, n.d.

33 J. Cobb, Jr., *Is It Too Late? A Theology of Ecology* (1972); C. Hartshorne, *Beyond Humanism: Essays in the Philosophy of Nature* (1937); Whitehead, *Science and the Modern World* (1925), chs. 5, 13; Griffin, "Whitehead's Contribution to the Theology of Nature", *Bucknell Rev.* 20 (1972): 95.

34 M. Heidegger, *The Question Concerning Technology and Other Essays* (W. Lovitt trans. 1977); G. Steiner, *Martin Heidegger* (1979); V. Vycinas, *Earth and Gods: An Introduction to the Philosophy of Martin Heidegger* (1961). On the approach taken by Heidegger and the contemporary ecological consciousness, see D. LaChapelle, *Earth Wisdom* (1978), ch. 9, "Martin Heidegger and the Quest for Being". The writings of Michael Zimmerman on Heidegger are also useful, including his "Beyond 'Humanism': Heidegger's Understanding of Technology," *Listening* 12 (1977): 74. See also Zimmerman, "Marx and Heidegger on the Technological Domination of Nature," *Philosophy Today* 12 (Summer 1979): 99.

35 Murdoch & Connell, "All About Ecology," in *Western Man and Environmental Ethics* (I. Barbour ed. 1973).

36 Donald Worster, epilogue, *Nature's Economy* (1977). The importance of the thinking of Aldo Leopold should be emphasized. There are many articles interpreting Leopold's message. See, e.g., Jung, "The Splendor of the Wild: Zen and Aldo Leopold," *Atlantic Naturalist* 29 (1974): 5.

37 Shepard, "Introduction: Man and Ecology," in *The Subversive Science* (P. Shepard & D. McKinley eds. 1969), 1. See also Everndon, "Beyond Ecology," *N. Am. Rev.* 263 (1978): 16.

38 D. Worster, *supra* note 36.

39 Huth, "Wilderness and Art," in *Wilderness: America's Living Heritage* (D. Brower ed. 1961), 60; Shepard, "A Sense of Place," *N. Am. Rev.* 262 (1977): 22.

40 Adams, "The Artist and the Ideals of Wilderness," in *Wilderness: America's Living Heritage* (D. Brower ed. 1961), 60.

41 M. Graves, *The Drawings of Morris Graves* (1974).

42 M. Sibley, *Nature and Civilization: Some Implications for Politics* (1977), 251 (ch. 7, "Nature, Civilization and the Problem of Utopia"). Sibley makes a case for more utopian visions from contemporary intellectuals. Although many people have been revolted by the visions of Marxism and Fascist dic-

tatorships, "the student of politics has an obligation not only to explain and criticize but also to propose and explicate ideals. We need more utopian visions, not fewer. For if politics be that activity through which man seeks consciously and deliberately to order and control his collective life, then one of the salient questions in all politics must be: Order and control for what ends? Without utopian visions these ends cannot be stated as wholes; and even a discussion of means and strategies will be clouded unless ends are at least relatively clear" (ibid., 47).

43 Jung, "To Save the Earth," *Philosophy Today* 8 (1975): 108.
44 Sessions, *supra* note 30.
45 Skolimowski, "Ecology Movement Re-examined," *Ecologist* 6 (1976): 298; Skolimowski, "Options for the Ecology Movement," *Ecologist* 7 (1977): 318.
46 Nash, "Do Rocks Have Rights?," *Centre Magazine* 10 (1977): 2.
47 Skolimowski, *supra* note 45.
48 J. Needleman, *A Sense of the Cosmos* (1975), 76–7, 100–2.
49 Sessions, *supra* note 30. Spinoza is one of the most important philosophers for deep ecology. The new translations of Spinoza's work are absolutely essential for understanding his thought. See P. Wienpahl, *The Radical Spinoza* (1979).
50 R. Ornstein, *supra* note 28.
51 Sessions, *supra* note 30; G. Snyder, *The Old Ways* (1977).
52 Capra, *supra* note 20; Sessions, *supra* note 30; Needleman, *supra* note 48.
53 Commoner, *supra* note 11; McHarg, *supra* note 12.
54 Needleman, *supra* note 48; Sessions, *supra* note 30.
55 E. Schumacher, *Small is Beautiful, Economics as if People Mattered* (1973). See as an example of this argument P. Ehrlich, A. Ehrlich, & J. Holdren, *Ecoscience* (1977).
56 W. Ophuls, *Ecology and the Politics of Scarcity* (1977). Schumacher, *supra* note 55.
57 S. Steiner, *supra* note 29; Snyder, *supra* note 51; P. Shepard, *The Tender Carnivore and the Sacred Game* (1974).
58 P. Berg, *Reinhabiting a Different Kind of Country* (1978).
59 R. Dasmann, *A Different Kind of Country* (1968); N. Myers, *The Sinking Ark: A New Look at the Problems of Disappearing Species* (1979).
60 R. Dasmann, *Ecological Principles for Economic Development* (1973).
61 W. Ophuls, *supra* note 56. On the "steady-state" see *Toward a Steady-State Economy* (H. Daly ed. 1973).
62 E. Schumacher, *supra* note 55.
63 T. Roszak, *supra* note 27.
64 J. Huizinga, *Homo Ludens: The Play Element in Culture* (1950). Collier, "The Fullness of Life Through Leisure," in *The Subversive Science* (P. Shepard & D. McKinley eds. 1969), 416.
65 W. Ophuls, *supra* note 56; *The Sustainable Society* (D. Pirages ed. 1977); P. Berg & R. Dasmann, *Reinhabiting a Separate Country* (1978).
66 "Editorial," *Greenpeace Chronicles* 3 (March, 1979), I, No. 14. On the suggestion for a "world wilderness police" see Iltis, "Wilderness, Can Man Do Without It?," in *Recycle This Book: Ecology, Society and Man* (J. Allan & A. Hanson eds. 1972), 167. On "earth festivals" and getting in touch with nature again, see D. LaChapelle, *supra* note 34.

The world wilderness police is a defensive force. The defense of nature against despoilers is the goal. I think the prototype of this policeman is Morel the hero of Romain Gary's novel *The Roots of Heaven* (1958). Only after all appeals to the United Nations, to all nations, to all reasonable men have failed does Morel turn to the force of arms to defend the elephants of central Africa from poachers.

41

Deeper than Deep Ecology: The Eco-Feminist Connection

Ariel Salleh

> ...beyond that perception of otherness lies the perception of pysche, polity and cosmos, as metaphors of one another....
>
> ***John Rodman***[1]

In what sense is eco-feminism "deeper than deep ecology"? Or is this a facile and arrogant claim? To try to answer this question is to engage in a critique of a critique, for deep ecology itself is already an attempt to transcend the shortsighted instrumental pragmatism of the resource-management approach to the environmental crisis. It argues for a new metaphysics and an ethic based on the recognition of the intrinsic worth of the nonhuman world. It abandons the hardheaded scientific approach to reality in favor of a more spiritual consciousness. It asks for voluntary simplicity in living and a non-exploitive steady-state economy. The appropriateness of these attitudes as expressed in Naess' and Devall's seminal papers on the deep ecology movement is indisputable.[2] But what is the organic basis of this paradigm shift? Where are Naess and Devall "coming from," as they say? Is deep ecology a sociologically coherent position?

The first feature of the deep ecology paradigm introduced by Naess is replacement of the Man/Nature dualism with a *relational total-field image*, where man is not simply "in" his environment, but essentially "of" it. The deep ecologists do not appear to recognize the primal source of this destructive dualism, however, or the deeply ingrained motivational complexes which grow out of it.[3] Their formulation uses the generic term *Man* in a case where use of a general term is not applicable. Women's monthly fertility cycle, the tiring symbiosis of pregnancy, the wrench of childbirth and the pleasure of suckling an infant, these things already ground women's consciousness in the knowledge of being coterminous with Nature. However tacit or unconscious this identity may be for many women, bruised by derogatory patriarchal attitudes to motherhood, including modern male-identified feminist ones, it is nevertheless "a fact of life." The deep ecology movement, by using the generic term *Man*, simultaneously presupposes the difference between the sexes in an uncritical way, and yet overlooks the significance of this difference. It overlooks the point that if women's lived experience were recognized as meaningful and were given legitimation in our culture, it could provide an immediate "living" social basis for the alternative consciousness which the deep ecologist is trying to formulate and introduce as an abstract ethical construct. Women already, to borrow Devall's turn of phrase, "flow with the system of nature."

The second deep ecology premise, according to Naess is a move away from anthropocentrism, a move toward *biological egalitarianism* among all living species. This assumption, however, is already cancelled in part by the implicit contradiction contained in Naess' first premise. The master-slave role which marks man's relation with nature is replicated in man's relation with woman. A self-consistent biological egalitarianism cannot be ar-

From *Environmental Ethics* 6/4 (1984): 339–45. Reprinted by permission of Center for Environmental Philosophy, University of North Texas, Denton, TX.

rived at unless men become open to both facets of this same urge to dominate and use. As Naess rightly, though still somewhat anthropocentrically, points out, the denial of dependence on Mother/Nature and the compensatory drive to mastery which stems from it, have only served to alienate man from his true self. Yet the means by which Naess would realize this goal of species equality is through artificial limitation of the human population. Now putting the merits of Naess' "ends" aside for the moment, as a "means" this kind of intervention in life processes is supremely rationalist and technicist, and quite at odds with the restoration of life-affirming values that is so fundamental to the ethic of deep ecology. It is also a solution that interestingly enough cuts right back into the nub of male dependence on women as mothers and creators of life – another grab at women's special potency, inadvertent though it may be.

The third domain assumption of deep ecology is the *principle of diversity and symbiosis*: an attitude of live and let live, a beneficial mutual coexistence among living forms. For humans the principle favors cultural pluralism, an appreciation of the rich traditions emerging from Africa, China, the Australian Aboriginal way, and so on. These departures from anthropocentrism, and from ethnocentrism, are only partial, however, if the ecologist continues to ignore the cultural inventiveness of that other half of the human race, women; or if the ecologist unwittingly concurs in those practices which impede women's full participation in his own culture. The annihilation of seals and whales, the military and commercial genocide of tribal peoples, are unforgivable human acts, but the annihilation of women's identity and creativity by patriarchal culture continues as a fact of daily existence. The embrace of progressive attitudes toward nature does little in itself to change this.

Deep ecology is an *anti-class posture*; it rejects the exploitation of some by others, of nature by man, and of man by man, this being destructive to the realization of human potentials. However, sexual oppression and the social differentiation that this produces is not mentioned by Naess. Women again appear to be subsumed by the general category. Obviously the feminist ecological analysis is not "in principle" incompatible with the anti-class posture of deep ecology. Its reservation is that in bypassing the parallel between the original exploitation of nature as object-and-commodity resource and of nurturant woman as object-and-commodity resource, the ecologist's anti-class stance remains only superficially descriptive, politically and historically static. It loses its genuinely deep structural critical edge. On the question of political praxis though, there is certainly no quarrel between the two positions. Devall's advocacy of loose activist networks, his tactics of nonviolent contestation, are cases in point.[4] Deep ecology and feminism see change as gradual and piecemeal; the violence of revolution imposed by those who claim "to know" upon those who "do not know" is an anathema to both.

The fight against *pollution and resource depletion* is, of course, a fundamental environmental concern. And it behooves the careful activist to see that measures taken to protect resources do not have hidden or long-term environmental costs which outweigh their usefulness. As Naess observes, such costs may increase class inequalities. In this context he also comments on the "after hours" environmentalist syndrome frequently exhibited by middle-class professionals. Devall, too, criticizes what he calls "the bourgeois liberal reformist elements" in the movement – Odum, Brower, and Lovins, who are the butt of this remark. A further comment that might be made in this context, however, is that women, as keepers of *oikos*, are in a good position to put a round-the-clock ecological consciousness into practice. Excluded as many still are from full participation in the social-occupational structure, they are less often compromised by the material and status rewards which may silence the activist professional. True, the forces of capitalism have targeted women at home as consumer par excellence, but this potential can just as well be turned against the systematic waste of industrialism. The historical significance of the domestic labor force in moves to recycle, boycott, and so on, has been grossly underestimated by ecologists.

At another level of analysis entirely, but again on the issue of pollution, the objectivist attitude of most ecological writing and the tacit mind-body dualism which shapes this, means that its comprehension of "pollution" is framed exclusively in external material terms. The feminist consciousness, however, is equally concerned to eradicate ideological pollution, which centuries of patriarchal conditioning have subjected us all to, women and men. Men, who may derive rather more ego gratification from the patriarchal status quo than women, are on the whole less motivated to change this system than women are. But radical women's consciousness-raising groups are continually engaging in an intensely reflexive political process; one that works on the psychological contamination produced by the culture of domination and helps women to build new and confident selves. As a foundation for social and political change, this work of women is a very thorough preparation indeed.

The sixth premise of Naess' deep ecology is the *complexity, not complication principle*. It favors the preservation of complex interrelations which exist between parts of the environment, and inevitably, it involves a systems theoretical orientation. Naess' ideal is a complex economy supported by division, but not fragmentation of labor; worker alienation to be overcome by opportunities for involvement in mental and manual, specialized and nonspecialized tasks. There are serious problems of implementation attached to this vaguely sketched scenario, but beyond this, the supporting arguments are also weak, not to say very uncritical in terms of the stated aims of the deep ecology movement. The references to "soft future research," "implementation of policies," "exponential growth of technical skill and intervention," are highly instrumental statements which collapse back into the shallow ecology paradigm and its human chauvinist ontology. What appears to be happening here is this: the masculine sense of self-worth in our culture has become so entrenched in scientistic habits of thought, that it is very hard for men to argue persuasively without resource to terms like these for validation. Women, on the other hand, socialized as they are for a multiplicity of contingent tasks and practical labor functions in the home and out, do not experience the inhibiting constraints of status validation to the same extent. The traditional feminine role runs counter to the exploitive technical rationality which is currently the requisite masculine norm. In place of the disdain that the feminine role receives from all quarters, "the separate reality" of this role could well be taken seriously by ecologists and reexamined as a legitimate source of alternative values. As Snyder suggests, men should try out roles which are not highly valued in society; and one might add, particularly this one, for herein lies the basis of a genuinely grounded and nurturant environmentalism. As one eco-feminist has put it:

> If someone has laid the foundations of a house, it would seem sensible to build on those foundations, rather than import a prefabricated structure with no foundations to put beside it.[5]

A final assumption of deep ecology described by Naess is the importance of *local autonomy and decentralization*. He points out that the more dependent a region is on resources from outside its locality, the more vulnerable it is ecologically and socially: for self-sufficiency to work, there must be political decentralization. The drive to ever larger power blocs and hierarchical political structures is an invariant historical feature of patriarchal societies, the expression of an impulse to compete and dominate the Other. But unless men can come to grips honestly with this impulse within themselves, its dynamic will impose itself over and over again on the anatomy of revolution. Women, if left to their own devices, do not like to organize themselves in this way. Rather they choose to work in small, intimate collectivities, where the spontaneous flow of communication "structures" the situation. There are important political lessons for men to learn from observing and participating in this kind of process. And until this learning takes place, notions like autonomy and decentralization are likely to remain hollow, fetishistic concepts.

Somewhat apologetically, Naess talks about his ecological principles as "intuitive formulations" needing to be made more "precise." They are a "condensed codification" whose tenets are clearly "normative"; they are "ecophilosophical," containing not only norms but "rules," "postulates," "hypotheses," and "policy" formulations. The deep ecology paradigm takes the form of "subsets" of "derivable premises," including at their most general level "logical and mathematical deductions." In other words, Naess' overview of ecosophy is a highly academic and positivized one, dressed up in the jargon of current science-dominated standards of acceptability. Given the role of this same cultural scientism in industry and policy formulation, its agency in the very production of the eco-crisis itself, Naess' stance here is not a rationally consistent one. It is a solution trapped in the given paradigm. The very term *norm* implies the positivist split between fact and value, the very term *policy* implies a class separation of rulers and ruled. Devall, likewise, seems to present purely linear solutions – "an objective approach," "a new psychology"; the language of cost-benefit analysis, "optimal human carrying capacity," and the language of science, "data on hunter gatherers," both creep back in. Again, birth "control programs" are recommended, "zoning," and "programming," the language of technocratic managerialism. "Principles" are introduced and the imperative *should* rides roughshod through the text. The call for a new epistemology is somehow dissociated in this writing from the old metaphysical presuppositions which prop up the argument itself.

In arguing for an eco-phenomenology, Devall certainly attempts to bypass this ideological noose – "Let us think like a mountain," he says – but again, the analysis here rests on what is called "a gestalt of person-in-nature": a conceptual effort, a grim intellectual determination "to care"; "to show reverence" for Earth's household and "to let" nature follow "its separate" evolutionary path. The residue of specular instrumentalism is overpowering; yet the conviction remains that a radical transformation of social organization and values is imminent: a challenge to the fundamental premises of the dominant social paradigm. There is a concerted effort to rethink Western metaphysics, epistemology, and ethics here, but this "rethink" remains an idealism closed in on itself because it fails to face up to the uncomfortable psychosexual origins of our culture and its crisis. Devall points by turn to White's thesis that the environmental crisis derives from the Judeo-Christian tradition, to Weisberg's argument that capitalism is the root cause, to Mumford's case against scientism and technics. But for the eco-feminist, these apparently disparate strands are merely facets of the same motive to control which runs a continuous thread through the history of patriarchy. So, it has been left to the women of our generation to do the theoretical housework here – to lift the mat and sweep under it exposing the deeply entrenched epistemological complexes which shape not only current attitudes to the natural world, but attitudes to social and sexual relations as well.[6] The accidental convergence of feminism and ecology at this point in time is no accident.

Sadly, from the eco-feminist point of view, deep ecology is simply another self-congratulatory reformist move; the transvaluation of values it claims for itself is quite peripheral. Even the Eastern spiritual traditions, whose authority deep ecology so often has recourse to – since these dissolve the repressive hierarchy of Man/Nature/God – even these philosophies pay no attention to the inherent Man/Woman hierarchy contained within this metaphysic of the Whole. The supression of the *feminine* is truly an all pervasive human universal. It is not just a supression of real, live, empirical women, but equally the supression of the feminine aspects of men's own constitution which is the issue here. Watts, Snyder, Devall, all want education for the spiritual development of "personhood." This is the self-estranged male reaching for the original androgynous natural unity within himself. The deep ecology movement is very much a spiritual search for people in a barren secular age; but how much of this quest for self-realization is driven by ego and will? If, on the one hand, the search seems to be stuck at an abstract cognitive level, on the other, it may be led full circle and sabotaged by the ancient compulsion to fabricate perfectability. Men's ungrounded restless search for the alienated Other part of themselves has led to a society where not life itself, but "change," bigger and better, whiter than white, has become the consumptive end. The dynamic to overcome this alienation takes many forms in the post-capitalist culture of narcissism – material and psychological consumption like karma-cola, clown workshops, sensitivity training, bio-energetics, gay lib, and surfside six. But the deep ecology movement will not truly happen until men are brave enough to rediscover and to love the woman inside themselves. And we women, too, have to be allowed to love what we are, if we are to make a better world.

Notes

1 John Rodman, "The Liberation of Nature?" *Inquiry* 20 (1977): 83–145; quoted by Bill Devall in "The Deep Ecology Movement," *Natural Resources Journal* 20 (1980): 317.

2 Arne Naess, "The Shallow and the Deep, Long-Range Ecology Movement," *Inquiry* 16 (1973): 95–100; Bill Devall, "The Deep Ecology Movement," pp. 299–322.

3 See Ariel Kay Salleh, "Of Portnoy's Complaint and Feminist Problematics," *Australian and New Zealand Journal of Sociology* 17 (1981): 4–13; "Ecology and Ideology," *Chain Reaction* 31 (1983): 20–21; "From Feminism to Ecology," *Social Alternatives* (1984): forthcoming.

4 And on this connection, see Ariel Kay Salleh, "The Growth of Eco-feminism," *Chain Reaction* 36 (1984): 26–28; also comments in "Whither the Green Machine?" *Australian Society* 3 (1984): 15–17.

5 Ann Pettitt, "Women Only at Greenham," *Undercurrents* 57 (1982): 20–21.

6 Some of this feminist writing is discussed in Ariel Kay Salleh, "Contributions to the Critique of Political Epistemology," *Thesis Eleven* 8 (1984): 23–43.

Part VI

Technology as Social Practice

Technology and the Lifeworld

Introduction

All the selections below make an explicit distinction, often overlooked, between very different ways of considering technological phenomena. On the one hand, there are the effects of technology *on* our lives as these appear to outside observers (e.g., social scientists or historians); on the other, there are the experiences of what happens *to* us in our living with the various technologies in everyday life. The authors in this section stress the importance of the latter, experiential perspective – the perspective, as Husserl called it, of the lifeworld. They all agree that it is both crucial for – and also routinely under-appreciated in – our attempts to understand the cultural transformation technology brings.

As the title of his essay indicates, Carl Mitcham begins with one of the key ideas from Heidegger's analysis of the human condition as "being-in-the-world." Though he admits he is removing the idea from its original context, Mitcham notes that his characterization of how it is to "be-with" technology still "takes off from" Heidegger's view that it is our practical-technical engagements, not our theoretical knowledge, that best illuminate and ordinarily predominate in our relationships with our surroundings and with other people. Mitcham identifies what he calls three historical-philosophical, ideal-typical attitudes toward technology – namely, ancient skepticism, Enlightenment optimism, and Romantic uneasiness. Each attitude expresses a fundamentally different sort of relation between ourselves and technology, and each, respectively, has been the most prominent in premodern, early modern, and (to a lesser extent) later modern times. Summarizing his results with a four-part comparative table, Mitcham explains how each outlook differently construes technology's essential character, its proper role in ethical and political affairs, the epistemological status of technological knowledge, and the metaphysical status of its artifacts. One salutary effect of his presentation is that it provides concise summaries of the outlook of a number of other thinkers who appear elsewhere in this anthology.

For Mitcham, however, the main point is a contemporary, not historical one. He thinks that the contrasts and conflicts among the three attitudes continue to echo in our own experiences with technology. So, regarding the first attitude, we (in the developed West) certainly can no longer embrace an "ancient skepticism" that appeals to myths about angered gods to express its sense that technology is "necessary but dangerous." We are also at least somewhat less likely to view manual labor as degrading, or hold out for a "higher wisdom" that makes practical inventiveness seem an unworthy expression of the genuinely "human." Yet who in our age of technical excess is not attracted to, say, Socrates' conviction that neither natural philosophy nor technical expertise will ever reveal the secrets of the good life?

Unlike the ancients, early modern philosophers like Bacon often construed invention and discovery as ordained by God, as providing us a way of reachieving the state of purity of the Garden of Eden. As Mitcham suggests, if one strips this thought of its specifically religious trappings, the core idea is certainly not dead that scientific knowledge – far from being a distraction to the pursuit of wisdom – is actually the way to the good life.

Further, as Mitcham notes, insofar as the test for scientific knowledge is success in manipulation and control of nature, the early modern view promotes the (certainly not unfamiliar) idea that experimentation – or more generally, technoscientific engagement with the world – supplants contemplation as the definitive expression of a true knower's outlook. Hence, from Bacon to Comte – and in some moods, still also for us – technology can seem to be a wholly beneficial force, promising not only dominion over nature but moral and political progress as well. Commerce and the crafts, more than the relatively less technological activity of agriculture, seem the practical ideals. And doesn't human inventiveness seem akin to divine creation? After all, if the world is God's artifact, and human artifacts are produced by the same "natural" principles, then they are just as "real" as Creation.

For all its initial appeal, however, there has been growing distaste for the sort of scientistic positivism just described. Yet as Mitcham notes, about this we have typically behaved more like ambivalent Romantics than outright opponents. Romantics, like early modern and Enlightenment thinkers, harbor a deep admiration for the tremendous possibilities of technological power. Yet Romanticism also expressed strong misgivings about the exercise of this power, for it seemed obvious that the industrial revolution which the New Science brought with it gave us not only new material wealth but environmental destruction and miserable working and living conditions for all but the few. Romantics thus reject a mechanical, rationalistic understanding of nature in favor of an organic and intuitive one; and like the ancients, they see technology as in need of restraint. While Enlightenment thinkers see nature in terms of human artifice, Romantics see human creativity as an outgrowth of nature. Mitcham closes with the provocative suggestion that we may already actually tend to be Romantics in our hearts, but that precisely by sharing Romanticism's ambivalent stance between the ancients' suspicion of technology and the Enlightenment's optimistic promotion of it, we keep losing whatever power we might otherwise presume to gain by explicitly adopting a Romantic outlook.

Don Ihde's phenomenological account of technology-enhanced perception, loosely based on Husserl's *Crisis* and Heidegger's *Being and Time*, makes use of the same idea that attracted Mitcham's attention – namely, that human beings, their technologies, and the surrounding world form a structural whole whose experientially recognizable features merit (but often fail to receive) careful analysis. Ihde says little about traditional and non-phenomenological interpretations of technology, but some of his criticisms of them are easy to detect between the lines. He rejects, for example, the very idea that relations with our surroundings are ever a matter of "bare" perception. There is, he insists, no "simple seeing" – no initial, contextless, foundational contact with an uninterpreted world. All our relations involve both "microperception" (immediate, focused, actual seeing, hearing, etc.) and "macroperception" (simultaneous sociocultural contextualizing); and these together constitute an existential relation whose meaning can always be elaborated but never needs to be "added" to the encounter itself. For Ihde, one reason this point is so frequently missed is that the usual model for human–world relations has been that of a mind standing over against a world full of things. The model, however, is inappropriate (i.e., unphenomenological) in several ways. For one thing, it ignores the fact that all I–world relations are "embodied" relations. I am engaged "with" my surroundings not just in terms of my thoughts, but through my body and my body's capacities. Moreover, the meaning of my relations with my surroundings is not simply a function of my directed, conscious awareness of this or that, but always involves "background relations" and sociocultural understanding that contextualizes and provides "cues" for dealing with whatever happens to be in focus. Finally, the model privileges direct, or what Ihde calls "unmediated," relations, perhaps in part out of epistemologically exaggerated fears of skepticism. Whatever the reasons, however, since technology relations are embodied, richly implicit, mediated relations, the traditional model cannot in principle do them justice.

Part of Ihde's alternative account – an analysis of the focal, microperceptual aspect of technology relations – appears below. Using the fiction of a "New Adam" in a technology-free Eden, Idhe contrasts the naïve, unmediated, "naked" apprehending of one's surroundings that would predominate in such an Eden with the mostly mediated, technoscientifically transformed relations one experiences today. Regarding these experiences specifically in terms of their "direct and focal" dimension, he identifies three sets of technology relations and distinguishes them in terms of the way our embodied selves, various technologies, and the world are meaningfully but differently configured in each. First, there are "embodiment relations," strictly so-called – relations in which technologies constitute and approximate the status of a "quasi-me" – as, for example, when we use eyeglasses or a hammer. In such cases, if the technologies do their job, we perceive *through* them, making them extensions of our own embodiment. Second, there are "hermeneutic relations" – relations in which "readable" (i.e., interpretable) technologies *with* which we encounter the world make it accessible in ways impossible for naked perception – as, for example, when written language presents us with what is otherwise remote from our time and place, or a thermometer gives us a scaled and enhanced measurement of what is otherwise simply some version of "hot," "cold," or too extreme to experience at all. Finally, there are "alterity relations" – relations where technologies themselves become "quasi-others" *to* which we relate – as, for example, in playing video games or interacting with robots. Though he does not use the world, Ihde makes it clear that alterity relations are sometimes a kind of fetishizing of otherwise useful technologies.

Finally, Ihde closes with a provocative critique of those who would see in our ever more technologically mediated experience the dream (or nightmare) of culture that has been utterly transformed into a kind of technological "cocoon." Could there ever be, he asks, a kind of "New Eve" in a "Spaceship" of Total Technology, where nothing remains "natural," and every human frailty and limitation is subject to a technoscientific fix? On this issue, Ihde sounds a much more optimistic note than most of those who have been influenced by Heidegger (e.g., Borgmann and Dreyfus). For him, an entirely artificial world – a sort of global recreational vehicle with all the amenities – is as much of a fiction as the idea of a pure, non-technological Eden. It only seems like a real possibility when one forgets the world's actual pluralism of cultures, that is, forgets that the idea of Spaceship Earth is a product of "our" distinctive history. Ihde says nothing further about the parameters of this "our," however, and in the end he leaves the issue unsettled by asking if there might actually be a "single trajectory" emerging toward a high-tech world culture. That, he says, would make "the Marcuses, Jonases, and Elluls ... the prophets of our times."

From the viewpoint of Jürgen Habermas, a phenomenology like Ihde's will seem both too happy about the effects of technology on the lifeworld and too willing to evaluate scientific and technological knowledge in their own terms. While Habermas agrees with Ihde that it is through technology that modern science makes its most powerful impact on our lives, he does not conceive science and technology together as the single phenomenon, "technoscience", whose influences on our experiences are best evaluated from within the circle of those experiences themselves. What is called for, he argues, is a kind of reflectively generated, critical social theory that, on the one hand, reveals how science and technology are themselves shaped by the (often undemocractic) political and economic interests that inform everyday life, and on the other hand, defends the domination-free discussion of the kind of conscious social planning (for which technological knowledge alone is insufficient) that would insure a genuinely democratic implementation of technology.

"Technical Progress and the Social Life-World" is an early expression of the philosophical position Habermas later elaborated somewhat differently in his *Theory of Communicative Action*. Here, referring specifically to Aldous Huxley's call for a literature that simply takes scientific phenomena for another of its objects, Habermas denies that the facts and theories of science can simply find a place in the imagery and thought patterns of the lifeworld. He rejects the apparent implication of postpositivistic philosophies of science like Toulmin's and Kuhn's that the paradigms and presuppositions of scientific theories share features with the ideologies and myths that impinge upon the social lifeworld of the time. (In this regard, however, one might want to consider, for example, the latitudinarian religion and politics of the Newtonians, or social Darwinism, or the impact of non-Euclidean geometry on early twentieth-century art, or the postmodernists' attraction to chaos theory.) Habermas also denies that scientific facts and theories can any longer be claimed to enter the lifeworld through traditional liberal or scientific education. Finally, he also challenges the Enlightenment habit of automatically associating instrumental control of technical matters with democratic freedom – a habit that has unfortunately tempted social theorists as diverse as Comte and Marx to expect that it would just be a matter of time before the technical control of productive forces was overseen by an essentially democratic decision-making process.

42

Three Ways of Being-With Technology

Carl Mitcham

In any serious discussion of issues associated with technology and humanity there readily arises a general question about the primary member in this relationship. On the one hand, it is difficult to deny that we exercise some choice over the kinds of technics with which we live – that is, that we control technology. On the other, it is equally difficult to deny that technics exert profound influences on the ways we live – that is, structure our existence. We build our buildings, Winston Churchill once remarked (apropos a proposal for a new Parliament building), then our buildings build us. But which comes first, logically if not temporally, the builder or the buildings? Which is primary, humanity or technology?

This is, of course, a chicken-and-egg question, one not subject to any straightforward or unqualified answer. But it is not therefore insignificant, nor is it enough to propose as some kind of synthesis that there is simply a mutual relationship between the two, that humanity and technology are always found together. Mutual relationship is not some one thing; mutual relationships take many different forms. There are, for instance, mutualities of parent and child, of husband and wife, of citizens, and so forth. Humanity and technology can be found together in more than one way. Rather than argue the primacy of one or the other factor or the cliché of mutuality in the humanity-technology relationship, I propose to outline three forms the relationship itself can take, three ways of being-with technology.

To speak of three ways of being-with technology is necessarily to borrow and adapt a category from Martin Heidegger's *Being and Time* (1927) in a manner that deserves some acknowledgment. In his seminal work Heidegger proposes to develop a new understanding of being human by taking the primordial human condition, being-in-the-world, and subjecting this given to what he calls an existential analysis. The analysis proceeds by way of elucidating three equiprimordial aspects of this condition of being human: *the world* within which the human finds itself, the *being-in* relationship, and the *being* who is in the relationship – all as a means of approaching what, for Heidegger, is the fundamental question, the meaning of Being.

The fundamental question need not, on this occasion, concern us. What does concern us is the central place of technics in Heidegger's analysis and the disclosure of being-with as one of its central features. For Heidegger the worldhood of the world, as he calls it, comes into view through technical engagements, which reveal a network of equipment and artifacts ready-to-hand for manipulation, and other human beings likewise so engaged. These others are neither just technically ready-to-hand (like tools) nor even scientifically present-at-hand (like natural objects); on the contrary, they are *like* the very human being who notices them in that "*they are there too, and there with it.*"[1]

From *From Artifact to Habitat: Studies in the Critical Engagement of Technology*, ed. Gayle L. Ormiston (Research in Technology series, vol. 3), Bethlehem, PA: LeHigh University Press, 1990, pp. 31–59. Reprinted by permission of Associated University Presses.

The being-with relationship thus disclosed through technical engagements is, for Heidegger, primarily social in character; it refers to the social character of the world that comes to light through technical practice. Such a world is not composed solely of tools and artifacts, but of tools used with others, and artifacts belonging to others. Technical engagements are not just technical but have an immediately and intimately social dimension. Indeed, this is all so immediate that it requires some labored stepping back even to recognize and state – the processes of distancing and articulation that are in part precisely what philosophy is all about.

The present attempt to step back and examine various ways of being-with technology rather than being-with others (through technology) takes off from but does not proceed in the same manner as Heidegger's social analysis of the They and the problem of authenticity in the technological world. For Heidegger, being-with refers to an immediate personal presence in technics. Social being-with can manifest itself, however, not just on the level of immediate or existential presence but also in ideas. Indeed, the social world is as much, if not more, a world of ideas as of persons. Persons hold ideas and interact with others and with things on the basis of them. These ideas can even enclose the realm of technics – that is, become a language or *logos* of technics, a "technology."

The idea of being-with technology presupposes this "logical" encompassing of technics by a society and its philosophical or protophilosophical articulation. For many people, however, the ideas that guide their lives may not be held with conscious awareness or full articulation. They often take the form of myth. Philosophical argument and discussion then introduces into such a world of ideas a kind of break or rupture with the immediately given. This break or rupture need not require the rejection or abandoning of that given, but it will entail the bringing of that given in to fuller consciousness or awareness – from which it must be accepted (or rejected) in a new way or on new grounds.

Against this background, then, I propose to develop historicophilosophical descriptions, necessarily somewhat truncated, of three alternative ways of being-with technology. The first is what may be called ancient skepticism; the second, Renaissance and Enlightenment optimism; and the third, romantic ambiguity or uneasiness. Even in the somewhat simplified form of ideal types in which they will be presented here, consideration of the issues that divide these three ways of being-with technology may perhaps illuminate the difficulties we face in trying to live with modern technology and its manifest problems.

Ancient Skepticism

The original articulation of a relationship between humanity and technics, an articulation that is in its earliest forms coeval with the appearance of recorded history, can be stated boldly as "technology is bad but necessary" or, perhaps more carefully, as "technology (that is, the study of technics) is necessary but dangerous." The idea is hinted at by a plethora of archaic myths, such as the story of the Tower of Babel or the myths of Prometheus, Hephaestos, or Daedalus and Icarus. Certainly the transition from hunting and gathering to the domestication of animals and plants introduced a profound and profoundly disturbing transition into culture. Technics, according to these myths, although to some extent required by humanity and thus on occasion a cause for legitimate celebration,[2] easily turn against the human by severing it from some larger reality – a severing that can be manifest in a failure of faith or shift of the will, a refusal to rely on or trust God or the gods, whether manifested in nature or in providence.[3]

Ethical arguments in support of this distrust or uneasiness about technical activities can be detected in the earliest strata of Western philosophy. According to the Greek historian Xenophon, for instance, his teacher Socrates (469–399 B.C.) considered farming, the least technical of the arts, to be the most philosophical of occupations. Although the earth "provides the good things most abundantly, farming does not yield them up to softness but...produces a kind of manliness....Moreover, the earth, being a goddess, teaches justice to those who are able to learn" (*Oeconomicus* 5.4 and 12). This idea of agriculture as the most virtuous of the arts, one in which human technical action tends to be kept within proper limits, is repeated by representatives of the philosophical tradition as diverse as Plato,[4] Aristotle,[5] St. Thomas Aquinas,[6] and Thomas Jefferson.[7]

Elsewhere Xenophon notes Socrates' distinction between questions about *whether* to perform an action and *how* to perform it, along with another distinction between scientific or technological questions concerning the laws of nature and ethical or political questions about what is right and wrong, good and bad, pious and impious, just and unjust. In elaborating on the first distinction, Socrates stresses that human beings must determine for themselves *how* to perform their actions – that they can take lessons in "construction (*tektonikos*), forging metal, agriculture, ruling human beings, and...calculation, economics, and military strategy," and therefore should not depend on the gods for help in "counting, measuring, or weighing"; the ultimate consequences of their technical actions are nonetheless hidden. His initial example is even taken from agriculture: the man who knows how to plant a field does not know whether he will reap the harvest. Thus whether we should employ our technical powers is a subject about which we must rely on guidance from the gods.[8]

At the same time, with regard to the second distinction, Socrates argues that because of the supreme importance of ethical and political issues, human beings should not allow themselves to become preoccupied with scientific and technological pursuits. In the intellectual autobiography attributed to him in the *Phaedo*, for instance, Socrates relates how he turned away from natural science because of the cosmological and moral confusion it tended to engender.[9] In the *Memorabilia* it is similarly said of Socrates that

> He did not like others to discuss the nature of all things, nor did he speculate on the "cosmos" of the sophists or the necessities of the heavens, but he declared that those who worried about such matters were foolish. And first he would ask whether such persons became involved with these problems because they believed that their knowledge of human things was complete or whether they thought they were obligated to neglect human things to speculate on divine things. (*Memorabilia* 1.1.11–12)

Persons who turn away from human things to things having to do with the heavens appear to think "that when they know the laws by which everything comes into being, they will, when they choose, create winds, water, seasons, and anything else like these that they may need" (*Memorabilia* 1.1.15).[10] As "the first to call down philosophy from the heavens and place it in the city and...compel it to inquire about life and morality and things good and bad,"[11] Socrates' own conversation, however, is described as always about human things: What is pious? What is impious? What is good? What is shameful? What is just? What is unjust? What is moderation? For, as Xenophon says on another occasion, Socrates "was not eager to make his companions orators and businessmen and inventors, but thought that they should first possess moderation [*sophrosune*]. For he believed that without moderation those abilities only enabled a person to become more unjust and to work more evil (*Memorabilia* 4.3). The initial distinction grants technical or "how-to" questions a realistic prominence in human affairs but recognizes their ambiguity and uncertainty; the subsequent distinction subordinates any systematic pursuit of technical knowledge to ethical and political concerns.

Such uneasiness before the immoderate possibilities inherent in technological powers is further elaborated by Plato. Near the beginning of the *Republic*, after Socrates outlines a primitive state and Glaucon objects that this is no more than a "city of pigs," Socrates replies

> The true state is in my opinion the one we have described – a healthy state, as it were. But if you want, we can examine a feverish state as well. ...For there are some, it seems, who will not be satisfied with these things or this way of life; but beds, tables, and other furnishings will have to be added, and of course seasonings, perfume, incense, girls, and sweets – all kinds of each. And the requirements we mentioned before can no longer be limited to the necessities of houses, clothes, and shoes; but [various *technai*] must also be set in motion....The healthy state will no longer be large enough either, but it must be swollen in size by a multitude of activities that go beyond the meeting of necessities....(372d–373b)

As this passage indicates, and as can be confirmed by earlier references to Homer and other poets, certain elements of classical Greek culture had a distrust of the wealth and affluence that the *technai* or arts could produce if not kept within strict limitations. For according to the ancients such wealth accustoms men to easy things. But *chalepa ta kala*, difficult is the beautiful or the perfect; the perfection of anything, including human nature, is the opposite of what is soft or easy. Under conditions of affluence human beings tend to become accustomed to ease, and thus tend to choose the less over the more perfect, the lower over the higher, both for themselves and for others.

With no art is this more prevalent than with medicine. Once drugs are available as palliatives, for instance, most individuals will choose them for the alleviation of pain over the more strenuous paths of physical hygiene or psychological enlightenment. The current (*techné*) of medicine, Socrates maintains to Glaucon later in the third book of the *Republic*, is an education in disease that "draws out death" (406b); instead of promoting health it allows the unhealthy to have "a long and wretched life" and "to produce offspring like themselves" (407d). That Socrates' description applies even more strongly to modern medical technology than it did to that in classical Athens scarcely needs to be mentioned.

Another aspect of this tension between politics and technology is indicated by Plato's observations on the dangers of technical change. In the words of Adeimantus, with whom Socrates in this instance evidently agrees, once change has established itself as normal in the arts, "it overflows its bounds into human character and activity and from there issues forth to attach commerical affairs, and then proceeds against the laws and political orders" (424d–e). It is desirable that obedience to the law should rest primarily on habit rather than force. Technological change, which undermines the authority of custom and habit, thus tends to introduce violence into the state. Surely this is a possibility that the experience of the twentieth century, one of the most violent in history, should encourage us to take seriously.

This wariness of technological activity on moral and political grounds can be supplemented by an epistemological critique of the limitations of technological knowledge and a metaphysical analysis of the inferior status of technical objects. During a discussion of the education of the philosopher-king in the seventh book of the *Republic*, Socrates considers what kind of teaching most effectively brings a student "into the light" of the highest or most important things. One conclusion is that it is not those *technai* that "are oriented toward human opinions and desires or concerned with creation and fabrication and attending to things that grow and are put together" (533b). Because it cannot bring about a conversion or emancipation of the mind from the cares and concerns of the world, technology should not be a primary focus of human life. The orientation of technics, because it is concerned to remedy the defects in nature, is always toward the lower or the weaker (342c–d). A physician sees more sick people than he does healthy ones. *Eros* or love, by contrast, is oriented toward the higher or the stronger; it seeks out the good and strives for transcendence. "And the person who is versed in such matters is said to have spiritual wisdom, as opposed to the wisdom of one with *technai* or low-grade handicraft skills," Diotima tells Socrates in the *Symposium* (203a).

Aristotle agrees, but for quite different, more properly metaphysical reasons. According to Aristotle and his followers, reality or being resides in particulars. It is not some abstract species Homo Sapiens (with capital H and capital S) that *is* in the primary sense, but Socrates and Xanthippe. However, the reality of all natural entities is dependent on an intimate union of form and matter, and the *telos* or end determined thereby. The problem with artifacts is that they fail to achieve this kind of unity at a very deep level, and are thus able to have a variety of uses or extrinsic ends imposed upon them. "If a bed were to sprout," says Aristotle, "not a bed would come up but a tree" (*Physics* 193b.10). Insofar as it truly imitates nature, art engenders an inimitable individuality in its products, precisely because its attempt to effect as close a union of form and matter as possible requires a respect for or deference to the materials with which it works. In a systematized art or technology, matter necessarily tends to be overlooked or relegated to the status of an undifferentiated substrate to be manipulated at will.[12] Indeed, in relation to this Aristotle suggests a distinction between the arts of cultivation (e.g., medicine, education, and agriculture, which help nature to produce more abundantly things that it can produce of itself) from those of construction or domination (arts that bring into existence things that nature would not).[13]

The metaphysical issue here can be illustrated by observing the contrast between a handcrafted ceramic plate and Tupperware dishes. The clay plate has a solid weight, rich texture, and explicit reference to its surroundings not unlike that of a natural stone, whereas Tupperware exhibits a lightness of body and undistinguished surface that only abstractly engages the environment of its creation and use. According to a Mobil Oil Company advertisement from the early 1980s, synthetic products are actually "better than the real thing," so the word "synthetic," which implies a "pallid imitation," ought to be discarded. But whether this is true or not depends heavily on a prior understanding of what is real in the first place. For Aristotle there is a kind of reality that can only be found in particulars and is thus beyond the grasp of mass-production, function-oriented, polymer technology.

For Plato and the Platonic tradition, too, artifice is less real than nature. Indeed, in the tenth book of the *Republic* there is a discussion of the making of beds (to which Aristotle's remarks from the *Physics* may allude) by god or nature, by the carpenter or *tekton*, and by painter or artist. Socrates' argument is that the natural bed, the one made by the god, is the primary reality; the many beds made in imitation by artisans are a secondary reality; and the pictures of beds painted by artists are a tertiary reality. *Techné* is thus creative in a second or "third generation" sense (597e) – and thus readily subject to moral and metaphysical guidance.

In moral terms artifice is to be guided or judged in terms of its goodness or usefulness. In metaphysical terms the criterion of judgment is proper proportion or beauty. One possible disagreement between Platonists and Aristotelians with regard to one or another aspect of making is whether the good or the beautiful, ethics or aesthetics, is the proper criterion for its guidance. Such disagreement should nevertheless not be allowed to obscure a more fundamental agreement, the recognition of the need to subject *poesis* and *technai* to certain well-defined limitations. Insofar as technical objects or activities fail to be subject to the inner guidance of nature (*phusis*), nature must be brought to bear upon them consciously, from the outside (as it were) by human beings. Again, the tendency of contemporary technical creations to bring about environmental problems or ecological disorders to some extent confirms the premodern point of view.

The ancient critique of technology thus rests on a tightly woven, fourfold argument: (1) the will to technology or the technological intention often involves a turning away from faith or trust in nature or providence; (2) technical affluence and the concomitant processes of change tend to undermine individual striving for excellence and societal stability; (3) technological knowledge likewise draws the human being into intercourse with the world and obscures transcendence; (4) technical objects are less real than objects of nature. Only some necessity of survival, not some ideal of the good, can justify the setting aside of such arguments. The life of the great Hellenistic scientist Archimedes provides us (as it did antiquity) with a kind of icon or lived-out image of these arguments. Although, according to Plutarch, Archimedes was capable of inventing all sorts of devices, he was too high minded to do so except when pressed by military necessity – yet even then he refused to leave behind any treatise on the subject, because of a salutary fear that his weapons would be too easily misused by humankind.[14]

Allied with the Judeo-Christian-Islamic criticism of the vanity of human knowledge and of worldly wealth and power,[15] this premodern distrust of technology dominated Western culture until the end of the Middle Ages, and elements of it can be found vigorously repeated in numerous figures since – from Samuel Johnson's neoclassicist criticism of Milton's promotion of education in natural science[16] to Norbert Wiener who in 1947, like Archimedes twenty-three hundred years before, vowed not to publish anything more that could do damage in the hands of militarists.[17] In one less well-known allusion to another aspect of the classical moral argument, John Wesley (1703–91), in both private journals and public sermons, ruefully acknowledges the paradox that Christian conversion gives birth to a kind of self-discipline that easily engenders the accumulation of wealth – which wealth then readily undermines true Christian virtue. "Indeed, according to the natural tendency of riches, we cannot expect it to be otherwise," writes Wesley.[18]

With regard to other aspects of the premodern critique, Lewis Mumford, for instance, has criticized the will to power manifested in modern technology, and Heidegger, following the lead of the poet Ranier Maria Rilke, has invoked the metaphysical argument by pointing out the disappearance of the thinghood of things, the loss of a sense of the earth in mass-produced consumer objects. From Heidegger's point of view, nuclear annihilation of all things would be "the mere final emission of what has long since taken place, has already happened."[19]

From the viewpoint of the ancients, then, being-with technology is an uneasy being-along-side-of and working-to-keep-at-arms-length. Phrased in terms of the contemporary discipline of technology assessment, this premodern attitude looks upon technics as dangerous or guilty until proven innocent or necessary – and in any case, the burden of proof lies with those who would favor technology, not those who would restrain it.

Enlightenment Optimism

A radically different way of being-with technology – one that shifts the burden of proof from those who favor to those who oppose the introduction of inventions – argues the inherent goodness of technology and the consequent accidental character of all misuse. Aspects of this idea or attitude are not without premodern adumbration. But in comprehensive and persuasive form arguments to this effect are first fully articulated in the writings of Francis Bacon (1561–1626) at the time of the Renaissance, and subsequently become characteristic of the Enlightenment philosophy of the eighteenth century.

Like Xenophon's Socrates, Bacon grants that the initiation of human actions should be guided by divine counsel. But unlike Socrates, Bacon maintains that God has given humanity a clear mandate to pursue technology as a means for the compassionate alleviation of the suffering of the human condition, of being-in-the-world. Technical know-how is cut loose from all doubt about the consequences of technical action. In the choice between ways of life devoted to scientific-technological or ethical-political questions, Bacon further argues that Christian revelation directs men toward the former over the latter.

> For it was not that pure and uncorrupted natural knowledge whereby Adam gave names to the creatures according to their propriety, which gave occasion to the fall. It was the ambitious and proud desire of moral knowledge to judge of good and evil, to the end that man may revolt from God and give laws to himself, which was the form and manner of the temptation.[20]

Contrary to what is implied by the myth of Prometheus or the legend of Faust, it was not scientific and technological knowledge that led to the Fall, but vain philosophical speculation concerning moral questions. Formed in the image and likeness of God, human beings are called upon to be creators; to abjure that vocation and to pursue instead an unproductive discourse on ethical quandaries brings about the just punishment of a poverty-stricken existence. "He that will not apply new remedies must expect new evils."[21] Yet "the kingdom of man, founded on the sciences," says Bacon, is "not much other than...the kingdom of heaven."[22]

The argument between Socrates and Bacon is not, it is important to note, simply one between partisans either against or in favor of technology. Socrates allows technics a legitimate but strictly utilitarian function, then points out the difficulty of obtaining a knowledge of consequences upon which to base any certainty of trust or commitment. Technical action is circumscribed by uncertainty or risk. Bacon, however, although he makes some appeals to a consequentialist justification, ultimately grounds his commitment in something approaching deontological principles. The proof is that he never even considers evaluating individual technical projects on their merit, but simply argues for an all-out affirmation of technology in general. It is right to pursue technological action, never mind the consequences. Intuitions of uncertainty are jettisoned in the name of revelation.

The uniqueness of the Baconian (or Renaissance) interpretation of the theological tradition is also to be noted. For millennia the doctrines of God as creator of "the heavens and the earth" (Gen. 1.1) and human beings as made "in the image of God" (Gen. 1.27) exercised profound influence over Jewish and later Christian anthropology, without ever being explicitly interpreted as a warrant for or a call to technical activity. Traditional or premodern interpretations focus on the soul, the intellect, or the capacity for love as the key to the *imago Dei*.[23] The earliest attribution to this doctrine of technological implications occurs in the early Renaissance.[24] The contemporary theological notion of the human as using technology to prolong creation or cocreate with God depends precisely on the reinterpretation of Genesis adumbrated by Bacon.

The Enlightenment version of Bacon's religious argument is to replace the theological obligation with a natural one. In the first place, human beings simply could not survive without technics. As D'Alembert puts it in the "Preliminary Discourse" to the *Encyclopedia* (1751), there is a prejudice against the mechanical arts that is a result of their accidental association with the lower classes.

> The advantage that the liberal arts have over the mechanical arts, because of their demands upon the intellect and because of the difficulty of excelling in them, is sufficiently counterbalanced by the quite superior usefulness which the latter for the most part have for us. It is their very usefulness which reduced them perforce to purely mechanical operations in order to make them accessible to a larger number of men. But while justly respecting great geniuses for their enlightenment, society ought not to degrade the hands by which it is served.[25]

In the even more direct words of Immanuel Kant, "Nature has willed that man should by himself, produce everything that goes beyond the mechanical ordering of his animal existence, and that he should partake of no other happiness or perfection than that which he himself, independently of instinct, has created by his own reason."[26] Nature and reason, if not God, command humanity to pursue technology; the human being is redefined not as *homo sapiens* but as *homo faber*. Technology is the essential human activity. In more ways than Kant explicitly proclaims, "Enlightenment is man's release from his self-incurred tutelage."[27]

Following a redirecting (Bacon) or reinterpreting (D'Alembert and Kant) of the will, Bacon and his followers explicitly reject the ethical-political argument against technological activities in the name of moderation. With no apparent irony, Bacon maintains that the inventions of printing, gunpowder, and the compass have done more to benefit humanity than all the philosophical debates and political reforms throughout history. It may, he admits, be pernicious for an individual or a nation to pursue power. Individuals or small groups may well abuse such power. "But if a man endeavor to establish and extend the power and dominion of the human race itself over the universe," writes Bacon, "his ambition (if ambition it can be called) is without doubt both a more wholesome and a more noble thing than the other two." And, of course, "the empire of man over things depends wholly on the arts and sciences" (*Novum Organum* 1.129).

Bacon does not expound at length on the wholesomeness of technics. All he does is reject the traditional ideas of their corrupting influence on morals by arguing for a distinction between change in politics and in the arts.

> In matters of state a change even for the better is distrusted [Bacon observes], because it unsettles what is established; these things resting on authority, consent, fame and opinion, not on demonstration. But arts and sciences should be like mines, where the noise of new works and further advances is heard on every side. (*Novum Organum* 1.90)

Unlike Aristotle and Aquinas, both of whom noticed the same distinction but found it grounds for caution in technology,[28] Bacon thinks the observation itself is enough to set technology on its own path of development.

Bacon's Enlightenment followers, however, go considerably further, and argue for the positive or beneficial influence of the arts on morals. The *Encyclopedia*, for instance, having identified "luxury" as simply "the use human beings make of wealth and industry to assure themselves of a pleasant existence" with its origin in "that dissatisfaction with our condition...which is and must be present in all men," undertakes to reply directly to the ancient "diatribes by the moralists who have censured it with more gloominess than light."[29] Critics of material welfare have maintained that it undermined morals, and apologists have responded that this is the case only when it is carried to excess. Both are wrong. Wealth is, as we would say today, neutral. A survey of history reveals that luxury "did not determine morals, but...it took its character rather from them."[30] Indeed, it is quite possible to have a moral luxury, one that promotes virtuous development.

But if a first line of defense is to argue for moderation, and a second for neutrality, a third is to maintain a positive influence. David Hume (1711–76), for instance, in his essay "Of Commerce," argues that a state should encourage its citizens to be manufacturers more than farmers or soldiers. By pursuit of "the arts of *luxury*, they add to the happiness of the state."[31] Then, in "Of Refinement in the Arts," he explains that the ages of luxury are both "the happiest and the most virtuous" because of their propensity to encourage industry, knowledge, and humanity."[32] "In times when industry and the arts flourish," writes Hume, "men are kept in perpetual occupation, and enjoy, as their reward, the occupation itself, as well as those pleasures which are the fruit of their labour."[33]

Furthermore, the spirit of activity in the arts will galvanize that in the sciences and vice versa; knowledge and industry increase together. In Hume's own inimitable words: "We cannot reasonably expect that a piece of woolen cloth will be wrought to perfection in a nation which is ignorant of astronomy."[34] And the more the arts and sciences advance, "the more sociable men become." Technical engagements promote civil peace because they siphon off energy that might otherwise go into sectarian competition. Technological commerce and scientific aspirations tend to break down national and class barriers, thus ushering in tolerance and sociability. In the words of Hume's contemporary, Montesquieu: "Commerce is a cure for the most destructive prejudices; for it is a general rule, that wherever we find tender manners, there commerce flourishes; and that wherever there is commerce, there we meet with tender manners."[35]

The ethical significance of technological activity is not limited, however, to its socializing influence. Technology is an intellectual as well as a moral virtue, because it is a means to the acquisition of true knowledge. That technological activity contributes to scientific advance rests on a theory of knowledge that again is first clearly articulated by Bacon, who begins his *Novum Organum* or "new instrument" with the argument that true knowledge is acquired only by a close intercourse with things themselves. "Neither the naked hand nor the understanding left to itself can effect much. It is by instruments and helps that the work is done, which are as much wanted for the understanding as for the hand" (*Novum Organum* 1.2). Knowledge is to be acquired by active experimentation, and ultimately evaluated on the basis of its ability to engender works. The means to true knowledge is what Bacon candidly refers to as the "torturing of nature"; left free and at large, nature, like the human being, is loath to reveal her secrets.[36] The result of this new way will be the union of knowledge and power (*Novum Organum* 2.3). Bacon is, quite simply, an epistemological pragmatist. What is true is what works. "Our only hope," he says, "therefore lies in a true induction" (*Novum Organum* 1.14).

The very basis of the great French *Encyclopedia or Dictionary of Sciences, Arts, and Crafts* is precisely this epistemological vision, a unity between theory and practice. Bacon is explicitly identified as its inspiration, and is praised for having conceived philosophy "as being only that part of our knowledge which should contribute to making us better or happier, thus...confining it within the limits of useful things [and inviting] scholars to study and perfect the arts, which he regards as the most exalted and most essential part of human science."[37] Indeed, in explicating the priorities of the *Encyclopedia*, the "Preliminary Discourse" goes on to say that "too much has been written on the sciences; not enough has been written well on the mechanical arts."[38] The article on "Art" in the *Encyclopedia* further criticizes the prejudice against the mechanical arts, not only because it has "tended to fill cities with...idle speculation,"[39] but even more because of its failure to produce genuine knowledge. "It is difficult if not impossible...to have a thorough knowledge of the speculative aspects of an art without being versed in its practice," although it is equally difficult "to go far in the practice of an art without speculation."[40] It is this new unity of theory and practice – a unity based in practice more than in theory[41] – that is at the basis of, for instance, Bernard de Fontenelle's eulogies of the practice of experimental science as an intellectual virtue as well as a moral one, and the Enlightenment reconception of Socrates as having called philosophy down from the heavens to experiment with the world.[42]

Bacon's true induction likewise rests on a metaphysical rejection of natural teleology. The pursuit of a knowledge of final causes "rather corrupts than advances the sciences," declares Bacon, "except such as have to do with human action" (*Novum Organum* 2.2). Belief in final causes or purposes inherent in nature is a result of superstition or false religion. It must be rejected in order to make possible "a very diligent dissection and anatomy of the world" (*Novum Organum* 1.124). Nature and artifice are not ontologically distinct. "All Nature is but Art, unknown to thee," claims Alexander Pope.[43] "Nature does not exist," declares Voltaire, "art is everything." The Aristotelian distinction between arts of cultivation and of construction is jettisoned in favor of universal construction.

With regard to Pope, although it is not uncommon to find comparisons of the God/nature and artist/artwork relationships in Greek and Christian, ancient and modern authors, there are subtle differences. For Plato (*Sophist* 265b and *Timaeus* 27c) and St. Augustine (*De civitate Dei* 11.21), for example, there is first a fundamental distinction to be made between divine and human *poiesis*, which is itself to be distinguished from *techné* and, second, the fact that even though made by a god the world is not to be looked upon as an artifact or something that functions in an artificial manner. Thomas Hobbes, Bacon's secretary, however, proposes to view nature not just as produced by a divine art but as itself "the art whereby God hath made and governs the world."[44] Indeed, so much is this the case that for Hobbes human art itself can be said to produce natural objects – or, to say the same thing in different words, the whole distinction between nature and artifice disappears.

This last point also links up with the first: metaphysics supports volition. If nature and artifice are not ontologically distinct, then the traditional distinction between technics of cultivation and technics of domination disappears. There are no technics that help nature to realize its own internal reality, and human beings are free to pursue power. If nature is just another form of mechanical artifice, it is likewise reasonable to think of the human being as a machine. "Man is a machine and...in the whole universe there is but a single substance variously modified," concludes LaMettrie.[45] "For what is the heart," wrote Hobbes a century earlier, "but a spring; and the nerves, but so many strings; and the joints, but so many wheeles."[46] But the activities appropriate to machines are technological ones; *homo faber* is yet another form of *l'homme machine*, and vice versa.

Like that of the anicents, then, the distinctly modern way of being-with technology may be articulated in terms of four interrelated arguments: (1) the will to technology is ordained for humanity by God or by nature; (2) technological activity is morally beneficial because, while stimulating human action, it ministers to physical needs and increases sociability; (3) knowledge acquired by a technical closure with the world is truer than abstract theory; and (4) nature is no more real than – indeed it operates by the same principles as – artifice. It is scarcely necessary to illustrate how aspects of this ideology remain part of intellectual discourse in Marxism, in pragmatism, and in popular attitudes regarding technological progress, technology assessment and public policy, education, and medicine.

Romantic Uneasiness

The premodern argument that technology is bad but necessary characterizes a way of being-with technology that effectively limited rapid technical expansion in the West for approximately two thousand years. The Renaissance and Enlightenment argument in support of the theory that technology is inherently good discloses a way of being-with technology that has been the foundation for a Promethean unleashing of technical power unprecedented in history. The proximate causes of this radical transformation were, of course, legion: geographic, economic, political, military, scientific. But what brought all such factors together in England in the mid-eighteenth century to engender a new way of life, what enabled them to coalesce into a veritable new way of being-in-the-world, was a certain kind of optimism regarding the expansion of material development that is not to be found so fully articulated at any point in premodern culture.[47]

In contrast with premodern skepticism about technology, however, the typically modern optimism has not retained its primacy in theory even though it has continued to dominate in practice. The reasons for this are complex. But faced with the real-life consequences of the Industrial Revolution, from societal and cultural disruptions to environmental pollution, post-Enlightenment theory has become more critical of technology. Romanticism, as the name for the typically modern response to the Enlightenment, thus implicitly contains a new way of being-with technology, one that can be identified with neither ancient skepticism nor modern optimism.

Romanticism is, of course, a multi-dimensioned phenomenon. In one sense, it can refer to a permanent tendency in human nature that manifests itself differently at different times. In another, it refers to a particular manifestation in nineteenth-century literature and thought. Virtually all attempts to analyze this particular historical manifestation interpret romanticism as a reaction to and criticism of modern science. Against Newtonian mechanics, the romantics propose an organic cosmology; in opposition to scientific rationality, romantics assert the legitimacy and importance of imagination and feeling. What is seldom appreciated is the extent to which romanticism can also be interpreted as a questioning – in fact, the first self-conscious questioning – of modern technology.[48] So interpreted, however, romanticism reflects an uneasiness about technology that is nevertheless fundamentally ambiguous; although as a whole the romantic critique may be distinct from ancient skepticism and modern optimism, in its parts it nevertheless exhibits differential affinities with both.

To begin with, consider the volitional aspect of technology. On the ancient view, technology was seen as a turning away from God or the gods. On the modern, it is ordained by God or, with the Enlightenment rejection of God, by Nature. With the romantics the will to technology either remains grounded in nature or is cut free from all extrahuman determination. In the former instance, however, nature is reconceived as not just mechanistic movement but as an organic striving toward creative development and expression. From the perspective of "mechanical philosophy," human technology is a prolongation of mechanical order; from that of *Naturphilosophie* it becomes a participation in the self-expression of life. When liberated from even such organic creativity, technology is grounded solely in the human will to power, but with recognition of its often negative consequences; the human condition takes on the visage of Gothic pathos.[49] The most that seems able to be argued is that the technological intention, that is the will to power, should not be pursued to the exclusion of other volitional options – or that it should be guided by aesthetic ideals.

William Wordsworth (1770–1850), for instance, the most philosophical of the English romantic poets, in the next-to-last book of his long narrative poem, *The Excursion* (1814), describes how he has "lived to mark / A new and unforseen creation rise" (bk. 8, lines 89–90).

Casting reserve away, exult to see
An intellectual mastery exercised
O'er the blind elements; a purpose given,
A perseverance fed; almost a soul
Imparted – to brute matter. I rejoice,
Measuring the force of those gigantic powers
That, by the thinking mind, have been compelled
To serve the will of feeble-bodied Man.
(bk. 8, lines 200–207)

Here the rejoicing in and affirmation of technological conquest and control is clearly in harmony with Enlightenment sentiments.

Yet in the midst of this exultation
I grieve, when on the dark side
Of this great change I look; and there behold
Such outrage done to nature.
(151–53)

And afterward he writes,

How insecure, how baseless in itself,
Is the Philosophy whose sway depends
On mere material instruments; – how weak
Those arts, and high inventions, if unpropped
By virtue.
(223–27)

Here Enlightenment optimism is clearly replaced by something approaching premodern skepticism.

Wordsworth clarifies his position in the last book of the poem. True, he has complained, in regard to the factory labor of children, that a child is

...subject to the arts
Of modern ingenuity, and made
The senseless member of a vast machine.
(bk. 9, lines 157–59)

Still, he is not insensitive to the fact that the rural life is also often an "unhappy lot" enslaved to "ignorance" "want" and "miserable hunger" (lines 163–65). Nevertheless, he says, his thoughts cannot help but be

...turned to evils that are new and chosen,
A bondage lurking under shape of good, –
Arts, in themselves beneficent and kind,
But all too fondly followed and too far.
(187–90)

In such lines Wordsworth no longer maintains with any equanimity the Enlightenment principle that the arts are "in themselves beneficent and kind." With his suggestion that the self-creative thrust has in technology been followed "too fondly" and "too far," and that bondage has been created under the disguise of good, a profound questioning is introduced. But unlike the ancients, who called for specific delimitations on technics, with the romantics there is no clear outcome other than a critical uneasiness – or a heightened aesthetic sensibility.

Later, in a sonnet on "Steamboats, Viaducts, and Railways" (1835), having observed contradictions between the practical and aesthetic qualities of such artifacts, Wordsworth concludes that

In spite of all that beauty may disown
In your harsh features, Nature doth embrace
Her lawful offspring in Man's art; and Time,
Pleased with your triumphs o'er his brother Space,
Accepts from your bold hands the proffered crown
Of hope, and smiles on you with cheer sublime.

Once again technology, in Enlightenment fashion, is viewed as an extension of nature, and even described in Baconian terms as the triumph of time over space.[50] The "lawful offspring" is nevertheless ugly, full of "harsh features" that beauty disowns. Yet from the "bold hands" of technology temporal change is given the "crown of hope... with cheer sublime" that things will work out for the good. In Wordsworth's own commentary on *The Excursion*, the problem "is an ill-regulated and excessive application of powers so admirable in themselves."[51] But it is precisely this ill-regulated and excessive technology that also gives birth to a new kind of admiration, that of the sublime.

With regard to the moral character of technology, ambiguity is even more apparent. Consider, for instance, the arguments of Jean-Jacques Rousseau

(1712–78), a man who is, in important respects, the founder of the romantic movement, and whose critique takes shape even before the inauguration of the Industrial Revolution itself, strictly in reaction to ideas expressed by the *philosophes*. In 1750, in a prize-winning "Discourse on the Moral Effects of the Arts and Sciences," critical of the kinds of ideas that would shortly be voiced by D'Alembert's "Preliminary Discourse," Rousseau boldly concludes that "as the conveniences of life increase, as the arts are brought to perfection, and luxury spreads, true courage flags, the virtues disappear."[52] "Money, though it buys everything else," he argues, "cannot buy morals and citizens.[53] "The politicians of the ancient world," he says, "were always talking of morals and virtue; ours speak of nothing but commerce and money."[54] In fact, from Rousseau's point of view, not only have "our minds...been corrupted in proportion as the arts and sciences have improved,"[55] but the arts and sciences themselves "owe their birth to our vices."[56] Action, even destructive action, particularly on a grand (or sublime) scale, is preferable to nonaction.[57]

What sounds, at first, like a straightforward return to the moral principles of the ancients, however, is made in the name of quite different ideals. Virtue, for Rousseau, is not the same thing it is for Plato or Aristotle – as is clearly indicated by his praise of Francis Bacon, "perhaps the greatest of philosophers."[58] In agreement with Bacon, Rousseau criticizes "moral philosophy" as an outgrowth of "human pride,"[59] as well as the hiatus between knowledge and power, thought and action, that he finds to be a mark of civilization; instead, he praises those who are able to act decisively in the world, to alter it in their favor, even when these are men whom the Greeks would have considered barbarians. Virtue, for instance, lies with the Scythians who conquered Persia, not the Persians; with the Goths who conquered Rome, not the Romans; with the Franks who conquered the Gauls, or the Saxons who conquered England.[60] In civilized countries, he says, "There are a thousand prizes for fine discourses, and none for good action."[61]

With Bacon, Rousseau argues the need for actions, not words, and approves the initial achievements of the Renaissance in freeing humanity from a barren medieval scholasticism.[62] But unlike Bacon, Rousseau sees that even scientific rationality, through the alienation of affection, can often weaken the determination and commitment needed for decisive action. Thus, in a paradox that will become a hallmark of romanticism, Rousseau turns against technology – but in the name of ideals that are at the heart of technology. He criticizes a particular historical embodiment of technology, but only to advance a project that has become momentarily or partially impotent.

It was in England, however, where the Industrial Revolution found its earliest full-scale manifestation, that this paradoxical critique achieved an initial broad literary expression. Such expression took a realistic turn, rejecting classical patterns in favor of the specific depiction of real situations often in unconventional forms. A poem such as William Blake's "London" (1794) or a novel like Charles Dickens's *Hard Times* (1854), in their presentation of the dehumanizing consequences of factory labor, equally well illustrate the force of this approach. Wordsworth, again, may be quoted to extend the issue of the alienation of affections to the social level. In a letter from 1801 he writes:

> It appears to me that the most calamitous effect which has followed the measures which have lately been pursued in this country, is a rapid decay of the domestic affections among the lower orders of society....For many years past, the tendency of society, amongst all the nations of Europe, has been to produce it; but recently, by the spreading of manufactures through every part of the country...the bonds of domestic feeling...have been weakened, and in innumerable instances entirely destroyed....If this is true,...no greater curse can befall a land.[63]

Romantic realism is, however, allied with visionary symbolism and, through this, epistemological issues. Consider, for instance, another aspect of Blake's genius, his prophetic poems. John Milton over a century before had in *Paradise Lost* (1667) already identified Satan with the technical activities of mining, smelting, forging, and molding the metals of hell into the city of Pandemonium.[64] Following this lead, Blake, in *Milton* (1804), identifies Satan with the abused powers of technology – and Newtonian science. Satan, "Prince of the Starry Hosts and of the Wheels of Heaven," also has the job of turning "the [textile] Mills day & night."[65] But in the prefatory lyric that opens this apocalyptic epic, Blake rejects the necessity of "these dark Satanic Mills" and cries out

> I will not cease from Mental Fight,
> Nor shall my Sword sleep in my hand

> Till we have built Jerusalem
> In England's green & pleasant Land.

This lyric, "And Did Those Feet in Ancient Time," is set to music and becomes the anthem of the British Socialism. A visionary, imaginative – not to say utopian – socialism is the romantic answer to the romantic critique of the moral limitations of technology. Mary Shelley's *Frankenstein* (1818), in another instance, likewise presents a love-hate relationship with technology in which that which is hated is properly redeemed not by premodern delimitation but by the affective correlate of an expansive imagination – namely, love.

Industrialization, then, undermines affection – that is, feeling and emotion, at both the individual and social levels. And this practical fact readily becomes allied with a more theoretical criticism of the Enlightenment emphasis on reason as the sole or principle cognitive faculty. The Enlightenment argued for the primacy of reason as the only means to advance human freedom from material limitations. According to the romantic reply, not only does such an emphasis on reason not free humanity from material bonds (witness the evils of the Industrial Revolution), but in itself it is (in the words of William Blake) a "mind-forged manacle." The focus on reason is itself a limitation that must be overcome; and through the consequent liberation of imagination the historical condition of technical activity can in turn be altered. In the "classic" epistemological defense and definition of Samuel Taylor Coleridge:

> The imagination...I consider either as primary, or secondary. The primary imagination I hold to be the living power and prime agent of all human perception, and as a repetition in the finite mind of the eternal act of creation in the infinite I AM. The secondary I consider as an echo of the former, co-existing with the conscious will, yet still as identical with the primary in the kind of its agency, and differing only in degree, and in the mode of its operation. It dissolves, diffuses, dissipates, in order to recreate; or where this process is rendered impossible, yet still, at all events, it struggles to idealize and to unify.[66]

Indeed, it is to this power that Blake also appeals as the source of his social revolution, when he proclaims, "I know of no other Christianity and of no other Gospel than the liberty both of body & mind to exercise the Divine Arts of Imagination, the real & eternal World of which this Vegetable Universe is but a faint shadow, & in which we shall live in our Eternal or Imaginative Bodies when these Vegetable Mortal Bodies are no more."[67]

Finally, with regard to artifacts, the romantic view is again both like and unlike that of the Enlightenment. It is Enlightenment-like in the belief that nature and artifice operate by the same principles. Contra the Enlightenment, however, the romantic view takes nature as the key to artifice rather than artifice as the key to nature. The machine is a diminished form of life, not life a complex machine. Furthermore, nature is no longer perceived primarily in terms of stable forms; the reality of nature is one of process and change. Wordsworth and other English romantics are taken with the "mutability" of nature. Lord Byron, for instance, at the conclusion of *Childe Harold's Pilgrimage* (1818), when he aspires "to mingle with the Universe, and feel / What I can ne'er express" (canto 4, stanza 177), describes nature as the

> ...glorious mirror, where the Almighty's form
> Glasses itself in tempests; in all time,
> Calm or convulsed – in breeze, or gale, or
> storm –
> Icing the Pole, or in the torrid clime
> Dark-heaving – boundless, endless, and
> sublime –
> The image of Eternity....
>
> (canto 4, stanza 183)

Nature, thus reconceptualized, reflects its new character into the world of artifice.

For the Enlightenment, nature and artifice both exhibit at their highest levels of reality various aspects of mechanical order, the interlocking of parts in a mathematical interrelation of the well-drafted lines of a Euclidean geometry. The metaphysical character of such reality is manifest to the senses through a "classical" vision of the beautiful – although there develops an Enlightenment excitement with the great or grandiose and the consequent projecting of art beyond nature that contradicts the models of harmonious stability within nature charateristic of classical antiquity and thus intimates romantic sensibilities. For romanticism, by contrast, the metaphysical reality of both nature and artifice is best denoted not by stable or well-ordered form but by process or change, especially as apprehended by the new aesthetic category of the sublime or the overwhelming and what Byron refers to as "pleasing fear" (canto 4, stanza 184).

As an aesthetic category, the idea of the sublime can be traced back to Longinus (third century A.D.) who departed from classical canons of criticism by praising literature that could provoke "ecstasy." But the concept received little real emphasis until Edmund Burke's *A Philosophical Enquiry into the Origin of Our Ideas of the Sublime and Beautiful* (1757). For Burke, beauty is associated with social order and is represented with harmony and proportion in word and figure; the sublime, by contrast, is concerned with the individual striving and is indicated by magnitude and broken line. "Whatever is fitted in any sort to excite the ideas of pain, and danger, whatever is in any sort terrible, or is conversant about terrible objects, or operates in a manner analogous to terror, is a source of the *sublime*," is Burke's famous definition.[68] Certainly modern technological objects and actions – from Hiroshima to Chernobyl – have tended to become a primary objective correlative of such a sentiment.

Like premodern skepticism and Enlightenment optimism the romantic way of being-with technology can thus be characterized by a pluralism of ideas that constitute a critical uneasiness: (1) the will to technology is a necessary self-creative act, which nevertheless tends to overstep its rightful bounds; (2) technology makes possible a new material freedom but alienates from the decisive strength to exercise it and creates wealth while undermining social affection; (3) scientific knowledge and reason are criticized in the name of imagination; and (4) artifacts are characterized more by process than by structure and invested with a new ambiguity associated with the category of the sublime. The attractive and repulsive interest revealed by the sublime expresses perhaps better than any other the uniqueness of the romantic way of being-with technology.

Coda

As the analysis of romantic being-with technology has especially tended to indicate, the ideas associated with the four aspects of technology as volition, as activity, as knowledge, and as object cannot be completely separated. Theology, ethics, epistemology, and metaphysics are ultimately aspects of a

Table 1 Historicophilosophical epochs

Conceptual elements	Ancient skepticism (suspicious of technology)	Enlightenment optimism (promotion of technology)	Romantic uneasiness (ambiguous about technology)
Volition or intention (religious)	Will to technology involves tendency to turn away from God or the gods.	Will to technology is ordained by God or by Nature.	Will to technology is an aspect of creativity – which tends to crowd out other aspects.
Action (ethics and politics)	Personal: Technical affluence undermines individual virtue.	Personal: Technical activities socialize individuals.	Personal: Technology engenders freedom but alienates from affective strength to exercise it.
	Societal: Technical change weakens political stability.	Societal: Technology creates public wealth.	Societal: Technology weakens social bonds of affection.
Knowledge (epistemology)	Technical information is not true wisdom.	Technical engagement with the world yields true knowledge (pragmatism).	Imagination and vision are more crucial than technical knowledge.
Objects (metaphysics and aesthetics)	Artifacts are less real than natural objects and thus require external guidance.	Nature and artifice operate by the same mechanical principles.	Artifacts expand the processes of life and reveal the sublime.

way of being in the world. Acknowledging this limitation, it is nevertheless possible to summarize the three ways of life in relation to technology by means of the matrix displayed in table 1.

At the outset, however, the argument of this essay was indicated to have some relation to Heidegger's early analysis of technology, although it has taken off in a trajectory not wholly consistent with Heidegger's own analysis or intentions. Yet there remains a final affinity worth noting. In Heidegger's existential analysis there is a paradox that the personal that is revealed through the technical is also undermined thereby. The use of tools is with others and in a world of artifacts owned by others, but the others easily become treated as all the same and thus become, as he calls it, a They – mass society.

> In utilizing public means of transport and in making use of information services such as the newspapers every Other [person] is like the next. The Being-with-one-another dissolves one's own *Dasein* [or existence] completely into the kind of Being of "the Others," in such a way, indeed, that the Others, as distinguishable and explicit, vanish more and more.[69]

With regard to the romantic way of being-with technology there is also a paradox. Not only is there a certain ambiguity built into this attitude, but the attitude itself has not been adopted in any wholehearted way by modern culture. Romanticism is, if you will, uneasy with itself. Indeed, this may be in part why romanticism has so far been unable to demonstrate the kind of practical efficacy exhibited by both premodern skepticism and Enlightenment optimism. The paradox of the romantic way of being-with technology is that, despite an intellectual cogency and expressive power, it has yet to take hold as a truly viable way of life. Given almost two centuries of active articulation, this impotence may well point toward inherent weaknesses. Perhaps the truth is that romanticism has been adopted, but that it is precisely its internal ambiguities, its bipolar attempt to steer a middle course between premodern skepticism and Enlightenment optimism, that vitiates its power.

Notes

1 Martin Heidegger, *Being and Time*, trans. John Macquarrie and Edward Robinson (New York: Harper and Row, 1962), p. 154.

2 One *locus classicus* of such celebration is Sophocles, *Antigone* 332.

3 For an interpretation of the specifically religious dimensions of this negative mythology, see Carl Mitcham, " 'The Love of Technology Is the Root of All Evil'," *Epiphany Journal* 8, no. 1. (Fall 1987): 17–28.

4 For Plato, see especially the fifth book of the *Laws* (743d), where agriculture is described as keeping production within proper limits and as helping to focus attention on the care of the soul and the body. Cf. also *Laws*, bk. 8 (842d–e) and bk. 10 (889d).

5 For Aristotle, see especially the *Politics* 1.8–11, and the distinction between two ways of acquiring goods, agriculture and business, the former of which is said to be "by nature" (1258a38), the latter "not by nature" (1258b41). In the *Politics* 6.2, agrarian based democracy is described as both "oldest" and "best" (1318b7–8).

6 Following Aristotle, St. Thomas Aquinas's commentary on the *Politics* terms farming "natural," "necessary," and "praiseworthy" (*Sententia libri Politicorum* 1, lectio 8), and again in *De regimine principum* 2.3, Thomas identifies farming as "better" than commercial activities for providing for material welfare. For Thomas Aquinas, however, farming tends to be spoken of in relation to all manual labor, and in consequence of the doctrine of the Fall takes on a certain ambiguity not found in Aristotle. For instance, in the *Summa theologiae* 2.2, quaestio 187, articulus 3, "Whether Religious Are Bound to Manual Labour," it is argued that all human beings must work with their hands for four reasons: to obtain food (as proof texts Thomas cites Gen. 3.19 and Ps. 128.2), to avoid idleness (Sirach 33.27), to restrain concupiscence by mortifying the body (2 Cor. 6.4–6), and to enable one to give alms (Eph. 4.28). Note that there is a subtle difference between the first two reasons (which cite the Hebrew Scriptures) and the second (which cite the Greek Scriptures). For a relevant interpretation of Thomas Aquinas's thought, which nevertheless fails to recognize the tensions alluded to here, see George H. Speltz, *The Importance of Rural Life according to the Philosophy of St. Thomas Aquinas* (Washington, D.C.: Catholic University of American Press, 1945). Cf. also Philo, "De agricultura," a commentary on Noah as farmer.

7 "Those who labour in the earth are the chosen people of God, if ever he had a chosen people, whose breasts he has made his peculiar deposit for substantial and genuine virtue....Corruption of morals in the mass of cultivators is a phaenomenon of which no age nor nation has furnished an example." Thomas Jefferson,

Notes on the State of Virginia (1782) (Chapel Hill: University of North Carolina Press, 1955), Query 19, "Manufactures." See also a letter to John Jay, 23 August 1785: "Cultivators of the earth are the most valuable citizens. They are the most vigorous, the most independent, the most virtuous, and they are tied to their country, and wedded to its liberty and interest, by the most lasting bonds. As long, therefore, as they can find employment in this line, I would not convert them into mariners, artisans or anything else."

8 Xenophon, *Memorabilia* (Oxford: Clarendon Press, 1890) 1.1.7, 1.1.9, and Cf. 4.7.10.

9 Cf. also *Memorabilia* 4.7.6–7.

10 Cf. Empedocles, frag. 111.

11 Cicero, *Tusculan Disputations* 5.4.10–11. See also *Academica* 1.4.15.

12 For a development of this argument, see "Philosophy and the History of Technology," *The History and Philosophy of Technology*, ed. George Bugliarello and Dean B. Doner (Urbana: University of Illinois Press, 1979), pp. 163–201.

13 Compare Aristotle, *Physics* 2.1.193a12–17; *Politics* 7.7.1337a2; and *Oeconomica* 1.1.1343a26–b2.

14 Plutarch, "Life of Marcellus," in *Plutarch's Lives*, ed. A. H. Clough (Boston: Little, Brown, and Company, 1859), 2.252–55.

15 On the inadequacy of human knowledge, see the Book of Job, Prov. 1.7, Isa: 44.25, and Col. 2.8. Power over the world, Satan says in the Gospel of Luke, has been given to him (Luke 4.6). The prince of this world, according to the Gospel of John, is to be cast out (John 12.31).

16 In his study of Milton (in "The Lives of the Poets," 1.99–100 [paragraphs 39–41]), Samuel Johnson criticizes a program of education which would concentrate on natural philosophy. "The truth is, that the knowledge of external nature, and the sciences which that knowledge requires or includes, are not the great or the frequent business of the human mind. Whether we provide for action of conversation,...the first requisite is the religious and moral knowledge of right and wrong....Physiological learning is of such rare emergence, that one may know another half his life, without being able to estimate his skill in hydrostatics or astronomy; but his moral and prudential character immediately appears....[And] if I have Milton against me, I have Socrates on my side. It was his labour to turn philosophy from the study of nature to speculations upon life; but the innovators whom I oppose...seem to think, that we are placed here to watch the growth of plants, or the motions of stars. Socrates was rather of the opinion that what he had to learn was, how to do good, and avoid evil." Cf. also "The Rambler" no. 24 (Saturday, 9 June 1750).

17 Norbert Wiener, "A Scientist Rebels," *Bulletin of the Atomic Scientists* 3, no. 1 (January 1947): 31.

18 John Wesley, *Works* (Grand Rapids, Mich.: Zondervan, n.d. [photomechanical reprint of the edition published by the Wesleyan Conference, London, 1872]), 7: 289.

19 Martin Heidegger, "The Thing," in *Poetry, Language, Thought*, trans. Albert Hofstadter (New York: Harper and Row, 1971), p. 166. See also Heidegger's essay on Rilke, "What Are Poets For?" esp. pp. 112–17.

20 Francis Bacon, "Preface," *The Great Instauration*, in *Selected Writings of Francis Bacon*, ed. Hugh G. Dick (New York: Modern Library, 1955).

21 Bacon, *Essays*, no. 24, "Of Innovations," in *Selected Writings of Francis Bacon*, p. 65.

22 Francis Bacon, *Novum Organum*, in *The Works of Francis Bacon* (Stuttgart-Bad Canstatt: Friedrich Fromann Verlag Gunther Holzboog, 1963), 1.68. Subsequent references to the *Novum Organum* will appear parenthetically within the text.

23 According to the Talmud, "As God fills the entire universe, so does the soul fill the whole body" (Berakhot 10a). According to the teachings of Jesus, "Love your enemies and pray for those who persecute you, so that you may be sons of your Father who is heaven; for he makes his sun rise on the evil and on the good, and sends rain on the just and on the unjust" (Matt. 5.44–45).

24 For a study of this transformation in the history of ideas, see C. E. Trinkhous, *In Our Image and Likeness: Humanity and Divinity in Italian Humanist Thought*, 2 vols. (London: Constable, 1970).

25 Jean Le Rond D'Alembert, *Preliminary Discourse to the Encyclopedia Of Diderot*, trans. Richard N. Schwab and Walter E. Rex (Indianapolis, Ind.: Bobbs-Merrill, 1963), p. 42.

26 Immanuel Kant, "Idea for a Universal History from a Cosmopolitan Point of View" (1784), Third Thesis, in *On History*, trans. Lewis White Beck (Indianpolis, Ind.: Bobbs-Merrill, 1963), p. 13.

27 Immanuel Kant, "What Is Enlightenment?" (1784), in *On History*, p. 3.

28 See Aristotle, *Politics* 1268b25–1269a25; and St. Thomas Aquinas, *Summa theologiae* 1–2, quaestio 97, articulus 2.

29 Charles-François de Saint-Lambert, "Luxury," in *Encyclopedia: Selections*, trans. Nelly S. Hoyt and Thomas Cassirer (Indianapolis, Ind.: Bobbs-Merrill, 1965), p. 204 (translation slightly altered).

30 Saint-Lambert, "Luxury," p. 231.

31 David Hume, *Essays* (London: Oxford University Press, 1963), p. 262.

32 Ibid., p. 276.

33 Ibid., p. 277.

34 Ibid., p. 277–78.

35 Charles Montesquieu, *The Spirit of the Laws*, trans. Thomas Nugent, ed. David Wallace Carrithers (Berkeley: University of California Press, 1977), vol. 1, bk. 20: 1.

36 See Francis Bacon, *The Great Instauration*, "The Plan of the Work," in *Selected Writings*.
37 D'Alembert, *Preliminary Discourse*, p. 75.
38 Ibid., p. 122.
39 Denis Diderot, "Art," in *Encyclopedia*, p. 5.
40 Ibid., p. 4.
41 For some discussion of this contrast, see Nicholas Lobkowicz, *Theory and Practice: History of a Concept from Aristotle to Marx* (Notre Dame, Ind.: University of Notre Dame Press, 1967).
42 On this interesting topic, see K. J. H. Berland, "Bringing Philosophy Down from the Heavens: Socrates and the New Science," *Journal of the History of Ideas* 47, no. 2 (April–June 1986): 299–308, a commentary on Amyas Busche's *Socrates: A Dramatic Poem* (1758). One point Berland does not consider is the extent to which this view of Socrates, which is also found in Aristophanes' *The Clouds* as well as other sources, might be legitimate; see, e.g., Leo Strauss, *Socrates and Aristophanes* (New York: Basic Books, 1966).
43 Alexander Pope, *An Essay on Man*, epistle 1, line 289.
44 Thomas Hobbes, *Leviathan, or the Matter, Forme, & Power of a Common Wealth Ecclesiasticall and Civill* (1651), edited by C. B. Macpherson (New York: Penguin Books, 1968), p. 81.
45 Julien Offray de La Mettrie, *Man a Machine* (*L'Homme Machine*, 1748) (La Salle, Ill.: Open Court, 1912), p. 148.
46 Hobbes, *Leviathan,* "Introduction," p. 23.
47 This is vividly demonstrated by the vicissitudes of development now taking place in countries of the Third World. Geographic advantage, scientific knowledge, imported hardware, political or economic decisions, piecemeal optimism, and envious desire cannot by themselves or even in concert effect industrialization. Despite the ideological rhetoric of Maoist China and Khomeni's Iran, modern technology does not seem to be adopted independently of certain key elements of Western culture. The westernization of Japan confirms the argument from the other side of the divide.
48 For one collection of texts that does begin to point in this direction, see Humphrey Jennings, *Pandaemonium: The Coming of the Machine as Seen by Contemporary Observers, 1660–1886*, ed. Mary-Lou Jennings and Charles Madge (New York: Free Press, 1985). Note that even Leo Marx in a later essay continues to speak of science as the focus of concern; see "Reflections on the Neo-Romantic Critique of Science," *Daedalus* 107, no. 2 (Spring 1978): 61–74. However, in *The Machine in the Garden: Technology and the Pastoral Ideal in America* (New York: Oxford University Press, 1964) he describes one aspect of the romantic critique of technology. Cf. also Wylie Sypher, *Literature and Technology: The Alien Vision* (New York: Random House, 1968).
49 Cf. Friedrich Nietzsche, *The Gay Science* (1882), section 12. For a mundane philosophy of gothic pathos, see Jean-Paul Sartre, *Being and Nothingness* (1943), trans. Hazel Barnes (New York: Washington Square Press, 1968), p. 784, the last sentence of the last chapter of which declares that "Man is a useless passion."
50 See Francis Bacon, "The Masculine Birth of Time," translation included in Benjamin Farrington, *The Philosophy of Francis Bacon* (Chicago: University of Chicago Press, 1966).
51 The note is to *The Excursion*, bk. 8, line 112, at the beginning of a passage describing the industrial transformation of the English landscape as one in which "where not a habitation stood before, / Abodes of men" are now "irregularly massed / Like trees in forests" (lines 122–24) and as a "trumph that proclaims / How much the mild Directress of the plough / Owes to alliance with these new-born arts!" (lines 130–32). "In treating of this subject," Wordsworth writes in his note, "it was impossible not to recollect, with gratitude, the pleasing picture…Dyer has given of the influences of manufacturing industry upon the face of this Island. He wrote at a time when machinery was first beginning to be introduced, and his benevolent heart prompted him to auger from it nothing but good." Wordsworth, as much as Sophocles (*Antigone*, 331), is capable of appreciating the benefits of technology. But, he adds, now "Truth has compelled me to dwell upon the baneful effects arising out of an ill-regulated and excessive application of powers so admirable in themselves."
52 Jean-Jacques Rousseau, "A Discourse on the Arts and Sciences," in *The Social Contract and Discourses*, trans. G. D. H. Cole (New York: Dutton, 1950), p. 164.
53 Ibid., p. 162.
54 Ibid., p. 161.
55 Ibid., p. 150.
56 Ibid., p. 158–59.
57 Cf. in this same regard, Niccolo Machiavelli's use of *virtú* as power in *The Prince* (1512).
58 Rousseau, "A Discourse on the Arts and Sciences," p. 173. The encyclopedics likewise praise Bacon above all other philosophers.
59 Ibid., p. 158.
60 Ibid., p. 152.
61 Ibid., p. 168.
62 Ibid., p. 146.
63 William Wordsworth, letter to Charles James Fox, 14 January 1801. In this commentary on his presentation of "domestic affections" in the poems "The Brothers" and "Michael," Wordsworth further remarks that "The evil [of the destruction of domestic affections] would be the less to be regretted, if these institutions [of industrialization] were regarded only as palliatives to a disease [in a manner

not unlike that associated with ancient skepticism]; but the vanity band pride of their promoters are so subtly interwoven with them, that they are deemed great discoveries and blessings to humanity [as per Enlightenment optimism]."

64 See John Milton, *Paradise Lost*, bk. 1, line 670. Milton also associates Satan's legions with engines and engineering at bk. 1, line 750 and bk. 6, line 553.

65 William Blake, *Milton* (1804), ed. Kay Pankhurst Esson and Roger R. Esson (Boulder, Colo.: Shambhala, 1978), bk. 1, sec. 4, lines 9–10.

66 Samuel Taylor Coleridge, *Biographia Literaria*, ed. George Watson (New York: Dutton, 1956), chap. 13, p. 167.

67 William Blake, *Jerusalem*, pt. 4, "To the Christians," introduction (London: George Allen & Unwin, 1964).

68 Edmund Burke, *A Philosophical Enquiry into the Origin of our Ideas of the Sublime and Beautiful* (1757) (London: R. & J. Dodsley, 1759). pt. 1, sec. 7, first sentence.

69 Heidegger, *Being and Time*, p. 164.

43

A Phenomenology of Technics

Don Ihde

The task of a phenomenology of human-technology relations is to discover the various structural features of those ambiguous relations. In taking up this task, I shall begin with a focus upon experientially recognizable features that are centered upon the ways we are bodily engaged with technologies. The beginning will be within the various ways in which I-as-body interact with my environment by means of technologies.

A Technics Embodied

If much of early modern science gained its new vision of the world through optical technologies, the process of embodiment itself is both much older and more pervasive. To embody one's praxis *through* technologies is ultimately an *existential* relation with the world. It is something humans have always – since they left the naked perceptions of the Garden – done.

I have previously and in a more suggestive fashion already noted some features of the visual embodiment of optical technologies. Vision is technologically transformed through such optics. But while the fact *that* optics transform vision may be clear, the variants and invariants of such a transformation are not yet precise. That becomes the task for a more rigorous and structural phenomenology of embodiment. I shall begin by drawing from some of the previous features mentioned in the preliminary phenomenology of visual technics.

Originally "Program One. A Phenomenology of Technics" in Don Ihde, *Technology and the Lifeworld*, Bloomington: Indiana University Press, 1990, pp. 72–108, abridged.

Within the framework of phenomenological relativity, visual technics first may be located within the intentionality of seeing.

I see–through the optical artifact–the world

This seeing is, in however small a degree, at least minimally distinct from a direct or naked seeing.

I see–the world

I call this first set of existential technological relations with the world *embodiment relations*, because in this use context I take the technologies *into* my experiencing in a particular way by way of perceiving *through* such technologies and through the reflexive transformation of my perceptual and body sense.

In Galileo's use of the telescope, he embodies his seeing through the telescope thusly:

Galileo–telescope–Moon

Equivalently, the wearer of eyeglasses embodies eyeglass technology:

I–glasses–world

The technology is actually *between* the seer and the seen, in a *position of mediation*. But the referent of the seeing, that towards which sight is directed, is "on the other side" of the optics. One sees *through* the optics. This, however, is not enough to specify this relation as an embodiment one. This is because one first has to determine *where* and *how*, along what will be described as a continuum of relations, the technology is experienced.

There is an initial sense in which this positioning is doubly ambiguous. First, the technology must be *technically* capable of being seen through; it must be transparent. I shall use the term *technical* to refer to the physical characteristics of the technology. Such characteristics may be designed or they may be discovered. Here the disciplines that deal with such characteristics are informative, although indirectly so for the philosophical analysis per se. If the glass is not transparent enough, seeing-through is not possible. If it is transparent enough, approximating whatever "pure" transparency could be empirically attainable, then it becomes possible to embody the technology. This is a material condition for embodiment.

Embodying as an activity, too, has an initial ambiguity. It must be learned or, in phenomenological terms, constituted. If the technology is good, this is usually easy. The very first time I put on my glasses, I see the now-corrected world. The adjustments I have to make are not usually focal irritations but fringe ones (such as the adjustment to backglare and the slight changes in spatial motility). But once learned, the embodiment relation can be more precisely described as one in which the technology becomes maximally "transparent." It is, as it were, taken into my own perceptual-bodily self experience thus:

(I–glasses)–world

My glasses become part of the way I ordinarily experience my surroundings; they "withdraw" and are barely noticed, if at all. I have then actively embodied the technics of vision. Technics is the symbiosis of artifact and user within a human action.

Embodiment relations, however, are not at all restricted to visual relations. They may occur for any sensory or microperceptual dimension. A hearing aid does this for hearing, and the blind man's cane for tactile motility. Note that in these corrective technologies *the same structural features of embodiment* obtain as with the visual example. Once learned, cane and hearing aid "withdraw" (if the technology is good – and here we have an experiential clue for the perfecting of technologies). I hear the world through the hearing aid and feel (and hear) it through the cane. The juncture (I-artifact)-world is through the technology and brought close by it.

Such relations *through* technologies are not limited to either simple or complex technologies. Glasses, insofar as they are engineered systems, are much simpler than hearing aids. More complex than either of these monosensory devices are those that entail whole-body motility. One such common technology is automobile driving. Although driving an automobile encompasses more than embodiment relations, its pleasurability is frequently that associated with embodiment relations.

One experiences the road and surroundings *through* driving the car, and motion is the focal activity. In a finely engineered sports car, for example, one has a more precise feeling of the road and of the traction upon it than in the older, softer-riding, large cars of the fifties. One embodies the car, too, in such activities as parallel parking: when well embodied, one feels rather than sees the distance between car and curb – one's bodily sense is "extended" to the parameters of the driver-car "body." And although these embodiment relations entail larger, more complex artifacts and entail a somewhat longer, more complex learning process, the bodily tacit knowledge that is acquired is perceptual-bodily.

Here is a first clue to the polymorphous sense of bodily extension. The experience of one's "body image" is not fixed but malleably extendable and/or reducible in terms of the material or technological mediations that may be embodied. I shall restrict the term embodiment, however, to those types of mediation that can be so experienced. The same dynamic polymorphousness can also be located in non-mediational or direct experience. Persons trained in the martial arts, such as karate, learn to feel the vectors and trajectories of the opponent's moves within the space of the combat. The near space around one's material body is charged.

Embodiment relations are a particular kind of use-context. They are technologically relative in a double sense. First, the technology must "fit" the use. Indeed, within the realm of embodiment relations one can develop a quite specific set of qualities for design relating to attaining the requisite technological "withdrawal." For example, in handling highly radioactive materials at a distance, the mechanical arms and hands which are designed to pick up and pour glass tubes inside the shielded enclosure have to "feed back" a delicate sense of touch to the operator. The closer to invisibility, transparency, and the extension of one's own bodily sense this technology allows, the better. Note that the design perfection is not one related to the machine alone but to the combination of machine and human. The machine is perfected along a bodily vector, molded to the perceptions and actions of humans.

And when such developments are most successful, there may arise a certain romanticizing of technology. In much anti-technological literature there are nostalgic calls for returns to simple tool technologies. In part, this may be because long-developed tools are excellent examples of bodily expressivity. They are both direct in actional terms and immediately experienced; but what is missed is that such embodiment relations may take any number of directions. Both the sports car driver within the constraints of the racing route and the bulldozer driver destroying a rainforest may have the satisfactions of powerful embodiment relations.

There is also a deeper desire which can arise from the experience of embodiment relations. It is the doubled desire that, on one side, is a wish for *total transparency*, total embodiment, for the technology to truly "become me." Were this possible, it would be equivalent to there being no technology, for total transparency would *be* my body and senses; I desire the face-to-face that I would experience without the technology. But that is only one side of the desire. The other side is the desire to have the power, the transformation that the technology makes available. Only by using the technology is my bodily power enhanced and magnified by speed, through distance, or by any of the other ways in which technologies change my capacities. These capacities are always *different* from my naked capacities. The desire is, at best, contradictory. I want the transformation that the technology allows, but I want it in such a way that I am basically unaware of its presence. I want it in such a way that it becomes me. Such a desire both secretly *rejects* what technologies are and overlooks the transformational effects which are necessarily tied to human-technology relations. This illusory desire belongs equally to pro- and anti-technology interpretations of technology.

The desire is the source of both utopian and dystopian dreams. The actual, or material, technology always carries with it only a partial or quasi-transparency, which is the price for the extension of magnification that technologies give. In extending bodily capacities, the technology also transforms them. In that sense, all technologies in use are non-neutral. They change the basic situation, however subtly, however minimally; but this is the other side of the desire. The desire is simultaneously a desire for a change in situation – to inhabit the earth, or even to go beyond the earth – while sometimes inconsistently and secretly wishing that this movement could be without the mediation of the technology.

The direction of desire opened by embodied technologies also has its positive and negative thrusts. Instrumentation in the knowledge activities, notably science, is the gradual extension of perception into new realms. The desire is to see, but seeing is seeing through instrumentation. Negatively, the desire for pure transparency is the wish to escape the limitations of the material technology. It is a platonism returned in a new form, the desire to escape the newly extended body of technological engagement. In the wish there remains the contradiction: the user both wants and does not want the technology. The user wants what the technology gives but does not want the limits, the transformations that a technologically extended body implies. There is a fundamental ambivalence toward the very human creation of our own earthly tools.

The ambivalence that can arise concerning technics is a reflection of one kind upon the *essential ambiguity* that belongs to technologies in use. But this ambiguity, I shall argue, has its own distinctive shape. Embodiment relations display an essential magnification/reduction structure which has been suggested in the instrumentation examples. Embodiment relations simultaneously magnify or amplify and reduce or place aside what is experienced through them.

The sight of the mountains of the moon, through all the transformational power of the telescope, removes the moon from its setting in the expanse of the heavens. But if our technologies were only to replicate our immediate and bodily experience, they would be of little use and ultimately of little interest. A few absurd examples might show this:

In a humorous story, a professor bursts into his club with the announcement that he has just invented a reading machine. The machine scans the pages, reads them, and perfectly reproduces them. (The story apparently was written before the invention of photocopying. Such machines might be said to be "perfect reading machines" in actuality.) The problem, as the innocent could see, was that this machine leaves us with precisely the problem we had prior to its invention. To have reproduced through mechanical "reading" all the books in the world leaves us merely in the library.

A variant upon the emperor's invisible clothing might work as well. Imagine the invention of perfectly transparent clothing through which we might technologically experience the world. We could see through it, breathe through it, smell and hear through it, touch through it. Indeed, it effects no changes of any kind, since it is *perfectly* invisible. Who would bother to pick up such clothing (even if the presumptive wearer could find it)? Only by losing some invisibility – say, with translucent coloring – would the garment begin to be usable and interesting. For here, at least, fashion would have been invented – but at the price of losing total transparency – by becoming that through which we relate to an environment.

Such stories belong to the extrapolated imagination of fiction, which stands in contrast to even the most minimal actual embodiment relations, which in their material dimensions simultaneously extend and reduce, reveal and conceal.

In actual human-technology relations of the embodiment sort, the transformational structures may also be exemplified by variations: In optical technologies, I have already pointed out how spatial significations change in observations through lenses. The entire gestalt changes. When the apparent size of the moon changes, along with it the apparent position of the observer changes. Relativistically, the moon is brought "close"; and equivalently, this optical near-distance applies to both the moon's appearance and my bodily sense of position. More subtly, every dimension of spatial signification also changes. For example, with higher and higher magnification, the well-known phenomenon of depth, instrumentally mediated as a "focal plane," also changes. Depth diminishes in optical near-distance.

A related phenomenon in the use of an optical instrument is that it transforms the spatial significations of vision in an instrumentally focal way. But my seeing without instrumentation is a full bodily seeing – I see not just with my eyes but with my whole body in a unified sensory experience of things. In part, this is why there is a noticeable irreality to the apparent position of the observer, which only diminishes with the habits acquired through practice with the instrument. But the optical instrument cannot so easily transform the entire sensory gestalt. The focal sense that is magnified through the instrument is monodimensioned.

Here may be the occasion (although I am not claiming a cause) for a certain interpretation of the senses. Historians of perception have noted that, in medieval times, not only was vision not the supreme sense but sound and smell may have had greatly enhanced roles so far as the interpretation of the senses went. Yet in the Renaissance and even more exaggeratedly in the Enlightenment, there occurred the reduction to sight as the favored sense, and within sight, a certain reduction *of* sight. This favoritism, however, also carried implications for the other senses.

One of these implications was that each of the senses was interpreted to be clear and distinct from the others, with only certain features recognizable through a given sense. Such an interpretation impeded early studies in echo location.

In 1799 Lazzaro Spallanzani was experimenting with bats. He noticed not only that they could locate food targets in the dark but also that they could do so blindfolded. Spallanzani wondered if bats could guide themselves by their ears rather than by their eyes. Further experimentation, in which the bats' ears were filled with wax, showed that indeed they could not guide themselves without their ears. Spallanzani surmised that either bats locate objects through hearing or they had some sense of which humans knew nothing. Given the doctrine of separate senses and the identification of shapes and objects through vision alone, George Montagu and Georges Cuvier virtually laughed Spallanzani out of the profession.

This is not to suggest that such an interpretation of sensory distinction was due simply to familiarity with optical technologies, but the common experience of enhanced vision through such technologies was at least the standard practice of the time. Auditory technologies were to come later. When auditory technologies did become common, it was possible to detect the same amplification/reduction structure of the human-technology experience.

The telephone in use falls into an auditory embodiment relation. If the technology is good, I hear *you* through the telephone and the apparatus "withdraws" into the enabling background:

(I–telephone)–you

But as a monosensory instrument, your phenomenal presence is that of a voice. The ordinary multidimensioned presence of a face-to-face encounter does not occur, and I must at best imagine those dimensions through your vocal gestures. Also, as with the telescope, the spatial significations are changed. There is here an auditory version of visual near-distance. It makes little difference whether you are geographically near or far, none at all whether you are north or south, and none with respect to anything but your bodily relation to the instrument. Your voice retains its partly irreal near-distance, reduced from the full dimensionality of direct perceptual situations. This telephonic distance is different both from immediate face-to-face encounters and from visual or geographical distance as normally taken. Its distance is a mediated distance with its own identifiable significations.

While my primary set of variations is to locate and demonstrate the invariance of a magnification/reduction structure to any embodiment relation, there are also secondary and important effects noted in the histories of technology. In the very first use of the telephone, the users were fascinated and intrigued by its auditory transparency. Watson heard and recognized Bell's *voice*, even though the instrument had a high ratio of noise to message. In short, the fascination attaches to magnification, amplification, enhancement. But, contrarily, there can be a kind of forgetfulness that equally attaches to the reduction. What is *revealed* is what excites; what is concealed may be forgotten. Here lies one secret for technological trajectories with respect to development. There are *latent telics* that occur through inventions.

Such telics are clear enough in the history of optics. Magnification provided the fascination. Although there were stretches of time with little technical progress, this fascination emerged from time to time to have led to compound lenses by Galileo's day. If some magnification shows the new, opens to what was poorly or not at all previously detected, what can greater magnification do? In our own time, the explosion of such variants upon magnification is dramatic. Electron enhancement, computer image enhancement, CAT and NMR internal scanning, "big-eye" telescopes – the list of contemporary magnificational and visual instruments is very long.

I am here restricting myself to what may be called a *horizontal* trajectory, that is, optical technologies that bring various micro- or macro-phenomena to vision through embodiment relations. By restricting examples to such phenomena, one structural aspect of embodiment relations may be pointed to concerning the relation to microperception and its Adamic context. While *what* can be seen has changed dramatically – Galileo's New World has now been enhanced by astronomical phenomena never suspected and by micro-phenomena still being discovered – there remains a strong phenomenological constant in *how* things are seen. All lenses and optical technologies of the sort being described bring what is to be seen into a normal bodily space and distance. Both the macroscopic and the microscopic appear within the same near-distance. The "image size" of galaxy or amoeba is the *same*. Such is the existential condition for visibility, the counterpart to the technical condition, that the instrument makes things visually present.

The mediated presence, however, must fit, be made close to my actual bodily position and sight. Thus there is a reference within the instrumental context to my face-to-face capacities. These remain primitive and central within the new mediational context. Phenomenological theory claims that for every change in what is seen (the object correlate), there is a noticeable change in how (the experiential correlate) the thing is seen.

In embodiment relations, such changes retain both an equivalence and a difference from non-mediated situations. What remains constant is the bodily focus, the reflexive reference back to my bodily capacities. What is seen must be seen from or within my visual field, from the apparent distance in which discrimination can occur regarding depth, etc., just as in face-to-face relations. But the range of what can be brought into this proximity is transformed by means of the instrument.

Let us imagine for a moment what was never in fact a problem for the history of instrumentation: If the "image size" of both a galaxy and an amoeba is the "same" for the observer using the instrument, how can we tell that one is macrocosmic and the other microcosmic? The "distance" between us and these two magnitudes, Pascal noted, was the same in that humans were interpreted to be between the infinitely large and the infinitely small.

What occurs through the mediation is not a problem *because our construction of the observation presupposes ordinary praxical spatiality*. We handle the paramecium, placing it on the slide and then under the microscope. We aim the telescope at the indicated place in the sky and, before looking through it, note that the distance is at least that of the heavenly dome. But in our imagination experiment, what if our human were *totally immersed* in a technologically mediated world? What if, from birth, all vision occurred only through lens systems? Here the problem would become more difficult. But in our distance from Adam, it is precisely the presumed difference that makes it possible for us to see both nakedly *and* mediately – and thus to be able to locate the difference – that places us even more distantly from any Garden. It is because we retain this ordinary spatiality that we have a reflexive point of reference from which to make our judgments.

The noetic or bodily reflexivity implied in all vision also may be noticed in a magnified way in the learning period of embodiment. Galileo's telescope had a small field, which, combined with early hand-held positioning, made it very difficult to locate any particular phenomenon. What must have been noted, however, even if not commented upon, was the exaggerated sense of bodily motion experienced through trying to fix upon a heavenly body – and more, one quickly learns something about the earth's very motion in the attempt to use such primitive telescopes. Despite the apparent fixity of the stars, the hand-held telescope shows the earth-sky motion dramatically. This magnification effect is within the experience of one's own bodily viewing.

This bodily and actional point of reference retains a certain privilege. All experience refers to it in a taken-for-granted and recoverable way. The bodily condition of the possibility for seeing is now twice indicated by the very situation in which mediated experience occurs. Embodiment relations continue to locate that privilege of my being here. The partial symbiosis that occurs in well-designed embodied technologies retains that motility which can be called expressive. Embodiment relations constitute one existential form of the full range of the human-technology field.

B Hermeneutic Technics

Heidegger's hammer in use displays an embodiment relation. Bodily action through it occurs within the environment. But broken, missing, or malfunctioning, it ceases to be the means of praxis and becomes an obtruding *object* defeating the work project. Unfortunately, that negative derivation of objectness by Heidegger carries with it a block against understanding a second existential human-technology relation, the type of relation I shall term *hermeneutic*.

The term hermeneutic has a long history. In its broadest and simplest sense it means "interpretation," but in a more specialized sense it refers to *textual* interpretation and thus entails *reading*. I shall retain both these senses and take hermeneutic to mean a special interpretive action within the technological context. That kind of activity calls for special modes of action and perception, modes analogous to the reading process.

Reading is, of course, a reading of____; and in its ordinary context, what fills the intentional blank is a text, something *written*. But all writing entails technologies. Writing has a product. Historically, and more ancient than the revolution brought about by such crucial technologies as the clock or the compass, the invention and development of writing was surely even more revolutionary than clock or compass with respect to human experience. Writing transformed the very perception and understanding we have of language. Writing is a technologically embedded form of language.

There is a currently fashionable debate about the relationship between speech and writing, particularly within current Continental philosophy. The one side argues that speech is primary, both historically and ontologically, and the other – the French School – inverts this relation and argues for the primacy of writing. I need not enter this debate here in order to note the *technological difference* that obtains between oral speech and the materially connected process of writing, at least in its ancient forms.

Writing is inscription and calls for both a process of writing itself, employing a wide range of technologies (from stylus for cuneiform to word processors for the contemporary academic), and other material entities upon which the writing is recorded (from clay tablet to computer printout). Writing is technologically mediated language. From it, several features of hermeneutic technics may be highlighted. I shall take what may at first appear as a detour into a distinctive set of human-technology relations by way of a phenomenology of reading and writing.

Reading is a specialized perceptual activity and praxis. It implicates my body, but in certain distinctive ways. In an ordinary act of reading, particularly of the extended sort, what is read is placed before or somewhat under one's eyes. We read in the immediate context from some miniaturized bird's-eye perspective. What is read occupies an expanse within the focal center of vision, and I am ordinarily in a somewhat rested position. If the object-correlate, the "text" in the broadest sense, is a chart, as in the navigational examples, what is represented retains a representational isomorphism with the natural features of the landscape. The chart represents the land- (or sea)scape and insofar as the features are isomorphic, there is a kind of representational "transparency." The chart in a peculiar way "refers" beyond itself to what it represents.

Now, with respect to the embodiment relations previously traced, such an isomorphic representation is both similar and dissimilar to what would be seen on a larger scale from some observation position (at bird's-eye level). It is similar in that the shapes on the chart are reduced representations of distinctive features that can be directly or technologically mediated in face-to-face or embodied perceptions. The reader can compare these similarities. But chart reading is also different in that, during the act of reading, the perceptual focus is the chart itself, a substitute for the landscape.

I have deliberately used the chart-reading example for several purposes. First, the "textual" isomorphism of a representation allows this first example of hermeneutic technics to remain close to yet differentiated from the perceptual isomorphism that occurs in the optical examples. The difference is at least perceptual in that one sees *through* the optical technology, but now one sees the chart as the visual terminus, the "textual" artifact itself.

Something much more dramatic occurs, however, when the representational isomorphism disappears in a printed text. There is no isomorphism between the printed word and what it "represents," although there is some kind of *referential* "transparency" that belongs to this new technologically embodied form of language. It is apparent from the chart example that the chart itself becomes the *object of perception* while simultaneously referring beyond itself to what is not immediately seen. In the case of the printed text, however, the referential transparency is distinctively different from technologically embodied perceptions. *Textual transparency is hermeneutic transparency, not perceptual transparency*.

Historically, textual transparency was neither immediate nor attained at a stroke. The "technology" of phonetic writing, which now is increasingly a world-wide standard, became what it is through a series of variants and a process of experimentation. One early form of writing was pictographic. The writing was still somewhat like the chart example; the pictograph retained a certain representational isomorphism with what was represented. Later, more complex ideographic writing (such as Chinese) was, in effect, a more abstract form of pictography.

Calligraphers have shown that even early phonetic writing followed a gradual process of formalizing and abstracting from a pictographic base (see figure). Letters often depicted a certain animal, the first syllable of whose name provided the sound for the letter in a simultaneous sound and letter. Built into such early phonetic writing was thus something like the way the alphabet is still taught to children: "C is for Cow." Most educated persons are familiar with the mixed form of writing, hieroglyphics. Although the writing is pictographic, not all pictographs stood for the entity depicted; some represented sounds (phonemes).

Egyptian Apis

Phoenician Aleph

Ionian Greek Alpha

Roman A

An interesting cross-cultural example of this movement from a very pictographic to a formalized and transformed ideographic writing occurs with Chinese writing. The same movement from relatively concrete representations in pictographs occurs through abbreviated abstractions – but in a different direction, non-phonetic and ideographic. Thus, for phonetic writing there is a double abstraction (from pictograph to letter and then reconstituting a small finite alphabet into represented spoken words), whereas the doubled abstraction of ideographic writing does not reconstitute to words as such, but to concepts.

In the most ancient Chinese writing in the period of the "Tortoise Shell Language" (prior to 2000 B.C.) and even in some cases through the later "Metal Language" period (2000–500 B.C.), if one is familiar with the objects as they occur within Chinese culture, one can easily detect the pictographic representation involved. For example, one can see in the figure below that the ideograph for boat actually abstractly represents the sampan-type boats of the riverways (still in use). Similarly, in the ideograph for gate one can still recognize the uniquely Oriental-type gate in the drawing. The modern variants – related but more abstracted – have clearly lost that instant representational isomorphism.

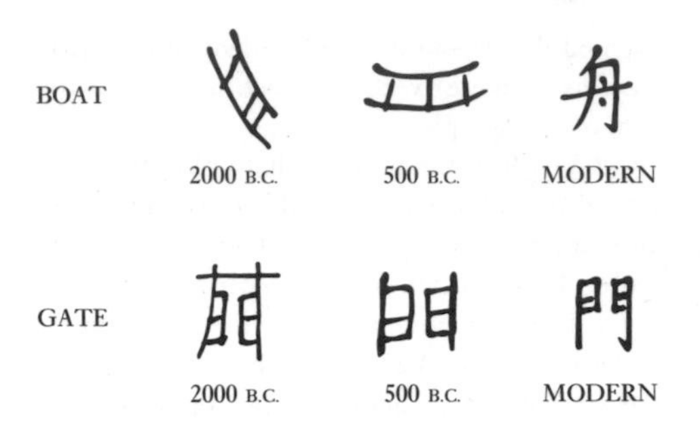

Implied in these transformations are changes of both technique and related technologies. Sergei Eisenstein, the film maker and one sensitive to such image technologies, has pointed to just such a transformation which arose out of the invention of the brush and India ink:

> But then, by the end of the third century, the brush is invented. In the first century after the "joyous event" (A.D.) – paper. And, lastly, in the year 220 – India ink.
>
> A complete upheaval. A revolution in draughtmanship. And, after having undergone in the course of history no fewer than fourteen different styles of handwriting, the hieroglyph crystallized in its present form. The means of production (brush and India ink) determined the form.
>
> The fourteen reforms had their way. As a result:
>
> 
>
> In the fierily cavorting hieroglyph *ma* (a horse) it is already impossible to recognize the features of the dear little horse sagging pathetically in its hindquarters, in the writing style of Ts'ang Chiech, so well-known from ancient Chinese bronzes.[1]

If this is an accurate portrayal of the evolution of writing, it follows something like a Husserlian origin-of-geometry trajectory. The trajectory was from the more concrete to the greater degrees of abstraction, until virtually all "likeness" to origins disappeared. In this respect, writing only slowly approximated speech.

Once attained, like any other acquisition of the lifeworld, writing could be read and understood in terms of its unique linguistic transparency. Writing becomes an embodied hermeneutic technics. Now the descriptions may take a different shape. What is referred to is referred by the text and is referred to *through* the text. What now presents itself is the "world" of the text.

This is not to deny that all language has its unique kind of transparency. Reference beyond itself, the capacity to let something become present through language, belongs to speech as well. But here the phenomenon being centered upon is the new embodiment of language in writing. Even more thematically, the concern is for the ways in which writing as a "technology" transforms experiential structures.

Linguistic transparency is what makes present the *world* of the text. Thus, when I read Plato, Plato's "world" is made present. But this presence is a *hermeneutic* presence. Not only does it occur *through* reading, but it takes its shape in the interpretative context of my language abilities. His world is linguistically mediated, and while the words may elicit all sorts of imaginative and perceptual phenomena, it is through language that such phenomena occur. And while such phenomena may be strikingly rich, they do not appear *as* word-like.

We take this phenomenon of reading for granted. It is a sedimented acquisition of the literate lifeworld and thus goes unnoticed until critical reflection isolates its salient features. It is the same with the wide variety of hermeneutic technics we employ.

The movement from embodiment relations to hermeneutic ones can be very gradual, as in the history of writing, with little-noticed differentiations along the human-technology continuum. A series of wide-ranging variants upon readable technologies will establish the point. First, a fairly explicit example of a readable technology: Imagine sitting inside on a cold day. You look out the window and notice that the snow is blowing, but you are toasty warm in front of the fire. You can clearly "see" the cold in Merleau-Ponty's pregnant sense of perception – but you do not actually *feel* it. Of course, you could, were you to go outside. You would then have a full face-to-face verification of what you had seen.

But you might also see the thermometer nailed to the grape arbor post and *read* that it is 28°F. You would now "know" how cold it was, but you still would not feel it. To retain the full sense of an embodiment relation, there must also be retained some isomorphism with the felt sense of the cold – in this case, tactile – that one would get through face-to-face experience. One could invent such a technology; for example, some conductive material could be placed through the wall so that the negative "heat," which is cold, could be felt by hand. But this is not what the thermometer does.

Instead, you read the thermometer, and in the immediacy of your reading you *hermeneutically* know that it is cold. There is an instantaneity to such reading, as it is an already constituted intuition (in phenomenological terms). But you should not fail to note that *perceptually* what you have seen is the dial and the numbers, the thermometer "text." And that text has hermeneutically delivered its "world" reference, the cold.[2]

Such constituted immediacy is not always available. For instance, although I have often enough lived in countries where Centigrade replaces Fahrenheit, I still must translate from my intuitive familiar language to the less familiar one in a deliberate and self-conscious hermeneutic act. Immediacy, however, is not the test for whether the relation is hermeneutic. A hermeneutic relation mimics sensory perception insofar as it is also a kind of seeing as____; but it is a referential seeing, which has as its immediate perceptual focus seeing the thermometer.

Now let us make the case more complex. In the example cited, the experiencer had both embodiment (seeing the cold) and hermeneutic access to the phenomenon (reading the thermometer). Suppose the house were hermetically sealed, with no windows, and the only access to the weather were through the thermometer (and any other instruments we might include). The hermeneutic character of the relation becomes more obvious. I now clearly have to know how to read the instrumentation and from this reading knowledge get hold of the "world" being referred to.

This example has taken actual shape in nuclear power plants. In the Three Mile Island incident, the nuclear power system was observed only through instrumentation. Part of the delay that caused a near meltdown was *misreadings* of the instruments. There was no face-to-face, independent access to the pile or to much of the machinery involved, nor could there be.

An intentionality analysis of this situation retains the mediational position of the technology:

I–technology–world
(engineer–instruments–pile)

The operator has instruments between him or her and the nuclear pile. But – and here, an essential difference emerges between embodiment and hermeneutic relations – what is immediately perceived is the instrument panel itself. It becomes the object of my microperception, although in the special sense of a hermeneutic transparency, I *read* the pile through it. This situation calls for a different formalization:

I–(technology–world)

The parenthesis now indicates that the immediate *perceptual* focus of my experience *is* the control panel. I read through it, but this reading is now dependent upon the semi-opaque connection between the instruments and the referent object (the pile). This *connection* may now become enigmatic.

In embodiment relations, what allows the partial symbiosis of myself and the technology is the capacity of the technology to become perceptually transparent. In the optical examples, the glassmaker's and lens-grinder's arts must have accomplished this end if the embodied use is to become possible. Enigmas which may occur regarding embodiment-use transparency thus may occur within the parenthesis of the embodiment relation:

(I–technology) → World
↑
enigma position

(This is not to deny that once the transparency is established, thus making microperception clear, the observer may still fail, particularly at the macroperceptual level. For the moment, however, I shall postpone this type of interpretive problem.) It would be an oversimplification of the history of lens-making were not problems of this sort recognized. Galileo's instrument not only was hard to look through but was good only for certain "middle range" sightings in astronomical terms (it did deliver the planets and even some of their satellites). As telescopes became more powerful, levels, problems with chromatic effects, diffraction effects, etc., occurred. As Ian Hacking has noted,

> Magnification is worthless if it magnifies two distinct dots into one big blur. One needs to resolve the dots into two distinct images.... It is

> a matter of diffraction. The most familiar example of diffraction is the fact that shadows of objects with sharp boundaries are fuzzy. This is a consequence of the wave character of light.[3]

Many such examples may be found in the history of optics, technical problems that had to be solved before there could be any extended reach within embodiment relations. Indeed, many of the barriers in the development of experimental science can be located in just such limitations in instrumental capacity.

Here, however, the task is to locate a parallel difficulty in the emerging new human-technology relation, hermeneutic relations. The location of the technical problem in hermeneutic relations lies in the *connector* between the instrument and the referent. Perceptually, the user's visual (or other) terminus is *upon* the instrumentation itself. To read an instrument is an analogue to reading a text. But if the text does not correctly refer, its reference object or its world cannot be present. Here is a new location for an enigma:

I → (technology–world)
↑
enigma position

While breakdown may occur at any part of the relation, in order to bring out the graded distinction emerging between embodiment and hermeneutic relations, a short pathology of connectors might be noted.

If there is nothing that impedes my direct perceptual situation with respect to the instrumentation (in the Three Mile Island example, the lights remain on, etc.), interpretive problems in reading a strangely behaving "text" at least occur in the open; but the technical enigma may also occur within the text-referent relation. How could the operator tell if the instrument was malfunctioning or that to which the instrument refers? Some form of *opacity* can occur within the technology-referent pole of the relation. If there is some independent way of verifying which aspect is malfunctioning (a return to unmediated face-to-face relations), such a breakdown can be easily detected. Both such occurrences are reasons for instrumental redundancy. But in examples where such independent verification is not possible or untimely, the opacity would remain.

Let us take a simple mechanical connection as a borderline case. In shifting gears on my boat, there is a lever in the cockpit that, when pushed forward, engages the forward gear; upward, neutral; and backwards, reverse. Through it, I can ordinarily feel the gear change in the transmission (embodiment) and recognize the simple hermeneutic signification (forward for forward) as immediately intuitive. Once, however, on coming in to the dock at the end of the season, I disengaged the forward gear – and the propeller continued to drive the boat forward. I quickly reversed – and again the boat continued. The hermeneutic significance had failed; and while I also felt a difference in the way the gear lever felt, I did not discover until later that the clasp that retained the lever itself had corroded, thus preventing any actual shifting at all. But even at this level there can be opacity within the technology-object relation.

The purpose of this somewhat premature pathology of human-technology relations is not to cast a negative light upon hermeneutic relations in contrast to embodiment ones but rather to indicate that there are different locations where perceptual and human-technology relations interact. Normally, when the technologies work, the technology-world relation would retain its unique hermeneutic transparency. But if the I-(technology-world) relation is far enough along the continuum to identify the relation as a hermeneutic one, the intersection of perceptual-bodily relations with the technology changes.

Readable technologies call for the extension of my hermeneutic and "linguistic" capacities *through* the instruments, while the reading itself retains its bodily perceptual location as a relation *with* or *towards* the technology. What is emerging here is the first suggestion of an emergence of the technology as "object" but without its negative Heideggerian connotation. Indeed, the type of special capacity as a "text" is a condition for hermeneutic transparency.

The transformation made possible by the hermeneutic relation is a transformation that occurs precisely through *differences* between the text and what is referred to. What is needed is a particular set of textually clear perceptions that "reduce" to that which is immediately readable. To return to the Three Mile Island example, one problem uncovered was that the instrument panel design was itself faulty. It did not incorporate its dials and gauges in an easily readable way. For example, in airplane instrument panel design, much thought has been given to pattern recognition, which occurs as a perceptual gestalt. Thus, in a four-engined aircraft, the four dials indicating r.p.m. will be coordinated so that a single glance will indicate which, if any, engine is out of synchronization. Such technical design accounts for perceptual structures.

There is a second caution concerning the focus upon connectors and pathology. In all the examples I have used to this point, the hermeneutic technics have involved material connections. (The thermometer employs a physical property of a bimetallic spring or mercury in a column; the instrument panel at TMI employs mechanical, electrical, or other material connections; the shift lever, a simple mechanical connection.) If reading does not employ any such material connections, it might seem that its referentiality is essentially different, yet not even all technological connections are strictly material. Photography retains representational isomorphism with the object, yet does not "materially" connect with its object; it is a minimal beginning of action at a distance.

I have been using contemporary or post-scientific examples, but non-material hermeneutic relations do not obtain only for contemporary humans. As existential relations, they are as "old" as post-Garden humanity. Anthropology and the history of religions have long been familiar with a wide variety of shamanistic praxes which fall into the pattern of hermeneutic technics. In what may at first seem a somewhat outrageous set of examples, note the various "reading" techniques employed in shamanism. The reading of animal entrails, of thrown bones, of bodily marks – all are hermeneutic techniques. The patterns of the entrails, bones, or whatever are taken to *refer* to some state of affairs, instrumentally or textually.

Not only are we here close to a familiar association between magic and the origins of technology suggested by many writers, but we are, in fact, closer to a wider hermeneutic praxis in an intercultural setting. For that reason, the very strangeness of the practice must be critically examined. If the throwing of bones is taken as a "primitive" form of medical diagnosis – which does play a role in shamanism – we might conclude that it is indeed a poor form of hermeneutic relations. What we might miss, however, is that the entire gestalt of what is being diagnosed may differ radically from the other culture and ours.

It may well be that as a focused form of diagnosis upon some particular bodily ailment (appendicitis, for example), the diagnosis will fail. But since one important element in shamanism is a wider diagnosis, used particularly as the occasion of locating certain communal or social problems, it may work better. The sometimes socially contextless emphasis of Western medicine upon a presumably "mechanical" body may overlook precisely the context which the shaman so clearly recognizes. The entire gestalt is different and differently focused, but in both cases there are examples of hermeneutic relations.

In our case, the very success of Western medicine in certain diseases is due to the introduction of technologies into the hermeneutic relation (fever/thermometer; blood pressure/manometer, etc.) The point is that hermeneutic relations are as commonplace in traditional and ancient social groups as in ours, even if they are differently arranged and practiced.

By continuing the intentionality analysis I have been following, one can now see that hermeneutic relations vary the continuum of human-technology-world relations. Hermeneutic relations maintain the general mediation position of technologies within the context of human praxis towards a world, but they also change the variables within the human-technology-world relation. A comparative formalism may be suggestive:

General intentionality relations
Human–technology–world

Variant A: embodiment relations
(I–technology) → world

Variant B: hermeneutic relations
I → (technology world)

While each component of the relation changes within the correlation, the overall shapes of the variants are distinguishable. Nor are these matters of simply how technologies are experienced.

Another set of examples from the set of optical instruments may illustrate yet another way in which instrumental intentionalities can follow new trajectories. Strictly embodiment relations can be said to work best when there is both a transparency and an isomorphism between perceptual and bodily action within the relation. I have suggested that a trajectory for development in such cases may often be a horizontal one. Such a trajectory not only follows greater and greater degrees of magnification but also entails all the difficulties of a technical nature that go into allowing what is to be seen as though by direct vision. But not all optical technologies follow this strategy. The introduction of hermeneutic possibilities opens the trajectory into what I shall call *vertical* directions, possibilities that rely upon quite deliberate hermeneutic transformations.

It might be said that the telescope and microscope, by extending vision while transforming it, remained *analogue* technologies. The enhancement and magnification made possible by such technologies remain visual and transparent to ordinary vision. The moon remains recognizably the moon, and the microbe – even if its existence was not previously suspected – remains under the microscope a beastie recognized as belonging to the animate continuum. Here, just as the capacity to magnify becomes the foreground phenomenon to the background phenomenon of the reduction necessarily accompanying the magnification, so the similitude of what is seen with ordinary vision remains central to embodiment relations.

Not all optical technologies mediate such perceptions. In gradually moving towards the visual "alphabet" of a hermeneutic relation, deliberate variations may occur which enhance previously undiscernible *differences*:

1) Imagine using spectacles to correct vision, as previously noted. What is wanted is to *return* vision as closely as possible to ordinary perception, not to distort or modify it in any extreme micro- or macroperceptual direction. But now, for snowscapes or sun on the water or desert, we modify the lenses by coloring or polarizing them to cut glare. Such a variation transforms *what* is seen in some degree. Whether we say the polarized lens removes glare or "darkens" the landscape, what is seen is now clearly different from what may be seen through untinted glasses. This difference is a clue which may open a new *telic direction* for development.

2) Now say that somewhere, sometime, someone notes that certain kinds of tinting reveal unexpected results. Such is a much more complex technique now used in infrared satellite photos. (For the moment, I shall ignore the fact that part of this process is a combined embodiment and hermeneutic relation.) If the photo is of the peninsula of Baja California, it will remain recognizable in shape. Geography, whatever depth and height representations, etc., remain but vary in a direction different from any ordinary vision. The infrared photo enhances the difference between vegetation and non-vegetation beyond the limits of any isomorphic color photography. This difference corresponds, in the analogue example, to something like a pictograph. It simultaneously leaves certain analogical structures there and begins to modify the representation into a different, non-perceived "representation."

3) Very sophisticated versions of still representative but non-ordinary forms of visual recognition occur in the new heat-sensitive and light-enhanced technologies employed by the military and police. Night scopes which enhance a person's heat radiation still look like a person but with entirely different regions of what stands out and what recedes. In high-altitude observations, "heat shadows" on the ground can indicate an airplane that has recently had its engines running compared to others which have not. Here visual technologies bring into visibility what was not visible, but in a distinctly now perceivable way.

4) If now one takes a much larger step to spectrographic astronomy, one can see the acceleration of this development. The spectrographic picture of a star no longer "resembles" the star at all. There is no point of light, no disk size, no spatial isomorphism at all – merely a band of differently colored rainbow stripes. The naive reader would not know that this was a picture of a star at all – the reader would have to know the language, the alphabet, that has coded the star. The astronomer-hermeneut does know the language and "reads" the visual "ABCs" in such a way that he knows the chemical composition of the star, its internal makeup, rather than its shape or external configuration. We are here in the presence of a more fully hermeneutic relation, the star mediated not only instrumentally but in a transformation such that we must now thematically *read* the result. And only the informed reader can do the reading.

There remains, of course, the *reference* to the star. The spectograph is *of* Rigel or *of* Polaris, but the individuality of the star is now made present hermeneutically. Here we have a beginning of a special transformation of perception, a transformation which deliberately enhances differences rather than similarities in order to get at what was previously unperceived.

5) Yet even the spectrograph is but a more radical transformation of perception. It, too, can be transformed by a yet more radical *hermeneutic* analogue to the *digital* transformation which lies embedded in the preferred quantitative praxis of science. The "alphabet" of science is, of course, mathematics, a mathematics that separates itself by yet another hermeneutic step from perception embodied.

There are many ways in which this transformation can and does occur, most of them interestingly involving a particular act of *translation* that often goes unnoticed. To keep the example as simple as possible, let us assume *mechanical* or *electronic* "translation." Suppose our spectrograph is read by a machine that yields not a rainbow spectrum but a set of numbers. Here we would arrive at the final hermeneutic accomplishment, the transformation of even the analogue to a digit. But in the process of hermeneuticization, the "transparency" to the object referred to becomes itself enigmatic. Here more explicit and thematic interpretation must occur.

Hermeneutic relations, particularly those utilizing technologies that permit vertical transformations, move away from perceptual isomorphism. It is the *difference* between what is shown and how something is shown which is informative. In a hermeneutic relation, the world is first transformed into a text, which in turn is read. There is potentially as much flexibility within hermeneutic relations as there are in the various uses of language. Emmanuel Mournier early recognized just this analogical relationship with language:

> The machine as implement is not a simple material extension of our members. It is of another order, an annex to our language, an auxiliary language to mathematics, a means of penetrating, dissecting and revealing the secret of things, their implicit intentions, their unemployed capacities.[4]

Through hermeneutic relations we can, as it were, *read* ourselves into any possible situation without being there. In science, in contrast to literature, what is important is that the reading retain *some* kind of reference or hermeneutic transparency to what is there. Perhaps that is one reason for the constant desire to reverse what is read back towards what may be perceived. In this reversal, contemporary technologically embodied science has frequently derived what might be called *translation technologies*. I mention two in passing:

(*a*) Digital processes have become *de rigueur* within the perceptual domain. The development of pictures from space probes is such a *double translation* process. The photograph of the surface of Venus is a technological analogue to human vision. It at least is a field display of the surface, incorporating the various possible figures and contrasts that would be seen instantaneously in a visual gestalt – but this holistic result cannot be transmitted in this way by the current technologies. Thus it is "translated" into a digital code, which can be transmitted. The "seeing" of the instrument is broken down into a series of digits that are radiographically transmitted to a receiver; then they are reasembled into a spatter pattern and enhanced to reproduce the photograph taken millions of miles away. It would be virtually impossible for anyone to read the digits and tell what was to be seen; only when the linear text of the digits has been retranslated back into the span of an instantaneous visual gestalt can it be seen that the rocks on Venus are or are not like those on the moon. Here the analogues of perception and language are both utilized to extend vision beyond the earth.

(*b*) The same process is used audially in digital recordings. Once again, the double translation process takes place and sound is reduced to digital form, reproduced through the record, and translated back into an auditory gestalt.

Digital and analogue processes blur together in certain configurations. Photos transmitted as points of black on a white ground and reassembled within certain size limits are perceptually gestalted; we see Humphrey Bogart, not simply a mosaic of dots. (Pointillism did the same in painting, although in color. So-called concrete poetry employs the same crossover by placing the words of the poem in a visual pattern so the poem may be both read and seen as a visual pattern.)

Such translation and retranslation processes are clearly transformations from perceptually gestalted phenomena into analogues of writing (serial translation and retranslation processes are clearly transformations from perceptual gestalt phenomena into analogues of writing serial transmissions along a "line," as it were), which are then retranslatable into perceptual gestalts.

I have suggested that the movement from embodiment relations to hermeneutic ones occurs along a human-technology continuum. Just as there are complicated, borderline cases along the continuum from fully haired to bald men, there are the same less-than-dramatic differences here. I have highlighted some of this difference by accenting the bodily-perceptual distinctions that occur between embodiment and hermeneutic relations. This has allowed the difference in perceptual and hermeneutic transparencies to stand out.

There remain two possible confusions that must be clarified before moving to the next step in this phenomenology of technics. First, there is a related sense in which perception and interpretation are intertwined. Perception is primitively already interpretational, in both micro- and macrodimensions. To perceive is already "like" reading. Yet reading is also a specialized act that receives both further definition and elaboration within literate contexts. I have been claiming that one of the distinctive differences between embodiment and heremneutic relations involves perceptual position, but in the broader sense, interpretation pervades both embodiment and hermeneutic action.

A second and closely related possible confusion entails the double sense in which a technology may be used. It may be used simultaneously both as something *through* which one experiences and as something *to* which one relates. While this is so, the doubled relation takes shapes in embodiment different from those of hermeneutic relations. Return to the simple embodiment relation illustrated in wearing eyeglasses. *Focally*, my perceptual experience finds its directional aim *through* the lenses, terminating my gaze upon the object of vision; but as a *fringe* phenomenon, I am simultaneously aware of (or can become so) the way my glasses rest upon the bridge of my nose and the tops of my ears. In this fringe sense, I am aware of the glasses, but the focal phenomenon is the perceptual transparency that the glasses allow.

In cases of hermeneutic transparency, this doubled role is subtly changed. Now I may carefully read the dials within the core of my visual field and attend to them. But my reading is simultaneously a reading through them, although now the terminus of reference is not necessarily a perceptual object, nor is it, strictly speaking, perceptually present. While the type of transparency is distinct, it remains that the purpose of the reading is to gain hermeneutic transparency.

Both relations, however, at optimum, occur within the familiar acquisitional praxes of the lifeworld. Acute perceptual seeing must be learned and, once acquired, occurs as familiarly as the act of seeing itself. For the accomplished and critical reader, the hermeneutic transparency of some set of instruments is as clear and as immediate as a visual examination of some specimen. The peculiarity of hermeneutic transparency does not lie in either any deliberate or effortful accomplishment of interpretation (although in learning any new text or language, that effort does become apparent). That is why the praxis that grows up within the hermeneutic context retains the same sense of spontaneity that occurs in simple acts of bodily motility. Nevertheless, a more distinctive presence *of* the technology appears in the example. My awareness of the instrument panel is both stronger and centered more focally than the fringe awareness of my eyeglasses frames, and this more distinct awareness is essential to the optimal use of the instrumentation.

In both embodiment and hermeneutic relations, however, the technology remains short of full objectiveness or *otherwise*. It remains the means through which something else is made present. The negative characterization that may occur in breakdown pathologies may return. When the technology in embodiment position breaks down or when the instrumentation in hermeneutic position fails, what remains is an obtruding, and thus negatively derived, object.

Both embodiment and hermeneutic relations, while now distinguished, remain basic existential relations between the human user and the world. There is the danger that my now-constant and selective use of scientific instrumentation could distort the full impact of the existential dimension. Prior to moving further along the human-technology-world continuum, I shall briefly examine a very different set of instrumental examples. The instrumentation in this case will be *musical instrumentation*.

In the most general sense, it should be easy to see that the use of musical instrumentation, in performance, falls into the same configurations as do scientific instruments:

I–musical instrument–world
I–scientific instrument–world

But the praxical context is significantly changed. If scientific or knowledge-developing praxis is constrained by the need to have a referential terminus within the world, the musical praxis is not so constrained. Indeed, if there is a terminus, it is a reference not so much to some thing or region of the environment as to the production of a musical event within that environment. The "musical object" is whatever sound phenomenon occurs through the performance upon the instrument. Musical sounds are produced, *created*. Whereas in the development of scientific instrumentation the avoidance of phenomena that would be artifacts of the instrument rather than of its referent are to be avoided or reduced as much as possible, the very discovery and enhancement of such instrumental artifacts may be a positive phenomenon in making music. There are interesting and significant differences in these two praxical contexts, but for the moment, I shall restrict myself to a set of observations about the similarities in the intentionality structures of both scientific and musical instrumentation.

It should be obvious that a very large use of musical instrumentation falls clearly into the embodiment relation pattern. The player picks up the instrument (having learned to embody it) and expressively produces the desired music:

Player–instrument–sound

In embodiment cases, the sound-making instrument will be partially symbiotically embodied:

(Player–instrument)–sound

Second, the previously noted amplification/reduction structure also occurs here. If our player is a trombonist, the "buzz" his lip vibrations produce can be heard without any instrument but, once amplified and transformed *through* the trombone, occur as the musical sound distinctive to the human-instrument pairing. Equally immediately, at least within the complex of contemporary instrumentation, one may detect that nothing like a restriction to human sound as such belongs to the contemporary musical context. Isomorphism to human sound, while historically playing a significant cultural role, now occupies only one dimension of musical sound.

This history, however, is interesting. There have been tendencies in Western musical history to restrict to or at least to develop precisely along horizontally variant ways. The restriction of musical sound to actual human voices (certain Mennonite sects do not allow any musical instrumentation, and all hymn singing is done a cappella) is a form of this tendency. Instrumentation that mimics or actually amplifies vocal sounds and their ranges is another example: woodwinds, horns, organs (even to the organ stop titles which are usually voice analogues) – all are ancient instruments that often deliberately followed a kind of vocal isomorphism. Medieval music was often doubly constrained. Not only must the music remain within the range of human similitude, but even the normatively controlled harmonics and chant lines were religio-culturally constrained. Later, one could detect a much more vocal model to much Italian (Renaissance through Baroque) music in contrast to a more instrumentally oriented model in German music.

The implicit valuational model of the human voice was also reflected in the music history of the West by the ranking of instruments by *expressivity*, with those instruments thought most expressive – the violin, for example – rated more highly than those farther from the vocal model.

The difference between embodiment and hermeneutic relations appears within this context as well. While embodiment relations in the most general existential sense need not be strictly constrained by isomorphism, hermeneutic variants occur very quickly along the musical spectrum. The piano retains little vocal isomorphism; yet when played, it falls into the embodiment relation, is expressive of the individual style and attainment of the performer, etc. Farther along the continuum, computer-produced music clearly occurs much more fully within the range of hermeneutic relations, in some cases with the emergence of random-sound generation very close to the sense of *otherness*, which will characterize the next set of relations where the technology emerges as *other*.

Instrumental music, as technics, may go in either embodiment or hermeneutic directions. It may develop its instrumentation in both vertical and horizontal trajectories. In either direction there are recognizable clear, technological transformations. If the Western "bionic" model of much early music was voice, in Andean music it was bird song (both in melody and in sound quality produced by breathy wood flutes). Contrarily, percussion instrumentation (drum music and communication) was, from the outset, a movement in a vertical and thus more hermeneutic direction. This exploration of possibility trees in horizontal and vertical directions belongs to the realm of musical praxis as much as to scientific, but is without any referentiality to a natural world.

The result of technological development in musical technics is also suggestively different from its result in scientific praxis. The "world" produced muscially through all the technical adumbrations is *not* that suggested either by the new philosophy of science or by a Heideggerian philosophy of technology. The closest analogy to the notion of standing reserve (resource well) that the musical "world" might take is that the realm of all possible sound may be taken and/or transformed musically. But the acoustical resources of musical technics are utilized through the creative sense of *play* which pervades musical praxis. The "musical object" is a created object, but its creation is not constrained by the same imperatives of scientific praxis. Yet the materialization of musical sound *through* instrumentation remains a fully human technological form of action.

What can be glimpsed in this detour into musical instrumentation is that while the human-technology structures are parallel with those found within scientific instrumentation, the "world" created does not at all imply the same reduction to what has been claimed as the unique Western view of the domination of nature. Here, then, is an opening to a different possible trajectory of development.

C Alterity Relations

Beyond hermeneutic relations there lie *alterity relations*. The first suggestions of such relations, which I shall characterize as relations *to* or *with* a technology, have already been suggested in different ways from within the embodiment and hermeneutic contexts. Within embodiment relations, were the technology to intrude upon rather than facilitate one's perceptual and bodily extension into the world, the technology's objectness would necessarily have appeared negatively. Within hermeneutic relations, however, there emerged a certain positivity to the objectness of instrumental technologies. The bodily-perceptual focus *upon* the instrumental text is a condition of its own peculiar heremeneutic transparency. But what of a positive or presentential sense of relations with technologies? In what phenomenological senses can a technology be *other*?

The analysis here may seem strange to anyone limited to the habits of objectivist accounts, for in such accounts technologies as objects usually come first rather than last. The problem for a phenomenological account is that objectivist ones are non-relativistic and thus miss or submerge what is distinctive about human-technology relations.

A naive objectivist account would likely begin with some attempt to circumscribe or define technologies by object characteristics. Then, what I have called the technical properties of technologies would become focal. Some combination of physical and material properties would be taken to be definitional. (This is an inherent tendency of the standard nomological positions such as those of Bunge and Hacking). The definition will often serve a secondary purpose by being stipulative: only those technologies that are obviously dependent upon or strongly related to contemporary scientific and industrial productive practices will count.

This is not to deny that objectivist accounts have their own distinctive strengths. For example, many such accounts recognize that technological or "artificial" products are different from the simply found object or the natural object. But the submergence of the human-technology relation remains hidden, since either object may enter into praxis and both will have their material, and thus limited, range of technical usability within the relation. Nor is this to deny that the objectivist accounts of types of technologies, types of organization, or types of designed purposes should be considered. But the focus in this first program remains the phenomenological derivation of the set of human-technology relations.

There is a tactic behind my placing alterity relations last in the order of focal human-technology relations. The tactic is designed, on the one side, to circumvent the tendency succumbed to by Heidegger and his more orthodox followers to see the otherness of technology only in negative terms or through negative derivations. The hammer example, which remains paradigmatic for this approach, is one that derives objectness from breakdown. The broken or missing or malfunctioning technology could be *discarded*. From being an obtrusion it could become *junk*. Its objectness would be clear – but only partly so. Junk is not a focal object of use relations (except in certain limited situations). It is more ordinarily a background phenomenon, that which has been put out of use.

Nor, on the other side, do I wish to fall into a naively objectivist account that would simply concentrate upon the material properties of the technology as an object of knowledge. Such an account would submerge the relativity of the intentionality analysis, which I wish to preserve here. What is needed is an analysis of the positive or presentential senses in which humans relate to technologies as relations *to* or with technologies, to technology-as-other. It is this sense which is included in the term "alterity."

Philosophically, the term "alterity" is borrowed from Emmanuel Levinas. Although Levinas stands within the traditions of phenomenology and hermeneutics, his distinctive work, *Totality and Infinity*, was "anti-Heideggerian." In that work, the term "alterity" came to mean the radical difference posed to any human by another human, an *other* (and by the ultimately other, God). Extrapolating radically from within the tradition's emphasis upon the non-reducibility of the human to either objectness (in epistemology) or as a means (in ethics), Levinas poses the otherness of humans as a kind of *infinite* difference that is concretely expressed in an ethical, face-to-face encounter.

I shall retain but modify this radical Levinasian sense of human otherness in returning to an analysis of human-technology relations. How and to what extent do technologies become other or, at least, *quasi-other*? At the heart of this question lie a whole series of well-recognized but problematic interpretations of technologies. On the one side lies the familiar problem of anthropomorphism, the personalization of artifacts. This range of anthropomorphism can reach from serious artifact-human analogues to trivial and harmless affections for artifacts.

An instance of the former lies embedded in much AI research. To characterize computer "intelligence" as human-like is to fall into a peculiarly contemporary species of anthropomorphism, however sophisticated. An instance of the latter is to find oneself "fond" of some particular technofact as, for instance, a long-cared-for automobile which one wishes to keep going and which may be characterized by quite deliberate anthropomorphic terms. Similarly, in ancient or non-Western cultures, the role of sacredness attributed to artifacts exemplifies another form of this pehnomenon.

The religious object (idol) does not simply "represent" some absent power but is endowed with the sacred. Its aura of sacredness is spatially and temporally present within the range of its efficacy. The tribal devotee will defend, sacrifice to, and care for the sacred artifact. Each of these illustrations contains the seeds of an alterity relation.

A less direct approach to what is distinctive in human-technology alterity relations may perhaps better open the way to a phenomenologically relativistic analysis. My first example comes from a comparison to a technology and to an animal "used" in some practical (although possibly sporting) context: the spirited horse and the spirited sports car.

To ride a spirited horse is to encounter a lively animal *other*. In its pre- or nonhuman context, the horse has a life of its own within the environment that allowed this form of life. Once domesticated, the horse can be "used" as an "instrument" of human praxis – but only to a degree and in a way different from counterpart technologies; in this case, the "spirited" sports car.

There are, of course, analogues which may at first stand out. Both horse and car give the rider/driver a magnified sense of power. The speed and the experience of speed attained in riding/driving are dramatic extensions of my own capacities. Some prominent features of embodiment relations can be found analogously in riding/driving. I experience the trail/road through horse/car and guide/steer the mediating entity under way. But there are equally prominent differences. No matter how well trained, no horse displays the same "obedience" as the car. Take malfunction: in the car, a malfunction "resists" my command – I push the accelerator, and because of a clogged gas line, there is not the response I expected. But the animate resistance of a spirited horse is more than such a mechanical lack of response – the response is more than malfunction, it is *dis*obedience. (Most experienced riders, in fact, prefer spirited horses over the more passive ones, which might more nearly approximate a mechanical obedience.) This life of the other in a horse may be carried much further – it may live without me in the proper environment; it does not need the *deistic* intervention of turning the starter to be "animated." The car will not shy at the rabbit springing up in the path any more than most horses will obey the "command" of the driver to hit the stone wall when he is too drunk to notice. The horse, while approximating some features of a mediated embodiment situation, never fully enters such a relation in the way a technology does. Nor does the car ever attain the sense of animation to be found in horseback riding. Yet the analogy is so deeply embedded in our contemporary consciousness (and perhaps the lack of sufficient experience with horses helps) that we might be tempted to emphasize the similarities rather than the differences.

Anthropomorphism regarding the technology on the one side and the contrast with horseback riding on the other point to a first approximation to the unique type of otherness that relations to technologies hold. Technological otherness is a *quasi-otherness*, stronger than mere objectness but weaker than the otherness found within the animal kingdom or the human one; but the phenomenological derivation must center upon the positive experiential aspects outlining this relation.

In yet another familiar phenomenon, we experience technologies as *toys* from childhood. A widely cross-cultural example is the spinning top. Prior to being put into use, the top may appear as a top-heavy object with a certain symmetry of design (even early tops approximate the more purely functional designs of streamlining, etc.), but once "deistically" animated through either stick motion or a string spring, the now spinning top appears to take on a life of its own. On its tip (or "foot") the top appears to defy its top-heaviness and gravity itself. It traces unpredictable patterns along its pathway. It is an object of *fascination*.

Note that once the top has been set to spinning, what was imparted through an embodiment relation now exceeds it. What makes it fascinating is this property of quasi-animation, the life of its own. Also, of course, once "automatic" in its motion, the top's movements may be entered into a whole series of possible contexts. I might enter a game of warring tops in which mine (suitably marked) represents me. If I-as-top am successful in knocking down the other tops, then this game of hermeneutics has the top winning for me. Similarly, if I take its quasi-autonomous motion to be a hermeneutic predictor, I may enter a divination context in which the path traced or the eventual point of stoppage indicates some fortune. Or, entering the region of scientific instrumentation, I may transform the top into a gyroscope, using its constancy of direction within its now-controlled confines as a better-than-magnetic compass. But in each of these cases, the top may become the focal center of attention as a quasi-other to which I may relate. Nor need the object of fascination carry either an embodiment or hermeneutic referential transparency.

To the ancient and contemporary top, compare briefly the fascination that occurs around video games. In the actual use of video games, of course, the embodiment and hermeneutic relational dimensions are present. The joystick that embodies hand and eye coordination skills extends the player into the displayed field. The field itself displays some hermeneutic context (usually either some "invader" mini-world or some sports analogue), but this context does not refer beyond itself into a worldly reference.

In addition to these dimensions, however, there is the sense of *interacting with* something other than me, the technological *competitor*. In competition there is a kind of dialogue or exchange. It is the quasi-animation, the quasi-otherness of the technology that fascinates and challenges. I must beat the machine or it will beat me.

In each of the cases mentioned, features of technological alterity have shown themselves. The quasi-otherness, the quasi-autonomy which appears in the toy or the game is a variant upon the technologies that have fascinated Western thinkers for centuries, the *automaton*.

The most sophisticated Greek (and similarly, Chinese) technologies did not appear in practical or scientific contexts so often as in game or theatrical ones. (War contexts, of course, have always employed advanced technologies.) Within these contexts, automatons were devised. From rediscovered treatises by Hero of Alexandria on pneumatics and hydraulics (which had in the second century B.C. already been used for humorous applications), the Renaissance builders began to construct various automata. The applications of Hero had been things like automatically opening temple doors and artificial birds that sang through steam whistles. In the Renaissance reconstructions, automata became more complex, particularly in fountain systems:

> The water garden of the Villa d'Este, built in 1550 at Tivoli, outside Rome, for the son of Lucrezia Borgia [was the best known]. The slope of the hill was used to supply fountains and dozens of grottos where water-powered figures moved and played and spouted.... The Chateau Merveilleux of Helbrun...is full of performing figures of men and women where fountains turn on and off unexpectedly or, operating in the intricate and quite amazing theatre of puppets, run by water power.[5]

The rage for automata was later to develop in a number of directions from music machines, of which the Deutsches Museum in Munich has a grand collection, to Vaucanson's automated duck which quacked, ate, drank, and excreted.[6] Much later, automation techniques were used in more practical contexts, although versions of partially automated looms for textiles did begin to appear in the eighteenth century (Vaucanson, the maker of the automated duck, invented the holed cylinder that preceded the punch-card system of the Jacquard loom).

Nor should the clock be exempted from this glance at automata fascination. The movements of the heavens. of the march of life and death, and of the animated figures on the clocks of Europe were other objects of fascination that seemed to move "autonomously." The superficial aspects of automation, the semblance of the animate and the similitude of the human and animal, remained the focus for even more serious concerns with automatons. That which is more "like" us seemed to center the fascination and make the alterity more quasi-animate.

Fascination may hide what is reductive in technological selectivities. But it may also hide, doubly, a second dimension of an instrumental intentionality, its possible dissimilarity direction, which may often prove in the longer run the more interesting trajectory of development. Yet semblance usually appears to be the first focus.

It was this *semblance* which became a worry for Modern (seventeenth and eighteenth century) Philosophy. Descartes's famous doubts also utilize the popular penchant for automata. In seeking to prove that it is the mind alone and not the eyes that know things, he argues:

> I should forthwith be disposed to conclude that the wax is known by the act of sight and not by the intuition of the mind alone were it not for the analogous instance of human beings passing on in the street below, as observed from a window. In this case, I do not fail to say that I see the men themselves, just as I say that I see the wax; and yet, what do I see from the window beyond hats and cloaks that might cover artificial machines, whose motions might be determined by springs?[7]

This can-I-be-fooled-by-a-cleverly-conceived-robot argument was to have an exceedingly long history, even into the precincts of contemporary analytic philosophies.

Were Descartes to become a contemporary of current developments in the attempt to mimic animal and human motions by automata, he might well rethink his illustration. Not only spring-run automata but also the most sophisticated computer-run automata look mechanical. These most sophisticated computer-run automata have difficulty maneuvering in anything like a lifelike motion. As Dreyfus has pointed out and as would be confirmed by many current researchers, bodily motion is perhaps harder to imitate than certain "mental" activities such as calculating.

To follow only the inclination towards similitude, however, is to reduce what may be learned from our relations with technologies. The current state of the art in AI research, for example, while having been partially freed from its earlier fundamentalistic state, remains primarily within the aim of creating similarities with human intelligence or modeling what are believed to be analogues to our intelligence. Yet it might well be that the *differences* that emerge from computer experimentation may be more informative or, at least, as informative as the similitudes.

There are what I shall call technological *intentionalities* that emerge from many technologies. Let us engage in a pseudo-Cartesian, imaginative construction of a humanoid robot, within the limits of easily combinable and available technologies, to take account of the similarity/difference structures which may be displayed. I shall begin with the technology's "perceptions" of sensory equipment: What if the robot were to hear? The inventor, perhaps limited by a humanist's budget, could install an omnidirectional microphone for ears. We could check upon what our robot would "hear" by adding a cassette player for a recorded "memory" of its "hearing." What is heard would turn out to be very differently structured, to have a very different form of intentionality than what any human listener would hear.

Assume that our robot is attending a university lecture in a large hall and is seated, as a shy student might be, near the rear. Given the limits of the mentioned technology, what would be heard would fail to have either the foreground/background pattern of human listening or the selective elimination of noise that even ordinary listening displays. The robot's auditory memory, played back, would reveal something much more like a sense-data auditory world than the one we are familiar with. The lecturer's voice, though recorded and within low limits perhaps detectable, would often be buried under the noise and background sounds that are selectively masked by human listening. For other purposes, precisely this differently structured technological intentionality could well be useful and informative. Such a different auditory selectivity could perhaps give clues to better architectural dampening of sounds precisely because what is repressed in human listening here stands out. In short, there is "truth" to be found in both the similarity and the difference that technological intentionalities reveal.

A similar effect could be noted with respect to the robot's vision. Were its eyes to be made of television equipment and the record or memory of what it has seen displayed on a screen, we would once again note the flatness of its visual field. Depth phenomena would be greatly reduced or would disappear. Although we have become accustomed to this flat field in watching television, it is easy to become reaware of the lack of depth between the baseball pitcher and the batter upon the screen. The technological shape of intentionality differs significantly from its human counterpart.

The fascination with human or animate similitude within the realm of alterity relations is but another instance of the types of fascination pervading our relations with technologies. The astonishment of Galileo at what he saw through the telescope was, in effect, the location of similitude within embodiment use. The magnification was the magnification *of* human visual capacity and remained within the range of what was familiarly visible. The horizontal trajectory of magnification that can more and more enhance vision is a trajectory along an already familiar praxis.

With the examples of fascination with automata, the fascination also remains within the realm of the familiar, now in a kind of mirror phenomenon for humans and the technology. Of all the animals in the earth's realm, it seems that the human ones are those who can prolong this fascination the most intensely. Paul Levinson, in an examination of the history of media technologies, has argued that there are three stages through which technologies pass. The first is that of technology as toy or novelty. The history of film technology is instructive:

> The first film makers were not artists but tinkers...."Their goal in making a movie was not to create beauty but to display a scientific curiosity." A survey of the early "talkies" like *The Jazz Singer*, first efforts in animation such as Disney's "Laugh-O-Gram" cartoons, and indeed the supposed debut of the motion picture in *Fred Ott's Sneeze* supports [this thesis] itself.[8]

The same observation could be made about much invention. But once taken more seriously, novelty can be transformed into a second stage, according to Levinson: that of technology as mirror of reality. This too happened in the history of film. Following the early curiosities at the onset of the film industry, the introduction of the Lumieres' presentation of "actualities" were, in part, fascinating precisely through the magnification/reduction selectivities that film technologies produce through unique film intentionalities. Examples could be as mundane as "workers leaving a factory, a baby's meal, and the famous train entering the station." What made such cinemas vérités dramatic were "in this case, a real train chugging into a real station, at an angle such that the audience could almost believe the train was chugging at *them*."[9]

This mirror of life, like the automaton, is not isomorphic with non-technological experience but is technologically transformed with the various effects that exaggerate or enhance some effects while simultaneously reducing others. Levinson is quite explicit in his analysis concerning the ways newly introduced technologies also enhance this development:

> The growth of film from gimmick to replicator was apparently in large part dependent upon a new technological component....The "toy" film played to individuals who peeked into individual kinetoscopes; but the "reality" film reached out to mass audiences, who viewed the reality-surrogate in group theatres. The connection between mass audiences and reality simulation, moreover, was no accident. Unlike the perception of novelties, which is inherently subjective and individualized, reality perception is a fundamentally objective, group process.[10]

Although the progression of the analysis here moves from embodiment and hermeneutic relations to alterity ones, the interjection of film or cinema examples is of suggestive interest. Such technologies are transitional between hermeneutic and alterity phenomena. When I first introduced the notion of hermeneutic relations, I employed what could be called a "static" technology: writing. The long and now ancient technologies of writing result in fixed texts (books, manuscripts, etc., all of which, barring decay or destruction, remain stable in themselves). With film, the "text" remains fixed only in the sense that one can repeat, as with a written text, the seeing and hearing of the cinema text. But the mode of presentation is dramatically different. The "characters" are now animate and theatrical, unlike the fixed alphabetical characters of the written text. The dynamic "world" of the cinema-text, while retaining many of the functional features of writing, also now captures the semblance of real-time, action, etc. It remains to be "read" (viewed and heard), but the object-correlate necessarily appears more "life-like" than its analogue – written text. This factor, naively experienced by the current generations of television addicts, is doubtless one aspect in the problems that emerge between television watching habits and the state of reading skills. James Burke has pointed out that "the majority of the people in the advanced industrialized nations spend more time watching television than doing anything else beside work."[11] The same balance of time use also has shown up in surveys regarding students. The hours spent watching television among college and university students, nationally, are equal to or exceed those spent in doing homework or out-of-class preparation.

Film, cinema, or television can, in its hermeneutic dimension, refer in its unique way to a "world." The strong negative response to the Vietnam War was clearly due in part to the virtually unavoidable "presence" of the war in virtually everyone's living room. But films, like readable technologies, are also *presentations*, the focal terminus of a perceptual situation. In that emergent sense, they are more dramatic forms of perceptual immediacy in which the presented display has its own characteristics conveying quasi-alterity. Yet the engagement with the film normally remains short of an engagement with an *other*. Even in the anger that comes through in outrage about civilian atrocities or the pathos experienced in seeing starvation epidemics in Africa, the emotions are not directed to the screen but, indirectly, through it, in more appropriate forms of political or charitable action. To this extent there is retained a hermeneutic reference elsewhere than at the technological instrument. Its quasi-alterity, which is also present, is not fully focal in the case of such media technologies.

A high-technology example of breakdown, however, provides yet another hint at the emergence of alterity phenomena. Word processors have become familiar technologies, often strongly liked by their users (including many philosophers who fondly defend their choices, profess knowledge about the relative abilities of their machines and programs, etc.). Yet in breakdown, this quasi-love relationship reveals its quasi-hate underside as well. Whatever form of "crash" may occur, particularly if some fairly large section of text is involved, it occasions frustration and even rage. Then, too, the programs have their idiosyncracies, which allow or do not allow certain movements; and another form of human-technology competition may emerge. (Mastery in the highest sense most likely comes from learning to program and thus overwhelm the machine's previous brainpower. "Hacking" becomes the game-like competition in which an entire system is the alterity correlate.) Alterity relations may be noted to emerge in a wide range of computer technologies that, while failing quite strongly to mimic bodily incarnations, nevertheless display a quasi-otherness within the limits of linguistics and, more particularly, of logical behaviors. Ultimately, of course, whatever contest emerges, its sources lie opaquely with other humans as well but also with the transformed technofact, which itself now plays a more obvious role within the overall relational net.

I have suggested that the computer is one of the stronger examples of a technology which may be positioned within alterity relations. But its otherness remains a quasi-otherness, and its genuine usefulness still belongs to the borders of its hermeneutic capacities. Yet in spite of this, the tendency to fantasize its quasi-otherness into an authentic otherness is pervasive. Romanticizations such as the portrayal of the emotive, speaking "Hal" of the movie *2001: A Space Odyssey*, early fears that the "brain power" of computers would soon replace human thinking, fears that political or military decisions will not only be informed by but also made by computers – all are symptoms revolving around the positing of otherness to the technology.

These romanticizations are the alterity counterparts to the previously noted dreams that wish for total embodiment. Were the technofact to be genuinely an other, it would both be and not be a *technology*. But even as quasi-other, the technology falls short of such totalization. It retains its unique role in the human-technology continuum of relations as the medium of transformation, but as a recognizable medium.

The wish-fulfillment desire occasioned by embodiment relations – the desire for a fully transparent technology that would *be* me while at the same time giving me the powers that the use of the technology makes available – here has its counterpart fantasy, and this new fantasy has the same internal contradiction: It both reduces or, here, extrapolates the technology into that which is not a technology (in the first case, the magical transformation is *into me*; in this case, *into the other*), and at the same time, it desires what is not identical with me or the other. The fantasy is for the transformational effects. Both fantasies, in effect, deny technologies playing the roles they do in the human-technology continuum of relations; yet it is only on the condition that there be some detectable differentiation within the relativity that the unique ways in which technologies transform human experience can emerge.

In spite of the temptation to accept the fantasy, what the quasi-otherness of alterity relations does show is that humans may relate positively or presententially *to* technologies. In that respect and to that degree, technologies emerge as focal entities that may receive the multiple attentions humans give the different forms of the other. For this reason, a third formalization may be employed to distinguish this set of relations:

1 → technology–(–world)

I have placed the parentheses thusly to indicate that in alterity relations there may be, but need not be, a relation through the technology to the world (although it might well be expected that the *usefulness* of any technology will necessarily entail just such a referentiality). The world, in this case, may remain context and background, and the technology may emerge as the foreground and focal quasi-other with which I momentarily engage.

This disengagement of the technology from its ordinary-use context is also what allows the technology to fall into the various disengaged engagements which constitute such activities as play, art, or sport.

A first phenomenological itinerary through direct and focal human-technology relations may now be considered complete. I have argued that the three sets of distinguishable relations occupy a continuum. At the one extreme lie those relations that approximate technologies to a quasi-me (embodiment relations). Those technologies that I can so take into my experience that through their semi-transparency they allow the world to be made immediate thus enter into the existential relation which constitutes my self. At the other extreme of the continuum lie alterity relations in which the technology becomes quasi-other, or technology "as" other *to* which I relate. Between lies the relation with technologies that both mediate and yet also fulfill my perceptual and bodily relation with technologies, hermeneutic relations. The variants may be formalized thus:

Human-technology–World Relations
Variant 1, Embodiment Relations
(Human-technology) → World
Variant 2, Hermeneutic Relations
Human → (technology–World)
Variant 3, Alterity Relations
Human → technology–(–World)

Although I have characterized the three types of human-technology relations as belonging to a continuum, there is also a sense in which the elements within each type of relation are differently distributed. There is a *ratio* between the objectness of the technology and its transparency in use. At the extreme height of embodiment, a background presence of the technology may still be detected. Similarly but with a different ratio, once the technology has emerged as a quasi-other, its alterity remains within the domain of human invention through which the world is reached. Within all the types of relations, technology remains artifactual, but it is also its very artifactual formation which allows the transformations affecting the earth and ourselves.

All the relations examined heretofore have also been focal ones. That is, each of the forms of action that occur through these relations have been marked by an implicated self-awareness. The engagements through, with, and to technologies stand within the very core of praxis. Such an emphasis, while necessary, does not exhaust the role of technologies nor the experiences of them. If focal activities are central and foreground, there are also fringe and background phenomena that are no more neutral than those of the foreground.

[...]

Notes

1 Sergei Eisenstein, *Film Form: Essays in Film Theory*, ed. and trans. Jay Leyda (New York: Harcourt, Brace and World, 1949), p. 29.

2 This illustration is my version of a similar one developed by Patrick Heelan in his more totally hermeneuticized notion of perception in *Space Perception and the Philosophy of Science* (Berkeley: University of California Press, 1983), p. 193.

3 Ian Hacking, *Representing and Intervening* (Cambridge: Cambridge University Press, 1983), p. 195. Hacking develops a very excellent and suggestive history of the use of microscopes. His focus, however, is upon the technical properties that were resolved before microscopes could be useful in the sciences. He and Heelan, however, along with Robert Ackermann, have been among the pioneers dealing with perception and

instrumentation in instruments. Cf. also my *Technics and Praxis* (Dordrecht: Reidel Publishers, 1979).

4 Emmanuel Mournier, *Be Not Afraid*, trans. Cynthia Rowland (London: Rockcliffe, 1951), p. 195.

5 James Burke, *Connections* (Boston: Little, Brown, 1978), p. 106.

6 Ibid., p. 107.

7 René Descartes, *A Discourse on Method*, trans. John Veitch (London: J. M. Dent, 1953), p. 92.

8 Paul Levinson, "Toy, Mirror and Art: The Metamorphosis of Technological Culture," in *Philosophy, Technology and Human Affairs*, ed. Larry Hickman (College Station: Ibis Press, 1985), p. 163.

9 Ibid., p. 165.

10 Ibid., p. 167.

11 Burke, *Connections*, p. 5.

44

Technical Progress and the Social Life-World

Jürgen Habermas

When C. P. Snow published *The Two Cultures* in 1959, he initiated a discussion of the relation of science and literature which has been going on in other countries as well as in England. Science in this connection has meant the strictly empirical sciences, while literature has been taken more broadly to include methods of interpretation in the cultural sciences. The treatise with which Aldous Huxley entered the controversy, however, *Literature and Science*, does limit itself to confronting the natural sciences with the belles-lettres.

Huxley distinguishes the two cultures primarily according to the specific experiences with which they deal: literature makes statements mainly about private experiences, the sciences about intersubjectively accessible experiences. The latter can be expressed in a formalized language, which can be made universally valid by means of general definitions. In contrast, the language of literature must verbalize what is in principle unrepeatable and must generate an intersubjectivity of mutual understanding in each concrete case. But this distinction between private and public experience allows only a first approximation to the problem. The element of ineffability that literary expression must overcome derives less from a private experience encased in subjectivity than from the constitution of these experiences within the horizon of a life-historical environment. The events whose connection is the object of the law-like hypotheses of the sciences can be described in a spatio-temporal coordinate system, but they do not make up a world:

> The world with which literature deals is the world in which human beings are born and live and finally die; the world in which they love and hate, in which they experience triumph and humiliation, hope and despair; the world of sufferings and enjoyments, of madness and common sense, of silliness, cunning and wisdom; the world of social pressures and individual impulses, of reason against passion, of instincts and conventions, of shared language and unsharable feelings and sensations...[1]

In contrast, science does not concern itself with the contents of a life-world of this sort, which is culture-bound, ego-centered, and pre-interpreted in the ordinary language of social groups and socialized individuals:

> ...As a professional chemist, say, a professional physicist or physiologist, [the scientist] is the inhabitant of a radically different universe – not the universe of given appearances, but the world of inferred fine structures, not the experienced world of unique events and diverse qualities, but the world of quantified regularities. (LS, p. 8)

Huxley juxtaposes the *social life-world* and the *worldless universe of facts*. He also sees precisely

From Jürgen Habermas, *Toward a Rational Society: Student Protest, Science and Politics*, trans. Jeremy J. Shapiro, Boston: Beacon, 1970, pp. 50–61.

the way in which the sciences transpose their information about this worldless universe into the life-world of social groups:

> Knowledge is power and, by a seeming paradox, it is through their knowledge of what happens in this unexperienced world of abstractions and inferences that scientists have acquired their enormous and growing power to control, direct, and modify the world of manifold appearances in which human beings are privileged and condemned to live. (LS, p. 10)

But Huxley does not take up the question of the relation of the two cultures at this juncture, where the sciences enter the social life-world through the technical exploitation of their information. Instead he postulates an immediate relation. Literature should assimilate scientific statements as such, so that science can take on "flesh and blood."

> ...Until some great artist comes along and tells us what to do, we shall not know how the muddled words of the tribe and the too precise words of the textbooks should be poetically purified, so as to make them capable of harmonizing our private and unsharable experiences with the scientific hypotheses in terms of which they are explained. (LS, p. 107)

This postulate is based, I think, on a misunderstanding. Information provided by the strictly empirical sciences can be incorporated in the social life-world only through its technical utilization, as technological knowledge, serving the expansion of our power of technical control. Thus, such information is not on the same level as the action-orienting self-understanding of social groups. Hence, without mediation, the information content of the sciences cannot be relevant to that part of practical knowledge which gains expression in literature. It can only attain significance through the detour marked by the practical results of technical progress. Taken for itself, knowledge of atomic physics remains without consequence for the interpretation of our life-world, and to this extent the cleavage between the two cultures is inevitable. Only when with the aid of physical theories we can carry out nuclear fission, only when information is exploited for the development of productive or destructive forces, can its revolutionary practical results penetrate the literary consciousness of the life-world: poems arise from consideration of Hiroshima and not from the elaboration of hypotheses about the transformation of mass into energy.

The idea of an atomic poetry that would elaborate on hypotheses follows from false premises. In fact, the problematic relation of literature and science is only one segment of a much broader problem: *How is it possible to translate technically exploitable knowledge into the practical consciousness of a social life-world?* This question obviously sets a new task, not only or even primarily for literature. The skewed relation of the two cultures is so disquieting only because, in the seeming conflict between the two competing cultural traditions, a true life-problem of scientific civilization becomes apparent: namely, how can the relation between technical progress and the social life-world, which today is still clothed in a primitive, traditional, and unchosen form, be reflected upon and brought under the control of rational discussion?

To a certain extent practical questions of government, strategy, and administration had to be dealt with through the application of technical knowledge even at an earlier period. Yet today's problem of transposing technical knowledge into practical consciousness has changed not merely its order of magnitude. The mass of technical knowledge is no longer restricted to pragmatically acquired techniques of the classical crafts. It has taken the form of scientific information that can be exploited for technology. On the other hand, behavior-controlling traditions no longer naively define the self-understanding of modern societies. Historicism has broken the natural-traditional validity of action-orienting value systems. Today, the self-understanding of social groups and their worldview as articulated in ordinary language is mediated by the hermeneutic appropriation of traditions as traditions. In this situation questions of life conduct demand a rational discussion that is not focused exclusively either on technical means or on the application of traditional behavioral norms. The reflection that is required extends beyond the production of technical knowledge and the hermeneutical clarification of traditions to the employment of technical means in historical situations whose objective conditions (potentials, institutions, interests) have to be interpreted anew each time in the framework of a self-understanding determined by tradition.

This problem-complex has only entered consciousness within the last two or three generations.

In the nineteenth century one could still maintain that the sciences entered the conduct of life through two separate channels: through the technical exploitation of scientific information and through the processes of individual education and culture during academic study. Indeed, in the German university system, which goes back to Humboldt's reform, we still maintain the fiction that the sciences develop their action-orienting power through educational processes within the life history of the individual student. I should like to show that the intention designated by Fichte as a "transformation of knowledge into works" can no longer be carried out in the private sphere of education, but rather can be realized only on the politically relevant level at which technically exploitable knowledge is translatable into the context of our life-world. Though literature participates in this, it is primarily a problem of the sciences themselves.

At the beginning of the nineteenth century, in Humboldt's time, it was still impossible, looking at Germany, to conceive of the scientific transformation of social life. Thus, the university reformers did not have to break seriously with the tradition of practical philosophy. Despite the profound ramifications of revolutions in the political order, the structures of the preindustrial work world persisted, permitting for the last time, as it were, the classical view of the relation of theory to practice. In this tradition, the technical capabilities employed in the sphere of social labor are not capable of immediate direction by theory. They must be pragmatically practiced according to traditional patterns of skill. Theory, which is concerned with the immutable essence of things beyond the mutable region of human affairs, can obtain practical validity only by molding the manner of life of men engaged in theory. Understanding the cosmos as a whole yields norms of individual human behavior, and it is through the actions of the philosophically educated that theory assumes a positive form. This was the only relation of theory to practice incorporated in the traditional idea of university education. Even where Schelling attempts to provide the physician's practice with a scientific basis in natural philosophy, the medical *craft* is unexpectedly transformed into a medical *praxiology*. The physician must orient himself to Ideas derived from natural philosophy in the same way that the subject of moral action orients itself through the Ideas of practical reason.

Since then it has become common knowledge that the scientific transformation of medicine succeeds only to the extent that the pragmatic doctrine of the medical art can be transformed into the control of isolated natural processes, checked by scientific method. The same holds for other areas of social labor. Whether it is a matter of rationalizing the production of goods, management and administration, construction of machine tools, roads, or airplanes, or the manipulation of electoral, consumer, or leisure-time behavior, the professional practice in question will always have to assume the form of technical control of objectified processes.

In the early nineteenth century, the maxim that scientific knowledge is a source of culture required a strict separation between the university and the technical school because the preindustrial forms of professional practice were impervious to theoretical guidance. Today, research processes are coupled with technical conversion and economic exploitation, and production and administration in the industrial system of labor generate feedback for science. The application of science in technology and the feedback of technical progress to research have become the substance of the world of work. In these circumstances, unyielding opposition to the decomposition of the university into specialized schools can no longer invoke the old argument. Today, the reason given for delimiting study on the university model from the professional sphere is not that the latter is still foreign to science, but conversely, that science – to the very extent that it has penetrated professional practice – has estranged itself from humanistic culture. The philosophical conviction of German idealism that scientific knowledge is a source of culture no longer holds for the strictly empirical scientist. It was once possible for theory, via humanistic culture, to become a practical force. Today, theories can become technical power while remaining unpractical, that is, without being expressly oriented to the interaction of a community of human beings. Of course, the sciences now transmit a specific capacity: but the capacity for control, which they teach, is not the same capacity for life and action that was to be expected of the scientifically educated and cultivated.

The cultured possessed orientation in action. Their culture was universal only in the sense of the universality of a culture-bound horizon of a world in which scientific experiences could be interpreted and turned into practical abilities, namely, into a reflected consciousness of the practically necessary. The only type of experience which is admitted as scientific today according to positivistic criteria is not capable of this transposition into practice. The capacity for *control* made possible by the empirical sciences is not to be confused with the capacity for *enlightened action*. But is science, therefore, completely discharged of this task of action-orientation, or does the question of academic education in the framework of a civilization transformed by scientific means arise again today as a problem of the sciences themselves?

First, production processes were revolutionized by scientific methods. Then expectations of technically correct functioning were also transferred to those areas of society that had become independent in the course of the industrialization of labor and thus supported planned organization. The power of technical control over nature made possible by science is extended today directly to society: for every isolatable social system, for every cultural area that has become a separate, closed system whose relations can be analyzed immanently in terms of presupposed system goals, a new discipline emerges in the social sciences. In the same measure, however, the problems of technical control solved by science are transformed into life problems. For the scientific control of natural and social processes – in a word, technology – does not release men from action. Just as before, conflicts must be decided, interests realized, interpretations found – through both action and transaction structured by ordinary language. Today, however, these practical problems are themselves in large measure determined by the system of our technical achievements.

But if technology proceeds from science, and I mean the technique of influencing human behavior no less than that of dominating nature, then the assimilation of this technology into the practical life-world, bringing the technical control of particular areas within the reaches of the communication of acting men, really requires scientific reflection. The prescientific horizon of experience becomes infantile when it naively incorporates contact with the products of the most intensive rationality.

Culture and education can then no longer indeed be restricted to the ethical dimension of personal attitude. Instead, in the political dimension at issue, the theoretical guidance of action must proceed from a scientifically explicated understanding of the world.

The relation of technical progress and social life-world and the translation of scientific information into practical consciousness is not an affair of private cultivation.

I should like to reformulate this problem with reference to political decision-making. In what follows we shall understand "technology" to mean scientifically rationalized control of objectified processes. It refers to the system in which research and technology are coupled with feedback from the economy and administration. We shall understand "democracy" to mean the institutionally secured forms of general and public communication that deal with the practical question of how men can and want to live under the objective conditions of their ever-expanding power of control. Our problem can then be stated as one of the relation of technology and democracy: how can the power of technical control be brought within the range of the consensus of acting and transacting citizens?

I should like first to discuss two antithetical answers. The first, stated in rough outline, is that of Marxian theory. Marx criticizes the system of capitalist production as a power that has taken on its own life in opposition to the interests of productive freedom, of the producers. Through the private form of appropriating socially produced goods, the technical process of producing use values falls under the alien law of an economic process that produces exchange values. Once we trace this self-regulating character of the accumulation of capital back to its origins in private property in the means of production, it becomes possible for mankind to comprehend economic compulsion as an alienated result of its own free productive activity and then abolish it. Finally, the reproduction of social life can be rationally planned as a process of producing use values; society places this process under its technical control. The latter is exercised democratically in accordance with the will and insight of the associated individuals. Here Marx equates the practical insight of a political public with successful technical control. Meanwhile we have learned that even a well-functioning planning bureaucracy with scientific control of the production of goods and services is not a sufficient condition for realizing the associated material and intellectual productive forces in the interest of the enjoyment and freedom of an emancipated society. For Marx did not reckon with the possible emergence at every level of a discrepancy between scientific control of the material conditions of life and a democratic decision-making process. This is the philosophical reason why socialists never anticipated the authoritarian welfare state, where social wealth is relatively guaranteed while political freedom is excluded.

Even if technical control of physical and social conditions for preserving life and making it less burdensome had attained the level that Marx expected would characterize a communist stage of development, it does not follow that they would be linked automatically with social emancipation of the sort intended by the thinkers of the Enlightenment in the eighteenth century and the Young Hegelians in the nineteenth. For the techniques with which the development of a highly industrialized society could be brought under control can no longer be interpreted according to an instrumental model, as though appropriate means were being organized for the realization of goals that are either presupposed without discussion or clarified through communication.

Hans Freyer and Helmut Schelsky have outlined a counter-model which recognizes technology as an independent force. In contrast to the primitive state of technical development, the relation of the organization of means to given or preestablished goals today seems to have been reversed. The process of research and technology – which obeys immanent laws – precipitates in an unplanned fashion new methods for which we then have to find purposeful application. Through progress that has become automatic, Freyer argues, abstract potential continually accrues to us in renewed thrusts. Subsequently, both life interests and fantasy that generates meaning have to take this potential in hand and expend it on concrete goals. Schelsky refines and simplifies this thesis to the point of asserting that technical progress produces not only unforeseen methods but the unplanned goals and applications themselves: technical potentialities command their own practical realization. In particular, he puts forth this thesis with regard to the highly complicated objective exigencies that in political situations allegedly prescribe solutions without alternatives.

> Political norms and laws are replaced by objective exigencies of scientific-technical civilization, which are not posited as political decisions and cannot be understood as norms of conviction or weltanschauung. Hence, the idea of democracy loses its classical substance, so to speak. In place of the political will of the people emerges an objective exigency, which man himself produces as science and labor.

In the face of research, technology, the economy, and administration – integrated as a system that has become autonomous – the question prompted by the neohumanistic ideal of culture, namely, how can society possibly exercise sovereignty over the technical conditions of life and integrate them into the practice of the life-world, seems hopelessly obsolete. In the technical state such ideas are suited at best for "the manipulation of motives to help bring about what must happen anyway from the point of view of objective necessity."

It is clear that this thesis of the autonomous character of technical development is not correct. The pace and *direction* of technical development today depend to a great extent on public investments: in the United States the defense and space administrations are the largest sources of research contracts. I suspect that the situation is similar in the Soviet Union. The assertion that politically consequential decisions are reduced to carrying out the immanent exigencies of disposable techniques and that therefore they can no longer be made the theme of practical considerations, serves in the end merely to conceal preexisting, unreflected social interests and prescientific decisions. As little as we can accept the optimistic convergence of technology and democracy, the pessimistic assertion that technology excludes democracy is just as untenable.

These two answers to the question of how the force of technical control can be made subject to the consensus of acting and transacting citizens are inadequate. Neither of them can deal appropriately with the problem with which we are objectively confronted in the West and East, namely, how we can actually bring under control the preexisting, unplanned relations of technical progress and the social life-world. The tensions between productive forces and social intentions that Marx diagnosed and whose explosive character has intensified in an unforeseen manner in the age of thermonuclear weapons are the consequence of an ironic relation of theory to practice. The direction of technical progress is still largely determined today by social interests that arise autochthonously out of the compulsion of the reproduction of social life without being reflected upon and confronted with the declared political self-understanding of social groups. In consequence, new technical capacities erupt without preparation into existing forms of life-activity and conduct. New potentials for expanded power of technical control make obvious the disproportion between the results of the most organized rationality and unreflected goals, rigidified value systems, and obsolete ideologies.

Today, in the industrially most advanced systems, an energetic attempt must be made consciously to take in hand the mediation between technical progress and the conduct of life in the major industrial societies, a mediation that has previously taken place without direction, as a mere continuation of natural history. This is not the place to discuss the social, economic, and political conditions on which a long-term central research policy would have to depend. It is not enough for a social system to fulfill the conditions of technical rationality. Even if the cybernetic dream of a virtually instinctive self-stabilization could be realized, the value system would have contracted in the meantime to a set of rules for the maximization of power and comfort; it would be equivalent to the biological base value of survival at any cost, that is, ultrastability. Through the unplanned sociocultural consequences of technological progress, the human species has challenged itself to learn not merely to affect its social destiny, but to control it. This challenge of technology cannot be met with technology alone. It is rather a question of setting into motion a politically effective discussion that rationally brings the social potential constituted by technical knowledge and ability into a defined and controlled relation to our practical knowledge and will. On the one hand, such discussion could enlighten those who act politically about the tradition-bound self-understanding of their interests in relation to what is technically possible and feasible. On the other hand, they would be able to judge practically, in the light of their now articulated and newly interpreted needs, the direction and the extent to which they want to develop technical knowledge for the future.

This *dialectic of potential and will* takes place today without reflection in accordance with interests for which public justification is neither demanded nor permitted. Only if we could elaborate this dialectic with political consciousness could we succeed in directing the mediation of technical progress and the conduct of social life, which until now has occurred as an extension of natural history; its conditions being left outside the framework of discussion and planning. The fact that this is a matter for reflection means that it does not belong to the professional competence of specialists. The substance of domination is not dissolved by the power of technical control. To the contrary, the former can simply hide behind the latter. The irrationality of domination, which today has become a collective peril to life, could be mastered only by the development of a political decision-making process tied to the principle of general discussion free from domination. Our only hope for the rationalization of the power structure lies in conditions that favor political power for thought developing through dialogue. The redeeming power of reflection cannot be supplanted by the extension of technically exploitable knowledge.

Note

1 Aldous Huxley, *Literature and Science*. New York: Harper & Row, 1963, p. 8 (henceforth LS).

Technology and Cyberspace

Introduction

Of all the technologies that shape contemporary life, computer technology is arguably the most powerful and pervasive. We commonly speak of "the computer revolution," the rise of a global "computer culture," and "the age of [computer-processed] information." Not everyone, of course, is happy about this; but it is hard to find anyone ready to assure us that the computer's influence matters little. At the same time, however, aside from those who are selling their own wares, it is difficult to find anyone who rejoices in the coming of cyberculture without at least some qualifications. Representative of the more celebratory commentators is Michael Heim, who is the author of three books on the issue. The other selections in this section present, in varying degrees, less sanguine reports.

In the two selections taken from his *The Metaphysics of Virtual Reality*, Heim agrees that cyber-technology is indeed omnipresent and needs to be analyzed in the broadest cultural terms. Yet he warns that we must not rely entirely upon studies undertaken from an objective or outside observer's perspective, for this only tells us of the effects of cyber-technology *on* us. Echoing the authors in the previous section of part VI, Heim argues that real understanding comes when we see "what happens *to* us as we collaborate" with computer technology. Thus, he praises Dreyfus's *What Computers [Still] Can't Do* for the "existential questions" it raises against artificial intelligence (AI) researchers, who see the mind as nothing but an information processor. Yet he also complains that a critique of AI research conceives the central issue of computerization too narrowly. Using Heidegger's terms and drawing on his own *Electric Language: A Philosophical Study of Word Processing* (1987, 1999), Heim contrasts his approach with Dreyfus's by calling the latter merely "ontic" and his, "existential–ontological." The question, he says, is "how *in general*" does our reality change when we everywhere live and work with computers? Heim thinks that on this question, social theorists like Marshall McLuhan and Walter Ong are more appropriately inclusive, especially insofar as they see the intimate connection between the development of computer technology and its impact on language and communication. Yet Heim complains that McLuhan and Ong display the opposite problem from Dreyfus – namely, their theories lack "a feel for the trade-offs" that are inevitably part of historical transformations as radical as that being effected by computerization.

For Heim, it is in Heidegger that one finds a perspective that is both sufficiently inclusive and existentially sensitive enough to evaluate the "gains and losses" of computerization. In "The Essence of VR," he gives a loosely Heideggerian summary of the sort of gain-and-loss analysis of virtual reality elaborated in his recent book, *Virtual Realism* (2000). Where Heidegger asks about the "essence" of technology (Part IV), Heim appears to settle for something more like family resemblances. Taking his cue from divergent tendencies within VR research itself, Heim tries to discern an "esoteric essence" of VR by examining goals set by its various developers. Quoting MIT Media Lab's David Zeltzer, Heim calls this essence the "Holy Grail" of VR research and likens it to something like the fictional "Holodeck" from TV's "Star

Trek: The Next Generation." In such a room, says Heim, one can create "the ideal human–computer interface," that is, a simulated, interactive, artificial, completely manipulable "world" – a world, on the one hand, in which there is both total, bodily immersion and also mental independence sufficient to give commands and a world, on the other hand, from which you can "come and go" at will. For elaboration, Heim appeals to Richard Wagner's last opera, *Parsifal*, in part because of its portrayal of the Grail legend but in part also because Wagner regarded it as his realization of a "total artwork," a created, alternate reality in terms of which "ordinary" reality might be transformed. Not everyone, of course, has been as willing as Heim to make positive use of Wagner or of his idea of a *Gesamtwerk*. For a disillusioned Nietzsche, *Parsifal* was the nadir of Wagner's operas, and for some media theorists, the Wagnerian *Gesamtwerk* is even a precursor to and possible inspiration for Nazi mass rallies. Indeed, in general, a number of other authors in this collection would undoubtedly criticize Heim for being impossibly apolitical. For example, although he mentions several times the crucial role played by both governmental and military agencies in the development of computer technology during the height of the Cold War, no conclusions are drawn from these references when it comes time to explain VR's Holy Grail.

In some ways, then, Andrew Ross's "Hacking Away at the Counterculture" provides an illuminating contrast to Heim's position. This selection exemplifies Ross's "cultural studies" approach to technology. The fact that Ross, a literary figure uneducated in science, writes about New Age and countercultural rejections of science that still unintentionally appropriate it, has been used against him in the so-called "Science Wars" (a term coined by Ross himself). In the present selection, Ross applies his general thesis to ironic and contradictory tendencies among computer hackers. For if they clearly appreciate the new technology and even develop a technological subculture, they just as clearly reject the established norms of the professional culture of technology, particularly when those involve corporate and military secrecy. Ross goes on to show that establishment culture is similarly ambivalent in its characterization of hackers. On the one hand, they are a "menace" to society. Ross compares the proposed quarantining and isolation of hackers with the old responses to cholera epidemics, and he notes the similarity of media accounts of computer viruses and AIDS – both imply that the problem is "uncontrolled," free interaction of (undesirable) individuals. Popular magazines demonize the more successful hackers, and the courts zealously prosecute them. Yet precisely because the talents of the hackers are the talents of prodigies and even potential employees of computerized corporate and governmental institutions, their talents are also applauded – even if in the meantime hacking itself serves as a "convenient vehicle" for giving law enforcement and corporate managers new investigatory powers and creates a lucrative market for anti-viral software, security experts, copyright lawyers, and Internet watchdogs. Ross's ironic postmodernism puts his social theory in interesting contrast with what might be called the existential personalist models of Dreyfus and Borgmann, as well as with Haraway's information network model. For Ross, the Internet is neither a dehumanizing masscult phenomenon that must be humanized nor the beginning of an inevitable cyber-future; rather, it is a site of conflict between expressions of anarchistic individualism and corporate bureaucracy, a conflict in which these antithetical impulses nevertheless nourish each another in surprising ways. Ross is thus no friend of social pictures like Foucault's seamless web of panoptical surveillance or Haraway's all-encompassing military command, control, and communication system. Ross emphasizes that hackers after all do have the ability to disrupt the system, and that the system itself is deeply inefficient; hence, there is a moral imperative not to encourage the sort of passive hopelessness about the cybercultural system that Foucault's and Haraway's models would appear, even if unintentionally, to imply.

Borgmann's article on "Information and Reality" employs Shannon and Weaver's classic post-World-War II work on communication theory to characterize "information" in the postmodern age. Shannon and Weaver have been criticized by humanists for positing a notion of "information" as something always analyzable into measurable bits that is quite different from the traditional idea of "information" as the meaning of a message. Borgmann, however, finds Shannon and Weaver's account, somewhat unfortunately, appropriate to our "information age." The technological developments of the last 150 years, he says, have given rise to a conception of information that rivals and undercuts the old idea that information is either "about" or "for" some actual reality and as thus contributing either to knowing something or to knowing how to do something. Traditionally, information has been conceived according to a three-part model in which there is a messenger/signal/symbol/vehicle, a recipient/reader/audience, and a message/content/news/information. According to Borgmann, the assumption was that the purpose of all this theoretical and practical information is to "mediate" between ourselves and a real outside world, where this mediation is basically understood in terms of a distinction between a focal center and an intelligible background – that is, between near and far. Today, however, technological developments are leading to information itself coming to be conceived "as" reality. Modern theories like Shannon and Weaver's, for example, "erase the distance between near and far" by conceiving all our relations with things and persons as inevitably represented "through" information. Moreover, by conceiving information as ultimately analyzable into measurable bits, the old distinction between an objectively identifiable *quantity* (the signal) and a subjective and elusive *quality* (the message) is also erased. The result of these theoretical innovations – so crucial for computerization – is that all information, reduced to its basic units or bits, becomes capable of being coded, stored, and transmitted. The Internet, concludes Borgmann sadly, can serve as an "emblem" for the results: for the Net provides "information" in a free-floating and completely indiscriminate fashion, not especially "for" or "about" anything in particular; and "communication" (as, e.g., via e-mail) is anonymous, much more copious than in any face-to-face relation, and it encourages people to create versions of themselves with virtues they actually lack. Thus does the Net tend to promote the flimsy, the bland, and the offensive, "enfeebling" persons and "attenuating" things. Borgmann closes with a call for regaining a sense of the "actuality" of people and a turn away from the false realism of information to the genuine realism of persons and natural surroundings. Partisans of the realism of virtual reality would resist this turn, of course, reveling in precisely the "erasure" of the difference between information and the supposed object of the information that Borgmann laments. Postmodernists such as Haraway would also reject Borgmann's traditionalist appeal to a human essence that takes priority over the new electronic self. Can Borgmann's feelings about the way electronic communication creates a new cyborg being with new rather than traditional humanist virtues do anything to change these "facts"?

In "Anonymity versus Commitment," Dreyfus ingeniously transfers to the Internet the critique of "the Press" developed by the nineteenth-century existentialist and religious thinker, Søren Kierkegaard. As Kierkegaard describes it, a popular press, in its effort to satisfy the insatiable demands of its readers, indiscriminately reports every new event, thus encouraging the sort of uncommitted attitude of curiosity in which everything is momentarily interesting but nothing seems worth knowing. In Kierkegaard's language, "the Press" creates "the Public." The trouble with being curious, of course, is that ultimately the only thing that can keep boredom at bay is yet another round of information about yet another "new" event. Here one sees similarities between the evaluations of Borgmann and Dreyfus concerning the availability of information on the Internet. In Dreyfus, however, the bulk of the discussion is directed to his provocative countersuggestion. With a contemporary reconsideration of Kierkegaard's "three stages on life's way," he argues, we might (1) cease to live *aesthetically*, like an undirected and idly curious Websurfer; (2) discover that even the "principled" conversion of information into knowledge from a "chosen" perspective as in the *ethical* stage is ultimately hollow, so long as one fails to see that (3) a genuine choice – the kind of choice that turns otherwise merely optional perspectives into a genuine "commitment" – is, as in Kierkegaard's depiction of commitment in the *religious* stage, unconditional.

45

Heidegger and McLuhan *and* The Essence of Virtual Reality

Michael H. Heim

5 Heidegger and McLuhan: the Computer as Component

Heidegger and Computers

An odd juxtaposition? No philosopher highlights the clash between technology and human values so sharply as Heidegger. Not only did he make technology central to metaphysics, but he came to see in it the root evil of the twentieth century, including the Nazi German catastrophe, which he described as "the confrontation of European humanity with global technology." In both his life and his writings, Heidegger felt technology to be an overwhelming force that challenges the reassuring maxims of traditional morality. Yet his death in 1976 did not permit him to see the century's most powerful technological revolution: the proliferation of the microcomputer. He saw only the first glimmerings of computerization, the mainframe dinosaurs of the computer age. But because his work spans the gap between the days before computers and the increasingly computerized present, Heidegger can become a springboard for understanding the new situation of the sciences and the humanities.

The images we have of Heidegger the thinker, both photographic and mythic, place him in another time, another generation. In posed photographs, we see him sitting in a hut on the quiet mountaintop of the Todtnauberg, surrounded by shelves of books as he bends intently over a wooden writing table. The sun pours in the window. Under his pen, the manuscripts bristle with marginalia and scrawled notations of every kind, his pages a palimpsest heaped with layers of minute revisions. Heidegger the thinker is Heidegger the scholar, and the scholar searches ancient texts for clues to the history of Being. He looks for hints about where our essence, our heart, is today and whither the pull of the future.

This image of Heidegger feeds on nostalgia. Even the Heidegger of the photos, seated in his hut a half-century ago, working with pen on paper, had a keen sense of just how faded this picture was soon to become, how quickly this image would turn antiquarian. Because he connected being with time, Heidegger knew that reality changes and with it the task of thinking. He sensed the pace of change in the twentieth century, and he seemed to foresee what librarians realize today: "The image of the humanist scholar in the book-crammed study, thinking deep thoughts, will continue to be less and less viable in professional scholarship."[1] This recent observation by the director of a great college library confirms what Heidegger in his writings surmised: our rapid technological advance challenges the legacy of human thinking. Who better than the contemporary librarian knows the inner trend of today's scholar? Bid adieu to the "hochgewölbtes, engen gotischen Studierzimmer" of Goethe's Faust. The *Schreibstube* is giving way to the computer workstation, and scholarship requires a cybersage.

Originally "Heidegger and McLuhan: The Computer as Component" and "The Essence of VR" in Michael H. Heim, *The Metaphysics of Virtual Reality*, Oxford: Oxford University Press, 1997, pp. 55–72, 109–28. Reprinted by permission of Oxford University Press.

Computerized libraries already exist today without paper books, and by the year 2000, nearly every text of human knowledge will exist in electronic form. Heidegger sensed, with anguish, that his works would one day come to light in a world of scholarship that had grown alien to the meditative pathways that nurtured his thoughts. In 1967, he saw a rising crest of information that, he suspected, might soon engulf his own writings: "Maybe history and tradition will fit smoothly into the information retrieval systems that will serve as resource for the inevitable planning needs of a cybernetically organized mankind. The question is whether thinking, too, will end in the business of information processing."[2]

In his essay "The Age of the World Picture," Heidegger unearthed seeds planted by seventeenth-century Cartesian philosophy that have blossomed today as science merges with computer science.[3] The computer began to appear indirectly in Heidegger's mid-century writings as he took up the theme of calculative versus meditative thinking, for the computer was to become the supreme calculator.

The first time I ran across the conjunction of Heidegger and computers was in 1977 when Joseph Kockelmans returned from giving seminars in Europe. While in Trier, he made the acquaintance of two graduate students, Rainer Bast and Heinrich Delfosse, who were at the time breaking new ground in Heidegger studies. Kockelmans showed me some work from these two students by handing me a stack of computer paper twenty centimeters thick. It was a series of computer printouts listing the textual discrepancies among the various German editions of *Being and Time*. Since the 1960s, the computer analysis of texts had been used occasionally by humanists, but mainly to detect stylistic differences in classical works like those of Homer and Shakespeare. There in my hands lay the first discomfiting conjunction of Heidegger and computers. That computer printout eventually became the book *Handbuch zum Textstudium von Martin Heideggers "Sein und Zeit."*[4]

Until then, Heidegger and computers had been for me an odd juxtaposition, an abstraction under the heading "the question of technology." What my hands held was not an abstract treatise but a concrete, oxymoronic fact. Heidegger speculated on an all-enframing *Gestell* (technological system), ominous and threatening, but an abstraction looming like a metaphysical sphinx, terrorizing thought with a puzzling lack of specificity. Now here was computer text concretely manifesting that abstraction. The stack of printouts highlighted both the inevitability of a technologically informed scholarship and the soundness of Heidegger's fears that his work would soon become an object of technological scrutiny. Heidegger was now on computer. The question of technology had become the question about how to go about studying Heidegger.

Just what were the specific dangers of computers? At that time, the main philosophical answer to this question was what I call the computer as opponent. In this approach, the computer appears as a rival intelligence that challenges the human being to a contest.

The computer as opponent

In 1972, Hubert Dreyfus called attention to the danger of computers. Applying phenomenological analysis, Dreyfus argued that we must delineate carefully what computers can and cannot do, lest we become unrealistic about computers and fall into a misunderstanding of the kind of beings we ourselves are. In *What Computers Can't Do*,[5] Dreyfus observed how mid-twentieth-century culture tended to interpret the human being as an information-processing system. Researchers spoke of the brain as a heuristically programmed digital computer. Because the brain is a physical thing, Dreyfus noted, and because we can metaphorically describe the brain as "processing information," we easily slip into the unexamined dogma that human thinking operates in formal patterns and that properly programmed computers might be able to replicate these patterns. If computers could replicate thought patterns, might we not then justifiably say that computers think or have an artificial intelligence (AI)? Research funds were flowing into AI when Dreyfus raised his doubts. Dreyfus argued then – and continued to argue in his book *Mind over Machine*[6] – that we are deluding ourselves if we believe that we can create machines to replicate human thought. Dreyfus sought to establish the limits of artificial intelligence, and he saw the computer as a metaphysical opponent.

Most philosophical reasoning about computers still moves within the narrow confines of artificial intelligence, the computer as opponent: Is it possible for computers to think? Can human mental and perceptual processes fit the formulas of digital programs? How far can computers advance in simulating or surpassing human reasoning? Such are the questions that held, and still hold, the attention of philosophers from Hubert Dreyfus to John Searle. This line of inquiry goes only a short distance in exploring the existential questions raised by the conjunction of Heidegger and computers. The computer opponent line takes for its paradigm the chess match. More combative than the Turing test, the chessboard places the human in a duel with the computer, the winner claiming superior intelligence. The game paradigm ensures that the relationship remains antagonistic. The combative paradigm still holds sway over the popular imagination, the human-versus-the-machine contest, with a winner/loser outcome.

Dreyfus first connected Heidegger with computers by working within this model. Observing an unbounded enthusiasm for artificial-intelligence research, Dreyfus drew on Heidegger's critique of technology to set limits on the kind of research that defines the human mind as an information processor. Dreyfus challenged the very idea that a chess-playing program "of any significance" could be built, and in 1965 he published a paper equating "Alchemy and Artificial Intelligence." This ruffled the AI researchers, and they took Dreyfus up on his challenge. In 1967, MIT researchers confronted Dreyfus with a computer chess program named Mac-Hack. To the delight of the AI community, Dreyfus lost to the computer in a public match.[7] Later, in *What Computers Can't Do*, Dreyfus explained his philosophical point – that he was concerned not with generic predictions but with the underlying comparison that hastily identifies intelligence with formal patterns or algorithms.

Dreyfus sharpened Heidegger's technology critique by focusing on the formal patterns that computers use. Because software programs run by explicitly stated instructions, the computer works on a level of intelligibility that Heidegger characterized as derived and not primordial. Formal patterns process reality but filter it through the screen of lucidity. What fails to fit the patterns gets lost in the process, even if we try to reintroduce the unknown into our interpretations. The tendency to interpret reality as essentially lucid or representable goes back to Plato, according to Heidegger's early reading of Plato.[8] Dreyfus sees in the computer, according to artificial intelligence researchers, the apotheosis of metaphysics. Plato postulated the Good as subsistent in itself. The Good, the *agathon*, energizes the forms of things, making them stable and self-consistent. So too artificial-intelligence research, at least in one of its early phases, postulates formal patterns as the be-all and end-all of intelligence. (Much recent AI research is turning away from the priority of formal algorithms and instead is looking to "fuzzy logic.")

Dreyfus applied Heidegger's critique of technology to computers, but he conceived the computer too narrowly as an artificial-intelligence device. He saw the computer only as opponent. Yet the opposition of computer and mind/brain remains, as Heidegger would say, ontic rather than ontological. The two terms *mind/brain* and *computer/program* refer to beings, to definite entities within the world. We can compare and research the nature of these entities. We can investigate the causes of their operations, sizing up their powers and limitations, but still we treat them as beings, as entities delimited by their respective natures. The mind-versus-computer question is neither ontological nor existential. Whether or not the computer could, in principle, outsmart the mind or simulate consciousness – however intriguing a question this might be – does not touch what is happening to us through computerization. The chess paradigm distracts us from the present issue, because it makes us construe our relationship to computers as confrontational rather than collaborative.

Very different from the computer as an opponent is the computer as a component. The computer has become an ingredient in human knowing. Instead of confronting a potential rival, we find ourselves interfacing with computers. Computers are woven into the fabric of everyday life, and they have become an important thread in the texture of Western civilization. Our daily reliance on computers affects the way our culture proceeds, in everything from architecture to zoology. Instead of regarding computers as opponents, we collaborate with them. Increasingly rare is a computer-free stance from which to regard the computer as a separate device. Even the research and development at major corporations is now moving away from artificial-intelligence research, in which the computer functions separately, to research on the human–computer symbiosis, including information environments that augment human bodily perception and create "virtual realities."[9] While we may legitimately inquire into the power wielded by computers independently of humans, the existential–ontological question really cuts in a different and deeper direction than does AI. As we now live and work with computers in our writing, building, banking, drawing, and so forth, how does our reality change? As Heidegger might put it: What is the meaning of this intimate connection of Being with computers? When he pondered technology as our destiny, Heidegger seemed to have had something in mind more intimidating than an external challenge to our dignity as human beings. What Heidegger saw was something even more sinister than a revolt of the machines.

What Heidegger called "the essence of technology" infiltrates human existence more intimately than anything humans could create. The danger of technology lies in the transformation of the human being, by which human actions and aspirations are fundamentally distorted. Not that machines can run amok, or even that we might misunderstand ourselves through a faulty comparison with machines. Instead, technology enters the inmost recesses of human existence, transforming the way we know and think and will. Technology is, in essence, a mode of human existence, and we could not appreciate its mental infiltration until the computer became a major cultural phenomenon.

Already in 1957 Heidegger noticed the drive for technological mastery pushing into the human interior where thought and reality meet in language. In his essay on Hebel, he wrote:

> The language machine regulates and adjusts in advance the mode of our possible usage of language through mechanical energies and functions. The language machine is – and above all, is still becoming – one way in which modern technology controls the mode and the world of language as such. Mean while, the impression is still maintained that man is the master of the language machine. But the truth of the matter might well be that the language machine takes language into its management and thus masters the essence of the human being.[10]

What did Heidegger mean when he referred to the "language machine" (*Sprachmaschine*)? He did not say "computer" – the only computers around then were huge mainframes like the UNIVAC that filled several rooms and performed only numerical calculations. Could we, twenty-five years later, translate what Heidegger meant by using the English term *computer* instead of *language machine*?

In the early 1980s, I ran into the meaning of language machine just as I was finishing the translation of another book by Heidegger, *The Metaphysical Foundations of Logic*. The translation required a lot of detailed organization, since the main body of the text includes extensive citations in Latin, French, and Greek. As a scholarly publication, the text had not only to render but also to preserve many of the references in their original languages. The translation required laborious cross-referencing with other texts and other English translations. Index cards and cut-and-paste scraps swamped my kitchen table. Perseverance was all. The work took me more than two years. Then, just as I finished typing the third and final draft of the translation, I discovered the language machine, the connection between Heidegger and computers. Not long after mailing the final draft of the translation, I installed my own personal computer for word processing. Imagine my mixed feelings when I came to realize that the two years of labor on the translation would have amounted to no more than one year if I had used a computer to handle the text and references. The meaning of language machine began to take shape in my mind.

The computer as component

Soon after trading in my electric typewriter for a portable computer in 1983, I came to believe that the machine in my hands was indeed the language machine of Heidegger's speculations. The "language machine" was Heidegger's groping term for the incipient phenomenon of word processing. Of course, word processing did not exist in Heidegger's lifetime, at least not as a cultural phenomenon. It existed only in the dreams of inventors like Doug Engelbart and Ted Nelson. Although he did not see the word processor, Heidegger did have a keen eye for the philosophical implications in the shift of writing technologies. He saw in writing technology a clue to the human relationship to language and to our awareness as beings embodied in the world:

> Not by chance does modern man write "with" the typewriter and "dictates" – the same word as "to invent creatively" [*Dichten*] – "into" the machine. This "history" of the kinds of writing is at the same time one of the major reasons for the increasing destruction of the word. The word no longer passes through the hand as it writes and acts authentically, but through the mechanized pressure of the hand. The typewriter snatches script from the essential realm of the hand – and this means the hand is removed from the essential realm of the word. The word becomes something "typed." Never-

> theless, mechanical script does have its own, limited importance in which mechanized script serves as a mere transcription for preserving handwriting or in which typewritten script substitutes for "print." When typewriters first became prevalent, a personal letter typed on a machine was regarded as a lapse of manners or as an insult. Today, handwritten letters slow down rapid reading and are therefore regarded as old-fashioned and undesirable. Mechanized writing deprives the hand of dignity in the realm of the written word and degrades the word to a mere means for the traffic of communication.[11]

Heidegger focuses on the increasing typification brought about by modern rationalist models of standardized intelligibility, models that underscore the advantages of repetition and instant recognition.

Heidegger's criticisms of the typewriter are somewhat off the mark now that the personal computer has replaced the mechanical typewriter. Unlike the typewriter, the word processor guides the hand into a nonmechanical process. The fingers on the keyboard might just as well be a voice that activates the information device, for the computer removes the writing activity from script and mechanical imprints. Word processing can also have a graphic interface that brings the hand back to bodily gestures like pointing and moving things around with a graphic pointing device or mouse. The actions are carried out by an already typified, digitized element. Unlike the typewriter, the computer does not simply replace direct hand movements with the industrial–mechanical action of springs, pulleys, and levers. The information environment allows gestures to work in ways that leave behind the industrial machine with its cumbersome but efficient mediation of human energy and attention. The electronic element shifts the quality of action to another level. The formulation of ideas on a word processor can establish impersonality while achieving a directness and flexibility undreamed of with the typewriter.

Heidegger sensed the power of the machine as an agent for changing our relationship to the word. In fact, the word processor changes our relationship to written language at least as much as the printing press does. Nor can scholarship go unchanged. Heidegger correctly feared that electronic digital text might absorb his own work. In 1967, he feared that a rising tide of information might soon obliterate his own writings: "Maybe history and tradition will fit smoothly into the information retrieval systems which will serve as resource for the inevitable planning needs of a cybernetically organized mankind. The question is whether thinking too will end in the business of information processing."[12]

If it has already transformed the epistemic stance of the natural sciences, the computer is transforming the humanities as well. The word processor is the calculator of the humanist, giving its users the power to manipulate written language in new ways. Just as the printing press altered culture and scholarship soon after its invention, so too the computer automates the composition, storage, and transmission of written words. And if the computer affects all written communication, will it not in turn affect the way in which we regard and use language in general – not only when we sit at the word processor but also, by aftereffect, whenever we speak and listen, perhaps even whenever we think?

Computer technology is so flexible and adaptable to our thought processes that we soon consider it less an external tool and more a second skin or mental prosthesis. Once acclimated to the technology, we play it much as a musician plays an instrument, identifying with it, becoming one with it. Writing on the language machine produces a new kind of writing and thinking. At our fingertips is the calculating machine dreamed of by Pascal and Leibniz, the fathers of modern metaphysics, but now this calculator operates on our language as we spontaneously produce it.

Heidegger sensed that the language machine belongs to our destiny. What did he mean when he said the language machine would "take language into its management and master the essence of the human being"? Was he simply reacting to change? Should we place him historically among the reactionaries of his time?[13] I think not. Political terms of reaction or progress are too crude here. Heidegger's statement invites us to insight, not political agendas. He was meditating on a technology still in the bud. Now that this technology is blossoming, we need to consider what he was getting at. Neither Luddite nor technophobe, Heidegger resisted every attempt to categorize his views as either optimistic or pessimistic. Whether the glass was half-empty or half-full, Heidegger was interested in the substance of its contents. He was a soft determinist, accepting destiny while studying the different ways of absorbing its impact. In this respect, he resembled the Canadian philosopher of communication Marshall McLuhan.

McLuhan and computers

Like McLuhan, Heidegger believed that he had grasped something unique and essential about the twentieth century. Both Heidegger and McLuhan felt an inner relationship to their epoch. Each believed that he was interpreting a destiny that the next generation would receive, and each believed that the legacy of his reflections on technology was far more important than his own personal value judgments about technology. McLuhan wrote that he held back his own value judgments from the public because they create "a smog in our culture." He wrote: "I have tried to avoid making personal value judgments about these processes [of technological transformation] since they seem far too important and too large in scope to deserve a merely private opinion."[14] Similarly, Heidegger held back statements of personal values from his philosophy, whether statements of self-justifications or of a moral agenda. The point was to reflect on the radical shifts brought about by an unprecedented development.

Both Heidegger and McLuhan saw intimate connections between information technology and the way the mind works. If Heidegger is the father of information anxiety, McLuhan is the child of the television medium of the 1960s. What synchronized their visions is the crucial role that technology plays in defining reality, in operating as an invisible backdrop within which the content or entities of the world appear. Behind the visible entities of the world, McLuhan glimpsed a hidden backdrop:

> To say that any technology or extension of man creates a new environment is a much better way of saying the medium is the message. Moreover, this environment is always "invisible" and its content is always the old technology. The old technology is altered considerably by the enveloping action of the new technology.[15]

For Heidegger, likewise, the question of technology was not an ontic one, not one about the proliferation of devices or even about the possible supremacy of the machine over human beings. His ontological question touches the world, the clearing or backdrop against which things appear. Ontology has to do with our understanding of the being of things, not with things as such. The ontological question probes the invisible background. As McLuhan saw it, "The content of the new environment is always the old one. The content is greatly transformed by the new technology.... Today the environment itself becomes the artifact."[16] Technology would not sweep the older things away but would transform them while placing them before us as though nothing had changed. Similarly, according to Heidegger, the future takes up the past while making it present, and the environment we live in quickly becomes an artifact in the omnivorous future of the technological system.

McLuhan helps us understand what the computer does specifically as a language machine, as a component of human knowledge. Both McLuhan and Heidegger considered the most awesome power of technology to reside in its newly achieved intimacy with language. McLuhan noted with approval Heidegger's treatment of language as a transcendental aspect of Being:

> The alphabet and kindred gimmicks have long served man as a subliminal source of philosophical and religious assumptions. Certainly Martin Heidegger would seem to be on better ground [than Kant was in assuming Euclidian space to be an a priori] in using the totality of language itself as philosophical datum. For there, at least in non-literate periods, will be the ratio among all the senses.... An enthusiasm for Heidegger's excellent linguistics could easily stem from naive immersion in the metaphysical organicism of our electronic milieu.... There is nothing good or bad about print but the unconsciousness of the effect of any force is a disaster, especially a force that we have made ourselves.[17]

McLuhan suggests that Heidegger's ideas have a greater appeal to a culture organized electronically because such a culture has already left behind the detached, linear, individualistic mentality of literate or print cultures. He agrees with Heidegger in asserting that language technology belongs to us more essentially than does any other tool. When a technology touches our language, it touches us where we live.

How can we philosophically reflect on the word processor? How can we get beyond the vague general talk about the dangers of the calculative mentality? McLuhan's work can help track the impact of word-processing technology more specifically and clearly. But for me it was not McLuhan but an illustrious student of his, Walter J. Ong, who provided a more precise conceptual angle from which I could better see the language machine. For specific insight into the way that the word processor alters our thought processes and even our sense of reality, I found help in the writings of Ong, who treats the psychodynamics – the shifts in mentality – that occur in Western history as new technologies for language storage come into prominence.

Ong traces two major shifts in knowledge storage: the oral-to-literate and the chirographic-to-print shifts. The first occurred when the culture moved from a predominantly oral-based society to a society increasingly based on the written word. The second shift moved from handwritten (chirographic) texts to the more widely disseminated, mechanically produced printed books. With more detail and coherence than his mentor McLuhan, Ong traced these shifts in writing technologies as they affected human awareness and in turn influenced interactive epistemology (knowledge as it occurs in relation to tools and to other persons). Unlike an absolute stance, this epistemological approach takes seriously the changes that mark the history of human knowledge. The studies by Ong and Eric Havelock (*Preface to Plato*) provide concrete material for distinguishing different historical epochs by their characteristic ways of symbolizing, storing, and transmitting truths. The patterns of psychic transformation they trace dovetail nicely with Heidegger's history of being.

According to Heidegger, we notice the eclipse of the truth of being occurring already in Plato's metaphysics. Once the truth of being becomes equated with the light of unchanging intelligibility, the nature of truth shifts to the ability of statements to reflect or refer reliably to entities. With the steadiness of propositional truth comes the tendency to relate to being as a type, a form, or an anticipated shape. With being as a steady form, entities gain their reality through their being typified. Already in Plato we see the seeds of the Western drive to standardize things, to find what is dependable and typical in them. Truth as the disclosure process, as the play of revealing/concealing disappears behind the scene in which the conscious mind grasps bright objects apprehended as clear, unwavering, rational forms. As humans develop the ability to typify and apprehend formal realities, the loss of truth as emergent disclosure goes unnoticed. All is light and form. Nothing hides behind the truth of beings. But this "nothing" finally makes an appearance after the whole world has become a rigid grid of standardized forms and shapes conceived and engineered by humans. As the wasteland grows, we see the devastation of our fully explicit truths. We see that there is, must be, more. The hidden extra cannot be consciously produced. Only by seeing the limits of standardization can we begin to respond to it. We have to realize that each advance in typifying and standardizing things also implies a trade-off. When we first reach forward and grasp things, we see only the benefits of our standardization, only the positive side of greater clarity and utility. It is difficult to accept the paradox that no matter how alluring, every gain in fixed intelligibility brings with it a corresponding loss of vivacity. Because we are finite, every gain we make also implies a lost possibility. The loss is especially devastating to those living in the technological world, for here they enjoy everything conveniently at their disposal – everything, that is, except the playful process of discovery itself.

The McLuhan-inspired theory of cultural transformation brings out even more sharply the impact of the word processor. But this theory lacks a poignant sense of loss or a feel for the trade-offs in finite historical transformations. Ong's version of cultural transformation has about it something of a grand Christian optimism, seeing in the global network of electronic radio, television, and film a way of reintegrating a fallen, fragmented humanity, creating a closer community. For Ong, the shift from a predominantly oral culture to a literate culture shattered the original tribal unity. In bringing about greater individualism and fostering the logical faculties, literacy cut into the psychic roots of belonging and severed the attachment to immediate interpersonal presence. The print culture even further reinforced literacy, spreading it ever more widely, lifting individualism to unprecedented heights. Then, in Hegelian fashion, Ong sees the electronic media sublating the earlier oppositions, the oral and the literate, so that electronics achieves an encompassing synthesis. Electronic visuals, supported by voices, re-creates human presence and reunites the individuated members of the community. Underneath, however, the electronic images still depend on the reading of scripts, prepared messages, and a print-informed society.

So the electronic media preserve individual literacy while surpassing it. Because of his hopeful Hegelian dialectic, Ong omits the critical evaluation that can take place only in the existential moment. Whereas McLuhan remained publicly silent on the adverse effects of the new media, Ong appears to have absorbed criticism in a larger picture based on the Christian narrative of Garden → Fall → Paradise Regained.

Heidegger, on the contrary, reminds us of the inevitable trade-offs in history. His philosophy does in fact proceed from the Hegelian sweep of historical epochs, but it denies the possibility of an integrative summation from one absolute standpoint. History is a series of ambiguous gains bringing hidden losses. The series of epochs that makes up the history of reality (*Seinsgeschichte*) expands or contracts with different hermeneutic projects but never permits a single cumulative narrative. Each moment of historical transformation brings a challenge of interpreting the losses and gains, the trade-offs in historical drift. The drift of history allows no safe haven from which to assess and collect strictly positive values once and for all.

In our era, Heidegger's notion of the intrinsic trade-offs of history can spark a critical analysis of computerized writing. Existential criticism can investigate the implications of a specific technology in all its ambiguity. Because it accepts historical drift, existential criticism proceeds without possessing a total picture of the whither and wherefore, without accepting the picture promoted by either technological utopians or dystopians. There is no need to enforce a closure of pro or con, wholesale acceptance or rejection. While recognizing the computer as a component in our knowledge process, we can attend to what happens to us as we collaborate with technology. Because human history is a path of self-awareness, as we deepen our understanding of computer interaction, we will also increase our self-understanding.

[...]

8 The Essence of VR

What is virtual reality?

A simple enough question.

We might answer: "Here, try this arcade game. It's from the Virtuality series created by Jonathan Waldern. Just put on the helmet and the datagloves, grab the control stick, and enter a world of computer animation. You turn your head and you see a three-dimensional, 360-degree, color landscape. The other players see you appear as an animated character. And lurking around somewhere will be the other animated warriors who will hunt you down. Aim, press the button, and destroy them before they destroy you. Give it a few minutes and you'll get a feel for the game, how to move about, how to be part of a virtual world. That's virtual reality!"

Suppose the sample experience does not satisfy the questioner. Our questioner has already played the Virtuality game. Suppose the question is about virtual reality in general.

Reach for a dictionary. *Webster's* states:

Virtual: "being in essence or effect though not formally recognized or admitted"
Reality: "a real event, entity, or state of affairs"

We paste the two together and read: "Virtual reality is an event or entity that is real in effect but not in fact."

Not terribly enlightening. You don't learn nuclear physics from dictionaries. We need insight, not word usage.

The dictionary definition does, however, suggest something about VR. There is a sense in which any simulation makes something seem real that in fact is not. The Virtuality game combines head-tracking device, glove, and computer animation to create the "effect" on our senses of "entities" moving at us that are "not in fact real."

But what makes VR distinctive? "What's so special," our questioner might ask, "about these computer-animated monsters? I've seen them before on television and in film. Why call them 'virtual realities'?"

The questioner seeks not information, but clarification.

Pointing to the helmet and gloves, we insist: "Doesn't this feel a lot different from watching TV? Here you can interact with the animated creatures. You shoot them down or hide from them or dodge their ray guns. And they interact with you. They hunt you in three-dimensional space just as you hunt them. That doesn't happen in the movies, does it? Here you're the central actor, you're the star!"

Our answer combines hands-on demonstration with a reminder of other experiences. We draw a contrast, pointing to something that VR is not. We still have not said what it is.

To answer what VR is, we need concepts, not samples or dictionary phrases or negative definitions.

OK, so what is it?

Our next reply must be more informed: "Go to the source. Find the originators of this technology; ask them. For twenty years, scientists and engineers have been working on this thing called *virtual reality*. Find out exactly what they have been trying to produce."

When we look to the pioneers, we see virtual reality going off in several directions. The pioneers present us with at least seven divergent concepts currently guiding VR research. The different views have built camps that fervently disagree as to what constitutes virtual reality.

Here is a summary of the seven:

Simulation

Computer graphics today have such a high degree of realism that the sharp images evoke the term *virtual reality*. Just as sound systems were once praised for their high fidelity, present-day imaging systems now deliver virtual reality. The images have a shaded texture and light radiosity that pull the eye into the flat plane with the power of a detailed etching. Landscapes produced on the GE Aerospace "visionics" equipment, for instance, are photorealistic real-time texture-mapped worlds through which users can navigate. These data-worlds spring from military flight simulators. Now they are being applied to medicine, entertainment, and education and training.

The realism of simulations applies to sound as well. Three-dimensional sound systems control every point of digital acoustic space, their precision exceeding earlier sound systems to such a degree that three-dimensional audio contributes to virtual reality.

Interaction

Some people consider virtual reality any electronic representation with which they can interact. Cleaning up our computer desktop, we see a graphic of a trash can on the computer screen, and we use a mouse to drag a junk file down to the trash can to dump it. The desk is not a real desk, but we treat it as though it were, virtually, a desk. The trash can is an icon for a deletion program, but we use it as a virtual trash can. And the files of bits and bytes we dump are not real (paper) files, but function virtually as files. These are virtual realities. What makes the trash can and the desk different from cartoons or photos on TV is that we can interact with them as we do with metal trash cans and wooden desktops. The virtual trash can does not have to fool the eye in order to be virtual. Illusion is not the issue. Rather, the issue is how we interact with the trash can as we go about our work. The trash can is real in the context of our absorption in the work, yet outside the computer work space we would not speak of the trash can except as a virtual trash can. The reality of the trash can comes from its handy place in the world woven by our engagement with a project. It exists through our interaction.

Defined broadly, virtual reality sometimes stretches over many aspects of electronic life. Beyond computer-generated desktops, it includes the virtual persons we know through telephone or computer networks. It includes the entertainer or politician who appears on television to interact on the phone with callers. It includes virtual universities where students attend classes on line, visit virtual classrooms, and socialize in virtual cafeterias.

Artificiality

As long as we are casting our net so wide, why not make it cover everything artificial? On first hearing the term *virtual reality*, many people respond immediately: "Oh, sure, I live there all the time." By this they mean that their world is largely a human construct. Our environment is thoroughly geared, paved, and wired – not quite solid and real. Planet Earth has become an artifice, a product of natural and human forces combined. Nature itself, the sky with its ozone layer, no longer escapes human influence. And our public life has everywhere been computerized. Computer analysis of purchasing habits tells supermarkets how high and where to shelve the Cheerios. Advertisers boast of "genuine simulated walnut."

But once we extend the term *virtual reality* to cover everything artificial, we lose the force of the phrase. When a word means everything, it means nothing. Even the term *real* needs an opposite.

Immersion

Many people in the VR industry prefer to focus on a specific hardware and software configuration. This is the model set for virtual reality by Sutherland, Fisher, Furness, and Brooks, before whom the term *virtual reality* did not exist, since no hardware or software claimed that name.

The specific hardware first called VR combines two small three-dimensional stereoscopic optical displays, or "eye-phones"; a Polhemus head-tracking device to monitor head movement: and a dataglove or hand-held device to add feedback so the user can manipulate objects perceived in the artificial environment. Audio with three-dimensional acoustics can support the illusion of being submerged in a virtual world. That is, the illusion is immersion.

According to this view, virtual reality means sensory immersion in a virtual environment. Such systems, known primarily by their head-mounted displays (HMD) and gloves, were first popularized by Jaron Lanier's VPL (Virtual Programming Language) Incorporated. The HMD cuts off visual and audio sensations from the surrounding world and replaces them with computer-generated sensations. The body moves through artificial space using feedback gloves, foot treadmills, bicycle grips, or joysticks.

A prime example of immersion comes from the U.S. Air Force, which first developed some of this hardware for flight simulation. The computer generates much of the same sensory input that a jet pilot would experience in an actual cockpit. The pilot responds to the sensations by, for instance, turning a control knob, which in turn feeds into the computer, which again adjusts the sensations. In this way, a pilot can get practice or training without leaving the ground. To date, commercial pilots can upgrade their licenses on certain levels by putting in a certain number of hours on a flight simulator.

Computer feedback may do more than readjust the user's sensations to give a pseudoexperience of flying. The feedback may also connect to an actual aircraft, so that when the pilot turns a knob, a real aircraft motor turns over or a real weapon fires. The pilot in this case feels immersed and fully present in a virtual world, which in turn connects to the real world.

When you are flying low in an F-16 Falcon at supersonic speeds over a mountainous terrain, the less you see of the real world, the more control you can have over your aircraft. A virtual cockpit filters the real scene and represents a more readable world. In this sense, VR can preserve the human significance of an overwhelming rush of split-second data. The heads-up display in the cockpit sometimes permits the pilot to view the real landscape behind the virtual images. In such cases, the simulation is an augmented rather than a virtual reality.

The offshoots of this technology, such as the Waldern arcade game, should not distract us – say the immersion pioneers – from the applications being used in molecular biology (docking molecules by sight and touch), airflow simulation, medical training, architecture, and industrial design. Boeing Aircraft plans to project a flight controller into virtual space, so that the controller floats thousands of feet above the airport, looking with an unobstructed view in any direction (while actually seated in a datasuit on the earth and fed realtime visual data from satellite and multiple camera viewpoints).

A leading model of this research has been the workstation developed at NASA – Ames, the Virtual Interface Environment Workstation (VIEW). NASA uses the VIEW system for telerobotic tasks, so that an operator on earth feels immersed in a remote but virtual environment and can then see and manipulate objects on the moon or Mars through feedback from a robot. Immersion research concentrates on a specific hardware and software configuration. The immersive tools for pilots, flight controllers, and space explorers are a much more concrete meaning of VR than is the vague generalization "everything artificial."

Telepresence

Robotic presence adds another aspect to virtual reality. To be present somewhere yet present there remotely is to be there virtually (!). Virtual reality shades into telepresence when you are present from a distant location – "present" in the sense that you are aware of what's going on, effective, and able to accomplish tasks by observing, reaching, grabbing, and moving objects with your own hands as though they were close up. Defining VR by telepresence nicely excludes the imaginary worlds of art, mathematics, and entertainment. Robotic telepresence brings real-time human effectiveness to a real-world location without there being a human in the flesh at that location. Mike McGreevy and Lew Hitchner walk on Mars, but in the flesh they sit in a control room at NASA–Ames.

Telepresence medicine places doctors inside the patient's body without major incisions. Medical doctors like Colonel Richard Satava and Dr. Joseph Rosen routinely use telepresence surgery to remove gall bladders without the traditional scalpel incisions. The patient heals from surgery in one tenth the usual time because telepresence surgery leaves the body nearly intact. Only two tiny incisions are needed to introduce the laparoscopic tools. Telepresence allows surgeons to perform specialist operations at distant sites where no specialist is physically present.

By allowing the surgeon to be there without being there, telepresence is a double-edged sword, so to speak. By permitting immersion, telepresence offers the operator great control over remote processes. But at the same time, a psycho-technological gap opens up between doctor and patient. Surgeons complain of losing hands-on contact as the patient evaporates into a phantom of bits and bytes.

Full-body immersion

About the same time that head-mounted displays appeared, a radically different approach to VR was emerging. In the late 1960s, Myron Krueger, often called "the father of virtual reality," began creating interactive environments in which the user moves without encumbering gear. Krueger's is come-as-you-are VR. Krueger's work uses cameras and monitors to project a user's body so it can interact with graphic images, allowing hands to manipulate graphic objects on a screen, whether text or pictures. The interaction of computer and human takes place without covering the body. The burden of input rests with the computer, and the body's free movements become text for the computer to read. Cameras follow the user's body, and computers synthesize the user's movements with the artificial environment.

I see a floating ball projected on a screen. My computerprojected hand reaches out and grabs the ball. The computer constantly updates the interaction of my body and the synthetic world that I see, hear, and touch.

In Krueger's Videoplace, people in separate rooms relate interactively by mutual body painting, free-fall gymnastics, and tickling. Krueger's Glowflow, a light-and-sound room, responds to people's movements by lighting phosphorescent tubes and issuing synthetic sounds. Another environment, Psychic Space, allows participants to explore an interactive maze in which each footstep corresponds to a musical tone, all produced with live video images that can be moved, scaled, and rotated without regard to the usual laws of cause and effect.

Networked communications

Pioneers like Jaron Lanier accept the immersion model of virtual reality but add equal emphasis to another aspect that they see as essential. Because computers make networks, VR seems a natural candidate for a new communications medium. The RB2 (Reality Built for Two) System from VPL highlights the connectivity of virtual worlds. In this view, a virtual world is as much a shared construct as a telephone is. Virtual worlds, then, can evoke unprecedented ways of sharing, what Lanier calls "post-symbolic communication." Because users can stipulate and shape objects and activities of a virtual world, they can share imaginary things and events without using words or real-world references.

Accordingly, communication can go beyond verbal or body language to take on magical, alchemical properties. A virtual-world maker might conjure up hitherto unheard-of mixtures of sight, sound, and motion. Consciously constructed outside the grammar and syntax of language, these semaphores defy the traditional logic of verbal and visual information. VR can convey meaning kinetically and even kinesthetically. Such communication will probably require elaborate protocols as well as lengthy time periods for digesting what has been communicated. Xenolinguists will have a laboratory for experiment when they seek to relate to those whose feelings and world views differ vastly from their own.

All right, enough!" shouts our questioner, bleary eyed with information overload.

"I've taken your virtual-reality tour, listened to the pioneers, and now my head is spinning. These pioneers do indeed explore in different directions. There's a general drift here but no single destination. Should I go home feeling that the real virtual reality does not exist?"

Let's not lose stamina now. We cannot let the question fizzle. Too much depends on searching for the true virtual reality.

We should not get discouraged because a mention of reality, virtual or otherwise, opens several pathways in the clearing.

Let us recall for a moment just how controversial past attempts were to define the term *reality*. Recall how many wars were fought over it.

People today shy away from the *R*-word. *Reality* used to be the key to a person's philosophy. As a disputed term, *reality* fails to engage scientific minds because they are wary of any speculation that distracts them from their specialized work. But a skeptical attitude will fall short of the vision and direction we need.

Here's a brief sidebar on how controversial the *R*-word has been throughout Western history:

> Plato holds out ideal forms as the "really real" while he denigrates the raw physical forces studied by his Greek predecessors. Aristotle soon demotes Plato's ideas to a secondary reality, to the flimsy shapes we abstract from the really real–which, for Aristotle, are the individual substances we touch and feel around us. In the medieval period, real things are those that shimmer with symbolic significance. The biblical–religious symbols add superreal messages to realities, giving them permanence and meaning, while the merely material aspects of things are less real, merely terrestrial, defective rubbish. In the Renaissance, things counted as real that could be counted and observed repeatedly by the senses. The human mind infers a solid material substrate underlying sense data but the substrate proves less real because it is less quantifiable and observable. Finally, the modern period attributed reality to atomic matter that has internal dynamics or energy, but soon the reality question was doomed by the analytical drive of the sciences toward complexity and by the plurality of artistic styles.

This reminder of metaphysics should fortify us for the long haul. If for two thousand years Western culture has puzzled over the meaning of reality, we cannot expect ourselves in two minutes, or even two decades, to arrive at the meaning of virtual reality.

The reality question has always been a question about direction, about focus, about what we should acknowledge and be concerned with. We should not therefore be surprised when VR proves controversial and elusive. Creating a new layer of reality demands our best shot, all our curiosity and imagination, especially since for us, technology and reality are beginning to merge.

When we look for the essence of a technology, we are engaging in speculation, but not in airy speculation. Our speculation involves where we plant our feet, who we are, and what we choose to be. Behind the development of every major technology lies a vision. The vision gives impetus to developers in the field even though the vision may not be clear, detailed, or even practical. The vision captures the essence of the technology and calls forth the cultural energy needed to propel it forward. Often a technological vision taps mythic consciousness and the religious side of the human spirit.

Consider for a moment the development of space technology. (Keep in mind that an inner connection exists between outer space and cyberspace, as I will point out later.) The U.S. space program enjoyed its most rapid development in the 1960s, culminating in the moon walk in 1969. What was the vision behind it?

The U.S. space program was a child of the cold war. The May 1961 speech by President John F. Kennedy that set NASA's goals incorporated traditional elements of myth: heroic struggle, personal sacrifice, and the quest for national prominence. Yet the impetus for Kennedy's speech came largely from without. What launched the U.S. space program was the fear of being surpassed by the Soviets, who had made a series of bold advances in human space travel. The goal of the moon landing was for the United States an attempt not to be overtaken by the Soviet developments in manned space exploration.

Few Americans know about the vision of their Russian competitors in space exploration. Everyone knows, of course, that the Communist revolution in 1917 froze Russian public goals in the hackneyed single-party language of a Marxist-Leninist agenda. Some historians know the name of the great Russian rocket pioneer Konstantin Tsiolkovsky (1857–1935), who stands with the American Robert H. Goddard (1882–1945) and the German-born Hermann Oberth (b. 1894). But less is known about the background of Tsiolkovsky's thinking and the visionary philosophy that influenced the first generation of Soviet space explorers.

What lay behind the energetic push to send human beings into outer space? The Russians to this day have gathered far more data on human survival in outer space. The need for information was more than curiosity or a vague lust for new frontiers; it was a moral mission, a complex and imaginative grasp of human destiny in the cosmos. The early Russian rocket pioneers, who gave the impetus to the program, felt there was an essence to their space technology, a deep inner fire that inspired and directed the research. They felt an existential imperative that drew on the religious and cultural traditions coming down through the main stream of Russian history. This essence was not itself technological, and so we might call it the esoteric essence of space technology, the hidden core of ideas that in themselves are not technological. In fact, the ideas behind the first space exploration were lofty, awe inspiring, and even mystical.

The visionary ideas fueling Tsiolkovsky and the early Russian explorers came from N. F. Fedorov. Nikolai Fedorovich Fedorov (1828–1903) was a powerful inspiration to Soloviev, Dostoevsky, Tolstoy, and a whole generation of Russians who sought to understand how modernization connects with traditional religion and culture. Even the engineers of the Trans-Siberian Railway came often to sit at the feet of the famous sage. Fedorov lived an intensely spiritual life, dedicated exclusively to ideas and learning. His profound vision applied certain strands of Russian Orthodox spirituality to the harnessing of modern technology.

Sketching a national vision, Fedorov drew large. He argued that Russia should marshal its military and national strength toward a single goal: the conquest of nature. Conquering nature meant regulating the earth as a harmonious system. It meant controlling the weather so that harvests would be plentiful. It meant balancing nature so that all life-forms could thrive together in harmony. In his vision, Fedorov saw armies producing solar energy and harnessing the electromagnetic energy of the earth, using the energy to regulate the earth's motion in space, turning the earth into a vessel for cosmic cruises. Overpopulation would cease to be a problem as humanity colonized other planets.

Unique to Fedorov's vision is its guiding moral spark. Instead of basing the conquest of nature on dominance, aggression, and egoism, Fedorov shunned the notion that humans should rule the cosmos out of a selfish desire for material wealth and abundance. Instead, he envisioned the conquest of nature as an act of altruism. But being generous to future generations can be less than purely altruistic, for they can return the favor by their acclaim of our deeds. We must regulate the forces of nature, he believed, so altruistically that we serve those who cannot possibly return our favors: we must conquer nature in order to resurrect our ancestors, the ultimate act of altruism.

The resurrection of all our dead ancestors, and it alone, provides a lofty enough ideal to mobilize humanity to explore the entire universe, including outer space. Fedorov found this thought in Russian Orthodox Christianity. According to Christian belief, the dead will rise again so that Christ, in a final judgment, will reorganize and completely redeem the world. The bodies of all human beings will one day rise again, and this resurrection, according to Fedorov, will take place through the work of human beings who carry out the divine plan. The long-range goal of human cooperation must be to discover the laws of nature to such a depth that we can eventually reconstitute the bodies of past human beings from their remaining physical particles still floating about in the universe.

Fedorov's strategy was to channel science and technology toward the reunion of all humanity. He decried the heartless positivism that builds on the sufferings and corpses of previous generations, instead seeking a purely idealistic motive. Without such a high aim, a heartless science would ultimately turn against society. For him, and for the many Soviet scientists inspired by him, the ultimate aim of the space program was, quite literally, nothing less than resurrecting the dead.

Contrast this sublime – and to us incredible and bizarre – vision of the space program with current U.S. public policy. "The commercialization of space," as promoted by administrations since the late 1970s, offers civilian entrepreneurs new opportunities for investment. To cover this naked self-interest, a mythic notion from U.S. history adds the sense of a new frontier. As a mere resource for commerce, space holds little allure, but a new frontier beyond earth adds adventure to the hope for personal gain. The vision even draws on the California gold rush in the nineteenth century, the spirit of enterprise.

In fact, this last word, *enterprise*, shows us where the commercialization of space falls short. Commercialization fails to touch the essence of space exploration, for commercial interests will neglect the long-range research needed for space science. Commercialization also drives up the cost of information derived from space exploration so that the data from space will not be available to small businesses, university scientists, farmers, state and local governments, and developing countries. In short, this kind of exploration envisions no future, only short-range profit.

But for NASA, for space enthusiasts, and for the Pentagon people, *enterprise* has a capital *E*. The word refers to a spirit of business adventure, but it also, in many minds, has another important meaning. Many technical people today also take *enterprise* to be the proper name in a science fiction myth, that of the starship *Enterprise* in "Star Trek," the popular science fiction television series about twenty-first-century space travelers. "Star Trek" contributed the code word, the handshake, the common inspiration for space exploration in the United States. (Shake hands informally with someone at the Pentagon or NASA and be prepared with an answer to the query "Are you a Trekkie?") For hundreds of technicians, the space program flies on the imaginative wings of Gene Roddenberry's brainchild, born on September 8, 1966, when the TV show was first aired. But Roddenberry was no Fedorov. The sage of Pasadena created no unifying vision to direct humanity "where no one has gone before." His fictional productions treated only a motley collection of profound moral questions pertaining to human behavior at any time, any place. But despite the limits of its lineage, "Star Trek" showed us more truly the esoteric essence, the real meaning, of space exploration than did government statements on the commercialization of space. The essence of the American space program, its heart and soul, comes from "Star Trek."

Where in VR is a counterpart to the space program's esoteric essence? What is the essence of VR, its inner spirit, the cultural motor that propels the technology? When the first conferences met on cyberspace and on virtual reality in 1989 and 1990, respectively, two threads of shared vision ran through the diverse groups of participants. One was the cyberpunk writings of William Gibson, known to both technical and literary types as the coiner of the term *cyberspace*. The other was the Holodeck from "Star Trek: The Next Generation."

Along with its cargo bay of imaginative treasures, the starship *Enterprise* brought the Holodeck. The Holodeck is familiar furniture in the vocabulary of virtual-reality pioneers. For most people, the Holodeck portrays the ideal human-computer interface. It is a virtual room that transforms spoken commands into realistic landscapes populated with walking, talking humanoids and detailed artifacts appearing so lifelike that they are indistinguishable from reality. The Holodeck is used by the crew of the starship *Enterprise* to visit faraway times and places such as medieval England and 1920s America. Generally, the Holodeck offers the crew rest and recreation, escape and entertainment, on long interstellar voyages.

While not every VR pioneer explicitly agrees on goals, the Holodeck draws the research onward. Publicly, researchers try to maintain cool and reasonable expectations about VR. Hyperbole from the media often stirs grandiose expectations in the public; when presented with actual prototypes, the public turns away with scorn. So researchers play down talk of the Holodeck. At the MIT Media Lab, leaders such as David Zeltzer avoid the term *virtual reality* not only because of the specter of metaphysics it evokes, but also because of the large promises it raises. The term seems to make greater claims than do terms like *virtual environments* (preferred at MIT and NASA) and *virtual worlds* (preferred at the universities of North Carolina and Washington). But when speaking at a VR conference for the Data Processing Management Association in Washington, D.C., on June 1, 1992, Zeltzer made an intriguing aside, one that touches, I think, on the highest possibilities of virtual reality, on its esoteric essence.

Did I say "esoteric essence"? How can we expect to give our young questioner an answer to "What is virtual reality?" when we have left the public, exoteric world of clear explanations and have embarked on a search for the esoteric essence of VR, its underlying vision? Well, our questioner seems to have gotten lost some time ago, most likely during the sidebar on the history of reality. I think I see someone off in the distance pulling avidly on the trigger of the Virtuality game. Maybe more time spent in VR will eventually deliver better answers than any verbal speculation. At any rate, on to the esoteric essence...

Zeltzer's remark went something like this: "True virtual reality may not be attainable with any technology we create. The Holodeck may forever remain fiction. Nonetheless, virtual reality serves as the Holy Grail of the research."

"Holy Grail?" Holy Grail!

Now when Zeltzer made this reference, he was not deliberately invoking a Jungian archetype. His remark expressed modesty and diffidence rather than alchemical arrogance. Still, archetypes do not have to hit us in the nose to wield their peculiar power. They work most powerfully at the back of the subconscious mind, and therein lies their magic. An effective archetype works its magic subtly.

David Zeltzer was calling up a mythic image far more ancient and infinitely more profound than "Star Trek." "Star Trek" has, after all, become the stuff of trivia: "Star Trek" ties and boxer shorts, "Star Trek" vinyl characters and mugs ("Fill them with a hot beverage and watch Kirk and Spock beam up to an unknown world"). "Star Trek" lost any sublimity it may have had when it came to occupy Kmart shelves along with electric flyswatters and noisemaker whoopee cushions.

The Holy Grail, though, sums up the aspirations of centuries. It is an image of the Quest. From Tennyson's romantic *Idylls of the King* to Malory's King Arthur and the Knights of the Round Table, the ancient Grail legend reaches back to Christian and pre-Christian times. The Grail has always been a symbol of the quest for a better world. In pre-Christian times, the Grail was the cup that holds a cure for an ailing king who, suffering from his own wounds, sees his country turning into a wasteland. Christians believed the Grail to be both the chalice of Jesus' Last Supper and the cup that caught the Savior's blood at the Crucifixion. Medieval legend links the spear that pierced Jesus' side on the cross with the sacred cup that held his blood. Later works of art, from T. S. Eliot's *The Waste Land* to Richard Wagner's *Parsifal*, have preserved the Grail story as a symbol of spiritual quest and lofty aspiration.

Perhaps the essence of VR ultimately lies not in technology but in art, perhaps art of the highest order. Rather than control or escape or entertain or communicate, the ultimate promise of VR may be to transform, to redeem our awareness of reality – something that the highest art has attempted to do and something hinted at in the very label *virtual reality*, a label that has stuck, despite all objections, and that sums up a century of technological innovation. VR promises not a better vacuum cleaner or a more engrossing communications medium or even a friendlier computer interface. It promises the Holy Grail.

We might learn something about the esoteric essence of VR by thinking about Richard Wagner's *Parsifal*. Wagner himself was searching for a Holodeck, though he did not know it. By the time he finished *Parsifal*, his final opera, Wagner no longer considered his work to be opera. He did not want it called opera or music or theater or even "art," and certainly not entertainment. By the time he finished his last work, Wagner realized he was trying to create another reality, one that would in turn transform ordinary reality. The term he came to use was "a total work of art," by which he meant a seamless union of vision, sound, movement, and drama that would sweep the viewer to another world, not to escape but to be changed. Nor could the viewer be a mere spectator. Wagner created a specially designed building in Bayreuth, Germany, well off the beaten track, where the audience would have to assemble after a long journey because he forbade the performance of *Parsifal* in any other building. The audience would have to prepare itself well ahead of time by studying the libretto, because *Parsifal* was long, mysterious, and full of complex, significant details. (Wagner's *Ring* cycle takes over fifteen hours to present a related myth.) Looking for the right terms to express his intent, Wagner called *Parsifal* "a festival play for consecrating the stage" (*ein Bühnenweihfestspiel*). The Bayreuth theater would become the site for a solemn, nearly liturgical celebration. The mythmaker would create a counter-reality, one reminiscent of the solemn mass of the Catholic church, which appeals to all the senses with its sights, sounds, touch, drama, even appealing to smell with incense and candles. The audiences at Bayreuth were to become pilgrims on a quest, immersed in an artificial reality.

The drama *Parsifal*, like a mysterious dream, resists easy summary, and it eludes interpretation. But the general story outline is clear. The protectors of "correct values" (the Knights) inevitably paint themselves into the corner of righteousness. Paralyzed, unable to act, their leadership suffers intense internal pain (Amfortas). They can regain the power of the Grail that they protect only through the intervention of someone who is still innocent of right and wrong, someone who is by all standards a fool. The innocent fool (Arabic, *fal parsi*) can clean out the sclerotic righteous society only after passing a test and learning to feel the sufferings of others. Once the innocent fool has acquired compassion for others and sensitivity to life's complexity, he can bring the power (the Spear) back to the righteous Knights of the Holy Grail. The Grail Knights then come to understand more deeply what the work of the Holy Grail, and their mission, means. The Grail grants its full power only to those who can be touched by compassion.

Wagner's Holodeck presents a Parsifal who mirrors the individual audience members at Bayreuth. Wagner shaped the drama with story and music so that strong sensations would engulf the audience and pierce them to the heart. Each listener begins as a naive spectator and is then gradually touched by the painful actions on stage until the listener becomes transformed into a more sensitive and compassionate member, ready to bring to a sick society some measure of healing and renewal.

Wagner hoped to do more than make music and theater; he believed that his music dramas could transform society by imparting new feelings and attitudes. This goal he shared with traditional religion; and religion returns the competition with distrust and the accusation of heterodoxy. For this reason, Wagner's work remains to this day controversial among religious people, including many artists and musicians who have strong religious faith.

How well did Wagner succeed? One of the most telling tributes to the success of Wagner's *Parsifal* comes from a Jesuit priest, Father Owen Lee, who in a radio broadcast intermission feature from the Metropolitan Opera in New York City said:

> I watched as usual from the least expensive seat under the roof, hovering there with an unearthly feeling for long half-hours floating in an immense space, suffused with a sense of what Baudelaire felt listening to Wagner: "A sense of being suspended in an ecstasy compounded of joy and insight." I can remember staggering out of theaters after *Parsifal*, hardly aware of people applauding, the music streaming through me, carried out of myself, seeing my experience – indeed, feeling that I was seeing all experience – at a higher level of awareness, put in touch with a power greater than myself, a kind of holy fool.[18]

Another holy fool was the Finnish composer Jan Sibelius, who wrote: "Heard *Parsifal*. Nothing else in all the world has made so overwhelming an impression on me. All my heartstrings throbbed." The German composer Max Reger wrote: "Heard *Parsifal*. Cried for two weeks, then decided to become a composer."

Someday VR will elicit similar rave reviews, not mere thrills, but insight into experience. As it evolves its art form, VR will have certain advantages over Wagner's "total work of art." Certain disadvantages might also plague it where Wagnerian solutions might help.

Activity/passivity

VR systems, as Jaron Lanier points out, can reduce apathy and the couch-potato syndrome simply by requiring creative decisions. Because computers make VR systems interactive, they also allow the artist to call forth greater participation from users. Whereas traditional art forms struggle with the passivity of the spectator, the VR artist finds a controlled balance between passivity and activity. The model of user navigation can be balanced by the model of pilgrimage and sacred awe.

Manipulation/receptivity

Some observers date the advent of VR to the moment when the dataglove appeared on the computer screen. At that moment, the user became visible as an active, involved force in the digital world. This implies that VR has a tilt toward manipulation, even a latent tendency toward aggressive, first-person attitudes. The VR artist will need strategies for inducing a more receptive atmosphere, so that the user can be open in all directions, receiving signals from and having empathy for other beings. The user must be able to be touched, emotionally moved, by non-first-person entities in the virtual world. The spear of manipulation must join the cup of sensitivity. If simulators serve to train hand-eye and other coordination skills, VR may take a further step and become a training tool to enhance receptivity.

Remote presence

The visual bias of current VR brings out a possible detachment in the user's sense of the world. Seeing takes place at a distance, whereas hearing and the other senses are more intimate to our organic life. The visual bias increases the detachment of telepresence. Some VR versions stress the "looking-at" factor, such as David Gelernter's Mirror Worlds, in which, in real-time, users can zoom in on miniature shoe-box worlds containing local homes, businesses, cities, governments, or nations. VR offers the opportunity to shift the Western philosophy of presence. From Pythagoras to Aristotle, from Berkeley to Russell, our philosophical sense of presence has relied on vision, consequently putting us in the position of spectators. To be touched, we need to introduce more sensory awareness. VR may develop a kind of feedback in which presence includes an openness and sensitivity of the whole body.

Augmented reality

VR will enhance the power of art to transform reality. The picture frame, the proscenium, the movie theater all limit art by blocking it off as a section of reality. VR, with its augmented reality, allows a smoother, more controlled transition from virtual to real and back. This capability, which may frighten psychologists, will offer artists an unprecedented power to transform societies.

These are a few of the differences that make virtual reality different from traditional art forms. They belong to the essence of VR, its Holy Grail. This goal means that we need a different breed of artist as well. And where will we find these new cybersages, these virtual-world makers? I see our young questioner smiling broadly now as yet another wounded pterodactyl drops from the pink sky of Waldern's arcade game. Plenty of fledgling enthusiasm here, and a society that needs healing and renewal.

Notes

1 Ralph Holibaugh, director of the Olin and Chalmers libraries at Kenyon College, in *The Kenyon College Annual Report 1988–90*, p. 5.

2 Martin Heidegger, Preface to *Wegmarken* (Frankfurt: Klostermann, 1967), p. i. [Author's translation]

3 Don Ihde, for one, sees science as merging with its instruments in *Technology and the Lifeworld: From Garden to Earth* (Bloomington: Indiana University Press, 1990).

4 Rainer Bast and Heinrich Delfosse, *Handbuch zum Text-studium von Martin Heideggers "Sein und Zeit"* (Stuttgart: Frommann-Holzboog, 1979).

5 Hubert Dreyfus, *What Computers Can't Do: The Limits of Artificial Intelligence*, rev. ed. (New York: Harper Colophon, 1979).

6 Hubert Dreyfus, *Mind over Machine: The Power of Human Intuition and Expertise in the Era of the Computer* (New York: Free Press, 1985).

7 The history of this chess match appears in Howard Rheingold, *Tools for Thought: The People and Ideas Behind the Next Computer Revolution* (New York: Simon and Schuster, 1985), pp. 161–62. Dreyfus explains what he takes to be the point of the match in *Mind over Machine* (p. 112).

8 Martin Heidegger, "Platons Lehre von der Wahrheit," in *Wegmarken*, pp. 109–44. This was a lecture series given in 1930 and 1931.

9 See Michael Benedikt, ed., *Cyberspace: First Steps* (Cambridge, Mass.: MIT Press, 1991). The term *cyberspace* originated with William Gibson, who uses science fiction to explore the symbiotic connection of humans and computers.

10 Martin Heidegger, *Hebel – Der Hausfreund* (Pfullingen: Günther Neske, 1957); reprinted as "Hebel – Friend of the House," trans. Bruce Foltz and Michael Heim, in *Contemporary German Philosophy*, ed. Darrel E. Christensen (University Park: Pennsylvania State University Press, 1983), vol. 3, pp. 89–101.

11 Martin Heidegger, *Parmenides*, vol. 54 of *Gesamtausgabe* (Frankfurt: Klostermann, 1982), pp. 119–19. [Author's translation] These were lectures given in the winter of 1942/1943. In this passage, Heidegger is commenting on the ancient Greek notion of "action" (*pragma*).

12 Heidegger, Preface to *Wegmarken*, p. ii. [Author's translation]

13 A recent study that locates Heidegger's theory of technology in the cultural reaction of the Weimar Republic is Michael Zimmerman, *Heidegger's Confrontation with Modernity: Technology, Politics, Art* (Bloomington: Indiana University Press, 1990).

14 Marshall McLuhan to Jonathan Miller, April 1970, in *The Letters of Marshall McLuhan*, comp. and ed. Matie Molinaro, Corinne McLuhan, and William Toye (New York: Oxford University Press, 1987), p. 405. In a letter to Jonathan Miller, April 1970, McLuhan wrote: "I take it that you understand that I have never expressed any preferences or values since *The Mechanical Bride*. Value judgments create smog in our culture and distract attention from processes. My personal bias is entirely pro-print and all of its effects" (ibid.). In other places, McLuhan was not so open about his stance. In writing to Eric Havelock, May 1970, for instance, he noted:

> My own studies of the effects of technology on human psyche and society have inclined people to regard me as the enemy of the things I describe. I feel a bit like the man who turns in a fire alarm only to be charged with arson. I have tried to avoid making personal value judgments about these processes since they seem far too important and too large in scope to deserve a merely private opinion. (ibid., p. 406)

15 McLuhan to John Culkin, September 1964, in ibid., p. 309.

16 McLuhan to Buckminster Fuller, September 1964, in ibid., p. 309.

17 H. Marshall McLuhan, *The Gutenberg Galaxy: The Making of Typographic Man* (Toronto: University of Toronto Press, 1962), p. 66, in a section entitled "Heidegger surf-boards along on the electronic wave as triumphantly as Descartes rode the mechanical wave."

18 Father Owen Lee, "Metropolitan Opera Broadcast Intermission Feature," March 28, 1992.

46

Hacking Away at the Counterculture

Andrew Ross

Ever since the viral attack engineered in November of 1988 by Cornell University hacker Robert Morris on the national network system Internet, which includes the Pentagon's ARPAnet data exchange network, the nation's high-tech ideologues and spin doctors have been locked in debate, trying to make ethical and economic sense of the event. The virus rapidly infected an estimated six thousand computers around the country, creating a scare that crowned an open season of viral hysteria in the media, in the course of which, according to the Computer Virus Industry Association in Santa Clara, California, the number of known viruses jumped from seven to thirty during 1988, and from three thousand infections in the first two months of that year to thirty thousand in the last two months. While it caused little in the way of data damage (some richly inflated initial estimates reckoned up to $100 million in downtime), the ramifications of the Internet virus have helped to generate a moral panic that has all but transformed everyday "computer culture."

Following the lead of the Defense Advance Research Projects Agency (DARPA) Computer Emergency Response Team at Carnegie-Mellon University, antivirus response centers were hastily put in place by government and defense agencies at the National Science Foundation, the Energy Department, NASA, and other sites. Plans were made to introduce a bill in Congress (the Computer Virus Eradication Act, to replace the 1986 Computer Fraud and Abuse Act, which pertained solely to government information) that would call for prison sentences of up to ten years for the "crime" of sophisticated hacking, and numerous government agencies have been involved in a proprietary fight over the creation of a proposed Center for Virus Control, modeled, of course, on Atlanta's Centers for Discase Control, notorious for its failure to respond adequately to the AIDS crisis.

Media commentary on the virus scare has run not so much tongue-in-cheek as hand-in-glove with the rhetoric of AIDS hysteria – the common use of terms like *killer virus* and *epidemic*; the focus on high-risk personal contact (virus infection, for the most part, is spread on personal computers, not mainframes); the obsession with defense, security, and immunity; and the climate of suspicion generated around communitarian acts of sharing. The underlying moral imperative being this: You can't trust your best friend's software any more than you can trust his or her bodily fluids – safe software or no software at all! Or, as Dennis Miller put in on *Saturday Night Live*, "Remember, when you connect with another computer, you're connecting to every computer that computer has ever connected to." This playful conceit struck a chord in the popular consciousness, even as it was perpetuated in such sober quarters as the Association for Computing Machinery, the president of which, in a controversial editorial titled "A Hygiene Lesson," drew comparisons not only with sexually transmitted diseases, but also with a cholera epidemic, and urged attention to "personal systems hygiene."[1] Some computer scientists who studied the symp-

From *Technoculture*, vol. 3, ed. Constance Penley and Andrew Ross, Minneapolis: University of Minnesota Press, 1991, pp. 107–34.

tomatic path of Morris's virus across Internet have pointed to its uneven effects upon different computer types and operating systems, and concluded that "there is a direct analogy with biological genetic diversity to be made."[2] The epidemiology of biological virus, and especially AIDS, research is being studied closely to help implement computer security plans, and, in these circles, the new witty discourse is laced with references to antigens, white blood cells, vaccinations, metabolic free radicals, and the like.

The form and content of more lurid articles like *Time's* infamous (September 1988) story, "Invasion of the Data Snatchers," fully displayed the continuity of the media scare with those historical fears about bodily invasion, individual and national, that are often considered endemic to the paranoid style of American political culture.[3] Indeed, the rhetoric of computer culture, in common with the medical discourse of AIDS research, has fallen in line with the paranoid, strategic mode of Defense Department rhetoric. Each language-repertoire is obsessed with hostile threats to bodily and technological immune systems; every event is a ballistic maneuver in the game of microbiological war, where the governing metaphors are indiscriminately drawn from cellular genetics and cybernetics alike. As a counterpoint to the tongue-in-cheek artificial intelligence (AI) tradition of seeing humans as "information-exchanging environments," the imagined life of computers has taken on an organicist shape, now that they too are subject to cybernetic "sickness" or disease. So too the development of interrelated systems, such as Internet itself, has further added to the structural picture of an interdependent organism, whose component members, however autonomous, are all nonetheless affected by the "health" of each individual constituent. The growing interest among scientists in developing computer programs that will simulate the genetic behavior of living organisms (in which binary numbers act like genes) points to a future where the border between organic and artificial life is less and less distinct.

In keeping with the increasing use of biologically derived language to describe mutations in systems theory, conscious attempts to link the AIDS crisis with the information security crisis have pointed out that both kinds of virus, biological and electronic, take over the host cell/program and clone their carrier genetic codes by instructing the hosts to make replicas of the viruses. Neither kind of virus, however, can replicate itself independently; they are pieces of code that attach themselves to other cells/programs – just as biological viruses need a host cell, computer viruses require a host program to activate them. The Internet virus was not, in fact, a virus, but a worm, a program that can run independently and therefore *appears* to have a life of its own. The worm replicates a full version of itself in programs and systems as it moves from one to another, masquerading as a legitimate user by guessing the user passwords of locked accounts. Because of this autonomous existence, the worm can be seen to be have as if it were an organism with some kind of purpose or teleology, and yet it has none. Its only "purpose" is to reproduce and infect. If the worm has no inbuilt antireplication code, or if the code is faulty, as was the case with the Internet worm, it will make already-infected computers repeatedly accept further replicas of itself, until their memories are clogged. A much quieter worm that that engineered by Morris would have moved more slowly, as one supposes a "worm" should, protecting itself from detection by ever more subtle camouflage, and propagating its cumulative effect of operative systems inertia over a much longer period of time.

In offering such descriptions, however, we must be wary of attributing a teleology/intentionality to worms and viruses that can be ascribed only, and, in most instances, speculatively, to their authors. There is no reason a cybernetic "worm" might be expected to behave in any fundamental way like a biological worm. So, too, the assumed intentionality of its author distinguishes the human-made cybernetic virus from the case of the biological virus, the effects of which are fated to be received and discussed in a language saturated with human-made structures and narratives of meaning and teleological purpose. Writing about the folkloric theologies of significance and explanatory justice (usually involving retribution) that have sprung up around the AIDS crisis, Judith Williamson has pointed to the radical implications of this collision between an intentionless virus and a meaning-filled culture:

> Nothing could be more meaningless than a virus. It has no point, no purpose, no plan; it is part of no scheme, carries no inherent significance. And yet nothing is harder for us to confront than the complete absence of meaning. By its very definition, meaninglessness cannot be articulated within our social language, which is a system *of* meaning: impossible to include, as

> an absence, it is also impossible to exclude – for meaninglessness isn't just the opposite of meaning, it is the end of meaning, and threatens the fragile structures by which we make sense of the world.[4]

No such judgment about meaninglessness applies to the computer security crisis. In contrast to HIV's lack of meaning or intentionality, the meaning of cybernetic viruses is always already replete with social significance. This meaning is related, first of all, to the author's local intention or motivation, whether psychic or fully social, whether wrought out of a mood of vengeance, a show of bravado or technical expertise, a commitment to a political act, or in anticipation of the profits that often accrue from the victims' need to buy an antidote from the author. Beyond these local intentions, however, which are usually obscure or, as in the Morris case, quite inscrutable, there is an entire set of social and historical narratives that surround and are part of the "meaning" of the virus: the coded anarchist history of the youth hacker subculture; the militaristic environments of search-and-destroy warfare (a virus has two components – a carrier and a "warhead"), which, because of the historical development of computer technology, constitute the family values of information technoculture; the experimental research environments in which creative designers are encouraged to work; and the conflictual history of pure and applied ethics in the science and technology communities – to name just a few. A similar list could be drawn up to explain the widespread and varied *response* to computer viruses, from the amused concern of the cognoscenti to the hysteria of the casual user, and from the research community and the manufacturing industry to the morally aroused legislature and the mediated culture at large. Every one of these explanations and narratives is the result of social and cultural processes and values; consequently, there is very little about the virus itself that is "meaningless." Viruses can no more be seen as an objective, or necessary, result of the "objective" development of technological systems than technology in general can be seen as an objective, determining agent of social change.

For the sake of polemical economy, I would note that the cumulative effect of all the viral hysteria has been twofold. First, it has resulted in a windfall for software producers, now that users' blithe disregard for makers' copyright privileges has eroded in the face of the security panic. Used to fighting halfhearted rearguard actions against widespread piracy practices, or reluctantly acceding to buyers' desire for software unencumbered by top-heavy security features, software vendors are now profiting from the new public distrust of program copies. So too the explosion in security consciousness has hyper-stimulated the already fast growing sectors of the security system industry and the data encryption industry. In line with the new imperative for everything from "vaccinated" workstations to "sterilized" networks, it has created a brand-new market of viral vaccine vendors who will sell you the virus (a one-time only immunization shot) along with its antidote – with names like Flu Shot +, ViruSafe, Vaccinate, Disk Defender, Certus, Viral Alarm, Antidote, Virus Buster, Gatekeeper, Ongard, and Interferon. Few of the antidotes are very reliable, however, especially since they pose an irresistible intellectual challenge to hackers who can easily rewrite them in the form of ever more powerful viruses. Moreover, most corporate managers of computer systems and networks know that by far the great majority of their intentional security losses are a result of insider sabotage and monkeywrenching.

In short, the effects of the viruses have been a profitable clamping down on copyright delinquency and the generation of the need for entirely new industrial production of viral suppressors to contain the fallout. In this respect, it is easy to see that the appearance of viruses could hardly, in the long run, have benefited industry producers more. In the same vein, the networks that have been hardest hit by the security squeeze are not restricted-access military or corporate systems, but networks like Internet, set up on trust to facilitate the open academic exchange of data, information, and research, and watched over by its sponsor, DARPA. It has not escaped the notice of conspiracy theorists that the military intelligence community, obsessed with "electronic warfare," actually stood to learn a lot from the Internet virus; the virus effectively "pulsed the system," exposing the sociological behavior of the system in a crisis situation.[5]

The second effect of the virus crisis has been more overtly ideological. Virus-conscious fear and loathing have clearly fed into the paranoid climate of privatization that increasingly defines social identities in the new post-Fordist order. The result – a psychosocial closing of the ranks around fortified private spheres – runs directly counter to the ethic that we might think of as residing at the architectural heart of information technology. In its basic assembly structure, information technology is a technology of processing, copying, replication, and simulation, and therefore does not recognize the concept of private information property. What is now under threat is the rationality of a shareware culture, ushered in as the achievement of the hacker counterculture that pioneered the personal computer revolution in the early seventies against the grain of corporate planning.

There is another story to tell, however, about the emergence of the virus scare as a profitable ideological moment, and it is the story of how teenage hacking has come to be defined increasingly as a potential threat to normative educational ethics and national security alike. The story of the creation of this "social menace" is central to the ongoing attempts to rewrite property law in order to contain the effects of the new information technologies that, because of their blindness to the copyrighting of intellectual property, have transformed the way in which modern power is exercised and maintained. Consequently, a deviant social class or group has been defined and categorized as "enemies of the state" to help rationalize a general law-and-order clampdown on free and open information exchange. Teenage hackers' homes are now habitually raided by sheriffs and FBI agents using strong-arm tactics, and jail sentences are becoming a common punishment. Operation Sundevil, a nationwide Secret Service operation conducted in the spring of 1990, involving hundreds of agents in fourteen cities, is the most recently publicized of the hacker raids that have produced several arrests and seizures of thousands of disks and address lists in the last two years.[6]

In one of the many harshly punitive prosecutions against hackers in recent years, a judge went so far as to describe "bulletin boards" as "hi-tech street gangs." The editors of *2600*, the magazine that publishes information about system entry and exploration indispensable to the hacking community, have pointed out that any single invasive act, such as trespass, that involves the use of computers is considered today to be infinitely more criminal than a similar act undertaken without computers.[7] To use computers to execute pranks, raids, fraud, or theft is to incur automatically the full repressive wrath of judges urged on by the moral panic created around hacking feats over the last two decades. Indeed, there is a strong body of pressure groups pushing for new criminal legislation that will define "crimes with computers" as a special category of crime, deserving "extraordinary" sentences and punitive measures. Over that same space of time, the term *hacker* has lost its semantic link with the journalistic *hack*, suggesting a professional toiler who uses unorthodox methods. So too its increasingly criminal connotation today has displaced the more innocuous, amateur mischief-maker-cum-media-star role reserved for hackers until a few years ago.

In response to the gathering vigor of this "war on hackers," the most common defenses of hacking can be presented on a spectrum that runs from the appeasement or accommodation of corporate interests to drawing up blueprints for cultural revolution. (a) Hacking performs a benign industrial service of uncovering security deficiencies and design flaws. (b) Hacking, as an experimental, free-form research activity, has been responsible for many of the most progressive developments in software development. (c) Hacking, when not purely recreational, is an elite educational practice that reflects the ways in which the development of high technology has outpaced orthodox forms of institutional education. (d) Hacking is an important form of watchdog counterresponse to the use of surveillance technology and data gathering by the state, and to the increasingly monolithic communications power of giant corporations. (e) Hacking, as guerrilla know-how, is essential to the task of maintaining fronts of cultural resistance and stocks of oppositional knowledge as a hedge against a technofascist future. With all of these and other arguments in mind, it is easy to see how the social and cultural *management* of hacker activities has become a complex process that involves state policy and legislation at the highest levels. In this respect, the virus scare has become an especially convenient vehicle for obtaining public and popular consent for new legislative measures and new powers of investigation for the FBI.[8]

Consequently, certain celebrity hackers have been quick to play down the zeal with which they pursued their earlier hacking feats, while reinforcing the *deviant* category of "technological hooliganism" reserved by moralizing pundits for "dark-side" hacking. Hugo Cornwall, British author of the best-selling *Hacker's Handbook*, presents a Little England view of the hacker as a harmless fresh-air enthusiast who "visits advanced computers as a polite country rambler might walk across picturesque fields." The owners of these properties are like "farmers who don't mind careful ramblers."

Cornwall notes that "lovers of fresh-air walks obey the Country Code, involving such items as closing gates behind one and avoiding damage to crops and livestock" and suggests that a similar code ought to "guide your rambles into other people's computers; the safest thing to do is simply browse, enjoy and learn." By contrast, any rambler who "ventured across a field guarded by barbed wire and dotted with notices warning about the Official Secrets Act would deserve most that happened thereafter."[9] Cornwall's quaint perspective on hacking has a certain "native charm," but some might think that this beguiling picture of patchwork-quilt fields and benign gentlemen farmers glosses over the long bloody history of power exercised through feudal and postfeudal land economy in England, while it is barely suggestive of the new fiefdoms, transnational estates, dependencies, and principalities carved out of today's global information order by vast corporations capable of bypassing the laws and territorial borders of sovereign nation-states. In general, this analogy with "trespass" laws, which compares hacking to breaking and entering other people's homes, restricts the debate to questions about privacy, property, possessive individualism, and, at best, the excesses of state surveillance, while it closes off any examination of the activities of the corporate owners and institutional sponsors of information technology (the almost exclusive "target" of most hackers).[10]

Cornwall himself has joined the lucrative ranks of ex-hackers who either work for computer security firms or write books about security for the eyes of worried corporate managers.[11] A different, though related, genre is that of the penitent hacker's "confession," produced for an audience thrilled by tales of high-stakes adventure at the keyboard, but written in the form of a computer security handbook. The best example of the "I Was a Teenage Hacker" genre is Bill (aka "The Cracker") Landreth's *Out of the Inner Circle: The True Story of a Computer Intruder Capable of Cracking the Nation's Most Secure Computer Systems*, a book about "people who can't 'just say no' to computers." In full complicity with the deviant picture of the hacker as "public enemy," Landreth recirculates every official and media cliché about subversive conspiratorial elites by recounting the putative exploits of a high-level hackers' guild called the Inner Circle. The author himself is presented in the book as a former keyboard junkie who now praises the law for having made a good moral example of him:

> If you are wondering what I am like, I can tell you the same things I told the judge in federal court: Although it may not seem like it, I am pretty much a normal American teenager. I don't drink, smoke or take drugs. I don't steal, assault people, or vandalize property. The only way in which I am really different from most people is in my fascination with the ways and means of learning about computers that don't belong to me.[12]

Sentenced in 1984 to three years' probation, during which time he was obliged to finish his high school education and go to college, Landreth concludes: "I think the sentence is very fair, and I already know what my major will be." As an aberrant sequel to the book's contrite conclusion, however, Landreth vanished in 1986, violating his probation, only to face later a stiff five-year jail sentence – a sorry victim, no doubt, of the recent crackdown.

Cyber-Counterculture?

At the core of Steven Levy's best-seller *Hackers* (1984) is the argument that the hacker ethic, first articulated in the 1950s among the famous MIT students who developed multiple-access user systems, is libertarian and crypto-anarchist in its right-to-know principles and its advocacy of decentralized technology. This hacker ethic, which has remained the preserve of a youth culture for the most part, asserts the basic right of users to free access to all information. It is a principled attempt, in other words, to challenge the tendency to use technology to form information elites. Consequently, hacker activities were presented in the eighties as a romantic countercultural tendency, celebrated by critical journalists like John Markoff of the *New York Times*, by Steward Brand of *Whole Earth Catalog* fame, and by New Age gurus like Timothy Leary in the flamboyant *Reality Hackers*. Fueled by sensational stories about phone phreaks like Joe Egressia (the blind eight-year-old who discovered the tone signal of the phone company by whistling) and Captain Crunch, groups like the Milwaukee 414s, the Los Angeles ARPAnet hackers, the SPAN Data Travellers, the Chaos Computer Club of Hamburg, the British Prestel hackers, *2600*'s BBS, "The Private Sector," and others, the dominant media representation of the hacker came to be that of the "rebel with a modem," to use Markoff's term, at least until the more recent "war on hackers" began to shape media coverage.

On the one hand, this popular folk hero persona offered the romantic high profile of a maverick though nerdy cowboy whose fearless raids upon an impersonal "system" were perceived as a welcome tonic in the gray age of technocratic routine. On the other hand, he was something of a juvenile technodelinquent who hadn't yet learned the difference between right and wrong – a wayward figure whose technical brilliance and proficiency differentiated him nonetheless from, say, the maladjusted working-class J.D. street corner boy of the 1950s (hacker mythology, for the most part, has been almost exclusively white, masculine, and middle-class). One result of this media profile was a persistent infantilization of the hacker ethic – a way of trivializing its embryonic politics, however finally complicit with dominant technocratic imperatives or with entrepreneurial-libertarian ideology one perceives these politics to be. The second result was to reinforce, in the initial absence of coercive jail sentences, the high educational stakes of training the new technocratic elites to be responsible in their use of technology. Never, the given wisdom goes, has a creative elite of the future been so in need of the virtues of a liberal education steeped in Western ethics!

The full force of this lesson in computer ethics can be found laid out in the official Cornell University report on the Robert Morris affair. Members of the university commission set up to investigate the affair make it quite clear in their report that they recognize the student's academic brilliance. His hacking, moreover, is described as a "juvenile act" that had no "malicious intent" but that amounted, like plagiarism – the traditional academic heresy – to a dishonest transgression of other users' rights. (In recent years, the privacy movement within the information community – a movement mounted by liberals to protect civil rights against state gathering of information – has actually been taken up and used as a means of criminalizing hacker activities.) As for the consequences of this juvenile act, the report proposes an analogy that, in comparison with Cornwall's *mature* English country rambler, is thoroughly American, suburban, middle-class, and *juvenile*. Unleashing the Internet worm was like "the driving of a golf-cart on a rainy day through most houses in the neighborhood. The driver may have navigated carefully and broken no china, but it should have been obvious to the driver that the mud on the tires would soil the carpets and that the owners would later have to clean up the mess."[13]

In what stands out as a stiff reprimand for his alma mater, the report regrets that Morris was educated in an "ambivalent atmosphere" where he "received no clear guidance" about ethics from "his peers or mentors" (he went to Harvard!). But it reserves its loftiest academic contempt for the press, whose heroization of hackers has been so irresponsible, in the commission's opinion, as to cause even further damage to the standards of the computing profession; media exaggerations of the courage and technical sophistication of hackers "obscures the far more accomplished work of students who complete their graduate studies without public fanfare," and "who subject their work to the close scrutiny and evaluation of their peers, and not to the interpretations of the popular press."[14] In other words, this was an inside affair, to be assessed and judged by fellow professionals within an institution that reinforces its authority by means of internally self-regulating codes of professionalist ethics, but rarely addresses its ethical relationship to society as a whole (acceptance of defense grants and the like). Generally speaking, the report affirms the genteel liberal ideal that professionals should not need laws, rules, procedural guidelines, or fixed guarantees of safe and responsible conduct. Apprentice professionals ought to have acquired a good conscience by osmosis from a liberal education rather than from some specially prescribed course in ethics and technology.

The widespread attention commanded by the Cornell report (attention from the Association of Computing Machinery, among others) demonstrates the industry's interest in how the academy invokes liberal ethics in order to assist in the managing of the organization of the new specialized knowledge about information technology. Despite or, perhaps, because of the report's steadfast pledge to the virtues and ideals of a liberal education, it bears all the marks of a legitimation crisis inside (and outside) the academy surrounding the new and all-important category of computer professionalism. The increasingly specialized design knowledge demanded of computer professionals means that codes that go beyond the old professionalist separation of mental and practical skills are needed to manage the division that a hacker's functional talents call into question, between a purely mental pursuit and the pragmatic sphere of implementing knowledge in the real world. "Hacking" must then be designated as a strictly *amateur* practice; the tension, in hacking, between *interestedness* and *disinterestedness* is different from, and deficient in relation to, the proper balance demanded by professionalism. Alternately, hacking can be seen as the amateur flip side of the professional ideal – a disinterested love in the service of interested parties and institutions. In either case, it serves as an example of professionalism gone wrong, but not very wrong.

In common with the two responses to the virus scare described earlier – the profitable reaction of the computer industry and the self-empowering response of the legislature – the Cornell report shows how the academy uses a case like the Morris affair to strengthen its own sense of moral and cultural authority in the sphere of professionalism, particularly through its scornful indifference to and aloofness from the codes and judgments exercised by the media – its diabolic competitor in the field of knowledge. Indeed, for all the trumpeting about excesses of power and disrespect for the law of the land, the revival of ethics, in the business and science disciplines in the Ivy League and on Capitol Hill (both awash with ethical fervor in the post-Boesky and post-Reagan years), is little more than a weak liberal response to working flaws or adaptational lapses in the social logic of technocracy.

To complete the scenario of morality play example-making, however, we must also consider that Morris's father was chief scientist of the National Computer Security Center, the National Security Agency's public effort at safeguarding computer security. A brilliant programmer and code breaker in his own right, he had testified in Washington in 1983 about the need to deglamorize teenage hacking, comparing it to "stealing a car for the purpose of joyriding." In a further Oedipal irony, Morris Sr. may have been one of the inventors, while at Bell Labs in the 1950s, of a computer game involving self-perpetuating programs that were a prototype of today's worms and viruses. Called Darwin, its principles were incorporated, in the 1980s, into a popular hacker game called Core War, in which autonomous "killer" programs fought each other to the death.[15]

With the appearance, in the Morris affair, of a patricidal object who is also the Pentagon's guardian angel, we now have many of the classic components of countercultural cross-generational conflict. What I want to consider, however, is how and where this scenario differs from the definitive contours of such conflicts that we recognize as having been established in the sixties; how the Cornell hacker Morris's relation to, say, campus "occupations" today is different from that evoked by the famous image of armed black students emerging from a sit-in on the Cornell campus; how the relation to technological ethics differs from Andrew Kopkind's famous statement, "Morality begins at the end of a gun barrel," which accompanied the publication of the "do-it-yourself Molotov cocktail" design on the cover of a 1968 issue of the *New York Review of Books*; or how hackers' prized potential access to the networks of military systems warfare differs from the prodigious Yippie feat of levitating the Pentagon building. It may be that, like the J.D. rebel without a cause of the fifties, the disaffiliated student dropout of the sixties, and the negationist punk of the seventies, the hacker of the eighties has come to serve as a visible public example of moral maladjustment, a hegemonic test case for redefining the dominant ethics in an advanced technocratic society. (Hence the need for each of these deviant figures to come in different versions – lumpen, radical chic, and Hollywood-style.)

What concerns me here, however, are the different conditions that exist today for recognizing countercultural expression and activism. Twenty years later, the technology of hacking and viral guerrilla warfare occupies a similar place in countercultural fantasy as the Molotov cocktail design once did. While such comparisons are not particularly sound, I do think they conveniently mark a shift in the relation of countercultural activity to technology, a shift in which a software-based technoculture, organized around outlawed libertarian principles about free access to information and communication, has come to replace a dissenting culture organized around the demonizing of abject hardware structures. Much, though not all, of the sixties counterculture was formed around what I have elsewhere called the *technology of folklore* – an expressive congeries of preindustrialist, agrarianist, Orientalist, antitechnological ideas, values, and social structures. By contrast, the cybernetic countercultures of the nineties are already being formed around the *folklore of technology* – mythical feats of survivalism and resistance in a data-rich world of virtual environments and posthuman bodies – which is where many of the SF-and technology-conscious youth cultures have been assembling in recent years.[16]

There is no doubt that this scenario makes countercultural activity more difficult to recognize and therefore to define as politically significant. It was much easier, in the sixties, to *identify* the salient features and symbolic power of a romantic preindustrialist cultural politics in an advanced technological society, especially when the destructive evidence of America's supertechnological invasion of Vietnam was being daily paraded in front of the public eye. However, in a society whose technopolitical infrastructure depends increasingly upon greater surveillance and where foreign wars are seen through the lens of laser-guided smart bombs, cybernetic activism necessarily relies on a much more covert politics of identity, since access to closed systems requires discretion and dissimulation. Access to digital systems still requires only the authentification of a signature or pseudonym, not the identification of a real surveillable person, so there exists a crucial operative gap between authentication and identification. (As security systems move toward authenticating access through biological signatures – the biometric recording and measurement of physical characteristics such as palm or retinal prints, or vein patterns on the backs of hands – the hackers' staple method of systems entry through purloined passwords will be further challenged.) By the same token, cybernetic identity is never used up – it can be re-created, reassigned, and reconstructed with any number of different names and under different user accounts. Most hacks, or technocrimes, go unnoticed or unreported for fear of publicizing the vulnerability of corporate security systems, especially when the hacks are performed by disgruntled employees taking their vengeance on management. So too authoritative identification of any individual hacker, whenever it occurs, is often the result of accidental leads rather than systematic detection. For example, Captain Midnight, the video pirate who commandeered a satellite a few years ago to interrupt broadcast TV viewing, was traced only because a member of the public reported a suspicious conversation heard over a crossed telephone line.

Eschewing its core constituency among white males of the pre-professional-managerial class, the hacker community may be expanding its parameters outward. Hacking, for example, has become a feature of young adult novel genres for girls.[17] The elitist class profile of the hacker prodigy as that of an undersocialized college nerd has become democratized and customized in recent years; it is no longer exclusively associated with institutionally acquired college expertise, and increasingly it dresses streetwise. In a recent article that documents the spread of the computer underground from college whiz kids to a broader youth subculture termed "cyberpunks," after the movement among SF novelists, the original hacker phone phreak Captain Crunch is described as lamenting the fact that the cyberculture is no longer an "elite" one, and that hacker-valid information is much easier to obtain these days.[18]

For the most part, however, the self-defined hacker underground, like many other protocountercultural tendencies, has been restricted to a privileged social milieu, further magnetized by the self-understanding of its members that they are the apprentice architects of a future dominated by knowledge, expertise, and "smartness," whether human or digital. Consequently, it is clear that the hacker cyberculture is not a dropout culture; its disaffiliation from a domestic parent culture is often manifest in activities that answer, directly or indirectly, to the legitimate needs of industrial R&D. For example, this hacker culture celebrates high productivity, maverick forms of creative work energy, and an obsessive identification with on-line endurance (and endorphin highs) – all qualities that are valorized by the entrepreneurial codes of silicon futurism. In a critique of the myth of the hacker-as-rebel, Dennis Hayes debunks the political romance woven around the teenage hacker:

> They are typically white, upper-middle-class adolescents who have taken over the home computer (bought, subsidized, or tolerated by parents in the hope of cultivating computer literacy). Few are politically motivated although many express contempt for the "bureaucracies" that hamper their electronic journeys. Nearly all demand unfettered access to intricate and intriguing computer networks. In this, teenage hackers resemble an alienated shopping culture deprived of purchasing opportunities more than a terrorist network.[19]

While welcoming the sobriety of Hayes's critique, I am less willing to accept its assumptions about the political implications of hacker activities. Studies of youth subcultures (including those of a privileged middle-class formation) have taught us that the political meaning of certain forms of cultural "resistance" is notoriously difficult to read. These meanings are either highly coded or expressed indirectly through media – private peer languages, customized consumer styles, unorthodox leisure patterns, categories of insider knowledge and behavior – that have no fixed or inherent political significance. If cultural studies of this sort have proved anything, it is that the often symbolic, not wholly articulate, expressivity of a youth culture can seldom be translated directly into an articulate political philosophy. The significance of these cultures lies in their embryonic or *protopolitical* languages and technologies of opposition to dominant or parent systems of rules. If hackers lack a "cause," then they are certainly not the first youth culture to be characterized in this dismissive way. In particular, the left has suffered from the lack of a cultural politics capable of recognizing the power of cultural expressions that do not wear a mature political commitment on their sleeves.

So too the escalation of activism-in-the-professions in the last two decades has shown that it is a mistake to condemn the hacker impulse on account of its class constituency alone. To cede the "ability to know" on the grounds that elite groups will enjoy unjustly privileged access to technocratic knowledge is to cede too much of the future. Is it of no political significance at all that hackers' primary fantasies often involve the official computer systems of the police, armed forces, and defense and intelligence agencies? And that the rationale for their fantasies is unfailingly presented in the form of a defense of civil liberties against the threat of centralized intelligence and military activities? Or is all of this merely a symptom of an apprentice elite's fledgling will to masculine power? The activities of the Chinese student elite in the prodemocracy movement have shown that unforeseen shifts in the political climate can produce startling new configurations of power and resistance. After Tiananmen Square, Party leaders found it imprudent to purge those high-tech engineer and computer cadres who alone could guarantee the future of any planned modernization program. On the other hand, the authorities rested uneasy knowing that each cadre (among the most activist groups in the student movement) is a potential hacker who can have the run of the communications house if and when he or she wants.

On the other hand, I do agree with Hayes's perception that the media have pursued their romance with the hacker at the cost of underreporting the much greater challenge posed to corporate employers by their employees. It is in the arena of conflicts between workers and management that most high-tech "sabotage" takes place. In the mainstream everyday life of office workers, mostly female, there is a widespread culture of unorganized sabotage that accounts for infinitely more computer downtime and information loss every year than is caused by destructive, "dark-side" hacking by celebrity cybernetic intruders. The sabotage, time theft, and strategic monkey-wrenching deployed by office workers in their engineered electromagnetic attacks on data storage and operating systems might range from the planting of time or logic bombs to the discrete use of electromagnetic Tesla coils or simple bodily friction: "Good old static electricity discharged from the fingertips probably accounts for close to half the disks and computers wiped out or down every year."[20] More skilled operators, intent on evening a score with management, often utilize sophisticated hacking techniques. In many cases, a coherent networking culture exists among female console operators, where, among other things, tips about strategies for slowing down the temporality of the work regime are circulated. While these threats from below are fully recognized in their boardrooms, corporations dependent upon digital business machines are obviously unwilling to advertise how acutely vulnerable they actually are to this kind of sabotage. It is easy to imagine how organized computer activism could hold such companies for ransom. As Hayes points out, however, it is more difficult to mobilize any kind of labor movement organized upon such premises:

> Many are prepared to publicly oppose the countless dark legacies of the computer age: "electronic sweatshops," military technology, employee surveillance, genotoxic water, and ozone depletion. Among those currently leading the opposition, however, it is apparently deemed "irresponsible" to recommend an active computerized resistance as a source of worker's power because it is perceived as a medium of employee crime and "terrorism."[21]

Processed World, the "magazine with a bad attitude," with which Hayes has been associated, is at the forefront of debating and circulating these questions among office workers, regularly tapping into the resentments borne out in on-the-job resistance.

While only a small number of computer users would recognize and include themselves under the label of "hacker," there are good reasons for extending the restricted definition of *hacking* down and across the caste hierarchy of systems analysts, designers, programmers, and operators to include all high-tech workers, no matter how inexpert, who can interrupt, upset, and redirect the smooth flow of structured communications that dictates their positions in the social networks of exchange and determines the pace of their work schedules. To put it in these terms, however, is not to offer any universal definition of hacker agency. There are many social agents, for example, in job locations who are dependent upon the hope of technological *reskilling*, and for whom sabotage or disruption of communicative rationality is of little use; for such people, definitions of hacking that are reconstructive, rather than deconstructive, are more appropriate. A good example is the crucial role of worker technoliteracy in the struggle of labor against automation and deskilling. When worker education classes in computer programming were discontinued by management at the Ford Rouge plant in Dearborn, Michigan, United Auto Workers members began to publish a newsletter called the *Amateur Computerist* to fill the gap.[22] Among the columnists and correspondents in the magazine have been veterans of the Flint sit-down strikes who see a clear historical continuity between the problem of labor organization in the thirties and the problem of automation and deskilling today. Workers' computer literacy is seen as essential not only to the demystification of the computer and the reskilling of workers, but also to labor's capacity to intervene in decisions about new technologies that might result in shorter hours and thus in "work efficiency" rather than worker efficiency.

The three social locations I have mentioned above all express different class relations to technology: the location of an apprentice technical elite, conventionally associated with the term *hacking*; the location of the female high-tech office worker, involved in "sabotage"; and the location of the shop-floor worker, whose future depends on technological reskilling. All therefore exhibit different ways of *claiming back* time dictated and appropriated by technological processes, and of establishing some form of independent control over the work relation so determined by the new technologies. All, then, fall under a broad understanding of the politics involved in any extended description of hacker activities.

The Culture and Technology Question

Faced with these proliferating practices in the workplace, on the teenage cult fringe, and increasingly in mainstream entertainment, where, over the last five years, the cyberpunk sensibility in popular fiction, film, and television has caught the romance of the popular taste for the outlaw technology of human/machine interfaces, we are obliged, I think, to ask old kinds of questions about the new silicon order that the evangelists of information technology have been deliriously proclaiming for more than twenty years. The postindustrialists' picture of a world of freedom and abundance projects a bright millenarian future devoid of work drudgery and ecological degradation. This sunny social order, cybernetically wired up, is presented as an advanced evolutionary phase of society in accord with Enlightenment ideals of progress and rationality. By contrast, critics of this idealism see only a frightening advance in the technologies of social control, whose owners and sponsors are efficiently shaping a society, as Kevin Robins and Frank Webster put it, of "slaves without Athens" that is actually the inverse of the "Athens without slaves" promised by the silicon positivists.[23]

It is clear that one of the political features of the new post-Fordist order – economically marked by short-run production, diverse taste markets, flexible specialization, and product differentiation – is that the postindustrialists have managed to appropriate not only the utopian language and values of the alternative technology movements but also the Marxist discourse of the "withering away of the state" and the more compassionate vision of local, decentralized communications first espoused by the anarchist. It must be recognized that these are very popular themes and visions (advanced most publicly by Alvin Toffler and the neoliberal Atari Democrats, though also by leftist thinkers such as André Gortz, Rudolf Bahro, and Alain Touraine) – much more popular, for example, than the tradition of centralized technocratic planning espoused by the left under the Fordist model of mass production and consumption.[24] Against the postindustrialists' millenarian picture of a postscarcity harmony, in which citizens enjoy decentralized access to free-flowing information, it is necessary, however, to emphasize how and where actually existing cybernetic capitalism presents a gross caricature of such a postscarcity society.

One of the stories told by the critical left about new cultural technologies is that of monolithic, panoptical social control, effortlessly achieved through a smooth, endlessly interlocking system of networks of surveillance. In this narrative, information technology is seen as the most despotic mode of domination yet, generating not just a revolution in capitalist production but also a revolution in living – "social Taylorism" – that touches all cultural and social spheres in the home and in the workplace.[25] Through routine gathering of information about transactions, consumer preferences, and creditworthiness, a harvest of information about any individual's whereabouts and movements, tastes, desires, contacts, friends, associates, and patterns of work and recreation becomes available in the form of dossiers sold on the tradable information market, or is endlessly convertible into other forms of intelligence through computer matching. Advanced pattern recognition technologies facilitate the process of surveillance, while data encryption protects it from public accountability.[26]

While the debate about privacy has triggered public consciousness about these excesses, the liberal discourse about ethics and damage control in which that debate has been conducted falls short of the more comprehensive analysis of social control and social management offered by left political economists. According to one Marxist analysis, information is seen as a new kind of commodity resource that marks a break with past modes of production and that is becoming the essential site of capital accumulation in the world economy. What happens, then, in the process by which information, gathered up by data scavenging in the transactional sphere, is systematically converted into intelligence? A surplus value is created for use elsewhere. This surplus information value is more than is needed for public surveillance; it is often information, or intelligence, culled from consumer polling or statistical analysis of transactional behavior, that has no immediate use in the process of routine public surveillance. Indeed, it is this surplus bureaucratic capital that is used for the purpose of forecasting social futures, and consequently applied to the task of managing the behavior of mass or aggregate units within those social futures. This surplus intelligence becomes the basis of a whole new industry of futures research that relies upon computer technology to simulate and forecast the shape, activity, and behavior of complex social systems. The result is a possible system of social management that far transcends the questions about surveillance that have been at the discursive center of the privacy debate.[27]

To challenge further the idealists' vision of post-industrial light and magic, we need only look inside the semiconductor workplace itself, which is home to the most toxic chemicals known to man (and woman, especially since women of color often make up the majority of the microelectronics labor force), and where worker illness is measured not in quantities of blood spilled on the shop floor but in the less visible forms of chromosome damage, shrunken testicles, miscarriages, premature deliveries, and severe birth defects. Semiconductor workers exhibit an occupational illness rate that by the late seventies was already three times higher than that of manufacturing workers, at least until the federal rules for recognizing and defining levels of injury were changed under the Reagan administration. Protection gear is designed to protect the product and the clean room from the workers, and not vice versa. Recently, immunological health problems have begun to appear that can be described only as a kind of chemically induced AIDS, rendering the T-cells dysfunctional rather than depleting them like virally induced AIDS.[28] In addition to the extraordinarily high stress patterns of VDT operators in corporate offices, the use of keystroke software to monitor and pace office workers has become a routine part of job performance evaluation programs. Some 70 percent of corporations use electronic surveillance or other forms of quantitative monitoring of their workers. Every bodily movement can be checked and measured, especially trips to the toilet. Federal deregulation has meant that the limits of employee work space have shrunk, in some government offices, below that required by law for a two-hundred-pound laboratory pig.[29] Critics of the labor process seem to have sound reasons to believe that rationalization and quantification are at last entering their most primitive phase.

These, then, are some of the features of that critical left position – or what is sometimes referred to as the "paranoid" position – on information technology, which imagines or constructs a totalizing, monolithic picture of systematic domination. While this story is often characterized as conspiracy theory, its targets – technorationality, bureaucratic capitalism – are usually too abstract to fit the picture of a social order planned and shaped by a small, conspiring group of centralized power elites.

Although I believe that this story, when told inside and outside the classroom, for example, is an indispensable form of "consciousness-raising," it is not always the best story to tell. While I am not comfortable with the "paranoid" labeling, I would argue that such narratives do little to discourage paranoia. The critical habit of finding unrelieved domination everywhere has certain consequences, one of which is to create a siege mentality, reinforcing the inertia, helplessness, and despair that such critiques set out to oppose in the first place. What follows is a politics that can speak only from a victim's position. And when knowledge about surveillance is presented as systematic and infallible, self-censoring is sure to follow. In the psychosocial climate of fear and phobia aroused by the virus scare, there is a responsibility not to be alarmist or to be scared, especially when, as I have argued, such moments are profitably seized upon by the sponsors of control technology. In short, the picture of a seamlessly panoptical network of surveillance may be the result of a rather undemocratic, not to mention unsocialist, way of thinking, predicated upon the recognition of people solely as victims. It is redolent of the old sociological models of mass society and mass culture, which cast the majority of society as passive and lobotomized in the face of the cultural patterns of modernization. To emphasize, as Robins and Webster and others have done, the power of the new technologies to transform despotically the "rhythm, texture, and experience" of everyday life, and meet with no resistance in doing so, is not only to cleave, finally, to an epistemology of technological determinism, but also to dismiss the capacity of people to make their own use of new technologies.[30]

The seamless "interlocking" of public and private networks of information and intelligence is not as smooth and even as the critical school of hard domination would suggest. In any case, compulsive gathering of information is no *guarantee* that any interpretive sense will be made of the files or dossiers, while some would argue that the increasingly covert nature of surveillance is a sign that the "campaign" for social control is not going well. One of the most pervasive popular arguments against the panoptical intentions of the masters of technology is that their systems do not work. Every successful hack or computer crime in some way reinforces the popular perception that information systems are not infallible. And the announcements of military-industrial spokespersons that the fully automated battlefield is on its way run up against an accumulated stock of popular skepticism about the operative capacity of weapons systems. These misgivings are born of decades of distrust for the plans and intentions of the military-industrial complex, and were quite evident in the widespread cynicism about the Strategic Defense Initiative. The military communications system, for example, worked so poorly and so farcically during the U.S. invasion of Grenada that commanders had to call each other on pay phones: ever since then, the command-and-control code of ARPAnet technocrats has been C^5 – Command, Control, Communication, Computers, and Confusion.[31] Even in the Gulf War, which has seen the most concerted effort on the part of the military-industrial-media complex to suppress evidence of such technical dysfunctions, the Pentagon's vaunted information system has proved no more, and often less, resourceful than the mental agility of its analysts.

I am not suggesting that alternatives can be forged simply by encouraging disbelief in the infallibility of existing technologies (pointing to examples of the appropriation of technologies for radical uses, of course, always provides more visibly satisfying evidence of empowerment), but technoskepticism, while not a *sufficient* condition for social change, is a *necessary* condition. Stocks of popular technoskepticism are crucial to the task of eroding the legitimacy of those cultural values that prepare the way for new technological developments: values and principles such as the inevitability of material progress, the "emancipatory" domination of nature, the innovative autonomy of machines, the efficiency codes of pragmatism, and the linear juggernaut of liberal Enlightenment rationality – all increasingly under close critical scrutiny as a wave of environmental consciousness sweeps through the electorates of the West. Technologies do not shape or determine such values. These values already preexist the technologies, and the fact that they have become deeply embodied in the structure of popular needs and desires then provides the green light for the acceptance of certain kinds of technology. The principal rationale for introducing new technologies is that they answer to already-existing intentions and demands that may be perceived as "subjective" but that are never actually within the control of any single set of conspiring individuals. As Marike Finlay has argued, just as technology is possible only in given discursive situations, one of which is the desire of people to have it for reasons of empowerment, so capitalism is merely the site, and not the source, of the power that is often autonomously attributed to the owners and sponsors of technology.[32]

There is no frame of technological inevitability that has not already interacted with popular needs and desires, no introduction of new machineries of control that has not already been negotiated to some degree in the arena of popular consent. Thus the power to design architecture that incorporates different values must arise from the popular perception that existing technologies are not the only ones, nor are they the best when it comes to individual and collective empowerment. It was this kind of perception – formed around the distrust of big, impersonal, "closed" hardware systems, and the desire for small, decentralized, interactive machines to facilitate interpersonal communication – that "built" the PC out of hacking expertise in the early seventies. These were as much the partial "intentions" behind the development of microcomputing technology as deskilling, monitoring, and information gathering are the intentions behind the corporate use of that technology today. The growth of public data networks, bulletin board systems, alternative information and media links, and the increasing cheapness of desktop publishing, satellite equipment, and international data bases are as much the result of local political "intentions" as the fortified net of globally linked, restricted-access information systems is the intentional fantasy of those who seek to profit from centralized control. The picture that emerges from this mapping of intentions is not an inevitably technofascist one, but rather the uneven result of cultural struggles over values and meanings.

It is in this respect – in the struggle over values and meanings – that the work of cultural criticism takes on its special significance as a full participant in the debate about technology. Cultural criticism is already fully implicated in that debate, if only because the culture and education industries are rapidly becoming integrated within the vast information service conglomerates. The media we study, the media we publish in, and the media we teach within are increasingly part of the same tradable information sector. So too, our common intellectual discourse has been significantly affected by the recent debates about postmodernism (or culture in a postindustrial world) in which the euphoric, addictive thrill of the technological sublime has figured quite prominently. The high-speed technological fascination that is characteristic of the postmodern condition can be read, on the one hand, as a celebratory capitulation on the part of intellectuals to the new information technocultures. On the other hand, this celebratory strain attests to the persuasive affect associated with the new cultural technologies, to their capacity (more powerful than that of their sponsors and promoters) to generate pleasure and gratification and to win the struggle for intellectual as well as popular consent.

Another reason for the involvement of cultural critics in the technology debates has to do with our special critical knowledge of the way in which cultural meanings are produced – our knowledge about the politics of consumption and what is often called the politics of representation. This is the knowledge that demonstrates that there are limits to the capacity of productive forces to shape and determine consciousness. It is a knowledge that insists on the ideological or interpretive dimension of technology as a culture that can and must be used and consumed in a variety of ways that are not reducible to the intentions of any single source or producer, and whose meanings cannot simply be read off as evidence of faultless social reproduction. It is a knowledge, in short, that refuses to add to the "hard domination" picture of disenfranchised individuals watched over by some scheming panoptical intelligence. Far from being understood solely as the concrete hardware of electronically sophisticated objects, technology must be seen as a lived, interpretive practice for people in their everyday lives. To redefine the shape and form of that practice is to help create the need for new kinds of hardware and software.

One of the aims of this essay has been to describe and suggest a wider set of activities and social locations than is normally associated with the practice of hacking. If there is a challenge here for cultural critics, then it might be presented as the obligation to make our knowledge about technoculture into something like a hacker's knowledge, capable of penetrating existing systems of rationality that might otherwise be seen as infallible; a hacker's knowledge, capable of reskilling, and therefore of rewriting the cultural programs and reprogramming the social values that make room for new technologies; a hacker's knowledge, capable also of generating new popular romances around the alternative uses of human ingenuity. If we are to take up that challenge, we cannot afford to give up what technoliteracy we have acquired in deference to the vulgar faith that tells us it is always acquired in complicity, and is thus contaminated by the poison of instrumental rationality, or because we hear, often from the same quarters, that acquired technological competence simply glorifies the inhuman work ethic. Technoliteracy, for us, is the challenge to make a historical opportunity out of a historical necessity.

Notes

1 Bryan Kocher, "A Hygiene Lesson," *Communications of the ACM*, 32 (January 1989), 3.

2 Jon A. Rochlis and Mark W. Eichen, "With Microscope and Tweezers: The Worm from MIT's Perspective," *Communications of the ACM*, 32 (June 1989), 697.

3 Philip Elmer-DeWitt, "Invasion of the Body Snatchers," *Time* (September 26, 1988), 62–67.

4 Judith Williamson, "Every Virus Tells a Story: The Meaning of HIV and AIDS," *Taking Liberties: AIDS and Cultural Politics*, ed. Erica Carter and Simon Watney (London: Serpent's Tail/ICA, 1989), 69.

5 "Pulsing the system" is a well-known intelligence process in which, for example, planes deliberately fly over enemy radar installations in order to determine what frequencies they use and how they are arranged. It has been suggested that Morris Sr. and Morris Jr. worked in collusion as part of an NSA operation to pulse the Internet system, and to generate public support for a legal clampdown on hacking. See Allan Lundell, *Virus! The Secret World of Computer Invaders That Breed and Destroy* (Chicago: Contemporary Books, 1989), 12–18. As is the case with all such conspiracy theories, no actual conspiracy need have existed for the consequences – in this case, the benefits for the intelligence community – to have been more or less the same.

6 For details of these raids, see *2600: The Hacker's Quarterly*, 7 (Spring 1990), #7.

7 "Hackers in Jail," *2600: The Hacker's Quarterly*, 6 (Spring 1989), 22–23. The recent Secret Service action that shut down *Phrack*, an electronic newsletter operating out of St. Louis, confirms *2600*'s thesis; a nonelectronic publication would not be censored in the same way.

8 This is not to say that the new laws cannot themselves be used to protect hacker institutions, however. *2600* has advised operators of bulletin boards to declare them private property, thereby guaranteeing protection under the Electronic Privacy Act against unauthorized entry by the FBI.

9 Hugo Cornwall, *The Hacker's Handbook*, 3rd ed. (London: Century, 1988), 181, 2–6. In Britain, for the most part, hacking is still looked upon as a matter for the civil, rather than the criminal, courts.

10 Discussions about civil liberties and property rights, for example, tend to preoccupy most of the participants in the electronic forum published as "Is Computer Hacking a Crime?" in *Harper's*, 280 (March 1990), 45–58.

11 See Hugo Cornwall, *Data Theft* (London: Heinemann, 1987).

12 Bill Landreth, *Out of the Inner Circle: The True Story of a Computer Intruder Capable of Cracking the Nation's Most Secure Computer Systems* (Redmond, Wash: Tempus, Microsoft, 1989), 10.

13 *The Computer Worm: A Report to the Provost of Cornell University on an Investigation Conducted by the Commission of Preliminary Enquiry* (Ithaca, NY.: Cornell University, 1989).

14 Ibid., 8.

15 A. K. Dewdney, the "computer recreations" columnist at *Scientific American*, was the first to publicize the details of this game of battle programs in an article in the May 1984 issue of the magazine. In a follow-up article in March 1985, "A Core War Bestiary of Viruses, Worms, and Other Threats to Computer Memories," Dewdney described the wide range of "software creatures" that readers' responses had brought to light. A third column, in March 1989, was written, in an exculpatory mode, to refute any connection between his original advertisement of the Core War program and the spate of recent viruses.

16 Andrew Ross, *No Respect: Intellectuals and Popular Culture* (New York: Routledge, 1989), 212. Some would argue, however, that the ideas and values of the sixties counterculture were only fullfilled in groups like the People's Computer Company, which ran Community Memory in Berkeley, or the Homebrew Computer Club, which pioneered personal microcomputing. So too the Yippies had seen the need to form YIPL, the Youth International Party Line, devoted to "anarcho-technological" projects, which put out a newsletter called *TAP* (alternately the *Technological American Party* and the *Technological Assistance Program*). In its depoliticized form, which eschewed the kind of destructive "dark-side" hacking advocated in its earlier incarnation, *TAP* was eventually the progenitor of *2600*. A significant turning point, for example, was *TAP*'s decision not to publish plans for the hydrogen bomb (the *Progressive* did so) – bombs would destroy the phone system, which the *TAP* phone phreaks had an enthusiastic interest in maintaining.

17 See Alice Bach's "Phreakers" series, which narrates the mystery-and-suspense adventures of two teenage girls: *The Bully of Library Place* (New York: Dell, 1988), *Double Bucky Shanghai* (New York: Dell, 1987), *Parrot Woman* (New York: Dell, 1987), *Ragwars* (New York: Dell, 1987), and others.

18 John Markoff, "Cyberpunks Seek Thrills in Computerized Mischief," *New York Times* (November 26, 1988), 1, 28.

19 Dennis Hayes, *Behind the Silicon Curtain: The Seductions of Work in a Lonely Era* (Boston: South End, 1989), 93. One striking historical precedent for the hacking subculture, suggested to me by Carolyn Marvin, was the widespread activity of amateur or "ham" wireless operators in the first two decades of the century. Initially lionized in the press as boy-inventor heroes for their technical ingenuity and daring adventures with the ether, this white middle-class subculture was increasingly demonized by the U.S. Navy (whose signals the amateurs prankishly interfered with), which was crusading for complete military control of the airwaves in the name of national security. The amateurs lobbied with democratic rhetoric for the public's right to access the airwaves, and, although partially successful in their case against the Navy, lost out ultimately to big commercial interests when Congress approved the creation of a broadcasting monopoly after World War I in the form of RCA. See Susan J. Douglas, *Inventing American Broadcasting 1899–1922* (Baltimore: Johns Hopkins University Press, 1987), 187–291.

20 "Sabotage," *Processed World*, 11 (Summer 1984), 37–38.

21 Hayes, *Behind the Silicon Curtain*, 99.

22 *The Amateur Computerist*, available from R. Hauben, P.O. Box 4344, Dearborn, MI 48126.

23 Kevin Robins and Frank Webster, "Athens without Slaves ... or Slaves without Athens? The Neurosis of Technology," *Science as Culture*, 3 (1988), 7–53.

24 See Boris Frankel, *The Post-Industrial Utopians* (Oxford: Basil Blackwell, 1987).

25 See, for example, the collection of essays edited by Vincent Mosco and Janet Wasko, *The Political Economy of Information* (Madison: University of Wisconsin Press, 1988), and Dan Schiller, *The Information Commodity* (Oxford University Press, forthcoming).

26 Tom Athanasiou and Staff, "Encryption and the Dossier Society," *Processed World*, 16 (1986), 12–17.

27 Kevin Wilson, *Technologies of Control: The New Interactive Media for the Home* (Madison: University of Wisconsin Press, 1988), 121–25.

28 Hayes, *Behind the Silicon Curtain*, 63–80.

29 "Our Friend the VDT," *Processed World*, 22 (Summer 1988), 24–25.

30 See Kevin Robins and Frank Webster, "Cybernetic Capitalism," *The Political Economy of Information*, ed. Vincent Mosco and Janet Wasko (Madison: University of Wisconsin Press, 1988), 44–75.

31 Barbara Garson, *The Electronic Sweatshop* (New York: Simon & Schuster, 1988), 244–45.

32 See Marike Finlay's Foucauldian analysis, *Powermatics: A Discursive Critique of New Technology* (London: Routledge & Kegan Paul, 1987). A more conventional culturalist argument can be found in Stephen Hill, *The Tragedy of Technology* (London: Pluto, 1988).

47

Information and Reality at the Turn of the Century

Albert Borgmann

Our ability to command information grows every day. We wake up to the news on the radio, read the paper during breakfast, are immersed in signs and advertisements as we make our way to the office, sit down and fire up our computers – that really opens the floodgate of information – return home, turn on the television set and let waves of information wash over us until we go to bed.

Many people say they get much pleasure from cruising and surfing on this sea of information. But, at times, we feel like the sorcerer's apprentice, unable to contain the powers we have summoned, and afraid of drowning in the flood we have loosened. Where does all this information come from and what is it doing to reality?

In earlier times, information arose from the interplay of three factors, a messenger, a recipient, and a message. Paul Revere was the messenger, the Patriots were the recipients, and "The English are coming," was the message. The dove with the olive branch was the messenger, Noah the recipient, and the message was "The waters are receding." Even today, smoke is the messenger, the pilot of the spotter plane the recipient, and "There is a fire in section 36" is the message. A rock-lined cavity can be the messenger, a historian the recipient, and the message: "This is the root cellar of an abandoned homestead."

The general pattern is always *x informs y about z*. X is the messenger, but we also call x the sign, the signal, the symbol, or the vehicle. Y is the recipient or the receiver, the audience, the listener, the reader, the viewer, the spectator, or the investigator. And z is the message, the information, the news, the intelligence, the meaning, or the content conveyed by the messenger or the signal. This basic triad of factors had an engaging sturdiness and intelligibility. It incorporated distinctions that may have been grist for abstruse philosophical mills; but they were clear and serviceable enough for lay people.

We can call this the core meaning of information.[1] It is *information about reality*. In its simplest form, it is a natural phenomenon relevant to animals as well as to humans. A scent carried by the wind is information to a bear that carrion is to be found down by the creek. Information is a difficult phenomenon, not because its origin is mysterious, but because, throughout human history, it has been evolving in tiny steps from something simple and natural to something exceedingly complex and technical.

The fire ring you left at a campsite a few years ago informs you, in a natural and incidental way, that this is a good spot to put up your tent. But if it took you a while to find it again and you want to avoid going in circles next time, you might pile up a few rocks where you have to leave the trail and head down to the creek. While the rocks around the fire ring compose an incidental and nearly natural sign, the rocks in the pile constitute an intentional and conventional sign.

The first conventional signs in human history were reminders, tallies on bones or sticks, and collections of pebbles or shells, to keep track of the phases of the moon or animal kills, or sheep,

From *Design Issues* 11/2 (1995): 21–30.

or goats. Pebbles were followed by clay tokens; these by incisions and impressions on clay tablets; and those, again, in succession by logographs, syllabaries and letters. These signs served, and some of them still serve, to set down information about the past. But they also can bring nearer what is distant in space, edicts from the king, reports of disasters, and news of life everlasting. Of course, information about things past and distant events does not literally convey those things and events to the here and now. A record of a bushel of grain delivered to the granary is not the same as the bushel of grain, itself. Nor is a royal edict the same as the majestic presence of the king. It takes comprehension to gather the message from the sign and to grasp the impact of the message on the here and now.

The ability to comprehend a message and integrate it into one's immediate world is, except for one's native language, an arduously acquired and ever incomplete skill. But once acquired, more or less, comprehension makes for an incomparably more comprehensive and comprehensible world. Reality no longer trails off rapidly in the mists of distance and past, but becomes perspicuous as far as information and comprehension can reach.

Information, however, not only can illuminate what is distant in time and space, but also what is remote in conception and imagination; and what would remain a distant possibility without the aid of conventional signs. Complex social arrangements, monumental buildings, and artful pieces of music would be inconceivable without the information laid down, e.g., in the texts of covenants, drawings of cathedrals, and scores of cantatas. This no longer is *information about reality* but *information for reality*; information for making a community, a building, or music. But here again, the steps from *information about* to *information for* reality and back are small. In the sketchbook of the medieval master mason, Villard de Honnecourt, plans that inform us about the existing cathedral of Laon are mingled with plans for churches that were never executed.[2] Gregorio Allegri wrote a score for the mass *Miserere*, to be performed in the Sistine Chapel only and, therefore, kept secret. But when the fourteen-year-old Mozart, on having heard the mass once, wrote down the score from memory, it was information about rather than for the performance of the music.

While *information about* reality chiefly requires comprehension and renders the world more perspicuous, *information for* reality calls for realization and makes for a more prosperous world. By realization, I mean the process of translating information into reality. To realize a covenant is to make it come alive in practices and celebrations. To realize a blueprint is to construct a building conforming to it. And to realize a musical score is to perform it.

Realization, too, requires the acquisition of demanding skills; the particular literacies of rabbis, artisans and musicians, among many others. But again, when these skills are mastered, the world becomes incomparably richer; adorned with festive customs, material structures and musical events that are intricate and magnificent beyond what human imagination, without information and notation, could have conceived and carried out.

Information about and for reality used to mediate between humanity and reality to produce a distinctive kind of world. The central feature of this world was a focal area of nearness that was understood against a comprehensible background of farness. Information about reality, even today, tells us of faraway things and events; but it does not transport them into our midst. They remain distant, yet they inform and illuminate what is present. If, on a hot and windy August afternoon in the Rockies, you see smoke billowing up like thunderheads behind the farthest ridge on the horizon; the smoke conveys information about a raging wildfire; but it does not carry the fire, itself, into your immediate world. The focal area of your life is still safe. The fire is distant though the fire danger is close.

Similarly, when you stumble upon a collapsed root cellar from early in this century, you learn of a way of life where perishables used to be entrusted to the coolness of the ground. Yet root cellars remain a thing of the past. They are part of the background that defines your way of life where food is stored in freezers and refrigerators. And so it is with letters from distant friends and colleagues; with journals and memoirs of homesteaders; and with tipi rings and cairns left by the Salish and Kootenai Indians. Information about reality keeps, or used to keep, the farness of space and time distinct from the nearness of those persons and things that make up the focal area of our lives. Yet information makes farness comprehensible, and provides an illuminating background for what is here and now.

While *information about reality* renders our world perspicuous in its order of nearness and farness, *information for reality* is the source of a distinctively prosperous culture. Such prosperity requires submission to a definite kind of discipline. Discipline is needed, first, to acquire the skills of reading texts, following plans, and playing musical scores; and then to submit continually to the text of a covenant, the specifications of a blueprint, and the notes of sheet music. The reward of such discipline is a franchise in a magnificent culture. If you are given the culture without being instructed in the requisite discipline, the culture is not truly yours. You are dependent on others for the realization of the founding information of the culture. But if you are skilled, you can fully and freely appropriate your inheritance.

In addition to discipline and competence, a world built on information for reality engenders a vigorous sense of continuity, community and intimacy. Jewish people trace their community back for some four thousand years. Their tradition would not be so astoundingly continuous and vigorous had it not been for a writer, now commonly called J, who set down the story of Abraham and Sarah a few hundred years later.[3] But to realize the promise God made to Abraham, one needs not only an unbroken tie to the founding event, but also a community of people with whom to enact the tradition.

Similarly, the music which Bach left us requires not only discipline and competence, but also fellow singers and players. And more is needed yet. Since information for reality is never more than an instruction or a recipe, we must gather from the tangible reality around us the materials and ingredients necessary for the realization of a text, a plan, or a score. To realize musical scores, we need instruments. To realize a blueprint, we need building materials. The realization of texts, particularly of poetry and fiction, is the most subtle. But it, too, calls for us to draw on our experiences with the tangible world. In any case, to realize information for reality, we have to be intimate with the visible, audible, palpable shapes of our immediate world; with the strings, mouthpieces and keys of instruments; with the stones and timbers of construction; and with the faces and gestures we recall when we read a poem.

Traditionally, information has been about and for reality. But through the technological developments of the past century and a half, information, though still about and for reality, also has begun to rival reality, itself; and has emerged virtually *as reality*. The culture of technology, in general, is animated by a pervasive and evident desire for more copious and refined consumption. Yet the implementation of this project looks more like bricolage than the execution of a grand design. Rather than follow the twists and turns of the transformation of information, I will trace the conceptual path that leads from information about and for reality to information as reality.

I begin with the landmark that has defined a crucial stretch of this path, viz., Claude Shannon's paper of 1948, "The Mathematical Theory of Communication," and Warren Weaver's comments of 1949.[4] Shannon's paper is strictly a technical accomplishment, but it was thought right away to be a cultural milestone. The excitement it generated is obvious in Weaver's remarks. Shannon's work served as a catalyst of powerful cultural elements that were ready to react. Weaver's reaction, in particular, is emblematic in what it had to say about the rising cultural force of information.

The rise of this force has inundated and begun to wash out the structure of the world that was built on the traditional information about and for reality. The threat to its central feature, the focal area of nearness surrounded by a comprehensible background of farness, was signaled by Weaver when he erased the distance between nearness and farness. Nearness used to be marked by the immediate presence of things and persons, as opposed to the farness of distant things and persons that were represented through the information we had about them. But, for Weaver, our access to persons, at any rate, is always through information; whether a person is near or far.

"In oral speech," Weaver said, "the information source is the brain, and the transmitter is the voice mechanism producing the varying sound pressure (the signal) which is transmitted through the air (the channel)."[5] Was Weaver simply making a technical point which, as one critic has charged, was not technically warranted?[6] Roughly half a century later, Deborah Tannen tells us about her colleague, Ralph:

> E-mail deepened my friendship with Ralph. Though his office was next to mine, we rarely had extended conversations because he is shy. Face to face, he mumbled, so I could barely tell he was speaking. But when we both got on e-mail, I started receiving long, self-revealing messages; we poured our hearts out to each other.[7]

Tannen calls e-mail "a souped-up conversation." Apparently, it brings people closer to each other than a traditional f2f (face to face) conversation can. The contrast between nearness and farness has evaporated. It is an irrelevant circumstance that Ralph's and Deborah's offices are on the same floor. They might as well be on different continents.

Underlying this particular illustration is the general view that humans are in touch with the world through information, and through information only. The mountain peak across the valley is visual information; the wind in the trees is acoustic information; and a caress from your loved one is tactile information. In this sense, everything out there, all of reality, has always been information.[8]

Though this view is a powerful undercurrent in our culture, there also is a more particular connection between information and reality, where reality is simply not revealed or claimed to be information, but information arises as a rival and pretender to the throne of traditional reality. The rise of this particular pretension is closely bound up with a threat to the discipline and competence the world of traditional information used to require.

On this issue, Weaver is divided. Like Shannon, he holds to a distinction between the signal and the message; between the vehicle and the content of information.[9] That seems to make eminently good sense, since the two traditionally have been so disparate from one another. A thin plume of smoke curling up from the Rocky Mountains on a breezy August afternoon is, all by itself, a slight and innocent signal. But the message it conveys is awful: within a few hours there might be a roaring wildfire.

It is similar with the messages that are conveyed by conventional signs or signals. There is a huge disparity between the mere 20 column inches of print that a report of a forest fire occupies in a newspaper, and the many acres of incinerated grass and charred trees the report informs us about. As e-mail illustrates, the austere signals of the alphabet can carry weighty messages even today. As long as signal and message, or vehicle and content of information, remain distinct; the need to learn how to gather the message from the signal or the content from the vehicle, and achieve competence in appropriating one's cultural heritage, will persist. But the traditional discipline is in jeopardy.

Signal and message contrast much like quantity and quality. Message and quality are subjective and elusive, while signal and quantity are objective and measurable.[10] Some people believe that the life of the spirit is essentially, and will forever be, beyond quantification and objectification. But a more aggressive and confident view of the matter holds that quality is simply unanalyzed quantity; and that all qualities, when analyzed all the way down to their basic constituents, will turn out to be objective and measurable quantities.

The crucial contribution Shannon made to technological culture was his definitive underscoring of information as a quantity measurable in bits, by showing that the basic unit of information could be used to demonstrate interesting theorems about the coding and transmission of information. Weaver sensed that differences in the quality of information are superficial and ultimately irrelevant; and that it really doesn't matter "what kinds of symbols are being considered – whether written letters or words, or musical notes, or spoken words, or symphonic music, or pictures."[11] In this spirit, he might have proposed to erase the distinction between the message and the signal, or the content and the vehicle of information, by analyzing the message or content into just so many bits of information. But he might as well have offered a thimble to someone trying to drain a pond. At the time, the technology available could not command and control of a sufficient number of bits.

Let me illustrate the problem as well as the solution that eventually emerged. Imagine I tell you the Baroque Consort will give a performance tonight and you ask, "What's the first piece on the program?" and I answer: "Bach's Cantata No. 10." The symbols I have strung together to reply carry about 175 bits of information, and so would the letters if I were to point silently to the first item on the program. "Bach's Cantata No. 10" is the signal or the vehicle. But what does it convey? In one sense, nothing less than an entire cantata of some 25 pages of sheet music and some 20 minutes of performance; a church cantata for four soloists, choir, trumpet, two oboes, basso continuo, and strings. But how many people could bring to mind Bach's music in all its details when they hear "Bach's Cantata No. 10"? Perhaps no more than a few dozen among the more than five billion people alive today. Of course, there is a gradation and division of competence when it comes to recalling music. To determine what "Bach's Cantata No. 10" should convey to an educated mind is a difficult and contentious task. E. D. Hirsch lists *Johann Sebastian Bach*, but not *cantata*, as an item literate Americans should know.[12] In any case, "Bach's Cantata No. 10" is an example both of the great disparity that is possible between signal and message; between an extremely austere vehicle of 175 bits and an abundantly rich content of baroque music; and of the inverse proportion between the magnitude of the signal and the magnitude of the competence needed fully to comprehend the message.

Now assume that, instead of pointing to an item on the program, I were to hand you the score and text of the cantata for you to peruse. This information comes to perhaps 164,000 bits, a vehicle richer than the entry on the program by a factor of about a thousand. The distance between signal and message has been decreased, and so have the demands on the competence needed to gather the content from the vehicle. All that is needed now are musical literacy and a command of German. But in contemporary culture, these are still forbidding requirements and, accordingly, the signal and message are still separated by a prohibitive abyss.

Moreover, even for competent readers of the score, there is a significant gap between it and the presence of the cantata in an actual performance. Much of that which still must be decided and filled in to have a performance is left blank in the score; including matters such as tempo, phrasing, pitch, the particular choice of instruments and singers, and more. How many bits would it take to have everything spelled out to the last detail? In a dubious but still enlightening sense, it would take about 1.2 billion bits, a signal about seven million times larger than the entry on the program, and seven thousand times larger than the score and text.[13]

You can find these billion-plus bits recorded in binary notation on a compact disk. Many would say that, at this level of richness, the signal has all but caught up with the message and perhaps even surpassed it. What you have on the disk is the music itself. And if the gap between vehicle and content has not shrunk to zero, the demands on the competence required to bring the music to life certainly have. It takes less than a minute to show someone how to use a CD player.

CDs not only suggest that, in contemporary culture, the line between information and reality has become questionable as Weaver had hinted; they demonstrate outright what Weaver had claimed explicitly, viz., that all information is fundamentally alike. The compact disk can contain all kinds of information indifferently; whether it consists, as Weaver has it, of "written letters or words, or musical notes, or spoken words, or symphonic music, or pictures." But this account still leaves us with a fuzzy picture of information as reality. Sound recordings, after all, have existed for more than a hundred years. Is it not a mere matter of technical convenience whether the sound is recorded in analog or digital form? The crucial point is that information in analog form, such as photographs, films, or vinyl records really isn't emancipated from reality and, like a satellite in low orbit, is in constant danger of falling back into reality. Moving analog information in and out of storage media is a relatively slow process; and it is irreversibly deteriorated as it moves through time and the mechanics of copying. Its internal structure is as viscous as honey, and difficult to manipulate and control.

Traditional writing and print, of course, are digital and therefore mobile, durable and crisp; where analog information is inert, corruptible and fuzzy. But writing also is an extremely austere vehicle of information. It was the wedding of binary digitality with electronics that, in time, provided the union of control and abundance of information.

A digitalized electronic recording of a cantata gives us a richer rendition of the music than an analog recording but, beyond that, digitalized electronic information can move at the speed of an electric current or of light. It retains its fidelity with immutable perfection from copy to copy; and, possessing a sharply articulated internal structure, it can easily and speedily be analyzed, sorted and modified any way you please. Digital electronic information has emancipated itself from the poverty of traditional digital information, and from the heaviness and caducity of reality. Thus, it hovers as an omnipresent and unblinking virtual reality above the slow and weary actual reality.

What has been the cultural effect of this new kind of information? Again, Weaver caught the rising wave of excitement. He underscored a phenomenon which is inherent in all information, but was not clearly grasped until the question of the amount of information was raised and answered in bits. In one sense, information signals behave like any normal vehicle or container. If you have two buckets, you can carry twice as much water as if you had one. On two pages, you can say twice as much as on one. But the vehicles or containers of information have the peculiarity that, when their capacity for amount is doubled, their capacity for variety is squared. So if on a page you can say a thousand different things, then on two pages you can say not just twice a thousand, (i.e., two thousand different things), but a thousand times a thousand (i.e., a million different things).

Weaver calls the availability of variety "freedom of choice" and, in his view, a steady increase in the vehicles of information leads to an explosive growth in freedom of choice.[14] Not very many people of the information age may be aware of this fine point of information theory. But the realization that the growth of information brings with it an explosion of possibilities is widespread. Of course, this development has simply benefited information about and for reality.

The Internet can serve as the emblem of the digital and electronic improvement of information about reality. By now, it is a truism that it furnishes us with information about persons and things that would be difficult to obtain without the net. Moreover, the Internet provides for the collective exchange of information by way of bulletin boards, lists, conferences, etc., that would be impossible without it. More generally, computers and databases furnish information that you would not think of looking for without them, e.g., information on how many places down you rank from the most frequently cited scholar in your field.

Less frivolous information, uniquely obtainable from computers, also is available. One example is information on precisely how air flow builds up drag around the pod of a jet engine, when the pod is placed under the wing or when it is ahead of the wing; and when it is round and when it is pear-shaped. Such information about airflow naturally becomes essential in designing of an airplane.[15]

The Boeing 777 can serve as an emblem of the power of digital electronic information for reality. Supercomputers are playing a crucial role in its development. Computerized information has largely replaced blueprints and mockups, and it has suffused the very thing in its final shape. The 777 no longer is guided by Newtonian levers and wires, but by information relayed from the pilot or a computer to servomechanisms.[16]

Such achievements are, undoubtedly, occasions for admiration with regard to the virtuosity that has gone into the conception and construction of the information technology. As for the end use this technology is put to, there again are occasions for gratitude. But when it comes to the overall cultural effect of it, the result lies somewhere between the trivial and the troubling.

To begin with the Internet, most of what flows through it, as far as I can tell, is overwhelmingly flimsy. There is much throat clearing, half-hearted criticism, throwing out of suggestions, crashing obviousness, and instruction by the moderately knowledgeable of the totally ignorant. The tone oscillates between the obsequiously laudatory and the rudely offensive, with much blandness in between. The Internet, for the most part, is a dump of wasted time.

Wasting time is not a uniquely late 20th century phenomenon, of course. Does electronic information technology have a specific impact on culture? It does, and its influence flows from information as reality down to information about and for reality. Information as virtual reality has inherited the enormous pliability and preternatural perfection of digital electronic information. The norm of pliability and perfection that is enshrined in virtual reality has cast a shadow on information about and for reality, and thereby rendered persons less personable and reality less real.

To return to Deborah Tannen's colleague as an example, Ralph has not actually been cured of his shyness nor has Deborah actually learned how to "break the ice." Both have used the Internet to create versions of themselves virtually possessing the virtues they actually lack and no longer have any reason actually to acquire. One hears anecdotes of virtual beginnings having actually happy endings. But veterans of virtual reality, as David Bennahum has well put it, "are wary – they've been burned too often by net friendships that collapsed once the stuff of real life came in."[17]

As for the bearing that the norm of pliability and perfection has on information for the shaping of reality, consider once more the Boeing 777. What is striking in historical perspective is the slackening of the power to shape things over the last thirty or forty years. To a lay person, the manifest differences between the 777 and the 707 of a generation ago are slight. The same must be said of buildings, highways and cars. The crucial changes have not been in shape, but in sophistication.

The latter is astounding and admirable in its own right, to be sure, and, other things being equal, beneficial to humanity and ecology. In the case of the 777, the end result will be cheaper, quieter and more commodious flights. But notice the end effect of this result. The sense of passage from place to place will be further attenuated, and so will be the distinction between nearness and farness.

Information as virtual reality has its own and direct effects on culture. It serves utility in design, manufacturing, medicine, and science. But it serves consumption as well, and saps everyone's vigor in dealing with the actual world. It is no respecter of class or education. Some it traps with keyboards and texts; others with buttons and pictures.

To put the drift of information technology briefly and starkly, information at the turn of the century tends to enfeeble persons and attenuate things. Luddism, of course, is not the answer. Information cannot and should not be wished away. But it has to be counterbalanced. We must regain the actuality of people and the nearness of things, and turn from information as reality to information and reality.

Notes

1 Fred I. Dretske, *Knowledge and the Flow of Information* (Cambridge: MIT Press, 1981), 41–47.
2 *The Sketchbook of Villard de Honnecourt*, ed. by Theodore Bowie (Westport, CT: Greenwood, 1982), 88–91, 106–7.
3 For background and a fanciful view of J see Harold Bloom, *The Book of J* (New York: Grove Weidenfeld, 1990).
4 Claude E. Shannon and Warren Weaver, *The Mathematical Theory of Communication* (Urbana: University of Illinois Press, 1949).
5 Ibid., 98.
6 David Ritchie, "Shannon and Weaver: Unraveling the Paradox of Information," *Communication Research* 13 (1986): 283–7.
7 Deborah Tannen, "Gender Gap in Cyberspace," *Newsweek* (16 May 1994): 52.
8 See Robert Wright, *Three Scientists and Their Gods* (New York: Harper, 1988), 21–38.
9 Shannon and Weaver, *Mathematical Theory*, 3 and 100. See also Seth Finn and Donald F. Roberts, "Source, Destination, and Eutropy: Reassessing the Role of Information Theory in Communication Research," *Communication Research* 11 (1984): 456.
10 Donald M. Mackay, *Information, Mechanism and Meaning* (Cambridge: MIT Press, 1969), 17.
11 Shannon and Weaver, 114.
12 E. D. Hirsch, *Cultural Literacy: What Every American Needs to Know* (Boston: Houghton, 1987).
13 There is no hard and fast way of measuring the amount of information of a given item in bits. For alphanumeric test, I have followed ASCII. For Bach's musical notation, I have assumed 12 bits per note. For the CD, I have assumed 54 million bits per minute.
14 Shannon and Weaver, 100–1.
15 Albert M. Erisman and Kenneth W. Neves, "Advanced Computing for Manufacturing," *Scientific American* (October 1987), 163–64.
16 See the April 1994 issue of *Aviation Week and Space Technology*, devoted to the Boeing 777.
17 David Bennahum, "Fly me to the MOO: Adventures in Textual Reality," *Lingua Franca* (May/June 1994), 26.

48

Anonymity versus Commitment: The Dangers of Education on the Internet

Hubert L. Dreyfus

Introduction

For two decades computers have been touted as a new technology that will revitalize education. In the eighties they were proposed as tutors, tutees, and drill masters, but none of those ideas seem to have taken hold. Now the latest proposal is that somehow the power of the World Wide Web will make possible a new approach to education for the twenty-first century.

In *School's Out*, Lewis J. Perelman, considering the future of high schools, announces with assurance:

> With knowledge doubling every year or so, "expertise" now has a shelf life measured in days; everyone must be both learner and teacher; and the sheer challenge of learning can be managed only through a globe-girdling network that links all minds and all knowledge.[1]

He continues:

> I call this new wave of technology *hyperlearning*...The *hyper* in hyperlearning refers not merely to the extraordinary speed and scope of new information technology, but to an unprecedented degree of connectedness of knowledge, experience, media, and brains – both human and nonhuman.... We have the technology today to enable virtually anyone who is not severely handicapped to learn anything, at a "grade A" level, anywhere, anytime.[2]

Speaking of higher education in *Transforming Education*, Michael Dolence and Donald Norris make the same point:

> Under the emerging vision for learning in the 21st century, information technology is a primary instrument of transformation. It is the key ingredient making feasible a network learning, distance-free, knowledge navigation-based vision for the Information Age learner.[3]

I propose to translate Kierkegaard's account of the dangers and opportunities of what he called "The Press" into a critique of "The Internet" so as to raise the question: what contribution – for good or ill – can the World Wide Web, with its capacity to deliver vast amounts of information to users all over the world, make to educators trying to pass on knowledge and to develop skills in their students? I will then use Kierkegaard's three-stage answer to the problem of anonymity and lack of involvement posed by the press – his claim that to have a meaningful life the learner must pass through the aesthetic, the ethical and the religious spheres of existence – to suggest that only the first two stages – the aesthetic and the ethical – can be implemented with Information Technology, while the religious stage, which alone makes meaningful learning possible, is undermined rather than supported by the tendencies of the Net.

From *Ethics and Information Technology* 1 (1999): 15–21. Reprinted by permission of Kluwer Academic Publishers, Dordrecht, The Netherlands.

How the Press and the Internet Undermine Commitment and Meaning

In his essay, *The Present Age*,[4] Kierkegaard claimed that his age was characterized by reflection and curiosity. People took an interest in everything but were not committed to anything. He attributed this growing cultivation of curiosity and the consequent failure to distinguish the important from the trivial to the press. Its new massive distribution of desituated information, he held, was making every sort of information immediately available to anyone thereby producing an anonymous, detached spectator. He wrote in his Journal: "...here ...are the two most dreadful calamities which really are the principle powers of impersonality – the press and anonymity."[5] Kierkegaard thought that, thanks to these powers, the press would complete the leveling of qualitative distinctions, distinctions of worthiness, a leveling that had been going on in the West since the Enlightenment.

What Kierkegaard envisaged as a consequence of the press's indiscriminate coverage and dissemination is now being realized on the World Wide Web. All qualitative distinctions are, indeed, being leveled. Relevance and significance have disappeared. And this is an important part of the attraction of the Web. Nothing is too trivial to be included. Nothing is so important that it demands a special place. In his religious writing, Kierkegaard criticized the implicit nihilism in the idea that God is equally concerned with the salvation of a sinner and the fall of a sparrow. On the Web, the attraction and the danger is that everyone can take this godlike point of view. One can view a coffee pot in Cambridge, the latest super-nova, read the latest news in Chile, look up references in a library in Alexandria, or direct a robot to plant and water a seed in Austria, not to mention plow through thousands of ads, all with equal ease and equal lack of any sense of what is important.[6]

When Kierkegaard attacks the Press, he attacks it for creating what he calls *the Public*. As Kierkegaard puts it, "In order that everything should be reduced to the same level, it is first of all necessary to procure a phantom, its spirit, a monstrous abstraction, and all-embracing something which is nothing, a mirage – and that phantom is the public."[7] According to Kierkegaard, the Press speaks for the Public but no one stands behind the views the public holds. Because of this absence of responsibility. Kierkegaard suggested as a motto for the Press: "Here men are demoralized in the shortest possible time on the largest possible scale, at the cheapest possible price."[8]

Kierkegaard would no doubt have been happy to transfer this same motto to the Internet, for, just as no individual assumes responsibility for the consequences of the information in "the Press", no one assumes responsibility for the accuracy of the information on the Web. The information has become so anonymous that no one knows or cares where it came from. Of course, in so far as one does not take action on the information, no one really cares if it is reliable. All that matters is that everyone passes the word along by forwarding it to other users. Moreover, in the name of protecting privacy, ID codes are being developed that will assure that even the sender's address can be kept secret. The Net is thus a perfect medium for slander and innuendo. Kierkegaard could have been speaking of the Internet when he said of the Press: "It is frightful that someone who is no one...can set any error into circulation with no thought of responsibility and with the aid of this dreadful disproportioned means of communication."[9]

Kierkegaard would surely see in the Net with its interest groups, which anyone in the world can join and where one can discuss any topic endlessly without consequences, the height of irresponsibility. Without rootedness in a particular problem, all that remains for the interest group commentator is endless gossip. In such groups anyone can have an opinion on anything and all are only too eager to respond to the equally deracinated opinions of other anonymous amateurs who post their views from nowhere. Such commentators do not take a stand on the issues they speak about. Indeed, the very ubiquity of the Net makes any such local stand seem irrelevant. As Kierkegaard puts it: "A public is neither a nation, nor a generation, nor a community, nor a society, nor these particular men, for all these are only what they are through the concrete; no single person who belong to the public makes a real commitment."[10]

The only alternative Kierkegaard saw to this anonymity and lack of commitment was to plunge into some kind of activity – any activity – as long as one threw oneself into it with passionate involvement. Towards the end of *The Present Age* he exhorts his contemporaries to make such a leap:

> There is no more action or decision in our day than there is perilous delight in swimming in

> shallow waters. But just as a grown-up, struggling delightedly in the waves, calls to those younger than himself: "Come on, jump in quickly" – the decision in existence...calls out. ...Come on, leap cheerfully, even if it means a lighthearted leap, so long as it is decisive. If you are capable of being a man, then danger and the harsh judgment of existence on your thoughtlessness will help you become one.[11]

The Aesthetic Sphere: Commitment to the Enjoyment of Sheer Information

Such a lighthearted leap into the deeper water is typified by the Net-surfer for whom information gathering has become a way of life. Such a surfer is curious about everything and ready to spend every free moment visiting the latest hot spots on the Web. He or she enjoys the sheer range of possibilities. Something interesting is only a click away.

Commitment to information as a boundless source of enjoyment puts one in what Kierkegaard calls *the aesthetic sphere of existence* – his anticipation of postmodernity. For such a person just visiting as many sites as possible and keeping up on the cool ones is an end in itself. The only qualitative distinction is between those sites that are interesting and those that are boring. Life consists in fighting off boredom by being a spectator at everything interesting in the universe and in communicating with everyone else so inclined. Such a life produces a self that has no defining content or continuity but is open to all possibilities and to constantly being drawn into new games.

But we have still to explain what makes this use of the Web attractive. Why is there a thrill in being able to find out about everything no matter how trivial? What motivates a commitment to curiosity? Kierkegaard thought that in the final analysis people were attracted to the Press, and we can now add the Web, because the anonymous spectator *takes no risks*. The person in the aesthetic sphere keeps open all possibilities and has no fixed identity that could be threatened by disappointment, humiliation or loss. Surfing the Web is ideally suited to such a life. On the Internet commitments are at best virtual commitments and losses only virtual losses.

But when he is speaking from the point of view of the next higher sphere of existence, Kierkegaard tells us that the self requires not "variableness and brilliancy" but "firmness, balance and steadiness."[12] So, according to Kierkegaard: "As a result of knowing and being everything possible, one is in contradiction with oneself."[13]

We would therefore expect the aesthetic sphere to reveal that it was ultimately unlivable, and, indeed, Kierkegaard held that if one threw oneself into the aesthetic sphere with total commitment it was bound to break down under the sheer glut of information and possibilities. Without some way of telling the relevant from the irrelevant and the significant from the insignificant, everything becomes equally interesting and equally boring. Writing under a pseudonym from the perspective of someone experiencing the melancholy that signals the breakdown of the aesthetic sphere he laments: "My reflection on life altogether lacks meaning. I take it some evil spirit has put a pair of spectacles on my nose, one glass of which magnifies to an enormous degree, while the other reduces to the same degree."[14]

This inability to distinguish the trivial from the important eventually stops being exciting and leads to the very boredom the aesthete and Net-surfer dedicate their lives to avoiding. Thus, Kierkegaard concludes: "every aesthetic view of life is despair, and everyone who lives aesthetically is in despair whether he knows it or not. But when one knows it...a higher form of existence is an imperative requirement."[15]

The Ethical Sphere: Turning Information into Knowledge

That higher form of life Kierkegaard calls *the ethical sphere*. In it one has a stable identity and one is committed to involved action. Information is not denigrated but is sought for serious purposes. Only if one makes a commitment to some perspective – chooses some interest or subject to learn about – can one have a sense of relevance and thus turn information into knowledge. To cite again *Transforming Higher Education*:

> Until recently, educators found it sufficient to distinguish between "data" and "information" – interpreted data that has a directed use. Today, a further value must be stipulated – knowledge, which is the perspective and insights that derive from the synthesis of information. Learners need to develop the capacity to search, select, and synthesize vast amounts of information to create knowledge.[16]

But if all one has is information, how is the student supposed to arrive at a perspective in terms of which to turn the information into knowledge? This is especially important since it turns out that for a beginner in any domain to become competent in that domain the beginner must take up a perspective from which to distinguish what is relevant from what is irrelevant.

Taking up such a perspective requires taking risks which involves the learner more and more in his or her tasks. While it might seem that this involvement would interfere with detached rule-testing and so would inhibit further skill development, in fact just the opposite seems to be the case. Only if the detached rule-following stance of the novice is replaced by involvement, can there be further advancement, while resistance to the frightening acceptance of risk and responsibility can lead to stagnation.

Patricia Benner has observed this phenomenon in the education of nurses. Nurses who protect themselves from becoming involved in their work of helping people get well or at least to have dignified deaths, and who therefore do not allow themselves to take up a perspective and then feel elation at their successes and sorrow at their failures, never get past this competence stage and burn out because there are too many facts and rules to keep track of.[17]

Thus studies of skill acquisition have shown that, unless the outcome matters and unless the person developing the skill is willing to accept the pain that comes from failure and the elation that comes with success, the learner will be stuck at the level of competence and never achieve mastery. Only those willing to take risks go on to become experts.

It follows that, since expertise can only be acquired through involved engagement with actual situations, the possibility of acquiring expertise is lost in the disengaged discussions and deracinated knowledge acquisition characteristic of the Net. Not only is the detached learner unable to acquire specific disciplinary skills, but, for the same reason, such a learner could not acquire the general skills for getting around in the world and getting on with others that Aristotle calls *phronesis* or *practical wisdom*.

The Net-enthusiast will presumably answer that all the learner has to do to turn information into knowledge, and rule-following into skilled behavior, is to choose a perspective – something that matters – and care about the outcome. But Kierkegaard would argue that the very ease of making choices on the Internet would ultimately lead to the inevitable breakdown of serious choice and so of the ethical sphere. Commitments that are freely chosen can and should be revised from minute to minute as new information comes along. But where there is no risk and every commitment can be revoked without consequences, choice becomes arbitrary and meaningless. Nothing really matters.[18]

To avoid the constant choice of new perspectives, one might, like Judge William, Kierkegaard's pseudonymous author of the description of the ethical sphere in *Either/Or*, turn to one's talents and one's social roles to limit one's choices. Judge William says that his range of possible relevant actions are constrained by his social roles as judge and husband. But Judge William admits, indeed he is proud of the fact, that, as an autonomous person, he is free to give whatever meaning he decides to his talents and his roles. He thus can choose which talents and therefore which commitments are the most important ones. This choice is based on a more fundamental choice of what is worthy and not worthy, what good and what evil. As Judge William puts it:

> The good *is* for the fact that I will it, and apart from my willing, it has no existence. This is the expression for freedom....By this the distinctive notes of good and evil are by no means belittled or disparaged as merely subjective distinctions. On the contrary, the absolute validity of these distinctions is affirmed.[19]

But, Kierkegaard argues, if everything is up for choice, including the standards on the basis of which one chooses, there is no reason to choose one set of standards rather than another. Choosing the guidelines for one's life never makes any serious difference since one can always choose to rescind one's previous choice.

The ethical breaks down because the power to make commitments and so to choose what information to seek out undermines itself. Any choice I make does not get a grip on me so it can always be revoked. It must be constantly reconfirmed by a new choice to take the previous one seriously. As Kierkegaard puts it:

> If the despairing self is *active*,...it is constantly relating to itself only experimentally, no matter what it undertakes, however great, however amazing and with whatever perseverance. It

> recognizes no power over itself; therefore in the final instance it lacks seriousness.... The despairing self is content with taking notice of itself which is meant to bestow infinite interest and significance on its enterprises, and which is exactly what makes them experiments.... The self can, at any moment, start quite arbitrarily all over again and, however far an idea is pursued in practice, the entire action is contained within an hypothesis.[20]

Thus the *choice* of perspective that was supposed to turn the glut of information into knowledge and provide the involvement necessary for skill acquisition only adds to the pool of possibilities, and one ends up in what Kierkegaard calls the despair of the ethical. Kierkegaard concludes that one cannot stop the proliferating of information and turn it into relevant knowledge by *deciding* what is worth knowing; one can only turn information into relevant and meaningful knowledge, and one can only care about one's performance and so develop skills, if one has a strong identity based on a serious, long-lasting commitment.

The Religious Sphere: Making One Unconditional Commitment

The view of commitments as choices open to being revoked does not seem to hold for those commitments that are most important. These special commitments are neither the ones that I arbitrarily choose nor the ones that I am obliged to keep because of my social role. Rather, these special commitments are experienced as grabbing my whole being. When I respond to such a summons by making an *unconditional* commitment, this commitment determines who I am and what will be the significant issue for me for the rest of my life. Political and religious movements can grab us in this way as can love relationships and, for certain people, such vocations as the law or music.

These unconditional commitments are different from the normal sorts of commitments. They define the world in which our everyday commitments are made. They thus determine which commitments really matter and why they do. Identities based on unconditional commitments, then, stop the proliferation of choices. They block nihilism by establishing qualitative distinctions between what, for the individual, is important and trivial, relevant and irrelevant, serious and playful.

But, of course, such a commitment is risky. One's cause may fail. One's lover may leave. And, since one has defined oneself in terms of them, one cannot just walk away. On the contrary, one's identity and world will collapse and one will experience grief. The detachment of the "Present Age", the flexibility of the aesthetic sphere, and the unbounded freedom of the ethical sphere are all ways of avoiding this risk. But it turns out, Kierkegaard claims, that for that very reason they level all qualitative distinctions and end in the despair of meaninglessness. There is no way to have a meaningful life and to develop particular skills and the skill of being a good human being without taking risks.

This leads to the perplexing question: What role can Information Technology play in encouraging and supporting such unconditional commitments? A first suggestion might be that the movement from the aesthetic stage to the religious stage will be facilitated by the Web just as flight simulators help one learn to fly. One would be solicited to throw oneself into Net-surfing and find that boring; then into making free choices and revocable commitments until they proliferated absurdly; and so finally be driven to let oneself be drawn into a risky irrevocable commitment as the only way of out despair. Indeed, at any stage from looking for all sorts of interesting web sites as one surfs the Net, to striking up a conversation in a chat room, to making commitments to interest groups, one might just get hooked by one of the ways of life opened up and find oneself drawn into a world-defining lifetime commitment. No doubt this might happen – people do meet in chat rooms and fall in love – but it is certainly infrequent.

Kierkegaard would surely argue that, while the Internet *allows* unconditional commitments, it does not *support* them. Like a simulator, it manages to capture everything but the risk. Our imaginations can be drawn in, as they are in playing games and watching movies. And no doubt game simulations sharpen our responses for non-game situations. But so far as games work by capturing our imaginations, they will fail to give us serious commitments. We teach or work in the laboratory day after day because these activities matter greatly to us. But we are unlikely to stay with either for long if we have only an imaginary ultimate commitment. The temptation is to live in a world of stimulating images and simulated commitments and thus to lead a simulated life. Far from encouraging unconditional commitments, the Net tends to turn all of life into a risk-free game. So, in the end, although Information Technology does not *prohibit* unconditional commitments, it does *inhibit* them.

The test as to whether one had acquired an unconditional commitment would come if one had the incentive and courage to transfer what one had learned to the concrete situation in the real world. There one would confront what Kierkegaard called "the danger and the harsh judgement of existence." And, precisely the attraction of the Net would inhibit that final plunge. Anyone using the Net that was led to risk his or her real identity in the real world would thus have to act against the grain of what attracted them to the Net in the first place.

If Kierkegaard is right, for the cyber-world to avoid despair, it would have to find a way of canceling its risk-free attraction by somehow supporting and encouraging unconditional commitments and strong identities in the real world where risk of failure and disappointment is inevitable.

Conclusion

If educators who teach in the world of hyper-information and hyper-connectivity want to impart knowledge and skills they will not only have to encourage their students to plunge in and swim in deep and dangerous waters, as Kierkegaard proposed, they will also have to encourage them to swim upstream. That is, they will have to foster the sort of unconditioned commitments and strong identities necessary for turning information into meaningful knowledge and the involvement necessary for developing the skills to use it. Only then can they develop in their students knowledge, skills and what Aristotle called practical wisdom.

But what will give these students the strength to resist the nihilistic pull of the new network culture? Their teachers will have to foster those social practices that support strong identities. Fortunately, there are still in our culture the narratives left from the Judeo-Christian tradition that Kierkegaard was drawing on and trying to preserve. We will have to go on preserving them. We also still have culture heroes like Martin Luther King, Jr. who show that it makes sense to die for one's commitments.

This shows us the basic problem with the Internet and explains why, despite wild promises, attempts to use it in education have so far failed. Since skills cannot be captured and transmitted in rules and the attraction of a life of risk and commitment cannot be fully portrayed by narratives, education at its best must be based on apprenticeship. Even science which starts out teaching rules and techniques ends up with the student as an apprentice in a successful scientist's laboratory. Computer games can teach hand-eye co-ordination, but where worldly expertise is concerned, one can only learn by imitation of the style and day-by-day responses to specific local situations of someone who already has the relevant mastery.

Only by working closely with students in a shared situation in the real world can teachers with strong identities ready to take risks to preserve their commitments pass on their passion and skill so their students can turn information into knowledge and practical wisdom. In so far as we want to teach skill in particular domains and practical wisdom in life, which we certainly do, we thus finally run up against the limits of the World Wide Web. As far as I can see, learning by apprenticeship can work only in the nearness of the classroom and laboratory; never in cyberspace.

Notes

1 Lewis J. Perelman, *School's Out*, New York: Avon/ Education, p. 22, 1993.
2 Ibid. p. 23.
3 Michael G. Dolence and Donald M. Norris, *Transforming Higher Education*, Ann Arbor: Society for College and University Planning, p. 36, 1995.
4 Translated separately by Alexander Dru as *The Present Age*, New York: Harper and Row, 1962.
5 *Pap*. IX A 378, 1848. The translations are from Søren Kierkegaard's Journals and Papers, tr. by Howard V. Hong and Edna H. Hong, Bloomington: Indiana University Press, vol. 2, p. 480, 1970 (indicated as *Journals and Papers*).
6 This leveling of differences is reflected in the way information is organized on the Web. When information is organized in a traditional hierarchical database, the user is forced to commit to a certain class of information before he can view more specific data that fall under that class. For example, I have to commit to an interest in animals before I can find out what I want to know about tortoises: and once having made that commitment to the animal line in the database, I can't then examine the data on problems of infinity without backtracking through my previous commitments. When information is organized in hypertext, as it is on the Web, however, instead of the privileged

relations between a class and its members, the organizing principle is the interconnectedness of all knowledge. The goal of hypertext is to allow the user easily to get from one data entry to any other, as long as they are related in at least some tenuous fashion. So, for instance, in examining the entry on tortoises, I might click on the bold text that reads "Tortoises – compared to hares," and be transported instantly to an entry on Zeno's paradox. In this way the user is encouraged to traverse a vast network of knowledge and information, all of which is equally accessible and none of which is privileged. Everything is linked to everything else on a single level. Moreover, the links are not based on a developing sense of relevance but on the canned, context-free relevance of key words. *Quantity* of connections has replaced any judgement as to the *quality* of those connections. [Ed.]

7 *The Present Age*, p. 59.
8 *Journals and Papers*, p. 489.
9 Ibid. p. 481.
10 *The Present Age*, p. 63.
11 Ibid. pp. 36–37.
12 *Either/Or*. tr. by Alastair Hannay, Harmondsworth: Penguin Books, p. 391, 1992.
13 *The Present Age*, p. 68.
14 *Either/Or* p. 46.
15 Ibid. p. 502.
16 *Transforming Higher Education*, p. 26.
17 Patricia Benner, *From Novice to Expert, Excellence and Power in Clinical Nursing Practice*, Reading: Addison-Wesley, 1984.
18 Enthusiasts try to make an advantage of this rapid turnover of ideas and commitments and use it as an excuse to throw out old-fashioned critical reflection and discussion. In *Transforming Higher Education* we read:

> Network scholarship increases the "bandwidth" of information that can be synthesized by an individual and shortens the timeframe....[But] some [faculties]...are stuck in yesterday's model of scholarship: print media and leisurely timeframes for debate and discussion. (p. 25)

But this call for keeping up with the accelerating rate of change is followed by the admission that this new form of constant adaptation is not working.

> The tools of network scholarship are revolutionizing discovery research and the synthesis of information in particular academic disciplines. They have not been applied as successfully...to other forms of...teaching/learning (Ibid.).

[Ed.]

19 *Either/Or*, p. 524.
20 *The Sickness unto Death*, tr. by Alastair Hannay, Harmondsworth: Penguin Books, p. 100, 1989.

References

Patricia Benner. *From Novice to Expert, Excellence and Power in Clinical Nursing Practice*. Reading, Addison-Wesley, 1984.

Michael G. Dolence and Donald M. Norris. *Transforming Higher Education*. Ann Arbor, Society for College and University Planning, 1995.

S. Kierkegaard. *The Present Age*. Translated separately by Alexander Dru, New York, Harper and Row, 1962.

S. Kierkegaard. *Journals and Papers*. Paper IX A 378, 1848. Translated by Howard V. Hong and Edna H. Hong. Bloomington, Indiana University Press, 1970.

S. Kierkegaard. *The Sickness unto Death*. Translated by Alastair Hannay. Harmondsworth, Penguin Books, 1989.

S. Kierkegaard. *Either/Or*. Translated by Alastair Hannay. Harmondsworth, Penguin Books, 1992.

Lewis J. Perelman. *School's Out*. New York: Avon/Education, 1993.

Technology, Knowledge, and Power

Introduction

This final section focuses on a number of approaches to the question of the relationship between (especially scientific) knowledge and (especially technological) power, the problem of their joint social impact, and the possibility of reorienting or transforming technology.

Michel Foucault, one of the most influential social philosophers of the last few decades, maintained a lifelong concern for the rise and growing dominance of scientific knowledge; but he is known above all for his analyses of the hidden as well as overt relations of power embedded in the "discourses of knowledge" and the political implications of these relations on human affairs. Thoroughly interdisciplinary in his procedures and choice of topics, Foucault is usually described as having gone through two intellectual stages. In earlier work, he undertakes an "archeological" and historical investigation of the emergence of the modern "episteme" (i.e., general conception of rationality), and analyzes its function as a paradigm for the sciences of both nature and human life, as well as for the more "dubious disciplines" of applied social science such as criminology, psychoanalysis and psychiatry, and education. In what is perhaps his most famous work of this period, *The Order of Things* (1966), the discontinuous history of the sciences in which epistemes succeed one another is described in a way that bears some resemblance to the works of Kuhn, Toulmin, and other postpositivist philosophers of science in the Anglo-American tradition.

In later works, Foucault's method is more "genealogical" and explicitly concerned with sociopolitical issues, and his central concern is with the largely unrecognized ways in which knowledge not only "is" power, as Bacon remarked, but also involves relationships of power that privilege some and dehumanize others. In *Discipline and Punish*, from which the excerpt below was taken, Foucault employs his general theory of knowledge and power in the production of "technologies of the self" to show how the scientific conception of rationality and irrationality gets utilized to classify, mold, isolate, and discipline "deviant" individuals. Foucault admits that, of course, part of the technology here is physical – for example, as it is found in the architecture of prisons and schools, including Bentham's famous panopticon described below. Yet he argues that to judge technology by its tools and productions is to miss its point. One gets closer to the truth by recognizing that aspect of technology which involves behavioral techniques – for example, the confession and the examination, whether in a clinical, academic, or juridical setting. Foucault's characterization of behavioral techniques bears a striking resemblance to the descriptions of Skinnerian behavioral technology in the United States. What is equally striking, however, is the contrast between the adulation heaped upon such techniques by its academic and therapeutic advocates and Foucault's deeply critical stance toward these same techniques. For him, too little concern has been shown for the fact that behavioral technologies need no tools and yet work powerfully on and often against people. In this emphasis on tool-free technology, Foucault clearly shares ground with Mumford, whose "megamachine" is a technological apparatus consisting in human beings, and with Ellul, who also refuses to tie "technique" to the involvement of any hardware but conceives it instead as involving the rationalization of all aspects of life for the purpose of maximizing efficiency.

With his critical conception of techniques as instruments of power in the background, Foucault argues against those who, like Guy Debord, would characterize modern society with its professional sports and mass media as a "society of the spectacle." For Foucault, on the contrary, ours is a society involving mutual surveillance. For example, in ancient times architecture was arranged so that a maximum number of spectators could observe a single thing or person. In the modern carcerial world, architecture is arranged so that the ideal is a single observer observing as many people as possible. Foucault shows how the apparently more "humane" treatment of criminals and the insane is really a more complete and insidious form of control. Under the older regimes of torture and inquisition, punishment was overt, physical, and administered by visible representatives of a state or royal power. Modern modes of punishment are at once less visible and more complete. By making criminals, the socially undersirable, and the insane objects of technoscientific "knowledge," they are more effectively and totally "understood," classified, and brought under control – not by physical force but by hidden forms of coercion – e.g., the modern juridical form of examination. One can also see this transition from ancient to modern practices of control reflected in changes in the characterization of scientific method. For Bacon, scientific experiment can verge on a sort of torture of nature. According to Kant in the *First Critique*, scientific inquiry is like a judicial examination, where the scientist is like a judge, putting the question to nature. Foucault was himself active in the prison liberation movement but denied that courses of action could be grounded ethically. His archeological and genealogical descriptions, like those of Marx in his more scientific moods, were supposedly value-free.

As both Chellis Glendinning and Langdon Winner know, the original Luddites were worker followers of a (possibly mythical) Ned Ludd who supposedly smashed weaving machines that were putting people out of work. Today, however, "Luddite" is typically used as a term of disparagement by technocrats and other supporters of high technology for anyone who opposes or criticizes new technologies. Glendinning takes advantage of this usage to wear the label, "Neo-Luddite," as a badge of honor. In her "Neo-Luddite Manifesto," she draws on Mumford's views concerning the social impact of technology and Winner's arguments for the necessity of participation by users in the design of technology in order to present her conception of a decentralized, renewable technology. Her critique of the worldview of modern technological society – which includes brief analyses of its notions of mechanism, rationality, efficiency, detachment, and ownership – resembles those of Rousseau and the Romantics.

In *Autonomous Technology*, Langdon Winner (to whom Glendinning refers with praise) combines partial concession to Ellul's view of this topic with arguments for opposition to the reigning technological order. In "Luddism as Epistemology," the selection from this book included below, Winner urges a piecemeal dismantling of technology as an experimental means of discovering what aspects of technology we would really miss, what effects the removal of those aspects would have on our life and society, and what parts of technological systems might actually be necessary for social functioning. His Luddism is therefore more systematic and analytical than that of Neo-Luddites like Glendinning.

In his exchange with Winner, Mark Elam criticizes Winner's "metaphysical" opposition to social constructivist conceptions of technology. Elam complains that from this traditional sort of stance, one merely embraces or opposes an idea by starting out from an ethical commitment based on presumed knowledge of how things really are. But that is just the problem. Such "knowledge" is merely presumptive. What is called for instead is an ironical, postmodern standpoint – which Elam sees himself deriving from Richard Rorty's case for the "liberal ironist" in *Contingency, Irony and Solidarity* – from which it is possible to defend, but without necessarily embracing, the unwillingness of social constructivist Steve Woolgar to "take a political stance" and evaluate technologies against Winner's criticism in Winner's metaphysical way. Winner replies that he is not defending metaphysical truth but rather community dialogue. (One might observe that Winner's position, though crudely sketched here, sounds *in nuce* like that of Habermas.) He seeks to undermine Elam's reliance on Rorty by pointing to Rorty's opposition to postmodern politics (or better, anti-politics) and to his explicit embrace of fairly traditional liberal positions. The detached, ironic stance, Winner suggests, may even serve as an excuse for economically motivated collaboration with that about which it is ironic. (Would he find significance in the fact that Woolgar has moved from his academic post in science studies to take up a business school position?)

Emmanuel Mesthene's upbeat, pioneering account of the social impact of technological change and John McDermott's acerbic response to him have come to constitute a classical exchange in the evaluation of modern technology. In "Technology and Wisdom" Mesthene, the director of the early and short-lived Harvard University Program on Technology and Society, presents with admirable openness and clarity the attitude that often lies less candidly acknowledged behind pro-technological writing. There are, says Mesthene, no longer any real limits to what we can do with nature or ourselves. The notion of physical impossibility has been overcome. Previous societies always labored under some notion of cosmic or divine impediment – for instance, a surd universe of matter, a chaotic cosmic substratum (e.g., Plato's receptacle), arbitrary but unavoidable fate as in Greek tragedy, or original sin as in Christian theology. Modern science, however, has made everything transparent and understandable. Nothing, as Mesthene sees it, is incomprehensible or uncontrollable. (One might note in passing that quantum mechanics, chaos theory, and anti-reductionistic claims concerning biology may all raise doubts about Mesthene's thesis, but the point to stress is that his attitude prereflectively guides the majority of advocates of technological progress and technocracy.) In the end, then, given the conceptual power of scientific understanding, there is nothing to stop unlimited technological development. In fact, says Mesthene, we have not and will not stop technological development because it is an expression of our humanity. Hence, those who urge technological caution are guilty of the same sort of "failure of nerve" that beset the Hellenistic world.

Mesthene's *Technological Change* is an expansion of his symposium contribution to the Fourth Annual Report (1967–8) of his Harvard program. In the book, he attempts to position himself between optimistic technological utopians and pessimistic technological dystopians, but also argues against those traditionalists who claim that modern technologies raise no new issues concerning the human condition. The simple fact is, he asserts, that new technologies open new possibilities and put within reach other possibilities that were previously too costly and difficult to realize. Thus, inevitably, technology changes the range and weightings of our possibilities; and the only proper response to this fact is to cease clinging to some given set of values and embrace instead the process of "valuation" called for by each new circumstance. As for the political consequences of this state of affairs, Mesthene claims that in some ways, it fosters democracy (e.g., with the promise of instant voting, or more importantly, by making closer and more organized forms of cooperation necessary). At the same time, he admits our understanding of what democracy is will have to undergo development. For example, democracy cannot be as "direct" as traditional theory depicts it. Technological change encourages us to produce an ever better-educated public, but nothing can stop the inevitably widening gap between technological experts who guide policy and citizens who are supposed to set it. "No amount of participation [in government] in the populist sense," says Mesthene, "can substitute for the expertise and decision-making technologies that modern government must use." Nevertheless, he adds, the presentation of facts by experts in, say, economics and international affairs will tend to increase consensus by removing issues from controversy.

It is Mesthene's view of the political consequences of technological development upon which McDermott seizes. In "Technology: The Opiate of the Intellectuals," he calls Mesthene's arguments – for all their soothing rhetoric about a "new democratic ethos" – the straightforward defense of a politically right-wing, technocratically elitist position. As Mesthene himself seems at times to realize, the society he envisions is fundamentally dependent on expert social scientific and business management opinion and is therefore increasingly without need of popular support or participation. McDermott coins the phrase, *laissez innover*, to describe Mesthene's kind of ideology of technocracy. Just as *laissez faire* was never a genuine description of real economies with their monopolies and government favors but rather an ideology to support entrepreneurs against employees, so the doctrine of *laissez innover* is designed, not to liberate technological creativity from old-fashioned doubts, but to free technocrats and their economic masters from popular restraint or regulation. In order to illustrate the increasingly undemocratic and expert-dependent character of the world Mesthene embraces, McDermott contrasts the ultra-rationality of strategic bombing during the Vietnam War with the myths created by uninformed and dismayed ground troops who could not understand why they were being beaten by a technologically inferior indigenous force. In a striking exemplification of the general argument of Marcuse and the critical social theorists, McDermott shows how, contrary to Mesthene's rosy picture of an ever better-educated population, technological "rationality" at the top breeds irrationality among the populace. In a brief retrospective, McDermott reaffirms his conviction that technology is still as central to our world and as problematically related to democratic interests as it was when he wrote "Technology."

Andrew Feenberg's "Democratic Rationalization" combines criticism of technological determinism with a call for the democratization of technology and concludes with a summary of the criticisms of Heidegger's account of technology elaborated in his selection in Part IV. Feenberg touches upon many of the themes in previous selections. He rejects the deterministic, "unilinear" picture of the course of technology (e.g., in the selection from Heilbroner, Part V, chapter 34). He also rejects Weber's view that technological rationalization makes industrial democracy impossible – a view which he claims to find in both Heidegger and Ellul and which is sketched in simple terms in Engels's "On Authority" (in Part I). Feenberg appeals to the social constructivism of Pinch and Bijker, arguing that it is only "decontextualized" conceptions of technology that it can make it appear to be the self-generating foundation of modern society. A closer look at technology in its actual social and cultural setting reveals that its development is in fact multidirectional, not unilinear, and its course is "overdetermined" by and variously adaptive to both technical and societal demands.

Feenberg argues that his non-deterministic position cannot help being political, since if there are no "technological imperatives," then technology needs to be interpreted and the choices it embodies and makes possible need to be made explicit and evaluated. Here, Feenberg gives the social constructivists' method of interpretive flexibility a hermeneutic twist. It is not how technology *functions* but what its objects are taken to *be* that requires illumination. Feenberg identifies "social meaning" and "cultural horizon" as the two hermeneutical dimensions of technical artifacts. Hackers of the sort discussed by Ross (Part VI, chapter 46) exemplify the way users of technical objects reinterpret and modify their social meaning. Similarly, the AIDS patients' movement transformed the way medical drug treatment and experimentation were understood. As regards the cultural horizon, Feenberg points out that in Foucault's theory of "power/knowledge," modern forms of oppression are shown to depend upon alleged "truths" of a technical frame of understanding that maintains its dominance precisely by not being recognized as the contingent, chosen cultural image of technical control that it is. Feenberg closes by complaining that from the level of abstraction maintained by people like Heidegger, the specific character of today's technology cannot appear – with the result that it can then seem to be a "metaphysical condition," instead of the result of a particular social hegemony that it really is.

49

Panopticism

Michel Foucault

The following, according to an order published at the end of the seventeenth century, were the measures to be taken when the plague appeared in a town.[1]

First, a strict spatial partitioning: the closing of the town and its outlying districts, a prohibition to leave the town on pain of death, the killing of all stray animals; the division of the town into distinct quarters, each governed by an intendant. Each street is placed under the authority of a syndic, who keeps it under surveillance; if he leaves the street, he will be condemned to death. On the appointed day, everyone is ordered to stay indoors: it is forbidden to leave on pain of death. The syndic himself comes to lock the door of each house from the outside; he takes the key with him and hands it over to the intendant of the quarter; the intendant keeps it until the end of the quarantine. Each family will have made its own provisions; but, for bread and wine, small wooden canals are set up between the street and the interior of the houses, thus allowing each person to receive his ration without communicating with the suppliers and other residents; meat, fish and herbs will be hoisted up into the houses with pulleys and baskets. If it is absolutely necessary to leave the house, it will be done in turn, avoiding any meeting. Only the intendants, syndics and guards will move about the streets and also, between the infected houses, from one corpse to another, the "crows", who can be left to die: these are "people of little substance who carry the sick, bury the dead, clean and do many vile and abject offices". It is a segmented, immobile, frozen space. Each individual is fixed in his place. And, if he moves, he does so at the risk of his life, contagion or punishment.

Inspection functions ceaselessly. The gaze is alert everywhere: "A considerable body of militia, commanded by good officers and men of substance", guards at the gates, at the town hall and in every quarter to ensure the prompt obedience of the people and the most absolute authority of the magistrates, "as also to observe all disorder, theft and extortion". At each of the town gates there will be an observation post; at the end of each street sentinels. Every day, the intendant visits the quarter in his charge, inquires whether the syndics have carried out their tasks, whether the inhabitants have anything to complain of; they "observe their actions". Every day, too, the syndic goes into the street for which he is responsible; stops before each house: gets all the inhabitants to appear at the windows (those who live overlooking the courtyard will be allocated a window looking onto the street at which no one but they may show themselves); he calls each of them by name; informs himself as to the state of each and every one of them – "in which respect the inhabitants will be compelled to speak the truth under pain of death"; if someone does not appear at the window, the syndic must ask why: "In this way he will find out easily enough whether dead or sick are being concealed." Everyone locked up in his cage, everyone at his window, answering to his name and showing himself when asked – it is the great review of the living and the dead.

From Michel Foucault, *Discipline and Punish: The Birth of the Prison*, trans. Alan Sheridan, New York: Pantheon, 1978, pp. 195–209, 216–28, 316–17, 326–33, abridged.

This surveillance is based on a system of permanent registration: reports from the syndics to the intendants, from the intendants to the magistrates or mayor. At the beginning of the "lock up", the role of each of the inhabitants present in the town is laid down, one by one; this document bears "the name, age, sex of everyone, notwithstanding his condition": a copy is sent to the intendant of the quarter, another to the office of the town hall, another to enable the syndic to make his daily roll call. Everything that may be observed during the course of the visits – deaths, illnesses, complaints, irregularities – is noted down and transmitted to the intendants and magistrates. The magistrates have complete control over medical treatment; they have appointed a physician in charge; no other practitioner may treat, no apothecary prepare medicine, no confessor visit a sick person without having received from him a written note "to prevent anyone from concealing and dealing with those sick of the contagion, unknown to the magistrates". The registration of the pathological must be constantly centralized. The relation of each individual to his disease and to his death passes through the representatives of power, the registration they make of it, the decisions they take on it.

Five or six days after the beginning of the quarantine, the process of purifying the houses one by one is begun. All the inhabitants are made to leave; in each room "the furniture and goods" are raised from the ground or suspended from the air; perfume is poured around the room; after carefully sealing the windows, doors and even the keyholes with wax, the perfume is set alight. Finally, the entire house is closed while the perfume is consumed; those who have carried out the work are searched, as they were on entry, "in the presence of the residents of the house, to see that they did not have something on their persons as they left that they did not have on entering". Four hours later, the residents are allowed to re-enter their homes.

This enclosed, segmented space, observed at every point, in which the individuals are inserted in a fixed place, in which the slightest movements are supervised, in which all events are recorded, in which an uninterrupted work of writing links the centre and periphery, in which power is exercised without division, according to a continuous hierarchical figure, in which each individual is constantly located, examined and distributed among the living beings, the sick and the dead – all this constitutes a compact model of the disciplinary mechanism. The plague is met by order; its function is to sort out every possible confusion: that of the disease, which is transmitted when bodies are mixed together; that of the evil, which is increased when fear and death overcome prohibitions. It lays down for each individual his place, his body, his disease and his death, his well-being, by means of an omnipresent and omniscient power that subdivides itself in a regular, uninterrupted way even to the ultimate determination of the individual, of what characterizes him, of what belongs to him, of what happens to him. Against the plague, which is a mixture, discipline brings into play its power, which is one of analysis. A whole literary fiction of the festival grew up around the plague: suspended laws, lifted prohibitions, the frenzy of passing time, bodies mingling together without respect, individuals unmasked, abandoning their statutory identity and the figure under which they had been recognized, allowing a quite different truth to appear. But there was also a political dream of the plague, which was exactly its reverse: not the collective festival, but strict divisions; not laws transgressed, but the penetration of regulation into even the smallest details of everyday life through the mediation of the complete hierarchy that assured the capillary functioning of power; not masks that were put on and taken off, but the assignment to each individual of his "true" name, his "true" place, his "true" body, his "true" disease. The plague as a form, at once real and imaginary, of disorder had as its medical and political correlative discipline. Behind the disciplinary mechanisms can be read the haunting memory of "contagions", of the plague, of rebellions, crimes, vagabondage, desertions, people who appear and disappear, live and die in disorder.

If it is true that the leper gave rise to rituals of exclusion, which to a certain extent provided the model for and general form of the great Confinement, then the plague gave rise to disciplinary projects. Rather than the massive, binary division between one set of people and another, it called for multiple separations, individualizing distributions, an organization in depth of surveillance and control, an intensification and a ramification of power. The leper was caught up in a practice of rejection, of exile-enclosure; he was left to his doom in a mass among which it was useless to differentiate; those sick of the plague were caught up in a meticulous tactical partitioning in which individual differentiations were the constricting effects of a power that multiplied, articulated and subdivided itself; the great confinement on the one hand; the correct training on the other. The leper and his separation; the plague and its segmentations. The first is marked; the second analysed and distributed. The exile of the leper and the arrest of the plague do not bring with them the same political dream. The first is that of a pure community, the second that of a disciplined society. Two ways of exercising power over men, of controlling their relations, of separating out their dangerous mixtures. The plague-stricken town, traversed throughout with hierarchy, surveillance, observation, writing; the town immobilized by the functioning of an extensive power that bears in a distinct way over all individual bodies – this is the utopia of the perfectly governed city. The plague (envisaged as a possibility at least) is the trial in the course of which one may define ideally the exercise of disciplinary power. In order to make rights and laws function according to pure theory, the jurists place themselves in imagination in the state of nature; in order to see perfect disciplines functioning, rulers dreamt of the state of plague. Underlying disciplinary projects the image of the plague stands for all forms of confusion and disorder; just as the image of the leper, cut off from all human contact, underlies projects of exclusion.

They are different projects, then, but not incompatible ones. We see them coming slowly together, and it is the peculiarity of the nineteenth century that it applied to the space of exclusion of which the leper was the symbolic inhabitant (beggars, vagabonds, madmen and the disorderly formed the real population) the technique of power proper to disciplinary partitioning. Treat "lepers" as "plague victims", project the subtle segmentations of discipline onto the confused space of internment, combine it with the methods of analytical distribution proper to power, individualize the excluded, but use procedures of individualization to mark exclusion – this is what was operated regularly by disciplinary power from the beginning of the nineteenth century in the psychiatric asylum, the penitentiary, the reformatory, the approved school and, to some extent, the hospital. Generally speaking, all the authorities exercising individual control function according to a double mode; that of binary division and branding (mad/sane; dangerous/harmless; normal/abnormal); and that of coercive assignment, of differential distribution (who he is; where he must be; how he is to be characterized; how he is to be recognized; how a constant surveillance is to be exercised over him in an individual way, etc.). On the one hand, the lepers are treated as plague victims; the tactics of individualizing disciplines are imposed on the excluded; and, on the other hand, the universality of disciplinary controls makes it possible to brand the "leper" and to bring into play against him the dualistic mechanisms of exclusion. The constant division between the normal and the abnormal, to which every individual is subjected, brings us back to our own time, by applying the binary branding and exile of the leper to quite different objects; the existence of a whole set of techniques and institutions for measuring, supervising and correcting the abnormal brings into play the disciplinary mechanisms to which the fear of the plague gave rise. All the mechanisms of power which, even today, are disposed around the abnormal individual, to brand him and to alter him, are composed of those two forms from which they distantly derive.

Bentham's *Panopticon* is the architectural figure of this composition. We know the principle on which it was based: at the periphery, an annular building; at the centre, a tower; this tower is pierced with wide windows that open onto the inner side of the ring; the peripheric building is divided into cells, each of which extends the whole width of the building; they have two windows, one on the inside, corresponding to the windows of the tower; the other, on the outside, allows the light to cross the cell from one end to the other. All that is needed, then, is to place a supervisor in a central tower and to shut up in each cell a madman, a patient, a condemned man, a worker or a schoolboy. By the effect of backlighting, one can observe from the tower, standing out precisely against the light, the small captive shadows in the cells of the periphery. They are like so many cages, so many small theatres, in which each actor is alone, perfectly individualized and constantly visible. The panoptic mechanism arranges spatial unities that make it possible to see constantly and to recognize immediately. In short, it reverses the principle of the dungeon; or rather of its three functions – to enclose, to deprive of light and to hide – it preserves only the first and eliminates the other two. Full lighting and the eye of a supervisor capture better than darkness, which ultimately protected. Visibility is a trap.

To begin with, this made it possible – as a negative effect – to avoid those compact, swarming, howling masses that were to be found in places of confinement, those painted by Goya or described by Howard. Each individual, in his place, is securely confined to a cell from which he is seen from the front by the supervisor; but the side walls prevent him from coming into contact with his companions. He is seen, but he does not see; he is the object of information, never a subject in communication. The arrangement of his room, opposite the central tower, imposes on him an axial visibility; but the divisions of the ring, those separated cells, imply a lateral invisibility. And this invisibility is a guarantee of order. If the inmates are convicts, there is no danger of a plot, an attempt at collective escape, the planning of new crimes for the future, bad reciprocal influences; if they are patients, there is no danger of contagion; if they are madmen there is no risk of their committing violence upon one another; if they are schoolchildren, there is no copying, no noise, no chatter, no waste of time; if they are workers, there are no disorders, no theft, no coalitions, none of those distractions that slow down the rate of work, make it less perfect or cause accidents. The crowd, a compact mass, a locus of multiple exchanges, individualities merging together, a collective effect, is abolished and replaced by a collection of separated individualities. From the point of view of the guardian, it is replaced by a multiplicity that can be numbered and supervised; from the point of view of the inmates, by a sequestered and observed solitude (Bentham, 60–64).

Hence the major effect of the Panopticon: to induce in the inmate a state of conscious and permanent visibility that assures the automatic functioning of power. So to arrange things that the surveillance is permanent in its effects, even if it is discontinuous in its action; that the perfection of power should tend to render its actual exercise unnecessary; that this architectural apparatus should be a machine for creating and sustaining a power relation independent of the person who exercises it; in short, that the inmates should be caught up in a power situation of which they are themselves the bearers. To achieve this, it is at once too much and too little that the prisoner should be constantly observed by an inspector: too little, for what matters is that he knows himself to be observed; too much, because he has no need in fact of being so. In view of this, Bentham laid down the principle that power should be visible and unverifiable. Visible: the inmate will constantly have before his eyes the tall outline of the central tower from which he is spied upon. Unverifiable: the inmate must never know whether he is being looked at at any one moment; but he must be sure that he may always be so. In order to make the presence or absence of the inspector unverifiable, so that the prisoners, in their cells, cannot even see a shadow, Bentham envisaged not only venetian blinds on the windows of the central observation hall, but, on the inside, partitions that intersected the hall at right angles and, in order to pass from one quarter to the other, not doors but zig-zag openings; for the slightest noise, a gleam of light, a brightness in a half-opened door would betray the presence of the guardian.[2] The Panopticon is a machine for dissociating the see/being seen dyad: in the peripheric ring, one is totally seen, without ever seeing; in the central tower, one sees everything without ever being seen.[3]

It is an important mechanism, for it automatizes and disindividualizes power. Power has its principle not so much in a person as in a certain concerted distribution of bodies, surfaces, lights, gazes; in an arrangement whose internal mechanisms produce the relation in which individuals are caught up. The ceremonies, the rituals, the marks by which the sovereign's surplus power was manifested are useless. There is a machinery that assures dissymmetry, disequilibrium, difference. Consequently, it does not matter who exercises power. Any individual, taken almost at random, can operate the machine: in the absence of the director, his family, his friends, his visitors, even his servants (Bentham, 45). Similarly, it does not matter what motive animates him: the curiosity of the indiscreet, the malice of a child, the thirst for knowledge of a philosopher who wishes to visit this museum of human nature, or the perversity of those who take pleasure in spying and punishing. The more numerous those anonymous and temporary observers are, the greater the risk for the inmate of being surprised and the greater his anxious awareness of being observed. The Panopticon is a marvellous machine which, whatever use one may wish to put it to, produces homogeneous effects of power.

A real subjection is born mechanically from a fictitious relation. So it is not necessary to use force to constrain the convict to good behaviour, the madman to calm, the worker to work, the schoolboy to application, the patient to the observation of the regulations. Bentham was surprised that panoptic institutions could be so light: there were no more bars, no more chains, no more heavy locks; all that was needed was that the separations should be clear and the openings well arranged. The heaviness of the old "houses of security", with their fortress-like architecture, could be replaced by the simple, economic geometry of a "house of certainty". The efficiency of power, its constraining force have, in a sense, passed over to the other side – to the side of its surface of application. He who is subjected to a field of visibility, and who knows it, assumes responsibility for the constraints of power; he makes them play spontaneously upon himself; he inscribes in himself the power relation in which he simultaneously plays both roles; he becomes the principle of his own subjection. By this very fact, the external power may throw off its physical weight; it tends to the non-corporal; and, the more it approaches this limit, the more constant, profound and permanent are its effects: it is a perpetual victory that avoids any physical confrontation and which is always decided in advance.

Bentham does not say whether he was inspired, in his project, by Le Vaux's menagerie at Versailles: the first menagerie in which the different elements are not, as they traditionally were, distributed in a park (Loisel, 104–7). At the centre was an octagonal pavilion which, on the first floor, consisted of only a single room, the king's *salon*; on every side large windows looked out onto seven cages (the eighth side was reserved for the entrance), containing different species of animals. By Bentham's time, this menagerie had disappeared. But one finds in the programme of the Panopticon a similar concern with individualizing observation, with characterization and classification, with the analytical arrangement of space. The Panopticon is a royal menagerie; the animal is replaced by man, individual distribution by specific grouping and the king by the machinery of a furtive power. With this exception, the Panopticon also does the work of a naturalist. It makes it possible to draw up differences: among patients, to observe the symptoms of each individual, without the proximity of beds, the circulation of miasmas, the effects of contagion confusing the clinical tables; among school children, it makes it possible to observe performances (without there being any imitation or copying), to map aptitudes, to assess characters, to draw up rigorous classifications and, in relation to normal development, to distinguish "laziness and stubbornness" from "incurable imbecility"; among workers, it makes it possible to note the aptitudes of each worker, compare the time he takes to perform a task, and if they are paid by the day, to calculate their wages (Bentham, 60–64).

So much for the question of observation. But the Panopticon was also a laboratory; it could be used as a machine to carry out experiments, to alter behaviour, to train or correct individuals. To experiment with medicines and monitor their effects. To try out different punishments on prisoners, according to their crimes and character, and to seek the most effective ones. To teach different techniques simultaneously to the workers, to decide which is the best. To try out pedagogical experiments – and in particular to take up once again the well-debated problem of secluded education, by using orphans. One would see what would happen when, in their sixteenth or eighteenth year, they were presented with other boys or girls; one could verify whether, as Helvetius thought, anyone could learn anything; one would follow "the genealogy of every observable idea"; one could bring up different children according to different systems of thought, making certain children believe that two and two do not make four or that the moon is a cheese, then put them together when they are twenty or twenty-five years old; one would then have discussions that would be worth a great deal more than the sermons or lectures on which so much money is spent; one would have at least an opportunity of making discoveries in the domain of metaphysics. The Panopticon is a privileged place for experiments on men, and for analysing with complete certainty the transformations that may be obtained from them. The Panopticon may even provide an apparatus for supervising its own mechanisms. In this central tower, the director may spy on all the employees that he has under his orders: nurses, doctors, foremen, teachers, warders; he will be able to judge them continuously, alter their behaviour, impose upon them the methods he thinks best; and it will even be possible to observe the director himself. An inspector arriving unexpectedly at the centre of the Panopticon will be able to judge at a glance, without anything being concealed from him, how the entire establishment is functioning. And, in any case, enclosed as he is in the middle of this architectural mechanism, is not the director's own fate entirely bound up with it? The incompetent physician who has allowed contagion to spread, the incompetent prison governor or workshop manager will be the first victims of an epidemic or a revolt. " 'By every tie I could devise', said the master of the Panopticon, 'my own fate had been bound up by me with theirs' " (Bentham, 177). The Panopticon functions as a kind of laboratory of power. Thanks to its mechanisms of observation, it gains in efficiency and in the ability to penetrate into men's behaviour; knowledge follows the advances of power, discovering new objects of knowledge over all the surfaces on which power is exercised.

The plague-stricken town, the panoptic establishment – the differences are important. They mark, at a distance of a century and a half, the transformations of the disciplinary programme. In the first case, there is an exceptional situation: against an extraordinary evil, power is mobilized; it makes itself everywhere present and visible; it invents new mechanisms; it separates, it immobilizes, it partitions; it constructs for a time what is both a counter-city and the perfect society; it imposes an ideal functioning, but one that is reduced, in the final analysis, like the evil that it combats, to a simple dualism of life and death: that which moves brings death, and one kills that which moves. The Panopticon, on the other hand, must be understood as a generalizable model of functioning; a way of defining power relations in terms of the everyday life of men. No doubt Bentham presents it as a particular institution, closed in upon itself. Utopias, perfectly closed in upon themselves, are common enough. As opposed to the ruined prisons, littered with mechanisms of torture, to be seen in Piranese's engravings, the Panopticon presents a cruel, ingenious cage. The fact that it should have given rise, even in our own time, to so many variations, projected or realized, is evidence of the imaginary intensity that it has possessed for almost two hundred years. But the Panopticon must not be understood as a dream building: it is the diagram of a mechanism of power reduced to its ideal form; its functioning, abstracted from any obstacle, resistance or friction, must be represented as a pure architectural and optical system: it is in fact a figure of political technology that may and must be detached from any specific use.

It is polyvalent in its applications; it serves to reform prisoners, but also to treat patients, to instruct schoolchildren, to confine the insane, to supervise workers, to put beggars and idlers to work. It is a type of location of bodies in space, of distribution of individuals in relation to one another, of hierarchical organization, of disposition of centres and channels of power, of definition of the instruments and modes of intervention of power, which can be implemented in hospitals, workshops, schools, prisons. Whenever one is dealing with a multiplicity of individuals on whom a task or a particular form of behaviour must be imposed, the panoptic schema may be used. It is – necessary modifications apart – applicable "to all establishments whatsoever, in which, within a space not too large to be covered or commanded by buildings, a number of persons are meant to be kept under inspection" (Bentham, 40; although Bentham takes the penitentiary house as his prime example, it is because it has many different functions to fulfil – safe custody, confinement, solitude, forced labour and instruction).

In each of its applications, it makes it possible to perfect the exercise of power. It does this in several ways: because it can reduce the number of those who exercise it, while increasing the number of those on whom it is exercised. Because it is possible to intervene at any moment and because the constant pressure acts even before the offences, mistakes or crimes have been committed. Because, in these conditions, its strength is that it never intervenes, it is exercised spontaneously and without noise, it constitutes a mechanism whose effects follow from one another. Because, without any physical instrument other than architecture and geometry, it acts directly on individuals; it gives "power of mind over mind". The panoptic schema makes any apparatus of power more intense: it assures its economy (in material, in personnel, in time); it assures its efficacity by its preventative character, its continuous functioning and its automatic mechanisms. It is a way of obtaining from power "in hitherto unexampled quantity", "a great and new instrument of government...; its great excellence consists in the great strength it is capable of giving to *any* institution it may be thought proper to apply it to" (Bentham, 66).

It's a case of "it's easy once you've thought of it" in the political sphere. It can in fact be integrated into any function (education, medical treatment, production, punishment); it can increase the effect of this function, by being linked closely with it; it can constitute a mixed mechanism in which relations of power (and of knowledge) may be precisely adjusted, in the smallest detail, to the processes that are to be supervised; it can establish a direct proportion between "surplus power" and "surplus production". In short, it arranges things in such a way that the exercise of power is not added on from the outside, like a rigid, heavy constraint, to the functions it invests, but is so subtly present in them as to increase their efficiency by itself increasing its own points of contact. The panoptic mechanism is not simply a hinge, a point of exchange between a mechanism of power and a function; it is a way of making power relations function in a function, and of making a function function through these power relations. Bentham's Preface to *Panopticon* opens with a list of the benefits to be obtained from his "inspection-house": "*Morals reformed – health preserved – industry invigorated – instruction diffused – public burthens lightened* – Economy seated, as it were, upon a rock – the gordian knot of the Poor-Laws not cut, but untied – all by a simple idea in architecture!" (Bentham, 39).

Furthermore, the arrangement of this machine is such that its enclosed nature does not preclude a permanent presence from the outside: we have seen that anyone may come and exercise in the central tower the functions of surveillance, and that, this being the case, he can gain a clear idea of the way in which the surveillance is practised. In fact, any panoptic institution, even if it is as rigorously closed as a penitentiary, may without difficulty be subjected to such irregular and constant inspections and not only by the appointed inspectors, but also by the public; any member of society will have the right to come and see with his own eyes how the schools, hospitals, factories, prisons function. There is no risk, therefore, that the increase of power created by the panoptic machine may degenerate into tyranny; the disciplinary mechanism will be democratically controlled, since it will be constantly accessible "to the great tribunal committee of the world".[4] This Panopticon, subtly arranged so that an observer may observe, at a glance, so many different individuals, also enables everyone to come and observe any of the observers. The seeing machine was once a sort of dark room into which individuals spied; it has become a transparent building in which the exercise of power may be supervised by society as a whole.

The panoptic schema, without disappearing as such or losing any of its properties, was destined to spread throughout the social body; its vocation was to become a generalized function. The plague-stricken town provided an exceptional disciplinary model: perfect, but absolutely violent; to the disease that brought death, power opposed its perpetual threat of death; life inside it was reduced to its simplest expression; it was, against the power of death, the meticulous exercise of the right of the sword. The Panopticon, on the other hand, has a role of amplification; although it arranges power, although it is intended to make it more economic and more effective, it does so not for power itself, nor for the immediate salvation of a threatened society: its aim is to strengthen the social forces – to increase production, to develop the economy, spread education, raise the level of public morality; to increase and multiply.

How is power to be strengthened in such a way that, far from impeding progress, far from weighing upon it with its rules and regulations, it actually facilitates such progress? What intensificator of power will be able at the same time to be a multiplicator of production? How will power, by increasing its forces, be able to increase those of society instead of confiscating them or impeding them? The Panopticon's solution to this problem is that the productive increase of power can be assured only if, on the one hand, it can be exercised continuously in the very foundations of society, in the subtlest possible way, and if, on the other hand, it functions outside these sudden, violent, discontinuous forms that are bound up with the exercise of sovereignty. The body of the king, with its strange material and physical presence, with the force that he himself deploys or transmits to some few others, is at the opposite extreme of this new physics of power represented by panopticism; the domain of panopticism is, on the contrary, that whole lower region, that region of irregular bodies, with their details, their multiple movements, their heterogeneous forces, their spatial relations; what are required are mechanisms that analyse distributions, gaps, series, combinations, and which use instruments that render visible, record, differentiate and compare: a physics of a relational and multiple power, which has its maximum intensity not in the person of the king, but in the bodies that can be individualized by these relations. At the theoretical level, Bentham defines another way of analysing the social body and the power relations that traverse it; in terms of practice, he defines a procedure of subordination of bodies and forces that must increase the utility of power while practising the economy of the prince. Panopticism is the general principle of a new "political anatomy" whose object and end are not the relations of sovereignty but the relations of discipline.

The celebrated, transparent, circular cage, with its high tower, powerful and knowing, may have been for Bentham a project of a perfect disciplinary institution; but he also set out to show how one may "unlock" the disciplines and get them to function in a diffused, multiple, polyvalent way throughout the whole social body. These disciplines, which the classical age had elaborated in specific, relatively enclosed places – barracks, schools, workshops – and whose total implementation had been imagined only at the limited and temporary scale of a plague-stricken town, Bentham dreamt of transforming into a network of mechanisms that would be everywhere and always alert, running through society without interruption in space or in time. The panoptic arrangement provides the formula for this generalization. It programmes, at the level of an elementary and easily transferable mechanism, the basic functioning of a society penetrated through and through with disciplinary mechanisms.

There are two images, then, of discipline. At one extreme, the discipline-blockade, the enclosed institution, established on the edges of society, turned inwards towards negative functions: arresting evil, breaking communications, suspending time. At the other extreme, with panopticism, is the discipline-mechanism: a functional mechanism that must improve the exercise of power by making it lighter, more rapid, more effective, a design of subtle coercion for a society to come. The movement from one project to the other, from a schema of exceptional discipline to one of a generalized surveillance, rests on a historical transformation: the gradual extension of the mechanisms of discipline throughout the seventeenth and eighteenth centuries, their spread throughout the whole social body, the formation of what might be called in general the disciplinary society.

A whole disciplinary generalization – the Benthamite physics of power represents an acknowledgement of this – had operated throughout the classical age. The spread of disciplinary institutions, whose network was beginning to cover an ever larger surface and occupying above all a less and less marginal position, testifies to this: what was an islet, a privileged place, a circumstantial measure, or a singular model, became a general formula; the regulations characteristic of the Protestant and pious armies of William of Orange or of Gustavus Adolphus were transformed into regulations for all the armies of Europe; the model colleges of the Jesuits, or the schools of Batencour or Demia, following the example set by Sturm, provided the outlines for the general forms of educational discipline; the ordering of the naval and military hospitals provided the model for the entire reorganization of hospitals in the eighteenth century.

[...]

A few years after Bentham, Julius gave this society its birth certificate (Julius, 384–6). Speaking of the panoptic principle, he said that there was much more there than architectural ingenuity: it was an event in the "history of the human mind". In appearance, it is merely the solution of a technical problem; but, through it, a whole type of society emerges. Antiquity had been a civilization of spectacle. "To render accessible to a multitude of men the inspection of a small number of objects": this was the problem to which the architecture of temples, theatres and circuses responded. With spectacle, there was a predominance of public life, the intensity of festivals, sensual proximity. In these rituals in which blood flowed, society found new vigour and formed for a moment a single great body. The modern age poses the opposite problem: "To procure for a small number, or even for a single individual, the instantaneous view of a great multitude." In a society in which the principal elements are no longer the community and public life, but, on the one hand, private individuals and, on the other, the state, relations can be regulated only in a form that is the exact reverse of the spectacle: "It was to the modern age, to the ever-growing influence of the state, to its ever more profound intervention in all the details and all the relations of social life, that was reserved the task of increasing and perfecting its guarantees, by using and directing towards that great aim the building and distribution of buildings intended to observe a great multitude of men at the same time."

Julius saw as a fulfilled historical process that which Bentham had described as a technical programme. Our society is one not of spectacle, but of surveillance; under the surface of images, one invests bodies in depth; behind the great abstraction of exchange, there continues the meticulous, concrete training of useful forces; the circuits of communication are the supports of an accumulation and a centralization of knowledge; the play of signs defines the anchorages of power; it is not that the beautiful totality of the individual is amputated, repressed, altered by our social order, it is rather that the individual is carefully fabricated in it, according to a whole technique of forces and bodies. We are much less Greeks than we believe. We are neither in the amphitheatre, nor on the stage, but in the panoptic machine, invested by its effects of power, which we bring to ourselves since we are part of its mechanism. The importance, in historical mythology, of the Napoleonic character probably derives from the fact that it is at the point of junction of the monarchical, ritual exercise of sovereignty and the hierarchical, permanent exercise of indefinite discipline. He is the individual who looms over everything with a single gaze which no detail, however minute, can escape: "You may consider that no part of the Empire is without surveillance, no crime, no offence, no contravention that remains unpunished, and that the eye of the genius who can enlighten all embraces the whole of this vast machine, without, however, the slightest detail escaping his attention" (Treilhard, 14). At the moment of its full blossoming, the disciplinary society still assumes with the Emperor the old aspect of the power of spectacle. As a monarch who is at one and the same time a usurper of the ancient throne and the organizer of the new state, he combined into a single symbolic, ultimate figure the whole of the long process by which the pomp of sovereignty, the necessarily spectacular manifestations of power, were extinguished one by one in the daily exercise of surveillance, in a panopticism in which the vigilance of intersecting gazes was soon to render useless both the eagle and the sun.

The formation of the disciplinary society is connected with a number of broad historical processes – economic, juridico-political and, lastly, scientific – of which it forms part.

1. Generally speaking, it might be said that the disciplines are techniques for assuring the ordering of human multiplicities. It is true that there is nothing exceptional or even characteristic in this: every system of power is presented with the same problem. But the peculiarity of the disciplines is that they try to define in relation to the multiplicities a tactics of power that fulfils three criteria: firstly, to obtain the exercise of power at the lowest possible cost (economically, by the low expenditure it involves; politically, by its discretion, its low exteriorization, its relative invisibility, the little resistance it arouses); secondly, to bring the effects of this social power to their maximum intensity and to extend them as far as possible, without either failure or interval; thirdly, to link this "economic" growth of power with the output of the apparatuses (educational, military, industrial or medical) within which it is exercised; in short, to increase both the docility and the utility of all the elements of the system. This triple objective of the disciplines corresponds to a well-known historical conjuncture. One aspect of this conjuncture was the large demographic thrust of the eighteenth century; an increase in the floating population (one of the primary objects of discipline is to fix; it is an anti-nomadic technique); a change of quantitative scale in the groups to be supervised or manipulated (from the beginning of the seventeenth century to the eve of the French Revolution, the school population had been increasing rapidly, as had no doubt the hospital population; by the end of the eighteenth century, the peace-time army exceeded 200,000 men). The other aspect of the conjuncture was the growth in the apparatus of production, which was becoming more and more extended and complex; it was also becoming more costly and its profitability had to be increased. The development of the disciplinary methods corresponded to these two processes, or rather, no doubt, to the new need to adjust their correlation. Neither the residual forms of feudal power nor the structures of the administrative monarchy, nor the local mechanisms of supervision, nor the unstable, tangled mass they all formed together could carry out this role: they were hindered from doing so by the irregular and inadequate extension of their network, by their often conflicting functioning, but above all by the "costly" nature of the power that was exercised in them. It was costly in several senses: because directly it cost a great deal to the Treasury; because the system of corrupt offices and farmed-out taxes weighed indirectly, but very heavily, on the population; because the resistance it encountered forced it into a cycle of perpetual reinforcement; because it proceeded essentially by levying (levying on money or products by royal, seigniorial, ecclesiastical taxation; levying on men or time by *corvées* of press-ganging, by locking up or banishing vagabonds). The development of the disciplines marks the appearance of elementary techniques belonging to a quite different economy: mechanisms of power which, instead of proceeding by deduction, are integrated into the productive efficiency of the apparatuses from within, into the growth of this efficiency and into the use of what it produces. For the old principle of "levying-violence", which governed the economy of power, the disciplines substitute the principle of "mildness-production-profit". These are the techniques that make it possible to adjust the multiplicity of men and the multiplication of the apparatuses of production (and this means not only "production" in the strict sense, but also the production of knowledge and skills in the school, the production of health in the hospitals, the production of destructive force in the army).

In this task of adjustment, discipline had to solve a number of problems for which the old economy of power was not sufficiently equipped. It could reduce the inefficiency of mass phenomena: reduce what, in a multiplicity, makes it much less manageable than a unity; reduce what is opposed to the use of each of its elements and of their sum; reduce everything that may counter the advantages of number. That is why discipline fixes; it arrests or regulates movements; it clears up confusion; it dissipates compact groupings of individuals wandering about the country in unpredictable ways; it establishes calculated distributions. It must also master all the forces that are formed from the very constitution of an organized multiplicity; it must neutralize the effects of counter-power that spring from them and which form a resistance to the power that wishes to dominate it: agitations, revolts, spontaneous organizations, coalitions – anything that may establish horizontal conjunctions. Hence the fact that the disciplines use procedures of partitioning and verticality, that they introduce, between the different elements at the same level, as solid separations as possible, that they define compact hierarchical networks, in short, that they oppose to the intrinsic, adverse force of multiplicity the technique of the continuous, individualizing pyramid. They must also increase the particular utility of each element of the multiplicity, but by means that are the most rapid and the least costly, that is to say, by using the multiplicity itself as an instrument of this growth. Hence, in order to extract from bodies the maximum time and force, the use of those overall methods known as time-tables, collective training, exercises, total and detailed surveillance. Furthermore, the disciplines must increase the effect of utility proper to the multiplicities, so that each is made more useful than the simple sum of its elements: it is in order to increase the utilizable effects of the multiple that the disciplines define tactics of distribution, reciprocal adjustment of bodies, gestures and rhythms, differentiation of capacities, reciprocal coordination in relation to apparatuses or tasks. Lastly, the disciplines have to bring into play the power relations, not above but inside the very texture of the multiplicity, as discreetly as possible, as well articulated on the other functions of these multiplicities and also in the least expensive way possible: to this correspond anonymous instruments of power, coextensive with the multiplicity that they regiment, such as hierarchical surveillance, continuous registration, perpetual assessment and classification. In short, to substitute for a power that is manifested through the brilliance of those who exercise it, a power that insidiously objectifies those on whom it is applied; to form a body of knowledge about these individuals, rather than to deploy the ostentatious signs of sovereignty. In a word, the disciplines are the ensemble of minute technical inventions that made it possible to increase the useful size of multiplicities by decreasing the inconveniences of the power which, in order to make them useful, must control them. A multiplicity, whether in a workshop or a nation, an army or a school, reaches the threshold of a discipline when the relation of the one to the other becomes favourable.

If the economic take-off of the West began with the techniques that made possible the accumulation of capital, it might perhaps be said that the methods for administering the accumulation of men made possible a political take-off in relation to the traditional, ritual, costly, violent forms of power, which soon fell into disuse and were superseded by a subtle, calculated technology of subjection. In fact, the two processes – the accumulation of men and the accumulation of capital – cannot be separated; it would not have been possible to solve the problem of the accumulation of men without the growth of an apparatus of production capable of both sustaining them and using them; conversely, the techniques that made the cumulative multiplicity of men useful accelerated the accumulation of capital. At a less general level the technological mutations of the apparatus of production, the division of labour and the elaboration of the disciplinary techniques sustained an ensemble of very close relations (cf. Marx, *Capital*, vol. 1, chapter XIII and the very interesting analysis in Guerry and Deleule). Each makes the other possible and necessary; each provides a model for the other. The disciplinary pyramid constituted the small cell of power within which the separation, coordination and supervision of tasks was imposed and made efficient; and analytical partitioning of time, gestures and bodily forces constituted an operational schema that could easily be transferred from the groups to be subjected to the mechanisms of production; the massive projection of military methods onto industrial organization was an example of this modelling of the division of labour following the model laid down by the schemata of power. But, on the other hand, the technical analysis of the process of production, its "mechanical" breaking-down, were projected onto the labour force whose task it was to implement it: the constitution of those disciplinary machines in which the individual forces that they bring together are composed into a whole and therefore increased is the effect of this projection. Let us say that discipline is the unitary technique by which the body is reduced as a "political" force at the least cost and maximized as a useful force. The growth of a capitalist economy gave rise to the specific modality of disciplinary power, whose general formulas, techniques of submitting forces and bodies, in short, "political anatomy", could be operated in the most diverse political régimes, apparatuses or institutions.

2. The panoptic modality of power – at the elementary, technical, merely physical level at which it is situated – is not under the immediate dependence or a direct extension of the great juridico-political structures of a society; it is nonetheless not absolutely independent. Historically, the process by which the bourgeoisie became in the course of the eighteenth century the politically dominant class was masked by the establishment of an explicit, coded and formally egalitarian juridical framework, made possible by the organization of a parliamentary, representative régime. But the development and generalization of disciplinary mechanisms constituted the other, dark side of these processes. The general juridical form that guaranteed a system of rights that were egalitarian in principle was supported by these tiny, everyday, physical mechanisms, by all those systems of micro-power that are essentially non-egalitarian and asymmetrical that we call the disciplines. And although, in a formal way, the representative régime makes it possible, directly or indirectly, with or without relays, for the will of all to form the fundamental authority of sovereignty, the disciplines provide, at the base, a guarantee of the submission of forces and bodies. The real, corporal disciplines constituted the foundation of the formal, juridical liberties. The contract may have been regarded as the ideal foundation of law and political power; panopticism constituted the technique, universally widespread, of coercion. It continued to work in depth on the juridical structures of society, in order to make the effective mechanisms of power function in opposition to the formal framework that it had acquired. The "Enlightenment", which discovered the liberties, also invented the disciplines.

In appearance, the disciplines constitute nothing more than an infra-law. They seem to extend the general forms defined by law to the infinitestimal level of individual lives; or they appear as methods of training that enable individuals to become integrated into these general demands. They seem to constitute the same type of law on a different scale, thereby making it more meticulous and more indulgent. The disciplines should be regarded as a sort of counter-law. They have the precise role of introducing insuperable asymmetries and excluding reciprocities. First, because discipline creates between individuals a "private" link, which is a relation of constraints entirely different from contractual obligation; the acceptance of a discipline may be underwritten by contract; the way in which it is imposed, the mechanisms it brings into play, the non-reversible subordination of one group of people by another, the "surplus" power that is always fixed on the same side, the inequality of position of the different "partners" in relation to the common regulation, all these distinguish the disciplinary link from the contractual link, and make it possible to distort the contractual link systematically from the moment it has as its content a mechanism of discipline. We know, for example, how many real procedures undermine the legal fiction of the work contract: workshop discipline is not the least important. Moreover, whereas the juridical systems define juridical subjects according to universal norms, the disciplines characterize, classify, specialize; they distribute along a scale, around a norm, hierarchize individuals in relation to one another and, if necessary, disqualify and invalidate. In any case, in the space and during the time in which they exercise their control and bring into play the asymmetries of their power, they effect a suspension of the law that is never total, but is never annulled either. Regular and institutional as it may be, the discipline, in its mechanism, is a "counter-law". And, although the universal juridicism of modern society seems to fix limits on the exercise of power, its universally widespread panopticism enables it to operate, on the underside of the law, a machinery that is both immense and minute, which supports, reinforces, multiplies the asymmetry of power and undermines the limits that are traced around the law. The minute disciplines, the panopticisms of every day may well be below the level of emergence of the great apparatuses and the great political struggles. But, in the genealogy of modern society, they have been, with the class domination that traverses it, the political counterpart of the juridical norms according to which power was redistributed. Hence, no doubt, the importance that has been given for so long to the small techniques of discipline, to those apparently insignificant tricks that it has invented, and even to those "sciences" that give it a respectable face; hence the fear of abandoning them if one cannot find any substitute; hence the affirmation that they are at the very foundation of society, and an element in its equilibrium, whereas they are a series of mechanisms for unbalancing power relations definitively and everywhere; hence the persistence in regarding them as the humble, but concrete form of every morality, whereas they are a set of physico-political techniques.

To return to the problem of legal punishments, the prison with all the corrective technology at its disposal is to be resituated at the point where the codified power to punish turns into a disciplinary power to observe; at the point where the universal punishments of the law are applied selectively to certain individuals and always the same ones; at the point where the redefinition of the juridical subject by the penalty becomes a useful training of the criminal; at the point where the law is inverted and passes outside itself, and where the counter-law becomes the effective and institutionalized content of the juridical forms. What generalizes the power to punish, then, is not the universal consciousness of the law in each juridical subject; it is the regular extension, the infinitely minute web of panoptic techniques.

3. Taken one by one, most of these techniques have a long history behind them. But what was new, in the eighteenth century, was that, by being combined and generalized, they attained a level at which the formation of knowledge and the increase of power regularly reinforce one another in a circular process. At this point, the disciplines crossed the "technological" threshold. First the hospital, then the school, then, later, the workshop were not simply "reordered" by the disciplines; they became, thanks to them, apparatuses such that any mechanism of objectification could be used in them as an instrument of subjection, and any growth of power could give rise in them to possible branches of knowledge; it was this link, proper to the technological systems, that made possible within the disciplinary element the formation of clinical medicine, psychiatry, child psychology, educational psychology, the rationalization of labour. It is a double process, then: an epistemological "thaw" through a refinement of power relations; a multiplication of the effects of power through the formation and accumulation of new forms of knowledge.

The extension of the disciplinary methods is inscribed in a broad historical process: the development at about the same time of many other technologies – agronomical, industrial, economic. But it must be recognized that, compared with the mining industries, the emerging chemical industries or methods of national accountancy, compared with the blast furnaces or the steam engine, panopticism has received little attention. It is regarded as not much more than a bizarre little utopia, a perverse dream – rather as though Bentham had been the Fourier of a police society, and the Phalanstery had taken on the form of the Panopticon. And yet this represented the abstract formula of a very real technology, that of individuals. There were many reasons why it received little praise; the most obvious is that the discourses to which it gave rise rarely acquired, except in the academic classifications, the status of sciences; but the real reason is no doubt that the power that it operates and which it augments is a direct, physical power that men exercise upon one another. An inglorious culmination had an origin that could be only grudgingly acknowledged. But it would be unjust to compare the disciplinary techniques with such inventions as the steam engine or Amici's microscope. They are much less; and yet, in a way, they are much more. If a historical equivalent or at least a point of comparison had to be found for them, it would be rather in the "inquisitorial" technique.

The eighteenth century invented the techniques of discipline and the examination, rather as the Middle Ages invented the judicial investigation. But it did so by quite different means. The investigation procedure, an old fiscal and administrative technique, had developed above all with the reorganization of the Church and the increase of the princely states in the twelfth and thirteenth centuries. At this time it permeated to a very large degree the jurisprudence first of the ecclesiastical courts, then of the lay courts. The investigation as an authoritarian search for a truth observed or attested was thus opposed to the old procedures of the oath, the ordeal, the judicial duel, the judgement of God or even of the transaction between private individuals. The investigation was the sovereign power arrogating to itself the right to establish the truth by a number of regulated techniques. Now, although the investigation has since then been an integral part of western justice (even up to our own day), one must not forget either its political origin, its link with the birth of the states and of monarchical sovereignty, or its later extension and its role in the formation of knowledge. In fact, the investigation has been the no doubt crude, but fundamental element in the constitution of the empirical sciences; it has been the juridico-political matrix of this experimental knowledge, which, as we know, was very rapidly released at the end of the Middle Ages. It is perhaps true to say that, in Greece, mathematics were born from techniques of measurement; the sciences of nature, in any case, were born, to some extent, at the end of the Middle Ages, from the practices of investigation. The great empirical knowledge that covered the things of the world and transcribed them into the ordering of an indefinite discourse that observes, describes and establishes the "facts" (at a time when the western world was beginning the economic and political conquest of this same world) had its operating model no doubt in the Inquisition – that immense invention that our recent mildness has placed in the dark recesses of our memory. But what this politico-juridical, administrative and criminal, religious and lay, investigation was to the sciences of nature, disciplinary analysis has been to the sciences of man. These sciences, which have so delighted our "humanity" for over a century, have their technical matrix in the petty, malicious minutiae of the disciplines and their investigations. These investigations are perhaps to psychology, psychiatry, pedagogy, criminology, and so many other strange sciences, what the terrible power of investigation was to the calm knowledge of the animals, the plants or the earth. Another power, another knowledge. On the threshold of the classical age, Bacon, lawyer and statesman, tried to develop a methodology of investigation for the empirical sciences. What Great Observer will produce the methodology of examination for the human sciences? Unless, of course, such a thing is not possible. For, although it is true that, in becoming a technique for the empirical sciences, the investigation has detached itself from the inquisitorial procedure, in which it was historically rooted, the examination has remained extremely close to the disciplinary power that shaped it. It has always been and still is an intrinsic element of the disciplines. Of course it seems to have undergone a speculative purification by integrating itself with such sciences as psychology and psychiatry. And, in effect, its appearance in the form of tests, interviews, interrogations and consultations is apparently in order to rectify the mechanisms of discipline: educational psychology is supposed to correct the rigours of the school, just as the medical or psychiatric interview is supposed to rectify the effects of the discipline of work. But we must not be misled; these techniques merely refer individuals from one disciplinary authority to another, and they reproduce, in a concentrated or formalized form, the schema of power-knowledge proper to each discipline (on this subject, cf. Tort). The great investigation that gave rise to the sciences of nature has become detached from its politico-juridical model; the examination, on the other hand, is still caught up in disciplinary technology.

In the Middle Ages, the procedure of investigation gradually superseded the old accusatory justice, by a process initiated from above; the disciplinary technique, on the other hand, insidiously and as if from below, has invaded a penal justice that is still, in principle, inquisitorial. All the great movements of extension that characterize modern penality – the problematization of the criminal behind his crime, the concern with a punishment that is a correction, a therapy, a normalization, the division of the act of judgement between various authorities that are supposed to measure, assess, diagnose, cure, transform individuals – all this betrays the penetration of the disciplinary examination into the judicial inquisition.

What is now imposed on penal justice as its point of application, its "useful" object, will no longer be the body of the guilty man set up against the body of the king; nor will it be the juridical subject of an ideal contract; it will be the disciplinary individual. The extreme point of penal justice under the Ancien Régime was the infinite segmentation of the body of the regicide: a manifestation of the strongest power over the body of the greatest criminal, whose total destruction made the crime explode into its truth. The ideal point of penality today would be an indefinite discipline: an interrogation without end, an investigation that would be extended without limit to a meticulous and ever more analytical observation, a judgement that would at the same time be the constitution of a file that was never closed, the calculated leniency of a penalty that would be interlaced with the ruthless curiosity of an examination, a procedure that would be at the same time the permanent measure of a gap in relation to an inaccessible norm and the asymptotic movement that strives to meet in infinity. The public execution was the logical culmination of a procedure governed by the Inquisition. The practice of placing individuals under "observation" is a natural extension of a justice imbued with disciplinary methods and examination procedures. Is it surprising that the cellular prison, with its regular chronologies, forced labour, its authorities of surveillance and registration, its experts in normality, who continue and multiply the functions of the judge, should have become the modern instrument of penality? Is it surprising that prisons resemble factories, schools, barracks, hospitals, which all resemble prisons?

Notes

1 Archives militaries de Vincennes, A 1,516 91 sc. Pièce. This regulation is broadly similar to a whole series of others that date from the same period and earlier.
2 In the *Postscript to the Panopticon*, 1791, Bentham adds dark inspection galleries painted in black around the inspector's lodge, each making it possible to observe two storeys of cells.
3 In his first version of the *Panopticon*, Bentham had also imagined an acoustic surveillance, operated by means of pipes leading from the cells to the central tower. In the *Postscript* he abandoned the idea, perhaps because he could not introduce into it the principle of dissymmetry and prevent the prisoners from hearing the inspector as well as the inspector hearing them. Julius tried to develop a system of dissymmetrical listening (Julius, 18).
4 Imagining this continuous flow of visitors entering the central tower by an underground passage and then observing the circular landscape of the Panopticon, was Bentham aware of the Panoramas that Barker was constructing at exactly the same period (the first seems to have dated from 1787) and in which the visitors, occupying the central place, saw unfolding around them a landscape, a city or a battle? The visitors occupied exactly the place of the sovereign gaze.

Bibliography

Archives militaires de Vincennes, A 1,516 91 sc.
Bentham, J., *Works*, ed. Bowring, IV, 1843.
Guerry, F. and Deleule, D., *Le Corps productif*, 1973.
Julius, N. H., *Leçons sur les prisons*, I, 1831 (Fr. trans.).
Loisel, G., *Histoire des ménageries*, II, 1912.
Marx, Karl, *The Eighteenth Brumaire of Louis Bonaparte*, ed. 1954. *Capital*, vol. I, ed. 1970.
Tort, Michel, *Q.I.*, 1974.
Treilhard, J. B., *Motifs du code d'instruction criminelle*, 1808.

50

Notes toward a Neo-Luddite Manifesto

Chellis Glendinning

I Why Neo-Luddism?

Most students of European history dismiss the Luddites of 19th century England as "reckless machine-smashers" and "vandals" worthy of mention only for their daring tactics. Probing beyond this interpretation, though, we find a complex, thoughtful, and little-understood social movement whose roots lay in a clash between two worldviews.

The worldview that 19th century Luddites challenged was that of *laissez-faire* capitalism with its increasing amalgamation of power, resources, and wealth, rationalized by its emphasis on "progress."

The worldview they supported was an older, more decentralized one espousing the interconnectedness of work, community, and family through craft guilds, village networks, and townships. They saw the new machines that owners introduced into their workplaces – the gig mills and shearing frames – as threats not only to their jobs, but to the quality of their lives and the structure of the communities they loved. In the end, destroying these machines was a last-ditch effort by a desperate people whose world lay on the verge of destruction.

The current controversy over technology is reminiscent of that of the Luddite period. We too are being barraged by a new generation of technologies – two-way television, fiber optics, biotechnology, superconductivity, fusion energy, space weapons, supercomputers. We too are witnessing protest against the onslaught. A group of Berkeley students recently gathered in Sproul Plaza to kick and smash television sets as an act of "therapy for the victims of technology." A Los Angeles business woman hiked onto Vandenberg Air Force Base and beat a weapons-related computer with a crowbar, bolt cutters, hammer, and cordless drill. Villagers in India resist the bulldozers cutting down their forests by wrapping their bodies around tree trunks. People living near the Narita airport in Japan sit on the tarmac to prevent airplanes from taking off and landing. West Germans climb up the smokestacks of factories to protest emissions that are causing acid rain, which is killing the Black Forest.

Such acts echo the concerns and commitment of the 19th century Luddites. Neo-Luddites are 20th century citizens – activists, workers, neighbors, social critics, and scholars – who question the predominant modern worldview, which preaches that unbridled technology represents progress. Neo-Luddites have the courage to gaze at the full catastrophe of our century: The technologies created and disseminated by modern Western societies are out of control and desecrating the fragile fabric of life on Earth. Like the early Luddites, we too are a desperate people seeking to protect the livelihoods, communities, and families we love, which lie on the verge of destruction.

II What Is Technology?

Just as recent social movements have challenged the idea that current models of gender roles,

From *Utne Reader* 38/1 (1990): 50–3. Reprinted by kind permission of the author.

economic organizations, and family structures are not necessarily "normal" or "natural," so the Neo-Luddite movement has come to acknowledge that technological progress and the kinds of technologies produced in our society are not simply "the way things are."

As philosopher Lewis Mumford pointed out, technology consists of more than machines. It includes the techniques of operation and the social organizations that make a particular machine workable. In essence, a technology reflects a worldview. Which particular forms of technology – machines, techniques, and social organizations – are spawned by a particular worldview depend on its perception of life, death, human potential, and the relationship of humans to one another and to nature.

In contrast to the worldviews of a majority of cultures around the world (especially those of indigenous people), the view that lies at the foundation of modern technological society encourages a mechanistic approach to life: to rational thinking, efficiency, utilitarianism, scientific detachment, and the belief that the human place in nature is one of ownership and supremacy. The kinds of technologies that result include nuclear power plants, laser beams, and satellites. This worldview has created and promoted the military-industrial-scientific-media complex, multinational corporations, and urban sprawl.

Stopping the destruction brought by such technologies requires not just regulating or eliminating individual items like pesticides or nuclear weapons. It requires new ways of thinking about humanity and new ways of relating to life. It requires the creation of a new worldview.

III Principles of Neo-Luddism

1) *Neo-Luddites are not anti-technology.* Technology is intrinsic to human creativity and culture. What we oppose are the *kinds* of technologies that are, at root, destructive of human lives and communities. We also reject technologies that emanate from a worldview that sees rationality as the key to human potential, material acquisition as the key to human fulfillment, and technological development as the key to social progress.

2) *All technologies are political.* As social critic Jerry Mander writes in *Four Arguments for the Elimination of Television*, technologies are not neutral tools that can be used for good or evil depending on who uses them. They are entities that have been consciously structured to reflect and serve specific powerful interests in specific historical situations. The technologies created by mass technological society are those that serve the perpetuation of mass technological society. They tend to be structured for short-term efficiency, ease of production, distribution, marketing, and profit potential – or for warmaking. As a result, they tend to create rigid social systems and institutions that people do not understand and cannot change or control.

As Mander points out, television does not just bring entertainment and information to households across the globe. It offers corporations a surefire method of expanding their markets and controlling social and political thought. (It also breaks down family communications and narrows people's experience of life by mediating reality and lowering their span of attention.)

Similarly, the Dalkon Shield intrauterine device did not just make birth control easier for women. It created tremendous profits for corporate entrepreneurs at a time when the largest generation ever born in the United States was coming of age and oral contraceptives were in disfavor. (It also damaged hundreds of thousands of women by causing septic abortions, pelvic inflammatory disease, torn uteruses, sterility, and death.)

3) *The personal view of technology is dangerously limited.* The often-heard message "but I couldn't live without my word processor" denies the wider consequences of widespread use of computers (toxic contamination of workers in electronic plants and the solidifying of corporate power through exclusive access to new information in data bases).

As Mander points out, producers and disseminators of technologies tend to introduce their creations in upbeat, utopian terms. Pesticides will increase yields to feed a hungry planet! Nuclear energy will be "too cheap to meter." The pill will liberate women! Learning to critique technology demands fully examining its sociological context, economic ramifications, and political meanings. It involves asking not just what is gained – but what is lost, and by whom. It involves looking at the introduction of technologies from the perspective not only of human use, but of their impact on other living beings, natural systems, and the environment.

IV Program for the Future

1) As a move toward dealing with the consequences of modern technologies and preventing further destruction of life, *we favor the dismantling of the following destructive technologies*:

- nuclear technologies – which cause disease and death at every stage of the fuel cycle;
- chemical technologies – which re-pattern natural processes through the creation of synthetic, often poisonous chemicals and leave behind toxic and undisposable wastes;
- genetic engineering technologies – which create dangerous mutagens that when released into the biosphere threaten us with unprecedented risks;
- television – which functions as a centralized mind-controlling force, disrupts community life, and poisons the environment;
- electromagnetic technologies – whose radiation alters the natural electrical dynamic of living beings, causing stress and disease; and
- computer technologies – which cause disease and death in their manufacture and use, enhance centralized political power, and remove people from direct experience of life.

2) *We favor a search for new technological forms.* As political scientist Langdon Winner advocates in *Autonomous Technology*, we favor the creation of technologies by the people directly involved in their use – not by scientists, engineers, and entrepreneurs who gain financially from mass production and distribution of their inventions and who know little about the context in which their technologies are used.

We favor the creation of technologies that are of a scale and structure that make them understandable to the people who use them and are affected by them. We favor the creation of technologies built with a high degree of flexibility so that they do not impose a rigid and irreversible imprint on their users, and we favor the creation of technologies that foster independence from technological addiction and promise political freedom, economic justice, and ecological balance.

3) *We favor the creation of technologies in which politics, morality, ecology, and technics are merged for the benefit of life on Earth*:

- community-based energy sources utilizing solar, wind, and water technologies – which are renewable and enhance both community relations and respect for nature;
- organic, biological technologies in agriculture, engineering, architecture, art, medicine, transportation, and defense – which derive directly from natural models and systems;
- conflict resolution technologies – which emphasize cooperation, understanding, and continuity of relationship; and
- decentralized social technologies – which encourage participation, responsibility, and empowerment.

4) *We favor the development of a life-enhancing worldview in Western technological societies.* We hope to instill a perception of life, death, and human potential into technological societies that will integrate the human need for creative expression, spiritual experience, and community with the capacity for rational thought and functionality. We perceive the human role not as the dominator of other species and planetary biology, but as integrated into the natural world with appreciation for the sacredness of all life.

We foresee a sustainable future for humanity if and when Western technological societies restructure their mechanistic projections and foster the creation of machines, techniques, and social organizations that respect both human dignity and nature's wholeness. In progressing towards such a transition, we are aware: We have nothing to lose except a way of living that leads to the destruction of all life. We have a world to gain.

51

Luddism as Epistemology

Langdon Winner

But what next? Following the normal pattern of twentieth-century writing, I should now rush forward with suggestions and recommendations for how things might be different. What good are analyses, criticisms, and perspectives, some might say, unless they point to positive courses of action?

In view of what we have seen, however, it is not easy simply to take a deep breath and begin spewing forth plans for a better world. The issues are difficult ones. It has not been my aim to make them seem any less difficult than they are. In my experience, virtually all of the remedies proposed are little more than tentative steps in uncertain directions. Goodman's plea for the application of moral categories to technological action, Bookchin's outlines for a liberatory technology, Marcuse's rediscovery of utopian thinking, and Ellul's call to the defiant, self-assertive, free individual – all of these offer us something.[1] But when compared to the magnitude of what is to be overcome, these solutions seem trivial. I could, I suppose, fudge the matter here and seem to be zeroing in on some useful proposals. Having gone this far, the reader can probably predict how it would look.

First, I could say that there is a need to begin the search for new technological forms. Recognizing the often wrong-headed and oppressive character of existing configurations of technology, we should find new kinds of technics that avoid the human problems of the present set. This would mean, presumably, the birth of a new sort of inventiveness and innovation in the physical arrangements of this civilization.

Second, I could suggest that the development of these forms proceed through the direct participation of those concerned with their everyday employment and effects. One major shortcoming in the technologies of the modern period is that those touched by their presence have little or no control over their design or operation. To as great an extent as possible, then, the processes of technological planning, construction, and control ought to be opened to those destined to experience the final products and full range of social consequences.

Third, I might point to the arguments presented here and offer some specific principles to guide further technological construction. One such rule would certainly be the following: *that as a general maxim, technologies be given a scale and structure of the sort that would be immediately intelligible to non-experts*. This is to say, technological systems ought to be intellectually as well as physically accessible to those they are likely to affect. Another worthy principle would be: *that technologies be built with a high degree of flexibility and mutability*. In other words, we should seek to avoid circumstances in which technological systems impose a permanent, rigid, and irreversible imprint on the lives of the populace. Yet another conceivable rule is this: *that technologies be judged according to the degree of dependency they tend to foster, those creating a greater dependency being held inferior*. This merely recognizes a situation we have seen again and again … . Those who must rely for their very existence upon

From Langdon Winner, *Autonomous Technology: Technics-out-of-Control as a Theme in Political Thought*, Cambridge, MA: The MIT Press, 1977, pp. 325–35, 372–3, abridged.

artificial systems they do not understand or control are not at liberty to change those systems in any way whatsoever. For this reason, any attempt to create new technological circumstances must make certain that it does not discover freedom only to lose it again on the first step.

Finally, I could suggest a supremely important step – that we return to the original understanding of technology as a means that, like all other means available to us, must only be employed with a fully informed sense of *what is appropriate*. Here, the ancients knew, was the meeting point at which ethics, politics, and technics came together. If one lacks a clear and knowledgeable sense of which means are appropriate to the circumstances at hand, one's choice of means can easily lead to excesses and danger. This ability to grasp the appropriateness of means has, I believe, now been pretty thoroughly lost. It has been replaced by an understanding which holds that if a given means can be shown to have a narrow utility, then it ought to be adopted straight off, regardless of its broader implications.[2] For a time, perhaps from the early seventeenth century to the early twentieth, this was a fruitful way of proceeding. But we have now reached a juncture at which such a cavalier disposition will only lead us astray. A sign of the maturity of modern civilization would be its recollection of that lost sense of appropriateness in the judgment of means. We would profit from regaining our powers of selectivity and our ability to say "no" as well as "yes" to a technological prospect. There are now many cases in which we would want to say: "After all a temptation is not very tempting."[3]

I am convinced that measures of this kind point to a new beginning on the problems we have seen.[4] At the same time, these proposals have overtones of utopianism and unreality, which make them less than compelling. It may be that the only innovation I have suggested is to use my hat as a megaphone. There are excellent reasons why *any* call for the taking of a new path or new beginning now falls flat.

Not the least of these is simply the fact that while positive, utopian principles and proposals can be advanced, the real field is already taken. There are, one must admit, technologies already in existence – apparatus occupying space, techniques shaping human consciousness and behavior, organizations giving pattern to the activities of the whole society. To ignore this fact is to take flight from the reality that must be considered. One finds, for example, that in the contemporary discussions those most sanguine about the prospects for tackling the technological dilemma are those who place their confidence in *new* systems to be implemented in the future. Their hope is not that the existing state of affairs will be changed through any direct action, only that certain superior features will be added. In this manner the mass of problems now at hand is skirted.[5]

Another barrier is this: even if one seriously wanted to construct a different kind of technology appropriate to a different kind of life, one would be at a loss to know how to proceed. There is no living body of knowledge, no method of inquiry applicable to our present situation that tells us how to move any differently from the way we already do. Mumford's suggestion that society return to an older tradition of small-scale technics and craftsmanship is not convincing. The world that supported that tradition and gave it meaning has vanished. Where and how techniques of that sort could be a genuine alternative is highly problematic. Certainly a technological revivalism could *add* things to the existing technological stock. But the kind of knowledge that would make a difference is not to be found in decorating the periphery.

In no place is the force of these considerations better exemplified than in the sorry fate of the counterculture of the late 1960s. The belief of those who followed the utopian dream was that by dropping out of the dominant culture and "raising one's consciousness," a better way of living would be produced. In several areas of social fashion – clothing, music, language, drug use – there were some remarkable innovations. But behind the facade of style, a familiar reality still held sway. The basic structures of life, many of them technological structures, remained unchallenged and unchanged. Members of the movement convinced themselves that with a few gestures they had transcended all of that. But all of the networks of practical connections remained intact. The best that was done was to give the existing patterns a hip veneer. Members of the management team began to wear bell-bottoms and medallions.

The lesson, I think, is evident. Even though one commits oneself to ends radically different from those in common currency, there is no real beginning until the question of means is looked straight in the eye. One must take seriously the fact that there are already technologies occupying the available physical and social space and employing the available resources. One must also take seriously the fact that one simply does not yet know how to go ahead to find genuinely new means appropriate to the new "consciousness." No doubt some faced with this realization will simply wish to stop. They will see the virtual necessity of co-optation and the impending disappointment for anyone who tries to resist one's technological fate. Some will find it impossible to do anything else than retreat into despair and blame their plight on "those in power." But if I am not mistaken, the logic of the problem admits at least one more alternative.

In many contemporary writings the response to the idea of autonomous technology reads something like this: "Technology is not a juggernaut; being a human construction it can be torn down, augmented and modified at will."[6] The author of this statement, Dr. Glenn I. Seaborg, would probably be the last person to suggest that any existing technology actually be "torn down." But in his mind, as in many others, the conviction that man still controls technology is rooted in the notion that at any time the whole thing could be taken apart and something better built in its place. This idea, for reasons we have seen all along, is almost pure fantasy. Real technologies do not permit such wholesale tampering. Changes here occur through "invention," "development," "progress," and "growth" processes in which more and more additions are made to the technological store while some parts are eventually junked as obsolete. The technologies generated are understood to be more or less permanent fixtures. That they might be torn down or seriously tinkered with is unthinkable.

But perhaps Seaborg's idea has some merit. As we have already noted, is not the fundamental business of technics that of taking things apart and putting them together? One conceivable approach to tackling whatever flaws one sees in the various sytems of technology might be to begin dismantling those systems. This I would propose not as a solution in itself but as a method of inquiry. The forgotten essence of technical activity, regardless of the specific purpose at hand, might well be revealed by this very basic yet, at the same time, most difficult of steps. Technologies identified as problematic would be taken apart with the expressed aim of studying their interconnections and their relationships to human need. Prominent structures of apparatus, technique, and organization would be, temporarily at least, disconnected and made unworkable in order to provide the opportunity to learn what they are doing for or to mankind. If such knowledge were available, one could then employ it in the invention of radically different configurations of technics, better suited to nonmanipulated, consciously, and prudently articulated ends.

None of this would be necessary if such information were obvious. But at present it is exactly this kind of awareness and understanding that is lacking. Our involvement in advanced technical systems resembles nothing so much as the somnambulist in Caligari's cabinet. Somewhat drastic steps must be taken to raise the important questions at all. The method of carefully and deliberately dismantling technologies, epistemological Luddism if you will, is one way of recovering the buried substance upon which our civilization rests. Once unearthed, that substance could again be scrutinized, criticized, and judged.

I can hear the outcry already. Isn't this man's Luddism simply an invitation to machine smashing? Isn't it mere nihilism with a sharp edge? How can anyone calmly suggest such an awful course of action?

Again, I must explain that I am only proposing a method. The method has nothing to do with Luddism in the traditional sense (the smashing and destroying of apparatus). The much-maligned original Luddites were, of course, merely unemployed workers with a flare for the dramatic. As they scrutinized the mechanization of the textile trade in the industrial revolution, they applied two interesting criteria. Does the new device enhance the quality of the product being manufactured? Does the machine improve the quality of work? If the answer to either question or both is "no," the innovation should not be permitted. Banned from lawful union activity, the Luddites did what they could and unwittingly brought upon themselves a lasting opprobrium.[7]

As best I can tell, there have never been any epistemological Ludites, unless perhaps Paul Goodman was one on occasion. I am not proposing that a sledge hammer be taken to anything. Neither do I advocate any act that would endanger anyone's life or safety. The idea is that in certain instances it may be useful to dismantle or unplug a technological system in order to create the space and opportunity for learning.

The most interesting parts of the technological order in this regard are not those found in the structure of physical apparatus anyway. I have tried to suggest that the technologies of concern are actually *forms of life* – patterns of human consciousness and behavior adapted to a rational, productive design. Luddism seen in this context would seldom refer to dismantling any piece of machinery. It would seek to examine the connections of the human parts of modern social technology. To be more specific, it would try to consider at least the following: (1) the kinds of human dependency and regularized behavior centering upon specific varieties of apparatus, (2) the patterns of social activity that rationalized techniques imprint upon human relationships, and (3) the shapes given everyday life by the large-scale organized networks of technology. Far from any wild smashing, this would be a meticulous process aimed at restoring significance to the question, What are we about?

One step that might be taken, for example, is that groups and individuals would for a time, self-consciously and through advance agreement, extricate themselves from selected techniques and apparatus. This, we can expect, would create experiences of "withdrawal" much like those that occur when an addict kicks a powerful drug. These experiences must be observed carefully as prime data. The emerging "needs," habits, or discomforts should be noticed and thoroughly analyzed. Upon this basis it should be possible to examine the structure of the human relationships to the device in question. One may then ask whether those relationships should be restored and what, if any, new form those relationships should take. The participants would have a genuine (and altogether rare) opportunity to ponder and make choices about the place of that particular technology in their lives. Very fruitful experiments of this sort could now be conducted with many implements of our semiconscious technological existence, such as the automobile, television, and telephone.

Other possibilities for Luddism as methodology can be found at virtually any point in which social and political institutions depend upon advanced technologies for their effective operation. Persons who, for any reason, wish to alter or reform those institutions – the factory, school, business, public agency – have an alternative open to them that they have previously overlooked. As preparation for changes one may later wish to make, one might try disconnecting crucial links in the organized system for a time and studying the results. There is no getting around the fact that the most likely consequences will be some variety of chaos and confusion. But it is perhaps better to have this out in the open rather than endure the subliminal chaos and confusion upon which many of our most important institutions now rest. Again, these symptoms must be taken as prime data. The effects of systematic disconnection must be taken as an opportunity to inquire, to learn, and seek something better. What is the institution doing in the first place? How does its technological structure relate to the ends one would wish for it? Can one see anything more than to plug the whole back together the way it was before? The Luddite step is necessary if such questions are to be asked in any critical way. It is, perhaps, not too farfetched to suppose that some positive innovations might result from this straightforward challenge to established patterns of institutional life.

By far the most significant of Luddite alternatives, however, requires no direct action at all: the best experiments can be done simply by refusing to repair technological systems as they break down. Many of society's biggest investments at present are those that merely prop up failing technologies. This propping up is usually counted as "growth" and placed in the plus column. We build more and more freeways, larger and larger suburban developments, greater and greater systems of centralized water supply, power, sewers, and police, all in a frantic effort to sustain order and minimal comfort in the sprawling urban complex. Perhaps a better alternative would be to let dying artifice die. One might then begin the serious search, not for something superficially "better" but for totally new forms of sociotechnical existence. …

In Mary Shelley's novel, Victor Frankenstein is portrayed as a "modern Prometheus." The young man's inevitable tragedy mirrors an ancient story in which the combined elements of ambition, artifice, pride, and power meet an unfortunate end.

Without doubt the most excellent of Promethean stories, however, is that written by Aeschylus 2,500 years ago. In *Prometheus Bound* we find in luminous, mythical outline many of the themes we have encountered in this essay, for an interesting feature of Aeschylus's treatment of the legend is that it emphasizes the importance of technology in Prometheus's crime against the gods. The fall of man is in Aeschylus's view closely linked to the introduction of science and the arts and crafts. Chained to a desolate rock for eternity, Prometheus describes his plight.

Prometheus I caused mortals to cease foreseeing doom.

Chorus What cure did you provide them with against that sickness?

Prometheus I placed in them blind hopes.
Chorus That was a great gift you gave to men.
Prometheus Besides this, I gave them fire.
Chorus And do creatures of a day now possess bright-faced fire?
Prometheus Yes, and from it they shall learn many crafts.
Chorus These are the charges on which –
Prometheus Zeus tortures me and gives me no respite.[8]

The theft of fire, Aeschylus makes clear, was in its primary consequence the theft of all technical skills and inventions later given to mortals. "I hunted out the secret spring of fire," Prometheus exclaims, "that filled the narthex stem, which when revealed became the teacher of each craft to men, a great resource. This is the sin committed for which I stand accountant, and I pay nailed in my chains under the open sky."[9] As the brash protagonist recounts the specific items he has bestowed upon the human race, it becomes evident that Aeschylus's tale represents the movement of primitive man to civilized society. "They did not know of building houses with bricks to face the sun; know how to work in wood. They lived like swarming ants in holes in the ground, in the sunless caves of the earth."[10] The fire enabled mankind to develop agriculture, mathematics, astronomy, domesticated animals, carriages, and a host of valuable techniques. But Prometheus ends his proud description on a sorry note.

> It was I and none other who discovered ships, the sail-driven wagons that the sea buffets. Such were the contrivances that I discovered for men – alas for me! For I myself am *without contrivance to rid myself of my present affliction* [emphasis added].[11]

Prometheus's problem is something like our own. Modern people have filled the world with the most remarkable array of contrivances and innovations. If it now happens that these works cannot be fundamentally reconsidered and reconstructed, humankind faces a woefully permanent bondage to the power of its own inventions. But if it is still thinkable to dismantle, to learn and start again, there is a prospect of liberation. Perhaps means can be found to rid the human world of our self-made afflictions.

Notes

1 Paul Goodman, *People or Personnel and Like a Conquered Province* (New York: Random House, Vintage Books, 1968), pp. 297–316; Murray Bookchin, *Post-Scarcity Anarchism* (Berkeley: Ramparts Books, 1971); Herbert Marcuse, *An Essay on Liberation* (Boston: Beacon Press, 1969); Jacques Ellul, *Autopsy of Revolution*, trans. Patricia Wolf (New York: Alfred A. Knopf, 1971), chap. 5.

2 Erich Fromm finds in this tendency the foremost principle of action in the technological society: "something ought to be done because it is technically possible to do so." *The Revolution of Hope* (New York: Bantam Books, 1968), p. 33. In some understandings this principle or the motive it expresses is taken to be "the technological imperative" itself. I have not adopted that definition here, preferring to employ the term in the context presented in chapters 2 and 6. The phenomenon Fromm and others have noticed is perhaps best called "technomania."

3 Gertrude Stein, *Look At Me Now and Here I Am, Writings and Lectures 1909–45*, ed. Patricia Meyerowitz (Baltimore: Penguin Books, 1971), p. 58.

4 See E. F. Schumacher, *Small is Beautiful: Economics As If People Mattered* (New York: Harper & Row, 1973); Ivan Illich, *Tools for Conviviality* (New York: Harper & Row, 1973); Wilson Clark, *Energy for Survival: The Alternative to Extinction* (Garden City: Doubleday & Company, Anchor Books, 1975). See also *The Journal of the New Alchemists* (Woods Hole, Mass.: The New Alchemy Institute, 1973, 1974).

5 One peculiar response of thinkers now worried about the technological society is to pretend in effect that one *already* lives in the future. A sophisticated technology has been directed toward more intelligent ends and given a more humane structure. Through the proper selection of new devices, the problems of the old order have been surmounted. But in these future fantasies, of which "postindustrialism" is now the most popular, there is almost no attempt to stipulate what will have happened to the technologies we will supposedly have "gone beyond." Since there are new technologies of information processing, for example, we are somehow entitled to assume that the world of industrial technology has vanished.

6 Glenn T. Seaborg and Roger Corliss, *Man and Atom* (New York: E. P. Dutton, 1972), p. 265.

7 See Malcolm I. Thomis, *The Luddites: Machine-Breaking in Regency England* (New York: Schocken Books, 1970); George Rudé, *The Crowd in History*

(New York: John Wiley & Sons, 1964), pp. 79–92. I owe the formulation of the Luddite criteria to Larry Spence.

8 Aeschylus, *Prometheus Bound*, in *Aeschylus II*, ed. David Grene and Richard Lattimore and trans. David Grene (New York: Washington Square Press, 1967), pp. 148–149.

9 Ibid., p. 144.

10 Ibid., p. 156.

11 Ibid.

52

Anti Anticonstructivism or Laying the Fears of a Langdon Winner to Rest

Mark Elam, with Langdon Winner's Reply

In a recent paper, Langdon Winner (1993) characterizes the new "constructivist" sociology of technology as an intellectual tragedy. Assuming high moral ground, he accuses everyone linked with "social constructivism" of being, among other things, elitist, implicitly conservative, blasé, and politically naive. Winner appears to be declaring war on something despicable and asking each of us to make our loyalties known: are you one of us or one of them? Although accepting that "some important differences" exist among the leading practitioners of social constructivism and that some of the writers he mentions might actually object to being classified as constructivists, Winner maintains that the "basic disposition and viewpoint" of a constructivist remains "fairly consistent" (p. 366) and that beyond dispute there has been "a concerted push to affirm social constructivism as a coherent mode of analysis and to include or exclude writers according to their degree of adherence to the new canonical standard" (p. 377).

After being confronted with Winner's call to arms against heinous constructivism, it is a relief to have some characteristically wise and instructive words from Clifford Geertz. According to Geertz, "a scholar can hardly be better employed than in destroying a fear" (1984, 263). The fear and loathing Winner expresses toward social constructivism strongly resembles the harmful pattern of intellectual dread that Geertz identifies under the rubric "antirelativism." Winner wishes to whip up something like "anticonstructivism." I feel no particular inclination to defend an extremely amorphous research program that, for the want of a better name, can be labeled "the social construction of technology"; and yet, I do feel compelled to condemn anticonstructivism. In the same way in which Geertz champions anti anti-relativism, I wish to encourage a commitment to anti anticonstructivism.

What does being anti anticonstructivist mean? It does *not* mean "sitting on the fence," because, as an intellectual attitude, it is well suited to combat the paranoia that leads to fences being raised in the first place. The logical analogy Geertz has in mind when he says he is anti antirelativist is what was described during the McCarthy era in the United States as anti anticommunism. This entailed a strong opposition to the obsession with the "Red Menace" without harboring any strong affection for communist ideals. In other words, an anti antiposition entails a concerted effort to counter a view without thereby committing oneself to what it opposes (Geertz 1984, 264).

To try to counter anticonstructivism in the very limited space available, I shall concentrate on just one of Winner's major charges against constructivism. I want to suggest that Winner's blind aversion to social constructivism prevents him from getting close enough to his target to come with an appropriate critique.

Good Liberals and Thoroughgoing Interpretivism

One of Winner's primary accusations against the constructivists is that they are rarely, if ever,

From *Science, Technology, & Human Values* 19/1 (1994): 101–9.

prepared to take a stand "on the larger questions about technology and the human condition that matter most in modern history" (1993, 372). Significantly, he singles out the recent work of Steve Woolgar as providing one of the clearest illustrations of the disturbing way in which the constructivists "side-step questions that require moral and political argument" (p. 373). In Winner's eyes, Woolgar is guilty of "blasé depoliticized scholasticism" (p. 376) because he refuses to offer a "solid, systematic standpoint or core of moral concerns from which to criticize or oppose any particular pattern of technological development" (p. 374). It is obvious from the way Winner presents his case against the "waggish" Woolgar that he lacks information about the identity of his foe. Woolgar's moral credentials appear to me as immaculate as the ones Winner pins to his own chest. In my view, Woolgar's refusal to offer us a "solid standpoint or core of moral concerns" reflects neither political naïveté nor scholastic frivolity on his part but, on the contrary, a very serious belief that "good liberals" just do not do that sort of thing.

In an important sense, Winner and Woolgar deserve each other. Of the approaches to the social study of technology that Winner wishes to tar with the same brush, Woolgar's program of "thoroughgoing interpretivism" (Woolgar 1991; Woolgar and Grint 1991; Grint and Woolgar 1992) is as "antirealist" as Winner's is anticonstructivist. Woolgar dreads realism in the same way that Winner dreads constructivism. The opposition between Woolgar's and Winner's ideas of what being a good liberal entails can be captured with reference to the basic distinction that Richard Rorty (1989, ch. 4) draws between the liberal "ironist" and the liberal "metaphysician." After commencing with the claim that each one of us possesses a set of words or a "final vocabulary" we use to justify our actions, beliefs, and lives, Rorty defines an ironist as someone who meets the following three conditions:

> (1) She has radical and continuing doubts about the final vocabulary she currently uses, because she has been impressed by other vocabularies, vocabularies taken as final by people and books she has encountered; (2) she realizes that argument phrased in her present vocabulary can neither underwrite nor dissolve these doubts; (3) insofar as she philosophizes about her situation, she does not think that her vocabulary is closer to reality than others, that it is in touch with a power not herself. (P. 73)

A metaphysician, on the other hand, tends to think of his or her final vocabulary as talking good common sense, and is typically cocksure about the nature of "reality" and highly confident that he or she can recognize the truth when he or she sees it. In addition, when a metaphysician discovers the right answer to a question, he or she does not simply equate this with his or her growing competence in a local final vocabulary, or with the initiation of a new one, but rather, with the stumbling upon "the real facts of the matter" (p. 76). Contrary to this, an ironist believes that truth is always contingent and that it is never possible to hold that one final vocabulary is superior to another. What is more, if an ironist is also a "good liberal," he or she will undoubtedly insist that privileging one final vocabulary over another is about the most illiberal thing one can do. As far as a good liberal ironist is concerned, any attempt to impress upon others the superiority of one's final vocabulary will always result in the serious humiliation of all those who happen to believe differently.

Therefore, when Winner, as a good liberal metaphysician, accuses Woolgar of political indifference and of sidestepping political and moral arguments, the latter, as a good liberal ironist, would no doubt respond that he is only fulfilling an obligation to respect other people's freedom to think for themselves. He would respond in this way because he believes that taking a firm public stance on moral and political issues can only lead to cruelty toward all those who do not share his point of view. And being cruel in this fashion is among the worst things we humans do (see Rorty 1989, 74).

Clearly, in his blanket criticism of constructivism, Winner fails to recognize that the kind of thoroughgoing interpretivism Woolgar practices is also, when viewed politically, libertarian interpretivism. It supports a strong negative view of freedom extended to encompass potentially conflicting ways of thinking about and interpreting reality. Because we can only ever know *differently* and never *better* than anyone else, we are never justified in thrusting our knowledge upon others. For Rorty, my final vocabulary is without a doubt my most valuable possession. It is what makes me the person I am. As a consequence, to be good liberals we must always strive to avoid making "lasting impressions" on each other. Although

Rorty readily accepts that such impressions will often be made – Dewey, Wittgenstein, and Nabokov, among others, have made lasting impressions on him – he also argues that we should never set out with the intention of making them. If we do happen to change someone's mind about something, this must always happen by pure chance. According to Rorty, the best way for us to avoid being illiberal by enforcing our views on others is to always maintain a strategic division between our private and public lives:

> For my private purposes, I may redescribe you and everybody else in terms which have nothing to do with my attitude to your actual or possible suffering. My private purposes, and that part of my final vocabulary which is not relevant to my public actions, are none of your business. But as I am a liberal, the part of my final vocabulary which is relevant to such actions requires me to become aware of all the various ways in which other human beings whom I might act upon can be humiliated. (Pp. 91–92)

In other words, we can only avoid cruel invasions of other people's privacy by never attempting to translate our private prejudices into public goods. Winner wants Woolgar to make exactly this translation. He wants Woolgar to come clean and take a stand on the big issues "which matter most." Woolgar would no doubt respond that his private opinions, to the extent that he possesses such things, belong to nobody but himself and that as a good liberal he would never wish to ram them down anybody else's throat. In perfect contrast to Winner, who wants to liberate us by converging on the truth, Woolgar is equally dedicated to protecting our liberty by escaping this truth.

By failing to recognize thoroughgoing interpretivism as libertarian interpretivism, Winner fails to provide us with an appropriate critique of the moral and political implications of (this particular breed of) constructivism. Crucially, what he is missing is that for Woolgar and his associates, emphasizing the "interpretative flexibility" of technology is not just a methodological imperative but a moral one as well. For Woolgar, arguing against the interpretative flexibility of technology is only something people with a predilection for cruelty do. Obviously, an extended critical discussion of the politics of constructivism, libertarian or otherwise, is urgently required.

Conclusion

The intention of this short commentary has not been to defend constructivism, but rather to counter anticonstructivism. I have indicated that I find it difficult to accept Woolgar's program of thoroughgoing interpretivism, but, at the same time, I see that this is not a justification for rejecting constructivism *tout court*. I have difficulties with Woolgar's program primarily because I believe it makes a virtue out of necessity. Unsettling dominant views about technology must be seen as one of the most important tasks of the social study of technology. But surely we do not need to be so sensitive about hurting the feelings of the bearers of these dominant views. The Rortyian charge of cruelty might well be justified if our ambition were to be a permanent "needle in the eye" of all those who happen to see things differently from us. But making a lasting impression on people with whom we disagree need not always be such a gruesome affair. What is wrong with some good old-fashioned straight talk anyway? Speaking our minds to others may be a way of encouraging them to experiment with a different point of view rather than necessarily a means of converting them to the truth. It all depends on who we think we are when we open our mouths and whether or not we are prepared to do our share of the listening when the time comes.

To elaborate on a distinction from Geertz (1984, 265), what ultimately confronts us is a tale of two fears. If you side with a Woolgar, you will always want us to be wildly troubled by our unwitting ability to both consume *and* produce the "overlearned and overvalued" interpretations of technology that pervade our society. If you side with a Winner, on the other hand, you will continually seek to warn us against the new nihilistic "value neutrality" that allegedly builds on the idea that every interpretation of technology is as significant, and thus as insignificant, as every other. Clearly, these two fears can successfully feed off and bolster each other. For this reason, no task could be more pressing than laying both of them quickly to rest. My ambition in this commentary has been to help to silence one without wishing to condone the other.

Reply to Mark Elam

My essay on the empty black box seems to have hit some raw nerves. I am grateful to Mark Elam for taking time to respond and I appreciate the passion in his words. Unfortunately, he seems more interested in applying ideological decals and passing peremptory judgments than in addressing any of the issues I posed. Presenting himself in the graceful image of a "good liberal ironist," Elam shoots some rather illiberal, nonironic shots at "fear and loathing," "predilection for cruelty," and such like. Oh my. It looks as though Baudelaire is wearing his Godzilla costume today.

Mr. Elam correctly notes that I urge scholars to take a position on the larger questions about technology and the human condition. I am convinced that important questions about moral and political life are answered within the artificial forms and processes of our technological world. For that reason it is incumbent upon a writer not only to describe, analyze, and interpret but also to let people know what he or she makes of these technologically embodied ways of living. This should not be the focus of every moment of scholarship and intellectual exchange. But, surely, at the end of the day, our readers and our students should know where we stand on such broad issues as the promise of "technological development" or on what we would advise on particular consequential social and technical patterns.

Elam mistakenly supposes that I insist on rooting positions of this kind in "truth." Wrong decal. As a matter of fact, my view is that both evaluations of technology and the cultivation of lasting virtues that concern technological choice must emerge from dialogue within real communities in particular situations. As regards technological decision making, I believe that the challenge is not that of how to impose universal standards of judgment clarified by "liberal metaphysics," but how to expand the social and political spaces where ordinary citizens can play a role in making choices early on about technologies that will affect them (as argued in Winner 1992).

From that point of view there is a great deal to be learned from social histories and sociological studies, including those written in a constructivist mode. Studying the rise of transit systems in New York City, for example, one can grasp how highway construction involved the reproduction of racial and economic divisions, divisions not easily removed once they were set in concrete. That observation is not derived from Immanuel Kant or from some bankrupt strand of Enlightenment metaphysics. It comes from listening to specific accounts of the lived experience of communities that suffered authoritarianism and inequality expressed in public works.

These are not concerns that should be "laid to rest." Drawing upon historical experience we start keeping an eye out for technical blueprints emerging today and we challenge those within corporations and government bureaucracies who are busily drawing them up. If we take notice of these matters only after the bulldozers arrive or fiber optic cables are strung, it is simply too late, too late for anything, except for Ph.D. dissertations, journal articles, or book series on social construction.

As he attacks the phantasm he mistakenly supposes is my persona – "A Langdon Winner" – it is interesting to see Mr. Elam cite Richard Rorty with such aplomb. Indeed, Rorty has done much to reveal the new demands on intellectual debate for a time in which there are no reliable, universal foundations. But Rorty does not carry his approach in quite the direction Elam suggests. Rorty has recently spoken out against the vain posturing that sometimes parades as "poststructural," or "postmodern," or "subversive" discourse. He argues that too much of contemporary scholarship is engaged with inward gazing, university-centered squabbles about language, method, and the self-indulgent "politics of difference." In plain words, Rorty exhorts intellectuals to be engaged in attempts to "use the mechanisms of government to help prevent the rich from ripping off the poor, the strong from trampling on the weak" (Rorty 1991). As self-proclaimed liberal ironist, Rorty is still willing to identify problems, take a stand, join with others, and seek practical remedies. Perhaps that aspect of Rorty's work has escaped Mr. Elam's attention.

Elam's most puzzling claim is that my standpoint "can only lead to cruelty towards all those who do not share his point of view." It would be fascinating to learn exactly which "predilection to cruelty" he thinks I harbor. But to get specific about that would require a level of concreteness Elam finds uncomfortable. It might even require a reasoned argument that could stand the light of critical scrutiny.

But let's lick that decal off as well. Thinkers like Gandhi and Martin Luther King have shown how moral and political commitment can be defined in ways that place the avoidance of cruelty at the very center of one's concerns. Affirmation of nonviolence as a way of acting and the recognition that means and ends are tightly intertwined suggest that brutality need not infect our politics, even when we are engaged in intense struggle. That lesson applies to technological politics and to more conventional political conflicts.

I understand Mr. Elam's fear that taking strong, inflexible positions is a dangerous game. But does the interpretivist or liberal ironist escape this problem? During the 1980s many of those most delightfully detached and ironic about the Strategic Defense Initiative were also accepting paychecks to work on it. Similar attitudes are still with us, for example, in the playful hermeneutics that surround "virtual reality" and "interactive media," while our populace is being maneuvered toward new frameworks of manipulation and exploitation.

Far from laying my misgivings to rest, Mr. Elam's response merely confirms them. One approach to inquiry in this field prefers niceties of analysis and interpretation to basic, widely recognized human concerns. This raises an interesting question. Will technology studies become a debate among meaningful positions or the display of fashionable postures?

References

Geertz, C. 1984. Anti anti-relativism. *American Anthropologist* 86: 263–78.

Grint, K., and S. Woolgar. 1992. Computers, guns and roses: What's social about being shot? *Science, Technology, & Human Values* 17: 366–80.

Rorty, R. 1989. *Contingency, irony and solidarity*. Cambridge: Cambridge University Press.

——. 1991. Intellectuals in politics. *Dissent* (Fall): 483–90.

Winner, L. 1992. Citizen virtues in a technological society. *Inquiry* 35 (3/4): 3412–61.

——. 1993. Upon opening the black box and finding it empty: Social constructivism and the philosophy of technology. *Science, Technology, & Human Values* 18: 362–78.

Woolgar, S. 1991. The turn to technology in social studies of science. *Science, Technology, & Human Values* 16: 20–50.

Woolgar, S., and K. Grint. 1991. Computers and the transformation of social analysis. *Science, Technology, & Human Values* 16: 368–78.

53

The Social Impact of Technological Change

Emmanuel G. Mesthene

Technology and Wisdom

My objective is to suggest some of the broader implications of what is new about our age. It might be well to start, therefore, by noting what is new about our age.

The fact itself that there is something new is not new. There has been something new about every age, otherwise we would not be able to distinguish them in history. What we need to examine is what in particular is new about our age, for the new is not less new just because the old was also at one time new.

The mere prominence in our age of science and technology is not strikingly new, either. A veritable explosion of industrial technology gave its name to a whole age two centuries ago, and it is doubtful that any scientific idea will ever again leave an imprint on the world so penetrating and pervasive as did Isaac Newton's a century before that.

It is not clear, finally, that what is new about our age is the rate at which it changes. What partial evidence we have, in the restricted domain of economics, for example, indicates the contrary. The curve of growth, for the hundred years or so that it can be traced, is smooth, and will not support claims of explosive change or discontinuous rise. For the rest, we lack the stability of concept, the precision of intellectual method, and the necessary data to make any reliable statements about the rate of social change in general.

I would, therefore, hold suspect all argument that purports to show that novelty is new with us, or that major scientific and technological influences are new with us, or that rapidity of social change is new with us. Such assertions, I think, derive more from revolutionary fervor and the wish to persuade than from tested knowledge and the desire to instruct.

Yet there is clearly something new, and its implications are important. I think our age is different from all previous ages in two major respects: first, we dispose, in absolute terms, of a staggering amount of physical power; second, and most important, we are beginning to think and act in conscious realization of that fact. We are therefore the first age who can aspire to be free of the tyranny of physical nature that has plagued man since his beginnings.

"Technology and Wisdom" from *Technology and Social Change*, ed. Emmanuel G. Mesthene, Indianapolis: Bobbs-Merrill Inc., 1967, pp. 109–15, with some revisions by the author. "Chapter I: Social Change," "Chapter II: Values," and "Chapter III: Economic and Political Organization" from Emmanuel G. Mesthene, *Technological Change: Its Impact on Man and* Society, Cambridge, MA: Harvard University Press, pp. 15–20, 34–65, 73–89, abridged.

I

The consciousness of physical impossibility has had a long and depressing history. One might speculate that it began with early man's awe of the bruteness and recalcitrance of nature. Earth, air, fire, and water – the eternal, immutable elements of ancient physics – imposed their requirements on men, dwarfed them, outlived them, remained indifferent when not downright hostile to them. The physical world loomed large in the affairs of men, and men were impotent against it. Homer celebrated this fact by investing nature with gods, and the earliest philosophers recognized it by erecting each of the natural elements in turn – water, air, earth, and fire – into fundamental principles of all existence.

From that day to this, only the language has changed as successive ages encountered and tried to come to terms with physical necessity, with the sheer "rock-bottomness" of nature. It was submitted to as fate in the Athenian drama. It was conceptualized as ignorance by Socrates and as metaphysical matter by his pupils. It was labeled evil by the pre-Christians. It has been exorcised as the Devil, damned as flesh, or condemned as illicit by the Church. It has been the principle of non-reason in modern philosophy, in the form of John Locke's Substance, as Immanuel Kant's formless manifold, or as Henri Bergson's pure duration. It has conquered the mystic as nirvana, the psyche as the Id, and recent Frenchmen as the blind object of existential commitment.

What men have been saying in all these different ways is that physical nature has seemed to have a structure, almost a will of its own, that has not yielded easily to the designs and purposes of man. It has been a brute thereness, a residual, a sort of ultimate existential stage that allowed, but also limited, the play of thought and action.

It would be difficult to overestimate the consequences of this recalcitrance of the physical on the thinking and outlook of men. They have learned, for most of history, to plan and act *around* a permanent realm of impossibility. Man could travel on the sea, by sail or oar or breast stroke. But he could not travel *in* the sea. He could cross the land on foot, on horseback, or by wheel, but he could not fly over it. Legends such as those of Daedalus and Poseidon celebrated in art what men could not aspire to in fact.

Thinking was similarly circumscribed. There were myriad possibilities in existence, but they were not unlimited, because they did not include altering the physical structure of existence itself. Man could in principle know all that was possible, once and for all time. What else but this possibility of complete knowledge does Plato name in his Idea of the Good? The task of thought was to discern and compare and select from among this fixed and eternal realm of possibilities. Its options did not extend beyond it, anymore than the chess player's options extend beyond those allowed by the board and the pieces of his game. There was a natural law, men said, to which all human law was forever subservient, and which fixed the patterns and habits of what was thinkable.

There was, occasionally, an invention during all this time that did induce a physical change. It thus made something new possible, like adding a pawn to the chess game. New physical possibilities are the result of invention; of technology, as we call it today. That is what *invention* and *technology* mean. Every invention, from the wheel to the rocket, has created new possibilities that did not exist before. But inventions in the past were few, rare, exceptional, and marvelous. They were unexpected departures from the norm. They were surprises that societies adjusted to after the fact. They were generally infrequent enough, moreover, so that the adjustments could be made slowly and unconsciously, without radical alteration of world views, or of traditional patterns of thought and action. The Industrial Revolution, as we call it, was revolutionary precisely because it ran into attitudes, values, and habits of thought and action that were completely unprepared to understand, accept, absorb, and change with it.

Today, if I may put it paradoxically, technology is becoming less revolutionary, as we recognize and seek after the power that it gives us. Inventions are now many, frequent, planned, and increasingly taken for granted. We were not a bit surprised when we got to the moon. On the contrary, we would have been very surprised if we had not. We are beginning to use invention as a deliberate way to deal with the future, rather than seeing it only as an uncontrolled disrupting of the present. We no longer wait upon invention to occur accidentally. We foster and force it, because we see it as a way out of the heretofore inviolable constraints that physical nature has imposed upon us in the past.

Francis Bacon, in the sixteenth century, was the first to foresee the physical power potential in scientific knowledge. We are the first, I am suggesting, to have enough of that power actually at hand to create new possibilities almost at will. By massive physical changes deliberately induced, we can literally pry new alternatives out of nature. The ancient tyranny of matter has been broken, and we know it. We found, in the seventeenth century, that the physical world was not at all like what Aristotle had thought and Aquinas had taught. We are today coming to the further realization that the physical world need not be as it is. We can change it and shape it to suit our purposes.

Technology, in short, has come of age, not merely as technical capability, but as a social phenomenon. We have the power to create new possibilities, and the will to do so. By creating new possibilities, we give ourselves more choices. With more choices, we have more opportunities. With more opportunities, we can have more freedom, and with more freedom we can be more human. That, I think, is what is new about our age. We are recognizing that our technical prowess literally bursts with the promise of new freedom, enhanced human dignity, and unfettered aspiration. Belatedly, we are also realizing the new opportunities that technological development offers us to make new and potentially big mistakes.

II

At its best, then, technology is nothing if not liberating. Yet many fear it increasingly as enslaving, degrading, and destructive of man's most cherished values. It is important to note that this is so, and to try to understand why. I can think of four reasons.

First, we must not blink at the fact that technology does indeed destroy some values. It creates a million possibilities heretofore undreamed of, but it also makes impossible some others heretofore enjoyed. The automobile makes real the legendary foreign land, but it also makes legendary the once real values of the ancient market place. Mass production puts Bach and Brueghel in every home, but it also deprives the careful craftsman of a market for the skill and pride he puts into his useful artifact. Modern plumbing destroys the village pump, and modern cities are hostile to the desire to sink roots into and grow upon a piece of land. Some values are unquestionably bygone. To try to restore them is futile, and simply to deplore their loss is sterile. But it is perfectly human to regret them.

Second, technology often reveals what technology has not created: the cost in brutalized human labor, for example, of the few cases of past civilization whose values only a small elite could enjoy. Communications now reveal the hidden and make the secret public. Transportation displays the better to those whose lot has been the worse. Increasing productivity buys more education, so that more people read and learn and compare and hope and are unsatisfied. Thus technology often seems the final straw, when it is only illuminating rather than adding to the human burden.

Third, technology might be deemed an evil, because evil is unquestionably potential in it. We can explore the heavens with it, or destroy the world. We can cure disease, or poison entire populations. We can free enslaved millions, or enslave millions more. Technology spells only possibility, and is in that respect neutral. Its massive power can lead to massive error so efficiently perpetrated as to be well-nigh irreversible. Technology is clearly not synonymous with the good. It *can* lead to evil.

Finally, and in a sense most revealing, technology is upsetting, because it complicates the world. This is a vague concern, hard to pin down, but I think it is a real one. The new alternatives that technology creates require effort to examine, understand, and evaluate them. We are offered more choices, which makes choosing more difficult. We are faced with the need to change, which upsets routines, inhibits reliance on habit, and calls for personal readjustments to more flexible postures. We face dangers that call for constant re-examination of values and a readiness to abandon old commitments for new ones more adequate to changing experience. The whole business of living seems to become harder.

This negative face of technology is sometimes confused with the whole of it. It can then cloud the understanding in two respects that are worth noting. It can lead to a generalized distrust of the power and works of the human mind by erecting a false dichotomy between the modern scientific and technological enterprises, on the one hand, and some idealized and static prescientific conception of human values, on the other. It can also color discussion of some important contemporary issues, that develop from the impact of technology on society, in a way that obscures rather than enhances understanding, and that therefore inhibits rather than facilitates the social action necessary to resolve them.

Because the confusions and discomfort attendant on technology are more immediate and therefore sometimes loom larger than its power and its promise, technology appears to some an alien and hostile trespasser upon the human scene. It thus seems indistinguishable from that other, older alien and hostile trespasser: the ultimate and unbreachable physical necessity of which I have spoken. Then, since habit dies hard, there occurs one of those curious inversions of the imagination that are not unknown to history. Our new-found control over nature is seen as but the latest form of the tyranny of nature. The knowledge and therefore the mastery of the physical world that we have gained, the tools that we have hewed from nature and the human wonders we are building into her, are themselves feared as rampant, uncontrollable, impersonal technique that must surely, we are told, end by robbing us of our livelihood, our freedom, and our humanity.

It is not an unfamiliar syndrome. It is reminiscent of the long-term prisoner who may shrink from the responsibility of freedom in preference for the false security of his accustomed cell. It is reminiscent even more of Socrates, who asked about that other prisoner, in the cave of ignorance, whether his eyes would not ache if he were forced to look upon the light of knowledge, "so that he would try to escape and turn back to the things which he could see distinctly, convinced that they really were clearer than these other objects now being shown to him." Is it so different a form of escapism from that, to ascribe impersonality and hostility to the knowledge and the tools that can free us finally from the age-long impersonality and hostility of a recalcitrant physical nature?

Technology has *two* faces: one that is full of promise, and one that can discourage and defeat us. The freedom that our power implies from the traditional tyranny of matter – from the evil we have known – carries with it the added responsibility and burden of learning to deal with matter and to blunt the evil, along with all the other problems we have always had to deal with. That is another way of saying that more power and more choice and more freedom require more wisdom if they are to add up to more humanity. But that, surely, is a challenge to be wise, not an invitation to despair.

An attitude of despair can also, as I have suggested, color particular understandings of particular problems, and thus obstruct intelligent action. I think, for example, that it has distorted the public debate about the effects of technology on work and employment.

The problem has persistently taken the form of fear that machines will put people permanently out of work. That fear has prevented recognition of a distinction between two fundamentally different questions. The first is a question of economic analysis and economic and manpower policy about which a great deal is known, which is susceptible to analysis by well-developed and rigorous methods, and on the dimensions and implications of which there is a very high degree of consensus among the professionally competent.

That consensus is that there is not much that is significantly new in the probable consequences of automation on employment. Automation is but the latest form of mechanization, which has been recognized as an important factor in economic change at least since the Industrial Revolution. What *is* new is a heightened social awareness of the implications of machines for men, which derives from the unprecedented scale, prevalence, and visibility of modern technological innovation. That is the second question. It, too, is a question of work, to be sure, but it is not one of employment in the economic connotation of the term. It is a distinct question, that has been too often confused with the economic one because it has been formulated, incorrectly, as a question of automation and employment.

This question is much less a question of whether people will be employed than of what they can most usefully do, given the broader range of choices that technology can make available to them. It is less a technical economic question than a question of the values and quality of work. It is not a question of what to do with increasing leisure, but of how to define new occupations that combine social utility and personal satisfaction.

I see no evidence, in other words, that society will need less work done on some day in the future when machines may be largely satisfying its material needs, or that it will not value and reward that work. But we are, first, a long way still from that day, so long as there remain societies less affluent than the most affluent. Second, there is a work of education, integration, creation, and eradication of disease and discontent to do that is barely tapped so long as most people must labor to produce the goods that we consume. The more machines can take over what we do, the more we can do what machines cannot do. That, too, is liberation: the liberation of history's slaves, finally to be people....

III

Such basically irrational fears of technology have a counterpart in popular fears of science itself. Here, too, anticipatory despair in the face of some genuine problems posed by science and technology can cloud the understanding.

It is admittedly horrible, for example, to contemplate the unintentional evil implicit in the ignorance and fallibility of man as he strives to control his environment and improve his lot.

What untoward effects might our grandchildren suffer from the drugs that cure our ills today? What monsters might we breed unwittingly while we are learning to manipulate the genetic code? What are the tensions on the human psyche of a cold and rapid automated world? What political disaster do we court by providing 1984's Big Brother with all the tools he will ever need? Better, perhaps, in Hamlet's words, to

> ...bear those ills we have
> Than fly to others that we know not of.

Why not stop it all? Stop automation! Stop tampering with life and heredity! Stop the senseless race into space! The cry is an old one. It was first heard, no doubt, when the wheel was invented. The technologies of the bomb, the automobile, the spinning jenny, gunpowder, printing, all provoked social dislocation accompanied by similar cries of "Stop!" Well, but why not stop now, while there may still be a minute left before the clock strikes twelve?

We do not stop, I think, for three reasons: we do not want to; we cannot, and still be men; and we therefore should not.

It is not at all clear that atom bombs will kill more people than wars have ever done, but energy from the atom might one day erase the frightening gap between the more and less favored peoples of the world. Was it more tragic to infect a hundred children with a faulty polio vaccine than to have allowed the scourge free reign forever? It is not clear that the monster that the laboratory may create, in searching the secret of life, will be more monstrous than those that nature will produce unaided if its secrets remain forever hidden. Is it really clear that rampant multiplication is a better ultimate fate for man than to suffer, but eventually survive, the mistakes that go with learning? The first reason we do not stop is that I do not think we would decide, on close examination, that we really want to.

The second reason is that we cannot so long as we are men. Aristotle saw a long time ago that "man by nature desires to know." He will probe and learn all that his curiosity prompts him to and his brain allows, so long as there is life within him. The stoppers of the past have always lost in the end, whether it was Socrates, or Christ, or Galileo, or Einstein, or Bonhoeffer, or Boris Pasternak they tried to stop. Their intended victims are the heroes.

We do not stop, finally, because we would not stop being men. I do not believe that even those who decry science the loudest would willingly concede that the race has now been proved incapable of coping with its own creations. That admission would be the ultimate in dehumanization, for it would be to surrender the very qualities of intelligence, courage, vision, and aspiration that make us human. "Stop," in the end, is the last desperate cry of the man who abandons man because he is defeated by the responsibility of being human. It is the final failure of nerve.

I am recalling that celebrated phrase, "the failure of nerve," in order to introduce a third and final example of how fear and pessimism can color understanding and confuse our values. It is the example of those who see the sin of pride in man's confident mastery of nature. I have dealt with this theme before, but I permit myself to review it briefly once more, because it points up the real meaning of technology for our age.

The phrase, "the failure of nerve," was first used by the eminent classical scholar, Gilbert Murray, to characterize the change of temper that occurred in Hellenistic civilization at the turn of our era. The Greeks of the fifth and fourth centuries B.C. believed in the ultimate intelligibility of the universe. There was nothing in the nature of existence or of man that was inherently unknowable. They accordingly believed also in the power of the human intelligence to know all there was to know about the world, and to guide man's career in it.

The wars and mixing of cultures that marked the subsequent period brought with them vicissitude and uncertainty that shook this classic faith in the intelligibility of the world and in the capacity of men to know and to do. There was henceforth to be a realm of knowledge and action available only to God, not subjected to reason or to human effort. Men, in other words, more and more turned to God to do for them what they no longer felt confident to do for themselves. That was the failure of nerve.

The burden of what I have been saying is that times are changing. We have the power and will to probe and change physical nature. No longer are God, the human soul, or the mysteries of life improper objects of inquiry. We are ready to examine whatever our imagination prompts us to. We are convinced again, for the first time since the Greeks, of the essential intelligibility of the universe: there is nothing in it that is in principle not knowable. As the sociologist Daniel Bell has put it,

"Today we feel that there are no inherent secrets in the universe, and this is one of the significant changes in the modern moral temper." That is another way of stating what is new about our age. We are witnessing a widespread recovery of nerve.

Is this confidence a sin? According to Gilbert Murray, most people "are inclined to believe that without some failure and sense of failure, without a contrite heart and conviction of sin, man can hardly attain the religious life." I would suspect that this statement is still true of most people, although it is clear that a number of contemporary theologians are coming to a different view. To see a sense of failure as a condition of religious experience is a historical relic, dating from a time when an indifferent nature and hostile world so overwhelmed men that they gave up thought for consolation. To persist in such a view today, when nature is coming increasingly under control as a result of restored human confidence and power, is both to distort reality and to sell religion short. It surely does no glory to God to rest his power on the impotence of man.

The challenge of our restored faith in knowledge and the power of knowledge is rather a challenge to wisdom – not to God.

Some who have seen farthest and most clearly in recent decades have warned of a growing imbalance between man's capabilities in the physical and in the social realms. John Dewey, for example, said: "We have displayed enough intelligence in the physical field to create the new and powerful instrument of science and technology. We have not as yet had enough intelligence to use this instrument deliberately and systematically to control its social operations and consequences." Dewey said this more than thirty years ago, before television, before atomic power, before electronic computers, before space satellites. He had been saying it, moreover, for at least thirty years before that. He saw early the problems that would arise when man learned to do anything he wanted before he learned what he wanted.

I think the time Dewey warned about is here. My more thoughtful scientific friends tell me that we now have, or know how to acquire, the technical capability to do very nearly anything we want. Can we...control our biology and our personality, order the weather that suits us, travel to Mars or to Venus? Of course we can, if not now or in five or ten years, then certainly in twenty-five, or in fifty or a hundred.

But if the answer to the question What can we do? is "Anything," then the emphasis shifts far more heavily than before onto the question What should we do? The commitment to universal intelligibility entails moral responsibility. Abandonment of the belief in intelligibility two thousand years ago was justly described as a failure of nerve because it was the prelude to moral surrender. Men gave up the effort to be wise because they found it too hard. Renewed belief in intelligibility two thousand years later means that men must take up again the hard work of becoming wise. And it is much harder work now, because we have so much more power than the Greeks. On the other hand, the benefits of wisdom are potentially greater, too, because we have the means at hand to *make* the good life, right here and now, rather than just to go on contemplating it in Plato's heaven.

The question What should we do? is thus no idle one but challenges each one of us. That, I think, is the principal moral implication of our new world. It is what all the shouting is about in the mounting concern about the relations of science and public policy, and about the impact of technology on society. Our almost total mastery of the physical world entails a challenge to the public intelligence of a degree heretofore unknown in history.

Chapter I Social Change

Three inadequate views about technology

Research aimed at discerning one or another of the particular effects of technological change on industry, government, or education is not new. Economists, social scientists, and other professional investigators have been engaged in it at least since Karl Marx, over a century ago. By contrast, systematic inquiry devoted to seeing all such effects together and to assessing their broader implications for contemporary society as a whole is relatively recent. Also, it calls for cooperation and joint effort among different academic disciplines, so that it lacks some of the methodological rigor and the richness of theory and data that mark the more established fields of scholarship.

As a result, efforts to understand and explicate the interaction between technological change and social change have to contend with a number of facile, one-dimensional or partial views about the nature of that interaction. It has always been so: Absence of knowledge encourages myth, or the comfortable illusion that there is nothing new to know.

One uncritical view that is prevalent at the present time holds that technology is a virtually unalloyed blessing for man and society. Technology is seen as the motor of all progress, as holding solutions for most of our social problems, as helping to liberate the individual from the clutches of a complex and highly organized society, and as the source of permanent prosperity – in short, as the promise of utopia in our time.

This view has its modern origins in the Baconian conception of knowledge as power, in the social philosophies of such nineteenth-century thinkers as Saint-Simon and Auguste Comte, and probably also, at least for Americans, in the pragmatic conviction that there is nothing a nation of doers cannot do. The view tends nowadays to be held by many scientists and engineers, by many military leaders and aerospace industrialists, by people who believe that man is fully in command of his tools and his destiny, by people who think he should be if he is not, and by many of the devotees of such modern techniques of "scientific management" as systems analysis and program planning and budgeting.

On its surface, this view exhibits the optimism that we associate with the rationalistic tradition in western intellectual history, as in the eighteenth-century Enlightenment in France, for example. It places great faith in the social efficacy of scientific methods and tools, and it by and large assumes a model of society according to which science is the principal determining element in the shape and destinies of men and their institutions.

Below the surface, one may detect traces of economic and political ideology in this view. Some vested interests find profit or other advantage in new technology, and many who would bend society to their purposes often see it as a means of bringing about deliberate social change.

A contrary view sees technology as an almost unmitigated curse. Technology is said to rob people of their jobs, their privacy, their participation in democratic government, and even, in the end, of their dignity as human beings. It is seen as autonomous and uncontrollable, as fostering materialistic values and as destructive of religion, as bringing about a technocratic society and bureaucratic state in which the individual is increasingly submerged, and as threatening, ultimately, to poison nature and blow up the world.

This view is heir to two different traditions. It is akin to historical "back-to-nature" attitudes toward the world, such as we associate with Jean-Jacques Rousseau and Henry Thoreau. It also derives from traditional socialist critiques of the appropriation of technology as capital. From one or another of these standpoints, this view is currently propounded by many artists and literary commentators, by popular social critics, and by existentialist philosophers and latter-day Marxists. It is attractive to many of our youth, and it tends to be held, understandably enough, by segments of the population that suffer dislocation as a result of technological change.

By contrast with the optimistic view of technology, this one is marked by the pessimism that has often been associated with the mystical tradition in the history of the West. Taken together, the two views may be seen as the latest stage of the eternal battle between God and the Devil. In the first view, technology is invested with an omnipotence heretofore reserved to the Almighty. In the second view, technology emerges as the modern counterpart of the Devil, responsible, as the Devil has traditionally been, for "man's eternal inhumanity to man." The psychological counterpart of this contrast, as the two views make clear, is that between optimists and pessimists. In sociological-political-economic terms, we are dealing with the contrast between those who have power and status and those who do not.

There is a third view, which is of a quite different sort, however, and the inadequacies of which derive from its being partial and hypercritical, rather than biased and uncritical. This view holds that technology as such is not worthy of special notice. The arguments advanced for this conclusion include the following: Technology is not new, and it has moreover been recognized as a factor in social change at least since the Industrial Revolution; it is unlikely that the social effects of modern technology – even of computers, for example – will be nearly so traumatic as the introduction of the factory system into eighteenth-century England; research has shown that technology has done little to accelerate the rate of economic productivity since the 1880s; and, there has been no significant change in recent decades in the time period between invention and widespread adoption of new technology.

Moreover, according to this argument, improved communications and higher levels of education make people today much more adaptable than ever before to new ideas and to the new social reforms required by technology. If anything, therefore, technological change is likely to be less upsetting than in the past, because its scope and rate are roughly in equilibrium with man's social and psychological development.

Although this view is supported by a good deal of empirical evidence, it tends to underemphasize a number of the characteristics of modern technology and of the modern world that deprive historical comparison of some of its force. Among these are the effects of sheer physical power, speed of communication, and population densities, which are less easy to identify and measure with precision than, say, more strictly economic changes. Such underemphasis of relevant factors reflects the difficulty of coming to grips with a new and broadened subject matter by means of concepts, methods, and intellectual categories that were designed to deal with older and different subject matters.

This third view tends to be held by historians, for whom continuity is an indispensable methodological assumption, and by many economists, who find that their techniques measure some things quite well while those of the other social sciences are as yet much less refined and reliable.

Stripped of the trace of caricature I have allowed to creep into my descriptions, each of the three views I have discussed contains a measure of truth and evokes a genuine aspect of the relationship of technology and society. Yet they are oversimplifications that do not really serve to advance understanding. One can find empirical evidence to support each of them – and especially so the third – without however gaining much knowledge about the mechanism by which technology leads to social change, or much insight about the implications of technology for the future. The first two views remain too uncritical, and the third too partial, to be reliable guides for inquiry. Recent research and analysis have been leading to more differentiated conclusions and revealing more subtle relationships, as I hope what follows will show.

[...]

On taking advantage of technological opportunities

The heightened prominence of technology in our society faces us with the interrelated tasks of profiting from its opportunities and containing its dangers. At the present time, much public and political concern is being devoted to the increasingly visible and distressing negative impacts of modern technology, on the environment as well as on man. The concern is deserved, but it is important to give at least equal attention to deriving the benefits for society that are also potential in technology.

This is not to suggest that we have not done so in many cases, of course: in space and in industrial production, as already noted, as well as in such areas as medicine and communications. In all such cases, we have been able to achieve a successful combination of technical inventiveness, adequate resources, and institutional adaptiveness. All three of those ingredients must be present. In areas in which they are not, there is a presumption that society has failed to take full advantage of its technological opportunities.

Failure of society to respond to the opportunities created by technological change means that much actual or potential technology lies fallow, that is, is not used at all or is not used to its full capacity. This can often mean that potentially solvable problems are left unsolved and potentially achievable goals unachieved, because we waste our technological resources or use them inefficiently.

The best hope of controlling the damaging effluent of the old technology of internal combustion, for example, lies with newer electric, chemical, or filtration technologies, and provision of adequate low- and medium-income housing could be moved off dead center if we would develop the political means of taking advantage of new materials and construction technologies. A society has at least as much stake in the efficient utilization of technology as in that of its natural or human resources.

There are often good reasons, of course, for not developing or utilizing a particular technology. The mere fact that it can be developed is not sufficient reason for doing so. The costs of development may be too high in the light of the expected benefits, as was the case a few years ago with the project to develop a nuclear-powered aircraft. Or, a new technological device may be so dangerous in itself or so inimical to other purposes that it is never developed, as in the cases of Herman Kahn's "Doomsday Machine" and the recent proposal to "nightlight" Vietnam by reflected sunlight.

But there are also cases where technology lies fallow because existing social structures are inadequate, and prevailing value systems poorly attuned, to exploiting the opportunities it offers. A clear example of this lack is provided in a recent examination of institutional failure in the urban ghetto by Richard S. Rosenbloom and Robin Marris.[1] At point after point this study confirms what has long been suspected, that is, that traditional institutions, attitudes, and approaches are by and large incapable of coming to grips with the new problems of our cities. Many of those problems were themselves caused by technological change, ... but existing social mechanisms seem unable to realize the possibilities for resolving them that are also inherent in technology.

Vested economic and political interests serve to obstruct adequate provision of low-cost housing. Community institutions wither for want of interest and participation by residents. City agencies are unable to marshal the skills and take the systematic approach needed to deal with new and intensified problems of education, crime control, and public welfare. Business corporations finally, which are organized around the expectation of private profit, are insufficiently motivated to bring new technology and management know-how to bear on urban projects where the benefits appear to be largely social. Business has yet to estimate the costs of racial discrimination in terms of decreased purchasing power and its consequent depressing effect on the private economy.

All these factors combine to dilute what could otherwise be translated into a genuine public desire to apply our best knowledge, our latest tools, and adequate resources to the resolution of urban tensions and the eradication of poverty in the nation.

There is also institutional failure of another sort. Government in general and agencies of public information in particular are not yet equipped for the massive task of public education that is needed if our society is to make full use of its technological potential, although the Federal government has been making significant strides in this direction in recent years. Thus, much potentially valuable technology goes unused, because the public at large is insufficiently informed about the possibilities and their costs to provide support for appropriate political action. To use a negative example, it was an informed and therefore aroused public opinion that finally led to governmental safety regulation of the automobile industry. More positively, it is education of the public in the potentialities of construction technology and in the barriers to its full utilization embodied in zoning ordinances and in the interests of the affected industries and labor unions that seems necessary for adequate formulation and implementation of urban housing programs.

The general point is that political action in a democracy depends on the availability of abundant, accurate, and widely disseminated political information. As noted, we have done very well with our technology in the face of what were or were believed to be crisis situations, as with our military technology in World War II and with our space efforts when beating the Russians to the moon was deemed a national goal of first priority. We have also done very well when the potential benefits of technology were close to home or easy to see, as in improved health care and better and more varied consumer goods and services.

We have done much less well in developing and applying technology where the need or opportunity has seemed neither so clearly critical nor so clearly personal as to motivate political action, as in the instance of urban policy discussed above, or in that of educational technology for that matter. Where technological possibility continues to lie fallow, it is to improved political information and to innovation in political institutions that society must attend if it is to use its tools to their full effectiveness.

Containing the negative effects of technology

The kinds and magnitude of the negative effects of technology are no more independent of the institutional structures and cultural attitudes of society than is realization of the new opportunities that technology offers. With few exceptions, technologies in our society are developed and applied as a result of individual decision. Individual entrepreneurs, individual firms, and individual government agencies are always on the lookout for technological opportunities: either new machines to reduce the costs of production, or new mechanical products to sell, or new technological systems to facilitate accomplishment of a mission. Technological innovation is so much sought after, in fact, that the Federal government puts $16 billion a year into research and development and big companies hire whole staffs of scientists and engineers to create new technologies.

In deciding whether to develop a new technology, the individual decision maker calculates the benefit that he can expect to derive from the new development and compares it with what it is likely to cost him. If the expected benefit to himself is greater than the cost he will have to pay, he goes ahead. In making this calculation, the individual decision maker does not pay very much attention to the probable benefits and costs of the development to others than himself or to society generally, except to the extent that he is required to do so indirectly by relevant laws or governmental regulations. These latter benefits and costs are characterized by economists as *external*.

The external benefits potential in new technology – what I referred to in the preceding section as technological opportunities – may thus not be fully realized by the individual developer. They will rather accrue to society as a result of deliberate social action, as pointed out above.

Similarly with the external costs. The direct costs of technological development include design and engineering, labor, raw materials, plant and equipment, packaging, marketing, distribution, and so forth. The individual developer pays for these. But he does not pay – or he pays very little and very indirectly – for the waste that he pours into the river, the smoke he discharges into the air, the job dislocations that he causes when he automates his plant, or the noise nuisance that may go with a new airplane design. In minimizing only expected costs to himself, in other words, the individual decision maker will tend to contain only some of the potentially negative effects of the new technology. The external costs, and therefore the negative effects on society at large, are not of principal concern to him, and in our society they are generally not expected to be.

Most of the consequences of technology that are causing concern at the present time – the proliferation of weapons technology, smog, water pollution, radioactivity, urban sprawl, sonic booms, threats to the beauty and balance of nature, social and psychological tensions and unrest, job dislocations, and encroachments on individual privacy – are negative externalities of this kind. They are with us in large measure because it has not been anybody's explicit business to foresee and anticipate them. They have fallen between the stools of innumerable individual decisions to develop individual technologies for individual purposes without anyone – any organization or agency – around to give explicit attention to what all these decisions add up to for society as a whole and for people as human beings.

This freedom of individual decision making is a value that we have cherished and that is built into the institutional fabric of our society. The negative effects of technology that we deplore are a measure of what this traditional freedom is beginning to cost us. They are traceable less to some mystical autonomy presumed to lie in technology and much more to the autonomy that our economic and political institutions grant to individual decision-making. Technology would seem much less autonomous if we could internalize all its costs and charge them directly to the producer. They would then be paid for either by the ultimate consumer (in the form of higher prices in the case of consumer goods or of some sort of fee, as in a toll-road) or by society as a whole when necessary (in one form or another of public subsidy), but only following a deliberate and explicit public decision to do so.

When the social costs of virtually unrestricted individual decision making in the economic realm achieved crisis proportions in the great depression of the 1930s, the Federal government introduced economic policies and measures many of which had the effect of abridging the freedom of individual decision. Now that some of the negative impacts of technology are threatening to become critical, the government is considering measures of control that will have the analogous effect of constraining the freedom of individual decision makers to develop and apply new technologies irrespective of social consequence.

[…]

All such attempts to assess, control, and mitigate the negative effects of technology appear to many to threaten freedoms that our traditions still take for granted as inalienable rights of men and good societies, however much such freedoms may have been tempered in practice by the social pressures of modern times. These include the freedom of the market, the freedom of private enterprise, the freedom of the scientist to follow the truth wherever it may lead, and the freedom of the individual to pursue his fortune and decide his fate.

There is thus set up a tension between the need to control technology and our wish to preserve our values, which leads some people to conclude that technology is inherently inimical to human values. The political effect of this tension takes the form of inability to adjust our decision making structures to the realities of technology so as to take maximum advantage of the opportunities it offers, and so that we can act to contain its potential ill effects before they become so pervasive and urgent as to seem uncontrollable.

To understand why such tensions are so prominent a social consequence of technological change, it is necessary to look explicitly at the effects of technology on social and individual values.

Chapter II Values

The challenge of technology to values

Despite the great importance of scientific techniques and of the processes and institutions of knowledge in contemporary society, it is clear that political decision making and the resolution of social problems are not dependent on knowledge and rational method alone. Many commentators have noted that ours is a "knowledge" society that is devoted to an "end of ideology" and that bases its decisions on the collection and analysis of data, but none would deny the role that values play in shaping the course of society and the decisions of individuals.[2]

On the contrary, questions of values can become more pointed and insistent in a society that organizes itself to control technology and that engages in deliberate social planning. Planning demands explicit recognition of value commitments and value priorities and often brings into the open value conflicts that remain hidden in the more impersonal and less conscious workings of the market system.

In economic planning, for example, we have to make explicit choices between the value we attach to leisure and the value we place on increased economic productivity, and we have to do so moreover without having a common measure on the basis of which the choice can be made. In planning education, we come face to face with the traditional American value dilemma of equality versus achievement: Do we opt for equality and nondiscrimination by giving all students the same basic education, or do we emphasize freedom and foster achievement by tailoring education to individual learning capacities, which are themselves often conditioned by social and economic background?

Current science-based decision making techniques also call for clarity in the specification of goals, thus serving to make value preferences explicit. The effectiveness of systems analysis, for example, requires either that objectives and criteria of evaluation be known in advance, or that alternative possible objectives be clearly enough formulated so that they can be compared; and criteria and objectives of specific actions obviously relate to a society's system of values. That, incidentally, is why the application of systems analysis meets with less relative success in educational or urban planning than in many facets of military planning.

The increased awareness of conflicts among our values that planning and rational decision making produce serves in part to explain the generally questioning attitude toward traditional values that appears to be endemic to a high-technology, knowledge-based society. Another part of the explanation lies in the discovery that values once considered immutable are often based on inadequate knowledge. As we learn more about the physiological and psychological bases of deviant behavior, for example, we begin to question ancient concepts of personal responsibility and the value of punishment as a deterent to crime. ...

To be sure, the values of a society change more slowly than do the realities of human experience; their persistence is inherent in their nature as values and in their function as criteria of judgment and action, which means that their own adequacy will tend to be judged later rather than earlier. But values do change, as a glance at history shows. Some are abandoned as circumstances change and new ones are formulated to deal with new situations. Most frequently, we make rearrangements in our value hierarchy; values once considered crucial become less relevant and therefore less important, while others, once relatively lower in our estimation, take on new importance. Values do not have to be eternal and unchanging in order to be values.

The fact that values come into question as our knowledge increases, therefore, and that some traditional values cease to function adequately when technology leads to changes in social conditions does not mean that values as such are being destroyed by knowledge and technology. What does happen is that values change through a process of accommodation between the system of existing values and the technological and social changes that impinge on it. Understanding of the effects of technological change on our values, therefore, depends on discovering the specific ways in which this process of accommodation occurs, on identifying the trends implicit in it, and on tracing its consequences for the value system of contemporary American society.

How technology leads to value change

... Robin Williams defines values as "those *conceptions of desirable states of affairs* that are utilized in selective conduct as *criteria* for preference or choice or as *justifications* for proposed or actual behavior," and he distinguishes "some fifteen major value-belief clusterings that are salient in American culture, as follows: (1) activity and work; (2) achievement and success; (3) moral orientation; (4) humanitarianism; (5) efficiency and practicality; (6) science and secular rationality; (7) material comfort; (8) progress; (9) equality; (10) freedom; (11) democracy; (12) external conformity; (13) nationalism and patriotism; (14) individual personality; (15) racism and related group superiority."[3]

It seems clear that values in this sense have their origins in the patterns of choice behavior that are characteristic of any given society. What we mean when we say that a society is committed to certain values is that the people in that society will typically make judgments and choose to act in ways that reveal and reinforce those values. It seems equally clear that choice behavior is determined, or at least circumscribed, by the options available to choose from at the time the choice is made. We can choose to go to the country or to go to the moon, but we cannot at this time choose to go on living for 150 years, because that option is not now available to us.

Available choice options do change over time, of course. Thirty years ago we could not have chosen to go to the moon; 30 years from now we may succeed in extending the human life span to 150 years. When options are thus changed or expanded, it is to be expected that choice behavior will change, too, and changed choice behavior can in turn be expected, given appropriate time lags, to be conceptualized or "habitualized" into a changed set of values.

Technology has a direct impact on values by virtue of bringing about just such changes in our available options. By literally creating new opportunities for action, it offers individuals and society new options to choose from. Space technology – to stay with the most conspicuous example – makes it possible for the first time to go to the moon or to communicate by satellite, and it thereby adds those two new options to the spectrum of choices available to society.

By adding new options in this way, technology can lead to changes in values in the same way that the appearance of new dishes on the heretofore standard menu of one's favorite restaurant can lead to changes in one's taste and choices of food. Specifically, technology appears to lead to value change either by bringing some previously unattainable goal within the realm of choice, or by making some values easier to implement than in the past, that is, by changing the costs associated with realizing them. ...

Thus, the economic affluence that industrial technology has helped to bring to American society makes possible fuller implementation of our traditional values of social and economic equality than was possible in the past. Unless or until it is acted upon, however, that very possibility gives rise to the tensions we associate with the rising expectations of the underprivileged and provokes both the activist response of the radical left and the modern hippie's rejection of society as hypocritical and unwilling to practice in reality what it preaches in its ideals.

[...]

Technology thus has the effect of facing us with contradictions in our own value system and of calling for deliberate attention to their resolution. The problem is aggravated by the regrettable but undisputed fact that our social and moral sciences are as yet far from being equipped, conceptually or methodologically, to deal with such issues in a concrete and rigorous way.

The value implications of economic change

The changes discussed and illustrated in the preceding section represent the relatively direct effects of technology on values. In addition, value change often comes about through the mediation of some more general social or cultural change produced by technology. In such cases, the effect of technological change on values can be thought of as being indirect. ...These are the value changes that result from the effects of technology on the economy and the value changes that result from the effects of technology on our religious belief systems. ...

Public goods and services differ from private consumer goods and services in that they are provided on an all-or-none basis and consumed in a joint way, so that more for one consumer does not mean less for another. For example, the clearing of a swamp or a flood-control project, once completed, benefits everyone in the vicinity. A meteorological forecast, once made, can be transmitted by word of mouth to additional users at no additional cost. Knowledge itself may be the prime example of a public good, since the research expenses needed to produce it are incurred only once, unlike consumer goods of which every additional unit adds to the cost of production.

As noted earlier, private profit expectation is an inadequate incentive for the production of public goods and services, because their benefit is "indiscriminate" and therefore not fully appropriable to the firm or individual that incurs the cost of producing them. Individuals are therefore motivated to dissimulate by understating their true preferences for such goods, in the hope of shifting their cost to others. This creates a "free-loader" problem, which skews the mechanism of the market. The market therefore provides no effective indication of the optimal amount of such public commodities from the point of view of society as a whole. If society got only as much public health care, flood control, or knowledge as individual profit calculations would generate, it would no doubt get less of all of them than it does now or than it expresses a desire for by collective political action.

This gap between collective preference and individual motivation imposes strains on a value system, such as ours, that is primarily individualistic rather than collective or "societal" in its orientation. That system arose out of a simpler, more rustic, and less affluent time than ours, when both benefits and costs were of a much more private sort than now. What public goods were needed, it was assumed, could be provided largely by private individual enterprise and action, given only a liberal constitution and a governmental guarantee of individual rights.

That assumption, and the individualistic value system it supported, are no longer fully adequate for our society, which industrial technology has made productive enough to be able to allocate significant resources to the purchase of public goods and services and in which modern transportation and communications as well as the sheer magnitude of technological effects lead to extensive ramifications of individual actions on other people and on the environment.

The response to this changed experience on the part of the public at large generally takes the form of increased government intervention in social and economic affairs to contain or guide these wider ramifications, as noted previously. The result is that the influence of values associated with the free rein of individual enterprise is put under strain. Society finds that it must strike a new balance between individual rights and the public interest and create the new social mechanisms needed to institutionalize it. This process will be accompanied, inevitably, by a change of orientation of the traditional American value system in the direction of legitimating the growing emphasis on collective, or society-wide, decision and action. To be sure, the tradition that ties freedom and liberty to a laissez-faire system of decision making remains very strong and the changes in social structures and cultural attitudes that can touch it at its foundations are still only on the horizon. But the trend seems clearly implicit in the imperatives of technological change.

Religion and values

Much of the unease that our society's emphasis on technology seems to generate among some sectors of society can perhaps be explained in terms of the impact that technology has on religion.

The formulations and institutions of religion are not immune to the influences of technological change, for they, too, tend toward an accommodation to changes in the social milieu in which they function. One of the ways in which religion functions, however, is as an ultimate belief system that provides legitimation, that is, a "meaning" orientation, to moral and social values. This ultimate meaning orientation, according to Harvey Cox, is even more basic to human existence than the value orientation. When the magnitude or rapidity of social change threatens the credibility of that belief system, therefore, and when the changes are moreover seen as largely the results of technological change, the meanings of human existence that we hold most sacred seem to totter, and technology appears to be the villain.

Religious change thus provides another mediating mechanism through which technology affects our values. That conditions are ripe for religious change at the present time is evident in the spectrum of events from the "Death-of-God" movement among Protestant theologians to the questioning of the authority of the Pope in moral matters among segments of the Catholic clergy. Such developments amount to asking whether our established religious syntheses and symbol systems are adequate any longer to the religious needs of people who are living in a generally scientific and secular age that is changing so fundamentally as to strain traditional notions of eternity. If they are not, how are they likely to change? ...

Not only must religion somehow meet the challenge of such "loss of faith," however; it needs also to come to terms with the pluralism of explicitly religious belief systems that is characteristic of the modern world. No religion that hopes to be influential today can be formulated in terms that ignore the parallel existence and competitive appeal of alternative faiths. This is so, partly because we have learned a good deal about how different religions function in different societies and partly because widespread knowledge of other peoples makes social and religious diversity seem less unnatural and more acceptable than was the case in earlier times.

This pluralism of belief systems poses serious problems for the ultimate legitimation or "meaning" orientation of moral and social values that religion seeks to provide, because it demands a religious synthesis that can integrate the fact of variant perspectives into its own symbol system. Western religions have been notoriously incapable of performing this integrating function and have rather gone the route of schism and condemnation of variance as heresy. The institutions and formulations of historical Christianity in particular, which once provided the foundations of Western society, carry the added burden of centuries of conflict with scientific world views as these competed for ascendancy in the same society. This makes it especially difficult for traditional Christianity to accommodate to a living experience so infused by scientific knowledge and attitudes as ours, and it helps explain why its adequacy is coming under serious question at the present time.

Cox has noted three major traditions in the Judeo-Christian synthesis and has found them inconsistent in their perceptions of the future: an "apocalyptic" tradition foresees imminent catastrophe and induces a negative evaluation of the world; a "teleological" tradition sees the future as the certain unfolding of a fixed purpose inherent in the universe itself; a "prophetic" tradition, finally, sees the future as an open field of human hope and responsibility and as becoming what man will make of it.[4]

Technology, as I have noted, creates new possibilities for human choice and action, but leaves their disposition uncertain. What its effects will be and what ends it will serve are not inherent in the technology but depend on what man will do with technology. Technology thus makes possible a future of open-ended options that seems to accord well with the presuppositions of the prophetic tradition. That tradition may therefore provide us with the context necessary to a new religious synthesis that is both adequate to our time and continuous with what is most relevant in our religious history.

Arriving at such a new synthesis would of course require an effort at deliberate religious innovation for which Cox finds insufficient theological ground at the present time, despite the attempts represented by Vatican II and other recent ecumenical movements. Although it is recognized that religions have changed and developed in the past, conscious innovation in religion has generally been condemned and is not provided for by the relevant theologies. The main task that technological change poses for theology in the next decades, therefore, is that of deliberate religious innovation and symbol reformulation to take specific account of religious needs in a technological age.

What consequences would such changes in religion have for values? Too little is known as yet about the relationship of religion and values to allow anything like a comprehensive answer to all aspects of so broad and fundamental a question. Considering the principal importance of the symbolic dimension of religious experience, however, examination of the effects of technology on specifically expressive or aesthetic values – rather than on social or moral values, in the first instance – seems particularly promising. Cox therefore attempts the beginning of an answer to the general question in the context of the familiar complaint that, since technology is principally a means, that is, a tool, it enhances merely instrumental values at the expense of expressive or consummatory values that are considered to be somehow more "real."

The appropriate distinction is not between technological instrumental values and nontechnological expressive values, however, but among different expressive values that attach to different technologies. The horse-and-buggy was a technology too, after all, and it is not self-evident that its charms were different in kind or superior to the sense of power and adventure, and the spectacular views, that go with jet travel.

In fact, technological advance is in many instances a condition for the emergence of new creative or consummatory values. Improved sound boxes in the past and structural steel and motion photography in the present have made possible the artistry of Jascha Heifetz, Frank Lloyd Wright, and Charles Chaplin. These artists have opened up wholly new ranges of expressive possibility without, moreover, in any way inhibiting a concurrent renewal of interest in medieval instruments and primitive art.

One concludes that the new religious synthesis we seek would similarly forge new symbols expressive of technological reality. It could thus show the way in which society's values might be brought into better accord with contemporary experience. Indeed, if religious innovation can provide a meaning orientation broad enough to accommodate the idea that new technology can be creative of new values – or can serve to enhance or provide new content for old values – a long step will have been taken toward providing a religious belief system adequate to the realities and needs of a technological age. And it might be quite enough if religion performed this function only in the realm of expressive values. The challenge to education in general and to the humanistic disciplines in particular would then be clear: to accomplish a similar transformation of values in the social and moral realms.

Chapter III Economic and Political Organization

The enhancement of the public sphere

When technology brings about social changes ... that impinge on our values ... it poses for society a number of problems that are ultimately political in nature.

The term "political" is used here in the broadest sense: It encompasses all of the decision making structures and procedures that have to do with the allocation and distribution of wealth and power in society. The political – or, better yet, the economic/political – organization of society thus includes more than the various bodies officially charged with public decision making; that is, it includes more than formal government. It also includes the market mechanism and other economic institutions, and it includes private decision making as well – by firms, or labor unions, or churches, or political parties, or professional or trade associations, or by individuals, for that matter – to the extent that such private decisions affect or have implications for public decisions (i.e., for public policy).

The term "political" is thus a bridging concept; it bridges public and private organizations, formal and informal procedures, small and large groups. It is particularly important to attend to the organization of the entire body politic in this way when technological change contributes to a blurring of once clear distinctions between the public and private sectors of society and to changes in the roles of its principal institutions.

What is the import of big science and of powerful technologies for the political organization of modern society? More generally, what happens when the new rationality, the proliferation of expertise, and the growing social importance of knowledge run head on into political structures, processes, attitudes, values, and practices that developed in the hundred years between Andrew Jackson and World War II, that is, during a time when science and technology were not so big and powerful as they have become since and when the social role of knowledge and its institutions was secondary at best or virtually nonexistent at worst?

If we look for an answer in terms of a basic and long-term trend, I think we must conclude that a major effect of an active science and technology and of a commitment to knowledge as an instrument of social action is a progressive enhancement of the range and influence of the public sector of society in general, and of public decision making in particular. A number of considerations lead to this conclusion.

For one thing, the development and application of technology seem increasingly to require large-scale and complex social concentrations, whether these be large cities, large corporations, big universities, or big government. In instances where technological advance appears to facilitate reduction of such first-order concentrations, it tends instead to enlarge the relevant *system* of social organization, that is, to lead to increased centralization. Thus, the physical dispersion made possible by transportation and communication technologies tends to enlarge the urban complex that must be governed as a unit.

A second characteristic of advanced technology is that its effects cover large distances, in both the geographical and the social senses of the term. Both its positive and negative effects are more extensive. For example, horse-powered transportation technology was limited in its speed and capacity, but its nuisance value was also limited, in most cases to the owner and to the local townspeople. The supersonic transport airplane would carry hundreds across long distances in minutes, but its noise and vibration damage would also be inflicted, willy-nilly, on everyone within the limits of a swath 3,000 miles long and several miles wide.

The concatenation of increased density, extended "distance," and multiplying population means that technological applications have increasingly wider ramifications, and that increasingly large concentrations of people and organizations become dependent on technological systems. A striking illustration of this was provided by the widespread effects of the power blackout in the northeastern part of the United States. The result, as noted in the discussion of technology assessment above, is that more and more decisions that could once be left to private decision makers because their effects were limited in impact and extent must now be taken in public ways, by society as a whole. Education, medicine, population policy, as well as the conduct of science and technology are additional examples of domains that were once largely private but are now increasingly coming into the public sphere.

[...]

The trend illustrated by such examples impinges on the American political tradition in a way that often makes it difficult for us to see what is happening. That tradition has produced two basic philosophies about the proper role of government in society. Depending on whether one has been a laissez-faire-type Republican or a socially-minded Democrat, government – in the popular phrase – was properly on tap or on top. Those were never pure types, of course, but they did define a spectrum in terms of which one could take a political stand.

Nowadays, however, this kind of typology is simply becoming irrelevant, precisely because of the new dominance of the public sphere I have been describing. To interpret this dominance as a victory of a Hamiltonian over a Jeffersonian concept of the proper role of government – as Senator Goldwater sought to do in his presidential campaign in 1964, for example – is to misunderstand and distort what is happening by trying to force it into old categories. For, as noted at the beginning of this chapter, the public or political sphere is much broader in scope and wider in extent than government in the formal sense.

What is actually happening is best described as a mixing-up of social institutions. Put more positively, what we are seeing is the forging of new partnerships between governmental and nongovernmental forms, as all institutions in the society become aware of the increasingly public character of the problems they face. In these new partnerships, government finds a new dimension, a new role, that we have not normally associated with it. No longer is government either the simple arbiter of conflicting interests between business, labor, farmers, or whatever, or the agent to whom all social action should be delegated. Instead, government – however haltingly as yet – is taking on the function of social pioneer and leader of a team; it seeks to identify opportunities over the horizon and problems before they are upon us and to marshal the forces, public and private, needed to deal with both. This helps explain the futuristic cast of many contemporary government programs. It also helps explain the intersectoral mechanisms that are being devised to enable cooperative efforts to deal with new developments as they emerge.

These changes are coming about in response to the growing need to make our social decisions deliberately and in public ways, rather than allowing them to "fall out," so to speak, of the impersonal interplay of innumerable private decisions. That need will continue to grow as technology, economic affluence, and increasing population combine to multiply both the opportunities and problems that society faces and to accelerate the changes with which it must come to terms. This means that allowing political change to come gradually and of its own accord may no longer be a viable strategy for contemporary society, as many of our youth are coming to insist. Instead, we face the problem of deliberately restructuring our political institutions and decision making mechanisms – including the system of economic decision making – to make them adequate to the enhanced social role of the public sphere. For example, the public goods problem alluded to earlier generates a need for such institutional innovation in the organization of the economy, as does the advent of modern computer-based information-handling techniques in the area of governmental decision making.

Private firms and public goods

[...]

The problem that these various findings and considerations point to can be restated in a manner that makes clear the issue of economic/political organization that is involved. Thus, corporations have proved to be highly efficient in exploring technological possibilities for new consumer products and services that can be marketed and sold for a profit. There are a number of reasons for this efficiency. The system of feedback and reward is speedy and highly visible; the corporation whose product does not sell quickly shifts its product line or goes bankrupt. It is this tension between the possibility of great success and quick failure, in fact, that creates the presumption that few technological possibilities for the private sector will go unexplored and that few of the wrong guesses will remain hidden for very long. Because of this kind of built-in efficiency, the corporate system has served us well – better than most, one is inclined to agree – when our greatest need as a society was feeding, clothing, and sheltering our population and raising our standard of living, that is, when our greatest need was for translating our technological progress into an abundance of private goods and services.

By contrast, there exist only relatively inadequate institutional mechanisms devoted to exploring technological possibilities for new socially desirable public goods and services. Traditionally, such mechanisms have included government agencies, corporations that serve public ends indirectly in the course of responding to a market or performing on government contract, mixed bodies such as port or river authorities, and combinations of public and private organizations.

As noted above, these institutional mechanisms tend to perform reasonably well in response to national crises or when the derivative personal benefit is clear, as in the provision of public health, for example. They have generally performed well also in the provision of the public counterparts of private goods, as in public roads for private automobiles. They have not been effective by and large, however, in guarding against the more general social costs of technology or in meeting needs – such as for adequate housing or environmental betterment – the satisfaction of which involves institutional rivalry and political conflict. Nor have they been aggressive in searching out technological possibilities for opportunities that society might well elect to choose if they were there to see. They have not been motivated to invent "new dishes" for the social menu.

A part of the reason for the latter failure, of course, is that such possibilities are difficult to see, because what we do not have and never have had is much more difficult to see or to miss than what we do have or what we may have had and lost. The full benefits of genuinely imaginative and creative uses of the mass media, for example, or the full potentials of public transportation technology cannot be appreciated until such time as they might be realized. Another reason is that public institutions do not have means equivalent to advertising and market research to inform the public about new possibilities and to test the potential demand for them. Finally, there is the difficulty of measuring the potential benefits of envisaged public goods and of deciding whether they are worth their costs. The discipline of the profit-and-loss statement is lacking, as are the clear feedback and the incentive of calculable rewards. Exactly analogous considerations, as we have seen, reduce the efficacy of these institutional mechanisms in containing the negative effects of technological change.

The need for institutional innovation that I cited above arises out of these inadequacies of present institutions. With the proper economic and political organization, we could derive greater benefits from our technology than we do. We might not then simply wait for them to fall out as incidental consequences of entrepreneurial activities but would be able to pursue them directly as a matter of deliberate public policy. That could stimulate development and application of technologies aimed primarily at producing social benefits and only secondarily at generating private profits. Such a reversal of our traditional priorities could help, not only to reduce pollution or rebuild our cities, but also to create new social opportunities in education, in health, or in cultural development.

Unfortunately, the knowledge and the ideas necessary to chart the necessary institutional innovation are largely lacking still, mainly because the recognition of the need is only very recent. In response to this lack, Dr. Marris is currently engaged in a study designed to formulate the relevant research questions and suggest some early answers.

Among the questions this study is addressing are: (1) the effects of corporate goals and organization on the provision of socially desirable goods, (2) the costs to society of a policy aimed principally at stimulating economic growth rather than at a broader and more balanced social development, (3) the incommensurability of individual incentive and public will, (4) the desirable balance between individual and social welfare when the two are inconsistent with each other, (5) changes in the roles of government and industrial institutions in the political organization of American society, and (6) the consequences of those changes for the functions of advertising and competing forms of communication in the process of public education.

In particular, attention is being directed to whether existing forms of company organization are adequate for marshaling technology to social purposes by responding to the demand for public goods and services, or whether new productive institutions will be required to serve that end. Other investigators are beginning to address one or another aspect of this problem too, of course, but it will be some time yet before these efforts begin to add up collectively to anything approximating a reasonable blueprint for the social innovation that we need.

Scientific decision making

A need for innovation in our political institutions and attitudes is implicit also in the new methods and tools of decision making that science and technology are making available to government.

One way in which this is coming about is through the proliferation of what might be called scientific or knowledge agencies in government itself: examples in the Federal government are the National Science Foundation, the National Aeronautics and Space Administration, and the Atomic Energy Commission (as specifically scientific or technological agencies) and the Departments of Health, Education, and Welfare, of Housing and Urban Development, and of Transportation (for what might be thought of as the more broadly knowledge-oriented agencies).

A second way in which science is affecting public decision making is by introduction of the professional scientist into the policy-making process, and in the development of greater technical sophistication on the part of the professional civil servant and government official. The broader face of this trend is the increasing use of experts of all sorts in the process of government, with the problems that this implies for the role of a less expert Congress and a less expert electorate in public decision making.

A third and increasingly important way in which governmental decision making is utilizing scientific techniques is by the introduction of computerized information-handling procedures and the adoption of some recently developed intellectual methodologies.

[...]

The imperatives of modern decision making thus appear to require greater and greater dependence on the collection and analysis of data and on the use of technological devices and scientific techniques. Indeed, many observers would be likely to agree that there is an "increasing relegation of questions which used to be matters of political debate to professional cadres of technicians and experts which function almost independently of the democratic political process."[5] In recent times that trend has been most noticeable in the areas of economic policy and national security affairs.

These imperatives of modern decision making, however, run counter to that element of traditional democratic theory that places high value on direct participation in the political process and thus generates the kind of discontent mentioned above. If it turns out on more careful examination that direct participation is becoming less relevant to a society in which the connections between causes and effects are long and often hidden – which is an increasingly "indirect" society, in other words – elaboration of a new democratic ethos and of new political institutions and procedures more adequate to the realities of a modern technological society will emerge as perhaps the major intellectual and political challenge of our time.

At a minimum, it would seem that such new political forms would have to embody a distinction between the expression of preferences and the prediction of consequences. The opportunity to express preferences cannot, in a democracy, be properly denied to any segment of the population. The more complex the society, in fact, and the more technical the issues it has to deal with, the more deliberate must be the effort to allow for the free expression of preferences. The simple institution of the quadrennial or biennial ballot on candidates presumed to represent all the issues does not any longer appear to be fully adequate to contemporary need; our society may be outgrowing it as it is outgrowing the early value system that attended its birth. A more differentiated system of electoral consultation needs to be devised that will be responsive at short intervals when necessary, that can distinguish among different kinds of issues, and that can inform those whom it consults about alternative options and their associated costs.

The need for a more refined system of electoral consultation does not however imply that the electorate at large, or any particular segment of it, has a role to play in the technical process of government itself. In a populous, modern, industrialized, and knowledge-oriented society, that process consists increasingly of adumbrating alternative policy options and calculating their probable consequences. It is clearly a job for experts and for all the sophisticated information-handling and management techniques that can be brought to bear on it. To be sure, the policy experts and decision makers must remain accountable to the electorate in the end. An adequate system of electoral consultation must therefore also provide effective mechanisms for monitoring and evaluating actual performance. But consultation and accountability remain distinct from the technical decision making process, and no amount or combination of them – that is, no amount of "participation" in the populist sense of the term – can substitute for the expertise and decision making technologies that modern government must use.

These considerations do no more than point to the general direction in which the needed restructuring of our political institutions must proceed. They do not begin to convey the difficulty of the intellectual and political problems that must be met before the actual restructuring can begin, nor do they evoke the strength of privilege and vested interest that will stand in the way. But institutional innovation here, as in the area of economic organization, is still in its early infancy. The hard work remains to be done.

Individual rights and public responsibilities

What do technological change and the social and political changes that it brings with it mean for the life of the individual and the responsibilities of citizenship? It is not clear that their effects are all one way.

We are often told by the pessimistic critics of technology ... that today's individual is alienated by the vast proliferation of technical expertise and complex bureaucracies, by a feeling of impotence in the face of "the machine," by a decline in personal privacy, and by the loss of an effective voice in the determination of public policy.

Although there is no way of proving the point, it is probably true that the social pressures placed on individuals today are more complicated than they were in earlier times. Increased geographical and occupational mobility can aggravate feelings of uncertainty in people by depriving them of the stabilities associated with living in one place or plying one trade over a lifetime. Also, the need to function in large organizations places difficult demands on individuals to conform or "adjust." This is often felt as degrading – by many of our youth, for example – because it is interpreted as a subordination of humanity to organizational efficiency.

It is also evident that individual privacy declines in a complex technological society, for a number of reasons. Many people voluntarily trade some of their privacy for benefits that they value more highly. For example, there is some abridgment of privacy entailed in having a telephone and being listed in the telephone book, in applying for social security benefits, or in enjoying the convenience of a credit card. There are also a number of ways in which individual privacy declines by involuntary invasion: either with our knowledge, as when we file an income tax return or register for selective service; or without our knowledge, as when we may be victims of wiretapping or any of the number of listening or visual surveillance devices that are available to anyone cheaply and by mail order. Such decline in individual privacy is probably an inevitable concomitant of the enhancement of the public sphere that was discussed earlier in this chapter.

Finally, it must certainly be conceded that the power, authority, influence, and scope of government are greater today than at any time in the history of the United States. The principal concerns of governments in the past have been to provide for national defense and to act as agents of social justice. Much more than that is demanded of modern government and much more is attempted by it. In today's highly industrialized mass societies, government takes on responsibility for education, for public health, for cultural development, for provision of housing, for the functioning of the economy, and for the support of scientific research and technological development. To the extent that government does so it encroaches on domains that were once exclusively private.

It might appear, therefore, that the prerogatives and political significance of the individual are in fact reduced if not submerged by big technology and complex social organization, as many contemporary social critics claim. But there is another, no less compelling side of the coin, which is currently being explored in a study by Edward Shils of the University of Chicago. For example, government may be more powerful and pervasive than in the past, but it also appears to be more lacking in confidence than ever before. As people become more educated and self-confident they yield less to, and demand more of, government. To the extent that government is not fully responsive to that demand it generates the kind of revolt against authority that we see in all parts of the world and thereby puts itself on the defensive. Where political rulers once relied on charisma and appealed to their traditional if not divine right to govern, they now consult opinion polls, accept criticism from press and public, and engage in dialogue with their constituents. If we compensate for the large populations and structural complexity of modern mass society, there is a strong presumption that the average citizen today can "reach" and influence his government more easily than could his counterparts in earlier societies.

There are also two sides to the privacy question. While privacy may be declining in the various ways indicated above, it also tends to decline in a sense that most of us are likely to approve. The average man in Victorian times, for example, probably "enjoyed" much more privacy than today. No one much cared what happened to him and he was free to remain ignorant, starve, fall ill, and die in complete privacy; that was "the golden age of privacy," as Professor Shils puts it. Compulsory universal education, social security legislation, and public health measures – that is, the very idea of a welfare state, which depends on availability of extensive and accurate information about the nation's population – are all antithetical to privacy in this sense, but it is the rare individual today who is loath to see that kind of privacy go.

It is not clear, finally, that technological and social complexity must inevitably lead to reducing the individual to "mass man" or "organization man." Economic productivity and modern means of communication allow the individual to aspire to more than he ever could before. Better and more easily available education not only provides him with skills and with the means to develop his individual potentialities, but also improves his self-image and his sense of value as a human being. This is probably the first time in two centuries that such high proportions of people have *felt* like individuals; few eighteenth-century English factory workers or nineteenth-century factory or farm workers in America, so far as we know, had the sense of individual worth that underlies the demands on society made by the average resident of the black urban ghetto today.

... [M]oreover, the scope of individual choice and action today are greater than in previous times, in the choice of consumer products, marital partner, occupation, place to live, objects of loyalty, and allegiance to religious, political, and other social groups. It should be emphasized that the computers and other information-processing and communication technologies that are said to submerge individuality may, on the contrary, be essential to its maintenance. ...

Even the much maligned modern organization may in fact "serve as a mediator or buffer between the individual and the full raw impact of technological change."[6] The lack of concern for the person that is cited as a failing of modern bureaucracy may thus have a silver lining after all. The very impersonality of the big organization – like the promise of anonymity that attracts so many people to a big city – can help to protect individuality and personal freedom. There can be such a thing as too much concern for the individual.

There are no doubt many times when the feeling that individuality is being sacrificed to organizational efficiency is justified, but much of it may also be another result of the application of old values and old criteria to essentially new situations. The fact is that technology and population growth do combine to reduce the scope and effectiveness of purely individual endeavor and to enhance the need for organized effort. But organized effort is not in itself demeaning, provided the participants in it understand their individual roles and learn to perform them well. I would doubt that the sense of achievement and quality of personal satisfaction of those who had a hand in the moon landing in 1969 were any less intense or different in kind from the feelings of Charles Lindbergh in 1927 or of Christopher Columbus and his band of adventurers in 1492. A good deal of the reason that many young people today suspect that they are, may be that we are failing to educate them properly about the complexity of the social roles they will be called upon to play and the opportunities that will be open to them when they are able to master those roles.

Recognition that the impact of modern technology on man as individual and as citizen has two faces, negative and positive, is consistent with the hypothesis about the double effect of technological change that was advanced in Chapter I. It also suggests that appreciation of that impact in detail may not be achieved in terms of old formulas, such as more or less privacy, more or less government, more or less freedom, or more or less individuality. ...

Given the inadequacy of old formulas, it may be that inquiry into the problem of human choice and effective citizenship in a technological society is more fruitfully conducted in terms of the kinds of commitment that individuals living in such a society are called upon to make. As at once individuals and social beings, all of us eventually come to some balance, appropriate to each of us, between the relative degree of commitment we are prepared to make to private and to public goals and values. Each of us must achieve a symbiotic relationship between our private and public selves, in other words, by deciding about the degree to which we will function as social beings and the degree to which we will pursue private satisfactions.

The ancient texts we have suggest that the free citizens of Periclean Athens were almost entirely public beings in this respect; their major rewards came from the public arena and even their entertainments, such as sports and theater, were public occasions. They literally lived in the market place. They only slept in their houses and their purely private pursuits and activities appear to have counted for little. By contrast, certain segments of today's youth seem by and large to turn their backs on society and on public life, to reject their values and responsibilities, and to devote themselves almost exclusively to the pursuit of private goals and gratifications, whether physical, chemical/physiological, or psychological.

The Athenian and the hippie of the sixties thus define the spectrum along which each of us will choose where he will stand. Few of us choose either extreme but rather settle for some point in between. The enhancement of the public sphere that I have discussed, the trend away from an individualistic and toward a more collective value orientation, and the growing complexity of society – that is, the principal ways in which technological change affects society – would seem to call for a shift of the commitment point along that spectrum, away from the private in the direction of the public. How, how fast, and to what degree that shift will occur may determine how successful modern society will be in dealing with its technology.

Notes

1 Richard S. Rosenbloom and Robin Marris, eds., *Social Innovation in the City: New Enterprises for Community Development*; a collection of working papers published by the Harvard University Program on Technology and Society. (Cambridge: 1969; distributed by Harvard University Press.)

2 For typical statements about our "knowledge" society, see Daniel Bell, *The End of Ideology* (New York: The Free Press, 1962), and "Notes on the Post-Industrial Society," I and II, *The Public Interest*, Nos. 6 and 7 (Winter and Spring 1967); and Robert E. Lane, "The Decline of Politics and Ideology in a Knowledgeable Society," *American Sociological Review*, Vol. 31, No. 5 (October 1966).

3 Robin Williams, "Individual and Group Values," *Annals of the American Academy of Political and Social Science*, Vol. 37 (May 1967), pp. 23 and 33.

4 Harvey Cox, "Tradition and the Future," I and II, *Christianity and Crisis*, Vol. XXVII, Nos. 16 and 17 (October 2 and 16, 1967).

5 Harvey Brooks, "Scientific Concepts and Cultural Change," in G. Holton, ed., *Science and Culture* (Boston: Houghton-Mifflin, 1965), p. 71.

6 Paul Lawrence and Jay Lorsch, *Organization and Environment: Managing Differentiation and Integration* (Boston: Division of Research, Harvard Business School, 1967), p. 241.

54

Technology: The Opiate of the Intellectuals, with the Author's 2000 Retrospective

John McDermott

Technology: The Opiate of the Intellectuals

If religion was formerly the opiate of the masses, then surely technology is the opiate of the educated public today, or at least of its favorite authors. No other single subject is so universally invested with high hopes for the improvement of mankind generally and of Americans in particular. …

These hopes for mankind's, or technology's, future, however, are not unalloyed. Technology's defenders, being otherwise reasonable men, are also aware that the world population explosion and the nuclear missiles race are also the fruit of the enormous advances made in technology during the past half century or so. But here too a cursory reading of their literature would reveal widespread though qualified optimism that these scourges too will fall before technology's might. Thus population (and genetic) control and permanent peace are sometimes added to the already imposing roster of technology's promises. What are we to make of such extravagant optimism?

[In early 1968] Harvard University's Program on Technology and Society, "…an inquiry in depth into the effects of technological change on the economy, on public policies, and on the character of society, as well as into the reciprocal effects of social progress on the nature, dimension, and directions of scientific and technological development," issued its Fourth Annual Report to the accompaniment of full front-page coverage in *The New York Times* (January 18). Within the brief (fewer than 100) pages of that report and most clearly in the concluding essay by the Program's Director, Emmanuel G. Mesthene, one can discern some of the important threads of belief which bind together much current writing on the social implications of technology. Mesthene's essay is worth extended analysis because these beliefs are of interest in themselves and, of greater importance, because they form the basis not of a new but of a newly aggressive right-wing ideology in this country, an ideology whose growing importance was accurately measured by the magnitude of the *Times*'s news report.

At the very beginning of Mesthene's essay, which attempts to characterize the relationships between technological and social change, the author is careful to dissociate himself from what he believes are several extreme views of those relationships. For example, technology is neither the relatively "unalloyed blessing" which, he claims, Marx, Comte, and the Air Force hold it to be, nor an unmitigated curse, a view he attributes to "many of our youth." (This is but the first of several reproofs Mesthene casts in the direction of youth.) Having denounced straw men to the right and left of him he is free to pursue that middle or moderate course favored by virtually all political writers of the day. This middle course consists of an extremely abstract and – politically speaking – sanitary view of technology and technological progress.

For Mesthene, it is characteristic of technology that it:

> …creates new possibilities for human choice and action but leaves their disposition uncertain. What its effects will be and what ends it will serve are not inherent in the technology, but depend on what man will do with technology. Technology thus makes possible a future of open-ended options.…

This essentially optimistic view of the matter rests on the notion that technology is merely "…the organization of knowledge for practical purposes…" and therefore cannot be purely boon or wholly burden. The matter is somewhat more complex:

> New technology creates new opportunities for men and societies and it also generates new problems for them. It has both positive and negative effects, and it usually has the two *at the same time and in virtue of each other*.

This dual effect he illustrates with an example drawn from the field of medicine. Recent advances there

> have created two new opportunities: (1) they have made possible treatment and cures that were never possible before, and (2) they provide a necessary condition for the delivery of adequate medical care to the population at large as a matter of right rather than privilege.

Because of the first, however,

> the medical profession has become increasingly differentiated and specialized and is tending to concentrate its best efforts in a few major, urban centers of medical excellence.

Mesthene clearly intends but does not state the corollary to this point, namely that the availability of adequate medical care is declining elsewhere.[1] Moreover, because of the second point, there have been

> …big increases in demand for medical services, partly because a healthy population has important economic advantages in a highly industrialized society. This increased demand accelerates the process of differentiation and multiplies the levels of paramedical personnel between the physician at the top and the patient at the bottom of the hospital pyramid.

…Mesthene believes there are two distinct problems in technology's relation to society, a positive one of taking full advantage of the opportunities it offers and the negative one of avoiding unfortunate consequences which flow from the exploitation of those opportunities. Positive opportunities may be missed because the costs of technological development outweigh likely benefits (e.g., Herman Kahn's "Doomsday Machine"). Mesthene seems convinced, however, that a more important case is that in which

> …technology lies fallow because existing social structures are inadequate to exploit the opportunities it offers. This is revealed clearly in the examination of institutional failure in the ghetto carried on by [the Program]. At point after point,…analyses confirm…that existing institutions and traditional approaches are by and large incapable of coming to grips with the new problems of our cities – many of them caused by technological change… – and unable to realize the possibilities for resolving them that are also inherent in technology. …

His diagnosis of these problems is generous in the extreme:

> All these factors combine to dilute what may be otherwise a genuine desire to apply our best knowledge and adequate resources to the resolution of urban tensions and the eradication of poverty in the nation.

Moreover, because government and the media "…are not yet equipped for the massive task of public education that is needed…" if we are to exploit technology more fully, many technological opportunities are lost because of the lack of public support. This too is a problem primarily of "institutional innovation."

Mesthene believes that institutional innovation is no less important in combatting the negative effects of technology. Individuals or individual firms which decide to develop new technologies normally do not take "adequate account" of their likely social benefits or costs. His critique is anti-capitalist in spirit, but lacks bite, for he goes on to add that

...[most of the negative] consequences of technology that are causing concern at the present time – pollution of the environment, potential damage to the ecology of the planet, occupational and social dislocations, threats to the privacy and political significance of the individual, social and psychological malaise – are *negative externalities of this kind*. They are with us in large measure because it has not been anybody's explicit business to foresee and anticipate them. [Italics added.]

Mesthene's abstract analysis and its equally abstract diagnosis in favor of "institutional innovation" places him in a curious and, for us, instructive position. If existing social structures are inadequate to exploit technology's full potential, or if, on the other hand, so-called negative externalities assail us because it is nobody's business to foresee and anticipate them, doesn't this say that we should apply technology to this problem too? That is, we ought to apply and organize the appropriate *organizational* knowledge for the practical purpose of solving the problems of institutional inadequacy and "negative externalities." Hence, in principle, Mesthene is in the position of arguing that the cure for technology's problems, whether positive or negative, is still more technology. This is the first theme of the technological school of writers and its ultimate First Principle.

Technology, in their view, is a self-correcting system. Temporary oversight or "negative externalities" will and should be corrected by technological means. Attempts to restrict the free play of technological innovation are, in the nature of the case, self-defeating. Technological innovation exhibits a distinct tendency to work for the general welfare in the long run. *Laissez innover!*

I have so far deliberately refrained from going into any greater detail than does Mesthene on the empirical character of contemporary technology (see below) for it is important to bring out the force of the principle of *laissez innover* in its full generality. Many writers on technology appear to deny in their definition of the subject – organized knowledge for practical purposes – that contemporary technology exhibits distinct trends which can be identified or projected. Others, like Mesthene, appear to accept these trends, but then blunt the conclusion by attributing to technology so much flexibility and "scientific" purity that it becomes an abstraction infinitely malleable in behalf of good, pacific, just, and egalitarian purposes. Thus the analogy to the laissez-faire principle of another time is quite justified. Just as the market or the free play of competition provided in theory the optimum long-run solution for virtually every aspect of virtually every social and economic problem, so too does the free play of technology, according to its writers. Only if technology or innovation (or some other synonym) is allowed the freest possible rein, they believe, will the maximum social good be realized.

What reasons do they give to believe that the principle of *laissez innover* will normally function for the benefit of mankind rather than, say, merely for the benefit of the immediate practitioners of technology, their managerial cronies, and for the profits accruing to their corporations? As Mesthene and other writers of his school are aware, this is a very real problem, for they all believe that the normal tendency of technology is, and ought to be, the increasing concentration of decision-making power in the hands of larger and larger scientific-technical bureaucracies. *In principle*, their solution is relatively simple, though not often explicitly stated.[2]

Their argument goes as follows: the men and women who are elevated by technology into commanding positions within various decision-making bureaucracies exhibit no generalized drive for power such as characterized, say, the landed gentry of preindustrial Europe or the capitalist entrepreneur of the last century. For their social and institutional position and its supporting culture as well are defined solely by the fact that these men are problem solvers. (Organized knowledge for practical purposes again.) That is, they gain advantage and reward only to the extent that they can bring specific technical knowledge to bear on the solution of specific technical problems. Any more general drive for power would undercut the bases of their usefulness and legitimacy.

Moreover their specific training and professional commitment to solving technical problems creates a bias against ideologies in general which inhibits any attempts to formulate a justifying ideology for the group. Consequently, they do not constitute a class and have no general interests antagonistic to those of their problem-beset clients. We may refer to all of this as the disinterested character of the scientific-technical decision-maker, or, more briefly and cynically, as the principle of the Altruistic Bureaucrat.

As if not satisfied by the force of this (unstated) principle, Mesthene like many of his schoolfellows spends many pages commenting around the belief that the concentration of power at the top of technology's organizations is a problem, but that like other problems technology should be able to solve it successfully through institutional innovation. You may trust in it; the principle of *laissez innover* knows no logical or other hurdle.

This combination of guileless optimism with scientific toughmindedness might seem to be no more than an eccentric delusion were the American technology it supports not moving in directions that are strongly antidemocratic. To show why this is so we must examine more closely Mesthene's seemingly innocuous distinction between technology's positive opportunities and its "negative externalities." In order to do this I will make use of an example drawn from the very frontier of American technology, the war in Vietnam.

II

At least two fundamentally different bombing programs [have been] carried out in South Vietnam. There are fairly conventional attacks against targets which consist of identified enemy troops, fortifications, medical centers, vessels, and so forth. The other program is quite different and, at least since March 1968, infinitely more important. With some oversimplification it can be described as follows:

Intelligence data is gathered from all kinds of sources, of all degrees of reliability, on all manner of subjects, and fed into a computer complex located, I believe, at Bien Hoa. From this data and using mathematical models developed for the purpose, the computer then assigns probabilities to a range of potential targets, probabilities which represent the likelihood that the latter contain enemy forces or supplies. These potential targets might include: a canal-river crossing known to be used occasionally by the NLF; a section of trail which would have to be used to attack such and such an American base, now overdue for attack; a square mile of plain rumored to contain enemy troops; a mountainside from which camp fire smoke was seen rising. Again using models developed for the purpose, the computer divides pre-programmed levels of bombardment among those potential targets which have the highest probability of containing actual targets. Following the raids, data provided by further reconnaissance is fed into the computer and conclusions are drawn (usually optimistic ones) on the effectiveness of the raids. This estimate of effectiveness then becomes part of the data governing current and future operations, and so on.

Two features must be noted regarding this program's features, which are superficially hinted at but fundamentally obscured by Mesthene's distinction between the abstractions of positive opportunity and "negative externality." First, when considered from the standpoint of its planners, the bombing program is extraordinarily rational, for it creates previously unavailable "opportunities" to pursue their goals in Vietnam. It would make no sense to bomb South Vietnam simply at random, and no serious person or air force general would care to mount the effort to do so. So the system employed in Vietnam significantly reduces, though it does not eliminate, that randomness. That canal-river crossing which is bombed at least once every eleven days or so is a very poor target compared to an NLF battalion observed in a village. But it is an infinitely more promising target than would be selected by throwing a dart at a grid map of South Vietnam. In addition to bombing the battalion, why not bomb the canal crossing to the frequency and extent that it *might* be used by enemy troops?

Even when we take into account the crudity of the mathematical models and the consequent slapstick way in which poor information is evaluated, it is a "good" program. No single raid will definitely kill an enemy soldier but a whole series of them increases the "opportunity" to kill a calculable number of them (as well, of course, as a calculable but not calculated number of nonsoldiers). This is the most rational bombing system to follow if American lives are very expensive and American weapons and Vietnamese lives very cheap. Which, of course, is the case.

Secondly, however, considered from the standpoint of goals and values not programmed in by its designers, the bombing program is incredibly irrational. In Mesthene's terms, these "negative externalities" would include, in the present case, the lives and well-being of various Vietnamese as well as the feelings and opinions of some less important Americans. Significantly, this exclusion of the interests of people not among the managerial class is based quite as much on the so-called technical means being employed as on the political goals of the system. In the particular case of the Vietnamese bombing system, the political goals of the bombing system clearly exclude the interests of certain Vietnamese. After all, the victims of the bombardment are communists or their supporters, they are our enemies, they resist US intervention. In short, their interests are fully antagonistic to the goals of the program and simply must be excluded from consideration. The technical reasons for this exclusion require explanation, being less familiar and more important, especially in the light of Mesthene's belief in the malleability of technological systems.

Advanced technological systems such as those employed in the bombardment of South Vietnam make use not only of extremely complex and expensive equipment but, quite as important, of large numbers of relatively scarce and expensive-to-train technicians. They have immense capital costs; a thousand aircraft of a very advanced type, literally hundreds of thousands of spare parts, enormous stocks of rockets, bombs, shells and bullets, in addition to tens of thousands of technical specialists; pilots, bombardiers, navigators, radar operators, computer programmers, accountants, engineers, electronic and mechanical technicians, to name only a few. In short, they are "capital intensive."

Moreover, the coordination of this immense mass of esoteric equipment and its operators in the most effective possible way depends upon an extremely highly developed technique both in the employment of each piece of equipment by a specific team of operators and in the management of the program itself. Of course, all large organizations standardize their operating procedures, but it is peculiar to advanced technological systems that their operating procedures embody a very high degree of information drawn from the physical sciences, while their managerial procedures are equally dependent on information drawn from the social sciences. We may describe this situation by saying that advanced technological systems are both "technique intensive" and "management intensive."

It should be clear, moreover, even to the most casual observer that such intensive use of capital, technique, and management spills over into almost every area touched by the technological system in question. An attack program delivering 330,000 tons of munitions more or less selectively to several thousand different targets monthly would be an anomaly if forced to rely on sporadic intelligence data, erratic maintenance systems, or a fluctuating and unpredictable supply of heavy bombs, rockets, jet fuel, and napalm tanks. Thus it is precisely because the bombing program requires an intensive use of capital, technique, and management that the same properties are normally transferred to the intelligence, maintenance, supply, coordination and training systems which support it. Accordingly, each of these supporting systems is subject to sharp pressures to improve and rationalize the performance of its machines and men, the reliability of its techniques, and the efficiency and sensitivity of the management controls under which it operates. Within integrated technical systems, higher levels of technology drive out lower, and the normal tendency is to integrate systems.

From this perverse Gresham's Law of Technology follow some of the main social and organizational characteristics of contemporary technological systems: the radical increase in the scale and complexity of operations that they demand and encourage; the rapid and widespread diffusion of technology to new areas; the great diversity of activities which can be directed by central management; an increase in the ambition of management's goals; and, as a corollary, especially to the last, growing resistance to the influence of so-called negative externalities.

Complex technological systems are extraordinarily resistant to intervention by persons or problems operating outside or below their managing groups, and this is so regardless of the "politics" of a given situation. Technology creates its own politics. The point of such advanced systems is to minimize the incidence of personal or social behavior which is erratic or otherwise not easily classified, of tools and equipment with poor performance, of improvisory techniques, and of unresponsiveness to central management.

For example, enlisted men who are "unrealistically soft" on the subject of civilian casualties and farmers in contested districts pose a mortal threat to the integral character of systems like that used in Vietnam. In the case of the soldier this means he must be kept under tight military discipline. In the case of the farmer, he must be easily placed in one of two categories; collaborator or enemy. This is done by assigning a probability to him, his hamlet, his village, or his district, and by incorporating that probability into the targeting plans of the bombing system. Then the enlisted man may be controlled by training and indoctrination as well as by highly developed techniques of command and coercion, and the farmers may be bombed according to the most advanced statistical models. In both cases the system's authority over its farmer subjects or enlisted men is a technical one. The technical means which make that system rational and efficient in its aggregate terms, i.e., as viewed from the top, themselves tend by design to filter out the "nonrational" or "nonefficient" elements of its components and subjects, i.e., those rising from the bottom.

To define technology so abstractly that it obscures these observable characteristics of contemporary technology – as Mesthene and his school have done – makes no sense. It makes even less sense to claim some magical malleability for something as undefined as "institutional innovation." Technology, in its concrete, empirical meaning, refers fundamentally to systems of rationalized control over large groups of men, events, and machines by small groups of technically skilled men operating through organizational hierarchy. The latent "opportunities" provided by that control and its ability to filter out discordant "negative externalities" are, of course, best illustrated by extreme cases. Hence the most instructive and accurate example should be of a technology able to suppress the humanity of its rank-and-file and to commit genocide as a by-product of its rationality. The Vietnam bombing program fits technology to a "T."

III

It would certainly be difficult to attempt to translate in any simple and direct way the social and organizational properties of highly developed technological systems from the battlefields of Vietnam to the different cultural and institutional setting of the US. Yet before we conclude that any such attempt would be futile or even absurd, we might consider the following story.

In early 1967 I stayed for several days with one of the infantry companies of the US Fourth Division whose parent battalion was then based at Dau Tieng. From the camp at Dau Tieng the well-known Black Lady Mountain, sacred to the Cao Dai religious sect, was easily visible and in fact dominated the surrounding plain and the camp itself. One afternoon when I began to explain the religious significance of the mountain to some GI friends, they interrupted my somewhat academic discourse to tell me a tale beside which even the strange beliefs of the Cao Dai sect appeared prosaic.

According to GI reports which the soldiers had heard and believed, the Viet Cong had long ago hollowed out most of the mountain in order to install a very big cannon there. The size of the cannon was left somewhat vague – "huge, fucking..." – but clearly the GI's imagined that it was in the battleship class. In any event, this huge cannon had formerly taken a heavy toll of American aircraft and had been made impervious to American counterattacks by the presence of two – "huge, fucking" – sliding steel doors, behind which it retreated whenever the Americans attacked. Had they seen this battleship cannon, and did it ever fire on the camp, which was easily within its range? No, they answered, for a brave flyer, recognizing the effectiveness of the cannon against his fellow pilots, had deliberately crashed his jet into those doors one day, jamming them, and permitting the Americans to move into the area unhindered.

I had never been in the army, and at the time of my trip to Vietnam had not yet learned how fantastic GI stories can be. Thus I found it hard to understand how they could be convinced of so improbable a tale. Only later, after talking to many soldiers and hearing many other wild stories from them as well, did I realize what the explanation for this was. Unlike officers and civilian correspondents who are almost daily given detailed briefings on a unit's situation capabilities and objectives. GI's are told virtually nothing of this sort by the army. They are simply told what to do, where, and how, and it is a rare officer, in my experience anyway, who thinks they should be told any more than this. Officers don't think soldiers are stupid; they simply assume it, and act accordingly. For the individual soldier's personal life doesn't make too much difference; he still has to deal with the facts of personal feelings, his own well-being, and that of his family.

But for the soldier's group life this makes a great deal of difference. In their group life, soldiers are cut off from sources of information about the situation of the group and are placed in a position where their social behavior is governed largely by the principle of blind obedience. Under such circumstances, reality becomes elusive. Because the soldiers are not permitted to deal with facts in their own ways, facts cease to discipline their opinions. Fantasy and wild tales are the natural outcome. In fact, it is probably a mark of the GI's intelligence to fantasize, for it means that he has not permitted his intellectual capacity to atrophy. The intelligence of the individual is thus expressed in the irrationality of the group.

It is this process which we may observe when we look to the social effect of modern technological systems in America itself. Here the process is not so simple and clear as in Vietnam, for it involves not simply the relations of today's soldiers to their officers and to the Army but the historical development of analogous relations between the lower and upper orders of our society. Moreover, these relations are broadly cultural rather than narrowly social in nature. It is to a brief review of this complex subject that I now wish to turn.

IV

Among the conventional explanations for the rise and spread of the democratic ethos in Europe and North America in the seventeenth, eighteenth, and nineteenth centuries, the destruction of the gap in political culture between the mass of the population and that of the ruling classes is extremely important. There are several sides to this explanation. For example, it is often argued that the invention of the printing press and the spread of Protestant Christianity encouraged a significant growth in popular literacy. In its earliest phases this literacy was largely expended on reading the Old and New Testaments, but it quickly broadened to include other religious works such as Bunyan's *Pilgrim's Progress*, and after that to such secular classics as *Gulliver's Travels*. The dating of these developments is, in the nature of the case, somewhat imprecise. But certainly by the middle of the eighteenth century, at least in Britain and North America, the literacy of the population was sufficient to support a variety of newspapers and periodicals not only in the larger cities but in the smaller provincial towns as well. The decline of Latin as the first language of politics and religion paralleled this development, of course. Thus, even before the advent of Tom Paine, Babeuf, and other popular tribunes, literacy and the information it carried were widely and securely spread throughout the population and the demystification of both the religious and the political privileges of the ruling classes was well developed. Common townsmen had closed at least one of the cultural gaps between themselves and the aristocracy of the larger cities.

Similarly, it is often argued that with the expansion and improvement of road and postal systems, the spread of new tools and techniques, the growth in the number and variety of merchants, the consequent invigoration of town life, and other numerous and familiar related developments, the social experiences of larger numbers of people became richer, more varied, and similar in fact to those of the ruling classes. This last, the growth in similarity of the social experiences of the upper and lower classes, is especially important. Social skills and experiences which underlay the monopoly of the upper classes over the processes of law and government were spreading to important segments of the lower orders of society. For carrying on trade, managing a commercial – not a subsistence – farm, participating in a vestry of workingmen's guild, or working in an up-to-date manufactory or business, unlike the relatively narrow existence of the medieval serf or artisan, were experiences which contributed to what I would call the social rationality of the lower orders.

Activities which demand frequent intercourse with strangers, accurate calculation of near means and distant ends, and a willingness to devise collective ways of resolving novel and unexpected problems demand and reward a more discriminating attention to the realities and deficiencies of social life, and provide thereby a rich variety of social experiences analogous to those of the governing classes. As a result not only were the processes of law and government, formerly treated with semireligious veneration, becoming demystified but, equally important, a population was being fitted out with sufficient skills and interests to contest their control. Still another gap between the political cultures of the upper and lower ends of the social spectrum was being closed.

The same period also witnesses a growth in the organized means of popular expression. In Britain, these would include the laboring people's organizations whose development is so ably described in Edward Thompson's *The Making of the English Working Class*. In America, the increase in the organized power of the populace was expressed not only in the growing conflict between the colonies and the Crown but more sharply and fundamentally in the continuous antagonism between the coastal areas and the backwoods, expressed, for example, in Shay's rebellion in western Massachusetts in 1786. Clearly these organizational developments were related to the two foregoing as both cause and effect. For the English workingmen's movement and the claims to local self-government in America spurred, and were spurred by, the growth in individual literacy and in social rationality among the lower classes. They were in fact its organizational expression.

These same developments were also reflected in the spread of egalitarian and republican doctrines such as those of Richard Price and Thomas Paine, which pointed up the arbitrary character of what had heretofore been considered the rights of the higher orders of society, and thus provided the popular ideological base which helped to define and legitimate lower-class demands.

This description by no means does justice to the richness and variety of the historical process underlying the rise and spread of what has come to be called the democratic ethos. But it does, I hope, isolate some of the important structural elements and, moreover, it enables us to illuminate some important ways in which the new technology, celebrated by Mesthene and his associates for its potential contributions to democracy, contributes instead to the erosion of that same democratic ethos. For if, in an earlier time, the gap between the political cultures of the higher and lower orders of society was being widely attacked and closed, this no longer appears to be the case. On the contrary, I am persuaded that the direction has been reversed and that we now observe evidence of a growing separation between ruling and lower-class culture in America, a separation which is particularly enhanced by the rapid growth of technology and the spreading influence of its *laissez-innover* ideologues.

Certainly, there has been a decline in popular literacy, that is to say, in those aspects of literacy which bear on an understanding of the political and social character of the new technology. Not one person in a hundred is even aware of, much less understands, the nature of technologically highly advanced systems such as are used in the Vietnam bombing program. People's ignorance in these things is revealed in their language. No clearer illustration of this ignorance is needed than the growing and already enormous difference between the speech of organizational and technical specialists and that of the man in the street, including many of the educated ones. To the extent that technical forms of speech within which the major business of American society is carried on are not understood or are poorly understood, there is a decline in one of the essentials of democracy.

This is not to say that the peculiar jargon which characterizes the speech of, say, aerospace technicians, crisis managers, or economic mandarins is intrinsically superior to the vocabulary of ordinary conversation, though sometimes this is indeed the case. What is important about technical language is that the words, being alien to ordinary speech, hide their meaning from ordinary speakers; terms like foreign aid or technical assistance have a good sound in ordinary speech; only the initiate recognizes them as synonyms for the old-fashioned, nasty word, imperialism. Such instances can be corrected but when almost all of the public's business is carried on in specialized jargon correction makes little difference. Like Latin in the past, the new language of social and technical organization is divorced from the general population. ...

Secondly, the social organization of this new technology, by systematically denying to the general population experiences which are analogous to those of its higher management, contributes very heavily to the growth of social irrationality in our society. For example, modern technological organization defines the roles and values of its members, not vice versa. An engineer or a sociologist is one who does all those things but only those things called for by the "table of organization" and the "job description" used by his employer. Professionals who seek self-realization through creative and autonomous behavior without regard to the defined goals, needs, and channels of their respective departments have no more place in a large corporation or government agency than squeamish soldiers in the army. ...

However, those at the top of technology's more advanced organizations hardly suffer the same experience. For reasons which are clearly related to the principle of the Altruistic Bureaucracy the psychology of an individual's fulfillment through work has been incorporated into management ideology. As the pages of *Fortune*, *Time*, or *Business Week* ... serve to show, the higher levels of business and government are staffed by men and women who spend killing hours looking after the economic welfare and national security of the rest of us. The rewards of this life are said to be very few: the love of money would be demeaning and, anyway, taxes are said to take most of it; its sacrifices are many, for failure brings economic depression to the masses or gains for communism as well as disgrace to the erring managers. Even the essential high-mindedness or altruism of our managers earns no reward, for the public is distracted, fickle, and, on occasion, vengeful. (The extensive literature on the "ordeal" of Lyndon Johnson is a case in point.) Hence for these "real revolutionaries of our time," as Walt Rostow has called them, self-fulfillment through work and discipline is the only reward. The managerial process is seen as an expression of the vital personalities of our leaders and the right to it an inalienable right of the national elite.

In addition to all of this, their lonely and unrewarding eminence in the face of crushing responsibility, etc., tends to create an air of mystification around technology's managers. When the august mystery of science and the perquisites of high office are added to their halos, they glow very blindingly indeed. Thus, in ideology as well as in reality and appearance, the experiences of the higher managers tend to separate and isolate themselves from those of the managed. Again the situation within the US is not so severe nor so stark as in the army in Vietnam but the effect on those who are excluded from self-management is very similar. Soldiers in Vietnam are not alone in believing huge, secret guns threaten them from various points; that same feeling is a national malady in the US.

It seems fundamental to the social organization of modern technology that the quality of the social experience of the lower orders of society declines as the level of technology grows no less than does their literacy. And, of course, this process feeds on itself, for with the consequent decline in the real effectiveness and usefulness of local and other forms of organization open to easy and direct popular influence their vitality declines still further, and the cycle is repeated.

The normal life of men and women in the lower and, I think, middle levels of American society now seems cut off from those experiences in which near social means and distant social ends are balanced and rebalanced, adjusted, and readjusted. But it is from such widespread experience with effective balancing and adjusting that social rationality derives. To the degree that it is lacking, social irrationality becomes the norm, and social paranoia a recurring phenomenon.

... With no great effort and using no great skill, Presidents Johnson and Nixon have managed to direct disorganized popular frustration over the continuation of the war and popular abhorrence over its unremitting violence on to precisely that element in the population most actively and effectively opposed to the war and its violence. ...

People often say that America is a sick society when what they really mean is that it has lots of sick individuals. But they were right the first time: the society is so sick that individual efforts to right it and individual rationality come to be expressed in fundamentally sick ways. Like the soldiers in Vietnam, we try to avoid atrophy of our social intelligence only to be led into fantasy and, often, violence. It is a good thing to want the war in Vietnam over for, as everyone now recognizes, it hurts us almost as much as the Vietnamese who are its intended victims. But for many segments of our population, especially those cut off from political expression because of their own social disorganization, the rationality of various alternatives for ending the war is fundamentally obscure. Thus their commendable desire to end the war is expressed in what they believe is the clearest and most certain alternative: use the bomb!

Mesthene himself recognizes that such "negative externalities" are on the increase. His list includes "...pollution of the environment, potential damage to the ecology of the planet, occupational and social dislocations, threats to the privacy and political significance of the individual, social and psychological malaise...." Minor matters all, however, when compared to the marvelous opportunities *laissez innover* holds out to us: more GNP, continued free world leadership, supersonic transports, urban renewal on a regional basis, institutional innovation, and the millenial promises of his school.

This brings us finally to the ideologies and doctrines of technology and their relation to what I have argued is a growing gap in political culture between the lower and upper classes in American society. Even more fundamentally than the principles of *laissez innover* and the altruistic bureaucrat, technology in its very definition as the organization of knowledge for practical purposes assumes that the primary and really creative role in the social processes consequent on technological change is reserved for a scientific and technical elite, the elite which presumably discovers and organizes that knowledge. But if the scientific and technical elite and their indispensable managerial cronies are the really creative (and hardworking and altruistic) element in American society, what is this but to say that the common mass of men are essentially drags on the social weal? This is precisely the implication which is drawn by the *laissez innover* school. Consider the following quotations from an article which appeared in *The New Republic* in December 1967, written by Zbigniew Brzezinski, one of the intellectual leaders of the school.

Brzezinski is describing a nightmare which he calls the "technetronic society" (the word like the concept is a pastiche of technology and electronics). This society will be characterized, he argues, by the application of "...the principle of equal opportunity for all but...special opportunity for the singularly talented few." It will thus combine

"...continued *respect* for the popular will with an increasing *role* in the key decision-making institutions of individuals with special intellectual and scientific attainments." (Italics added.) Naturally, "The educational and social systems [will make] it increasingly attractive and easy for those meritocratic few to develop to the fullest of their special potential."

However, while it will be "...necessary to require everyone at a sufficiently responsible post to take, say, two years of [scientific and technical] retraining every ten years...," the rest of us can develop a new "...interest in the cultural and humanistic aspects of life, *in addition to purely hedonistic preoccupations*." (Italics added.) The latter, he is careful to point out, "would serve as a social valve, reducing tensions and political frustration."

Is it not fair to ask how much *respect* we carefree pleasure lovers and culture consumers will get from the hard-working bureaucrats, going to night school two years in every ten, while working like beavers in the "key decision-making institutions"? The altruism of our bureaucrats has a heavy load to bear.

Stripped of their euphemisms these are simply arguments which enhance the social legitimacy of the interests of new technical and scientific elites and detract from the interests of the rest of us; that is to say, if we can even formulate those interests, blinded as we will be by the mad pursuit of pleasures (and innovation??!) heaped up for us by advanced technology. Mesthene and his schoolfellows try to argue around their own derogation of the democratic ethos by frequent references, as we have seen, to their own fealty to it. But it is instructive in this regard to note that they tend, with Brzezinski, to find the real substance of that democratic ethos in the principle of the equality of opportunity. Before we applaud, however, we ought to examine the role which that principle plays within the framework of the advanced technological society they propose.

As has already been made clear the *laissez innover* school accepts as inevitable and desirable the centralizing tendencies of technology's social organization, and they accept as well the mystification which comes to surround the management process. Thus equality of opportunity, as they understand it, has precious little to do with creating a more egalitarian society. On the contrary, it functions as an indispensable feature of the highly stratified society they envision for the future. For in their society of meritocratic hierarchy, equality of opportunity assures that talented young meritocrats (the word is no uglier than the social system it refers to) will be able to climb into the "key decision-making" slots reserved for trained talent, and thus generate the success of the new society, and its cohesion against popular "tensions and political frustration."

The structures which formerly guaranteed the rule of wealth, age, and family will not be destroyed (or at least not totally so). They will be firmed up and rationalized by the perpetual addition of trained (and, of course, acculturated) talent. In technologically advanced societies, equality of opportunity functions as a hierarchical principle, in opposition to the egalitarian social goals it pretends to serve. To the extent that it has already become the kind of "equality" we seek to institute in our society, it is one of the main factors contributing to the widening gap between the cultures of upper- and lower-class America.

V

Approximately a century ago, the philosophy of laissez-faire began its period of hegemony in American life. Its success in achieving that hegemony clearly had less to do with its merits as a summary statement of economic truth than with its role in the social struggle of the time. It helped to identify the interests of the institutions of entrepreneurial capitalism for the social classes which dominated them and profited from them. Equally, it sketched in bold strokes the outlines of a society within which the legitimate interests of all could supposedly be served only by systematic deference to the interests of entrepreneurial capitalists, their institutions, and their social allies. In short, the primary significance of laissez-faire lay in its role as ideology, as the cultural or intellectual expression of the interests of a class.

Something like the same thing must be said of *laissez innover*. As a summary statement of the relationship between social and technological change it obscures far more than it clarifies, but that is often the function and genius of ideologues. *Laissez innover* is now the premier ideology of the technological impulse in American society, which is to say, of the institutions which monopolize and profit from advanced technology and of the social classes which find in the free exploitation of *their* technology the most likely guarantee of their power, status, and wealth.

This said, it is important to stress both the significance and limitations of what has in fact been said. Here Mesthene's distinction between the positive opportunities and negative "externalities" inherent in technological change is pivotal; for everything else which I've argued follows inferentially from the actual social meaning of that distinction. As my analysis of the Vietnam bombing program suggested, those technological effects which are sought after as positive opportunities and those which are dismissed as negative externalities are decisively influenced by the fact that this distinction between positive and negative within advanced technological organizations tends to be made among the planners and managers themselves. Within these groups there are, as was pointed out, extremely powerful organizational, hierarchical, doctrinal, and other "*technical*" factors, which tend by design to filter out "irrational" demands from below, substituting for them the "rational" demands of technology itself. As a result, technological rationality is as socially neutral today as market rationality was a century ago.

Turning from the inner social logic of advanced technological organizations and systems to their larger social effect, we can observe a significant convergence. For both the social tendency of technology and the ideology (or rhetoric) of the *laissez-innover* school converge to encourage a political and cultural gap between the upper and lower ends of American society. As I have pointed out, these can now be characterized as those who manage and those who are managed by advanced technological systems.

This analysis lends some weight (though perhaps no more than that) to a number of wide-ranging and unorthodox conclusions about American society today and the directions in which it is tending. It may be useful to sketch out the most important of those conclusions in the form of a set of linked hypotheses, not only to clarify what appear to be the latent tendencies of America's advanced technological society but also to provide more useful guides to the investigation of the technological impulse than those offered by the obscurantism and abstractions of the school of *laissez innover*.

First, and most important, technology should be considered as an institutional system, not more and certainly not less. Mesthene's definition of the subject is inadequate, for it obscures the systematic and decisive social changes, especially their political and cultural tendencies, that follow the widespread application of advanced technological systems. At the same time, technology is less than a social system per se, though it has many elements of a social system, viz., an elite, a group of linked institutions, an ethos, and so forth. Perhaps the best summary statement of the case resides in an analogy – with all the vagueness and imprecision attendant on such things: today's technology stands in relation to today's capitalism as, a century ago, the latter stood to the free market capitalism of the time.

The analogy suggests, accurately enough I believe, the likelihood that the institutional links and shared interests among the larger corporations, the federal government, especially its military sector, the multiversity and the foundations, will grow rather than decline. It suggests further a growing entanglement of their elites, probably in the neo-corporations of technology, such as urban development corporations and institutes for defense analysis, whose importance seems likely to increase markedly in the future.

Finally, it suggests a growing convergence in the ethos and ideology of technology's leading classes along lines which would diminish slightly the relative importance of rhetoric about "property" and even about "national security," while enhancing the rhetoric of *laissez innover*. This does not necessarily imply any sacrifice in the prerogatives of either the private sector or of the crisis managers and the military, for one can readily understand how the elite strictures of *laissez innover* may be applied to strengthen the position of the corporate and military establishments.

[...]

Galbraith's concept of an "educational and scientific" elite class overlooks the peculiar relationship which the members of that supposed class have to advanced technology. Specifically, it overlooks the fact that most technical, scientific, and educational people are employed at relatively specialized tasks within very large organizations whose managing and planning levels are hardly less insulated from their influence than from the influence of the technically unskilled.

The obvious growth in status and, I, think, power of such men as Ithiel de Sola Pool, Herman Kahn, Samuel Huntington, Daniel Patrick Moynihan, Henry Kissinger, Charles Hitch, and Paul Samuelson hardly represents the triumph of wisdom over power – an implication not absent from Galbraith's analysis. An examination of the role which these men now play in our national life should emphasize that they are scientific and technical entrepreneurs whose power is largely based on their ability to mobilize *organized* intellectual, scientific, and technical manpower and other resources, including foundation grants and university sponsorship, in behalf of the objectives of going institutions. They are much more like managers than like intellectuals, much more like brokers than like analysts. ...

A second major hypothesis would argue that the most important dimension of advanced technological institutions is the social one, that is, the institutions are agencies of highly centralized and intensive social control. Technology conquers nature, as the saying goes. But to do so it must first conquer man. More precisely, it demands a very high degree of control over the training, mobility, and skills of the work force. The absence (or decline) of direct controls or of coercion should not serve to obscure from our view the reality and intensity of the social controls which are employed (such as the internalized belief in equality of opportunity, indebtedness through credit, advertising, selective service channeling, and so on).

Advanced technology has created a vast increase in occupational specialties, many of them requiring many, many years of highly specialized training. It must motivate this training. It has made ever more complex and "rational" the ways in which these occupational specialties are combined in our economic and social life. It must win passivity and obedience to this complex activity. Formerly, technical rationality had been employed only to organize the production of rather simple physical objects, for example, aerial bombs. Now technical rationality is increasingly employed to organize all of the processes necessary to the utilization of physical objects, such as bombing systems. For this reason it seems a mistake to argue that we are in a "postindustrial" age, a concept favored by the *laissez innover* school. On the contrary, the rapid spread of technical rationality into organizational and economic life and, hence, into social life is more aptly described as a second and much more intensive phase of the industrial revolution. One might reasonably suspect that it will create analogous social problems.

Accordingly, a third major hypothesis would argue that there are very profound social antagonisms or contradictions not less sharp or fundamental than those ascribed by Marx to the development of nineteenth-century industrial society. The general form of the contradictions might be described as follows: a society characterized by the employment of advanced technology requires an ever more socially disciplined population, yet retains an ever declining capacity to enforce the required discipline.

[...]

Politically, the advance of technology tends to concentrate authority within its managing groups in the ways I have described. But at the same time the increasing skill and educational levels of the population create latent capacities for self-management in the work place and in society.

[...]

These are brief and, I believe, barely adequate reviews of extremely complex hypotheses. But, in outline, each of these contradictions appears to bear on roughly the same group of the American population, a technological underclass. If we assume this to be the case, a fourth hypothesis would follow, namely that technology is creating the basis for new and sharp class conflict in our society. That is, technology is creating its own working and managing classes just as earlier industrialization created its working and owning classes. Perhaps this suggests a return to the kind of class-based politics which characterized the US in the last quarter of the nineteenth century, rather than the somewhat more ambiguous politics which was a feature of the second quarter of this century. I am inclined to think that this is the case, though I confess the evidence for it is as yet inadequate.

This leads to a final hypothesis, namely that *laissez innover* should be frankly recognized as a conservative or right-wing ideology. This is an extremely complex subject for the hypothesis must confront the very difficult fact that the intellectual genesis of *laissez innover* is traceable much more to leftist and socialist theorizing on the wonders of technical rationality and social planning than it is to the blood politics of a De Maistre or the traditionalism of a Burke. So be it. Much more important is the fact that *laissez innover* is now the most powerful and influential statement of the demands and program of the technological impulse in our society, an impulse rooted in its most powerful institutions. More than any other statement, it succeeds in identifying and rationalizing the interests of the most authoritarian elites within this country, and the expansionism of their policies overseas. Truly it is no accident that the leading figures of *laissez innover*, the Rostows, Kahn, Huntington, Brzezinski, to name but a few, are among the most unreconstructed cold warriors in American intellectual life.

The point of this final hypothesis is not primarily to reimpress the language of European politics on the American scene. Rather it is to summarize the fact that many of the forces in American life hostile to the democratic ethos have enrolled under the banner of *laissez innover*. Merely to grasp this is already to take the first step toward a politics of radical reconstruction and against the malaise, irrationality, powerlessness, and official violence that characterize American life today.

Author's Retrospective (2000): Atavism and Modernism

"Technology" is embedded in a very distinctive way in modern American institutions, a way which has tended to be copied in other modern societies. In fact, much of what is most attractive and unattractive, "modern" and "backward" in our society stems from this distinctive social embeddedness.

We can adapt the French concept of *cadre* to our purposes here. Its ambit is wider than, say, Galbraith's older concept of a "technostructure" since it incorporates not only "technical" specialties but also all those professional and managerial specialties which, together, innovate, deploy, maintain, and direct those industrial, sales, communications, military and other systems in which modern, up-to-date technology is employed. This usage emphasizes that a director for the TV show, a sales manager for frozen food products, or a line manager presiding over a shipping department represents a contemporary, changing, university-based "technology", "technique," and credentialling not radically different in its practical and institutional character from an engineer designing new radar components or the maintenance manager for the latest jet engines. That is, if we avoid sacralizing "hardware" or the results of the purely physical sciences as on a higher plane than the sorts of things studied in business schools or design departments, if instead we look to the ensembles of relations which connect different employees to the physical world through their professions and their employing institutions, to the latters' hierarchies, to the way in which different modern professions are recruited, trained, acculturated, deployed, and paid, we can see a quite remarkable convergence of all sorts of once different "fields" to the common patterns exhibited across today's *cadre*.

Typically these *cadre* are located within a management echelon in which their authority, job tenure, personal emoluments, and technological authority are markedly greater – carefully distinguished in institutional practice and even in law – from those of their subordinates in the workforce and, of course, from persons located outside of the main technology-employing institutions.[3]

This double action – the convergence of the different professions to a common pattern, and especially their separation from "the workforce" into a special managerial echelon – deeply affects the way technology is deployed in a modern society. What in fact we call "technology" is increasingly what these *cadre* learn, understand, do, and effect, and the technology's social characteristics are deeply marked by that dual convergence and separation. We have, essentially, a "technology" of a special minority whose embeddedness within very large, hierarchically organized institutions profoundly insulates the direction and magnitude of technological change from the experience, knowledge, and authority of the great majority.

Obviously, this sort of social organization has certain advantages. The classic "Market" of Economics orthodoxy would never have had the financial patience, scale of physical resources, and access to multiple technologies by as many different co-operating institutions, including the marketing resources and skills, to have brought about the mass usage of the LP, the cassette tape and the CD. If late Beethoven is within easy reach of Everyman, it must be accounted a cultural triumph of our distinct way of socially embedding technology. Its other achievements, in industrial productivity, in medicine, in travel and financial services, in PCs and the Internet, are equally triumphs for the special autonomy we accord our technological *cadre* and the scale of inter-institutional resources made available to them.

Of course, there is a social price to this. As I suggested in my old essay, there has been an emerging divide between *cadre* and the mass. I think the increasing maldistribution of income, not only in the advanced countries but everywhere, is subtly linked to the perhaps excessive social and economic esteem accorded *cadre* and the idea that the others, the masses of "the mass market," enjoy an undeserved free ride on the technology so laboriously created by others.[4]

There is also a certain irony in the fact that our most modern of societies is arguably re-producing a premodern kind of politics. The concept of "the political classes" seemed to have disappeared for good at the turn of the nineteenth to the twentieth century, especially given the rise of the mass electorate, liberalism and social democracy, and increasing literacy. That only a minority – the titled, the connected, the propertied, and the educated – would exert a semi-monopoly of influence on the political process, others being overlooked and excluded, was then seen as an anachronism being wiped out by the march of Progress. Whatever our expectations in this regard, I do think that the ranks of the politically significant have shrunk; in the US and the UK, for example, both the Democrats and the Labour Party increasingly value and seek the votes of the same "middle class" (= *cadre*) that the Republicans and the Tories vie for. The French, the German and other European Socialists have gone much the same route. In my youth this historical tendency to privilege the *cadre* against the rest was defended by an Economic Growth narrative. In the 1950s that changed to a Cold War narrative. In my essay of the late 1960s more or less the same social privileging was championed by "Laissez innover!" Today it's "the Market" which demands that we especially privilege the interests and the autonomy of the *cadre* and, of course, of the elite who call our larger institutions their own. It is this very plasticity of ideology, its frequent change of form around a stable core of social interests, that leads me to think that the way we socially embed our technology is, whatever its other merits, a main source of the deeply political atavism of what we deem "technologically advanced society."

Notes

1 This is almost certainly true of persons living in rural areas or in smaller towns and cities. However, a New York-based New Left project, the Health-Policy Advisory Center, has argued with considerable documentation, that roughly half of New York City's population is now medically indigent and perhaps 80 percent of the population is indigent with respect to major medical care.

2 For a more complete statement of the argument which follows, see Suzanne Keller, *Beyond the Ruling Class* (New York: Random House, 1963).

3 I analyzed the rise and significance of the managerial echelon in *Corporate Society: Class, Property and Contemporary Capitalism* (Boulder: Westview, 1991).

4 A notable difference between the older and the modern "professions" is that in the latter the purely professional ethos has been radically de-rated. For example, scores of Ford Motors' *cadre* had to have known that certain Ford ignition systems, like the Ford/Firestone tires, were a source of multiple deaths and injuries (*New York Times*, September 12, 2000, p.Al). It would be facetious to say that we need more "whistle-blowers". Most modern professions are definitively acculturated to give their first, almost exclusive loyalty to their employers when in fact the social/technological impact of modern *cadre* and their relative autonomy from the public's scrutiny and intervention really demand far more. All of the modern professions, on the model of medicine, should be under both the moral/professional obligation – and the legal requirement – to practice their specialties with the interests of the public, not their employers, foremost in mind. A radical resocialization of the modern professions, including all kinds of management, seems a utopian idea, but it is imperatively called for.

55

Democratic Rationalization: Technology, Power, and Freedom

Andrew Feenberg

I The Limits of Democratic Theory

Technology is one of the major sources of public power in modern societies. So far as decisions affecting our daily lives are concerned, political democracy is largely overshadowed by the enormous power wielded by the masters of technical systems: corporate and military leaders, and professional associations of groups such as physicians and engineers. They have far more to do with control over patterns of urban growth, the design of dwellings and transportation systems, the selection of innovations, our experience as employees, patients, and consumers, than all the governmental institutions of our society put together.

Marx saw this situation coming in the middle of the nineteenth century. He argued that traditional democratic theory erred in treating the economy as an extra-political domain ruled by natural laws such as the law of supply and demand. He claimed that we will remain disenfranchised and alienated so long as we have no say in industrial decision-making. Democracy must be extended from the political domain into the world of work. This is the underlying demand behind the idea of socialism.

Modern societies have been challenged by this demand for over a century. Democratic political theory offers no persuasive reason of principle to reject it. Indeed, many democratic theorists endorse it.[1] What is more, in a number of countries socialist parliamentary victories or revolutions have brought parties to power dedicated to achieving it. Yet today we do not appear to be much closer to democratizing industrialism than in Marx's time.

This state of affairs is usually explained in one of the following two ways.

On the one hand, the common-sense view argues that modern technology is incompatible with workplace democracy. Democratic theory cannot reasonably press for reforms that would destroy the economic foundations of society. For evidence, consider the Soviet case: although they were socialists, the communists did not democratize industry, and the current democratization of Soviet society extends only to the factory gate. At least in the ex-Soviet Union, everyone can agree on the need for authoritarian industrial management.

On the other hand, a minority of radical theorists claim that technology is not responsible for the concentration of industrial power. That is a political matter, due to the victory of capitalist and communist elites in struggles with the underlying population. No doubt modern technology lends itself to authoritarian administration, but in a different social context it could just as well be operated democratically.

In what follows, I will argue for a qualified version of this second position, somewhat different from both the usual Marxist and democratic formulations. The qualification concerns the role of technology, which I see as *neither* determining nor as neutral. I will argue that modern forms of he-

A revised version of "Subversive Rationality: Technology, Power, and Democracy," in *Inquiry* 35/3–4 (1992): 301–22, reprinted by permission of Taylor & Francis, Oslo, Norway.

gemony are based on the technical mediation of a variety of social activities, whether it be production or medicine, education or the military, and that, consequently, the democratization of our society requires radical technical as well as political change.

This is a controversial position. The common-sense view of technology limits democracy to the state. By contrast, I believe that unless democracy can be extended beyond its traditional bounds into the technically mediated domains of social life, its use value will continue to decline, participation will wither, and the institutions we identify with a free society will gradually disappear.

Let me turn now to the background to my argument. I will begin by presenting an overview of various theories that claim that insofar as modern societies depend on technology, they require authoritarian hierarchy. These theories presuppose a form of technological determinism which is refuted by historical and sociological arguments I will briefly summarize. I will then present a sketch of a non-deterministic theory of modern society I call "critical theory of technology." This alternative approach emphasizes contextual aspects of technology ignored by the dominant view. I will argue that technology is not just the rational control of nature; both its development and impact are intrinsically social. I will then show that this view undermines the customary reliance on efficiency as a criterion of technological development. That conclusion, in turn, opens broad possibilities of change foreclosed by the usual understanding of technology.

II Dystopian Modernity

Max Weber's famous theory of rationalization is the original argument against industrial democracy. The title of this essay implies a provocative reversal of Weber's conclusions. He defined rationalization as the increasing role of calculation and control in social life, a trend leading to what he called the "iron cage" of bureaucracy.[2] "Democratic" rationalization is thus a contradiction in terms.

Once traditionalist struggle against rationalization has been defeated, further resistance in a Weberian universe can only reaffirm irrational life forces against routine and drab predictability. This is not a democratic program but a romantic anti-dystopian one, the sort of thing that is already foreshadowed in Dostoevsky's *Notes from Underground* and various back-to-nature ideologies.

My title is meant to reject the dichotomy between rational hierarchy and irrational protest implicit in Weber's position. If authoritarian social hierarchy is truly a contingent dimension of technical progress, as I believe, and not a technical necessity, then there must be an alternative way of rationalizing society that democratizes rather than centralizes control. We need not go underground or native to preserve threatened values such as freedom and individuality.

But the most powerful critiques of modern technological society follow directly in Weber's footsteps in rejecting this possibility. I am thinking of Heidegger's formulation of "the question of technology" and Ellul's theory of "the technical phenomenon."[3] According to these theories, we have become little more than objects of technique, incorporated into the mechanism we have created. As Marshall McLuhan once put it, technology has reduced us to the "sex organs of machines." The only hope is a vaguely evoked spiritual renewal that is too abstract to inform a new technical practice.

These are interesting theories, important for their contribution to opening a space of reflection on modern technology. I will return to Heidegger's argument in the conclusion to this essay. But first, to advance my own argument, I will concentrate on the principal flaw of dystopianism, the identification of technology in general with the specific technologies that have developed in the last century in the West. These are technologies of conquest that pretend to an unprecedented autonomy; their social sources and impacts are hidden. I will argue that this type of technology is a particular feature of our society and not a universal dimension of "modernity" as such.

III Technological Determinism

Determinism rests on the assumption that technologies have an autonomous functional logic that can be explained without reference to society. Technology is presumably social only through the purpose it serves, and purposes are in the mind of the beholder. Technology would thus resemble science and mathematics by its intrinsic independence of the social world.

Yet unlike science and mathematics, technology has immediate and powerful social impacts. It would seem that society's fate is at least partially dependent on a non-social factor which influences it without suffering a reciprocal influence. This is what is meant by "technological determinism." Such a deterministic view of technology is commonplace in business and government, where it is often assumed that progress is an exogenous force influencing society rather than an expression of changes in culture and values.

The dystopian visions of modernity I have been describing are also deterministic. If we want to affirm the democratic potentialities of modern industrialism, we will therefore have to challenge their deterministic premises. These I will call the thesis of unilinear progress, and the thesis of determination by the base. Here is a brief summary of these two positions.

1 Technical progress appears to follow a unilinear course, a fixed track, from less to more advanced configurations. Although this conclusion seems obvious from a backward glance at the development of any familiar technical object, in fact it is based on two claims of unequal plausibility: first, that technical progress proceeds from lower to higher levels of development; and second, that that development follows a single sequence of necessary stages. As we will see, the first claim is independent of the second and not necessarily deterministic.

2 Technological determinism also affirms that social institutions must adapt to the "imperatives" of the technological base. This view, which no doubt has its source in a certain reading of Marx, is now part of the common sense of the social sciences.[4] Below, I will discuss one of its implications in detail: the supposed "trade-off" between prosperity and environmental values.

These two theses of technological determinism present decontextualized, self-generating technology as the unique foundation of modern society. Determinism thus implies that our technology and its corresponding institutional structures are universal, indeed, planetary in scope. There may be many forms of tribal society, many feudalisms, even many forms of early capitalism, but there is only one modernity, and it is exemplified in our society for good or ill. Developing societies should take note: as Marx once said, calling the attention of his backward German compatriots to British advances: *De te fabula narratur* – of you the tale is told.[5]

IV Constructivism

The implications of determinism appear so obvious that it is surprising to discover that neither of its two theses can withstand close scrutiny. Yet contemporary sociology of technology undermines the first thesis of unilinear progress while historical precedents are unkind to the second thesis of determination by the base.

Recent constructivist sociology of technology grows out of new social studies of science. These studies challenge our tendency to exempt scientific theories from the sort of sociological examination to which we submit non-scientific beliefs. They affirm the "principle of symmetry," according to which all contending beliefs are subject to the same type of social explanation, regardless of their truth or falsity.[6] A similar approach to technology rejects the usual assumption that technologies succeed on purely functional grounds.

Constructivism argues that theories and technologies are underdetermined by scientific and technical criteria. Concretely, this means two things: first, there is generally a surplus of workable solutions to any given problem, and social actors make the final choice among a batch of technically viable options; and second, the problem-definition often changes in the course of solution. The latter point is the more conclusive but also more difficult of the two.

Two sociologists of technology, Trevor Pinch and Wiebe Bijker, illustrate it with the early history of the bicycle.[7] The object we take to be a self-evident "black box" actually started out as two very different devices, a sportsman's racer and a utilitarian transportation vehicle. The high front wheel of the sportsman's bike was necessary at the time to attain high speeds, but it also caused instability. Equal-sized wheels made for a safer but less exciting ride. These two designs met different needs and were in fact different technologies with many shared elements. Pinch and Bijker call this original ambiguity of the object designated as a "bicycle," "interpretative flexibility."

Eventually the "safety" design won out, and it benefited from all the later advances that occurred in the field. In retrospect, it seems as though the high-wheelers were a clumsy and less efficient stage in a progressive development leading through the old "safety" bicycle to current designs. In fact, the high-wheeler and the safety shared the field for years and neither was a stage in the other's development. The high-wheeler represents a possible alternative path of bicycle development that addressed different problems at the origin.

Determinism is a species of Whig history which makes it seem as though the end of the story was inevitable from the very beginning by projecting the abstract technical logic of the finished object back into the past as a cause of development. That approach confuses our understanding of the past and stifles the imagination of a different future. Constructivism can open up that future, although its practitioners have hesitated so far to engage the larger social issues implied in their method.[8]

V Indeterminism

If the thesis of unilinear progress falls, the collapse of the notion of determination by the technological base cannot be far behind. Yet it is still frequently invoked in contemporary political debates.

I shall return to these debates later in this essay. For now, let us consider the remarkable anticipation of current attitudes in the struggle over the length of the workday and child labor in mid-nineteenth-century England. The debate on the Factory Bill of 1844 is entirely structured around the deterministic opposition of technological imperatives and ideology. Lord Ashley, the chief advocate of regulation, protests in the name of familial ideology that "The tendency of the various improvements in machinery is to supersede the employment of adult males, and substitute in its place, the labour of children and females. What will be the effect on future generations, if their tender frames be subjected, without limitation or control, to such destructive agencies?"[9]

He went on to deplore the decline of the family consequent upon the employment of women, which "disturbs the order of nature," and deprives children of proper upbringing. "It matters not whether it be prince or peasant, all that is best, all that is lasting in the character of a man, he has learnt at his mother's knees." Lord Ashley was outraged to find that "females not only perform the labour, but occupy the places of men; they are forming various clubs and associations, and gradually acquiring all those privileges which are held to be the proper portion of the male sex...they meet together to drink, sing, and smoke; they use, it is stated, the lowest, most brutal, and most disgusting language imaginable...."

Proposals to abolish child labor met with consternation on the part of factory owners, who regarded the little worker as an "imperative" of the technologies created to employ him. They denounced the "inefficiency" of using full-grown workers to accomplish tasks done as well or better by children, and they predicted all the usual catastrophic economic consequences – increased poverty, unemployment, loss of international competitiveness – from the substitution of more costly adult labor. Their eloquent representative, Sir James Graham, therefore urged caution: "We have arrived at a state of society when without commerce and manufactures this great community cannot be maintained. Let us, as far as we can, mitigate the evils arising out of this highly artificial state of society; but let us take care to adopt no step that may be fatal to commerce and manufactures."

He further explained that a reduction in the workday for women and children would conflict with the depreciation cycle of machinery and lead to lower wages and trade problems. He concluded that "in the close race of competition which our manufacturers are now running with foreign competitors...such a step would be fatal...." Regulation, he and his fellows maintained in words that echo still, is based on a "false principle of humanity, which in the end is certain to defeat itself." One might almost believe that Ludd had risen again in the person of Lord Ashley: the issue is not really the length of the workday, "but it is in principle an argument to get rid of the whole system of factory labour." Similar protestations are heard today on behalf of industries threatened with what they call environmental "Luddism."

Yet what actually happened once the regulators succeeded in imposing limitations on the workday and expelling children from the factory? Did the violated imperatives of technology come back to haunt them? Not at all. Regulation led to an intensification of factory labor that was incompatible with the earlier conditions in any case. Children ceased to be workers and were redefined socially as learners and consumers. Consequently, they entered the labor market with higher levels of skill and discipline that were soon presupposed by technological design. As a result no one is nostalgic for a return to the good old days when inflation was held down by child labor. That is simply not an option (at least not in the developed capitalist world).

This example shows the tremendous flexibility of the technical system. It is not rigidly constraining but, on the contrary, can adapt to a variety of social demands. This conclusion should not be surprising given the responsiveness of technology to social redefinition discussed previously. It means that technology is just another dependent social variable, albeit an increasingly important one, and not the key to the riddle of history.

Determinism, I have argued, is characterized by the principles of unilinear progress and determination by the base; if determinism is wrong, then technology research must be guided by the following two contrary principles. In the first place, technological development is not unilinear but branches in many directions, and could reach generally higher levels along more than one different track. And, secondly, technological development is not determining for society but is overdetermined by both technical and social factors.

The political significance of this position should also be clear by now. In a society where determinism stands guard on the frontiers of democracy, indeterminism cannot but be political. If technology has many unexplored potentialities, no technological imperatives dictate the current social hierarchy. Rather, technology is a scene of social struggle, a "parliament of things," on which civilizational alternatives contend.

VI Interpreting Technology

In the next sections of this essay, I would like to present several major themes of a non-determinist approach to technology. The picture sketched so far implies a significant change in our definition of technology. It can no longer be considered as a collection of devices, nor, more generally, as the sum of rational means. These are tendentious definitions that make technology seem more functional and less social than in fact it is.

As a social object, technology ought to be subject to interpretation like any other cultural artifact, but it is generally excluded from humanistic study. We are assured that its essence lies in a technically explainable function rather than a hermeneutically interpretable meaning. At most humanistic methods might illuminate extrinsic aspects of technology, such as packaging and advertising, or popular reactions to controversial innovations such as nuclear power or surrogate motherhood. Technological determinism draws its force from this attitude. If one ignores most of the connections between technology and society, it is no wonder that technology then appears to be self-generating.

Technical objects have two hermeneutic dimensions that I call their *social meaning* and their *cultural horizon*.[10] The role of social meaning is clear in the case of the bicycle introduced above. We have seen that the construction of the bicycle was controlled in the first instance by a contest of interpretations: was it to be a sportsman's toy or a means of transportation? Design features such as wheel size also served to signify it as one or another type of object.[11]

It might be objected that this is merely an initial disagreement over goals with no hermeneutic significance. Once the object is stabilized, the engineer has the last word on its nature, and the humanist interpreter is out of luck. This is the view of most engineers and managers; they readily grasp the concept of "goal" but they have no place for "meaning."

In fact the dichotomy of goal and meaning is a product of functionalist professional culture, which is itself rooted in the structure of the modern economy. The concept of "goal" strips technology bare of social contexts, focusing engineers and managers on just what they need to know to do their job.

A fuller picture is conveyed, however, by studying the social role of the technical object and the lifestyles it makes possible. That picture places the abstract notion of "goal" in its concrete social context. It makes technology's contextual causes and consequences visible rather than obscuring them behind an impoverished functionalism.

The functionalist point of view yields a decontextualized temporal cross-section in the life of the object. As we have seen, determinism claims implausibly to be able to get from one such momentary configuration of the object to the next on purely technical terms. But in the real world all sorts of unpredictable attitudes crystallize around technical objects and influence later design changes. The engineer may think these are extrinsic to the device he or she is working on, but they are its very substance as a historically evolving phenomenon.

These facts are recognized to a certain extent in the technical fields themselves, especially in computers. Here we have a contemporary version of the dilemma of the bicycle discussed above. Progress of a generalized sort in speed, power, and memory goes on apace while corporate planners struggle with the question of what it is all for.

Technical development does not point definitively toward any particular path. Instead, it opens branches, and the final determination of the "right" branch is not within the competence of engineering because it is simply not inscribed in the nature of the technology.

I have studied a particularly clear example of the complexity of the relation between the technical function and meaning of the computer in the case of French videotex.[12] Called *Teletel*, this system was designed to bring France into the Information Age by giving telephone subscribers access to databases. Fearing that consumers would reject anything resembling office equipment, the telephone company attempted to redefine the computer's social image; it was no longer to appear as a calculating device for professionals but was to become an informational network for all.

The telephone company designed a new type of terminal, the *Minitel*, to look and feel like an adjunct to the domestic telephone. The telephonic disguise suggested to some users that they ought to be able to talk to each other on the network. Soon the *Minitel* underwent a further redefinition at the hands of these users, many of whom employed it primarily for anonymous on-line chatting with other users in the search for amusement, companionship, and sex.

Thus the design of the *Minitel* invited communications applications which the company's engineers had not intended when they set about improving the flow of information in French society. Those applications, in turn, connoted the *Minitel* as a means of personal encounter, the very opposite of the rationalistic project for which it was originally created. The "cold" computer became a "hot" new medium.

At issue in the transformation is not only the computer's narrowly conceived technical function, but the very nature of the advanced society it makes possible. Does networking open the doors to the Information Age where, as rational consumers hungry for data, we pursue strategies of optimization? Or is it a postmodern technology that emerges from the breakdown of institutional and sentimental stability, reflecting, in Lyotard's words, the "atomisation of society into flexible networks of language games?"[13] In this case technology is not merely the servant of some predefined social purpose; it is an environment within which a way of life is elaborated.

In sum, differences in the way social groups interpret and use technical objects are not merely extrinsic but make a difference in the nature of the objects themselves. *What* the object *is* for the groups that ultimately decide its fate determines what it *becomes* as it is redesigned and improved over time. If this is true, then we can only understand technological development by studying the sociopolitical situation of the various groups involved in it.

VII Technological Hegemony

In addition to the sort of assumptions about individual technical objects we have been discussing so far, that situation also includes broader assumptions about social values. This is where the study of the cultural horizon of technology comes in. This second hermeneutic dimension of technology is the basis of modern forms of social hegemony; it is particularly relevant to our original question concerning the inevitability of hierarchy in technological society.

As I will use the term, hegemony is a form of domination so deeply rooted in social life that it seems natural to those it dominates. One might also define it as that aspect of the distribution of social power which has the force of culture behind it.

The term "horizon" refers to culturally general assumptions that form the unquestioned background to every aspect of life.[14] Some of these support the prevailing hegemony. For example, in feudal societies, the "chain of being" established hierarchy in the fabric of God's universe and protected the caste relations of the society from challenge. Under this horizon, peasants revolted in the name of the king, the only imaginable source of power. Rationalization is our modern horizon, and technological design is the key to its effectiveness as the basis of modern hegemonies.

Technological development is constrained by cultural norms originating in economics, ideology, religion, and tradition. We discussed earlier how assumptions about the age composition of the labor force entered into the design of nineteenth-century production technology. Such assumptions seem so natural and obvious they often lie below the threshold of conscious awareness.

This is the point of Herbert Marcuse's important critique of Weber.[15] Marcuse shows that the concept of rationalization confounds the control of labor by management with control of nature by technology. The search for control of nature is generic, but management only arises against a specific social background, the capitalist wage system. Workers have no immediate interest in output in this system, unlike earlier forms of farm and craft labor, since their wage is not essentially linked to the income of the firm. Control of human beings becomes all-important in this context.

Through mechanization, some of the control functions are eventually transferred from human overseers and parcellized work practices to machines. Machine design is thus socially relative in a way that Weber never recognized, and the "technological rationality" it embodies is not universal but particular to capitalism. In fact, it is the horizon of all the existing industrial societies, communist as well as capitalist, insofar as they are managed from above. (In section X, I discuss a generalized application of this approach in terms of what I call the "technical code.")

If Marcuse is right, it ought to be possible to trace the impress of class relations in the very design of production technology, as has indeed been shown by such Marxist students of the labor process as Harry Braverman and David Noble.[16] The assembly line offers a particularly clear instance because it achieves traditional management goals, such as deskilling and pacing work, through technical design. Its technologically enforced labor discipline increases productivity and profits by increasing control. However, the assembly line only appears as technical progress in a specific social context. It would not be perceived as an advance in an economy based on workers' cooperatives in which labor discipline was more self-imposed than imposed from above. In such a society, a different technological rationality would dictate different ways of increasing productivity.[17]

This example shows that technological rationality is not merely a belief, an ideology, but is effectively incorporated into the structure of machines. Machine design mirrors back the social factors operative in the prevailing rationality. The fact that the argument for the social relativity of modern technology originated in a Marxist context has obscured its most radical implications. We are not dealing here with a mere critique of the property system, but have extended the force of that critique down into the technical "base." This approach goes well beyond the old economic distinction between capitalism and socialism, market and plan. Instead, one arrives at a very different distinction between societies in which power rests on the technical mediation of social activities and those that democratize technical control and, correspondingly, technological design.

VIII Double Aspect Theory

The argument to this point might be summarized as a claim that social meaning and functional rationality are inextricably intertwined dimensions of technology. They are not ontologically distinct, for example, with meaning in the observer's mind and rationality in the technology proper. Rather they are "double aspects" of the same underlying technical object, each aspect revealed by a specific contextualization.

Functional rationality, like scientific–technical rationality in general, isolates objects from their original context in order to incorporate them into theoretical or functional systems. The institutions that support this procedure, such as laboratories and research centers, themselves form a special context with their own practices and links to various social agencies and powers. The notion of "pure" rationality arises when the work of decontextualization is not itself grasped as a social activity reflecting social interests.

Technologies are selected by these interests from among many possible configurations. Guiding the selection process are social codes established by the cultural and political struggles that define the horizon under which the technology will fall. Once introduced, technology offers a material validation of the cultural horizon to which it has been pre-formed. I call this the "bias" of technology: apparently neutral, functional rationality is enlisted in support of a hegemony. The more technology society employs, the more significant is this support.

As Foucault argues in his theory of "power\ knowledge," modern forms of oppression are not so much based on false ideologies as on the specific technical "truths" which form the basis of the dominant hegemony and which reproduce it.[18] So long as the contingency of the choice of "truth" remains hidden, the deterministic image of a technically justified social order is projected.

The legitimating effectiveness of technology depends on unconsciousness of the cultural–political horizon under which it was designed. A recontextualizing critique of technology can uncover that horizon, demystify the illusion of technical necessity, and expose the relativity of the prevailing technical choices.

IX The Social Relativity of Efficiency

These issues appear with particular force in the environmental movement today. Many environmentalists argue for technical changes that would protect nature and, in the process, improve human life as well. Such changes would enhance efficiency in broad terms by reducing harmful and costly side effects of technology. However, this program is very difficult to impose in a capitalist society. There is a tendency to deflect criticism from technological processes to products and people, from a priori prevention to a posteriori clean-up. These preferred strategies are generally costly and reduce efficiency under the horizon of the given technology. This situation has political consequences.

Restoring the environment after it has been damaged is a form of collective consumption, financed by taxes or higher prices. These approaches dominate public awareness. This is why environmentalism is generally perceived as a cost involving trade-offs, and not as a rationalization increasing overall efficiency. But in a modern society, obsessed by economic well-being, that perception is damning. Economists and businessmen are fond of explaining the price we must pay in inflation and unemployment for worshipping at Nature's shrine instead of Mammon's. Poverty awaits those who will not adjust their social and political expectations to technology.

This trade-off model has environmentalists grasping at straws for a strategy. Some hold out the pious hope that people will turn from economic to spiritual values in the face of the mounting problems of industrial society. Others expect enlightened dictators to impose technological reform even if a greedy populace shirks its duty. It is difficult to decide which of these solutions is more improbable, but both are incompatible with basic democratic values.[19]

The trade-off model confronts us with dilemmas – environmentally sound technology vs. prosperity, workers' satisfaction and control vs. productivity, etc. – where what we need are syntheses. Unless the problems of modern industrialism can be solved in ways that both enhance public welfare and win public support, there is little reason to hope that they will ever be solved. But how can technological reform be reconciled with prosperity when it places a variety of new limits on the economy?

The child labor case shows how apparent dilemmas arise on the boundaries of cultural change, specifically, where the social definition of major technologies is in transition. In such situations, social groups excluded from the original design network articulate their unrepresented interests politically. New values the outsiders believe would enhance their welfare appear as mere ideology to insiders who are adequately represented by the existing designs.

This is a difference of perspective, not of nature. Yet the illusion of essential conflict is renewed whenever major social changes affect technology. At first, satisfying the demands of new groups after the fact has visible costs and, if it is done clumsily, will indeed reduce efficiency until better designs are found. But usually, better designs can be found, and what appeared to be an insuperable barrier to growth dissolves in the face of technological change.

This situation indicates the essential difference between economic exchange and technique. Exchange is all about trade-offs: more of A means less of B. But the aim of technical advance is precisely to avoid such dilemmas by elegant designs that optimize several variables at once. A single cleverly conceived mechanism may correspond to many different social demands, one structure to many functions.[20] Design is not a zero-sum economic game, but an ambivalent cultural process that serves a multiplicity of values and social groups without necessarily sacrificing efficiency.

X The Technical Code

That these conflicts over social control of technology are not new can be seen from the interesting case of the "bursting boilers."[21] Steamboat boilers were the first technology regulated in the United States. In the early nineteenth century the steamboat was a major form of transportation similar to the automobile or airplane today. Steamboats were necessary in a big country without paved roads and lots of rivers and canals. But steamboats frequently blew up when the boilers weakened with age or were pushed too hard. After several particularly devastating accidents in 1816, the city of Philadelphia consulted with experts on how to design safer boilers, the first time an American governmental institution interested itself in the problem. In 1837, at the request of Congress, the Franklin Institute issued a detailed report and recommendations based on rigorous study of boiler construction. Congress was tempted to impose a safe boiler code on the industry, but boilermakers and steamboat owners resisted and government hesitated to interfere with private property.

It took from that first inquiry in 1816 to 1852 for Congress to pass effective laws regulating the construction of boilers. In that time 5,000 people were killed in accidents on steamboats. Is this many casualties or few? Consumers evidently were not too alarmed to continue traveling by riverboat in ever-increasing numbers. Understandably, the shipowners interpreted this as a vote of confidence and protested the excessive cost of safer designs. Yet politicians also won votes demanding safety.

The accident rate fell dramatically once technical changes such as thicker walls and safety valves were mandated. Legislation would hardly have been necessary to achieve this outcome had it been technically determined. But in fact boiler design was relative to a social judgment about safety. That judgment could have been made on strictly market grounds, as the shippers wished, or politically, with differing technical results. In either case, those results *constitute* a proper boiler. What a boiler "is" was thus defined through a long process of political struggle culminating finally in uniform codes issued by the American Society of Mechanical Engineers.

This example shows just how technology adapts to social change. What I call the "technical code" of the object mediates the process. That code responds to the cultural horizon of the society at the level of technical design. Quite down-to-earth technical parameters such as the choice and processing of materials are *socially* specified by the code. The illusion of technical necessity arises from the fact that the code is thus literally "cast in iron," at least in the case of boilers.[22]

Conservative anti-regulatory social philosophies are based on this illusion. They forget that the design process always already incorporates standards of safety and environmental compatibility; similarly, all technologies support some basic level of user or worker initiative. A properly made technical object simply *must* meet these standards to be recognized as such. We do not treat conformity as an expensive add-on, but regard it as an intrinsic production cost. Raising the standards means altering the definition of the object, not paying a price for an alternative good or ideological value as the trade-off model holds.

But what of the much discussed cost/benefit ratio of design changes such as those mandated by environmental or other similar legislation? These calculations have some application to transitional situations, before technological advances responding to new values fundamentally alter the terms of the problem. But all too often, the results depend on economists' very rough estimates of the monetary value of such things as a day of trout fishing or an asthma attack. If made without prejudice, these estimates may well help to prioritize policy alternatives. But one cannot legitimately generalize from such policy applications to a universal theory of the costs of regulation.

Such fetishism of efficiency ignores our ordinary understanding of the concept which alone is relevant to social decision-making. In that everyday sense, efficiency concerns the narrow range of values that economic actors routinely affect by their decisions. Unproblematic aspects of technology are not included. In theory one can decompose any technical object and account for each of its elements in terms of the goals it meets, whether it be safety, speed, reliability, etc., but in practice no one is interested in opening the "black box" to see what is inside.

For example, once the boiler code is established, such things as the thickness of a wall or the design of a safety valve appear as essential to the object. The cost of these features is not broken out as the specific "price" of safety and compared unfavorably with a more efficient but less secure version of the technology. Violating the code in order to lower costs is a crime, not a trade-off. And since all further progress takes place on the basis of the new safety standard, soon no one looks back to the good old days of cheaper, insecure designs.

Design standards are only controversial while they are in flux. Resolved conflicts over technology are quickly forgotten. Their outcomes, a welter of taken-for-granted technical and legal standards, are embodied in a stable code, and form the background against which economic actors manipulate the unstable portions of the environment in the pursuit of efficiency. The code is not varied in real-world economic calculations but treated as a fixed input.

Anticipating the stabilization of a new code, one can often ignore contemporary arguments that will soon be silenced by the emergence of a new horizon of efficiency calculations. This is what happened with boiler design and child labor; presumably, the current debates on environmentalism will have a similar history, and we will someday mock those who object to cleaner air as a "false principle of humanity" that violates technological imperatives.

Non-economic values intersect the economy in the technical code. The examples we are dealing with illustrate this point clearly. The legal standards that regulate workers' economic activity have a significant impact on every aspect of their lives. In the child labor case, regulation helped to widen educational opportunities with consequences that are not primarily economic in character. In the riverboat case, Americans gradually chose high levels of security and boiler design came to reflect that choice. Ultimately, this was no trade-off of one good for another, but a non-economic decision about the value of human life and the responsibilities of government.

Technology is thus not merely a means to an end; technical design standards define major portions of the social environment, such as urban and built spaces, workplaces, medical activities and expectations, life patterns, and so on. The economic significance of technical change often pales beside its wider human implications in framing a way of life. In such cases, regulation defines the cultural framework *of* the economy; it is not an act *in* the economy.

XI Heidegger's "Essence" of Technology

The theory sketched here suggests the possibility of a general reform of technology. But dystopian critics object that the mere fact of pursuing efficiency or technical effectiveness already does inadmissible violence to human beings and nature. Universal functionalization destroys the integrity of all that is. As Heidegger argues, an "objectless" world of mere resources replaces a world of "things" treated with respect for their own sake as the gathering places of our manifold engagements with "being."[23]

This critique gains force from the actual perils with which modern technology threatens the world today. But my suspicions are aroused by Heidegger's famous contrast between a dam on the Rhine and a Greek chalice. It would be difficult to find a more tendentious comparison. No doubt modern technology is immensely more destructive than any other. And Heidegger is right to argue that means are not truly neutral, that their substantive content affects society independent of the goals they serve. But I have argued here that this content is not *essentially* destructive; rather, it is a matter of design and social insertion.

However, Heidegger rejects any merely social diagnosis of the ills of technological societies and claims that the source of their problems dates back at least to Plato, that modern societies merely realize a *telos* immanent in Western metaphysics from the beginning. His originality consists in pointing out that the ambition to control being is itself a way of being and hence subordinate at some deeper level to an ontological dispensation beyond human control. But the overall effect of his critique is to condemn human agency, at least in modern times, and to confuse essential differences between types of technological development.

Heidegger distinguishes between the *ontological* problem of technology, which can only be addressed by achieving what he calls "a free relation" to technology, and the merely *ontic* solutions proposed by reformers who wish to change technology itself. This distinction may have seemed more interesting in years gone by than it does today. In effect, Heidegger is asking for nothing more than a change in attitude toward the selfsame technical world. But that is an idealistic solution in the bad sense, and one which a generation of environmental action would seem decisively to refute.

Confronted with this argument, Heidegger's defenders usually point out that his critique of technology is not merely concerned with human attitudes but with the way being reveals itself. Roughly translated out of Heidegger's language, this means that the modern world has a technological form in something like the sense in which, for example, the medieval world has a religious form. Form is no mere question of attitude but takes on a material life of its own: power plants are the Gothic cathedrals of our time. But this interpretation of Heidegger's thought raises the expectation that he will offer criteria for a reform of technology. For example, his analysis of the tendency of modern technology to accumulate and store up nature's powers suggests the superiority of another technology that would not challenge nature in Promethean fashion.

Unfortunately, Heidegger's argument is developed at such a high level of abstraction he literally cannot discriminate between electricity and atom bombs, agricultural techniques and the Holocaust. In a 1949 lecture, he asserted: "Agriculture is now the mechanized food industry, in essence the same as the manufacturing of corpses in gas chambers and extermination camps, the same as the blockade and starvation of nations, the same as the production of hydrogen bombs."[24] All are merely different expressions of the identical enframing which we are called to transcend through the recovery of a deeper relation to being. And since Heidegger rejects technical regression while leaving no room for a better technological future, it is difficult to see in what that relation would consist beyond a mere change of attitude.

XII History or Metaphysics

Heidegger is perfectly aware that technical activity was not "metaphysical" in his sense until recently. He must therefore sharply distinguish modern technology from all earlier forms of technique, obscuring the many real connections and continuities. I would argue, on the contrary, that what is new about modern technology can only be understood against the background of the traditional technical world from which it developed. Furthermore, the saving potential of modern technology can only be realized by recapturing certain traditional features of technique. Perhaps this is why theories that treat modern technology as a unique phenomenon lead to such pessimistic conclusions.

Modern technology differs from earlier technical practices through significant shifts in emphasis rather than generically. There is nothing unprecedented in its chief features, such as the reduction of objects to raw materials, the use of precise measurement and plans, the technical control of some human beings by others, and large scales of operation. It is the centrality of these features that is new, and of course the consequences of that are truly without precedent.

What does a broader historical picture of technology show? The privileged dimensions of modern technology appear in a larger context that includes many currently subordinated features that were defining for it in former times. For example, until the generalization of Taylorism, technical life was essentially about the choice of a vocation. Technology was associated with a way of life, with specific forms of personal development, virtues, etc. Only the success of capitalist deskilling finally reduced these human dimensions of technique to marginal phenomena.

Similarly, modern management has replaced the traditional collegiality of the guilds with new forms of technical control. Just as vocational investment in work continues in certain exceptional settings, so collegiality survives in a few professional or cooperative workplaces. Numerous historical studies show that these older forms are not so much incompatible with the "essence" of technology as with capitalist economics. Given a different social context and a different path of technical development, it might be possible to recover these traditional technical values and organizational forms in new ways in a future evolution of modern technological society.

Technology is an elaborate complex of related activities that crystallizes around tool making and using in every society. Matters such as the transmission of techniques or the management of its natural consequences are not extrinsic to technology per se but are dimensions of it. When, in modern societies, it becomes advantageous to minimize these aspects of technology, that too is a way of accommodating it to a certain social demand, not the revelation of its preexisting "essence." In so far as it makes sense to talk about an essence of technology at all, it must embrace the whole field revealed by historical study, and not only a few traits ethnocentrically privileged by our society.

There is an interesting text in which Heidegger shows us a jug "gathering" the contexts in which it was created and functions. This image could be applied to technology as well, and in fact there is one brief passage in which Heidegger so interprets a highway bridge. Indeed, there is no reason why modern technology cannot also "gather" its multiple contexts, albeit with less romantic pathos than jugs and chalices. This is in fact one way of interpreting contemporary demands for such things as environmentally sound technology, applications of medical technology that respect human freedom and dignity, urban designs that create humane living spaces, production methods that protect workers' health and offer scope for their intelligence, and so on. What are these demands if not a call to reconstruct modern technology so that it gathers a wider range of contexts to itself rather than reducing its natural, human, and social environment to mere resources?

Heidegger would not take these alternatives very seriously because he reifies modern technology as something separate from society, as an inherently contextless force aiming at pure power. If this is the "essence" of technology, reform would be merely extrinsic. But at this point Heidegger's position converges with the very Prometheanism he rejects. Both depend on the narrow definition of technology that, at least since Bacon and Descartes, has emphasized its destiny to control the world to the exclusion of its equally essential contextual embeddedness. I believe that this definition reflects the capitalist environment in which modern technology first developed.

The exemplary modern master of technology is the entrepreneur, singlemindedly focused on production and profit. The enterprise is a radically decontextualized platform for action, without the traditional responsibilities for persons and places that went with technical power in the past. It is the autonomy of the enterprise that makes it possible to distinguish so sharply between intended and unintended consequences, between goals and contextual effects, and to ignore the latter.

The narrow focus of modern technology meets the needs of a particular hegemony; it is not a metaphysical condition. Under that hegemony technological design is unusually decontextualized and destructive. It is that hegemony that is called to account, not technology per se, when we point out that today technical means form an increasingly threatening life environment. It is that hegemony, as it has embodied itself in technology, that must be challenged in the struggle for technological reform.

XIII Democratic Rationalization

For generations faith in progress was supported by two widely held beliefs: that technical necessity dictates the path of development, and that the pursuit of efficiency provides a basis for identifying that path. I have argued here that both these beliefs are false, and that furthermore, they are ideologies employed to justify restrictions on opportunities to participate in the institutions of industrial society. I conclude that we can achieve a new type of technological society that can support a broader range of values. Democracy is one of the chief values a redesigned industrialism could better serve.

What does it mean to democratize technology? The problem is not primarily one of legal rights but of initiative and participation. Legal forms may eventually routinize claims that are asserted informally at first, but the forms will remain hollow unless they emerge from the experience and needs of individuals resisting a specifically technological hegemony.

That resistance takes many forms, from union struggles over health and safety in nuclear power plants to community struggles over toxic waste disposal, to political demands for regulation of reproductive technologies. These movements alert us to the need to take technological externalities into account and demand design changes responsive to the enlarged context revealed in that accounting.

Such technological controversies have become an inescapable feature of contemporary political life, laying out the parameters for official "technology assessment."[25] They prefigure the creation of a new public sphere embracing the technical background of social life, and a new style of rationalization that internalizes unaccounted costs borne by "nature," i.e., some-thing or -body exploitable in the pursuit of profit. Here respect for nature is not antagonistic to technology but enhances efficiency in broad terms.

As these controversies become commonplace, surprising new forms of resistance and new types of demands emerge alongside them. Networking has given rise to one among many such innovative public reactions to technology. Individuals who are incorporated into new types of technical networks have learned to resist through the Net itself in order to influence the powers that control it. This is not a contest for wealth or administrative power, but a struggle to subvert the technical practices, procedures, and designs structuring everyday life.

The example of the Minitel can serve as a model of this new approach. In France, the computer was politicized as soon as the government attempted to introduce a highly rationalistic information system to the general public. Users "hacked" the network in which they were inserted and altered its functioning, introducing human communication on a vast scale where only the centralized distribution of information had been planned.

It is instructive to compare this case to the movements of AIDS patients.[26] Just as a rationalistic conception of the computer tends to occlude its communicative potentialities, so in medicine, caring functions have become mere side effects of treatment, which is itself understood in exclusively technical terms. Patients become objects of this technique, more or less "compliant" to management by physicians. The incorporation of thousands of incurably ill AIDS patients into this system destabilized it and exposed it to new challenges.

The key issue was access to experimental treatment. In effect, clinical research is one way in which a highly technologized medical system can care for those it cannot yet cure. But until quite recently access to medical experiments has been severly restricted by paternalistic concern for patients' welfare. AIDS patients were able to open up access because the networks of contagion in which they were caught were parallelled by social networks that were already mobilized around gay rights at the time the disease was first diagnosed.

Instead of participating in medicine individually as objects of a technical practice, they challenged it collectively and politically. They "hacked" the medical system and turned it to new purposes. Their struggle represents a counter-tendency to the technocratic organization of medicine, an attempt at a recovery of its symbolic dimension and caring functions.

As in the case of the Minitel, it is not obvious how to evaluate this challenge in terms of the customary concept of politics. Nor do these subtle struggles against the growth of silence in technological societies appear significant from the standpoint of the reactionary ideologies that contend noisily with capitalist modernism today. Yet the demand for communication these movements represent is so fundamental that it can serve as a touchstone for the adequacy of our concept of politics to the technological age.

These resistances, like the environmental movement, challenge the horizon of rationality under which technology is currently designed. Rationalization in our society responds to a particular definition of technology as a means to the goal of profit and power. A broader understanding of technology suggests a very different notion of rationalization based on responsibility for the human and natural contexts of technical action. I call this "democratic rationalization" because it requires technological advances that can only be made in opposition to the dominant hegemony. It represents an alternative to both the ongoing celebration of technocracy triumphant and the gloomy Heideggerian counterclaim that "Only a God can save us" from technocultural disaster.[27]

Is democratic rationalization in this sense socialist? There is certainly room for discussion of the connection between this new technological agenda and the old idea of socialism. I believe there is significant continuity. In socialist theory, workers' lives and dignity stood for the larger contexts modern technology ignores. The destruction of their minds and bodies on the workplace was viewed as a contingent consequence of capitalist technical design. The implication that socialist societies might design a very different technology under a different cultural horizon was perhaps given only lip service, but at least it was formulated as a goal.

We can make a similar argument today over a wider range of contexts in a broader variety of institutional settings with considerably more urgency. I am inclined to call such a position socialist and to hope that in time it can replace the image of socialism projected by the failed communist experiment.

More important than this terminological question is the substantive point I have been trying to make. Why has democracy not been extended to technically mediated domains of social life despite a century of struggles? Is it because technology excludes democracy, or because it has been used to suppress it? The weight of the argument supports the second conclusion. Technology can support more than one type of technological civilization, and may someday be incorporated into a more democratic society than ours.

Notes

This paper expands a presentation of my book, *Critical Theory of Technology* (New York: Oxford University Press, 1991), delivered at the American Philosophical Association, December 28, 1991, and first published in an earlier version in *Inquiry* 35/3–4 (1992).

1 See, for example, Joshua Cohen and Joel Rogers, *On Democracy: Toward a Transformation of American Society* (Harmondsworth: Penguin, 1983); Frank Cunningham, *Democratic Theory and Socialism* (Cambridge: Cambridge University Press, 1987).

2 Max Weber, *The Protestant Ethic and the Spirit of Capitalism*, T. Parsons, trans. (New York: Scribners, 1958), pp. 181–2.

3 Martin Heidegger, *The Question Concerning Technology*, W. Lovitt, trans. (New York: Harper & Row, 1977); Jacques Ellul, *The Technological Society*, J. Wilkinson, trans. (New York: Vintage, 1964).

4 Richard W. Miller, *Analyzing Marx: Morality, Power and History* (Princeton: Princeton University Press, 1984), pp. 188–95.

5 Karl Marx, *Capital* (New York: Modern Library, 1906), p. 13.

6 See, for example, David Bloor, *Knowledge and Social Imagery* (Chicago: University of Chicago Press, 1991), pp. 175–9. For a general presentation of constructivism, see Bruno Latour, *Science in Action*

(Cambridge, Mass.: Harvard University Press, 1987).

7 Trevor Pinch and Wiebe Bijker, "The Social Construction of Facts and Artefacts: or How the Sociology of Science and the Sociology of Technology Might Benefit Each Other," *Social Studies of Science* 14 (1984).

8 See Langdon Winner's blistering critique of the characteristic limitations of the position, "Upon Opening the Black Box and Finding it Empty: Social Constructivism and the Philosophy of Technology," *The Technology of Discovery and the Discovery of Technology: Proceedings of the Sixth International Conference of the Society for Philosophy and Technology* (Blacksburg, Va.: Society for Philosophy and Technology, 1991). [This essay is reprinted as chapter 22 in this volume.]

9 *Hansard's Debates, Third Series: Parliamentary Debates 1830–1891*, vol. LXXIII, 1844 (Feb. 22–Apr. 22). The quoted passages are found between pp. 1088 and 1123.

10 A useful starting point for the development of a hermeneutics of technology is offered by Paul Ricoeur, "The Model of the Text: Meaningful Action Considered as a Text," in P. Rabinow and W. Sullivan, eds., *Interpretive Social Science: A Reader* (Berkeley: University of California Press, 1979).

11 Michel de Certeau used the phrase "rhetorics of technology" to refer to the representations and practices that contextualize technologies and assign them a social meaning. De Certeau chose the term "rhetoric" because that meaning is not simply present at hand but communicates a content that can be articulated by studying the connotations technology evokes. See the special issue of *Traverse* 26 (October, 1982), entitled *Les Rhétoriques de la Technologie*, and, in that issue, especially Marc Guillaume's article, "*Téléspectres*," pp. 22–3.

12 See chapter 7, "From Information to Communication: The French Experience with Videotex," in Andrew Feenberg, *Alternative Modernity* (Berkeley and Los Angeles: University of California Press, 1995).

13 Jean-François Lyotard, *La Condition Postmoderne* (Paris: Editions de Minuit, 1979), p. 34.

14 For an approach to social theory based on this notion (called, however, *doxa* by the author), see Pierre Bourdieu, *Outline of a Theory of Practice*, R. Nice, trans. (Cambridge: Cambridge University Press, 1977), pp. 164–70.

15 Herbert Marcuse, "Industrialization and Capitalism in the Work of Marx Weber," in *Negations*, J. Shapiro, trans. (Boston: Beacon Press, 1968).

16 Harry Braverman, *Labor and Monopoly Capital* (New York: Monthly Review Press, 1974); David Noble, *Forces of Production* (New York: Oxford University Press, 1984).

17 Bernard Gendron and Nancy Holstrom, "Marx, Machinery and Alienation," *Research in Philosophy and Technology* 2 (1979).

18 Foucault's most persuasive presentation of this view is *Surveiller et Punir* (Paris: Gallimard, 1975).

19 See, for example, Robert Heilbroner, *An Inquiry into the Human Prospect* (New York: Norton, 1975). For a review of these issues in some of their earliest formulations, see Andrew Feenberg, "Beyond the Politics of Survival," *Theory and Society* 7 (1979).

20 This aspect of technology, called "concretization," is explained in Gilbert Simondon, *Du Mode d'Existence des Objets Techniques* (Paris: Aubier, 1958), chap. 1.

21 John G. Burke, "Bursting Boilers and the Federal Power," in M. Kranzberg and W. Davenport, eds., *Technology and Culture* (New York: New American Library, 1972).

22 The technical code expresses the "standpoint" of the dominant social groups at the level of design and engineering. It is thus relative to a social position without for that matter being a mere ideology or psychological disposition. As I will argue in the last section of this essay, struggle for sociotechnical change can emerge from the subordinated standpoints of those dominated within technological systems. For more on the concept of standpoint epistemology, see Sandra Harding, *Whose Science? Whose Knowledge?* (Ithaca: Cornell University Press, 1991).

23 The texts by Heidegger discussed here are, in order, "The Question Concerning Technology," *op. cit.*; "The Thing," and "Building Dwelling Thinking" in *Poetry, Language, Thought*, A. Hofstadter, trans. (New York: Harper & Row, 1971).

24 Quoted in T. Rockmore, *On Heidegger's Nazism and Philosophy* (Berkeley: University of California Press, 1992), p. 241.

25 Alberto Cambrosio and Camille Limoges, "Controversies as Governing Processes in Technology Assessment," *Technology Analysis & Strategic Management* 3/4 (1991).

26 For more on the problem of AIDS in this context, see Andrew Feenberg, "On Being a Human Subject: Interest and Obligation in the Experimental Treatment of Incurable Disease," *The Philosophical Forum* 23/3 (1992).

27 "Only a God Can Save Us Now," Martin Heidegger interviewed in *Der Spiegel*, D. Schendler, trans. *Graduate Philosophy Journal* 6/1 (1977).

Index

Notes: Entries in **bold** refer to text of articles